7th Research in Engineering Education Symposium (REES 2017)

Research in Engineering Education

Bogota, Columbia
6-8 July 2017

Volume 1 of 2

ISBN: 978-1-5108-4941-9

Printed from e-media with permission by:

Curran Associates, Inc.
57 Morehouse Lane
Red Hook, NY 12571

Some format issues inherent in the e-media version may also appear in this print version.

Printed by Curran Associates, Inc. (2017)

For permission requests, please contact Research in Engineering Education Network
at the address below.

Research in Engineering Education Network
c/o Bill Williams
Escola Sup. De Tecno Barreiro
Rua Americo Da Silva Marinho 1
2239-001 Lavaradio, Portugal

www.reen.co

bwbillwilliamsbw@gmail.com

Additional copies of this publication are available from:

Curran Associates, Inc.
57 Morehouse Lane
Red Hook, NY 12571 USA
Phone: 845-758-0400
Fax: 845-758-2633
Email: curran@proceedings.com
Web: www.proceedings.com

TABLE OF CONTENTS

VOLUME 1

VOLUME 2

2017 Research in Engineering Education Symposium. REES2017 will be held in Bogotá, Colombia from July 6 to July 8 2017

IMPORTANCE EVALUATION OF PHYSICS EDUCATION IN THE PRODUCTION ENGINEER TRAINING AND PROFESSIONAL PRACTICE

Roberta Matta de Araújo
Instituto Federal Fluminense, Bom Jesus do Itabapoana, Brazil
rmaraujo@iff.edu.br

Henrique Rego Monteiro da Hora
Instituto Federal Fluminense, Campos dos Goytacazes, Brazil
henrique.dahora@iff.edu.br

Eduardo Shimoda
Universidade Candido Mendes, Campos dos Goytacazes, Brazil
shimoda@ucam-campos.br

Currently there is a great demand for professionals in Production Engineering in the market. Nevertheless, the number of graduates per year in this area lags far behind the number of incoming students*. One of the factors causing this disparity is the high rate of retention and drop-out of students due to poor performance in subjects of the basic cycle of the course, such as* Physics*. In this context, the main goal of this research is to identify the* Physics *content considered essential in the perception of the* Physics *faculty members of a Production Engineering course in the municipality of Campos dos Goytacazes, through the use of questionnaires application and text mining in public contests for Production Engineer position to check what content is needed to follow this career. Research has shown that, according to the respondent professors, there are few* Physics *contents considered essential for the formation of a Production Engineer. Therefore, this research suggests a discussion of the curriculum related to the* Physics *education specifically for this course.*

Keywords: Physics *education, Production Engineering, Essential contents, Questionnaire, Text-mining.*

Introduction

To obtain an economical and productive development and to guarantee the competitiveness of a country, it is necessary that technology is absorbed in the necessary speed and, consequently, Brazil needs to graduate, with quality, an expressive quantity of engineers who are able to bring themselves up to date constantly (Tozzi & Tozzi, 2011). Salerno *et al.* (2014) corroborate with this affirmation declaring that the engineer is of basic importance for the innovation of a country, making clear the interest for the quantity and quality of these professionals. For this purpose, Brazil is investing in important public policies, trying to increase the number of opportunities and the access to the higher education in the country.

However, in accordance with Canto *et al.* (2012), the increasing number of opportunities is not accompanied by the interest of the students for the engineering career, causing a decrease in the candidate/opportunity relation and making it easier for students with bigger gaps in their education to ingress. According to the authors, the facilitated entry into the course causes most of the students to present difficulties of permanence due to the contents disparity to complete basic subjects.

The disproportion between the number of incoming students and alumini is, for the most part, due to the number of retention and evasion in the Engineering courses. According to Oliveira *et al.* (2013) and Tozzi and Tozzi (2011), the average evasion rate for each 5 years is of 50%, being able to reach 60% in the private Higher Education Institutions. For Reis, Cunha and Spritzer (2012), evasion represents an investment without return for the public sector and a loss of revenue for the private sector. In agreement with Martins *et al.* (2014) evasion takes place for countless reasons, being in its majority, "low performance in subjects, non-adaptation to the university and bad education in the elementary and high school, contributing to a bad performance in the courses of the technological line". Oliveira (2009) declares that the explanation of the students failure is, in part, in the elementary formation. In this case, the teacher tries to teach the content in textual and linear ways, making the learning discouraging and without meaning to the student (SILVA; CECÍLIO, 2007). Oliveira *et al.* (2013) argue that in order to increase the number of graduating students, first it is necessary to develop projects and mechanisms to reduce the high rates of quitting the engineering programs.

In accordance to the National Curriculum Guidelines, the professionals from engineering programs must have a profile with strong scientific, technological, humanist and enterprising formation (BRAZIL, 2002). According to Oliveira and Passos (2014), the scientific culture is inserted in the basic cycle of the course, where the Mathematics and Physics subjects are comprised and. Szajnberg and Zakon (2001) affirm that the students of the basic cycle of engineering programs receive, along two years, teachings referring to all areas of Physics, contemplating theoretical and practical contents. For the authors, the function of the basic cycle is to prepare the student for the professionalizing cycle. Nevertheless, the Physics teaching for Engineering programs is offered, in great part of the universities, in the basic teaching centers, by professors who do not know the continuity of the program, generating a segmented teaching and without integration between the basic and professionalizing learning cycles (MELLO; MELLO, 2003).

In general, the basic cycle holds the highest rates of evasion and retention of the students, more precisely in the first year of course, being Physics one of the greatest factors of this evasion/retention (OLIVEIRA ; PASSOS, 2014). Barbeta and Yamamoto (2002) affirm that the students reach the higher education without mastery of the basic concepts of Physics, causing bad performance, since in the first moment, these concepts are reintroduced, however in a deeper way.

Based on the education of the engineer, Physics plays a key role in the process. However, some Engineering programs only present Physics in the basic cycle, as it is the case of the Production Engineering (PE) in many higher education institutions. According to Faria and Souza Junior (2007), the PE is a generalist modality, allowing the engineer to integrate the production with human resources, finances and market. The authors mention that the great areas of knowledge in Production Engineering are: production management, quality management, financial management, ergonomics and work safety, product management, operational research, strategic and organizational management, knowledge management, environmental management and PE education.

The Brazilian Association of Production Engineering (ABEPRO, short in Portuguese), in a document on the curricular references of the Production Engineering, emphasizes that this modality "when returning to its emphasis on characteristics of goods and/or services and productive systems, links strongly with the ideas of projecting and enabling products and productive systems, planning the production, producing and distributing products" (CUNHA, 2002). Having in mind the importance of Production Engineering for the success of the Brazilian businesses and consequently of the country, the great demand for this course, and the demand for these professionals in the labor market, their education must be taken into account.

Understanding the need of the country for versatile professionals like the production engineer, it is relevant to know the importance of Physics for the engineering teaching in Production Engineering programs. The present work aims at presenting the data obtained from a survey applied to Physics professors in Production Engineering programs based on a questionnaire besides a text mining in public contests for production engineer job opportunities.

Metodology

The first step of the research was the elaboration of a questionnaire, followed by its application to professors who teach Physics subjects in the courses of PE in the municipality of Campos dos Goytacazes in the period corresponding to the months of October and November of 2014. In it was presented a relation of Physics contents referring to high school and higher education, aiming at revealing the perspective of observation and understanding of the teacher regarding the listed contents, in reference to the PE students.

The municipality has 6 (six) higher education institutions offering the PE course, in which 18 (eighteen) professors of interest of the research are distributed. However, the questionnaires were not all returned. As a result, the total collected was around 72.2%.

The questionnaire was elaborated from a compilation of Physics contents taught in High School and Higher Education. Having the State sphere predominance over high school enrollments, the selection of Physics contents for the elaboration of the questionnaire was based on the minimum Physics curriculum of the state of Rio de Janeiro. In addition, the content of Physics required in the National High School Examination (ENEM, short in Portuguese) was also considered because it is the main means of applying to higher education. For the compilation of the Physics contents of the higher education course in Production Engineering, were used Volumes 1, 2, 3 and 4 of Physics Fundamentals by Halliday Resnick Walker's, as it is the most commonly indicated in the course curriculum.

Each item exposed had as answer options: not important (1), important but not essential (2), essential (3) and I do not know (N). Therefore, the respondent teacher should point out one of the alternatives, referring initially to the importance or not of the content as prerequisites for the student entering the course of Physics in the higher education of Production Engineering and later, to indicate the importance degree of the same subject for the graduate student of the course, or rather, for the formation of a production engineer.

The purpose of the questionnaire was to identify high school Physics contents that are essential as prerequisites for studying higher education Physics subjects and to identify higher education Physics contents considered essential for the training of the production engineer in the view of the researched professors. To do this, after tabulation of the data in an Excel spreadsheet, a descriptive statistical evaluation was used, using the mean and frequency distribution, as well as a statistical method used to validate the items, the Lawshe method (see more details in Lawshe, 1975).

The second step of the research consisted of a search, through the internet, of the tests of public contests for the position of production engineer. The site <https://www.questoesdeconcursos.com.br/> was chosen to capture these tests. For the sorting of the questions was chosen the option "Questions". From then on, a filter was used to obtain only the contents pertinent to the research, where was selected to "Post" the option "Production Engineer" and "Subject", the option "Production Engineering". The questions were organized in a separated database with 582 specific questions of the area, contemplating the numerous contests.

Next, a second file with key words for each content was elaborated, and these were divided into thematic blocks. A text mining was then carried out in the file of questions, using text mining software (INTEXT, 2015). To do this, the keyword file was inserted into the mining tool, and the classification tool listed the number of questions in which the key words appeared, per block. The software used listed each keyword with the question file. Each time a given keyword is identified in the text (questions file), it is counted in a certain block of contents, which can be part of more than one block, causing it to be computed more than once.

The contents considered essential for entry and exit from the perspective of the responding professors will be presented below. The contents were separated into eight distinct blocks, each one with a set of contents. Table 1 shows the contents pertaining to each block.

Content Block	Content Relation
Basic and Fundamental Knowledge	1. Notions of order of magnitude and scientific notation; 2. International System of Units; 3. Basic tools: graphics and vectors; 4. Conceptualization of vector and scalar quantities.
The movement, the balance and the discovery of the laws of Physics	5. Fundamental quantities of Mechanics: time, space, velocity and acceleration; 6. Historical relation between force and movement; 7. Descriptions of movement and its interpretation: quantification of movement and its mathematical and graphic description; 8. Special cases of movements and their observable regularities; 9. Concept of inertia; 10. Notion of systems of inertial and non-inertial references; 11. Force and variation of the amount of movement; 12. Newton's laws and their applications; 13. Friction force, force weight, normal force of contact and traction; 14. Center of mass, center of gravity and the idea of material point; 15. Force diagrams; 16. Identification of the forces that act in the circular movements; 17. Notion of centripetal force and its quantification; 18. Concept of external and internal forces; 19. Dynamic notion of mass and amount of movement (linear momentum); 20. Law of conservation of momentum (linear momentum) and impulse theorem; 21. Moment of a force (torque); 22. Conditions of static equilibrium of material point and rigid bodies; 23. Moment of inertia, rolling and angular momentum; 24. Hydrostatic; 25. Hydrodynamics
Energy, work and power	26. Conceptualization of work, energy and power; 27. Concept of potential energy and kinetic energy; 28. Conservation of mechanical energy and energy dissipation; 29. Work of a force; 30. Conservative and dissipative forces; 31. Different forms of energy

Table 1: List of Physics contents. Source: Data from ENEM and the collection "Fundamentals of Physics".

Content Block	Content Relation
Mechanics and the operation of the Universe	32) Gravitational acceleration and the Law of Universal Gravitation; 33) Kepler's Laws; 34) Movements of celestial bodies; 35) Influence on the Earth: tides and climatic variations; 36) Historical conceptions about the origin of the universe and its evolution

Heat and thermal phenomena	37) Concepts of heat and temperature; 38) Thermometry; 39) Heat transfer, thermal equilibrium, heat capacity and specific heat; 40) Heat propagation; 41) Thermal expansion; 42) Changes of physical state, latent heat of transformation and phase diagrams; 43) Ideal gas behavior; 44) Laws of Thermodynamics; 45) Applications and thermal phenomena of everyday use; 46) Understanding climatic phenomena related to the water cycle
Oscillations, waves, optics and radiation	47) Fundamental concepts and principles of geometrical optics; 48) Reflection and Refraction of light; 49) Geometric optics: lenses, mirrors and imaging; 50) Human eye: vision anomalies; 51) Simple optical instruments; 52) Simple harmonic motion; 53) Wave concept: Period, frequency, cycle, pulses and nature; 54) Beams and wave fronts; 55) Propagation: relationship between speed, frequency and wavelength; 56) Wave phenomena: reflection, refraction, diffraction, interference, resonance and polarization; 57) Waves in different propagation media; 58) Acoustics: velocity and propagation of sound, acoustic interval, sound intensity, reflection, sound tubes and Doppler effect; 59) Radiation, its interactions and its technological applications
Electrical and magnetic phenomena	60) Electric charge, electrification and Coulomb law; 61) Electrical field, work and electric potential; 62) Capacitors and electrostatic shielding; 63) Electrical current and resistance; 64) Relations between electrical magnitudes: voltage, current, power and energy;
Electrical and magnetic phenomena	65) Single electric circuits; 66) Circuit of Multiples Meshes; 67) Special circuits; 68) Alternating current; 69) Electrical meters; 70) Graphical representation of circuits and conventional symbols; 71) Electric generators and receivers; 72) Power and power consumption in electrical devices; 73) The imams and their properties; 74) Magnetic field and magnetic force; 75) Electromagnetic Induction; 76) Inductance; 77) Electromagnetic oscillations; 78) Maxwell equations; 79) Electric motors, alternators, dynamos and transformers
Modern Physics	80) Electromagnetic waves; 81) Notions of special relativity; 82) Notions of general relativity; 83) Qualitative knowledge of Quantum Physics: quantum theory, photoelectric effect, Bohr atom, dual nature of light and uncertainty principle; 84) Quantum mechanics: quantitative studies; 85) Principles of nuclear Physics

Table 1 (continued): List of Physics contents. Source: ENEM Data and the collection "Fundamentals of Physics".

The first block addresses the contents related to "Basic and fundamental knowledge". Table 2 shows the importance of these contents as prerequisites for the student in the course of production engineering and the importance of these contents for the graduates of the course according to the Lawshe method. Consider Ne as the number of professors who considered the essential item and N the number of respondents.

Table 2: The essentiality of the contents of the "Basic and fundamental knowledge" in the viewpoint of professors, according to the Lawshe method.

Question	Stratum	Ne	N	%Ne	CVRcalc	CVRcrit	Conclusion
Q1	Incoming	9	13	69.2%	0.385	0.544	
	Alumni	12	13	92.3%	0.846	0.544	essential
Q2	Incoming	11	13	84.6%	0.692	0.544	essential
	Alumni	13	13	100.0%	1.000	0.544	essential
Q3	Incoming	7	13	53.8%	0.077	0.544	
	Alumni	13	13	100.0%	1.000	0.544	essential
Q4	Incoming	9	13	69.2%	0.385	0.544	
	Alumni	11	13	84.6%	0.692	0.544	essential

Source: Own elaboration

It may be noticed that by the method, all content is essential for graduates, but only item 2 is essential for admission. The other items, although considered essential for alumni by most respondents, their calculated CVRs were lower than the critical ones.

The second block surveys on the subjects related to "The movement, the balance and the discovery of laws of Physics". Table 3 shows the essentiality of this block for incoming and alumni students according to the method used.

Table 3: The essentiality of related content to "The movement, balance and discovery of the laws of Physics" in the view of professors, according to the Lawshe method.

Question	Stratum	Ne	N	%Ne	CVRcalc	CVRcrit	Conclusion
Q5	Incoming	11	13	84.6%	0.692	0.544	essential
	Alumni	10	13	76.9%	0.538	0.544	
Q6	Incoming	3	13	23.1%	-0.538	0.544	
	Alumni	5	13	38.5%	-0.231	0.544	
Q7	Incoming	6	13	46.2%	-0.077	0.544	
	Alumni	9	13	69.2%	0.385	0.544	
Q8	Incoming	2	13	15.4%	-0.692	0.544	
	Alumni	4	13	30.8%	-0.385	0.544	
Q9	Incoming	8	13	61.5%	0.231	0.544	
	Alumni	10	13	76.9%	0.538	0.544	
Q10	Incoming	2	13	15.4%	-0.692	0.544	
	Alumni	6	13	46.2%	-0.077	0.544	
Q11	Incoming	5	13	38.5%	-0.231	0.544	
	Alumni	10	13	76.9%	0.538	0.544	
Q12	Incoming	12	13	92.3%	0.846	0.544	essential
	Alumni	13	13	100.0%	1.000	0.544	essential
Q13	Incoming	10	13	76.9%	0.538	0.544	
	Alumni	13	13	100.0%	1.000	0.544	essential
Q14	Incoming	5	13	38.5%	-0.231	0.544	
	Alumni	11	13	84.6%	0.692	0.544	essential
Q15	Incoming	6	13	46.2%	-0.077	0.544	
	Alumni	11	13	84.6%	0.692	0.544	essential
Q16	Incoming	4	13	30.8%	-0.385	0.544	
	Alumni	7	13	53.8%	0.077	0.544	
Q17	Incoming	4	13	30.8%	-0.385	0.544	
	Alumni	7	13	53.8%	0.077	0.544	
Q18	Incoming	3	13	23.1%	-0.538	0.544	
	Alumni	8	13	61.5%	0.231	0.544	
Q19	Incoming	2	13	15.4%	-0.692	0.544	
	Alumni	6	13	46.2%	-0.077	0.544	
Q20	Incoming	4	13	30.8%	-0.385	0.544	
	Alumni	6	13	46.2%	-0.077	0.544	
Q21	Incoming	4	13	30.8%	-0.385	0.544	
	Alumni	10	13	76.9%	0.538	0.544	
Q22	Incoming	4	13	30.8%	-0.385	0.544	
	Alumni	11	13	84.6%	0.692	0.544	essential
Q23	Incoming	1	13	7.7%	-0.846	0.544	

	Alumni	7	12	58.3%	0.167	0.566	
Q24	Incoming	4	13	30.8%	0.385	0.544	
	Alumni	9	13	69.2%	0.385	0.544	
Q25	Incoming	3	13	23.1%	-0.538	0.544	
	Alumni	9	13	69.2%	0.385	0.544	

Source: own elaboration

The statistical method used shows that only items 5 and 12 are essential in the professors' opinion, however, we should not ignore the relevance of item 13, considering that 76.9% of professors classify it as essential and the other 23.1% consider it important. In relation to the essential contents for the student who is getting a degree in the higher education institution, the method classifies items 12, 13, 14, 15 and 22 as essential, however, items 5, 9, 11 and 21 presented a significant percentage, considering that of the thirteen respondents, ten (76.9%) considered it essential and the other three considered them important. Therefore, such items should not be totally discarded.

In the third block, subjects related to the theme "Energy, work and power" are discussed. The result of the Lawshe method with respect to each subject of this set is revealed by table 4, referring to incoming and alumni.

Table 4: The essentiality of the contents referring to "Energy, work and power" in the professors' perspective, according to the Lawshe method.

Question	Stratum	Ne	N	%Ne	CVRcalc	CVRcrit	Conclusion
Q26	Incoming	5	13	38.5%	-0.231	0.544	
	Alumni	12	13	92.3%	0.846	0.544	essential
Q27	Incoming	7	13	53.8%	0.077	0.544	
	Alumni	11	13	84.6%	0.692	0.544	essential
Q28	Incoming	5	13	38.5%	-0.231	0.544	
	Alumni	11	13	84.6%	0.692	0.544	essential
Q29	Incoming	6	13	46.2%	-0.077	0.544	
	Alumni	9	13	69.2%	0.385	0.544	
Q30	Incoming	3	13	23.1%	-0.538	0.544	
	Alumni	7	13	53.8%	0.077	0.544	
Q31	Incoming	7	13	53.8%	0.077	0.544	
	Alumni	11	13	84.6%	0.692	0.544	essential

Source: own elaboration.

According to table 4, with regards to the incoming students, no content was considered essential, while for the alumini, items 26, 27, 28 and 31 were classified as essential in the view of the professors studied.

The fourth block presents the contents concerning "Mechanics and the workings of the Universe". Table 5 presents the result obtained by Lawshe, regarding the assessment of the professors as to the importance of each subject of the set for the incoming student and alumni, respectively.

Table 5: The essentiality of the contents referring to "Mechanics and the workings of the Universe" according to the Lawshe method.

Question	Stratum	Ne	N	%Ne	CVRcalc	CVRcrit	Conclusion
Q32	Incoming	0	13	0.0%	-1.000	0.544	
	Alumni	1	13	7.7%	-0.846	0.544	
Q33	Incoming	0	13	0.0%	-1.000	0.544	
	Alumni	0	13	0.0%	-1.000	0.544	
Q34	Incoming	0	13	0.0%	-1.000	0.544	
	Alumni	0	13	0.0%	-1.000	0.544	
Q35	Incoming	0	13	0.0%	-1.000	0.544	
	Alumni	0	13	0.0%	-1.000	0.544	
Q36	Incoming	0	13	0.0%	-1.000	0.544	

	Alumni	0	13	0.0%	-1.000	0.544	

Source: own elaboration.

It is observed in the table that no content was considered essential. The Lawshe method uses the frequency of responses to calculate the CRV and compare it with the critical CRV. To obtain an at least positive calculated CRV, the frequency of responses must be greater than 50%. In this case all frequencies are below the minimum, which leads to a result with negative CRVs and no essential items.

The fifth block refers to the theme "Heat and thermal phenomena". The result obtained regarding the importance of the subject is presented in table 6.

Table 6: The essentiality of the contents referring to "The heat and the thermal phenomena" in the professors' perspective, according to the Lawshe method.

Question	Stratum	Ne	N	%Ne	CVRcalc	CVRcrit	Conclusion
Q37	Incoming	7	13	53.8%	0.077	0.544	
	Alumni	12	13	92.3%	0.846	0.544	essential
Q38	Incoming	7	13	53.8%	0.077	0.544	
	Alumni	12	13	92.3%	0.846	0.544	essential
Q39	Incoming	6	13	46.2%	-0.077	0.544	
	Alumni	11	13	84.6%	0.692	0.544	essential
Q40	Incoming	5	13	38.5%	-0.231	0.544	
	Alumni	10	13	76.9%	0.538	0.544	
Q41	Incoming	5	13	38.5%	-0.231	0.544	
	Alumni	10	13	76.9%	0.538	0.544	
Q42	Incoming	7	13	53.8%	0.077	0.544	
	Alumni	9	13	69.2%	0.385	0.544	
Q43	Incoming	3	13	23.1%	-0.538	0.544	
	Alumni	7	13	53.8%	0.077	0.544	
Q44	Incoming	7	13	53.8%	0.077	0.544	
	Alumni	10	13	76.9%	0.538	0.544	
Q45	Incoming	7	13	53.8%	0.077	0.544	
	Alumni	9	13	69.2%	0.385	0.544	
Q46	Incoming	3	13	23.1%	-0.538	0.544	
	Alumni	6	13	46.2%	-0.077	0.544	

Source: own elaboration

It is observed that no content was classified as essential for the participants in the PE course, while for the graduates, items 37, 38 and 39 were essential. It should be pointed out the possible importance of items 40, 41e 44, due to their high percentage of responses.

The sixth block presents the theme "Oscillations, waves, optics and radiation." Table 7 shows the data obtained.

Table 7: The essentiality of the related content "Oscillations, waves, optics and radiation" from the viewpoint of professors, according to the Lawshe method.

Question	Stratum	Ne	N	%Ne	CVRcalc	CVRcrit	Conclusion
Q47	Incoming	4	12	33.3%	-0.333	0.566	
	Alumni	6	12	50.0%	0.000	0.566	
Q48	Incoming	5	12	41.7%	-0.167	0.566	
	Alumni	7	12	58.3%	0.167	0.566	
Q49	Incoming	3	12	25.0%	-0.500	0.566	
	Alumni	4	12	33.3%	-0.333	0.566	
Q50	Incoming	1	13	7.7%	-0.846	0.544	
	Alumni	0	12	0.0%	-1.000	0.566	
Q51	Incoming	3	13	23.1%	-0.538	0.544	
	Alumni	3	13	23.1%	-0.538	0.544	
Q52	Incoming	6	13	46.2%	-0.077	0.544	
	Alumni	10	12	83.3%	0.667	0.566	essential
Q53	Incoming	7	13	53.8%	0.077	0.544	
	Alumni	9	13	69.2%	0.385	0.544	
Q54	Incoming	2	12	16.7%	-0.667	0.566	
	Alumni	4	12	33.3%	-0.333	0.566	
Q55	Incoming	6	13	46.2%	-0.077	0.544	

	Alumni	8	12	66.7%	0.333	0.566	
Q56	Incoming	4	13	30.8%	-0.385	0.544	
	Alumni	7	13	53.8%	0.077	0.544	
Q57	Incoming	3	3	23.1%	-0.538	0.544	
	Alumni	7	13	53.8%	0.077	0.544	
Q58	Incoming	3	13	23.1%	-0.538	0.544	
	Alumni	7	13	53.8%	0.077	0.544	
Q59	Incoming	4	12	33.3%	-0.333	0.566	
	Alumni	7	12	58.3%	0.167	0.566	

Source: own elaboration

It is observed that for the Incoming students, no content is considered essential and for the alumni, although several contents are considered essential by the majority, only item 52 is classified as such.

The seventh block addresses contents on "Electrical and magnetic phenomena". Table 8 shows the percentage of importance of the abovementioned contents regarding the incoming students and alumni of the course of PE.

Table 8: The essentiality of the contents referring to the "Electrical and magnetic phenomena" for the incoming and alumni in the perspective of the professors.

Question	Stratum	Ne	N	%Ne	CVRcalc	CVRcrit	Conclusion
Q60	Incoming	8	13	61.5%	0.231	0.544	
	Alumni	9	13	69.2%	0.385	0.544	
Q61	Incoming	7	13	53.8%	0.077	0.544	
	Alumni	8	13	61.5%	0.231	0.544	
Q62	Incoming	6	13	46.2%	-0.077	0.544	
	Alumni	8	13	61.5%	0.231	0.544	
Q63	Incoming	8	13	61.5%	0.231	0.544	
	Alumni	9	13	69.2%	0.385	0.544	
Q64	Incoming	9	13	69.2%	0.35	0.544	
	Alumni	11	13	84.6%	0.692	0.544	essential
Q65	Incoming	8	13	61.5%	0.231	0.544	
	Alumni	9	13	69.2%	0.385	0.544	
Q66	Incoming	5	13	38.5%	-0.231	0.544	
	Alumni	5	13	38.5%	-0.231	0.544	
Q67	Incoming	3	13	23.1%	-0.538	0.544	
	Alumni	5	13	38.5%	-0.231	0.544	
Q68	Incoming	5	13	38.5%	-0.231	0.544	
	Alumni	9	13	69.2%	0.385	0.544	
Q69	Incoming	4	13	30.8%	-0.385	0.544	
	Alumni	9	13	69.2%	0.385	0.544	
Q70	Incoming	4	13	30.8%	-0.385	0.544	
	Alumni	7	13	53.8%	0.077	0.544	
Q71	Incoming	3	13	23.1%	-0.538	0.544	
	Alumni	6	13	46.2%	-0.077	0.544	
Q72	Incoming	7	13	53.8%	0.077	0.544	
	Alumni	11	13	84.6%	0.692	0.544	essential
Q73	Incoming	5	13	38.5%	-0.231	0.544	
	Alumni	8	13	61.5%	0.231	0.544	
Q74	Incoming	6	13	46.2%	-0.077	0.544	
	Alumni	7	13	53.8%	0.077	0.544	
Q75	Incoming	3	13	23.1%	-0.538	0.544	
	Alumni	6	13	46.2%	-0.077	0.544	
Q76	Incoming	3	13	23.1%	-0.538	0.544	
	Alumni	4	3	30.8%	-0.385	0.544	
Q77	Incoming	2	13	15.4%	-0.692	0.544	
	Alumni	6	13	46.2%	-0.077	0.544	
Q78	Incoming	1	13	7.7%	-0.846	0.544	
	Alumni	5	13	38.5%	-0.231	0.544	
Q79	Incoming	2	13	15.4%	-0.692	0.544	
	Alumni	6	13	46.2%	-0.077	0.544	

Source: own elaboration

Note that by the applied method only items 64 and 72 were classified as essential for the alumni. Finally, the eighth block concerns the theme "Modern Physics". The result obtained regarding the essentiality of the contents listed above, for the incoming and alumni, in the view of the professors studied is presented in table 9.

Table 9: The essentiality of content related to "Modern Physics" in the view of professors, according to the Lawshe method.

Question	Stratum	Ne	N	%Ne	CVRcalc	CVRcrit	Conclusion
Q80	Incoming	2	13	15.4%	-0.692	0.544	
	Alumni	5	12	41.7%	-0.167	0.566	
Q81	Incoming	0	13	0.0%	-1.000	0.544	
	Alumni	1	12	8.3%	-0.833	0.566	
Q82	Incoming	0	13	0.0%	-1.000	0.544	
	Alumni	2	12	16.7%	-0.667	0.566	
Q83	Incoming	0	13	0.0%	-1.000	0.544	
	Alumni	3	12	25.0%	-0.500	0.566	
Q84	Incoming	0	13	0.0%	-1.000	0.544	
	Alumni	1	12	8.3%	-0.833	0.566	
Q85	Incoming	0	13	0.0%	-1.000	0.544	
	Alumni	1	13	7.7%	-0.846	0.544	

Source: own elaboration

According to the applied method, no content was considered essential, neither for the incoming students nor for the alumni. It should be noted that, as the fourth block which refers to "Mechanics and the workings of the Universe", all frequencies of responses were below 50%, obtaining negative calculated CRV.

It is observed that, although professors complain about the poor background of incoming students to a Higher Education Institution, they point out as prerequisites for those students who are leaving high school only the subjects referring to the International System of Units, mechanics quantities fundamentals and Newton's laws and their applications. According to respondent professors, what most affects students' performance in the Physics subjects is the poor mathematical base, since the course starts from the most basic subjects approach.

It is also noteworthy that for these professors, the essential contents are for the most part those referring to the Basic Knowledge and some topics of Classical Mechanics, which are presented in volume 1 of the collection " Fundamentals of Physics ", corresponding to the Physics course 1, generally offered in the first and/or second term in Production Engineering graduation program.

It is important to note that for the statistical method adopted, the critical CRV calculation depends on the number of respondents, that is, the larger the sample, the lower this value (critical CRV). In this case, for 13 respondents, as in the case of the present study, the critical CRV, for the most part, was equal to 0.544, making an item with a frequency of responses greater than 77.2% to be characterized as essential. If the entire population corresponding to the 18 professors had participated in the research, the critical CRV, according to Ayre and Scally (2014), would pass to 0.444, and could slightly shift the scenario of the essential contents.

Below are presented the contents considered necessary for the exercise of the profession according to the public contests tests for the position of engineer of production, obtained from a text mining done in a collection of tests of public contests for such position. The contents were grouped by themes, totaling eight blocks. Figure 1 shows the percentage related to each block.

Figure 1: Contents collected in public contests for the pursuit of the production engineer profession.

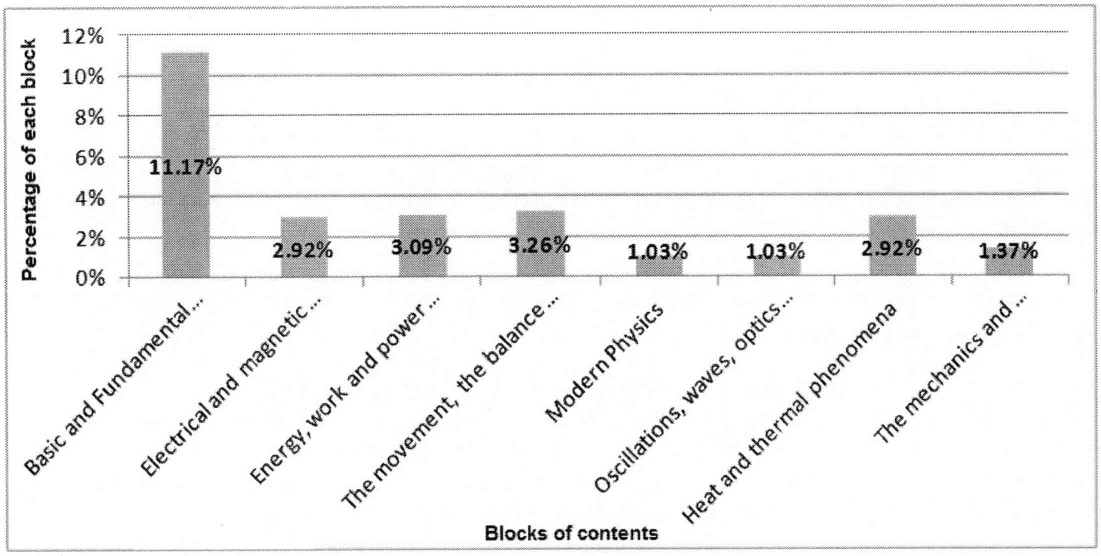

For each block presented there is a list of contents addressed. It should be noted that the percentage relative to each block is very low, considering the number of issues. It can be observed from the graph that the contents that appear in greater proportion are those related to the "Basic and fundamental knowledge" where the contents are inserted: Notions of order of magnitude and scientific notation, International System of Units, Basic Tools: Graphs and vectors; and Conceptualization of vector and scalar quantities. Such contents are essential for all courses in the area of science and technology. They are considered as tools, being used not only by Physics, but also by mathematics, which causes them to appear in a larger quantity of questions. Note that the other blocks show a percentage below 3.5%. Considering the fact that a certain keyword of the ontology used in the text mining can be cataloged in more than one block, this percentage becomes insignificant, that is, the physical contents collected in public contests for the position of production engineer are irrelevant.

The contents referring to "The movement, the balance and the discovery of the laws of Physics", as well as "Heat and thermal phenomena" and "Energy, work and power", were considered by the respondents as essential in their totality or great majority, but presented percentages approximately equal to 3.3%, 2.9% and 3.1%, respectively. Whereas the theme "Oscillations, waves, optics and radiation", also understood for the most part as essential, obtained an even lower percentage, of about 1%.

The subjects "Modern Physics" and "Mechanics and the workings of the Universe" presented about 1% and 1.4%, respectively, of the questions collected in the last public contests. In the perception of the researched professors, both themes are not considered essential for neither the incoming student nor the alumni.

Finally, the block "Electrical and magnetic phenomena", where few contents presented were understood as essential, obtained a percentage equal to or close to themes considered essential.

Table 10 presents a summary of the data obtained through the research carried out, revealing the percentage of the essentiality of each block of contents from the respondent professors point of view and the questions of public contest for the production engineer.

Block of contents	Incoming	Alumni	Profession Practice
Basic and fundamental knowledge	25.00%	100.00%	11.17%

Movement, Balance and the discovery of the laws of Physics	9.52%	23.81%	3.26%
Energy. work and power	0.00%	66.66%	3.09%
Mechanics and the workings of the Universe	0.00%	0.00%	1.37%
Heat and thermal phenomena	0.00%	30.00%	2.92%
Oscillations. waves. optics and radiation	0.00%	7.69%	1.03%
Electric and magnetic phenomena	0.00%	10.00%	2.92%
Modern Physics	0.00%	0.00%	1.03%

Table 10: Percentage referring to the essentiality of contents according to the responding professors and the questions of the public contests for PE. Source: Own elaboration.

Final Considerations

The higher education institutions, in general, that offer the degree program in Production Engineering offer, for the basic cycle of engineering, mixed classes, that is, they do not provide specific classes to each modality. This reality does not allow the teacher to carry out a work oriented to production engineering; in other words, there is no basic Physics teaching specifically directed to the needs of a student of production engineering.

It is known that it is in the basic cycle that the highest evasion rate occurs and the low performance in Physics is a big factor. According to Silva and Cecílio (2007), there is a great lack of motivation among students in relation to the basic subjects, among them Physics, because they do not understand why they are learning certain contents and where it will be applied in the professionalizing subjects. The vast majority of professors who teach Physics for PE have major in Physics and all of them teach their classes without distinction based on the program the students are enrolled in.

According to Silva and Cecílio (2007), it is necessary that the professors of engineering courses are acquainted with this profession practice for their students to prepared in the best possible way.

Regarding production engineering, Physics is decisive to ground the subjects of the professionalizing cycle, such as resistance of materials, mechanical/thermal and fluid systems, technical processes, etc., besides enabling and stimulating critical professional practice. This paper does not question the importance of the subject, however it recognizes that it does have a high time load in the basic cycle and, for PE, as it can be noticed, few of its contents are considered relevant in the education and professional practice of the production engineer.

For that matter, it is suggested that the teaching of Physics be directed to this course, giving emphasis to the contents that are in fact important while suppressing the ones that do not add to the education of the production engineer. According to the National Curriculum Guidelines for Engineering courses, the Higher Education Institutions must reserve a minimum of 30% of the total time load for the subjects specified in the core basic contents listed by it. However, within the determinations, the institution has total freedom to elaborate the curriculum and program for each subject, thus it is crutial that the specificity of each modality be respected.

References

Ayre, C., & Scally, A. J. (2014). Critical Values for Lawshe's Content Validity Ratio Revisiting the Original Methods of Calculation. *Measurement and Evaluation in Counseling and Development*, *47*(1), 79–86.

BRASIL. Institui Diretrizes Curriculares Nacionais do Curso de Graduação em Engenharia., Pub. L. No. RESOLUÇÃO CNE/CES 11/2002 (2002). Retrieved from http://portal.mec.gov.br/cne/arquivos/pdf/CES112002.pdf

Cunha, G. D. da. (2002). Um Panorama Atual da Engenharia da Produção. ABEPRO. Retrieved from http://www.abepro.org.br/arquivos/websites/1/PanoramaAtualEP4.pdf

do Canto Filho, A. B., Ferreira, L. F., Bercht, M., Tarouco, L. M. R., & de Lima, J. V. (2012). Objetos de aprendizagem no apoio à aprendizagem de engenharia: explorando a motivação extrínseca. *RENOTE, 10*(3). Retrieved from http://www.seer.ufrgs.br/renote/article/view/36390/0

Faria, A. F., & Souza Junior, A. C. R. (2007). Propostas de melhoria do projeto pedagógico através do acompanhamento dos egressos. *Revista GEPROS*, (1), Pag–33.

Intext. (2015). Text mining. Retrieved February 25, 2015, from http://www.intext.com.br/

Lawshe, C.H. (1975). A quantitative approach to content validity. *Personnel Psychology, 28*, 563–575. doi:10.1111/j.1744-6570.1975.tb01393.x

Martins, L. M., Neto, J. M. M., dos Santos, F. F. P., & dos Santos, M. do S. F. (2014). Experiências adquiridas com o ensino da disciplina de Processos Químicos no curso de Engenharia de Produção. *Revista de Ensino de Engenharia*, *32*(1). Retrieved from http://www.abenge.org.br/revista/index.php/abenge/article/view/135

Mello, J. C. C. B. S. de, & Mello, M. H. C. S. de. (2003). Integração entre o ensino de cálculo e o de pesquisa operacional. *Production*, *13*(2), 123–129. https://doi.org/10.1590/S0103-65132003000200012

Oliveira, H. P., & Passos, W. de A. C. (2014). Ensino de Física Básica para as Engenharias: O caso da Univasf. *Revista de Ensino de Engenharia*, *33*(2). Retrieved from http://www.abenge.org.br/revista/index.php/abenge/article/viewFile/213/187

Oliveira, P. J. C. de. (2009). *Ensino da física num curso superior de engenharia: na procura de estratégias promotoras de uma aprendizagem activa*. Retrieved from https://ria.ua.pt/handle/10773/1480

Oliveira, V. F., Almeida, N. N., Carvalho, D. M., & Pereira, F. A. A. (2013). Um Estudo Sobre a Expansão da Formação em Engenharia no Brasil. *Revista de Ensino de Engenharia*, *32*(3), 37–56.

Reis, V. W., Cunha, P. J., & Spritzer, I. (2012). Evasão no ensino superior de engenharia no Brasil: um estudo de caso no CEFET/RJ. In *XL Congresso Brasileiro de Ensino de Engenharia*. Belém, PA: Abenge. Retrieved from http://www.abenge.org.br/CobengeAnteriores/2012/artigos/103734.pdf

Salerno, M. S., Lins, L. M., de Vasconcelos Gomes, L. A., & Bottan, T. (2014). Tendências e Perspectivas da Engenharia no Brasil. *Observatório Da Inovação E Competitividade-OIC*, 52.

Silva, L. P., & Cecílio, S. (2007). A mudança no modelo de ensino e de formação na engenharia. *Educação Em Revista*, *45*, 61–80.

Szajnberg, M., & Zakon, A. (2001). A ampliação e a readequação do ensino de física para a engenharia do terceiro milênio. *COBENGE, Porto Alegre*.

Tozzi, M. J., & Tozzi, A. R. (2011). Escassez de Engenheiros no Brasil: Mito ou realidade? In *XXXIX Congresso Brasileiro de Educação em Engenharia*. Blumenau,SC. Retrieved from http://www3.fsa.br/LocalUser/cobenge2011/sessoestec/art1579.pdf

Human-Robot Scaffolding, a cognitive architecture to foster the metacognitive and emotional support using the Baxter robot.

John Jairo Páez Rodríguez, Enrique González Guerrero
jjpaezr@udistrital.edu.co, egonzal@javeriana.edu.co

Abstract

In education, robots have been considered as learning tools. They have evolved from simple tools that only followed students' instructions to complex cognitive artificial systems which allow robots to behave as tutor, peer, or learner. This research proposal suggests the Human-Robot Scaffolding concept as a path to explore new strategies in the learning process which is supported by Bruner's work on scaffold instruction. The expected results foresee the acknowledgement of robots as learning tools which through artificial cognitive conditions can support the mediation process. In this sense, robots will be learning tools that recognize and test the mental models to support schema constructions, influence the zone of proximal development and assist students during their learning process.

Keywords: Human-Robot Scaffolding, Human-Robot Interaction, Problem-Solving Strategies, Constructionism.

Introduction

Robots can give physical, emotional and cognitive support based on the learner's characteristics. In the physical aspect, robot can assist the child during the mechanical assembly process by manipulating and reorganizing the blocks. They can also give emotional support showing icons representing three emotional states like sadness, anger, surprise to encourage the user during the learning process. And finally, they can provide cognitive support through direct explanations, guided instructions and formulation of questions according to the user needs. In conclusion, robots as learning tools represent physical, emotional and cognitive support to the learners.

This document is divided in five sections: Introduction, Human-robot scaffolding, Artificial cognitive architecture and Conclusions. The Introduction presents the main objective of the thesis's doctoral proposal and describes a brief analysis of differences between both tutors (human and robot). The next section named Human-Robot Scaffolding describes the concept and presents different research works from Human-Robot Interaction field. Later, the artificial cognitive architecture is presented. In this section, there are special attention in the scaffolder module. Finally, The Conclusions are presented.

The main objective aims to develop an artificial cognitive architecture to anthropomorphic robot support the students' learning processes of the Means-Ends Analysis strategy. In a typical case, the user demonstrates how to solve a problem by way of moving of mechanical parts and showing emotional and verbal expressions. With this information, the cognitive, meta-cognitive and emotional states are identified and the scaffolding strategy is done by the robot. The behaviour's robot mimics three roles: tutor, peer and learner to support the learning process through five actions; general verbal or non-verbal encouragement, specific verbal or non-verbal Instruction, assists choice of material, arrange material for assembly and show an operation.

The evolution and responsibility of tools during learning process has been related to cognitive, emotional and social aspects. For example, in early 80's the LEGO blocks were used as external-representation systems to show the thought process because they did not have cognitive or emotional skills. But today, new social-robot's characteristics allows them to be used as tools that complement human work in aspects where robots have strengths. For example, during the learning problem-solving strategies through the Hanoi towers, robot could reorganize the blocks to make explanations and to return to the state of the problem. The robot could return more than seven states to problem and promote cognitive and metacognitive learner's process.

From robotics and artificial intelligence, the research process continues the exercise of inquiry how the cognitive science models the cognitive systems to explain and support the learning process related to problem-solving on the similar social conditions but with social robots. This document showing the advances of thesis's research proposal, which opens a way of analysis about implications of robots as learning tools.

Human-Robot Scaffolding

The *Human-Robot Scaffolding* concept merge two big disciplinary fields: Human-Robot Interaction and Education. From HRI field, the cognitive, emotional and social aspects are highlighted. Several works show different frameworks, artificial cognitive architectures, and mathematic models which make contributions in the naturalization of the interaction process which lessens the users cognitive load. From Education field, process to support the learning which is named Scaffolding is presented. As educative strategy, scaffolding was proposed by Jerome Brunner who highlighted the role of tutoring in the problem-solving process. The regulation process about these interventions changed intentionally according to the learning needs. During this process, the student's cognitive load is increased while the responsibility of robot or agent responsible for carrying out the process of scaffolding is diminished. When the interaction is intentional, the specific learning takes place Fisher and Frey [2010]. The intersection between the two fields of knowledge raises the question: How do scaffolding with social robots? According to the literature review, there are not experiences in which the anthropomorphic robots develop scaffolding processes to learn problem-solving strategies. Moreover, some robotics' experiences grounded in scaffolding theory are presented.

Below the investigations in which the robot supports the user in making decisions are presented. Reardon et al. [2015] use the System of Least Prompts methodology for support robot's instructor-behavior and its present the *fading* concept, which is accord with the scaffolding theory to explain the rate of intervention that robot doing in the range of student-learning responsibility. The author describes four increasingly intrusive prompts levels which are related to the feedback verbal conditions, but the robot always has the same role. Giuliani et al. [2010] present a human-robot interaction cooperative strategy, to assemble an artefact. The robot identifies the user mistake and suggests how to correct this. The robot uses the situated reference generation module, which recognizes the environment task properties and analyses the verbal and non-verbal of human behavior. In this experience, the robot and user know the same path to assembly the artifact, but the user was induced to do the mistake to validate the robot behavior. Alves-Oliveira et al. [2014] use a Nao robot which is combined with an intelligent tutor system increase dialogue options. In this regard, the Nao robot has the skill to present intentions to interact with the user thought of gaze, mutual eye contact, and ostensive signals. It has a multi-modal behavior to support the process learning. The robot language was developed through of Behavior Markup Language *BML* this is useful to specify gaze, animations, deictic gestures and speech behavior. Vouloutsi et al. [2016] use the Zeno robot to implement the Distributed Adaptive Control architecture (DAC), a bio-inspired framework that explain how the robot can interact with children between 8-11 years to resolve the *balance beam problem* using the Flipper Dialog Manager. The system uses a tablet device and augmented reality strategy, for doing suggest and support the learning process. Although the problem involves physical movements, the robot does not interact manually with the problem. Reidsma et al. [2016] use the same problem and evaluate for topics: impact of robot behavior and characteristics on perception of robot by student, potential for longer term relation between robot and student, impact of robot on learning outcome and impact of Robot on Learning Process. But there highlights the challenge to evaluate the child-robot interaction. Köse et al. [2015] in some tasks the guide instruction is a *pseudo* guide instruction because the robot only is reactive to human behavior during learning process. Ramacliandran and Scassellati [2014] highlight student's knowledge and learning preferences as foundations for the robot to develop an adaptive support. Although this paper doesn't speak about of scaffold, recognize in two levels the mediation process. Serholt et al. [2015] compare children attitudes when they interact with two tutors, human a robot, while they build a LEGO house; students ask for more help human tutor and when working with the robot, have more eager to finish the task. They learn with robots but is not necessary that the robot act as tutor. Malik et al. [2014] uses a NAO robot with children that have especial educational needs. NAO stimulus the motor learning by imitation. For example, Thien et al. [2016] uses of KUKA robot as learning tool to explain the "Tower of Hanoi" problem.

Different to the previous method by which the robot makes suggestions; now investigations which have evidence of collaborative work between humans and robots are presented. Foster et al. [2008b] presents one strategy to generate multimodal expression from robot while is working with a human partner in a physical task. The method had been named *Haptic Ostensive Reference*. They use an incremental algorithm to describe objects around the environment. This kind of strategy is to improve the collaborative work and looking to solving the communication problems in the cooperative task in HRI field Foster et al. [2008a] recognizes the importance of gestural-deictic reference in human-robot cooperation, and they present the haptic-ostensive reference concept as a strategy to improve the communications process. The research presented has complemented the before proposal, in which the object description process was related to using more semantic descriptors. In the same project, Ziaie et al. [2008] presented a method for recognizing hand expressions through Bayesian models. Rickert et al. [2007] highlight the importance of multimodal communication. In this case, the robot architecture considers modules as face tracker, gaze tracker, hand tracker, speech recognition and object recognition. With this information, the robot represents, interprets and reasons about conceptual strategies to assembly the 3D parts, which are present to the user through physical movements and speech. Chidambaram et al. [2012] present a method to robots can persuade people while both are interacting. The author considers aspect as gaze, proximity, gestures: *iconic, metaphoric, deictic, and vocal cues.* Foster et al. [2009] the task assembly process has two strategies: pre-order or post-order. In the first strategy, the robot labelled the assembly before they gave the instructions to the user. In the second, the robot gave the instructions and when the user finished the task, the robot labelled the assembly. The results show that the first strategy to improve the collaborative work. Kanda et al. [2004] presents the partner robot concept and recognize how this concept have been winning interest. He says as several papers speak about the partner robots as health assister, companion, and assistant. The author also presents the Interactive behaviors concept which considers the hugging, shaking hands, greeting, pointing to an object, as an action to foster the interactive process with children. Burghart et al. [2006] the authors recorded a tutor to a child interaction while they were assembling a puzzle. With this information, the tutor behavior was modeled in the ART-R architecture and tested with an anthropomorphic robot arm with 7 DOF. Additionally, the systems learned to identify human hands, puzzle pieces, deictic gestures, and the context and environment of the task. During the development of the task the robot to user suggestions were accepted or rejected, which allowed to create a base of knowledge about the process of assembling the puzzle. In the same path, Chandra et al. [2015] present a research which Nao has a facilitator role, in this work two children and one robot work collaboratively without a human teacher, the result suggests that if a child has a teacher role and the robot has a facilitator, he act with more responsibility. Jarrassé et al. [2014] presents the motor interaction concept, for speaking about different roles that define the human-robot motor interaction: robotic slaves and partner. In this field is common to find the *Theory of Mind* concept, to explain as people can work cooperatively. Peternel et al. [2014] present the cooperative dynamic manipulation concept to develop a control strategy support in the human sensorimotor learning which the robot can work with the human in the same manipulation task. Here, the learning by demonstration was useful.

The examples associated with guided instruction and collaborative work just presented. Now, experiences that correspond to the final stage of the scaffolding process in which the student assumes greater responsibility for their learning process, are presented. Students can apply the knowledge and skills which they have acquired in the other states to solve problems with the same characteristics. Hood et al. [2015], Okazaki et al. [2015] presented a research in which children uses the learning by teaching strategy for learning about handwriting process. Okazaki et al. [2016] use de Wizard-of-Oz method to research about the user conditions while is teaching a robot. In this case, the robot has a behavior that showed interest in learning which is presented by voice and movements on the tablet device. The author analyzes the voice sound, gaze direction, and hand motion of the subjects while they are explaining a learning task. The *Motionese* phenomenon is highlighted in relation with the user-behavior combinations.

The robot's actions which have been described above, do not consider students' needs and its do not merge information from student, task, and strategy as recommend the scaffolding fundaments which can be considered as essential conditions of artificial cognitive systems that assume the role of a tutor or learner during the learning process, Mateos [2001]. Also, Pea [2004] remark that this kind of frameworks should consider the following features: recognition of the processes in which the student identifies their progress, distinguish different levels of student knowledge, development a path which is characterized by performance levels, establish the relation between the student's status and adaptive support. Winne and Nesbit [2009] also analyzes the technological systems in relation to the information that must be recognized by the tool, aspects in which it beats the student, conditions to foster meta-comprehension and metacognition, strategies that using the error as learning mechanism, strategies to foster the self-regulation and self-efficacy.

Conceptual layer of the proposed architecture

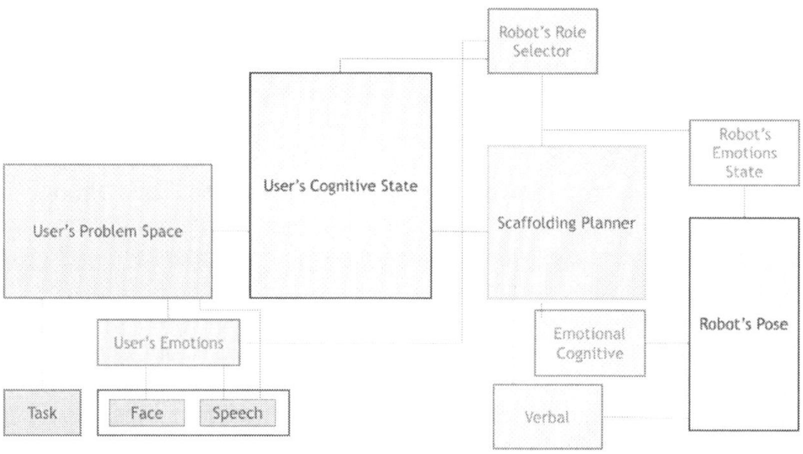

Figure 1: *Human-Robot Scaffolding Architecture.*

The figure 1 describes the conceptual layer of the architecture named BDI-HRS and is grounded by Vernon [2014] who has proposed a methodology to developed artificial cognitive systems, which considers four aspects: the computational/bio-inspired spectrum, the level of abstraction in the biological model, the mutual dependence of brain, body, and environment, and the ultimate-proximate distinction *(why vs how)*. Later proposals about multi-agent systems and BDI (Believe, Desire, Intentions) architecture will be used. The architecture will be implement in the Baxter Robot, figure 2.

Input: in this level, the scaffolding process rises when a mechanical problem is presented. The path in which robot can recognize the student's mental states is through of three characteristics: task, face and speech. The task movements are joined to mechanical blocks as velocity, turns, delays, blocks exchange, grasp and drop the blocks. Emotions linked to the learning process *(eureka, satisfaction, curiosity, confusion, disappointment, frustration and dispiritedness)* and specific words speech which are associated with cognitive judgments.

Interpretation: the interpretation module is grounded in the processing information theory. In this regard, the *memory state* is an attribute of an information processing system which can store and preserving symbols structures, which is composed of symbols and connectors. For example (Block:X, Emotional-State: Happy, Verbal-Judgment: Change-Position). With this data, the robot can recognize the *Memory States*. According to cognitive theories is possible estimating cognitive and metacognitive model's behaviors which allow creating metaphorical and mathematical interpretations. The metaphorical interpretation is known as *descriptive models* and the mathematics how *predictive models*. For example, Guiard developed a protocol to analyze the hand's movement within HCI field.

Literature review shows different methods to cognitive task analysis as think-aloud protocols, content analysis, process isolation and situated studies, hierarchical task analysis, link analysis, operational diagram sequence, time line analysis and GOMS *(Goals, Operators, Methods, Selection Rules)* that describes the *"how to do it"* or procedural knowledge which has been inspired by the information processing theory.

Reasoning: The memory state sequence can show the *thinking states* and the doubts and pauses *(d,p)* correspond to changes in the direction of thought. Each memory state *(Ms)* is joined to a knowledge *(Kw)* about the problem solution process and its are useful to determine the knowledge's student needs about Mean-End Analysis strategy *(transform, reduce, apply, evaluate)* and emotional conditions to interact with him by the robot's role (tutor or peer). The scaffolding process not only depends on student characteristics, also depends on the strategy and the task's characteristics. When the task is developed in technological environments, conditions scaffolding change because the technology offers new opportunities to get the data on which the decisions of support are taken and it can adapt the environment to meet the learning needs. In this sense aspects as details about action, knowledge assessment, time and previous interaction acquires other dynamics. The above aspects open possibilities of generating levels of granularity in the scaffolding process.

Action: The planner coordinates the kinematic, dynamic, visual and audio issues, to present the required conditions to do the pose emotional and physical movements to intervene physically in the mechanical task. The actions of the robot are two: cognitive and emotional. The cognitive actions of the robot guide the development of the task according to the user's learning needs and can be verbal or physical. Physics actions involve the movement of blocks of the robot or the robot's intervention to rearrange the block with the aim of promoting learning. The robot control integrates two platforms: ROS and RobotAct González [2007]. ROS is the native program to control the Baxter robot. RobotAct is a proposal developed at the Pontificia Universidad Javeriana inspired by multi-agent systems paradigm. In RobotAct, robots are considered as a social actor with beliefs, desires, intentions and emotions.

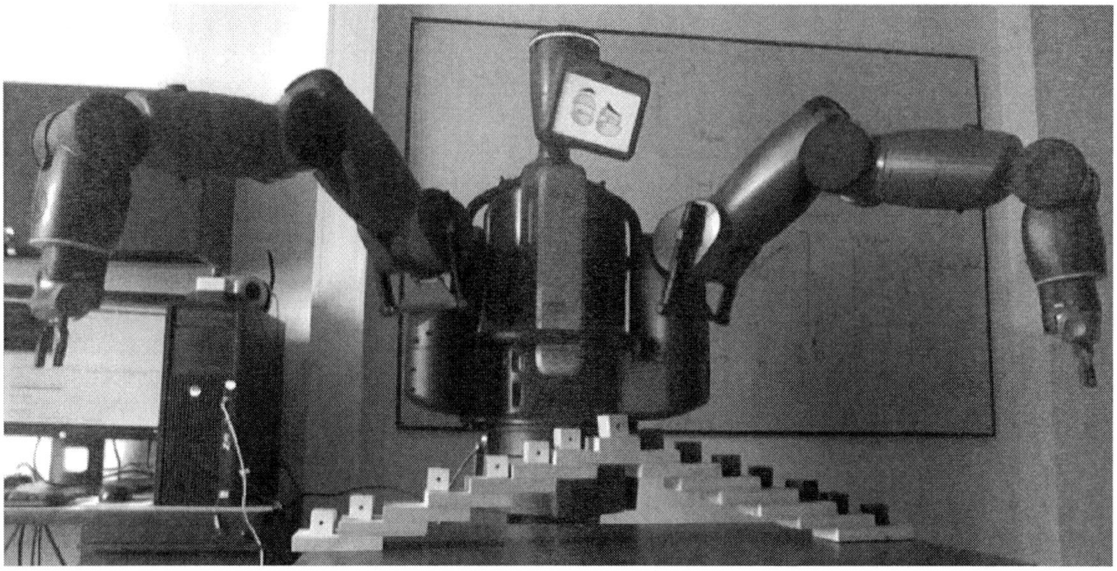

Figure 2: Baxter Robot – Robotics Lab at the Pontificia Universidad Javeriana.

The scaffold module estimates the assertive actions to foster learning. The previous module named the user's cognitive state estimates the student's knowledge about the mechanical task assembly and the strategy as Mean-End Analysis. With the information collected, the scaffolding module decides one of the four intervention possibilities: focus lesson, guided instruction, cooperative work and independent work.

Conclusions

This thesis proposal aims to develop an artificial cognitive system which supports the learning process of the Mean-End Analysis strategy. The architecture joins three aspects: emotional interaction joints to the learning process, physical interaction, and adaptive scaffolding. The proposed plan is divided into three milestones; the theoretical construction about Human-Robot Scaffolding, the developing of an artificial cognitive architecture according to Human-Robot Scaffolding principles and the validation process which use the mixed-method to acknowledge the robot performance relate to technical and psychological issues.

This paper is a brief of doctoral thesis proposal which show the *Human-Robot Scaffolding* concept as a strategy which converge pedagogical and technological issues. The aim is that anthropomorphic robots with handling capacity can participate explicitly while learning process. The novel concept is according to new research tendencies related to robot and education which had been showed as educative field to build electronic and mechanical devices, to check reasoning process and nowadays as physical agents that take over the rol of virtual intelligent tutors.

References

A.-L. Vollmer, M. Mühlig, J.J. Steil, K. Pitsch, J. Fritsch, K.J. Rohlf- ing, and B. Wrede. Robots show us how to teach them: Feedback from robots shapes tutoring behavior during ac- tion learning. *PLoS ONE*, 9(3), 2014. doi: 10.1371/journal. pone.0091349. cited By 7.

Aníbal Colón. Filosofía de la técnica. *Editorial de la Universidad de Puerto Rico, Puerto Rico*, 1992.

Astrid M Rosenthal-Von Der Putten, Frank P Schulte, Sabrina C Eimler, Sabrina Sobieraj, Laura Hoffmann, Stefan Maderwald, Matthias Brand, and Nicole C Kramer. Investigations on empathy towards humans and robots using fmri. *Comput- ers in Human Behavior*, 33:201–212, 2014.

Catherina Burghart, Christian Gaertner, and Heinz Woern. Co-operative solving of a children's jigsaw puzzle between human and robot: First results. In *Cognitive Robotics: Papers from the AAAI Workshop: Papers from the 2006 AAAI Workshop, ed. M. Beetz, K. Rajan, M. Thielscher, and RB Rusu*, pages 33–39, 2006.
David Vernon. *Artificial cognitive systems: A primer.* MIT Press, 2014.
Douglas Fisher and Nancy Frey. *Better learning through structured teaching: A framework for the gradual release of responsibility.* 2013.
Douglas Fisher and Nancy Frey. *Guided instruction: How to develop confident and successful learners.* ASCD, 2010.
Enrique González and César Bustacara. *Desarrollo de aplicaciones basadas en sistemas multiagentes.* Editorial Universidad Javeriana, 2007.
Guanglong Du and Ping Zhang. A markerless human–robot in- terface using particle filter and kalman filter for dual robots. *IEEE Transactions on Industrial Electronics*, 62(4):2257–2264, 2015.
H. Köse, P. Uluer, N. Akaln, R. Yorganc?, A. Özkul, and G. Ince. The effect of embodiment in sign language tutoring with as- sistive humanoid robots. *International Journal of Social Robotics*, 7(4):537–548, 2015. doi: 10.1007/s12369-015-0311-1.
H. Okazaki, K. Hasegawa, Y. Kanai, K. Ishii, M. Ogata, and M. Imai. Building pedagogical relationships between hu- mans and robots in natural interactions. pages 115–120, 2015. doi: 10.1145/2814940.2814941.
H. Okazaki, Y. Kanai, M. Ogata, K. Hasegawa, K. Ishii, and M. Imai. Toward understanding pedagogical relationship in human-robot interaction. *Journal of Robotics and Mechatronics*, 28(1):69–78, 2016. doi: 10.20965/jrm.2016.p0069.
L. Brown and A.M. Howard. Engaging children in math ed- ucation using a socially interactive humanoid robot. volume 2015-February, pages 183–188, 2015. doi: 10.1109/ HUMANOIDS.2013.7029974.
L. Peternel, T. Petri, E. Oztop, and J. Babi? Teaching robots to cooperate with humans in dynamic manipulation tasks based on multi-modal human-in-the-loop approach. *Autonomous Robots*, 36(1-2):123–136, 2014. doi: 10.1007/ s10514-013-9361-0. cited By 9.
M.E. Foster, E.G. Bard, M. Guhe, R.L. Hill, J. Oberlander, and A. Knoll. The roles of haptic-ostensive referring expressions in cooperative, task-based human-robot dialogue. pages 295– 302, 2008b. doi: 10.1145/1349822.1349861. cited By 11.
Manuel Giuliani and Alois Knoll. Using embodied multimodal fusion to perform supportive and instructive robot roles in human-robot interaction. *International Journal of Social Robotics*, 5(3):345–356, 2013.
Manuel Giuliani, Mary Ellen Foster, Amy Isard, Colin Matheson, Jon Oberlander, and Alois Knoll. Situated reference in a hybrid human-robot interaction system. In *Proceedings of the 6th international natural language generation conference*, pages 67–75. Association for Computational Linguistics, 2010.
Mar Mateos. *Metacognición y educación.* Aique Buenos Aires, 2001.
Markus Rickert, Mary Ellen Foster, Manuel Giuliani, Tomas By, Giorgio Panin, and Alois Knoll. Integrating language, vision and action for human robot dialog systems. In *International Conference on Universal Access in Human-Computer Interaction*, pages 987–995. Springer, 2007.
Mary Ellen Foster, Ellen Gurman Bard, Markus Guhe, Robin L Hill, Jon Oberlander, and Alois Knoll. The roles of haptic- ostensive referring expressions in cooperative, task-based human-robot dialogue. In *Proceedings of the 3rd ACM/IEEE international conference on Human robot interaction*, pages 295– 302. ACM, 2008a.
Minchi C Kim and Michael J Hannafin. Scaffolding problem solving in technology-enhanced learning environments (teles): Bridging research and theory with practice. *Computers & Education*, 56(2):403–417, 2011.
N. Jarrassé, V. Sanguineti, and E. Burdet. Slaves no longer: Re- view on role assignment for human-robot joint motor action. *Adaptive Behavior*, 22(1):70–82, 2014. doi: 10.1177/ 1059712313481044. cited By 4.
N.A. Malik, H. Yussof, and F.A. Hanapiah. Development of imitation learning through physical therapy using a humanoid robot. volume 42, pages 191–197, 2014. doi: 10.1016/j.procs. 2014.11.051. cited By 1.
N.D. Thien, A. Terracina, L. Iocchi, and M. Mecella. Robotic teaching assistance for the "tower of hanoi" problem. *International Journal of Distance Education Technologies*, 14(1):64–76, 2016. doi: 10.4018/IJDET.2016010104.
P. Alves-Oliveira, S. Janarthanam, A. Candeias, A. Deshmukh, T. Ribeiro, H. Hastie, A. Paiva, and R. Aylett. Towards dialogue dimensions for a robotic tutor in collaborative learning scenarios. volume 2014-October, pages 862–867, 2014. doi: 10.1109/ROMAN.2014.6926361.
Philip H Winne and John C Nesbit. 14 supporting self-regulated learning with cognitive tools. *Handbook of metacognition in education*, page 259, 2009.
Pujan Ziaie, Thomas Muller, and Alois Knoll. A novel approach to hand-gesture recognition in a human-robot dialog system. In *2008 First Workshops on Image Processing Theory, Tools and Applications*, pages 1–8. IEEE, 2008.
R. Aylett, A. Kappas, G. Castellano, S. Bull, W. Barendregt, A. Paiva, and L. Hall. I know how that feels - an empathic robot tutor. 2016. doi: 10.1109/eCHALLENGES.2015. 7441088.
Ramacliandran and B. Scassellati. Adapting difficulty lev- els in personalized robot-child tutoring interactions. volume WS-14-07, pages 56–59, 2014.
Reardon, H. Zhang, R. Wright, and L.E. Parker. Response prompting for intelligent robot instruction of students with intellectual disabilities. volume 2015-November, pages 784– 790, 2015. doi: 10.1109/ROMAN.2015.7333651.
Reidsma, V. Charisi, D. Davison, F. Wijnen, J. van der Meij, V. Evers, D. Cameron, S. Fernando, R. Moore, T. Prescott, D. Mazzei, M. Pieroni, L. Cominelli, R. Garofalo, D. de Rossi, V. Vouloutsi, R. Zucca, K. Grechuta,

M. Blancas, and P. Ver- schure. The easel project: Towards educational human- robot symbiotic interaction. *Lecture Notes in Computer Sci- ence (including subseries Lecture Notes in Artificial Intelligence and Lecture Notes in Bioinformatics)*, 9793:297–306, 2016. doi: 10.1007/978- 3- 319- 42417- 0_27.
Richard E Richard E Mayer et al. *Pensamiento, resolución de prob- lemas y cognición*. Number 153.43 M3. 1986.
Roy D Pea. The social and technological dimensions of scaffold- ing and related theoretical concepts for learning, education, and human activity. *The journal of the learning sciences*, 13(3): 423–451, 2004.
S. Chandra, P. Alves-Oliveira, S. Lemaignan, P. Sequeira, A. Paiva, and P. Dillenbourg. Can a child feel responsible for another in the presence of a robot in a collaborative learn- ing activity? volume 2015-November, pages 167–172, 2015. doi: 10.1109/ROMAN.2015.7333678.
S. Serholt, C.A. Basedow, W. Barendregt, and M. Obaid. Com- paring a humanoid tutor to a human tutor delivering an in- structional task to children. volume 2015-February, pages 1134–1141, 2015. doi: 10.1109/HUMANOIDS.2014.7041511.
Takayuki Kanda, Takayuki Hirano, Daniel Eaton, and Hiroshi Ishiguro. Interactive robots as social partners and peer tutors for children: A field trial. *Human-computer interaction*, 19(1): 61–84, 2004.
V Nitsch and M Popp. Emotions in robot psychology. *Biological cybernetics*, 108(5):621–629, 2014.
V. Vouloutsi, M. Blancas, R. Zucca, P. Omedas, D. Reidsma, D. Davison, V. Charisi, F. Wijnen, J. van der Meij, V. Evers, D. Cameron, S. Fernando, R. Moore, T. Prescott, D. Mazzei, M. Pieroni, L. Cominelli, R. Garofalo, D. de Rossi, and P.F.M.J. Verschure. Towards a synthetic tutor assistant: The easel project and its architecture. *Lecture Notes in Computer Sci- ence (including subseries Lecture Notes in Artificial Intelligence and Lecture Notes in Bioinformatics)*, 9793:353–364, 2016. doi: 10.1007/978- 3- 319- 42417- 0_32.
Vijay Chidambaram, Yueh-Hsuan Chiang, and Bilge Mutlu. Designing persuasive robots: how robots might persuade people using vocal and nonverbal cues. In *Proceedings of the sev- enth annual ACM/IEEE international conference on Human-Robot Interaction*, pages 293–300. ACM, 2012.

Economic expectations and their effects on college career choice in Colombia

Juan Carlos Morales Piñero

Universidad Sergio Arboleda, Bogotá, Colombia

juancar.morales@correo.usa.edu.co

Nestor Orlando Cordero Saenz

Universidad Sergio Arboleda, Bogotá, Colombia

nocordero@yahoo.com

José Daniel Ramírez Camargo

Universidad Sergio Arboleda, Bogotá, Colombia

danielramirezcam@gmail.com

***Abstract**: In this research, we analyze the playing role by growth in the vocational careers supply, salary expectations and reputation of the university on the decision to choose Industrial Engineering as a professional alternative. The study was conducted using secondary information on wages, population and enrollment information in Colombia; this analysis covers the years 2007-2015. The main conclusions reveal that the average wage received by university graduates tends to increase in a greater proportion than that of graduates of vocational careers. Although there are not significant differences in the wages of the recent graduates in industrial engineering, graduates from the most expensive universities tend to earn significantly higher wages than graduates of other universities. However, the high cost of education involves rates of return much lower than those of universities cheaper.*

Introduction

The transition from secondary education to higher education involves for students a decision that defines their occupational future in their adolescence (Villada et al., 2002). Given the psychological insecurity that the student faces, this decision may be influenced by factors such as family beliefs, social reputation, salary expectations, among others (Ghansah et al., 2016; Martin, Simmons, & Yu, 2014; Taylor, 2007).

In this sense, (Carnasciali, Thompson, & Thomas, 2013) analyze the influence of socializers, self-identified competence, and media sources on choosing a career and indicate that gender and parental educational achievement levels affect students' choice of engineering. As well, (Piñero Ramírez, 2015) in his research on the Rational Action Theory concludes that individual characteristics (e.g., social origin, sex and school history) are the main factor that determines the career choice. Specifically in Bogota, we find the study of (Pineda Barón, 2015) who using a Logit Multinomial Model, concludes that this the career choice depends on the genre, mother's education, family income, and the results of Saber 11 Test and wage expectations.

On the other hand, when the option of studying at a public university is not available, families need financial assistance to pay the cost of higher education; this involves the uncertainty of whether or not such investment will be worthwhile. In addition, there is a latent question—

tacitly or explicitly—of how many years the student will be attending higher education before getting into the job market.

Likewise, choosing a specific university can be an important decision for the candidate. In this context, we find the study of (Dunnett, Moorhouse, Walsh, & Barry, 2012) who examines the impact of fee changes on how students weigh up their university choices. They affirm that "students from families where there is no history of attending university will experience more disutility from the higher fees" (p.199)

Specifically, in the case of young people who wish to study Industrial Engineering, several questions can be arising such as follows: should they pursue a vocational or a university career? Which university should they choose? Will become a worthwhile investment? In this research, we analyze the influence of an increase in the offer of vocational careers, student´s salary expectations and the reputation of the university on the decision to choose Industrial Engineering as a professional alternative in Colombia.

Methodology

The study was made using secondary information of private Colombian universities obtained from (Observatorio Laboral -MEN, 2015), looking for salary information, from (DANE, 2015) for population statistics, and from (SNIES- MEN, 2015) for enrollments information. We analyzed the available data for the period 2007 – 2015.

The variables analyzed were: number of enroll students, career, number of industrial engineering programs, universities, affordability, graduate incomes, wage increase per year of experience, employability rate.

The first stage consisted of evaluating the growth of vocational careers in relation to university careers during the study period in order to evaluate the trend in the preference of high school graduates.

Later, the study focused on the area of industrial engineering and we compare the average wage (in current Colombian pesos COP) of a graduate in vocational career with the average wage of a university graduate. In this case, the variable wage was built on the average monthly salary reported each year and it was adjusted based on the years of experience of the graduate.

The next stage of the analysis was focused on the students enrolled in Industrial Engineering in 51 private Colombian universities for which complete information was available. For this purpose, the universities were grouped by cluster analysis based on the tuition cost of the year 2014. The cluster analysis was applied using the SPSS software and three groups/clusters were defined: 12 universities were grouped as the most expensive, 22 universities at intermediate cost and 17 universities as economic.

Subsequently, students' preference for these three types of universities was analyzed according to the number of students who applied, were admitted and enrolled in Industrial Engineering by universities over the study period. This was done to compare the salary differences in relation to the cost of the tuition of the institution where students attended the course.

Finally, an economic analysis of the alternative of studying Industrial Engineering was carried out in a horizon of 10 years since the student enrolls in the career until he completes 5 years working as a graduate; that is, we assume a scenario of 5 years studying and 5 years working. The investment in the studies was calculated on the basis of the tuition fee reported by the universities for the year 2014 and this value was adjusted annually assuming an annual inflation rate of 5%. Revenues were calculated since the sixth year on the basis of the average salary of a graduate of the year 2014 adjusted for inflation in the previous 5 years. From year 7, the salary was annually adjusted based on the growth rate of the experience curve. The monetary values are expressed in Colombian Pesos (COP).

The discount rate used was the lowest market charged by the (ICETEX, 2017) that was 3.66% for the year 2014. This rate only applies to poor students, who technically could only apply to affordable universities. However, to carry out the study under equal conditions, in all cases ICETEX rate was applied. After calculating the cash flows for each year and university, the net present value and the internal rate of return on investment were calculated.

Results

To carry out the study, the first step was to examine the evolution of access to higher education in Colombia. For that, the population enrolled in universities and in vocational careers in the period 2007-2015 was evaluated. This population was contrasted with the population of the country between the ages of 17 and 21, which are the age at which young people enter higher education (see Table 1).

Table 1: Coverage of higher education in Colombia

	Enrolled per year				
Year	Vocational education	University Education	Total Enrolled per year	Population between 17 and 21 years	Coverage rate
2007	395.437	910.228	1.305.665	4.125.881	31,65%
2008	463.980	963.167	1.427.147	4.187.317	34,08%
2009	498.382	1.015.608	1.513.990	4.241.585	35,69%
2010	542.627	1.045.133	1.587.760	4.284.916	37,05%
2011	586.471	1.159.512	1.745.983	4.319.415	40,42%
2012	593.684	1.218.816	1.812.500	4.342.603	41,74%
2013	670.930	1.296.123	1.967.053	4.354.649	45,17%
2014	711.291	1.369.149	2.080.440	4.356.453	47,76%
2015	717.521	1.431.983	2.149.504	4.349.823	49,42%
Source: SNIES - MEN, DANE					

With this information was calculated the rate of coverage of higher education in Colombia differentiated by vocational and university education. The results shown in figure 1 indicate that in the group of people who opt for higher education, the university remains in time as the principal option.

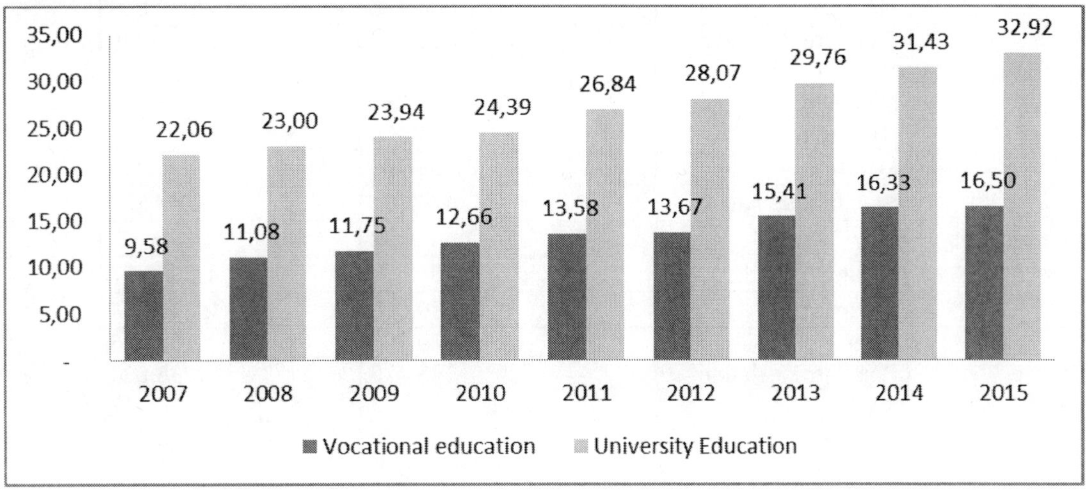

Figure 1: Coverage rate by Institution

This information allowed the analysis of the evolution in average salaries of graduates of vocational careers and of universities in areas related to industrial engineering during the study period (see Figure 2). The analysis shows that not only the salaries of university graduates are much higher than that of vocational career graduates, but that this tends to increase in a greater proportion over time. This can be an element that discourages the preference for vocational careers.

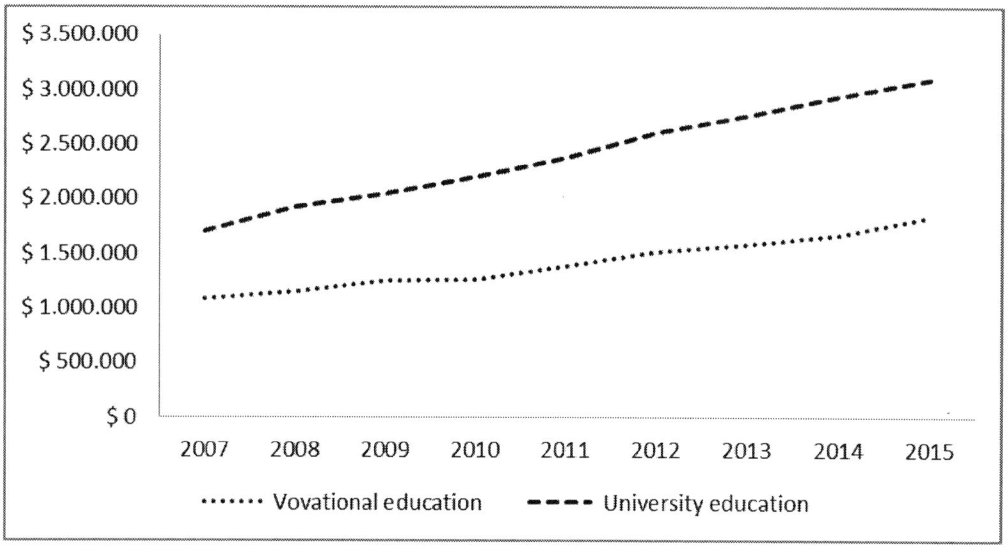

Figure 2: Undergraduate average wage in COP

After analyzing the differences between vocational and university graduates, the focus was on evaluating the behavior of the market segment that is inclined to study Industrial Engineering in Colombia in private universities. To do this, a cluster analysis was initially carried out to classify universities according to the value of tuition for 2014 (see Table 2).

Table 2: Industrial engineering career by university tuition fee

Cluster Number		**Average Tuition fee in 2014 ***	**Average wage in 2014 ***
More Expensive	Mean	7.064.476	2.033.856
	Standard deviation	2.302.656	290.245
	N	12	12
	Minimum	5.057.350	1.573.126
	Maximum	13.144.000	2.419.651
Intermediates	Mean	3.788.760	1.698.348
	Standard deviation	470.704	267.856
	N	22	22
	Minimum	3.184.000	1.300.667
	Maximum	4.780.000	2.440.914
Lower Cost	Mean	2.479.698	1.766.226
	Standard deviation	348.473	288.684
	N	17	17
	Minimum	1.599.950	1.205.750
	Maximum	2.910.000	2.147.820
* Values in COP			

On the basis of the classification, the more expensive group of universities stands out the most because they have the greatest dispersion in the data, since the value of the standard deviation represents more than one-third of the mean. Once the classification was made, the evolution of the Industrial Engineering stakeholders in each type of university was analyzed, separated by phases of the process. It is important to stress the fact that in Colombia for applicants to enter a university must submit their application to be evaluated by examinations and / or interviews. Once that phase passes, the candidate is advised of the admission to the program, but not all those who are admitted eventually enroll because it usually happens that they apply to several universities.

Table 3: Number of Industrial Engineering students by process stage

University by tuition fee	Students by process	2007	2008	2009	2010	2011	2012	2013	2014	2015
More Expensive	Students who applied	4924	5169	5458	4877	5276	5598	5495	5465	6385
	Acknowledged Students	3014	3632	3747	3180	3947	4047	4174	4174	4473
	Enrolled Students	2320	2126	2384	2209	2290	2289	1995	2106	2476
Intermediates	Students who applied	4482	6218	6014	6091	6968	7247	9082	8601	8487
	Acknowledged Students	3742	5170	4688	4619	5872	5179	6054	6133	6620
	Enrolled Students	3483	4024	3694	3569	4873	4123	4733	4799	4836
Lower Cost	Students who applied	2070	1933	2648	2738	3428	3803	3707	4371	4202
	Acknowledged Students	1943	1659	2340	2354	2903	2871	3031	3309	3108
	Enrolled Students	1411	1800	1549	1608	1873	2114	2496	2782	2740

In figure 3 the evolution of the number of Industrial Engineering students during the study period classified by the tuition fee is showed. It is clear that universities at a cost of intermediate tuition cost capture the largest proportion of the market and maintain a positive trend. It is also observed that the cheaper universities maintain a growing trend, and 2013 began to capture a greater proportion than the more expensive universities. Besides, the most expensive universities have been stable in the time managing between 2000 and 2500 students per year.

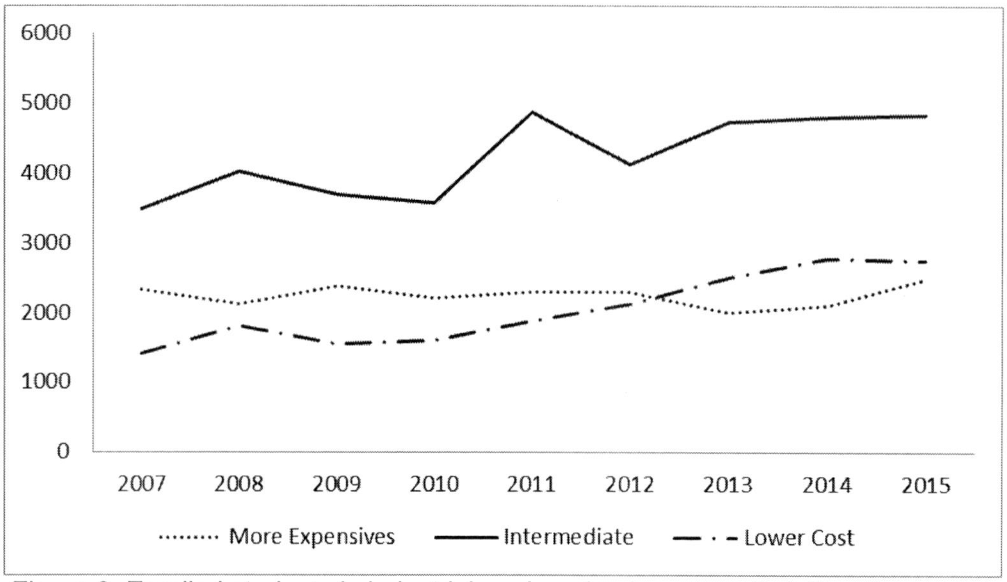

Figure 3: Enrolled students in industrial engineering by university tuition fee

On the other side, from the information previously shown in Table 2, it is observed that the average salary received by the new graduate for the year 2014 is not necessarily directly related to the cost of enrollment. Also, there is no significant difference in the average salary of the recent graduate. To evaluate the behavior over time of this variable, we performed this analysis during the study period but comparing the average salary for the year 2015 (Y axis) depending on the year of graduation (X axis) (see Figure 4).

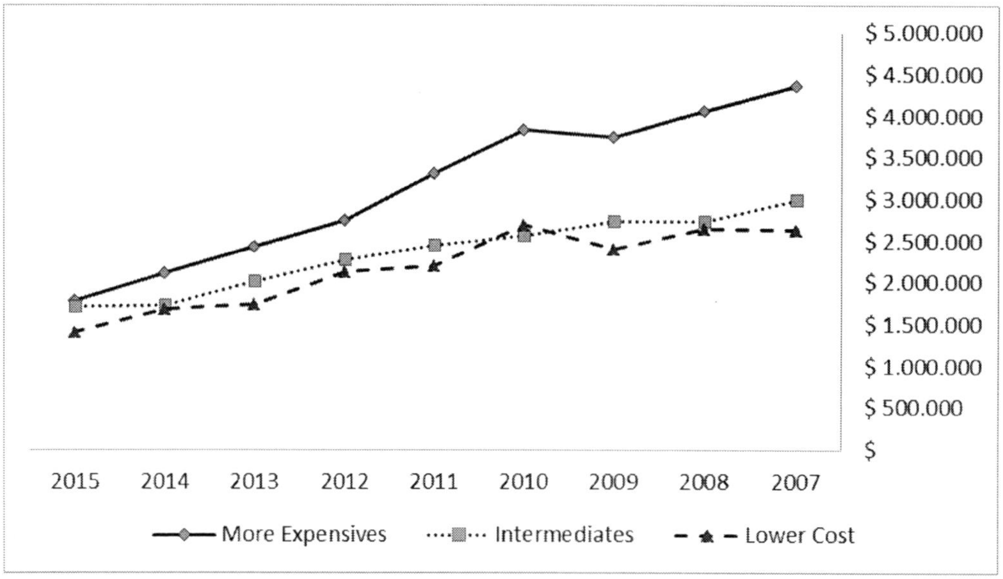

Figure 4: Wage in 2015 by year of graduation in COP

As showed in figure 4, there are no significant differences in the salaries of the recent graduates, but as time goes on, graduates from more expensive universities tend to earn significantly higher salaries than graduates of other universities. On the other hand, graduates of universities with intermediate or cheap tuition fees tend to receive equivalent salaries as the years pass.

In order to evaluate the economic scenario that has a candidate of Industrial Engineering in Colombia based on the information analyzed so far, an economic analysis of the investment that a person has to make in his career was made, taking as a temporal horizon 10 years from the beginning of study. The cash flows obtained, the net present value of the investment and the internal rate of return were calculated as showed in table 4.

Table 4: Economic expectation for an industrial engineering in Colombia

		University by tuition fee		
		More Expensive	**Intermediates**	**Lower Cost**
Average tuition fee in COP (2014)		7.064.476	3.788.760	2.479.698
Tuition fee*		12,07	6,15	4,02
Average wage in COP (2014)		2.033.856	1.698.348	1.766.226
Employability rate		87,51	84,03	82,27
NPV** in COP (10 years)		79.167.418	86.999.412	104.934.732
IRR *** in COP (10 years)		20,98%	30,98%	43,51%
Tuition fees during the career in COP	Year 1	-14.128.952	-7.577.520	-4.959.396
	Year 2	-14.835.399	-7.956.396	-5.207.366
	Year 3	-15.577.169	-8.354.216	-5.467.734
	Year 4	-16.356.028	-8.771.927	-5.741.121
	Year 5	-17.173.829	-9.210.523	-6.028.177
Income in the first graduate years in COP	Year 6	31.149.279	26.010.848	27.050.425
	Year 7	35.487.594	29.633.509	30.817.872
	Year 8	40.712.756	33.996.720	35.355.469
	Year 9	44.491.848	37.152.408	38.637.280
	Year 10	48.235.981	40.278.904	41.888.732
* In number of minimum wage in 2014; ** Net Present Value; *** Internal Rate of Return				

As can be noted, although the people who study industrial engineering in the most expensive universities earn significantly higher wages over time than the rest of the graduates, the high investment involved in the studies makes the return rates much lower than the rest.

It is important to note here that if a differentiated rate is applied to the market, the gap would be much higher, perhaps leading to economic advice do not recommend to study in an expensive university. Finally, it can be observed that the most economically profitable option is offered by studies in the lower costs universities. Besides, it is observed that the employability rate is equivalent to any of the universities.

Conclusions

From the research carried out, it can be concluded that although the number of students enrolled in vocational careers tends to grow in proportional form to those enrolled in universities, the average wage received by university graduates tends to increase in a higher proportion in the long-term.

It is also observed that universities with an intermediate tuition cost capture the largest proportion of the market and hold an increasing trend throughout the study.

Although there are no significant differences in the wages of just industrial engineering graduates, graduates of more expensive universities tend to earn significantly higher wages than graduates of other universities over time. However, the high cost of studies in those universities makes the investment have rates of return much lower than the rest of universities.

References

Carnasciali, M.-I., Thompson, A. E., & Thomas, T. J. (2013). Factors influencing students' choice of engineering major. In ASEE Annual Conference and Exposition, Conference Proceedings (p. 36).

DANE. (2015). Proyecciones de población. Retrieved February 4, 2017, from http://www.dane.gov.co/index.php/estadisticas-por-tema/demografia-y-poblacion/proyecciones-de-poblacion

Dunnett, A., Moorhouse, J., Walsh, C., & Barry, C. (2012). Choosing a University: A conjoint analysis of the impact of higher fees on students applying for university in 2012. Tertiary Education and Management, 18(3). http://doi.org/10.1080/13583883.2012.657228

Ghansah, B., Benuwa, B.-B., Ansah, E. K., Ghansah, N. E., Magama, C., & Ocquaye, E. N. N. (2016). Factors that influence students' decision to choose a particular university: A conjoint analysis. International Journal of Engineering Research in Africa, 27, 147–157. http://doi.org/10.4028/www.scientific.net/JERA.27.147

ICETEX. (2017). Crédito a largo plazo - 0%. Retrieved March 1, 2017, from https://www.icetex.gov.co/dnnpro5/es-co/créditoeducativo/pregrado/cortoplazo100porciento.aspx

Martin, J. P., Simmons, D. R., & Yu, S. L. (2014). Family roles in engineering undergraduates' academic and career choices: Does parental educational attainment matter. International Journal of Engineering Education, 30(1).

Observatorio Laboral -MEN. (2015). Consultas Avanzadas - liferay.com. Retrieved February 4, 2017, from http://bi.mineducacion.gov.co:8380/eportal/web/men-observatorio-laboral/consultas-avanzadas

Pineda Barón, L. A. (2015). Factores que afectan la elección de carrera: caso Bogotá. Vniversitas Económica, 15(3), 1–35.

Piñero Ramírez, S. L. (2015). Factores asociados a la selección de carrera: una aproximación desde la Teoría de la Acción Racional. Revista de Investigación Educativa, 20(enero-junio), 72–99.

SNIES- MEN. (2015). Sistemas de información. Retrieved February 4, 2017, from http://www.mineducacion.gov.co/sistemasdeinformacion/1735/w3-article-212400.html

Taylor, D. (2007). Employment Preferences and Salary Expectations of Students in Science and Engineering. BioScience. http://doi.org/10.1641/B570212

Villada, C. A., Lopez, L. M., Patiño, C. D., Ramirez, W., Jurado, C. M., & Ossa, J. (2002). Factores asociados a la elección de carrera y universidad. Unipluriversidad, 2(2), 23–30.

A previous version of this paper was accepted to be published in the Revista Espacios

Copyright statement

Faculty Perception Before and After Implementation of Standards-Based Grading

Eunsil Lee & Adam Carberry
Arizona State University, Mesa, AZ, United States
elee78@asu.edu; adam.carberry@asu.edu

Heidi Diefes-Dux
Purdue University, West Lafayette, IN, United States
hdiefes@purdue.edu

Sara Atwood
Elizabethtown College, Elizabethtown, PA, United States
atwoods@etown.edu

Matthew Siniawski
Loyola Marymount University, Los Angeles, CA, United States
Matthew.siniawski@lmu.edu

Abstract: *An effective transition to a new grading system requires a process of reflecting on experiences to understand and reframe perceptions of grading. Standards-based Grading (SBG), an alternative to the traditional, summative score-based grading, was introduced to higher education engineering faculty either through a workshop or as a direct participant of a larger research project. These faculty were surveyed to better understand the perceptions faculty hold before and after implementation of SBG. Perceived student gains, barriers, and best practices were solicited using an open-ended survey of faculty who had attended the workshop on SBG, were preparing to implement SBG, or had implemented SBG for the first time. Themes among the different groups emerged, providing insights into faculty perceptions at different phases of SBG implementation. The study also demonstrated the necessity of faculty support structures to establish an environment favorable for widespread grading reformation.*

Context

Grading systems have been used for assessing students since the late 1700s (Postman, 1992). Such systems involve the collection and evaluation of evidence for the purposes of showing how well students achieve the established learning objectives over a certain period of time, communicating achievement to other interested parties, providing information to students for self-evaluation, providing incentives for students to learn, documenting student performance to evaluate the effectiveness of instructional programs, and to select, identify, or group students for certain educational pathways or programs (Guskey & Pollio, 2012; Muñoz & Guskey, 2015). Use of a single grading method (e.g., percentage, letter, norm-referenced, mastery-based, or narrative) does not achieve all purposes (Brookhart, 2010). Consequently, educators have been asked to select and develop the most adequate approach based on identified purposes for grading in their educational contexts (Guskey & Pollio, 2012).

Summative score-based grading models have long been the tradition and remain widely used internationally in higher education (Suskie, 2004). Such an approach relies on assigning an end-of-semester letter grade based on how well students performed on completing separate assignments rather than how well they learned specific course objectives (Angelo & Cross, 1993; Sadler, 2005). The resulting grade fails to communicate what is intended due to a lack of richness in explaining students' learning progress relative to specific learning criteria (Brookhart & Nitko, 2008; Guskey, 2011). A reason for using this approach may have roots in one of the oldest misunderstandings about grading – that grades should serve to differentiate students based on demonstrated performance (Guskey, 2011). This idea disagrees not only with the primary function of grading, but also with the goal of educators to develop, not to select, talent by assessing students' learning progress (Guskey & Bailey, 2001).

Efforts made to reform traditional grading practices have suggested applying grading systems with greater focus on monitoring students' development of articulated standards (Guskey & Bailey, 2001) and providing appropriate feedback (Armacost & Pet-Armacost, 2003). Systems that monitor student learning of what students have and have not learned, paired with regular and specific feedback, enables a dynamic relationship between students and educators that can lead to improved sound assessment of student learning (Bloom, 1968; Guskey, 1997). One alternative approach that integrates these actions is to directly measure the quality of students' proficiency with well-defined course objectives (Sadler, 2005), heretofore referred to as ***Standards-based grading (SBG)***.

SBG was first developed during the 1990s when all US states reformed public K-12 education by setting academic standards (Marzano, 2011; Reeves, 2003). The system is based on backward course design, which establishes clearly articulated course objectives at the beginning of the course with a system of well-developed rubrics (Post, 2014). Student strengths and weaknesses toward the objectives are monitored and assessed using student work throughout the duration of a course. Personalized and meaningful feedback is provided to each student allowing them to track individual learning and development. The system has been shown to lead students toward a learning-orientation, rather than a grade-orientation, by giving them the opportunity to self-evaluate and self-learn (Atwood, Siniawski & Carberry, 2014). This action redefines their role as a learner in the learning process (Heflebower & Hoegh, 2014).

Although SBG is used frequently in K-12 education, higher education has been slow to adopt this grading approach due to cultural attitudes and grading system limitations. Studies of SBG in K-12 educational environments suggest that a transition to SBG requires a process of reflection, new learning, and changes in practice for educators and students to reframe held beliefs about grading (Heflebower & Hoegh, 2014). Our overall work is an effort to better understand the perceptions held by students and faculty in order to inform extended use of this non-traditional system into higher education, specifically engineering classrooms. Previous investigations based on available theory and evidence obtained in K-12 learning environments (Sadler, 2005) have revealed a positive influence of SBG on students' perceptions and behaviors toward the SBG system (Atwood, et al., 2014; Siniawski, Carberry & Dionisio, 2012; Vaishnav & Carberry, 2017). SBG also improved students' domain-specific self-efficacy and helped students develop more sophisticated beliefs about STEM knowledge (Atwood, et al., 2014; Carberry, Siniawski & Dionisio, 2012; Siniawski, et al., 2012). An identified barrier to propagation of SBG is buy-in from faculty practitioners. We first investigated expert faculty perceptions of SBG from among those who had piloted SBG in their courses for multiple years (Carberry, Siniawski, Atwood & Diefes-Dux, 2016).

This study extends our previous work by investigating faculty who are interested in, preparing for, or conducting their first adoption of this new grading system into their classrooms. The goal for this analysis is to identify how these novice SBG implementers perceive student gains, barriers for implementation, and identified best practices for implementing SBG. We aim to discuss the similarities and differences among faculty at different points in their adoption by comparing their perceptions.

Research Methods

Sample/Data Collection

Faculty participating in a variety of SBG interventions were surveyed to investigate their perceptions of SBG. Three faculty groups were included in this study. Group 1 included faculty who participated in a 2016 ASEE workshop (labeled as *After Workshop Group*). An open-ended paper survey was administered immediately after the workshop to 39 faculty. The survey examined perceived student gains, expected barriers for implementing SBG, and their perceived ability to implement SBG after attending the workshop. All 39 faculty came from institutions across the United States and voluntarily participated in the workshop. This group had little, if any, prior experience with SBG implementation.

Groups 2 and 3 included faculty who were part of an initial cohort tasked with preparing and implementing SBG within their engineering courses during the Fall 2016/Spring 2017 academic year. The cohort consisted of 20 faculty from 19 institutions who applied and were selected during Summer 2016. Selection was based on the applicant's reasons for wanting to implement SBG within their courses regardless of past experiences with SBG. The cohort has participated in activities such as an online community (Standard-based Grading Hub/ www.sbghurb.lmu.build), support groups, and webinars provided by the research team who share their experiences using SBG. An open-ended electronic survey was administered to this cohort of faculty in Fall 2016 using Qualtrics; 18 of the 20 faculty responded. Those faculty who had not yet implemented SBG, but were preparing for implementation in the Spring (n = 10) were placed in Group 2 (labeled as *Pre-Implementation Group*). The remaining faculty (n = 8) who implemented SBG prior to the survey were placed in Group 3 (labeled as *Post-Implementation Group*). The survey included questions regarding faculty perceptions of student gains, barriers for implementation, and best practices of SBG, based on their experiences and/or expectations. Course types where SBG was planned or implemented varied, including design-based project courses and technical courses in diverse STEM disciplines.

Data Analysis

An emergent open-coding approach was taken to identify themes in the data (Miles & Huberman, 1994). Direct quotes from the raw data investigating perceived student gains, barriers, and best practices were grouped and compiled into overarching summary statements. The *After Workshop Group* was not coded for best practices since they had no previous experience preparing for or applying SBG in a course. Instead, this cohort was coded simply on whether they felt they would be able to implement SBG following the workshop. Two raters independently analyzed responses and created a succinct list of emergent themes in each category by grouping the salient utterances. A final discussion among the raters was made to establish inter-rater reliability and determine agreement for a final list of emergent themes in each category. Similarities and differences in faculty concerns or expectations at different implementation points were investigated by comparing the emergent themes within and across the groups.

Results

After Workshop Group

Emergent themes from the surveys included eight areas of perceived student gains and seven barriers to implementation.

Perceived student gains

1. Opportunity for students to self-evaluate/assess personal achievement by identifying strengths and weakness throughout the course.

2. Clear understanding of what was learned in the course and what needs to be learned beyond the course.
3. Transparency in the grading system that can provide a better connection between learning and the grade.
4. Effective, clear, and timely feedback that can motivate student action.
5. Rigorous evaluation of student progress.
6. Lower student anxiety toward their grade.
7. Alignment of learning objectives with ABET criteria.

Perceived barriers

1. Time and effort required of faculty to prepare and implement SBG (e.g., defining and developing learning objectives and rubrics, grading assignments, providing detailed feedback, and keeping a gradebook).
2. 'Buy-in', acceptance and understanding from students, faculty, departments and/or other administrators.
3. Impact on course-specific features (e.g., department-guided courses, co-teaching, large class sizes, and open-ended courses).
4. Integration with learning management systems.
5. Lack of resources (e.g., teaching assistants) for data entry and evaluation.
6. Training others (e.g., faculty and teaching assistants) to use SBG.

The majority of workshop participants (77%) felt they would be able to implement SBG following the workshop; 15% believed they could not and 8% were unsure.

Pre-Implementation Group

Emergent themes from the surveys included six areas of perceived student gains, four barriers to implementation, and five areas of best practices.

Perceived Student Gains

1. Clear and effective understanding of why, how, and what they are learning.
2. Opportunity to shift from being merit-based to competency-oriented, allowing students to be authentic in their learning and to contextualize their knowledge.
3. Meta-learning enabled by the opportunity to self-assess/evaluate.
4. Improved learning habits and preparation for future learning (e.g., life-long learning).
5. Trust afforded to students that they are being evaluated on their development.
6. Leveling the playing field for underrepresented minorities by focusing on the quality of learning.

Perceived Barriers

1. Initial workload to prepare (e.g., developing rubrics and embedding within a learning management system) and time to implement or grade.
2. Faculty concerns on the uncertainty and lack of practice with implementing such a system.
3. 'Buy-in' from others including students, faculty, and administrators.
4. Consistency of grading across multiple graders in a course.

Best Practices

1. Adjust and align learning objectives with assignments and rubrics to be applicable to the specific course.
2. Prepare SBG with the aid of the SBG community activities (e.g., asking for advice,

reviewing best practices, and sharing practical tips).

3. Provide rubrics to students prior to assignment submissions.
4. Train teaching assistants in how to use SBG.
5. Seek support from other faculty, departments, or other administration.

Post-Implementation Group

Emergent themes from the surveys included eight areas of perceived student gains, six barriers to implementation, and seven areas of best practices.

Perceived Student Gains

1. Effective student self-evaluation and an opportunity to be self-directed through clear learning objectives and progress updates.
2. Better understanding of course learning objectives with a focus on the quality of work.
3. Faculty possess a better understanding of student learning and areas of struggle.
4. Meaningful, clear, and transparent grading that allows students to identify areas for improvement.
5. In-depth feedback provided to students on their progress.
6. Gained experience with performance management criteria, i.e., learning objectives, used in industry
7. Reduction in student anxiety about grades.
8. Increased enjoyment toward the course.

Perceived Barriers

1. Difficulties in grading management (e.g., weighting, conversions, late penalties, extra credit).
2. Challenges in differentiating between barrier levels of achievement and different levels of mastery.
3. Difficulties in data management (e.g., data tracking) needed for students to view progress on learning objectives.
4. Identifying how to implement the system in different types of courses.
5. Time to check consistency among multiple graders.
6. Infrastructure support for SBG at small institutions.

Best Practices

1. Establish detailed learning objectives, build rubrics, and assess assignments to properly demonstrate each learning objective.
2. Provide learning objectives to students at the beginning of the course and rubrics when activities are assigned.
3. Grade performance based on the quality of work through the use of rubrics.
4. Develop and modify the system to be applicable to specific courses using prior experiences.
5. Identify tools (e.g., learning management systems) to compile, track, and provide scores and feedback to students.
6. Gradually expand the use of SBG in a course, i.e., start with assignments and work up to exams.

Discussion and Implications

This study built on our previous work investigating the impact of SBG on student learning in higher education, specifically within engineering education environments. The goal of this study was to expand our understanding of faculty perceptions to include those interested in, preparing for, or implementing SBG for the first time. Results of this study are intended to be a useful resource for designing faculty support mechanisms to propagate SBG with the intention to improve higher education assessment.

Perceived Student Gains

The perceived student gain commonly identified among all three groups was students' self-regulated learning enabled by evaluating their own learning process, which is also noted in research conducted on K-12 environments (Guskey, 2011; Postman, 1992; Wormeli, 2006). This finding is likely due to the explicit set of learning objectives and timely feedback on the objectives required for SBG (Wiggins & McTighe, 1998). These components allow students to clearly understand why, how, and what they are learning and provide the opportunity to monitor their personal progress against each objective (Atwood et al,, 2014). Meta-learning associated with this process can intentionally establish habits of learning, i.e., fail early and improve later, which may guide students to take control of their own learning (Iran-Nejad & Chissom, 1992; Paris & Paris, 2001). This result aligns with our previous data from expert faculty (Carberry et al., 2016) and potentially implies that faculty acknowledge a positive impact of SBG toward student learning, regardless of their implementation of SBG.

Respondents across the three groups reported a number of different student gains not identified by experts. These gains were mostly presented with a positive expectation toward students, but in some cases were reported with uncertainty. Some of these insights are recognitions that SBG can shift student and faculty perceptions of how grading can impact learning, attitudes, and classroom culture. The transparent nature of SBG affords students with an opportunity to connect to the grading system, which can lead to attitude shifts and lowered anxiety. Additional gains for students came in the way of faculty awareness. Faculty reported that SBG helped them in identifying areas where students were struggling and allowed for better alignment of learning objectives with ABET. This overall approach was believed to benefit all students, particularly underrepresented minorities, and allowed students to experience an assessment system that more closely aligns with that of industry. These new insights demonstrate the growing potential of SBG.

Perceived Barriers

Respondents across the three groups expressed concerns about implementing SBG despite perceived student gains. Most notably were the expected time or initial workload to prepare and implement the new system, 'buy-in' from others including students, faculty, and administrators, and methods to maintain consistency of grading. These concerns were also noted by expert faculty surveyed (Carberry et al, 2016). This is likely due to the initial investment that is required for start-up, including pre-determining learning objectives and aligning learning objectives. These barriers indicate the need to study methods to alleviate such concerns to lower the entry barrier for SBG.

Different concerns emerged for faculty at different stages of SBG implementation. The *Post-Implementation Group* identified barriers that were largely practical and connected to challenges faced during implementation. Emerging themes include difficulties in grading, grading management, data management, and how to operate the new grading system effectively. The *Pre-Implementation Group* cited barriers faced during the preparation phase. These concerns focused on a lack of resources and previous experience with SBG impacting normal classroom flow. Finally, the *After Workshop Group*'s concerns related mainly to student acceptance of SBG and lack of resources (e.g., teaching assistants). Teaching assistant concerns were observed across the groups; however, concerns took on different aspects based on experience with SBG. Less experienced SBG faculty were likely to focus on recruitment/training, while those with experience focused more on the time and methods for maintaining consistency among their teaching assistants. Overall, the results indicate that faculty in different phases of SBG implementation perceive different aspects as barriers.

Best Practices

Best practices identified by all groups include establishing well-aligned learning objectives, assignments, and rubrics and providing these materials to students at the appropriate time.

Additional best practices identified by the *Post-Implementation Group* included identifying tools for data management (e.g., learning management systems) and gradually expanding the use of SBG within and across courses. These best practices may reflect how faculty new to SBG cope with perceived barriers. Such practices may be a part of expert SBG implementers' general knowledge of how to implement SBG and are no longer considered best practices to them. It also highlights the significance of understanding the perceived barriers for potential implementers in order to find solutions that can become best practices. This is crucial for those planning to implement SBG. Those in the *Pre-Implementation Group* are keen on understanding how to overcome perceived barriers with best practices and recommended preparing for SBG implementation using the newly established SBG online community, teaching assistant training programs, and support from those who have experienced implementing SBG.

These recommendations have been taken into consideration and are being addressed through the online community intended to assist new implementers in acquiring the content knowledge needed prior to implementation. The existence of the SBG community demonstrates the necessity of developing support mechanisms that can assist faculty as they advance through the different phases of implementation.

Conclusions & Future Work

Changing grading practices is considered the most expedient way to improve student achievement because of its significant role in facilitating learning and teaching (Guskey & Bailey, 2001; Heflebower & Hoegh, 2014). Reformation of grading can be a challenging task because it takes time for students and educators to become familiar and comfortable with a new system. The extensive effort necessary to shape an environment favorable for good practice requires the support of educators, understanding of perceptions, time for learning, and opportunities to practice, so that educators can become empowered to push the reformation.

We have demonstrated through a comparison of faculty perceptions that this new system provides benefits to students, which include self-regulated learning and clear connections between grading and learning. Emergent barriers include initial workload, 'buy-in' from others, grading consistency, and a lack of resources. Best practices such as well-aligned learning objectives, assignments, and rubrics can facilitate use of the grading system. Strategies addressed in different stages of implementation, including identification of a data management method, gradual expansion of SBG, and utilization of available resources, indicate that solutions faculty develop to cope with perceived barriers can become best practices to share with others. Of utmost importance was the necessity of support structures including the online SBG community.

Changing grading systems is challenging for educators, institutions, and students, but can result in worthwhile benefits for students and faculty. Our future work aims for large-scale implementation of SBG in higher STEM education by developing supports based on the results from this study, including better integration with learning management systems, alignment with ABET assessment procedures, and resources to train new graders. Another aim is to generate systematic knowledge on how engagement with SBG feedback enhances student learning.

References

Angelo, T.A., & Cross, K.P. (1993). *Classroom Assessment Techniques: A Handbook for College Teachers* (2nd ed.). San Francisco, CA: Jossey-Bass.

Armacost, R. L., & Pet-Armacost, J. (2003). Using mastery-based grading to facilitate learning. *Proceedings of the ASEE/IEEE Frontiers in Education Conference, Boulder, CO.*

Atwood, S., Siniawski, M., & Carberry, A. (2014). Using standards-based grading to effectively assess project-based design courses. *Proceedings of the ASEE Annual Conference & Exposition, Indianapolis, IN.*

Bloom, B. S. (1968). *Learning for Mastery. Instruction and Curriculum*. Durham: Regional Education Laboratory for the Carolinas and Virginia.

Brookhart, S. M. (2010). *How to assess higher-order thinking skills in your classroom*. Alexandria, VA: ASCD.

Brookhart, S. M., & Nitko, A. J. (2008). *Assessment and grading in classrooms*. Upper Saddle River, NJ: Prentice Hall.

Carberry, A., Siniawski, M., Atwood, A., & Diefes-Dux, H. (2016). Best practices for using standards-based grading in engineering courses. *Proceedings of the ASEE Annual Conference & Exposition, New Orleans, LA.*

Carberry, A. R., Siniawski, M., & Dionisio, J. (2012). Standards-based grading: Preliminary studies to quantify changes in student affective and cognitive behaviors. *Proceedings of the ASEE/IEEE Frontiers in Education Conference, Seattle, WA.*

Guskey, T. R. (1997). Research needs to link professional development and student learning. *Journal of staff development*, *18*, 36-41.

Guskey, T. R. (2011). Five obstacles to grading reform. *Educational Leadership, 69*(3), 17–21.

Guskey, T. R., & Bailey, J. M. (2001). *Developing grading and reporting systems for student learning*. Thousand Oaks, CA: Corwin Press.

Guskey, T. R. & Pollio, H. R. (2012). *Grading systems – school, higher education.* Thousand Oakes, CA: Sage Publications.

Heflebower, T., & Hoegh, J. K. (2014). *A School Leader's Guide to Standards-Based Grading*. Bloomington, IN: Solution Tree Press.

Iran-Nejad, A., & Chissom, B. S. (1992). Contributions of active and dynamic self-regulation to learning. *Innovative Higher Education*, *17*(2), 125-136.

Marzano, R. J. (2011). *Formative assessment and standards-based grading*. Bloomington, IN: Marzano Research Laboratory.

Miles, M. & Huberman, M. (1994). *Qualitative data analysis: A source book for new methods.* Thousand Oaks, CA: Sage Publications.

Muñoz, B. M. A., & Guskey, T. R. (2015). Standards-based grading and reporting will improve education. *Phi Delta Kappan, 96*(7), 64-68.

Paris, S. G., & Paris, A. H. (2001). Classroom applications of research on self-regulated learning. *Educational psychologist*, *36*(2), 89-101.

Post, S. L. (2014). Standards-based grading in a fluid mechanics course. *Proceedings of the ASEE Annual Conference & Exposition*, *Indianapolis, IN.*

Postman, N. (1992). Technopoly: The Surrender of Culture to Technology. New York, NY: Alfred A. Knopf.

Reeves, D. B. (2003). *Making standards work: how to implement standards-based assessments in the classroom, school, and district* (3rd ed.). Englewood, CO: Advanced Learning Press.

Sadler, D. R. (2005). Interpretations of criteria-based assessment and grading in higher education. *Assessment & Evaluation in Higher Education, 30*(2), 175-194.

Siniawski, M., Carberry, A., & Dionisio, J. (2012). Standards-based grading: An alternative to score-based assessment. *Proceedings of the ASEE Pacific Southwest Section Conference, San Luis Obispo, CA.*

Suskie, L. (2004). *Assessing student learning: A common sense approach*. Boston, MA: Anker.

Vaishnav, S. & Carberry, A. (2017). Student perspectives on standards-based grading used in engineering project-based courses. American Educational Research Association Annual Conference, San Antonio, TX.

Wiggins, G., & McTighe, J. (1998). *Understanding by design*. Alexandria, VA: ASCD.

Wormeli, R. (2006). *Fair isn't always equal: Assessing & grading in the differentiated classroom.* Portland, ME: Stenhouse Publishers.

Acknowledgements

This work was made possible by support from the National Science Foundation (NSF DUE-1503794). Any opinions, findings, conclusions or recommendations expressed in this material are those of the author and do not necessarily reflect the views of the National Science Foundation.

Interaction between engineering program and academic skills on academic performance in mathematics and physics: a case study

Guillermo Mejía A.
Universidad Industrial de Santander, Bucaramanga, Colombia
gmejia@uis.edu.co

Tulia E. Rivera F.
Universidad Industrial de Santander, Bucaramanga, Colombia
trivera@uis.edu.co

***Abstract**: Studies show that students with weak academic skills in the early semesters of their higher educational programs demonstrate poor academic performance in later semesters; hence, engineering students must enhance their basic academic skills in order to ensure better performance at higher levels. Three of the basic academic skills required by an engineering student are: applying knowledge in practice, abstraction and synthesis, and problem-solving. These skills enable them to engage with their professional context. Basic sciences such as mathematics and physics provide engineering students with the sound knowledge and academic skills required to understand the natural phenomena that characterize their professional environment. The present study found significant interaction between engineering program and academic skills on academic performance in mathematics and physics of higher education.*

Introduction

Statistical evidence shows that quality of education, measured as citizens' cognitive skills, correlates with national long-run economic growth (Hanushek and Woessmann, 2010). Indeed, quality of education impacts more significantly on economic growth than quantity of schooling (Barro, 2013). Likewise, there is evidence of positive correlation between investment in higher education and economic growth (Aghion, Boustan, Hoxby, and Vandenbussche, 2009). Therefore, national education policies ought to promote quality of higher education, and drive improvement of students' basic academic skills.

Specifically, findings from Barro's study (2003) indicate that the math and sciences scores of a nation's students are significant predictive variables of its national economic growth, and that, furthermore, science is somewhat more predictive than mathematics. Congruently, the National Education Office of Colombia, in its Resolution No. 2773 of 2003 (MEN 2003), states that fundamental knowledge in engineering must be provided by mathematics and sciences (e.g. physics) courses, since these disciplines provide the knowledge and academic skills necessary to understand and explain the physical phenomena that engineers face in their daily activities.

Because of its importance, it is imperative to undertake studies of academic performance, especially that which relates to mathematics and physics in engineering education. However, in Colombia, there is little information available to assess the effectiveness of learning and teaching practices in mathematics and physics within higher education, aside from a few isolated studies which show evidence of poor performance among engineering students in their first year of study (Tutalcha et al, 2013; Valencia et al., 2007).

Accordingly, in order to gather evidence with a view towards improving learning and teaching practices, this study aimed to analyze students' academic performance at the College of Engineering and Physical-Mechanical Sciences at the Universidad Industrial de Santander (UIS in Spanish), in Bucaramanga, Colombia. Specifically, this study tested the interaction of a given discipline (mathematics and physics) and a given engineering program upon the academic skills of the students, based on their scores on a standardized test.

Theoretical framework

Bloom's Taxonomy

Academic performance, or achievement, is understood as the level of learning achieved by students within a formal education system. This achievement is measured through the application of standardized tests with respect to a specific body of knowledge and academic skills.

Several taxonomies have been developed to understand the level of learning achieved by students, with Bloom's taxonomy being the most commonly used in engineering. This taxonomy serves as a benchmark of academic skills acquired throughout the cycle of study (Bell and Floger, 1995; Howard, Carver and Lane, 1996).

Bloom's taxonomy establishes a hierarchical ranking of six thinking processes: 1) retention and reproduction of information (knowledge level); 2) reproduction of information based on understanding (understanding level); 3) use of learned knowledge in different contexts (application level); 4) understanding problems through their components (analysis level); 5) comprehension of an entire concept from its components (synthesis level); and 6) assessment of processes and information (assessment level) (Bloom et al., 1956).

According to this taxonomy, the present study assumes the following hierarchical-ascending order for three academic skills that will be measured and assessed in order to attain the objectives proposed: "*Apply knowledge in practice*", "*Abstract-analyze-synthesize*", and "*Identify and solve problems*".

Mathematics and Physics in Higher Education

There is a generalized idea that an engineer is a scientist who uses physics and mathematics to promote the welfare of mankind. Therefore, mathematics and physics are two basic disciplines recognized as pivotal knowledge for the first years of engineering curricula. Indeed, measures of achievement in these disciplines are considered meaningful predictors of success in engineering education. Good performance in early mathematics and physics courses is thought to positively predict performance in subsequent engineering courses (Simpson and Fernandez, 2014).

The importance of mathematics in engineering has long been recognized, as well as its criticality as a tool of progress in science and technology (Gnedenko and Khalil, 1979). It therefore follows that institutions of higher education should seek to improve the teaching practices of mathematics using solid evidence.

Physics, too, has become one of most influential disciplines in our world, especially in such fields as electromagnetism, the electron, and quantum mechanics (Brinkman and Lang, 199). According to the UNESCO-Physics Action Council, physics will play an important role in understanding the global environmental and energy problems of the coming years (Mackintosh, 1999); therefore, physics teaching must provide approaches relevant to the prevailing scientific circumstances at specific periods in time.

This research is a case study conducted in Colombia to analyze the academic performance of engineering students based on their mathematics and physics scores on the EXIM standardized test. The study measured students' academic performance at the College of Engineering and Physical-Mechanical Sciences at the UIS, in Bucaramanga, Colombia. It is hoped that the study's findings will provide insights, based on statistical evidence, to guide the design of improvement strategies for engineering curricula, and for learning and teaching practices.

Methodology

This case study is of a non-experimental quantitative research approach, using a cross-sectional design to compare the academic performance of students. The cross-sectional design was implemented using a data set collected between 2009 and 2015, which included mathematics and physics scores achieved on the Colombian standardized test EXIM. The study primarily sought to analyze interactions through the categorization of variables, controlling for engineering program and discipline.

Sample

The study defined the individual student as the unit of analysis. The corresponding population is the whole set of engineering students of the College of Engineering and Physical-Mechanical Sciences at the UIS between 2009 and 2015. The sample was made up of students who were randomly chosen to take the EXIM test. During the specified period, EXIM had been applied to a total of 800 engineering students at the college (see Table 1).

Department	# Students	%
Civil Engineering	127	15.9%
Electrical Engineering	114	14.3%
Electronic Engineering	133	16.6%
Industrial Engineering	135	16.9%
Systems Engineering	143	17.9%
Mechanical Engineering	148	18.5%
Total	800	100.0%

Table 1. Sample composition by engineering program (N = 800)

The sampled individuals were third-year students who had finished the basic levels of the engineering curricula, including mathematics and physics courses. For each year of study, there are two academic semesters of 16 weeks each. The mathematics and physics content for each of the first three years of study for each engineering program is given in Table 2.

	1st Year				2nd Year				3rd Year			
	Mathematics		Physics		Mathematics		Physics		Mathematics		Physics	
	Courses	Hours	Courses	Hours	Courses	Hours	Courses	Hours	Courses	Hours	Courses	Hours
Civil Eng.	3	192	1	96	3	192	4	320	1	64	1	96
Electrical Eng.	3	192	1	96	3	192	4	352	2	128	3	192
Electronic Eng.	3	192	1	96	3	192	4	352	1	64	1	64
Industrial Eng.	3	192	1	96	2	128	4	320	3	192	1	80
Systems Eng.	3	192	1	96	3	192	3	288	2	128	0	0
Mechanical Eng.	3	192	1	96	2	128	5	384	0	0	5	352

Table 2. Number of mathematics and physics courses and hours per year during the first three years of the engineering curricula at the UIS

In the first year of study, all engineering programs include the same mathematics courses: Calculus I, Linear Algebra, and Calculus II; and the same physics course: Physics I. Variation among the curricula begin to arise in the second and third years of study (see Table 3).

	2nd Year		3rd Year	
	Mathematics	Physics	Mathematics	Physics
Civil Eng.	Calculus III	Physics II	Statistics	Fluid Mechanics
	Differential Equations	Engineering Statics		
	Numerical Methods	Physics III		
		Solid Mechanics		
Electrical Eng.	Calculus III	Physics II	Numerical Methods	Electromagnetism
	Linear Algebra II	Electrical Circuits I	Statistics	Analytical Mechanics
	Differential Equations	Physics III		Thermodynamics
		Electrical Circuits II		
Electronic Eng.	Calculus III	Physics II	Statistics	Electromagnetism
	Linear Algebra II	Electrical Circuits I		
	Differential Equations	Physics III		
		Electrical Circuits II		
Industrial Eng.	Calculus III	Physics II	Statistics I	Electronics
	Differential Equations	Analytical Mechanics	Statistics II	
		Physics III	Operations Research	
		Thermodynamics		
Systems Eng.	Calculus III	Physics II	Numerical Methods	
	Differential Equations	Physics III	Statistics I	
	Discrete Mathematics	Electronics		
Mechanical Eng.	Calculus III	Physics II		Thermodynamics I
	Differential Equations	Engineering Statics		Fluid Mechanics
		Physics III		Machine Mechanics
		Dynamics		Thermodynamics II
		Strength of Materials		Electronics

Table 3. Mathematics and physics courses of the engineering curricula at the UIS

The instrument

Each student's academic achievement was measured using the EXIM standardized test. This test is applied to engineering students in Colombia by the Colombian Association of Engineering Faculties (ACOFI in Spanish).

EXIM assesses four basic disciplines through 140 questions: mathematics (50), physics (40), chemistry (30), and biology (20). Each area is assessed along two axes: fundamental knowledge and academic skills (ACOFI, 2015). For benchmarking purposes, EXIM establishes an achievement scale from 0 to 100 points, which is categorized as follows: from 0 to 49 points, low achievement; from 50 to 70 points, moderate achievement; and from 71 to 100 points, high achievement.

This study assumes that the EXIM exhibits acceptable indices of reliability and validity, based on those found in Rodríguez's study (2013); moderate difficulty (0.33 for mathematics and 0.30 for physics); acceptable discrimination (0.26 for mathematics and 0.32 for physics); and moderate reliability (α = 0.53 for mathematics and α = 0.55 for physics).

Variables

The principal variables analyzed were defined according to the EXIM criteria as follows:

- *Mathematics*: the student's overall average in mathematics, based on scores of five indicators of fundamental knowledge and three academic skills.
- The three academic skills in mathematics are: *MAS1*: "*Abstract-analyze-synthesize*", *MAS2*: "*Apply knowledge in practice*", and *MAS3*: "*Identify-and-solve problems*".
- *Physics*: the student's overall average in physics, based on scores of four indicators of fundamental knowledge and three academic skills.
- The three academic skills in physics were: *PAS1*: "*Abstract-analyze-synthesize*", *PAS2*: "*Apply knowledge in practice*"; and *PAS3*: "*Identify-and-solve problems*".

Statistical techniques

Descriptive statistics were computed for the aforementioned variables, also evaluating normality and outlier patterns. Because of the large sample, N = 800, and invoking the central limit theorem, this study used parametric procedures to achieve the objectives stated. A Grubbs' test evaluated outlier cases for each variable and found no outliers at level of significance of $\alpha = 0.05$.

Regression models were used to test if, according the hierarchy of Bloom's taxonomy, the lower-level academic skills "*Abstract-analyze-synthesize*" and "*Apply knowledge in practice*" are significant predictors of the more advanced academic skill "*Identify-and-solve problems*". A t-test assuming equal variances was used to compare the students' performance on academic skills in mathematics and physics, and the results were adjusted according to the Bonferroni criterion. General linear models were used to analyze both the interaction of a given discipline (mathematics or physics) and of a given engineering program upon academic skills.

Results

Academic performance in mathematics and physics

Statistical summaries of the main variables of the students' scores in mathematics and physics are given in Table 4 and Table 5.

	Mathematics	MAS1	MAS2	MAS3
Mean	55.1	53.8	54.6	53.5
Mean error	0.37	0.35	0.37	0.36
SD	10.4	10.0	10.3	10.3
CV	18.8	18.6	18.9	19.2
Minimum	16.0	24.0	24.0	25.0
Q1	49.0	47.0	47.0	46.0
Median	56.0	53.0	55.0	53.0
Q3	62.0	61.0	62.8	61.0
Maximum	84.0	82.0	84.0	83.0
Skewness	-0.17	0.01	-0.15	-0.03
Kurtosis	0.03	-0.19	-0.25	-0.40

Table 4. Statistics of academic skills in mathematics (N = 800)

Based on Table, the statistics show that the best average performance for academic skills in mathematics was "*Apply knowledge in practice*" (MAS2), though there are only slight differences among the three academic skills assessed. A regression model showed that, although the lower-level academic skills "*Abstract-analyze-synthesize*" and "*Apply knowledge in practice*" are significant predictors ($p < 0.001$) of the more advanced skill "*Identify-and-solve problems*", they are weak predictors, as they explain only 30% of its variance. These scores show a low-to-moderate overall performance in mathematics based on the EXIM achievement scale.

	Physics	PAS1	PAS2	PAS3
Mean	53.5	52.4	52.5	52.4
Mean error	0.41	0.38	0.39	0.38
SD	11.6	10.6	11.1	10.8
CV	21.7	20.3	21.1	20.6
Minimum	26.0	29.0	26.0	28.0
Q1	45.0	44.0	44.0	45.0
Median	53.0	52.0	51.0	51.0
Q3	61.0	59.0	61.0	59.0
Maximum	95.0	82.0	89.0	86.0
Skewness	0.31	0.16	0.27	0.23
Kurtosis	-0.08	-0.42	-0.31	-0.24

Table 5. Statistics of academic skills in physics (N = 800)

Likewise, for the results in physics, Table shows that the average performance for academic skills in physics is quite similar among the three academic skills, however "*Abstract-analyze-synthesize*" (PAS1) showed the tightest dispersion (SD = 10.6). A regression model showed that, although the lower-level academic skills "*Abstract-analyze-synthesize*" and "*Apply knowledge in practice*" are significant predictors ($p < 0.001$) of the more advanced skill "*Identify-and-solve problems*", they are weak predictors, as they only explain 25% of its variance. As with mathematics, the results show a low-to-moderate overall performance in physics, according to the EXIM achievement scale.

Interaction of discipline and program upon academic skills

The study tested the differences in academic skills between mathematics and physics, and found statistical evidence to state that UIS students' perform better in mathematics than in physics. This finding implies a possible interaction of discipline (mathematics, physics) and academic skills achievement (see Table 6).

	Mean	St. Dev.	SE mean	Lower limit 95%	Upper limit 95%	t-value	p-value	d Cohen
MAS1	53.8	10.0	0.35					
PAS1	52.4	10.6	0.38					
MAS1-PAS1	1.4	12.9	0.46	0.48	2.27	3.02	<0.01	0.11
MAS2	54.6	10.3	0.37					
PAS2	52.5	11.1	0.39					
MAS2-PAS2	2.1	13.7	0.49	1.14	3.04	4.3	<0.01	0.15
MAS3	53.5	10.3	0.36					
PAS3	52.4	10.8	0.38					
MAS3-PAS3	1.1	13.5	0.48	0.13	2.00	2.24	0.03	0.08

Table 6. Paired t-test for differences (adjusted $\alpha = .017$) (N=800)

To maintain the appropriate significance level **α** for the comparisons in the table using a Bonferroni adjustment, a p-value must be less than $p = 0.017$ (i.e., 0.05/3) to be considered statistically significant. Congruently, the difference for the academic skill "*Identify-and-solve problems*" is not significant, while the differences for the other two skills are significant; although the effect sizes of these differences are insignificant, with Cohen's d being less than 0.2 (see Table 6).

The interaction plot in Figure 1 shows a possible effect for discipline upon the three academic skills. Students achieved better scores for academic skills in mathematics than in physics, with the skill "*Apply knowledge in practice*" (MAS2) showing the best average score in mathematics.

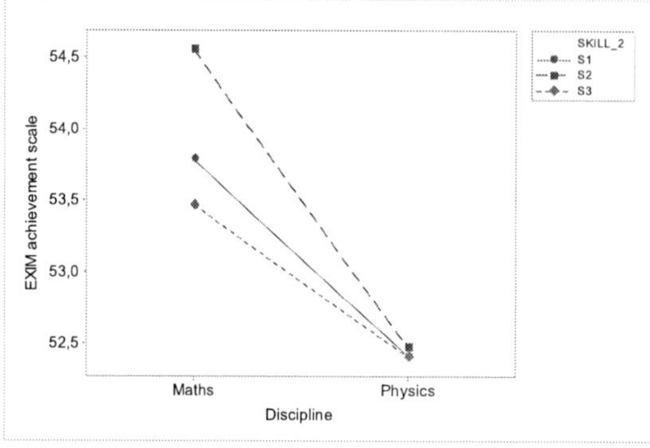

Figure 1. Interaction plot between discipline and academic skills

The interaction plot in Figure 2 shows a possible effect for engineering program upon the three academic skills. Students in the mechanical engineering program achieved better scores for academic skills compared with the other five engineering programs.

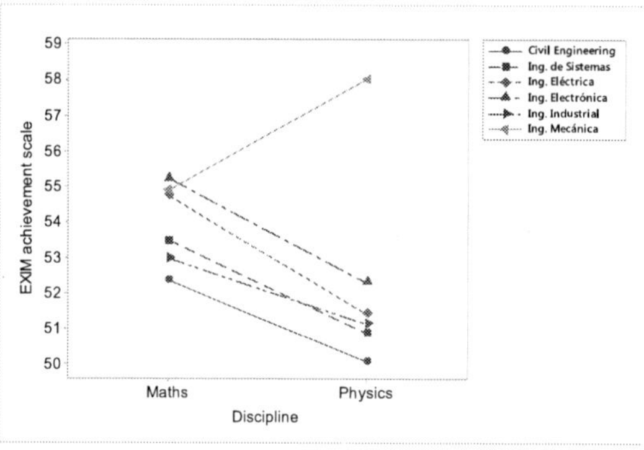

Figure 2. Interaction plot between discipline and academic skills

This study used a Mauchly's test to assess the assumption of sphericity and, when necessary, adjusted the degrees of freedom using Greenhouse-Geisser estimates. The study found a significant interaction effect between engineering program and discipline, $F(4.65, 524.93) = 7.82$, $P < 0.001$; and at a level of significance of $\alpha = 0.10$, the study found significant interaction between discipline and academic skill, $F(2, 226) = 2.52$, $P = 0.085$. These results indicate that the engineering program had different effects on students' academic skills depending upon which discipline, mathematics or physics, was assessed. Also, there is statistical evidence to state that discipline had different effects on students' achievement depending on which academic skills were assessed. The contrast analysis revealed significant interactions when comparing the scores of the mechanical program students to those in other programs, and when comparing "*Apply knowledge in practice*" achievement to the other two skills for mathematics, compared to physics. However, these contrasts did yield small effect sizes (Cohen's $d < 0.2$).

Discussion

The findings indicate that engineering students at the UIS achieve better performance in mathematics than in physics. Nevertheless, these students spend more time attending physics classes than mathematics classes during their first three years. On average, per engineering program, students attend 563 hours of physics classes and 459 of mathematics classes. Therefore, it is necessary to undertake additional research into physics teaching in engineering education at the UIS in order to boost academic performance in this discipline. It is important to take into account that, according to Barro (2013), mathematics performance exhibits less of an influence upon economic growth than does science performance.

Additionally, the results showed that students achieved better performance in MAS2, "*Apply knowledge in practice*", in mathematics, which seems to indicate that mathematics teaching during the first three years of the engineering curricula boosts this academic skill. Likewise, in physics, there is a similar effect for the three academic skills, even though the dispersion measures seem to indicate that physics teaching boosts, to a somewhat greater degree, the academic skills related to PAS1, "*Abstract-analyze-synthesize*".

Additionally, this study shows that the lower-level academic skills "*Abstract-analyze-synthesize*" and "*Apply knowledge in practice*" are weak predictors of performance in the more advanced skill "*Identify-and-solve problems*". This finding appears similar to those of Zewdie's study (2014), where students were found to concentrate more on basic mathematical procedures than on scientific approaches while solving problems in physics courses. Nonetheless, other studies have found that in physics, basic skills such as knowledge of individual facts within a system are significant and excellent predictors of problem-solving abilities (Friege and Lind, 2006). Due to the ambiguity of findings in other studies, it is imperative to know in depth how the teaching practices at the UIS influence the academic skills of engineering students.

Finally, the findings suggest that the teaching of mathematics and physics may be strongly influenced by the choice of engineering program. The results show that mechanical engineering students who attended more physics-related courses achieved better scores in physics. While this study shows a significant difference in physics achievement between the mechanical and electrical engineering programs, other studies found no significant differences between them (Hari, Anuar, Mokhtar, and Mokhtar, 2016). Based on the present study, it seems that mathematics and physics teaching practices are dissimilar within engineering education.

Conclusion

The results of this study provide insights, based on statistical evidence, to guide the design of improvement strategies for curricula and teaching practices in engineering at the UIS. First, the results show that current learning and teaching practices in mathematics have a greater effect on the academic skill of applying knowledge in comparison with the other two skills studied. The statistical evidence suggests that engineering students achieve better academic performance in mathematics than in physics. Second, in both disciplines, mathematics and physics, the evidence shows that current learning and teaching practices weakly influence the academic skill for problem solving. Third, the findings suggest that the teaching of mathematics and physics may strongly vary among engineering programs. These results indicate that it is necessary to undertake more research into physics teaching in engineering education at the UIS in order to boost the academic performance in this discipline.

References

Aghion, P., Boustan, L., Hoxby, C. M., & Vandenbussche, J. (2009). The causal impact of education on economic growth: Evidence from the U.S. (mimeo) Department of Economics: Harvard University

Asociación Colombiana de Facultades de Ingeniería (ACOFI) (2015). Resultados generales EXIM 2015 (pdf). Accessed at http://www.acofi.edu.co/wp-content/uploads/2016/09/RESULTADOS-2015.pdf on 1 feb., 2017.

Barro, R. J. (2013). Education and economic growth. *Annals of Economics and Finance, 14,* 301-328.

Bell, J. T., & Fogler, H. (1995). Investigation and application of virtual reality as an educational tool. *ASEE Annual Conference Proceedings, 2* (Investing in the Future), 1718-1728.

Bloom, B. S., Engelhart, M. D., & Committee of College and University Examiners. (1956). *Taxonomy of educational objectives: The classification of educational goals.* London: Longmans.

Brinkman, W. F., & Lang, D. V. (January 01, 1999). Physics and the communications industry. *Reviews of Modern Physics, 71.*

Gnedenko, B. V., & Khalil, Z. (1979). The mathematical education of engineers. Educational Studies In Mathematics, 10(1), 71.

Friege, G., & Lind, G. (2006). Types and Qualities of Knowledge and their Relations to Problem Solving in Physics. *International Journal of Science and Mathematics Education, 4,* 3, 437-465.

Hanushek E A and Wößmann L (2010). Education and economic growth in: Penelope Peterson, Eva Baker, Barry McGaw, (Editors), *International Encyclopedia of Education.* Volume 2, pp. 245-252. Oxford: Elsevier.

Hari, Z., Anuar, A., Mokhtar, M. Z., and Mokhtar, R. (2016) "Evaluating engineering students performance in physics courses during foundation studies," *2016 IEEE 8th International Conference on Engineering Education (ICEED)*, Kuala Lumpur, Malaysia, 2016, pp. 214-217. doi: 10.1109/ICEED.2016.7856074

Howard, R. A., Carver, C. A., & Lane, W. D. (1996). Felder's learning atyles, Bloom's taxonomy, and the Kolb Learning Cycle: Tying it All Together in the CS2 Course. *Sigcse Bulletin, 28,* 1, 227-231.

Mackintosh, Raymond S. (1999). *International Workshop on the Future of Physics and Society, Debrecen, Hungary, 4-6 March 1999, Workshop Summary.* Accessed at http://www.unesco.org/science/wcs/meetings/eur_debrecen_99.htm#part_2, on feb 20, 2017.

Ministerio de Educación Nacional -MEN (2003). Resolución Núm. 2773 de 2003 sobre la calidad para los programas de formación profesional de pregrado en Ingeniería. Accessed at http://www.mineducacion.gov.co/1621/articles-86417_Archivo_pdf.pdf on feb. 20, 2017.

Rodriguez, Olga R. (2013). La evaluación objetiva en ingeniería aportes en procesos de evaluación y mejora curricular. World Engineering Education Forum WEEF Cartagena Colombia 2013. Accessed at http://www.acofipapers.org/index.php/acofipapers/2013/paper/view/367/186 on 20 feb., 2017.

Simpson, J. and Fernandez, E. (2014) "Student performance in first year, mathematics, and physics courses: implications for success in the study of electrical and computer engineering", conference proceedings, *Frontiers in Education Conference* (FIE), August 2014.

Tutalcha, Martha C. et al. (2013) Preocupa bajo nivel de matemáticas en bachilleres. Accessed at http://www.agenciadenoticias.unal.edu.co/ndetalle/article/preocupa-bajo-nivel-de-matematicas-en-bachilleres.html on 10 feb. 2017.

Valencia G., A., Mejía V., L. F., Restrepo G., G., Parra M., C. M., Muñoz O., L. D. y Ochoa Á., J. (2007). Primer año en ingeniería: más allá del rendimiento académico. Revista Educación en Ingeniería, (2007) (4), 55.

Zewdie, Z. M., (2014). An investigation of students' approaches to problem solving in physics courses. *International Journal of Chemical and Natural Science.* Vol. 2, No. 1 (2014): 77-89

Problem Formulation in Engineering Design

Douglas Lecorchick
North Carolina State University, Raleigh, NC USA
dlecorc@ncsu.edu

Matthew Lammi
North Carolina State University, Raleigh, NC USA
mdlammi@ncsu.edu

***Abstract**: Problem solving is an essential ability and skill for life and engineering is no exception. Although there has been research on problem solving, there is little known about problem formulation, particularly within engineering design and in educational settings. The purpose of this study was to generate a substantive grounded theory on high school student problem formulation within engineering design. Thirty-two high school students grouped together in teams of two (dyads) were given an engineering design problem and asked to solve the open-ended problem using sketch paper, a whiteboard, and access to a computer. Each dyad was video recorded during the entire design session. The data were analyzed from a classic grounded theory methodological perspective including substantive and theoretical coding, field note annotations, and memo annotations toward a substantive grounded theory of high school students' experience of and navigation through problem formulation within engineering design.*

Introduction

Problem solving is an essential ability and skill for life. In STEM settings, engineering is no exception; the ability to understand and solve open-ended technical problems is paramount. Problem solving in engineering design includes more than devising a solution, but typically involves handling ambiguity; understanding and communicating constraints, variables, and specifications; formulating, defining, and scoping problems; and implementing a solution (Dym, 1992; Jonassen, Strobel, & Lee, 2006).

Although research has been done on problem solving, little is known about problem formulation, particularly within engineering design and in educational settings. This is problematic mainly because problem formulation is crucial to the problem solving process (Eierman & Philip, 2003).

It is generally accepted that problem-solving skills are highly valued and that skillset continues throughout most people's careers regardless of what formal training they have had or what career path they have chosen (Sweller, 1988). However, a problem typically needs to begin formulation before problem solving can occur (Baer, Dirks, & Nickerson, 2013). Problem formulation, for this study, is described as the process of exploring, identifying, and discovering a problem driven by the given constraints and one's own personal beliefs.

Knowing how people perceive information is important to understanding how they solve problems (Bardwell, 1991; Eierman & Philip, 2003). By exploring how people experience and engage in problem formulation, we can further understand how they approach the problem solving process. This starts with a working theory of how students experience and engage in problem formulation. With a better understanding of how individuals formulate problems researchers can then suggest best practices to increase problem solving effectiveness (Eierman & Philip, 2003).

Research Questions

The purpose of this study was to generate a theory of high school students' problem formulation experience within engineering design. The aim of this study was to understand how high school students experience and engage in the process of formulating an open-ended engineering design problem.

The primary focus of this research study was the observed problem formulation experience of high school students during an engineering design challenge and the characteristics of problem formulation within that design challenge. The following research questions framed the data analysis for this study:

1. How is problem formulation characterized during an engineering design challenge?
2. How do the participants navigate through problem formulation?

Methodology - Classic Grounded Theory

Grounded theory is the generation or discovery of a theory from systematically collecting and analyzing the data (Glaser & Strauss, 1967). The researchers of this study aimed to generate a theory of how high school students formulate problems during an engineering design challenge. Thus, grounded theory is the most influential qualitative method of use when researchers aim to generate theory from the data (Chism, Douglas, & Hilson, 2008; Strauss & Corbin, 1997).

Furthermore, grounded theory data analysis focuses more on identifying patterns than describing the data (Griffiths, 2013). Through meaning, action, and interaction the researchers begin to see the fit and relevance within the context of the study (Giske & Artinian, 2007; Glaser, 2014). This allows the researchers to see things as they currently are and not how one could preconceive them to be (Glaser, 2014).

A grounded theory conceptually presents the relationship between social behavior concepts which emerge from the systematically collected and analyzed data (Glaser & Strauss, 1967; Holton & Walsh, 2016). For this study, the student's navigation through the problem formulation experience is the social behavior that is being explored and the grounded theory is a presentation of the relationships between and among the emerged concepts. Holton and Walsh (2016, p. 10) stated,

"To understand the nature of the classic GT methodology, one must understand the distinction between conceptualization and description. Grounded theory is not about accurate, descriptive reporting of findings in a study, nor is it an act of interpreting meaning as ascribed by the participants in a study; rather, it is an act of conceptual abstraction. For a classic grounded theorist, what matters are the concepts, the relationships among those concepts, and their power to offer the reader a conceptual study, free of the particularistic details of specific empirical incidents (indicators)."

Below are terms and definitions to facilitate a better understanding of the grounded theory process:

Field Notes are the annotations written by the researcher during the observation (for this study) and are specific to the dyad being observed (Charmaz, 2006; Holton & Walsh, 2016).

Substantive Coding is the process of fragmenting the data by coding incidents and concepts toward identifying a core category that explains the main concern of the study (Glaser & Strauss, 1967; Holton & Walsh, 2016).

Theoretical Coding is the process of conceptualizing the relationship of substantive codes toward identifying the emergent substantive theory (Glaser, 1978; Holton & Walsh, 2016).

Memos are the written theoretical ideas of the researcher occurring sporadically and continually throughout the research study that assists in elevating the data analysis toward a conceptual level (Charmaz, 2006; Holton & Walsh, 2016).

Theoretical Sorting is the process of hand sorting memos to toward outlining the fit in the emerging theory (Glaser, 1978; Glaser & Strauss, 1967; Holton & Walsh, 2016).

Constant Comparative Analysis begins as the analytical process of comparing fragments (incidents) of data to other fragments (incidents) of data toward identifying recurring patterns (concepts) and later assists with category development, theory delimiting, and the actual writing up of the theory (Charmaz, 2006; Glaser & Strauss, 1967; Holton & Walsh, 2016).

Participants

This study consisted of thirty-two participants, sixteen dyads, of high school juniors enrolled in one of three rural mid-Atlantic school districts. The high schools were similar in size and course offerings and there was no separation of dyads with or without prior engineering course experience and each dyad was student self-selected.

Engineering design problems are usually solved by a collection of individuals forming a team where each member contributes their skills and knowledge to form an acceptable solution (Jonassen et al., 2006). For this study, each dyad was viewed as a problem solving team and the data collection of each dyad was regarded as a single unit of data.

Design Challenge

Student self-selected dyads were asked to design a device to assist physically impaired elderly nursing home residents in opening a stuck double-hung window without the use of an existing electrical energy source. Each dyad was given a design challenge sheet that outlined the challenge requirements. In addition to the design sheet the students had access to a computer with a video describing the function and parts of a double hung window. Each dyad was given a timeframe of no longer than forty-five minutes to develop the design and were instructed to provide a detailed sketch of their design solution on the whiteboard.

Data Collection

The researchers aimed to capture both the interaction of the dyad while reading and discussing the challenge (before advancing to the whiteboard) and the whiteboard sketch by ways of unstructured observations. Unstructured observations restrict the interaction and interference of the observer, which allows the participants being observed to act as naturally as possible. Recording these observations allows the researcher to capture most everything that occurs and then an analysis of that data is performed at a later time (Chism et al., 2008). This was done with two cameras and lapel microphones; one recording the interaction of the participants and one recording the whiteboard, as shown below in Figure 1.

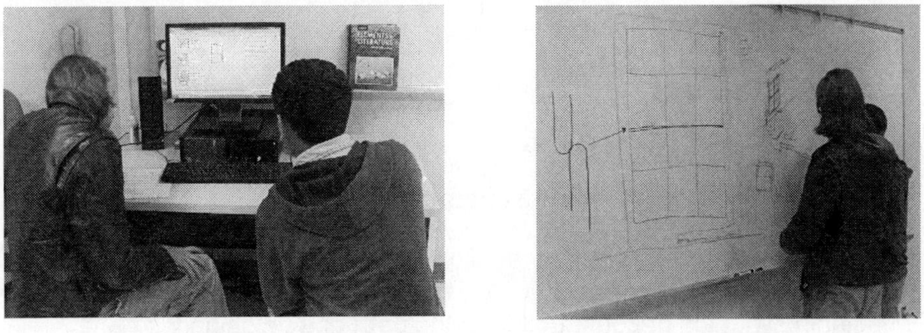

Figure 1. Vantage points captured by recording.

The audio from each design session was transcribed and broken down into utterances as lines of text guided by the verbal protocol method of analysis. The researcher performed analysis while simultaneously viewing the videos and reading through the transcripts. Verbal protocol analysis (concurrent) is when objects of study verbalize their thoughts while they are performing a task without being prompted by the observer (Trickett & Trafton, 2007). For this study, the dyads are the objects of study and their verbal utterances were transcribed and analyzed.

Verbal protocol analysis has been used in many scientific research studies. Trickett and Trafton (2007) state that verbal protocol analysis is a valuable tool stretching across several domains including both scientific and analogical reasoning. This study uses a modified verbal protocol analysis in that the participants are in teams of two and are not prompted to share their thoughts as this was viewed as natural interactions between the participants (Wells et al., 2016).

There is immense value in using observations as a data collection technique when the researcher takes field notes on the behavior and activities of the participants in the study (Creswell, 2014; Griffiths, 2013). From an observation, the researcher can transcribe text and actions of the participants to better analyze the data (Fernández, 2003). From the transcribed text there are usually important data that would have been missed if not recorded (Giske & Artinian, 2007). Therefore, observations and transcribed texts provided the necessary platform for the researchers to both collect and analyze the experience of the students' problem formulation during the engineering design challenge.

Data Analysis

The coding scheme for this study occurred in two phases beginning with substantive coding (1) and ending with theoretical coding (2). Substantive coding has two levels: open coding (1a) and selective coding (1b). Open coding is where the researcher looks to the raw data to identify conceptual codes; selective coding is where the researcher looks for the relationship between those open codes to identify relevant core concepts. Theoretical coding is where the researcher looks for the relationship between the core concepts, previously identified in the substantive coding phase, to establish theoretical codes or phases (Glaser & Strauss, 1967; Holton & Walsh, 2016). A visual representation of the coding scheme is shown below in Figure 2 that shows the two stages of coding and the primary purpose of those stages.

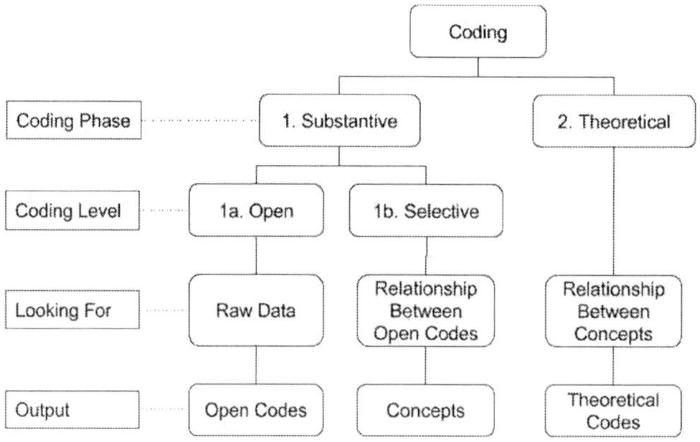

Figure 2. **Coding scheme progression.**

The coding of data is a multi-level process that allows the researcher to start defining a theory from the data collected (Charmaz, 2006). Each coding level has a different analytical focus usually beginning with the deconstruction of the data and finishing with a reconstruction of that data framing core categories and concepts into a substantive theory. As the researcher moves from one coding level to the next, the current level builds upon the previous levels providing clarity regarding emergent categories and themes. A unique characteristic within grounded theory is that the researcher collects and analyzes the data simultaneously (Chism et al., 2008). This highly active coding process requires the researcher to remain open and aware of what emerges from the data while actively coding (Charmaz, 2006).

Research Study Overview

This study was conducted in four phases as the researchers attempted to generate a theory as to how high school student's problem formulate during an engineering design challenge. Figure 3, shown below, shows the overall study approach and a detailed explanation of each phase follows. It is important to note that this study started with the general interest of how problem formulation and engineering design are experienced by problem solvers.

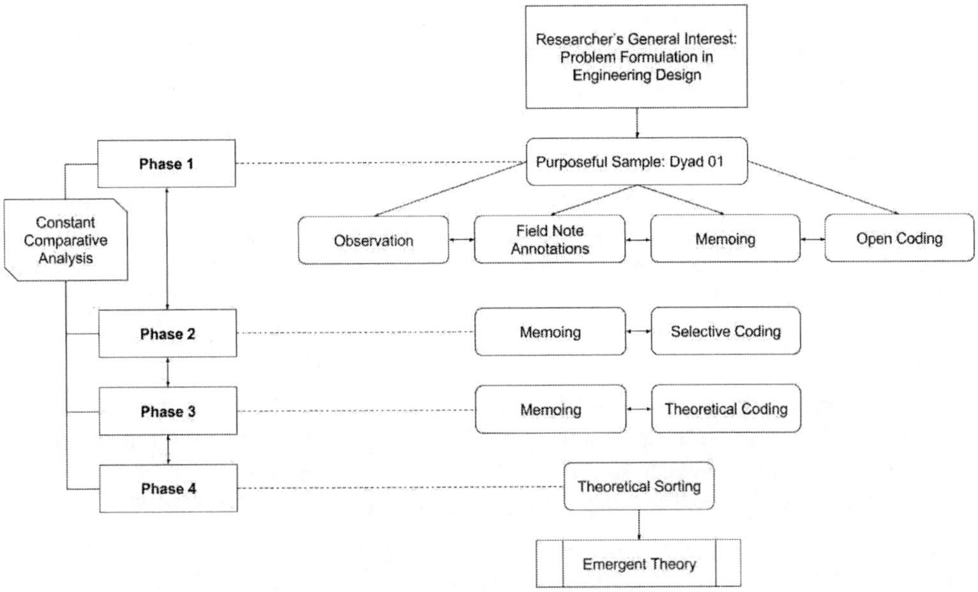

Figure 3. **Overall study approach.**

Phase 1

Phase 1 began with establishing a purposeful sample and selection of an initial dyad. The researchers then observed this dyad's entire attempt at solving the engineering design challenge problem. During the observation the researchers annotated both field notes and memos capturing the researcher's conceptual thoughts. After observing the dyad, the researcher began open coding the transcribed text of the dyad aiming to identify core categories by assigning codes to lines of raw text. During this phase the constant comparative method of analysis was primarily a comparison between open codes toward identifying relationships among them concerning the primary concern of problem formulation within engineering design.

Phase 2

Phase 2 began with the selective coding of the open coded transcripts from phase 1. During this selective coding process, the researchers used the constant comparative method of analysis to compare the emerging core category and assigned open codes from phase 1 to begin identifying relatedness among them concerning the primary concern of problem formulation within engineering design. During this phase theoretical concepts began to emerge and through theoretical saturation of the assigned selective codes the researchers were guided into phase 3. The researchers also annotated memos during this phase capturing the researcher's conceptual ideas as they occurred.

Phase 3

Phase 3 began with the researchers reconstructing the pieces of text initially deconstructed in phases 1 and 2 by using the constant comparative method of analysis between the previously identified theoretical concepts. During this process the researchers looked for relatedness or patterns among the selective codes. Through theoretical saturation of these newly formed theoretical codes the researcher stopped the reconstruction of the data and began to generate theory that began phase 4. The researchers also annotated memos during this phase capturing the researcher's conceptual ideas as they occurred.

Phase 4

Phase 4 began with the theoretical sorting of field note and memo annotations toward identifying patters or relatedness among them using the constant comparative method of analysis. A substantive grounded theory was generated as a result of the systematic data collection and analysis of the study.

Grounded Theory

This grounded theory is an exposition of the identified problem formulation concepts and phases, the participant's navigation through those concepts and phases, and the participant's overall problem formulation experience. This theory is grounded in the data and applicable to other populations within similar contexts (Holton & Walsh, 2016). Figures are used throughout this presentation to provide clarity of the emergent concepts and themes (Kennedy-Lewis, 2016).

Problem formulation is a multiplex experience that was present in varying degrees for these participants during this engineering design challenge. The problem formulation process began with the participant's awareness of an unmet need. For this study, that was a device to assist elderly nursing home patients with opening a stuck double-hung window.

The navigation through problem formulation continued with an introduction of a solution idea. This solution idea was usually followed by an exploration of both solution idea constraints and the overall design challenge limitations and requirements. The participants then proceeded to discuss the implementation of the solution idea, modifications needed of both the solution idea and the implementation process, and finally a decision to accept or reject the solution idea.

The navigation through these concepts was generally consistent. Figure 4 below shows the typical navigation through the core concepts of problem formulation and the cyclical relationship between three of them: *Explore*, *Implement*, and *Modify*.

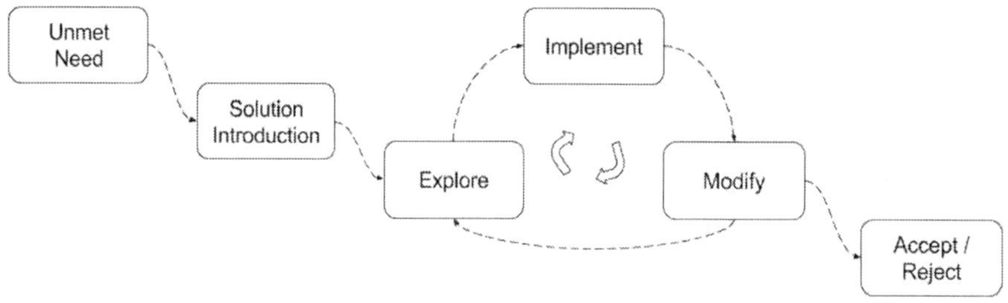

Figure 4. **Typical navigation through problem formulation for this study.**

Participants engaged in problem formulation by first becoming aware of an unmet need and then identifying a possible solution toward satisfying that need. This initial engagement, or relationship between the unmet need and a solution idea is an intuitive process and theoretical phase that the researchers labeled *Tacit Formulation*. There was usually no verbal communication that took place until after a solution idea had been suggested. However, the researcher still observed cognitive activity regarding mental constructs to facilitate the suggestion of such ideas.

After a solution idea was introduced a cyclical relationship followed between exploring that solution idea, implementing the solution idea, and making modifications to both the solution idea and the implementation process. This cyclical relationship is identified by the researchers as *Problem Formulation Framing* and is a theoretical phase that emerged from data analysis during theoretical coding. This phase is where most of the student's involvement in the problem formulation process occurred. It is during this phase that the solution idea and implementation process are both being applied and evaluated. The researcher also identified two supporting categories within this theoretical phase: *Applied Discourse* and *Evaluative Discourse. Applied* Discourse is where most of the problem formulation activity occurred as constraint analysis and solution idea implementation. *Evaluative Discourse* is where the dyad discussed the implementation process and suggested modifications to overcome issues discovered by the exploration of the constraints and the implementation process.

The entire problem formulation process is an implicit navigation activity toward the decision of either acceptance or rejection of the solution idea. This theoretical phase of *Implicit Formulation* is an approach and pathway taken through the problem solving process. Situated within this phase are *Intuitive Formulation* and *Formulation Framing*, the 2 supporting categories of *Applied Discourse* and *Evaluative Discourse*, and the core concepts of problem formulation as shown below in Figure 5.

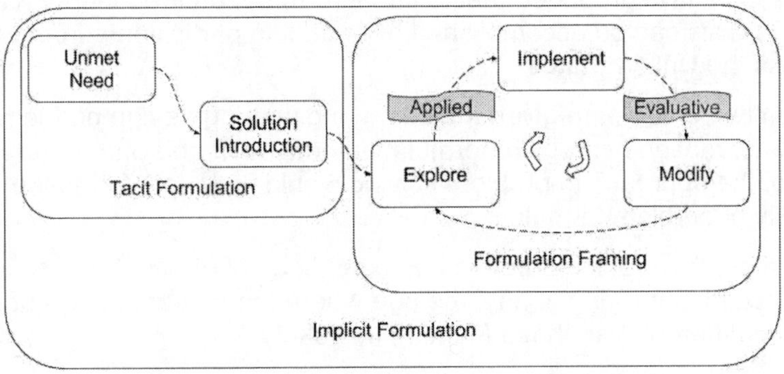

***Figure 5.* Implicit formulation.**

Implications for Engineering Education

The following are implications for engineering education action following the presentation of the findings of this research study:

1. Toward an accepted definition of problem formulation within the K 12 setting. "Professional language is replete with jargon, loaded words, easily misunderstood words, or words that have different meanings to different people" (Nathaniel, 2006, p. 38). This is also true in K12 settings where the problem solving terminology varies in meaning, thus becoming even more confusing for students (Woods, Felder, Rugarcia, & Stice, 2000).
2. Establishing an agreed upon and widely recognized process model of K12 problem formulation within engineering design as outlined in this research study. There are various models that aim to capture and predict the engineering design process. These models however are not in alignment with the observed engineering design process of the participants in this study. Howard, Culley, and Dekoninck (2008) compared different engineering design literature finding most models suggest an investigative or exploratory

phase before the introduction of a solution. However, the findings of this study show that the exploration and investigative approach within problem formulation happens after a solution idea has been introduced. The current engineering design models do not accurately predict or outline the problem solving process that was observed in this study, specifically regarding problem formulation as it occurred for these participants.

3. Further the ability for academic exercises to more closely represent problems one experiences in work and life. This is achieved by providing curriculum designers with accurate information of how high schools students navigate through problem formulation within engineering design and the actual observed characteristics of problem formulation within engineering design, as presented in this study.
4. To expand and clarify the existing body of problem formulation literature. Specifically, at the intersection of problem formulation, K12 education, and engineering design.

Future Research

The following are recommendations for further research in engineering education following the presentation of the findings of this research study:

1. A study using Verbal Protocol Analysis in addition to observations similar to this study would provide the researcher with information about why decisions were made during the design session. This would be particularly useful in probing the thought process of each participant during the Tacit Formulation phase as described in this study.
2. A study increasing the sample size and widening the sample selection is needed to develop the substantive grounded theory into a formal grounded theory (Glaser & Strauss, 1967). This can be accomplished by including participants from various school districts across the United States.
3. A study where two groups are given the same engineering design problem. One group is provided an intervention of problem formulation education and one group is not. A comparison of the approach to problem solving would provide information regarding the teaching ability of problem formulation.
4. A study to explore the problem formulation experience of elementary school aged participants toward identifying, if any, the age where the characteristics of problem formulation are different than those found in this study.
5. A study to explore creativity within problem formulation using Verbal Protocol Analysis and observations where the researcher focuses on the impact, if any, of a student's creativity on the navigating the problem formulation process.

References

Baer, M., Dirks, K. T., & Nickerson, J. A. (2013). Microfoundations of strategic problem formulation. *Strategic Management Journal*, *34*(2), 197–214. https://doi.org/10.1002/smj.2004

Bardwell, L. V. (1991). Problem-Framing: A perspective on environmental problem-solving. *Environmental Management*, *15*(5), 603–612. https://doi.org/10.1007/BF02589620

Charmaz, K. (2006). *Constructing grounded theory: a practical guide through qualitative analysis. Book* (Vol. 10). https://doi.org/10.1016/j.lisr.2007.11.003

Chism, N. V. N., Douglas, E., & Hilson, W. J. (2008). Qualitative Research Basics: A Guide for Engineering Educators. *Engineering Education*, 1–65. Retrieved from http://cleerhub.org/resources/8

Creswell, J. W. (2014). *Research design: qualitative, quantitative and mixed methods approaches. Research Design: qualitative, quantitative and mixed methods approaches.*

Dym, C. L. (1992). Representation and Problem-Solving - the Foundations of Engineering Design. *Environment and Planning B-Planning & Design*, *19*(1), 97–105.

Eierman, M. a., & Philip, G. C. (2003). The task of problem formulation. *International Journal of Information Technology & Decision Making*, *2*, 353–372. Retrieved from http://www.worldscinet.com/abstract?id=pii:S0219622003000690

Fernández, W. D. (2003). Using the glaserian approach in grounded studies of emerging business practices. *Electronic Journal of Business Research Methods*, *2*(2), 83–94.
Giske, T., & Artinian, B. (2007). A Personal Experience of Working with Classical Grounded Theory: From Beginner to Experienced Grounded Theorist. *International Journal of Qualitative Methods*, *6*(4), 67–80. Retrieved from http://offcampus.lib.washington.edu/login?url=http://search.ebscohost.com/login.aspx?direct=true&db=a9h&AN=28333758&site=ehost-live%5Cnhttp://content.ebscohost.com/ContentServer.asp?T=P&P=AN&K=28333758&S=R&D=a9h&EbscoContent=dGJyMNHX8kSeqLE4zOX0OLCmr0uep7
Glaser, B. (2014). Applying grounded theory. *The Grounded Theory Review*, *13*(1), 46–50.
Glaser, B. G. (1978). *Theoretical sensitivity: Advances in the methodology of grounded theory. Sociology Pr.* https://doi.org/Casa
Glaser, B. G., & Strauss, A. L. (1967). *The Discovery of Grounded Theory: Strategies for Qualitative Research. Aldine* (Vol. 1). https://doi.org/10.2307/2575405
Griffiths, C. (2013). Using grounded theory to analyze qualitative observational data that is obtained by video recording. *The Grounded Theory Review*, *12*(1), 26–36.
Holton, J. A. (2008). Grounded theory as a general research methodology. *Grounded Theory Review*, *7*(2), 22–27. Retrieved from http://groundedtheoryreview.com/2008/06/30/grounded-theory-as-a-general-research-methodology/
Holton, J. A., & Walsh, I. (2016). *Classic grounded theory, Applications with qualitative and quantitative data*. Los Angeles, CA: SAGE.
Howard, T. J., Culley, S. J., & Dekoninck, E. (2008). Describing the creative design process by the integration of engineering design and cognitive psychology literature. *Design Studies*, *29*(2), 160–180. https://doi.org/10.1016/j.destud.2008.01.001
Jonassen, D. H., Strobel, J., & Lee, C. B. (2006). Everyday Problem Solving in Engineering : *Journal of Engineering Education*, 139–151.
Kennedy-lewis, C. B. L. (2016). Using Diagrams to Make Meaning in Grounded Theory Data Collection and Analysis Using Diagrams to Make Meaning in Grounded Theory Data Collection and, 1–23. https://doi.org/10.4135/978144627305014534926
Nathaniel, A. (2006). Thoughts on the Literature Review and GT | Grounded Theory Review. *Grounded Theory Review*, *5*(2/3), 35–42. Retrieved from http://groundedtheoryreview.com/2006/06/30/1416/
Strauss, A., & Corbin, J. (1997). *Grounded theory in practice*. Retrieved from http://books.google.com/books?hl=fr&lr=&id=TtRMolAapBYC&pgis=1
Sweller, J. (1988). Cognitive load during problem solving: Effects on learning. *Cognitive Science*, *12*(2), 257–285. https://doi.org/10.1016/0364-0213(88)90023-7
Trickett, S. B., & Trafton, J. G. (2007). A primer on verbal protocol analysis. *Handbook of Virtual Environment …*, 332–346. Retrieved from http://ammonwiemers.com/IdetPortfolio/articles/Assessment/A Primer on verbal protocol analysis.pdf
Wells, J. G., Lammi, M., Gero, J., Grubbs, M., Paretti, M., & Williams, C. (2016). Characterizing design cognition of high school students: Initial analyses comparing those with and without pre-engineering experiences. *Journal of Technology Education*.
Woods, D. R., Felder, R. M., Rugarcia, A., & Stice, J. E. (2000). The Future of Engineering Education III. Developing Critical Skills. *Chemical Engineering Education*, *34*(2), 108–117. https://doi.org/10.1080/030437900308562

Copyright statement

Defining entrepreneurial behavior as a learning outcome of engineering entrepreneurship programs

Prateek Shekhar
University of Michigan, Ann Arbor, USA
pshekhar@umich.edu

Aileen Huang-Saad
University of Michigan, Ann Arbor, USA
aileenhs@umich.edu

Julie Libarkin
Michigan State University, East Lansing, USA
libarkin@msu.edu

***Abstract**: Entrepreneurship is receiving increasing attention in engineering education. As more universities expose engineering students to entrepreneurship through formal and informal programs, it is important to develop a coherent understanding of their learning outcomes and develop assessment measures. Our work focuses on entrepreneurial behaviour, which is a commonly intended learning outcome of engineering entrepreneurship programs. In this work-in-progress paper, we present our on-going instrument development work involving identification of constructs using existing literature and expert feedback; development of assessment instrument items and expert response to generated items. Our preliminary work presents a 12-item open-ended instrument for assessing six sub-constructs describing entrepreneurial behaviour - opportunism, innovativeness, networking, risk management, adaptability and perseverance.*

Introduction

In recent years, engineering colleges and universities in the US have witnessed a significant growth in engineering entrepreneurship programs (Besterfield-Sacre, Ozaltin, Shartrand, Shuman, & Weilerstein, 2011; Shartrand, Weilerstein, Besterfield-Sacre, & Golding, 2010). Similar to business schools, engineering entrepreneurship programs initially focused on teaching and supporting student venture creation (Standish-Kuon & Rice, 2002). Yet, as more programs have launched, their focus has broadened beyond business fundamental knowledge and skills to address the more contemporary global and economic demands on engineering graduates. Such demands include the national calls for engineering curricular reform to meet changes in technology and society (National Academy of Engineering, 2004, 2005) and the addition of professional skills to accreditation standards (Shuman, Besterfield-Sacre, & McGourty, 2005).

With the advent and growth of diverse engineering entrepreneurship programs, comes the need for more tools to assess the impact of entrepreneurship education (Duval-Couetil, 2013). This is particularly challenging for engineering entrepreneurship education, as it extends beyond the dissemination of business knowledge and seeks to impact students' entrepreneurial attitude and behaviour. In addition, engineering entrepreneurship programs have evolved to be diverse programs with varied objectives, programs and pedagogical approaches (Duval-Couetil, Shartrand, & Reed, 2016; Gilmartin, Chen, & Estrada, 2016). A recent review of engineering entrepreneurship program models (Gilmartin et al., 2016) explored the breadth of engineering entrepreneurship programs through in-depth interviews with 12 different program directors. They found that program goals varied from 'starting a business' to 'teaching business fundamentals' to 'developing an entrepreneurial mindset'. In addition, they noted that program evaluation was limited across the programs and metrics tended to focus on output metrics, e.g. enrolment and degrees, as opposed to student learning or behaviour. These variations can be attributed to the intersectionality of the academic culture and individual perception of entrepreneurial behaviour.

Limited program evaluation and assessment work in engineering entrepreneurship education and recent calls to improve productivity measures for higher education (Massy, Sullivan, & Mackie, 2013), clearly identify a significant gap and need to examine the impact of engineering entrepreneurship education on student learning. Our presented work focuses on assessing entrepreneurial behaviour as a learning outcome of engineering entrepreneurship programs. Here we present our on-going work on developing an entrepreneurial behavioural inventory that can be used to assess student learning in engineering entrepreneurship programs.

Background

The lack of quality assessment measures in entrepreneurship education has been of significant concern for the last several years (Carey & Matlay, 2010; Duval-Couetil, 2013; Pittaway & Edwards, 2012). Recognizing that entrepreneurship education is more than business knowledge, but should also include impacting entrepreneurial behaviours, attitudes and skills (Herrmann, Hannon, Cox, Termouth, & Crowley, 2008; Pittaway & Edwards, 2012), there has been calls for more innovative approaches to assessment (Pittaway & Edwards, 2012). To date, there has been limited rigorous work in the assessment of engineering entrepreneurship education (Purzer, Fila, & Nataraja, 2016). A recent review of current assessment methods in engineering entrepreneurship education identified 29 articles that utilized 52 assessment instruments (Purzer et al., 2016). Of those 52 assessment instruments, only 10 reported evidence of instrument reliability and validity. Instrument reliability and validity are critical when choosing an appropriate instrument to collect data (Creswell, 2009). The 10 instruments measure a range of constructs covering student perceptions of knowledge gained including, business principles (Besterfield-Sacre et al., 2012), leadership skills (Gerhart & Grunow, 2009) and personal professional competencies (Pistrui, Layer, & Dietrich, 2013); and affect measures include self-efficacy (Chen, Greene, & Crick, 1998; Duval-Couetil, Reed-Rhoads, & Haghighi, 2012) and attitudes (Duval-Couetil et al., 2012) towards entrepreneurship and leadership (Wielkiewicz, 2000).

In addition, five of the 10 instruments attempt to measure some aspect of behaviour. While one study directly measured behaviour through participation in entrepreneurship related activities (Duval-Couetil et al., 2012), others followed an indirect approach by using entrepreneurial intention or motivation (Kuckertz & Wagner, 2010; Liñán & Chen, 2009; Pistrui et al., 2013). To date, entrepreneurial intent has widely been used as a predictor of entrepreneurial behaviour. According on Ajzen's (1991) Theory of Planned Behaviour (TPB) (Ajzen, 1991), intent is a fundamental cognitive determinant of behaviour; the stronger an individual's intent to engage in a specific behaviour, the more likely s/he is to perform that behaviour.

However, there are two fundamental challenges with using entrepreneurial intent as a measure of engineering entrepreneurship education assessment. First, while TPB has been shown to be a predictor of entrepreneurial behaviour, it has been narrowly studied in the context of the intent to start a business. Second, these studies also rely on participant self-report responses, which may be more distorted than observed or measured behaviour (Fredricks & Dossett, 1983). Given that engineering entrepreneurship programs seek to motive entrepreneurial behaviours beyond new business development and the limitations of self-reported data, engineering education researchers should consider exploring different methods of measuring engineering entrepreneurial behaviour. To address this gap, we chose to develop a new measure of engineering entrepreneurial behaviour by identifying constituents of entrepreneurial behaviour and subsequently developing an instrument for assessing entrepreneurial behaviour as a learning outcome in engineering entrepreneurship programs.

Methodology

We followed a three-step process for developing an instrument for assessing entrepreneurial behaviour: identifying constructs, generating items, and gathering expert responses. For *identifying constructs*, we first conducted a literature review to examine the way in which entrepreneurial behaviour has been defined. Our review identified several constructs that constitute entrepreneurial behaviour. Second, experts were presented with the definition of 5 different constructs associated with entrepreneurs as defined by the literature review and were asked to review the list and answer - 1) Do these five constructs cover all the important behaviours that entrepreneurs exhibit? If not, what else should be included? 2) Which constructs, if any, would you remove from the list? Why? and 3) Are the definitions of the 5 listed constructs complete? If not, what should be changed? The 10 experts were either entrepreneurs and/or entrepreneurship course instructors in universities that teach undergraduate and graduate engineering entrepreneurship courses. The feedback was sought through an online survey from 9 experts. One expert preferred answering the survey questions verbally over the phone. Third, one researcher performed a simple analysis of the expert's responses. The common recommendations emerging from analysis were utilized to modify and expand the definition of entrepreneurial behaviour and its constituent sub-constructs.

For *generating items*, the final revised definitions were used. Two questions were developed for each construct. These open-ended questions asked how one would respond to a given hypothetical scenario. To *gather expert response*, the resulting survey was administered to 15 experts such that 5 expert responses were collected for each question. The responses were open coded by four researchers.

Findings

Identifying Constructs

Our review of literature identified Lau, Shaffer, Fai Chan, & Wing Yan Man (2012)'s work on entrepreneurial behaviour as our guiding framework. Lau, Shaffer, Fai Chan, & Wing Yan Man (2012) conceptualized entrepreneurial behaviour as a multi-faceted construct and identified five individual entrepreneurial behaviours: innovativeness, risk taking, networking, opportunism and change orientation. These five constructs were used as the basis for our work. However, we modified the definitions of these constructs to incorporate other relevant work in entrepreneurship education (Chen et al., 1998; McGee, Peterson, Mueller, & Sequeira, 2009). In addition to these published studies, resources from two additional sources - Kern Entrepreneurial Engineering Network (KEEN) and LeanLauch Curriculum were used to refine the definitions of the constructs. For the past several years, KEEN has been building an engineering entrepreneurship higher education network with the purpose of supporting the development of entrepreneurial engineers (KEEN, 2017). As part of their initiative, they have defined specific attributes (curiosity, connections, creating value) of an entrepreneurial engineer that extend beyond new venture creation. These attributes were used to inform this study. In addition, the recent widespread adoption of the Lean Launch curriculum for entrepreneurship has also influenced the scope of entrepreneurship education (Osterwalder & Pigneur, 2010). Thus, fundamental principles of the Lean Launch curriculum were also used to inform constructs examined.

Overall, all the experts' agreed that the five listed constructs were the most relevant entrepreneurial behaviours. However, there were several recommendations to modify the definitions and include an additional construct in the list. First, several experts suggested that we include an additional behaviour related to 'perseverance'. In multiple instances, experts explicitly stated 'perseverance' as an additional characteristic. For example, responding to the question asking about additional constructs that are part of entrepreneurial behaviour, one expert stated, "Perseverance: You need to stick to something for a long period of time to really make a difference". In other instances, expert responses described perseverance in the context of passion, determination and resilience. For example, describing a true entrepreneur, one respondent wrote, “A true entrepreneur can ‘not’ pursue his/her vision until all ways of advancing it have been completely exhausted...and sometimes beyond that point as well”. Similarly, describing an entrepreneur as a passionate and obstinate believer, another expert stated - “They have heard their whole lives ... ‘that would be too hard’, ‘that's been done before’, ‘what makes you think you can do this’, ‘shouldn't you just settle for a job’, ‘everyone is telling you this won't work, why aren't you listening?’, ‘are you crazy’ ... somewhere in there is this passionate, obstinate believer who believes when everyone else doubts”.

Second, there was an overall disagreement of the definition of ‘Risk Taking’. While one expert suggested removing it from the list, others recommended changing the definition of risk taking to “reframing the magnitude of risk”. For example, one expert recommended, “I would adjust ‘risk taking’ as I think entrepreneurs are often the most risk adverse people I know. What they are is ‘risk aware’”. In other words, experts suggested that the entrepreneurs are more risk aware and able to manage risks often by taking actions based on informed judgements about the associated risk. Third, experts frequently commented on the need to incorporated “relationship building” in the definition of ‘Networking’. In particular, several experts expressed that in the context of entrepreneurship, networking involves building relationships rather than just interacting formally and informally. In one such remark, an expert mentioned, “The word ‘relationships’ needs to be in there. Networking is not about interactions, it's about relationships”. Focussing on team building, one expert described networking as “the ability to build and maintain relationships in teams”. Another echoed, “For networking, nothing about person to person connection is mentioned in the process. It reads very transactional, which sometimes it is, but tends to work better when it's relational.”

Fourth, other recommendations included minor changes to the definitions. For ‘opportunism’, experts suggested removing ‘mundane’ from the definition. For ‘Innovativeness’, several experts’ responses reported that entrepreneurship involves “new ways of thinking, new ways of describing [and] reframing old problems in new ways”. Also, they underscored the importance of addressing customer and market needs in entrepreneurship and recommended adding "that meet customer/market needs" in the definition of innovativeness. Lastly, for ‘adaptability’, experts recommended not using ‘precautionary’ in describing the proactive approach. In sum, expert feedback provided directions for modifying and refining our initial definitions of entrepreneurial behaviour that were derived from existing literature. Experts’ recommendations were addressed by making changes accordingly. The revised constructs defining entrepreneurial behaviour are presented in Table 1.

Table 1: Entrepreneurial Behaviour Constructs

	Definition	**Informed by**
Opportunism	*Recognizing a need to develop new products, services or processes to* • Improve existing operations, and/or develop new marketing approaches; • Discover opportunities that enhance efficiency or quality; • Evaluate opportunities that involve balancing inadequate commitment of resources and the potential for return	• Opportunism (Lau et al., 2012) • Marketing (Chen et al., 1998) • Creating Value/Curiosity (KEEN, 2017) • Product market fit (Osterwalder & Pigneur, 2010)

Innovativeness	*Developing new and/or new ways of improving existing* • products, • services or • processes *that meet customer/market needs*	• Innovativeness (Lau et al., 2012) • Innovation (Chen et al., 1998) • Searching (McGee et al., 2009) • Value proposition (Osterwalder & Pigneur, 2010)
Networking	*Purposefully establishing informal and formal relationships with other people with the intent of recruiting physical and monetary resources, and exchanging information, experiences and advice.* • Internal (within an organization) • External (outside an organization)	• Networking (Lau et al., 2012) • Management (Chen et al., 1998) • Marshalling (McGee et al., 2009) • Customer discovery (Osterwalder & Pigneur, 2010)
Risk Management	*Take actions that reduce the probability of risk occurring or reduce the potential impact if the risk were to occur (Morris, 2013).* • Financial • Resources • Time	• Risk Taking (Chen et al., 1998; Lau et al., 2012) • Financial Control (Chen et al., 1998) • Connections (KEEN, 2017) • Lean Launch (Osterwalder & Pigneur, 2010)
Adaptability	*Altering oneself or the environment to the changed circumstances to one's advantage* • Proactive approach - involves seeking knowledge about the environment and taking necessary actions • Reactive approach - involves responding to changes rather than exploiting and initiating change	• Change Orientation (Lau et al., 2012) • Pivot (Osterwalder & Pigneur, 2010) • Proactive (Bateman & Crant, 1993)
Perseverance	*Ability to sustain goal-directed action and energy when confronting difficulties and obstacles that impede goal achievement.*	• Perseverance (Morris, Webb, Fu, & Singhal, 2013)

Item Generation

We used the final constructs and their definitions presented in Table 1 to develop an instrument for assessing student entrepreneurial behaviour. For each of the six constructs, two items were created resulting in a 12-item assessment instrument (Table 2). Considering the lack of prior instruments and consistency in terminology, we created open-ended rather than close-ended items at this stage of instrument development. These items asked 'what one would do' in response to the given scenarios that intended to assess the six identified constructs of entrepreneurial behaviour. These scenarios were created such that they were applicable from an undergraduate students' perspective. The use of open-ended response type allowed us to capture a wide range of responses to the developed items.

Expert Response

Overall, the responses provided insights on how experts will respond to the presented scenarios. In regard to instrument development, two findings were particularly noteworthy. First, although we received a wide range of responses for opportunism and innovativeness, the responses for the two constructs were similar.

In particular, similar to innovativeness, responses for opportunism proposed different solutions to improvise the existing operation. This is because our items for opportunism asked, 'would you use the same approach? If not, *what would you change*?' A logical solution is to only ask whether one would use the same approach? However, a yes/no response would have not provided detailed insights. A plausible implication of this finding is that while theoretically opportunism and innovativeness may be two separate constructs, they may be difficult to delineate when designing assessment items. This is primarily because their definitions are intertwined with opportunism (recognizing a need to improve existing products) being the precursor of innovativeness (developing new ways to improve existing products).

Second, in contrast with other items, the second item for perseverance construct received only two response types - 'take elective' and 'seek other opportunities for interaction'. Furthermore, the number of responses was also highly skewed with 'take elective' noted in 5 expert responses and the other in one expert response.

Table 2: Expert response to behaviour items

	Item	Response Codes	Responses
Opportunism	This year you have been asked to recruit students to new interdisciplinary undergraduate courses introduced by the College of Engineering. Typically, students learn about new classes through registrar emails sent through the department list serv. Would you use the same approach? If not, what would you change?	No Change	2
		In person approach (e.g. Information sessions)	3
		Distribution (e.g. flyers)	3
		Virtual Approach (e.g. personalized emails)	1
	You are organizing a career fair for the College of Engineering. For the last ten years, the fair has been a one-day event that included companies from different engineering disciplines. Students usually walk around different booths to find potential jobs for their engineering specialization and skillset. Would you use the same approach? If not, what would you change?	No Change	5
		Increase efficiency	2
		In person approach (e.g. Workshops and student booths)	4
		Distribute student information (e.g. resume books)	2
Innovativeness	You recently learned that many online class assignments take extra time because of the slow response time of computers in the residence halls. The problem causes a great deal of inconvenience to your classmates and also has led to a decline in homework grades. What would you do?	Leave problem (e.g. drop course)	3
		Fix technical problem	1
		Talk to person in-charge	2
		Remove electronic submission	1
	Something is wrong with the operating system of your computer. You seem to recall that your friend who has now graduated had a similar problem and was eventually able to solve it. You know it will take a long time to have the problem fixed by the technical support department of your college. What would you do?	Identify someone else to fix the problem	3
		Online help	3
		Fix yourself	1
		Contact friend	3

Networking	You are part of a design team developing medical devices for people outside your home country. To better design a product, you need to understand the potential users in that country. How will you learn about them?	Talk to users	5
		Talk to stakeholders and/or experts	5
		Online search	1
		Run experiments	1
	As an engineer, you want to pursue a career with financial firms. But, you do not know the different roles engineers can play in financial firms. What would you do?	Connect virtually	2
		Use online sources	4
		Use print media	4
		In-person approach (e.g. events)	4
Risk Management	Your design team has been working with a non-profit organization in South America to deliver a low-cost solar refrigerator. Your advisor says you cannot book the flights to deliver the refrigerator unless you have positive results on the validation tests. If you do not buy the tickets right now, the price will go up and all the team members will not be able to go due to limited funding. What would you do?	Depends on the situation	2
		Perform validation	3
		Purchase refundable tickets	1
	The Dean of your college has offered the Student Government funds to host a weeklong series of events for alumni during homecoming. Unfortunately, the Student Government has committed to organizing a career fair during that same period. Not hosting the homecoming event would damage your relationship with the Dean and alumni. What would you do?	Communicate with stakeholders	2
		Merge events	3
		Run more valuable event	1
Adaptability	A friend has told you that the course you want to take next semester requires extra time and effort due to multiple projects. You want to graduate next semester. This course will be beneficial in finding a job after graduation, but you are taking an overwhelming amount of credits next semester to graduate on time. What would you do?	Take the beneficial course	2
		Evaluate priorities (e.g. delay graduation)	2
		Adapt (e.g. audit course)	3
		Do not take the beneficial course	2
	You have organized a student-led event on campus. Two days before the event, you are notified that the venue you reserved will be unavailable due to a building repair emergency. What would you do?	Find alternate venue	5
		Evaluate cost	1
		Find alternate time	2
Perseveranc e	You conceptualized an efficient prototype in your freshman design course. You are confident that if you spend more time on it, you can develop the prototype and it can be sold. But, you have a lot of courses to take to complete your degree requirements. What would you do?	Take course later	2
		Devise a plan	1
		Pursue design (e.g. develop prototype)	4

	You are trying to get a job at a local automotive company. You have an opportunity to enrol in an elective course that includes a mentorship opportunity with employees from this automotive company. This course is an elective course and you have a lot of required courses to take to complete your degree requirements. What would you do?	Take elective	5
		Seek other opportunities for interaction	1

Conclusion and Future Work

Entrepreneurial behavior can be expressed in multiple ways and should not be limited to starting a company. Leveraging existing literature and expert feedback, this work identifies six sub-constructs describing entrepreneurial behavior – opportunism, innovativeness, networking, risk management, adaptability and perseverance. Also, we present items generated for these constructs and preliminary expert response data. While the expert response data provides insights about the validity of the constructs and developed items, reliance on only expert response to modify the instrument will not be the best approach considering the instrument ultimately aims to assess entrepreneurial behavior as a student learning outcome. Our next steps include, collecting students' responses to the behavior items to revise the instrument, establishing validity of the items in measuring the constructs and developing a reliable coding schema for analysis of open-ended items.

References

Ajzen, I. (1991). The theory of planned behavior. *Organizational Behavior and Human Decision Processes*, *50*(2), 179–211.

Besterfield-Sacre, M. E., Robinson, A. M., Ozaltin, O., Shuman, L. J., Shartrand, A. M., & Weilerstein, P. (2012). Essential factors related to entrepreneurial knowledge in the engineering curriculum. *119th ASEE Annual Conference and Exposition*. Retrieved from http://www.scopus.com/inward/record.url?eid=2-s2.0-84864995680&partnerID=40&md5=f1b4ccdc00a2a13d51ed7f0abb5d197b

Besterfield-Sacre, M., Ozaltin, N. O., Shartrand, A., Shuman, L. J., & Weilerstein, P. (2011). Understanding the technical entrepreneurship landscape in engineering education (CCLI-Phase 2). In *ASEE Annual Conference and Exposition, Conference Proceedings*.

Carey, C., & Matlay, H. (2010). Creative disciplines education: a model for assessing ideas in entrepreneurship education? *Education + Training*, *52*(8/9), 694–709. https://doi.org/10.1108/00400911011088999

Chen, C. C., Greene, P. G., & Crick, A. (1998). Does entrepreneurial self-efficacy distinguish entrepreneurs from managers? *Journal of Business Venturing*, *13*(4), 295–316.

Creswell, J. W. (2009). *Research design: Qualitative, quantitative, and mixed methods approaches*. Thousand Oaks, CA: Sage Publishing.

Duval-Couetil, N. (2013). Assessing the impact of entrepreneurship education programs: Challenges and approaches. *Journal of Small Business Management*, *51*(3), 394–409. https://doi.org/10.1111/jsbm.12024

Duval-Couetil, N., Reed-Rhoads, T., & Haghighi, S. (2012). Engineering students and entrepreneurship education: Involvement, attitudes and outcomes. In *International Journal of Engineering Education* 28(2), 425–435). Retrieved from https://uspfodc.us/resources/41/download/Engeering_Students_and_EShip_Education_-_Involvement_Attitudes_and_Outcomes.pdf

Duval-Couetil, N., Shartrand, A., & Reed, T. (2016). The role of entrepreneurship program models and experiential activities on engineering student outcomes. *Advances in*

Engineering Education, *5*(1), 1–27.

Fredricks, A. J., & Dossett, D. L. (1983). Attitude-behavior relations: A comparison of the Fishbein-Ajzen and the Bentler-Speckart models. *Journal of Personality and Social Psychology*, *45*(3), 501–512. https://doi.org/10.1037/0022-3514.45.3.501

Gerhart, A., & Grunow, M. (2009). Leadership models and practices course - Student perceptions and development of leadership skills and incorporation of a new leadership course. In *ASEE Annual Conference and Exposition, Conference Proceedings*.

Gilmartin, S. K., Chen, H. L., & Estrada, C. (2016). Investigating Entrepreneurship Program Models in Undergraduate Engineering Education, *International Journal of Engineering Education 32*(5), 2048–2065.

Herrmann, K., Hannon, P., Cox, J., Termouth, P., & Crowley, T. (2008). *Developing Entrepreneurial Graduates*. London.

KEEN. (2017). Kern Entrepreneurial Engineering Network. Retrieved from http://engineeringunleashed.com/keen/

Kuckertz, A., & Wagner, M. (2010). The influence of sustainability orientation on entrepreneurial intentions - Investigating the role of business experience. *Journal of Business Venturing*, *25*(5), 524–539. https://doi.org/10.1016/j.jbusvent.2009.09.001

Lau, T. L. M., Shaffer, M. A., Chan, K. F., & Man, T. W. Y. (2012). The entrepreneurial behaviour inventory: A simulated incident method to assess corporate entrepreneurship. *International Journal of Entrepreneurial Behavior & Research*, *18*(6), 673–696.

Liñán, F., & Chen, Y. W. (2009). Development and cross-cultural application of a specific instrument to measure entrepreneurial intentions. *Entrepreneurship: Theory and Practice*, *33*(3), 593–617. https://doi.org/10.1111/j.1540-6520.2009.00318.x

Massy, W. F., Sullivan, T. a., & Mackie, C. (2013). *Improving Measurement of Productivity in Higher Education. Change: The Magazine of Higher Learning* (Vol. 45). National Academies Press. https://doi.org/10.1080/00091383.2013.749140

McGee, J. E., Peterson, M., Mueller, S. L., & Sequeira, J. M. (2009). Entrepreneurial self-efficacy: refining the measure. *Entrepreneurship Theory and Practice*, *33*(4), 965–988.

Morris, M. H., Webb, J. W., Fu, J., & Singhal, S. (2013). A competency-based perspective on entrepreneurship education: Conceptual and empirical insights. *Journal of Small Business Management*, *51*(3), 352–369. https://doi.org/10.1111/jsbm.12023

National Academy of Engineering. (2004). *The Engineer of 2020: Visions of Engineering in the New Century*. Washington DC: The National Academies Press. Retrieved from http://www.nap.edu/openbook.php?record_id=10999

National Academy of Engineering. (2005). *Educating the engineer of 2020: Adapting engineering education to the new century* . Washington, D.C: The National Academies Press.

Osterwalder, A., & Pigneur, Y. (2010). *Business model generation: a handbook for visionaries, game changers, and challengers*. John Wiley & Sons.

Pistrui, D., Layer, J. K., & Dietrich, S. (2013). Mapping the behaviors, motives and professional competencies of entrepreneurially-minded engineers in theory and practice: An empirical investigation. *Journal of Engineering Entrepreneurship*, *4*(1), 39–54.

Pittaway, L., & Edwards, C. (2012). Assessment: examining practice in entrepreneurship education. *Education + Training*, *54*(8/9), 778–800. https://doi.org/10.1108/00400911211274882

Purzer, S., Fila, N. D., & Nataraja, K. (2016). Evaluation of Current Assessment Methods in

Engineering Entrepreneurship Education. *Advances in Engineering Education*, 1–13. Retrieved from http://advances.asee.org/mission.cfm

Shartrand, A., Weilerstein, P., Besterfield-Sacre, M., & Golding, K. (2010). Technology Entrepreneurship Programs in U.S. Engineering Schools: Course and Program Characteristics at the Undergraduate Level. In *American Society for Engineering Education*.

Shuman, L. J., Besterfield-Sacre, M., & McGourty, J. (2005). The ABET "Professional Skills" - Can They Be Taught? Can They Be Assessed? *Journal of Engineering Education*, *94*(1), 41–55. https://doi.org/10.1002/j.2168-9830.2005.tb00828.x

Standish-Kuon, T., & Rice, M. P. (2002). Introducing engineering and science students to entrepreneurship: Models and influential factors at six American universities. *Journal of Engineering Education*, *91*(1), 33.

Wielkiewicz, R. M. (2000). The leadership attitudes and beliefs scale: An instrument for evaluating college students' thinking about leadership and organizations. *Journal of College Student Development*, *41*(3), 335–347.

Acknowledgements

This project is funded by the U.S National Science Foundation through grant number 1504257. The opinions are those of the authors and do not necessarily represent the National Science Foundation. We are thankful to Laura Hirshfield for her help in this work.

Copyright statement

Work in Progress—Another Step Toward an Internationally-Inclusive Framework Characterizing the Impact of Engineering Education Research

Jeremi London
Arizona State University, Mesa, AZ, USA
Jeremi.London@asu.edu

Parth Patel
Arizona State University, Mesa, AZ, USA
Parth@asu.edu

***Abstract**: It is important for engineering education researchers to be able to effectively communicate the impact of their work to colleagues and stakeholders in and outside of the engineering education community. Unfortunately, there is little scholarship on how to do so. The purpose of this study is to develop a conceptual framework that characterizes the impact of engineering education research and is inclusive of international perspectives. The process for developing the framework is informed by Messick's Unified Theory for Instrument Development and Validation. To date, engineering education researchers around the world have provided input on the contents of the framework; this has resulted in a draft that is ready for review. The workshop surrounding this work in progress is designed to gather input on the generalizability of the draft. Because of this work, the field of engineering education is one step closer to having an internationally-inclusive research impact framework.*

Context

In an environment of increased accountability and shrinking support for research, engineering education researchers and practitioners are being asked to identify the tangible impacts of their work. More broadly, the benefits of research are under scrutiny because its importance is not well understood by society, and is often considered to have no practical usefulness (Bornmann & Marx, 2014; Petit, 2004; Salter & Martin, 2001; Smith, 1997; Ziman, 2000). Unfortunately, there is a dearth of scholarship on how research impact should be defined and evaluated in any field; and the engineering education community lacks a shared understanding and language around the topic of impact (London, 2014). As a result, many struggle to articulate the impact of our work to diverse audiences, and are unable to advocate for engineering education using consistent messages. It is imperative that the global engineering education research community coalesces on a shared meaning of what impact looks like in this context– or else an inability to clearly communicate the impact of our work may lead to a loss of public support for our research and development (R&D). Because this community spans many geographic regions, it is important that the conversation around this topic engages members of the worldwide engineering education community.

Purpose & Research Questions

The purpose of this study is to develop a framework that characterizes the impact of engineering education research and is inclusive of international perspectives.

The guiding research question is as follows: *What is a meaningful and shared description of the impact of engineering education research that is internationally inclusive and according to engineering education researchers' perspectives?*

Theoretical Lens & Methodological Approach

Messick's unified theory for instrument development and validation (Messick, 1995) serves as the methodological framework for this study. Based on Messick's theory, validity includes six facets: substantive, content, generalizability, consequential, structural, and external. Insights from a literature review, card sorting activities among engineering education researchers, and hierarchical cluster analysis provide evidence for the first two aspects of validity. The next step is to evaluate the extent to which the draft framework is representative of the construct domain and generalizes to other groups and settings.

Results

The content of frameworks developed to characterize research impact in the medical sciences (Dembe et al., 2014), informal science education (Allen et al., 2008), arts and humanities (Levitt et al., 2020), and science, in general (Godin & Doré, 2005) served as the starting place for the content of the engineering education research impact framework. Through manual and electronic card sorting activities, engineering education researchers (n=116) around the world provided input on whether the contents were relevant to engineering education, not relevant, or relevant if adapted. The participants also provided feedback on which of the "relevant" ideas should be grouped together, and proposed new ideas that were not reflected among the existing frameworks. Based on the cluster analysis results, members of the community perceive that nearly three-quarters of the 125 impact descriptors from existing frameworks are relevant or relevant if adapted; and there are clusters of ideas within each group. Such insights have led to a draft framework that needs additional input from the international engineering education research community.

Conclusion

While there is little scholarship on the impact of research, existing research impact frameworks serve as a meaningful starting place for developing a valid research impact framework for the field of engineering education. A sound methodological lens and a combination of methods help guide the exploration of this unfamiliar territory. Because of this work, the field of engineering education is one step closer to having an internationally-inclusive research impact framework.

References

Allen, S., Campbell, P.B., Dierking, L.D., Flagg, B.N., Friedman, A.J., Garibay, C., . . . Ucko, D.A. (Eds.). (2008). *Framework for Evaluating Impacts of Informal Science Education Projects*. (Available at: http://informalscience.org/evaluations/eval_framework.pdf).

Bornmann, Lutz, & Marx, Werner. (2014). How should the societal impact of research be generated and measured? A proposal for a simple and practicable approach to allow interdisciplinary comparisons. *Scientometrics, 98*(1), 211-219.

Dembe, A.E., Lynch, M.S., Gugiu, P.C., & Jackson, R.D. (2014). The Translational Research Impact Scale: Development, Construct Validity, and Reliability Testing. *Evaluation & the Health Professions, 37*(1), 50-70.

Godin, B., & Doré, C. (2005). Measuring the Impacts of Science: Beyond the Economic Dimension. http://www.csiic.ca/PDF/Godin_Dore_Impacts.pdf

Levitt, R., Claire, C., Diepeveen, S., Chonaill, S.N., Rabinovich, L., & Tiessen, J. (2010). Assessing the impact of arts and humanities research at the University of Cambridge: Rand Corporation. Technical Report.

London, Jeremi S. (2014). *The Impact of National Science Foundation Investments in Undergraduate Engineering Education Research: A Comparative, Mixed Methods Study.* (Doctor of Philosophy Dissertation), Purdue University, ProQuest Dissertations and Theses database. (3687797). http://docs.lib.purdue.edu/open_access_dissertations/327

Messick, S. (1995). Validity of psychological assessment: Validation of inferences from persons' responses and performances as scientific inquiry into score meaning. *American psychologist, 50*(9), 741-749.

Petit, J C. (2004). Why do we need fundamental research? *European Review, 12*(2), 191-207.

Salter, Ammon J, & Martin, Ben R. (2001). The economic benefits of publicly funded basic research: a critical review. *Research Policy, 30*, 509-532.

Smith, Christopher Llewellyn. (1997). What's the use of basic science? http://wwwnew.jinr.ru/section.asp?sd_id=94

Ziman, J. (2000). *Real Science: What it is, and what it means*. Cambridge, UK: Cambridge University Press.

Acknowledgements

This work is supported by the U.S. National Science Foundation award EEC-1564629. Any opinions, findings, and conclusions or recommendations expressed in this material are those of the authors and do no necessarily reflect the views of the National Science Foundation.

Shifting the Paradigm in the Teaching/Learning Evaluation

Engelberth Soto-Estrada
EIA University, Envigado, Colombia
engelberth.soto@eia.edu.co

Jairo Gómez-Lizarazo
EIA University, Envigado, Colombia
jairo.gomez@eia.edu.co

***Abstract**: Student learning is influenced by many factors, from the student's own motivation to the curriculum structure. Despite this, teaching/learning evaluation in higher education institutions remains focused on teaching performance, while the results of such instruments are used as an indicator (and in some cases the only indicator) of quality and effectiveness of learning. This article argues that teacher-centered evaluation provides insufficient and unreliable information to improve the teaching/learning process. Instead, a process-based evaluation model (PBEM) is proposed. PBEM is intended to deliver reliable information about teaching effectiveness, but also about other factors influencing the teaching/learning process such as institutional organization, curriculum structure, and content. PBEM was designed in accordance with the EIA University's pedagogical model. The implementation of the instrument started in 2015. Since then, more than 2,000 courses have been evaluated. This article addresses some key issues and challenges related to the implementation of such a model.*

Introduction

The EIA University (known until recently as Antioquia School of Engineering) is located 20 km east of Medellin, in the Colombian department of Antioquia. The university offers undergraduate and postgraduate degrees covering a range of engineering disciplines such as civil, mechanical, systems, environmental and industrial engineering.

Until 2014, all courses were evaluated by the use of an electronic questionnaire designed to provide information on students' perceptions of teaching effectiveness. Questions were raised about the suitability of the lecturer-centered evaluation method as the information delivered by this instrument was inconsistent with the characteristics of well-known and well-qualified lecturers and courses. Furthermore, the evaluation items were unrelated to the Institutional Educational Project (IEP). IEP is a mandatory document for Higher Education Institutions (HEI) in Colombia. It explains how education is conceived by the HEI in terms of principles and pedagogical strategies (Congreso de Colombia, 1994).

Considering the above mentioned, the EIA University Academic Council decided to modify the evaluation procedure. Two main goals were established for the new mechanism: to reduce bias and to assess the progress made towards the adoption of the IEP in the teaching and learning process. The implementation of the new instrument started in 2015 and is currently in use.

This paper is intended to explain the characteristics of the EIA University new evaluation instrument (section 3) and to discuss the results of its implementation (section 4). Section 2 provides background information about significant research findings considered in this project. Finally, in the last section conclusions and future work are outlined.

Background

Student Evaluation of Teaching (SET) is a common practice in HEI (Muñiz & Fonseca-Pedrero, 2008). Many instruments and procedures for collecting and analyzing information as the main and "in some cases, the sole indicator of teaching quality" have been developed (Spooren, et al., 2013). SET serves three purposes: improving teaching quality, support faculty decisions, and providing evidence on institution accountability. However, the implementation, the reliability, and the validity of the different methods and instruments have received considerable faculty and administrators' criticism and concern, which makes SET an active research field (McCarthy, 2012).

SET research areas include reliability, validity, and the analysis of factors that influence students' ratings (Addison & Stowell, 2012). For almost a century, since the pioneering work carried out at the universities of Pennsylvania (Freyd, 1923) and Purdue (Remmers & Brandenburg, 1927), thousands of publications on SET have been issued. The structure of the instruments has evolved, from the consideration of the two factors (professional maturity and rapport) proposed by Remmers & Brandenburg (1927) to the six factors identified by Isaacson et al. (1964): overloading, skill, structure, feedback, group interaction, and rapport. Marsh (1991) proposed nine factors of teaching quality: breadth of coverage, organization/clarity, learning/value, examinations/grading, enthusiasm, rapport, group interaction, assignments/readings, and workload/difficulty. Addison & Stowell (2012) concluded that teaching quality can be explained by few factors related to how a course is designed and taught and to the instructor ability to establish a good connection with the students.

While several variables have been studied as potential influencers of students' response, the most common is the connection with academic performance. However, this argument requires further examination and discussion. Aleamoni (1987) was unable to find evidence of this relationship. Zabaleta (2007) concluded that "SET results are not transferable and cannot be used to compare teachers". The size of the class was also investigated to validate SET instruments. Hobson & Talbot (2001) proposed a minimum size class of 15 for quality results. Meng et al. (2013) pointed out another factor affecting SET, the lack of support to carry out internal validity in relation to the psychometric value of the instrument.

In order to improve the reliability of a SET instrument, Lemos et al. (2011) suggested to start with the clarification of the pedagogical model where the SET mechanism is going to be used, then to establish the objectives of the evaluation, and finally to define the dimensions that represent teaching quality.

The EIA University pedagogical model, described in its IEP, belongs to the social development theory, which was first proposed by Vygotsky. The central objective of the university's model is to develop students' intellectual abilities and skills, followed by profession-specific knowledge. Four social competencies are included: creativity, teamwork, communication, and systemic thinking. In addition, the university promotes the values of honesty, respect, responsibility, and solidarity. The institution fosters the intellectual and character development of students in key transverse training areas, including information management, research, sustainability, and entrepreneurship.

Methodology and PBEM design

A four-step methodology was used, similar to the strategy proposed by Lemos et al. (2011). The first step was to assess the preceding evaluation instrument. The assessment was based on an online survey. Lecturers' responses pointed out weakly formulated or redundant questions, as well as relevant aspects that were omitted in the evaluation procedure.

Next, the new evaluation instrument was outlined. As previously mentioned, two fundamental goals were defined: the new mechanism should reduce bias and it had to provide information on the adoption of the university's IEP during the teaching/learning process.

To achieve these objectives, the design concept of the institutional SET instrument was modified, from the lecturer-centered to a process evaluation model. In a process evaluation, four broad aspects must be considered: inputs, activities, outputs, and the integration between these components (Unite for Sight, 2015). Table 1 shows how process evaluation was used as a framework for teaching/learning assessment at the EIA University.

Table 1: Process evaluation vs. teaching/learning at the EIA University

Process evaluation	Teaching/learning factors	Responsible
Inputs	Students' previously acquired knowledge	Academic staff involved in curriculum design / former lecturers
	Course content and schedule	Program director / support staff / lecturer
	Physical facilities and resources	Physical facilities' manager / academic director
Activities	Methodological strategies and didactic resources	Lecturer
Outputs	Course learning outcomes	Lecturer / academic staff involved in curriculum design / program director and support staff / physical facilities' manager
Integration between components	Teaching/learning process carried out according to IEP, course content, and schedule	Lecturer

As indicated in Table 1, a process-based evaluation approach encompasses more aspects and actors involved in the teaching/learning process than a traditional SET instrument. Therefore, a process-based method is suitable to examine many of the factors that influence student success. The Organization for Economic Co-operation and Development summarizes these factors (OECD, 2009): "Student learning is influenced by many factors. These include the student's own skills, expectations, motivation, and behavior along with the support they receive from their families and the influence of their peer group. In addition to the quality of teachers, other factors include school organization, resources, and climate; and curriculum structure, and content. The effect of teachers is also cumulative, i.e. at a given moment in time student learning is influenced not only by the current teachers but also by former teachers."

A number of studies have explored the influence of the organizational performance of HEI on student learning outcomes, focusing on the contribution of lecturers and non-academic staff. Non-academic staff, such as the physical facilities' manager, is decisive for teaching and learning. The influence of facility conditions on learning has been extensively documented. Earthman (2002) studied the influence of school facility conditions on student academic achievement. Peccolo (1962) analyzed the effect of thermal environment on learning. Connolly et al. (2015) and Woolner & Hall (2010) examined the impact of noise on the learning process.

The evaluation questionnaire was also improved. To reduce students' response bias, the teacher-centered questioning was modified to process-centered. Table 2 contains two examples of process-centered questions.

Table 2: Comparison between teacher-centered and process-centered questions

Item	Teacher-centered	Process-centered
Learning outcomes	How much have you learned from this teacher?	How do you rate the overall value of this course?
Instructional aids	Did the teacher use instructional aids to enhance learning?	How do you rate the instructional aids used to enhance learning in this course?

The final PBEM-questionnaire is composed of 27 items that provide information on four factors of the teaching/learning process: planning, teaching effectiveness, adoption of the IEP and student self-assessment. A simple factor structure was proposed, i.e., each item has a relatively strong loading on one factor (target-loading) and relatively small loading on other factors (cross-loadings). However, a cross-loading analysis is also used to examine latent variables that can be related to observed variables (or manifest variables), based on questionnaire items, such as the students' perception of learning. In the PBEM the perception of learning is evaluated by a single item, and by the manifest variables shown in Table 3.

Table 3: Manifest variables used to examine the students' perception of learning

Latent variable	Manifest variable	Primary target factor
Perception of learning	Previous knowledge	Planning
	Motivation	Teaching effectiveness / Student self-assessment
	Peer group	Student self-assessment
	Course content covered	Teaching effectiveness
	Teaching quality (5 items)	Teaching effectiveness
	Curriculum structure	Planning
	Condition of university facilities	Planning

Implementation of PBEM was the third phase of this project. After the Academic Council's authorization, PBEM started to be applied during the first semester (spring-summer) 2015. Since then, more than 2,000 courses have been evaluated. During the last two years (2016-2017), the opinion of lecturers, students and university staff have been gathered. The project is currently in its fourth phase where PBEM results are under analysis and the evaluation questionnaire adjusted.

Results and discussion

As mentioned previously, more than 2,000 courses were evaluated so far. Figure 1 shows the number of courses evaluated per semester, Figure 2 displays the students' response rate. These two figures evidence that, while the number of courses in the university has increased, the average response rate has declined. This trend may be due to various factors such as the introduction of the new evaluation instrument, but is mainly the result of the voluntary character of teaching/learning evaluation at the EIA

University and to the lack of support from the academic community to fully implement PBEM.

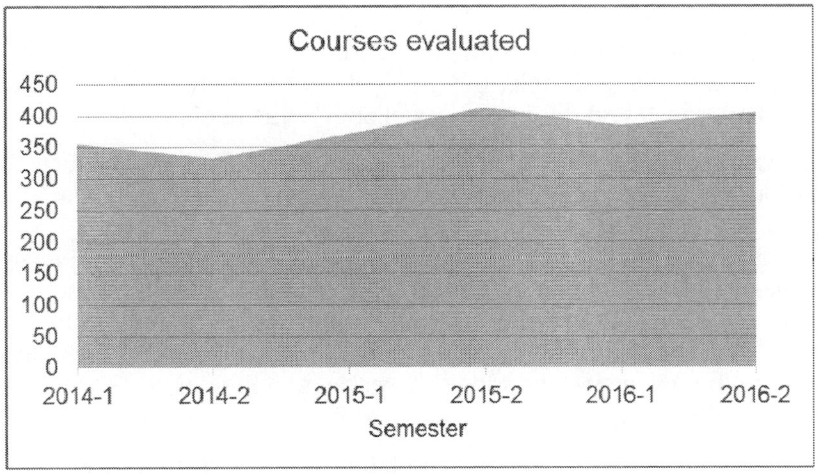

Figure 1: Number of courses evaluated through PBEM

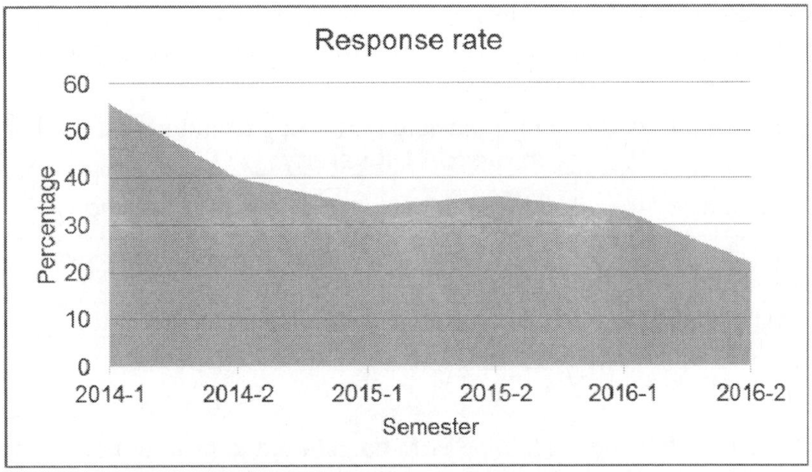

Figure 2: Students' response rate

After the introduction of the new evaluation procedure, questions were raised about its validity. Both students and university staff questioned the decision to choose a process-based approach.

Students were unfamiliar with process questioning and asked for more items related directly to teaching effectiveness. They (those students who expressed any opinion about the evaluation procedure) showed little motivation to evaluate the process, on the contrary, they wanted to continue focusing on teaching performance. It would seem that the students were unable to recognize that in the process of teaching and learning many factors and actors are involved. Conversely, they considered the teacher as the person responsible for the overall quality of learning and therefore the only factor that had to be improved by using SET. However, the significance and usefulness of students' feedback were extensively debated among university staff (and was, in fact, one of the reasons to modify teacher-centered SET), as most of the comments were superficial or influenced by lecturers' grading (for good and for bad).

Despite the previously mentioned, the fact that PBEM is a voluntary activity at the EIA University remains the major challenge to increasing students' response rate.

In the same way, the new instrument generated concerns and doubts among academic staff. The vast majority of lecturers agreed that the information of the previous evaluation procedure was unreliable, but many of them were unable to understand the process-based approach and requested "traditional" evaluation.

Practical and conceptual reasons are behind this reaction. Part-time lecturers require teaching performance records that can be used as experience evidence in other universities, which means that the records received at the EIA University had to be comparable with those used in other institutions (and the comparison between PBEM and teacher-centered evaluation was challenging, at least in the first version of the questionnaire, since the instrument was tailored to the needs of the institution). On the other hand, many lecturers, both part-time and full-time, were unable to recognize the relation between the university's pedagogical model and their courses, which made PBEM implementation a laborious case-by-case process.

The Academic Council agreed that the dimensions that represent teaching effectiveness at the EIA University had to include traditional variables such as those suggested by Marsh (1991): breadth of coverage, organization/clarity, examinations/grading, etc., and variables related to the university's pedagogical model. As mentioned previously, the EIA University's IEP includes values, competencies, and key transverse areas of training (see Table 4). These variables were included in the PBEM questionnaire.

Table 4: Institutional values, competencies, and key transverse areas of training at the EIA University

Competencies	Values	Key areas of training
Creativity	Honesty	Information management
Teamwork	Respect	Research
Communication	Responsibility	Sustainability
Systemic thinking	Solidarity	Entrepreneurship

Despite the multiple challenges faced, PBEM has started to provide useful information to improve planning, teaching effectiveness, and the adoption of the IEP at the university. Table 5 resumes average students' ratings (on a scale of 1-5 with 5 being excellent) per academic program, corresponding to the courses taught during the spring semester 2016.

Table 5: Average students' ratings per program (2016-1)

Factor	Undergraduate academic program										UEIA mean value
	1	2	3	4	5	6	7	8	9	10	
Planning	4,0	4,2	4,3	4,2	4,3	4,3	4,3	4,1	4,1	4,3	4,3
Teaching effectiveness	3,8	4,2	4,4	4,2	4,3	4,3	4,4	4,1	4,1	4,3	4,2
IEP adoption	4,0	4,3	4,4	4,3	4,1	4,2	4,4	4,2	4,1	4,1	4,1
Student self-assessment	4,4	4,2	4,6	4,4	4,5	4,4	4,6	4,2	4,3	4,6	4,3

During the spring semester 2016, the overall students' rating was 4.21, which can be considered satisfactory. None of the four factors were perceived negative (below 3) in any of the 10 undergraduate academic programs evaluated. However, further analysis is required for specific items that consistently received lower ratings (two of them belonging to IEP adoption: sustainability and entrepreneurship) and for the academic program 1, which, in comparison to others, received lower ratings.

Conclusions and future work

Despite the difficulties encountered during PBEM implementation, the evaluation instrument started to show its potential to improve teaching effectiveness and curriculum design and to provide information to understand how learning and intellectual development of students are taking place at the EIA University. That was possible due to the incorporation of the institutional competencies, values, and key transverse training areas to the evaluation procedure.

However, in order to achieve the full and effective implementation of PBEM two fundamental tasks have to be accomplished: to improve students and lecturers' understanding of the institutional pedagogical model, and to shift the way teaching/learning evaluation has been conceived by the academic community at the EIA University.

Understanding the pedagogical model is crucial to clarify the content of the evaluation instrument. To address this issue students and lecturers are being informed about the academic objectives that the university is trying to achieve, and consequently, evaluate. PBEM represents an opportunity to enhance the coherence between the institutional educational project and teaching/learning practices at the EIA University.

Finally, to increase the acceptance of the process-based evaluation more data about its validity is required. A confirmatory factor analysis is expected to take place this year to determine the reliability of the PBEM.

References

Addison, W. E., & Stowell, J. R. (2012). Conducting Research on Student Evaluations of Teaching. In M. E. Kite (Ed.), *Effective Evaluation of Teaching: A guide for faculty and administrators.* Retrieved from the Society for the Teaching of Psychology website: http://teachpsych.org/ebooks/evals2012/index.php

Aleamoni, L. M. (1987). Student rating myths versus research facts. *Journal of Personnel Evaluation in Education,* Issue 1, pp. 111-119.

Congreso de Colombia, (1994). *Ley general de educación, Decreto 115 de 1994.* Bogotá.

Connolly, D. et al. (2015). Students' Perceptions of School Acoustics and the Impact of Noise on Teaching and Learning in Secondary Schools: Findings of a Questionnaire Survey. *Energy Procedia,* Volumen 78, pp. 3114-3119.

Earthman, G. I. (2002). School Facility Conditions and Student Academic Achievement, California: UCLA´s Institute for Democracy, Education, and Access.

Freyd, M. (1923). A Graphic Rating Scale for Teachers. *The Journal of Educational Research,* 8(5), pp. 133-139.

Hobson, S. M., & Talbot, D. M. (2001). Understanding student evaluations: What all faculty should know. *College Teaching,* Issue 49, pp. 26-31.

Isaacson, R. L., et al. (1964). Dimensions of student evaluations of teaching. *Journal of Educational Psychology,* Issue 55, pp. 344-351.

Lemos, M.; Queiros, C.; Teixeira, P., & Menezes, I. (2011). Development and validation of a theoretically based, multidimensional questionnaire of students´ evaluation of university teaching. *Assessment & Evaluation in Higher Education,* Issue 36, pp. 843-864.

Marsh, H. W. (1991). A multidimensional perspective on students' evaluation of teaching effectiveness: A reply to Abrami and d'Apollonia. *Journal of Educational Psychology,* Issue 83, pp. 416-421.

McCarthy, M. A. (2012). Using Student Feedback as One Measure of Faculty Teaching Effectiveness. In M. E. Kite (Ed.), *Effective Evaluation of Teaching: A guide for faculty and administrators.* Retrieved from the Society for the Teaching of Psychology website: http://teachpsych.org/ebooks/evals2012/index.php

Meng, L. T.; Chea, C., & Nooi, P. S. (2013). Questionnaire design and data analysis: an alternative approach in student evaluation of Teaching. *Sunway Academic Journal,* Volumen 9, pp. 1-15.

Muñiz, J., & Fonseca-Pedrero, E. (2008). Construcción de instrumentos de medida para la evaluación universitaria. *Revista de Investigación en Educación,* Volumen 5, pp. 13-25.

OECD (2009). Teacher Evaluation. A Conceptual Framework and examples of Country Practices, Mexico: OECD.

Peccolo, C. (1962). *The Effect of Thermal Environment on Learning. A Pilot Study,* Iowa City: U.S. Department of Health, Education, and Welfare.

Remmers, H. H., & Brandenburg, G. C. (1927). Experimental data on the Purdue Rating Scale for Instructors. *Educational Administration and Supervision,* Issue 13, pp. 519-527.

Spooren, P.; Brockx, B., & Mortelmans, D. (2013). On the Validity of Student Evaluation of Teaching: The State of the Art. *Review of Education Research,* Volumen 83, pp. 598-642.

Unite for Sight (2015). *Process Evaluation.* [Online]
Available at: http://www.uniteforsight.org/evaluation-course/module5#_ftn7
[Last access: 2 18 2017].

Woolner, P., & Hall, E. (2010). Noise in Schools: A Holistic Approach to the Issue. *International Journal of Environmental Research and Public Health,* 7(8), pp. 3255-3269.

Zabaleta, F. (2007). The use and misuse of student evaluation of teaching. *Teaching in Higher Education,* Issue 12, pp. 55-76.

Acknowledgments

The authors would like to acknowledge the valuable contributions of the academic and support staff who participated during the design and implementation of the evaluation instrument. We express our sincere gratitude to Carlos Felipe Londoño Álvarez, rector of the EIA University, and the Academic Council for their support during this process.

Copyright statement

Contributions and reflections of the STS studies in the education of engineers in the Colombian case

Javier Jiménez
Universidad de los Andes, Bogotá, Colombia
ja.jimenez911@uniandes.edu.co

Jorge Rojas-Alvarez
Universidad de los Andes, Bogotá, Colombia
j.rojas46@uniandes.edu.co

***Abstract**: This contribution presents the experience of the research group Technology and Society of the School of Engineering at Universidad de los Andes (Colombia), which since 2002 has made a proposal of research and education for engineering relying on Science, Technology and Society (STS) studies. This paper emphasizes the time when STS was first introduced to the School of Engineering, the educational experience from the active learning perspective as well as the investigative for design in context.*

Introduction

Since June 2002 the Initiative of Research and Teaching Technology, Engineering and Society (ITIS in Spanish) of the School of Engineering at Universidad de los Andes offers courses, as part of the Basic Undergraduate Cycle, promoting a reflection on the role of technology today. Currently some of those courses are Technology and Society, Technology and Globalization, Techno-cultures in Latin America and History of Technology. Each semester more than 280 undergraduate students from different programs attend the courses, especially from engineering, industrial design, economics and related programs.

The courses are inspired by a worldwide effort that some people denominate Engineering, Technology and Society Education (Brown, Downey, & Diogo, 2009; Lucena & Schneider, 2008; Johri, 2011). Many authors involved in that field of study agree that an understanding of engineering and its professional practice is the path to comprehend technology's role in our society because that profession is responsible of the development and maintenance of technological systems (Hughes, 1993). It is the profession where social, organizational and physical components of an innovative process come together (Callon & Law, 1998) and through it the boundaries of technical rationality are analyzed because, in relation with technology, our distinctive feature as a society is an engineering issue (Mitcham, 1994).

Throughout the courses it is recognized that as our society becomes more technological, engineers take a part as core social agents because their professional function will organize, design, create, manage and implement societal changing projects (Downey, 2009). That is the reason why engineering education should be put in context, so that any decision taken will not leave the social dimension aside.

The courses are closely linked with the criteria of the ABET accreditation. They especially meet the criteria: 3 Student Outcomes, items: c) an ability to design a system, component, or process to meet desired needs within realistic constraints such as economic, environmental, social, political, ethical, health and safety, manufacturability, and sustainability; 3 d) an ability to function on multidisciplinary teams: 3 h) the broad education necessary to understand the impact of engineering solutions in a global, economic, environmental, and societal context ; and j) a knowledge of contemporary issues. The School abides by them - about good teaching practices and the promotion of their innovation. It is also required that engineering teaching matches social context needs encouraging multidisciplinary work and providing students with different perspectives to the problems that their profession will positively influence.

Likewise, ITIS courses contribute to the Development Plan of the School of Engineering, particularly to the objectives about Quality and Differentiation. Particularly to the objectives. “1. To keep updated academic qualifications profiles of the programs and to guarantee that they are achieved” and “4. To promote research activities.” (Behrentz Valencia, 2013; Mariño Drews, 2013).

Course design

During the first semester of 2014, the course length of Technology and Society was 15 weeks, 3 hours a week. There were 65 students enrolled, 67% of engineering and 33% of other programs. The general objectives, methodology, and its relation to expected learning are detailed below.

General Objective

The general objective of Technology and Society was the analysis of the main controversies caused by the inclusion of Information and Communication Technologies (ICT) in Higher Education in Colombia. The reference of this analysis was the theoretical framework systemic vision of technology (Vinck, 2012; Osorio Marulanda, 2003; Pacey, 1990) and its methodology was the Social Construction of Technology - SCOT (Bijker, 2005).

Through a systemic vision of technology students problematized in context ICT in Higher Education in Colombia, considering its main cultural, technical and organizational features, finding its identity and mutual influences; and the way possible characteristics of technology in accordance with national reality are prefigured.

The purpose of the SCOT was to discuss about the relevance and effectiveness of such technologies in addition to examining their significant actors, interests, strategies, motivations and major challenges to adopt them in the national context related to the previous systemic vision.

Specifically, there was an analysis of the main controversies caused by the possible introduction of Massive Open Online Courses (MOOC) at Universidad de los Andes, which could bring access to Higher Education to the population in remote areas whose institutions and programs are mainly on campus.

Course methodology

The course structure had two linked activities: on the one hand, it was to study the context and the problematic relationship between technology and society; on the other hand, it was the construction of possible scenarios to implement a solution for a technological controversy through design in context. The general activities of the course included:

- Theoretical sessions. The Science, Technology and Society (STS) field is introduced, offering examples of systemic vision of technology and the methodology SCOT.
- Case study. The methodology combines the study made by the students of the role of MOOC at Universidad de los Andes with the recognition of social and technological actors. In order to find out how the system works, groups of students research about each actor's interests, strategies and motivations. The case study had the following dynamics:
    - Discussion, personal and group reflection on what was said in class.

- o Real time videoconferencing with experts in MOOC
- o Learning off-campus sessions with related videos and short writing exercises.

Expected contributions of the group Technology and Society to the education of engineers

Analyzing the articles written by the ITIS and the reflections of different authors regarding the importance of social role when reflecting on engineering (Schneider, 2010; Trevelyan, 2010; Johri, 2011; Valderrama Pineda, Burbano Valdés, Escobar Gutiérrez, & García Rozo, 2007) it can be concluded that the contribution of the group activities to the education of engineers was focused on pedagogy, the vision of technology, theoretic base, and interdisciplinary proposal

Pedagogical Contribution

Pedagogical contribution refers to the emphasis on STS that the ITIS courses bring to the student to comprehend, appropriate and recognize the existing relationships between technology and society, and the relevance of the courses to create an interdisciplinary culture and reflection on teaching and research at the university. This includes the connection between teaching-learning of conceptual contents of the courses as well as the development of skills that provide students in both university and professional/working fields, to assume a responsible, critical and proactive position on social and technological realities when faced with the decision-making and implementation of technology in Colombia.

Vision of Technology

Regarding the vision of technology, students who have taken the courses and according to the final survey of the courses conducted by the School of Engineering, say that they had access to a nontraditional vision or perspective of technology. A vision that seems to fit to a great extent the one intended by the group, that is to say, a critical contextualized non-linear one. It is also very satisfactory that students can identify by contrast different approaches with elements of the traditional vision of technology. The understanding of such elements by the students may be summarized as follows:

- Dynamic vision vs. static vision: there is a traditional vision that shows technological tasks as something linear, as a "natural" outcome of a series of chained events. From this point of view, the main technological problem would be related to the implementation of certain advances in determined contexts. The courses show another perspective of technology: a social process, where technology is created by society and it is basic that it stays under its control.
- Technology as human phenomenon vs. artefactual vision: in conjunction with the last item, the courses extend the general and traditional vision of technology that includes only artifacts and tools. The courses show technology as an evident social phenomenon. The proof is that almost every activity – if not all of them – implies or is related to it. Men depend on technology and that is why a reflection on its role, boundaries and extents in the society is so important and pertinent along with an analysis on the ethical, moral and political dimension of the activities imbued with technology.
- Technology as a social construction tool: the courses show that the question about technology is also a question about the intended society and the influence that a type of technology has in a certain type of society. Different aspects such as equality, equity, opportunities and future development are determined depending on the technology a

society chooses for supply (water), education (internet), and the people responsible to manage it (multinationals, the Estate, the communities, etc.).

- Technology adaptation vs. technology neutrality: future engineers consider that courses contents confirm that every technology is neither beneficial nor the most suitable. There is a need to contextualize technology according to where it will be implemented. Advantageous technologies for the first world may be damaging or useless in developing countries when solving a certain problem.
- Technology as a means vs. technology as an end: the courses confirm the relevance of considering technology as a means to solve social problems with a responsible attitude. Conceiving technology as an end may be a naïve attitude that believes that social problems will be solved by technological advances (one should never look further) and the solution of those problems don't depend on people but on the implementation of one available up-to-date technology. This implies that problems of our surroundings, our city and country are not due to a technological setback.

Theoretical Foundation

It is important to have integration when developing subjects, showing their progressive relationship and connecting theoretical aspects taught throughout the courses.

That is why there is a need to know the how and why of the theoretical foundation and its relationship to the courses. To this extent, concrete information on the subjects with theoretical perspectives or approaches should be formulated in a clear and differentiated way.

Besides, the involvement of experts, researchers and decision-makers to illustrate the way technology and society act together, has awoken students' sensibility about the importance of reflecting on the interactions between technology and society not only at a macro level but also in ordinary life because of the wide incorporation of technology.

Interdisciplinary Proposal

One of the core elements of the group is the construction of an interdisciplinary and integrated perspective of the relationship between technology and society. The already mentioned reflections consider that it can only be achieved when that relationship, as a generic module, includes the next levels: philosophical-epistemological, dynamic and pragmatic. The first one is necessary to understand the other but specially to consolidate academic fundaments. This is where should appear the epistemological assumptions and fundaments on why it is important to resort to an interdisciplinary and paradigmatic knowledge different from the traditional ways of understanding and analyzing technology and society. The second one, the dynamic, is about making evident the difference between a static vision of technology, where it is defenseless, centered on the artifacts and follows linear courses of action; and the vision of it as a social process, logically dynamic and discontinuous. It is important to show how and where such process is formed, to identify production, maintenance and legitimatization methods of technological knowledge, different uses and conceptions proving until what point and why this process should be considered a social one.

As for the pragmatic relationship, it is, if you will, the least abstract level between technology and society. The former elements are articulate and concrete cases are analyzed for understanding specific knowledge so the students can discern how to make knowledge operational.

Strategy of evaluation and findings

By the end of the course at the first semester in 2014, there was an evaluation with the students to reflect on the contribution of the STS studies in the education of engineers in the Colombian case.

There was a survey with focus heterogeneous groups of eight students. An advantage of it was that participants could interact with one another and come up with ideas as a group (Smith & Leith, 2015). A focus group allows accumulating experiences, reactions and attitudes in respect to the theme, making the discussion more meaningful (Gibbs, 1997; Kitzinger, 1995).

Although the organizers of the survey were afraid of the influences in the interventions from some participants on others, strategies of confirmation of the student's opinions along the interview were implemented to mitigate this issue. Also, these strategies tried that students felt more comfortable and enjoy more being in group than having individual interviews.

The focus groups offered a result on these core subjects:

- Pedagogical contribution
- The vision of technology and theoretic base of the course
- Interdisciplinary proposal

Out of the 55 students, 48 took part in the heterogeneous focus groups. The groups were heterogeneous regarding the semester they were in and their program – 65% engineering and the remaining percentage social sciences, economics and related programs.

As for the first core subject, 79% of students agreed that the course shows a wider vision of certain goals of the University. Videoconferencing clearly demonstrates the interest of the University in being part of virtual education, by implementing MOOC, blended courses, etc. They also indicate that along with audiovisual material some people could clarify the objectives, so students were informed about the position of the University on the subject of new technologies.

In relation to a systemic vision of technology, 80% of students affirmed that the course helped them to understand the educational model learning/teaching of MOOC thanks to the conversations with guests who participated in this type of courses and even had been or are part of the design and development of MOOC. The following figure states the distribution of the topics in the vision of the technology:

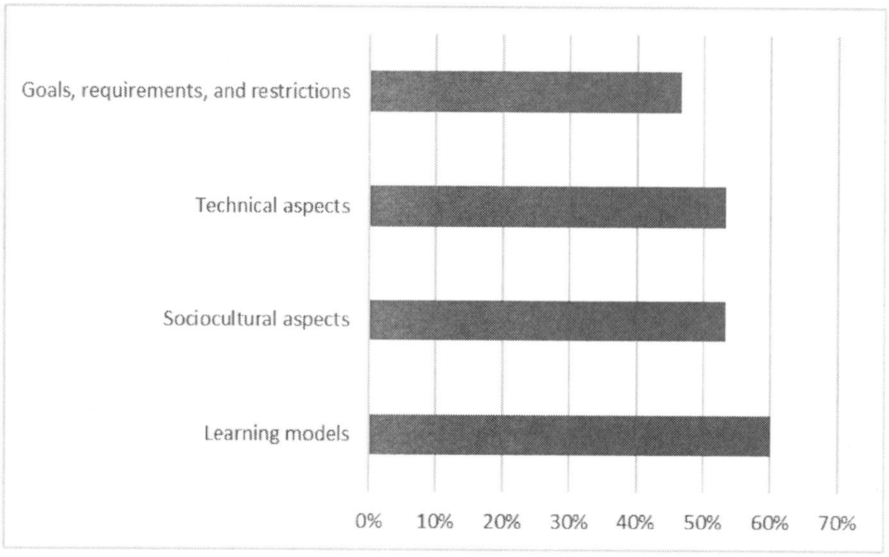

Figure 1: Perception of the topic distribution in the systemic vision of technology

The students believe that during the course a strategy was designed in which through videos, conferences, essays and special guests, an attempt was made to look from different points of view at the key elements for the implementation of a MOOC. The material never tries to focus on a single specific point, but we study different components, whether economic, social or cultural; Which allowed us to design strategies for the future introduction of MOOCs at the University of Los Andes.

With respect to technical aspects, 53% of the students indicated how significant these aspects are for the design and management of MOOC. They recognized the importance that behind any platform there are always people in charge of technical aspects to ensure projects are carried out, i.e. one of the special guests show how the public seem to think technological tools appear "out of thin air," but at the end it was clear that there are human beings providing such service.

Students highlighted a better understanding and appropriation of the relationships between technology and society due to the course. They realized the relevance of an interdisciplinary vision of technological problems in our context. Thus, they have tools to take a responsible, critical and proactive stand on technological and social reality regarding decision-making and implementation of technology in Colombia.

Participants of the classes pointed out how this type of courses encourages them to understand the importance of a systematic vision of technology. Through exercises they realize that asking about technology also implies the plans, purposes and values of our society. Therefore, any technological decision should meet the criteria of scientists and engineers along with community participation to help determine its equality, equity, opportunities and future development.

Conclusions

The reflections elaborated in the contribution, can help to identify strategies so that teaching-learning of conceptual contents of the courses as well as the development of skills that provide students in both university and professional/working fields. Course's teaching staff reflections on the proposal of STS in engineering educations are as follows:

- Different teaching-learning of conceptual contents of the courses as well as the development of skills that provide students in both university and professional/working fields can harmoniously coexist at Technology and Society course according to the objectives already set, i.e. a deeper reflection on technological artifacts, their consequences, history, impact, possible solutions, etc. are complemented by reflexive short texts, later by longer descriptive collaborative texts, real time videoconferencing and video observation. Also, depending on the kind of work to be developed with the student - individual or group -writing strategies are enriched by a detailed planning of their scope and learning.
- Concrete information on the subjects with theoretical perspectives or approaches should be formulated in a clear and differentiated way. This requires planning and reflection on their sense and connection to the course's instances. Too many or too few perspectives may become a restriction to achieve learning objectives.
- Previously to practical design in context exercises, there must be a clarification for students on what tools will be used, the reason they are chosen, how they complement writing strategies and what are their scope and limitations. There must be all the time clear directions on their use and what vision of technology is emphasized.
- The fact that these courses perfect argumentative tools and analytical capacity linked to a wider perspective for design and management technology, make it possible to enhance situated knowledge on the practice of engineering with conscious social impact.
- From the perspective of ABET skills, in the area of technological system design regarding contextual problems and at a multidisciplinary team, students emphasized how strategies used helped them to better apply a systemic vision of technology. They think that class exercises showed them how the question about technology is also a question about plans, purposes and values of our society, so criteria mentioned by scientists and engineers can interact with community participation for every technological decision, determining equality, equity, opportunities and future development in general terms.
- According to strategic objectives of the School of Engineering, the complement of the STS reflections activities in class strengthens and deepens students' results when they worked in group. Since real time videoconferencing and video observation were activities mostly made in an individual way, there was a need of making other activities to create a meeting point for discussion of their perspectives in and out of the classroom. In this way, theoretical reflections about a certain case were quickly enriched by precise exercises and groups were focused consciously on an objective.

We think this type of courses create an active and dialogic space so that students can reach significant agreements on the world they want. Consequently, there is a promotion to develop some required skills for decision-making on the design and socio-technical implementation, for the analysis of their implications and impact on society and the role of engineering to build the world.

The of STS studies in the education of engineers in the Colombian case, has created flexible and autonomous approaches useful for students' learning by visualizing different standpoints of a technological problem and formulating contextual reflections on the relationship between technology and society.

This approach to STS contributions to engineering education is an invitation to researchers for approaching to questions about what changes experience students in their own disciplines after taking this kind of courses, how they give propose to design specific tools, and in what way they broad its context of appliance.

References

Behrentz Valencia, E. (2013). Palabras de Instalación Foro 'Fortalecimiento de la enseñanza de la ingeniería con las nuevas tecnologías de la información y las comunicaciones'. *Revista de Ingeniería*, (39), 50–52.

Bijker, W. E. (2005). ¿Cómo y por qué es importante la tecnología? *Redes*, *11*(21), 19–53.

Brown, J. K., Downey, G. L., & Diogo, M. P. (2009). The Normativities of Engineers: Engineering Education and History of Technology. *Technology and Culture*, *50*(4), 737–752.

Callon, M., & Law, J. (1998). De los intereses y su transformación. Enrolamiento y contraenrolamiento. In M. Domènech & F. J. Tirado (Eds.), *Sociología simétrica: ensayos sobre ciencia, tecnología y sociedad* (pp. 1–2). Barcelona: Gedisa.

Downey, G. L. (2009). What is engineering studies for? Dominant practices and scalable scholarship. *Engineering Studies*, *1*(1), 55–76. https://doi.org/10.1080/19378620902786499

Gibbs, A. (1997). Focus Groups. *Social Research Update*, *5*(2), 1–8.

Hughes, T. P. (1993). *Networks of Power: Electrification in Western Society, 1880-1930*. Baltimore: JHU Press.

Johri, A. (2011). Global, technological, and environmental challenges for engineering professionals. *Engineering Studies*, *3*(2), 71–77. https://doi.org/10.1080/19378629.2011.613571

Kitzinger, J. (1995). Qualitative Research: Introducing focus groups. *BMJ*, *311*(7000), 299–302. https://doi.org/10.1136/bmj.311.7000.299

Lucena, J., & Schneider, J. (2008). Engineers, development, and engineering education: From national to sustainable community development. *European Journal of Engineering Education*, *33*(3), 247–257. https://doi.org/10.1080/03043790802088368

Mariño Drews, O. (2013). Fortalecimiento de la enseñanza de la ingeniería con las tecnologías de información y comunicaciones. *Revista de Ingeniería*, (39), 46–49.

Mitcham, C. (1994). *Thinking Through Technology: The Path Between Engineering and Philosophy*. Chicago: University of Chicago Press.

Osorio Marulanda, C. A. (2003). *Aproximaciones a la Tecnología desde los enfoques en CTS*. Presented at the Cátedra de Ciencia, Tecnología, Sociedad e Innovación, Panamá. Retrieved from www.oei.es/ctsiparaguay/osoriotec.pdf

Pacey, A. (1990). La práctica tecnológica. In *La cultura de la tecnología* (pp. 2–13). Buenos Aires: Fondo de Cultura Económica.

Schneider, J. (2010). Engineering and the values of social justice. *Engineering Studies*, *2*(1), 1–4. https://doi.org/10.1080/19378621003615975

Smith, R., & Leith, K. (2015). Interviewing. In *D-Lab Scale-Ups User Research Framework version 1.2* (pp. 22–31). Massachusetts Institute of Technology. Retrieved from https://d-lab.mit.edu/research-about

Trevelyan, J. (2010). Reconstructing engineering from practice. *Engineering Studies*, *2*(3), 175–195. https://doi.org/10.1080/19378629.2010.520135

Valderrama Pineda, A., Burbano Valdés, A. E., Escobar Gutiérrez, J. A., & García Rozo, A. (2007). STS Problems in Colombia. In *Memories of the Annual Meeting of the Society for the Social Studies of Science*. Montreal, Canadá.

Vinck, D. (2012). Pensar la técnica. *Universitas Philosophica*, *29*(58), 17–37.

Acknowledgements

The authors would like to give special thanks to professor Ángel Alonso Gutiérrez Pérez of the School of Engineering at Universidad de los Andes for his collaboration to write this contribution.

Work-In-Progress: Development of the Environmental Engineering Issues Test (E2IT): Process and Theoretical Framework.

Andrea Mazzurco
The University of Queensland, Brisbane, Australia
a.mazzurco@uq.edu.au

Homero Murzi
The University of Queensland, Brisbane, Australia
h.murzi@uq.edu.au

Melanie Fleming
The University of Queensland, Brisbane, Australia
melanie.fleming@uq.edu.au

Abstract: *Engineers are increasingly expected to develop solutions that can preserve our environment and meet sustainability goals. Consequently, it is becoming increasingly important that engineering students develop a high level of moral judgment when it comes to environmental engineering issues. However, while there are instruments to measure broad engineering ethical development, the literature still lacks an instrument to measure the development of engineering students' moral judgement related to environmental issues. To address this gap, we have undertaken the development of the Environmental Engineering Issues Test (E2IT), a scenario-based assessment instrument. In this work-in-progress paper, we present best practices to develop scenario-based instruments to assess moral judgment, the process we are following to develop the E2IT, and the result of the first step of our process: a theoretical framework that comprises several dimensions of environmental ethics.*

Introduction

Environmental awareness and sustainable practices are becoming extremely important for practicing engineers. Many engineering educators and scholars have designed curriculum and modules to prepare engineering students to develop solutions that foster environmental sustainability (Nair, Jones, & White, 2002; Murzi, Fleming, Mazzurco, & Pikaar, 2017). Many scholars have especially focused on ethics education with a focus on environmental engineering. For instance, Santi (2000) developed several ethics exercises for environmental, civil, and geological engineers, while others have used recent environmental scandals, like the Volkswagen Diesel Scandal (Warford, 2016), to create case-studies for instruction. However, properly assessing the effectiveness of these types of educational interventions remain a key challenge. While there are a few instruments to assess gains in environmental knowledge (e.g., Azapagic, Perdan, & Shallcross, 2005), there is still a lack of instruments that effectively measure moral judgment in the context of environmental engineering.

To address this, we undertook the development of the Environmental Engineering Issues Test (E2IT), a scenario-based instrument that is being modelled after the Defining Issues Test (DIT), which has been used in hundreds of studies to assess general moral judgment skills (Center for the Study of Ethical Development, 2016). Specifically, in this work-in-progress paper we report on the first stage of the development of the E2IT. First, we provide a literature review on DIT and on other DIT-like instruments. In this literature review, we discuss specifically issues and best practices related to develop and validate DIT-like instruments. Second, we provide a through overview of our plan to develop and validate the

E2IT, which is based on the best practices discussed in the literature review. Finally, we present the theoretical framework for the E2IT, which is the outcome of our first step in developing and validating the E2IT.

Measuring moral judgment

The most commonly used instrument to measure moral judgment is the Defining Issues Test (DIT) that was developed in the early 70's by Rest and colleagues (see Rest et al., 1999). The DIT is grounded in Neo-Kohlbergian cognitive moral development theory. Kolbergh (1984) proposed that individuals go through six self-contained stages of moral development. In contrast, the Neo-kohlbergian theory substituted the six stages with three schemas (Rest et al, 1999), thus conceiving moral development in terms of shifting distributions of schemas rather than a stepwise progression (for more details see Rest et al., 1999). The three schemas of the Neo-Kohlbergian theory are pre-conventional or personal-interest, conventional or maintaining norms, and post-conventional (see table 1). Individuals who predominantly use the pre-conventional schema will make decisions based on self-interest when faced with ethical dilemmas. Individuals who rely mostly on the conventional schema will make decisions based on laws and norms. Individuals who rely mostly on the post-conventional schema will make decisions based on ethical ideals (e.g., universal rights and social justice). The DIT measures the extent to which each schema contributes to an individual's moral decision-making thinking when faced with ethical dilemmas. The higher an individual's score on the DIT the more the post-conventional schema contributes to the individual moral decision-making choice.

Table 1. Schemas of the DIT (adapted from Narvaez and Bock (2002))

Schema	**Features of schema**
Pre-Conventional / Personal-Interest Schema	Self-focused Advantage to self is primary Survival orientation Scope includes others who are known
Conventional / Maintaining Norms Schema	Need for norms Society-wide view Uniform categorical application Duty orientation
Post-Conventional Schema	Appeal to an Ideal Shareable ideals Primacy of moral ideal Rights orientation

The DIT consists of five scenarios that present moral dilemmas followed by two rating tasks. In the first task, respondents are asked to rate the importance of 12 questions on a 5-point scale, in which each question reflect one of the three schema. In the second task, respondents are asked to choose the four most important questions and rank them in order of importance. The combination of the two ratings tasks provides scores that indicate the level of moral development of the respondent (Rest et al., 1999). The most appealing characteristic of the DIT is that it has been validated by over 400 studies and has been used in multiple disciplines, including accounting (Abdolmohammadi & Ariail, 2009), veterinary science (Batchelor, Creed, & McKeegan, 2015), medicine (Bohm et al., 2014), counselling (Eriksen & McAuliffe, 2006), pharmacy (Gallagher, 2011) and others (see Center for the Study of Ethical Development (2006) for more). Additionally, it has been used as a model to create discipline-specific DIT-like instruments. For instance, Doyle, Frecknall-Hughes, and Summers (2009) developed a DIT for a taxation-specific scenario, Verrinder and Phillips (2014) created the VetDIT for veterinary professionals, and Welton, Lagrone, and Davis (1994) created a DIT-like instrument for accounting-specific scenarios.

The DIT has been used in engineering education to evaluate the effect of professional ethics courses on students' development of moral judgment. Self and Ellison (1998) used the DIT to evaluate a course on engineering ethics. They gave the DIT to students at the beginning and at the end of the course and found significant increases in students' scores, suggesting that the course was able to improve students' moral reasoning abilities. Similarly, Staehr and Byrne (2003) used the DIT to evaluate a course on computer ethics. They gave the DIT to students in the course and to a control group (i.e., students who were not attending a course on ethics). Like Self and Ellison's (1998) study, Staehr and Byrne's (2003) analysis of the pre-post data showed higher increase of moral reasoning ability in the computer ethics students than in the control group. Drake et al (2005) used the DIT to evaluate the effect of a short module on engineering ethics and found no significance in pre-post results. Among the reasons for the lack of significant results, Drake et al (2005) observes that the DIT focus on general, non-engineering situations and may not capture changes of moral judgment of engineering-specific ethical dilemmas. Thus, they concluded by suggesting that "it might be beneficial to develop a new instrument, perhaps modelled on the DIT-2, incorporating ethical dilemmas likely to be faced by engineers" (Drake et al., 2005, p. 229).

The Engineering and Science Issues Test (ESIT) was developed in response to Drake et al (2005) call for the creation of a DIT-like instrument specifically focused on engineering professional ethics (Borenstein et al., 2010) The ESIT was created based on the hypothesis that "a measure of moral reasoning more precisely tailored to the practical context of engineers and researchers would be more sensitive to the effects of professional ethics education" (Borenstein et al., 2010 p. 389). Their validation study showed that the ESIT was able to effectively capture students' changes in moral judgment abilities of engineering-specific ethical dilemmas (Borenstein et al., 2010).

To develop and validate the ESIT, Borenstein et al., (2010) the same procedure used for the DIT and other DIT-like instruments. They developed six scenarios based on realistic situation in the context of the engineering workplace. To develop the questions, they used an inter-rater reliability process. The four authors independently created a subset of questions for each scenario that covered the three schemas of Neo-Kohlbergian theory (i.e., pre-conventional, conventional, post-conventional). Then, each author rated the other authors' questions based on the three schemas and they kept the questions were they achieved the highest results of inter-rater reliability. Finally, the last step of their validation process was to administer the ESIT to students in the professional ethics course they had previously evaluated using the DIT and to a control group who was not receiving any overt ethics instruction. The results showed that ESIT scores of students in the professional ethics course improved as compared to the scores of the control group, suggesting that ESIT is more sensitive than the DIT in capturing change in levels of moral judgment in the context of engineering ethical dilemmas

In addition to the ESIT, another DIT-like instrument specific to engineering context is the Engineering Ethical Reasoning Instrument (EERI) (Zhu et al., 2014). In contrast with the ESIT that uses scenarios situated in the engineering workplace, the EERI presents scenario focused on students-led, project-based team in community engagement. The EERI was specifically developed to measure the changes of moral judgment in project-base programs like Engineering Projects In Community Service (EPICS). Similar to the ESIT, the EERI contains 12 questions and a ranking task. In contrast with the DIT and DIT-like instruments, upon completion of the scenario, participants are asked whether or not they would proceed with an action related to the scenario. An example scenario from the EERI is the "Soap Box Derby" case reported below, which was inspired from a real project that students worked on at Purdue University:

> Your student design team has designed a new Soap Box Derby car that allows children with physical and cognitive disabilities to race by allowing an adult to ride in a backseat and maintain full control of the car. Based on suggestions from the adults, you have added spring tension to the child's steering wheel in front in

order to simulate the feeling of driving and make the child's experience more realistic and fun. The child will not have the ability to control the car, only the illusion of control. Before the first test run with an adult and a 14-year-old child on board you hear the child's parent tell the child to "be careful" and to "drive safety." The parent turns to you, explains that because of a cognitive disability the child likely won't understand the difference anyway, and asks you to tell the child that the front steering wheel is actually functional. The request that you lie to the child would take advantage of the child's disability and it creates the possibility that the child would feel responsible if they were to lose the race or have an accident. (Zhu et al., 2014, p. 5).

This scenario is evidently different from the ESIT one as it is clearly focused on students' project-based design teams and therefore is not well suited to be used in the same context of the ESIT.

To validate the EERI, Zhu et al (2014) used a mixed, qualitative and quantitative, approach. For the first version of the EERI, they created eight scenarios with 12 questions each. They asked engineering experts to review the scenarios and questions from a technical perspective and asked 15 Ph.D. students from English and Philosophy departments to review the language used in the instruments to make sure it was clear and well-articulated. They also gave the EERI to three different groups of students and each time they updated the scenarios, items, and instructions. The final group of students comprised 800 respondents and the instrument presented satisfactory values of internal consistency, measured by Cronbach's alpha (between 0.72 and 0.85 for five scenarios and 0.57 for another one). Two of the eight scenarios were dropped because did not present satisfactory properties. Finally, they also interviewed students who took the survey and observed them while working in teams. The combination of interview data and observations provided further evidence that the students' performance on the EERI was indicative of their level of moral judgment.

Both the ESIT and EERI are valid, robust, and provide with effective ways to measure levels of moral judgement in engineering-specific contexts. The ESIT is best suited for professional ethics in industrial settings and the EERI is more appropriate for students' project-based design work. However, they are limited in terms of assessing moral judgment in the context of environmental engineering. Therefore, we decided to undertake the development of the E2IT to provide a practical and effective way to assess moral judgment related to ethical issues in the context of environmental engineering. The E2IT will be especially well suited to assess gains from courses on professional ethics in environmental engineering like the one presented in another paper (Murzi et al., 2017).

Development and validation of the E2IT

To develop and validate the E2IT, we will follow three sequential steps that bring together the best practices for developing DIT-like instruments that we reviewed in the previous sections. The first step is to develop a solid theoretical framework which informs all aspects of the ESIT. The theoretical framework was developed by reviewing and integrating literature on moral judgment and environmental issues. The theoretical framework is presented in the results section of this paper.

The second step focuses on creating the scenarios and questions for the ESIT. To create the scenario, we will use three main strategies:

1. We will search the literature for cases-studies focused on environmental issues that present ethical dilemmas likely faced by engineers. For instance, some of the scenarios could be adapted from Harris et al's (2005) *Engineering Ethics: Concepts and Cases.*

2. We will interview practicing environmental engineers to gather real-life ethical dilemmas that could be used as adapted to scenarios.
3. We will ask experts in the field to provide ideas for and examples of realistic scenarios

Like Zhu et al (2014), we will ask PhD students from English and Philosophy departments to review the scenarios to make sure that the language used is clear. We also ask environmental engineering experts to review the scenarios and rate them based on how realistic they are. Scenarios that receive low realistic ratings will be revised or discarded.

To develop the 12 questions for each scenario, we will follow the same process used by Bornestein et al (2010) to create the questions for their ESIT. Each researcher will create independently a subset of questions for each scenario. Then we will rate each other's questions and keep the ones that receive higher scores of inter-rater reliability. Like for the scenarios, we will ask PhD students and experts to review the questions and provide feedback on clarity and accuracy of the questions.

To assess the validity of the E2IT, we will test it against the seven criteria used to demonstrate the validity of the DIT (Rest et al, 1999; Thoma, 2006):

1. Differentiation of various age/education groups. Studies using the DIT with thousands of subjects show that 30% to 50% of variance of DIT score can be attributed to difference in educational background (i.e., high school students score worse than PhD's). Therefore, we will administer the E2IT to first year engineering students and graduate environmental engineering students to check that E2IT is able to capture differences between the two groups.
2. Longitudinal gains: administering the DIT to the same group of people over multiple consecutive years, showed gains in DIT scores. Therefore, we will administer the E2IT to the same group of students once a year for four years to check that the E2IT is sensitive to longitudinal changes.
3. DIT scores were found to be correlated with multiple measures of moral comprehension and cognitive development. Therefore, we will test to what extent the E2IT is correlated with such types of measures.
4. DIT is able to capture gains in moral judgment due to moral education interventions. Therefore, we will use the E2IT in a pre-post study design to evaluate students' changes in moral judgment as a result of attending a course on environmental ethics.
5. DIT scores are correlated with multiple measures of prosocial behaviours. Similarly, we will test E2IT correlation with measures of prosocial behaviours.
6. DIT scores is correlated with measures of cultural ideology and political orientation. We will test such types of correlation for the E2IT as well.
7. For E2IT's reliability to be satisfactory, Cronbach's alpha and test-retest reliability must be at least in the upper .70s.

Each time one of the above criteria is tested we will revise the instrument as necessary.

Furthermore, in line with modern theories of validity, we will evaluate the instrument in terms of its use (Douglas & Pruzer, 2015). Specifically, we suggest the E2IT should be use to evaluate effectiveness of courses focused on environmental ethics. Therefore, all our validation efforts will be focused on courses evaluation and development. Finally, we also acknowledge that "[v]alidation is never quite over" (Douglas & Pruzer, 2015, p. 111). Consequently, we will continuously collect evidence that can "inform future improvements to the instrument and to provide evidence of appropriate use" (Douglas & Pruzer, 2015, p. 111).

Step 1: Establishing the theoretical framework

The E2IT will be informed by two complementary theoretical frameworks. The first focus on environmental issues relevant to environmental engineers and provides a conceptual map of the issues that the E2IT scenario will focus on. The E2IT scenarios will focus on issues defined in the environmental issues core knowledge that was used in developing curricula for environmental engineers (e.g, Nair et al., 2002; Murzi et al., 2017). Specifically, the scenarios will focus on ethical dilemmas in the context of the following environmental issues: 1) urbanization and population growth, 2) water scarcity and quality, 3) waste management, 4) air and noise pollution, 5) biodiversity, and 6) life cycle analysis. The cited topics represents some of the most significant environmental issues for engineers, which are also typically addressed in environmental engineering curricula, making the E2IT relevant especially to this specific discipline. This framework will help us search for and select relevant case-studies that will inspire the creation of the scenarios for the E2IT.

The second theoretical framework integrates Neo-Kohlbergian cognitive moral development theory (Rest et al., 1999) and McCuen's (1979) professional stages for engineers. Building upon Kohlberg's six stages model, McCuen's (1979) developed six stages describing the moral development of engineers that go from a "pre-professional" stage to a "principled professional" stage (see table 2). While we recognize the value of using a moral development model focused on engineering professionalism, we also agree with Neo-Kohlbergian theorists that a stage model is not accurate and we choose to adopt the Neo-Kohlbergian schema with an emphasis on professionalism (see table 2). Specifically, the E2IT will be based on three moral schemas. The first schema is the "pre-professional" schema which is very similar to the DIT pre-conventional schema and aligns with the first two stages of McCuen's model. It suggests that engineers make decisions based on self-interest. The second schema is the "professional" schema which is very similar to the DIT conventional schema and aligns with McCuen's third and fourth schema. Engineers relying on this schema will make decisions out of a sense of loyalty to their firm or based on established professional norms (e.g., code of ethics). Finally, the third schema for the E2IT is the "post-professional" which is similar to the DIT "post-conventional" schema and aligns with McCuen's fifth and sixth stage. Individuals relying on this schema make decision by drawing on universal norms and values (e.g., justice and human rights). The 12 questions of each scenario will be phrased to reflect these three schemas.

Table 2. Comparison of Kohlberg's and McCuen's stages and DIT and E2IT schemas

Kohlberg's stages	McCuen's stages	DIT schema	E2IT schemas
Stage 1: Obedience	Stage1 : individual gains	Pre-conventional schema*:* survival orientation / advantage to self	Pre-professional: self-interest
Stage 2: Purposeful exchange	Stage 2: self-advancement		
Stage 3: Conforming to group norms	Stage 3: loyalty to firm	Conventional schema: need for norms / duty orientation	Professional: conforming to professional norms
Stage 4: Law and order	Stage 4: loyalty to the profession		
Stage 5: Societal consensus	Stage 5: service to society	Post-conventional: primacy of moral ideal / rights orientation.	Post-professional: adherence to universal norms and values.
Stage 6: Universal morality	Stage 6: universal justice, fairness, and caring for humans		

Conclusions and future work

In this work-in-progress paper, we presented the theoretical foundation and the development and validation plan for the Environmental Engineering Issues Test (E2IT). The E2IT will be a DIT-like instrument which will allow engineering educators and scholars to evaluate the effects attending a courses on environmental issues and professional ethics on engineering students' development of moral judgment. The next steps in the development of the E2IT include the creation of scenarios presenting real-life moral dilemmas in the context of environmental engineering and the development of 12 questions for each scenario. Afterwards, we will proceed with the validation of the instrument. It is important to note that moral judgment is one of the key components of ethical development. Therefore, the E2IT is still limited in terms of measuring overall ethical development and we look forward to other scholars to lead the way in developing instruments focused on moral sensitivity, moral motivation and identity formation, and ethical implementation.

References

Abdolmohammadi, M. J., & Ariail, D. L. (2009). A test of the selection-socialization theory in moral reasoning of CPAs in industry practice. *Behavioral Research in Accounting*, 21(2), 1-12.

Azapagic, A., Perdan, S., & Shallcross, D. (2005). How much do engineering students know about sustainable development? The findings of an international survey and possible implications for the engineering curriculum. *European Journal of Engineering Education, 30*(1), 1-19.

Batchelor, C., Creed, A., & McKeegan, D. (2015). A preliminary investigation into the moral reasoning abilities of UK veterinarians. *Veterinary Record: Journal of the British Veterinary Association*, 177(5), 124.

Bohm, K. C., Van Heest, T., Gioe, T. J., Agel, J., Johnson, T. C., & Van Heest, A. (2014). Assessment of moral reasoning skills in the orthopedic surgery resident applicant. *The Journal of Bone and Joint Surgery, 96A* (17), 1-6.

Borenstein, J., Drake, M. J., Kirkman, R., & Swann, J.L. (2010). The Engineering and Science Issues Test (ESIT): A discipline-specific approach to assessing moral judgment. *Science and Engineering Ethics, 16,* 387-407.

Center for the Study of Ethical Development (2016). Searchable database of DIT usage. Retrieved on 03 March 2017 from http://ethicaldevelopment.ua.edu/database-of-dit-usage.html

Eriksen, K. P., & McAuliffe, G. J. (2006). Constructive development and counselor competence. *Counselor Education & Supervision, 45*, 180-192.

Douglas, K. A., & Purzer, S. (2015). Validity: Meaning and relevancy in assessment for engineering education research. *Journal of Engineering Education, 104*(2), 108-118.

Doyle, E., Frecknall-Hughes, J., & Summers, B. (2009). Research methods in taxation ethics: Developing the Defining Issues Test (DIT) for a tax-specific scenario. *Journal of Business Ethics, 88*, 35-52.

Drake, M. J., Griffin, P. M., Kirkman, R., & Swann, J. L. (2005). Engineering ethical curricula: Assessment and comparison of two approaches. *Journal of Engineering Education, 94*(2), 223.

Gallagher, C. T. (2011). Assessment of levels of moral reasoning in pharmacy students at different stages of the undergraduate curriculum. *International Journal of Pharmacy Practice, 19,* 374-380.

Harris, C. E., Davis, M., Pritchard, M. S., & Rabins, M. J. (2005). *Engineering ethics: Concepts and cases* (3rd ed.). Belmont, CA: Wadsworth.

Kohlberg, L. (1984). *The Psychology of Moral Development: The Nature and Validity Of Moral Stages.* San Fransisco: Harper and Row.

Murzi, H., Fleming, M., Mazzurco, A., & Pikaar, I. (2017). Work-in-progress: Environmental course redesign: Learning ethics from the environmental engineering perspective. *Proceeding of the 2017 Research in Engineering Education Symposium.*

McCuen, R. (1979). The ethical dimensions of professionalism. Journal of Professional Activities, 105 (12), 89-105.

Nair, I., Jones, S., & White, J. (2002). A curriculum to enhance environmental literacy. *Journal of Engineering Education, 91*(1), 57.

Narvaez, D., & Bock, T. (2002). Moral schemas and tacit judgement or how the defining issues test is supported by cognitive science. *Journal of Moral Education, 31*(3), 297-314.

Rest, J., Narvaez, D., Bebeau, M.J., & Thoma, S.J. (1999). A neo-Kohlbergian approach: The DIT and schema theory. *Educational Psychology Review, 11* (4), 291- 324.

Santi, P. M. (2000). Ethics exercises for civil, environmental and geological engineers. *Journal of Engineering Education, 89*(2), 151-160.
Self, D.J., & Ellison, E.M. (1998). Teaching engineering ethics: Assessment of its influence on moral reasoning skills. *Journal of Engineering Education*, *87(*1), 29–34.
Staehr, J.L., & Byrne, G.J. (2003). Using the defining issues test for evaluating computer ethics teaching. *IEEE Transactions on Education, 46*(2), 229-234.
Thoma, S. J. (2006). Research on the defining issues test. In M. Killen, & J. G. Smetana, *Handbook of moral development* (pp. 67-91). Mahwah, NJ: Lawrence Erlbaum Associates Publishers.
Verrinder, J. M., & Phillips, C. J. (2014). Development of a moral judgment measure for veterinary education. *Journal of Veterinary Medical Education*, 41(3), 258-264.
Warford, E. L. (2016). Ethics in the classroom: The Volkswagen diesel scandal. *Proceedings of the 2016 ASEE Conference & Exposition, New Orleans, LA, June 26-29.*
Welton, R.E., Lagrone, R. M., & Davis, J. R. (1994). Promoting the moral development of accounting graduate students: An instructional design and assessment. *Accounting Education, 3*(1), 35-50.
Zhu, Q., Zoltowski, C. B., Feister, M. K., Buzzanell, P. N., Oakes, W. C., Mead, A. D. (2014). The development of an instrument for assessing individual ethical decision-making in project-based design teams: Integrating quantitative and qualitative methods. *Proceedings of the 2014 ASEE Conference and Exposition, Indianapolis, IN, June 15-18.*

Acknowledgements

We would like to thank the Faculty of Engineering, Architecture, and Information Technology of the University of Queensland to support this study and the anonymous reviewers who helped us improve the quality of the paper.

Copyright statement

Work-in-Progress: Environmental Course Redesign: Learning Ethics from the Environmental Engineering Perspective.

Homero Murzi
The University of Queensland, Brisbane, Australia
h.murzi@uq.edu.au

Melanie Fleming
The University of Queensland, Brisbane, Australia
melanie.fleming@uq.edu.au

Andrea Mazzurco
The University of Queensland, Brisbane, Australia
a.mazzurco@uq.edu.au

Ilje Pikaar
The University of Queensland, Brisbane, Australia
i.pikaar@uq.edu.au

***Abstract**: Ethics represent a very important component of the civil engineering degree. In this paper, we present our experiences redesigning an environmental and sustainability engineering course to include and ethical component. We explain the theoretical framework derived from engineering education literature that we used as a reference for the restructure, we also provide details about assessments that we created, and on the different sources of data that we are collecting to analyse the effectiveness of the proposed interventions.*

Introduction

Environmental issues are a major pillar of the engineering profession. Most activities engineers do are affected by, or will affect the environment to some degree (Nair, Jones, & White, 2002). Santi (2000) affirms that civil and environmental engineering share many common dilemmas that are required to be considered by engineering students when learning about environmental issues. Furthermore, Drake, Griffin, Kirkman, and Swann (2005) explain that engineers not only need to learn environmental issues as part of their education, there is also a need for the increase of the ethical perspective in environmental engineering courses.

Because of the nature of the field, engineers will be faced with unpredictable situations that involve materials, design parameters, and the constant interaction between clients and regulatory agencies, therefore the engineering profession requires the utmost ethical standards and ethical understanding (Drake et al., 2005; Finelli et al., 2012; Haws, 2001; Santi, 2000). However, although there has been an interest in the engineering education field to promote ethical awareness, engineering undergraduate education has been slow in develop effective learning interventions that allow students strong ethical development and ethical thinking. Furthermore, teaching and learning topics related to environmental issues are complex by nature; when including an ethical perspective, the complexity is higher and becomes very challenging (Alpay, 2013; Nair et al., 2002). The problem is that students bring preconceived opinions about environmental issues (Nair et al., 2002) and have negative perceptions about the ambiguity and subjectivity of ethics (Alpay, 2013; Newberry, 2004; Perlman & Varma, 2001a). This situation provides room for improving the way environmental engineering is taught in Civil Engineering.

Civil engineering students are required to have a thorough understanding of environmental problems. In the school of Civil Engineering at the University of Queensland one of the most important compulsory courses that the program has is "*Environmental Issues and Sustainability in Engineering*," this course has been historically offered as a 1 credit course that focused on the assessment and monitoring of environmental issues. Considering the previous explained opportunity to improve how we teach environmental issues in civil engineering, we used educational research to restructure the environmental engineering course and expand its scope to include –in addition to other topics- an ethical component to it.

Research Purpose and rationale

In this work-in-progress paper, we present our research-driven approach to restructure a course to integrate environmental issues and professional ethics. Specifically, we first discuss the theoretical framework that guided our approach to restructure the course, second we present information about the course in detail, and finally we close by presenting our reflections through the process.

The re-structuring responds to several needs in our civil engineering program. First, a recommendation from our accreditation body was to incorporate more ethics in our engineering courses. In addition, the size of engineering students coming into second year is increasing therefore part of the restructuring of the course responds to being able to satisfy a larger demand of students and make sure we are providing them with a high quality educational experience. Finally, by developing reflective learning activities and assessments -as part of the restructuring of the course, we are providing students with the opportunity to approach issues of sustainability, environmental engineering, and ethical development in a different way that requires more active reflection, which also provides the students with a deeper learning experience by following best educational practices (Benedict, Napper, & Guice, 2000; Boyd & Fales, 1983; Brockbank & McGill, 2007; Felder & Silverman, 1988).

Theoretical background

Engineering education related to course redesign

We based part of our redesign process on Finelli, Daly, and Richardson (2014) "Institutional Change Plan Using Local Evidence." The authors developed an institutional change plan to bridge the research-to practice gap in engineering education. The plan was proposed based on data collected from faculty members, students, and classroom observations. They provide several recommendations for the adoption of effective teaching practices in engineering.

According to Finelli et al. (2014) research has demonstrated that students' success is highly influenced by the use of effective teaching practices and development of effective learning environments. There are educational practices that can improve student learning, engagement, and interest in engineering, therefore Finelli et al. (2014) encourage engineering faculty members to accelerate the adoption of their model. The model is based on theories from organizational change, faculty motivation, and instructional development, and its goal is to propose a change in teaching practices that drive institutional transformation. Some of the suggestions Finelli et al. (2014) propose that are considered in our process of course re-design are:

- The use of experienced educational developers that provide critical specialized support.
- Inclusion of credible research evidence such as active learning and rapport building.
- Because time is a key barrier, there should be a focus on pedagogies that can be easy to learn and implement.

- Because improving the student experience is a powerful enabler, highlight practices that increase student learning and engagement.
- Because student reaction is a key theme, it is important to include an opportunity for participants to review and act on midterm student feedback.
- Provide demonstrations of class principles, giving students good feedback, talking to students if they perform poorly, and praising students for good comments and answers.
- Provide frequent student feedback, understand student motivation, student misconceptions and preconceptions, and incorporate instructional technology
- Because community building it is an important enabler, community should be included in the scope of the course both in the use of examples for the materials being taught and considered in the development of assignments and projects.

Previous studies in similar courses

While Finelli's work provided the framing for our approach to restructure the course, we also made changes based on previous studies and best practices. Alpay (2013) presented a novel approach in an ethics engineering course offered in several engineering departments at Imperial College in London. In the course, first-year undergraduate engineering students are responsible for proposing activities of ethical relevance to their peers in their specific disciplinary area. In the study, Alpay (2013) provides recommendations on the importance of students' awareness when learning ethics. His proposals provide methods for introducing an ethics component into conventional engineering courses, and promote teaching methods which are relevant to student needs and interests in ethical development. We used his focus on contextualized reflection to address the problem of students' negative preconceptions.

Perlman and Varma (2001a) conducted another study focused on teaching ethics in engineering. The authors concluded that when teaching ethics in engineering, one of the most important things is to provide contextualized relevance of the topics being discussed. As part of our new course proposal, we plan to bring several industry speakers and asked them to include –in addition to their technical presentation- an example of an ethical dilemma that they have faced. Students, as one of their assignments must write an essay on the ethical lessons they learned from the industry speakers.

Similarly, Drake et al. (2005) found that to improve students moral reasoning and sensitivity to ethical issues, engineering ethics must incorporate specific discipline context. We considered their study when developing course materials and class assignments, making them relevant and contextualized to the civil engineering practice in Queensland, Australia.

Regarding sustainability, Azapagic, Perdan, and Shallcross (2005) conducted a study on how to better teach sustainability to engineering students. The authors propose to include sustainability as a core course in the curriculum. As part of our course redesign, we changed the name of this course from "Environmental Issues, Monitoring & Assessment" to "Environmental Issues and Sustainability in Engineering" to make explicit that the focus of the course is mainly on sustainability.

Finally, we also used several studies that focused on how to measure ethical development in engineering (Drake et al., 2005; Finelli et al., 2012; Nair et al., 2002; Self & Ellison, 1998; Walling, 2015), and the effective assessment of sustainability issues in engineering (Azapagic et al., 2005; Beever & Hess, 2016; Blackwelder, Coleman, Colunga-Santoyo, Harrison, & Wozniak, 2016; Nair et al., 2002).

Development of the Environmental Issues and Sustainability in Engineering Course

Course Overview

The Environmental Issues and Sustainability in Engineering course was recently redesigned and developed and the new version will be delivered in the first semester of 2017. Several changes were made to improve the overall instructional design of the course. Some of the changes can be seen in Table 1. The new course description is as follows:

> Engineers, professionally and as members of society, encounter environmental issues either directly or by way of regulations and workplace practices. They are also at the forefront of designing and implementing many of the changes needed to manage impacts and lead to sustainable development. This course is an introduction to the more significant environmental issues and associated sustainability challenges with a focus on implications for engineers in practice. Topics include: population & urbanisation issues, climate change; waste management, biodiversity; air, water & noise pollution and principles of sustainable and water sensitive urban design. Through these topics students are introduced to the monitoring & assessment of environmental data; and project-based investigation of issues is used to encourage deeper exploration of several topics. (https://www.uq.edu.au/study/course.html?course_code=CIVL2135)

The course is offered as a compulsory course for second year civil engineering students that are joining the civil engineering program after a general first-year engineering experience. The course now is 2 credit units with 3 hours of lecture every week, and 1 hour of tutorials. During the Lecture, students are presented with different information regarding environmental, sustainability, and ethics topics. The lectures have been designed using several active learning strategies to make sure that students actively engaged in the sessions. For example, we included formative assessments via real-time online surveys where students are provided with quick response questions during class with no associated grade, to foster active processing and provide formative feedback about understanding of course content.

We also developed a pre-test to understand students' initial pre-conceptions on the most relevant topics of the course, this information will allow us to make effective decisions on the focus of the information provided during lecture. The pre-test will be taken by every student in the first week of class during the tutorial time.

Lectures are conducted by a teaching team that includes several experts in the school of civil engineering. In addition, there are 5 lectures conducted by industry experts. Industry experts were asked not only to provide their traditional technical presentation, but to bring examples about decisions that they had to made in their daily job that had complex ethical implications. In addition to the 3 hours of lecture, students are provided with a 1 hour tutorial session per week where they can apply the information learned from the lecture into more active sessions conducted by several tutors that are advanced undergraduate students, PhD students, and faculty members.

Tutors were provided with guidance on the new structure of the course, and understood the importance of emphasizing sustainability and ethics issues throughout the semester. They are also included in the continuous analysis of students' data during the semester to be able to focus the tutorial time on the aspects that students are finding more challenging.

Table 1: Course main changes

	Previous version	Current semester (Semester 1-2017)
Couse Title	Environmental Issues, Monitoring & Assessment	Environmental Issues and Sustainability in Engineering
Students	300 average per/semester	256 students enrolled
Learning objectives	**1** Explain the key processes that cause contemporary environmental issues **2** Explain modern concepts of sustainability **3** Source, process, evaluate and effectively communicate various environmental monitoring data **4** Assess environmental monitoring data against relevant standards and guidelines to evaluate the state of a given environmental system **5** Apply simple calculations to examine different options for minimizing the environmental impact of various engineering scenarios	**1** Explain the key drivers and processes that cause contemporary environmental issues. **2** Explain modern concepts of sustainability and the potential contributions that engineers can make to shift human societies towards or away from a more sustainable way of operating. **3** Source, process, evaluate and effectively communicate various environmental monitoring data **4** Assess environmental monitoring data against relevant standards and guidelines to evaluate the state of a given environmental system **5** Apply simple calculations to examine different options for minimizing the environmental impact of various engineering scenarios **6** Identify the ethical issues associated with a range of different situations in professional engineering practice and discuss potential solutions to address these ethical issues
Graded requirements	Project 1 - Case study assessment (35%) Final Exam (65%)	Report - Individual assignments (10%) Project 1 - Case study assessment (30%) Project 2 - Case study Ethics in engineering (20%) Final Exam (40%)
Units credit	1	2
Hours per week	1.5 Hours Lecture <1 Hours Tutorials	3 Hours Lecture 1 Hour Tutorial

Course Assessment

Bailey, Floersheim, and Ressler (2002) suggest that assessment methods must be consistent with the purpose of the course, and should be used as instruments to promote positive change and help the students achieve superior performance. We developed an assessment plan based on recommendations from different engineering education studies (Azapagic et al., 2005; Blackwelder et al., 2016; Drake et al., 2005; Finelli et al., 2012; Nair et al., 2002; Self & Ellison, 1998; Walling, 2015). The focus of the assessment plan was to be able to effectively evaluate students' gains in terms of ethical thinking, and sustainability awareness. In this section, we explain in more detail the different assessments we created.

Reflection assessment: Individual assignment

According to Beever and Hess (2016), case studies are one of the most prominent means for ethical development in engineering students, since they can reach higher stages of moral development. Discipline-specific case studies are also useful to overcome one of the barriers engineering students have to learn ethics and sustainability (i.e. the preconceptions that these topics are subjective and ambiguous) because students are required to be involved in decision making and problem solving approaches that involve some of the issues that they are currently dealing with (Alpay, 2013). We developed part of the assessment to be individual reflection assessments, where students need to reflect on their own learning. Students are asked to explain an important topic from the class (e.g. sustainability, biodiversity, climate change, and air pollution) in their own words, and then reflect how the topic is relevant and important with reference to (i) them personally, (ii) the global society, and (iii) their future career as professional engineers.

Project 1 - Case study assessment

In order to integrate relevant research and theory from the course contextualized in the students' own environment, for the main project of the course students will conduct engineering research on their own houses and will evaluate the quality of their water. This project not only contextualize and engage the students in the real application of the course topics but also poses a critical and complex ethical situation. The project also helps to enable students to feel comfortable about using their knowledge in a known situation and develop methods for recognizing commonalities and differences in classes of problems so that the transfer of learning to a new problem can occur (Nair et al., 2002).

Project 2 - Case study Ethics in engineering

As previously mentioned, one of the main focus of this course redesign was the inclusion of a high focus on ethical development. As suggested byMcGinn (2003) it is necessary to implement serveral metodhs to complement the variety of learning experiences in engineering ethics. Thus, ethics education need to be based on activities, experimentation, and reflection on the understanding of information. We approached the learing on ethics from several perspectives. First we included an ethical component during the lectures provided by the industry speakers. They were asked to reflect and show the studetns real examples that invovled complex ethical situations. We also explained the preference if the ethical dilemma is also a sustainability issue.

Project 3 – Report on industry speakers.

As suggested by Perlman and Varma (2001b), we decided to invite industry experts and asked them to talk about ethical dilemas that they have faced. Students need to reflect on the industry speakers ethical dilemmas presented and create a report based on their findings. This provides students with the opportunity of reflect on a real issue and think about how they –as practicing engieers- will react on a similar situation. We believe students will actively engage in this activity since they will be able to relate to the experts in the field that they could consider role models.

Course evaluation plan

In order to be able to evaluate the effectiveness of our actions to restructure the course, we developed an evaluation plan in which we use two sources of evidence: 1) a pre-post test, and 2) student's end-of-semester evaluation. The analysis of these two sources of data will allows us to see what worked best and what still needs to be improved.

The pre-post test assesses two important objectives of the course: 1) awareness of topics related to the enviroment and sustainable development, and 2) moral judgment. We will also collect demographic information to analyse pre-post data across groups. To measures gians in students' awareness of enviromental issues, we used an adapted version of Azapagic et al. (2005) survey. In this survey, respondents are asked to self-assess how much they know about a topic on scale from 1 (not heard of) to 5 (know a lot). The survey includes four main topics: 1) environmental issues (e.g., acid, rain, deforestation, etc), 2) environmental legislation, policy, and standards (e.g., The Florence Convention, Kyoto Protocol, Rio Declaration, etc), 3) environmental tools, technologies and approaches (e.g., clean technology, eco-labelling, renewable energy), and 4) sustainable development (e.g., precautionary principles, stakeholders' participation, social responsibility, etc). While the four main topics remained, some of the specific items were dropped and others were added in order to better align with the course content. For instance, we dropped the item EU EMAS from the environmental legislation, policy, and standard because this is not as relevant in the

Australian context and would probably not have been encountered by the class. Example items are available in table 2.

Table 2. Example items for environmental awareness survey.

1. How do you rate your knowledge of the following *environmental issues*?				
	Not heard of	Heard of but could not explain	Have some knowledge	Know a lot
Acid rain	☐	☐	☐	☐
Air pollution	☐	☐	☐	☐
Biodiversity	☐	☐	☐	☐
Climate change	☐	☐	☐	☐
Deforestation	☐	☐	☐	☐

The second part of the pre-post test is the assessment of students' moral judgment development. To assess this, we decided to use the Engineering and Science Issues Test (ESIT) (Borenstein, Drake, Kirkman, & Swann, 2010). The ESIT is an instrument that was modelled after the famous Defining Issues Test (DIT) and measures the level of moral judgment development in the context of ethical dilemmas faced by professional engineers. Specifically, the ESIT contains six scenarios (one paragraph of length) that present ethical dilemmas in real-life engineering work situations. An example of a scenario from the ESIT is as follow:

> Engineer Jameson owns stock in RJ Industries, which is a vendor for Jameson's employer, Modernity, Inc., a large manufacturing company. Jameson's division has been requested by management to cut one vendor: either RJ Industries or Pandora Products, Inc. Pandora Products makes a component that is slightly higher in quality and slightly more expensive than that made by RJ Industries. Management and the other engineers in her division do not know that Jameson has a financial interest in one of the two vendors. Jameson is unsure whether she should participate in the decision. (Borestein et al., 2010, p. 391)

After reading the scenario, respondents are asked to complete two rating tasks. In the first task, respondents rate 12 questions on the importance to solve the ethical dilemma. In the task, respondents pick the top four questions and rank them in order of importance. Examples of questions for the above scenario is "Will Jameson's decision potentially cause harm to the public?". The rating on each task is then used to calculate an index that indicates the developmental level of moral judgment of the respondent.

In addition to the pre-post test, we will use end-of-the-semester students' evaluation survey responses to evaluate the learning experience of the students. This evaluation survey contains eight rating questions with items like "the course was well structured" and "assessment requirements were made clear to me", and two open-ended questions "what were the best aspects of this course?" and "what improvements would you suggest?". We will use the rating questions to look at trends and compare it to previous years and the two open-ended questions to look for work best and what needs improvements in the eye of the students.

Reflections on the course redesign process

Redesigning the Environmental Issues and Sustainability in Engineering was a demanding process that required an interdisciplinary team to work constantly on the course every week for the last 7 months. The redesign team consisted of two faculty members with technical expertise on the course, two engineering education researchers, and one expert in instructional design. There were several challenges that the team faced to be able to include a high ethical component, develop effective teaching pedagogies and active learning, create an effective assessment plan, include qualified industry speakers, design data collection procedures, align with accreditation needs, but not to the expense of the main technical knowledge that is required to be provided in the course. We needed to find a balance between the innovation regarding new activities, assignments, and experiences, and making sure the students were receiving all the information that was expected from the traditional lecture. To make this process successful, engineering education literature was crucial in helping us understand what was important in the redesign process and learning similar experiences from other engineering faculty members.

We focused our redesign process on the work done by Finelli et al. (2014) and our main goals as part of the redesign were (i) to improve the student experience during the lectures, (ii) to make an assessment plan relevant to not only improving the student experience of the assignment but furthermore to provide valuable information for both the instructors and the students, and (iii) to have sources of valid information (i.e. data) that help us measure the effectiveness of the proposal we are testing during this semester. In this work-in-progress paper, we explained the entire redesign process with the purpose of providing faculty members teaching sustainability and ethics in engineering with a glimpse of our own process. We believe that part of our success in the redesign of the course, was based on the considerations we took during the process from the engineering education literature. For example there were several actions that we took into account based on Finelli et al. (2014) recommendations:

- We used an experienced educational developer and engineering education researchers that provided specialized support.
- We included evidence-based active learning strategies that have been proven to be effective in the past.
- We developed pedagogies and assessment with the support of technology (online) that can be easy to implement.
- We designed rubrics for each assignment and activity, so the students can receive consistent, detailed, and constant feedback. In addition, tutors were trained to maintain open communication channels with the students so they feel that they can always have a voice.
- With the implementation of the pre-test we will be able to understand the misconceptions and preconceptions that students bring.
- By assigning students with a project based on their own houses, we are creating community building and providing local context to the students.

Conclusion

In this work-in-progress, we described how we designed a course on environmental and sustainability issues in engineering with a focus on ethical development. The course was developed by integrating engineering education research and theories into the traditional environmental engineering curriculum. The course will help the School of Civil Engineering meet the need for inclusion of ethical thinking as part of the accreditation process with Engineers Australia. In addition, the course will help with the increase in engineering student's enrolment by providing with more capacity since the course was expanded.

As part of this research project, we will collect data during the semester and analyse it and present the complete results of this study in different academic venues once we have enough relevant data. We also plan to identify the impact of the redesign of this course in the program of civil engineering.

One problem that we have identified so far regarding data collection is the need for a valid and reliable instrument that will allow us to measure ethical development –based on moral judgment- of engineering students. Therefore the second step of this research will be to develop, pilot, and validate an instrument focus specifically on ethical dilemmas in the context of environmental engineering.

In conclusion, our evidence-based redesign process of this course and evaluation plan can function as a case-study of how to use engineering education research to help improve normal tasks that engineering instructors need to do as part of their job, closing the research-to-practice gap. We hope that engineering educators interested in teaching courses that merge environmental issues with engineering ethics will find this paper helpful to guide their endeavours.

References

Alpay, E. (2013). Student-inspired activities for the teaching and learning of engineering ethics. *Science and engineering ethics, 19*(4), 1455-1468.

Azapagic, A., Perdan, S., & Shallcross, D. (2005). How much do engineering students know about sustainable development? The findings of an international survey and possible implications for the engineering curriculum. *European Journal of Engineering Education, 30*(1), 1-19.

Bailey, M., Floersheim, R. B., & Ressler, S. J. (2002). Course assessment plan: A tool for integrated curriculum management. *Journal of Engineering Education, 91*(4), 425-434.

Beever, J., & Hess, J. (2016). Deepwater Horizon oil spill: An ethics case study in environmental engineering: American Society for Engineering Education, New Orleans, LA.

Benedict, B., Napper, S., & Guice, L. (2000). Restructuring for strategic outcomes. *Journal of Engineering Education, 89*(2).

Blackwelder, B., Coleman, K., Colunga-Santoyo, S., Harrison, J. S., & Wozniak, D. (2016). *The Volkswagen Scandal*. Paper presented at the American Society for Engineering Education ASEE, New Orleans, LA.

Borenstein, J., Drake, M. J., Kirkman, R., & Swann, J. L. (2010). The engineering and science issues test (ESIT): A discipline-specific approach to assessing moral judgment. *Science and Engineering Ethics, 16*(2), 387-407.

Boyd, E. M., & Fales, A. W. (1983). Reflective learning: Key to learning from experience. *Journal of Humanistic Psychology, 23*(2), 99-117.

Brockbank, A., & McGill, I. (2007). *Facilitating reflective learning in higher education*: McGraw-Hill Education (UK).

Drake, M. J., Griffin, P. M., Kirkman, R., & Swann, J. L. (2005). Engineering ethical curricula: Assessment and comparison of two approaches. *Journal of Engineering Education, 94*(2), 223.

Felder, R. M., & Silverman, L. K. (1988). Learning and teaching styles in engineering education. *Engineering education, 78*(7), 674-681.

Finelli, C. J., Daly, S. R., & Richardson, K. M. (2014). Bridging the Research-to-Practice Gap: Designing an Institutional Change Plan Using Local Evidence. *Journal of Engineering Education, 103*(2), 331-361.

Finelli, C. J., Holsapple, M. A., Ra, E., Bielby, R. M., Burt, B. A., Carpenter, D. D., . . . Sutkus, J. A. (2012). An Assessment of Engineering Students' Curricular and Co-Curricular Experiences and Their Ethical Development. *Journal of Engineering Education, 101*(3), 469-494.

Haws, D. R. (2001). Ethics instruction in engineering education: A (mini) meta-analysis. *Journal of Engineering Education, 90*(2), 223.

McGinn, R. E. (2003). "Mind the gaps": An empirical approach to engineering ethics, 1997–2001. *Science and Engineering Ethics, 9*(4), 517-542.

Nair, I., Jones, S., & White, J. (2002). A curriculum to enhance environmental literacy. *Journal of Engineering Education, 91*(1), 57.

Newberry, B. (2004). The dilemma of ethics in engineering education. *Science and Engineering Ethics, 10*(2), 343-351.

Perlman, B., & Varma, R. (2001a). *Teaching engineering ethics*. Paper presented at the American Society for Engineering Education Annual Conference and Symposium, Albuquerque, NM.

Perlman, B., & Varma, R. (Eds.). (2001b). *Teaching engineering ethics* (Vol. 6). Proceedings of American Society for Engineering Education Annual Conference and Symposium.: ASEE.

Santi, P. M. (2000). Ethics exercises for civil, environmental and geological engineers. *JOURNAL OF ENGINEERING EDUCATION-WASHINGTON-, 89*(2), 151-160.

Self, D. J., & Ellison, E. M. (1998). Teaching engineering ethics: Assessment of its influence on moral reasoning skills. *JOURNAL OF ENGINEERING EDUCATION-WASHINGTON-, 87*, 29-34.

Walling, O. (2015). Beyond ethical frameworks: Using moral experimentation in the engineering ethics classroom. *Science and engineering ethics, 21*(6), 1637-1656.

Copyright statement

Work in progress: Approach for assessing learning outcomes in K-12 Engineering Education

Delby Catalina Bello Guzmán

Universidad Pedagógica y Tecnológica de Colombia, *Research Group in Automation and Robotics*
GIRA, Sogamoso, Colombia
delby.bello01@uptc.edu.co

Liliana Fernández - Samacá

Universidad Pedagógica y Tecnológica de Colombia, *Research Group in Automation and Robotics*
GIRA, Sogamoso, Colombia
liliana.fernandez@uptc.edu.co

***Abstract**: This paper presents a review and analysis of the state of the art about the contributions related to robotics approaches in K-12 engineering education. The results show as consulted research works focus mainly on the description of approaches where educational robotics appears as a strategy to encourage the study of technology as a resource for digital literacy, facilitate the learning of other areas as an element mediator in the teaching and learning process, and promote the development of skills as a strategy for innovation and creativity. Currently, given the growth of approaches and research in educational robotics, evaluating the effectiveness of approaches developed for the learning of engineering concepts becomes a subject of great interest to researchers in this field, especially when the Robotics approaches are developed for children.*

Introduction

Currently, the robotics appears as motivational and playful tool for engaging children in technological topics and encouraging engineering subjects. Moreover, robotics stimulates the development of transversal skills, especially those related to digital or computing thinking; thus, robotics has become a powerful tool for the technological literacy. This work is based on previous experiences in which the Robotics was used to motivate the learning of engineering topics in preschool and first courses of elementary school (Fernandez-Samaca, Barrera, Ariel Mesa, & Javier Perez-Holguin, 2017).Taking into account the age of the participants, one of the most important challenges in these experiences is to find alternatives to follow up the students' performance in the topics of engineering by using suitable assessment approaches. The research has taken as context elementary schools in which technological resources or commercial robotic platforms are not affordable. Therefore, the researchers have developed an approach that takes advantage of robotics by using the prototypes of open technology.

This paper devotes to describing a work in progress, which focuses on the evaluation of learning outcomes for four robotics concepts considered in an educational robotics approach, these concepts are circuit, control unit, sensor and actuator. This work is oriented to answer the research question: How to evaluate engineering concepts and learning outcomes in preschool and elementary school when robotics is used in the learning process?

To contribute to answer the research question, this work proposes a methodological framework that considers four stages, such as: the (i) design of the approach based on a documental review by using contribution from data bases like IEEEexplore, SCOPUS and

ScienceDirect, among others, (ii) the construction of the protocol from the design of rubrics to evaluate emotions (Russell & Bullock, 1985), this stages involves the design of questionnaires for children and definition of assessment activities, some qualitative data gathering also will be taken into account (Pekrun, Goetz, Frenzel, Barchfeld, & Perry, 2011); (iii) application of the assessment protocol in a robotic experience in a K-3 grade in which the designed K-12 approach was used; (iv) development of a descriptive study and a correlational statistical analysis based on the information obtained directly from the learning activities developed by the children. The work presented herein shows the main results of the first stage.

The rest of the paper is organized as follows: Section 1 has the theoretical framework with the most important concepts considered into the research, Section 2 describes the literature review analysis and finally, Section 3 presents some concluding remarks.

Theoretical Framework

The following are some theories and definitions that underpin the research work in K-12 Engineering Education.

The National Academy of Sciences ("National Academy of Sciences," 2017) describes three general principles that should characterize K-12 engineering education, the first of which refers to the emphasis (Katehi, Pearson, & Feder, 2009). This means the engineering design must itself be a tool for identifying and solving problems; taking into account that a problem can have more than one possible solution, and it must be a significant environment to learn technological, mathematical and scientific concepts, in turn, should stimulate systems thinking, modeling and analysis. The second principle, presented by the authors (Katehi et al., 2009) mentions that it must include appropriate knowledge in areas such as mathematics, science, and technology that should foster the development of skills in students. Finally, as a third principle, authors propose the promotion of "Habits of the mind," which are consistent with some of the skills of the citizen of the twenty-first century must develop, among them, systemic thinking, creativity, optimism, collaboration, communication and attention to ethical considerations.

In other words, the engineering must identify and solve problems of the context based on scientific knowledge through the construction of models or technology suitable to the needs of the society, whose development imposes interesting challenges. Facing these challenges must allow not only deepening the knowledge about in engineering per se but also developing transversal competencies in a holistic way.

Educational Robotics

Educational Robotics is based on the description of the use of robotics as a tool for learning; it basically handles two learning objectives. The first one is the use of robots to create interest in children to learn about the world of technology incorporating them into the class and teaching activities. The second objective is to integrate children into activities with robots while they learn concepts which are complex to teach through conventional approaches. In this second instance, the robot is seen as a learning support tool.

In educational robotics, it is common to use commercial kits or platforms, which allow to the user a friendly graphical programming environment (Alb, Salle, & Ramon, 2010; Kiss, 2010; Martin, 1995). One of the most popular robotic platforms on the world market is the LEGO® Mindstorms kit.

Constructivism and constructionism

According to (Serrano GonzC!lez-Tejero & Pons Parra, 2011), the constructivist theory of Piaget is one of the bases that support the learning through the education in robotics since this theory proposes that the new knowledge in the child is an active construction since its interaction with the world. Piaget emphasizes that children retain their perception of the world and have very good arguments of this, so they rarely leave this alone because someone tells them that it is not right. Piaget links the 'Learning by doing ' (Papert, 1993) as an approach to education (Bers, Flannery, Kazakoff, & Sullivan, 2014). Based on this argument, Seymour Papert extends the conception of the internal construction of the knowledge and indicates that these are supported by the constructions of the world, referring to the use of computers and robots.

The constructivist teaching approach allows children to explore their own interests while they are investigating a specific topic, solve problems and acquire reasoning skills. Papert mentions that well-designed constructionist activity can generate "powerful ideas" which can be defined as those ideas that are useful in the construction and extension of future knowledge (Papert & Harel, 1991).

Literature review analysis

In the state of the art, authors found several contributions about K-12 engineering education based on robotics; these describe diverse approaches, which are applied in different context and case studies. These experiences describe a direct relationship between the learning of concepts in STEM areas and the strengthening of communication skills, analysis, teamwork, critical thinking and analytical development (Fontenot et al.; Kimmel et al., 2014; Stubbs & Yanco, 2009).

The studies cases are usual in the literature related to K-12 engineering education; it has become a valuable tool for the analysis and characterization of the application of the approach and the presentation of results. In (Bers et al., 2014) , for example, a school experiment is presented in three classrooms, whose target population are pre-school children, the goal is to manipulate robots, develop fine motor skills, hand-eye coordination, social skills, and teamwork,(Bers et al., 2014). The authors, in turn, describe the influence of play on learning. The purpose of this research is to adapt and re-evaluate the curriculum from the TangibleK robotics program to attract children from early childhood to the development of computational thinking (Kimmel et al., 2014). The experience used commercial robot platforms (LEGO) and the language CHERP (Creative Hybrid Environment for Robotics Programming). The evaluation of the approach is mainly oriented to review seven programming concepts and to compare the results of the three classrooms of preschool, taking into account the variability in the teacher's instruction. The data analysis is presented on a Likert scale between 0 -5 as a product of the application of tests.

The STEM K-12 education offers an environment so students can see the interdependence between different subjects of the curriculum, through integrated contexts or "units" as described in 'the high school science teachers at Eagle' (Fa, Kelly, & Gardne, 2003). There, experiences are developed based on four general concepts that allow to students to interconnect different fields and science, such as matter, energy, force, and charge. These concepts were applied successfully to earth sciences (9th grade), chemistry (10th grade) and general physics in (11th grade). However, the class of earth sciences is developed isolated from chemistry, biology, and other related areas.

In (Karahoca, Karahoca, & Uzunboylub, 2011) authors present a case study in the school with children aged between 10 and 15 years to develop skills and abilities in areas such as science and technology from the design and construction of a line follower robot. The building of this robot integrates the contents of a course of electronic circuits in a collaborative learning environment by project-based learning (PBL). Students, in addition to the construction and testing of the robot, must interact each other, discuss, analyze, predict, make decisions, analyze data and reach conclusions. It is a worth noting that this is one of the few contributions in which a robotic platform is developed in its entirety by student teams, in other words, they do not use low-cost commercial robotic platforms. The evaluation of the approach is made from the weekly assessment of the activities and the achievement of objectives. Authors present a quantitative analysis of data.

In the documental review, researchers have found not only on the design of student-oriented approaches but also descriptions of experiences of programs for the training of teachers who lead the teaching-learning processes. In these practices, the teacher is oriented towards the development of skills in the design of Hands-on activities and the implementation of STEM lessons based on robotics, a good example of this kind of work is presented in (Fontenot et al.; Franz & Elmore, 2009; Karp et al., 2010; Kim et al., 2015)

In (Kim et al., 2015)), authors use robotic kits (called 'My robot time' and 'RoboRobo') and expose multiple instruments and a protocol for the evaluation of the proposed approach. This protocol was designed based on previous studies and the observation of student participation. In this case study, researchers develop a video in order to evaluate several topics, among which, the cognitive commitment and use of self-regulation questionnaire instrument (SRQL_L). These instruments correspond to those suggested by (Black & Deci, 2000) Who describe" an evaluation method that integrates the collection of quantitative and qualitative data prior to and after the development of the activities, in products such as a) student reports on their STEM commitment and knowledge, b) class plans, c) videos of robotics activities and c) interview with students".

Among works oriented to show a timely application of K-12 education tools in the classroom, it is the case of the software 'Dolitle,' (Kanemune, Nakatani, Mitarai, Fukui, & Kuno, 2004), a programming language used as a strategy of the Japanese government to promote the education in information technologies in primary and secondary school. In (Kanemune et al., 2004), authors describe an experience using this object-oriented software. The software was specially designed to learn computer programming at early ages given its easy syntax. In this experience, researchers use as evaluation instrument a questionnaire, which is applied to students to evaluate the understanding of concepts.

Commonly in engineering education literature about STEM approaches, there are case studies that describe outreach programs and trends in robotics, such as the creation of robotic platforms for participation in competitions and competitions at local and global level. These competitions seek to promote the creativity and develop skills like leadership and innovation. In recent years, the growth of the research in robotics has been encouraged by the access and availability of robotic platforms and programs that facilitate work with adolescents and school-age children (Cruz-Martín et al., 2012). In (Eguchi, 2016) for example, authors describe an experience of this type, the competition 'RoboCupJunior (RCJ),' a competition organized by Robocup ("rcj.robocup.org," 2017), with participants from more than 30 countries.

Robocup is an initiative characterized by maintaining the same competition objectives with rules adjusted annually to improve participants' experience and learning. The challenges of the tests are soccer, rescue and dance. Furthermore, it gives an emphasis more educative than competitive (Eguchi, 2016). Due to the strong interest in these types of events and the degree of commitment and innovation of the RCJ participating teams, the author presents two technological contributions resulting from these developments; on one hand a vision sensor for the soccer competition and on the other a low-cost Robot Arduino open robotic platform.

A widely worked and mentioned theme in research in education in engineering K-12; it is the effect or linkage of the game with activities and learning environments. In (Lindh & Holgersson, 2007), a study is presented to determine the impact of LEGO Dacta on students' ability to solve problems of logic and mathematics. According to the constructionist theory mentioned by (Lindh & Holgersson, 2007) citing to (Gustafsson & Lindh, 2001; Harel & Papert, 1991) the knowledge is not only transmitted from teacher to student but must be built in her mind by active learning. This research raises a null hypothesis that states that H: 'LEGO robots have no impact, neither positive nor negative in solving problems of logic and mathematics'; based on this hypothesis, (Lindh & Holgersson, 2007) proposes to divide students into two study groups as an alternative of comparison; one is called the control group (CG), composed of students who have not been linked to an experimental work with LEGO Dacta and another called experimental group (EG) represented by students who have participated pedagogically with LEGO Dacta proposals. One of the remarkable results obtained after the application of the ANOVA test (variance analysis) and a rigorous statistical analysis is that LEGO training can be used only for some groups of students, among other factors because the students present different styles of learning (Papert & Harel, 1991).

Regarding to the environments in which these researches are carried out, authors observed that usually there are many extra class activities and these are developed in a different environment than the classroom, such as the experience described in (Yilmaz, Ren, Custer, & Coleman, 2010) that uses a summer camp to attract students to the different fields of engineering, by using hands-on activities.

Results

The contributions analyzed were obtained through searching in a database by using the following five keywords: 1) Learning Outcomes, 2) Assessment, 3) K-12 Engineering Education, 4) Robotics and 5) STEM. Table 1 summarizes of the most relevant aspects regarding studied contributions related directly to the scope of the research proposal presented herein. The studied works have been classified according to six aspects; namely, i) Type of contribution (journal article, conference paper, others), ii) Ages of the participants (children, teenagers, preschool kids), iii) Used platform (If the research used or not commercial platforms), iv) Assessment (If use or not rubrics), v) Subject (knowledge field in which the approach is being applied) and vi) STEM curriculum (If the contribution focuses on STEM education syllabus).

The sample for the analysis corresponds to journal articles (68.4%), followed by conference proceeding (21.5%). Regarding student evaluation, 58% of the research experiences do not use rubrics, and those that use rubrics do not necessary perform a statistical analysis. In short, the approaches' assessment is focused more on the application of the approach than on the determination of knowledge level or learning outcomes, the rubrics appear like a mandatory part of the approach design but not like an instrument to assess it, see Figure 1.

In the most of the cases presented in journal papers (68%), authors describe experiences and activities designed for children over 12 years; generally, in diverse learning environments like summer camps and competitions, and whose purpose is engaging young people in STEM sciences. The analysis of state of the art shows that the application of robotics in early childhood is still very limited, only about 16% of the contributions consulted describe experiences for kids younger than six years. Moreover, 63% of works use commercial robotic platforms for the development of research learning activities, the most of these using LEGO® Minstorm; only a lower percentage (21%) build their robotic prototype, which usually is an open-source and low-cost platforms, see Figures 1 and 2.

According to the topics dealt in consulted articles, researchers observed that authors describe experiences in areas such as engineering and technology (30%), in science and mathematics, corresponds (15%), and teaching and learning process (11%), see Figure 2.

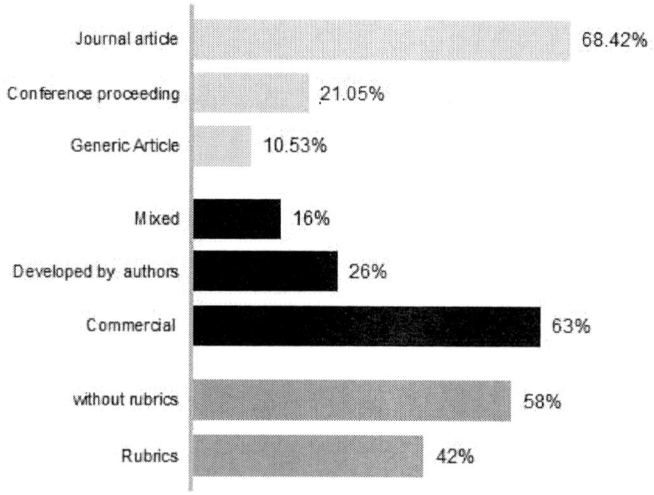

Figure 1. Types of contributions, platforms and use of rubrics.

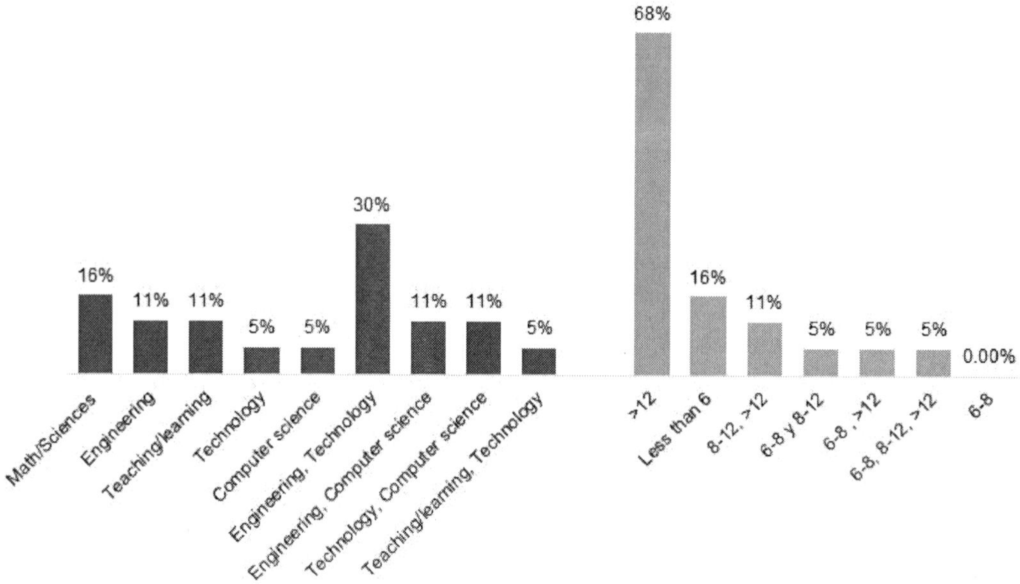

Figure 2. STEM areas of interest to researchers and age of the children participating in the approaches.

Although the information presented in Table 1 along with the analysis of each of the aspects found in the state of the art evaluation of K-12 engineering education approaches allows establishing that there is a broad field of research related to Robotics Education, there is a lower development of robotic approaches focused on early childhood. Likewise, the documental review shows a low percentage of approaches are oriented to use or design open-source and low-cost platforms.

Table 1. Classification of the main contributions found in the literature review.

Authors	Type of contribution			Ages of the participants				Platforms		STEM curriculum	Assessment		Subject				
	Journal article	Generic article	Conference papers	Less than 6	6-8	8-12	Older than 12	Commercial	Developed by authors		Rubrics	Without rubrics	Math/Sciences	Engineering	Teaching/ learning	Technology	Computer science
Fernández-Samacá, Barrera, Mesa, & Pérez-Holguín. (2017)	X			X				X	X		X			X			
Kim, C., Kim, D., Yuan, J., Hill, & R. B., Doshi, P.(2015)	X						X	X		X	X				X		
(Ospennikova, Ershov, & Iljin, 2015)	X				X		X	X		X		X		X			
Barrera-Lombana, & Mesa-Mesa. (2007)		X			X			X	X	X		X				X	
Bers, Flannery, Kazakoff, & Sullivan, 2014	X			X					X			X					X
(Oppliger, 2002)	X					X	X	X				X		X		X	
(Klassner & Anderson, 2003)		X					X	X		X		X				X	X
Kanemune,Nakatani , Mitarai, Et al 2004			X			X	X		X	X	X					X	X
Ruiz-del-Solar, 2010	X						X	X				X			X	X	
Jojoa, Bravo, & Cortés, 2010	X				X	X	X	X		X	X		X				
(Lindh & Holgersson, 2007)	X						X	X				X	X				
(Williams, 2007)	X						X	X				X	X				
Sullivan & Bers, 2017	X			X				X		X	X			X			X
Eguchi, 2015	X						X	X	X		X			X			X
(Gómez-De-Gabriel, Mandow, Fernández-Lozano, & García-Cerezo, 2011)	X						X	X			X				X		X
Yilmaz, Ren, Custer, & Coleman, 2010	X						X		X	X	X			X		X	
Karp & Schneider, 2011			X			X	X	X		X				X		X	
Fontenot, Dean Talkmitt, Susan Et al, 2009			X			X	X	X		X		X			X		
(Connaughton & Modlin, 2009)			X				X		X	X		X		X	X		
This research proposal				☑	☑				☑		☑			☑			

☑ Possible fields of action for research in K-12 engineering education. X Fields worked by researchers.

Concluding remarks

From the review of the state of the art, researchers identified that most of the contributions are focused on presenting learning approaches instead of assessment approaches. Likewise, the most of these learning approaches are related to Engineering and Technology and describe experiences developed for children older than 12 years by using commercial platforms, especially LEGO® Mindstorm. Although in the learning approaches, authors have considered the evaluation of the impact of their learning approaches, these are not devoted specifically to learning outcomes of engineering topics. Therefore, it is necessary to strengthen the research about how to design assessment approaches, especially when the learning approach is developed for children, and in this particular instance, in K-12 engineering education approaches based on Robotics.

REFERENCES

Alb, J., Salle, L., & Ramon, U. (2010). DTCNN Implementation in a LEGO Mindstorm NXT for Infrared and Ultrasonic Sensor Data Processing. 8-9.

Bers, M. U., Flannery, L., Kazakoff, E. R., & Sullivan, A. (2014). Computational thinking and tinkering: Exploration of an early childhood robotics curriculum. *Computers & Education, 72*, 145-157. doi:http://doi.org/10.1016/j.compedu.2013.10.020

Black, A. E., & Deci, E. L. (2000). The effects of student self-regulation and instructor autonomy support on learning in a college-level natural science course: A self-determination theory perspective. *Science Education, 84*(6), 740-756.

Connaughton, R., & Modlin, M. (2009). *A modular and extendable robotics platform for education.* Paper presented at the Frontiers in Education Conference, 2009. FIE'09. 39th IEEE.

Eguchi, A. (2016). RoboCupJunior for promoting STEM education, 21st century skills, and technological advancement through robotics competition. *Robotics and Autonomous Systems, 75*, 692-699. doi:10.1016/j.robot.2015.05.013

Fa, S., Kelly, J., & Gardne, J. (2003). THE DOTS. 83725-83725.

Fernandez-Samaca, L., Barrera, N., Ariel Mesa, L., & Javier Perez-Holguin, W. (2017). Engineering for Children by using Robotics. *International Journal of Engineering Education, 33*(1), 389-397.

Fontenot, D., Talkmitt, S., Morse, A., Marcy, B., Chandler, J., & Stennett, B. (2009). *Providing an engineering design model for secondary teachers.*

Franz, D., & Elmore, B. B. (2009). Work in Progress - Collaborative Outreach to " at- risk " Middle School Students using LEGO Robotics. 1-2.

Gómez-De-Gabriel, J. M., Mandow, A., Fernández-Lozano, J., & García-Cerezo, A. J. (2011). Using LEGO NXT mobile robots with Lab VIEW for undergraduate courses on mechatronics. *IEEE Transactions on Education, 54*(1), 41-47. doi:10.1109/TE.2010.2043359

Kanemune, S., Nakatani, T., Mitarai, R., Fukui, S., & Kuno, Y. (2004, 29-30 Jan. 2004). *Dolittle - experiences in teaching programming at K12 schools.* Paper presented at the Proceedings. Second International Conference on Creating, Connecting and Collaborating through Computing, 2004.

Karahoca, D., Karahoca, A., & Uzunboylub, H. (2011). Robotics teaching in primary school education by project based learning for supporting science and technology courses. *Procedia Computer Science, 3*, 1425-1431. doi:10.1016/j.procs.2011.01.025

Karp, T., Member, S., Gale, R., Member, S., Lowe, L. A., & Medina, V. (2010). Generation NXT : Building Young Engineers With LEGOs. *53*(1), 80-87.

Katehi, L., Pearson, G., & Feder, M. (2009). *Copyright © National Academy of Sciences. All rights reserved. Unless otherwise indicated, all materials in this PDF File are copyrighted by the National Academy of Sciences. Distribution, posting, or copying is strictly prohibited without written permiss.*

Kim, C., Kim, D., Yuan, J., Hill, R. B., Doshi, P., & Thai, C. N. (2015). Robotics to promote elementary education pre-service teachers' STEM engagement, learning, and teaching. *Computers & Education, 91*, 14-31. doi:10.1016/j.compedu.2015.08.005

Kimmel, H. S., Burr-Alexander, L. E., Hirsch, L., Rockland, R. H., Carpinelli, J. D., & Aloia, M. (2014, 22-25 Oct. 2014). *Pathways to effective K-12 STEM programs.* Paper presented at the 2014 IEEE Frontiers in Education Conference (FIE) Proceedings.

Kiss, G. (2010). USING THE LEGO-MINDSTORM KIT IN GERMAN COMPUTER SCIENCE. 101-104.

Klassner, F., & Anderson, S. D. (2003). LEGO MindStorms: Not just for K-12 anymore. *IEEE Robotics and Automation Magazine, 10*(2), 12-18. doi:10.1109/MRA.2003.1213611

Lindh, J., & Holgersson, T. (2007). Does lego training stimulate pupils' ability to solve logical problems? *Computers & Education, 49*(4), 1097-1111. doi:10.1016/j.compedu.2005.12.008

Martin, F. G. (1995). The Art of LEGO Design. *1*(2), 1-19.

National Academy of Sciences. (2017). Retrieved from http://www.nasonline.org/

Oppliger, D. (2002). Using FIRST LEGO League to enhance engineering education and to increase the pool of future engineering students (work in progress). *32nd Annual Frontiers in Education, 3*, 11-15. doi:10.1109/FIE.2002.1158731

Ospennikova, E., Ershov, M., & Iljin, I. (2015). Educational Robotics as an Inovative Educational Technology. *Procedia - Social and Behavioral Sciences, 214*, 18-26. doi:10.1016/j.sbspro.2015.11.588

Papert, S. (1993). The children's machine. *Technology Review-Manchester Nh-, 96*, 28-28. doi:Article

Papert, S., & Harel, I. (1991). Situating constructionism. *Constructionism, 36*(2), 1-11.

Pekrun, R., Goetz, T., Frenzel, A. C., Barchfeld, P., & Perry, R. P. (2011). Measuring emotions in students' learning and performance: The Achievement Emotions Questionnaire (AEQ). *Contemporary Educational Psychology, 36*(1), 36-48. doi:10.1016/j.cedpsych.2010.10.002

Russell, J. A., & Bullock, M. (1985). Multidimensional scaling of emotional facial expressions: Similarity from preschoolers to adults. *Journal of Personality and Social Psychology, 48*(5), 1290-1298. doi:10.1037/0022-3514.48.5.1290

Saxton, E., Burns, R., Holveck, S., Kelley, S., Prince, D., Rigelman, N., & Skinner, E. A. (2014). A Common Measurement System for K-12 STEM education: Adopting an educational evaluation methodology that elevates theoretical foundations and systems thinking. *Studies in Educational Evaluation, 40*, 18-35. doi:10.1016/j.stueduc.2013.11.005

Serrano GonzC!lez-Tejero, J. M., & Pons Parra, R. M.-a. (2011). El Constructivismo hoy: enfoques constructivistas en educaciC3n. *Revista electrC3nica de investigaciC3n educativa, 13*, 1-27.

Stubbs, K., & Yanco, H. (2009). STREAM: A workshop on the use of Robotics in K--12 STEM education [Education. *IEEE Robotics & Automation Magazine, 16*(4), 17-19. doi:10.1109/mra.2009.934830

Williams, D. C. M. Y. P. L. F. M. J. (2007). Acquisition of Physics Content Knowledge and Scientific Inquiry Skills in a Robotics Summer Camp. *Journal of Research on Technology in Education, 40*(2), 201-216. doi:10.1080/15391523.2007.10782505

Active collaborative learning experience: a course of engineering students at the half point of their studies

Julia Díaz
Universidad de los Andes, Bogotá, Colombia
jh.diaz84@uniandes.edu.co

Catalina Ramírez
Universidad de los Andes, Bogotá, Colombia
mariaram@uniandes.edu.co

Abstract: *A permanent challenge in engineering schools is to develop skills and capabilities related to sustainability. In developing countries engineers will play managerial roles, where they are engaged with social problems requiring solutions with ethical character. The course Proyecto Intermedio - Ingenieros sin Fronteras has been developed between professors and students from two universities in Colombia, addressed to establish a bridge between the academy and vulnerable communities immersed in entangled problematics. Collaborative learning is useful for learning in context in engineering. Although collaborative learning has been studied in university context, we know less about experiences involving engineering students and green micro-entrepreneurs. Exploring conditions that enable or prevent collaborative learning, we found that active listening, continuous learning and concreteness enable the collaborative learning. While fear, lack of knowledge and believes about partners inhibit it. Finally, we learned that these initiatives influence university´s institutional answers related to issues of social responsibility.*

KEYWORDS: Collaborative learning, sustainability, developing country.

Introduction

It is increasingly necessary that the Colombian engineers commit to work in solution for the complex problems that affect the urban and rural areas and the country in general. To that end they must not only be very good in the technical aspects, but also understand the value of working along with other professionals and with the communities that will have a conjoint participation in the engineering solutions. That is framed in the proposal of Johri & Olds (2011) about situated learning giving to the context (physical and social elements) a central role in engineering learning process.

This implies that the engineer must be an acute observer of the reality that surrounds him, that he realizes the difficult situation of several Colombian communities. To observe the complexity of these contexts allows the engineering students to conceive and design sustainable solutions from any of their respective disciplines, as long as the persons which they are working with are being taken into account. Additionally, as Trevelyan (2010, p. 388) states: "engineers work as much or more by leveraging and coordinating the expertise of others as they contribute themselves. To do this, engineers need to be good listeners, learners, teachers and leaders".

Those are the reasons why the team of Ingenieros sin Fronteras-Colombia, students and professors of several universities are dedicated to bring closer teaching and researching to engineering participation-observation-conception-design and solution implementation activities concerning contexts of vulnerability or of development opportunity. This paper presents the design of a teaching-research proposal in which the students of the Ingenieros sin Fronteras Intermediate Project course work collaboratively with the communities, designing engineering proposals that generate knowledge in both directions. Thus facilitating the articulation between theoretical and the practical components of the course. The specific working context is the collaboration between engineering students and entrepreneurs that seek to develop projects that have a positive socio-environmental impact. The context defined above gives its sense to the research question that is laid out below: Which factors favor or prevent the collaborative learning between engineering students and entrepreneurs of green businesses?

In order to address the research question a process of participant observation of the dynamics of the course Intermediate project was undertaken. Afterwards, a phase of qualitative analysis of the observations took place, leading to the findings proposed in the final part of this document. Looking to establish the context of the study, the motivation that originated the design of the course is described; followed by the manner in which the course itself is carried out. In a subsequent section the research methodology for addressing the research question is presented, within the framework of the course and of the findings, learnings and projections derived.

Context, the role of Engineering Education research before the vulnerable communities' situation

The engineering students are increasingly interested in contributing to solving effectively complex problem situations of Colombia. It is therefore important that they understand how Engineering has a fundamental impact in the viability of the Colombian regions. To understand that bad engineering decisions lead to compromising resources and generating corruption is a heartfelt necessity for professors, researchers and students. Additionally, in the stream of university-business cooperation had been found that universities have the potential to contribute to business innovation particularly if those business to collaborate with local universities Hewitt-Dundas (2013). In our case we explore basic elements of this relationship with a less common focus: green microenterprises.

That is the reason why the group of Ingenieros sin Fronteras – Colombia intends to bring closer the students to realities that they are unaware of. Difficult realities such as persons that lack potable water, children in poor health conditions due to the polluted water they drink, persons forced to move to the cities due to the lack of proper opportunities in the regions, among others. The purpose is that, confronted to these situations, the students will be able to evaluate the viability of implementing engineering projects for these contexts. Furthermore, to find and observe these situations promotes in the student a civic position of engagement and sustainability towards the country.

This work in teaching and education is also an exercise that can give an insight on the social responsibility of the universities in the Colombian context. The involvement of professors, researchers and students in these processes has helped in giving these issues a more visible position in the University's institutional discussion agenda. Currently a dynamic of positive reinforcement has been generated, supported by the priority position that has been given to the regions and communities subject in the Institutional Development Program (2010-2016) of the Universidad de los Andes.

Based on the above, Ingenieros sin Fronteras – Colombia seeks concrete objectives for promoting the development of communities through collaborative projects that are also interesting for Engineering students and professors. The development of the initiatives promotes:

For entrepreneurs

- Active participation in solutions planning and design.
- Acquisition of technical and managerial knowledge that can be applied in multiple contexts.

- Development of sustainable autonomy regarding engineering solutions.
- Internal union and group work.

For the Students

- Development of interdisciplinary work, group work, communication and research skills in real Colombian contexts.
- Innovation and solutions proposal working together with professors and communities.
- Carrying out research with students and professors with prominent international organizations and engineering schools.

For teachers

- Development and innovation through technically viable and culturally appropriable solutions.
- Design and implementation of relevant models for joint work between University-Community-Government.
- Carrying out applied research along with colleagues from prominent international organizations and engineering schools.

For the higher education institution

- Achievement, through concrete actions, of its priorities, namely helping its students to become social transformation agents.
- Having a concrete program that allows it to articulate teaching, research and presence in problem situations that affect vulnerable communities in the Colombian context.

Theoretical approach to course design

The methodologies used by Ingenieros sin Fronteras – Colombia are based in its work principles, moreover they are designed to accompany applied research processes in social contexts and work for and by the community. In this course we used in an integrated way the collaborative learning process and some elements from sustainability education. We adopt the Dillenbourg's (1990) definition of collaborative learning as "a situation in which particular forms of interaction among people are expected to occur, which would trigger learning mechanisms" (p.5). Additionally, we agree with González & Díaz (2005) that collaborative learning is useful in learning processes because it contribute as a deliberation tool and given it help to recognize peers in order to face topics of responsibility, as is the case of our course and the projects in which its students are involved. Sustainability education is another edge in our approach. Universities have been starting to include in their objectives topics in sustainability. But it needs to go beyond including a shift in education oriented to social change and ecological thinking (Moore, 2005). Considering that entrepreneurs in our course are or are trying to be green, sustainability emerges as a necessary topic to focus our attention in this study.

Also we include elements from other methodologies used in our projects of Ingenieros sin Fronteras - Colombia to date as: OCDIO, PAR and gamification. In first place, OCDIO Methodology is an engineering intervention methodology (Ramírez et al., 2016), that aims to provide the students with the required tools to address complex social problems in an innovative and flexible way. So our students developed group projects addressing problem situations in six phases: Observe, Conceive, Design, Evaluate, Implement and Operate. Through the application of this methodology the students generate prototypes arising from a systemic analysis of the problem situation, from the integration of the knowledge, and from innovative teamwork.

In second place, *Participatory Action Research (PAR)* is a project development methodology in which intervene the community and the institution in charge of the research. The community participates in the assessment of their problem and in the implementation of the projects, which are usually focused on natural resources and its sustainability. This is a suitable method for working with rural communities because it encourages the development of sustainable solutions (Ramírez et al., 2010). It generates a progressive transformation in the society, and it enhances community participation through feedback and adjustments to the proposals. Furthermore, PAR allows the educational institutions to open up to real problems and to seek solutions integrated with teaching and research processes that involve all the stakeholders

Finally, G*amification* is the last methodological approach used. It's the process of using a game, group exercise or activity in which different mechanics are employed for attracting the participants (Kapp K.M., 2012) and thus generating in them desired behaviors that result in solutions to specific problems. The objective is that the participant incorporates these behaviors in his/her daily activities through the game experience designed by the researchers.

Course's operating methodology

Based on the theoretical contexts described above, the Ingenieros sin Fronteras Colombia group developed the collaborative research frame based in seven engineering phases: Participate, Observe, Conceive, Design, Evaluate, Implement, and Operate. The EWB-COL group promotes the collaborative participation among the different groups of the community and observes the vulnerability of the different communities with the purpose of creating improvements cooperatively. During this stage it approaches the specific problem situation through the active participation of the persons that are potentially affected and through the exchange of ideas with experts and researchers.

Once this preliminary observation has been executed, the participants conceive the formulation, contextualization and a possible solution to the reality observed. This conception requires a strong emphasis in the development of creativity and innovation exercises that enables the technological proposal to be suitable for the community intervened (it's in this moment that gamification dynamics are included. After at least one evaluation phase, comes the phase of the preliminary participative design of the prototype and a proposal for an implementation of the technology that will improve the conditions of the problem situation that's being dealt with, along with the community. Afterwards, the counseling by the professors is intensified as well as the detailing of the development of the problem situation; in this sense the design of the prototype is carried out with more precision, more detail is given to the implementation and the operation phase begins.

Research methodology

The course Proyecto Intermedio - Ingenieros sin Fronteras is an initiative developed between professors and students from Engineering Schools of Corporación Universitaria Minuto de Dios and Universidad de los Andes. Aiming to achieve a better understanding of the aspects that enable or prevent the collaborative learning between students and communities, for this case entrepreneurs of green businesses (Pacheco, Ramírez, González, 2013), a process of observation and information gathering was undertook during two academic semesters in this course. The articulation between the actors is illustrated in Figure 1.

A convenience sample was made in this course based on the collaborating teams, composed by 12 entrepreneurs and the students of the course. This research process was carried out during the first and second semesters of 2015 with the objective of examining systematically the impact of the work model on the initiatives. 15 interviews were conducted with entrepreneurs and a survey was applied to 15 entrepreneurs and 26 students all of them over the age of majority. The interviews had a semi-structured format, they were preceded by the informed consent protocol, and they had an average duration of 15 minutes; the surveys were conducted in a similar manner. Both the interviews and the surveys focused on three aspects: learnings, conditions surrounding the process and prospective. During the second semester of 2015, 4 entrepreneurs that had participated in the first phase continued with the process, now with a new group of students.

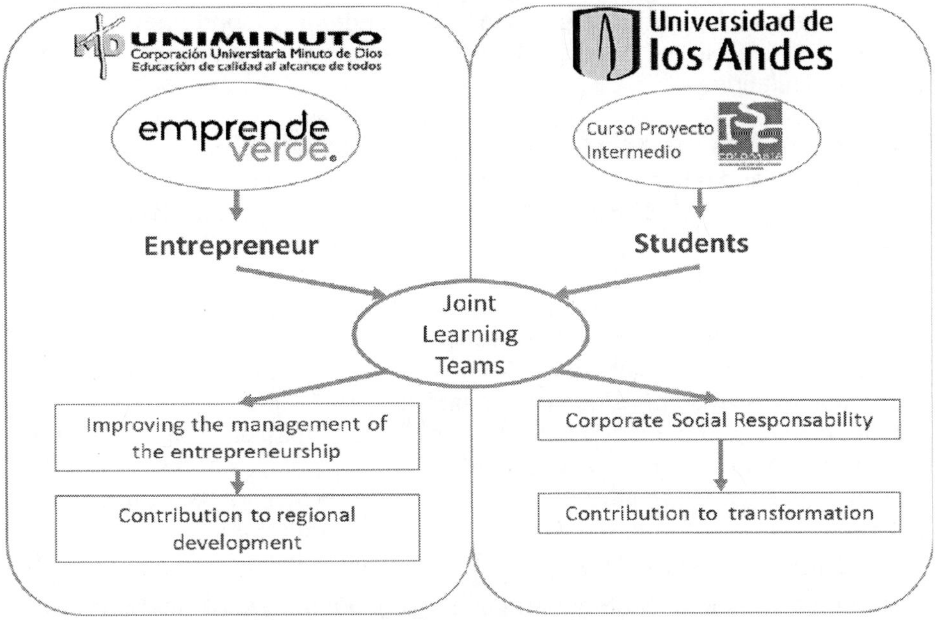

Figure 1. Stakeholders articulation
Source: Ramírez, Díaz, Blanco. Ascolfa Conference Presentation 2016.

The research was carried out with an exploratory analytic, reflexive and interpretative intention (Flyvbjerg, 2006), and with some inductive elements linked to the grounded theory (Glaser and Strauss, 1967). A qualitative approach methodology was used in order to identify those factors that enable or prevent the collaborative learning in blended teams (students and entrepreneurs). A set of 15 teams composed by engineering students and green entrepreneurs of Bogotá and Cundinamarca region in Colombia, were the participants in the study. Participative observation during teaching lessons and 4 workshops, surveys and semi-structured interviews were the tools employed for data gathering. The data were analyzed using a qualitative emergent categories approach.

The analysis of the interviews allowed a deeper understanding of some of the aspects mentioned in the surveys by the entrepreneurs, especially those referring to the preliminaries of the process, with questions such as: Did you have any reservations regarding the process when it began? What were your biggest concerns regarding your work as entrepreneur, before beginning this process?

Regarding the process itself: Tell me about your experience in the process with students: Do you think that it helped you to learn? What did you learn through this experience? Did the process enrich your entrepreneurship? There was also an inquiry regarding the internal and external conditions that had an influence such as communication, time and commitment; the possibility of maintaining the work with the student was explored; as well as what the entrepreneurs conceive as University Social Responsibility.

Findings

Some of the business ventures that participated in the study were: a Girardot food shop; two catering companies from Girardot and Soacha; a gastronomy line bakery from Madrid; three services companies: recycled paper products, product recovery for livestock exploitation, and pedagogical counseling in Soacha, Suesca and Madrid respectively; and a Sustainable Construction company.

Entrepreneurs: 89% said that collaborative learning with students helped them to redesign their green entrepreneurships, to include new ideas in planning and design and to learn how to work in a collaborative team. New proposals coming from students support and engagement influenced positively in the results. However, they pointed out lack of time and geographic distances as limitations.

Students: 92% said that conjoint work with entrepreneurs influenced their professional formation, in their knowledge and reflection about what implies to be an entrepreneur and about the social impact of the use of their knowledge in engineering serving social initiatives.

In terms of the aspects that favor or prevent the learning the following was found:
In the context of the theoretical research made and of the information gathered some findings and elaborations regarding the research questions are presented. According to the coordinators of the trading company (Velásquez, 2015), the main concerns identified in the entrepreneurs regarding their own administrative performance are: fear of taking risks, difficulty for delegating, high regard of local products, fear of public speaking for advertising the company, lack of product innovation. It was found that due to its nature and size, the management of the small business is highly determined by the decisions of its owner, who acts as general manager. This is corroborated by the statements of the participants whose perception influence the possibility of partnerships with new institutions:

"As a young man you see a lot of it, a lot of that prevention, because we know that our university corporation has a highly social approach; therefore the people that go to study there, go with the bare minimum but filled with desire for improvement, and having that approach generates some preventions on our side" (Sustainable construction producer)

The statement shows that prevention and fear of new processes can become an obstacle, but their entrepreneurial nature allows them to face this in such a way that they can benefit from these new opportunities. The global analysis of the occurrence of concepts in the interviews shows that fear and prevention are some of the terms with the highest incidence in the perception of the entrepreneurs regarding their own processes.

Another participant said: "Let's say that thanks to this process that we have undertaken, I've begun to have 360 degree view, I am looking to different horizons and that's what's important" (Food business manager). The statement confirms the data gathered in the survey, as the work optimized his capacity for generating networks, interacting with interested people, and blending the business with the concept of green business.

Finally, a participant explained his process for overcoming certain difficulties: "Talking, they changed completely, they changed by talking about them, maybe the main fears and negative perceptions that I had were psychological, it's always the psychological aspect; but all of that psychological stuff comes out by talking, then, as you talk about it, you solve it, and all those perceptions go away (Sustainable construction producer).

It was found that the interaction with the students has a positive effect in the entrepreneurs' management, mainly because of the overcoming of the preventions related to their own management attitude and the relation with the interested people.
Other concerns of the entrepreneurs identified by coordinators of trading company were related to aspects in the context that they consider hard to control such as: production capacity, technological equipment, single provider, logistic planning, orders fulfillment, use of family financial resources (Velásquez, C. 2015). These findings also were found in the field work as is illustrated following: "I'm worried for the low output capacity, very worried, also for

the few technological equipment, for lack of product innovation and for the lack of logistic planning, very worried" (Sustainable construction producer). That means that the capabilities of the entrepreneur to acquire resources, innovate and to interact with actors in relevant enterprise´s environment is crucial to assure the stability of his/her entrepreneurship.

In accordance with the results of the surveys, the tools provided by the students, related to planning, technological management and marketing, show how the joint work can favor the use of tools for having more control over these variables. One of the entrepreneurs manifested: "I've learned that the entrepreneur must improve himself constantly, each order is an opportunity for re-inventing oneself each day, a way for improving my logistic planning" (Recycled materials products).

However, some of these variables generate incertitude because they cannot be controlled exclusively with managerial tools, and the joint work doesn't guarantee completely their control. "This improvement process is a path that requires a longer time. In order to continue, more dedication and deepening are needed" (Dairy producer). Hence, it was found that the collaboration with the students has a variable effect in the entrepreneurs' management, especially regarding the variables highly dependent of the context.

This agrees with the surveys that acknowledge the role of the university as a knowledge generator, that provides analysis and that must come closer to the activities of the local entrepreneurs. Regarding this topic the following was stated: "To me, RSU means the students thinking beyond their careers, understanding that in Bogota, in Cundinamarca there are other things, there are other things outside; we're all young and entrepreneurs and what we do is beneficial for the same country" (Food industry entrepreneur). Another supporting statement was: "The support they gave us was very positive. We learned things that we wouldn't be able to learn otherwise" (Dairy producer).

As a result of the experience and of the initiatives and projects developed by OSAL, the course throughout multiple semesters, and the Verde-Ando (Going Green) project, three aspects that foster the RSU have been identified: the generation of institutional mechanisms that establish bridges between research and the local context; funding research projects that favor joint work with regional stakeholders; and playing an active role in regional projects.

It must be taken into account that the analyses of these perspectives of the RSU cannot be isolated, and therefore in the strategies developed by each university they may coexist as suggested by Quezada (2011). As the transformational perspective establishes the guidelines of the socially responsible behavior, the managerial and normative perspectives establish the manners in which the universities can express this behavior. That also influences the way in that university assumes its business-university collaborations.

We found therefore that, as the RSU goes from being nominal to becoming a part of the institutional action, the associated strategy goes from being normative or managerial to having a multiple perspective, including the transformational one. This is related to the prevalence of the RSU in the institutional development plan.

Conclusions and future research

Overcoming fears and distrust was a first step for those teams that had better results, also the conjoint motivation and commitment. The learnings, obtained both from at least 10 semesters of deliver the course and from this focalized research, allowed to carry out a deep reflection that, along with the results of the research itself, will enable to incorporate some aspects in the design of the course. And also to continue with the research activities, which will now focus in future designs that will allow maintaining the positive impact and possibly to develop more ambitious initiatives in an institutional level.

Firstly, the focus of the projects regarding water conservation and green businesses will be maintained. Second, a new aspect that is already in process to be included is the integration of an additional actor to the model. A non-profit organization dedicated to posing concrete challenges arising from specific necessities of communities in the Cundinamarca region. This organization has been incorporated to the course and the semester project for the student groups is currently being oriented towards these challenges.

Additionally, teams from two universities have been integrated to the project; these universities, one private and one public, are giving courses with a similar orientation. The goal is to generate two stages: one of collaborative work, in which teams from the three universities propose joint solutions. And another one in which they propose several approaches which will allow address the challenge in the best way. It has been planned to include funding for the implementation of the best project, so the groups will have the motivation of seeing their proposal implemented through these resources. The Green business concept is still under construction. Indeed, participant entrepreneurs have an intention to be green in practice, and part of what they do with students is to build strategies for "greening" their businesses.

Coincident with the work of Velásquez (2016), active listening, continuous learning, concreteness, membership and overcome fears are key ideas that form the basis for a methodological model that allows students to properly impact the management of enterprises. Situated learning is a crucial approach in engineering learning process, it give to students the possibility of be aware of real difficulties and real perception of those involved in his interventions. Engineers needs to develop capabilities to leverage and to facilitate coordination between several actors in their projects, collaborative learning and partnerships with entrepreneurs set the stage for engineering students to develop those capabilities.

Through a retrospective look of the course, of its evolution throughout the semesters, and of the related research process, it can be concluded that the course is very productive for all the actors involved, but especially for the students and their learning process. Even though they find some difficulties to engaging with process in the beginning of the course, a constant throughout the semesters has been that the students realized, as they undertake the project, that the knowledge they have acquired during their studies makes sense and become relevant. Additionally, that they are really applicable as they put them into use for addressing the challenges posed by the communities, and realize their potential as future engineers of being participants in transformation processes in their regional contexts.

References

Arias, J., Ramirez C., Duarte D., Florez M., Sanabria, J. (2016). *poCDIO: A Methodological Proposal for Promoting Active Participation in Social Engineering .Projects* Journal Systemic Practice and Action Research, 29,(4), p 379-403

Dillenbourg, P. (1999). *Collaborative learning: Cognitive and computational approaches.* Advances in learning and instruction series. Elsevier Science, New York, NY 10160-0757.

Flyvbjerg, B. (2006). Five misunderstandings about case-study research. Qualitative inquiry, 12(2), 219-245.

Glaser, B., y Strauss, A. (1967). The discovery grounded theory: strategies for qualitative inquiry. London, England: Wiedenfeld and Nicholson.

González, G., Díaz, L. (2005). Aprendizaje colaborativo: una experiencia desde las aulas universitarias. Educación y educadores, 8:21-44.

Hewitt-Dundas, N. (2013). The role of proximity in university-business cooperation for innovation. The Journal of Technology Transfer, 38(2), 93-115.

Johri, A., & Olds, B. M. (2011). Situated engineering learning: Bridging engineering education research and the learning sciences. Journal of Engineering Education, 100(1), 151-185.

Kapp K.M. (2012). The Gamification of Learning and Instruction: Gamed-based methods and strategies for training and education. California, USA. ASTD.

Moore, J. (2005). Is higher education ready for transformative learning? A question explored in the study of sustainability. Journal of transformative education, 3(1), 76-91.
Pacheco, J.F., Ramírez, M.C., González, M.A. (2013) Transformación de Unidades Productivas Tradicionales en Negocios Verdes. *Revista Digital Desarrollo Regional.* Edición 1. 27 de julio de 2013. ISSN 2389-7821. Bogotá: Corporación Universitaria Minuto de Dios.
Ramírez, C., Plazas, J., Torres, C., Pacheco, J.F. (2008). Proyecto de innovación comunitaria-PIC: Una experiencia de trabajo de ingeniería multidisciplinar e interuniversitario para la comunidad. Memorias de la Reunión Anual ACOFI.
Ramírez, C., Mereu, R. Bengo, I. Bejarano, A. & Silva, J. (2011). Participative Methodology for local development: The contribution of Engineers without Borders Italy and Colombia towards the improvement of water quality in vulnerable communities. *Systemic Practice and Action Research*, *24*, (1), p. 45-66
Zichermann,G. (2010, Octubre). Fun is the Future: Mastering Gamification. Consultado el 10 de mayo de 2012 en Google Tech Talk.
Ramírez, C., Diaz-Ramírez, J., Blanco G. Proceedings Conferencia Ascolfa Mayo 2016. Barranquilla. Ponencia: Apoyo a la gestión de emprendimientos verdes en Cundinamarca. Un ejercicio de responsabilidad social universitaria.
Shaffer. D. & Kipp. K. (2007). Psicología del Desarrollo Infancia y Adolescencia. México, D.F., pp. 516- 523.
Trevelyan, J. (2010). Mind the gaps: engineering education and practice. In Proceedings of the 21st Annual Conference for the Australasian Association for Engineering Education (p. 383). Engineers Australia.
Velásquez, C. (2015) Gestión de Casos, emprendimientos verdes. Comercializadora Social Emprendeverde.
Velásquez, C. (2016) Presentación sobre negocios verdes. Curso Proyecto Intermedio.
http://www.youtube.com/watch?v=6O1gNVeaE4g Recuperado Febrero 10 de 2017
https://planeacion.uniandes.edu.co/pdi/pdi-2016-2020/programa-de-desarrollo-integral-pdi-2016-2020 Recuperado Febrero 20 de 2017

Acknowledgements

The research is framed in the research project: Verde-ando, un ejercicio de responsabilidad social universitaria, mediante el aprendizaje compartido con emprendedores de negocios verdes en la region de Cundinamarca. It was developed with support of Centro de Ética Aplicada de Uniandes. Also the conjoint work with Uniminuto Sede Cundinamarca particularly Camilo Velásquez y Juan Fernando Pacheco

Copyright statement

The following copyright statement should be included at the end of your paper. Substitute authors' names in final copy only.

Enhancing conceptual learning through flipped classroom modality in electrical engineering education

Michaël Canu
Universidad de Los Andes, Bogotá, Colombia
m.canu134@uniandes.edu.co

Mauricio Duque
Universidad de Los Andes, Bogotá, Colombia
maduque@uniandes.edu.co

***Abstract**: Conceptual understanding is one of the core objectives of any course. In fact, many courses address only declarative or procedural learnings or assess only those kinds of learning. Flipped or Inverted classroom is an approach that allows the use of real conceptual teaching- learning activities with students during classroom sessions. The declarative/procedural part of the course, which is of a great importance lies on the on-line videos sessions and some tasks developed in the classroom and homework. This article proposes an inverted classroom architecture in which the classroom activities are oriented to conceptual understanding. Roughly speaking, we claim that the declarative and procedural part of the course is not a problem and could be provided to students through short videos outside the classroom supplemented with homework and short activities in classroom time. The maximum amount of time is devoted to conceptual problem solving. The question is, can this kind of modality improve the conceptual understanding and, at the same time, preserve a sufficient level of declarative and procedural knowledge? Declarative and procedural knowledges are provided through short videos, exercises, and assess through classical multiple-choice questions (MCQs) at the end of the semester. Students 'conceptual understanding is promoted through specific problem resolution activities during the classroom time. In order to measure the students' conceptual understanding improvement, a quasi-experimental design research was used within the Concept Inventory framework. Therefore, the "DIRECT" Electrical Circuit Concept Inventory from Engelhardt and Beichner was adapted to fit our constraints and preserving its internal validity. The results show an improvement of electric circuit conceptual understanding in students during the course period. The use of short theoretical videos seem to have resulted in no declarative declarative and procedural knowledge weakening according to last year evaluations. Inverted or flipped classroom allows a time gain – classroom time - which can be used to provide students with specific conceptual activities. All the declarative and procedural content of the course can be proposed on-line using short videos and exercises without loss of students' performance.*

Introduction

"Flipping the classroom" has become something of a buzzword in the last several years, driven in part by high profile publications in The New York Times (Fitzpatrick, 2012) or The Chronicle of Higher Education (Berrett, 2012). However, nearly ten years after the famous article in Science by Mazur (2009), some aspects of flipped classroom are still debated issues (Vliet et al., 2015; Faillet, 2014 or Strayer, 2012). Briefly, "flipping the classroom" means that students gain first exposure to new material outside of class, usually via reading or lecture videos, and then use classroom time to do the upper level work of assimilating that knowledge, through problem-solving, discussion, or debates. In terms of Bloom's revised taxonomy (Anderson et al., 2001), this means that students are doing the lower levels of cognitive work (gaining knowledge and comprehension) outside the classroom, and focusing on the higher forms of cognitive work (application, analysis, synthesis, and/or evaluation) in the classroom, where they have the support of their peers and instructor. This model contrasts from the traditional model in which "first exposure" occurs via lecture in classroom, with students assimilating knowledge through homework; thus, the term "flipped classroom" (Mazur et al., 2015).

According to some authors, this methodology allows some pedagogical advantages, for example, students could be more open to cooperative learning and innovative teaching methods (Strayer, 2012). But one of the main issue is about the kind of disciplinary knowledge (declarative, procedural, conceptual, strategical, etc.) the students will develop in this context and the resulting gain/loss related to those knowledge using this methodology.

Students' conceptual understanding is one of those kinds of knowledge that is the central objectives of any courses. However, many courses address only declarative or procedural knowledges or assess only those kinds of knowledge. Flipped classroom (Mazur et al, 2015) is a modality that allows the use of real conceptual teaching-learning activities with students during classroom sessions. Most of the declarative/procedural part of the course, which is of a great importance, lies on the on-line material (videos, animations, documents, etc.) supplemented with some specific sessions and tasks developed in the classroom and homework.

Following the proposition from Bishop and Vergeler (2013), we intend to study an inverted classroom architecture in which the classroom activities are mainly oriented to conceptual understanding. Like Mazur, we claim that the declarative and procedural part of the course is not a problem for the learners and could be provided to them through short time videos outside the classroom, supplemented with some class and home tasks. The maximum amount of classroom time is then devoted to conceptual problems solving. The question is, can this kind of modality improve or preserve conceptual understanding and, at the same time, not weaken the level of declarative and procedural knowledge?

The structure of our work is similar to the study by Papadopoulos (2010): the impact of the flipped classroom methodology on student conceptual understanding is assesses by a Conceptual Inventory test within a quasi or experimental design framework (pre and post-test with or without control group).

The focus of the article is on the assessment of conceptual knowledge and not specifically on flipped classroom structure or design, for more details and definitions, see for example Abeysekera and Dawson (2014) or Bishop and Verleger (2013).

The Conceptual Inventory assessment frameworks

Concept Inventory provides good examples of the use of conceptual models of understanding in instructional domains to generate sets of test questions systematically (the cognition and observation vertices of the assessment triangle, Pelegrino *et al.*, 2013; Jorion *et al.*, 2015). In general, concept inventories are intended to measure students' deep conceptual understanding within a relatively narrow domain and they share several common features. The questions are frequently based on science and engineering education research, including research on misconceptions and common student errors. Typically, they consist of multiple-choice questions whose distractor options were developed from misconceptions research (by asking students open-ended questions and analyzing the responses obtained, it is possible to find one of those distractors). Figure 1 presents such an item where the focus is on a typical mistaken reasoning called "linear causal reasoning" (Viennot and Rainson, 1992). All of the items call for a qualitative reasoning and do not contain any calculus. Many inventories have developed explicit links between incorrect multiple-choice distractors and specific misconceptions. A wide array of inventories is now freely available in STEM education (e.g. http://ciHUB .org).

Methodology

The study took place in the framework of a Basics Electrical course of a Colombian Engineer School. It is the first course about this topic in the student career (4TH to 5th semester). As mentioned above, the course is made up of theoretical short videos (between 3 and 6 min) that provide, at the same time, some short tests (quizzes) and classroom problem solving sessions. Videos are embedded in the university learning management system (LMS) which allows some automatic feedback from those on-line tests to students.

Declarative and procedural knowledges are then provided through short time videos, exercises, and assessments through "classical" multiple-choice questions (MCQs) exam at the end of the semester. Students 'conceptual understanding is promoted through specific problem solving activities during the classroom time, supervised by the professor. In order to measure the students' conceptual understanding improvement, a quasi-experimental design research was used (through pre and post-test MCQs) within the Concept Inventory framework. Two semesters of a school of engineering Fundamentals of Electronic Circuits course were used to collect representative data. Since there was no randomization of students' participation, it cannot be called an "experimental design study". Moreover, during the first semester only two sections of the same flipped-classroom course were used (without control group). One section accounted for 39 students and the other one for 26 (total n=65), two professors were involved with the same flipped-classroom curricula. Two sections from different courses were used for the second semester study: the two sections accounted for 36 students each (total n=72). Two professors were involved with two different curricula: the first with a a flipped-classroom and the other with a more "classical" one (lectures/exercises/laboratories).

The first semester's final test was made of a conceptual (a concept inventory) part and a more classical declarative/procedural part (from last year tests, see example fig. 3). For the two studies, the "DIRECT" Electrical Circuit Concept Inventory from Engelhardt and Beichner (2004) was adapted in order to fit our exam time constraints, preserving its internal validity. In particular, some of the DIRECT original questions were removed for the first and second semester test, for example the ones with low discrimination and difficulty (10, 11, 16, 20 and 28 items for example). This gave us a 22 items test instead of 29 (example in fig. 1).

For the second semester test, few items were added to the first semester test in order to include some questions about electrical voltage/current circuit measures or electrical conservation flow (example in fig. 2). This gave us a final 30 items test for the conceptual part. Old and new items were analysed in order to verify their discrimination and difficulty and the overall internal coherence or reliability of the test with alpha's Cronbach coefficient (Cronbach, 1951). The declarative/procedural part of the second semester test was different for the two sections.

26) If you increase the resistance C, what happens to the brightness of bulbs A and B?

(A) A stays the same, B dims
(B) A dims, B stays the same
(C) A and B increase
(D) A and B decrease
(E) A and B remain the same

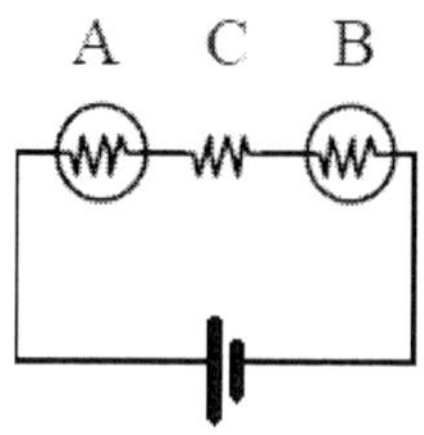

Figure 1: Example of DIRECT item

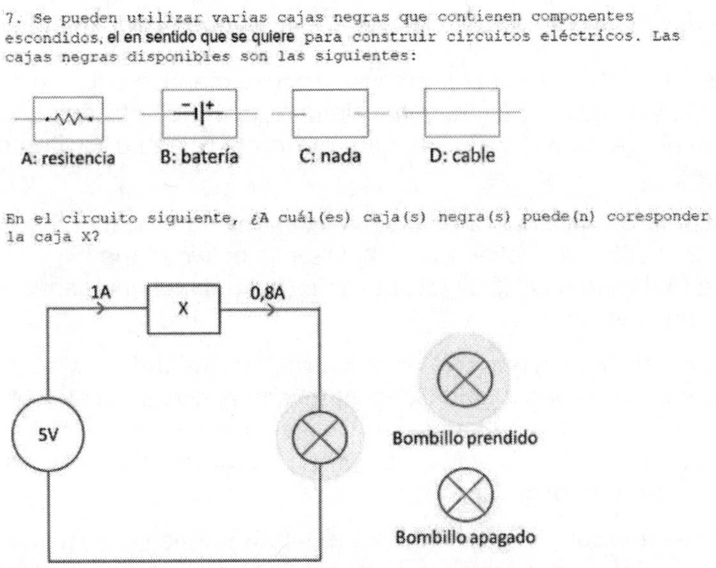

Figure 2: Example of new item about linear causal reasoning

Teniendo en cuenta el siguiente circuito. Seleccionar la ecuación de nodos correspondiente para Va.

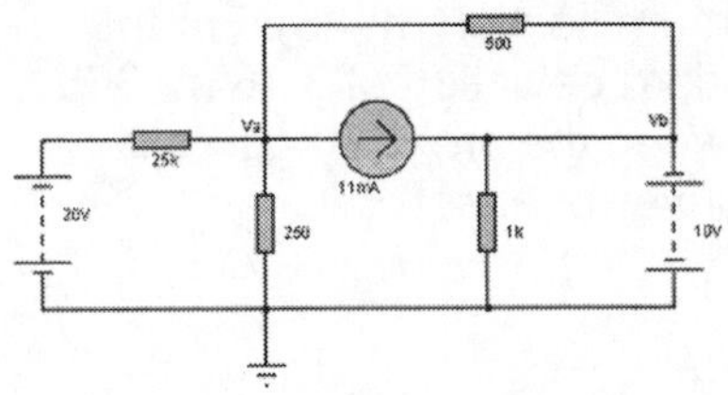

a)

$$\frac{Va-20}{25K}=\frac{Va}{250}+11mA+\frac{Va+10}{500}$$

b)

$$\frac{20-Va}{25K}=\frac{Va}{250}+11mA+\frac{Va+Vb}{500}$$

c)

$$\frac{Va-20}{25K}=\frac{Va}{250}-11mA+\frac{10-Va}{500}$$

d)

$$\frac{20-Va}{25K}=-\frac{Va}{250}-11mA+\frac{Va-Vb}{500}$$

Figure 3: Example of “classical” procedural item (Kirchhoff ley aplication)

Results

The first semester experimental design allowed us to test our university student population and compare it to the DIRECT data from Engelhardt and Beichner (university and high school data). The second cohort (second semester) allowed us to compare two student sub populations from two sections of the same course but under the flipped and non-flipped classroom scenario with separated procedural/declarative and conceptual components.

First semester analysis

Figure 3 shows a global comparison between DIRECT ECI test, our Pre and post-test. As showed in figure 4, there is no statistical difference between the two populations with respect to the same items. The DIRECT overall correct response mean is equal to 51% with 3.95 standard error (n=1135) and 0.71 Cronbach's alpha reliability coefficient, and the pre-test overall mean is equal to 48% with 3.99 standard error (n=65) and 0.75 Cronbach's alpha reliability coefficient.

This result is confirmed by a Chi squared test between the item correct responses distribution (p-value 0.002). Moreover, our population is between the high school one (41%) and university one (52%) from DIRECT overall data (but a direct comparison is not accurate since the number of items differs).

As showed in figure 5 (by items) and 6 (global score), the Pre and Post- test comparison reveals a significate improvement of the Concept Inventory score, on average, from 10.7 (sd=4.17) to 17.13 (sd=3.73) for the sample. A Chi squared test prove that this difference cannot be due to a random error (p-value < 0.00001). The Cohen's *d* effect size coefficient is 1.63; which can be considered as very large.

The students' final exam score is around 3.64/5 (sd=0.58), which is a little bit higher to the previous year score (2015, same professor) without flipped-classroom (3.46/5, sd=0.42).

Second semester analysis

Figure 7 shows the global comparison between Pre and post-test (frequency repartition). The pre-test mean is equal to 15.65 with 4.11 for the flipped-classroom section (n=36). For the non-flipped-classroom, the mean is equal to 13.03 with a standard deviation of 3.16 (n=36). As showed in this table, both sections improve their score between the pre and post-test. The difference is out of the error limits so this exclude a random error effect.

Table 1: Second semester results

	Total		Flipped-classroom		Classical modality	
	Pre	**Post**	**Pre**	**Post**	**Pre**	**Post**
Average	14,34	20,46	15,65	21,14	13,03	19,67
SD	3,90	4,49	4,11	4,22	3,16	4,65

The students' final exam score is around 3.61/5 (sd=0.58), which is very similar to the previous semester and match to the average of the three ultimate years for this professor.

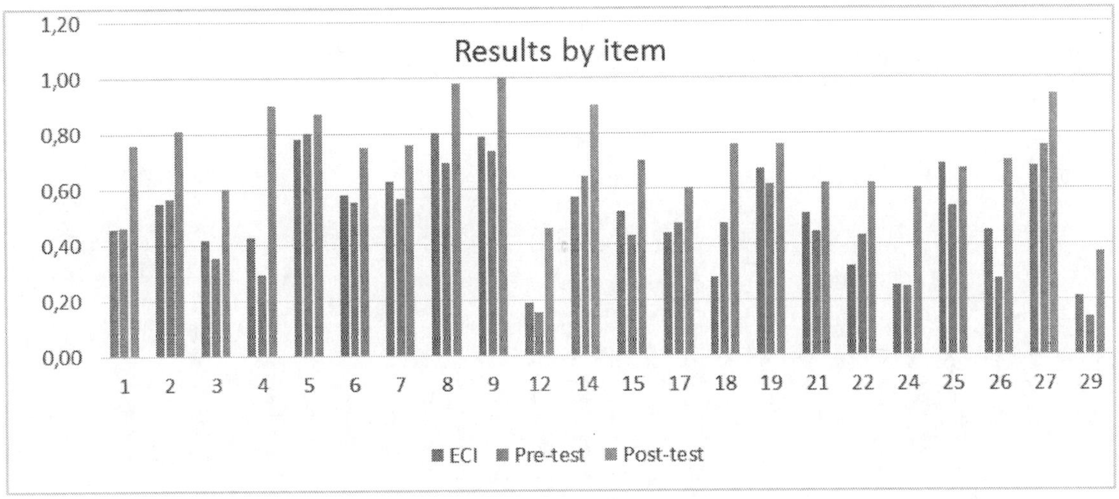

Figure 4: ECI comparison between DIRECT original application (n=1135), our pre and post-test (n= 65, first semester of 2016).

The students' final exam score of the non-flipped-classroom is around 4.26/5 (sd=0.55).

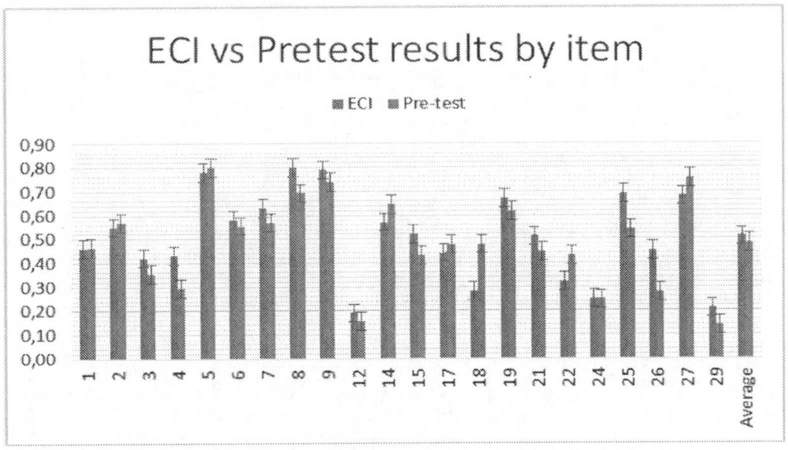

Figure 5: Comparison between original DIRECT (n=1135) application and our pre-test (n=65, first semester of 2016).

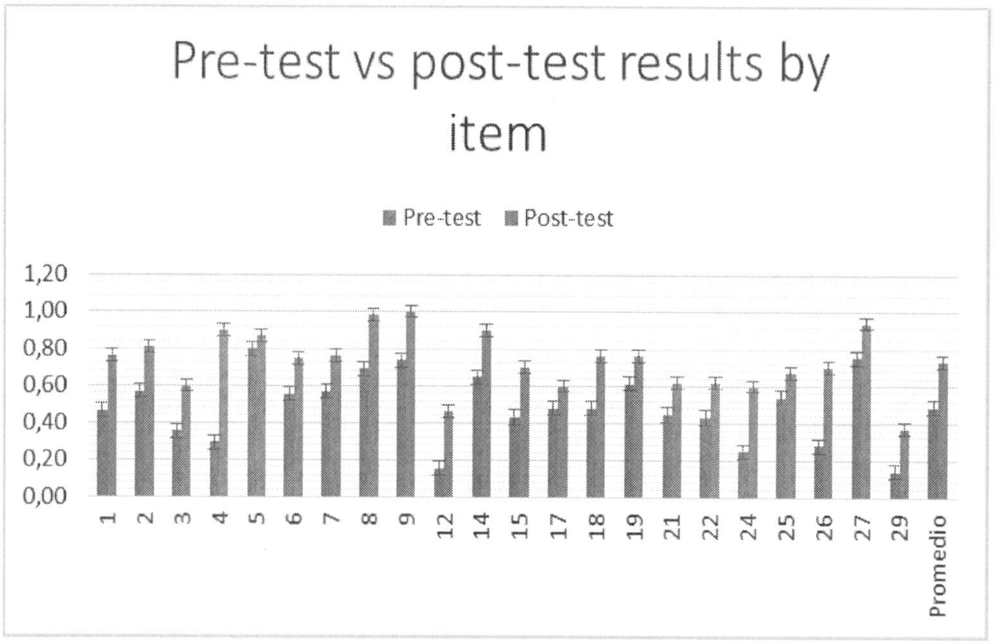

Figure 6: Comparison between our pre and post-test (n=65, first semester of 2016)

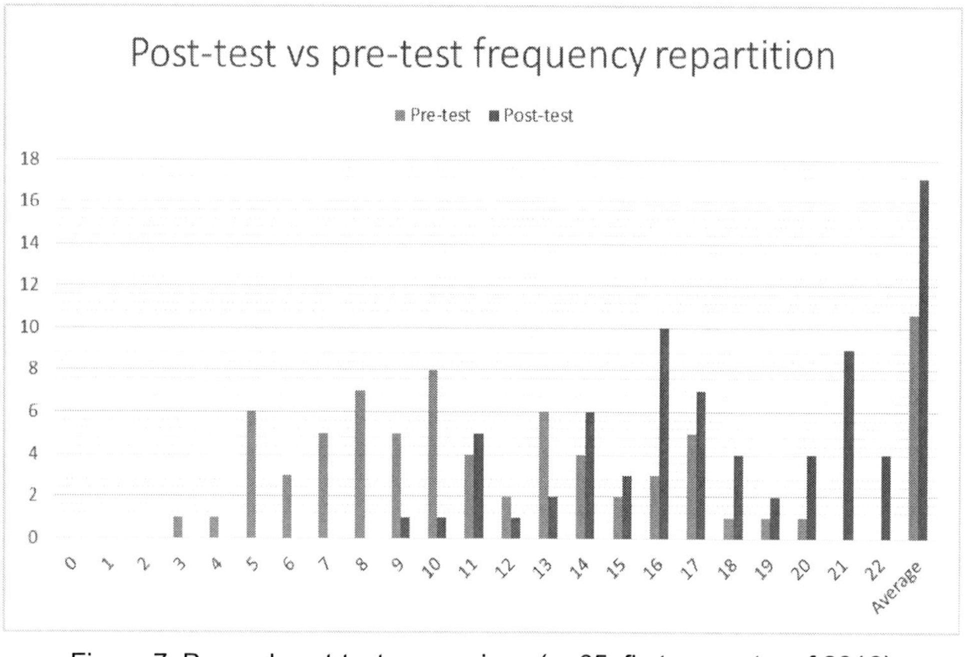

Figure 7: Pre and post-test comparison (n=65, first semester of 2016)

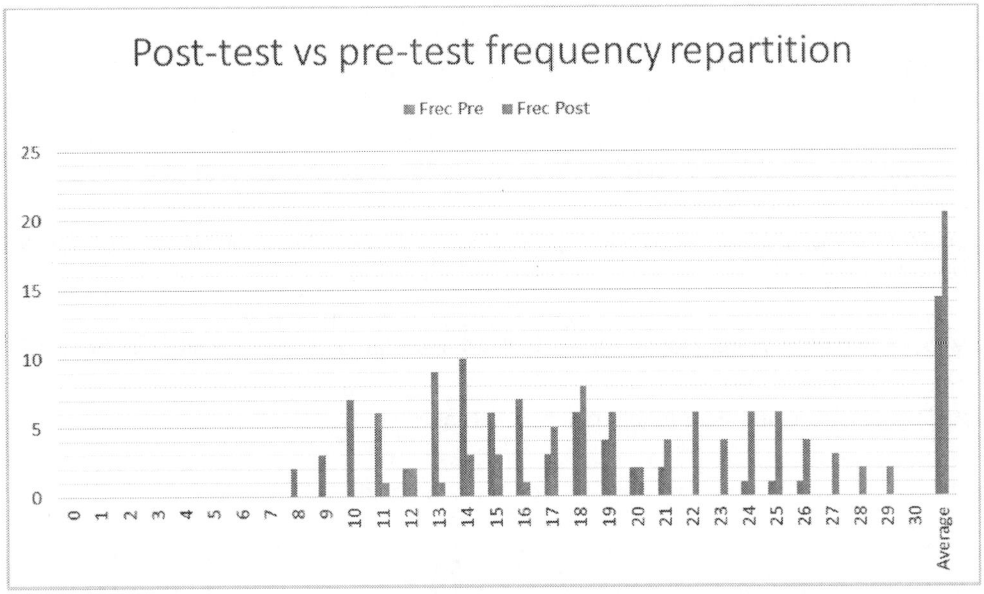

Figure 8: Pre and post-test comparison (n=72, for second semester of 2016)

Discussion

The use of short theoretical videos seems to have resulted in no declarative and procedural knowledge weakening according to last year evaluations. At the same time, the results show an improvement in electric circuit conceptual understanding in students during the course period, for the two periods. As data shows, the improvement in student's conceptual understanding occurs for the flipped and non-flipped-classroom in the same magnitude. Even if there is a difference between the Electric Circuit Conceptual Inventory scores for flipped and non-flipped classroom, it is not possible to prove that this difference is due to the use of this methodology. Indeed, the professor effect is hard to cancel in this kind of quasi-experimental design and the students' section selection process is not so clear. A Cronbach's alpha analysis between the two sections of the second semester show important differences between pre and post-testing. This suggests that the student sample here is not representative of all the student population and that the section selection process is linked to a section dependent factor. Moreover, the global exam score for the two different sections (and modalities) indicates a significant difference (3.61 - 4.26), which is impossible to directly link to the experiment, given the lack of an appropriate common exam (with an appropriate design methodology). Indeed, each professor can design its exam with different learning goals: mathematical/procedural and/or conceptual assessment objectives, for example. However, the relative stability of the students' exam score across the ultimate three years for the same professor indicates that the modality hasn´t lowered their global performance.

Conclusion

Inverted or flipped classroom allows a time gain – classroom time - that can be used to provide students with specific conceptual activities. All the declarative and procedural content of the course can be proposed on-line by the use of short videos and exercises without loss of students' performance. The mix between videos and problem solving session seem to have no negative effect upon global conceptual knowledge.

References

Abeysekera and Dawson (2015). *Motivation and cognitive load in the flipped classroom: definition, rationale and a call for research.* Higher Education Research & Development, 34(1), 1-14.

Anderson, Lorin W., Krathwohl, David R., Bloom and Benjamin S. (2001). A taxonomy for learning, teaching, and assessing: a revision of Bloom's taxonomy of educational objectives. Complete Ed. New York: Longman. 352p. ISBN 0321084055

Berrett, D (2012). How 'flipping' the classroom can Improve the traditional lecture. *The Chronicle of Higher Education*, February 19.

Bishop, J., & Verleger, M. A. (2013), The Flipped Classroom: A Survey of the Research Paper presented at 2013 ASEE Annual Conference & Exposition, Atlanta, Georgia. Retrieved from https://peer.asee.org/22585

Cronbach, L. J. (1951). Coefficient Alpha and the Internal Structure of Tests. Psychometrika, 16(3), 297–334.

Engelhard, V. P., & Beichner, R. J. (2004). Students' understanding of direct current resistive electrical circuits. Igarss 2014.

Faillet,V. (2014). La pédagogie inversée : recherche sur la pratique de la classe inversée au lycée. Sticef Journal, 21, 651-665, Retrieved from www.sticef.org

Fitzpatrick, M (2012). Classroom Lectures Go Digital. *The New York Times*, June 24.

Jorion, N., Gane, B. D., James, K., Schroeder, L., DiBello, L. V., & Pellegrino, J. W. (2015). An Analytic Framework for Evaluating the Validity of Concept Inventory Claims. *Journal of Engineering Education*, 104(4), 454–496. http://doi.org/10.1002/jee.20104

M.J. Lage, G.J. Platt, and M. Treglia. Inverting the classroom: A gateway to creating an inclusive learning environment. The Journal of Economic Education, 31(1):30–43, 2000.

Mazur, A.D., Brown, B. and Jacobsen, M. (2015). Learning designs using flipped classroom instruction. *Canadian Journal of Learning and Technology*, 41(2). 2-26.

Mazur, E. (2009). Farewell, Lectures. *Science*, 323(5910), 50-51. DOI: 10.1126/science.1168927

Papadopoulos, C., & Santiago Roman, A. (2010), Implementing An Inverted Classroom Model In Engineering Statics: Initial Results Paper presented at 2010 Annual Conference & Exposition, Louisville, Kentucky. https://peer.asee.org/16768

Pellegrino, J. W., Dibello, L. V, & Brophy, S. P. (2013). The science and design of assessment in engineering education. Cambridge Handbook of Engineering Education Research, 571–598.

Strayer, J. F. (2012). How learning in an inverted classroom influences cooperation, innovation and task orientation. Learning Environments Research, 15(2), 171-193. doi: 10.1007/s10984-012-9108-4

Viennot, L. and S. Rainson (1992). Students' reasoning about the superposition of electric fields, *International Journal of Science Education*, 14,475-487.

von Vliet, E. A., Winnips, J. C. and Brouwer, N. (2015). *CBE Life Sciences Education*, 14(3). DOI: 10.1187/cbe.14-09-0141

Copyright statement

The following copyright statement should be included at the end of your paper. Substitute authors' names in final copy only.

A maturity model for curriculum management

Juan Bernardo Quintero
Facultad de Ingeniería, Universidad de Medellín
Email address juquintero@udem.edu.co

Bell Manrique-Losada
Facultad de Ingeniería, Universidad de Medellín
Email address bmanrique@udem.edu.co

***Abstract**: In the academic context, a curriculum is mainly conceived as a process instead of as an object. For its design, it is possible to identify its maturity level to define the roadmap in which strategic practices that improve the whole teaching-learning process can be implemented. We propose a maturity model for curriculum management in higher education, based on curriculum design practices, the Feature Oriented Domain Analysis (FODA) approach, and IT support elements. As fundamentals, we are considering the phases of curriculum life cycle (design, reviewing, implementing, monitoring, and evolution) and the structural curriculum layers: macro curriculum, middle curriculum, and micro curriculum, for defining maturity levels by features, in people, processes, and tools dimensions.*

Introduction

There are conceptions where the curriculum is considered a process rather than an object (Gimeno Sacristán, 2007). In this context, the curriculum maturity is defined by the process maturity applied for building it. Frequently, in the higher education institutions exist problems in the curriculum management processes, *i.e.* professors don't follow the curriculum guidelines, or the political documents of the institutions lacking the necessary tools for aiding such processes. A maturity model for curriculum management is useful in defining the roadmap for implementing practices to improve the whole teaching-learning process.

In this approach, we present a maturity model for curriculum management in higher education, based on curriculum design practices, Feature Oriented Domain Analysis (FODA) (Kang, Cohen, Hess, Novak, & Peterson, 1990) technique elements, and IT support. In this model, we are taking into consideration the phases of curriculum life cycle (design, reviewing, implementing, monitoring, and evolution) and the structural curriculum layers: macro curriculum, middle curriculum, and micro curriculum (Runge & Garcés, 2011), in order to define maturity levels in curriculum management. Such levels are described by features in dimensions of people, processes, and tools.

Maturity in curriculum process is achieved with the integration of different disciplines *i.e.* institutional regulation, information systems, and infrastructure services among others. However, is very important that higher education institutions have guidelines to identify the "*as is*" and define the roadmap to reach the "*to be*" and therefore improve the curriculum management process. The main outcome of this work is a framework to evaluate and improve a curriculum management process: from the assessment viewpoint, the specific features of each curriculum maturity level are presented; from the process improvement viewpoint, we propose guidelines for integrating best practices and IT support in the curriculum process. An implementation of some practices in this framework was carried out in the Computer Sciences program at University of Medellín.

This paper is structured as follows: in the first section, we contextualize our work from two perspectives: curriculum life cycle and structural layers, and aspects of Feature Oriented

Domain Analysis. In the next section, we review related key concepts and work on maturity levels and software systems in curriculum management. In the following section, we propose a framework of Curriculum Maturity Levels. Then, we present an evaluation of the proposal in a case study. Finally, in the last section we discuss conclusions and future work.

Background

Curriculum life cycle and Structural curriculum layers

Given the diverse conceptions about curriculum, it is correct to consider it as a cultural and collective construction, mixing a documentary/object and practical/interactive dimension. Thus, and according to Gimeno Sacristán (2007), in this work we refer to a curriculum as a *process* instead of an *object*. Therefore, the curriculum is recognized as a conjunction between a substantive dimension and a practical-experiential dimension (Pinar *et al.*, 2008)

The curriculum follows a cyclical process through activities or steps in a learning context. Such a recurring lifecycle comprises to assess, evolve, and evaluate an entire life. According to JISC (2008) there is overlap and interplay between the processes called "curriculum design" and "curriculum delivery" and include the following moments: Initiate or review, Develop or redevelop, Approve, Communicate, Resource, Deliver, Support, Assess, and Evaluate. McKimm (2007) distinguishes among several kinds of curriculums, mainly official or formal curriculum. The official curriculum can be distinguished from the hidden, unofficial, or counter curriculum. Such kinds are implemented by means of stages of curriculum development, comprised in three stages:

- Initial stage: Determine educational or professional context in which the program is to be developed; define the learners needs aligned with the requirements of professional bodies; determine the aims and broad learning outcomes; Identify ideas and constraints.
- Core stage: Agree the broad structure and framework of the program, the main areas of teaching and learning, the topics sequence and key assessments; Detail the development of each topic with objectives and learning outcomes; Define assessment methods; and Implement and refine the program.
- Final stage: Develop an appropriate deliverable evaluation strategy; Review, and revise the course based on feedback.

Mooney and Mausbach (2008) describe how the curriculum development should imply in-depth revision and update processes for the curriculum, determined by a curriculum revision cycle. This development cycle is a critical piece to managing the whole work of curriculum design and include the following steps: (i) Establish foundations based on standards; (ii) Data Analysis including test data, surveys, and instructional practices; (iii) Benchmark assessments; (iv) Writing the scope and sequence in a curriculum map; (v) Resources Review; (vi) Pilot Process with teachers; (vii) Board Approval of education reviews curriculum; (viii) Staff development and training with new curriculum and materials; and (ix) Implementation through curriculum maps. Such development stages are the foundation to define assessments, resources review, and implementation in a cycle. Based on the previous references and with practical purposes, in this work we consider a curriculum life cycle comprising design, reviewing, implementing, monitoring, and evolution phases.

Following Gimeno Sacristán (2007) and Runge and Garcés (2011), we define structural layers of curriculum in terms of macro, middle, and micro curriculum. The differentiation among such levels is useful for implementing curricular activities (van den Akke, 2010), and some of the main distinctions among them are:

- **Macro curriculum**: specify elements in a macro level, considering the system, society, nation, state or institution (*e.g.* national syllabi or core objectives from university). Involves the socio-cultural system and the educational system (educational and curricular policies, laws, and rules). Thus, includes an interaction between the authority and the practice within which it is implemented (Carl, 2008).

- **Middle curriculum**: define elements in intermediate level (meso level), by according to the transversal competences and common curricular areas into faculty or university (*e.g.* school-specific curriculum). Includes the organizing structures and institutional approaches for orienting the curriculum.
- **Micro curriculum**: specify detailed elements at classroom level (*e.g.* methodological strategies by courses, syllabus, sequence of thematic units, and instructional materials). In practical terms, comprise the development and implementation of classroom-based activities and the guide to the teacher-student interaction.

Feature Oriented Domain Analysis (FODA)

FODA is a method for identifying prominent or distinctive features of systems in a specific domain. These features are aspects or characteristics of the domain and are used to define the domain in terms of the mandatory, optional, or alternative characteristics of the related systems (Kang *et al.*, 1990). The FODA method has been applied in several domains, looking for contributing to feature-oriented development approaches. These use the concept of a feature to structure design and code of a system, where features are the primary units of reuse and the variants of system vary in the features they provide. It stems from the different uses of features.

Some approaches have contributed to integrate FODA elements with Learning Management Systems (Duong, 2014) trying to make teaching and learning valuable, once teachers are supported by a tool, they can integrate everyday classroom practice and subject-matter curricula.

FODA is carried out in the framework of domain analysis, gathering and representing information on systems that share a common set of capabilities and data. According to Kang et al. (1990) there are three basic phases in the domain analysis process: (i) Context analysis, for defining the extent of a domain; (ii) Domain modeling, for describing the problems within the domain that are addressed by system; and (iii) Architecture modeling: creating the architecture that implements a solution to the problems in the domain. The participants in the domain analysis process are: sources (end user, uses system in the domain and domain expert who provide information about system in the domain), producer (analyst who gathers information and perform analysis), consumers (requirements analyst, end users, and designers). The previous phases and participants considered by FODA method are referents in our proposal in this paper.

Related Work

We identified relevant studies related to maturity levels in curriculum design, as well as applications and information systems for curriculum management, as follows:

Maturity levels in curriculum management

Thong *et al.* (2012) propose a process improvement model for curriculum design called CDMM in higher education institutions. Such a model comprises guidelines for self-assessment and a method with instructions for designing a solid curriculum. The proposed levels in the model are Initial, Repeatable, Defined, Managed, and Optimized. Such levels are evaluated with a Likert scale based on the adequacy grade (not adequate, partially adequately, moderately adequate, adequate, fully adequate).

Neuhauser (2004) recommends maturity levels for designing online courses (Online Course Design Maturity Model-OCDMM). He introduces best practices to virtual education faculties and provides an integrated system for facilitating to designer the alignment among best practices, learning principles, technologies, student and faculty objectives, performances, and changing needs. Based on such a system, a faculty could select improvement actions in course design based on their own available technology. The design maturity of a course shifts the student focus from passive to active learning, and based on this the student performance can be continuously measured. The assessment process involves a five-point scale: 1-Initial, 2-Exploring, 3-Awakening, 4-Strategizing, and 5-Integrating Best Practices.

Dennis and Minnie (2008) define a method for curriculum redesign with a process improvement model, comprising experiences in implementing a curriculum redesign from one based on a traditional course-based design to a project-centric design using the CMM as a process improvement tool. The entire curriculum was redesigned to have a project-centric focus with each course and the goal is enhancing the experiential learning of students by focusing it on a community-based system development project.

Chen *et al.* (2011) propose a Teaching Capability Maturity Model (T-CMM) with an assessment five-scale initial, managed, defined, quantitatively managed, and optimizing. Marchewka (2013) discusses how CMM can be applied to Assurance of Leaning (AoL) projects in order to assess, plan, and improve a business school. Each institution can follow such approach and assess its current level for then developing a plan to reach the next higher level of AoL process maturity: Initial, Repeatable, Defined, Managed and Optimized.

Table 1: maturity models for curriculum management

	Maturity Model Focus			**Approach Goal**		**Measurement scale**		
Approach	**General**	**Course**	**Professor**	**Curriculum involvement**	**Learning**	**Likert scale**	**Max Diff**	**Not specified**
Thong *et al.* (2012)	X			X		X		
Neuhauser, C. (2004)		X		X	X	X		X
Dennis & Minnie (2008)		X			X		X	
Chen *et al.* (2011)			X			X		
Marchewka (2013)	X				X	X		
Lutteroth *et al.* (2007)		X				X		
Kushtina *et al.* (2009)	X	X		X				X

Lutteroth *et al.* (2007) propose a conceptual model (CEMM) to be used in rating educational organizations according to their competence to deliver high quality education on a five-level scale. Only maturity levels are delivered in the work.

A conceptualization of the curriculum modification decision model is proposed by Kushtina *et al.* (2009). They describe the following sequential knowledge processing: i) syllabus preparation, ii) providing education services, iii) developing didactic materials, iv) the process of acquiring competences based on a specified knowledge model, and v) statistical evaluation of student progress. Such processes are interesting for carrying out curriculum modification decision and similar approaches from curricula designer point of view.

Curriculum management supported in technology

Curriculum management implies the selection and provision of teaching programs, plans, methods, resources, and materials for teachers in an academic domain. In many education institutions, this is done digitally through disparate systems working alongside other online tools, such as wikis, Learning Management Systems (LMSs), academic social networks, and document sharing tools. Consequently, the curriculum management is not fully integrated into a teaching workflow, and so fails to deliver the inherent benefits that an integrated curriculum management system can provide.

For learning management, most of the technological, digital, or online solutions/systems are oriented to the LMSs. Such LMSs connect the people, content, and systems in an academic context. LMSs have evolved in creating courses and organizing them with all instructional content toward innovative curricular and academic practices, including: Linking assessments and classroom learning; sharing successful practices and encouraging participants to engage in global collaboration; supporting professional learning communities; creating real-world learning experiences; sharing projects or portfolios with community leaders and business experts; faculty accountability; and, community functionality.

Several higher education institutions have implemented projects for curriculum management supported in information systems, architectures, and Enterprise Resource Planning (ERP). Such projects include the implementation of new student information systems and dependent technologies, as well as activities pertaining to the re-engineering of institutional policies and processes. Some of them are as follows:

- *The Open Campus Website* at University of the West Indies (www.open.uwi.edu/)
- *The Ilios Project*, Curriculum Management for the Health Professions (www.iliosproject.gitbooks.io). The Ilios workflow is based on a curriculum structure which includes several interconnected levels (displayed below starting with the highest level)
- *Entrada Project* team at Queen's University, University of Calgary, UCLA, University of Ottawa, Rush University, and University of British Columbia (http://www.entrada-project.org). Entrada is an Integrated Teaching and Learning Platform and consists of 3 independent apps: Entrada Admissions, Entrada ME, and Entrada CPD. The major pillars that make up the Entrada ME ecosystem are Curriculum Mapping and Reporting; Academic and Clinical Scheduling and Calendaring; Learning Management, Assessment and Evaluation, and Learner ePortfolios.
- *ItsLearning* (http://www.itslearning.mx/) includes in-built curriculum management capabilities that enable you to connect courses, resources and assessments with common standards and learning objects.

Proposal of Curriculum Maturity Levels

The maturity model for curriculum management has two main components: structure and process. The *structure component* describes the levels and dimensions to organize features. The *process component* explains how to improve the level of maturity.

Structure of maturity model

The structure has three dimensions: first, *people* involved in curricular process; second, considerations for *curriculum design*; and third, *tools* used in design and implementation processes. Each dimension has three features defined by FODA analysis. In the features the main elements regarding to each dimension are considered. In order to define the maturity level, each feature has six possible values associated with a question, as shown on Table 2.

Table 2: Dimensions, features, questions and features values

Dimension 1 – People		
Feature 1.1 - Designer	**Feature 1.2 – Reviewer**	**Feature 1.3 - Implementer**
Who is responsible of micro-curriculum design?	***Who is responsible of micro-curriculum inspection?***	***What action does the course leader (professor) take?***
0. Nobody **1. Manager**: anybody with an administrative role **2. Professor**: course responsible **3. Consultant**: external expert **4. Committee**: group formed by teachers and experts **5. Collaborative**: with students	**0. Nobody** **1. Person of the institution** **2. Person of the program** **3. Internal committee** **4. Committee with external staff** **5. Committee including students**	**0. Use** **1. Comment**: professor opinions are considered **2. Modify**: professor can modify the micro-curriculum **3. Propose**: professor proposes new elements and changes **4. Evolve reactively**: professor proposes changes reactively **5. Evolve proactively**: professor proposes changes proactively
Dimension 2 – Design		
Feature 2.1 - Access levels	**Feature 2.2 - Formative activities**	**Feature 2.3 - Evaluative activities**
What access level do participants have on the micro-curriculum documents?	***What activities to educate are planned in the curriculum?***	***What evaluative activities are planned in the curriculum?***
0. Restricted singled: unshared	**0. None**: just thematic units	**0. None**: just thematic units

file **1. Restricted public**: shared file with restricted access **2. Public**: shared file with everybody **3. Unsynchronized**: changes no need to be updated **4. Synchronized**: changes are updated online **5. Collaborative**: documents are built with all online participants	**1. General**: activities are proposed without considering differences among thematic units **2. Detailed**: activities are proposed by thematic unit **3. Supported**: professor has tools that support educational activities **4. Automated**: educational activities are automated in the web **5. Inter-course**: educational activities are planned considering several courses	**1. General**: activities are proposed without considering differences among thematic units **2. By learning unit**: activities are proposed by thematic unit **3. Feedback process**: feedback activities are expressly considered for evaluating **4. Intra-course with rubrics**: courses have instruments to score every evaluation activity **5. Inter-course**: evaluation activities are planned considering several courses
	Dimension 3 – Tools	
Feature 3.1 - Curricular referents	**Feature 3.2 - Thematic references**	**Feature 3.3 - Information technology tools**
How the curricular referents are used to design the curriculum?	***How the thematic references are used to design the micro-curriculum?***	***How information systems are used to design the curriculum?***
0. None **1. Intuitive**: curriculum designer knows the referents but don't use them **2. Non-formal**: curriculum designer uses the referents but don't use methodology **3. Formalized explicit**: curriculum designer uses the referents and methodology **4. Explicit evaluated**: curriculum designer uses the referents and the methodology including an evaluation process **5. Explicit collaborative**: curriculum designer is supported by internal or external expert	**0. None or obsolete** **1. General**: references are proposed without considering differences among thematic units **2. By learning units**: references by thematic unit are proposed **3. By topic**: references for each topic are proposed—learning units are composed by topics **4. Structured**: the appendix for references uses standard—like APA or ISO—and hyperlinks by topics **5. References manager**: references are automated with a references manager like Zotero or Mendeley	**0. None or obsolete**: tools are used only to manage files **1. Templates**: there are files with all necessary items in micro-curriculum documents **2. Repository**: each micro-curriculum is stored in a repository **3. Versioned repository**: each micro-curriculum arise versioned and audit actions are available **4. Portal**: a website with the micro-curriculum documents is available **5. Portal 2.0**: the website allows collaboration in design process and course activities

The average of the correspondent value to each response, defines the maturity level evaluated. The ranking of level is defined as follows:

- *Level 0 – Chaotic:* no formal curricular process
- *Level 1 – Initial:* first step to define a curricular process
- *Level 2 – Defined:* curricular process is defined, but it can be improved
- *Level 3 – Agreed:* curricular process is defined and socialized
- *Level 4 – Optimized:* curricular process is defined and adjusted
- *Level 5 – Collaborative:* curricular process is defined by all process participants

In Figure 1 we show the graphic representation of structure for maturity model for curriculum management.

Process of maturity model

The way of using this model of curricular maturity is defined by four main phases:

- *Phase 1*: evaluate current maturity level in curriculum management
- *Phase 2*: select dimensions and features to strengthen
- *Phase 3*: define actions to improve in feature values
- *Phase 4*: implement actions and evaluate the improvement

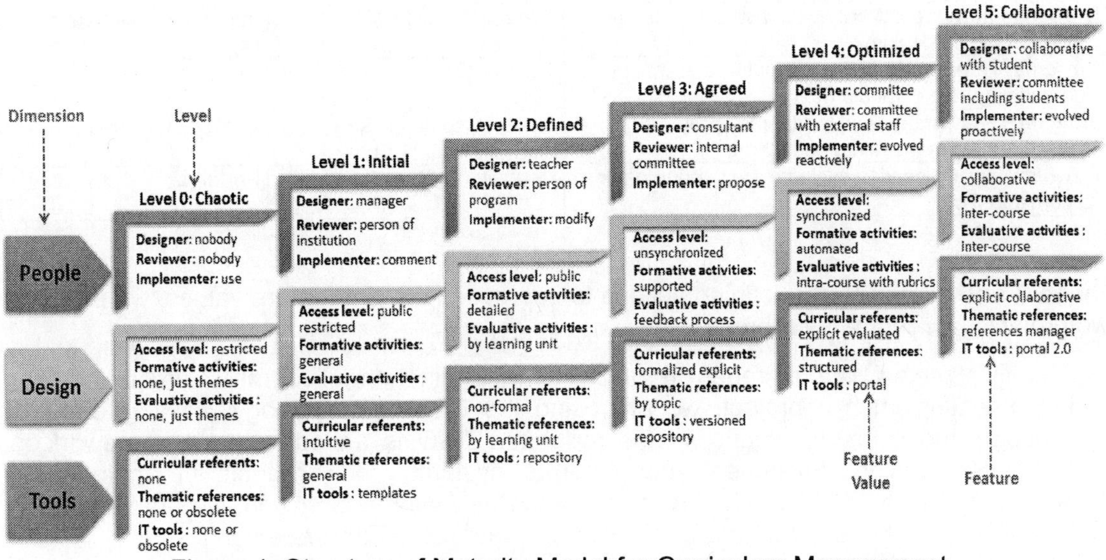

Figure 1: Structure of Maturity Model for Curriculum Management

In Figure 2 we show the graphic representation of the process of maturity model for curriculum management. In the validation chapter of this article we present how this process is applied in a particular case.

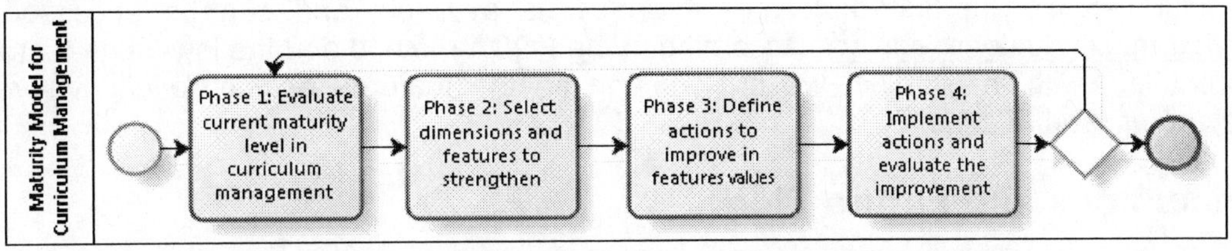

Figure 2: Process of Maturity Model for Curriculum Management

Evaluation

In order to validate the maturity model for curriculum management, we use our approach in the Computer Sciences program at University of Medellin. Following the phases presented in Figure 2, we obtain a current characterization of the program under study, as follows:

Phase 1 and 2: We develop an evaluation of the current situation in the academic program and identify the target feature values by dimension. An example of such evaluation is presented in Table 3. As we show in the 'current feature value' and 'target feature value' columns, we define the average of current features (1.78) which generates the current maturity level in such academic program between *initial* and *defined.* Meanwhile, the average of target features values is 3.33, which generates the future maturity level (target) between agreed and optimized.

Table 3: Dimension 3 - Tools

Feature: Question	**Current feature value**	**Target feature value**
Designer: Who is responsible of micro curriculum design?	Professor, sometimes manager **(1.5)**	Committee **(4)**
Reviewer: Who is responsible of micro curriculum inspection?	Person of program **(2)**	Internal committee **(3)**
Implementer: What action does the course leader (professor) take?	Propose **(3)**	- -
Access level: What access level do participants have on the micro curriculum documents?	Public restricted **(1)**	Synchronized **(4)**
Formative activities: What activities to educate are planned in the curriculum?	Detailed **(2)**	- -

Evaluative activities: What evaluative activities are planned in the curriculum?	By program units **(2)**	Intra course with rubrics **(4)**
Curricular referent: How the curricular referents are used to design the curriculum?	Intuitive **(2)**	- -
Thematic references: How the thematic references are used to design the micro curriculum?	General, sometimes by learning units **(1.5)**	Explicit evaluated **(4)**
IT Tools: How information systems are used to design the curriculum?	Repository **(2)**	Portal **(4)**

Phase 3 and 4: looking for defining actions to improve in current feature values, we propose two projects, which are in progress:

- The first project is the 'redesign of Systems Engineering Program Curriculum.' For implementing such a project, we are using the CDIO methodology, as a referent to redesign the courses, practices, and syllabus. CDIO is an educational framework to **C**onceive, **D**esign, **I**mplement and **O**perate systems and products in engineering. According to the FODA method and the low-value features identified, we are working in this project on the features: *designer*, *reviewer,* and *curricular referents*.
- The second project is called OCMS (Open Curricular Management System), as a transactional portal with functionalities based on *The Open Campus Website*, the *Ilios Project, Entrada Project.* and *ItsLearning* (analyzed in the previous section). We are implementing such OCMS with the following functional modules: (i) curriculum management, (ii) project bank, (iii) academic schedule, (iv) academic advisor, and (v) social media. The first prototype of project is available and can be accessed at http://macondo.udem.edu.co/. According to the FODA method and the low-value features identified, with this project we are working on the features *access level*, *evaluative activities*, and *IT tools*.

Conclusions and Future Work

In this work, we present a framework to evaluate and improve the curriculum management process, from two perspectives: *assessment*, with the specific features of each curriculum maturity level, in terms of a structure; and, improvement, proposing guidelines for integrating best practices and IT support in the curriculum process for changing a low level into a high level of curriculum maturity.

Maturity in curriculum processes are achieved by integrating different disciplines *i.e.* institutional regulation, information systems, infrastructure services, among others. However, it is very important that higher education institutions have guidelines to identify the "as is" and define the roadmap to reach "to be" and consequently, to improve its curriculum management process.

As an evaluation approach, we have implemented some practices from the proposed framework in the Computer Science Program at University of Medellín, reaching interesting advances in academic projects and overcoming the low-valued features in the model. Such projects are oriented to the curriculum redesign and a management system for supporting the whole process.

By following the phases presented in this process, a higher education institution can obtain a characterization of an academic program (evaluation of the current situation) and identify the target feature values by dimension, according to the structure of maturity levels. Based on the collected results, actions can be defined to improve current feature values.

References

Arend, C. (2008). Teacher Empowerment Through Curriculum Development. South Africa: Juta and Company Ltd.

Cao, F. Bryant, B.R. Burt, C.C., Huang, Z., Raje, R., Olson, A.M., & Auguston, M. (2003) Automating Feature-Oriented Domain Analysis. Department of Computer and Information Sciences. *International Conference on Software Engineering Research and Practice (SERP'03)* (pp. 944–949). Birmingham, AL.

Chen, C.Y., Kuo, C.Y., & Chen P.C. (2011). The Teaching Capability Maturity Model for Teachers in Higher Education: a Preliminary Study. Available at https://pdfs.semanticscholar.org/f6b5/489bc3de842e4960e6147737706b586e4183.pdf

Dennis, D., & Yen, M. (2008). Controlling Curriculum Redesign with a Process Improvement Model. *Journal of Information Systems Education, 19(1),* 331-342.

Duong, H.D. (2014). A Feature-Oriented Software Engineering Approach to Integrate ASSISTments with Learning Management Systems. Master Thesis, Faculty of the Computer Science, Worcester Polytechnic Institute.

Gimeno Sacristán, J. (2007). El curriculum: una reflexión sobre la práctica. Madrid: Morata.

Kang, K. C., Cohen, S. G., Hess, J. A., Novak, W. E., & Peterson, A. S. (1990). Feature-Oriented Domain Analysis (FODA) Feasibility Study. Pittsburgh, PE: Software Engineering Institute.

Kushtina, E., Zaikin, O., Rózewski, P., & Małachowski, B. (2009). Cost estimation algorithm and decision-making model for curriculum modification in educational organization. *European Journal of Operational Research 197(1),* 752–763.

Marchewka, J.T. (2013). Applying the Capability Maturity Model to Assurance of Learning. *Communications of the IIMA, 13(1).*

McKimm, J. (2007). Curriculum design and development. project led by the London Deanery. Imperial College Centre for Educational Development. Association for Supervision and Curriculum Development. Available at http://www.faculty.londondeanery.ac.uk/e-learning/setting-learning-objectives/Curriculum_design_and_development.pdf

Mooney, N.J. & Mausbach, A.T. (2008). Align the Design. Chapter *1: Developing Curriculum Leadership and Design*. Hershey, PA: ASCD organization.

Neuhauser, C. (2004). A maturity model: Does it provide a path for online course design?. *The Journal of Interactive Online Learning, 3(1).*

Pinar W. (2008) Understanding curriculum. An introduction to the study of historical and contemporary curriculum studies. New York: Peter Lang.

Runge Peña, A., & Garcés, J. (2011). Educabilidad, formación y antropologia pedagógica. *Revista Cientifica Guillermo de Ockham, 9(2)*, 13-25.

Thong. C. L., Yusmadi Y. J., Member, IAENG, Rusli A., and Hayati, N. (2012). Applying Capability Maturity Model to Curriculum Design: A Case Study at Private Institution of Higher Learning in Malaysia A. *Proceedings of the World Congress on Engineering 2012,* London, U.K.: WCE.

van den Akker, J. (2010). An introduction to educational design research. In Tjeerd Plomp & Nienke Nieveen (Eds.) *Curriculum Design Research*. Germany, EA: Netherlands institute for curriculum development.

Witthaus, G. (2012) Managing Curriculum Change. JISC e-Learning program. Available at www.jisc.ac.uk/elearningprogramme

Acknowledgements

This work has been partly sponsored by the 'Vicerrectoría de Investigaciones' and 'Faculty of Engineering' from Universidad de Medellín, under the project "Open Curricular Management System: Prototype I".

Copyright statement

Further Along the Engineering Pathway: The Impact of Design Spaces on the Development of Engineering Student Outcomes

Daniel W. Knight
University of Colorado Boulder, USA
Knightdw@Colorado.edu

Daria A. Kotys-Schwartz
University of Colorado Boulder, USA
Daria.Kotys@Colorado.edu

***Abstract**: A culminating engineering design experience, known in the US as Capstone Design, has become ubiquitous across accredited engineering programs in a number of countries. This design experience is thought to be a key milestone for preparing students for professional practice in industry settings. There is ongoing debate on the best practices for producing engineers who are ready for the workforce. Previous research has explored the impact of student and instructor characteristics and the role of curriculum design. The present study expands upon this investigation to explore the impact of dedicated Capstone design facilities housed in a customized design space known as the Idea Forge on the development of student outcomes including design, communication, teamwork, and project management skills. This investigation occurred during a time when the department of mechanical engineering under investigation experienced significant growth. During this period, Capstone design moved into a new space. Glimpses of these findings will be explored.*

Context

The present study investigates the impact of dedicated design facilities housed in a customized design space on the development of student engineering outcomes. The desired outcomes in the current investigation include skills in design, use of engineering tools, communication, teamwork, and project management in a mechanical engineering Capstone Design course.

A culminating design experience has become ubiquitous across accredited engineering programs in a number of countries (Rasul, et al. 2014) and an important milestone along the engineering pathway (Atman, et al. 2010). Capstone design courses emerged in the 1990's to balance the focus on theory and applied science in US engineering education that emerged after the Second World War (Dutson, et al. 1997). Motivations to bring back Capstone design include: to integrate technical concepts learned in previous years, to satisfy accreditation requirements, and to prepare students for professional practice (Rasul, et al. 2014, Ohland, et al. 2014). Given the link between Capstone design and accreditation, assessment of student outcomes is an important factor (Rasul, et al. 2014) that provides valuable assessment data on outcomes that are challenging to measure in other engineering courses (Stwalley, 2016).

There is ongoing debate as to best practices for producing engineers who are workforce ready. Previous research has explored the impact of student and instructor characteristics and the role of curriculum design (Howe, Rosenbauer, & Poulos, 2016). The present study focuses on a newer area of investigation concerning the impact of facilities and the associated human resources. In the past, many Capstone design programs have had minimal support facilities. But, progress has been made since the turn of the new century.

First, learning spaces in general have more recently incorporated learning principals such as active learning and the integration of learning technology to improve outcomes (Brown & Long, 2006). Also, the advent of university makerspaces, influenced by the Maker movement, has inspired the planning of traditional engineering design spaces to include rapid prototyping and 3D printing capabilities, along with electronics support and design studios. (Barrett, et al. 2015, Carlson & Sullivan, 1999).

Research Setting

The present study addresses a mechanical engineering Capstone design course that has been housed in two design spaces across six years of investigation. The course is a two-semester, industry-sponsored offering. Team-based design projects with specified roles are the foundation of the course, and the curriculum offers a high fidelity industry simulation with an employee handbook instead of a syllabus and industry style performance reviews to provide feedback (Kotys-Schwartz & Knight, 2016, Dutson, et al. 1997). Figure 1 provides an overview of the organizational structure of the course. Each team meets weekly with the sponsoring companies to refine the project scope, develop design alternatives; and manufacture, assemble and test products to meet client specification. Project teams are supervised by university faculty Directors, industry clients and course instructors. The project culminates in a year-end design expo (Kotys-Schwartz, et al. 2014).

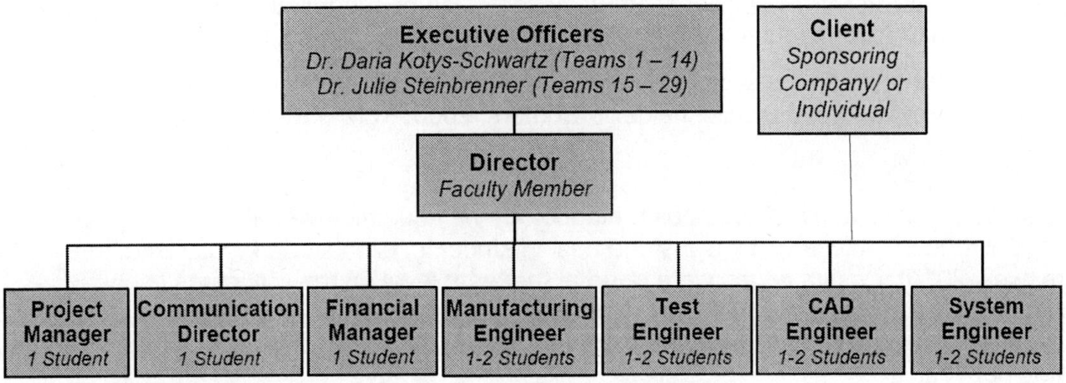

Figure 1: Organization structure of the Capstone Design Course

The course is housed in a 22,000 sq. foot design space known as the Idea Forge. The space has a variety of resources includes machining and electronics shops, rapid prototyping and 3D printing capability, wood-working, sewing support, an assembly, storage and testing area and support staff for each lab area. This facility is currently in its second full year of supporting the Capstone course (Kotys-Schwartz & Knight, 2016).The course was previously housed in a smaller, 6,000 sq. foot facility with machining and assembly capabilities that was closed upon the opening of the Idea Forge.

Capstone Design outcomes investigated for this study are as follows:

1. Skill at **Engineering Methods**: Students will have the necessary technical background to complete a long term, industry-sponsored project and be able to apply a knowledge of relevant manufacturing skills and design software,
2. **Design**: Students will be able to demonstrate an understanding of the iterative design process and contextual considerations of design,
3. **Communication**: Students will produce high quality communications including oral presentations and written reports,
4. **Teamwork**: Students will develop skill at building group cohesion and resolution of conflict, and
5. **Project Management**: Students will demonstrate project management skills including planning, organizing, and goal accomplishment.

Barrett, et al. (2017) suggested a need for future research into university makerspaces to investigate the impact on students who use the facilities, and that research is needed for Capstone design spaces as well. The present investigation builds upon this with the following research question: ***What is the impact of Capstone design spaces on the development of student outcomes in a mechanical engineering Capstone design course?***

Theoretical framework

General Systems Theory is employed as a theoretical lens for investigating the impact of design facilities on students. Growing out of biology and the social sciences, systems theory has been employed across multiple disciplines and specifically to explain organizational behaviour (Melão, Nuno, & Pidd, 2000). The theory provides a useful analytical framework for viewing an organization holistically and understanding its integration and flow of resources. A few of the important systems theory concepts that will be applied to the results include:

1. **Entropy**: the tendency for systems to become chaotic and disorganized,
2. **Subsystems**: each system is composed of interrelated subsystems,
3. **Open vs. Closed Systems**: the variability in systems with respect to the transfer of information and material between systems,
4. **Input-Transformation-Output Model**: systems can be modelled as a process in which inputs are transformed into outputs via movement through the system,
5. **Steady State**: when the system reaches a state of dynamic equilibrium within the flow of materials and information (Schneider & Somers, 2006, Katz & Kahn, 1978).

Methodology

This study employs a mixed-methods methodology with quantitative data collected via ratings of student outcomes by Capstone design students, clients and faculty Directors (Creswell, 2014). Design teams were rated at the team level on the outcomes (Kotys-Schwartz & Knight, 2016, Kristof-Brown, et al. 2014). Team level ratings are also more efficient and pragmatic than individual member ratings with multiple groups of raters and ~30 teams per year.

These ratings are collected via a course post-assessment survey. This survey also includes an open-ended feedback question for comments on skills. Ratings were collected from 2011-2016 totalling 847 team-level ratings. Table 1 breaks down the responses by rater. Note the large number of student responses. Table 2 depicts a breakdown of student raters by year. Note the large jump in student enrolment between 2013 and 2014, the last year of the old design facility.

An additional question was added for the past two years asking raters for feedback on project work at the Idea Forge. Interviews were scheduled with three members of the class of 2015 about the transition between the two design spaces. The means for each of the five skills ratings were compared against a program assessment goal of 4.0/5.0. One-way analyses of variance tested for significant changes across years for each skill. Qualitative data analysis procedures drawn from grounded theory established themes to support the quantitative results (Creswell, 2014).

Findings

Results by outcomes are presented across 2011-2016 in Figures 2-4. Figure 2 displays results across all three raters. Lines have been placed on each graph to indicate the post-assessment goal (4.0/5.0) and date of opening of the new design facility.

Table 1: Capstone Design Ratings by Rater

Raters (2011-2016)	n
Capstone Design Students	692
Faculty Directors	76
Industry Clients	79

Table 2: Student Capstone Design Ratings by Year

Student Raters	Number	Enrolment	Response Rate
2011	99	104	95%
2012	67	119	56%
2013	90	118	76%
2014	110	164	67%
2015	156	186	84%
2016	170	188	90%

In Figure 2, course enrolment is plotted to illustrate the relationship between skills decline and increasing enrolment. Mean scores in Figure 2 are composed of 82% student ratings. Figure 3 and Figure 4 provide a break out of faculty and client ratings respectively. Across all three figures, student outcomes are generally above the 4.0 target indicating that students were able to meet course skills objectives in either facility. The exception is skill at project management. Also, the range of industry client ratings from the last year of the old design facility, 2014, were below the desired outcome rating goal.

To understand the differences across years, a one-way analysis of variance was conducted for each outcome in Figure 2 (all raters) along with post-hoc comparisons of differences between years. Significant differences ($p<.05$) were found for each outcome. The strongest differences were found for skill at *engineering methods* ($F=8.10$, $p<.01$). Post hoc tests established that the *highest* skills ratings were in 2012 and *lowest* in 2015 for 4 out of 5 outcomes including *design*, *engineering methods*, *communication* and *teamwork*. All of these outcomes dropped significantly ($p<.05$) between 2012 and 2015.

Project management achieved its lowest numerical ratings in 2014, but these ratings did not differ significantly from ratings made before that time. We note the pattern of ratings for outcomes changes after the first full year of operation in the new design facility with rating averages increasing across all outcomes and significant increases ($p<.05$) for engineering methods and teamwork. Project management ratings gained significantly from 2014 to 2016.

Analysis of Figure 3, faculty Director and Figure 4, industry client ratings support the statistical analysis performed on the data from Figure 2, covering all raters. The numbers by year for these two data sets are too low for statistical analysis, but a pattern analysis demonstrates a drop in ratings for faculty during the opening of the new facility and for industry clients during the final year of the old facility. This analysis indicates a general rise in scores during the most recent year for faculty and clients.

Qualitative analysis of the open-ended data and student interviews from the 2015 class revealed three emerging themes around resources, integration and capacity for access to the design space.

Theme 1: Resources for the Design Projects

For the theme related to resources, participants described the process of using resources necessary for completion of the design project. Resources could be broken out into three categories: *space resources*, *equipment resources*, and *human resources*. Resources were also internal and housed within the design facility or external to the facility.

Space resources refer to the overall design facility and the spaces contained within. One client described the overall space as follows: *"The Idea Forge is a fantastic resource for the students pursuing their technical degrees…The space is very configurable to meet the ever changing demands of innovation by the various student teams." – Industry Client 2016"*

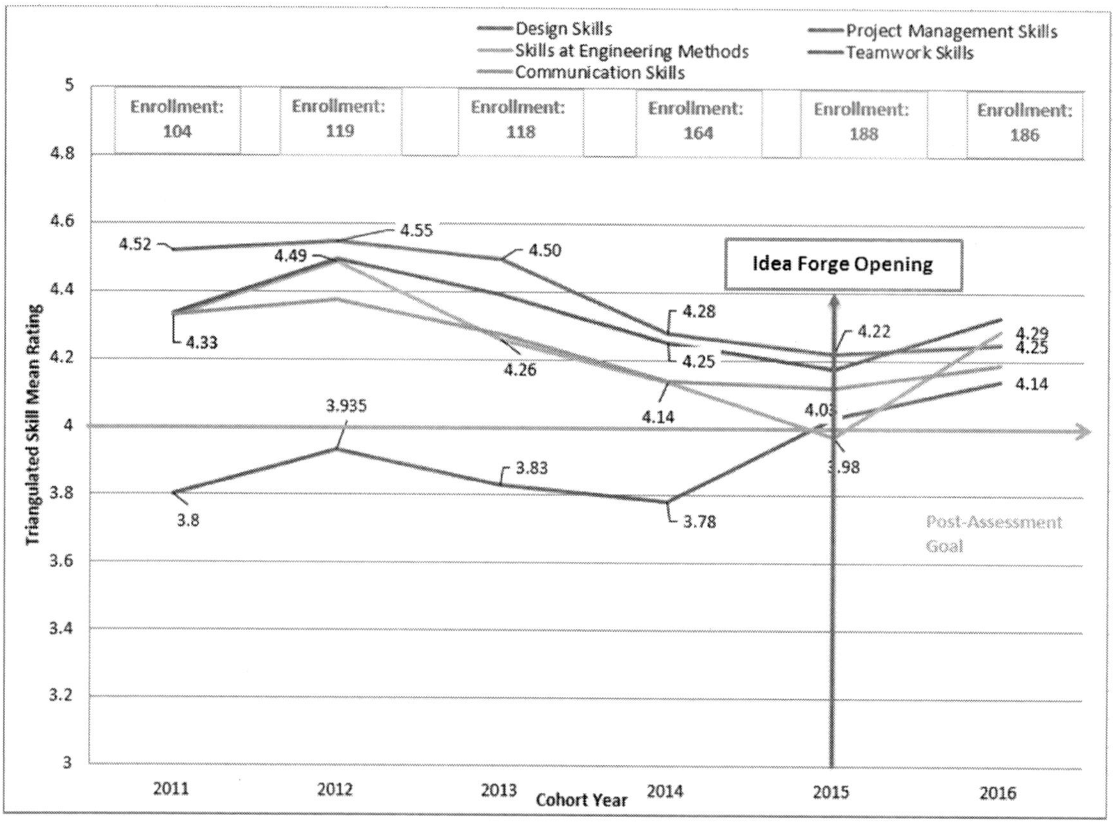

Figure 2: Capstone Design Student Outcomes across All Raters

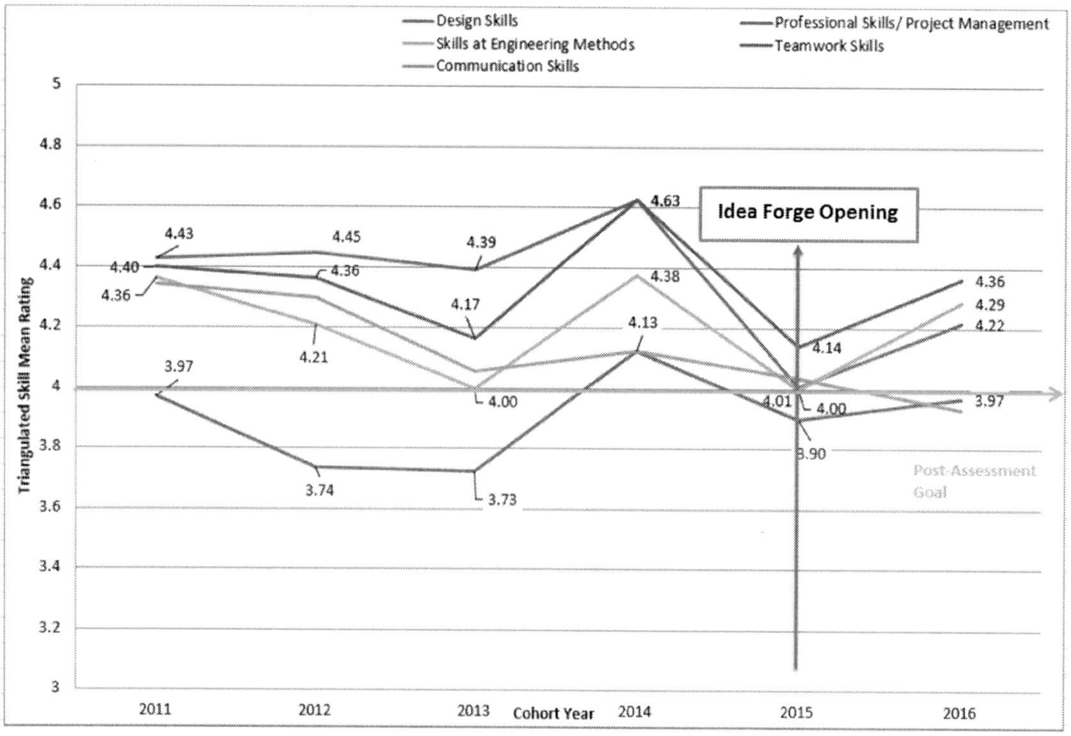

Figure 3: Capstone Design Student Outcomes Rated By Faculty Directors Only

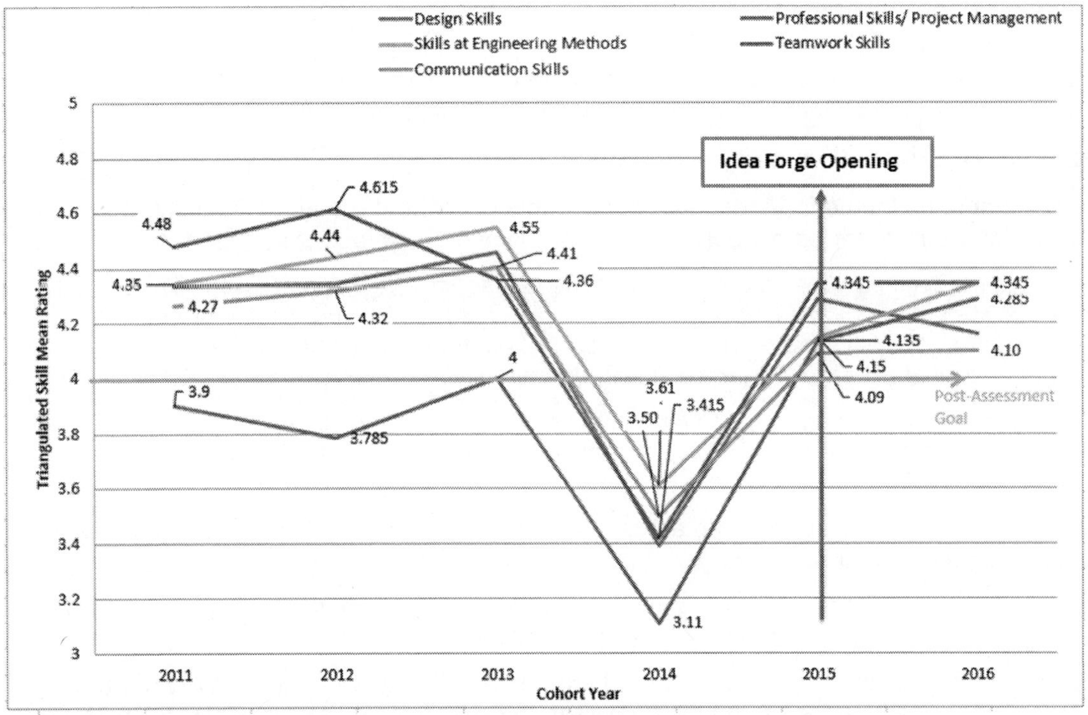

Figure 4: Capstone Design Student Outcomes Rated By Industry Clients Only

Equipment resources refer to the machines, computers, and other equipment used to support project development. For example, one student mentioned:

> *"Machining resources in the Idea Forge have been great, plenty of machines and knowledgeable staff." – Capstone student 2015*

One type of resource mentioned more often was the ***human resources*** with comments as follows:

> *"The Idea Forge personnel have been excellent during the course of the semester." – Faculty Director 2015*

External resources that served the projects but were housed outside the design facility were described in relation to ***internal resources*** housed within the design facility:

> *"Our client wanted us working at their company for the semester so we didn't get a chance to fully utilize the Idea Forge for our project. This is changing during the spring semester because we are moving our project to the Idea Forge to have better access to it."- Capstone student 2015*

Theme 2: Integration of Resources into a Cohesive Design Environment

Research participants also described integrating resources into a cohesive design environment to support design projects. One student interviewee described integration of different spaces within the design facility:

> *"I like that the machine shop is so close and helpful." – Capstone student 2015*

There were also descriptions of the integration of internal and external resources:

> *"...And one of the things I liked best was Daria (Design Facility Director). She is very keen on getting Idea Forge to work more with other parts of the university, like (other design facility)...This project couldn't have happened without her support." – Industry Client 2016*

Integration of external and internal resources could be a challenge as well with the necessity of integration support from external facilities:

> *"The Idea Forge is great for later stages of the project, however it lacks the rapid prototyping materials that the (other design facility) has, also rather difficult to transport our project from one location to the other." – Capstone student 2015*

Theme 3: Resource Capacity = Access

Another emerging theme from research participants was about resource capacity. Specifically, would there be enough capacity for teams to have access to needed resources?

> *"The Idea Forge has been a great place for us to meet as a team and to work. There has been plenty of storage for our project components (we don't need to store any large items). We have been using the soldering / electronics lab and have always had access to what we needed with plenty of help from staff." – Capstone student 2016*

> *"The Idea Forge is a great space for student meetings and working alongside the students. I enjoy having an open space that is not a pre-scheduled meeting room to work and learn." – Industry Client - 2016*

Challenges to access were reported across participants as well. For example, some wanted more resources for their projects:

> *"I think the Idea Forge needs more 3D printers. Iterating through 3D printing is a vital part of the modern design process." – Faculty Director 2016*

One of the capacity challenges is increased enrolment and impact on availability of the resources. This issue arose during the final year of the first design facility (2014). One faculty director commented,

> *"The resources provided by the University were the single biggest limitation/hurdle encountered by the team. While having a machine shop is nice… it isn't nearly big enough to support so many teams…– Faculty Director 2014*

As the enrolment in the course is expected to continue to rise, one faculty director commented,

> *"I think the teams have a great place to work here…I worry when we have more students/ teams." – Faculty Director 2016*

Discussion

Triangulated results across raters and years point to sufficient resources to meet design outcomes against the targeted post-assessment goal. Using a General Systems Theory perspective, these results can be viewed as an input-transformation-output model. Seniors are an input into the system (i.e., the design facility) and transformed via access to resources that are integrated into a design environment, which outputs the expected student skill outcomes (Melão, Nuno, & Pidd, 2000).

Project management was the one outcome that performed better in the new facility than the old. To succeed with their projects, students need to integrate multiple course aspects such as the technical resources, team members, multiple supervisors, and curriculum requirements. Systems theory views these interlocking elements as subsystems or components of the course and encourages integration to produce the expected output from the transformation process (Schneider & Somers, 2006). Perhaps having more course components better integrated under the same roof with increased capacity supports better project management for student teams.

Achievement of outcomes becomes problematic when access is decreased. This is seen when course enrollment increased starting in 2014, creating a bottleneck with respect to resource capacity. With the opening of the new facility, space was available, but equipment resources and human resources were not online until 2016. General Systems Theory offers the concept of entropy signifying the natural tendency for systems to break down and become chaotic over time. Here, the increasing numbers of students generated entropy in the system as teams lost access to resources, which affected the transformation process. Similarly, the opening of the new facility during the semester created additional chaos as new resources were coming on-line (Schneider & Somers, 2006).

Systems theory distinguishes between open systems, which exchange energy, information, and materials through boundaries with their environment, and closed systems, which do not (Katz & Kahn, 1978). Capstone design teams were invariably accessing resources outside the primary design facility, such as those associated with other suppliers, design spaces, and the facilities of their industry clients. There is a challenge inherent in maintaining an internally integrated design environment running at capacity while remaining open to design resource support from outside the system.

Implications

These results imply that integrated design facilities, with strong resource support and capacity for reasonable access to resources are instrumental in producing the desired outcomes in mechanical engineering Capstone design. This supports previous research that these design spaces impact retention in freshman design courses (Fortenberry, Sullivan, Jordan & Knight, 2007, Carlson & Sullivan, 1999).

These results further imply that an integrated design environment necessary to produce outcomes can be compromised by resource capacity as happened in the first design facility, and there are fears that increased enrolments will overwhelm the new facility. Huddleston (2000) discusses enrolment management as follows, "Optimally, an institution's enrolment is comprehensively developed and is based on a strategic, integrative plan that includes the identification, attraction, selection, encouragement, registration, retention, and graduation of targeted student segments." Caution is advised when increasing enrolments in a hands-on course such as Capstone design with an eye towards resource capacity to produce outcomes.

In systems thinking, a steady state is reached when the system reaches a state of equilibrium within the input-transformation-output cycle (Melão, & Pidd, 2000) This lens suggests caution when moving a Capstone course into a new facility before the facility has reached this steady state as four of five outcomes registered their lowest ratings during the first year of operation of the new facility. This movement to the new facility was necessary due to the closing of the old facility and it should be noted that outcomes were still largely maintained against targets.

Future Research

This investigation is limited by its setting in one university design space and could be expanded to include other Capstone design courses in other facilities. Also, only skills outcomes in a Capstone course associated with ABET criteria were investigated. This investigation could be expanded to include the development of attitudes or identity in a design space. In addition, the Capstone course instructor and curriculum changed during the course of the investigation and future research could disaggregate the impact of these factors from the change in facilities.

Future research could also gather more qualitative data on the design facilities as only two years of survey data specific to the facility and three interviews were gathered. Presented themes should be considered as emerging. Finally, this research could be expanded to investigate the impact of different subsystems within the design space on course outcomes such as the impact of the machine shop on skill at engineering methods, which showed the greatest positive changes after opening the new facility. Future research will delve more deeply into these issues.

References

ABET (2017) Criteria for Accrediting Engineering Programs, 2016 – 2017. ABET.org.

Atman, C. J., Sheppard, S. D., Turns, J., Adams, R. S., Fleming, L. N., Stevens, R., Streveler, R.A., Smith, K.A., Miller, R.L., Leifer, L.J. & Yasuhara, K. (2010). Enabling Engineering Student Success: The Final Report for the Center for the Advancement of Engineering Education. CAEE-TR-10-02. *Center for the Advancement of Engineering Education (NJ1).*

Barrett, T., Pizzico, M., Levy, B.D., Nagel, R.L., Linsey, J.S., Talley, K.G., Forest, C.R. & Newstetter, W.C. (2015). A Review of University Maker Spaces. *Proceedings of the 2015 ASEE Annual Conference.* June, Seattle, WA.

Brown, M. & Long, P. (2006). Trends in Learning Space Design. *Learning Spaces, Ch. 9, 1-11. www.educause.edu/learningspaces.*

Carlson, L. E., & Sullivan, J. F. (1999). Hands-on Engineering: Learning By Doing in the Integrated Teaching and Learning Program. *International Journal of Engineering Education*, 15(1), 20-31.

Creswell, J. W. (2014). *A Concise Introduction to Mixed Methods Research.* Sage Publications.

Dutson, A. J., Todd, R. H., Magleby, S. P., & Sorensen, C. D. (1997). A Review of Literature on Teaching Engineering Design through Project-Oriented Capstone Courses. *Journal of Engineering Education, 86*(1), 17-28.

Fortenberry, N.L., Sullivan, J.S., Jordan, P.N. & Knight, D.W. "Engineering Education Research Aids Instruction, "*Science* (317) 31, 1175-1176.

Huddleston Jr, T. (2000). Enrollment Management. *New Directions for Higher Education, 2000*(111), 65-73.

Howe, S., Rosenbauer, L., & Poulos, S. (2016). 2015 Capstone Design Survey–Initial Results. In *Proceedings of the 2016 Capstone Design Conference,* June, Columbus, OH.

Katz, D., & Kahn, R. L. (1978). *The Social Psychology of Organizations (Vol. 2).* New York: Wiley.

Kotys-Schwartz, D. A., Daly, S. R., Yilmaz, S., Knight, D.W., & Polmear, M.R. (2014). Evaluating The Implementation Of Design Heuristic Cards In An Industry Sponsored Capstone Design Course. In *Proceedings of the 2014 Capstone Design Conference,* June, Indianapolis, IN, .

Kotys-Schwartz, D.A. & Knight, D.W. (2016). Designing the Engineer: Curriculum and Infrastructure to Create Professionally Prepared Engineers. 2016 IE3 Summit, University of Florida.

Kristof-Brown, A. L., Seong, J. Y., Degeest, D. S., Park, W. W., & Hong, D. S. (2014). Collective Fit Perceptions: A Multilevel Investigation Of Person–Group Fit With Individual-Level And Team-Level Outcomes. *Journal of Organizational Behavior*, 35(7), 969-989.

Melão, N., & Pidd, M. (2000). A Conceptual Framework For Understanding Business Processes And Business Process Modelling. *Information Systems Journal*, 10(2), 105-129.

Ohland, M. W., Giurintano, D., Novoselich, B., Brackin, P., & Sangelkar, S. (2015). Supporting Capstone Teams: Lessons from Research on Motivation. *International Journal of Engineering Education, 31*(6), 1748-1759.

Rasul, M. G., Lawson, J. J., Howard, P., Martin, F., & Hadgraft, R. (2015). Learning and Teaching Approaches and Methodologies of Capstone Final Year Engineering Projects. *International Journal of Engineering Education*, 31(6), 1727-1735.

Schneider, M., & Somers, M. (2006). Organizations As Complex Adaptive Systems: Implications Of Complexity Theory For Leadership Research. *The Leadership Quarterly*, 17(4), 351-365.

Stwalley, R.M. (2016) Capstone Communications. *Resource Magazine.* 23(5): 15-15.

Copyright statement

Beyond superficiality: The Relationship Between Motivation and Deep Learning in Electronic Engineering Students

Juan M. Cruz
Universidad Javeriana, Bogota, Colombia. Virginia Tech, Blacksburg, USA
juan.cruz@javeriana.edu.co

Indhira M. Hasbun
Virginia Tech, Blacksburg, USA
Imhasbun@vt.edu

Brett D. Jones
Virginia Tech, Blacksburg, USA
brettjones@vt.edu

Stephanie G. Adams
Old Dominion University, Norfolk, USA
sgadams@odu.edu

Abstract*: Research indicates that a deep approach to learning is highly and positively related with students' development of expertise and transfer abilities, making the fostering of such deep learning approaches extremely relevant within higher education. Motivation plays a key role in fostering these learner experiences because it influences the activities that a learner chooses to engage in, as well as the level of engagement in such activities (i.e, deep approach to learning). This cross-sectional survey research employed Spanish versions of the MUSIC® Model of Academic Motivation Inventory (Jones, 2016), and a subset of the math and science engagement development scales (Wang, Fredricks, Ye, Hofkens, & Linn, 2016) to better understand which motivational factors correlate to deep learning in the context of an undergraduate electronic engineering program in Colombia. The relationships were determined by correlation and regression analyses of 146 student responses. All elements of the MUSIC model moderately correlated to deep learning.*

KEYWORDS

MUSIC Model of Motivation, cognitive engagement, behavioral engagement

CONTEXT

Higher education should encourage the types of learning approaches that lead to deep understanding of the concepts and principles of a domain – defined by Craik and Lockhart as deep approaches to learning (Hattie, Biggs, & Purdie, 1996) – because they are highly related to the development of expert levels of performance and transfer abilities. Recent studies in higher education claim that strengthening skills for developing expertise, and increasing transfer abilities should be the main purposes of all professional education programs (Bransford, Brown, & Cocking, 2000; Jamieson & Lohmann, 2009). Research also indicates that learning experiences that lead to deep understanding of domain concepts and principles are highly related to the attainment of these purposes of education programs. For example, in their thorough literature review about the implications on the development of expertise Litzinger, Lattuca, Hadgraft, and Newstetter (2011) concluded that "engineering education should encompass a set of learning experiences that allow students to construct deep conceptual knowledge to develop the ability to apply key technical and professional skills fluently" (p.1).

Litzinger et al. (2011) based their conclusions on studies such as Wierstra, Kanselaar, Van der Linden, Lodewijks, and Vermunt (2003) who stated that deep learning approaches correlated with learning experiences that emphasize "conceptual connections between the content of the learning domain (conceptual understandings) and that in which students were active and exercise a high degree of self-regulation"(p. 514). There are also other studies that suggest that students who take a deep approach to learning show superior learning outcomes relative to students who take a surface approach (Meyer, Parsons, & Dunne, 1990; Ramsden, 2003, as cited in Johri and Olds (2014), Ch. 21). Although these authors do not necessarily address the concept of engagement, research describes that the approaches to learning have a close relationship with academic involvement and in particular, the deep approach to learning as characteristic of higher levels of engagement (Kong, Wong, & Lam, 2003).

The latter suggests that engineering teachers should encourage a deep learning approach in their students, rather than a surface learning approach, to foster the development of expertise and the transfer of knowledge. However, how can teachers encourage a deep approach to learning? Traditionally, the study of learning approaches are related to the analysis of study skills (Biggs, Kember, & Leung, 2001). However, as Hattie et al. (1996) described, it is necessary to analyse other contextual factors beyond the study skills because they fail "to take account of the interaction between students' intentions and the context of learning" (p. 122). Moreover, research indicates that motivation plays a key role in student learning because motivation influences the activities that a learner chooses to engage in, as well as the level of engagement (i.e, deep approach to learning) (Ormrod & Jones, 2015). Similarly, Robbins et al. (2004) emphasized that the real question to predict college outcomes should be "how study skills combine with social and motivational factors to ensure positive student action" (P. 276). Hence research suggests the importance of understanding the relationship between academic motivation and study skills that are representative of a deep approach to learning. It is expected, then, that internal and external motivational factors influence the students' decision to have a deep approach to learning.

In summary, research suggests: 1) a deep approach to learning is highly related to the development of expertise, 2) there are motivational factors that encourage students to select a deep approach to learning; and, 3) the approach to learning is manifested in the students' engagement in their courses as well as in their study skills.

Although the relationship between motivation and deep learning has been studied (Biggs et al., 2001; Laird & Shoup, 2005; Wang & Eccles, 2013), its context has been less oriented to Spanish speaking cultures and engineering. Hence, to better understand which motivational factors correlate to deep learning in the context of an undergraduate engineering course, the purpose of this quantitative study is to determine the relationship between academic motivation, deep learning and subsequent performance within the Electronics Engineering department in a particular school of engineering in Colombia.

RESEARCH QUESTIONS

To reach its goals, this study started by asking the following question: What is the relationship between academic motivation, deep learning, and performance within courses in the Electronics Engineering Department in a particular school of engineering in Colombia?.

This study is important to engineering education because by establishing a relationship between motivation and deep learning in an engineering context, engineering teachers could use tools suggested by motivational models to more effectively foster in students the development of expertise and knowledge transfer to new contexts. As it will be explained in the following sections, to operationalize and measure the academic motivational factors this study employed the MUSIC® Model of Motivation developed by (Jones, 2009, 2015). Because the constructs of deep learning are closely related to cognitive engagement (Kong, Wong, & Lam, 2003; Wang et al., 2016; Zimmerman, 1990) to operationalize the deep learning concept, this study employed the model of engagement developed by Wang et al. (2016).

THEORETICAL FRAMEWORK

MUSIC® Model of Motivation. The MUSIC® Model of Motivation (abbreviated as the MUSIC model) is a model developed by Jones (2009) that encompasses several other motivation theories and their implications on students' motivation in academic settings. The motivation theories this model is based on have been extensively studied in multiple contexts by leading researchers in the field and have been proven to have implications for student engagement and subsequently student learning (Bandura, 1997; Eccles, 1983; Hidi & Renninger, 2006; Wigfield & Eccles, 2000)

The research upon which the model is based was drawn from a variety of fields and with students of all ages and cultural backgrounds (Jones, 2009, 2015).The constructs that are included in the MUSIC model are: 1) eMpowerment – the extent to which students feel in control of how to meet specific course objectives; 2) Usefulness –students' perceptions of how useful they feel the course content is or will be for their short or long term goals; 3) Success – students' perceptions of their ability to succeed in the course by exercising the required effort; 4) Interest – the extent to which students are interested in the course content and delivery methods and 5) Caring –students' perceptions that the instructor and/or their fellow classmates care about them as individuals and their learning in the course.

The instrument to measure motivation employed in this article, the MUSIC® Model of Academic Motivation Inventory (Jones, 2016), consists of 26 items divided into five scales. Students respond to the scale items by choosing one of six options on a Likert-scale format: 1 = *Strongly disagree*, 2 = *Disagree*, 3 = *Somewhat disagree*, 4 = *Somewhat agree*, 5 = *Agree*, 6 = *Strongly agree*. The items in each scale are averaged to create a mean scale score. This inventory has shown high validity, with Cronbach alphas ranging between 0.88 to 0.95 with samples of undergraduate students in Colombia (Jones, Li, & Cruz, 2017). The complete inventory is available at Jones (2016).

Engagement and approaches to learning. The concept of student engagement is also founded on the ideas of C. Robert Pace and Alexander Astin (Lattuca & Stark, 2011; Smith, Sheppard, Johnson, & Johnson, 2005). On one hand, Pace describes engagement as quality of effort; he made the assertion that the more quality of effort a student puts in an academic task, the more he or she will learn. On the other hand, Astin developed the concept of involvement as a student's engagement with his or her academic environment. Although the concept of engagement appears related to motivation, Pace and Astin distinguished them by acknowledging that motivation is a psychological state whereas involvement implies behavior (Lattuca & Stark, 2011).

Research suggests that the outcomes of learning are highly related to the approaches to learning chosen by students (Biggs, 1987; Laird & Shoup, 2005). The most common approaches are surface and deep. Students using a surface approach emphasize repetition and memorization, focus their study to pass the tests and tend not to relate new information with previous knowledge. Students who use a deep approach tend to focus on key concepts, meaning, application, and relate the new information with previous knowledge (Biggs, 1987; Entwistle & McCune, 2004; Laird & Shoup, 2005). In other words, a deep approach to learning is represented by higher levels of engagement (Laird & Shoup, 2005).

Measurements of the approaches to learning have traditionally employed the Motivated Strategies for Learning Questionnaire (MLSQ) developed by Entwistle and McCune (2004) or the Revised Two-Factor Study Process Questionnaire (R-SPQ-2F) developed by Biggs et al. (2001). Both take into account strategies and motivators that are related to the approach to learning. Although they are highly used, other researchers argue that its reliability is questionable (i.e, Cronbach alphas below 0.6), hence, they have developed more reliable measurements (Kong et al., 2003; Wang et al., 2016). Other measurements of deep approaches to learning use the National Survey of Student Engagement (NSSE) (Kuh, 2009), but research suggest that it is unlikely to function as a replacement for other, more in-depth measurements (Laird & Shoup, 2005).

We used the 16-item behavioral and cognitive engagement scale of the math and science development scales developed by Wang et. al. (2016). All items included a 5-point Likert-scale format (1 = *Not at all like me*, 2 = *Not much like me*, 3 = *Somewhat like me*, 4 = *Mostly like me*, 5 = *Very much like me*). The alpha reliability estimate for this scale was good for behavioral engagement (α = .81) and acceptable for cognitive engagement (α = .75) with samples of U.S. high school students.

METHODOLOGY

As mentioned earlier, this research was a cross-sectional survey study measuring students' motivation and engagement on a particular engineering school in Bogota, Colombia. We obtained a random, self-selected sample from students at the intended department. We identified a gatekeeper (Creswell, 2009) who invited all 579 students from the department to participate. The survey was administered online through the Qualtrics software. This electronic survey offered a variety of advantages, such as reduced cost, time savings, increased availability for large sample sizes, increased interactivity and customization for respondents, and ease of data analysis by eliminating the need to enter data manually (Singleton Jr., Straits, & Straits, 1993). To offset the coverage error and limited sampling frames, through our gatekeeper we had access to a reliable sampling frame and she assured students had access to computers and internet connections through the schools' computer labs. We obtained a response rate of 25% (146 out of 579 students). Data were analysed using the Statistical Package for the Social Sciences (SPSS) software. This procedure received Institutional Review Board (IRB) approval.

Given that the Spanish version of the MUSIC Inventory was validated elsewhere (Jones et al., 2017), our validation process was focused on the Spanish translation of the cognitive and behavioral components of Wang et al. (2016). The research team developed a Spanish version which was reviewed by peers and students in Colombia to check for clarity. Two independent collaborators back-translated it from Spanish to English and the research team reviewed the newly worded English version looking for language consistency. Once approved, the team proceeded to revise certain Spanish-wording to make it applicable to vocabulary utilized in Colombia, while still maintaining the same meaning. The reliability results for the Spanish version for cognitive engagement and for behavioral engagement were α=.74 and α=.70 respectively, which are considered acceptable values (George & Mallery, 2003) and consistent with the reliability of the original versions. Additionally, these instruments asked students to answer motivation and engagement questions based on the class of their choosing that they were currently studying. They also included additional questions regarding students' demographics information, GPA and expected grade of the selected class.

To identify the relationship between academic motivation and deep learning we started by computing some descriptive statistics, followed by correlation and regression analyses between the measurements of motivation and cognitive and behavioral engagement. ANOVA and t-test analyses were used to determine differences by gender, age, or class rank.

FINDINGS AND CONCLUSIONS

Our sample constituted 146 electronic engineering students. As shown in Figures 1 and 2, this sample was representative of the population of the intended department. Correlations across each construct were also calculated and are presented in Table 1. Even though each Motivation factor loaded individually, they were highly correlated with one another at the $p<0.01$ significance level (coefficients range between 0.55 to 0.83 for across all Motivation factors). All Motivation factors and Engagement variables were correlated, as verified by Pearson's correlation coefficients ranging between 0.396 to 0.520 at the $p<0.01$ level. To analyse cognitive engagement, we conducted multiple regression analysis to establish if there is statistical evidence to suggest that higher levels of motivation lead to deep learning. We checked and met assumptions of linearity, normally distributed errors, and collinearity. Three outliers were removed that did not meet the criteria for a sample of 146 respondents (i.e, Dffit>0.37, Standard residual error >3, Cook's distance>0.4) (Field, 2013). As a result of these manipulations, the model was run with a total of 141 resultant observations. As shown in Table 2, the combination of variables of motivation moderately predicted cognitive engagement $F(5,136)=6.523$, $p<.001$; only Empowerment $t(5,136)=1.977$, $p<.05$ was significant to the prediction. The adjusted R squared value was .164, indicating that 16.4% of the variance in the cognitive engagement was explained by the model.

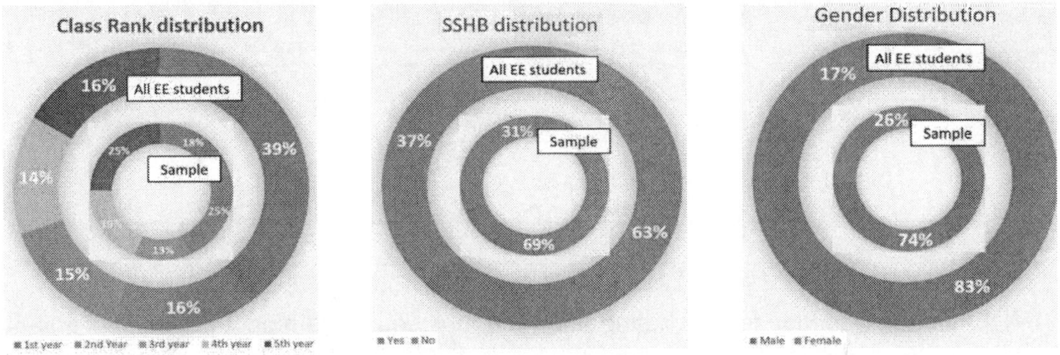

Figure 1: Descriptive Statistics

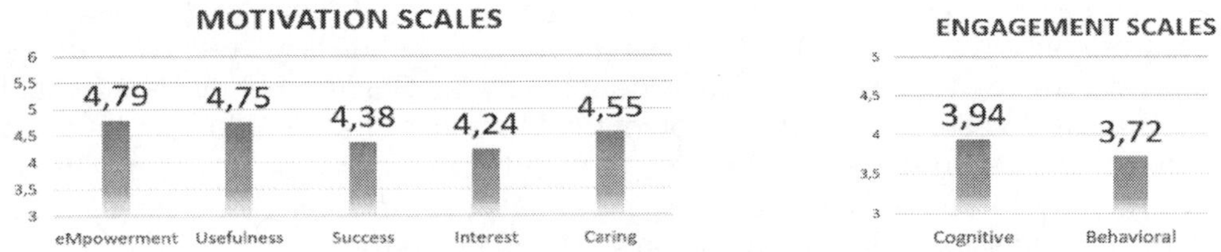

Figure 2: Motivation and Engagement Average Results

Table 1: Correlations

	Usefulness	Success	Interest	Caring	Cognitive Engagement	Behavioral Engagement
eMpowerment	,686**	,583**	,741**	,734**	,397**	,418**
Usefulness		,634**	,832**	,692**	,367**	,423**
Success			,596**	,519**	,324**	,336**
Interest				,777**	,388**	,430**
Caring					,288**	,296**
Cognitive Engagement						,602**

**. Correlation is significant at the 0.01 level (2-tailed).

Table 2: Regression model for Cognitive Engagement

	Sum of Squares	df	Mean Square	F	Sig.
Regression	6,610	5	1,322	6,523	,000b
Residual	27,561	136	0,203		
Total	34,171	141			

Model Summary[a]

R	R Square	Adjusted R Square	Std. Error of the Estimate	Durbin-Watson
,440[a]	,193	,164	,45017	2,311

a. Predictors: (Constant), Caring, Success, Usefulness, eMpowerment, Interest

We conducted a similar process to analyse Behavioral engagement. All responses met the assumptions of linearity, normally distributed errors and collinearity and no outliers were removed. As shown in Table 3, the combination of variables of motivation moderately predicted behavioral engagement $F(5,140)=8.53$, $p<.001$; only Empowerment $t(5,140)=2.101$, $p<.05$ was significant to the prediction. The adjusted R squared value was .206, indicating that 20.6% of the variance in the cognitive engagement was explained by the model

Table 3: Regression model for Behavioral Engagement

	Sum of Squares	df	Mean Square	F	Sig.
Regression	9,824	5	1,965	8,530	,000b
Residual	32,247	140	0,230		
Total	42,072	145			

Model Summary[a]

R	R Square	Adjusted R Square	Std. Error of the Estimate	Durbin-Watson
,483	,234	,206	,47994	2,068

a. Predictors: (Constant), Caring, Success, Usefulness, eMpowerment, Interest

We then turned our attention to completing analyses that would examine if the motivation and engagement scales discriminate between students' various demographics. An analysis of covariance (ANOVA) or t-test were run to establish whether there were statistically significant differences between gender, age, class rank and high school origin. Within our sample, 26% of respondents were female (n=38) and 74% males (n=108); almost 66% fall within the range of 18 to 21 years old (n=96). High school origin is denoted by the Student Senior High School in Bogota (SSHB) classification. It is generally of interest to administrators to understand the differences between effects on students who came from the capital city (Bogota) versus those who did not, as a way to understand the decisions´ impact on diversity. For our sample, students from Bogota constituted 69% of the sample (101 out of 146).

A one-way ANOVA was conducted to compare the effect of age on motivation and engagement in the 9 age groups (i.e, ages from 17 to 25). No significant effects of age in motivation and engagement were noted. We conduct a one-way ANOVA to compare the effect of Class Rank on motivation and engagement in 1st year, 2nd year, 3rd year, 4th year and 5th year. No significant effects of class rank in eMpowerment, Success, and engagement were noted. However, as show in Table 4 there was a significant effect on Usefulness and Interest at the $p<.01$ level and on Caring at the $p<.01$ for the five groups of class rank [$F(4, 141) = 7.5$, $p< .01$]. [$F(4, 141) = 5.2$, $p< .01$] and $F(4, 141) = 3.2$, $p< .05$ respectively. Post hoc comparisons using the Games-Howell test indicated that the mean score of Usefulness for students of 2nd year (M = 5.34) and 3rd year (M = 5.18) was significantly higher than 4th year (M = 4.08) and 5th year (M = 4.42). The mean score of Interest for students of 2nd year (M = 4.84) was significantly higher than 4th year (M = 3.63) and 5th year (M = 3.79). Finally, the mean score of Caring for students of 2nd year (M = 4.94) was significantly higher than students of 5th year (M = 4.14). Taken together, these results suggest that Class Rank does affect motivation but not engagement. Specifically, our results suggest that 2nd year students have higher levels of Usefulness, Interest, and/or Caring than students of the 4th and 5th year.

An independent-samples t-test was conducted to compare mean results in respondents who studied their Senior High School in Bogota (SHSB) and outside of Bogota, and female and male respondents. There was a significant difference in the scores for Usefulness in SHSB (M=4.6, SD=1.2) and no SHSB (M=5.03, SD=1.0) conditions; $t(144)=-2.0$, $p < 0.05$. In the same way, there was a was a significant difference in the scores for Success in female students (M=3.9, SD=1.3) and male students (M=4.55, SD=1.15); $t(144)=-2.9$, $p < 0.05$.These results suggest that people who attended senior high school outside of Bogota find their classes more useful than those who studied SHSB. Similarly, male students are more confident of their potential to succeed than female students.

Because the sample is representative of the population, we can infer some relationships between motivation and engagement with the statistical evidence presented. All the motivational factors moderately correlate to cognitive and behavioral engagement. These results in addition to the regression analysis suggest that **increasing empowerment will drive more actions that indicate cognitive and behavioral engagement.** From the remaining 4 factors, usefulness and interest showed higher correlations to engagement. Based on the definitions of these constructs we could infer that students who feel they have choices in the learning process, see the usefulness of the material they are learning, and are interested in the material, will adopt constructive conduct in class and more likely enact deep learning strategies. While elements of success and caring are also related to engagement, especially given that they are correlated with the other factors and may affect them, their influence is not as strong on engagement as the other components.

Table 4: Comparison of Means

Construct	Gender (t-test)	Class Rank (ANOVA)	Region of Origin (t-test)
	t(144)	**F(4,141)**	**t(144)**
eMpowerment	-1.5	2.2	-.1.2
Usefulness	-1.3	7.5** [a]	-2.0*
Success	-2.9*	2.5*	-1.25
Interest	-.91	5.2** [a]	-1.5
Caring	-.79	3.2* [a]	-1.25
Cognitive Engagement	1.1	.42	-.64
Behavioral Engagement	-.30	1.3	-.87

*at p<0.05, **at p<0.01 [a] Results of Welch robust test of equality of means

Table 5: Significant difference of Means

Construct	2nd year	3rd year	4th year	5th year
Usefulness	5.34	5.18[a]	4.08	4.42
Interest	4.84		3.63	3.79
Caring	4.94			4.14

a. No significant difference between the means of 2nd year and 3rd year students

Each of the motivational constructs were also highly correlated with each other, as expected by literature suggestions that indicate that motivation is explained by multiple factors and that these factors are often interrelated (Jones & Wilkins, 2013)

The difference in success by gender is consistent with other research that has found that, traditionally, women have lesser perceptions of self-efficacy and lower expectancy for success than their male counterparts (Bong & Skaalvik, 2003; Hocevar, Hagedorn, & Vogt, 2007; Wigfield & Eccles, 2000). For example, Eccles (1983) found that students' with high levels of self-efficacy need to be supported and developed to maintain interest in the academic context; the conditions needed to support the development of the students' interest often include allowing the student the autonomy to pursue those interests (Csikszentmihalyi, Rathunde, & Whalen, 1997; Hidi & Renninger, 2006), as well as fulfilling a student's need for relatedness either by interacting with peers or with an instructor (Deci & Ryan, 2000; Sosniak, 1990).

Inconsistent with theory, 4th and 5th year students means of Usefulness and Interest were lesser than 2nd or 3rd year. This is surprising because as students' progress through the program, content in engineering classrooms evolves from foundational knowledge to applications. With persistence, literature suggests, students have likely advanced through different states of interest, moving from a more situational interest towards an internalized stage of emerging or well developed individual interest (Hidi & Renninger, 2006). Additionally, results suggest that Class Rank does not impact the level of engagement, probably because our results did not show that class rank has an effect change in the perceived empowerment. It was also found that empowerment predicted engagement with a variance explanation of 16%. Although this prediction is low, this study presents evidence that there is a relationship between certain motivational constructs and engagement. Moreover, this relationship is positive, consistent with theoretical underpinnings and other empirical data in different contexts.

The results of this study suggest that both instruments can be successfully applied within the context of engineering in Colombia. The motivation instrument was also generally able to make differentiations by gender and Class Rank, consistent with literature suggestions. Lastly, correlation coefficients were also as expected across MUSIC constructs.

RECOMMENDATIONS AND IMPLICATIONS

Engineering programs may want to consider best practices to empower students, address the usefulness of their coursework, and prompt interest in the material. As posited by self-determination theory, one of the significant ways we can empower students is to give them a sense of autonomy (Deci & Ryan, 2000). Through autonomy, students will have greater psychological capacity to be at ease in learning environments and feel intrinsically motivated. To address aspects of usefulness, teacher should follow a course of action to help students perceive the value of their coursework towards the fulfilment of the students´ goals. Finally, interest can be achieved in stages that do not necessarily require students to be inherently attuned to the topic or learning activities presented to them (Hidi & Renninger, 2006). To prompt interest, teachers can expose students to novel experiences, which may involve industry visitations, new design projects based on contemporary or popular subjects, or simply more in-class demonstrations to keep student attention. More ideas for engaging students can be found in (Jones, 2009, 2015)

ACKNOWLEDGEMENT

This study started as a motivation and cognition class project at Virginia Tech, we appreciate the efforts and insights provided from our colleague Joshua Garcia-Sheridan who, during the class, helped on the development of the idea.

REFERENCES

Bandura, A. (1997). *Self-efficacy: The exercise of control*: Macmillan.

Biggs, J. B. (1987). *Student Approaches to Learning and Studying. Research Monograph*: ERIC.

Biggs, J. B., Kember, D., & Leung, D. Y. (2001). The revised two-factor Study Process Questionnaire: R-SPQ-2F. *The British journal of educational psychology, 71*, 133-149.

Bong, M., & Skaalvik, E. M. (2003). Academic self-concept and self-efficacy: How different are they really? *Educational Psychology Review, 15*(1), 1-40.

Bransford, J. D., Brown, A. L., & Cocking, R. R. (2000). *How People Learn: Brain, Mind, Experience, and School*. Washington, D.C.: National Academy Press.

Csikszentmihalyi, M., Rathunde, K., & Whalen, S. (1997). *Talented teenagers: The roots of success and failure*: Cambridge University Press.

Deci, E. L., & Ryan, R. M. (2000). The "what" and "why" of goal pursuits: Human needs and the self-determination of behavior. *Psychological Inquiry, 11*(4), 227-268.

Eccles, J. S. (1983). Expectancies, values, and academic behaviors.

Entwistle, N., & McCune, V. (2004). The conceptual bases of study strategy inventories. *Educational Psychology Review, 16*(4), 325-345.

Field, A. (2013). *Discovering statistics using IBM SPSS statistics*: Sage.
George, D., & Mallery, M. (2003). Using SPSS for Windows step by step: a simple guide and reference.
Hattie, J., Biggs, J., & Purdie, N. (1996). Effects of learning skills interventions on student learning: A meta-analysis. *Review of Educational Research, 66*, 99-136.
Hidi, S., & Renninger, K. A. (2006). The four-phase model of interest development. *Educational psychologist, 41*(2), 111-127.
Hocevar, D., Hagedorn, L. S., & Vogt, C. M. (2007). A social cognitive construct validation: Determining women's and men's success in engineering programs. *The Journal of Higher Education, 78*(3), 337-364.
Jamieson, L. H., & Lohmann, J. R. (2009). *Creating a Culture for Scholarly and Systematic Innovation in Engineering Education: Ensuring U.S. Engineering has the right people with the right talent for a global society*. Retrieved from Washington, D.C: American Society of Engineering Education:
Johri, A., & Olds, B. M. (2014). *Cambridge handbook of engineering education research*: Cambridge University Press.
Jones, B. D. (2009). Motivating Students to Engage in Learning: The MUSIC Model of Academic Motivation. *International Journal of Teaching and Learning in Higher Education, 21*, 272-285.
Jones, B. D. (2015). *Motivating students by design: Practical strategies for professors*. Charleston, SC: CreateSpace.
Jones, B. D. (2016). The MUSIC Model of Academic Motivation. Retrieved from www.themusicmodel.com
Jones, B. D., Li, M., & Cruz, J. M. (2017). A Cross-Cultural Validation of the MUSIC® Model of Academic Motivation Inventory: Evidence from Chinese- and Spanish-Speaking University Students. *International Journal of Educational Psychology, 6*(1), 25-44. doi:10.17583/ijep.2017.2357
Jones, B. D., & Wilkins, J. L. (2013). Testing the MUSIC Model of Academic Motivation through confirmatory factor analysis. *Educational Psychology, 33*(4), 482-503. doi:10.1080/01443410.2013.785044
Kong, Q.-P., Wong, N.-Y., & Lam, C.-C. (2003). Student engagement in mathematics: Development of instrument and validation of construct. *Mathematics Education Research Journal, 15*(1), 4-21.
Kuh, G. D. (2009). The national survey of student engagement: Conceptual and empirical foundations. *New directions for institutional research, 2009*(141), 5-20.
Laird, T. F. N., & Shoup, R. (2005). Measuring Deep Approaches to Learning Using the National Survey of Student Engagement. *Annual Meeting of the Association for Institutional Research*, 1-28.
Lattuca, L. R., & Stark, J. S. (2011). *Shaping the college curriculum: Academic plans in context*: John Wiley & Sons.
Litzinger, T., Lattuca, L. R., Hadgraft, R., & Newstetter, W. (2011). Engineering Education and the Development of Expertise. *Journal of Engineering Education, 100*(1), 123-150. doi:10.1002/j.2168-9830.2011.tb00006.x
Ormrod, J. E., & Jones, B. (2015). *Essentials of educational psychology: Big ideas to guide effective teaching* (4th ed.). Upper Saddle River, NJ: Pearson.
Robbins, S. B., Lauver, K., Le, H., Davis, D., Langley, R., & Carlstrom, A. (2004). Do psychosocial and study skill factors predict college outcomes? A meta-analysis. *Psychological bulletin, 130*(2), 261.
Singleton Jr., R. A., Straits, B. C., & Straits, M. M. (1993). *Approaches to social research (2nd ed.).* New York, NY: Oxford University Press.
Smith, K. A., Sheppard, S. D., Johnson, D. W., & Johnson, R. T. (2005). Pedagogies of engagement: Classroom-based practices. *Journal of Engineering Education, 94*, 87-101.
Sosniak, L. A. (1990). The tortoise, the hare, and the development of talent. *Encouraging the development of exceptional skills and talents*, 149-164.
Wang, M.-T., & Eccles, J. S. (2013). School context, achievement motivation, and academic engagement: A longitudinal study of school engagement using a multidimensional perspective. *Learning and Instruction, 28*, 12-23.
Wang, M.-T., Fredricks, J. A., Ye, F., Hofkens, T. L., & Linn, J. S. (2016). The Math and Science Engagement Scales: Scale development, validation, and psychometric properties. *Learning and Instruction*.
Wigfield, A., & Eccles, J. S. (2000). Expectancy–value theory of achievement motivation. *Contemporary educational psychology, 25*(1), 68-81.
Zimmerman, B. J. (1990). Self-regulated learning and academic achievement: An overview. *Educational psychologist, 25*(1), 3-17.

Methodology for assessing, evaluating and improving two student competencies. A case for effective communication and life-long learning in the Graduation Project course

Katherine Ortegon
Universidad Icesi, Cali, Colombia
kortegon@icesi.edu.co

Angelica Burbano
Universidad Icesi, Cali, Colombia
aburbano@icesi.edu.co

***Abstract**: The appropriate assessment and evaluation process of learning outcomes continue being a challenge for academic programs. The variety of performance indicators, techniques, and rubrics available might be overwhelming when starting the path to an accreditation process. In this study, a practical methodology for assessing, evaluating, and following up two student competencies in an undergraduate program of Industrial Engineering (IE) is presented. The competencies of effective communication and life-long learning are discussed as examples. The Graduation Project I and II courses were used for collection and analysis of student performance data on these two specific competencies. Results from the step-by-step implementation of the methodology as well as the underlying improvement process are explained in detail.*

Background

Entering an accreditation process for the first time creates a lot of uncertainty about how to start and about the right path towards creating a sustainable program improvement process integrated with course assessment. Since only four universities in Colombia are internationally accredited (ABET, 2016), there are plenty opportunities for other higher educational institutions (HEIs) of entering the accreditation arena. A variety of studies have focused their attention in the assessment process and rubrics (Allahverdi & Aldowaisan, 2015; Shaeiwitz, 1996; Walvoord, 2004) letting out of the scope the proper way to carry out the evaluation and follow-up process. Without a follow-up process, the development and attainment of learning outcomes could not be guaranteed.

Although there is no standard formula, this study constitutes a roadmap for beginners in the process of ensuring quality through the assessment, evaluation, and improvement of learning outcomes. It is expected this study will raise awareness about a practical methodology to create an improvement process that can be sustainable in the long term.

Research Questions

How to assess, evaluate, and improve student effective communication and life-long learning? How to make the proper follow up to improve students' performance?

Theoretical Framework

Assessment, evaluation, and improvement have their basis on the quality assurance (QA) concept. According to Harman and Meek (Harman & Meek, 2000), QA is defined as: "*systematic management and assessment procedures adopted by a higher educational institution or system to monitor performance and to ensure achievement of quality outputs program assessment and course assessment.*" This definition could be seen from the institutional culture perspective, the program perspective, or the course perspective. From the program perspective, the Input-Process-Output (IPO) framework suggested by Gray et.al. (2009), highlights that the most important activity for an academic cycle is the feedback loop or in other words, closing the assessment loop from outputs (learning outcomes or competencies) to inputs (student intake). However, this closing activity is typically misunderstood with the use of rubrics. Although rubrics are a successful tool for identifying the level of knowledge or competence acquired by students, rubrics are not the end of the closing loop for the program assessment. A complete assessment framework requires an underlying process including specific improvement actions across the formative learning process, stating the personnel responsible, allocating resources and the re-assessing students.

Methodology

The methodology proposed for assessing, evaluating and improving two competencies (i.e., effective communication and life-long learning) in undergraduate engineering students includes five steps and it is depicted in Figure 1. This methodology was implemented during the Graduation Project I and II courses (GPI and GPII). During the GPI course, students develop a research proposal and during GPII, the research project is developed. The Graduation project challenges students to integrate the knowledge acquired during previous semesters, to identify knowledge gaps, learn by themselves, present results, give oral presentations and deliver a written report at the end of the semester.

The GPI and GPII courses were reviewed and aligned which means that the syllabus was adjusted and the assessment plan and course card (i.e., containing the competencies to develop) for each course was developed. The courses in the proposed format were delivered and data on student performance was collected and analyzed by the *Assessment and Evaluation Committee*. A total of 32 students were assessed and evaluated in GPI and 37 students in GPII during 2015. This methodology could be also replicated for the assessment, evaluation, and improvement of any other student competencies.

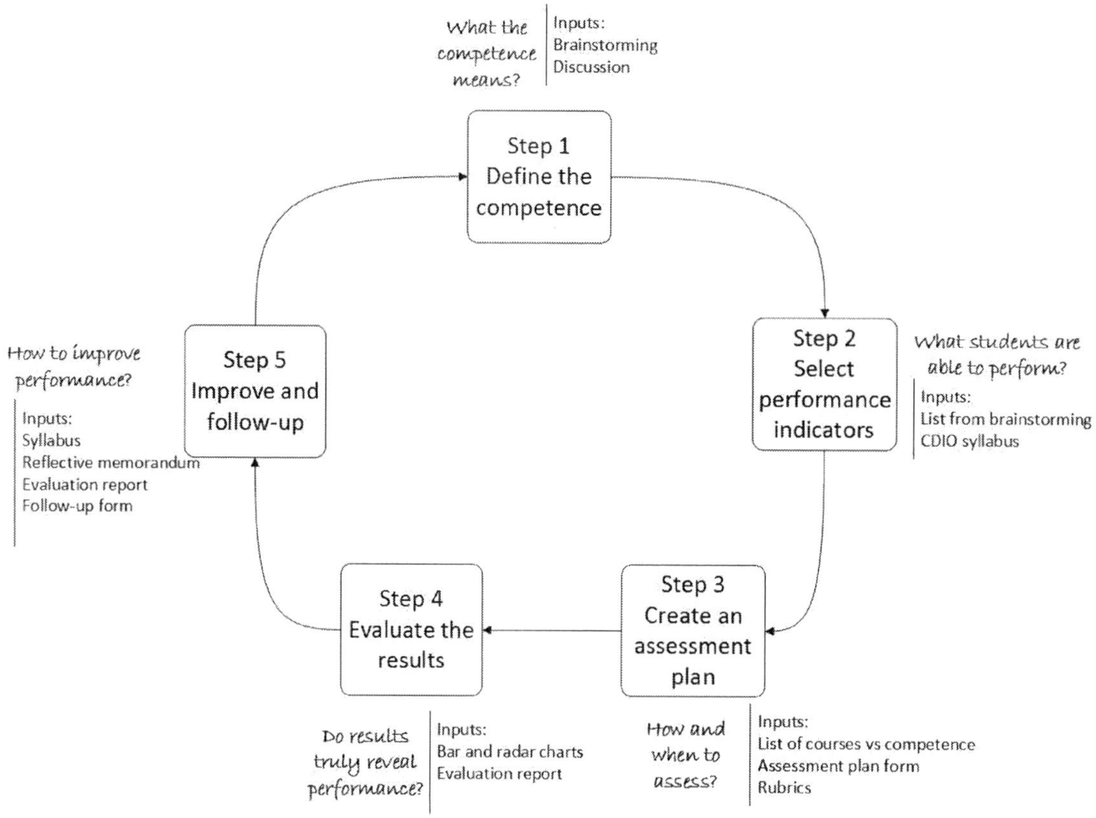

Figure 1. Methodology for the assessment, evaluation, and improvement of student competencies

Step 1. In this phase, it is important to question **what the competence means**. What could be an indicator or evidence that the student has acquired this competency? After a brainstorming and discussion within the assessment and evaluation team, a list of definitions and possible indicators of evidence should come up. In ABET terms, the a – k definitions could be explored.

Step 2. Once the competence is defined, it is critical to **select key performance indicators (PIs).** Performance indicators are the evidence of the competence acquired or in other words, what students are able to perform (ABET, 2015). The CDIO syllabus (conceive-design-implement-operate) could facilitate this process since it offers a detailed description of the knowledge, skills, and attitudes students must develop to perform the essential functions as an engineer (Edward F. Crawley, 2001). The proposed definition selected in step 1 and the indicators are then contrasted with the CDIO syllabus. This process will serve to validate if the team agrees in the definition and indicators or if those should be adjusted. It is recommended to have no more than four performance indicators per competence.

Step 3. Create an **assessment plan**. An assessment plan serves to identify which courses during the formative process contribute to the acquirement of the performance indicators selected, identify which course should be selected as a source of assessment, the method and mechanism to use, the rubric, date for data collection and responsible. It is important to clarify that the *assessment method* could be direct (i.e., measured directly through the student performance, for instance, a final course project) or indirect (i.e., measured by an external source, for instance, a survey). The *assessment mechanism* refers to the specific activity or instrument to be used to collect data (e.g., survey, final course project, video, questionnaire, national standard test). Rubrics should have a triangulation process to ensure that all members of the team agree on the evaluation levels.

Step 4. Develop the assessment plan and **evaluate the results**. In this phase, data is collected from students and the results are aggregated. Results should be presented in a graphical, simple, and clear manner so professors and students can understand easily the performance achieved. The evaluation report should contain the performance indicators assessed, the course source of assessment, the methodology carried out, the population, performance achieved with its possible causes, alternatives for improvement actions, and responsible. Lastly, it is important that the assessment committee reflects on the evaluation results obtained, in regards to the real performance observed. In other words, discuss if the method, mechanisms, and rubrics used truly reveal what students are able to do and what the acquirement of the competence actually is.

Step 5. In this phase, the team must question **how can the students improve their performance and progress towards the highest level in the rubric?** Improvement actions come from different sources. For instance, the results obtained from the evaluation process, the comparison between the results and the current syllabus, the reflective memorandum (i.e., a faculty self-assessment on course goals, class strategies, and achievements), from faculty involved during the formative process or in charge of the assessment process. The improvement actions are then grouped and displayed in the follow-up report. This report contains the performance indicators to be improved, the assessment results obtained, the cause analysis, actions, activities planned for improvement, responsible, the expected period for implementation and detailed follow-up deliverables. Once the improvement actions are implemented and the assessment cycle is repeated, the team must reflect on the effectiveness of the plan implemented in regards to the acquirement of the assessed competence. A plan-do-check-act (PDCA) approach is suggested in this step.

Findings

A practical methodology for the assessment, evaluation, and improvement of any student competence has been proposed. In this section, the step-by-step implementation and results are presented for two competencies as an example: effective communication and life-long learning in the context of the Graduation Project course delivered to senior Industrial Engineering (IE) undergraduate students.

Step 1. Define the competence. An Assessment and Evaluation committee formed by ten faculty members came up with the following definitions regarding the graduating profile defined for IE students. *Life-long learning* is the ability to learn independently identifying its own learning gaps and conducting a self-learning experience. *Effective communication* is the ability to communicate assertively in a written and oral manner, considering the structure, and style as well as recognizing other´s work.

Step 2. Select performance indicators (PIs). The list of initial PIs suggested by faculty and the final ones used for the assessment process are presented in Table 1. A discussion-review process was carried out in the light of the CDIO syllabus. According to the syllabus, the life-long learning skill includes motivation for self-learning, self-education in own´s learning style, and relationships with mentors. Likewise, oral communication skills include appropriate structure and style as well as non-verbal language and ability to argument properly (Edward F. Crawley, 2001).

Table 1 Performance Indicators for Life-long learning and Effective Communication

Performance Indicators Suggested	Competence	Performance Indicators Selected
1. To devise a strategy for searching and organizing information. 2. To devise a learning strategy using critical thinking. 3. Identify the effectiveness of the devised learning strategy.	Life-long learning	1. To recognize the importance of lifelong learning by assuming the learning experience as a self-learning process. 2. To search/identify, organize and analyse relevant information in an independent learning opportunity. 3. To demonstrate the ability to engage in lifelong learning by effectively conducting self-learning experiences that allow them to achieve the expected results.
1. To communicate properly in a written manner. 2. To communicate properly in an oral manner. 3. To build and defend arguments.	Effective Communication	1. To produce documents with the appropriate structure, grammar, and style. 2. To produce oral presentations with the appropriate structure, language, flow, and style. 3. To build and defend arguments. 4. To demonstrate proficiency in English (i.e., B2 level) by the time of graduation.

Step 3.Create an assessment plan. In this paper, only the assessment plans for the performance indicators related to the Graduation Project courses are displayed. In the case of life-long learning, PIs No. 1 and 3 were assessed through an employer survey during the professional internship. An example of an assessment plan is shown in Table 2

Table 2. Example of an Assessment Plan for PIs No. 2

	Performance Indicator (PI)	Source of assessment	Assessment Method	Date of data collection	Responsible
Life-long Learning	PI No. 2 To search/identify, organize and analyse relevant information in an independent learning opportunity.	GP I	Direct, Rubric applied to the research proposal final report.	Fall 2015	Professor GP I
Effective Communication	PI No. 2 To produce oral presentations with the appropriate structure, language, flow, and style.	GP II	Direct, Rubric applied to the final project oral presentation.	Spring 2015	Professor GP II

Step 4 and 5. Evaluate the results, improve and follow-up. The reflection process was focused on the appropriateness of the mechanisms used (i.e., final written report and oral presentation) to assess if students have *learned to learn* individually and are able to develop any oral presentation in a structured manner.

The research proposal final report delivered in GPI had been improved over time and perhaps, student advisor´s guidance and format refinement could have influenced the writing style during the class. Thus, a change in the instrument used was suggested and during the following cycle, students were asked to write one page summarizing their projects and the sources used for research. An example of the evaluation reports is presented in Figure 2 and Figure 3. The improvement plan summary including the basics of the PDCA approach is shown in Table 3.

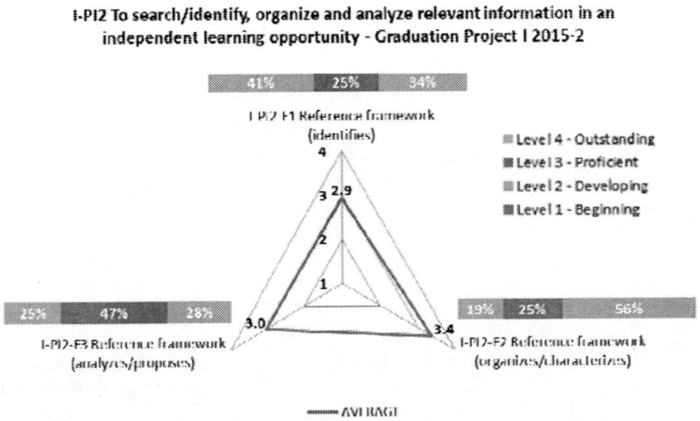

. Results: Groups performed at the developing level (41%) when asked to identify relevant information. Cause: students do not make an intensive use of Library resources.

Groups performed at an outstanding level (56%) when asked to organize information; this is one of the strengths. Groups were assessed at the proficient level (47%) when asked to analyse the gathered information. Cause: formative classes do not include activities that involve information gathering.

Improvement Actions: promote self-learning linked to the library initiative of information gathering and analysis skills. Strength the reference framework construction in Graduation Project I and Graduation Project II. Responsible: Professor GP I

Figure 2 Evaluation report for life-long learning, PI No, 2, Course GP I, population: 32 students

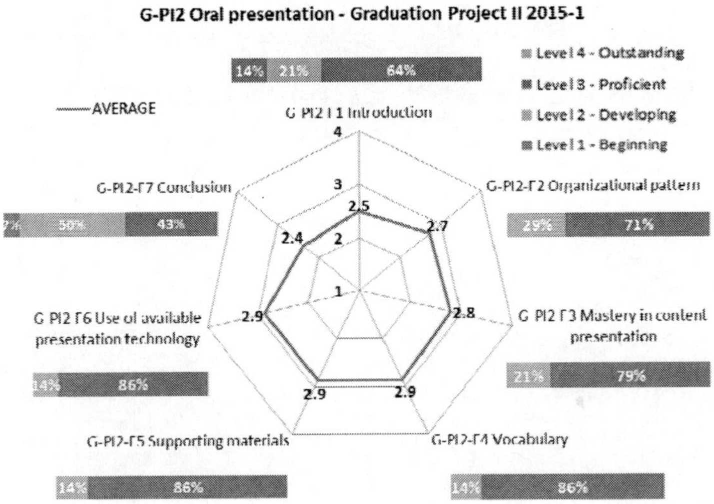

Results: Students were assessed at the outstanding and proficient level (64% and 43% respectively) in the aspects of Introduction and Conclusion associated with the oral presentation.

Most groups performed at a proficient level above 70% for the main aspects of the presentation. Cause: since it was accomplished, it is recognized the need of having consistent results over time.

Improvement Actions: During the formative process, at least three courses could be adjusted to expose students more frequently to oral presentations. Responsible: Professor GP I

Figure 3 Evaluation report for effective communication, PI No, 2, Course GP II, population: 37 students

Table 3. Example of Improvement and Follow-up Plan

	Aspect to be improved (PLAN)	Detailed Improvement Activities (DO)	Follow-up actions (CHECK)	Progress (%)
Life-long Learning PI No. 2, course GP I	Plan a use of library resources and to strength the reference framework construction.	1. Library resources are explored by the first time in the course of Introduction to Engineering. Their use should be reinforced in other formative courses. 2. The use of library resources during the reference framework construction should be accompanied by the development of online support material (e.g., videos or MOOCs). This will be discussed with the Library director.	1. Meeting with librarians in charge of training to identify potential courses to develop this competence and improvement of online resources. 2. Meeting with faculty involved in the formative process to identify course activities that could include searching, organizing, and analysing information. 3. Library training sessions should be included in the class schedule for the course of Graduation project I.	80%
Effective Communication PI No. 2, Course GP II	To strength oral communication skills during the formative process.	1. Review with faculty from formative courses where oral communication aspects might be strengthened. 2. Review the course syllabus to figure out which learning strategies or activities might be implemented (e.g., a training session with an expert in oral communication).	1. Meeting with faculty involved in the formative process to review the syllabus and include oral presentations across their courses (four courses selected) 2. A specialized workshop in oral presentations delivered by a field professional has been included in the class schedule of the course GPI and GPII. 3. A standard rubric to assess oral presentations has been created, validated, and distributed to faculty.	100%
Do this improvement worked? Yes, the problem was completely solved_____ Partially solved _____ Did not work _____ (ACT)				

To answer the original question: How to assess, evaluate, and improve student effective communication and life-long learning? A simple and practical step-by-step methodology is offered. This methodology could be followed by beginners in the practice of assessment reducing the time of implementation, and increasing the effectiveness of results. Two examples assessing effective communication and life-long learning in undergraduate Industrial Engineering students were presented using the proposed methodology. Even though the CDIO syllabus serves as a helpful guidance during the definition of learning outcomes or the selection of performance indicators, it is critical that each institution takes the time to conceptualize their own competencies considering faculty, students, and university resources, and culture.

As to the second question: How to improve and follow-up over time? A format based on the plan-do-check-act (PDCA) approach was proposed to facilitate the detail of activities, their tracking, and evaluation of effectiveness of the actions taken. The complete format can be provided upon request. Certainly, in assessing engineer's competencies there is much work left to be done. However, the proposed methodology could be helpful in facilitating the selection of performance indicators, presenting an evaluation report simple and concise, and more importantly, having the underlying process for continuous improvement in place.

Recommendations and Future research

Main questions raised within the assessment and evaluation team about how to reduce variability in results from one sample of students to another. Thus, additional research is recommended to expand the assessment and follow-up process so it can be comprehensive and allow a longitudinal assessment of student competencies. A longitudinal assessment will display not only the current state of a competence at the end-of-pipe but also, its evolution over time. As a result, improvement actions could be anticipated and higher performance levels might be achieved.

References

ABET. (2015). ABET. Retrieved from http://www.abet.org/

ABET. (2016). Abet-accredited programs. Retrieved from http://main.abet.org/aps/accreditedprogramsearch.aspx

Allahverdi, A., & Aldowaisan, T. (2015). Assessment of student outcomes of the Industrial and Management Systems Engineering programme at Kuwait University. Global Journal of Engineering Education, 17(3).

Edward F. Crawley. (2001). The CDIO Syllabus. Retrieved from http://cdio.org/files/CDIO_Syllabus_Report.pdf

Gray, P. J., & Patil, A. S. (2009). Engineering Education Quality Assurance. Springer. http://doi.org/10.1007/978-1-4419-0555-0

Harman, G., & Meek, V. L. (2000). Repositioning Quality Assurance and Accreditation in Australian Higher Education. Australia. Retrieved from https://www.researchgate.net/profile/V_Meek/publication/247391085_Repositioning_Quality_Assurance_and_Accreditation_in_Australian_Higher_Education/links/02e7e53b4d3f4e944c000000.pdf

Shaeiwitz, J. A. (1996). Outcomes Assessment in Engineering Education. Journal of Engineering Education, 85(3), 239–246. http://doi.org/10.1002/j.2168-9830.1996.tb00239.x

Walvoord, B. E. (2004). Assessment Clear and Simple: A Practical Guide for Institutions, Departments, and General Education (Jossey-Bass Higher and Adult Education) (2nd ed., Vol. (0)). San Francisco, CA: Jossey Bass.

Copyright statement

Work in Progress: Predictive analysis of conceptual understanding based on self-reported student engagement with resources in a blended engineering class

Taylor Prebel
Purdue University, West Lafayette, IN, United States
tprebel14@gmail.com

Nick A. Stites
Purdue University, West Lafayette, IN, United States
nstites@purdue.edu

Edward Berger
Purdue University, West Lafayette, IN, United States
bergere@purdue.edu

Jeffrey F. Rhoads
Purdue University, West Lafayette, IN, United Stated
jfrhoads@purdue.edu

Jennifer DeBoer
Purdue University, West Lafayette, IN, United Stated
deboerj@purdue.edu

***Abstract**: In an attempt to increase student retention and improve academic performance within science, technology, engineering, and mathematics disciplines, educators have begun to foster the use of active, blended, and collaborative pedagogical approaches within traditional classroom settings. Developed for a core mechanical engineering course, Freeform was among the first of these classroom environments to incorporate all three unique strategies and has consistently shown trends of increased student performance. This work in progress further investigates the Freeform environment to more specifically determine how it benefits student learning. It is among the first efforts to examine the extent to which engagement with online and collaborative resources predicts students' conceptual understanding of the course content. The results suggest that engagement with Freeform's online resources may not contribute to the students' overall conceptual understanding or gains in their conceptual understanding of dynamics. However, the results also suggest that peer collaboration may foster more engagement with online resources.*

Introduction

Higher-education institutions worldwide are working to address the continued high rates of student attrition in science, technology, engineering, and mathematics (STEM) majors. An estimated 60% of all students pursuing a STEM-related major decide to switch to a non-STEM field or end up not graduating altogether, making STEM courses a high priority for pedagogical adjustment (Christensen, Knezek, & Tyler-Wood, 2014). To improve the retention of students in STEM, many instructors are turning to active (Freeman et al., 2014), blended (Bernard, Borokhovski, Schmid, Tamim, & Abrami, 2014; Francis & Shannon, 2013; Means, Toyama, Murphy, & Baki, 2013), and collaborative (Johnson, Johnson, & Smith, 1991; Smith, Johnson, & Johnson, 1981) learning environments which have been shown to increase the students' engagement and performance. In addition, in the field of engineering, prominent works have called for an increased emphasis on students' conceptual understanding to better prepare them for the ill-defined problems they will face in professional practice (Litzinger, Lattuca, Hadgraft, & Newstetter, 2011; Sheppard, Macatangay, Colby, & Sullivan, 2009, Chapters 4-5). This work in progress study is positioned at the intersection of introducing novel pedagogical strategies into a traditional, lecture-based classroom environment and the emphasis on conceptual understanding. Specifically, this paper studies the extent to which students' engagement with online and collaborative resources predicts their conceptual understanding of material taught in an active, blended, and collaborative learning environment. We anticipate that the results of this research will add to the growing body of evidence documenting the benefits of implementing non-traditional educational strategies into a university-level STEM course.

Freeform: An Active, Blended, and Collaborative Learning Environment

In 2008, three professors at Purdue University developed an active, blended, and collaborative classroom environment, which they called *Freeform*, with the goals of accommodating the diverse learning styles of engineering students, improving student academic achievement, and increasing student retention within the course (Rhoads, Nauman, Holloway, & Krousgrill, 2014). *Freeform* was developed for two core, introductory-level mechanical engineering courses (Statics and Dynamics) at Purdue University, but has since expanded to five different institutions worldwide. Dynamics is the most mature instantiation of *Freeform* and will be the focus in this paper.

Five fundamental elements form the *Freeform* core: student-centered in-class instruction, a custom-written textbook (known as a "lecturebook", which the developers describe as a hybrid between a workbook and a textbook), an interactive online course discussion forum (referred to as the course blog), extensive multimedia content (such as online example and homework-solution videos), and refined student assessment tools, including an abbreviated Dynamics Concept Inventory (Rhoads et al., 2014; Stites et al., 2016). As a part of *Freeform*, instructors are encouraged to facilitate student collaboration in class (e.g., through group quizzes) and online (e.g., the course blog). Most *Freeform* instructors choose to solve several example problems from the lecturebook during class, whereas video solutions for these problems can be found on the course website. In addition, the homework problems, interactive simulations, and videos that demonstrate various concepts covered during class are also available on the course website for the students to view at their own convenience.

One of the defining characteristics of the *Freeform* environment is its emphasis on the students' conceptual understanding of the material. Every chapter of the lecturebook has a series of short-answer or multiple choice questions that the students may use to assess their own conceptual understanding. Quizzes often consist of multiple-choice conceptual questions, and typically one third of the content on each of the four exams (three midterm exams and the final exam) is devoted to assessing conceptual knowledge. Overall, *Freeform* is a flexible learning environment with extensive resources to scaffold and enhance the students' conceptual understanding and problem-solving abilities, with the overarching goal of improving student success and persistence in engineering.

Since the inception of *Freeform*, student performance in Dynamics has markedly increased. Prior to *Freeform*, the rate of students who earned a D, F, or withdrew from Dynamics (on an A-B-C-D-F grading scale), also known as the "DFW" rate, was above 20% (Rhoads et al., 2014). Most students who withdrew from the course or earned an F had to retake the class because Dynamics is a required class for the majority of those enrolled. Students who earned a D in the class likely lacked sufficient understanding of the material on which future courses build. Since implementing *Freeform*, the DFW rate has steadily decreased to approximately 10% (DeBoer et al., 2016; Stites et al., 2016). Previous research investigating this trend suggested that a student's incoming grade point average (GPA, which is measured on a 4.0-scale), international status, grade in Statics (which is a pre-requisite for Dynamics), year since *Freeform* inception (a proxy measure for the quantity and quality of *Freeform* resources), and instructor experience in the *Freeform* environment were significant predictors of overall student performance (DeBoer et al., 2016; Stites et al., 2017). Among these studies, none have begun to analyze the impact of the students' engagement with specific components of *Freeform*—namely the blended and collaborative resources—on student learning or performance in the class. Furthermore, no studies have used the students' conceptual understanding as the outcome variable; only the students' overall grade in Dynamics has been used as the outcome. Therefore, this work contributes to the body of knowledge for active, blended, and collaborative learning environments by examining the relationship between the students' usage of resources and their conceptual understanding of dynamics.

Conceptual Framework

Freeform is founded on a combination of findings from the literature and multiple theoretical frameworks supporting the benefits of implementing active, blended, and collaborative pedagogies. Research has highlighted the advantages of active and collaborative learning for decades now (e.g., Smith et al., 1981). More recently, Freeman et al. (2014) conducted a meta-analysis of the literature on active learning (including a substantial number of articles about collaborative learning) and concluded that students enrolled in an active classroom were 1.5 times more likely to pass when compared to students enrolled in a passive classroom. These results corroborated earlier claims that collaborative learning enhances knowledge construction (Jeong and Chi, 2007).

A blended course, as defined by Allen and Seaman (2013), is when a "substantial portion of the content is delivered online" (p. 7). A meta-analysis of blended learning in K-12 and higher education (including graduate and professional programs) found that students in blended courses significantly outperform those in face-to-face (only) courses (Means et al., 2013). A similar meta-analysis, focused solely on higher education, also concluded that students in blended classes have significantly higher achievement outcomes than those in classes without online components (Bernard et al., 2014). In the field of engineering specifically, Francis and Shannon (2013) found that the performance of architectural engineering students was higher for those who engaged with online resources. These prior findings illustrate the advantages of blended instruction and directly inform our hypothesis that the students' engagement with *Freeform* resources positively impacts their conceptual understanding of dynamics.

Research Question

This work in progress study is the first to examine how student engagement with online videos and a course blog relates to conceptual learning outcomes in a second-year, undergraduate engineering course. Specifically, our research question is: to what extent does engagement with online videos and a collaborative peer-to-peer discussion forum within the *Freeform* environment predict students' conceptual understanding of dynamics? We hypothesize that student engagement with these online resources is a significant positive predictor for attained conceptual understanding, which would provide further support for implementing blended and collaborative learning strategies in core engineering classes.

Methods

Participants

The data for this study were collected from students enrolled in Dynamics at Purdue University during the semesters of Fall 2015, Spring 2016, and Fall 2016. Of the 602 students enrolled in Dynamics during the three semesters, only the data from students who had completed an optional post-course survey were included in the final analyses of this study. The final sample consisted of 163 students (77% male and 23% female) of which 74% were domestic students (~60% white, ~7% Asian, and ~8% underrepresented minorities/other) and 26% were international students (any student with non-US citizenship).

Measures

During the final weeks of the semester, the students enrolled in Dynamics were asked to complete an online survey, which included questions about their study behaviors (e.g., group collaboration), their engagement with the course blog and online solution videos, as well as their attitudes towards and perceptions of the course. Four of the original fifty survey questioned were analyzed in this preliminary study. Due to the extensive length of the survey, only students completing 50% of the analyzed questions were included in the final sample. The four survey questions of interest included one study behavior question and three resource engagement questions. To measure in-person, collaborative interactions, students were asked to report the average number of hours they spent studying with at least one other student each week. This study-behavior metric was used as a control variable because students have told us during post-course interviews that their study group would utilize the online resources when working on homework assignments. In this study, we wanted to focus on the individualized usage of online resources rather than the resource usage that occurs during group collaboration.

Engagement with the course blog and the online solution videos was determined by self-reported frequency of use, with the responses ranging from "never" to "at least once per day." The students reporting engagement with the resources 1-2 times per week or more were categorized as "frequent" users, whereas the students reporting usage of less than 1-2 times per week were categorized as "infrequent" users.

To measure conceptual understanding, an abbreviated, 12-question version of the Dynamics Concept Inventory was administered to all consenting students at the beginning and end of the semester (Gray et al., 2005; Stites et al., 2016). The abbreviated Dynamics Concept Inventory (aDCI) is made up of multiple-choice questions that require minimal calculations (if any) and targets the assessment of the students' conceptual understanding of 12 subtopics of dynamics (one question per subtopic). The students' scores on the aDCI post-test (M = 64%, SD = 19%) and their normalized gain (M = 0.41, SD = 0.27) were used as measures of conceptual understanding in this study. The aDCI was scored out of 100%, while the normalized gain (G) for each student ranged from zero to one and was calculated by the following established metric (Coletta, Phillips, & Steinert, 2007):

$$G = \frac{(postscore\ \%) - (prescore\ \%)}{100 - (prescore\ \%)},$$

which is an individualized version of the normalized gain popularized by Hake (1998). The post-test score measures the students' conceptual understanding at the end of the course, and the normalized gain accounts for the fact that some students may enter the course with significant prior knowledge of dynamics. Therefore, the normalized gain attempts to measure how much conceptual knowledge the students gain specifically from Dynamics.

A Kolmogorov-Smirnov test for the post-test scores (p = 0.999) and the normalized gains (p = 0.835) suggested that the students who completed the survey and those who did not take the survey were likely to be derived from the same population. Additionally, the proportion of students in each of the ethnicity and gender categories were compared for the students who did and did not complete the survey, and no difference in proportions exceeded 6%. Therefore, the performance and demographic comparisons suggested that the sample of students who completed the survey were representative of the population of students who enrolled in Dynamics at Purdue University, which minimizes the bias in our results.

Statistical Analysis

A sequential linear regression analysis was conducted to determine if engagement with the course blog or online solution videos significantly predicted conceptual understanding of dynamics. The correlations between the continuous variables were used to check for collinearity. The predictor variables included in base models for the aDCI post-test scores (Model 1) and the normalized gains (Model 3) were demographic data (gender and ethnicity), prior academic performance (aDCI pre-test scores, Statics grade, and cumulative GPA at the start of the academic period in which they enrolled in Dynamics), and study behaviors (the number of hours studying in a group). The full models (Models 2 and 4) for each outcome variable added the self-reported number of videos watched and the frequencies of using the course blog and online solution videos. All variables except gender, ethnicity, and self-reported frequencies were considered continuous, and all numeric predictors were centered on their mean for the regression analysis. Multiple imputation (m = 5) using chained equations (Rubin, 1987) was used to estimate any missing values.

Results

The results of the correlations, as shown in Table 1, suggest that there is a positive correlation between aDCI post-scores and both Statics grade and incoming GPA. This may suggest that students who have a better academic history are more likely to score higher on the aDCI post-test. The post-test scores were also significantly correlated with both the pre-test scores and normalized gain (which is not surprising given that the post-test scores are part of the normalized gain equation). However, the correlation results suggest the lack of a significant relationship between aDCI pre-test scores and normalized gain. This may indicate that regardless of the level of conceptual understanding at the beginning of the semester, all students are equally likely to improve their conceptual understanding of the material throughout the semester. Additionally, both incoming GPA and Statics grade have only a slightly positive correlation with the pre-test scores. This relationship was expected due to the fact that majority of the aDCI assesses topics that are unfamiliar to students who have never taken Dynamics previously.

Table 1. Correlation Coefficients (Bottom Left) and Their Associated *p*-Values (Upper Right) for Numeric Variables in the Regression Analysis

	Post-test Score (%)	Pre-test Score (%)	Normalized Gain (*G*)	Incoming GPA	Statics Grade	Study Group Hours	Number of Videos Watched
Post-test Score (%)	-	<0.001	0.000	<0.001	<0.001	0.29	0.26
Pre-test Score (%)	**0.52**	-	0.13	0.01	0.009	0.23	0.40
Normalized Gain (*G*)	**0.83**	0.12	-	<0.001	<0.001	0.34	0.11
Incoming GPA	**0.47**	**0.21**	**0.47**	-	0.000	0.80	0.50
Statics Grade	**0.44**	**0.21**	**0.46**	**0.75**	-	0.12	0.19
Study Group Hours	-0.09	-0.10	-0.08	-0.02	-0.129	-	<0.001
Number of Videos Watched	-0.09	-0.07	-0.13	-0.05	-0.11	**0.53**	-

Note: Incoming GPA and Statics Grade were measured on a scale of 0.0 – 4.0.

A positive relationship exists between the average weekly hours spent studying with a group and the number of videos watched. This may suggest that students choose to utilize the online resources when studying with other students. Although these variables are significantly correlated to one another, both group study hours and the number of online videos watched are not significantly related to post-test score or normalized gain.

Finally, the students' incoming GPAs and their Statics grades are highly correlated. This is partially due to the part-whole relationship between the two variables (the Statics grade contributes to the overall GPA). To reduce variance inflation due to multicollinearity, we only included incoming GPA in our regression model.

Table 2. Summary of Sequential Linear Regression Results Predicting aDCI Post-test Score and Normalized Gain

	Post-test Scores (%)		Normalized Gain (G)	
	Model 1	Model 2	Model 3	Model 4
	B (SE)	B (SE)	B (SE)	B (SE)
Gender: Female	**-7.35**** (2.58)	**-6.95**** (2.59)	**-0.12**** (0.05)	**-0.11*** (0.05)
Ethnicity:				
Domestic: Asian	-7.47 (4.36)	-7.76 (4.37)	-0.07 (0.08)	-0.08 (0.08)
Domestic: URM/Other	-4.93 (4.05)	-5.31 (4.13)	-0.06 (0.07)	-0.07 (0.07)
International	0.69 (2.56)	0.89 (2.66)	0.02 (0.05)	0.02 (0.05)
Pre-test Scores (%)	**0.61***** (0.07)	**0.62***** (0.08)	0.00 (0.00)	0.00 (0.00)
Incoming GPA	**15.01***** (2.78)	**14.16***** (2.84)	**0.25***** (0.05)	**0.24***** (0.05)
Study Group Hours	-0.06 (0.19)	0.06 (0.24)	0.00 (0.00)	0.00 (0.00)
Number of Videos Watched		-0.07 (0.07)		0.00 (0.00)
Frequency of Video Usage		-0.03 (3.27)		-0.01 (0.06)
Frequency of Blog Usage		4.10 (2.73)		0.08 (0.05)
Constant	42.02*** (3.18)	40.56*** (3.24)	0.44*** (0.06)	0.41*** (0.06)
Observations	163	163	163	163
adj. R^2	0.49	0.50	0.18	0.19
ΔR^2	-	0.01 (p = 0.356)	-	0.01 (p = 0.368)

*$p < 0.05$. **$p < 0.01$. ***$p < 0.001$
Note: The comparison group for gender is males, and the comparison group for ethnicity is domestic, white students.

The results of the sequential linear regression analysis for both the post-test scores and normalized gains can be seen in Table 2. In Model 1, where the post-test score was the dependent variable, the initial set of variables accounted for 49% of the total variance in the post-test scores. In this model, gender, pre-test scores, and incoming GPA were significant predictors of aDCI post-test scores. Regarding the significance of the pre-test scores, it is reasonable to expect that students who come into the class with a higher conceptual understanding of dynamics also leave with a higher understanding. In Model 2, the group of engagement predictor variables (the total number of videos watched, frequency of online video usage, and frequency of blog usage) were not significant predictors of aDCI post-test scores, collectively or individually. Model 2 had a nonsignificant change in R^2, and none of the individual coefficients for the additional variables had p-values less than 0.05.

The variables included in Model 3 for the normalized gain accounted for 18% of the total variance. Unlike in Model 1 and 2 for the post-test scores, the aDCI pre-test scores did *not* have a significant effect on normalized gain. This might suggest that all of the students enrolled in Dynamics have an equal opportunity to increase their conceptual understanding (proportionally), no matter how they performed on the pre-test. In Model 4, the engagement variables explained an additional 1% of the total variance in the normalized gains, but this change in R^2 was not statistically significantly different from that of Model 3. Similar to the models for post-test scores, the variables for online video and course blog engagement, both individually and as a group, were not significant predictors of normalized gain.

Discussion

The results of the regression analyses for the aDCI post-test scores and normalized gains suggest that engagement with the *Freeform* course blog and online videos does not significantly predict the students' overall conceptual understanding or their proportional increase in conceptual understanding of dynamics. However, the students' self-reported average weekly hours spent studying with a group significantly correlated with the number of videos watched throughout the course. Therefore, the online resources may still be impacting the students' understanding of dynamics via group settings, but our model does not examine this relationship. The possibility of an interaction effect between study behaviors and resource usage deserves further investigation.

The regression coefficients suggest that for every 10% increase in a student's pre-test score, their post-test score increases by approximately 6%. The pre-test scores and incoming GPA, however, do not significantly explain the variance in Model 3 or Model 4 where normalized gain was the dependent variable. The insignificance of the pre-test scores influencing the normalized gains suggest that students with various levels of conceptual understanding at the beginning of the semester increase their conceptual understanding of dynamics proportionally. However, because the students with lower pre-test scores had more conceptual knowledge to gain, their absolute gain is higher than that of students with higher pre-test scores. In other words, the students who entered the class with a higher conceptual understanding of dynamics also exit Dynamics with a higher understanding, but the differences in conceptual understanding across all students have been reduced.

The results of all four models also indicate that, all other factors being equal, female students are predicted to have lower post-course conceptual understanding and lower gains as compared to males. Models 1 and 2 suggest that females score about 7% lower on their post-test aDCI than males, when the values for all other variables are held constant. Models 3 and 4 also estimate that females' normalized gains are approximately 11% lower than the normalized gains for similar males. The significance of gender as a predictor for conceptual understanding contrasts with the insignificance of gender in predicting a student's success in the class, as measured by the student not earning a D, F, or withdrawing from the course (DeBoer et al., 2016; Stites et al., 2017). Similarly, the significance of a student's residency status in this work contradicts results that have been published for predicting a student's overall success in the course. *As such, one of the main conclusions of this study is the need to further investigate the relationship between gender, residency status, resource usage, and study behaviors on conceptual understanding and overall grade in the course.*

Conclusion

This work in progress study is among the first to investigate how engagement with online and collaborative resources affects student conceptual understanding of material taught in an introductory-level mechanical engineering course. The results of the sequential linear regression analysis suggest that students' self-reported engagement with *Freeform*'s online videos and blog does not significantly predict their overall conceptual understanding of dynamics or their proportional gain of conceptual understanding.

Although the preliminary results of this study indicate that the collective effect of online-resource engagement on conceptual understanding of dynamics is not significant, there are several opportunities beyond those mentioned in the Discussion section to improve and expand on this work. For example, the predictor variables measuring engagement with the course blog and online videos were self-reported averages for the entire semester and may have been inaccurate reports of usage. Additionally, the regression models treated the aDCI post-test and normalized gain as continuous variables, but a similar analysis should treat the outcome variables as ordinal measures to see if the results change.

Future work may also focus on how engagement with these blended and collaborative resources affects the development of problem-solving skills. Although conceptual information makes up a sizable portion of the Dynamics curriculum, instructors in *Freeform* often devote the majority of class time to the development of students' problem-solving skills. Thus, it would be useful to investigate how resource engagement affects problem-solving skills.

In conclusion, this work in progress study indicates that the students' engagement with online and collaborative resources in the *Freeform* environment does not significantly predict their conceptual understanding in a core, undergraduate engineering class. However, the average weekly hours spent studying with a group and the number of videos watched are positively correlated which may indicate an interaction between the blended and collaborative resource utilization. Furthermore, this study identifies interesting paths for future work, including the effect of gender and residency status on the conceptual understanding of dynamics.

Acknowledgements

We would like to thank David Evenhouse and Craig Zywicki, as well as the many other faculty, staff, and students that make up the *Freeform* team. This project was funded by the National Science Foundation, grant number: DUE-1525671. Any opinions, findings, and conclusions or recommendations expressed in this material are those of the author(s) and do not necessarily reflect the views of the National Science Foundation.

References

Allen, I. E., & Seaman, J. (2013). *Changing course: Ten years of tracking online education in the United States*. Oakland, CA: Babson Survey Research Group and Quahog Research Group, LLC.

Bernard, R. M., Borokhovski, E., Schmid, R. F., Tamim, R. M., & Abrami, P. C. (2014). A meta-analysis of blended learning and technology use in higher education: From the general to the applied. *Journal of Computing in Higher Education*, *26*(1), 87-122. doi:10.1007/s12528-013-9077-3

Christensen, R., Knezek, G., & Tyler-Wood, T. (2014). Student perceptions of Science, Technology, Engineering and Mathematics (STEM) content and careers. *Computers in Human Behavior*, *34*(2014), 173-186. doi:10.1016/j.chb.2014.01.046

Coletta, V. P., Phillips, J. A., & Steinert, J. J. (2007). Interpreting force concept inventory scores: Normalized gain and SAT scores. *Physical Review Special Topics-Physics Education Research*, *3*(1), 10101-10106. doi:10.1103/PhysRevSTPER.3.010106

DeBoer, J., Stites, N., Berger, E. J., Rhoads, J. F., Krousgrill, C. M., Nelson, D. B., Zywicki, C., & Evenhouse, D. A. (2016). *Work in Progress: Rigorously Assessing the Anecdotal Evidence of Increased Student Persistence in an Active, Blended, and Collaborative Mechanical Engineering Environment*. Paper presented at ASEE Annual Conference & Exposition, New Orleans, LA. doi: 10.18260/p.27032

Francis, R., & Shannon, S. J. (2013). Engaging with blended learning to improve students' learning outcomes. *European Journal of Engineering Education*, *38*(4), 359-369. doi:10.1080/03043797.2013.766679

Freeman, S., Eddy, S. L., McDonough, M., Smith, M. K., Okoroafor, N., Jordt, H., & Wenderoth, M. P. (2014). Active learning increases student performance in science, engineering, and mathematics. *Proceedings of the National Academy of Sciences*, *111*(23), 8410-8415. doi:10.1073/pnas.1319030111

Gray, G. L., Costanzo, F., Evans, D., Cornwell, P., Self, B., & Lane, J. L. (2005). *The dynamics concept inventory assessment test: A progress report and some results introduction.* Paper presented at the ASEE Annual Conference & Exposition, Portland, OR. https://peer.asee.org/15378

Hake, R. R. (1998). Interactive-engagement versus traditional methods: A six-thousand-student survey of mechanics test data for introductory physics courses. *American Journal of Physics*, *66*(1), 64-74. doi:10.1119/1.18809

Jeong, H. & Chi, M.T.H. (2007). Knowledge convergence and collaborative learning. *Instructional Science*, *35*(4), 287 - 315. doi: 10.1007/s11251-006-9008-z

Johnson, D. W., Johnson, R. T., & Smith, K. A. (1991). Cooperative learning: Increasing college faculty instructional productivity. *ASHE-ERIC Higher Education Report No. 4.* Washington, D.C.: The George Washington University, School of Education and Human Development.

Litzinger, T., Lattuca, L. R., Hadgraft, R., & Newstetter, W. (2011). Engineering education and the development of expertise. *Journal of Engineering Education*, *100*(1), 123-150. doi:10.1002/j.2168-9830.2011.tb00006.x

Means, B., Toyama, Y., Murphy, R., & Baki, M. (2013). The effectiveness of online and blended learning: A meta-analysis of the empirical literature. Teachers College Record, *115*(3), 1-47.

Rhoads, J. F., Nauman, E., Holloway, B. M., Krousgrill, C. M. (2014), *The Purdue Mechanics Freeform Classroom: A New Approach to Engineering Mechanics Education* Paper presented at ASEE Annual Conference & Exposition, Indianapolis, Indiana. https://peer.asee.org/23174

Rubin, D. B. (1987). Multiple imputation for nonresponse in surveys. New York, NY: Wiley.

Sheppard, S. D., Macatangay, K., Colby, A., & Sullivan, W. M. (2009). *Educating engineers: Designing for the future of the field.* San Francisco, CA: Jossey-Bass.

Smith, K. A., Johnson, D. W., & Johnson, R. T. (1981). Structuring learning goals to meet the goals of engineering education. *Engineering Education*, *72*(3), 221-226.

Stites, N., Evenhouse, D. A., Tafur, M., Krousgrill, C. M., Zywicki, C., Zissimopoulos, A. N., Nelson, D. B., DeBoer, J., Rhoads, J. F., Berger, E. J. (2016), *Analyzing an Abbreviated Dynamics Concept Inventory and Its Role as an Instrument for Assessing Emergent Learning Pedagogies* Paper presented at ASEE Annual Conference & Exposition, New Orleans, LA. doi: 10.18260/p.26242.

Stites, N. A., Zywicki, C., Berger, E., Krousgrill, C., Rhoads, J. F., & DeBoer, J. (2017). *The impact of instructor experience on student success for a blended, undergraduate engineering class.* Paper presented at the AERA Annual Meeting, San Antonio, TX.

Student Self-Reported Use of Standards-Based Grading Feedback and Resources

Heidi A. Diefes-Dux
Purdue University, West Lafayette, IN, USA
hdiefes@purdue.edu

***Abstract**: Standards-based grading tightly links student assessment to course learning objectives, providing students with meaningful feedback. If students fail to access and use this feedback to guide their learning, the effort to construct an SBG system is for naught. In an end-of-semester survey, first-year engineering students in a course using SBG self-reported their access to and use of their feedback and related SBG resources to guide their work on assignments and prepare for exams. Students' reported access and use was as low as 25% at the start of the semester and improved around exam time, reaching a maximum of 73% mid-semester. Insights on students' engagement related to course management decisions are offered.*

Introduction

Summative score-based grading systems are the tradition in higher education and continue to be widely used and promoted internationally (Freeman & Lewis, 1998; Huba & Freed, 2000; Morgan, Dunn, Parry, & O'Reilly, 2004; Stevens & Levi, 2004; Suskie, 2004; Walvoord & Anderson, 2010). Within these systems, points are typically assigned or deducted based on a variety of performance indicators (e.g., correctness, completeness, and effort) without a direct link to explicit course learning objectives. The resulting grade provides students with insufficient feedback on their learning (Brookhart & Nitko, 2008), leaving students focused on their scores as an indicator of their learning rather than their progress towards meeting specific course learning objectives.

A standards-based grading (SBG) system, on the other hand, provides students with feedback explicitly on their achievement of course learning objectives. The system uses clearly articulated course learning objectives to guide course design and ground assessment of student learning (Marzano, 2010). This backward design approach entails clearly articulating learning objectives at the beginning of a course and linking assessments (e.g., rubrics) to the learning objectives (Post, 2014; Wiggins & McTighe, 1998), thus establishing greater validity, reliability, and fairness of grading (Muñoz & Guskey, 2015; Wiggins & McTighe, 1998). A well-designed SBG system provides actionable formative feedback to both students and instructors on performance with regards to stated learning objectives.

SBG is being adopted in engineering courses in the United States (e.g., Atwood, Siniawski, & Carberry, 2014, Post, 2014). In a pilot cross-institutional study, Carberry, Siniawski, and Dionisio (2012) found that SBG increased engineering students' self-efficacy and sophistication in epistemological beliefs. In another pilot cross-intuitional study, Vaishnav and Carberry (2017) found that engineering students viewed SBG as an effective form of grading. That said, one barrier to implementing SBG is students' lack of familiarity with SBG (Carberry, Siniawski, Atwood, & Diefes-Dux, 2016, Lee, Carberry, Atwood, Diefes-Dux, & Siniawski, 2017). This may lead to frustration for some students (Vaishnav & Carberry, 2017) that can manifest itself in students failing to engage with their feedback. SBG may provide directed feedback, but it will not help students focus on their learning if students do not review their feedback, are unable to interpret it, or cannot devise a plan to act on it. As SBG gains traction in undergraduate engineering education, instructors need to be aware of the student supports necessary for the learning benefits of SBG to be realized and for ensuring that the learning benefits are commensurate with the effort to design an SBG system.

Prior research has shown that first-year engineering students unfamiliar with SBG did not reflect deeply on the feedback provided on the first assignments of the semester (Diefes-Dux, 2016). What is not known is whether and how these same students used SBG feedback and related resources to guide their learning throughout the semester, particularly in preparation for exams. The question this paper investigates is whether, over the course of the semester, students improved their use of SBG resources and feedback. Insights into student behaviour within an SBG system will guide the design of courses and student instruction on the use of SBG resources and feedback for greater learning.

Methods

Setting and Participants

This study was conducted by the author in her single section (*n* = 85 initially enrolled) of a large required second-semester first-year engineering (FYE) course offered in Spring 2016 at a Midwest United States RU/VH institution. The overall course enrolment (N~1600) was divided among 15 sections (classes), each led by an instructor who follows a common course curriculum. This course emphasizes the development of skills associated with programming, data analysis and representation, and mathematical modelling. The computer tool students learn to use is MATLAB®, though some assignments are done for comparative purposes in Excel®. The course also supports students' development of teaming skills and professional habits. The use of SBG began in this course in 2010, with a number of aspects of its implementation being investigated over time (Hicks & Diefes-Dux, 2017; Hylton, & Diefes-Dux, 2016; Marbouti, Diefes-Dux, & Madhavan, 2016; Marbouti, 2017).

Learning Objectives and Assessment of Student Work

Each topic of the course has stated learning objectives (LOs) with multiple sub-LOs. For instance, the LO for the topic of selection structures (i.e., MATLAB's if-end) is: students will be able to create and troubleshoot a selection structure. The sub-LOs state that students will be able to: (1) convert between these selection structure representations: English, a flowchart, and code, (2) code a selection structure, (3) track execution of a selection structure, and (4) create test cases to evaluate a selection structure.

Students' complete problem sets, project milestones, and exams assessed using rubrics consisting of 5-10 items. All rubric items are LOs or sub-LOs (heretofore collectively called LOs). Each LO is assessed on a four-level achievement scale: fully achieved, partially achieved, underachieved, and no evidence. For instance, consider a problem in which a student is given a selection structure with three branches. To demonstrate full achievement of the sub-LO concerning the creation of test cases to evaluate a selection structure, the student must create three different and appropriate sets of variable assignments to test each branch of the structure.

As students work on a problem set or project milestone, they have access to a list of associated LOs. Students can and are encouraged to review their performance on each LO assessed after grading of an assignment has been complete. Similarly, students have access to a list of relevant LOs when preparing for exams. Students can review their performance on each LO assessed following grading of an exam. As all grading for this course is done through Blackboard® (online course management system) rubrics, students can review their performance on the LOs at any time.

Student Self-Report

During the semester, students were periodically asked to reflect on what they believed they had and had not learned, whether the learning-objective based assessment reflected their learning, and what actions they planned to take as a result of their feedback. Students' responses to the first two of these reflections were discussed by Diefes-Dux (2016). During the last week of class, students responded to a survey in which they were asked to reflect on their use of learning objective resources (5 items), their feedback (9 items), and their beliefs about various aspects of the SBG system (5 items). A four point Likert-scale (i.e. strongly agree to strongly disagree) with options of 'I do not recall' or 'No opinion' was used to rate the degree of agreement with each statement. Validity of the survey items was addressed through careful wording of the items, using language that was consistently and repeatedly used throughout the course documents and by the author in conversations with her students when referring to learning objectives (LOs), learning objective resources, and feedback. Reliability, in terms of consistent test-retest results, was checked by comparing students' responses to open-ended reflection prompts following the second and third exams and their responses to the end-of-semester survey items concerning their use of learning objective resources and their feedback around those exams.

Data Analysis

Sixty-three (63) students completed the survey. Pair-wise comparisons of proportions of students' agreement (strongly agree and agree) to the survey items were done with Pearson's Chi-squared test when appropriate. All other analyses of results focused on descriptive statistics, primarily percentages of students within each agree-disagree response category.

Results

Students were asked to consider whether they referred to the LO lists while completing problem sets or preparing for exams. The results can be seen in Table 1. Students more often agreed (strongly agree and agree) that they used the LO lists to prepare for exams (65-73%) than for completing the problem sets (25-33%). This was significant for each paired comparison between each of the exams and early problem sets ($p < 0.0001$) and between each of the exams and the later problem sets ($p \leq 0.0004$). There was no significant differences in student agreement on whether they referred to the LO lists while completing early and later problem sets nor while preparing for Exams 1, 2, and 3.

Table 1: Students' self-reported use of learning objective lists (*n*=63)

..., I referred to the LO lists posted on Bb (n=63)	Strongly Agree	Agree	Disagree	Strong Disagree	I Do Not Recall
While completing problem sets early in the semester (PS 1-4)	5	11	26	16	5
While completing problem sets later in the semester (PS 5-8)	3	18	25	17	0
While preparing for Exam 1	9	32	10	9	3
While preparing for Exam 2	14	32	8	6	3
While preparing for Exam 3	13	28	15	7	0

Students were asked whether they reviewed their LO feedback soon after it was released. Only 40% of students agreed that they looked at their early problem set feedback soon after it was graded (Table 2). More (60-65%) students agree that they looked at their feedback on later problem sets and Exams 1 and 2 soon after grading. Fewer (51%) students agreed that they reviewed their feedback on Exam 3. There was a significant difference between students reviewing their LO feedback soon after grading for the early problem sets and Exam 2 ($p = 0.0043$) and between early and later problem sets ($p = 0.0075$). There were no significant differences in the responses to any of the other items at a level of $p \leq 0.01$.

Forty percent of students agreed that they used their feedback on a previous problem sets to guide their work on the next problem set (Table 3). They were split between agreeing and disagreeing on whether they used their feedback to prepare for Exam 1. They agreed to a higher extent that they used their performance feedback to prepare for Exams 2 and 3 (68% and 62%, respectively). There was a significant difference between students use of their LO feedback as a guide on problem sets versus Exam 2 (p = 0.0013). There were no significant differences in the responses to any of the other items at a level of $p \leq 0.01$.

Table 2: Students' self-reported review of learning objective feedback (*n*=63)

..., I reviewed my performance on the LOs ...	Strongly Agree	Agree	Disagree	Strong Disagree	I Do Not Recall
Early in the semester, ... soon after feedback on a problem set was released (PS 1-4 Blackboard rubric feedback)	9	16	29	7	2
Later in the semester, ... soon after feedback on a problem set was released (PS 5-8 Blackboard rubric feedback)	11	29	19	4	0
associated with Exam 1 soon after feedback was released (Exam 1 Bb rubric feedback).	9	29	17	3	5
associated with Exam 2 soon after feedback was released (Exam 2 Bb rubric feedback).	13	28	14	2	6
associated with Exam 3 soon after feedback was released (Exam 3 Bb rubric feedback).	9	23	26	4	1

Table 3: Students' self-reported use of their feedback (*n*=63)

I used the assessment of my performance ...	Strongly Agree	Agree	Disagree	Strong Disagree	I Do Not Recall
on one problem set's LOs to guide my work on the next problem set.	6	19	31	2	5
on the LOs to guide my preparations for Exam 1	3	28	27	0	5
on the LOs to guide my preparations for Exam 2	9	34	17	2	1
on the LOs to guide my preparations for Exam 3	12	27	21	2	1

Figure 1 arranges the students' responses to the items in Tables 1 through 4 in an approximate timeline. The single item concerning students use of their performance on one problem set's LOs to guide work on the next problem set is repeated in each problem set (PS) category as a reminder that this type of event was ongoing across problem sets. In this representation, it can be seen more readily that students' self-reported engagement with SBG resources and feedback was low at the start of the semester (as low as 25%), and their highest engagement centred on the exams, particularly Exam 2 (at 73%).

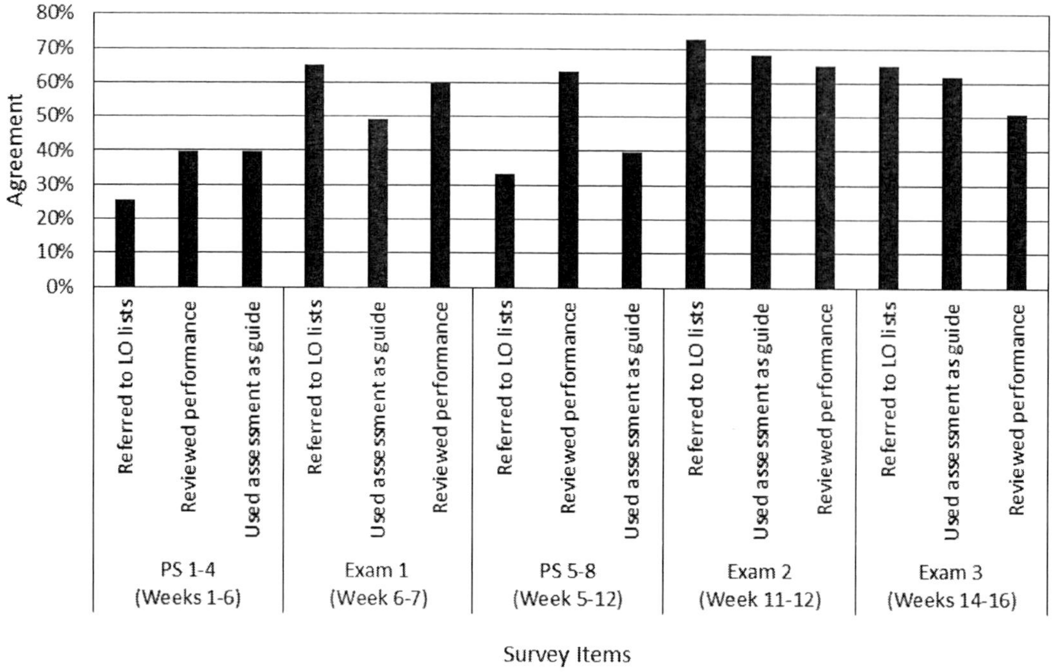

Figure 1: Students' responses to survey items arranged approximately in time order of events

Table 4 shows that students tended to agree that having a list of LOs available with an assignment made them aware of what they needed to learn (73%), though they were less likely to agree that the lists guided their study habits (54%). Similarly, students agreed that having their work assessed based on learning objectives made them aware of what they needed to learn (65%), though they were less likely to agree that the learning objective-based assessment of their work guided their study habits (56%). Despite the generally favourable beliefs about what SBG affords, only 40% of students believed that all courses should use learning objectives as the basis for grading.

Table 4: Students' self-reported beliefs about the SBG system (*n*=63)

I believe that ...	Strongly Agree	Agree	Disagree	Strong Disagree	No Opinion
having access to a list of LOs associated with an assignment makes me aware of what I need to learn.	7	39	10	0	7
that having access to a list of LOs associated with an assignment guides my study habits.	7	27	17	5	7
having my work assessed based on LOs keeps me aware of what I have and have not learned.	9	32	13	4	5
having my work assessed based on LOs guides my study habits.	6	29	20	5	3
all courses should use LOs as the basis for assessment of my work.	4	21	21	8	9

Discussion

The research question analysed in this study is: Did students improve their use of SBG resources and feedback over the course of the semester? The results indicate that towards the middle of the semester, there was a shift in students' use of the LO lists and their SBG feedback, particularly when preparing for exams; however, a large fraction of the students reported not paying much attention to the LO lists or their SBG feedback. In looking at these results, we have to think carefully about the manner in which SBG is being implemented in the course and how the implementation may be contributing to students' self-reported behavior.

There are issues around the ease of students' access to their SBG feedback via Blackboard. It is time consuming to look up feedback on Blackboard, requiring a multistep drill down into the rubric for each problem set and exam. This alone is a deterrent to accessing feedback. Further, Blackboard provides no means for tracking learning objective performance across assignments. These hurdles make tracking performance improvement, or lack thereof, difficult for students to follow across assignments and exams. This is another reason why course learning management systems, not just Blackboard, have been noted as a barrier to SBG implementation (Carberry, Siniawski, Atwood, & Diefes-Dux, 2016).

Beyond Blackboard issues, the size of this course constrains some management decisions around the implementation of SBG. Grading is done by undergraduate teaching assistants (UGTAs) and overseen by graduate teaching assistants (GTAs). In Spring 2016, the UGTAs and supervising GTAs, were given one week to grade a problem set and release the grades to the students; this was intended to give the UGTAs time to manage their own coursework alongside their grading responsibilities. That decision meant that one assignment was being returned at the same time as another was being submitted, leaving no time for students to review their feedback before the next assignment was due. The result may have been that many students submitted an assignment and forgot about it until a review of the feedback was useful, like prior to an upcoming exam. This suggest that there was no real need to look at the SBG feedback soon after it was returned.

In a study of UGTA feedback in Spring 2016, it was found that grading, while guided by a detailed solution and rubric, was inconsistent and that written feedback was often missing or unclear on learning objectives marked as partially achieved, underachieved, and no evidence (Hicks & Diefes-Dux, 2017). Low quality feedback may have resulted in students forgoing reviewing their feedback or using it as a guide due to a lack of faith in the grades. This contradicts how most students reported the assessment reflecting their learning in the beginning of the semester (Diefes-Dux, 2016) and suggests that this belief may have deteriorated over the course of the semester.

Some section level decisions by the author may have resulted in the general upward trend in student engagement with SBG. First, she did ask students to reflect on their learning with respect to the LOs periodically across the semester. So, students' attention was constantly drawn to the LOs. Second, the large positive shift around Exam 2 in favor of students using the LO lists and their SBG feedback may be the result of better access to LO performance data. Based on the students' ongoing reflections, the author was aware that her students were not critically reviewing their feedback as much as she had intended. She attempted to make the students more aware of their performance on the LOs prior to Exams 2 and 3 by printing individual reports of student performance and handing them out in the week prior to exams and having students reflect on their reports. It cannot be overlooked, however, that poor performance on Exam 1 may have also inspired more attention to feedback and the use of LO lists.

It is noted that the limitations of this study reduce generalizability of the findings. One limitation is that the author's section met at 7:30 am twice a week. Early morning classes do negatively impact student performance (Marbouti & Diefes-Dux, 2014) and likely their overall level of engagement. Sections of the course offered later in the day might engage with the

SBG feedback and resources quite differently. A second limitation is that the students were first-year engineering students with unknown prior exposure to SBG. While this group provides insight into how students with perhaps little SBG exposure might engage with SBG, it does not shed light on how students with prior and repeated exposure to SBG might behave.

The overall findings, despite the limitations of the study, do provide direction for next steps to encourage greater student engagement with SBG resources and feedback. First, the lack of ease of access to SBG feedback inherent to the design of the course management system needs to be overcome. One way of doing this is to raise the value of the SBG feedback. In Spring 2017, a weekly TA training with the problem set rubric items (i.e., the LOs) was launched. This training has the TAs practice assessing two pieces of sample student work for each LO graded and demonstrates to the TAs the written feedback that should accompany all non-proficient assessments. It is hoped that more consistent grading with more written feedback will draw students into reviewing their feedback even though it is somewhat tedious to access. A long term approach to the course management system problem is to initiate open conversations with the system providers to help them better understand assessment strategies and user needs.

At the end of the survey, there was a question asking the students, If you could have the LO lists associated with problem sets or exams provided to you differently, what would it look like and why? A number of students discussed how the language used in the learning objectives themselves was vague and that more description would make them better understood. With this in mind, in Spring 2017, a detailed list of evidence pertaining to what proficiency entails for each learning objective was provided to students. For instance, the proficiency evidence for the learning objective about creating test cases to evaluate a selection structure entails: (1) creating a thorough set of test cases to test all possible paths in the selection structure, (2) using English to completely describe each test and the intended path through the selection structure, (3) listing the test values in an appropriate format, and (4) selecting test values that are consistent with the test description. The LO lists with their proficiency evidence is made available with every assignment and prior to each exam. This evidence is now what is looked for and what given feedback addresses. Further, identifying the evidence in student work is the focus of TA training. It is hoped that the more detailed LO lists will better encourage students to use them as a guide when working on assignments and preparing for exams.

Three additional actions were also taken in Spring 2017 to encourage students to review their feedback in a more timely fashion. First, the turnaround time on grading has been shortened. Students are provided with a two-day window between receiving feedback on one assignment and the due date of the next assignment. This gives them time to review their feedback and use that feedback to guide improvements on their next submission. Second, the assessment of a greater number of LOs are repeated on successive problem sets to provide ongoing feedback on particularly difficult LOs. This repetition may encourage students to review their feedback before the next assignment submission. Finally, in a few pilot sections, weekly reflection on the LOs most recently assessed or practiced is being used to help students attend to their progress on the LOs.

Conclusion

First-year engineering students enrolled in a course using a SBG system were assessed to better understand their use of SBG resources and feedback. Students self-reported greater use of both components across the semester. The percentage of students engaged in the use of SBG related course elements was found to be low at the start of the course. While self-reported usage improved, particularly around exams, a large fraction of the students self-reported not being engaged at various points across the semester. Students' lack of familiarity with SBG could be a contributing factor to low engagement, though various course management decisions may factor in more significantly.

It is recommended that instructors identify and reduce course management decisions that may present barriers to students fully utilizing SBG resources and feedback to benefit their learning. It is also recommended that instructors consider ways to keep students attention on their learning progress, such as through ongoing reflection.

References

Atwood, S., Siniawski, M., & Carberry, A. (2014). Using standards-based grading to effectively assess project-based design courses. *Proceedings of the 121st American Society for Engineering Education Annual Conference & Exposition*, Indianapolis, IN: ASEE.

Brookhart, S. M., & Nitko. A. J. (2008). *Assessment and grading in classrooms*. Upper Saddle River, NJ: Pearson Education.

Carberry, A. R., Siniawski, M., & Dionisio, J. (2012). Standards-based grading: Preliminary studies to quantify changes in student affective and cognitive behaviors. *Proceedings of the 24nd ASEE/IEEE Frontiers in Education Conference*, Seattle, WA: ASEE/IEEE.

Diefes-Dux, H. A. (2016). Student reflections on standards-based graded. *Proceedings of the 46th ASEE/IEEE Frontiers in Education Conference,* Erie, PA: ASEE/IEEE.

Freeman, R., & Lewis, R. (1998). *Planning and implementing assessment*. London, U.K.: Kogan Page.

Carberry, A., Siniawski, M., Atwood, S., & Diefes-Dux, H. A. (2016). Best practices for using standards-based grading in engineering courses. *Proceedings of the 123rd American Society for Engineering Education Annual Conference and Exposition, New Orleans*, LA: ASEE.

Huba, M. E., & Freed, J.E. (2000). Learner-centered assessment on college campuses: Shifting the focus from teaching to learning. Needham Heights, MA: Allyn & Bacon.

Hicks, N. M., & Diefes-Dux, H. A. (2017). Grader consistency in using standards-based rubrics. *Proceedings of the 124th ASEE Annual Conference and Exposition,* Columbus, OH: ASEE.

Hylton, J. B., & Diefes-Dux, H. A. (2016). A standards-based assessment strategy for written exams. *Proceedings of the 123rd American Society for Engineering Education Annual Conference and Exposition*, New Orleans, LA: ASEE.

Lee, E., Carberry, A., Atwood, S., Diefes-Dux, H. A., & Siniawski, M. (2017). Faculty perception before and after implementation of standard-based grading. *Proceedings of the Research in Engineering Education Symposium 2017*, Bogota, Colombia: REEN.

Marbouti, F., & Diefes-Dux, H. A. (2014). You may be able to teach early classes, but students may not be awake yet! *Proceedings of the 121st American Society for Engineering Education Annual Conference and Exposition*, Indianapolis, IN: ASEE.

Marbouti, F., Diefes-Dux, H. A., & Madhavan, K. (2016). Models for early prediction of at-risk students in a course using standards-based grading. *Computers & Education*, 103(December 2016), 1-15.

Marbouti, F. (2017). A standards-based grading model to predict students' success in a first-year engineering course (Unpublished doctoral dissertation). Purdue University, West Lafayette, IN, USA.

Marzano, R. (2010). *Formative assessment and standards-based grading*. Bloomington, IN: Marzano Research Laboratory.

Morgan, C., Dunn, L., Parry, S., & O'Reilly, M. (2004). *The student assessment handbook: New directions in traditional and online assessment*. London, U.K.: Routledge Falmer.

Muñoz, B. M. A., & Guskey, T. R. (2015). Standards-based grading and reporting will improve education. *Phi Delta Kappan*, 96(7), 64-68.

Post, S. L. (2014). Standards-based grading in a fluid mechanics course. *Proceedings of the 121st American Society for Engineering Education Annual Conference & Exposition*, Indianapolis, IN: ASEE.

Suskie, L. (2004). Assessing student learning: A common sense approach. Boston, MA: Anker.

Stevens, D. D., & Levi, A. J. (2004). *Introduction to rubrics: An assessment tool to save grading time, convey effective feedback and promote student learning*. Sterling, VA: Stylus.

Vaishnav, S., & Carberry, A. (2017). Student perspectives on standards-based grading used in engineering project-based courses. *Submitted to American Educational Research Association Annual Conference*, San Antonio, TX: AERA.

Walvoord, B. E., & Anderson, V. J. (2010). *Effective Grading: A Tool for Learning and Assessment in College, 2nd Edition*. San Francisco, CA: Jossey-Bass.

Wiggins, G., & McTighe, J. (1998). *Understanding by design*. Alexandria, VA: ASCD.

Acknowledgements

This work was made possible by a grant from the National Science Foundation (NSF DUE-1503794). Any opinions, findings, and conclusions or recommendations expressed in this material are those of the author and do not necessarily reflect the views of the National Science Foundation.

The Tensions Measuring Instructional Practices

Kevin A. Nguyen
University of Texas at Austin, Austin, TX, USA
kevin.a.nguyen@utexas.edu

Matt DeMonbrun
University of Michigan, Ann Arbor, MI, USA
mdemonbr@umich.edu

Maura Borrego
University of Texas, Austin, TX, USA
maura.borrego@austin.utexas.edu

Jenefer Husman
University of Oregon, Eugene, OR, USA
jhusman@uoregon.edu

Michael Prince
Bucknell University, Lewisburg, PA, USA
prince@bucknell.edu

Cynthia Finelli
University of Michigan, Ann Arbor, MI, USA
cfinelli@umich.edu

Prateek Shekhar
University of Michigan, Ann Arbor, MI, USA
pshekhar@umich.edu

Charles Henderson
Western Michigan University, Kalamazoo, USA
charles.henderson@wmich.edu

Cynthia Waters
North Carolina A&T State University, Greensboro, USA
kwaters@ncat.edu

Abstract: *This study explores the challenges of measuring instructional practices in the context of an ongoing research project on student resistance to active learning. We will discuss how our research team handled the issue when our survey data did not match previous constructed theoretical frameworks. Survey data was collected in two phases: the first phase included four courses with 179 students, and the second phase included 18 courses with 961 students. Survey data was studied using exploratory and confirmatory factor analysis, but results for the first and second phases of data collection conflicted, and students' reports of instructional practices did not readily map into our existing frameworks. We present a new framework for instructional practices that is still emerging. Researchers should take care to note the potential tensions when applying theoretical and lab-based frameworks to real classrooms.*

Introduction

The United States (U.S.) has been actively calling for improvements in the nations' STEM and engineering education programs (Board, 2007, 2010; Council on STEM Education, 2013). Heeding this call, the National Science Foundation (NSF) and other U.S. agencies have advocated the need to disseminate effective research-based instructional practices across programs (Borrego, Cutler, Prince, Henderson, & Froyd, 2013; Elrod & Kezar, 2014; Freeman et al., 2014). However, adoption of these effective instructional practices in the U.S. has been slow, and there is a need for research to identify instructional practices that have effectively entered the nation's engineering classrooms and are embraced by engineering faculty and instructors (Finelli, Daly, & Richardson, 2014).

Supported by the NSF, we developed the Student Response to Instructional Practices (StRIP) survey instrument to measure the various types of instructional practices being used in undergraduate engineering classrooms across the U.S. We specifically focused on nontraditional instructional practices, which we define as any teaching methods that involve student engagement in the classroom, and we use the term active learning as an umbrella term to talk about these types of instructional practices (Prince, 2004). In contrast to nontraditional instructional practices, traditional instructional practices involve listening to the instructor lecture or watching the instructor solve problems on the board. (Borrego et al., 2013; Prince & Felder, 2006; Smith, Sheppard, Johnson, & Johnson, 2005). Specifically, we focus on in-class activities in normal classroom sessions and not in lab or recitation sections.

In addition to measuring types of instructional practice, we were also interested in students' engagement with these instructional practices in the classroom. Of particular interest was students' resistance to more nontraditional instructional practices, as faculty and instructors have noted their reluctance to adopt nontraditional instructional practices was due to fear of students' resistance (Felder, 2011; Gaffney, Gaffney, & Beichner, 2010; Hendersen & Dancy, 2007; Prince, Borrego, Henderson, Cutler, & Froyd, 2013). Exploring students' engagement or lack of engagement with nontraditional instructional practices was important for us to address.

In the context of this study and survey development, we provide a research case (Ragin & Becker, 1992) to demonstrate the tensions of measuring instructional practices and student engagement. As we moved from survey development to pilot study to full scale dissemination, we experienced both successes and drawbacks to incorporating a lab-based psychological theoretical model, the Interactive-Constructive-Active-Passive (ICAP) framework (Chi & Wylie, 2014), in our research design. The ICAP framework provided a lens to examine both instructional practice and student engagement, but real-world classrooms result in limitations not experienced in a theoretical laboratory setting. We present in this paper empirical data to highlight the ways we navigated through the survey development process, and we especially focus on the problems we encountered when our proposed theoretical framework became incompatible with our final dissemination results. We hope that this paper will benefit fellow researchers working in survey development, quantitative research, and mixed methods research designs.

Survey Development

Early in our research project, we needed to categorize the instructional practices most frequently experienced in the engineering classroom to understand how students would respond to different characteristics of instruction. Because the original motivation for the project was to assess the level of student resistance to nontraditional instructional practices, we needed to clearly differentiate traditional and non-traditional practices. In addition, since we use active learning as an umbrella term encompassing a range of distinct practices (Prince, 2004), we also needed to differentiate among different types of active learning instructional practices.

Our initial framing for the instructional practices scale of our StRIP instrument drew heavily from Chi and Wylie's (2014) Interactive-Constructive-Active-Passive (ICAP) framework. Chi and Wylie's (2014) framework had features that led us to use it as a major influence for our initial instructional practices scale, as it differed instructional practices based on students' cognitive engagement ranging from interactive, constructive, active, and passive. As we desired to distinguish between traditional and non-traditional types of instruction, the framework recognized a clear distinction between "passive" and "active" pedagogies, where "passive" was a traditional instructional practice, and "interactive," "constructive," and "active" differentiated three different categories nontraditional instructional practices. These distinctions aligned with factors we considered important for influencing student resistance. For example, the distinction between "active" strategies of students working alone and "interactive" strategies that involved student groups aligned loosely with literature that suggested that the use of student teams tended to generate student resistance (Bacon, Stewart, & Silver, 1999; Donohue & Richards, 2009; Lake, 2001; Oakley, Hanna, Kuzmyn, & Felder, 2007; Powell & Kalina, 2009).

However, there were also features of the ICAP framework that needed modification for our study and were carried through in future iterations of the survey. First, the ICAP framework focused on cognitive processes, and though these processes may be an important factor in student learning, resistance is an affective outcome, rather than a cognitive one. More importantly, the specific scales of the ICAP framework were inconsistent with the active learning terminology found in the engineering education literature. For example, according to the ICAP model, students taking notes would be considered "active," however, few instructors would consider this to be an example of active learning (Prince, 2004; Felder & Brent, 2009). Consequently, while our initial subscale for instructional practices borrowed heavily from Chi and Wylie's (2014) ICAP model, this model was significantly modified for our study on student resistance.

Specifically, we defined a categorization scheme for Research Based Instructional Strategies (RBIS) which modified the ICAP framework to focus on a grouping for the items that captured the most common active learning methods found in the engineering education literature (Borrego et al., 2013; Prince et al., 2013). There four categories were passive lecture, active learning lecture, grouped based activities, and self-directed activities (K. Nguyen et al., 2017).

The items that comprised our initial instructional practices subscale are shown in Table 1, and the categorization within both the ICAP and RBIS frameworks is listed. The question stem for each of items as well as response options are provided below Table 1.

Initial Phase of Survey Testing

We pilot tested our instrument using data collected from 4 courses with 179 students (DeMonbrun et al., 2017; K. A. Nguyen, Borrego, et al., 2016). Table 2 provides Cronbach alpha (α) values for both categorization schemes, illustrating the internal reliability of the these items (Cronbach, 1951). We needed a measure of how students felt about the type of instruction in order to assess student their response or level of resistance. For this, we decided to look at how items held together for student responses on the "ideal scale," or response options "b" underneath Table 1. This question asked students if they would ideally like more or less of a given type of instruction. The rationale for this approach is that if students "feel" the same about certain sorts of instruction, they will want either more or less of that sort of instruction in the class. This should happen reliably, regardless of the frequency of that instruction in the class or whether they want more or less of that type of instruction. With this in mind, the consistency of student response on the "ideal" instruction section was used to determine reliability for purposes of instrument validation and these are the values shown in Table 2.

Table 1: Individual Items for our Initial Instructional Practices Subscale

Survey Item*	Categorization	
	ICAP Framework	RBIS Framework
Listen to the instructor lecture during class.	Passive	Passive Lecture
Get most of the information needed to solve the homework directly from the instructor.	Passive	Passive Lecture
Watch the instructor demonstrate how to solve problems.	Passive	Passive Lecture
Solve problems individually during class.	Active	Active Learning Lecture
Answer questions posed by the instructor during class.	Active	Active Learning Lecture
Ask the instructor questions during class.	Active	Active Learning Lecture
Be graded on my class participation	Active	Group Based Activities
Make individual presentations to the class	Active	Self-Directed Activities
Preview concepts before class by reading, watching videos, etc.	Active	Dropped
Work in assigned groups to complete homework or other projects.	Interactive	Group Based Activities
Study course content with classmates outside of class.	Interactive	Group Based Activities
Discuss concepts with classmates during class.	Interactive	Group Based Activities
Solve problems in a group during class.	Interactive	Group Based Activities
Do hands-on group activities during class.	Interactive	Group Based Activities
Be graded based on the performance of my group	Interactive	Dropped
Brainstorm different possible solutions to a given problem.	Constructive	Self-Directed Activities
Find additional information not provided by the instructor to complete assignments.	Constructive	Self-Directed Activities
Assume responsibility for learning material on my own.	Constructive	Self-Directed Activities
Make and justify assumptions when not enough information is provided.	Constructive	Self-Directed Activities
Solve problems that have more than one correct answer.	Constructive	Self-Directed Activities
Take initiative for identifying what I need to know.	Constructive	Self-Directed Activities

*Question Stem: For each of the following things, please indicate how often you did each thing in this course[a] and how often you would like to do each in your ideal course[b].

[a]Response options for how often you did were: 1 = Never; 2 = Seldom (1-5 times/semester); 3 = Sometimes (5-10 times/semester); 4 = Often (once a week); 5 = Very often (more than once/week).

[b]Response options for how often you would like were: 1 = Much less; 2 = Slightly less; 3 = About the same; 4 = Slightly more; 5 = Much more.

Reliability values for the classification scheme derived for both the ICAP and RBIS framework were acceptable (Table 2). The research team decided to continue modifying the instructional practices subscale to reflect more of the RBIS literature as well as to focus more on instructional practices that occurred in the classroom. We dropped items such as "Work in assigned groups to complete homework or other projects" that did not focus on in-class instructional practices. We also dropped items to reduce the overall number of items in the subscale and to reduce survey fatigue.

Second Phase of Survey Testing

After instrument modification, we conducted a final testing of the instrument by administering our StRIP survey to students enrolled in 18 courses across the United States with 961 students. After removing all responses with any missing values across any of the instructional practices subscale, our final sample consisted of 940 students.

Table 2: Internal Reliability for Factors within each of the Categorization Schemes for our Initial Instructional Practices Subscale (N=179)

Framework and Related Factor	Cronbach Alpha
ICAP Framework	
Passive (3 items)	0.67
Active (6 items)	0.73
Interactive (6 items)	0.80
Constructive (6 items	0.77
RBIS Framework	
Passive Lecture (3 items)	0.67
Active Learning Lecture (3 items)	0.78
Group Based Activities (5 items)	0.81
Self-Directed Activities (6 items)	0.79

Since we made considerable changes to our instrument between the pilot testing and this final phase of data collection, we conducted an exploratory factor analysis (EFA) to determine what factors or constructs emerged from our data. We split the full sample of 940 students into two random samples of 470 students each. First, we specified a model with no limitations on the number of factors extracted from the data on the first random sample. After determining that there were three factors with eigenvalues above 1.0, we respecified the model with a three-factor solution to maximize the loadings across these three factors.

Table 3: Individual Items and Exploratory Factor Analyses Results for our Final Instructional Practices Subscale (Sample 1, N=470)

Survey Item	Factor Loading			Communality
	1	2	3	
Be called on personally by the instructor to answer a question during class.	**0.75**	-0.12	-0.07	0.51
Be graded on my class participation.	**0.72**	-0.04	-0.07	0.52
Be asked to answer a question during class on material not previously covered in class.	**0.60**	0.11	-0.04	0.59
Be graded based on the performance of my group.	**0.61**	0.12	-0.11	0.59
Preview concepts before class by reading, watching videos, etc.	**0.52**	0.03	0.17	0.64
Solve problems during class that have more than one correct answer	**0.57**	-0.01	0.05	0.66
Be given time to think or discuss before answering a question posed by the instructor during class.	0.14	**0.48**	0.08	0.65
Discuss concepts with classmates during class.	-0.13	**0.78**	0.04	0.43
Solve problems in a group during class.	0.10	**0.73**	-0.12	0.47
Solve problems individually during class.	0.00	**0.50**	0.17	0.65
Listen to the instructor lecture during class.	-0.11	-0.01	**0.59**	0.68
Watch the instructor demonstrate how to solve problems during class.	-0.11	0.04	**0.61**	0.62
Eigenvalues	2.84	2.34	1.55	
Construct Reliability	0.80	0.75	0.61	
Percent of Variance Explained	60%	49%	33%	

Table 3 presents final survey items and the EFA analysis which includes factor loadings for a three-factor solution and overall indices of fit. The EFA suggested a three-factor solution, which is different from either of the two frameworks we used for our pilot data (Table 1). All factors had eigenvalues above 1.0 (Kaiser, 1958) and each of the three factors had a construct reliability above the recommended benchmark of 0.60 (Bagozzi & Yi, 1988). We dropped two items ("Receive an individual grade for group work" and "Be graded based on the performance of my group.") because of low factor loadings across all three factors (below 0.32) (Comrey & Lee, 1992).

Next, using the three factors identified during the EFA process, we conducted a confirmatory factor analysis (CFA) using this factor structure on the second random sample of students. We analyzed model fit indices for single factor, two factor, and three factor structures to determine if the three-factor structure identified in the EFA process was the best fit for this subsample of our data. The fit indices for each of these solutions is presented in Table 4.

The statistics in Table 4 suggest improvement in the models from the one-factor to the three-factor model, but the fit indices indicate mixed results. While the chi-square statistic is significant, this test is generally oversensitive to sample size. The root-mean-square error of approximation was 0.08, with the lower bound of our 90% confidence interval at 0.07 and an upper bound at 0.09. Additionally, while the standardized root-mean-square residual is acceptable (<0.08), the comparative fit index statistic is not acceptable (<0.95) (Hu & Bentler, 1999).

Although the model fit indices are not optimal, they do show marked improvement in the three-factor model over a one-factor solution. Furthermore, pattern coefficients (factor loadings) in the CFA suggest that all loadings are at or above 0.50, which is acceptable for each of the factors (Hair, Anderson, Tatham, & Black, 1992).

Table 4: Confirmatory Factor Analysis Results for our Final Instructional Practices Subscale (Sample 2, N=470)

	df	χ^2	χ^2/df	RMSEA (90% CI)	CFI	SRMR
Single Factor	54	442.99	8.20	0.12 (0.11, 0.14)	0.66	0.090
Two Factor	53	262.13	4.95	0.09 (0.08, 0.10)	0.82	0.069
Three Factor	51	192.40	3.77	0.08 (0.07, 0.09)	0.88	0.059

After closely examining the three factor solution, a new framework emerged, and we categorized the factors as: (1) high anxiety active learning, (2) low anxiety active learning, and (3) passive. High anxiety instructional practices are active learning pedagogies that either involve an element of individual accountability (being graded or called upon individually by the instructor) or place higher expectations on the students to be more self-directed learners or to confront high levels of ambiguity. As a result, these practices may be linked to anxiety. In contrast, low anxiety active learning includes other types of active learning pedagogies that involve either problem solving (something we anticipate is comfortable for an audience of engineering students) or simple discussion. We note that there are low levels of individual accountability associated with these practices, and they do not place an extraordinary burden on students. And finally, passive lecture includes instructional practices that do not require students to actively engage in class. Table 5 includes the CFA results for the second set of data in our final instrument administration. The factor loadings are high, but we do not yet have high confidence in the validity of these scales, as the original survey was not designed with this theoretical framework in mind.

Discussion

Overall, our main goal was to develop a scale that was sensitive to different types of instruction as well as student engagement. ICAP allowed us to design a scale with different types of instruction but attempts to categorize students' preferences for these four categories of instructional practices were problematic. The second phase of survey testing presented a three factor solution that was more closely aligned to an anxiety framework instead of the original ICAP framework. The lack of success of the ICAP framework with this subscale may point to a need to clarify the application of the ICAP framework to educational settings. The foundational research on the ICAP framework is based on experimental and observational research (Chi, Roy, & Hausmann, 2008). It may not be the case that students would have preference for their own classroom activities in the ways that Chi and her colleagues have described them. Rather, the strength of the ICAP framework lies in its ability to tease out cognitive processes displayed by students during class. More work should be done to tease apart or combine cognitive processes with affective processes. It appears that a pedagogical framework utilizing RBIS or anxiety or other forms of affect may be closer to modelling students' affective engagement with instructional practices. The anxiety framework, especially, touches on how instruction can be tied to affect. However, the anxiety framework presented here is an evolving theory that was not part of our original survey design and needs additional testing and research.

Table 5: Individual Items and Exploratory Factor Analyses Results for our Final Instructional Practices Subscale (Sample 2, N=470)

Item	High anxiety active learning	Low anxiety active learning	Passive
Be called on personally by the instructor to answer a question during class.	0.55		
Be graded on my class participation.	0.56		
Be asked to answer a question during class on material not previously covered in class.	0.54		
Be graded based on the performance of my group.	0.50		
Preview concepts before class by reading, watching videos, etc.	0.59		
Solve problems during class that have more than one correct answer	0.70		
Be given time to think or discuss before answering a question posed by the instructor during class.		0.49	
Discuss concepts with classmates during class.		0.71	
Solve problems in a group during class.		0.60	
Solve problems individually during class.		0.59	
Listen to the instructor lecture during class.			0.68
Watch the instructor demonstrate how to solve problems during class.			0.66
Covariance			
High anxiety active learning	1.00		
Low anxiety active learning	0.48	1.00	
Passive	0.19	0.49	1.00

Scaling up in survey design is an important step that researchers should consider carefully. Instrument modification from the first phase to the second phase were slight but enough to drastically change the final categorization for the second phase of the study. Moving from 179 students and 4 classrooms to 940 students and 18 classrooms presented its own difficulties and interesting implications. It may be the case that students' preference for categories of instructional practices across the United States do not fall into pedagogical frameworks commonly found in engineering education literature, such as active/passive or lecture/non-lecture. Rather, much of our data suggests that even though we surveyed the courses of faculty who self-identified as nontraditional instructors, many of them still employ significant periods of lecture (K. Nguyen et al., 2017; K. A. Nguyen, Shekhar, et al., 2016). We would give caution then to quickly labelling an instructor's practices into a few different categorizations throughout the semester.

Conclusion

We provided a research case of developing a survey from the initial phase to final survey dissemination. In our presentation of our survey development, we discussed how our theoretical frameworks changed and were modified during initial and final survey development. There are not many candid discussions of researchers' experiences with survey development, and we hope that ours serve as a case to talk about the application of theory in education research. In terms of measuring and connecting pedagogical theoretical frameworks to classrooms and students' experiences, there is still much research work to be done. Moving from the ICAP framework to the RBIS framework and then to an anxiety framework highlights how theory can be applied and reconfigured in light of data. Further exploring the role of student anxiety in engagement or resistance to active learning is a potentially fruitful direction for future work.

References

Bacon, D. R., Stewart, K. A., & Silver, W. S. (1999). Lessons from the best and worst student team experiences: How a teacher can make the difference. *Journal of Management Education, 23*(5), 467-488.

Bagozzi, R. P., & Yi, Y. (1988). On the evaluation of structural equation models. *Journal of the academy of marketing science, 16*(1), 74-94.

Board, N. S. B. (2007). *A national action plan for addressing the critical needs of the US science, technology, engineering, and mathematics education system*: National Science Foundation.

Board, N. S. B. (2010). *Preparing the next generation of stem innovators: Identifying and developing our nation's human capital*: National Science Foundation.

Borrego, M., Cutler, S., Prince, M., Henderson, C., & Froyd, J. E. (2013). Fidelity of Implementation of Research-Based Instructional Strategies (RBIS) in Engineering Science Courses. *Journal of Engineering Education, 102*(3), 394-425.

Chi, M. T., Roy, M., & Hausmann, R. G. (2008). Observing tutorial dialogues collaboratively: Insights about human tutoring effectiveness from vicarious learning. *Cognitive science, 32*(2), 301-341.

Chi, M. T., & Wylie, R. (2014). The ICAP framework: Linking cognitive engagement to active learning outcomes. *Educational Psychologist, 49*(4), 219-243.

Comrey, A., & Lee, H. (1992). Interpretation and application of factor analytic results. *Comrey AL, Lee HB. A first course in factor analysis, 2.*

Council on STEM Education. (2013). Federal Science Technology, Engineering, and Mathematics (STEM) Education 5-Year Strategic Plan: A Report from the Committee on STEM Education National Science and Technology Council: Executive Office of the President National Science and Technology Council.

Cronbach, L. J. (1951). Coefficient alpha and the internal structure of tests. *psychometrika, 16*(3), 297-334.

DeMonbrun, M., Finelli, C. J., Borrego, M., Shekhar, P., Prince, M., Hendersen, C., & Waters, C. (2017). Creating an Instrument to Measure Student Response to Instructional Practices. *Journal of Engineering Education, 106*(2).

Donohue, S. K., & Richards, L. G. (2009). *Factors affecting student attitudes toward active learning activities in a graduate engineering statistics course.* Paper presented at the Frontiers in Education Conference, 2009. FIE'09. 39th IEEE.

Elrod, S., & Kezar, A. (2014). Developing leadership in STEM fields: the PKAL summer leadership institute. *Journal of Leadership Studies, 8*(1), 33-39.

Felder, R. M. (2011). Hang in there! Dealing with student resistance to learner-centered teaching. *Chemical Engineering Education, 43*, 131-132.

Finelli, C. J., Daly, S. R., & Richardson, K. M. (2014). Bridging the Research-to-Practice Gap: Designing an Institutional Change Plan Using Local Evidence. *Journal of Engineering Education, 103*(2), 331-361.

Freeman, S., Eddy, S. L., McDonough, M., Smith, M. K., Okoroafor, N., Jordt, H., & Wenderoth, M. P. (2014). Active learning increases student performance in science, engineering, and mathematics. *Proceedings of the National Academy of Sciences, 111*(23), 8410-8415.

Gaffney, J. D., Gaffney, A. L. H., & Beichner, R. J. (2010). Do they see it coming? Using expectancy violation to gauge the success of pedagogical reforms. *Physical Review Special Topics-Physics Education Research, 6*(1), 010102.

Hair, J., Anderson, R. E., Tatham, L., & Black, W. (1992). Multivariate data analysis with regarding (Vol. 6, pp. 1-15): MacMillan Publishing Company New York.
Hendersen, C., & Dancy, M. H. (2007). Barriers to the Use of Research-Based Instructional Strategies: The Influence of Both Individual and Situational Characteristics. *Physical Review Special Topics: Physics Education Research, 3*(2), 020102-020101 to 020102-020114.
Hu, L. t., & Bentler, P. M. (1999). Cutoff criteria for fit indexes in covariance structure analysis: Conventional criteria versus new alternatives. *Structural equation modeling: a multidisciplinary journal, 6*(1), 1-55.
Lake, D. A. (2001). Student performance and perceptions of a lecture-based course compared with the same course utilizing group discussion. *Physical Therapy, 81*(3), 896-902.
Nguyen, K., Husman, J., Borrego, M., Shekhar, P., Prince, M., DeMobrun, M., . . . Waters, C. (2017). Students' Expectations, Types of Instruction, and Instructor Strategies Predicting Student Response to Active Learning. *International Journal of Engineering Education, 33*(1(A)), 2-18.
Nguyen, K. A., Borrego, M., Finelli, C. J., Shekhar, P., DeMonbron, M., Hendersen, C., . . . Waters, C. (2016). Measuring Student Response to Instructional Practices (StRIP) in Traditional and Active Classrooms. *Paper presented at 2016 ASEE Annual Conference and Exposition, New Orleans, Louisiana.*
Nguyen, K. A., Shekhar, P., Husman, J., Borrego, M., Prince, M., Finelli, C., . . . Henderson, C. (2016). Extended Abstract-Students' Expectations and Responses to Active Learning in Undergraduate Engineering Courses. *Paper presented at 2016 Mid Years Engineering Education Conference at College Station, Texas.*
Oakley, B. A., Hanna, D. M., Kuzmyn, Z., & Felder, R. M. (2007). Best practices involving teamwork in the classroom: Results from a survey of 6435 engineering student respondents. *Education, IEEE Transactions on, 50*(3), 266-272.
Powell, K. C., & Kalina, C. J. (2009). Cognitive and social constructivism: Developing tools for an effective classroom. *Education, 130*(2), 241.
Prince, M. (2004). Does active learning work? A review of the research. *Journal of Engineering Education, 93*, 223-232.
Prince, M., Borrego, M., Henderson, C., Cutler, S., & Froyd, J. (2013). Use of research-based instructional strategies in core chemical engineering courses. *Chemical Engineering Education, 47*(1), 27-37.
Prince, M., & Felder, R. (2006). Inductive teaching and learning methods: Definitions, comparisons, and research bases. *Journal of Engineering Education, 95*(2), 123-138.
Ragin, C. C., & Becker, H. S. (1992). *What is a case?: exploring the foundations of social inquiry*: Cambridge university press.
Smith, K. A., Sheppard, S. D., Johnson, D. W., & Johnson, R. T. (2005). Pedagogies of engagement: Classroom-based practices. *Journal of Engineering Education, 94*(1), 87-101.

Acknowledgements

This project is funded by the U.S. National Science Foundation through grant numbers 1347417, 1347482, 1347580, 1347718, and 1500309. The opinions are those of the authors and do not necessarily represent the National Science Foundation. The authors would like to thank the instructors and students who agreed to be part of the initial and final phases of the study, as well as project advisory board members and fellow researchers who helped us along the survey development process.

Using online collaborative exercises to promote the development of critical analysis skills in a large first year class

Jacqueline Bridge
The University of the West Indies, St. Augustine, Trinidad
Jacqueline.Bridge@sta.uwi.edu

***Abstract**: An important requirement of engineering education is that students develop skills in problem identification and definition, organising information, selection of appropriate solution strategies, proposing solutions, critically analysing proposed solutions and drawing conclusions from the results. In large classes, it becomes difficult to monitor the development of these skills using traditional methods. This article examines how the development of these skills may be facilitated using online collaborative exercises. Groups of students were asked to collaboratively solve an online open-ended problem based on an engineering concept taught within an Engineering Dynamics module. All contributions were input to an online (Moodle) discussion board. The form of the questions and the necessity of verbalising their thought processes to explain the logic behind their suggestions forces participants to engage in metacognitive activities, which facilitate the enhancement of critical analysis skills.*

Introduction

Problem solving and critical analysis are very important facets of the engineering profession. Traditionally, undergraduate engineering programs place heavy emphasis on the solution of exercises, with students solving about 2500 questions over their undergraduate education (Mourtos, DeJong-Okamoto & Rhee, 2004). Yet many employers report that recent engineering graduates are not sufficiently prepared to apply their knowledge to the solution of real-world engineering problems (IET, 2015). This disconnect arises because the problems that students are exposed to in the typical engineering program are mostly well-defined - the problem statement is given, assumptions are prescribed, the principles to be used are indicated and all the information required to solve the problem is in the statement. Conversely, real world problems are ill-defined. The solution of these problems requires skills in problem identification and definition, organizing information, selection of appropriate solution strategies, proposing solutions and analyzing proposed solutions and drawing conclusions from the results. These skills form part of the accepted domain of critical thinking (CT).

Douglas (2012) points out that about 90% of academics believe critical thinking is a major learning objective of higher education. However, only 9% explicitly include lectures on the development of CT skills in their courses (Ralston & Bays, 2010). Tiruneh, De Cock, Weldeslassie, Elem & Janssen (2016) argue that educators must aim to foster the development of domain specific critical thinking skills which enable their students to appropriately perform critical analysis tasks in their area of study. Alfrey and Cooney (2009) note that the two most successful methods for teaching critical thinking are writing for reflection and problem-based learning. Tsui (1998) posits that open-ended problem solving is especially effective in developing critical thinking skills. Given the nature of their proposed profession, this suggests that engineering students would benefit from having opportunities to apply their CT skills and domain-specific knowledge in a variety of open-ended contexts. Furthermore, these efforts must be carefully evaluated with coherent and consistent feedback on the quality of thinking being demonstrated (Ralston and Bays, 2010). However, attempts to provide this in-depth assessment are stymied by increasing class size. In this article, we examine how online open-ended collaborative exercises in Engineering Dynamics can be used to improve the problem solving/critical analysis skills of engineering students.

Theoretical Framework

The theoretical underpinnings for this study come from the literature on critical thinking, the development of critical thinking skills using open-ended questions and the enhancement of the learning experience through the use of collaborative exercises.

There is no universally accepted definition of critical thinking. Cosgrove (2010) suggests that it is the "analysis and evaluation of thinking with a view to improving it", focusing on the process itself. Halpern (2003) describes it as "the use of those cognitive skills or strategies that increase the probability of a desirable outcome", thereby seeming to focus on critical thinking as a tool for decision making. However, regardless of the definition of choice, engineering educators acknowledge that it is an important skill in the profession and should therefore be an outcome of good engineering education. But how can critical thinking be taught?

Ralston and Bays (2010) and Welch, Hieb & Graham (2015) present the Paul-Elder framework as a formal structure for teaching critical analysis skills. This framework decomposes critical thinking into the Standards, the Elements, and the Intellectual Traits. The components are interrelated as "the Standards are applied to the Elements as we learn to develop Intellectual Traits" (Paul & Elder, 2009).

Alfrey and Cooney (2009) note that an individual's critical thinking skills can be judged by their approach to open-ended problems. However, assessment of the finished product alone is not enough to indicate the attainment of critical thinking skills. Students must also report what their thought processes were. Previous research (Beckmann & Weber, 2016; Lovelace, Eggers & Dick, 2016; Boyer, Maher & Kirkman, 2006) has shown the suitability of online collaborative activities for this purpose using different tasks such as simulations and joint essays. A key aspect of these activities was the method used to analyze the critical thinking content. In the current work, the Newman, Webb & Cochrane (1995) content analysis tool is used to assess discussion board contributions.

Methods

Research Goals

This work explores the extent to which the development of critical analysis and problem solving skills can be facilitated in an online open-ended collaborative exercise. The research hypotheses that govern this study are:

1. Open-ended problems solved in an online collaborative discussion board facilitate the development of critical thinking skills more readily than the same exercise done solo.
2. Demonstration of critical analysis skills on the discussion board is positively correlated to subsequent performance in open-ended examination questions.

Participants

The study used data collected from two cohorts of students enrolled in a course in Engineering Dynamics which the author taught at a regional university. The first cohort of 117 students (Cohort A) were introduced to the concept of critical thinking in a 1-hr lecture and were assigned individual take-home critical thinking problems (CTPs). These were discussed in the weekly recitations. The following academic year, the second cohort of 129 students (Cohort B) had the same lecture introducing critical thinking. However, they were required to complete their CTPs in an online collaborative discussion board. The exercise contributed 10% of the overall final mark and all students were required to complete the exercise. All students signed informed consent forms. Table 1 shows the distribution of the students in Cohorts A and B, respectively. All the students in the study had completed (or been exempted from) a set of six required courses including engineering mathematics and mechanics (statics and elementary strength of materials). The difference in the mean GPA of the cohorts (2.87 (0.70) and 2.78(0.75)) in these six courses is statistically insignificant. The final Engineering Dynamics examinations of both cohorts were written by the same lecturer. In particular, the critical thinking questions required the students to demonstrate the same level of critical analysis and were marked using the same rubric. This data suggests that the level of student preparation and assessment was comparable over the two cohorts thus strengthening the case for comparison of performance between the two cohorts as a measure of the effectiveness of the intervention.

Table 1: Demographic Information for Participants in the Study

	Cohort A		Cohort B	
	n	%	n	%
Total	117	100	129	100
Gender				
Male	84	72	99	77
Female	32	27	30	23
Unreported	1	1		
Attempt				
First	97	83	100	78
Second	12	10	17	13
Third or higher	8	7	12	9
Major				
Mechanical	102	87	106	82
Industrial	7	6	17	13
Biosystems	8	7	6	5

Structure of the exercise

The aim of the open-ended online collaborative exercise is to foster the development of critical analysis skills. Thus the problems presented to the students typically are ill-defined. There is some vagueness in the context provided. This forces the students to progressively refine the problem for themselves. They must make assumptions about the information given and provide valid estimates for any missing data. The student also has to decide on the engineering principle to be used in the solution since there are no verbal cues in the problem context. Consequently, there are generally multiple valid approaches to finding an answer for the problem. There are also various levels of refinement that can be applied to the problem analysis. The choice of the content of the exercises is guided by the syllabus for the course (kinematics and kinetics (sum of forces, work-energy principles, impulse-momentum principles) of particles and rigid body in planar motion). As a result, the students have several opportunities to develop domain-specific critical thinking skills. Figure 1 shows a representative CTP.

CRITICAL THINKING PROBLEM *Engineering Dynamics* **Situation** According to the IAAF website, the Trinidadian athlete, Cleopatra Borel has a personal best of 19.42 m in the outdoor shot-putt (achieved in Paris Saint-Denis) and a personal best of 19.48 m in the indoor shot-putt (Blacksburg, Virginia, USA). (a) Estimate her initial shot putt release conditions which could achieve these results, neglecting air resistance. What could contribute to the difference between the indoor achievement and the outdoor achievement? (b) Redo the estimate done in (a), now accounting for air resistance. Investigate typical forms of air resistance on shot-putts and their effects on the kinematic equations for the shot-putt. Use the problem solving methodology discussed in Engineering Dynamics. State any assumptions used with appropriate justification, the engineering principles involved and also how you would verify your results. If you need any further information, state its source.

Figure 1: A representative Critical Thinking Problem

Procedure

Introductory Session

An introductory session on critical thinking skills and open-ended problem solving was delivered to the participants in the fifth week of teaching (in a 13 week semester). The session began with the class being asked to solve a simple open-ended problem individually; the students were then asked to discuss how they had approached the problem. The instructor used this as an example to point out that:

- Open-ended problems could have several correct solutions,
- Each student had constructed his/her own problem statement from the available information.
- Several assumptions had been made in the analysis.
- They relied on subject specific principles to solve the problem.

This formed the entry point into a lecture on open-ended problem solving and critical analysis.

For Cohort A, this lecture was followed by the distribution of the first CTP. Students had three days to complete the task. The exercises were graded by the instructor and teaching assistants. While specific comments were made on each submission, more general comments were addressed in the tutorial session immediately after the grading was completed.

For Cohort B, the lecture was followed by a discussion of the procedure for completing the open-ended CTP on the online discussion board. The instructor described the different types of comments that could be made and highlighted the level of critical discourse demonstrated in each. These include comments on the structure of the response, how coherently the response is presented, the adequacy of the information and principles used in the solutions and the validity of any assumptions made.

Collaborative groups in Cohort B

Using the automatic group creator within the MOODLE environment, the students in Cohort B were randomly distributed into groups of five or six participants. On each Tuesday of the CTP weeks, two students from each group were assigned the task of developing initial proposals to the solution of the critical thinking problem. The discussion board opened at 12:00 noon and the initial proposers had 24 hours to upload their submissions. Each discussion board was only visible to the members of the specific group.

All members of the group were then required to critique the submissions using the criteria previously discussed in the introductory lecture; (the two proposers critiqued the other's work). All contributions had to be made through the online discussion board. The group then had to come to a consensus about how they wanted to present the solution under the following categories: problem statement, givens, assumptions, additional information obtained and their sources, principles to be used in the solution of the problem. The group was given four days to complete this task. The instructor and the teaching assistants then viewed the discussion boards, pointing out any unsound arguments therein and commending any exceptionally cogent arguments presented. The group then had 24 hours to revise their submission, present a solution to the CTP and discuss the implications/limitations of their findings. Three CTPs were assigned during the semester in Weeks 7, 9 and 11, respectively.

Data Analysis

The performance on the final examination open-ended problem was used as the basis for determining how similar the average skill of the two cohorts were in solving CTPs i.e. to analyze the first research question. The distribution of the marks awarded in each cohort was assumed to be normal with approximately equal variances. A pooled t-test, at the 0.05 level of significance, was then used to investigate the hypothesis that the online collaborative forum led to the development of better critical thinking skills.

The evaluation of both hypotheses was enabled by analyzing each discussion board using the content analysis method developed by Newman, Webb and Cochrane (1995). Contributions were categorized as being either a positive or negative indicator of critical thinking skills in the domains of *Relevance, Importance, Novelty, Ambiguity, Outside Knowledge, Linking, Justification, Critical Assessment, Practical Utility* or *Width of Understanding.* The reader is referred to Newman, Webb, Cochrane (1995) for more information on each category. The difference between the number of positive and negative indicators for a particular category was normalised by their sum to produce a measure of the quality of the contribution ranging from -1 to +1. This measure was used in several ways:

- There was a breakdown by category for each CTP discussion board which gave the authors an overview of the general quality of the group effort.
- There was an overall measure of the quality of contribution from each participant to each CTP discussion board.

The data were examined for significant trends. They were also compared with a similar assessment of critical thinking skills on an open-ended final examination question.

Findings and Discussion

The first hypothesis that directed this research was that open-ended problems solved in an online collaborative discussion board facilitate the development of critical thinking skills more readily than the same exercise done solo. Cohorts A and B were assumed to come from populations with equal variances. This assumption was supported by the ratio of the standard deviations of the samples (1.16) and an F-test which showed that at the 95% confidence level, the variances were statistically the same. A pooled t-test was done using a significance level of 5%. The results show that there is a statistically significant difference (p=2.1e-6) in the performance of the two groups of students on the critical thinking question on the final examination. This supports the hypothesis that online collaborative discussion boards are better at facilitating the development of critical thinking skills.

Table 2: Averages of Discussion Board Critical Thinking Ratios

Scoring Criteria	*CTP1*	*CTP2*	*CTP3*	*Significant increase at 0.05 level*
Relevance	0.533 (0.092)	0.553 (0.113)	0.619 (0.069)	Yes
Importance	0.472 (0.119)	0.581 (0.106)	0.571 (0.106)	Yes
Novelty	0.421 (0.190)	0.558 (0.171)	0.600 (0.150)	Yes
Ambiguity	0.643 (0.149)	0.620 (0.130)	0.651 (0.171)	No
Outside Knowledge	0.454 (0.217)	0.581 (0.223)	0.520 (0.199)	No
Linking	0.600 (0.174)	0.690 (0.110)	0.664 (0.135)	No
Justification	0.465 (0.162)	0.569 (0.106)	0.584 (0.110)	Yes
Critical Assessment	0.521 (0.171)	0.578 (0.125)	0.583 (0.180)	No
Practical Utility	0.404 (0.158)	0.451 (0.132)	0.489 (0.125)	Yes
Width of Understanding	0.437 (0.162)	0.487 (0.130)	0.514 (0.241)	No

Table 2 shows that the discussion boards provided evidence of critical thinking in all areas. In each of the three critical thinking problems, the mean scores in each category was positive. The *Practical Utility* and *Novelty* domains scored the lowest. This is not surprising as the Engineering Dynamics students are typically in their first year at the university; they are reluctant to volunteer their opinion. The increase in the Novelty category is promising and may be attributed to growing maturity. We note that there was a significant increase in the critical thinking ratios of five categories as the semester progressed; this might be interpreted as being indicative of growing competence in critical thinking.

Table 3: Combined Averages of Student Critical Thinking Ratios

	CTP1	*CTP2*	*CTP3*	*Final Exam Open-ended Question*
Minimum CT ratio	0.059	0.053	0.170	-0.3
Maximum CT ratio	0.814	0.815	0.961	1
Average CT ratio	0.498	0.550	0.555	0.509
Std. Dev. of CT ratio	0.153	0.143	0.139	0.336

Table 3 shows the average of the individual participants' critical thinking ratio for each critical thinking problem and the ratio for the open-ended question in the final examination. We note that there was a slight increase in the mean values as students progressed through CTP1 to CTP3. Further examination of the data shows that this is due to the submission of less "uncritical" contributions. The author notes that as each student gains more practice in critical thinking, they (amongst other things) no longer make entries to the discussion board which either repeat something already said without further elaboration or offer judgments without explanation or accept the first solution without exploring other options.

The form of the CTP is crucial to the student development of the desired cognitive skills. Since the questions are open-ended and have no single correct answer, group members must use concise and logically presented arguments to convince their teammates of the validity of their contributions. The collaborative aspect of the activities benefits from the fact that each participant is exposed to multiple perspectives of the same problem (So & Brush, 2008). Critical thinking skills are honed as students judge, support/oppose and value these different viewpoints. When this is done on an electronic discussion board, the participant has to write clear statements as they cannot use nonverbal cues/charisma to win others over to their point of view. This too fosters critical thinking. Hence the student engaged in an open-ended problem on a discussion board has more opportunities to participate in critical thinking activities than a student working alone.

The second hypothesis addressed in this study was that demonstration of critical analysis skills on the discussion board is positively correlated to subsequent performance in open-ended examination questions.

Table 3 shows there is a decrease in the average of the critical thinking ratio on the open-ended examination question compared to the average ratio obtained per discussion boards. The coefficient of determination (r^2) between this paired set of data is 0.544. This suggests that there is a moderate correlation between performance on the discussion boards and subsequent performance on the open-ended final examination question. The drop in the average between discussion board contribution and final examination questions may be due to the individual nature of the examination. Within the discussion board, multiple perspectives are posed, leading to a wider discussion, brain-storming and logically stated objections/agreement. Because this rich influx of information cannot be replicated in the examination setting, the solutions are less developed. Furthermore, the participant has time to think through and craft a coherent discussion board input while the final examination question must be completed within a specified time.

Implications and Limitations

A major contribution of this study is the demonstration that participation in online collaborative discussion boards results in better performance on open-ended examination questions than working similar problems solo. This may be generalized to suggest that the participant develops transferrable critical thinking skills which may reduce the reported gap between employers' expectations and recent graduates' ability. Previous research (Beckmann & Weber, 2016; Lovelace, Eggers & Dick, 2016; Silva, Lopes, Dominguez, Payan-Carreira, Morais, Nascimento, & Morais, 2016) suggests that online collaboration is useful in developing the skills of participants for the solution of realistic problems. The current work extends these principles to engineering domain-specific CT skills. It is conjectured that this approach can be adapted for use in other engineering science courses (such as Statics, Mechanics of Solids and Thermodynamics).

The use of an actual course results in some limitations of the study. First, since the online collaboration CTPs were assigned 10% of the mark, it was not possible to have a true control group. Secondly, the two cohorts used in the study were from different year groups. While they had similar exposure to the critical thinking/problem solving techniques, there would typically be some differences in how information is presented from one year to another; for example, shortcomings observed in the offering to Cohort A would be referred to and corrected in the next offering. Furthermore, since the context of the problems is supposed to be new to the students, the CTP exercises and the open-ended question on the final examination were different for Cohort A and Cohort B. Thus the possibility that there was a difference in the cognitive load of the tasks assigned to the two sets of students cannot be definitively ruled out. Finally the gains in critical thinking for individual students were difficult to assess because there was not a formal analysis of these skills prior to the start of the intervention.

Conclusions

This study provides evidence that appropriate collaborative online exercises can assist engineering students in developing critical analysis via open-ended problem solving skills.

References

Alfrey, K. & Cooney, E. (2009). AC 2009-653: Developing a rubric to assess critical thinking in assignments with an open-ended component. . In *American Society for Engineering Education*.

Beckmann, J. & Weber, P. (2016). Cognitive Presence in Virtual Collaborative Learning: Assessing and Improving Critical Thinking in Online Discussion Forums, *Interactive Technology and Smart Education*, 13(1), 52 -70.

Boyer, N., Maher, P. & Kirkman, S. (2006). Transformative Learning in Online Settings: The use of self-directed metacognition and collaborative learning, *Journal of Transformative Education*, 4(4), 335 – 361.

Cosgrove, R. (2011). Critical thinking in the Oxford tutorial: a call for an explicit and systematic approach, *Higher Education Research and Development*, 30(3), 343 – 356.

Douglas, E. (2012). Defining and measuring Critical Thinking in Engineering, *Procedia – Social and Behavioral Sciences*, 56, 153 – 159.

Halpern, D. F. (2003).Thought and knowledge: An introduction to critical thinking. Mahwah, NJ: Erlbaum.

Institution of Engineering & Technology (2015). *Skills and Demand in Industry – 2015 Survey*. IET Publications.

Lovelace, K.J., Eggers,F. & Dyck, L.R. (2016). I Do And I Understand: Assessing The Utility Of Web-Based Management Simulations To Develop Critical Thinking Skills, *Academy of Management Learning & Education*, 15(1), 100-121.

Mourtos, N., DeJong-Okamoto, N. & Rhee, J. (2004). Open-ended problem-solving in thermal fluids engineering. *Global Journal of Engineering Education*, 8(2).

Newman, D.R., Webb, B. & Cochrane, C. (1995). A content analysis method to measure critical thinking in face-to-face and computer supported group learning. *Interpersonal Computing and Technology*, 3(2), 56-77.

Ralston, P. & Bays, C. (2010). AC2010-1518: Refining a Critical Thinking Rubric for Engineering. In *American Society for Engineering Education*.

Silva, H., Lopes, J., Dominguez, C., Payan-Carreira, R., Morais, E., Nascimento, M. & Morais, F. (2016). Fostering critical thinking through peer review between cooperative learning groups, *Revista Lusófona de Educação,* 32, 31-45.

So, H. & Brush T. (2008). Student perceptions of collaborative learning: social presence and satisfaction in a blended learning environment: Relationships and critical factors, *Computers and Education* 51, 318 – 336.

Tiruneh, D.T., De Cock, M., Weldeslassie, A.G., Elen, J. & Janssen, R. (2017). Measuring Critical Thinking in Physics: Development and Validation of a Critical Thinking Test in Electricity and Magnetism, *International Journal of Science and Mathematics Education,* 15, 663–682.

Tsui, L. (1998). *A Review of Research on Critical Thinking.* Paper presented to the 23rd National Conference, Association for the Study of Higher Education, Miami, Nov. 5-8, 1998. (ERIC Documentation Reproduction Services ED 427572).

Welch, K., Hieb, J. & Graham, J. (2015). A Systematic Approach to teaching Critical Thinking skills to Electrical and Computer Engineering Undergraduates, *American Journal of Engineering Education,* 6(2), 113.

Copyright statement

Community Learning Model for Promoting Entrepreneurship POCDILD

María Catalina Ramírez
Universidad de los Andes, Bogotá, Colombia
mariaram@uniandes.edu.co

Libia Alejandra Navas
Universidad de los Andes, Bogotá, Colombia
la.navas251@uniandes.edu.co

***Abstract**: This article aims to expose a community entrepreneurial learning model formulated by the research group Engineers without Borders Colombia for providing communities with tools to create sustainable productive solutions, relevant for their members and for customers in a potential market. As the goal of this proposal is to carry out community learning processes, guided by members' decisions, to produce sustainable and replicable solutions, a bibliographical review with methodological approaches sustaining this purpose is presented. Subsequently, the learning model methodological construction is described and its applicability is evidenced through a case study project devoted to strengthening community green businesses in Guavio Province, Colombia. Finally, to replicate this learning model in other similar rural communities, conclusions and main lessons of the methodological application are explained.*

Introduction

Entrepreneurship is fundamental for both regional and community development (Hindle, 2010) and previous efforts to obtain a better theoretical understanding of the phenomenon of social entrepreneurship within developing areas have been done before, but in many cases, these efforts are limited to ontological and epistemological discussions. Likewise, community development remains more of a social movement than an applied but under researched behavioral science; it is not based upon well-articulated models, (Blakely, 1980).

Consequently, the research group Engineers without Borders Colombia, an alliance between two Colombian universities, is concerned about developing models to promote community entrepreneurial systems with potential to thrive, involving learning and investigation processes, supported on organizational learning and applied research methodological foundations. Therefore, this paper describes the methodological development of a model applied by Engineers without Borders Colombia in the execution of a community project, taken here as case study. This project took place in the context of the Guavio Region, articulating members of government entities, academia and rural communities. Organizational learning processes assessment and the main lessons learned from this methodological application will contribute to its replication in communities with similar conditions.

Context

Close to Bogota, Colombian capital city, we can find Guavio Province, a natural region that offers about 72% of the water demand to supply Bogota and 9.88% of total national hydroelectric production. Nevertheless, despite life quality of about 15% of the country's population, settled in Bogota and surrounding areas, depends on Guavio environmental sustainability, 27.2% of its population has unfulfilled basic needs and 75% of the corporate structure is represented by micro-companies and family productive units (Cámara de Comercio de Bogotá, 2006).

Therefore, Engineers without Borders Colombia, an alliance between Universidad de los Andes and Corporación Universitaria Minuto de Dios, seeks to develop entrepreneurial models, meaningful and replicable for this and similar contexts, combining community project management and academic training to strengthen people skills for boosting a social change, focused on triggering progress of rural communities, based on equity and social, economic, and environmental sustainability.

As far as Guavio problems, Engineers without Borders Colombia has engaged entrepreneurs, interested on adequate environmental practices, efficient use of natural resources, social ecological awareness, among others, in a community green businesses strengthening project focused on generating population financial freedom in an environmentally and socially sustainable way (Engineers without Borders Colombia, n.d.). Besides Engineers without Borders members, the Science, Technology, and Innovation Secretariat of Cundinamarca Governor's Office, the Social Innovation Scientific Park, 35 productive units, and 350 high school students have participated in this project.

Theoretical Framework

Community entrepreneurial projects can be executed as applied research processes, since their purpose is the application and use of knowledge while other knowledge is gained simultaneously; knowledge use and research results allow analyzing reality in a rigorous, organized, and systematic manner (Vargas, 2009).Three applied research approaches are considered here for community entrepreneurial project development: Participatory Action Research (PAR), Design Thinking, and a holistic technological thinking.

PAR purpose is empowerment, defined by the World Bank as "the expansion of assets and capabilities of people to participate in, negotiate with, influence, control, and hold accountable institutions that affect their lives" (Narayan, 2002). This approach is focused on transferring researcher capacities and empowering people to assume actions to improve their life conditions. Then, community members are transformed into researchers and their decisions guide the process. (Park, 1993).

Besides, as entrepreneurship survival depends on successful introduction of products or services into a market, Design Thinking approach is relevant, seeking to build not just functional but significant products or services for users, incorporating their needs to prototyping. Considering this approach, after identifying a problem or opportunity and a brainstorming process, a prototype or test is developed to transform ideas into products or services (Brown & Wyatt, 2010).

In terms of technology, a non-artefactual but holistic vision is taken into consideration, indicating that functionality depends on relation building within the context. For a technology to be functional, it is not only necessary determining whether solutions are viable, but also how technology functionality is perceived by a social group, whether people know how it works and is able to appropriate it.

On the other hand, this methodological development contemplates that survival of entrepreneurship results of learning processes, being effective learning capacity one of the main tasks of organizational management (Espejo, Schuhmann, Schwaninger, & Bilello, 1996). Specifically, this methodology is supported on two learning models.

OADI-SMM Organizational Learning Model

A learning process, described by Kofman (1992) through an observation, assessing, designing and implementation cycle, arises from individual learning definition: the potential improvement for an effective action (Espejo et al., 1996). This type of learning comprises two levels; operational learning (know-how) is related to skill acquisition while conceptual learning (know-why) is linked to conceptual understanding of an experience (Kim, 1993).

According to Espejo et al. (1996), it is possible to differentiate between mental models including conceptual learning (frameworks) and those including operational learning (routines). Operational learning is connected to observation and implementation while assessment and design to conceptual learning. In general, individual learning consists in improving individual mental models; individuals yield frameworks which allow design improvements, as well as routines and procedures to implement those improvements.

From individual learning definition, organizational learning can be defined as an organizational capacity to boost learning processes aiming at an effective action. However, it is pertinent to determine when individual becomes organizational learning. March & Olsen (1975) linked individual and organizational action by describing a cycle where beliefs and convictions lead to problem solving through individual actions, and then organizational actions trigger environmental answers and reactions.

Consequently, organizational learning results from individual actions supported on shared mental models. The Espejo et al. (1996) OADI-SMM organizational learning integrated model describes how individuals may learn independently from the organization and how double-cycle learning is fundamental for individual mental model transformation, influencing organizational shared mental models.

In contrast, although learning may not be separated from action, actions are not enough to generate learning; considerations must be made on what was learned from them. Therefore, learning results from planning, action and consideration processes (Johanessen, Olaisen, & Olsen, 1999). Also, it may casually result but it requires an effective structural design, setting forth practices to generate conditions for human communication systems (Reyes, 2012).

Garzón and Fisher Organizational Learning Model

Garzón & Fisher (2008) define learning as the capacity to create, organize, and process information from its sources to generate new knowledge at different organizational levels, creating a culture that promotes learning and conditions to develop new capacities, to design products and services, to increase existing offer, and to improve processes focused on durability. These authors characterize learning capacity in a static dimension made up by knowledge stocks embedded in the organization and a dynamic dimension determined by knowledge generation, absorption and diffusion flows, making stock evolution possible.

Garzón and Fisher model not only incorporates individual and organizational but also team and inter-organizational levels, representing much better the execution of community projects that require multidisciplinary teamwork. On the other hand, as the model considers learning sources, conditions and culture, it allows determining organizational durability factors.

POCDILD Methodological Proposal

Community entrepreneurial projects must be devoted to generating sustainable productive and technological solutions that are both relevant to the community as functional for consumers in a potential market. Hence, these solutions must be formulated by community members, responding to their own requirements and as community projects involve various actors beyond community members, including actors from universities and governmental entities, it is crucial to promote their participation along the entire project through participatory spaces for brainstorming and prototyping. Besides interaction and participation, these spaces must guarantee organizational learning conditions and culture consolidation as well as learning assessment mechanisms and knowledge production tools. To increase the efficacy of the participatory spaces, community projects are proposed to be developed in three phases; the initials of the phases, along with those of "learning diagnostic", give name to the methodology.

Preparation and Observation

Participatory spaces require a unifying language to simplify communication among all participants, being language a symbolic interpretation mechanism to build social realities

(Romero, Matamoros, & Campo, 2013). Also, participants must be familiarized with participatory spaces development methodologies. Therefore, there must be a stage devoted to establishing networks between participants and conduct workshops for methodological training before participatory spaces development.

It is highlighted that although this stage is just focused on individual learning, since there is no concrete organizational action, it generates conditions required for organizational learning. Relations among participants as well as methodological preparation workshops promote commitment, competencies and learning communities.

Conception

In this stage, brainstorming participatory spaces are developed for the first collective actions, from which learning is an expected result. These actions, associated to formulation and selection of ideas to be developed as prototypes, depend on conditions, organizational culture and communication mechanisms established from the previous stage. In this stage, individuals are expected to participate ideas construction and the selection of ideas at group and organizational levels. Also, teamwork must be documented to conduct knowledge management processes.

Since this stage is meant not only to share knowledge and improving skills but also to formulate solutions to previously identified needs, learning becomes the Argyris & Schön (1978) single-cycle, where individuals, under their mental models, detect inconsistencies in terms of expectations and respond to them by modifying strategies and assumptions in constant organizational bylaws. Also, since there is not yet a product or service prototyping process, from Garzón & Fisher (2008) learning sources, competition and customers are not yet considered.

Design and implementation

Again, actions conducted in this stage are possible due to organizational culture and conditions previously developed and the establishment of communication mechanisms. These actions, focused on prototyping, take place in participatory design spaces, proposed by Engineers without Borders Researcher Diana Duarte.

Prototyping requires feedback to validate bylaws and assumptions. This is consistent with the Argyris & Schön (1978) double-cycle learning, an episode with a feedback double cycle connecting mistake detection of effective performance strategies and assumptions with the guidelines that define such performance. Furthermore, since prototypes involve products or services for commercial activities, it is important to consider customers and competitors and every learning source proposed by Garzón & Fischer (2008) must be taken into consideration.

Learning Diagnostic

Through the implementation of qualitative and quantitative methodologies, it is recommended to conduct and organizational learning analysis for brainstorming and prototyping stages, where organizational actions occur, considering the following Garzón and Fisher (2010) criteria:

- **Considerations to assess conditions**

- **Competencies**
    - Learning processes must strengthen a capacity to change behaviors to perform activities in a better way.

- **Communities**
    - Participatory spaces must support the conformation of a community, as well as learning communities and commitment communities.

❖ The implementation stage must contribute to practice community conformation, as people may participate on the development of prototypes appropriate to their domain areas.

- **Considerations to assess culture**

➢ **Leader's concept of the man in the organization.**

❖ Leaders must propitiate constructive relationships and must be capable to upgrade learning processes.

➢ **Cultural system**

❖ Symbolism, philosophy and decision making must be aligned.

➢ **Organizational climate**

❖ A stimulating atmosphere must be generated, facilitating learning and promoting discussion, dialog, free flow of ideas, and knowledge utilization.

The following diagram describes the methodological proposal POCDILD.

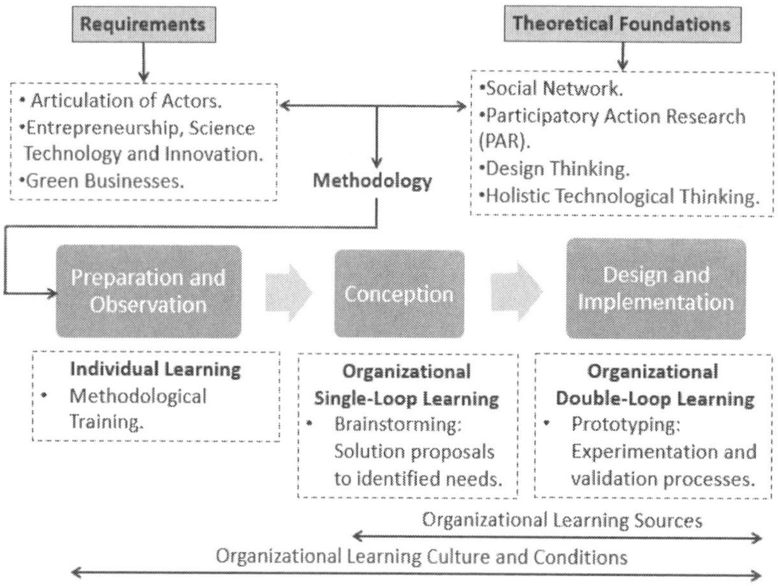

Figure 1: Community Entrepreneurial Learning Methodology

Proposal Case Study: Guavio Region Community Green Business Strengthening Project

The project "Fortalecimiento de las Capacidades de Innovación Social en Bogotá-Cundinamarca", which means Strengthening of Social Innovation Capacities in Bogotá-Cundinamarca, was raised through the agreement No. SCTel 019 of 2013 in charge of the Scientific Park of Corporación Universitaria Minuto de Dios and Cundinamarca Governor's Office and it was financed by the Science, Technology and Innovation Fund of the Colombian General System of Royalties.

Part of this project was devoted to consolidating a network of 35 or more productive units (companies or entrepreneurships), located in Gachetá, Junín, and Guasca municipalities, strengthening their green profile through a community learning process. Besides, 4 high schools participated in the project, with the purpose of inspiring entrepreneurial interest among

350 students. Likewise, it also encourages the participation of college students taking courses at Universidad de los Andes and Corporación Universitaría Minuto de Dios, reinforcing their academic education.

In accordance with the methodological formulation, this process of community learning is described below.

Preparation and Observation

After perceiving a discouragement of Guavio community members to perform regional development projects and identifying all potential project participants, Engineers without Borders Colombia established as a priority the construction of solid relations, a task that required two years of interaction between the group and the community. Subsequently, it was clear that to generate conditions for future participatory spaces it was necessary to foment interactions and to develop a unifying language between participants, as well as training them methodologically; considering participant features, it was proposed to develop a relation process with entrepreneurs, methodological preparation workshops with students and an event with entrepreneurs and students interacting for the first time.

First of all, expecting to communicate project objectives and network importance for regional economic dynamics, entrepreneur integration events were performed, involving two types of assistance motivation:

- Business expected benefits communication.
- Food was offered supplied by project participants.

In parallel with the entrepreneur integration events development, Innovation-Actions Workshops were executed with the students from the four schools: El Carmen Departmental Educational Institution and Mariano Ospina Rodríguez Technical – Commercial Departmental Educational Institution at Guasca municipality, Gachetá Normal High School, and Junín Normal High School. These training sessions provided students practical skills to become part of a social innovation process and to recognize their environment as a potential ecosystem for green entrepreneurship.

To close the preparation phase of the project, the Sustainable Entrepreneurship Seminar: "Quinoa Farming and Other Green Businesses" was held, congregating all participants. During the seminar, from the academy and practical experience, eight speakers illustrated the thematic of the event around sustainable entrepreneurships and successful quinoa plantation and commercialization cases.

Conception

In a brainstorming event, named Green Solution Laboratory, 30 teams were constituted by productive unit representatives and high-school and college students. Organized in three areas, depending on the main activity of the productive unit, each team proposed a solution idea. Each area selected three ideas and a panel with eight judges selected another idea. The criteria for formulating and selecting the ideas was their potential to provide technological solutions and strengthen productive units in a social, economic and environmentally responsible way.

Design and Implementation

In this stage, seven interdisciplinary teams experimented to transform the ideas into productive solutions, through participatory spaces focused on encouraging the community to perform autonomous activities. Therefore, Participatory Design Workshops were configured for generating business plans, conceptual models and functional prototypes, considering theoretical and practical limitations. The ideas were transformed into the following four prototypes, associated to seven research documents.

✓ **Touristic plan: A healthy journey:** A journey around four farms was designed, allowing the tourist to interact with the environment and try delicious and healthy regional cuisine.

✓ **Quinoa: Everyone's for everyone:** A quinoa producers' association was constituted.

✓ **Honey byproducts:** This green solution had the objective to formulate a business model for a portfolio of honey byproducts, starting by soap in bar and liquid presentations.

✓ **Research programs for the network from the schools:** Two solution ideas were developed, a material collection for a magazine and a photography gallery and a Smartphone application focused on regional visibility.

Organizational Learning Assessment in Community Entrepreneurial Learning Processes

Learning assessment in brainstorming and prototyping phases, is conducted through the implementation of performance indicators for participatory spaces.

With respect to brainstorming events, assistance is taken as Key Perform Indicator. It evidences commitment and learning desire from participants, being an approximation of learning and commitment community consolidation. For the case study project, the brainstorming event was the Green Solution Laboratory and from it assistance, high participation of students is highlighted; attending and completing tasks associated with the laboratory had an academic evaluation. On the other hand, although entrepreneur participation was lower than expected, possibly by not considering that the day of the event was appropriate for their commercial activities, it was enough to achieve the objectives of the event.

Table 1. Green Solution Laboratory Assistance

	College Student	High School Student	Productive Unit Representative	Total
Potential	46	66	51	**163**
Current	34	63	23	**120**
Assistance percentage	73.91%	95.45%	45.10%	**73.62%**

Regarding the design and implementation phase, success depends on the community articulation capacity to develop participatory spaces. Therefore, the number of participatory spaces is taken as indicator for community conformation. For the case study project, four solution ideas were developed, each one with more than two participatory spaces and at least one prototype.

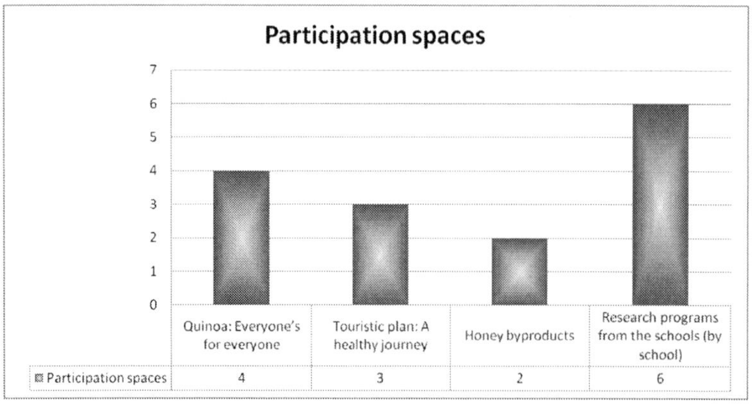

Figure 2. Number of Participation Spaces Held by Green Solution

Conclusions

Community projects must provide pertinent solutions for community members; if projects don't offer solutions to their relevant issues, by considering their interests, they won´t be committed with project execution. Participatory methodologies allow community members to identify their critical issues and foment their commitment with solutions proposing and developing. Therefore, this kind of methodologies must be incorporated in community project management to empower community members of the entire process.

On the other hand, students' participation in community projects strengthen their academic learning processes from knowledge, experiences and problematics shared by community members. Reciprocally, students' knowledge and perspectives reinforce solution construction processes.

The effective execution of participatory methodologies in community projects requires a previous process of preparation and observation, concerned with comprehension of community dynamics, establishment of participant's relations and methodological training. This process is crucial for transferring research capabilities to community members and generating conditions to develop effective brainstorming and prototyping spaces.

Brainstorming spaces, where all community members can participate to discuss common issues and possible ways to solve them, are fundamental to offer pertinent solutions for communities; at this point, participants are empowered and their own decisions begin to guide the project. Once established possible ways to solve community problems, learning process starts and community members can experiment in prototyping spaces for creating conditions and improving capabilities focused on ensuring solutions' viability in the long term. Particularly, this learning process is focused on promoting entrepreneurial solutions and experimentation spaces must employ methodologies for developing products or services relevant for potential customers.

As a learning process, learning assessment and knowledge management documenting is fundamental not only to improve current project processes but also to make it easier their replication, helping to prevent unintended consequences or repetition of possible errors.

As seen before, this methodological proposal for community project management focuses on achieving three fundamental purposes. The first is the generation of empowerment by community members, encouraging the development of their own ideas. The second is the sustainability of entrepreneurial solutions through a methodic development of functional prototypes. The third is to conduct replicable processes by documenting the lessons derived therefrom.

References

Argyris, C., & Schön, D. (1978). Organizational Learning: A Theory of Action Perspective. Reading, Mass: Addison-Wesley.

Blakely, E. (1980). Building theory for CD practice. Ames: Iowa State University Press.

Brown, T., & Wyatt, J. (2010). Design Thinking for Social Innovation. Stanford Social Innovation Review. Retrieved from http://ssir.org/articles/entry/design_thinking_for_social_innovation

Cámara de Comercio de Bogotá. (2006). Descripción de la provincia del Guavio. Retrieved November 5, 2013, from http://aulas.alianzaporelguavio.net/pluginfile.php/89/mod_resource/content/1/Descripcion%20guavio.pdf

Engineers without Borders Colombia. (n.d.). Fortalecimiento en Negocios Verdes en el Guavio. Retrieved July 25, 2016, from https://isfcolombia.uniandes.edu.co/index.php/proyectos/negocios-verdes

Espejo, R., Schuhmann, W., Schwaninger, M., & Bilello, U. (1996). Organizational Transformation and Learning. New York: John Wiley.

Garzón, M. A., & Fisher, A. L. (2008). Modelo teórico de aprendizaje organizacional. Pensamiento & Gestión(24), 195-224.

Hindle, K. (2010). How community context affects entrepreneurial process: a diagnostic framework. Entrepreneurship & Regional Development, 22(7), 599-647.

Johanessen, J.-A., Olaisen, J., & Olsen, B. (1999). Systemic thinking as the philosophical foundation for knowledge management and organizational learning. Kybernetes, 28(1), 24-46.

Kim, D. H. (1993). The Link between Individual and Organizational Learning. Sloan Management Review.

Kofman, F. (1992). Lecture slides, Cambridge Massachusetts: MIT Sloan School of Management.

March, J., & Olsen, J. (1975). The Uncertainty of the Past: Organizational Learning Under Ambiguity. European Journal of Political Research, 147-171.

Narayan, D. (2002). Empowerment and Poverty Reduction: A Sourcebook. Washington, DC: World Bank. © World Bank. https://openknowledge.worldbank.org/handle/10986/15239 License: CC BY 3.0 IGO. Retrieved from https://openknowledge.worldbank.org/handle/10986/15239 License: CC BY 3.0 IGO

Park, P. (1993). What is participatory research? A theoretical and methodological perspective. Voices of Change: Participatory Research in the United States and Canada.

Reyes, A. (2012). Organizational learning and the effective management of complexity. Kybernetes, 41(3/4), 318-326.

Vargas, Z. (2009). La investigación aplicada: una forma de conocer las realidades con evidencia científica. Revista Educación, 33(1), 155-165.

Copyright statement

Measuring active learning at the student level

John Chen,[1] David Janzen,[2] James Widmann,[3]
Karen McGaughey,[4] Jennifer Teramoto Pedrotti[5]
California Polytechnic State University, San Luis Obispo, USA
[1]jchen24@calpoly.edu, [2]djanzen@calpoly.edu, [3]jwidmann@calpoly.edu,
[4]kmcgaugh@calpoly.edu, [5]jpedrott@calpoly.edu

Abstract: *This research paper describes a method for measuring active learning at the student level. The quality of active learning is based on the ICAP framework, which delineates active learning into the four observable levels of Interactive, Constructive, Active and Passive. The method relies on a smartphone app that students use to measure active learning not only during class time, but also during their out-of-class learning times. In this paper we describe results from two studies using this method. The first measures active learning among students taking the same course from two different instructors, and compares how their measurements are similar and different. These results are further compared with in-class measurements by trained observers to validate their accuracy. The second study examines data from a larger sample of students taking an engineering mechanics course, but with several instructors.*

Introduction

Over the past few decades, there has been rising interest across higher education in active-learning pedagogies and, more recently, in research on active learning and its benefits for student learning. Across most academic disciplines, active learning is treated as a binary – a class is either active learning or not – and is driven by what the instructor does in the class. We propose that active learning should be considered on a continuum, both in its *quantity* and its *quality*. The quality of active learning is determined by the level of cognitive engagement that the student puts forth during a learning activity. Furthermore, while the instructor primes the class for active learning (or not), what the student does and thinks in response to that could vary, from low engagement to high. Finally, since students spend more time out-of-class than in-class when learning the content of a course, their quantity and quality of active learning during these times must also be measured in order to arrive at a true measure of each student's active learning. In this paper we describe a method that attempts to accurately measure active learning and its variability from student to student.

Background

Active learning

Active learning encompasses a broad range of classroom pedagogies that seek to transform students into participants in their learning, as opposed to merely transcribers of what the instructor says or writes at the front of the class. Clearly, the instructor plays a large role in setting up the class for this type of learning but, really, it is what the student *thinks* – about the problem or topic presented, the process undertaken to achieve a goal, and the result or solution – that drives the learning. According to Prince (2004), active learning requires "students to do meaningful learning activities and think about what they are doing" while engaging in the designed activities. Prince (2004) cites extensive evidence that such activities improve many aspects of learning, including information recall, concept mastery, critical thinking, and the correcting of misconceptions.

So, an instructor can prime a class for active learning, but do all students engage with the learning to the same extent? Surely the level of cognitive engagement could vary from student to student in such a class. Furthermore, it is also plausible that a skilled instructor can, purely through lecturing, induce his or her students to cognitively engage with the material, perhaps even deeply. This point emphasizes that perhaps active learning could be achieved without reliance on active-learning pedagogies. Finally, researchers and practitioners, when discussing active learning, usually only refer to the in-class environment when, in actuality, students spend significantly more time outside of class to learn course content. A complete measure of active learning, therefore, needs to consider all of these variables.

Framework for quantifying active learning

Engineering studies involve learning in a variety of contexts and activities. Students experience various levels of active learning through attending lectures, completing homework assignments, preparing for class, studying for quizzes and examinations, and seeking additional help. To describe these experiences for a complete measure of each student's *quality and quantity of active learning*, this project will use a framework developed by Chi and coworkers (Chi, 2009; Fonseca and Chi, 2011): the interactive-constructive-active-passive (ICAP) differentiated learning activities.

The ICAP framework classifies learning activities by observable, overt actions of the learner. A **P**assive learning activity is one in which the learner essentially engages in no overt actions. Listening to a lecture, watching a video, and reading text are examples. By contrast, an **A**ctive learning activity is characterized by overt actions that demonstrate paying attention. Examples include note-taking or highlighting of text. (Note that at this point "active" learning has taken on a definition that is quite different than the general use of the term in education, which would classify note-taking as a "passive" learning activity.) If the learner goes one step further and generates additional knowledge or information beyond that which is provided, she is engaging in **C**onstructive learning. Solving homework problems alone or resolving questions while reviewing notes alone are examples. The final category of **I**nteractive learning requires learners to interact with someone (e.g., a peer or expert) or something (e.g., a computer tutor) in order to build on the provided information. There must be an exchange of information between the members, such as defending one's responses, responding to questions, or correcting noted errors. The conventionally accepted active-learning techniques (Prince, 2004; Felder and Brent, 2009; Prince and Felder, 2006; Prince and Felder, 2007) would be classified as either constructive or interactive in the ICAP framework. Furthermore, based on the possible underlying cognitive mechanisms being activated by each kind of activity, the expected learning gains should increase in the order of passive < active < constructive < interactive; this is supported by the studies cited by Fonseca and Chi (2011).

This project adopts the ICAP framework to measure the "quality" of active learning experienced by students, with passive learning being lowest quality and interactive learning as the highest. The "quantity" of active learning is then simply the amount of total time students expend under each of the four ICAP categories for each course. Students will no doubt experience learning across this range of activities for a single course, and across the curriculum as they traverse it with various instructors.

Method for capturing student data

What remains is a method to capture the quantity and quality of active learning experienced by each student. Retrospective self-reports are known to be highly inaccurate due to recall bias (Bradburn, Rips and Shevell, 1987; Tourangeau, 2000) and observer reporting would require an inordinate effort and cost for any sizable study population. Instead, this project relies on the use of ecological momentary assessment (EMA) (Shiffman, Stone and Hufford, 2008) built in the form of a software application (or "app") for smartphones. EMA has a long history in psychological studies (Shiffman, Stone and Hufford, 2008). These brief, in-the-moment, self-reported data entries may take the form of pen-paper diaries, telephone calls, "beeper" reminders to complete a diary card or, most recently, digital data entries occurring at scheduled or random times throughout a day. This method for collecting data is superior to retrospective recalls because it occurs in real-time (or nearly so) for a specific event of interest, thereby minimizing recall bias. The data collection also occurs in the subject's natural ecology, which provides the data with context and ecological validity (Shiffman, Stone and Hufford, 2008). Another advantage is that EMA can provide detailed data on behaviors as they occur over time and in varying contexts, which provides for more nuanced interpretations (Shiffman, Stone and Hufford, 2008).

An EMA smartphone app is desirable since it is convenient and familiar to college students and facilitates data collection and storage. The app sends reminders to each participant after every scheduled class, as well as a few other times throughout the day to capture other learning experiences (e.g., study or homework time, office-hour visit). The participant then records the quality and quantity of each learning experience within the app, which uploads the data to a cloud-based server.

Methodology

Study context and population

This project is based in the engineering college of a large, western U.S., state-supported university. Two studies are reported here, with one study involving undergraduate engineering students of various majors taking an introductory thermodynamics course. The second study, still on-going and thus only preliminary findings are presented here, involves mechanical engineering undergraduates taking their first engineering mechanics (statics) course. Both studies were undertaken in September-December of 2016, which corresponds to the 10-week-long fall term of the university. Nearly all participants are in their second or third year of engineering studies. All participants were or are being compensated monetarily for contributing data, but their participation had no impact on their course grade.

Data recording and description

Study participants were recruited early in the fall term and each was provided with the smartphone app (called "Actively Learning") on either the iOS or Android platform. The participants were trained on using the app for data entry and also to discern the different levels of cognitive engagement in the ICAP framework. They next selected their specific section of either the thermodynamics (Study 1) or statics (Study 2) course from the database in the app; this allows for notifications to be set up to remind participants to make data entries.

The app sends reminders (notifications) to students to record both the quantity and quality of active learning throughout a day in order to minimize errors due to memory recall. Notifications are sent immediately after a class meeting (Fig. 1(a) and otherwise hourly, from 10 am to 10 pm (Fig. 1(b)). Figure 1(c) shows the user interface for making a data entry in the app. The app is connected via the cellular network or Wi-Fi to a server that stores all of the data (and also displays prior data, which can be edited or deleted if necessary). What is stored on the server, therefore, is a database containing all users, the course or courses being tracked for each participant, and the ICAP data. The latter contains time spent under each of the I, C, A or P levels, the learning event (e.g., class or homework or office hour) being recorded, and the date and time of each set of ICAP entries. The database can be exported from the server and imported into common software for analysis.

Results

Study 1

The first study involved 15 students taking an introductory thermodynamics course, and lasted for seven weeks (out of ten total weeks in the academic term). The primary purposes of this study were to gain user testing of the app (not discussed here) and serve as a validation study for the data gathered. The participants took the course from one of two instructors, with seven enrolled with one instructor (Instructor A) in the same class, while the remaining eight were in one of three possible sections of the other instructor (Instructor B), who conducted all sections in the same format and in succession on the same days of instruction. Instructor A is a veteran faculty member who relies almost exclusively on the lecture method of instruction. Instructor B has been teaching for six years and employs the Team-Based Learning (TBL) pedagogy (Michaelsen & Sweet, 2008). TBL is considered an active-learning pedagogy and requires students to do individual work before each class meeting that may include, for example, reading the textbook or watching a video. In class, students first take an individual quiz based on the pre-class work and then, working in term-long groups, they complete all other activities, which include a group-based quiz and solving one or more instructor-designed problems. The problem-solving activities are interspersed with mini-lectures (<10 min) to set up the problem or provide background information.

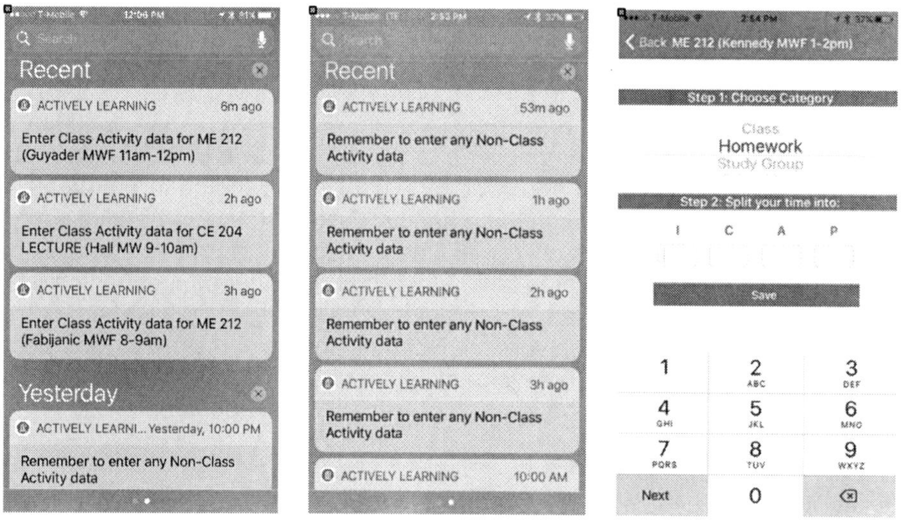

(a) (b) (c)

Fig. 1: Screenshots from an iOS-based smartphone of (a) post-class notifications, (b) non-class notifications, and (c) the user interface for data entry.

Figure 2 shows data recorded by individual participants in each instructor's class(es) on one particular day near the start of the study. The difference in pedagogy is strikingly clear: Students in the lecture class (Fig. 2(a)) recorded nearly all Active and Passive engagement, while the TBL class (Fig. 2(b)) recorded a mix of all four engagement levels, with most being at the Constructive or Interactive level. These are expected results since a lecture style requires students to listen and take notes, while TBL requires students to interact with peers to achieve the requested outcomes (quiz or problem solution).

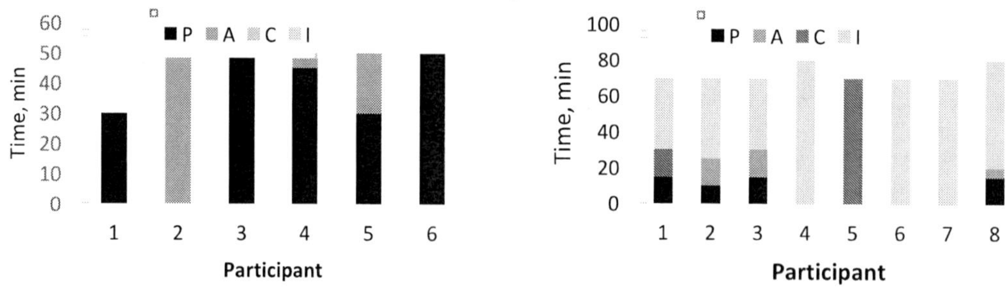

(a) (b)

Fig. 2: ICAP data from each participant in a (a) lecture-based class and (b) TBL class. Note that one participant in the lecture-based class did not record data this day.

In each mode of instruction shown in Fig. 2, there is still substantial student-to-student variation among each student's perceived level of engagement. The question is whether these are true variations or poor training on the ICAP framework, which leads students to assign times to incorrect categories. To test this, and to validate the accuracy of the data entries, two of the investigators attended randomly selected classes and measured the ICAP times of some study participants through direct observation (recall that the ICAP framework relies on overt, observable behaviors). The participants were told beforehand that the investigators were attending class to observe "a typical class," but not that any of them were being directly observed and measured. Figure 3 shows comparisons between the investigators' and each student's measurements from each class. In general, the agreements between the two investigators and between the investigators and each student are very good. All measurements correctly capture the predominant instruction mode of each class. Figure 3 further demonstrates that, indeed, there are variations between students in their level of engagement during the same class. Furthermore, these variations are largely confirmed by the investigators' observations.

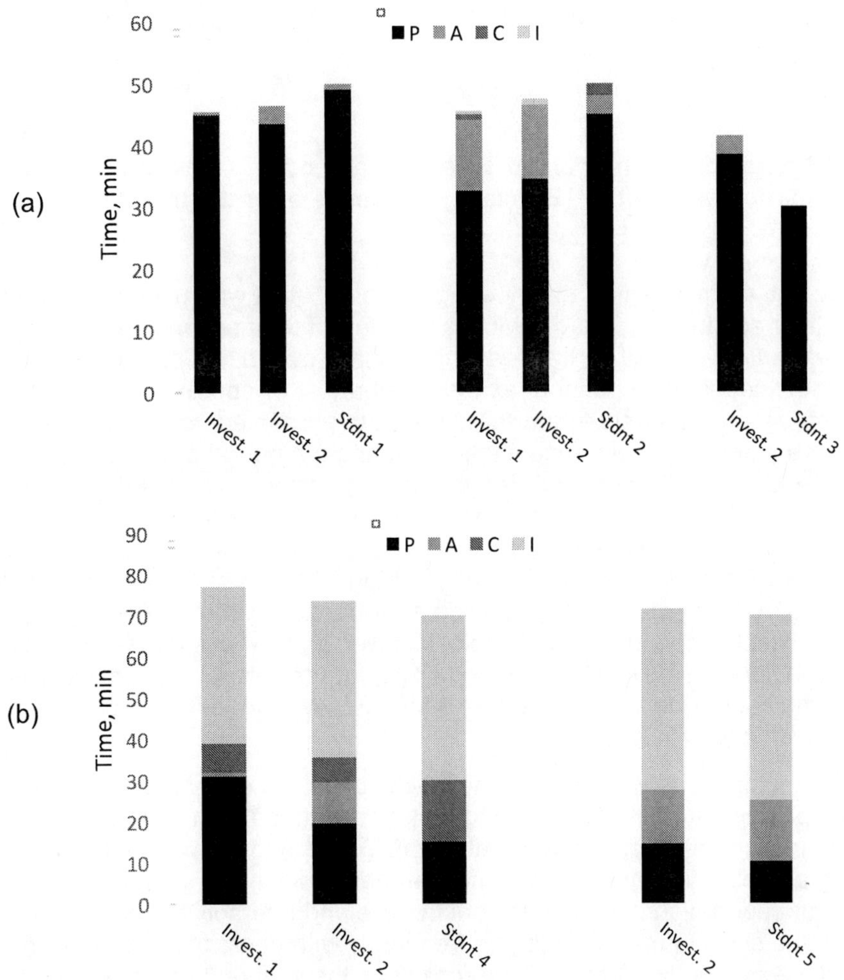

Fig. 3: ICAP data comparison between two investigators observing individual students in (a) a lecture-based class and (b) a TBL class.

The data shown in Fig. 3 were taken during the first week of data collection, which means that the students' training on the ICAP framework was very recent. We were concerned with students straying from this training after some time and therefore repeated the classroom observation six weeks later, and the results are shown for the TBL class in Fig. 4. As with the earlier validation measurement, the agreement between the investigators was very good, while that between the students and the investigators was reasonably good. A repeat observation was also done for the lecture-based class and the agreements were as good or better than that shown in Fig. 3(a).

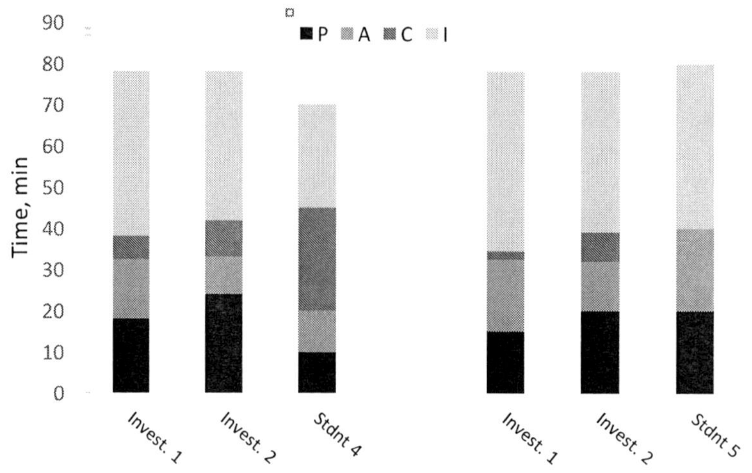

Fig. 4: ICAP data comparing measurements made by two investigators observing two participants in the TBL class, six weeks after the initial training.

Study 2

This study involved 42 mechanical engineering students, all of whom were at roughly the same point in their academic careers, which is the start of their second year of engineering studies. These participants will be tracked longitudinally over a five-course engineering-mechanics sequence that will span four academic quarters or approximately 15 months (including a non-academic summer quarter). The data presented here was taken during the first course in the mechanics sequence, statics. During the period of this study, which lasted the final four weeks of the 10-week quarter, the participants were enrolled in one of 14 possible sections of statics, taught by six instructors with wide-ranging teaching experience and varied prior student evaluations of teaching. Most instructors relied on the traditional lecture method but some also used questioning during lecture to induce critical thinking or deep reflection. One instructor, while not using a formal active learning pedagogy, employed a strategy of brief lectures to introduce a topic, followed by having students work in ad hoc pairs to solve problems as he walked around to monitor the progress or answer questions. All sections of the course were conducted on a Monday-Wednesday-Friday format with each class meeting lasting 50 minutes.

Figure 5 shows the distribution of an average 50-minute class time across the four ICAP levels for each participant. It is clear that students, even those in the same class, do not cognitively experience the class in very similar ways. Instructor C, who uses the informal active-learning approach in class, has the most students within the sample (28 of 42). Many of these students report high amounts of Constructive and Interactive times, as would be consistent with the class format of engaging frequently with peers to solve new problems. The remaining five instructors, while relying mostly on the lecture method, which would result in almost exclusively Passive (listening) and Active (notetaking) times, still were able to induce some of their students to work at the cognitively higher end (C & I) of the ICAP framework.

Next, we focus on data recorded by participants for learning outside of class. Figure 6(a) shows the total number of non-class entries made by each participant during the study period. The variation in study habits among students is striking: The entries range from three (slightly less than once per week) to over 40 (or about 10 per week). Just as interesting is that the total time spent (or the total ICAP time) for this out-of-class learning is much less varied, and seems to bear no relationship to the total number of non-class entries (black columns, Fig. 6(b)). For example, participants 1-3 recorded 47, 40 and 29 entries, while their total ICAP times were 1530, 1745 and 1525 min, respectively. Several other participants, who each had fewer entries, exceeded these total ICAP times. Also plotted in Fig. 6(b) are the sum of I and C times for each participant. This confirms, as would be expected and hoped for, that the majority of the out-of-class learning times is spent at the high-quality Interactive and Constructive levels, where students are gaining new knowledge while, for example, completing homework or attending office hours.

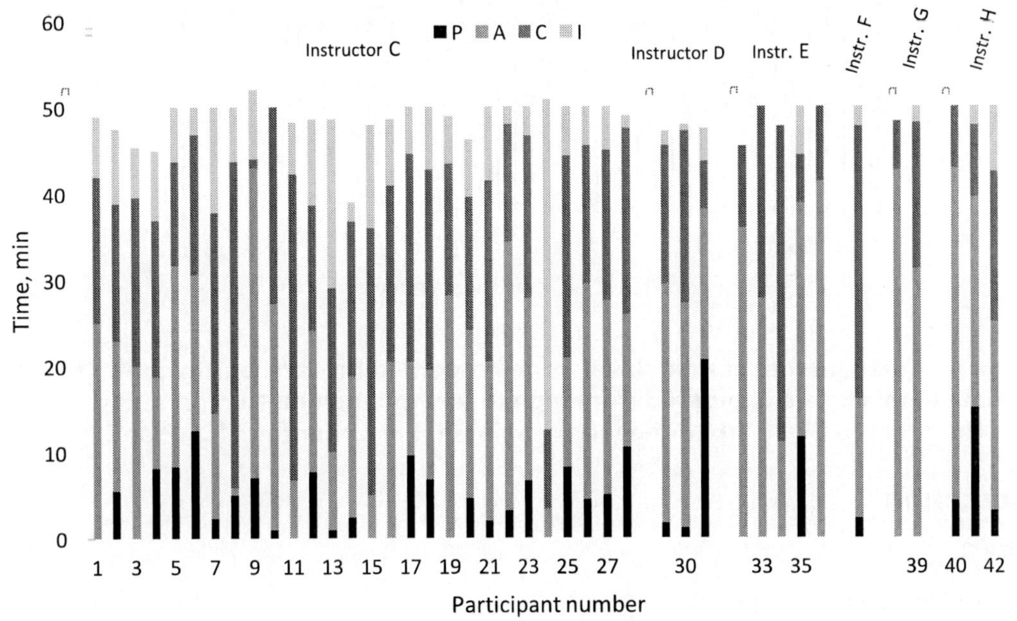

Fig. 5: Average ICAP time distribution for each participant for a 50-min class.

(a)

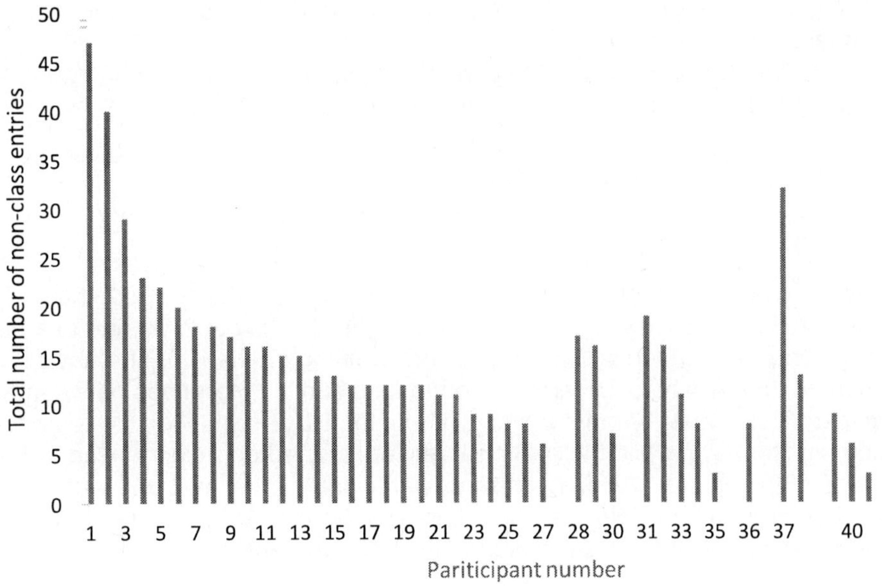

(b)

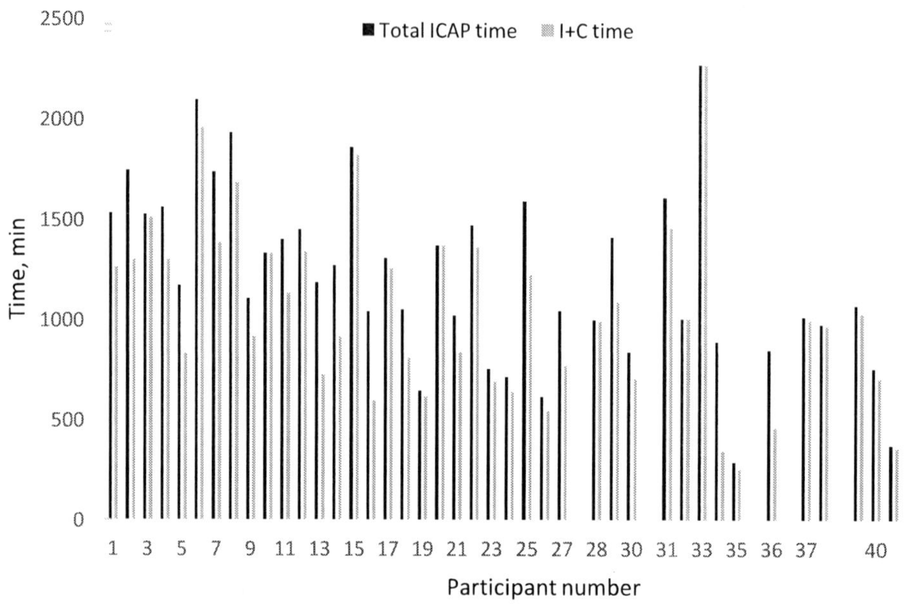

Fig. 6: (a) Total number of non-class entries recorded by each participant. [Note that the data in this figure is ordered from high to low for each instructor, and rearranged from the order in Fig. 5.] (b) For each participant, his or her total ICAP and I+C times.

Discussion

Study 1

Study 1 established several inferences:

- ICAP data recorded by participants of a class reliably differentiate between lecture and an active-learning mode of instruction (Fig. 2).
- Within a single class, the level of engagement (in reference to the ICAP framework) with the learning experience can vary widely from student to student (Fig. 3)
- These student-to-student variations are largely confirmed by observer measurements of individual students (Fig. 3).
- The current ICAP training provided to participants appears to be durable, with reliable measurements lasting six weeks after the initial training (Fig. 4).

At this preliminary stage of the overall project, we have not yet performed detailed statistical analyses on the data to establish, for example, inter-rater reliability between the two investigators' measurements of student engagement or the intraclass correlation between participants within the same class.

While measurements between the two investigators were generally very good, the comparison with each participant was only reasonably good. This could be due to a number of speculative factors. Strengthening the participants' training on the ICAP framework could perhaps improve on their ability to discern the sometimes subtle differences between the levels. The investigators certainly have a firm grasp of the ICAP framework based on reading the primary literature, while the participants were trained on a condensed translation of this framework in 30 minutes. Also, since study participants must wait until after a class ends, or later, to enter data, this delay may result in slight errors during recall. This is in contrast to the investigators who were measuring the engagement in real-time, down to half-minute intervals.

Study 2

This study reconfirmed that students' cognitive engagement can vary significantly between students within the same class (Fig 5). Although this study involved several instructors, with only one that employs active learning strategies, the differences between the classes are not obvious at first glance. Even in this case, however, the question could still be asked: Is there a measurable difference between an active-learning class and a mostly lecture-based class? We probe this question by performing a t-test between the data from Instructor C (for 28 participants) against that from the remaining 5 instructors (14 participants). The test compares the means of the average I plus C times per class between these two groups, and the difference is statistically significant (p=0.002; Instructor C: mean=27.1 min, std. dev.=9.14; all others: mean=16.9 min, std. dev.=9.89). This result should be viewed with care, however, given the sample sizes and data aggregation undertaken.

Study 2 also demonstrated that students' study habits outside of class vary widely in the frequency of study but far less so in the total time spent studying (Fig. 6). This would suggest, therefore, that those students who study frequently do so in shorter time increments than those who study less frequently. The true question, perhaps, is whether one study approach is more effective or efficient than the other. We cannot answer this question at present.

Summary

In this paper we presented early findings from an ongoing longitudinal study of active learning in engineering students. The data included students' engagement with learning during class, as well as their learning events out of class. As the data set grows from both additional data from the current participants plus additional future participants, the possible research questions that could be answered through these data are intriguing. For example, correlations between the quantity and quality of active learning with performance measures (such as exam scores or grades) or affective factors could be investigated.

References

Bradburn, N., Rips, L., & Shevell, S. (1987). Answering autobiographical questions: the impact of memory and inference on surveys. *Science*, 236, 157–161.

Chi, M.T.H. (2009). Active-Constructive-Interactive: A conceptual framework for differentiating learning activities. *Topics in Cognitive Science*, 1, 73-105.

Felder, R.M. & Brent, R. (2009). Active Learning: An Introduction. *ASQ Higher Education Brief*, 2(4).

Fonseca, B.A. & Chi, M.T.H. (2011). Instruction based on self-explanation. Ch. 15 in *Handbook of Research on Learning and Instruction*, R.E. Mayer and P.A. Alexander (Eds.). New York, NY: Routledge.

Michaelsen, L.K. & Sweet, M. (2008). The essential elements of team-based learning. *New Directions for Teaching and Learning*, 116, 7-27.

Prince, M. (2004). Does active learning work? A review of the research. *Journal of Engineering Education*, 93(3), 223-231.

Prince, M. & Felder, R.M. (2006). Inductive teaching and learning methods: definitions, comparisons, and research bases. *Journal of Engineering Education*, 95(2), 123-138.

Prince, M. & Felder, R.M. (2007). The many faces of inductive teaching and learning. *Journal of College Science Teaching*," 36(5), 14-20.

Shiffman, S., Stone, A.A., & Hufford, M.R. (2008). Ecological momentary assessment. *Ann. Rev. Clin. Psychol.*, 4, 1-32.

Tourangeau, R. (2000). Remembering what happened: memory errors and survey reports. In *The Science of Self-Report: Implications for Research and Practice*, A.A. Stone et al. (Eds.), Erlbaum, Mahwah, NJ, pp. 29-48, 2000.

Acknowledgements

This material is based in part upon work supported by the U.S. National Science Foundation under Grant No. DUE-1561623 and DUE-1431879. Any opinions, findings, and conclusions or recommendations expressed in this material are those of the authors and do not necessarily reflect the views of the National Science Foundation.

Copyright statement

Refining the WLR: Quantifying the difference in learning success between courses

Julie Direnga
Hamburg University of Technology, Hamburg, Germany
julie.direnga@tuhh.de

Dion Timmermann
Hamburg University of Technology, Hamburg, Germany
dion.timmermann@tuhh.de

Ferdinand Kieckhäfer
Hamburg University of Technology, Hamburg, Germany
ferdinand.kieckhaefer@tuhh.de

Christian Kautz
Hamburg University of Technology, Hamburg, Germany
kautz@tuhh.de

Abstract: *Pre- and posttests are often used in Engineering Education Research to assess teaching and learning. In some cases, the pretest and the posttest are different instruments, making it difficult to analyze the data with traditional methods such as learning gain. Previously, we proposed a statistical method, the so-called Weighted Linear Regression (WLR), to analyze such data. The WLR is a tool for teachers and researchers to effectively compare two or more courses or cohorts based on data from non-identical pre- and posttests (NIPPs). So far, however, it has not been used to statistically quantify the difference between courses or cohorts. This work refines the current WLR method so that the difference in learning success between courses using NIPPs can be quantified. Consequently, informed conclusions can be drawn if courses differ in effectiveness.*

Introduction

Researchers in Engineering Education Research often use pre- and post-tests to assess the effect of a course or intervention on student learning success (e. g. Hake (1998)). Sometimes, non-identical pre- and post-tests (NIPPs) are being used, i.e. the pre-test and the post-test are different instruments. This might be the case when an instrument valid for use as both pre- and posttest does not exist (e. g. Brose and Kautz (2011)). Consequently, the result of the analysis is not a learning gain as used by Hake (1998) and others. In Direnga et al. (2014), we proposed a statistical method, the so-called Weighted Linear Regression (WLR), to analyze such data. The WLR is a tool for teachers and researchers to effectively compare two or more courses or cohorts based on data from NIPPs. So far, however, it has not been laid out how to statistically quantify the difference between courses or cohorts with this method.

After stating our research question and theoretical framework, we will give a short introduction to the WLR and how to interpret it. We will then lay out the theory of the statistics used to quantify any differences between two courses or cohorts, which we will from now on refer to as different treatment groups. The statistics involve linear regression confidence bounds, Analysis of Covariance (ANCOVA), and effect size. In the Results section, we will present two examples to illustrate the quantification methods and how to interpret the statistics. Finally, we will discuss the results and finish with a short conclusion.

Research Question

How can we quantify the significance of a difference between treatment groups by means of the WLR?

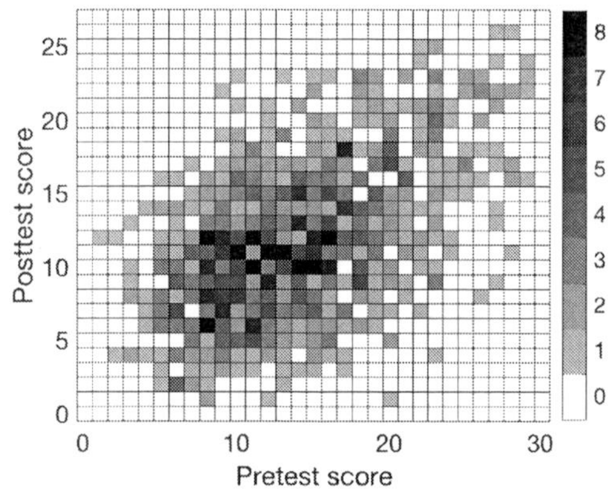

Figure 1: Distribution of individual pre- and posttest score pairs, N = 828.

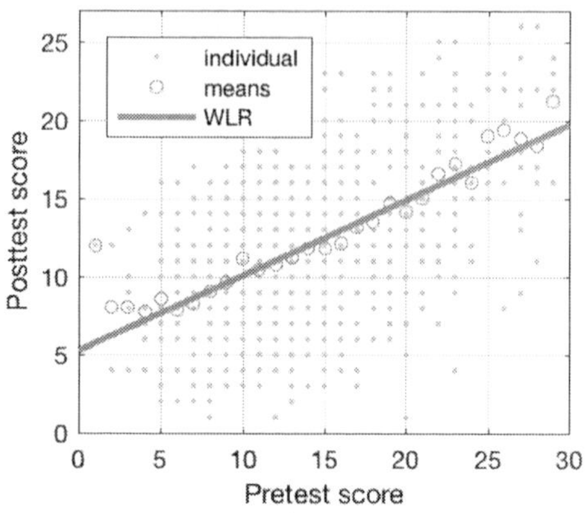

Figure 2: Mean posttest scores for each pretest score and resulting WLR. (Multiple occurrences of identical pre-/posttest score pairs are not represented here, see Figure 1 instead.)

Theoretical Framework

The statistical method we call WLR can be applied to many types of data. The purpose for which we want to use and promote it is to evaluate effectiveness or quality of teaching. Conceptual understanding is a critical achievement on the way from novice to expert thinking (Mazur (1997)). Furthermore, conceptual understanding can be measured by means of validated instruments, e. g. concept inventories (CIs). We can thus measure and compare the quality of teaching by multiple choice questions and CI scores (Adams and Wieman (2011)). In doing so, we do not focus on individuals but on aggregated data from entire cohorts. These scores are taken at face value, i. e. measurement errors will not be discussed here.

In case there is no CI that is valid as both pre- and posttest, but it is reasonable to assume different baseline understanding for the different treatment groups, NIPPs are a useful alternative, if chosen carefully.

The Weighted Linear Regression (WLR)

The WLR was developed as a tool to compare the effectiveness of teaching based on pre- and posttest score pairs. It is based on a regression line with the n (discrete) pretest scores X_i as covariate, the posttest scores as outcome, and the different groups or teaching methods as predictor variable. An example of the raw data, namely the distribution of pre- and posttest scores, is shown in Figure 1 (for one treatment group only). Even though the variation is rather large for the individual data, we found that the mean posttest scores $\bar{Y}_i$ for each possible pretest score X_i are quite well described by a general linear model (see Figure 2). We use the *mean* posttest score for two reasons: (1) to stress that we rather make inferences about the entire cohort instead of individuals, and (2) to visualize that the linear model is indeed appropriate.

The linear model is given by

$$\bar{Y}_i = \beta_0 + \beta_1 X_i + \varepsilon_i. \tag{1}$$

It assumes that all $\bar{Y}_i$ and the error terms ε_i are independent and normally distributed. The latter are also assumed to have an expected value of zero (Neter et al., 1985, p. 49 and 60).

The line parameters of the model are estimated by a weighted least squares approach by minimizing the following term with respect to β_0 and β_1

$$\Sigma_{i=1}^{n} w_i \left(\bar{Y}_i - \beta_0 - \beta_1 X_i\right)^2 \tag{2}$$

(Neter et al., 1985, p. 167) which results in the estimated points on the line

$$\hat{Y}_h = b_0 + b_1 X_h \tag{3}$$

(Neter et al., 1985, p. 72). The weights should take into account the distribution of the sampled individual pre- and posttest score pairs (see example data in Figure 1). We generally have a higher confidence in $\bar{Y}_i$ if the sample size n_i (the number of students with pretest score X_i) is large. In Direnga et al. (2014), we take the sample size into account by defining the weights as

$$w_i = n_i. \tag{4}$$

Note that with these weights, the resulting line parameters b_0 and b_1 are actually identical to regression parameters on the unweighted individual data. Thus, once linearity has been confirmed, one can also perform a standard linear regression on the unweighted individual data to determine the line parameters. The residuals will of course be much greater in the latter case, but they will not be of any importance for the quantification of the difference between courses.

How to interpret the WLR

The line is best interpreted as the expected posttest score for "model students" with the respective pretest scores, given that linearity is assumed and the best fit to the data is applied. What the WLR generally *cannot* do is predict individual students' posttest scores from their pretest scores (see Figure 4). Apart from that, the WLR provides us with three valuable pieces of information:

1. A high degree of linearity of the average posttest scores can be interpreted as constant sensitivity of the posttest over the entire score ranges given constant sensitivity of the pretest.
2. The general post-instruction level is represented by $\hat{Y}_{50\%}$, the estimated posttest score at a pretest score of 50 %.
3. The discriminating effect of the instruction for students with different pretest score levels is represented by the slope, b_1.

A high general post-instruction level as well as a non-negative slope are of course favorable. Whether the ideal course should have a strong positive slope (i. e. model students with a significantly higher pretest score should also have a significantly higher posttest score compared to low scoring model students) or not (i. e. model students should reach more or less the same level after instruction independent of their pretest scores) is on the other hand rather a matter of personal opinion and/or education politics. The results are most meaningful when there is a certain correlation between the pretest and posttest scores, so that the assumed linear model is appropriate. If this correlation cannot be found, it is questionable if the test pair is an adequate combination to use.

NIPPs generally do not allow to identify the pre- and posttest score pairs that are the equivalent to "no learning". Also, the resulting line parameters depend strongly on the test pair. Therefore, a single regression line does not tell us much. Instead, it must be compared to other datasets using the same test pair.

The WLR is based on sampled data. Therefore, we have to consider statistical uncertainties while comparing different courses. What tells us that the lines do not only differ due to sampling error? One way to answer this question is to determine the confidence bands around the regression lines.

Refining the WLR

Confidence Bounds

As we are dealing with sampled data, the resulting regression parameters, both slope b_1 and intercept b_0, can only be estimated with a certain confidence $1 - \alpha$. By using a joint estimation technique, we make use of the fact that β_1 and β_0 are not independent. The inequality

$$\frac{N\left(b_0-\beta_0\right)^2+2\left(\Sigma X_i\right)\left(b_1-\beta_1\right)+\left(\Sigma X_i^2\right)\left(b_1-\beta_1\right)^2}{2\ MSE} \leq F\left(1-\alpha; 2, N-2\right) \tag{5}$$

describes an elliptical $1-\alpha$ confidence region in the β_0, β_1-space (Neter et al., 1985, p. 147f). Here, MSE is the mean squared error (see Equation 8), N is the sample size, and $F\left(1-\alpha; 2, N-2\right)$ denotes the inverse of the F-distribution with 2 degrees of freedom in the numerator and $n-2$ degrees of freedom in the denominator for the percentile $1-\alpha$ (Neter et al., 1985, p. 8f). All β_0, β_1-pairs contained in this region determine the set of possible regression lines. From this, a hyperbolically shaped confidence band around the estimated regression line delimited by the upper and lower confidence bounds can be derived:

$$c_{upper,lower}|_h = \hat{Y}_h \pm W \cdot s\left(\hat{Y}_h\right). \tag{6}$$

$\hat{Y}_h$ is defined by Equation 3, W is a constant given by

$$W^2 = 2F\left(1-\alpha; 2, N-2\right), \tag{7}$$

and the variance of the estimated points is

$$\begin{aligned} s^2\left(\hat{Y}_h\right) &= MSE \cdot \left[\frac{1}{N} + \frac{\left(X_h-\bar{X}\right)^2}{\Sigma\left(X_i-\bar{X}\right)^2}\right] \\ &= \frac{\Sigma\left(Y_i-\hat{Y}_i\right)^2}{N-2} \cdot \left[\frac{1}{N} + \frac{\left(X_h-\bar{X}\right)^2}{\Sigma\left(X_i-\bar{X}\right)^2}\right] \end{aligned} \tag{8}$$

(Neter et al., 1985, p. 154 and 74). Here, $\bar{X}$ is the overall average pretest score. Note that the values of the variance for any h also depends on data from other pretest scores.

If the experiment were repeated again and again with an assumed constant population, then 95 % of the times (for $\alpha = 0.05$), the confidence band would include the entire regression line. We can thus conclude that if the confidence bands of two regression lines do not overlap in a large part of the pretest score range, it is highly unlikely that the data stem from the same population. In other words, we are most likely observing an effect of the intervention to the experiment group.

Significance

Can't we use Analysis of Covariance (ANCOVA) to test for statistically significant difference in the data? The answer depends on whether or not there is a difference in slopes. And yes, we can test for this difference with an ANCOVA model that includes an interaction factor between the predictor variable (treatment groups) and the covariate (pretest score). The statistical significance of the interaction term tells us whether or not the slopes of the regression line should be considered different or equal. When the slopes are not the same across treatment groups, one of the prior assumptions of the ANCOVA model is violated (Neter et al., 1985, p. 850f). We must therefore pay close attention to which aspects of the interpretation of the data are independent of the assumption of equal slopes. Unequal slopes prevent us from reporting "a single, overall effect of the independent variable on the dependent variable, at every value of the covariate" (Grace-Martin (2014)). This is not what we want to do. Instead, we acknowledge a statistically significant difference in slope by interpreting it as the discriminating effect of the instruction as mentioned above.

Effect Size

As often critiqued (e. g. Coe (2002)), statistical significance only shows whether or not there is a measurable effect, but it does not report the size of the effect. Furthermore, significance depends heavily on the sample size. Effect size, on the other hand, is a statistic that quantifies

ANCOVA Table					
Source	d.f.	Sum Sq	Mean Sq	F	Prob>F
treatment	1	3871.4	3871.44	216.99	0
prescore	1	7528.1	7528.1	421.95	0
treatment*prescore	1	197.4	197.37	11.06	0.0009
Error	1174	20945.7	17.84		

Figure 3: Interaction term between treatment and pretest score is statistically significant at a significance level of p $<$ 0.001.

the effect and is independent of sample size. It is generally defined as the relation between difference in group means and the common standard deviation (Coe (2002)).

To determine the effect of one treatment compared to another, we take the difference in the overall post-instuction levels $\hat{Y}_{50\%}$ and divide it by the group average sample standard deviation $\bar{s}$. To be on the safe side, we take the larger one of the two groups i and j.

$$d_{i,j} = \frac{|\hat{Y}_{50\%,i} - \hat{Y}_{50\%,j}|}{\max(\bar{s}_{i,j})} \tag{9}$$

Results

To demonstrate the suggested refinements, we calculate the confidence bounds on data previously published in Direnga et al. (2015). Figure 4 shows the resulting regression lines of two courses labelled treatment 1 and treatment 2. The maximum possible scores on the test instruments are reflected by the ranges of the axes: 30 points on the pretest and 27 points on the posttest. The dotted lines represent the respective 95 % confidence bounds calculated by Equation 6. Compared to the variation in individual test scores (see Figure 1 and 2) the band seems quite narrow. This is because the band does not serve as a prediction for the next individual measurement, but instead as a prediction for the next regression line in case the experiment were repeated and a new sample of the same size was generated from the same population.

The confidence bands in Figure 4 do not overlap except for the very low pretest scores between 0 and 2 points where very few measurements exist, so this region is of low relevance. At $x = 15$, the narrowest part of the confidence bands, we have a gap of 4.0 points between the means of the two treatments, 3.1 points between the inner confidence bounds, and 4.9 points between the outer confidence bounds. Because we have to account for two uncertainties of α, the total confidence is $(1-\alpha)^2$. For $1-\alpha = 95\%$, this results in a total confidence of 90.3 %. If a larger total confidence is required, $1-\alpha$ must be chosen accordingly.

The resulting interpretation would be that with 90.3 % confidence, a model student in the treatment group 1 with a pretest score of 15 points will score between 3.1 and 4.9 points higher on the posttest compared to the same model student in the treatment group 2. Equation 9 results in an effect size of $d_{1,2} = 0.9$ (Coe, 2002, p. 4 presents multiple ways to interpret this number). Furthermore, the ANCOVA shows that the interaction term between treatment group and pretest score is statistically significant at $p < 0.001$ (see Figure 3), i. e. the slopes are different. Therefore, we can conclude that the different treatments actually affect students differently depending on their pretest score. To be specific, although treatment 1 results in superior expected posttest scores for all model students independent of pretest score, students with higher pretest scores tend to benefit even more from treatment 1 compared to students with lower pretest scores.

Another example is presented in Figure 5 which shows a comparison of again treatment 1 and a third treatment. The values of the regression lines are identical in the low pretest score range and diverge slightly towards the high pretest scores. Looking only at the regression lines, it is impossible to decide whether the slightly higher values for treatment 1 should be

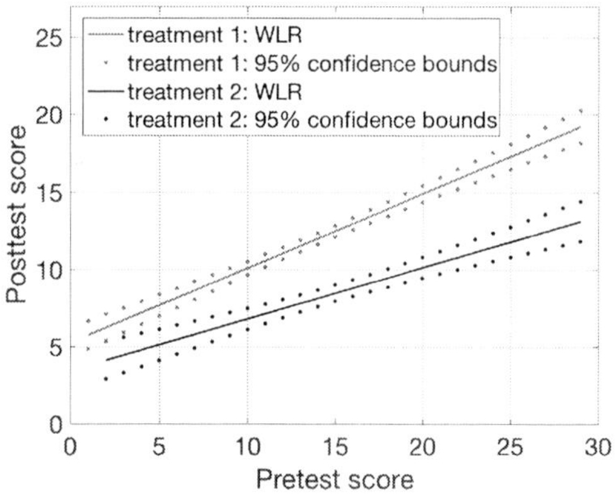

Figure 4: 95 % confidence bounds on the regression lines. Comparing treatment 1 (N = 828) and treatment 2 (N = 350).

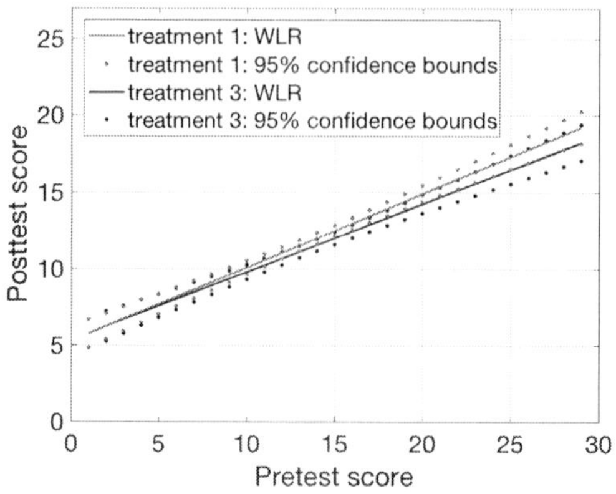

Figure 5: 95 % confidence bounds on the regression lines. Comparing treatment 1 (N = 828) and treatment 3 (N = 674).

interpreted as a measurable effect or as sampling error. By means of the strongly overlapping confidence intervals, however, we can conclude that we cannot see a difference between treatments 1 and 3 in the data.

The ANCOVA shows no significant difference in slopes. In this case, we can assume a parallel slope model in the ANCOVA. This results in the intercepts differing by 0.44 points at a significance level of $p < 0.05$. Different approaches can thus lead to different results in interpreting the data. Here, the difference between significance and effect size also becomes clear. Even though the difference might be statistically significant, the effect size is negligible at only $d_{1,3} = 0.1$.

The original idea of the WLR is to take the weighted means as a basis for the regression line. The confidence bounds, on the other hand, are calculated based on the individual data because by taking into account only the weighted means, we would lose valuable information about the confidence. In other words, the number of points the confidence bounds would be based on would never exceed the number of different pretest scores, and this would result in wider bounds than necessary.

Discussion and Conclusions

In this paper, we have provided an answer to the question how to quantify differences between two regression lines to assess the effectiveness of different courses on student conceptual understanding. Confidence bounds on the regression lines give us an insight how likely the lines and therefore the effectiveness of a treatment or course differ. In case of parallel lines, an ANCOVA can provide additional insight to determine the existence of an effect. Calculating the effect size at 50 % pretest score can furthermore quantify this effect also for non-parallel slopes.

It should be noted that the WLR works best with larger samples sizes. With small N, the assumption of linearity is often more difficult to support by means of the data and the confidence bounds of the WLR are wider. Furthermore, the linear model used by (Neter et al., 1985, p. 60) is assumed to have independent errors ε_i, normally distributed $N(0, \sigma)$, where σ is unknown but equal for all ε_i. Working with large data sets helps with this assumption. Another prerequisite is that the tool works on *linked* pre- and posttests of individual students (see Direnga et al. (2016) for a proposal of how to link test scores). The linking procedure must be taken care of when administering the tests.

With all these aspects taken into account, the WLR is a very valuable tool that provides us with two parameters to characterize the effect of a course or treatment. The data can furthermore be easily visualized and the confidence bands add extra value to the interpretation of the data. Introducing the confidence bounds also confirms that $\hat{Y}_{50\%}$ together with the slope b_1 is a better characteristic parameter for a course or treatment than e. g. the intercept b_0. The "bottleneck" of the confidence band shows that of all values $\hat{Y}_h$, $\hat{Y}_{50\%}$ is the one most precisely predicted and most stable.

Engineering instructors and researchers who wish to assess their teaching my means of a pair of pre- and posttests that are not necessarily the same instrument or who want to use another predictor instead of a pretest can use the WLR as a statistical tool to evaluate, interpret, and compare their data.

Ackknowledgements

We would like to thank Maarten Pinxten for giving valuable inspiration and feedback.

References

Adams, W. K. and Wieman, C. E. (2011). Development and Validation of Instruments to Measure Learning of Expert-Like Thinking. *International Journal of Science Education*, 33(9):1289–1312.

Brose, A. and Kautz, C. (2011). Identifying and Addressing Student Difficulties in Engineering Statics. In *Proceedings of the 2011 ASEE Annual Conference and Exposition*, Vancouver.

Coe, R. (2002). It's the Effect Size, Studid; What effect size is and why it is important. Exeter.

Direnga, J., Brose, A., and Kautz, C. (2015). Auswirkung verschiedener Lehrformate auf das konzeptionelle Verständnis im Fach Statik. In *Tagungsband zum 2. HD-MINT Symposium 2015*, pages 210–217, Nürnberg.

Direnga, J., Timmermann, D., Brose, A., and Kautz, C. (2014). A statistical method for assessing teaching effectiveness based on non-identical pre-and post-tests. In *Proceedings of the SEFI 2014 Annual Conference*, Birmingham, UK.

Direnga, J., Timmermann, D., Lund, J., and Kautz, C. (2016). Design and Application of Self-Generated Identification Codes (SGICs) for Matching Longitudinal Data. In *Proceedings of the 44th SEFI Annual Conference*, Tampere, Finland.

Grace-Martin, K. (2014). ANCOVA Assumptions When Slopes are Unequal. http://www.theanalysisfactor.com/ancova-assumptions-when-slopes-are-unequal/, last visited Feb 2017. (We dated this post to 2014 by the date of the first comment).

Hake, R. R. (1998). Interactive-engagement versus traditional methods: A six-thousand-student survey of mechanics test data for introductory physics courses. *American Journal of Physics*, 66:64–74.

Mazur, E. (1997). *Peer Instruction: A User's Manual.* Prentice Hall.

Neter, J., Wasserman, W., and Kutner, M. H. (1985). *Applied linear statistical models.* Irwin, second edition.

Methodology of Problematization and Teaching Strategy Interactive: teaching proposal in the course of Production Engineering

Silvana Rodrigues Quintilhano
UTFPR, Londrina, Brasil
squintilhano@utfpr.edu.br

Rogerio Tondato
UTFPR, Londrina, Brasil
rogeriotondato@utfpr.edu.br

***Abstract**: The current context has required the educator a different capacity to meet contemporary demands, In which student training and teacher planning are closely linked. However, in Brazil, the vast majority of undergraduate courses in Production Engineering have a vision where the teaching-learning relationship is based on the mere transmission of knowledge, leading students to a fragmented training. Thinking in this context, the National Curricular Guidelines (NCGs) of the Undergraduate Engineering Courses bring in their body the intention to change the philosophical basis of the course, focusing on developing skills for skills training and educational student-centered approach with emphasis on synthesis and transdisciplinary. This implies the search for new interactional practices, effectively structured and planned. Therefore, the objective of this article will be to analyze the Active Methodology as a new perspective for teaching in Production engineering at the Federal Technological University - Campus Londrina, Paraná, Brazil.*

Introduction

For Mezzadri et al. (2010), the current context has required of the educator a differentiated capacity to meet contemporary demands, where the student's formation and the teaching structure of a university are strictly linked. In the field of engineering education, for example, it is no different. The changes that have been taking place must be absorbed by the training of students. There are so many transformations in the field of engineering that impact on a new profile of this professional led by a new profile of the engineer engineer.

Therefore, the teacher engineer needs to be updated on technological advances and other methodological intervention techniques that the profession requires reviewing his university teaching practice so that future professionals are not labeled with that formation of an era that no longer applies to reality. The authors cite Bogossian (2000) classifies the following complementary skills and attitudes of the engineering profession: ability to lead, communicate and operate (with teams), understand economics, industrial organization and management (to understand the place of work and its relationship with the external environment); And finally, the relationship between technology and man, society and nature (the impact of technology on its environment).

The National Curriculum Guidelines of the Engineering Undergraduate Program (NCGs) bring in its body the intention to change the philosophical basis of the course, focusing on the development of skills considered as basic for the formation of competences and the student-centered pedagogical approach with emphasis on synthesis and transdisciplinarity. But even today, the great majority of undergraduate courses in engineering have a vision where the teaching-learning relationship is based on the transmission of knowledge and the student as a passive agent.

On the subject, Rezende (1999) comments that, the teaching-learning process is something complex where the movement between the democratic educator must be focused on the subject that learns and implies in the search of an interactionist practice and strictly methodical, leaving aside a practice traditional. This traditional practice, based on the teacher who privileges the transmission and assimilation of knowledge as a short-term result, generates a fragmented knowledge, held within watertight disciplines that are far from being grounded and complement each other, making it difficult to construct a body- Theoretical-practical where, in fact, their interrelationships and laws are perceived, understood, discussed and appropriated by the student.

The effort to change this conception comes from the Law of Guidelines and Bases of National Education and DCNs, which establish the focus of the learner-centered teaching-learning process as an active subject of its formation. This view is in line with that dictated by science education, aimed at the formation of a student with critical autonomy in the intellectual, political and economic spheres. Along the same lines, the need arises for the elaboration of new pedagogical proposals capable of accounting for the new professional profiles. A professional who should be able to solve individual and collective problems, positioning himself critically on issues related to society, exercising leadership, administration and management (CNE, 2002).

For Bassoli (2014), the absence of epistemological reflections and uncritical acceptance of a teaching of knowledge transmission already elaborated, is caused by the presence of deformed visions on science. Inertia and epistemological obstacles need to be overcome. If teachers are impregnated with a strong environmental and cultural training, made up of when they were students and this training permeates their pedagogical practices, it is necessary to lead teachers to experience new experiences that present them with the desirable aspects of planning and organization Of teaching, which are also opportunities for didactic and pedagogical reflection (CHINELLI et al., 2010).

In this sense, it is necessary to stimulate all agents involved in education so that they can streamline their processes and thus reach the relevant educational quality with the demands of the connected world. The changes in professional profiles lead all those involved to question the way of the university, on the traditional model of the school centered only on pre-established contents and governed by a curricular structure, non-adaptive current perspectives (MENESTRINA and MORAES, 2011).

Faced with such reflections, the ABEPRO - Brazilian Association of Production Engineering through ENEGEP - National Meeting of Production Engineering (2008) and ENCEP - National Meeting of the Coordinators of the Production Engineering Course (2008)), Proposed, throughout several discussions, references for the formation of the engineer-teacher to act in the courses of production engineering, based on the resolution of the National Council of Education 11/2002, aiming the qualitative advance of the specific professional contents, creating the Area of Education in Production Engineering, with the intention of improving the teaching, research and extension activities; Having as one of its sub-areas the "Pedagogical Practices and Evaluation of Teaching-Learning Process in Production Engineering".

Therefore, the objective of this study will be to present the Active Methodologies as a new perspective for teaching and learning in Engineering. Once it is up to teachers aware of their key role in the formation of the cognitive structure to be developed in its students. Thus, they must use their sensitivity and propose other types of activities choose methods by which students acquire mastery of concepts and, with this, develop real models of the world in which they live. (MASSONI and MOREIRA, 2014).

Theoretical Framework

The teacher needs to have a range of pedagogical strategies to create and support discussions and arguments, so that different visions can be explained, compared and contrasted, allowing students to make decisions about their positions (FARIA et al., 2014) . For that, the Active Methodologies and their application in the Production Engineering course are discussed in more depth in the next sessions.

Active Methodologies

The modern world requires man-centered education, not learning to learn with an emphasis on human development. In this new conception, with the adaptations of the new professional profiles, the teaching must start from what the student already knows and a flexible curricular structure that recognizes in the student the agent of his own knowledge. For Paulo Freire (1996 apud BORGES and ALENCAR, 2014), Higher Education Institutions should rethink their pedagogical model, considering learning by overcoming challenges, solving problems and building new knowledge based on previous knowledge and the experiences of individuals , As recommended by the Active Methodologies.

In general terms, the Active Methodologies (BERBEL, 2011) are interactive processes of knowledge, analysis, studies, research and individual or collective decisions, with the purpose of finding solutions to a specific problem. Through this pedagogical path, the teacher acts as a facilitator or adviser so that the student does research, reflect and decide for himself, what to do to achieve the established goals.

The use of Active Methodologies promotes autonomy as a motivational strategy in the school. The potential of this pedagogical proposal is to awaken the curiosity of the students, as they are inserted in the theory and bring new elements, not yet considered in class. Thus, the students' sense of commitment, competence and belonging are stimulated, as well as demonstrating their persistence in studies, accentuating their autonomy, Berbel (2011) adds.

Alberti et al. (2015), point out that the fragmentation and isolation of the disciplines taught in university education are a problem that deserves attention, and that active teaching methodologies stimulate the student to deepen in world problems, relating the sciences and the model experience Activity-experience. By this model the contents of learning are inserted in significant contexts helping to overcome the dichotomy between academic formation and reality - theory and practice. In a meaningful learning the cognitive structures are altered in those who learn, the preexisting concepts are modified and new connections between concepts are promoted, leading to the differentiated formation of the professional.

Borges and Alencar (2014) commented that, it is time to review the university pedagogical practice and suggest a constructivist proposal for higher education that consists of an education with autonomy through innovative methodologies, based on research, based on research And the participation of the students, work in group and deepening in the knowledge. The university reality evolves, advances and needs to be met through the development of more effective didactic abilities that involve science, world view, human being and current reality, where the role of the teacher is mediation in the processes of elaboration of knowledge.

Bastos (2010 apud BORGES and ALENCAR, 2014) explains that the Operative Groups have the purpose of generating a critical reading of reality, an investigative attitude among the students, with openness to doubts and new concerns. A relationship between the student and the environment as a source of mutual transformations, where social circumstances strongly influence human evolution. Where the construction of knowledge is based on the interaction between peers and the construction of the self that this relationship generates.

Gemignani (2012), also emphasizes the active methodologies and emphasizes that the university can contribute to this paradigm transition, from a pedagogical model that can provide a collective and interactive construction of working with knowledge. Freire (2006 apud GEMIGNANI, 2012) emphasizes that the active methodology is based on the challenging problems that demand research and discover solutions applicable to reality. Among them, the method of problematization, this is based on the Arch of Maguerez and problem-based learning (PBL).

The methodology of the problematization with the Arch of Maguerez has been well explained and disseminated. It was proposed by Charlez Maguerez and explained by Bordenave and Pereira (1982) who also applied it (Figure 1).

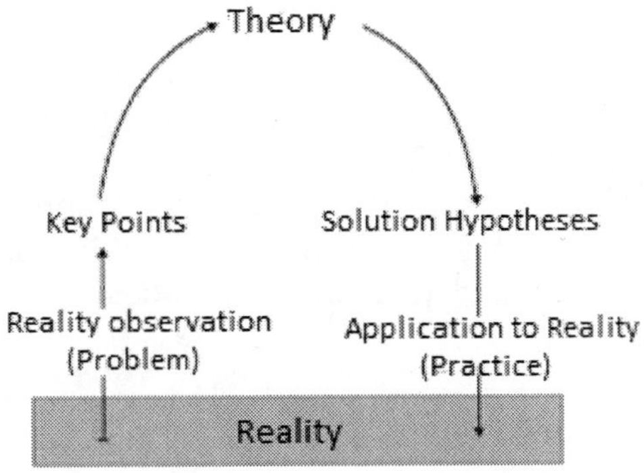

Figure 1: Maguerez Arch

In order to clarify how it is developed, Berbel et al. (2011), emphasize that the Arc is composed of five stages:

1. It begins with a clipping of reality (observation of reality) and definition of the problem to be studied;
2. The identification of the key points of the problem, for example possible causes of the existence of the problem under study, because the problem exists, that is, raising questions that make students reflect on the complexity of the problem and its multi-determiners;
3. Next, theorizing, the actual study stage;
4. With base, the definition of solution hypotheses, as fruit of the deep understanding that was obtained on the problem and,
5. Finally; The application to reality, which goes beyond the intellectual exercise, the students' true commitment to the environment.

Berbel et al. (2011) describe that the role of the teacher in the methodology of problematization is to stimulate this new learning to the students, completely changing the tradition of the teachers in presenting the problems for the students to solve. By means of this methodology, the focus of the study leads the students, who problematize thinking about aspects of living reality; this is an undeniably rich pedagogical fact. In (PBL) problem-based learning, the role of the teacher is to allow the student, independently and autonomously, to appropriate the problem and, in this context, it is the responsibility of the teacher to generate and provide Models and teaching scenarios capable of conducting work and learning at levels of appropriate complexity and relevance (ROCHA and LEMOS, 2014).

The PBL, according to Rocha and Lemos (2014), despite being the best known and used method of the active methodologies in the world, for Rocha (2012), is a proposal that still needs studies in Brazil in view of the deficient process And the current educational culture of the passive student.

Berbel (2011), emphasizes that this methodology requires the preparation of a list of situations that the student should master. There will be several themes that, in the cognitive sphere, are transformed into a problem. Every cognitive sphere of PBL should be designed in a way that ensures that the student studies the amount and depth of sufficient situations, enabling him to seek knowledge by himself when faced with a problem situation. Gemignani (2012) cites the BPA tutorial process in seven steps:

1. The reading of the group, of the presented problem;

2. The little-known conceptual terms and generators of doubts are clarified;
3. The problem is discussed and synthesized, its areas and relevant points identified;
4. A brainstorming is carried out, based on the group's previous knowledge of the problem;
5. Hypotheses to explain the problem are developed and, also, knowledge gaps are identified;
6. Then, define the learning objectives and what appropriate learning resources;
7. By means of individual study the search for information is carried out.
8. In this sense, it is perceived that, in these dynamic teaching-learning relationships, higher education should aim at the reconstruction of its social role, making the student capable of self-managing his training process.

Methodology

The research method adopted is Research-Action, a qualitative basis of approach to problem-oriented action. According to Turrioni and Mello (2012), "it is a type of empirical social research that is conceived and carried out in close association with an action." In this case, the action of the applicability of the Methodology in the course of Production Engineering, seeks the possibility of resolution in the teaching-learning process, for which researchers (teachers) and participants (students) are involved in a participatory manner.

As Turrioni and Mello (2012) describe: "In action research, the term research refers to the production of knowledge and the term action refers to an intentional modification of a given reality." Therefore, methodologically, for the application of this new pedagogical approach, a theoretical reference was made on the Active Methodologies, then a lesson plan for the Quality of Life Management in the Work discipline of the Production Engineering course was applied, considering Aspects of the interactionist practice and autonomy as a motivational strategy.

Freire (2006 apud GEMIGNANI, 2012) points out that the active methodologies are based on the challenging problems that demand research and discover solutions applicable to reality. In this study, we opted for the methodology of problematization with the Arch of Maguerez (Bordenave and Pereira, 1989), forming Operative Groups, which, based on diagnoses, allowed students to develop Quality of Life at Work programs (QLW) in a real company. The practice obviated the 5 steps: 1. Reality cut and problem definition; 2. Causes of the problem; 3. Theory of Quality of Life at Work; 4. Definition of solution hypotheses; 5. Proposition of the Quality of Life at Work Management (QLWM) program for the company.

Application of problem-based learning (PBL) in the Production Engineering Course

The methodology of the problematization was applied in the discipline of Work Quality of Life Management (GQVT) during the 2nd semester of 2016 in the course of Production Engineering of the Federal Technological University, Campus Londrina, Paraná, Brazil. The discipline composes the model of Professional Formation, totaling 25 students. The results of this application of the Problematization methodology will be compared to the application of the traditional classroom held in the same discipline during the first semester of 2016.

Before starting the work, the teacher responsible for the WQLM discipline established a partnership with a large Metalworking Equipment industry in the city of Londrina. The company, for 52 years in the Market, has approximately 600 employees, divided into three factories in the industrial park, all in the metallurgical branch. After confirming the proposal of the students' stage, it was applied in the classroom, theoretical lectures on Human Relations Theory and the emergence of Quality of Life at Work, as well as the models and programs of WQLM.

From theory, practice began. The practical application was based on the Arch of Maguerez, therefore methodologically, in five stages:

Step I - Observation of Reality

The students, in groups, applied a Satisfaction Survey, based on the model proposed by Walton (1973), which brings humanistic and social aspects, prioritizing the need for employee satisfaction. They were evaluated eight conceptual categories, described by Freitas (2009): 1. Fair Compensation and Adequate; 2. Working Conditions; 3.Use and Capacity Development; 4. Opportunity for Growth and Security; 5. Social Integration; 6.Constitutionalism; 7. Work and Total Living Space; And 8. The Social Relevance of Work in Life. The purpose of this activity was to detect points of employee dissatisfaction, so that we could define the problem to be studied.

Step II - Key Points

At this stage, the students interviewed the company's Human Resources manager, in order to identify possible causes of the existence of "dissatisfaction" under study. After this data collection, classroom discussions were promoted, leading students to question about human behavior correlating with Maslow's "Theory of Needs", seeking to reflect on the multi-determinants of company employees' dissatisfaction.

The results obtained through the satisfaction survey were inserted in Microsoft Excel, from the Radarchart chart, allowing the visualization of all the indicators evaluated, totaling the analysis of 452 employees. It is noteworthy that the indicators were 1 to 5 represented respectively by: Totally Dissatisfied, Dissatisfied, Reasonably Satisfied, Satisfied and Totally Satisfied.

As shown in figure 2, reflections generated in the classroom were guided by the lower degree of satisfaction that was summarized in three aspects: a) Fair Remuneration; B) Health and Fatigue; C) Training.

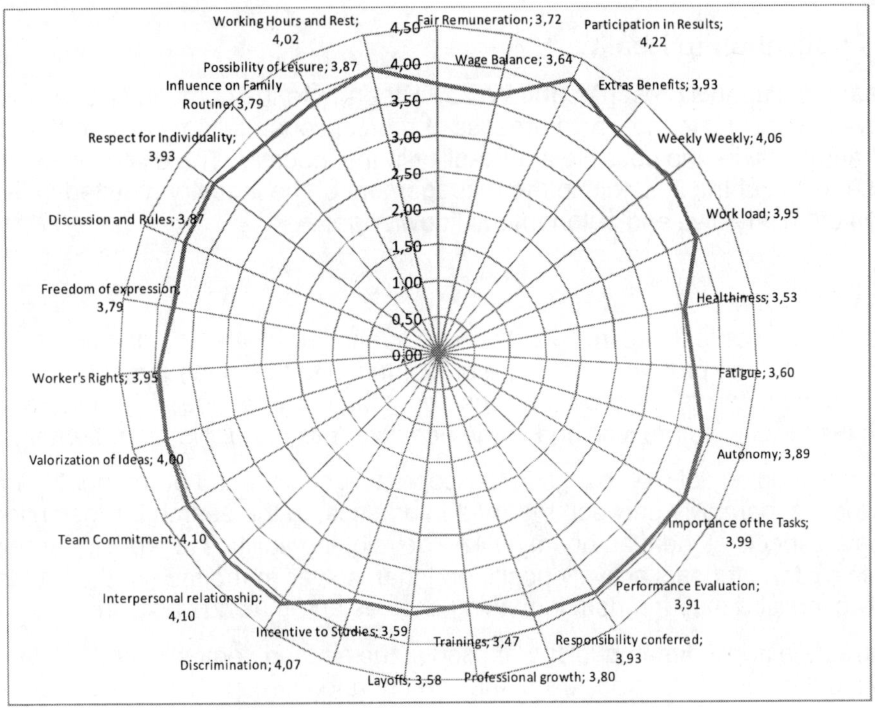

Figura 2: Radarchart Nível de Satisfação

Step III - Theorization

In this stage, in addition to the lectures on models and program implementation, the students sought to deepen the theoretical foundation on the WQLM programs. Leading them to consider that for the implementation of the program they would need institutional responsibility, social responsibility, commitment of leader's at all hierarchical levels, Inter-sectoral partnerships and effective employee participation. For a simulation exercise, the teacher brought to the classroom a problem situation so that the students, in operative groups, could simulate resolutions, based on WQLM actions.

Step IV – Solution Hypotheses

At this stage, the operative groups produced a WQLM Program for the industry under study. The program called for fruitful action on the three smallest indices of employee dissatisfaction. To elaborate the actions, the operating groups should answer the following questions:

- What is the proposed action?
- How to make?
- How to measure the results?
- What are the benefits to the company?
- What are the benefits to employees?

The actions proposed for the category "Fair and Adequate Remuneration" were geared towards drawing up a plan of positions and salaries. For the category of "Working Conditions", healthiness and fatigue, they focused on the employee's ergonomics and health, recommending the individual protection equipment (IPE). Lastly, the "Opportunities" category included an incentive to study and train in partnership with higher education institutions of the private network for discounts on tuition fees in undergraduate courses and partnerships with companies in the area of entrepreneurship, such as SEBRAE and SENAI for vocational courses and / or further training.

Step V – Application to Reality

Lastly, a team of students presented the WQLM Program to the company's board of directors so that they could evaluate the whole process of project construction, validate the research developed and consider the possible applicability in the industry. The discussions promoted by the presentation of the students to the management of the industry resulted in the construction of an effective and autonomous knowledge.

Results

The validity of the excellence of this practice was evaluated by the acumen of the academics in the elaboration of the program and in the presentation to the management of the industry under study, bringing up relevant discussions about the performance of companies and the engineer in the productive process and in the effectiveness of people management.

This active methodology of PBL brings in its scope a significant and permanent learning, promoting the inseparability between theory and practice. In the search for resolutions, this methodology respects student autonomy, while developing abilities to work in groups. From the concrete reality, students actively positioned themselves in the face of the problem and consciously exercised their freedom in making choices and making decisions.

Table 1 shows the applied methodology applied in the second semester of 2016 in comparison to the traditional exposure of the content and its respective evaluation.

Table 1: Student Revenue Results

	Traditional Classroom		Active Methodology	
	1st Sem/2016	%	2nd Sem/2016	%
Total Students	25	100	24	100
Average greater than 6.0	12	48,0	15	62,5
Average less than 6.0	13	52,0	9	37,5

It can be observed that there was an increase of 30.2% in the percentage result of students who obtained the above average score. However, despite the improvement of the evaluation results, a greater involvement of the students in the development of the work and the consequent learning in practice, which is indeed important for the training in Production Engineering, can be observed. One can also verify the quality of the work developed by the students, which can be observed the difficulties and challenges during the development of the work.

Below average performance by the students had a reduction, however, despite being a practical work, students still does not comply with the grade. This can be explained by the lack of engagement of these students in the development of the work, considering the greater need of application and dedication to it.

It can be concluded that the actions of the Problematization Methodology enabled the development of reflective and critical skills of students, mobilizing knowledge and acquisition of new knowledge in different contexts.

Recommendations and Implication for Education in Production Engineering

The experience with new practices leads the teacher to reflect on his role as mediator of the teaching-learning process. The study of the application of these new methodological perspectives brings, for the scope of the academic research, reflections on the area of "Education in Production Engineering", proposed by ABEPRO (2008), defined with "a systemic approach encompassing the management of educational systems in All its aspects: the training of people (faculty and administrative staff); The pedagogical didactic organization, especially the pedagogical course project; Methodologies and means of teaching / learning "; With the purpose of consolidating a "Pedagogical Engineering" for the improvement of teaching, research and extension activities.

References

ABEPRO (2016). *Associação Brasileira de Engenharia de Produção. Engenharia de Produção.* Accessed at: http://www.abepro.org.br/interna.asp?p=399&m=399&ss=1&c=408. on 05 Apr 2016.

Alberti, R. A; Furtado, J. C; Kipper, L. M. (2015). Simulação como ferramenta no ensino de engenharia: problematização e promoção da vivência em processos produtivos. *Revista de Ensino de Engenharia.* 83 v. 34, n. 1.

Bassoli, F. (2014). Atividades práticas e o ensino-aprendizagem de Ciência (s): mitos, tendências e distorções. *Ciência & Educação.* v. 20, n.3, p.579-593.

Berbel, N. A. N. (2011). As metodologias ativas e a promoção da autonomia de estudantes. *Semana: Ciências Sociais e Humanas.* Universidade Estadual de Londrina v. 32, n. 1.

Bordenave, J. D.; Pereira, A. M. (1982). Estratégias de ensino-aprendizagem. 4. ed. Petrópolis: Vozes.

Borges, T. S. Alencar, G. (2014). Metodologias Ativas na promoção da formação crítica do estudante: o uso das metodologias ativas como recurso didático na formação crítica do estudante do ensino superior. *Cairu em Revista.* Ano 03, n° 04.

CNE (2002) Conselho Nacional de Educação. Resolução CNE/CES 11, de 11 de março de 2002: DCNs do Curso de Graduação em Engenharia. Accessed at: http://portal.mec.gov.br/cne/arquivos/pdf/CES112002.pdf. on 20 Jan 2017.

Faria, C.; Freire, S.; Galvão, C.; Reis, P.; Figueiredo, O. (2014). "Como trabalham os cientistas?": potencialidades de uma atividade de escrita para a discussão acerca da natureza da ciência nas aulas de ciências. *Ciência & Educação.* v. 20, n.1, p.1-22.

Gemignani, E. Y. M. Y. (2012). Formação de Professores e Metodologias Ativas de Ensino-Aprendizagem. *Revista Fronteira da Educação* [online], Recife, v. 1, n. 2.

Massoni, N. T. Moreira, M. A. (2014). Uma análise cruzada de três estudos de caso com professores de física: a influência de concepções sobre a natureza da ciência nas práticas didáticas. *Revista Ciência & Educação.* .Baurú-SP, vol.20, n.3, p. 595-616. Accessed at:: http://www.scielo.br/scielo.php?pid=S1516-73132014000300595&script=sci_arttext. on 30 jun 2016.

Menestrina, T. C.; Moraes, A. F. (2011). Alternativas para uma aprendizagem Significativa em Engenharia: Curso de Matemática Básica. *Revista Brasileira de Ensino de Engenharia,* v.30, n.1, p.52-60. Accessed at:: http://www.abenge.org.br/revista/index.php/abenge/article/view/97/77. on 30 jun 2016.

Mezzadri, A. L. Farago, A. D. Horst, K. M. Picinin, C. Pilatti, L. A. (2010). O engenheiro-professor: um novo perfil docente. Congresso Internacional de Administração. Accessed at: http://www.admpg.com.br/2010/selecionados.php. on 02 Feb 2017.

Rezende, L.A. (1999). O processo ensino-aprendizagem: reflexões. *Semena: Ciências Sociais e Humanas.* Universidade Estadual de Londrina, v. 19/20, n. 3, p. 51-56.

Rocha, H.M. Lemos, W. M. (2016). Metodologias Ativas do que estamos falando? Base Conceitual e Relato de Pesquisa em Andamento. *I Simpósio Pedagógico e Pesquisas em Comunicação.* Accesses at: http://www.aedb.br/wp-content/uploads/2015/05/41321569. on 03 jun 2016.

Turrioni, J.B & Mello, C. H. P. (2012). Pesquisa-ação na Engenharia de Produção. In: Metodologia de Pesquisa em Engenharia de Produção e Gestão de Operações. 2.ed. Rio de Janeiro: Elsevier: ABEPRO.

Copyright statement

 This article was supported by the PROPPG of the UTFPR Londrina campus.

Engaging Freshers to Final Years': Embedding Student Engagement in Engineering Education

Robin Clark
Aston University, Birmingham, UK
r.p.clark@aston.ac.uk

Jane Andrews
Aston University, Birmingham, UK
j.e.andrews@aston.ac.uk

Abstract: *Based upon the findings of an Action Research study in which students' experiences were analysed at two key stages of their career, Foundation and Graduate levels, this paper sets out to discuss the issues around student engagement. Set within a context where students are paying more for their education than ever before, the paper discusses how to promote student engagement in what seems to be an 'culture of self-entitlement' wherein students expect to simply 'pass'. Aimed at academic and support staff alike, the paper describes how survey findings were used to inform decision making and ultimately impact the student experience. The conclusion argues that through the application of evidence-based practice it is possible to promote student engagement at all levels of the engineering learning cycle.*

Introduction & Background

Building upon previous work by the Project Leaders which saw the development of the RVS model of Engineering Education [1] and adopting an Action Research approach [2] the *Embedding Engagement in Engineering* Project focused on student engagement for two very different cohorts of students; Foundation level students, the majority of whom were from the UK, 18 or 19 years of age and had just left School or College; and Graduate level students, all of whom had a first Degree in an Engineering discipline, with the majority originating from outside the UK, and who were aged 23 to 42 years. Set at a time where undergraduate education is generally more expensive than graduate education, the responsibility for financing Higher Education for UK Students has shifted from the State to individual students themselves [3]. Undergraduate fees are currently set by Government and range from $12000 per year for UK students to $18000 per year for international students (making the total cost for a three or four year-long Bachelor's Degree between $36,000 and $72,000). Graduate levels, which are set institutionally, generally range from between $8700 to $18000 for a one year Master's Degree. The systems is operationalised through a mixture of government owned 'Student Loans' (at undergraduate level), bank loans and individual debt / payment.

Since the introduction of fees for undergraduate level UK students some 12 years ago, there has been a defined change in what students expect from their education. With a significant minority, having 'paid for' their education, appearing to simply expect to pass with minimum effort; the majority, perhaps understandably, simply expect more individualised support than ever before. This manifests itself in unprecedented pressures being put on academic staff to provide a high quality, almost individually bespoke, educational experience; with 'student success' and the concept of an 'enhanced student experience' central to the whole system [4,5,6]. At the same time more and more students are being offered places at university creating a 'pressure cooker' environment fuelled by debt, increasing demands, highly stressed academic staff and 'customer savvy' demanding students. On the positive side, more young people are now able to access Higher Education than ever before, although some argue that such widening participation is at a cost, and admissions standards have lowered [7, 8].

Within this somewhat pressured environment, previous work by the paper authors has highlighted the importance in promoting a *sense of belonging* to a particular discipline and institution. Such *belonging* is crucial to creating a sense of identity and a central part of promoting student engagement and success. It is particularly important for those students from 'non-traditional' backgrounds, which in engineering includes those from BME groups (Black and Minority Ethnic), young women and those from 'working class' backgrounds or without the 'traditional' academic qualifications [9, 10]. At the university where the study took place the vast majority of students fit one or more of the above categorisations, with the majority coming from working class, BME backgrounds.

Whilst previous studies have sought to examine student 'satisfaction' with their university experience [11,12], literature on how to embed student engagement within an organisational culture is harder to find, particularly when it comes to engineering education in the UK. Subsequently, whilst much effort is put into promoting student engagement, most of this is done on a 'trial and error' basis, with support staff particularly seeming to struggle in an ever-demanding culture. In seeking to address this, the *Embedding Engagement in Engineering* project set out to look at the issues around the student experience, focusing on the student experience and beginning with the research question "How can student engagement be promoted and embedded across the Engineering School?".

The Case-Study Organisation

Located in the UK's second largest city, Gosta University is one of the UK's most diverse Higher Education Institutions with over 60% of its student population originating from Black and Minority Ethnic (BME) backgrounds. The University prides itself on the employability of its graduates the majority of whom go on to have professional careers. The *Embedding Engagement in Engineering Project* took place in the University's School of Engineering & Applied Science. With a current student population of just over 3,300 enrolled upon a range of programmes in 6 different 'subject areas' the sample included students studying one of the following engineering disciplines: Mechanical Engineering & Design: Computing Sciences: Chemical Engineering & Chemistry: Information Engineering & Maths: Electronic, Electrical Engineering & Physics: Engineering Systems and Engineer Management.

Methodology: Project Approach

Building on previous studies and adopting an Action Research approach the study commenced by taking a School wide view. Descriptive data was accessed from two internal surveys which focused on why undergraduate students' withdraw and on final year undergraduate student satisfaction. Validity and reliability were addressed through a process of piloting and, having undertaken an initial analysis of the emergent data the decision was taken to narrow the study down to two cohorts of students. In order to capture the full breadth of views, it was decided to sample students on one year programmes at either end of the undergraduate 'scale'. Hence, students enrolled on a one year 'Foundation' Programme in General Engineering were sampled alongside students enrolled on a 12 month long Master's in Engineering Management were sampled. For this part of the survey previous research was used to develop a survey in which students were asked about their expectations and experiences of university [13, 14]. Focusing primarily on two different thematic areas, drivers and barriers to university: preferences and expectations of study [15, 16], the survey questions were built around "Agreement Likert Scales" and presented in a matrix fashion. It should be noted that this paper does not intend to provide a detailed statistical analysis of student outcomes; instead it uses data to inform and enhance the student experience and decision making within the institution. The findings from both stages of the project are presented in the findings section.

Having conducted the study the third, and most important part of the project involved introducing a number of changes to the School. These changes are reviewed in the discussion section of the paper.

Project Phase 1: The Wider Picture

In seeking to identify how to better embed engagement in engineering education the first part of the study involved a data trawling exercise whereby the reasons for student attrition in the previous academic year were examined. The purpose of focusing in this area was to determine whether there are any particular areas of concern which the project needed to specifically target. This part of the study sampled students from across the undergraduate cohort within the School. The purpose in taking a 'wider view' in the first instance was to gain some baseline data upon which both ground subsequent research questions and to ground any School wide changes. The first task was to look at why students 'drop out' of engineering. Using 'post-attrition data' students' reasons were accessed and thematically ordered. This data is accessed from national statistics which are used to benchmark universities. The results are shown below in Figure 1.

Figure 1: Reasons Why Students Withdrew From Aston University School of Engineering 2015 / 2016

Reason for withdrawing	N	%
Programme unsuitable	23	21.7
Transferred to another university	19	17.92
No reason given	14	13.21
Non-attendance	11	10.38
Other	11	10.38
Work commitments	7	6.6
Found full time employment	6	5.66
Financial issues	4	3.77
Health reasons	3	2.83
Academic failure	3	2.83
Family reasons	2	1.89
Personal reasons	2	1.89
Unknown	1	0.94
Total	106	100

The above table shows that the majority of students leave having found either the programme or the university unsuitable for their needs. Since October 2015 the Project Leads have been responsible for another project focusing on attrition and progression. It was therefore something of a relief to see that academic failure represented a relative low cause of individual student's withdrawal.

Having examined why a small number of students drop out of their engineering courses, the next task was to look at the degree to which students' views of their course. This part of the study focused on final year students' experiences; a total of 204 students from across the School were surveyed using an Agreement Likert Scale. Figure 2 provides an overview of the results, with the data presented using a three point scale. The purpose in looking at students' views of their course was to try and gauge the academic validity of undergraduate engineering education from the perspectives of one of the 'end users'. Figure 2 provides an overview of answers to five questions:

Figure 2: Final Year Engineering Students Views of their Engineering Course

	Disagree %	Neither Agree or Disagree %	Agree %
The subject is interesting	9	8	83
The course is intellectually stimulating	6	12	82
The quality of teaching on my course is of a high standard	11	16	73
I have opportunities to ask questions on my course	6	9	85
The course has developed my subject knowledge	6	4	90

Project Phase 2: Focus on 2 Programmes

- **Sampling**

The first part of Phase 2 of the Project involved identifying a suitable sample. In selecting two programmes at opposite end of undergraduate education the Project Leaders managed to acquire some data from students at every level. Figure 3 shows that 70% of the sample were from the Foundation Year whilst the other 30% were enrolled on Graduate Programmes, whilst Figure 4 provides an overview of the sample demographics.

Figure 3: Phase 2: Sample Level of Study

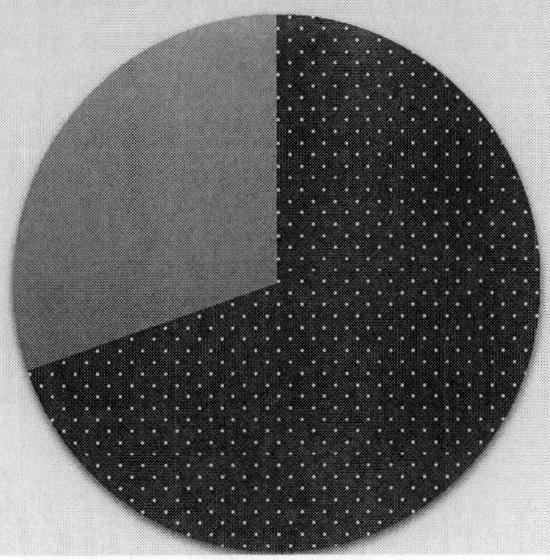

Figure 4: Phase 2 Sample Demographics

Character	ALL %	PG %	FY %
Male	82	74	83
Female	18	26	17
Overseas	15	67	5
Disability	3	0	4
Prior work experience	53	53	48
Under 21 years	78	0	93
Over 22 years	22	100	7

As can be seen from Figure 4, the vast majority of the graduate students originated from overseas, whereas only 15% of foundation level students were from outside the UK. This is somewhat different from the national picture where 46% of graduate students are from outside of the UK / EU compared to only 3% of undergraduates [17]. Such a notable difference is perhaps best explained by the location in which the project occurred, Gosta University is situated in one of the UK's most ethnically diverse cities, it is a 'minority-majority' university meaning that more than 60% of its students are from Black & Minority Ethnic (BME) backgrounds.

Survey Findings

- **A Sense of Belonging**

The main of the Embedding Engagement in Engineering Project was to promote a sense of belonging across the study body within the School of Engineering. Grounded in the Project Leader's previous work and taking into account the findings of the university-wide data analysis 8 questions were articulated. These questions focused on friendship, the availability of staff and the importance of being 'known' to others. Figure 5 provides an overview of the findings, showing the percentage of students in both courses, and across the sample, who either agreed or strongly agreed with each question statement.

Figure 5: A Sense of Belonging

	ALL (% of total sample)	PG%	FY%
The opportunity to make friends at university an important part of study	90	85	92
Being able to build friendships with people studying on different courses is important	86	85	86
It is important that a member of academic staff knows me by name	90	88	90
It is important that the Programme Director is available for me to go and talk to should I run into difficulty	94	92	95
It is important that the support staff are readily available to answer any queries I have	96	96	97
It is important that I 'fit in' with my class	69	81	67
I prefer academic staff who are approachable and friendly	92	99	92
I would rather not make friends with my classmates	8	0	10

In looking at the above table it appears evident that both graduate and foundation level students value the opportunity to build friendships with individuals on their programme. 'Fitting in' however, seemed to be more important for the older than the younger students, although this may be indicative of the fact that the majority of the graduate cohort were overseas and so had no support network available outside of the classroom. That 10% of the foundation year students indicated they would rather not make friends with their classmates was something of surprise, particularly given previous study findings relating to this.

- **Learning Preferences**

Students' learning preferences and expectations of learning were identified as being key to embedding engagement and promoting student success were examined as part of two different studies being conducted by the Project Leaders, hence this theme was included in two parallel studies. In terms of the *Embedding Engagement in Engineering Project,* the purpose of looking at learning preferences and expectations related to the School aim of providing a synergetic learning experience [1]. Two main sub-themes were examined within this area: study support: learning in the classroom & lab and an Agreement Likert Scale developed along with a number of question statements. Figure 6 provides an overview of the percentage of students in each group who either agreed or strongly agreed with each *question statement.*

Figure 5: Students' Learning Preferences and Expectations

Student Preferences & Expectations	ALL %	PG %	FY %
Student Support			
I prefer to be given pre-reading so that I can prepare for lectures in advance	72	61	85
I expect there to be a high level of academic support available should I experience difficulties	93	81	95
It is important that I have access to recordings of lectures and tutorials when I am studying	91	87	92
Learning Preferences			
I find working in groups beneficial to learning	71	84	68
I prefer traditional lectures where I sit and take notes and the lecturer talks	48	44	49
I prefer to be in a learning environment where a range of different teaching methods are used	70	60	74
I find role-play a good way of acquiring and practising 'people management' skills	52	77	48
I prefer active learning approaches such as model making and group work	72	77	71
I prefer problem based learning approaches	83	89	83

In looking at the data presented above it would appear that foundation students expect to receive far higher levels of academic support than do graduate students. This is perhaps not surprising given the different in ages of the two cohorts and the fact that many of the younger students were straight out of a school culture where independent learning does not seem to be encouraged and students receive overly intensive, support throughout their education [21, 22]. This has created a generation of young people many of whom appear to have little or no idea how to proactively learn; but instead expect to be 'drip-fed' knowledge and almost learn automatically without taking any responsibility for themselves [23]

Age, however, is not the only factor behind the differences between the two cohorts learning preferences and expectations. Other possible variables relate to culture and prior knowledge of engineering. That both groups indicated a leaning towards problem-based learning is perhaps not surprising given that the sample comprised solely engineering students.

Discussion: Embedding Engagement in Engineering

This paper builds on basic survey findings which have been used to inform a series of interventions, each of which is framed by one of the key concepts identified within the RVS approach to Engineering Education [1]. This concept is *Relationships,* with the interventions purposefully constructed so as to promote engagement across the School of Engineering.

Within the engineering classroom the need to engage students has possibly never been quite so important. Many academics find themselves torn between organizational requirements to provide recordings of live lecturers (or to pre-record lectures) and to make available either on line, or within the lab, an unprecedented depth and breadth of learning resources so that students can 'teach themselves'. However, amongst a generation of students which for all intents and purposes appears to view the concept of independent learning as something of an anathema, those students who are particularly "poor attenders" are often the most demanding; expecting individual attention and tutoring to make up for missed lecture time. This results in a somewhat impenetrable oxymoron that is, as attendance drops so student expectations that they will succeed rise.

One way of overcoming poor attendance and overly demanding 'un-engaged students' is for lecturers to develop positive learning relationships within the classroom. Such relationships can be between students (encouraged by small group working) and between lecturers and students, with interactive learning and class discussion forming an important part of learning. Previous studies have found that relationships are central to addressing high levels of attrition, and that by promoting a sense of belonging, individual academics can do much to engage students and prevent failure and promote attendance [24, 25, 26, 27].

In considering how to encourage students to develop scholarly relationships with each other and with academic staff, the Project Leaders purposefully constructed a foundation level curriculum that involves scientifically arranging students into small groups. This particular cohort was identified as the most in need of direct attention as, as a group, the foundation students originate from the widest social and educational backgrounds. They bring with them the lowest a priori qualifications and so have the greatest learning needs. Having organized students into small groups, project and problem based learning approaches were introduced across the curriculum and an emphasis on active learning and team working adopted.
In other areas of the School colleagues were supported in adopting a more 'hands on' and active approach to teaching. The Project Leaders have put in place a series of Staff Development workshops, each focused on a different key indicator for learning and teaching with the expectation that student engagement will become embedded into the organizational culture central.

Binding all of the interventions together synergetically is the concept of 'positive working relationships'. Encapsulated by the concept of ***Synergetic Configuration*** [28] which is "*the requirement that university level Engineering Education should equip students with the technical capabilities required by industry, whilst providing them with a sound theoretical knowledge base, within a supportive curriculum in which 'softer skills' are embedded alongside technical competencies and independent learning is the norm*", this project has highlighted the need for a cohesive approach to embedding engagement within engineering education.

Conclusion

In conclusion, the promotion of engagement in engineering education is not an easy goal to achieve. Younger students bring with them a learning style based on a culture of 'bottle feeding' whilst all students simply appear to 'expect to learn' – with little or no individual effort or shift towards independent learning. Yet, despite the numerous barriers and pitfalls, by aligning students' expectations and experiences with their learning preferences, and by taking into account different ontological, cultural and epistemological backgrounds, it is possible to make a difference. Positive learning relationships are key to success; by taking positive and direct action this project has shown that throughout the university engineering education learning cycle it is possible to promote student engagement through the implication of relatively non-complicated evidence-based practice.

References

1. Clark R. & Andrews, J. (2014). 'Relationships, Variety & Synergy: The Secret to Success in Engineering Education?'. *European Journal of Engineering Education*. 39. 6. pp 585-600.
2. Norton, L.S., (2009). *Action research in teaching and learning: A practical guide to conducting pedagogical research in universities*. London. Routledge.
3. Gibbs, P. (2001) "Higher education as a market: a problem or solution?." *Studies in Higher Education* 26. 1. pp. 85-94.
4. Tricker, T., (2005) *Student Expectations- How do we measure up*. University of Sheffield. Available from: http://inter-disciplinary.net/ati/education/ioe/ioe2/tricker%20paper.pdf Accessed 7/2/17
5. Arambewela, R. (2010). Student experience in the globalized higher education market: Challenges and research imperatives. *Globalization and internationalization in higher education: Theoretical, strategic and management perspectives*, 155-173.
6. Gurney-Read, J. (2014) Meeting students high expectations. *The Telegraph*. 3/6/14. http://www.telegraph.co.uk/education/educationopinion/10872594/Meeting-students-high-expectations.html Accessed 7/2/17
7. Paton, G. & Carter, C. (2014). Universities lowering entry grades to fill places this year. *The Telegraph*. 14/8/14. http://www.telegraph.co.uk/education/educationnews/11035385/Universities-lowering-entry-grades-to-fill-places-this-year.html Accessed 10/2/17.
8. Coughlan, S. (2016). University lowers entrance grades for disadvantage. *BBC*. 15/12/17. http://www.bbc.co.uk/news/education-38301844 Accessed 10/2/17.
9. PAPER AUTHORS - ANONYMISED
10. PAPER AUTHORS - ANONYMISED
11. PTES (2016) Postgraduate Taught Experience Survey https://www.heacademy.ac.uk/institutions/surveys/postgraduate-taught-experience-survey Accessed 7/2/17
12. NSS (2016) National Student Survey http://www.hefce.ac.uk/lt/nss/results/2016/ Accessed 7/2/17
13. Hernandez-Martinez, P., (2016). "Lost in transition": Alienation and drop out during the transition to mathematically-demanding subjects at university. *International Journal of Educational Research*. https://www.researchgate.net/profile/Paul_Hernandez-Martinez/publication/297225891_Lost_in_transition_Alienation_and_drop_out_during_the_transition_to_mathematically-demanding_subjects_at_university/links/56ddbcba08ae628f2d24ae30.pdf Accessed 7/2/17

14. Pampaka, M., Williams, J. and Hutcheson, G., (2012). Measuring students' transition into university and its association with learning outcomes. *British Educational Research Journal*, 38. 6. pp. 1041-1071.
15. Tinto, V., (1987). *Leaving college: Rethinking the causes and cures of student attrition*. University of Chicago Press, 5801 S. Ellis Avenue, Chicago, IL 60637.
16. Tinto, V., (2010). From theory to action: Exploring the institutional conditions for student retention. In Smart, J.C. (Ed) *Higher Education: Handbook of Theory and Research*. Volume 25. pp. 51-89. Springer. Netherlands.
17. HESA (2016) *HE student enrolments at HE providers in the UK by level of study, mode of study and domicile 2015 / 16*. Higher Education Statistics Authority. http://institutions.ukcisa.org.uk//Info-for-universities-colleges--schools/Policy-research--statistics/Research--statistics/International-students-in-UK-HE/ Accessed 8/2/17.
18. Pereira, D., 2014. Improving female participation in professional engineering geology to bring new perspectives to ethics in the geosciences. *International journal of environmental research and public health*, *11*(9), pp.9429-9445.
19. Hartman, H. and Hartman, M., 2008. How undergraduate engineering students perceive women's (and men's) problems in science, math and engineering. *Sex roles*, *58*(3-4), pp.251-265.
20. Phillips, J.A. and Head, N.W., 2000. Role Models in Engineering and Technology. *age*, *4*, ASEE. PEER. https://peer.asee.org/role-models-in-engineering-and-technology Accessed 8/2/17
21. Paton, G. (2012) Spoon fed students given tuition in basic skills at university. *The Telegraph*. 3/4/12. http://www.telegraph.co.uk/education/educationnews/9180982/Spoon-fed-students-given-tuition-in-basic-skills-at-university.html Accessed 9/2/17
22. Ramsey C. (2017) If we are serious about raising standards in universities improving the transition from the sixth form must be a priority. *The Telegraph*. 10/1/17. http://www.telegraph.co.uk/education/2017/01/10/serious-raising-standards-universities-improving-transition/ Accessed 9/2/17
23. Blessinger, P. and Carfora, J.M., 2015. Innovative Approaches in Teaching and Learning: An Introduction to Inquiry-Based Learning for STEM Programs. In *Inquiry-Based Learning for Science, Technology, Engineering, and Math (Stem) Programs: A Conceptual and Practical Resource for Educators* (pp. 3-19). Emerald Group Publishing Limited.
24. Read, B., Archer, L., & Leadwood, C. (2007). "Challenging Cultures? Student Conceptions of 'Belonging' and 'Isolation' at a Post-1992 University". *Studies in Higher Education*. 28. 3. pp. 261-277.
25. Quinn, J., (2005). "Belonging in a Learning Community: The Re-Imagined University and Imagined Social Capital". *Studies in the Education of Adults*. 37. 1. pp 4-17
26. Pitman, L.D. & Richmond, A. (2008). "University Belonging, Friendship Quality, and Psychological Adjustment During the Transition to College". *Journal of Experimental Education*. 76. 4. pp 343-362.
27. Clark, R., Andrews, J. & Gorman, P. (2013). "Tackling Transition: The Value of Peer Mentoring". What Works? Special Edition of the *Journal of Widening Participation and Lifelong Learning*. 14. pp. 57-75. [1]
28. Clark, R., Andrews, J., Thomas, L. & Aggarwal, R. (2013). *Compendium of Effective Practice in Higher Education Retention & Success: Vol II*. York. Higher Education Academy. ISB: 978 1 85449 447 4

From Trailing & Failing to Learning & Progressing: A bespoke approach to failure in engineering education

Robin Clark
Aston University, Birmingham, UK
r.p.clark@aston.ac.uk

Jane Andrews
Aston University, Birmingham, UK
j.e.andrews@aston.ac.uk

Abstract: *Starting with the research question "How can we reverse the negative impact of failure on engineering students' futures?" the 'Changing Futures Project' aims to identify and address the pedagogy of failure and in doing so make a positive difference to students' educational outcomes and progress. It builds on previous work [1, 2] to look at the issues behind 'failure' from the perspectives of individual students. In looking at the issues through the eyes of the students themselves this paper makes a distinctive contribution to current debates to the field of engineering education in general, but particularly in the areas of attrition, retention and student support. The paper ends with a total of 10 recommendations for institutions, colleagues and students.*

Introduction & Background: Brexit, the EU & Higher Education in the UK

Brexit is arguably the most recent political turbulence to impact Higher Education in the UK; resulting in an uncertain future for European and UK students alike, a number of unanswered questions have arisen with regards to continued funding and support of individual students [3]. Following over two decades of widening participation, massification and marketization, Brexit has resulted in Higher Education unexpectedly finding itself at something of an impasse, with unprecedented numbers of students facing an uncertain future as the country moves to disentangle itself from the laws, policies and procedures indicative of membership of the European Union [4]. Within this context, the issue of student attrition across Europe remains somewhat controversial issue, with data suggesting that, at best, around 85% of students complete their studies, whilst at the lower end of the scale the figure is just over 50% [5]. Likewise, in the USA the percentage of students graduating after six year of undergraduate study is just over 55% [6].

One of the inevitable impacts of attrition is the effect it can have on individual students; many of whom leave Higher Education with nothing to show for their efforts. Whilst the impact of attrition on individual student's futures and life-chances may well be significant, when considered from a wider perspective the implications of attrition are potentially much more far-reaching. Indeed, the cumulative effect of attrition combined with current and predicted skills shortages within the Engineering Sector, means that as the impact of Brexit begins to be realised, the UK economy is at risk of being damaged by scarcities of engineering talent [7]. In discussing this, Peter Finegold, Head of Education & Skills at the Institute of Mechanical Engineers, London, stated "*Without the development of these [engineering] skills, the UK will be unable to complete the vital infrastructure projects in the transport and energy sectors the country so desperately needs. We need enough people with the right skills to be confident of the country's economic growth*" (www.imeche.org 2016) [8]. Give the potential dire consequences of attrition for the future prosperity of British society, the subject has been much debated across policy, professional body and educational circles alike [9, 10, 11, 12, 13].

Within this context this paper focuses on how one, relatively well placed university located in the centre of the University is dealing with attrition. A university since the 1960's, and a Technical College well before that, Gosta University has an attrition rate of 6% overall. In its School of Engineering this rate doubles to around 12% [14]. Whilst this means that 88% of the

School's students go on to complete their education and graduate, the two project leads are both very concerned about the impact that 'failure' can have on individual students. It was this concern, together with institutional drivers, that prompted the *Changing Futures Project.*

Literature Review: What Determines Failure or Success in Higher Education?

Under the auspices of 'the student experience' the question of what determines failure or success at university has been the subject of much discussion across the Higher Education Sector; resulting in a number of socio-psychological, economic and pedagogic related variables being linked with the student experience [15, 16, 17, 18]. Yet despite this, no clear determinants of success or failure have been identified. Indeed, much of the literature seems to ponder upon the intransigence of attrition, noting that it is a complex issue indicative of a number of factors which vary in character from the learning environment through to a lack of individual engagement with studies [19, 20, 21].

With regards to engineering education in particular, one of the key influences on student success is arguably the nature of Science teaching in High School; with some studies suggesting that limitations in the content and context of the secondary school science curriculum is a key driver impacting students' poor performance once they get to university [22, 23, 24]. Whilst a lack of positive role-models in engineering is arguably relevant in terms of inspiring young people to embark on a study of engineering, a number of studies suggest that once students enrol upon an engineering programme, demographic factors, including gender, social class and ethnicity, come into play and may indeed influence individual student's chances of success or failure [25, 26].

From a pedagogical perspective, the way in which students learn can have a significant impact on success at university. Indeed, research suggests that engineering students are less likely to succeed when subjected to traditional pedagogic approaches, such as didactic 'traditional' lectures, and are more likely to benefit from active, hands-on and blended learning approaches [27, 28, 29, 30].

The Changing Futures Project at Gosta University.

Located in the UK's second largest city, Gosta University comprises a forward-thinking diverse institution where more than 60% of the students originate from Black and Minority Ethnic (BME) backgrounds. The university prides itself on delivering high quality learning and teaching, with an emphasis on social mobility and graduate employment.

Studying within one of six distinctive Faculty areas, Gosta University School of Engineering & Applied Science students study a range of BSc, BEng, MSc & MEng degree programmes in the following areas: Chemical Engineering & Chemistry: Computing Sciences: Electronic, Electrical Engineering & Physics: Engineering Systems and Engineer Management: Information Engineering & Maths: Mechanical Engineering & Design. Additionally the School houses internationally recognised research expertise in, amongst other areas, Bio-Energy: Engineering Education Research: Logistics: Photonics: and, Systems Analytics; meaning that students are have access to world class researchers and cutting edge knowledge. In year 1 of the Project the School had just over 2,500 students. In year 2 this number had risen to 3,200. This rise was indicative in an increase in full time undergraduate students at Bachelor's and Foundation levels, as well as an increase in Part Time students.

Despite such expertise, like many UK Schools of Engineering, Gosta University experiences some difficulties in attracting new students onto its engineering and applied science programmes. Moreover, as elsewhere, once students enrol on engineering programmes at the university they are more likely to drop out that students from the other subject areas [7, 8] a situation that has not been helped in the short term by Brexit; in the *aftershock* of the referendum applications to universities in general have dropped [31].

Within this somewhat tumultuous environment, the question of ***why*** engineering students are more likely to drop out of university than any other students is becoming increasingly important. Yet despite numerous studies and considerable investment, the answer to this question remains largely unknown. What is evident however, is the negative impact that attrition can have both on the individual student and on the institution in which they are enrolled. From an institutional perspective each student who "drops out", for whatever reason, means a loss of income of between £9000 to £12,000 per year (including fees, housing accommodation costs and other money spent at university). From an individual perspective, leaving university without a graduating qualification leaves a large 'hole' in the CV! It can subsequently result in an individual missing out on future career opportunities and all of the associated benefits with having a graduate level job. It is this second, individual impact of attrition, that prompted the two project leaders into action. Both were, and are, acutely aware of the individual externalities associated with 'failure' and 'dropping out'. Hence, in putting together the proposal for the project the decision was made to consider in some depth the individual perspective in terms of both failure and attrition.

Adopting an 'Action Research' methodology, the Changing Futures Project began in November 2015. The first cycle of the project ran until September 2015 whereupon the second cycle, which is now mid-way through, began. This paper follows the first cycle of the project through to completion and, where appropriate, provides comparative data with year 2 of the project. It focuses on student failure, considering in depth the issues and drivers behind the student experience, whilst looking at a number of interventions and actions taken to address the problem. In doing so the paper makes a notable contribution to current discussions in this area.

Project Aim & Objectives

The underlying aim of the Changing Futures Project is:

> "To directly tackle the issue of attrition and student failure in engineering and applied science at Gosta University".

A number of objectives were articulated in year 1 of the project:

- Using data-analytics, identify students at risk of attrition.
- Investigate students' perceptions of the reasons behind their failure.
- Identify modules with high levels of student failure
- Develop cost-effective and academically appropriate interventions to support failing students including Study Skills Packages
- Work with individual 'at risk' students to identify bespoke 'Study Pathways' with the aim of engendering success.
- Work with individual 'at risk' students to identify and, where necessary make referrals to appropriate counselling, medical, financial or other support services both within the university and externally.

Midway through year 2 of the project, following an insertion of cash to enable the development of further resources and interventions, a number of additional objectives were introduced:

- Work with the Student's Union to identify how better to support at risk students
- Put in place a series of student-focused interventions including *Attrition Advocates* and Peer Mentoring
- Further expand the Study Packs to encapsulate a wider-range of discipline specific resources

Project Approach

Defining "At Risk" Students: Sample Identification

In seeking to articulate exactly what is meant by the term 'At Risk Student' and hence identify the sample, the first stage of the project involved an examination of the demographic make-up of the student body. Figures 1 shows the current breakdown of the study body within the School of Engineering by ethnicity.

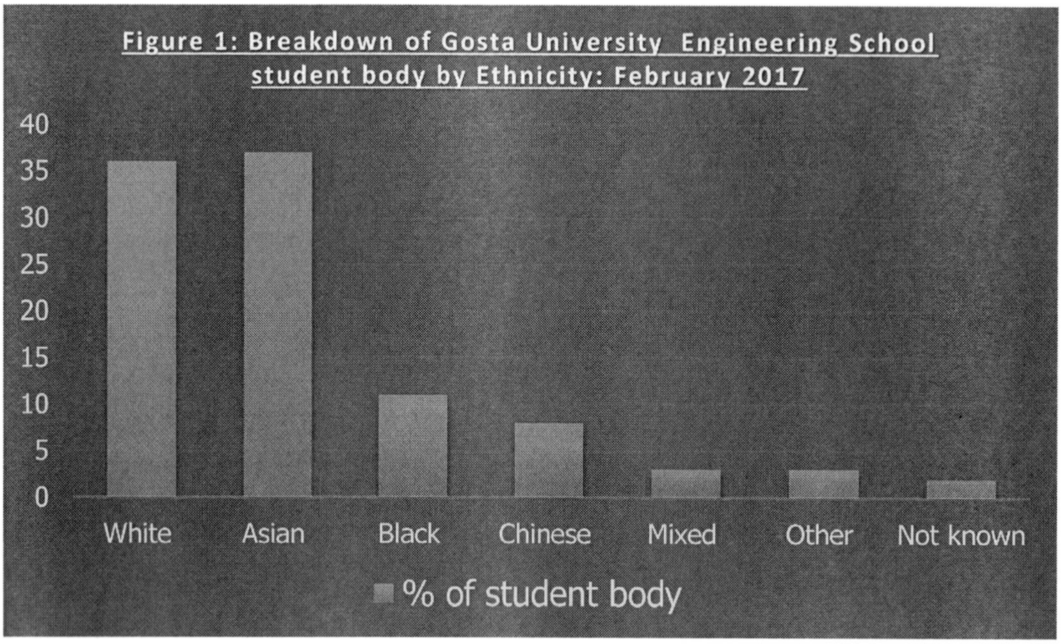

The above table shows that the largest ethnic group originate from an Asian background (37%) whereas White students make up 36% of the student body in the School, Black students represent 11% and Chinese students, 10%. This demography reflects that of the wider University, which is one of only a handful of 'minority-majority' universities in the UK.

Gender on the other hand is a completely different matter. Figure 2 shows that 23% of the population of the School of Engineering are female; whereas the university female population comprises 52% females (as shown in Figure 3).

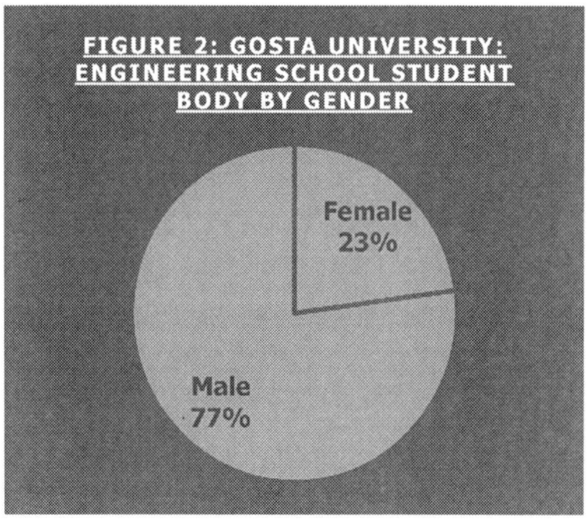

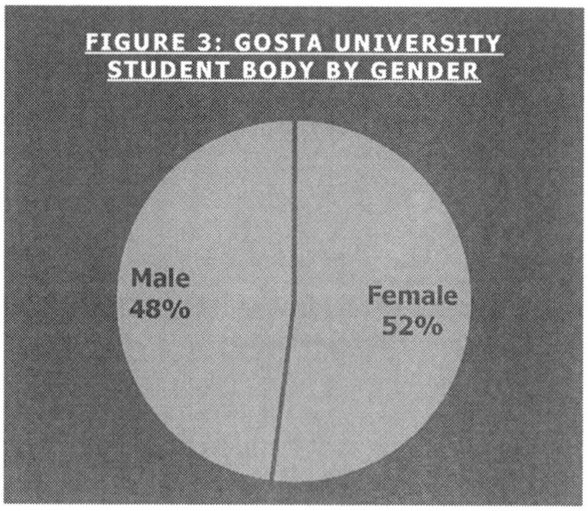

Having identified the ethnic and gender constitution of the study-body in the School the next stage of the project was to try and access data about attrition and cross-tabulate this with ethnicity. However, attrition data by ethnicity was not available at School level, hence a wider perspective was adopted and university data examined. A cross-tabulation of ethnicity and attrition at a university level for full-time undergraduate programmes (excluding pre and sub degree programmes) was undertaken. This analysis indicated that more students leave the university for non-academic reasons than actually fail.

In examining the attrition data at the University, what became apparent was the complex nature of the student body. Unlike elsewhere in the UK, British Asian and British Black students are slightly more likely to succeed than other ethnic groups at Gosta University; a phenomena that may be explained by considering the students' place of residence. British White students are more likely to live on campus than British Asian & British Black students; the majority of whom live locally and remain at home with their parents during their studies. However, the difference between different ethnic groups and likelihood of attrition was not found to be significant, a factor that reinforced the project leaders belief that no single ethnic group at the university are more at risk of failing or dropping out than any other.

That ethnicity is not a determinant of 'risk of attrition' is not surprising in an institution where the majority of students are classified as being from 'non-traditional' backgrounds, with most originating from a BME (Black & Minority Ethnicity) background. In seeking to put aside demographic indicators, the decision was taken to focus on academic achievement and the term 'at risk' initially defined as *"Those students who have on at least **two** occasions failed one or more modules".* This was amended in the second year of the project to *"Those students who have on at least **one** occasion failed one or more modules".*

The UK is part of the European Higher Education Area, which, under the Lisbon Convention and Bologna Agreement, has three distinctive stages of study: Bachelor's: Master's: Doctoral [32]. The Changing Futures Project is focused upon students enrolled on undergraduate (Bachelor's) programmes only. This means that the above definition includes all of those students who achieved less than 40% in one or more modules.

In the first year of the project, the Sampling Field comprised 1368 students. This figure excluded Foundation Degree[1] students who, whilst registered as 'fulltime', attend university under the auspices of a 'blended / block learning' arrangement. In year 2 the decision was made to include the 159 Foundation Degree students who were enrolled within the School as it became apparent that a small number of these students were 'at risk' of failing. Including Foundation Degree students, the sampling field four the second year of the project was 3,384, the significant rise on the previous year being due mainly to an increase in the numbers of students accepted into the first year of study.

Changing Students' Futures through Positive Actions & Interventions

Data Trawling & Analysis

The first task within the 'Action Research Cycle' was to identify those students 'at risk' of attrition. A data-trawling exercise was taken in October and November in both year 1 and 2 of the project. In year 1 this resulted in a total of 96 students being identified as 'at risk' of attrition; equivalent to 7% of the sampling field. In year 2 a similar number of students, 94, were identified as being 'at risk' as attrition; however, this figure only represented 2.8% of the total sampling field, much lower than the previous year. The drop in the percentage of students

[1] Foundation Degree: A blended learning programme which runs over three years. A Foundation Degree is a UK specific qualification similar to a College Degree in the USA. At Gosta University each Foundation Degree has a 'progression pathway' to enable students to top up their qualification to Bachelor's level. All Gosta Foundation Degree students are in full time employment.

being identified as being 'at risk' of attrition is somewhat difficult to explain, other than to suggest it may be indicative of the success of the first year of the project

In considering the students' demographic and background characteristics one of the unexpected characteristics identified in the initial data trawling exercise conducted each year was that a number of students had an 'AAB indicator' attached to their file. This indicator is linked with achievement and given to those who are in the top percentile of the intake in terms of academic achievement at the pre-university stage. In year 1 of the project 5 students fell into this category, whereas in year 2 the number more than tripled to 18 students. The reasons for this increase is not known, although it may be that the university is, year on year, recruiting 'higher achieving' students.

Having identified which students should be identified as being 'at risk' of attrition, the next action was to clarify which discipline groups the students were part of and to contact their subject group heads and individual tutors. Figure 4 provides an overview of the students' subject groups in both years. The data indicates that in year 1 of the project, Chemical Engineering & Chemistry had the highest number of 'at risk' students enrolled upon their programmes, whereas in year 2 it was Mechanical Engineering & Design where most of the sample students were housed. It should be noted, however, that both Chemical Engineering (Chem Eng) & Mechanical Engineering and Design (MED) represent the biggest groups in the School.

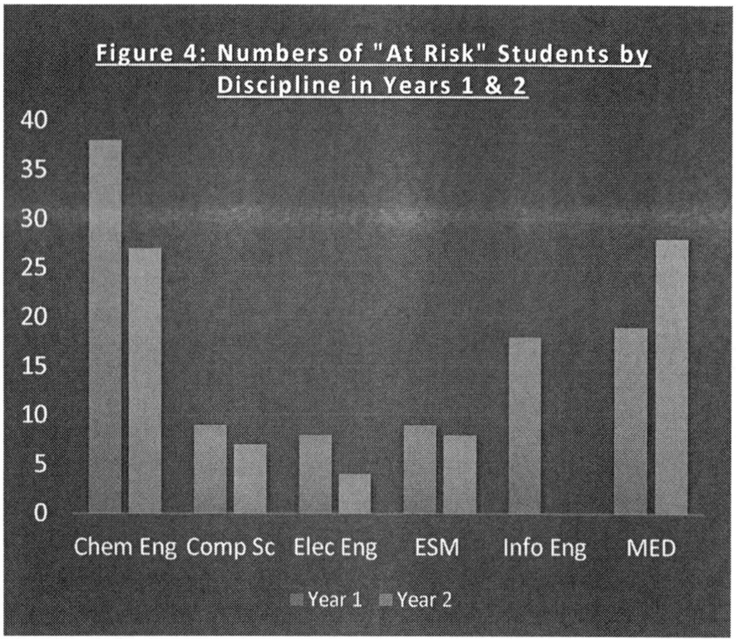

Key: All Programmes are UG level: MEng: BEng: BSc (with Honours)

Chem Eng: Chemical Engineering & Chemistry
Comp Sc: Computing Sciences
Elec Eng: Power, Electrical & Electronic Engineering
ESM: Engineering Systems & Management including Logistics and Construction Management
Info Eng: Maths & Physics
Mech Eng: Mechanical Engineering & Design

The final demographic characteristic which was examined was the students' prior qualifications. Figure 5 shows that the two different cohorts of students brought with them very different academic backgrounds. In the first year of the project the majority of those identified as being 'at risk' had undertaken a university Foundation Year programme; whereas in the second year most of those included within the project had entered university with traditional GCE 'A' levels grades B and / or C in Maths and at least one other science (including Physics, Chemistry, ICT).

Figure 5: Entrance qualifications of 'At Risk' Students

ENTRANCE QUALIFICATIONS	Yr 1	Yr 2
University Based Foundation Programme	46	15
Traditional GCE 'A' Levels Grades B,C in Maths, Physics and / or Chemistry	29	25
Traditional GCE'A' Levels Grades 'A' / 'A*' in Maths, Physics and / or Chemistry (High Achievers)	5	16
Vocational Qualifications	12	19
College Based Access Course	3	8
Undergraduate Bachelor's Degree	1	1
Other	0	10
Total	96	94

To check validity the in both cycles of the project the data was disaggregated and then individual students' records accessed and cross-checked using content analysis techniques. This process, which took place in November and December in year 1, and October and November in year 2 was extremely time consuming. It did however enable the researchers to confirm that all those within the sample were correctly identified as being 'at risk'.

The content analysis of each individual student record involved making a note of the following data:

- Gender
- Ethnicity
- Programme
- Failing / Trailing Modules
- Exceptional Circumstances recorded (such as illness or family bereavement)
- Physical or learning disabilities
- Personal Tutor
- Qualifications on admission
- Address of work-placement (where relevant)
- Email

Analysing the data at an individual level allowed for errors in student records to be identified and corrected. It also meant that the Project Leaders were able to identify patterns of behaviour in terms of engagement and academic achievement.

The purpose of undertaking the content analysis was not only to make sure that the correct students were targeted to be included in the project, but was also to allow for the Project Leaders to develop appropriately targeted resources and interventions. The following section provides an overview of the interventions and actions specifically aimed at supporting those students identified as being 'at risk' of failure in both year 1 and 2. The first year of the project is followed through to its end and the discussion shows how the second year of the project is built upon the first. Emergent data from the second year of the project, which is still very much underway, is used to frame some discussion. The conclusion builds on the findings and outcomes of the first year of the project to provide three key sets of recommendations.

Academic Study Skills Pack & Tutorial Support

Having undertaken the data-trawling exercise and content analysis the Project Leaders set about developing an Academic Skills Study Pack. In year 1 the Skills Study Pack was divided into two main sections: Learning at University: Studying Engineering & Applied Science.

Written specifically for Engineering and Applied Science students, the first part of the pack contained the following:

- Learning Styles: Information about learning styles and approaches including a 'learning styles' quiz for students to identify their own approach and guidance on how to maximise one's own approach.
- Academic Writing: Advice and guidance for STEM students on how to write reports: essays: lab reports.
- Referencing: Advice and guidance mainly focused on the Harvard System (which is the Universities preferred style).
- Self-Management: Including advice and guidance on; time management; maintaining a work-life balance; and, being a Professional.
- Where to go for help: Referral points within the University relating to both academic and personal matters listed and identified
- Virtual Resources: Useful websites

The second part of the pack was much smaller and provided virtual resources about:

- Maths
- Physics
- Chemistry

At the time of writing the paper the Study Skills Pack is under review. It will be relaunched in October 2017 for the third year of the project and will include far more Engineering-Applied Science content.

Individual & Bespoke Support

One of the key aims of the data trawling exercise was to identify each student's Personal Tutors. The 'massification' of Higher Education has resulted in a university system whereby each academic member of staff has up to around between 20 and 60 students for whom they are responsible for in terms of being the individual's appointed Personal Tutor (the number varies so much dependent on the discipline and area of study). Given such numbers, it is somewhat difficult for academics to be fully cognisant of each individual's profile and circumstances. Hence, in linking each trailing student with their Personal Tutor, the project acted as an 'additional' safety-net, making tutors aware of trailing students' situations and reinforcing the need for individual tutors to make contact with students and vice-versa.

Having made contact with the Personal Tutors, each 'at risk' student was contacted individually and invited to a one-to-one meeting with one of the Project Leaders. In the first year of the project ninety six students were contacted and over the course of terms 2 and 3, fifty-two attended face to face meetings. The project is now in its second year. To date (midpoint of term 2) ninety-four students have been contacted this year and 22 have attended meetings thus far.

The purpose of meeting with students on a one to one basis is to identify any issues the students feel may have effected their academic performance. Given the confidential nature of the meetings no 'formal' recording mechanism was utilised during the discussions. Instead, a contemporaneous hand-written note was made of students' general responses to three key questions:

- *How are you?*
- *What went wrong last year with regards to your studies?*
- *What can we do to help you get things back on track?*

During the course of the ensuing conversations the students are encouraged to discuss why they believe they failed. Figure 6 provides an overview of their answers (given in percentages) for both years of the project (bearing in mind that at the time of writing the second cycle of the project is only midway through – hence the findings will change).

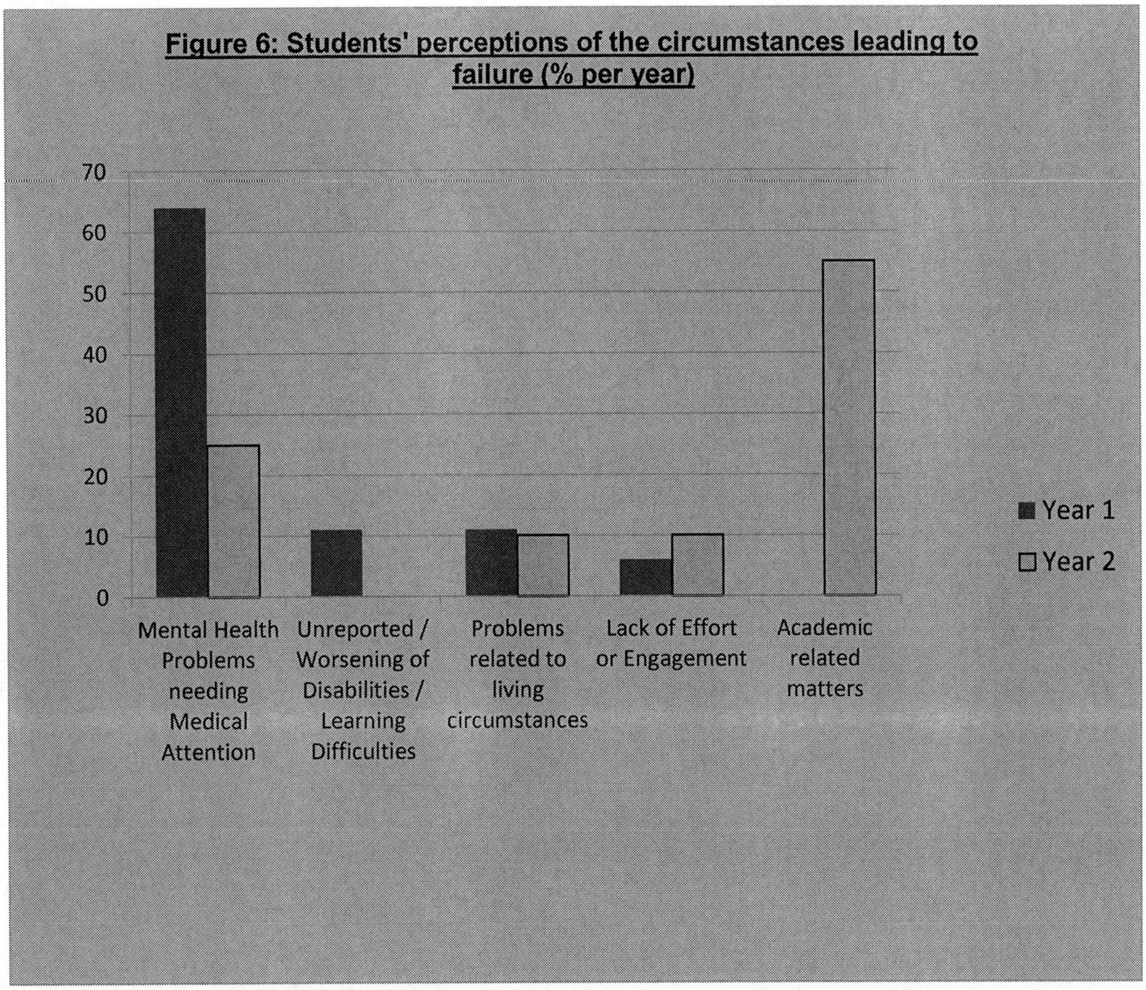

Within a few weeks of the first round of one-to-one meetings in year 1, it quickly became evident that many of the students had experiencing a serious problem with their mental health. Conversely, at the point of writing this paper, it would seem that mental health is slightly less of an issue in year 2 (this year). However it should be noted that the second cycle of the project has another one and half terms to run (5 months). This period includes several assessment deadlines and an examination period. In the first year of the project three students were identified during term 2 who were experiencing a serious 'mental health crisis'. All three were quickly referred to appropriate mental health support, and in the case of one student, his parents notified. Midway through term 2 in the second cycle of the project, one student has been identified as currently experiencing a mental health crisis. This individual is currently in hospital and one of the Project Leaders has taken responsibility for liaising with his family and sponsors. Although not taught by either of the Project Leaders, this particular student has been allocated a new Personal Tutor from within the project team.

In looking at the data in Figure 6 it would seem that the incidence of mental health problems reported in the second cycle is typical of the wider picture in terms of mental health problems in the UK Higher Education Sector. A recent study undertaken by the National Union of Students found that 25% of students experience mental health problems whilst at university. Conversely, most do not report such problems to their university [33] (possibly because of the stigma still attached to mental illness within the UK). Whilst 'pastoral care' is an important part of the 'Changing Futures Project', identifying the aetiology of students' mental health problems is not. Indeed, neither of the Project Leaders are mental health experts, although one is a trained counsellor. Hence, in seeking to try and make sense of the findings a literature search was undertaken in an attempt to identify key 'crisis' points and 'triggers'. This confirmed that, with regards to the student population, the causes of mental health crises are extremely complicated. However, amongst the most frequently reported 'triggers' are deadlines and exams. Figure 7 provides an overview of the findings of a national survey conducted to examine this issue amongst the UK student population.

Figure 7: Reasons why students experience mental health problems (NUS, 2016) [33]

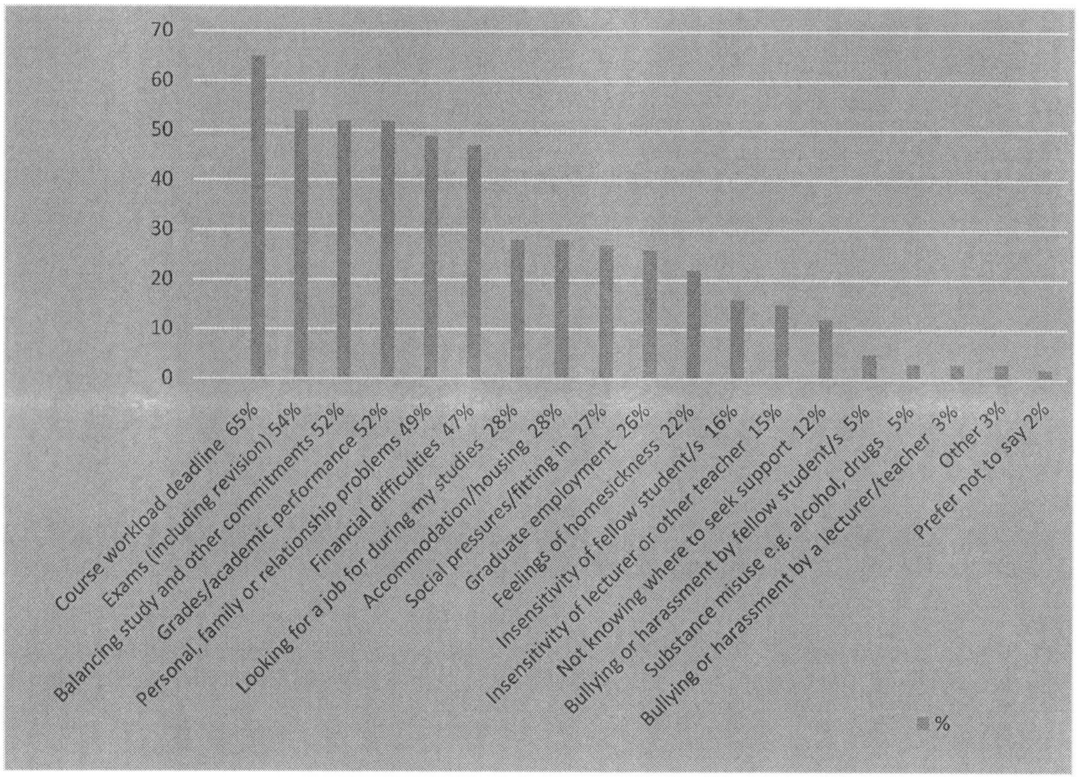

Support & Study Pathways.

In the first year of the project a number of non-academic issues were identified by students as being a factor they believed influenced their performance. Bearing this in mind, the two Project Leaders worked together to develop and put in place appropriate support mechanisms to provide students with individual 'Support Pathways'. Whilst acute mental health problems were dealt personally by the Project Leaders; the Project Leaders worked with individual students to work out bespoke Support Pathways on a one-to-one basis. Having done this, responsibility for accessing other support service was firmly placed on the student.

Figure 8 provides an overview of the various components of the support services used to develop individual pathways for students.

Figure 8: Support Pathways: Problems and Solutions for Pastoral Care

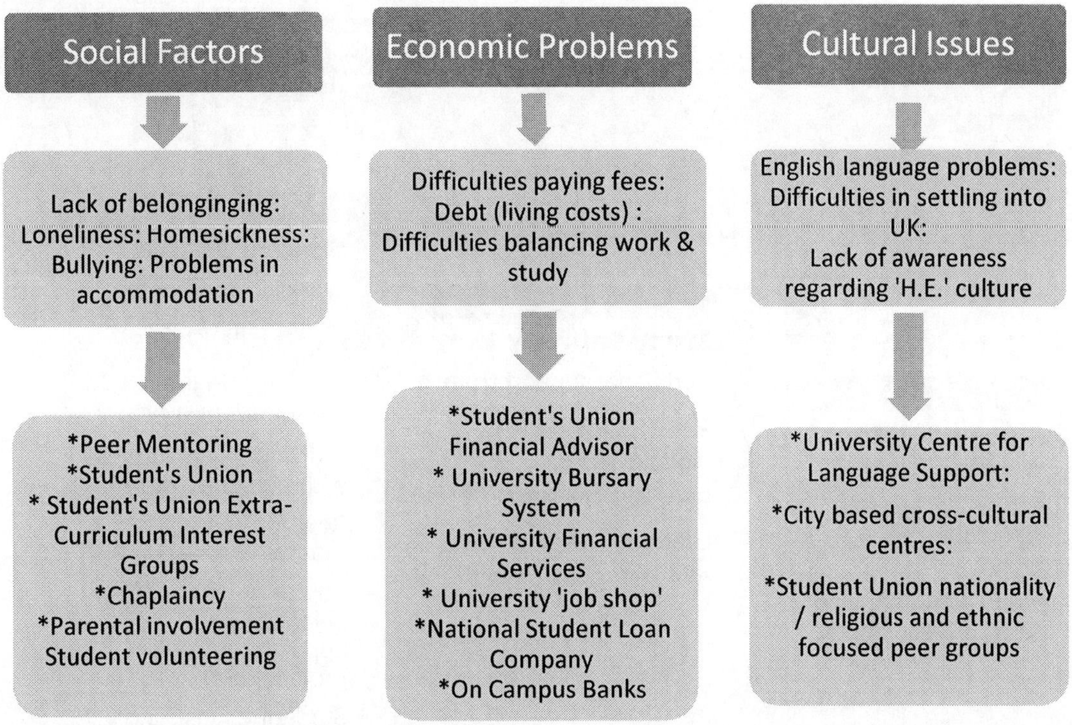

Whilst student support represents an important aspect of the project, in both years of the project cycle, academic issues have been prioritised. During the first cycle, the majority of the students indicated that they felt that the causes of failure were due to matters generally outside of their study. However, at the time of writing this has not been the case during the second cycle whereupon over half of the students have identified 'problems at the modular level' as being a factor determining their failing.

Such problems vary in nature from difficulties understanding the course content, misunderstandings regarding what is required in terms of the learning outcomes, issues around the type of assessment and, from the students' perspectives, a lack of constructive alignment in one particular assessment [34]. Following a similar approach to that described above, the Project Leaders worked with individual students to develop bespoke Study Pathways. Figure 9 below provides an example of one such pathway

Figure 9: A typical study pathway

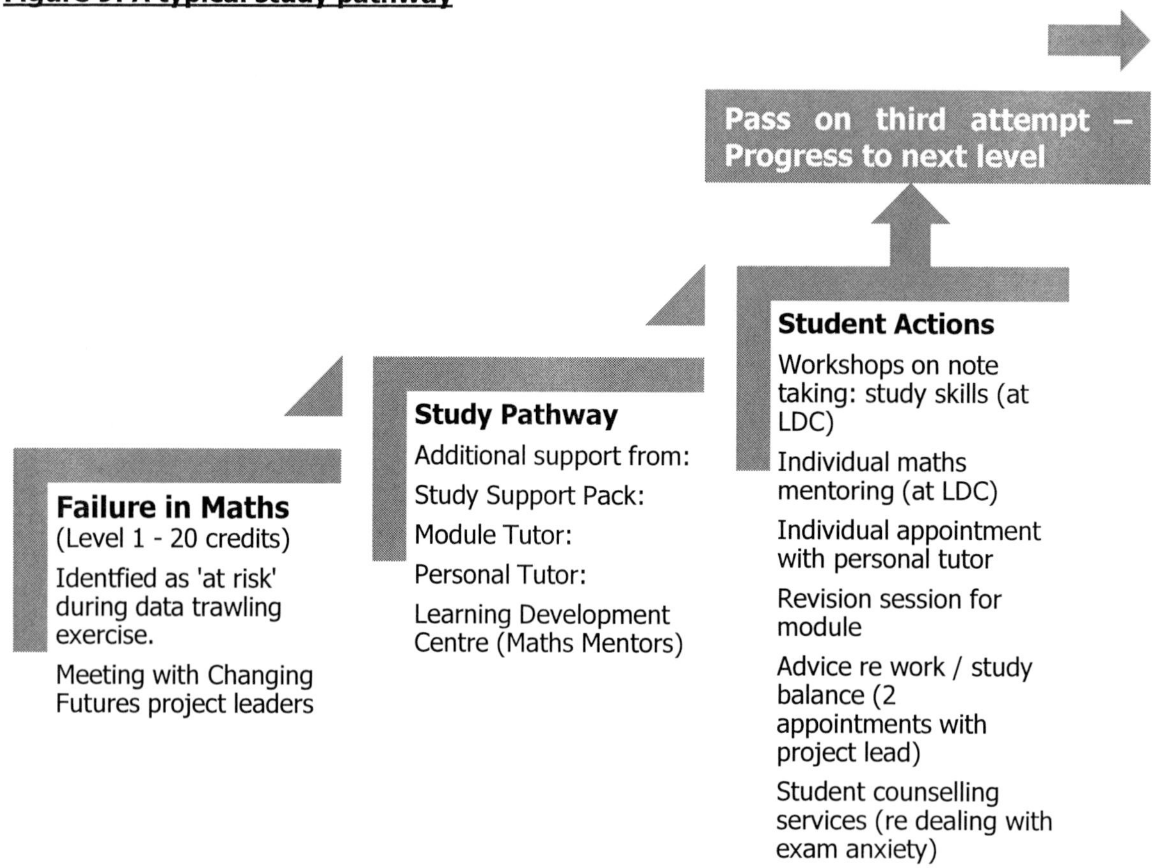

Project Outcomes and Next Steps

At the end of the first cycle of the project, the attrition rate within the School dropped markedly, resulting in a higher than predicted number of students registering to study this academic year. That the percentage of students identified as being 'at risk' dropped from 7% in year one to 2.8% in year 2 is perhaps the best indicator of the success of the project.

Within the School and University the most significant unpredicted outcome of the *Changing Futures Project* has been a raised awareness of the issues around student Mental Health. Grounded in the emergent findings, a number of symposia and training events have been held whereupon colleagues have been encouraged to make themselves aware of the various processes and services available for students experiencing acute or chronic mental health problems. The Exceptional Circumstances Application Process has been simplified and a top-level drive made to encourage tutors and students to make sure that the process is used when appropriate.

In conclusion, whilst the first cycle of the *Changing Futures Project* was very successful, the intensive nature of the support given to individual students has demanded a lot of commitment from the Project Leaders; with much of the work being done on top of usual academic duties, and resulting in the working week being extended considerably! Yet, in reflecting upon the project thus far both agree that the positive impact on each individual student's prospects and life-chances makes the effort worthwhile.

Ten key recommendations for institutions, colleagues and students are made:

Recommendations for Institutions:

1. **Financial Resources:** Funds needs to be made available to employer an 'Attrition Project Worker'.
2. **Staff Training**: Mental Health Awareness and Personal Tutor Training needs to be made available.
3. **Academic Support Services:** Need to be effectively marketed internally so students are aware of what support is available and how to access it.
4. **Social & Cultural Support Services**: There is a need for a more cohesive approach to student support, with all agencies working together

Recommendations for Colleagues

1. **Personal Tutors:** Need to meet with all tutees at least once a term, with contact maintained as frequently as possible. Training regarding mental health should be accessed. All tutors need to make themselves aware of what support services are available.
2. **Support Staff:** Non-academic colleagues are more likely to develop close working relationships with students and as such need to access mental health training also make sure they are aware of where and when to refer students in all circumstances.

Recommendations for Students

1. **Individual Students:** Need to take responsibility for proactively seeking academic and social support
2. **Mental Health Awareness:** The Student's Union needs to consider how to better support individual students with mental health problems
3. **Access to Tutors:** Individual students should make sure they meet with their Personal Tutor when needed, and access help from Subject Tutors when needed.

References

1. Bernold, L.E., Spurlin, J.E., Anson, C.M. (2007). "Understanding Our Students: A Longitudinal Study of Success and Failure in Engineering: With Implications for Increased Retention". *Journal of Engineering Education.* 96. 3. 263-273
2. Murray, M.L. (2016). *3 Reasons for High Engineering School Dropout Rates.* Types of Engineering Degrees. http://typesofengineeringdegrees.org/engineering-school-dropout-rates/ Accessed 3/3/16
3. Universities UK (2016) *Brexit FAQs. http://www.universitiesuk.ac.uk/policy-and-analysis/brexit/Pages/brexit-faqs.aspx* Accessed 14/2/17
4. House of Commons (2017). *The impact of exiting the European Union on Higher Education.* Oral Evidence to House of Commons Evidence Committee. http://data.parliament.uk/writtenevidence/committeeevidence.svc/evidencedocument/education-committee/the-impact-of-exiting-the-european-union-on-higher-education/oral/45114.html Accessed 14/2/17
5. EC (2013). *Drop Out and Completion Rates in Higher Education in Europe. http://ec.europa.eu/dgs/education_culture/repository/education/library/study/2015/dropout-completion-he_en.pdf* Accessed 27/1/17.
6. Weissman, J. (2014). *America's Awful Drop Out Rates in Four Charts.* MoneyBox. *http://www.slate.com/blogs/moneybox/2014/11/19/u_s_college_dropouts_rates_explained_in_4_charts.html* Accessed 27/1/17
7. Engineering UK (2016) *State of Engineering in the UK* http://epc.ac.uk/state-of-engineering-2016/ Accessed 14/2/17
8. Finegold P. (2016). *Engineering industry hindered by skills shortage.* https://www.imeche.org/news/news-article/strong-engineering-industry-hindered-by-skills-shortage Accessed 14/2/17

9. King, L.C. (1967) "High Student Failure Rates a Serious Problem". In *Chem. Eng. News*, **1967**, *45* (8), p 44. Available from http://pubs.acs.org/doi/pdf/10.1021/cen-v045n008.p044. Accessed 3/3/16
10. Felder, R.M., K.D. Forrest, L. Baker-Ward, E.J. Dietz & Mohr, P.H. (1993) "A Longitudinal Study of Engineering Student Performance and Retention: I. Success and Failure in the Introductory Course," *Journal of Engineering Education*, 82.1. pp. 15–21.
11. Seymour, E., & Hewitt, N.M. (1994), "Talking about Leaving-Factors Contributing to High Attrition Rates Among Science, Mathematics, and Engineering Undergraduate Majors," Final Report to the Alfred P. Sloan Foundation on an Ethnographic Inquiry at Seven Institutions, University of Colorado, Boulder, CO: Bureau of Sociological Research.
12. Bernold, L.E., Spurlin, J.E., Anson, C.M. (2007). "Understanding Our Students: A Longitudinal-Study of Success and Failure in Engineering: With Implications for Increased Retention". *Journal of Engineering Education.* 96. 3. pp. 263-273
13. Murray, M.L. (2016). *3 Reasons for High Engineering School Dropout Rates.* Types of Engineering Degrees. http://typesofengineeringdegrees.org/engineering-school-dropout-rates/ Accessed 3/3/16
14. HESA (2016) *Non Continuation Rates.* Higher Education Statistics Authority. Available from: https://www.hesa.ac.uk/data-and-analysis/performance-indicators/non-continuation Accessed 27/1/17.
15. Thomas, L. (2002). Student Retention In Higher Education: The Role Of Institutional Habitus. *Journal of Education Policy*, 17. 4. pp. 423-442.
16. Yorke, M., & Longden, B. (2004). *Retention & Student Success in Higher Education.* McGraw-Hill Education (UK).
17. Tinto, V. (2006). Research And Practice Of Student Retention: What Next?. *Journal of College Student Retention: Research, Theory & Practice*, 8. 1. pp. 1-19.
18. Crosling, G., Heagney, M., & Thomas, L. (2009). Improving Student Retention in Higher Education: Improving Teaching & Learning. Available from: http://www.universityworldnews.com/filemgmt_data/files/AUR_51-02_Crosling.pdf Accessed 4/3/16
19. Mann, S.J. (2001) Alternative Perspectives on the Student Experience: Alienation & Engagement. *Studies in Higher Education*. 26. 1. pp. 7–19.
20. Kuh, G.D. (2009) Afterword. In: Harper, S.R. & Quaye, S.J. (Eds.) *Student Engagement in Higher Education.* New York and London: Routledge, pp. 313–318.
21. Marton, F. & Säljö, R. (1976) On Qualitative Differences in Learning - I: Outcomes & Process. *British Journal of Educational Psychology.* 46, pp. 4–11.
22. Tytler, R., & Symington, D. (2006). Science in School & Society. *Teaching science: the journal of the Australian Science Teachers Association*, 52. 3.
23. Bøe, M. V., Henriksen, E. K., Lyons, T., & Schreiner, C. (2011). Participation in Science & Technology: Young People's Achievement-Related Choices in Late-Modern Societies. *Studies in Science Education*, 47. 1. pp. 37-72
24. Tan, D. L. (2002). Majors in Science, Technology, Engineering, and Mathematics: Gender & Ethnic Differences in Persistence and Graduation. *Norman, Okla: Department of Educational Leadership and Policy Studies.*
25. Prince, M. (2004). Does Active Learning Work? A Review of the Research. *Journal of engineering education*, 93.3. pp. 223-231.
26. Crawley, E., Malmqvist, J., Ostlund, S., & Brodeur, D. (2007). Rethinking Engineering Education. *The CDIO Approach*, 302.
27. Ohland, M. W., Sheppard, S. D., Lichtenstein, G., Eris, O., Chachra, D., & Layton, R. A. (2008). Persistence, Engagement and Migration in Engineering Programs. *Journal of Engineering Education*, 93. 3. pp. 259-278
28. Ohland, M. W., Brawner, C. E., Camacho, M. M., Layton, R. A., Long, R. A., Lord, S. M., & Wasburn, M. H. (2011). Race, Gender & Measures Of Success in Engineering Education. *Journal of Engineering Education*, 100. 2. pp. 225-235.
29. Tinto, V. (1993). Building Community. *Liberal Education*, 79. 4. pp. 16-21.

30. Fry, H., Ketteridge, S., & Marshall, S. (2008). *A Handbook for Teaching & Learning in Higher Education: Enhancing Academic Practice*. Routledge.
31. Collier, S. (2017). *UK Forecasting Shortages of Engineers.* Top Universities, London. https://www.topuniversities.com/student-info/university-news/eu-applications-uk-universities-down-post-brexit Accessed 14/2/17
32. EU (2017) *The Bologna Process and the European Higher Education Area.* http://ec.europa.eu/education/policy/higher-education/bologna-process_en Accessed 17/2/17
33. NUS (2016) *Mental Distress Survey.* National Union of Students. https://www.nus.org.uk/global/campaigns/20130517%20mental%20distress%20survey%20%20overview.pdf Accessed 27/1/17
34. Biggs, J., 1996. Enhancing teaching through constructive alignment. *Higher education*, *32*(3), pp.347-364.

Peers, the Internet, and Professors: Academic-help-seeking Behaviours Among Engineering Undergraduates

Edward Berger[a], Elizabeth Briody[b], Gireesh Guruprasad[a], Ryan Senkpeil[a], Elizabeth Wirtz[a]

[a]**Purdue University, West Lafayette IN USA (bergere@purdue.edu, ggurupra@purdue.edu, rsenkpei@purdue.edu, wirtz@purdue.edu)**

[b]**Cultural Keys, LLC, Detroit MI USA (Elizabeth.briody@gmail.com)**

***Abstract**: Undergraduate engineering students enrol in rigorous and challenging undergraduate programs that often require them to seek help for conceptual and problem-solving understanding. The local organizational culture around help seeking plays an important role in shaping how students secure help, including accessibility of support resources, their locations, and the general social acceptability of asking for certain kinds of help. This study focuses on academic-help-seeking behaviours of undergraduate mechanical engineers, specifically by probing their habits, the sources from which they receive help, and the alignment between their actions around help seeking and the larger organizational culture. Students generally have an overwhelming preference to seek help from peers rather than instructors. These behaviours are a complex mix of students' goals of task performance and task efficiency, as well as the prevailing culture of the department.*

Introduction

Student Academic-Help-Seeking Behaviours

Modern approaches to educating engineering students rightly focus on the collaborative nature of engineering practice. Engineering curricula include many opportunities for both formal and informal teaming among students. In the informal teaming space, students often form study groups as one manifestation of 'help seeking' to support both their learning and their academic performance (i.e., their grades). However, academic help seeking encompasses much more than engaging peers in study groups. Students can consult with their professors, talk to their advisor, use internet resources, study the course textbook, ask course teaching assistants (TAs), and access many other forms of help.

Moreover, students today see both performance and efficiency arguments in favor of collaborative work. Today's engineering curricula are quite rigorous, with both high performance expectations and high workload. Students seek to optimize their experience by achieving the highest performance with the most efficient effort. Therefore, seeking help may involve learning material that would otherwise have been very difficult to understand (improving performance), or seeking help might simply accelerate the student's path to understanding (improving efficiency). Either way, students' motivations for seeking help are a complex mix of performance and efficiency criteria.

Context for This Study

This study explores student academic-help-seeking behaviour at a large research-intensive university in the Midwest of the United States. We study a population of mechanical engineering undergraduate students engaged in a rigorous undergraduate program that requires 120 semester-credit hours to complete the degree. The university overall is viewed as 'selective' from an admission standpoint. The undergraduate profile in mechanical engineering includes about 14% female students and over 20% international students (mostly from China and India), but otherwise there are relatively few students from other under-represented minority (URM) groups. The undergraduate curriculum is viewed as technically very rigorous, and the workload is acknowledged by faculty and students alike to be very high. The enrolment is about 1,400 undergraduate students in mechanical engineering, and the student-to-faculty ratio is about 20:1. The academic culture in the department has been described by mechanical engineering community members (faculty, staff, students) as 'traditional' in its pedagogy and approach (Guruprasad, Berger, & Senkpeil, 2017). Students routinely cite the high workload as detrimental to their sleep habits and social relationships (Briody et al., 2017).

Research Questions

This study is part of a broader research program about student success in mechanical engineering, and their achievement of specific professional learning outcomes such as communication skills, teaming, and creativity. As part of this broader project, we are interested in how students navigate the rigors of the undergraduate program, and specifically how, when, and from whom they seek academic help. The specific ***research questions*** posed in this paper are: (i) what is the relative role of peers and faculty in student help-seeking? and (ii) to what extent are these patterns/relationships a function of the larger culture in the department? The working hypothesis is that the student academic-help-seeking behaviours are consistent with the larger academic culture within the department that emphasizes peer collaboration rather than professor contact, a preliminary characterization of which is presented elsewhere (Briody et al., 2017).

Theoretical Framework: Identity Congruence

The literature on student study groups has examined the role of group size (Barkley, Major, & Cross, 2014; Wood, 2009), homogeneity/diversity (Lazar, 1993; Li, Remedios, & Clarke, 2010; Sokolove & Marbach-Ad, 1999; Takeda & Homberg, 2014), virtual and physical presence (Arnold & Reeves, 2014; Johnson & Johnson, 2006), academic performance (Springer, Stanne, & Donovan, 1999; Vogt, 2008), and students' general propensity to study in groups versus studying alone (Fullilove & Treisman, 1990). The general conclusions from this broad view of the literature are that single metrics or characteristics of individuals do not adequately predict study group preferences or performance. For instance, a brief replication study (Robbins & Fredendall, 2001) confirmed that study groups were aware of and sensitive to their own 'team cohesiveness', and that team homogeneity—not diversity—correlated positively with team motivation and success. In these studies, homogeneity indicates one or more shared identity elements among the group, including country of origin, race, language, or gender, for instance. This notion of homogeneity was reinforced in several studies of Chinese students in higher education (Lazar, 1993; Li et al., 2010), which generally concluded that Chinese students prefer to study with other Chinese students. Gender dynamics in study groups have also been studied (Sokolove & Marbach-Ad, 1999; Takeda & Homberg, 2014), with observed differences in both team dynamics and performance based upon gender balance.

In our consideration of help seeking and group work in this study, we use a theoretical framework called Identity Congruence Theory (Hughes, 2010), which posits that three categories of congruence affect study group dynamics and performance:

- *Social identity congruence* characterizes the study group members' affiliation with each other around identity constructs, including demographics, over-lapping interests in non-academic subjects, social activities, hobbies, and so forth.
- *Operational identity congruence* describes group members' alignment in their philosophy around how to learn, including approaches, preferences, technology usage, etc.
- *Knowledge-related identity* congruence captures the relative stage of expertise across members of the study group, including prior knowledge and past learning experiences.

Hughes' study was conducted at a large university in the UK, comprising of students from varied backgrounds collaborating in informal online study groups. She found that the social aspect is least important for membership in learning groups, and that knowledge-related aspects are fundamental to engagement and learning. Although social identity congruence often led to positive reflections on collective working, social incongruence was not necessarily an indicator of operational problems.

Methods

Sampling Frame and Data Collection

The sampling frame for both our data collection was the entire undergraduate student population in a mechanical engineering (ME) program at a large research-intensive university in the Midwest United States, about 1400 students in total. The interview sample included 14 students heavily weighted toward upper-class students, while related focus groups included 37 additional students, all sophomores. Participants were recruited through broad email announcements, direct, in-person recruitment in ME classes, and through direct contact with several ME student organizations. We gathered over 100 hours of interview and focus group data audio. Our questions to both groups were semi-structured, with a common set of questions related to their experiences in the ME program. There were about 150 survey respondents, although not all of them responded to each question. Responses were all collected using the Qualtrics online software tool. In all cases, students were consented for the purpose of their participation in this research according to our approved IRB protocols.

Quantitative Data Analysis

We used basic descriptive statistics to understand underlying trends in the survey data. For Likert-scale questions, we employed two approaches to data summary. We aggregated responses from multiple scales into a single measure, for instance by combining 'agree' and 'strongly agree' into a single estimate of general agreement with a particular statement. We also used an odds ratio approach as follows. The odds of a student giving a particular response on the survey were simply the number of students who submitted that response divided by the total number of responses. For some data, we calculated the odds of each response occurring. Then, we normalized those odds by the metric with the lowest odds, creating an *odds ratio* that expressed how much more or less likely one response was as compared to another.

Qualitative Data Analysis

Interview and focus group audio files were transcribed, and we performed content analysis on the resulting transcripts. We coded themes and patterns using a grounded theory approach, with the resulting code structure iterated upon by multiple researchers on our team until a consensus emerged.

Findings

All Students Study in Groups; Some Study Alone First

Essentially all participants indicated they actively study with their peers on a routine basis for many different reasons (e.g., to complete homework, prepare for exams). However, students also generally acknowledged the role of self-reliance and individual work in their understanding of course material. As a result, more than 60% of survey respondents reported they they study in peer groups only *after* they have attempted to complete the assignment on their own. About 80% of students reported routinely studying in groups to complete homework assignments and prepare for exams, suggesting that they recognize the value of peer support for both learning and meeting deadlines. They also reported studying for a longer period of time each week alone (almost 10 hours) than in study groups (about 6 hours). Notably, less than half of survey respondents (44%) indicated they study in groups to 'enjoy the social interactions'.

Students also reported their typical times and locations of study sessions. When they study alone, students are typically studying in their residence (dorm room, apartment) in the evening (defined as 17:00-22:00) or later at night (defined as after 22:00). Students also indicated that when they study in a group, they meet in the evening and to a lesser extent later at night, but they tend to meet in academic spaces (vacant classrooms, the library) rather than residences.

Taken together, these survey responses suggest that students take a largely operational and knowledge-based orientation within the identity congruence framework, with social relationships taking a secondary role as suggested by Hughes (2010). Overall, the vast majority of participants—across the interviews, focus groups, and the survey—indicated that they routinely engage in group work and recognize its value. We note that many students indicated they began their studying alone, and later engaged in group work when faced with either a performance (they do not understand how to complete an assignment) or efficiency (they need to complete the assignment quickly) constraint. The social element of identity congruence was also minimized, as illustrated by the general preference to work in groups in non-residential spaces (which we interpret to mean creating a separation between personal life and academic life) and an overall trend of working in groups during the day and evening, rather than late at night.

The survey results also add depth to the knowledge-based and operational dimensions of identity congruence. We asked students to respond to six questions about their peers with whom they study in a group. We used a five-point Likert scale for each ranging from 1 = 'strongly disagree' to 5 = 'strongly agree'; the results are shown in Table 1.

Table 1. Student survey responses about their study partners. Each question indicates the identity congruence dimension it probes in parenthesis (S = social, O = operational, K = knowledge).

The students with whom I study...	*SD, D, N*	*A, SA*
1. ...earn grades similar to mine in ME courses. (K)	33.1%	66.9%
2. ...prefer the same style of studying and learning that I do. (O)	42.0%	58.0%
3. ...have similar class schedules to mine. (O)	24.2%	75.8%
4. ...live in the same dorm or apartment complex that I do. (S)	76.4%	23.6%
5. ...share the same interests outside of academics that I do. (S)	57.3%	42.7%
6. ...contribute equally to the work of the group. (K)	23.6%	76.4%

Note: 'SD, D, N' = strongly disagree, disagree, and neutral responses; 'A, SA' = agree and strongly agree responses.

These results clearly support the notion that our survey respondents placed less value on the social dimension of identity congruence (items 4 and 5), in favor of the operational and knowledge dimensions. As a result, in help seeking we conclude that students generally possess a utilitarian view that emphasizes the performance and efficiency benefits of collaboration with peers, rather than the social outcomes.

Students Seek Help from Peers and the Internet Before Professors

Results from the interviews and focus groups, consistent with the survey findings, extend our understanding of student academic help-seeking strategies. First, they offer details about getting help with their engineering course work:

- "Sometimes I look on the internet for ideas of how people have approached similar problems, but it's mostly check my notes and then ask friends."

- "Your friends are your co-workers. You walk into the ME building, second floor, you see your friends there and you just sit down and do your homework. It's a really fun environment."
- "Everybody's in the same boat. So, if you walk in somewhere by yourself and plan to work by yourself, odds are the person next to you, or across from you, is working on the same thing. And, you'll end up working with those people because you can't do this homework alone. You have to have other people there to help you make sure you're on the right track understanding these concepts. It's nice that random people can come together over this horrible homework."
- "[I use the] internet sometimes...it doesn't really have the specific problems you're looking for. But sometimes you'll find some corner of the internet, some random piece of information, and you're like, 'Oh, that helps out'."

Second, they provide study-participant explanations for why they use certain strategies and in particular why essentially all participants study in groups:

- "[It's important to] find a good group of people to work with because it's infinitely easier and better [than working alone]."
- "You have to do everything in groups because there is no way you could survive on your own unless you never leave your house!" (laughter follows)
- "We meet in the help room because that's a convenient place for all of us...when you run into something that you don't immediately understand, rather than staring at it for 15 or 20 minutes, you can say, 'Hey, did you get this?' And generally somebody else understands that part. Then you also have the opportunity to help other people with their 'Oh, I don't quite understand how to do this.' When you explain that, you form a better understanding of it as well."

Students find that they can learn from each other. Typically, someone understands a given concept or how to solve a homework problem, and they willingly share with their peers. Collaborating in study groups has the added benefit of encouraging students to develop their communication skills by explaining a challenging topic or concept to a peer.

Third, the qualitative results also suggest that students do not routinely approach their professors for academic help:

- "I don't think I ever went and talked to a professor about homework until probably this [senior] year. I went through the first six semesters of my college career without really communicating with professors a whole lot, which I probably shouldn't have done. I probably should have gone in and done a little more."
- "Office hours aren't necessarily in line with when I'm available."
- "The professor hasn't been the best to ask. They (sic) haven't been very communicable (sic) or haven't had the time."

Our intent here is not to explain fully why this pattern exists since we are in the process of continued data analysis. However, these statements are consistent with the survey results, as demonstrated in the questions about frequency of use and the usefulness of 10 academic support resources. Frequency was rated on a six-point scale ranging from 'once per day' to 'never', and the responses here have been categorized as either 'at least once per week' (encompassing the three most frequent points on the scale) or 'less than once per week' (encompassing the three least frequent points on the scale). We chose one week as the relevant unit of time measurement, because in most classes students have assignments to be completely at least weekly. Usefulness was rated on a five-point scale ranging from 'very useful' to 'completely useless' (with 3 = 'neutral'); a further option of 'no opinion' was also available to survey respondents.

Table 2 summarizes this data using an odds ratio (OR) calculation that expresses the frequency of use or the usefulness *as compared to professor office hours.* We believe that consultation with professors is a key component of a student's education, because it supports academic, professional, and social goals inherent in the overall undergraduate experience. As such, *we use student responses about professor office hours as the reference category for the calculations.* It is also conspicuous the professor office hours were the least used resource from among all the choices for the respondents to this survey, despite being viewed as useful. The odds that a student respondent visited office hours at least once per week were 0.24, meaning that just 19.5% of respondents reported attending office hours at least once per week (19.5%/80.5% = 0.24). We make a similar calculation for usefulness. The odds that a student respondent viewed professor office hours as either 'somewhat useful' or 'very useful' were 2.63, meaning that 72.5% of respondents reported that professor office hours were 'somewhat' or 'very' useful (72.5%/27.5% = 2.63).

The ORs were calculated using professor office hours as a reference as follows:

- For frequency of use, the OR is the ratio of the odds that a student reported using the resource at least once per week, compared to the odds that a student reported using professor office hours at least once per week. Example: the odds that a student respondent used the textbook at least once per week were 2.46, so the $OR_{freq,textbook}$ = 2.46/0.24 = 10.27. The interpretation is that it was about 10.27 times more likely that student respondents used the textbook at least once per week than it was that they visited professor office hours at least once per week.
- For usefulness, the OR represents the ratio of the odds that a student reported the usefulness of the resource as 'somewhat' or 'very' useful, compared to students who rated the professor office hours as 'somewhat' or 'very' useful. Example: the odds that a student respondent viewed the textbook to be either 'somewhat' or 'very' useful were 4.73, so $OR_{useful,textbook}$ = 4.73/2.63 = 1.80. The interpretation is that it was about 1.80 times more likely that student respondents rated the textbook 'somewhat' or 'very' useful than it was that they rated professor office hours 'somewhat' or 'very' useful.

Table 2. Odds ratios for frequency and usefulness of various support resources, ordered by descending frequency OR.

Resource	OR_{freq}	OR_{useful}
Online course portal (e.g., Blackboard, course blog)	26.88	0.83
Peers/currently enrolled in course	17.24	2.20
Online resources specifically created for the course	13.08	1.80
Textbook	10.27	0.75
Other online resources not created specifically for the course (e.g., YouTube videos)	8.51	0.88
Teaching assistants	3.40	0.67
Professor/in-class questions	2.44	0.83
Other peers/not currently enrolled in course	1.43	0.14
Other supplemental instruction options (e.g., tutors)	1.23	0.24
Professor/office hours	1.00	1.00

Table 2 shows quite dramatically just how infrequent consultation with professors during office hours was for our survey respondents. Indeed, visiting professors during office hours is the *last academic support option students act on*, and there are four items on the list that students were at least 10 times more likely to use on a weekly basis than professor office hours. It was more than 8.5 times as likely that students used 'other online resources not created specifically for the course' (e.g., YouTube videos) at least once per week than it was that they went to professor office hours at least once per week. *Moreover, it was more than*

17 times more likely that students consulted with their peers for academic help at least once per week than it was that they went to faculty office hours at least once per week.

This frequency pattern exists despite student responses that visiting faculty office hours was more useful than all but two other resources on the list. Peer interactions and the online resources specifically created for the course were rated 'somewhat' or 'very' useful by survey respondents about twice as often as professor office hours were rated 'somewhat' or 'very' useful. These ORs also reinforce the rather alarming notion that students spent a great deal of time seeking out academic help resources that were not particularly useful.

It is likely true that students use the online portal for their classes essentially daily, without receiving much academic support from it. Blackboard and similar learning management system (LMS) tools are typically used in very transactional ways by faculty (e.g., to assign homework). Perhaps this observation explains why students use it so frequently yet rate its usefulness comparatively lower. As a result, these data suggest that the most useful academic support resources are peers and the online resources specifically created for their courses, whose ORs for both frequency and usefulness are quite high compared to the other resources.

Once again, student interviews shed light on this behaviour. Students reported a variety of reasons for not attending faculty office hours largely revolving around convenience. Students often reported time conflicts with professor office hours, either due to their schedule of classes or the basic fact that they complete much of their work in the evening (Table 1) when professors are not available in their office. Students also reported that they sometimes complete their work in their residence (especially when working alone), and it was inconvenient to travel across campus to their professor's office. It is much more convenient to phone or text a friend, or meet up in person with a peer to discuss the assignment.

In the interviews, a small but non-zero number of students also cited explicit professor avoidance as a reason they do not seek help during office hours. Students described being intimidated by their professors, whom they perceive to be both brilliant and very busy; one commented: "I just feel like weird just going to them [the professors] for like basic homework questions I guess...I don't want to waste [their] time with something super trivial." Another student expressed this same sense of intrusion: "There is also the occasional faculty who is either never there, or does not appreciate you barging in, which is understandable if they are busy." Both these quotations capture the sense among (at least some) students that their professors are not approachable, or that they are not welcoming of visitors to their office hours.

Social congruence theory does not neatly fit into the narrative around professor avoidance revealed here. However, it does suggest that the undergraduate culture in mechanical engineering promotes peer interactions over faculty interactions, perhaps because students perceive more operational, knowledge-based, or social benefits (more 'usefulness') to their peer interactions as compared to their professor interactions. It is also possible that cultural barriers within the department result in the low priority placed on faculty interactions reported by interviewees, focus group participants, and survey respondents..

Implications for Engineering Education Practice

The Role of Culture in Shaping Student Behaviours

These data confirm a specific instance (not going to office hours) of a broader characteristic of the departmental culture that we have examined in more detail elsewhere (Briody et al., 2017). Student self- and peer-reliance are endemic to the departmental culture, and seem to be linked to the department's culture as well as the massive scale of its educational enterprise. The department faculty sometimes insulate themselves from time-consuming engagement with the undergraduate program because of the incentive structures that reward research productivity and graduate education. And the scale of the program reduces the number of interactions between faculty and students. Class sizes routinely exceed 100 students per section, and in many lower-division courses it is not uncommon to have multiple sections offered in a single semester (meaning that 600 or more students might be taking the same class in a given semester). In this scenario, it is difficult for a faculty member to learn the names of the students (although valiant efforts are often made), let alone get to know students in a more personal way. This creates a rather impersonal environment that does not encourage student-faculty interaction, nor does it provide the relationship-building opportunities that might be presented to faculty and students in smaller programs. The result of this situation is that students do not prioritize interaction with the faculty to support their learning outside of class, and therefore tend to not engage with faculty during office hours. Instead, peer reliance is very strong. Students seem to view peer collaboration as almost mandatory to manage the intense workload successfully and meet the high expectations set for students in the program.

Many students commented on this situation to us. Student often made an efficiency argument about completing the large workload in the program, clearly suggesting that the peer collaborations are a time-efficient approach to academic success. Other students emphasized the academic performance imperative to develop productive working relationships with their peers, again reinforcing the efficiency argument but also hinting at performance and, potentially, social benefits. Interviewees and focus group participants commonly described their work habits and help seeking in the context of their peers; their quotes clearly connect to the identity congruence framework. These observations on culture provide an important context to the interview, focus group, and survey responses, and further emphasize the peer-driven academic-help-seeking behaviours in the program.

Recommendation: Building Relationships Between Professors and Students

This research elevates an important finding that is surely generalizable to many engineering departments: faculty and students do not spend much time together, and students most often turn to their peers for academic help. When students and faculty do not spend time together, students miss crucial opportunities for mentorship, guidance, advice, and general networking with engineering professionals. As such, the primary recommendation borne out by this research is to design intentional ways to build stronger relationships between faculty and students, one component of which would be to address the office hours disconnect revealed here. We have a number of on-going experiments in which office hours are held in neutral locations (e.g., campus coffee shop) or at different times (e.g., early evening, during the dinner hour). Through this research we are also convening 'happy hour' gatherings among faculty, staff, and students, arranging for other social gatherings throughout the day, and encouraging students to become more involved in key departmental activities including undergraduate research or departmental/student organization governance. All these potential opportunities have the goal of fostering closer relationships between faculty and students, which may have the outcome of re-shaping help seeking by students to include their professors more prominently.

Conclusions

This study has revealed the academic-help-seeking patterns of undergraduate engineering students in a large, research-intensive mechanical engineering department. Students prioritize collaboration with peers and with online resources over their engagement with the faculty. Students at this institution appear to be quite self-reliant, and a significant portion of the survey respondents spent time studying alone before seeking help from peers. Professors reside at the bottom of the list of help resources sought out by students, for reasons that become clearer when the departmental culture is considered. The program's expectations for students, and their workload, drive students toward each other and away from the program faculty, and students make both efficiency and performance arguments about this process. Identity congruence theory helps us understand many student choices about how they complete their work, and the issue of faculty-student interactions has emerged as an important action to re-shape student academic-help-seeking behaviours.

References

Arnold, B., & Reeves, J. (2014). Translating Best Practices for Student Engagement to Online STEAM Courses. In *Proceedings of the 2014 American Society for Engineering Education Zone IV Conference* (pp. 1–15). Long Beach CA: AMERICAN SOCIETY FOR ENGINEERING EDUCATION.

Barkley, E., Major, C. H., & Cross, K. P. (2014). *Collaborative learning techniques: A handbook for college faculty* (2nd ed.). Jossey-Bass.

Briody, E., Wirtz, E., Berger, E., Morrison, E., Ramos, A., & Guruprasad, G. (2017). Ritual as work strategy: a window into organizational culture. *Journal of Higher Education*, (in review).

Fullilove, R. E., & Treisman, P. U. (1990). Mathematics Achievement Among African American Undergraduates at the University of California, Berkeley: An Evaluation of the Mathematics Workshop Program Mathematics Achievement Among African American Undergraduates at the University of California , Berk. *Journal of Negro Education*, *59*(3), 463–478.

Guruprasad, G., Berger, E., & Senkpeil, R. (2017). Characterizing the alignment between faculty and student beliefs. In *Proceedings of the 124th American Society for Engineering Education Annual Conference and Exposition.*

Hughes, G. (2010). Identity and belonging in social learning groups: the value of distinguishing the social, operational and knowledge-related dimensions. *British Educational Research Journal*, *36*(1), 47–63. http://doi.org/10.1080/01411920902834167

Johnson, G. M., & Johnson, J. A. (2006). Learning Style and Preference for Online Learning Support: Individual Quizzes versus Study Groups. In *Proceedings of the Annual World Conference on Educational Media, Hypermedia, and Telecommunications* (pp. 1861–1868). Orlando. FL.

Lazar, A. M. (1993). *The construction of two college study groups. Doctoral dissertation.* University of Pennsylvania.

Li, D., Remedios, L., & Clarke, D. (2010). Chinese Students' Perception of Out-of-class Groupwork in Australia. *The Australian Educational Researcher*, *37*(3), 95–112.

Robbins, T. L., & Fredendall, L. D. (2001). Correlates of Team Success in Higher Education. *Journal of Social Psychology*, *141*(1), 135–136. http://doi.org/10.1080/00224540109600534

Sokolove, P. G., & Marbach-Ad, G. (1999). The Benefits of Out-of-Class Group Study for Improving Student Performance on Exams: A Comparison of Outcomes in Active-Learning and Traditional College Biology Classes. *Journal on Excellence in College Teaching*, *10*(3), 49–67.

Springer, L., Stanne, M. E., & Donovan, S. S. (1999). Effects of Small-Group Learning on Undergraduates in Science, Mathematics, Engineering, and Technology: A Meta-Analysis. *Review of Educational Research*, *69*(1), 21–51.

Takeda, S., & Homberg, F. (2014). The effects of gender on group work process and achievement: an analysis through self- and peer-assessment. *British Educational Research Journal*, *40*(2), 373–396. http://doi.org/10.1002/berj.3088

Vogt, C. M. (2008). Faculty as a Critical Juncture in Student Retention and Performance in Engineering Programs. *Journal of Engineering Education*, *97*(1), 27–36.

Wood, B. B. S. (2009). Learning Science While Constructing Learning Teams. *Journal of College Science Teaching*, *38*(5), 28–33.

Acknowledgements

This material is based upon work supported by the National Science Foundation under Grant No. EEC-1519412. Any opinions, findings, and conclusions or recommendations expressed in this material are those of the authors and do not necessarily reflect the views of the National Science Foundation.

Copyright statement

Self-concept and psycho-pedagogical accompaniment: Relationship analysis with the academic performance of first-semester students at Faculty of Engineering of Universidad de Antioquia

Abstract: *In Colombia, dropout rates in university studies are around 57% of the students that enroll undergraduate programs. The Faculty of Engineering at Universidad de Antioquia (FI-UdeA) offers 12 undergraduate programs, and it registers a dropout rate of 51%. A high proportion of the dropout rate occurs during the first semester, and therefore, the FI-UdeA has implemented a set of actions that includes academic and economic accompaniment for students, which has impacted positively on retention and graduation rates. The purpose of this work is to identify the relationship between the self-concept and the psycho-pedagogical accompaniment with the academic performance of first-semester students at FI-UdeA. Results show that self-concept has a direct-positive relation with the academic performance, and psycho-pedagogical accompaniment has a positive effect on the academic performance of students that presented a low self-concept.*

Introduction

Higher education is related to the economic development of a nation, and at the end of the 1990s it was included in the indicators of sustainable development (United Nations, 2000). Part of the goals for economic development is to increase student rates in higher education, reducing therefore dropout rates. Education in engineering is one of the areas related to technological development, but at the same time, one of the areas with the higher dropout rates. In Colombia, according to the SPADIES Program from the Ministry of Education (SPADIES: *Sistema para la Prevención de la Deserción en Educación Superior*), dropout rates in university studies are around 57% of the students that enroll undergraduate programs. The Faculty of Engineering at Universidad de Antioquia (FI-UdeA) offers 12 undergraduate programs, and it registers a dropout rate of 51%. A high proportion of the dropout rate occurs during the first semester, and therefore, the FI-UdeA has implemented a set of actions that include academic and economic accompaniment for students, which has impacted positively retention and graduation rates.

Multiple factors affect dropout rates. Braxon et al. (2000) group dropout factors in economic, organizational, psychological and societal. Works related to dropout rates in universities studies from the Ministry of Education in Colombia, have included the group of academic factors in order to identify how previous academic conditions and high-school educational background can influence the academic performance during university studies (Ministerio de Eduación Nacional, 2009). According with the Ministry of Education, the two set of factors that have received the major interventions in the last decades are the economic and the academic ones. Within the economic factors, university students have received financial support for feeding and transportation, whereas in academic factors, students have received preparatory courses and training before entering universities with the aim of prepare them to assume academic life and responsibilities (Valencia et al., 2015).

One of the key factors in students' academic life, is the personal one, which contains socio-economic and psychological variables. Despite there is a considerable amount of works in the literature that study such factors, little has been done to study the organizational and psychological variables (Valencia et al., 2015). Psychological variables are related to the student's identity, i.e., the way in which students perceive the world and how they perceive themselves (Lerner, 2010; Villarroel, 2000; Covington and Beery, 1976; Watson, 1970). Such identity includes the so-called self-schemes, and within those schemes, the self-concept constitutes one of the variables of interest of this research (Valencia et al., 2015, Lent et al., 2007; Abouserie, 1995; Chapman and Boersma, 1982; Kifer, 1975). In this work, we focus on the psychological aspects that influence academic performance.

According to Musitu, García and Gutiérrez (1997), the self-concept can be understood as the individual's perception of himself, based on his experiences with others and on attributions of his own behavior. The self-concept is multidimensional, as it involves the individual's emotional, social, family, physical, and academic perceptions about himself. Garcia and Musitu (2014) suggest that the self-concept is an organized configuration of self-perceptions. González and Tourón (1992) indicate that the self-concept is constructed based on experiences that the subject has with the world that surrounds it. Costa and Tabernero (2012) indicate that students' academic performance is correlated to their self-concept. Moreover, the authors specify that the self-concept is strongly related to the family and physical perceptions, which evidences the importance of these factors to students' academic success. Authors such as Musitu et al. (1997) and Purkey (1970), confirm the effect of self-concept on academic performance. Furthermore, Kifer (1975), Covington and Omelich (1979), Byrne (1984), Hamachek (1987), Markus, Cross and Wurf (1990) and Leondari (1996) demonstrated that the perception and the feelings about itself are determinants in the academic success of the individuals (Urquijo, 2002). A negative self-concept generates a lack of self-confidence, a distorted view of self, feelings of disability and incapacity, which can later be reflected in poor academic and social performance.

In the last decades, several works have studied the correlations between self-concept and academic performance through the AF5 test (*Autoconcepto Forma* 5, AF5), (García and Musitu, 2001). In this line of research, the study carried out in Valencia, Spain by Gargallo (2006), confirmed the influence of self-concept on the academic performance in university students. In their study, the authors worked with a sample of 1298 students from three universities, and the analysis showed a moderate relationship between three of the five dimensions of self-concept (academic, family and physical) with the academic performance. In the final recommendations of the study, the authors propose to explore the role of teachers in the student's self-perceptions, and to support the development of good self-concepts in university students in order to promote better academic results.

Given the relevance of the self-concept in the academic performance and student retention, in this work we analyze the relationship of the self-concept and the psycho-pedagogical accompaniment with the academic performance of first-semester students at Faculty of Engineering of Universidad de Antioquia (FI-UdeA). The evaluation tool used to measure self-concept is the quantitative questionnaire called *Autoconcepto Forma* 5 (AF5), by García and Musitu (2001). This questionnaire has been validated internationally and can be applied to individuals from 9 to 62 years-old. This 30-question test addresses the study of self-concept from a multidimensional perspective and considers five factors or dimensions: Academic, Social, Emotional, Family and Physical. Each question is answered using a numerical scale from 1 to 99, where 1 is the lowest score and 99 is the highest.

The paper is organized as follows. First, we present the methodology to evaluate the self-concept and the academic performance, as well as a description of the psycho-pedagogical accompaniment completed. Second, we analyze the results obtained in each of the variables. Lastly, we conclude and present some future work opportunities.

Methodology

The application of the AF5 instrument allowed identifying students level in each of the five dimensions proposed in the test. During the intervention sessions (screening and workshops), students' opinions and feelings were recorded, such as perception about courses and teachers, construction of identity and body image based on cultural models, vocational and professional orientations, and imaginaries about the institution. Source of data included results of the students in the AF5 test and their academic records at the end of the semester 2015-01, which were consulted in the System of Registration and Systematized Registration (MARES, for its acronym in Spanish). For data analysis, the following quantitative and qualitative techniques were used: exploratory analysis, cross tabulation, analysis of variance, direct observation and categorization.

Students population and sampling

A population of 1231 students entered in January 2015 to the first academic semester in the 12 undergraduate academic programs of FI-UdeA. From this population, 90% entered through the admission exam that the University applies to all applicants; a 2% entered in the modality of indigenous students, a 2% in the afro-american modality, and the remaining 6% entered through a re-entry process. Moreover, from this population 72% are male, 28% female, and 86% has ages between 16 and 25 years old. Students enrolled in engineering programs offered in the virtual modality and in regions out of Medellín, Antioquia, where the main campus of the UdeA is located, are not included in the students population.

A proportionated stratified random sampling was completed in two directions: academic program and gender. A reliability of 90% and an error of 7% were used to estimate parameter *P*: Proportion of students with high self-concept in five dimensions. In order to calculate the sample, such parameter was fixed in 0.5, obtaining a theoretical sample number of 125 students. The final sample obtained through the study was 131 students.

Self-concept and academic performance evaluation

Scores obtained with the application of the self-concept evaluation through the AF5 procedure were transformed into percentiles. These values were the base input for the estimation of statistical indicators that allowed identifying the students' status in the five dimensions, considering the academic program and gender. Percentiles were also the base to classify students into two groups: high and low self-concept. It can be said that a student has a high self-concept in a given dimension, when his score is superior to the mean of the international scale for the AF5 test in the university population. A student was said to have a low self-concept if in at least three dimensions he showed a score inferior to the mean of its scale.

Psycho-pedagogical accompaniment

A psycho-pedagogical accompaniment program was offered by psychologists from the research team to a random sample of 30 students, and 22 of them enrolled in the program. This group of students is called *treatment group*. The accompaniment consisted of an individual psycho-orientation and four group workshops. The psycho-orientation included a work in each of the five dimensions evaluated by the AF5. The workshops were focused to improve self-awareness, positive attitudes towards oneself, auto-control techniques, goals establishment, decision making, problems and conflict solving, and leadership. Students with a low self-concept that did not participated of the psycho-pedagogical accompaniment were part of the *control group*. Furthermore, a third group was formed by students with a high self-concept that did not participated of the accompaniment. This group was called *reference group* since they are expected to obtain better academic performance.

Results Analysis

Validity and data processing

To evaluate the consistency of the instrument, the Alpha-Cronbach index was calculated using two methods: the analysis of items variance, and the correlation matrix. Index values obtained by each of the methods were 0.86 and 0.87, thus confirming the validity of the instrument reported in the literature.

To calculate the average scores, each of the six questions related to the five dimensions were used. Then, each average score was transformed into a percentile using the international scale table, which provides the distribution for dimension and gender. The higher the percentile, the better the location of the student with respect to the scale. The average of the percentiles for dimension and gender allows identifying the condition of students that enroll into an academic program.

Due to the fact that there is no scale independent of gender, and with the aim to identify the conditions of students in each dimension, a global weighted average was calculated, using the averages obtained by gender in each academic program. This result express the average percentile for each academic program and for each dimension. Percentile 50 (P50) was the reference point that students presented when they enter the academic program, i.e., it is assumed that if the average percentile of the students of an academic program is higher than P50, then the students of such program entered with a high self-concept.

Self-concept and academic performance

To evaluate academic performance three variables were used: grade point average (GPA), academic condition, and academic continuity. The first variable has a range from 0 to 5, in which 5 represents the highest grade that a student can achieve. The second variable represents the qualitative levels that evidence the academic situation of a student at the end of the semester, and is a function of the GPA. The levels that comprise this variable are presented in Table 1. The third variable indicates if a student can or cannot continue in the academic program that first entered. Regarding the academic condition, the values of "not calculated" and "insufficient" refer respectively to students that cancelled the academic semester, and to students that obtained an insufficient academic performance that bans them the possibility to enroll in the University for five years.

Table 1. Variables associated to Academic Performance

GPA	Academic Condition	Academic Continuity
No register	Nor calculated	Cannot continue
Less than 2.5	Insufficient	
From 2.5 to 2.8	Probation Period	Can continue
From 2.8 to 3.99	Normal	
From 4.0 to 5.0	Outstanding	

From the 131 students in the sample, five of them cancelled the academic semester and therefore, registers of 126 students were processed. The GPA values obtained for each group of self-concept are shown in Table 2, as well as the variance analysis. As it can be observed, the mean GPA in the high self-concept group is 0.6 higher than in the low self-concept group.

This difference is statistically significant given a *P* value of 0.0002 when completing the *Snedecor f*. A similar conclusion was obtained when the multiple range test was applied.

Table 2. Variance analysis for the GPA for each group of Self-concept

Self-concept group	Number of Students	Mean	Statistical error (grouped)	Inferior Limit	Superior Limit
High	69	3.3	0.1025	3.107	3.394
Low	57	2.7	0.1128	2.517	2.833
Total	**126**	**3,0**			

Source	Sum of Squares	Gl	Average Square	F Ratio	P Value
Between groups	10.34	1	10.34	14.27	**0.0002**
Intra groups	89.9	124	0.725		
Total (Corr.)	100.2	125			

Table 3 presents the analysis of students for academic condition and each group of self-concept. For each academic condition, the percentage of students that corresponds to the groups of self-concept are presented as "percentage by row". Thereby, from the 21 students that had an insufficient condition, 81% are part of the low self-concept. Similarly, the "percentage by column" shows the percentage of students that obtained a specific academic condition for each group of self-concept. Thereby, from the 69 students that are part of the high self-concept group, a 6% obtained an insufficient academic condition.

Table 3. Analysis of Students by Academic Condition and Self-concept

Academic Condition	Self-concept		Total by Row
	High	Low	
Insufficient	**4**	**17**	**21**
Percentage by row	19%	81%	16%
Percentage by column	6%	27%	
Normal	**52**	**27**	**79**
Percentage by row	66%	34%	60%
Percentage by column	75%	44%	
Probation Period	**11**	**11**	**22**
Percentage by row	50%	50%	17%
Percentage by column	16%	18%	
Not calculated	**0**	**5**	**5**
Percentage by row	0%	100%	4%
Percentage by column	0%	8%	
Outstanding	**2**	**2**	**4**
Percentage by row	50%	50%	3%
Percentage by column	3%	3%	
Total by Column	**69**	**62**	**131**
Percentage by row	53%	47%	100%

When observing the Normal academic condition, it can be noticed that 75% of the students presented a high self-concept, whereas a 44% presented a low self-concept. Although there are values with frequencies of less than five, the value of $P = 0.0004$ in the independency tests shows the existence of a statistical relation between the academic condition and the self-concept. As an additional data, the five students that cancelled the academic semester presented a low self-concept.

To solve the limitation of the application of the Chi-square test, the relation analysis of the self-concept with the academic continuity was completed. Results are presented in Table 4. As it can be observed, from the 32 students that cannot continue, 75% present have a low self-concept. Likewise, from the 69 students with high self-concept, 88% can continue, while from the 62 with low self-concept, only 61% can continue, i.e. 39% of the students with low self-concept cannot continue in their academic program. The hypothesis of independency between the two variables is significantly rejected ($P = 0.0003$), which gives more statistical arguments for the positive relation between the self-concept and the academic performance.

Table 4. Analysis of Students by Academic Continuity and Self-concept

Academic Continuity	**Self-concept**		**Total by Row**
	High	**Low**	
Can continue	**61**	**38**	**99**
Percentage by row	62%	38%	76%
Percentage by column	88%	61%	
Cannot continue	**8**	**24**	**32**
Percentage by row	25%	75%	24%
Percentage by column	12%	39%	
Total by Column	**69**	**62**	**131**
Percentage by row	53%	47%	100%

Psycho-pedagogical accompaniment

As explained in the methodology section, a psycho-pedagogical accompaniment program was offered to a random sample of students, called the *treatment group*. Students with a low self-concept that did not participated of the psycho-pedagogical accompaniment were part of the *control group*. Students with a high self-concept that did not participated of the accompaniment were part of the *reference group*. Consequently, the psycho-pedagogical accompaniment had three levels: *control*, *reference* and *treatment*. For each level, its effect was evaluated on the GPA using the variance analysis technique. Results are presented in Table 5.

Tabla 5. Variance analysis for the GPA for each level of Psycho-pedagogical accompaniment

Accompaniment Level	**Number of Students**	**Mean**	**Statistical error (grouped)**	**Inferior Limit**	**Superior Limit**
Control	39	2.3	0.1266	2.154	2.509
Reference	66	3.2	0.09735	3.112	3.385
Treatment	18	3.4	0.1864	3.158	3.68
Total	**123**	**3.0**			

Source	**Sum of Squares**	**Gl**	**Average Square**	**F Ratio**	**P Value**
Between groups	24.65	2	12.32	19.70	**0.0000**
Intra groups	75.06	120	0.6255		
Total (Corr.)	99.71	122			

From Table 5 it can be observed that students with a low self-concept that received psycho-pedagogical accompaniment (*treatment group*) had a mean GPA of 1.1 points higher than those that in the same self-concept group that did not received it (*control group*). When completing the variance analysis tests, it can be concluded that there is a significant effect of the psycho-pedagogical accompaniment on the GPA, given a value of $P = 0.000$. Additionally, although the GPA mean in the students with *treatment* surpasses the GPA mean in students of *reference*, the multiple range test does not show significant differences between them.

Results obtained when analyzing the academic condition for each level of psycho-pedagogical accompaniment are presented in Table 6. Results show that none of the students that were in the *treatment group* had an insufficient academic condition. Moreover, it should be noticed that from the 21 students in such academic condition, 81% are part of the *control group*. Despite there are values with less than five cases, the value of P = 0.0000 in the independency test, gives indications of the existence of a statistical relationship between academic status and levels of the psycho-pedagogical accompaniment level, where treatment has a positive effect on academic performance.

Table 6. Analysis of Students by Academic Condition and Psycho-pedagogical accompaniment level

Academic Condition	Accompaniment Level			Total by Row
	Control	Reference	Treatment	
Insufficient	**17**	**4**	**0**	**21**
Percentage by row	81%	19%	0%	16%
Percentage by column	40%	6%	0%	
Normal	**17**	**49**	**10**	**76**
Percentage by row	22%	65%	13%	59%
Percentage by column	40%	74%	53%	
Probation Period	**5**	**11**	**6**	**22**
Percentage by row	23%	50%	27%	17%
Percentage by column	11%	17%	32%	
Not calculated	**4**	**0**	**1**	**5**
Percentage by row	80%	0%	20%	4%
Percentage by column	9%	0%	5%	
Outstanding	**0**	**2**	**2**	**4**
Percentage by row	0%	50%	50%	3%
Percentage by column	0%	3%	10%	
Total by Column	**43**	**66**	**19**	**128**
Percentage by row	34%	51%	15%	100%

Given the fact that the academic continuity has fewer levels than the academic condition, a Chi-square test was completed. Table 7 presents the analysis of the academic continuity and the psycho-pedagogical accompaniment group. As observed, from the 32 students that cannot continue, only one had a accompaniment treatment. Furthermore, from the 19 students in such level 95% can continue, while in the control level only 47% can continue, i.e. 53% of these students cannot continue in their academic program. The hypothesis of independency between the two variables is significantly rejected (P = 0.0000), which gives more statistical arguments for the positive relation between the psycho-pedagogical accompaniment level and the academic performance.

Table 7. Analysis of Students by Academic Continuity and Psycho-pedagogical accompaniment level

Academic Continuity	Accompaniment Level			Total by Row
	Control	Reference	Treatment	
Can continue	**20**	**58**	**18**	**96**
Percentage by row	21%	60%	19%	75%
Percentage by column	47%	88%	95%	
Cannot continue	**23**	**8**	**1**	**32**
Percentage by row	72%	25%	3%	25%
Percentage by column	53%	12%	5%	
Total by Column	**43**	**66**	**19**	**128**
Percentage by row	34%	51%	15%	100%

Conclusions and Future Work

We have analyzed the relationship of the self-concept and the psycho-pedagogical accompaniment with the academic performance of first-semester students at Faculty of Engineering of Universidad de Antioquia. Results allow concluding that there is a positive relation between the self-concept and the academic performance. Moreover, the psycho-pedagogical accompaniment had a positive effect on the academic performance of the students that were part of this study.

Future work includes three main axis of work. First, the identification of the role of teachers during the first semester. In this axis, the role of teachers is relevant to the development of students' academic and emotional self-concept, as they can create a positive view of students and their ability to solve academic assignments. It is recommended that teachers take modules on self-concept in the pedagogical foundation courses, so they can consider the consequences of their attitudes and language in their relationship with students.

Second, the inclusion of psycho-pedagogical support in basic courses for students can strengthen the self-concept in its different dimensions. An example in this axis is found at Harvard University, where the course "Greater Happiness" is offered in first semester, aimed at strengthening this student-subject conscious relationship with its context.

Third, the students' family inclusion and the strengthening social skills can improve students' self-concept and therefore increase the probabilities of student retention. This work can be done through programs such as "*Encuentros de familia en la Universidad*" and workshops of extracurricular activities developed in the Faculty of Engineering at Universidad de Antioquia (FI-UdeA).

References

Abouserie, R. (2006). Self-esteem and achievement motivation as determinants of students approaches to studying. *Studies in Higher Education*, 20, 19-26.

Braxton, J., Milen, J. y Sullivan, A. (2000). The influence of active learning on the collage student departure process: toward a revision of Tinto's Theory. *Journal of Higher Education*, 17(5), 569-590.

Byrne, B. (1984). The general/academic self-concept nomological network: A review of construct validation research. *Review of educational research*, 54, 427-456

Chapman, J. and Boersma, F. J. (1982). Affective correlates of learning disabilities. *Australia and New Zealand Journal of Developmental Disabilities*, 8 (1), 51-53.

Costa, S. and Tabernero, C. (2012). Rendimiento académico y autoconcepto en estudiantes de educación secundaria obligatoria según el género. Infocop. Accessed at http://www.infocop.es/view_article.asp?id=4212 on 18 May 2017.

Covington, M. and Omelich, C. (1979) Are causal attributions causal?: A path analysis of the cognitive model of achievement motivation. *Journal of Personality and Social Psychology*, 37, 1487-1504.

Covington, M.V., and Beery, R.G. (1976). Self worth and school learning, New York: Holt, Rinehart & Winston.

García, F. and Musitu, G. (2001). AF5 Autoconcepto Forma 5. (2ª ed.). Madrid: TEA Ediciones.

García, F. and Musitu, G. (2014). AF5 Autoconcepto Forma 5. (4ª ed.). Madrid: TEA Ediciones.

Gargallo, B. (2006) Estrategias de aprendizaje, rendimiento y otras variables relevantes en estudiantes universitarios. *Revista de psicología general y aplicada*, 59, 1-2.

González, M. C. and Tourón, J. (1992). Autoconcepto y rendimiento escolar: sus implicaciones en la motivación y en la autorregulación del aprendizaje. Pamplona: EUNSA.

Hamachek, D. (1987). Encounters with the self. New York. Holt, Rinehart & Wiston.

Kifer, E. (1975). Relationships between academic achievement and personality characteristics: a quasi longitudinal study. *American Educational Research Journal*, 12, 191-210.

Lent, R.W., Singley, D., Sheu, H.B., Schmidt, J.A. and Schmidt, L. C.. (2007). Relation of social cognitive factors to academic satisfaction in engineering student. *Journal of Career Assessment*, 15, 87-97.

Leonadari, A. (1996). Comparability of self-concept among normal achievers, low achievers and children with learning difficulties. *Educational Studies*, 19(3), 357-371.

Lerner, J.M. (2010). Rendimiento académico en la perspectiva del desarrollo y bienestar del estudiante. Una visión clínico psicoanalítica. *Revista Universidad EAFIT*, 46, 77.

Markus, H., Cross, S., and Wurf, E. (1990). The role of the self-system in competence. In R.J. Stenberg & J. Kolligian (Eds.), Competence considered. (205-225) New Haven, CT Yale University Press.

Ministerio de Educación Nacional. Deserción Estudiantil en la Educación Superior Colombiana. Ministerio de Educación (2009). Accessed at http://www.mineducacion.gov.co/sistemasdeinformacion/1735/articles-254702_libro_desercion.pdf on 18 May 2017.

Musitu, G. and Allatt, P. (1994). Psicosociología de la familia. Valencia. Editorial: Albatros.

Musitu, G., García, F. and Gutiérrez, M. (1997). AFA (Autoconcepto Forma-A) (3ª ed.). Madrid: TEA Ediciones.

Purkey, W. (1970). Self-cpncept and school achievement. Englewood Cñiffs, NJ: Prentice-Hall.

United Nations (2000). Sustainable Development Goals. The United Nations Development Programme (UNDP). Accessed at http://www.undp.org/content/dam/undp/library/corporate/brochure/SDGs_Booklet_Web_En.pdf on 18 May, 2017.

Urquijo, S. (2002). Auto-concepto y desempeño académico en adolescentes. Relaciones con sexo, edad e institución. *Psico-USF*. 7 (2), 211-218

Valencia, A, Mejía, L, Parra-Mesa, C, Restrepo, G, Castañeda, E, and Mendoza, R. (2015). Observar para Decidir. *Revista Ingeniería y Sociedad*, 10, 23-25.

Villarroel, V. (2000). Relación entre autoconcepto y rendimiento académico. *Psykhe*, 10 (31). 3, 18.

Watson, W. (1970). Self concept and school achievement, Minnesota, USA. Prentice-Hall.

Acknowledge

Authors would like to express their gratitude to the Department of Industrial Engineering and the Faculty of Engineering at Universidad de Antioquia who partially funded this project. We also thank to Psychologists Erika Uñates and Alejandra Gil, and to Professor Olga C. Úsuga and students María Alejandra Lopera and Laura López for their support in the statistical analysis.

What is the theoretical basis of entrepreneurial learning in engineering education? A case study analysis using constructivism.

Isabel Hilliger
Pontificia Universidad Catolica de Chile, Santiago, Chile
ihillige@ing.puc.cl

Mar Perez-Sanagustin
Pontificia Universidad Catolica de Chile, Santiago, Chile
mar.perez@uc.cl

Constanza Miranda
Pontificia Universidad Catolica de Chile, Santiago, Chile
csmirand@uc.cl

Catalina Berner
Pontificia Universidad Catolica de Chile, Santiago, Chile
cdberner@uc.cl

Manuela de la Vega
Pontificia Universidad Catolica de Chile, Santiago, Chile
mdelavega@uc.cl

Abstract: *Entrepreneurial learning has gained ground in engineering education. Several studies have described a wide variety of entrepreneurship courses offered to engineering students. However, most of them have no theoretical approach to understand the learning process within these type of courses. Thus, education theories are needed to create shared meaning of entrepreneurial learning as it expands and evolves. This study is part of ongoing research to evaluate student learning in core-engineering courses focused on technology-based entrepreneurship. It uses mixed methods to explore the approach adopted by the School of Engineering in Pontificia Universidad Catolica de Chile. We collected qualitative information about course methods to analyze it by using the theory of constructivism. Different constructive elements were identified as problem-based or project-based learning. Then, quantitative data was used to address the learning benefits obtained from these course components. Finally, we discuss implications for the research site and future work.*

Introduction

Nowadays, there is a wide variety of entrepreneurial learning opportunities for engineering students. By entrepreneurial learning, academics typically refer to the acquisition of an entrepreneurial mind-set, knowledge, skills and attitudes (Duval-Couetil, Shartrand, & Reed 2016; Rideout & Gray, 2013). Over the last decades, in the U.S., engineering students had accessed mainly to entrepreneurship minors and certificates (Shartrand, Weilerstein, Besterfield-Sacre & Golding, 2010). Nowadays, the offer has expanded to core-curriculum courses, majors and campus-wide initiatives (Duval-Couetil et al., 2016). Meanwhile, in developing countries, the entrepreneurial learning opportunities have also risen, so policy makers expect positive effects on economic growth and employment (Moreno Zacarías, Moreno Meza & Santa Ana Escobar, 2016; Odora, 2015; Papayannakis, Kastelli, Damigos, & Mavrotas, 2008; Satheesh Raju, Suman Kumar, & Ramgopal, 2015).

In literature, several case studies of engineering entrepreneurship have been documented. Researchers have described elective courses targeted not only to engineering students (Sullivan, Carlson, & Carlson, 2001), but also from other majors (Sidhu, Singer, Suoranta, & Johnson, 2014). Articles have also analysed the implementation of sequences of more than one course (Creed, Suuberg, Crawford, & Gregory, 2002; Davis & Rose, 2007; Hazelwood, Valdevit, & Ritter, 2010), and even engineering minors (Bilén, Kisenwether, Rzasa, & Wise, 2005). There are also product design courses adopted as an alternative to capstone experiences (Bousaba & Conrad, 2015; Ochs, Lennon, Watkins, & Mitchell, 2006; Ohland, Sharron, Zhang, Brawner, & Miller 2004). Regardless of the varied publications along these lines, few of them refer to theories of learning.

Most studies on the effects of entrepreneurship education have based their analysis on psychological theories rather than learning principles. One of these theories is Bandura's self-efficacy (Bandura, 1977), which refers to 'the strength of a person belief that he or she is capable of performing the various roles and tasks of entrepreneurship' (Chen, Greene & Crick, 1998, p. 295). Other studies have alluded to the theory of planned behavior (Kautonen, van Gelderen, & Fink, 2015; Maresch, Harms, Kailer, & Wimmer-Wurm, 2016; Miralles, Giones, & Riverola, 2016; Rauch & Hulsink, 2015; Souitaris, Zerbinati, & Al-Laham 2007; von Graevenitz, Harhoff, & Weber, 2010). This theory was developed by Azjen (1991) to explain how entrepreneurship education influences individuals' beliefs about their entrepreneurial capabilities, in addition to attitude towards entrepreneurship and subjective norms. Under both theories, researchers have explored mostly psychosocial outcomes (Rideout & Gray, 2013), so the extent and nature of learning outcomes remains unexplored (Duval-Couetil, 2013).

Consequently, theories of learning are needed to explain the underlying mechanisms by which students acquire entrepreneurship knowledge, attitudes and skills. So far, assessment methods have lacked features to differentiate levels of competencies, having limited utility for formative purposes (Purzer, Fila, & Naraja, 2015). By applying learning theories, a holistic approach could be adopted to measure student outcome attainment of different educational opportunities in engineering entrepreneurship. Learning sciences enable a systematic understanding on how students' progress, besides informing the design of better learning environments (Johri & Olds, 2011). Concerning entrepreneurial learning, a better understanding is required to ensure students' development of a wide variety of skills, such as the ability to address a real-world problem, to work in multidisciplinary teams, to communicate effectively, and to manage uncertainty, risk and failure (Duval-Couetil, Kisenwether, Tranquilo, & Wheadon, 2014).

This study explores two core-engineering courses implemented in Pontificia Universidad Catolica de Chile (UC-Engineering): a first-semester course on product design and a third-year course on technology-based entrepreneurship. Both courses include problem-finding activities in a defined real-world context. Mixed methods were used to analyze their methodology using the theory of constructivism. First, a qualitative approach was adopted to categorize course elements. Second, quantitative data was collected from final course surveys to analyze learning benefits. Among different theories, we chose constructivism as a theoretical basis. This theory conceptualizes learning as knowledge acquisition facilitated by authentic tasks (Ertmer & Newby, 2013). Finally, we discuss different aspects to be improved in UC-Engineering instructional design, besides suggesting directions for future research.

Context

Over the last five years, policy makers in Europe and North America have promoted entrepreneurship education to boost employment and economic growth (Maresch et al., 2016; Oosterbeek, van Praag, & Ijsselstein, 2010; Rauch & Hulsink 2015; von Graevenitz et al., 2010). Researchers have documented different training strategies as the number of entrepreneurship courses rises (see Table 1). Rideout and Gray (2013) describe three education modalities regarding different entrepreneurship settings: small business management, small business start-up, and technology venture start-up. The first modality makes use of case study teaching and lectures, whereas the other modalities integrate practical applications and project-based learning. The third modality also involves simulations, prototyping, mentoring and even starting real venture (Rideout & Gray, 2013). Concerning varied types of delivery, Celis and Huang-Saad (2015) distinguish among an individual course, a certificate, a specific program (e.g. Minor or Major), and/or co-curricular entrepreneurship experiences. Furthermore, Duval-Couetil, Shartrand and Reed (2016) refer to different models depending on the institution imparting the training, the targeted students, the expected outcomes, and the teaching methodologies.

Table 1: Characteristics of Different Types of Entrepreneurship Training Strategies (adapted from Rideout and Gray (2013), and Duval-Couetil, Shartrand and Reed (2016))

Setting	Delivery	Leading Institution	Methodologies
Small business Startup New venture	Elective course Mandatory course Certificate Minor Major Co-curricular	Business school Engineering school Multi-school	Lectures Case studies Project-based learning Mentoring Prototyping Simulation and role play Fundraising New venture learning

In the U.S., entrepreneurship programs have gained ground in engineering schools (Celis & Huang-Saad, 2015; Duval-Couetil et al., 2016). Developing countries, such as Chile, Mexico, South Africa, India, and Greece have followed this trend (Grose, 2015; Moreno et al. 2016; Odora, 2015; Papayannakis et al., 2008; Satheesh et al., 2015). In Chile, since 2013, the National Agency for Innovation and Development (CORFO) has been co-financing strategic plans of engineering schools to reinforce entrepreneurship, innovation and applied research. Schools participating in this program are expected to prepare engineers to thrive in new technology markets, to work collaboratively, to think creatively and to communicate their ideas effectively (Grose, 2015). In UC-Engineering, this governmental initiative has become an opportunity to boost prior efforts to reform the engineering curriculum (Celis & Hilliger 2016). In the early 2000s, a cornerstone course on engineering design was implemented to expose first-year students to problem formulation (ING1004 Engineering Challenges). Then, in 2013, there was a curriculum overhaul to offer students multiple academic pathways, which included a core-engineering course on technology-based entrepreneurship in the third year (ING2030 Research, Innovation and Entrepreneurship).

Both ING1004 and ING2030 are mandatory courses in which students develop different entrepreneurial skills, such as the ability to address real-world problems, to work in multidisciplinary teams, and to communicate effectively (Hilliger, Perez-Sanagustín, & De la Llera 2016). Table 2 shows the characteristics of these two courses regarding the types of entrepreneurship-training strategies described in Table 1.

Table 2: Common Characteristics of ING1004 and ING2030 according to Different Types of Entrepreneurship Education Modalities

Course	Setting	Delivery	Leading Institution	Teaching and Assessment Methods
ING1004	Product design	Mandatory course	Engineering school	Lectures Case studies Project deliverables Prototyping Oral presentations
ING2030	New venture			

Research Questions

The research questions addressed in this article are:

1. What elements of constructivist learning are present in the ING1004 and ING2030 courses?
2. What learning benefits students have obtained from these constructivist elements?

Theoretical Framework

Concerning higher education, researchers have criticized the excessive lecturing and the low practical component within teaching and learning experiences (De los Rios, Cazorla, Díaz-Puentes, & Yagüe 2009). Consequently, the use of the constructivism theory has strengthen teaching towards bridging theoretical and practical aspects (Parsons, Caylor, & Simmons, 2005). Constructive approaches reinforce learning through problems and projects that enable the acquisition of knowledge and skills (Jones & Monaco, 2005). Additionally, they support the development of non-technical skills such as communication, information management, problem solving, and teamwork (Parsons et al., 2005). Constructive learning environments motivate students to adapt to a new situation and apply what they have learned (Wijni, Loyens, & Derous, 2010; Brundiers & Wiek, 2013). Therefore, constructivism is conceived as a student-centered approach that builds on prior students' learning. By introducing this approach to university teaching, students become more aware of the practical reality of their future practice (Parsons et al., 2005).

Constructivism learning could take two forms: problem-based learning and project-based learning (Parson et al., 2005). These two forms are characterized by being student-centered (Brundiers & Wiek, 2013; Wijni, Loyens & Derous, 2010; Hung, 2011; Liztinger, Lattuca, Hadgraft, & Newstetter, 2011). Students build small groups to work in real-world situations, undertaking tasks that simulate future professional situations. These learning experiences usually integrate peer-based assessments to encourage collaborative learning (Brundiers & Wiek, 2013; Jones & Monaco, 2005).Therefore, students' work actively in their teamwork skills, besides enhancing the social aspect of learning (Hung, 2011; Wijni, Loyens & Derous, 2010).

Despite common elements, there are differences between problem-based learning and project-based learning (see Table 3). At a first glance, problem-based learning could be conceived as a subset of project-based learning. This approach focuses on framing a specific problem rather than conducting multiple problem solving activities (Jones & Monaco, 2005). Additionally, problem-based learning is commonly more theoretical than project-based learning (Brundiers & Wiek, 2013). Students generate deep understanding from the theoretical construction of a problem that exists in the real world. Thus, students assume responsibility for their learning process. They are required to think critically about real-life situations in an iterative exercise of question and answer (Brundiers & Wiek, 2013). In contrast, project-based learning seeks to generate an understanding of a specific context with a practical emphasis. To produce something practical, students are required to apply knowledge towards project management and outcomes attainment. A predefined framework of action limits students' autonomy (Brundiers & Wiek, 2013). Thus, the role of the tutor or instructor is to guide the students towards the creation of a better outcome or product (Brundiers & Wiek, 2013; Hung, 2011; Wijni, Loyens & Derous, 2010).

Concerning entrepreneurship education, learning opportunities encourage stimulating activities that integrate theory with practice (Taksa, Tynjäläb & Kukemelka, 2016). These courses commonly consider a team-based project that demands different tasks through a specific course period. Students assume high responsibility during this process by finding a problem, and then, developing a project to solve it within time constraints. Instructors have to provide constant feedback to support the generation of ideas and the detection of good business opportunities. Peer-review processes are also needed through different methods and spaces, such as debates and presentations (Täksa et al., 2016).

Table 3: Differences of Problem-based and Project-based Learning

Aspects	Problem-based Learning	Project-based Learning
Outcome	Deep and theoretical	Practical and case-based
Students' Main Activity	Iterative exercise of questioning a real-world situation	Produce an expected outcome from knowledge application
Guiding Principle	Deep learning	Project management
Students' Autonomy	Full student-centered approach	Limited Student-centered

Therefore, it is possible to find constructive elements in the entrepreneurship courses. In empirical terms, there is evidence that subjects taking entrepreneurship courses are more confident about their abilities compared to subjects who do not take these courses (Duval-Couetil, Reed-rhoads & Haghighi, 2012). This is in part due to the benefits learning benefits students obtained from opportunities to apply content into practice (Täksa, Tynjälä, Toding, Kukemelk, & Venesaar, 2014).

Methodology

This study used mixed methods in an exploratory sequential design (Creswell, 2012). We collected and analyzed qualitative information to explore courses' constructive elements, followed by a quantitative phase to explain the learning benefits students obtained from these course components.

Data Collection and Analysis

To answer the first research question, we gathered qualitative information from courses' syllabus and materials. Under the theory of constructivism, we classified different course aspects into problem-based or project based.

In order to answer the second reseach question, we collected quantitative data in a course evaluation survey. This survey was applied online by the end of the semester, and students' participation was voluntary. We used a 5-point Likert scale to ask students about the learning benefits obtained from different course elements (1: not beneficial, 5: extremely beneficial). Then, we estimated the percentage of respondents per element that reported high learning benefits (i.e. 4: very beneficial and 5: extremely beneficial). Additionally, we asked students about the essential learning activities of each course in an open-ended question.

Participants and Sample

The ING1004 and ING2030 courses are both recommended for the first semester of freshmen and junior year correspondingly. Thus, we collected and analyzed information obtained from courses' versions implemented in the first semester of 2016 (see Table 4).

Table 4: Sample obtained from ING1004 and ING2030 Versions in 1st semester of 2016

Participants and Sample	ING1004	ING2030
N° of students enrolled	752	469
N° of sections offered	9	7
N° of responses in course evaluation survey	194	324
Response rates	26%	69%

Findings

Regarding ING1004, students must work in defined groups to design a product that solves a real-world problem (~7-person teams). Each semester, there is a specific theme in order to provide a framework for problem finding. In the course syllabi, the main learning outcome expected in the ability to identify and solve a real problem regarding diverse contextual factors, such as social, economic and environmental. Additionally, students are expected to develop the ability to work in a team, besides recognizing safe and responsible behaviors in the use of prototyping technologies. The course considers lectures and peer-tutoring activities in which students share the learnings obtained from participating in different workshops. At the end of the course, there is a technological fair were teams exhibit their final prototypes and present their findings to contextual stakeholders. There are other assessment methods during the course, such as individual homework, exams, peer assessments and oral presentations where each team has to demonstrate progress in their problem solving process.

The ING2030 course engages students with the process of creating a new venture. Each semester, a specific theme serves as a framework for recognizing a business opportunity. Students have to conform 5-person teams to work collaboratively in the generation of a technology-based idea that is both practical and feasible. Among learning outcomes, students are expected to develop entrepreneurial skills, such as the ability to function effectively on teams, the capacity to achieve product/market fit, and the ability to pitch an idea effectively. Course instruction techniques include guest lectures, section lectures, prototyping workshops and mentoring sessions. During mentoring, students get feedback from entrepreneurs on their venture and their business model. Assessment activities consist in reflection blogs, group homework, peer assessments, team presentations, final prototype submission, and a final pitch.

Both courses encourage students to perform an iterative question-answer process. However, students have limited autonomy to investigate problems and gain theoretical knowledge. Students' autonomy diminishes even more throughout the course because the tutor requires team to converge on a specific product. By classifying the elements of ING1004 and ING2030, we identified both project-based and problem-based elements (see Table 5). Still, we also identified some aspects from traditional teaching strategies.

In order to answer the second research question, we estimated the percentage of respondents that considered different course elements as very beneficial or extremely beneficial (see Table 6). A larger percentage of respondents obtained benefits from constructive elements compared to traditional teaching methods. By traditional methods, we mean lectures, testing and homework. Additionally, assessment methods report higher benefit percentages than teaching methodologies. ING1004 students obtained more benefits from mid-term project presentations, whereas ING2030 students obtained more benefits from the final pitch.

Regarding teaching methods, the most beneficial activities of both courses involved stakeholders. In ING1004, the teaching method with the largest percentage of students that reported learning benefits was the technological fair. Key agents are invited to this activity in order to assess teams' final prototypes and presentations. These agents represent stakeholders from real-life challenge given to the students. During the first semester of 2016, the challenge was waste minimization. Meanwhile, in ING2030, the methodology with the largest percentage of students obtaining learning benefits were the mentoring sessions. In these sessions, real entrepreneurs are invited, so they provide students valuable feedback on their ventures. Since 2015, entrepreneurs are recruited from the Endeavor network, a non-profit organization that supports high impact entrepreneurs through networking activities and connections with enterprise and governmental leaders.

Table 5: Classification of Course Elements of ING1004 and ING2030 According to Problem-based or Project based

Characteristics	ING1004	ING2030	Classification
Focus	Product design	Venture creation	Project-based learning
Outcome	Practical and context-based		Project-based learning
Team Work	7-person	5-person	Problem and Project based
Students' Main Activity	Iterative exercise of questioning a real-world problem followed by solution prototyping		Problem and Project based
Guiding Principle	Engineering Design Process	Technological Innovation Process	Project-based learning
Teaching Methods	Lectures Workshops Peer Tutoring Presentations in front of stakeholders (Technological Fair)	Lectures Workshops Sessions with Entrepreneurs Mentoring	Traditional Problem and Project based
Assessment Methods	Individual Exams Individual Homework Project Presentations Final Prototype Peer Assessments	Reflection Blog Group Homework Final Prototype Final Presentation Peer Assessments	Traditional Problem and Project-based
Students' Autonomy	Limited student-centered		Project-based learning

Table 6: Classification of Course Elements of ING1004 and ING2030 According to Problem-based or Project based

Teaching Methodologies			
ING1004 (N=194)		**ING2030 (N=324)**	
Technological Fair	34%	Mentoring	24%
Workshops	18%	Guest Lectures	15%
Peer Tutoring	12%	Section Lectures	14%
Lectures	9%	Workshops	13%
Assessment Methods			
ING1004 (N=194)		**ING2030 (N=324)**	
Mid-term Presentations	51%	Final Pitch	50%
Final Prototype	46%	Final Prototype	43%
Peer Assessments	21%	Peer Assessments	27%
Individual Exams	7%	Group Homework	11%
Individual Homework	1%	Reflection Blog	3%

Concerning the key learning activities declared in the open-ended question, ING1004 students alluded to the proximity with engineering practice by undertaking project presentations, workshops and peer assessments. Meanwhile, ING2030 students valued learning opportunities to develop professional skills, such as teamwork and oral communication. Students also mentioned instructors' feedback and the entrepreneurship experience to a lesser degree.

Discussion, Conclusions and Implications

Regarding the first research question, we identified different constructive elements in UC-Engineering courses. First, both courses are focused in a real-life setting, so students become aware of the practical reality of future engineering practice (Parson et al., 2015), In ING1004, students have to solve a problem in a given real-world context. While, in ING2030, students are required to create a new venture to fulfil an existing need. Furthermore, teamwork and peer-assessments are used to emphasize social construction of learning (Brundiers & Wiek, 2013). Additionally, instructors and stakeholders play an essential role, as they provide feedback in formative and summative assessment.

Concerning the second research question, higher percentages of students obtained significant benefits from constructive elements, declaring these course components as key learning activities. By reinforcing learning through team-based projects, students have more opportunities to develop professional skills (Jones & Monaco, 2005). This effect is particularly observed in the ING2030 course, whose students describe the acquisition of non-technical skills in the respective open-ended question. Consequently, engineering programs that have embedded entrepreneurial learning within the core-engineering curriculum might have improved the profession training of their students.

Nevertheless, the courses addressed in this study are more oriented to project-based rather than problem-based learning. UC-Engineering students are highly encouraged to produce practical results (Brunders & Wiek, 2013), so deeper learning efforts are restrained to a short phase at the beginning of each course. Therefore, UC-Engineering might need to balance project-based approaches with problem-based learning, so students have more opportunities to think critically about real-life situations (Brunders & Wiek, 2013). At an instructional level, traditional lecturing included in ING1004 and ING2030 courses might have to be replaced by flipped approaches where students work in their project in class hours. This improvement action could increase the number of positive interactions among and within teams. Furthermore, instructors would have also have more opportunities to provide feedback and make these entrepreneurial experiences more meaningful.

In conclusion, entrepreneurship courses have grown rapidly over the past decades under the motivation of policy makers and accreditation agencies. Learning theories could facilitate the revision of existing entrepreneurial learning experiences. The analysis of course elements under a theoretical basis facilitates the understanding of learning mechanisms within an entrepreneurship course. Thus, improvement actions could be formulated to ensure student outcome attainment. These actions might help students to develop the attitudes and skills they need to succeed in employment and self-employment careers. Therefore, leaders of schools of engineering should consider learning theories to reflect about the approaches that they have adopted towards entrepreneurial learning, so they make sure that future generations of engineers have deeper understanding of real-world issues.

References

Ajzen, I. (1991). The theory of planned behavior. *Organizational Behavior and Human Decision Processes, 50*, 179-211.

Bandura, A. (1977). Self-Efficacy: Toward a Unifying Theory of Behavioral Change. *Psychological Review, 84* (2), pp. 191-215

Bilen, S. G., Kisenwether, E. C., Rzasa, S. E., & Wise, J. C. (2005). Developing and assessing students' entrepreneurial skills and mind-set. *Journal of Engineering Education, 94* (2), 233–243.

Bousaba, N. A., & Conrad, J. M. (2014, June), Recent Graduates' Perspectives on Innovation and Entrepreneurship and the Creation of New Entrepreneurship Course. *Paper presented at 2014 ASEE Annual Conference & Exposition*. Indianapolis, IN: ASEE.

Brundiers, K & Wiek, A. (2013). Do we teach what we preach? An international comparison of problem and project-based learning courses in sustainability. *Sustainability, 5,* 1725-1746.

Chen, C. C., & Greene, P. G. (n.d.). Distinguish Entrepreneurs From Managers ? *Journal of Business Venturing, 9026* (97), 295–316.

Celis, S., & Huang-Saad, A. (2015, June), Students Seeking Different Paths to Entrepreneurial Education. *Paper presented at 2015 ASEE Annual Conference & Exposition*. Seattle, WA: ASEE.

Celis, S., & Hilliger, I. (2016, June), Redesigning Engineering Education in Chile: How Selective Institutions Respond to an Ambitious National Reform. *Paper presented at 2016 ASEE Annual Conference & Exposition*. New Orleans, LA: ASEE

Creed, C. J., Suuberg, E. M., & Crawford, G. P. (2002). Engineering Entrepreneurship: An Example of A Paradigm Shift in Engineering Education. *Journal of Engineering Education, 91(2)*, 185–195.

Creswell, J. W. (2012). Educational research: Planning, conducting, and evaluating quantitative and qualitative research. Boston, MA: Pearson Education Inc.

Davis, D., & Rose, J. (2007). Entrepreneurial engineering capstone course with research-based outcomes assessment. Paper presented at 2007 ASEE Annual Conference & Exposition. Honolulu, HI: ASEE.

De los Ríos, I., Cazorla, A., Díaz-Puentes, J.M., & Yagüe, J. L. (2009). Project-based learning in engineering higher education: two decades of teaching competences in real environments. *Procedia – Social and Behavioral Sciences, 2* (2), 1368-1378.

Duval-Couetil, N., Reed-rhoads, T. & Haghighi, S. (2012). Engineering Students and Entrepreneurship Education: Involvement, attitudes and outcomes. *International Journal of Engineering Education, 28* (2), 425.

Duval-Couetil, N. (2013). Assessing the impact of entrepreneurship education programs: Challenges and approaches. *Journal of Small Business Management, 51*(3), 394–409.

Duval-Couetil, N., & Kisenwether, E. C., & Tranquillo, J., & Wheadon, J. D. (2014, June), Catalyzing the Adoption of Entrepreneurship Education in Engineering by Aligning Outcomes with ABET *Paper presented at 2014 ASEE Annual Conference & Exposition*. Indianapolis, IN: ASEE.

Duval-Couetil, N., Shartrand, A., & Reed, T. (2016). The Role of Entrepreneurship Program Models and Experiential Activities on Engineering Student Outcomes. *Advances in Engineering Education. Winter 2016*, 1-27.

Ertmer, P. A., & Newby, T. J. (2013). Behaviorism, Cognitivism, Constructivism: Comparing Critical Features from an Instructional Design Perspective. *Performance Improvement Quarterly, 26 (2)*, 43-71.

Hazelwood, V., Valdevit, A., & Ritter, A. (2010). A Model for a Biomedical Engineering Senior Design Capstone Course, with Assessment Tools to Satisfy ABET "Soft Skills. *Paper presented at the Capstone Design Conference*. Boulder, CO.

Hilliger, I., Perez-Sanagustín, M. & de la Llera, J.C. (2016). "Preparing all engineerings to become entrepreneurial: Core entrepreneurship education in Chile". *Global Engineering: IFEES-GEDC Bulletin, 1* (3).

Hung, W. (2011) Theory to reality: A few issues in implementing problem-based learning. *Education Technology and Research Development, 59*, 529-552.

Johri, A., & Olds, B. (2011). Situated Engineering Learning: Bridging Engineering Education Research and the Learning Sciences. *Journal of Engineering Education, 100(1),* 151–185.

Jones & Monaco. (2005) Project-based learning/problem-based learning. Laureate Education.

Kautonen, T., van Gelderen, M., & Fink, M. (2015). Robustness of the theory of planned behavior in predicting entrepreneurial intentions and actions. *Entrepreneurship: Theory and Practice, 39* (3), 655–674.

Liztinger, T., Lattuca, L., Hadgraft, R. & Newstetter, W. (2011) Engineering education and the development of expertise. *Journal of Engineering Education, 1,* 123-150.

Maresch, D., Harms, R., Kailer, N., & Wimmer-Wurm, B. (2016). The impact of entrepreneurship education on the entrepreneurial intention of students in science and engineering versus business studies university programs. *Technological Forecasting and Social Change, 104*, 172–179.

Miralles, F., Giones, F., & Riverola, C. (2016). Evaluating the impact of prior experience in entrepreneurial intention. *International Entrepreneurship and Management Journal, 12* (3), 791–813.

Moreno Zacarías, H.M., Moreno Meza, R. de J., Santa Ana Escobar, M.B. (2016). "Las Universidades Públicas Autónomas Mexicanas: licenciaturas en ingeniería y su relación con el fomento académico de ser emprendedor". *In Vestigium Ire,10*, 118-138.

Ochs, J., Lennon, G., Watkins, T., & Mitchell, G. (2006). A Comprehensive Model for Integrating Entrepreneurship Education and Capstone Projects while Exceeding ABET Requirements. *Paper presented at the American Society for Engineering Education*. Chicago, IL: ASEE.

Odora, R. J. (2015). Integrating Product Design and Entrepreneurship Education: A Stimulant for Enterprising Design and Engineering Students in South Africa. *Procedia Technology, 20*, 276-283.

Ohland, M. W., Frillman, S. A., Zhang, G., Brawner, C. E., & Miller, T. K. I. (2004). The effect of an entrepreneurship program on GPA and retention. *Journal of Engineering Education, 93* (4), 293–301.

Oosterbeek, H., van Praag, M., & Ijsselstein, A. (2010). The impact of entrepreneurship education on entrepreneurship skills and motivation. *European Economic Review, 54* (3), 442–454.

Papayannakis, L., Kastelli, I., Damigos, D. & Mavrotas, G. (2008). Fostering entrepreneurship education in engineering curricula in Greece. Experience and challenges for a Technical University. *European Journal of Engineering Education, 33* (2), 199-210.

Parsons, Caylor & Simmons (2005). Cooperative Education Work Assignments: The Role of Organizational and Individual Factors in Enhacing ABET Competencies and Co-op workplace Well-Being. *Journal of Engineering Education, 94* (3), 309-318.

Purzer, S., Fila, N., & Nataraja, K. (2016). Evaluation of Current Assessment Methods in Engineering Entrepreneurship Education. *Advances in Engineering Education, 5* (1), 1–41.

Rauch, A., & Hulsink, W. (2015). Putting entrepreneurship Education where the intention to Act lies: An investigation into the impact of entrepreneurship education on entrepreneurial behavior. *Academy of Management Learning and Education, 14* (2), 187–204.

Rideout, E. C., & Gray, D. O. (2013). Does entrepreneurship education really work? A review and methodological critique of the empirical literature on the effects of university-based entrepreneurship education. *Journal of Small Business Management, 51* (3), 329–351.

Satheesh Raju G., Suman Kumar N. & Ramgopal N. Ch. (2015). Entrepreneurship and Innovation: A Study on Factors Affecting Engineering Graduates towards Entrepreneurship and Innovation. *Journal of Engineering Education Transformations, Special Issue*.

Shartrand, A., Weilerstein, P., Besterfield-sacre, M., & Golding, K. (2010). Technology Entrepreneurship Programs in U.S. Engineering Schools: Course and Program Characteristics at the Undergraduate Level. *American Society for Engineering Education Annual Conference and Exposition, 6*, 4527–4536.

Sidhu, I., Singer, K., Suoranta, M., & Johnsson, C. (2014). SUBMITTED TO THE 74 th ANNUAL MEETING OF THE ACADEMIA OF MANAGEMENT 2014. Introducing Berkeley Method of Entrepreneurship - a game-based teaching approach.

Souitaris, V., Zerbinati, S., & Al-Laham, A. (2007). Do entrepreneurship programmes raise entrepreneurial intention of science and engineering students? The effect of learning, inspiration and resources. *Journal of Business Venturing, 22* (4), 566–591.

Sullivan, J. F., Carlson, L. E., & Carlson, D. W. (2001). Developing Aspiring Engineers into Budding Entrepreneurs: An Invention and Innovation Course. *Journal of Engineering Education, 90* (4), 571-576.

Täks, M., Tynjälä, P. & Kukemelk, H. (2016). Engineering students' conceptions of entrepreneurial learning as part of their education. *European Journal of Engineering Education, 41* (1), 53-69.

Täks, M.,Tynjälä, P., Toding, M., Kukemelk, H. & Venesaar, U. (2014). Engineering Students' Experiences in studying entrepreneurship. *Journal of Engineering Education, 103* (4), 573-598.

von Graevenitz, G., Harhoff, D., & Weber, R. (2010). The effects of entrepreneurship education. *Journal of Economic Behavior and Organization, 76* (1), 90–112.

Wijinia, L., Loyens, S. M. M., & Derous, E. (2011). Effects of problem-based versus lecture based learning environments on student motivation. *Contemporary Educational Psychology, 36*, 101-113.

Acknowledgements

This work was supported by CORFO under grant n°: 14ENI2-26862, and partially supported by the Fondecyt Project nº: 11150231 and the EU Project MOOC Maker nº: 561533-EPP-1-2015-1-ES-EPPKA2-CBHE-JP.

Copyright statement

Applied Mathematics: An Innovative Course in Applied Mathematics and Science for an Undergraduate Curriculum in Computing Disciplines

Pedro Guillermo Feijóo García
Universidad El Bosque, Bogotá D.C., Colombia
pfeijoo@unbosque.edu.co

Orlando López-Cruz
Universidad El Bosque, Bogotá D.C., Colombia
orlandolopez@unbosque.edu.co

***Abstract**: A common denominator in systems engineering programs, is the disconnection between mathematics and science with the general professional guideline, circularly proposed for their students. To reduce the existing gap, we designed, developed and implemented an Applied Mathematics course at Universidad El Bosque, Colombia, aiming to educate our students on deciding when to apply analytical tools based on mathematics or physics of Descartes, Leibniz and Newton, or when to apply, instead, other mathematical and scientific tools best suited to address complex environments, where information systems are to be developed and deployed. This paper describes the purpose, the focus and the students' perception, corresponding to the course's development and the topics covered in it, supplementing them with relative findings per the role it exposes within the academic curriculum.*

Introduction

The pervasive use and permanent advances in information and communication technologies (ICT's) have evidenced the gap between mathematics taught in engineering curricula and mathematics needed by those who graduate in computing related disciplines, to be able to succeed in an engineering career. This gap is wider and more evident when people is involved in the actual operation of engineering artifacts produced by disciplines such as information systems, information technology, software engineering, computer science and industrial engineering. Thus, the mentioned gap is wider in knowledge disciplines where the materiality or physical dimension of artifacts being engineered is minimum or null at all.

The high cohesion between the artifact and people being served by it, demands design skills to build models including people dynamics and their interactions. Far from the environment in which artifacts must obey to physical laws in nature, artifacts of engineering systems, tackled in the course being reported, are in the domain of social phenomena uncertainty, which emerges from the interaction of people and artifacts by means of information and information devices.

To try to reduce this gap, we designed and developed an "Applied Mathematics" course aiming to educate students to become professionally able to apply criteria to decide when to apply analytical tools based on mathematics or physics of Descartes, Leibniz and Newton, and when to apply, instead, other mathematical and scientific tools best suited to address complex environments where information systems are to be developed and deployed.

Throughout this document, we present the corresponding findings and results, looking forward to answer how an undergraduate course of a systems engineering program can teach students how to use the best suited mathematical tools when developing information systems deployed in complex and real social environments. Latter in this paper, we present the results analysis that conducts us to the conclusions arose from this experience.

Context

The widely extended belief during the Middle Ages was that all things lived only by virtue of a corporate existence (Francke, et al., 1913). Therefore, learning existed only as incorporated in a university. When the holy father Pope Alexander IV stated in a constitution addressed to the University of Paris in 1256 "Quasi lignum vitae in Paradiso Dei, et quasi lucerna fulgoris in Domo Domini, est in Sancta Ecclesia Parisiensis Studii disciplina" (i.e. "As the tree of life in God's Paradise and the lamp of glory in the house of God, such in the Holy Church is the place of the Parisian corporation of learning"), he did not know that his statement would define the place of science in modern civilization also (Veblen, 1906). This is to say, science is considered the quintessence of knowledge in the western world, and the scientific method, the valid road to discover knowledge. This has led to the axiomatic belief that engineering is applied science. This belief implies that engineering does not produce knowledge or it is not as relevant as scientific knowledge.

This unfounded belief ignores that the steam engine –for instance- was engineered about a century before thermodynamics were created by Nicolas Sadi Carnot. This historical fact, as well as many others, implies that scientific knowledge does not necessarily precedes engineering. Engineering differs from science in many respects. One of them is that while science studies and explain phenomena, even predict them, engineering transforms reality (Olaya, 2013). Engineering transforms one situation into another –desired- situation (Petroski, 2011) by using specific artifacts in specific contexts (Petroski, 1992). If not the only one, engineering is the science of the artificial par excellence (Simon, 1998). The artificial world (i.e. the world with things not created by nature) is created by human beings (Simon, 2008).

Besides mechanical artifacts, that may be designed by analytical mathematics and physics, there are some artifacts such as organizations, organizational changes, which are essentially social systems because of the concurrence of interactions among human beings that are complex artifacts, complex systems as well the systems advised by the creator of systems theory (Bertalanffy, 2007), who focused special attention to those systems not analysable, systems that may not be reduced to a contingent set of components (López-Cruz, 2002).

To study those systems, furthermore, to engineer those complex systems, the engineer needs a different mathematical toolbox. Tools that enables not only to apply analytical methods to explain and predict, such as those of advanced engineering mathematics (Kreyszig, 2010): Frobenius series, Bessel equation, Laplace transform, convolution, Fourier analysis, Z transform, complex numbers and functions, partial differential equations, and so on; but also tools that allow to engineer complex dynamic adaptive systems such as the information systems required by the 21th century knowledge-based economics (Hasenclever & Silva, 2007).

To tackle this challenge, and avoid widely known problems in teaching mathematics (Laguna,et al., 2016) that may be applied when engineering social and socio-technical systems (Cano Ibarra,et al., 2007), a different course of Applied Mathematics was designed. The course was designed to deal with the need of new knowledge related to the concepts of emergence, dynamic self-regulation, self-organization, dynamic adaptive systems, which recall complexity as thinking method (Delgado Díaz, 2007).

Applied Mathematics in Universidad El Bosque addresses cellular automata as a mean to design emergent behaviour in complex systems (Wasserman, 1994), network analysis (Newman, 2001), an introduction to artificial neural networks, and entropy optimization principles (Kapur and Kesayan, 1995).

A Social Approach from Multi-Agent Systems Modelling

Considering the importance of mathematics and science comprehension, and further appliance in the systems engineering curriculum, the activity explained through this section was implemented successfully to take students towards problems arising from the Colombian social, political and cultural context. This activity included as main topic, multi-agent systems modelling and design, using innovative modelling frameworks such as NetLogo (Caillou, et al., 2017).

Students were asked to do brainstorming about social problems existing in Colombia's actuality, identifying the context, the actors and the relations looking towards to model, explain and even propose a solution for them. This, by understanding how the society could be abstracted as a general system, constituted by autonomous microsystems or agents, explained by relations occurred from and within a specific context. In other words, we conducted the students to think and visualize society as an Actor-Network (Nieto Olarte, 2013).

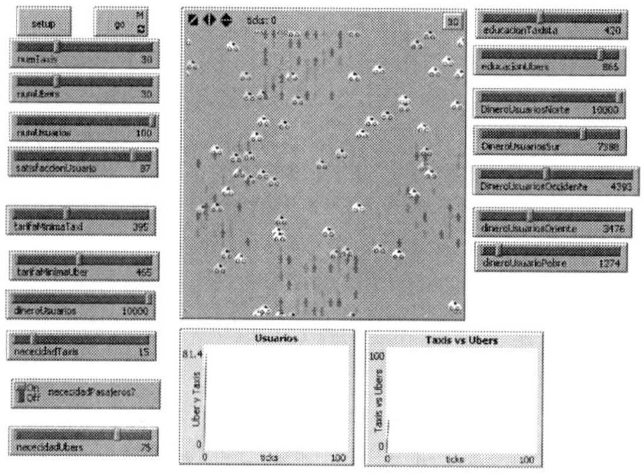

Figure 1: "Uber vs Taxicabs" NetLogo Model

A first example of our students' works, was a modelling proposal towards to illustrate interactively the current conflict between Uber and Taxicab drivers in Bogotá D.C. Students identified three main actors: Taxicab driver, Uber driver and the citizen; which interacted towards variables of security, trust and payment capacity (see Figure 1).

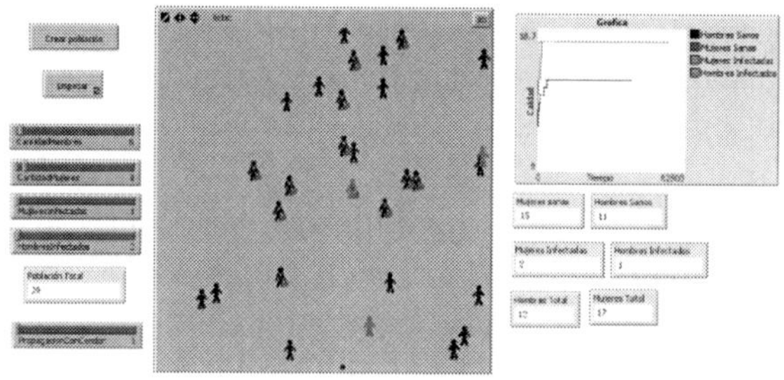

Figure 2: "HIV Propagation" NetLogo Model

On a second approach on the results corresponding to this same activity, a group of our students modelled HIV propagation in the national territory, considering four main actors involved within this propagation schema: female citizen, male citizen, infected female citizen and infected male citizen. The model exposed successfully the virus propagation through citizens' reproduction, measuring variables according to the number of infected individuals, using and omitting the condom as prevention mechanism (see Figure 2).

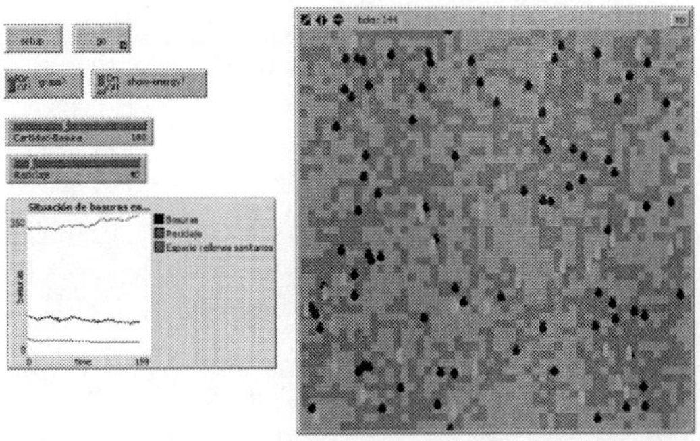

Figure 3: “Garbage Distribution” NetLogo Model

As a last example in this activity’s development, another group of our students modelled an approach regarding garbage distribution in Bogotá D.C. They considered for their model a total of three actors: garbage, recycling and landfills. Their approach presented interesting results, taking into consideration numerical variables which correlated the actors identified and included (see Figure 3).

All the students were asked to explain their model in a poster session, with their classmates as peer reviewers. The activity exposed to be an excellent academic approach for the appliance of mathematics, science and data science, in complex social scenarios where high levels of abstraction were required for the models proposal and development.

Students Perception and Satisfaction

Looking towards to justify the role of the proposed “Applied Mathematics” course, we conducted a survey to get and analyze the students’ perception towards the topics, purpose and course approach. A total of 21 students participated in this survey, with the following gender and age distribution, illustrated in Figure 4.

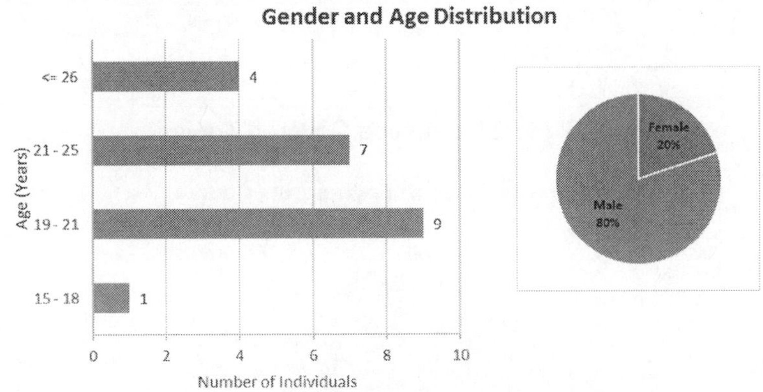

Figure 4: Gender and Age Distribution

The first question asked corresponded to which courses, from the mathematics courses taken by systems engineers, they consider more alike to *Applied Mathematics*. As it is shown in Figure 5, for a list of six courses, more than the 50% of the sample population indicated that the *Applied Mathematics* was not alike to any of the courses previously listed, followed by a 24% who related it to *Differential Calculus*, and a 14% who found similarities with the *Discrete Mathematics* course.

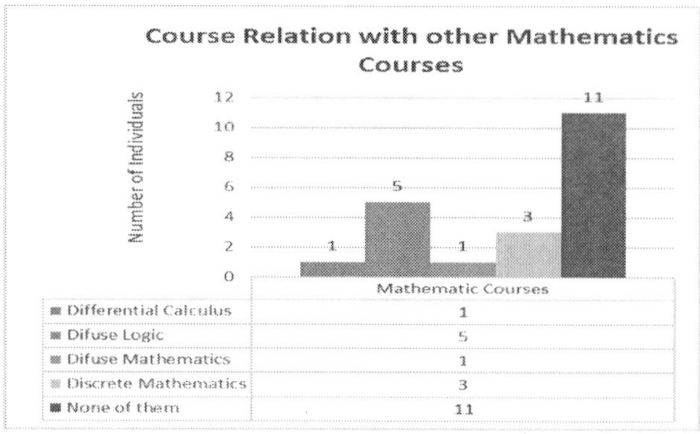

Figure 5: Course Relation with other Mathematics Courses

Following the results illustrated with Figure 5, corresponding to which affirmation described better the students' experience with the *Applied Mathematics* course, we can observe that more than the 90% of the sample population stated that *"Applied Mathematics emphasizes in the use of mathematics for real world problems modelling"*. Furthermore, as it is exposed with Figure 6, most of the students indicated as most relevant course's topics: Agent-based systems (57,1%), graphs and network analysis (19%), and evolutionary algorithms (19%). Result that are favourable for the course proposal and objectives.

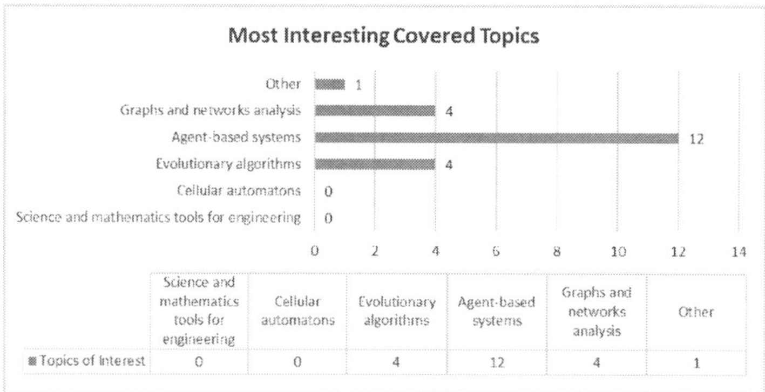

Figure 6: Most Interesting Covered Topics

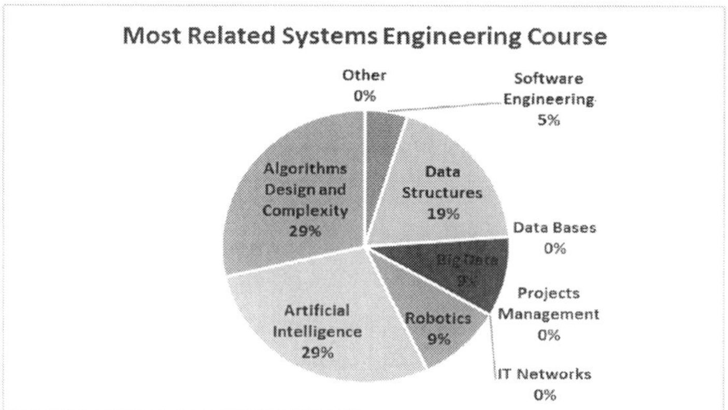

Figure 7: Course Relation with other Systems Engineering Courses

Complementing the results detailed before, and considering Figure 7, we can visualize that most of the participant students related the course with other courses of the systems engineering curriculum, such as: Algorithms design and complexity (28,6%), artificial intelligence (28,6%), data structures (19%), big data (9,5%) and robotics (9,5%).

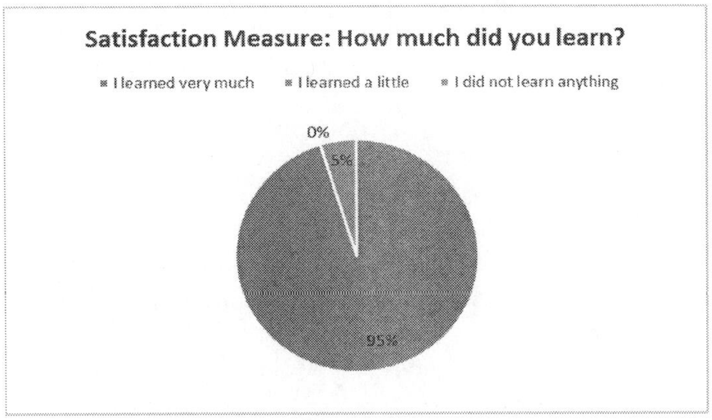

Figure 8: Satisfaction Measure: How much did you learn?

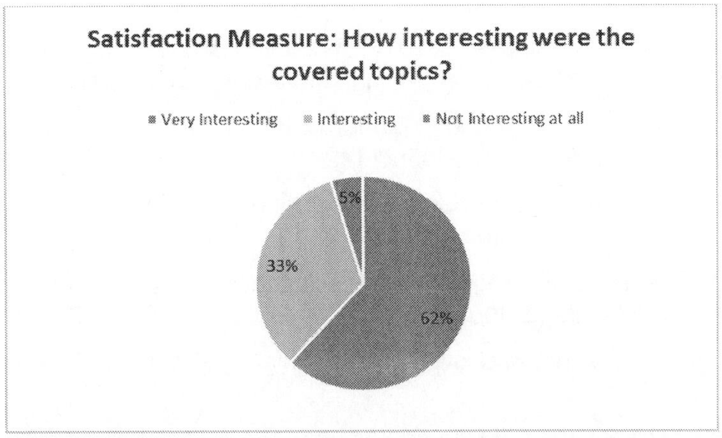

Figure 9: Satisfaction Measure: How interesting do you consider the topics covered?

Complementarily, and looking to analyse students' satisfaction, we asked how much they considered to have learned in the course, and how interesting were the topics covered with it. As it can be seen with Figures 8 and 9, the results are satisfactory, with a 95.2% of the population who considered to have learned very much, and a 61,9% who indicated that the course's topics were very interesting, in contrast to a 33.3% who stated that they were just interesting. Only 4,8% of the sample population stated that the course's topics were not interesting.

Findings and Conclusions

Per the obtained results, a first conclusion is that the design and development of this course has helped to reduce the existing gap between fundamental and professional cycles of the systems engineering program at Universidad El Bosque, teaching students how to articulate mathematics and physics when designing and developing information systems to be deployed in complex social environments.

Furthermore, as it was shown, the Applied Mathematics course at Universidad El Bosque was engineered to deal with a specific context and need to develop competences in individuals responsible to change present situations converting them in desired situations. The characteristics of this course respond to the biopsychosocial and cultural approach at Universidad El Bosque: a significant learning pedagogical model (Fink, 2013) which requires a student-centered approach. Therefore, this is not a static course and slightly evolves according to results that have been obtained each semester.

Finally, we conclude that the design and implementation of the Applied Mathematics course at Universidad El Bosque, is a significant approach to bring closer the systems engineering program to other scientific and social disciplines. Interdisciplinary is enriched with this type of courses, guaranteeing a healthy articulation of topics from scientific disciplines to society through technology.

Acknowledgements

Our sincere gratitude to students who actively and voluntarily participated in this study. Their help and collaboration were essential for this academic production, and furthermore, for the designed course improvement.

References

Francke, K., Howard, W. G., & Singer, I. (1913). The German classics of the nineteenth and twentieth centuries; masterpieces of German literature. Albany, NY: J.B. Lyon Co.

Veblen, T. (1906). The Place of Science in Modern Civilization. American Journal of Sociology, 11(5), 585-609. doi:10.1086/211429

Olaya, C. (2013). Más ingeniería y menos ciencia por favor. XI Congreso Latinoamericano de Dinámica de Sistemas, Monterrey, México.

Petroski, H. (2011). The essential engineer: why science alone will not solve our global problems. New York: Vintage Books.

Petroski, H. (1992). Engineering: The Evolution of Artifacts. American Scientist, vol. 80, pp. 416-420.

Simon, H. A. (1988). The Science of Design: Creating the Artificial. Design Issues, 4(1/2), 67. doi:10.2307/1511391

Simon, H. A. (2008). The sciences of the artificial. Cambridge, Mass.: MIT Press.

Bertalanffy, L. V. (2007). Teoría general de los sistemas: fundamentos, desarrollo, aplicaciones. Buenos Aires: Fondo de Cultura Económica.

López-Cruz, O (2002). Sobre lo sistemático y lo sistémico. Revista de Tecnología, vol. 1, p. 47.

Kreyszig, E. (2010). Advanced engineering mathematics. John Wiley & Sons.

Hasenclever, L., & Silva, E. H. (2007). The Knowledge-based Economy: modeled, measured, simulated, Loet Leydesdorff. Reciis, 1(1). doi:10.3395/reciis.v1i1.36en

Laguna Cortés, J.Ó., Alvarado Arellano, M., & Santacurz Vázquez V. (2016). Problemática de la enseñanza y evaluación de las matemáticas en la formación para Ingenieros. ANFEI Digital, p 1-9.

Cano Ibarra, S.T., Ramos Beltrán, J.A., & Medina Torres, M.G. (2007). Análisis del estrés académico en estudiantes de ingeniería como estrategia para el aprendizaje significativo. ANFEI Digital, p 1-8.

Delgado Díaz, C.J. (2007). Hacia un Nuevo saber: La bioética en la revolución contemporánea del saber. Vol. 2. Bogotá D.C., Universidad El Bosque.

Wasserman, S. (1994). Social network analysis: Methods and applications. Cambridge University Press, vol 8.

Newman, M. E. (2001). Scientific collaboration networks. II. Shortest paths, weighted networks, and centrality. Physical Review E, 64(1). doi:10.1103/physreve.64.016132

Kapur, J. N., & Kesavan, H. (1995). Entropy optimization: principles with applications. Boston, Mass.: Academic press.

Fink, L. D. (2013). Creating significant learning experiences, revised and updated: an integrated approach to designing college courses. San Francisco, CA: Jossey-Bass.

Caillou, P., Coyrehourq, S. R., Marilleau, N., & Banos, A. (2017). Exploring Complex Models in NetLogo. Agent-Based Spatial Simulation with Netlogo, Volume 2, 173-208. doi:10.1016/b978-1- 78548-157- 4.50006-6

Nieto Olarte, M. (2013). Las máquinas del imperio y el reino de Dios: reflexiones sobre ciencia, tecnología y religión en el mundo atlántico del siglo XVI. Bogotá, D.C., Colombia: Universidad de los Andes, Facultad de Ciencias Sociales, Departamento de Historia, 1 – 18.

Copyright statement

Enhancing the cultural and global awareness of STEM undergraduates through research abroad experiences

Stacie LeSure, Ph.D.
Howard University, Washington, D.C., United States of America
stacie.lesure@howard.edu

Lorraine Fleming, Ph.D.
Howard University, Washington, DC, United States of America
lfleming@howard.edu

Inez Moore
Howard University, Washington, DC, United States of America
inezmoore@gmail.com

Abstract*: The Global Education, Awareness and Research Undergraduate Program (GEAR UP) is a five-year, mixed-method investigation aimed to better understand the experiences and perspectives of undergraduate STEM students who participate in international research experiences. An overarching goal of this study is to discover effective strategies to prepare engineering and science undergraduates for global engagement and leadership. This paper shares preliminary data analysis that reveals positive impacts for STEM undergraduate students who participate in research abroad experiences including: strengthen career identity/awareness, increase desire to pursue graduate school and careers in research and deeper commitment to service.*

Background

Globalization efforts in engineering education are among the dominant critical initiatives emerging in 21stcentury post-secondary education. Endeavors to increase the pool of globally engaged engineers address both documented industrial needs and the apparent lack of students who have international engineering experience (National Science Board, 2010; Smerdon, 2000). The industry's need for globally competent engineers stems from the multinational climate in which many engineering and science firms operate today. Engineering is consistently ranked among the top industries where globalization efforts have produced widespread gains in economic growth and employment (Ferraro, 2006; NSB, 2010). Engineering products are oftentimes manufactured for world-wide use. This intuitively suggests that engineering students may benefit from a more complete global perspective in their educational training. Other examples that further illustrate the demand for globally engaged engineers are: (i) increases in international collaborations in publications/patents; and (ii) more peer-reviewed literature co-authorship among science and engineering research facilities. Unlike those from prior eras, current and future engineering students must be equipped to collaborate internationally (NSB, 2010).

Unfortunately, too few engineering graduates are prepared to work in a global marketplace. To address this shortcoming, researchers at a midsized, private institution located in the United States, designed an investigation to better understand how to prepare globally engaged STEM undergraduates. This research initiative is known as *The Global Education, Awareness and Research Undergraduate Program (GEAR UP)*. The primary purposes of GEAR UP are two-fold: (i) to assess if (and how) research abroad experiences enhance the cultural and global awareness of STEM undergraduate students; and (ii) to evaluate if (and how) research abroad experiences help prepare STEM undergraduate students to develop and maintain international research collaborations.

To accomplish these goals, groups of four STEM undergraduate students travel to international locations to conduct research under the supervision of faculty researchers at various universities. For this mixed method study, researchers collect quantitative and qualitative data to identify the impact of conducting research abroad. Over the past 4 years, 120 STEM undergraduate students from the US travelled to a total of 13 different countries and spent four weeks conducting research with international STEM graduate students and professors. These students have written research papers and presented their findings at various research conferences. Some have maintained professional relationships with their international colleagues. In this paper, the authors share preliminary findings. The two **research questions** investigated are:

RQ1: What are the factors that attract STEM undergraduate students to apply for research abroad opportunities?

RQ2: What impact does a research abroad experience have on STEM undergraduate students' career interests and plans after college?

Theoretical framework

The theoretical framework used in this study is *Kolb's Model of Experiential Learning* (Kolb, 1984). This framework supports the study's underlying hypothesis that GEAR-UP prepares STEM undergraduate students to be globally engaged leaders upon graduation. The fundamental principle of Kolb's model is that learning is established and enhanced through experience. In other words, information transmission can be enriched when accompanied by an experiential learning component or strategy. Kolb's strategies for enhanced experiential learning include: concrete experiences (e.g., field trips); observation and reflection (e.g., discussion groups); and abstract conceptualization (e.g. sharing content, journaling).

Methodology

GEAR UP is a five-year, mixed-method investigation aimed to better understand the experiences and perspectives of undergraduate STEM students who participate in international research experiences. Selected students participate in a summer research experience in one of 13 developing countries in Southeast Asia, Africa, and North and South America. GEAR UP is open to undergraduate students majoring in STEM (engineering, computer science, biology, chemistry, math, and physics) disciplines. The student selection process is thorough. Interested students complete an online application and participate in an interview. They also submit academic transcripts and two letters of recommendations from STEM professors. After gaining provisional acceptance, a selection committee assigns students to teams of four. Each team travels to a different country. The committee considers several factors when assigning students to teams, including academic majors, students' preferences, and attempts to balance gender and maturity level. The selection committee nominates provisionally accepted students for admission into a host country's research laboratory and sends application packets (online application, transcripts, and letters of recommendations) for review. Ultimately, the host country decides which students are officially selected to participate in GEAR UP. To date, all participating host countries have accepted all students nominated by the selection committee.

Participants

In this paper, the authors focus on students who participated in GEAR UP during the summer of 2013. The eight host countries, the universities and the specific STEM majors the researchers were willing to accept at each international location are highlighted in **Table 1**. In 2013, four students travelled to each of the eight countries except Kenya where there were only three. The academic status of student participants ranged from those who have just completed their first year of undergraduate school to those who had finished their fourth year, but have not completed all requirements to graduate. An equal number of males and females travelled to Cameroon, Mexico, Senegal and Turkey. More females than males travelled to Kenya, Philippines, South Africa and Thailand. In Mexico, Thailand, and Turkey only engineering majors travelled to conduct research. Researchers in Cameroon and the

Table 1: 2013 Host Countries

COUNTRY	UNIVERSITY	MAJORS
Cameroon	*University of Buea*	Mathematics, Physics, Electrical Engineering, Mechanical Engineering
Kenya	*University of Nairobi*	Biology, Chemistry, Chemical Engineering
Mexico	*Universidad Popular Autonoma del Estado de Puebla*	Civil Engineering, Computer Science, Electrical Engineering, Biology, Chemistry, Chemical Engineering
Philippines	*Ateneo de Manila University*	Computer Science
Senegal	*Université Cheikh Anta Diop de Dakar*	Chemistry, Physics, Chemical Engineering, Biology
South Africa	*University of the Western Cape*	Computer Science, Mathematics, Physics, Chemistry
Thailand	*Chulalongkorn University*	Biology, Chemistry, Electrical Engineering
Turkey	*Istanbul Technical University*	Civil Engineering, Mechanical Engineering

Philippines hosted engineering and computer science majors. In addition to engineering, biology and chemistry majors travelled to Senegal and Kenya. Undergraduates studying math, computer engineering and electrical engineering travelled to South Africa.

Data Collection

For this study, researchers employed survey instruments coupled with multiple qualitative research methods (i.e., semi-structured interviews, focus groups, and journaling) to triangulate results (**Table 2**). This mixed-method approach allows for the generation of new, comprehensive knowledge that fills a void in the literature with findings that target a population that is rarely studied.

Participants were required to complete surveys both prior to travelling abroad and upon their return. They were also expected to participate in focus group interviews after returning home from their research abroad experience. During the month that they were in the host country, participants maintained a reflective journal to share their perspectives on how well they were navigating and adapting to new and unfamiliar cultural settings.

Table 2: Data Collection

Data Collected	Instrument	When
Demographic information and responses to questions about the students' research interests, academic and career goals, and expectations of the research abroad experience	Application Packet	Prior to acceptance
Reflective data regarding students' academic and career goals and research abroad expectations	Survey	Prior to research abroad experience
Reflective data regarding research abroad experience and the impact of the experience on students' personal development and intellectual growth	Survey	Post research abroad experience
Comparative data on the cultural similarities and differences and student experience at the various site	Focus group	Post research experience
Longitudinal, comprehensive data on students' navigation and adaption to the cultural setting from their perspectives	Reflective journal	During research abroad experience

Data Analysis

For this paper, the authors report findings from two data sources, pre- and post- surveys. One hundred percent of GEAR UP participants completed the pre-travel survey and 61% completed the post-travel survey. The process the researchers employed to analyse data is straightforward. For the questions with answer choices provided, the researchers simply counted how many participants selected a specific response. For example, participants were asked if they ever travelled internationally and if they participated in study or research abroad experiences in the past (**Chart 1**). Out of 31 participants, 22 answered *"yes"* they have travelled internationally and 24 responded *"no"* to prior study/research experiences. In the case of open-ended questions, two researchers employed content analysis in search of emerging themes. After individually reviewing the responses, they compared findings and mutually agreed upon key themes.

Chart 1: Prior International Travel and Research Abroad Experiences

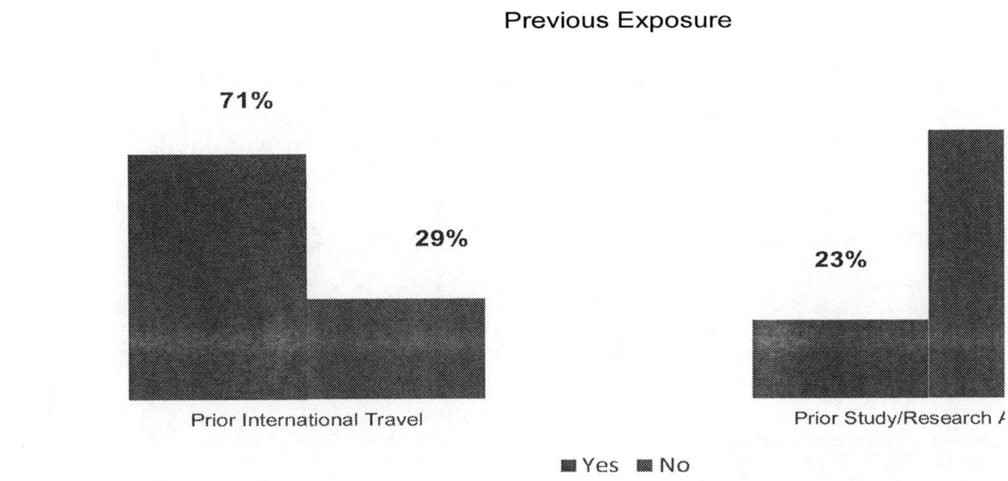

Findings

Typical international experiences for undergraduate students focus more on immersing students in the culture and improving their language skills. Study abroad experiences are more common for students majoring in the liberal arts than STEM disciplines. Little is known about the impact of placing STEM students in research laboratories in foreign locations.

RQ 1: What are the factors that attract STEM undergraduate students to apply for research abroad opportunities?

The researchers wanted to learn more about the typical profile of STEM students who applied to participate in a research abroad opportunity and specifically what attracted these students to apply to GEAR UP. Although 71% of the 2013 GEAR UP participants had traveled abroad, only 23% had engaged in study/research experiences (**Chart 1**). More than half of the participants reported an inability to speak the dominant language spoken in their host countries. Fourteen out of 31 confessed to not being able to speak the language at all. Six reported their ability to speak the native language as "*very little*" and five could speak it "adequately". Only 19% believed they could speak the dominant language of the host country "*exceptionally well*". These were the participants who travelled to countries in which the dominant language spoken was the same at the students' first language, English.

Regarding the students' plans immediately after college, the majority stated a desire to attend graduate school (**Table 3**).

Table 3: GEAR UP 2013: Participants plans after college

Plans immediately after college	*Number of respondents*
Industry/Private Sector	5
Service Program	4
Graduate School	12
Government/Public Sector	3
Post-Baccalaureate Program	2
Medical School	3
Military	1
Not Sure	1

On the pre-travel survey, students were asked *"What were the most important factors in accepting a GEAR UP assignment?"* (**Chart 2**). It is not surprising that 97% of STEM undergraduates selected *"research experience"* as an attraction. However, the fact that 90% also chose "exposure to a different culture" suggests that STEM students are as interested in conducting research as they are gaining exposure to different cultures. This finding is encouraging since one of the objectives of GEAR UP is to develop globally engaged STEM professionals.

Chart 2: Factors that attract students to GEAR UP

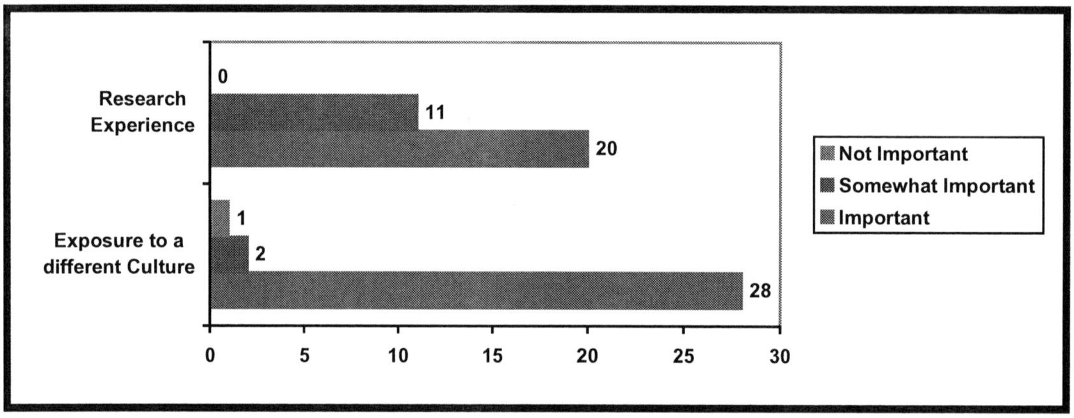

RQ2: What impact does a research abroad experience have on STEM undergraduate students' career interests and plans after college?

The researchers used content analysis to identify emerging themes revealed from responses to two open ended questions from the post-travel surveys:

1. Did GEAR UP impact your STEM-Career interests? If so, how? and
2. Did GEAR UP impact your post baccalaureate plans? If so, how?

During data analysis, emerging themes revealed that GEAR UP had a positive impact in four areas— career identity/awareness, graduate school and research interests, and service.

The following comments from students support these findings.

Increase in STEM career identity/awareness

Many participants expressed how GEAR UP increased their career identify and enhanced their awareness of the diverse career options for STEM professionals. A female engineering student who travelled to South Africa clearly outlined how her career path was changed. *She stated, "GEAR UP increased my interest in civil engineering because it exposed me to real world applications of the field that I find interesting. Also, the research prepared me for my internship with the Water Enterprise in San Francisco."*

Another female engineering student who travelled to Mexico remarked, *"Before my Gear Up experience I felt torn between environmental and water resource engineering. However, after working with a project directly related to environmental engineering and ecotechnology, I know this is what I want to explore further."*

In both cases, it is evident that the participants' career identity as STEM professionals and their awareness of STEM career options were influenced by GEAR UP.

Increase desire to attend graduate school and pursue a career in research

Majority of the participants emphasized an increased desire to attend graduate school. Some even realized how research is instrumental in innovation.

A female, civil engineering major who conducted research in Thailand declared that prior to GEAR UP, graduate school was not even a consideration. She shared, *"Although I do want to work soon after graduation, I never thought about attending graduate school until after my experience with GEAR UP. Observing the graduate students at the university inspired me to continue my learning, which will help me achieve more of my career goals."*

Two other female, civil engineering students who travelled to South Africa made similar comments regarding their newfound desires to attend graduate school after their GEAR UP experiences.

One remarked, *"Before this summer I wasn't proactive in considering graduate school. Now I have begun looking at graduate schools and their requirements for admissions."*

The other stated, *"Before GEAR UP, I had no interest in grad school, but I am now considering it."*

Beyond their desire to attend graduate school, participants described how GEAR UP made them more aware of the diverse activities that take place in a research laboratory.

A female, mechanical engineering students who worked in a laboratory in Cameroon discussed how GEAR UP made her realize that research is more than testing. She offered, *"It [GEAR UP] helped me see that research can be incorporated into inventing and not just boring lab testing."*

Deeper commitment to service

GEAR UP influenced participants to have a deeper commitment to service both abroad and domestically.

A female engineering student who travelled to Mexico revealed how participating in GEAR UP confirmed her interest in the Peace Corps. She stated, *"For about a year, I've known that I wanted to become a Peace Corps Volunteer but I wasn't sure if I should apply and strive for that opportunity. After my GEAR UP experience, I am now 100% sure that I want to become a Peace Corps volunteer."*

A male, mechanical engineering student described how the relationships he formed in Cameroon strengthened his ties and desire to help develop the country. He expressed, *"When I establish myself in my career, I will spend some of my free time or vacation days both visiting my friends in Cameroon and helping to develop the country that I feel has adopted me."*

Some students were even more motivated to the serve people in their own communities. A female, civil engineering student described the impact of her time in South Africa. She offered, *"My own community in the Mississippi Delta is lacking so much. It would be more rewarding to see change in a community that I grew up in as opposed to one that is thousands of miles away."*

Implications

Researchers have studied *"global learning and development, intercultural competence, intercultural maturity, and intercultural sensitivity"* of students studying abroad, (Braskamp, Braskamp, & Merrill, 2009, p. 101). However, the work presented in this paper is the first to focus specifically on STEM students as researchers.

This work lays the foundation for the critical need to build STEM professionals equipped with global awareness. Research abroad experiences have the potential to reward STEM undergraduates with a greater appreciation for different cultures, increase their desire to solve world challenges and remove stereotypical barriers that can inhibit productivity, economic prosperity and international collaborations. Conducting research in an international community exposes students to the perspectives and norms of others while simultaneously introducing people in other countries to the student researchers' diverse cultures. This mutual awareness allows for increase opportunities to earn shared respect of unfamiliar culture norms and an exchange of creative and innovative ideas to solve global challenges.

References

Braskamp, L. A., Braskamp, D. C., & Merrill, K. (2009). Assessing progress in global learning and development of students with education abroad experiences. *Frontiers: The Interdisciplinary Journal of Study Abroad, 18*, 101-118.

Ferraro, G. P. (2006). The cultural dimension of international business (5th ed.). New Jersey: Pearson Education.

Kolb, D. (1984) Experiential Learning: Experience as the Source of Learning and Development. Englewood Cliffs, NJ: Prentice-Hall.

National Science Board. 2010. Science and Engineering Indicators 2010. Arlington, VA: National Science Foundation (NSB 10-01).

Smerdon, E. T. (2000, June). Challenges in Engineering Education–The Impact of Internationalization of Engineering. In *6th International Conference on CAD/CAM, Robotics and Factories of the Future.*

Acknowledgements

The authors wish to thank the National Science Foundation for supporting this work: Grant Number 1238466. ACE Implementation Grant: STEM Global Undergraduate Research Initiative. Special thanks to each of the eight universities that hosted GEAR UP participants and the researchers that advised students during their four-week research abroad experiences.

Copyright statement

The following copyright statement should be included at the end of your paper. Substitute authors' names in final copy only.

Preparing Industrial Engineers through Project-based Learning using ICT: An Exploratory Analysis

Andres Esteban Acero

Universidad de los Andes, Bogotá, Colombia

ae.acero539@uniandes.edu.co

Luisa Fernanda Payan-Duran

Universidad de los Andes, Bogotá, Colombia

lf.payan890@uniandes.edu.co

Edier Ernesto Espinosa-Diaz

Universidad de los Andes, Bogotá, Colombia

ee.espinosa10@uniandes.edu.co

***Abstract**: Project-based learning (PBL) has proven to be a powerful tool for developing professional skills by creating a formative identity through active learning. In addition, in a changing and increasingly ICT-oriented world, educators should be able to include these tools in their practices. However, few studies had explored the relationship between PBL and the use of technologies within and outside the classroom. Therefore, this paper seeks to understand the integration of ICT tools in the classroom through PBL and how they can support the development of professional skills, in the case of Engineers without Borders Colombia and their courses offered to industrial engineering students at Universidad de los Andes. Results show that students prefer the courses with a higher technological usage, and that they are motivated to change their learning processes if ICT tools are involved. In terms of PBL, results show that technology promotes flexibility, autonomy, initiative and active participation.*

Introduction

Information and Communication Technologies (ICT's) importance in education has been widely researched finding for example, that it is normally related with a higher efficiency and better educational outcomes (Figuères & Eugelink, 2014) and it can create a powerful learning environment that enhances teaching methods and improves students learning experience (Malinovski, Vasileva, & Trajkovik, 2014). Therefore, many governments have sought to embrace ICT on their education policies. Colombian government did, arguing that this tools will enhance the archiving of the country's educational goals and will improve quality, pertinence, coverage and development of the educative contents (DNP Colombia, 2011). For Kolo and Breiter (2010), ICT-based innovations in education, as the one under study, are important mainly because "they prepare for a future workplace in a knowledge society increasingly dependent on ICT and furthermore, they support student-centered learning processes" (p. 89).

When a professional in engineering is performing his work, tackling a problem to solve it, the technical knowledge he has obtained through his training in the academy is not enough to be competent in the work teams he takes part of, he also needs to develop certain professional skills. The engineering programs of high-quality educational institutions are committed to supporting the development of such professional skills, which according to Litchfield, Amy, & Andrew (2016) are: (1) an ability to function on multidisciplinary teams, (2) an understanding of professional and ethical responsibility, (3) an ability to communicate effectively, (4) the broad education necessary to understand the impact of engineering solutions in a global and societal context, (5) a recognition of the need for, and ability to engage in, lifelong learning, (6) a knowledge of contemporary issues, (7) an ability to manage a project, including a familiarity with business, market-related, and financial matters, (8) a multidisciplinary systems perspective, (9) an understanding of and appreciation for the diversity of students, faculty, staff, colleagues, and customers (10) a strong work ethic. The above-mentioned professional skills match the criteria required by accreditation bodies for engineering programs such as ABET - Student Outcomes (ABET, 2016) and, consequently, meet the requirements of employers (Daud, 2015).

Although there are studies of the professional and technical learning outcomes of engineers involved and not involved in engineering services, these are anecdotal or low-scaled (Litchfield et al., 2016). The large-scale study proposed by those authors allowed us to identify that engineers involved in engineering services, such as professional practices or a course where they had to propose a solution to a not necessarily academic problem, perceived higher professional skills than those that were not (Mohan, Merle, Jackson, Lannin, & Nair, 2010; Reich et al., 2015). Studies like these inspired the realization of this paper that seeks to analyze the implications and relationship of professional skills (Markes, 2006) inside the courses of Ingenieros Sin Fronteras Colombia (Engineers without Borders Colombia – EWB Colombia).

Research questions

Given the background presented above, this study want to explore two research questions:

- Do PBL courses enhance the development of professional skills?
- Does ICT support the development of professional skills through PBL?

The hypothesis is that one or more of the professional skills are positively affected for those students that participated in the PBL courses. Additionally, the effects of these skills on professionals and university students is proposed as a measure of short-term and long-term effects. Finally, a hypothesis is raised about the effect of the use of ICT on the relationship between professional skills. The goal of the latter question, in addition, is to determine which technologies are useful to promote certain professional skills in students.

Theoretical framework

Professional skills

Professionals with a degree in engineering are increasingly compelled to demonstrate higher competences on professional skills. Different accreditation bodies for engineering programs, such as ABET (Accreditation Board for Engineering and Technology), include the development of some professional skills: lifelong learning, understanding then need for high ethical standards in engineering practice, the ability to work effectively as an individual and in multidisciplinary teams and the ability to communicate effectively with the engineering community and with the society (Duffy, 2011). Professional skills development is strongly linked to the intellectual development growth (Lewis & Bonollo, 2002), learning autonomy (Tlhoaele, Suhre, & Hofman, 2015), ethics commitment (Lynch & Kline, 2000) and the ability to present and lead an initiative (Lam, Cheng, & Ma, 2009). Engineers who do not develop these skills during their training and tend to wait for the lead of an authority figure to make decisions when addressing and solving a problem, are less wanted by employers (Markes, 2006). Therefore, the engineering curriculum should promote the development of professional skills from a student-centered learning process (Duffy, 2011).

At the international level, professional skills in higher education institutions are a requirement for employers shaping their work teams. According to Daud (2015), professional skills are aligned to a conceptual model that seeks to evaluate them in engineering education with the achievement of active teaching and learning, reflected in the expected learning outcomes:

verbal communication, problem solving and teamwork. This model seeks to ensure that through the interaction between higher education, labor market and employer feedback, the teacher can implement the student-centered education system (Koh, Herring, & Hew, 2010). However, engineering faculty face the challenge of assessing professional skills within active learning environments with the need to make them more rigorous and quantifiable, to respond to market needs and accreditation quality standards (Ramírez Cajiao, Carvajal Diaz, & Peñaloza, 2010).

Project-based learning

Deep changes within the classical teaching of engineering are necessary to face the problems of the 21st century. Engineering programs, since their consolidation after the Second World War, have been taught in a deductive way (Goldberg, 2012). This has privileged the sciences within engineering (Petroski, 2011), where teachers approach first to the required theory, followed by typical problems of textbooks and finally, sometimes, real-world applications.

As King (2012) points out, this structure needs profound changes that allow an engineering education focused on professional practice, autonomy and a deep and experiential learning. As a response to these dynamics, various methodologies from social constructivism have been applied in the engineering curriculum, including active learning (Oliveira & Oliveira, 2014) and project-based Learning (PBL) (Alves et al., 2016). While with active learning teaching focuses on the development of technical skills and attracting student attention in the learning process, PBL seeks to connect theory with practice by incorporating knowledge into real-life situations.

One of the central aspects on PBL research has been to define the characteristics that delimit its application. There is a consensus in the literature that PBL incorporates the following elements:

1. Problem solution or task completion that requires students to complete a series of educational activities that drive learning (Frank, Lavy, & Elata, 2003).
2. The project is not trivial and it is often multidisciplinary, requiring work over an extended period of time (Palmer & Hall, 2011).
3. The project normally involves the development of a concrete artefact - a design, a model, a thesis, a computer simulation, etc. (Puente, Eijck, & Jochems, 2014; Simões, Relvas, & Moreira, 2004).
4. Teachers assume a more advisory than authoritarian role (Alves et al., 2016; Lam et al., 2009).

Other aspect that has been included in the PBL research processes is the impact on the personal and social development of the students involved (Lou & Macgregor, 2004; Seet & Quek, 2010). However, research on the effects that the application of PBL on technical and professional skills are still inconclusive. In this article, we will investigate the effect that PBL implementation has on the professional and employability skills of students.

Study context

Course overview

Engineers without Borders (EWB) Colombia designed a learning space where engineering students' work becomes relevant by interacting directly with vulnerable communities. The projects are based on guidelines that students, professors, practitioners and volunteers on EWB Colombia must understand, develop and share. These guidelines or objectives point important characteristics of the social-responsible engineering, solution-based thinking and technological positivism for development. These objectives are:

- To recognize the contribution of engineering in improving life quality of communities.
- To identify the specific problems of vulnerable communities and the opportunities of intervention from engineering.

- To apply science and technology knowledge in projects that address issues in vulnerable communities.
- To work in multidisciplinary teams for the conception, design and implementation of innovative solutions to social problems.

This section provides a brief description of the most relevant aspects concerning the implementation of PBL and ICT at EWB Colombia courses.

Project structure

During 2012 to 2016 several kinds of projects were developed using the objectives previously mentioned. Each one of these projects followed the same methodological structure based on the PBL structure, namely:

- Recognition phase: Students were assigned to groups from 2 to 5 students. This division was based on the preferences and interest of each student. After that, each group meet and introduced themselves. In addition, the groups were assigned to a partner (a school, business, social organization or community) with whom they should work to co-create a solution to the problematic situation. Finally, the professor, the teaching assistants and the students visited their partner to get an overview of the situation of the region and the problem they will be working on.
- Problem identification: Each group should meet several times with their partner. Before these meetings, each group proposed a series of activities that allowed them to collect information about possible problematic situations and the perspectives of the actors involved, while creating bonds of trust between students and community. The proposals received feedback from other groups and the teaching staff, and each group finally designed a guide for every meeting.
- Problem quantification: The students and the communities, working together, imagined possible impacts and consequences derived from the problematic situation. These scenarios were tools, too, to create idealized design, where community played the most important role in the transformation process. Using this information, a complete picture of the situation and the expectation was accomplished, to be used lately in the design process.
- Co-design process: Using the same structure from the problem identification phase, activities were proposed by students for the co-design of possible solutions for the problematic situations. This design process was also iterative because students and partners made changes to fulfill the EWB Colombia solutions criteria. These criteria seek for culturally and technically viable, economically possible, and socially and environmentally sustainable solutions, and it is used as a tool to assure real social impact.
- Prototyping and solution implementation: In this phase, groups and community created prototypes of the possible solutions that were later presented to an expert panel. This panel evaluated each solution and gave advices and constructive critiques to students seeking for a successfully implementation of the solution. Finally, a written document was prepared with the solution and some guidance in the implementation process.

Using this methodology, based on the PBL approach, students developed a project with vulnerable communities and provided engineering solution to specific problems that could be implemented by the people.

Technologies used in the course

During the period that was considered for the current study, five technologies were considered. These technologies were selected for their relevance in the development of the course activities or their functionality in the process with the community.

First, the students and the teaching staff of the EWB course maintained in-person communication, in addition to exchanging e-mails during field trips. Second, given that summer courses were taught to people who might not be a student of the University of the Andes, the page of EWB Colombia was used. On this page were uploaded the contents, the qualifications and the guides on the activities of the course. During the regular semester courses, the learning management system Blackboard Learn was used. Both platforms used forums and videos to support the course content. Fourth, in the face-to-face classes, videoconferences with international experts were made through skype, in addition to having meetings with the community using this technology. Finally, students and their partners in the community communicated through phone calls or instant messaging.

Methods

Subjects

The participants of the study were 56 students who enrolled in the engineering courses "Proyecto Intermedio Ingenieros sin Fronteras" (Engineers Without Borders mid-career project) or "Ingeniería e Innovación Social" (Engineering and Social Innovation) at Universidad de los Andes, a large private university in Colombia. In this course, students were involved in service learning, active learning activities and project-based learning to co-create with a community a solution to a problematic situation or social innovation challenge. Accordingly, this course was selected for this study considering that it uses ICT and project-based learning aiming to foster a social impact on vulnerable communities. From 2012 to 2016, students from Universidad de los Andes, Pontificia Universidad Javeriana, Universidad Nacional de Colombia, Corporación Universitaria Minuto de Dios and Universidad Sergio Arboleda participated in these courses. Each course under study is considered as an illuminative case that would provide information about the impact of technology usage in (i) professional skills and (ii) involvement in PBL.

Procedure

This article uses an sequential mixed method approach, suitable for engineering education research, as suggested by Borrego, Douglas and Amelink (2009). The first phase of the study, a quantitative study of the experience of the students involved in the EWB courses were carried out by Acero, Payán and Ramírez (2017). According to the findings of this previous study, an exploratory analysis of the experience with the technologies and the impact on their professional skills is proposed in this article. The first step of this study requires understanding how being a student of the courses of EWB (Engineers without borders) affects self-perception of technology usage and professional development. To accomplish this purpose, the researchers design a survey where they can determine:

- Student's perceptions of the course methodology and how this course was useful for their academic or professional lives.
- Student's perceptions about the contribution of EWB Colombia courses to their professional skills. The chosen professional skills are based on Markes' (2006) research on the skills that employers value in engineering students, the assessments of Mohan, Merle, Jackson, Lannin, and Nair (Mohan et al., 2010) for professional skills within engineering curricula and the professional skills derived from ABET's evaluation of engineering programs (Reich et al., 2015). Therefore, fourteen professional skills were selected and evaluated using a Likert-scale perception questionnaire[1]. The selected professional skills were:

    1. Verbal communication
    2. Writing communication
    3. Teamwork
    4. Working on interdisciplinary teams
    5. Solving engineering problems
    6. Creative problem-solving

[1] A translated version of the questionnaire is available in the following link: https://goo.gl/Uh2tcy.

7. Negotiation
8. Adaptation
9. Time management
10. Autonomy
11. Planning
12. Initiative
13. Networking
14. Ethic commitment

- The diverse ICT that were used by the students during the course, both inside and outside of the classroom.
- Student's perceptions about the contribution to their professional skills of using ICT in the EWB Colombia courses. The same fourteen professional skills were evaluated using a Likert-scale perception questionnaire.
- Different socio-demographic characteristics of students participating in the study (gender, age and occupation).

In addition, academic level of the students when they enrolled in the course (undergraduate or graduate) were considered. Furthermore, an informed consent was provided at the beginning of the survey, where participants know about the objective of the survey and the possible risks of answering it. The survey had 46 questions, with three open questions[2]. Finally, survey was upload to Google Forms© to be available on an online format. The survey was send by email to 360 students who took the courses between 2012 and 2016. The response rate was 15.5%, with a total amount of 56 responses[3].

Data Analysis

The responses were collected and organized in a database using SPSS and a progressive focusing approach on data organization and collection. A contingency table analyzing associations between professional skills, ICT used and project-based learning characteristics was built. In addition, Pearson's correlation coefficient between variables were calculated to determine possible association between effects of technologies and spurious correlations. Finally, an analysis of variance (ANOVA) between technologies and professional skills were carried out.

Results

From the 56 engineers and engineering students that answer the survey, 38.2% were female and 61.8% male. 81.8% of the respondents are between 21 and 26 years, while 5.4% are between 30 and 35 years and 1.8% are outliers with 19 and 49 years. 76.7% of the respondents stated that they are currently studying and / or working; from them, 23.3% are only studying, while 31.7% only work and the remaining 21.7% perform both activities.

Regarding the perceptions of the different activities developed in the courses, it was found that 92.7% of the respondents perceived the EWB courses as important or very important for engineering students training. In addition, more than the half of the participants agreed that the theoretical concepts on which the courses are based are useful for their professional practice. According to the courses curriculum, students should develop a project along the course, working in teams. 61.9% of them say that was relevant for their professional development. Overall, 92% of respondents said the EWB Colombia course in which they participated was interesting, 72.7% consider that the course is useful for the professional life and 83.6% added that courses like this should be included in the curriculum of the engineering programs.

During the development of the courses some ICT were used. 90.9% of the respondents stated that they used the EWB Colombia website (isfcolombia.uniandes.edu.co), since most of the courses' content is hosted there. 96.4% stated that they used the e-mail as a means of

[2] The original version of the survey in Spanish is available in the following link: https://goo.gl/tgD3jd.
[3] The data should be available by request.

communication between classmates, teachers and course assistants, while 72.7% used the Learning Management System (LMS) (Blackboard Learn©) to upload some of the tasks. However, 67.3% of the respondents stated that they had not used the virtual portfolio where the project deliverables should be uploaded, 70.9% said they had not seen the videoconferences in the classes and 74.5% stated that they had not participated in Forums during the courses.

A very relevant component of EWB Colombia courses is the work with the communities, where the engineering students must face the challenge of using the acquired knowledge to propose engineering solutions to problems that these communities face. Therefore, students use some ICTs to maintain communication with the partner community. The result of the survey on this topic, shows that 47.3% of respondents used telephone calls, 60% used WhatsApp and 67.3% used email as a means of communication. On the other hand, 65.5% of respondents did not use social networks, 89.1% did not use videoconferences and 74.5% did not use forums to maintain communication with the community.

Below, some results presented and analyzed regarding the relation between the contribution of the courses to the development of professional skills and the use of ICT within the courses for communication with the communities, as well as the current dedication of the respondents.

Table 1: The differences between graduated engineers and students according to the professional skills

Variable	Mean (SD)	F	Significance
Oral communication * working	1.58 (0.692)	7.89	0.00
Writing communication * working	1.42 (0.60)	7.69	0.00
Engineering problems * working	1.53 (0.51)	7.58	0.00
Adaptation * working	1.37 (0.68)	3.92	0.05
Time management * working	1.21 (0.78)	4.49	0.03
Interdisciplinary * studying	0.36 (1.21)	6.56	0.01

When performing a variance analysis (ANOVA), among the graduated respondents and the engineering students, can be deduced that graduated engineers privilege several professional skills within their professional development. These skills were the communication skills (both written and oral expression), the ability to solve engineering problems, the ability to adapt to changes in their work environments and the time management. On the other hand, engineering students privilege the ability to work in interdisciplinary groups over the graduated group.

Table 2: Learning Management System (LMS) and professional skills

Variable	Mean (SD)	F	Sig
Ethics * Course Management	0.83 (1.25)	4.95	0.03
Autonomy * Course management	1.05 (1.06)	6.25	0.01
Negotiation * Course management	0.75 (1.17)	3.99	0.05

Creativity in problems * Course management	1.27 (0.87)	4.52	0.03
Engineering problems * Course management	1.10 (0.90)	5.13	0.02

On Table 2, it can be seen the differences between groups that used SicuaPlus© (The Learning Management System of Uniandes) and groups how doesn´t. This table suggest that this ICT had a mediation effect on the abilities of understanding the ethical considerations of the profession, working autonomously, negotiation, solving problems creatively and solving engineering problems. Only 72.7% of the respondents used the LMS. Some important remarks about these results are that 83% of the users of this technology stated that allowed them to understand the ethical considerations of the profession due to the higher levels of control this system applies. Otherwise, 67% of the participants stated that the ability to solve engineering problems is not due to the use of the Learning Management System.

Table 3: Phone calls with the community and professional skills

Variable	Mean (SD)	F	Sig
Ethics * Phone calls	1.00 (1.47)	6.27	0.01
Initiative *Phone calls	1.14 (1.02)	7.77	0.00
Planning * Phone calls	1.38 (0.82)	4.03	0.05
Autonomy * Phone calls	1.24 (1.15)	8.46	0.00
Adaptation * Phone calls	1.21 (1.01)	8.07	0.00
Creativity in problems * Phone calls	1.38 (0.72)	5.06	0.02
Engineering problems *Phone calls	1.17 (0.92)	4.00	0.05

A valuable tool for the development of the activities with interest groups (partner communities), is the constant communication through telephone calls that allow contact with the group and avoid displacement. 52.7% of the respondents said they used a phone call to communicate with the community. According to the analysis of variances of Table 3, it can be evidenced that there is an effect between making or not a call to the community with respect to the development of the ability to plan and coordinate and the ability to solve engineering problems. In addition, differential effect between making a call to the community and developing the ability to understand the ethical considerations of the profession, the ability to take the initiative, the ability to work autonomously, the ability to adapt to complex situations and ability to solve problems creatively are presented. This results suggest that talking actively with the community (even through a phone call) allow students to share more than information and help them to manage their time throughout the project.

Conclusions

The use of a social perspective in engineering education is not new (Abaté, 2011; Al Lily, 2013). This perspective of engineering teaching has been focused on the ethics implications of engineering practice and the inclusion of the social justice (Baillie, Ko, Newstetter, &

Radcliffe, 2011; Kabo & Baillie, 2009; Leydens & Lucena, 2014). However, this approaches and the use of project-based learning inside courses and its relationship with service learning has a big opportunities to learn about it. Most of the engineers and engineering students that participated on the survey stated that the courses where interesting and useful for their professional development. Furthermore, a high proportion indicated that social-oriented PBL courses as the ones offered by EWB Colombia should be part of engineering curriculum. It was probed that sharing experiences with the communities affect the way respondents evaluate the courses' contributions towards the development of professional skills. Further evaluation in this idea show us that undergraduate students tend to have a short-termed perception of the utility of this social approach, so they didn't find the courses had such a significant contribution otherwise than working in interdisciplinary groups. Engineers that are working or doing post-graduate student, on the other hand, valued more positively the contribution of the courses to their professional skills, especially those related with work management, working on groups and creative problem-solving. Additionally, regardless of the context, most of the respondents pointed out the EWB Colombia courses foster their ability to solve engineering problems and the participants pointed out this professional skill development is a response to the PBL approach of the courses. Compared with our first hypothesis, this results can confirm that PBL approaches have some positive impact on the development of professional skills in the students. Furthermore, the differences between students and professional are notorious according with our results. Further studies must explore deeply how this abilities are valuable at a professional level.

Other important characteristic of this study was to understand if the inclusion of technologies can mediate positively in the development of this abilities. In consequence, this study explores the same professional skills with interesting results per the use and perception of the mediation of two technologies on the learning process within PBL courses. First, results suggest that phone calls and the Learning Management Systems are two technologies with the potential to foster the development of professional skills inside PBL courses. However, this results suggest that these technologies are useful not only as a tool to organize and manage the information of the project, but also to give them opportunities to understand the implications of an ethical behavior.

This final results match with several result in the literature. First, the use of phone calls with the partners of the community are useful to develop highly valuable skills (Markes, 2006) for employability and communication skills (Leavitt, 1951; Mohamed, Murray, & Mohamed, 2010). In addition, the use of course management tools will be recommended due to their positives implication on problem solving and structuring (El-Zein & Hedemann, 2016). Employers and academia must recognize the importance of this results to prepare engineers with the needed abilities to face the task that 21st century proposes. Finally, this study provides results that can be valuable in the design of the curricula in engineering programs, where major changes reside.

Limitations and Future Work

The results and findings presented in this research paper could be generalizable if the study's limitations are considerate. The limited possibilities to contact the students and its low response rate is a major opportunity. Because this study includes both practitioners and students, authors encourage a further additional research with bigger samples of PBL courses in Uniandes in the last 5 years. The inclusion of this information will be useful to understand if the use of a social perspective has different implications on the learning of professional skills. In addition, the use of self-report scales for the perception of support on professional skills needs further validation. The exploration of this assessment and more and complete studies about the scale could lead to confirm the results presented in this paper and to create a standard method to measure the relationship between technologies usage and professional skills. Furthermore, further research in three results is required: the relationship between gender and technology usage, the ethical dimension of being part of a social-oriented PBL course and the implications of technology usage on practitioners and professionals.

Finally, this study focused in the role of the students on the PBL course. Further research must include the teaching perspective and the community perspective on the learning of technical and professional skills (Litchfield et al., 2016). These future studies will acknowledge the relationships created with the community, and the activities and facilitation process that teaching staff should accomplish to have high levels of the participation and engagement.

Acknowledgements

The authors wish to acknowledge the financial assistance of Conecta-TE Uniandes through the call for publications on ICT innovations in higher education. The authors also acknowledge the support of Nicolás Pacheco during the research and writing process.

References

Abaté, C. J. (2011). Should Engineering Ethics be Taught? *Science and Engineering Ethics*, *17*(3), 583–596. http://doi.org/10.1007/s11948-010-9211-9

ABET. (2016). *Criteria for Accrediting Engineering Programs*. Baltimore, MD: ABET.

Acero López, A. E., Payán Durán, L. F., & Ramirez, C. (2017). *Active Learning and Action Research Outside Classroom: Engineering with Social Impact* (No. 2965030). Retrieved from https://papers.ssrn.com/sol3/papers.cfm?abstract_id=2965030

Al Lily, A. (2013). Social Change and Educational Technologies: By Invitation or Invasion. *Journal of Organisational Transformation & Social Change*, *10*(1), 42–63. http://doi.org/10.1179/1477963312Z.0000000004

Alves, A. C., Sousa, R. M., Fernandes, S., Cardoso, E., Carvalho, M. A., Figueiredo, J., & Pereira, R. M. S. (2016). Teacher's experiences in PBL: implications for practice. *European Journal of Engineering Education*, *41*(2), 123–141. http://doi.org/10.1080/03043797.2015.1023782

Baillie, C., Ko, E., Newstetter, W., & Radcliffe, D. F. (2011). Advancing Diverse and Inclusive Engineering Education Practices through Interdisciplinary Research and Scholarship. *Journal of Engineering Education*, *100*(1), 6–13.

Borrego, M., Douglas, E. P., & Amelink, C. T. (2009). Quantitative, Qualitative, and Mixed Research Methods in Engineering Education. *Journal of Engineering Education*, (January).

Daud, M. F. (2015). Lecturers' Experiences of Assessing Generic Skills in an Active Learning Environment for Engineering Education in Malaysia – Emergent Findings. In *6th Research in Engineering Education Symposium: Translating Research into Practice, REES 2015*.

DNP Colombia. (2011). Plan Nacional de Desarrollo 2010-2014 Colombia. *Departamento Nacional de Planeación. Imprenta Nacional de Colombia.*, *Tomo 1*, 541.

Duffy, G. (2011). Facilitating Intellectual and Personal Skills Development in Engineering Programmes. In *Proceedings of the Research in Engineering Education Symposium*. Madrid. Retrieved from http://arrow.dit.ie/engscheleart/187/

El-Zein, A. H., & Hedemann, C. (2016). Beyond problem solving: Engineering and the public good in the 21st century. *Journal of Cleaner Production*, *137*, 692–700. http://doi.org/10.1016/j.jclepro.2016.07.129

Figuères, C. M., & Eugelink, H. (2014). The Role of ICTs in Poverty Eradication: More Than 15 Years' Experience from the Field. In H. Kaur & X. Tao (Eds.), *ICTs and the Millennium Development Goals: A United Nations Perspective* (pp. 199–222). Boston,

MA: Springer US. http://doi.org/10.1007/978-1-4899-7439-6_12

Frank, M., Lavy, I., & Elata, D. (2003). Implementing the project-based learning approach in an academic engineering course. *International Journal of Technology and Design Education*, *13*(3), 273–288. http://doi.org/10.1023/A:1026192113732

Goldberg, D. E. (2012). A Moribund Profession? *Communities Education*, 16.

Kabo, J., & Baillie, C. (2009). Seeing through the lens of social justice: a threshold for engineering. *European Journal of Engineering Education*, *34*(922781082), 317–325. http://doi.org/10.1080/03043790902987410

King, C. J. (2012). Restructuring Engineering Education: Why, How And When? *Journal of Engineering Education*, *101*(1), 1–5. http://doi.org/10.1002/j.2168-9830.2012.tb00038.x

Koh, J. H. L., Herring, S. C., & Hew, K. F. (2010). Project-based learning and student knowledge construction during asynchronous online discussion. *Internet and Higher Education*, *13*(4), 284–291. http://doi.org/10.1016/j.iheduc.2010.09.003

Kolo, C., & Breiter, A. (2010). An Integrative Model for the Dynamics of ICT-based Innovations in Education. *Digital Culture & Education (DCE)*, *1*(2), 89–103.

Lam, S., Cheng, R. W., & Ma, W. Y. K. (2009). Teacher and student intrinsic motivation in project-based learning. *Instructional Science*, *37*, 565–578. http://doi.org/10.1007/s11251-008-9070-9

Leavitt, H. J. (1951). Some effects of certain communication patterns on group performance. *Journal of Abnormal and Social Psychology*. http://doi.org/10.1037/h0057189

Lewis, W. ., & Bonollo, E. (2002). An analysis of professional skills in design: implications for education and research. *Design Studies*, *23*(4), 385–406. http://doi.org/10.1016/S0142-694X(02)00003-0

Leydens, J. A., & Lucena, J. C. (2014). Social Justice: A Missing, Unelaborated Dimension in Humanitarian Engineering and Learning Through Service. *International Journal for Service Learning in Engineering, Humanitarian Engineering and Social Entrepreneurship*, *9*(2), 1–28. Retrieved from http://library.queensu.ca/ojs/index.php/ijsle/article/view/5447%5Cnhttp://queens.scholarsportal.info/ojs/index.php/ijsle/article/view/5447

Litchfield, K., Javernick-Will, A., & Maul, A. (2016). Technical and Professional Skills of Engineers Involved and Not Involved in Engineering Service. *Journal of Engineering Education*, *105*(1), 70–92. http://doi.org/10.1002/jee.20109

Lou, Y., & Macgregor, S. K. (2004). Enhancing Project-Based Learning Through Online Between-Group Collaboration. *Educational Research and Evaluation*, *10*(4–6), 419–440. http://doi.org/10.1080/13803610512331383509

Lynch, W. T., & Kline, R. (2000). Engineering practice and engineering ethics. *Science, Technology, & Human Values*, *25*(2), 195–225. http://doi.org/10.1177/016224390002500203

Malinovski, T., Vasileva, M., & Trajkovik, V. (2014). Integrating computer games in primary education for increased students' QoE. In *Advances in Intelligent Systems and Computing* (Vol. 231, pp. 35–44). http://doi.org/10.1007/978-3-319-01466-1_3

Markes, I. (2006). A review of literature on employability skill needs in engineering. *European Journal of Engineering Education*, *31*(6), 637–650. http://doi.org/10.1080/03043790600911704

Mohamed, M., Murray, A., & Mohamed, M. (2010). The role of information and communication technology (ICT) in mobilization of sustainable development knowledge: a quantitative evaluation. *Journal of Knowledge Management*, *14*, 744–758.

http://doi.org/10.1108/13673271011074872

Mohan, a., Merle, D., Jackson, C., Lannin, J., & Nair, S. S. (2010). Professional Skills in the Engineering Curriculum. *IEEE Transactions on Education*, *53*(4), 562–571. http://doi.org/10.1109/TE.2009.2033041

Oliveira, P. C., & Oliveira, C. G. (2014). Integrator element as a promoter of active learning in engineering teaching. *European Journal of Engineering Education*, *39*(2), 201–211. http://doi.org/10.1080/03043797.2013.854318

Palmer, S., & Hall, W. (2011). An evaluation of a project-based learning initiative in engineering education. *European Journal of Engineering Education*, *36*(4), 357–365. http://doi.org/10.1080/03043797.2011.593095

Petroski, H. (2011). Softening the Curriculum. *ASEE Prism*, *21*(3), 25.

Puente, S. M. G., Eijck, M. Van, & Jochems, W. (2014). Professional development for design-based learning in engineering education: a case study. *European Journal of Engineering Education*, *40*(1), 14–31. http://doi.org/10.1080/03043797.2014.903228

Ramírez Cajiao, M. C., Carvajal Diaz, J. a, & Peñaloza, J. T. H. (2010). Innovation and teamwork training in undergraduate engineering education: A case of a computing engineering course. *International Journal of Engineering Education*, *26*(19), 1536–1549.

Reich, A., Rooney, D., Gardner, A., Willey, K., Boud, D., Fitzgerald, T., ... Fitzgerald, T. (2015). Engineers' professional learning: a practice-theory perspective. *European Journal of Engineering Education*, *40*(4), 366–379. http://doi.org/10.1080/03043797.2014.967181

Seet, L. Y. B., & Quek, C. L. (2010). Evaluating students' perceptions and attitudes toward computer-mediated project-based learning environment: A case study. *Learning Environments Research*, *13*(2), 173–185. http://doi.org/10.1007/s10984-010-9073-8

Simões, J. A., Relvas, C., & Moreira, R. (2004). Project-based teaching-learning computer-aided engineering tools. *European Journal of Engineering Education*, *29*(1), 147–161. http://doi.org/10.1080/0304379032000129223

Tlhoaele, M., Suhre, C., & Hofman, A. (2015). Using technology-enhanced, cooperative, group-project learning for student comprehension and academic performance. *European Journal of Engineering Education*, *3797*(September), 1–16. http://doi.org/10.1080/03043797.2015.1056102

Copyright statement

Understanding Student Use of Worked Examples to Support Problem Solving via Real-Time Measurements of Problem-Solving Events

Michael Crawford and Edward Berger

Purdue University, West Lafayette IN USA (crawfo20@purdue.edu, bergere@purdue.edu)

***Abstract**: Worked-example videos are ubiquitous and freely-available to students, and they are often used to support problem solving. The research suggests that learning by worked examples can be both effective and efficient for students. However, very few studies have examined the real-time problem-solving processes used by students. We conducted laboratory experiments in which participants solved a problem on paper, with access to a worked example video described as "potentially helpful". We collected real-time records of student problem solving events via a Livescribe system, video usage via an eye gaze capture system, and student thought processes via an audio recorded think-aloud protocol. Although few of their strategies for worked example use were systematic, many students nonetheless completed the problem with reasonable success. Worked-example video usage is a complex function of student sophistication and confidence as a learner, comfort with the subject matter, and individual goals about task performance and efficiency.*

Introduction

Multimedia Learning

The past two decades have seen enormous strides in all manner of digital technologies, as well as the personalization of learning based upon many of those technologies. The ubiquity of multimedia content through YouTube and other channels provides students with countless free resources for learning. The scholarship around multimedia learning has also developed, led by Mayer (Mayer, 2009; Mayer & Moreno, 2003) and others (Berger & Krousgrill, 2012; Brunken, Plass, Leutner, & Brünken, 2003) who developed important frameworks to help us understand the appropriate role of multimedia resources for learning, as well as how to construct multimedia resources effectively. The challenge today is one of scale and strategy: given a specific learning task, how do students (i) select an appropriate, useful video from the massive library of available videos to help them with that task, (ii) how exactly do they use that video while executing the task, and (iii) in what ways can their professors coach them in optimizing their effort and maximizing their performance on that task? These are critical questions, and this paper focuses squarely on the second.

Research Question

Our focus on the ways in which students use worked-example videos to support their problem solving generates our **research question**: to what extent are there detectable patterns to student usage of a worked-example video during problem solving? The working hypothesis is that there exist a small number of usage patterns for worked-example videos, and that our approach to the experiment will allow us to observe those patterns.

Theoretical Frameworks

Worked-Examples and Learning Trajectory

The worked-example effect has been widely studied for teaching and training in many disciplines (Paas & van Gog, 2006; Rourke & Sweller, 2009; Smith, Mestre, & Ross, 2010;

Sweller & Cooper, 1985; Ward & Sweller, 1990), as well as a particular instantiation of multimedia learning (Moreno, 2006; Moreno, Reisslein, & Delgoda, 2006; Moreno, Reisslein, & Ozogul, 2009). The underlying construct of worked-example research is cognitive load theory (CLT), which posits that humans have finite cognitive capacity that is the summation of three dimensions: germane (related to schema automation and commitment to long-term memory), intrinsic (related to the inherent difficulty of the material to be learned), and extraneous (related to the way the material is presented). The idea behind worked examples is that if they are well designed, they can help learners optimize cognitive load by minimizing the extraneous component and diverting available cognitive capacity to the germane load of schema formation/automation. Video-based worked examples provide the added benefit of being replayable, sharable, portable, and device-independent (Berger & Krousgrill, 2012).

However, worked-examples as learning tools have diminishing returns as the learner becomes more sophisticated, and this is called the expertise-reversal effect (Kalyuga, Ayres, Chandler, & Sweller, 2003; Salden, Aleven, Schwonke, & Renkl, 2009). Broadly speaking, this phenomena suggests that learners who are new to a subject can benefit substantially from watching experts solve problems related to that subject (i.e., watching well-constructed video worked examples). However, as students become more competent problem solvers, the relative utility of worked examples decreases, and their learning improves more readily when they simply practice solving problems themselves. This notion of a learning trajectory, and the placement of worked-examples along it, helps us understand the research as well as it implications for engineering education practice.

Illusion of Explanatory Depth

To understand the implications of this research for engineering education practice, the illusion of explanatory depth (IOED) provides a useful framework. The seminal IOED study (Rozenblit & Keil, 2002), through a clever series of experiments, illuminated the severely limited ability of test subjects to accurately estimate their own level of knowledge about reasonably complicated topics, such as devices (ex.: a toaster) or natural phenomena (ex.: lightening or earthquakes). The typical trend was for a test subject to wildly over-estimate their level of knowledge until confronted with an assessment and/or an expert explanation of the topic, at which point a more realistic (although still inaccurate) self-rating of knowledge emerged. This systematic prediction inaccuracy was not valid for fact-driven topics, such as narratives of a movie or world capitals; subjects generally had a good sense of their own knowledge of, say, the capital of Colombia. But for complex, nuanced topics—and engineering problem solving falls into this category—the typical behavior of test subjects was to provide very poor predictions of their actual knowledge until confronted with an assessment. *The essence of IOED is that people often believe that they understand or can do that which they have merely seen*, and the potential relationship to worked-example videos is clear. The implication in this research, expanded upon later in the paper, involves the way in which we coach students to use worked-example videos to support problem solving or test preparation so that they can avoid the pitfalls of IOED.

Methods

Experimental Apparatus and Data Collection

We collected data from 24 experiment participants, all undergraduate engineering students, and have analyzed the data in detail for 10 of them so far. The full experimental setup has been described in detail elsewhere (Berger & Wilson, 2016), but several of the salient details are summarized here. Figure 1 shows a screenshot of the outcome of the experiment, including the three data collection modalities used. Each participant, after having been consented for the purposes of this research according to our approved IRB protocol, completes a single engineering dynamics problem on paper using a Livescribe notebook (which captures a real-time record of everything they write). The problem focused on particle kinetics and had three sub-parts: (a) requiring a work-energy formulation, (b) utilizing

Newton's law, and (c) using a different work-energy equation from part (a). Simultaneously, participants have access to a worked-example video of an expert solution to a dynamics problem (duration: 04:23) that shares several, but not all, features of the problem they are trying to solve. An eye gaze capture system plus video screen capture software records, in real time, the participant's interactions with the video including play, pause, stop, and gaze events. Finally, participants are asked to think aloud and narrate their thought process throughout the experiment; this audio is capture via the Livescribe pen. These data streams are merged into a single time-synchronized video (see the screenshot of Figure 1).

Participants also completed a pre-survey about their comfort with technology for learning, and a post-survey based upon the NASA-TLX workload measurement (Hart & Staveland, 1988). Participants also consented to allow us to access their academic transcript, so we know their grade in the dynamics course and all its pre-requisites.

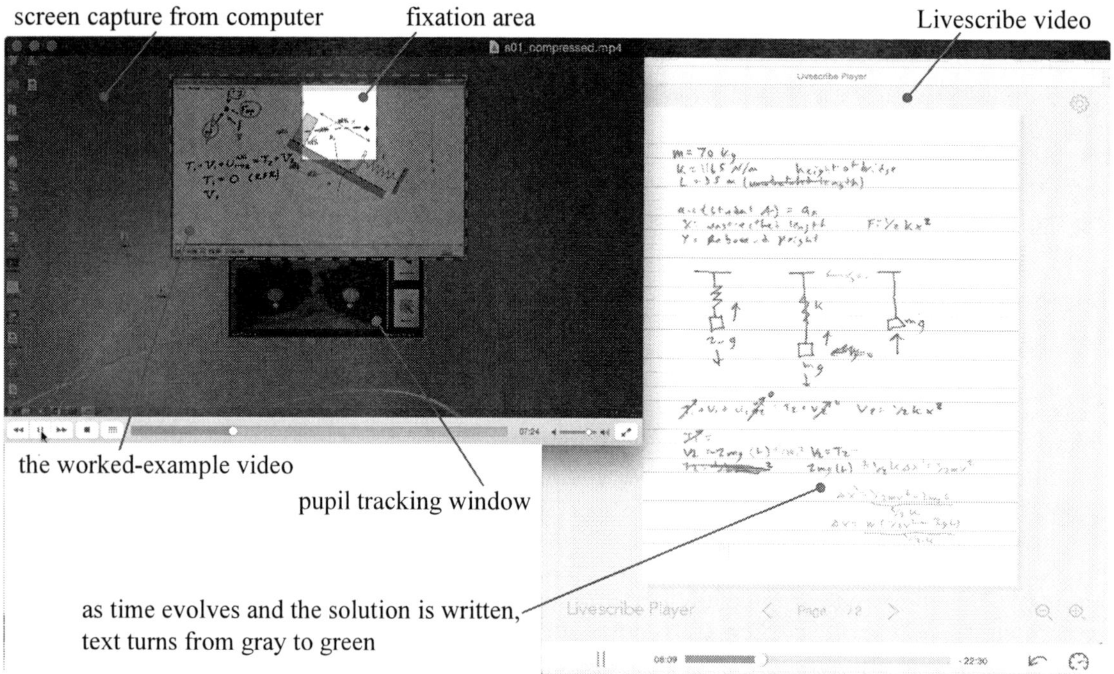

Figure 1: Still frame from experimental composite video showing both eye gaze capture panels and Livescribe video [from (Berger & Wilson, 2016)**].**

Data Analysis

The resulting video recording of each experiment was transcribed by a dynamics content expert to capture the following features: (i) the think-aloud narration from the participant, (ii) the specific solution events for the problem (ex.: drawing a free body diagram), and (iii) interaction events with the video. This experimental transcript is non-trivial to compose, and particularly solution-specific events really do require the coder to be a content expert. The transcripts created and coded for this research were processed by the authors (a faculty member who teaches the dynamics course, and an advanced undergraduate mechanical engineering honors student).

Each transcript was then coded according to a previously-developed code structure, with themes and sub-themes related to solution events, video events, and affect events. Affect events were generally audible expressions by participants that expressed confidence, frustration, confusion, or any other emotion throughout the experiment. This coding structure has also been presented in detail elsewhere (Berger & Wilson, 2016). We highlight here the category of 'forward reasoning' events, which capture participants advancing their solution in one of three ways: (i) using verbal or written information, (ii) using sketches or drawings, or

(iii) using algebraic operations. Forward reasoning is one of the key facets of the solution approach and one important metric we consider characteristics when comparing participants.

After coding, the data were processed to determine code counts, frequencies, time durations, sequencing, and other metrics related to the solution process. An example outcome is the coding stripe output of Figure 2, which shows the categories of events, their duration, and their sequencing. The coding stripe figure reports on each of the broad themes observed in the data; from top to bottom on the figure, they are: (i) which part of the problem they were solving, (ii) their engagement with the video, (iii) their forward reasoning events, (iv) their other solution events, and (v) their affect events. On the right side of the figure, the frequency (F) and duration (D, in minutes) of each event is shown. Frequency and duration for each event was calculated and summarized across subjects using descriptive statistics.

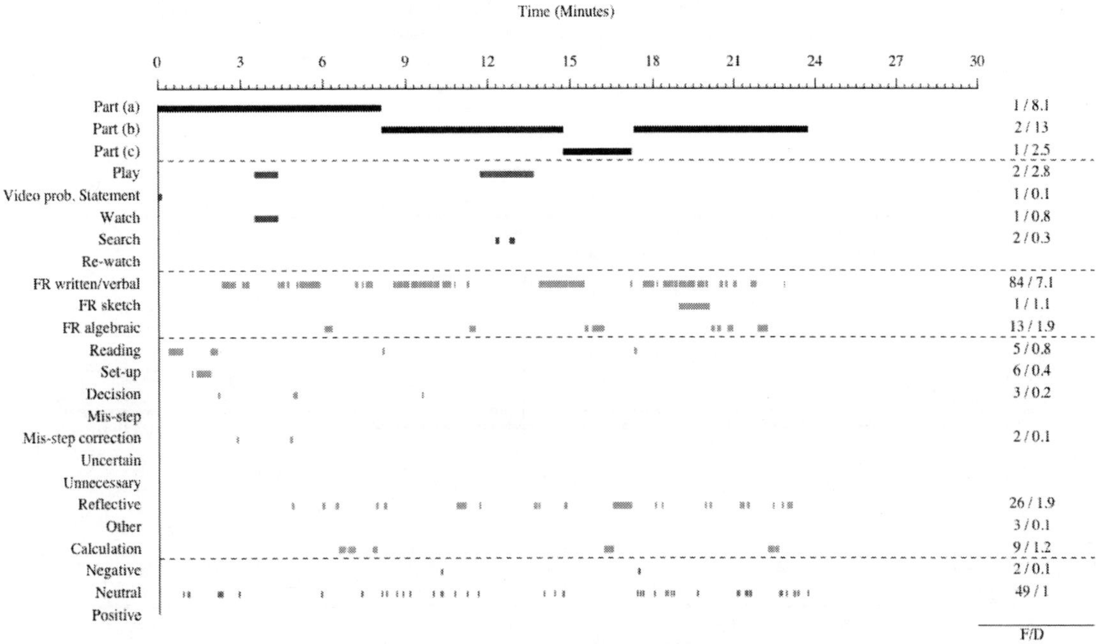

Figure 2: Coding stripe output of experimental coding for subject #7. The coding themes are listed on the left side, while the frequency (F, number of instances) and duration (D, in minutes) for each event/theme are on the right side.

Findings

Data Analysis Results

We can use the coding stripe figures, such as Figure 2, as a way to understand the temporal patterns used by participants during the problem solving experiment. Figure 3 shows a coding stripe comparison for five test subjects across a subset of the themes described above and shown in Figure 2. These sub-themes were identified in our exploratory data analysis as capturing much of the variation across participants. In Figure 3, we focus on time devoted to each part of the problem (black stripes), time devoted to playing and watching the video (blue stripes), and time devoted to the three categories of forward reasoning (orange stripes).

The distribution of time spent on each part of the problem generally favors part (a), likely because part (a) contains many of the start-up portions of the problem (such as drawing a free body diagram, defining a coordinate system, and understanding/modeling the overall dynamics of the problem). Many of these achievements in part (a) can be directly applied to parts (b) and (c), meaning that those parts of the problem are likely to take less time to complete.

The variation in video usage is instructive in two respects. First, there was wide variation in the amount of time each participant played the video during the experiment. Moreover, it is clear that video 'play' time and 'watch' time are not necessarily the same—or even similar. During the experiment, we saw many cases in which students played the video in the background, listening to the audio without actually watching the video, while they were thinking about the problem. It is conspicuous that generally the forward reasoning events took place when the video was not playing or being watched.

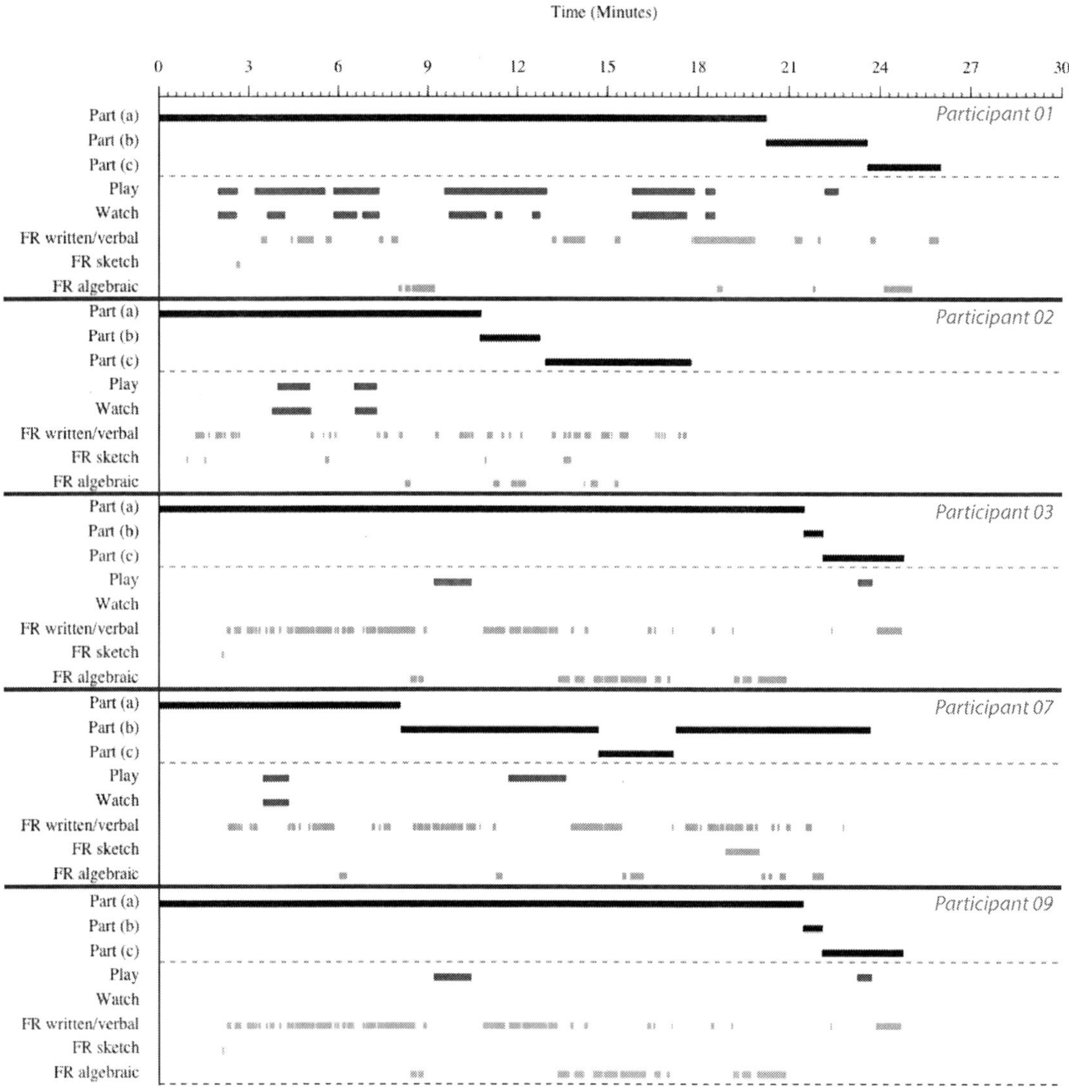

Figure 3: Code stripe composite showing several of the most important themes for five different participants. Parts (a), (b), and (c) correspond to the three parts of the problem. 'Play' and 'Watch' refer to participant interaction with the video. The 'FR' categories describe the three types of forward reasoning.

Table 1 summarizes some of the key time metrics across all 10 participants in this phase of our study. Participants generally used almost 2/3 of the total experiment time to complete part (a) of the problem, for the reasons described above. As observed from the coding stripe figure, there is a dramatic difference between the play and watch time for the video, and the ratio of the percentage of time spent playing to time spent watching was 19/7 = 2.7. Although not shown on the table, the percentage of time participants spent searching for content within the video was very small, just 0.6% of the total time for the experiment. Table 1 also makes it clear that these participants generally favor written/verbal modes of forward reasoning, rather than sketching or mathematical (algebraic) approaches.

Table 1: Time duration summary for specific codes across all 10 participants. Metrics are listed as mean and standard deviation of time (in minutes), and as a percentage of the total time taken to complete the problem.

Code	*Time (minutes; mean/SD)*	*% of total time*
Part (a)	12.1/5.6	62.0
Part (b)	3.7/3.8	18.0
Part (c)	4.0/3.2	20.0
Video play	3.5/3.7	19.0
Video watch	1.5/2.2	7.0
FR—written/verbal	5.9/1.9	32.3
FR—sketch/drawing	0.7/0.8	4.2
FR--algebraic	1.7/1.1	8.1

Participant verbal accounts of their work generally did not give the impression of intentional use of the video. We often heard speculative comments such as ("[the video honestly doesn't look that helpful" (Participant 3), or "video is kind of useful" (Participant 1). Several participants simply started the video at its beginning and watched the video to assess its content and potential usefulness for their problem solution. We did observe several instances of pattern matching, in which equations from the video were mimicked and used in the participant's own solution to the experimental problem.

Discussion

Two general patterns emerged from our research results.

1. Students did not know how to use the video efficiently. The ratio of 'play' time to 'watch' time was a surprising 2.7, clearly suggesting that many students played the video while they were working without paying very close attention to it. Moreover, very little time was spent on intentional searching within the video for any participant (and only about 1 minute of searching in the aggregate, across all 10 participants), suggesting that they: (i) searched highly efficiently, (ii) did not know what to search for, or (iii) did not employ a study strategy that included searching videos for specific pieces of information (i.e., their 'usual' approach was to watch large chunks of video). Based upon our observations during the experiment, option (i), highly efficient searching, simply did not occur and we discount this as a possible explanation for the very low search time. More likely, and consistent with the high play/watch ratio, is that *participants did not possess the required insights of abilities to enable efficient search within the video or recognize portions of the video that could have helped them solve the experimental problem more efficiently or with higher performance*. Their choice to play the video, often in large chunks, suggests that they were attempting to engage in pattern matching with the video's overall problem solving approach, rather than picking out the specific features of the video that would allow them to advance their own solution.

2. Most video-inspired forward reasoning seems to be pattern matching. The other observation is that forward reasoning and video watch were rarely synchronous in our data. Video watching and forward reasoning were often sequential, but our observations from each experiment as well as the audio of the experiment do not suggest that

watching the video directly led to a 'breakthrough' or induced specific conceptual understanding by the participant. The sequence of video watching and then forward reasoning was more likely related to pattern matching (say, using the equation used in the video) than it was to any improvement in conceptual understanding by the participant. *The conclusion here is that for these participants, video usage did not induce improved conceptual understanding, so much as it triggered a pattern matching behaviour.* This observation highlights differences between the current study (about using worked-example videos as an aid to problem solving) and past worked-example research (which often focused on teaching and training via worked examples). It appears that the sequence of video playing or watching and forward reasoning evident in the code stripes (Figures 2 and 3) are more likely the result of an ad-hoc approach to solving the experimental problem. Participant usage of the video seems to have been more hopeful than strategic, in which they used the video to identify a solution approach and equations that they could use in the experimental problem, rather than attempting to improve their fundamental understanding of dynamics.

Taken together, we believe both of these patterns have serious consequences in light of the IOED. The evidence presented here suggests that *students work with videos for the purpose of 'doing' rather than 'learning'*. They do not seem to view the video as an expert source from which they can learn, so much as they see it as a resource to help them achieve an objective (completing an assignment). The danger in this strategy is obvious: students can convince themselves that they understand a topic by simply matching a pattern observed in a video to solve a problem, when in fact their understanding of the topic might be highly superficial. This is the essence of the IOED and has important implications for how we coach students to use worked-example videos, as described later.

Limitations

There are two important limitations to this study. First, we recognize the small sample size in the data so far—just 10 participants. While the experiment itself takes about 45 minutes, including the pre- and post-surveys, the data analysis of all the video evidence, transcript generation, and coding is an extremely time-intensive process. We continue to analyze the experimental data and build our sample size.

The other limitation relates to the measurement of watch time. We suspect that our eye gaze equipment is producing a lower-bound estimate of the actual watch time, for the following reasons. 'Watch' time was only tabulated when the participant's pupils were visible in the pupil window of Figure 1, meaning that their head position and angle were aligned with the initial set-up and calibration of the system at the beginning of the experiment. In the presence of misalignment in head position, the eye gaze system still measures a gaze location, but we believe that location estimate to be in error based upon the geometric differences in participant position from the calibration condition. As a result, given this strict definition of 'watch' time, we believe the actual watch time is larger than that reported here (resulting in an actual play/watch ratio of less than 2.7). However, based upon our observations during the experiment, we do not believe that the reported watch times were dramatically different from the actual watch times. Nonetheless, addressing this experimental set-up issue will be an important refinement to future experiments.

Implications for Engineering Education Practice

Student Coaching

One important implication of this research is that students do not seem to make optimal use of the videos, in two respects. First, their pattern matching behaviour is not desirable from either a pedagogical or IOED perspective, and we would rather they take insight and meaning from the video. Second, their lack of efficient search processes suggests that they cannot articulate or recognize that which they do not know. That is, they do not know what they are looking for in the video, and perhaps they do not recognize it when they see it. Taken together, these two observations suggest that we must do a better job of explicitly coaching students on how to make the best use of the video resources.

The current generation of college students has grown up with YouTube, and it is natural to assume that their sophistication in using videos is very high. However, using worked-example videos for problem solving requires two important actions: (i) select the appropriate video that has the strongest alignment with the problem to be solved, and (ii) find the specific elements/concepts/processes within the selected video that most effectively enlighten the viewer about proper solution procedures. Coaching students about this process could be invaluable in both their performance (they earn a better score) and their efficiency (they know how to quickly identify useful information, and therefore can complete the assignment more quickly). The evidence presented here strongly suggests that students would benefit from more intentional coaching about how to use worked-example videos.

Video Creation

This research also holds value for video producers. Given what we know about student approaches to pattern matching, perhaps video creators can explicitly indicate the contents of the video using more extensive metadata, chapter markers, or other guideposts to help students evaluate the utility of a given video. Other options include explicit links to existing taxonomies or tools such as concept inventories for the video contents. Moreover, the voiceover within the video could include more than the simple mechanics of the solution (first we do this, then we do that) and instead include expert insights about why the video proceeds as it does. For instance, the video creator could intentionally describe the thought process behind a particular approach: "I'm using a work-energy approach because I'm attempting to solve for a change in position rather than a change in time…" These expert insights would certainly enlighten student users of the videos, and also may shift their mindset from one of pattern matching to a more intentional approach of video usage for learning in the service of problem solving.

Conclusions

This research reports the results of 10 experiments that help us understand student usage of worked-example videos to support their problem solving. Our careful transcripting and coding of the experimental data suggest that students engage in strong pattern matching efforts rather than a sophisticated search approach when using worked-example videos. The results also suggest that video watching does not trigger improvements in 'understanding' so much as it triggers pattern matching and mimicry of the processes used in the video. The implication for engineering education is that explicit coaching for students about how to most effectively use worked-example videos seems beneficial, as do improved video creation guidelines that include more detailed metadata and fully articulated expert insights. Our on-going data analysis continues to enhance our understanding of student use of worked examples, with further advances to be reported in future publications.

References

Berger, E. J., & Krousgrill, C. M. (2012). HigherEd 2.0: Web 2.0 in Higher Education. In I. Deliyannis (Ed.), *Interactive Multimedia* (pp. 67–92). Rijeka, Croatia: InTech Europe.

Berger, E. J., & Wilson, M. (2016). A Laboratory Study of Student Usage of Worked-example Videos to Support Problem Solving. In *Proceedings of the 123rd American Society for Engineering Education Annual Conference and Exposition* (p. 17 pages/Paper ID #16199). Seattle, Washington.

Brunken, R., Plass, J. L., Leutner, D., & Brünken, R. (2003). Direct Measurement of Cognitive Load in Multimedia Learning Direct Measurement of Cognitive Load in Multimedia Learning. *Educational Psychologist*, *38*(1), 53–61. http://doi.org/10.1207/S15326985EP3801

Hart, S. G., & Staveland, L. E. (1988). Development of NASA-TLX (Task Load Index): Results and Empirical and Theoretical Research. In P. A. Hancock & N. Meshkati (Eds.), *Human Mental*

Workload. Amsterdam: North Holland Press.
Kalyuga, S., Ayres, P., Chandler, P., & Sweller, J. (2003). The expertise reversal effect. *Educational Psychologist*, *38*(1), 23–31. http://doi.org/10.1207/S15326985EP3801
Mayer, R. E. (2009). *Multimedia Learning* (2nd ed.). New York: Cambridge University Press.
Mayer, R. E., & Moreno, R. (2003). Nine ways to reduce cognitive load in multimedia learning. *Educational Psychologist*, *38*(1), 43–52.
Moreno, R. (2006). When worked examples don't work: Is cognitive load theory at an Impasse? *Learning and Instruction*, *16*(2), 170–181. http://doi.org/10.1016/j.learninstruc.2006.02.006
Moreno, R., Reisslein, M., & Delgoda, G. (2006). Toward a fundamental understanding of worked example instruction: Impact of means-ends practice, backward/forward fading, and adaptivity. *Proceedings. Frontiers in Education. 36th Annual Conference*, 5–10. http://doi.org/10.1109/FIE.2006.322285
Moreno, R., Reisslein, M., & Ozogul, G. (2009). Optimizing Worked-Example Instruction in Electrical Engineering: The Role of Fading and Feedback during Problem-Solving Practice. *Journal of Engineering Education*, *98*(1), 83–92.
Paas, F., & van Gog, T. (2006). Optimising worked example instruction: Different ways to increase germane cognitive load. *Learning and Instruction*, *16*(2), 87–91. http://doi.org/10.1016/j.learninstruc.2006.02.004
Rourke, A., & Sweller, J. (2009). The worked-example effect using ill-defined problems: Learning to recognise designers' styles. *Learning and Instruction*, *19*(2), 185–199. http://doi.org/10.1016/j.learninstruc.2008.03.006
Rozenblit, L., & Keil, F. (2002). The misunderstood limits of folk science: an illusion of explanatory depth. *Cognitive Science*, *26*(5), 521–562. http://doi.org/10.1207/s15516709cog2605_1
Salden, R. J. C. M., Aleven, V., Schwonke, R., & Renkl, A. (2009). The expertise reversal effect and worked examples in tutored problem solving. *Instructional Science*, *38*(3), 289–307. http://doi.org/10.1007/s11251-009-9107-8
Smith, A. D., Mestre, J. P., & Ross, B. H. (2010). Eye-gaze patterns as students study worked-out examples in mechanics. *Physical Review Special Topics - Physics Education Research*, *6*(2), 1–9. http://doi.org/10.1103/PhysRevSTPER.6.020118
Sweller, J., & Cooper, G. A. (1985). The Use of Worked Examples as a Substitute for Problem Solving in Learning Algebra. *Cognition and Instruction*, *2*(1), 59–89. http://doi.org/10.1207/s1532690xci0201
Ward, M., & Sweller, J. (1990). Structuring effective worked examples. *Cognition and Instruction*, *7*(1), 1–39. http://doi.org/10.1207/s1532690xci0701

Acknowledgment

This material is based upon work supported by the National Science Foundation under Grant DUE-1525671. Any opinions, findings, and conclusions or recommendations expressed in this material are those of the authors and do not necessarily reflect the views of the National Science Foundation.

Copyright Statement

Student Engagement in Assessment through Group Work in a Mathematics Course for Bioengineers

Carrere L. Carolina
Math Department, Engineering Education Research and Development Group, Faculty of Engineering, National University of Entre Ríos, Oro Verde, Argentina
ccarrere@ingenieria.uner.edu.ar

Ilardo Juan
Math Department, Faculty of Engineering, National University of Entre Ríos, Oro Verde, Argentina
juancarlos.ilardo1@gmail.com

Ruiz Joaquín V.
Math Department, Faculty of Engineering, National University of Entre Ríos, Oro Verde, Argentina
joaquinruiz@hotmail.com

Iván Lapyckyj
Math Department, Faculty of Engineering, National University of Entre Ríos, Oro Verde, Argentina
ilapyckyj@ingenieria.uner.edu.ar

Escher Leandro
Math Department, Faculty of Engineering, National University of Entre Ríos, Oro Verde, Argentina
lgescher@ingenieria.uner.edu.ar

Waigandt Diana
Humanities and Languages Department, Faculty of Engineering, National University of Entre Ríos, Oro Verde, Argentina,
dwaigandt@ ingenieria.uner.edu.ar

***Abstract**: Assessment for learning in engineering programs is a central aspect when trying to improve the teaching-learning process of future engineers who need to meet the new challenges of society. In this regard, we reformulated the objectives of the Differential Equations (DE) course to suit the desired professional profile. Our general objective is that "students work effectively in teams." The purpose of this paper is to describe a research study which was carried out in an undergraduate Bioengineering program by answering the following questions: Which activities could engage students in the assessment process and provide information about group dynamics? Laboratory assignments were redesigned to provide personal participation opportunities. Individual contribution progress report instances and self-assessment and peer assessment activities were incorporated. The results allowed us to perceive the individual situation of students and, as a result of this, to improve the way we support them during their learning process.*

Introduction

Assessment for learning in engineering programs is a central aspect when trying to improve the teaching-learning process of future engineers who need to meet the new challenges of society. Numerous investigations indicate that the assessment systems implemented by teachers influence the learning process of the students (Biggs and Tand, 2011). Recognizing the complexity and importance of this issue was the starting point for an Action Research (Elliot, 2015) project centered on the assessment of the learning of mathematics for future bioengineers. The project is carried out at the Faculty of Engineering belonging to the National University of Entre Rios (FI-UNER). The research team is made up of all the mathematics teachers and engineers who teach DE, the auxiliary teaching students, a teacher from the Department of Humanities and Languages and the Pedagogical Advisor of the institution. This interdisciplinary group allows to approach assessment from different perspectives.

In order to begin our investigation, seek informed answers, and to promote the necessary changes, it was natural to ask ourselves the following questions: Why do we assess our students? What do we assess? The first led us to the two senses of assessment in higher education: to accredit knowledge and to contribute to the process of vocational training. The second led us to revise the objectives of the course. We also analyzed the bioengineering professional profile established in the Bioengineering 2008 curriculum, the "Valparaiso Agreement on Generic Competences for the Argentine Engineer", a document published by the Federal Council of Engineering Deans, and the demands of the labor market through the job offers published periodically by our Extension Secretariat in online "Newsletters" (CONFEDI, 2014).

In this way, the importance of forming a professional who can be part of multidisciplinary groups with team work skills was identified (Carrere et.al, 2016).Thus, we agreed to incorporate "that the student manages to work effectively in a team" as a new general objective of our DE course. It is worthwhile mentioning that the development of these skills is rarely mentioned explicitly in the syllabi of math courses in engineering programs. In ED we propose group activities that are developed in groups formed by three students. The process of group conformation is carried out with freedom according to affinity between the students. However, group-work assessment of laboratory assignments has not provided sufficient information about the learning process nor about group interaction dynamics.

The purpose of this paper is to describe a study which was carried out with an Action Research methodology in the 2016 DE course of an undergraduate Bioengineering program belonging to the FI-UNER. As a starting point, we wanted to answer the following questions:

- How can we contribute to the development of teamwork skills in our students?
- Which activities could engage students in the assessment process and provide information about group dynamics?
- How is it possible to foster individual contributions to ~~the~~ teamwork?
- How can we get information about the learning process during group task accomplishment?

Theoretical framework

Group Work is often used in engineering education to achieve a number of pedagogical goals and make students more active participants in the learning process than in traditional coursework (Gibbs, 2009). In particular, it has been shown that group work activities could help students to internalize new contents and bring them closer to professional life. Nevertheless assessing individual contributions to the success of the group is a hard task. Even more when a significant part of the engineer's degree is determined by the results of a teamwork project (Pung, 2011).

Due to the complex nature of the students' interactions in a collaborative learning environment, accurately assessing the contributions of a student to his or her group is a difficult task. However, inaccurate assessment of collaborative contributions gives rise to a variety of issues like: free-riding, the sucker-effect, student apathy toward collaboration, etc (Roberts and McInnerney, 2007).

All of these issues reduce the quality of the students' collaborations and thereby reduces the learning outcome of the collaborating students. Students become dissatisfied when they perceive that members of their team do not contribute sufficiently to the team project, forcing them to work harder or get a lower grade than they want (Oakley,Felder,Brent and Elhajj, 2004). In this sense numerous researches suggest that peer evaluations and self-evaluations can be used to develop students' team skills and to assign grades fairly based upon students' contributions (Falchikov, 2005; Kaufmann and Felder, 2000; Valverde, 2006). It has also been proposed that one way to evaluate the contributions of a student towards his or her group is by tracking all students' interactions with their group members in details and then using the quantity as well as the quality of the contributions of those interactions to create a model that represents a student's contributions toward his or her group (Khandaker and Soh, 2010).

On the other hand, assessment should be designed in such a way that it recognizes the importance of the learning process and its results. In particular formative assessment of this process during the group work task allows the teacher to understand the group dynamics and to give proper feedback on the activity (Willey and Freeman, 2006). Furthermore the assessment design should involve students, as some studies point this could improve the students' performance and enhance personal or intellectual development, or social competencies (Falchikov, 2005). Based on these conceptual aspects, changes in the design of the group work task were introduced and new actions to engage the student in the assessment were implemented.

The starting point: the original group task proposal

In the DE course, two non-face-to-face group tasks called Laboratory Practical Work (LPW) were given to the students. Through these activities, students were encouraged to explore the relationship between theoretical concepts and their applications in different contexts, while incorporating the numerical, graphical and computational advantages of using mathematical software. The LPW consisted of three integrative problems proposed by teachers. Sometimes, these integrative problems included concepts from other subjects (Electricity and Magnetism, Mechanical Physics and Advanced Computing) that allowed to explore mechanical or electrical systems mathematically modeled by differential equations and related to biomedical technology. The LPW were presented through a guide, which explained the objectives, bibliography and other relevant indications. For the accomplishment of each LPW, students counted on an average time of four weeks and the support of the teacher. The group was composed of three students and the conformation process was done freely according to affinity.

The teacher tried to accompany the learning process by providing feedback that reoriented the students to solve the problems. Students could choose to submit a draft paper stating the doubts and obstacles they encountered. The teacher, through the virtual platform, made written observations to improve the preliminary work presented and indicated those aspects that should be reviewed. Through this written feedback, the teacher provided information about specific errors or misconceptions avoiding to explicitly present the correct answer. This activity was optional and students participation rate had been decreasing since 2012, as it is shown in Figure 1. As a final product of the work of the team, the students presented a written report (WR).

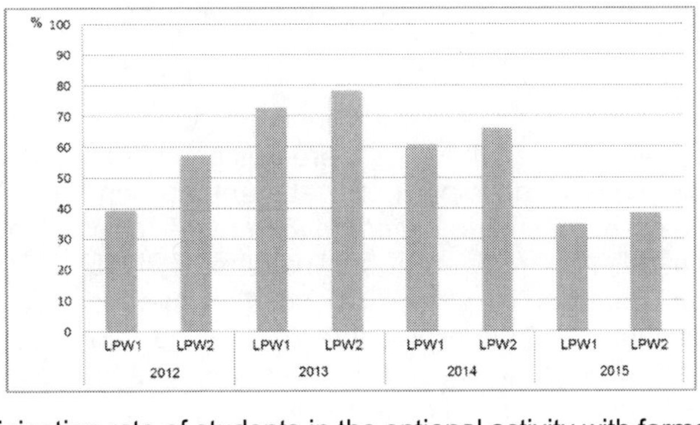

Figure 1. Participation rate of students in the optional activity with formative feedback

In this proposal, the two senses of assessment were considered: the formative one, since the teacher could reorient the process of learning; and the summative one, quantitatively evaluating the WR. However, this assessment method did not allow to recognize the interaction of the students and did not provide sufficient information about the learning process during group work.

Methodology

Participants

The activities described were developed in a DE undergraduate course for future Bioengineers during 2016 with the participation of 37 students (9 females & 28 males). DE is a weekly 8-hour subject taught in the second term of the second year at FIUNER. Its minimum contents include: Ordinary Differential Equations (ODE). Linear ODE systems. Equations and non-linear ODE systems. Fourier series. Equations in Partial Derivatives. These contents are specified in the 2008 Bioengineering curriculum.

The new group task proposal

According to the theoretical framework above mentioned, the new proposal tries to favor the dynamics of group work and to give opportunity of individual participation to the members. In this sense, the LPW was redesigned incorporating activities that had to be done by each one of the members, but whose results would be used in a later group discussion in order to solve the problems. Figure 2 shows the rubric of a LPW problem in which the individual work of the students and the subsequent discussion to reach conclusions as a team are evidenced.

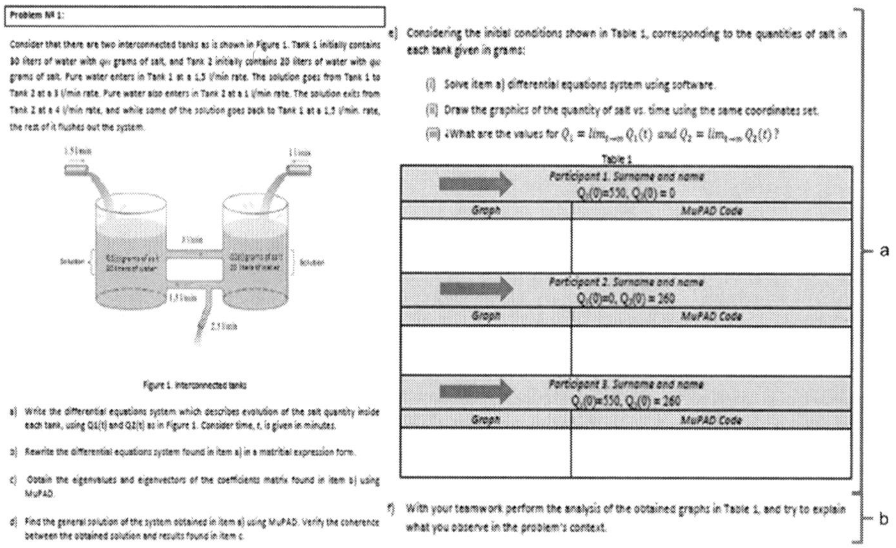

Figure 2. Example of a rubric corresponding to a LPW problem. a) Individual work of the students. b) Proposal of group activity

Furthermore, the problems raised in the LPW were designed considering the competence to solve problems that we seek to develop in our students during the DE course. Thus, the rubrics included the assessment criteria based on the stages for problem solving that experts perform and that are mentioned in different publications (CONFEDI, 2014): Identification of data and goals; Model; Selection of strategy or solution method; Procedure; Calculations, Results and communication; Graphic representation. Taking into account the objectives of the LPW, the use of software was incorporated as another assessment criterion. Moreover, instances of group dynamics observation were generated as they served as windows to provide feedback during the learning process. In addition, self- and peer-assessment activities were designed in order to involve the students in the group assessment process (Falchikov, 2005). The instances are described below:

1. Group dynamics observation instances

- Workshop classes:

Workshop classes were held in the computer lab. In each workshop class there were between 15 and 20 students who worked in groups of 2 or 3 together with 2 teachers (a head of practical assignments and a teaching assistant). During the workshops, students explored the relationship between theoretical concepts and their applications in different contexts, while learning about the numerical, graphical and computational advantages of using mathematical software. In these classes, the teachers played the role of tutor and assessor as they provided advice and guidance and helped students clarify ideas. This methodology turned the classes into a window of observation of the learning process, allowing teachers to incorporate formative assessment through feedback based on dialogue. The teachers, by means of guiding questions, tried to understand the difficulties the students had, and were able to offer suggestions and different alternatives so that students could find the answers by themselves.

In the new group task proposal, the workshop classes were planned as a meeting instance for the members of the groups. During the meeting, students were given the opportunity to discuss the strategy to solve the LPW problems, to put forward alternative solutions, to simulate them using software and to discuss results in order to reach conclusions. In this way, the workshop classes became a window of observation not only of the learning process, but also of group dynamics. The teachers were able not only to recognize the individual contributions to the assignment, but also to obtain information about the difficulties of each of the members in relation to the understanding of the concepts involved in the proposed problems.

- Progress reports:

The new proposal incorporated the individual presentation of progress reports of the group task. The students were given two possibilities to make this presentation. One of the options consisted of the presentation of drafts using the virtual classroom. Students could submit the number of drafts they deemed necessary and received written feedback from the teachers. In some cases, the teachers suggested the revision of some of the stages of the solution process, such as the analysis of the data of the problem, the selected procedures, the hypotheses considered, the computational simulation, the calculations; in others, teachers guided the students through questions or examples, encouraged the students to perform verification processes and analyze the consistency of the results or even requested the expansion of a justification or interpretation.

The second option consisted of the possibility to present progress reports in face-to-face meetings with the teaching assistants. They took place in the computer lab. In this case, the students presented their work, consulted about the proposed solution or the results interpretation, and made queries about the use of software. Teachers provided guidance through questions without directly responding to students' concerns.

In each LPW, the students had to present three Progress Reports. In this way, teachers had the possibility to monitor the individual work and obtained information on the contributions of each member of the group to the final product of the task. An indicator of the individual participation of the students in this activity was obtained, which corresponded to a score between 0 and 100, distributed as follows:

- 100: presented the total of the Progress Reports completely, solving the task with the software, attaching the corresponding code and obtaining particular conclusions to the task.
- 75: presented the total of the Progress Reports but incompletely.

- 25: presented only one Progress Report
- 0: did not submit any Progress Reports

The indicator was used to calculate the individual final grade, with the aim of promoting the participation of students in the proposed activities and their commitment to the group task. In each case, teachers recorded the difficulties observed and the questions they were asked to obtain information about the stages of problem solving that represented a greater obstacle to the students.

2. *Instances of student participation in the assessment process*

- Workshop to elaborate the assessment criteria of the Group Work

In order to carry out the new group task proposal, a workshop was held to elaborate the criteria to be used for Group Work assessment. It took place after the presentation of the Written Report corresponding to LPW 1.The temporal sequence of the workshop is shown in Figure 3.

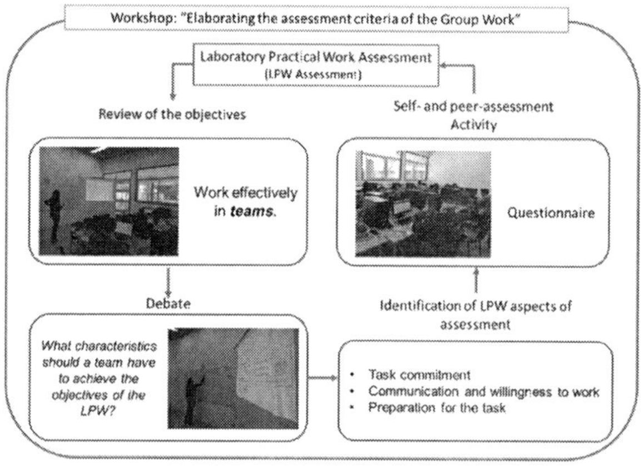

Figure 3. Temporal sequence of the workshop dynamics for the elaboration of criteria for the assessment of team work.

In the first place, the teacher explained the activity to be carried out and its importance in the context of LPW assessment, then asked the students the following question: *Why is group work important?* In order to find an answer to this question, the objectives of the LPW were reviewed in the subject planning and presented to the students at the beginning of the course:

- Identify, formulate and solve ***problems***.
- Use ***technological resources*** effectively.
- Work effectively in ***teams***.
- ***Communicate*** effectively.
- Act with ***ethics*** and professional responsibility.
- ***Learn*** in a continuous and autonomous way, and carry out self-assessment.

The teacher then made remarks on the importance of being able to "measure", that is, to assess, how the goals were reached in order to reorient the activities. Then the teacher raised the following question: *What characteristics should a team have to achieve the objectives of the LPW?* This question triggered a debate, and the various responses were written on the board. The students' answers highlighted "... *the importance of organization*" and that "*debate and proper communication of the members of the group were important*". This activity allowed the students to recognize the characteristics of an effective and efficient team, and consensus was reached in the sense that these corresponded to the aspects to be assessed. During this activity, the following aspects were identified:

- Task commitment
- Communication and willingness to work

- Preparation for the task

Finally, the teacher emphasized self-assessment and peer-assessment as tools for assessing the contribution of each of the members to the team, and used the identified aspects to elaborate the self- and peer-assessment questionnaire shown in Figure 4.

- Self-assessment and Peer-assessment

For each LPW the students carried out a self- and peer-assessment activity, in which the students worked individually in the presence of a teacher. The questionnaire shown in Figure 4 was used. It was elaborated using the assessment criteria that had been previously agreed upon and it was presented through the virtual campus.

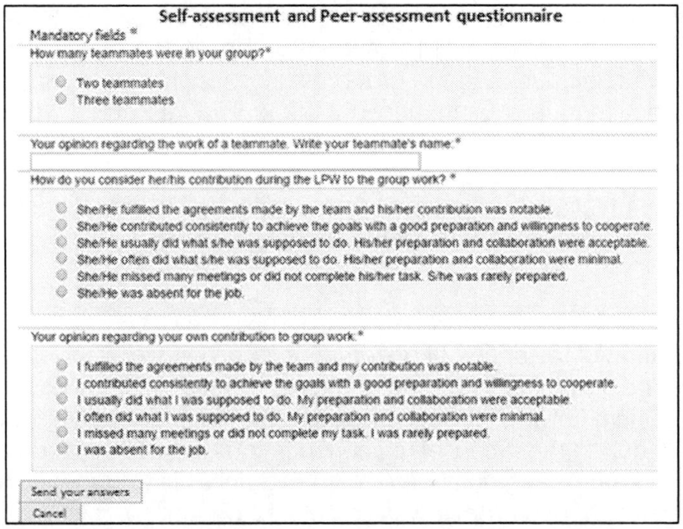
Self-assessment and Peer-assessment questionnaire

Mandatory fields *

How many teammates were in your group?*

- Two teammates
- Three teammates

Your opinion regarding the work of a teammate. Write your teammate's name:*

How do you consider her/his contribution during the LPW to the group work? *

- She/He fulfilled the agreements made by the team and his/her contribution was notable.
- She/He contributed consistently to achieve the goals with a good preparation and willingness to cooperate.
- She/He usually did what s/he was supposed to do. His/her preparation and collaboration were acceptable.
- She/He often did what s/he was supposed to do. His/her preparation and collaboration were minimal.
- She/He missed many meetings or did not complete his/her task. S/he was rarely prepared.
- She/He was absent for the job.

Your opinion regarding your own contribution to group work.*

- I fulfilled the agreements made by the team and my contribution was notable.
- I contributed consistently to achieve the goals with a good preparation and willingness to cooperate.
- I usually did what I was supposed to do. My preparation and collaboration were acceptable.
- I often did what I was supposed to do. My preparation and collaboration were minimal.
- I missed many meetings or did not complete my task. I was rarely prepared.
- I was absent for the job.

Send your answers

Cancel

Figure 4. Self-Assessment and Peer-assessment questionnaire

The questionnaire consisted of closed-ended questions. Each one corresponded to a literal descriptor that was converted to a numerical value to obtain the self-assessment and peer-assessment mark. Table 1 shows the equivalence used to perform the conversion. To obtain a free and sincere opinion from the students, the assigned numerical value was not informed in the questionnaire.

Table 1. Literal descriptors of self and peer-assessment questionnaire

Descriptor	Numerical conversion value
The student fulfilled the agreements made by the team and his/her contribution was notable.	100
The student contributed consistently to achieve the goals with a good preparation and willingness to cooperate.	80
The student usually did what s/he was supposed to do. His/her preparation and collaboration were acceptable.	60
The student often did what s/he was supposed to do. His/her preparation and collaboration were minimal.	40
The student missed many meetings or did not complete his/her task. She/he was rarely prepared.	20
She/he was absent for the job.	0

Group Work Assessment

Group work assessment contemplated the two meanings of assessment: the formative aspect, in the instances of observation of the process of learning previously described; and the summative one, through an individual mark in the assessment of group work.

First, the assessment of the WR was carried out taking into account the assessment criteria mentioned before (in *The new group task proposal*) and the Report Mark was determined, which turned out to be the same for all the members of the group. To obtain the Individual Mark, a modification was made to the methodology proposed by Valverde (2006). The author calculates the mark of each member taking into account a correction factor that contemplates self- and peer-assessment. In this research, the student's commitment to the group task is incorporated as a variable for the calculation of the individual grade. In this way, an Individual Factor was defined. It integrates self-assessment, peer-assessment and the student's commitment with the task. The calculation of the individual mark was made using a formula presented below and takes into account two terms:

$$Individual\ Mark = w\% * Report\ Mark + (100 - w\%) * Individual\ Factor * Report\ Mark$$

The first term allows a percentage (w%) of the grade of the Written Report to be assigned directly to the student. The value w% is selected by the teacher, and in this research it was 40% to give greater relevance to the assessment of group work process. The second term reflects the cooperative contribution to the group task (for each student and for the rest of the team). This term is calculated taking into account the Individual Factor, which is obtained for each member as the ratio between the arithmetic mean of self-assessment, peer-assessment and individual participation marks and the average of the means of all the members of the group. In this way, this factor represents the individual contribution of each member in relation to the other members of the group.

Results

As a result of the creation of instances of observation of group work, information was obtained about the learning process during the task and the difficulties encountered in the different stages of problem solving. In each instance, the teacher registered the type of query made by the students during the face-to-face consultations or in the presentations of drafts through the virtual classroom. The data were classified according to the stages of problem solving and the frequency of occurrence was calculated (Figure 5). It was noticed that in LPW 1 the Procedure step, which is related to the development of the solution step by step, using the appropriate mathematical notation and expressing the theoretical justification when applying a property or theorem, was the stage in which the students needed more orientation. However, the frequency decreased in LPW 2. One reason for this decrease could be that during the course the students learnt the mathematical tools that they need to solve the problem and that they used them with greater security.

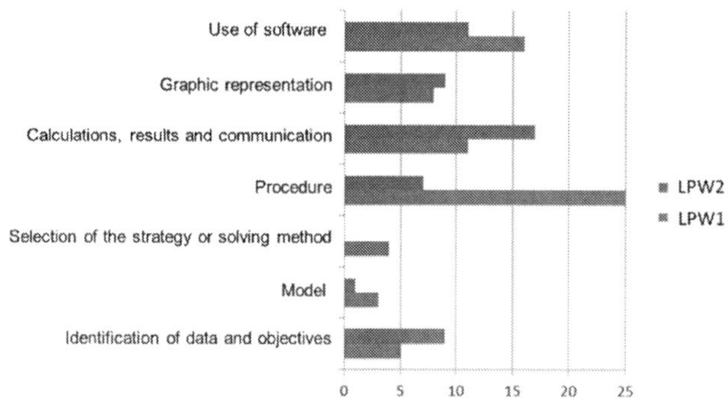

Figure 5. Frequency of feedback provided at each stage

Table 2. Percentage of students and participation indicator in each LPW

Individual participation indicator	% of students	
	TPL 1	TPL 2
0	15	8
25	15	5
75	8	19
100	62	68

With regards to individual contributions to teamwork Table 2 shows, for LPW 1 and 2, the value of the individual participation indicator and the percentage of students that each indicator obtained. The participation rate of the students in the instances to observe the individual contribution was obtained from these results. This rate was calculated as the sum of the percentages of students that obtained an individual participation indicator other than zero. In the first LPW, the participation rate of the students was 85%, while in LPW 2 this rate increased to 92%. In this way, an evolution in the participation of the students in this type of activities is observed.

On the other hand, in LPW 1, 70% of the students obtained an individual participation indicator equal to or greater than 75, and in LPW 2 this percentage increased to 87%. Contrasting these results with the marks of the Written Report shown in Figure 6.a, it can be inferred that a greater and better participation of the students in the proposed activities allowed the teachers to provide feedback that had a positive effect on the learning process. This reorientation had a direct impact on the report marks, since 93% of the students obtained a grade higher than 50 in LPW 2.

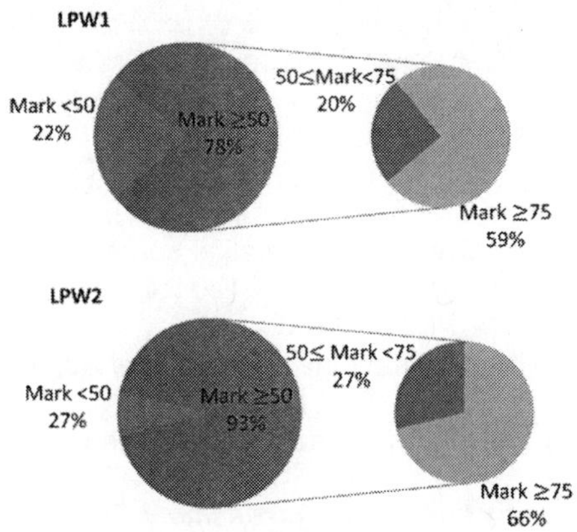

(a)

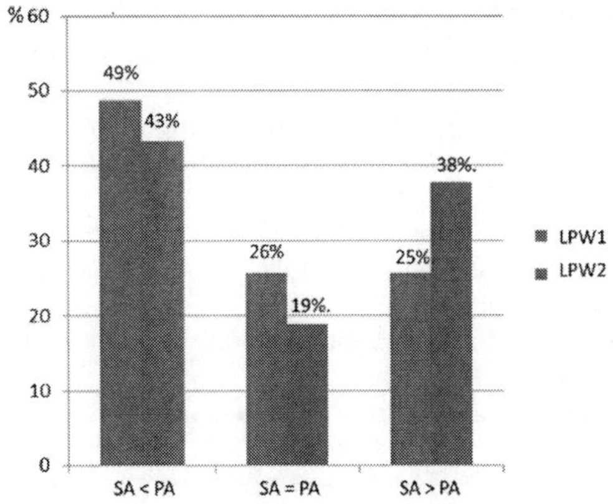

(b)

Figure 6. a) Marks corresponding to the Written Report. b) Comparison between self-assessment and peer-assessment marks.(SA: Self-assessment, PA: Peer-Assessment)

In relation to the activities proposed to get information about group dynamics, Figure 6.b shows the percentage of students that were self-assessed with a grade that was lower, equal to or greater than the average of the scores obtained in peer-assessment. The observed differences in the results show the interaction between the members of the group. In the first place, it is emphasized that in both LPWs, the highest number of students was self-assessed with a lower grade than their peers (49% and 43%, respectively). This may indicate that students have a low personal perception of their contribution to the group task.

Thus, it can also be observed that in LPW 2 the percentage of students that were self-assessed with a higher grade than the average peer-assessment grade increased with respect to the first LPW. From this result it is evident that the students were stricter when assessing their peers. Furthermore, the results in Figure 6.b suggest that despite knowing the assessment instrument there was no speculation in the second activity when carrying out self- and peer- assessment. This result could be an indicator of the students' engagement in the assessment process, which might have been promoted by their participation in the assessment criteria elaboration and the self- and peer assessment.

Conclusions

From this research, individual participation indicators were constructed, which allowed the observation of the commitment of each one of the members performing team work. Moreover, it is worth highlighting that the methodology proposed to obtain the individual mark is a contribution to the approach of the reference authors mentioned in this paper. Furthermore, the teachers who participated expressed greater agreement with the assessment methodology to obtain the individual grade regarding the previous methodology. In addition, the activities developed in this research showed the obstacles that arose in the different stages of the resolution of the problems proposed, allowing teachers to get information of the learning process and provide adequate and personalized feedback that helped redirect learning. Likewise, the differences in the results obtained from self- and peer-assessment allowed to know the interaction between the members of the group.

Although the results in this work correspond to a single cohort, we can say that this inquiry improved our understanding of the problem and allowed us to better understand the process undergone by the students when working as a group. In addition, it allowed us to perceive the individual situation of each student and, as a result, to improve our way of accompanying them in their professional training process. We plan to carry out a more complete and deep analysis and also inquire about how our students perceived this innovation in relation to the development of their teamwork skills.

References

Biggs, J. B. (2011). *Teaching for quality learning at university: What the student does.* McGraw-Hill Education.

CONFEDI (2014) *Competencias en Ingeniería.* REDI. Universidad FASTA, 2014. Accessed at http://redi.ufasta.edu.ar:8080/xmlui/handle/123456789/409 on 01 Mar 2017.

Carrere, C., Milesi, S., Lapyckyj, I., Ravera, E., Escher, L., Miyara, A., ... & Añino, M. (2016). Formative Assessment and Professional Training: Reflections from a Mathematics course in Bioengineering. In *Journal of Physics: Conference Series* (Vol. 705, No. 1, p. 012043). IOP Publishing.

Elliott, J. (2015). Educational action research as the quest for virtue in teaching. *Educational Action Research*, *23*(1), 4-21.

Falchikov, N. (2005). Improving assessment through student involvement: Practical solutions for aiding learning in higher and further education. Routledge.

Gibbs, G. (2009). The assessment of group work: lessons from the literature. *Assessment Standards Knowledge Exchange.*

Kaufman, D. B., Felder, R. M., & Fuller, H. (2000). Accounting for individual effort in cooperative learning teams. *JOURNAL OF ENGINEERING EDUCATION-WASHINGTON-*, *89*(2), 133-140.

Khandaker, N., & Soh, L. K. (2010). Assessing individual contributions to groupwork by a summary of tracked computer-supported collaborative activities. In *Frontiers in Education Conference (FIE), 2010 IEEE* (pp. S1E-1). IEEE.

Oakley, B., Felder, R. M., Brent, R., & Elhajj, I. (2004). Turning student groups into effective teams. *Journal of student centered learning*, *2*(1), 9-34.

Pung, C. (2011). Assessment of the CATME Perr Evaluation Tool Effectiveness.In *Proceedings of the 2011 American Society for Engineering Education Annual Conference & Exposition.*

Roberts, T. S., & McInnerney, J. M. (2007). Seven problems of online group learning (and their solutions). *Educational Technology & Society*, *10*(4), 257-268.

Valverde, G. J. (2006). Obtención de notas individuales a partir de una nota de grupo mediante una evaluación cooperativa. *Revista Iberoamericana de Educación*, *38*(5), 3.

Willey, K., & Freeman, M. (2006). Completing the learning cycle: The role of formative feedback when using self and peer assessment to improve teamwork and engagement. In *Proceedings of the 17th Annual Conference of the Australasian Association for Engineering Education: Creativity, Challenge, Change; Partnerships in Engineering Education* (p. 751). Australasian Association for Engineering Education.

Copyright statement

Learning Engineering Skills in a First-Year Course

Farshid Marbouti
San Jose State University, San Jose, CA, U.S.A.
farshid.marbouti@sjsu.edu

Jale Ulas
Middle East Technical University, Ankara, Turkey
jaleulas@gmail.com

Heidi A. Diefes-Dux
Purdue University, West Lafayette, IN, U.S.A.
hdiefes@purdue.edu

Abstract: *Teamwork and problem solving skills are important for engineering students to develop. These skills, in addition to communication and critical thinking skills, are necessary for tackling engineering-based design and open-ended problems. From an educational perspective, it is important to know what types of learning activities lead to acquiring these skills. In this study, in which we used structural equation modeling (SEM), we found that in-class learning (measured by quiz grades) and outside-class learning (measured by homework grades) were significantly related to students' demonstration of engineering skills (measured by engineering-based design project and open-ended problem solving activities grades). However, overall learning in the course (measured by exam grades) did not relate to engineering skills. The conclusion that this study confirms that not all the assessments are measuring the same student skills and knowledge is muddied by a number of limitations linked to a traditional score-based grading strategy. This is a clear call for the adoption of standard-based grading as means of generating grading result that are actionable.*

Moving from high school to college is considered a big change in students' educational life and considerable efforts have focused on improving students' success in their first year of college. Adding first-year seminar courses (e.g., Miller, Janz, & Chen, 2007), orientation classes (e.g., Starke, Harth, & Sirianni, 2001), and summer programs (e.g., Vinson, 2008) are some of the efforts to help first-year students succeed. Creating living/learning communities (e.g., Purdie & Rosser, 2011), assistant programs such as a program for socially/economically disadvantaged students (Braunstein, Lesser, & Pescatrice, 2008), and remedial English programs (Leake & Lesik, 2007) are examples of programs from which all college students may benefit. While the creators of these programs are optimistic, the research results on whether or not these programs improve students' educational success are controversial. In other words, some studies show that these programs are beneficial for students while some conclude there is no difference in the academic success of students who participated in these programs and those who did not.

One way to improve student learning is to use active learning strategies. Over the past decade, an increasing number of instructors are opting to employ active learning strategies (e.g., flipped classroom, group discussions) over the traditional extensive lecturing (Hurtado, Eagan, Pryor, Whang, & Tran, 2012). Collaborative (temporarily formed group) learning methods along with project-based learning and teamwork are often key components of new active learning strategies. The new pedagogies are usually implemented along with new assessment methods. For example, in first-year engineering program courses at a large mid-western United States university, in addition to weekly quizzes and homework, students solve design and open-ended problems in teams.

Teamwork and problem solving are important skills for engineering students. The completion of engineering-based design and open-ended problems requires communication, teaming, problem formulation, and critical thinking skills, which align with the skills required in STEM education (ABET, 2014; CCSSM, 2010; NAE, 2004; NCATE, 2008). Development of all these skills fulfills accreditation requirements and needs for future student success by giving them a foundation in problem solving in realistic contexts.

Project-based learning is an approach that "organizes learning around projects" (Thomas, 2000, p. 1). The projects in this approach pose authentic nontrivial problems that require students to engage in complex tasks such as asking and refining questions, discussing their ideas, making predictions and decisions, designing solution plans and/or experiments, gathering and analyzing data, making conclusive judgments, and coming up with realistic products (Blumenfeld et al., 1991; Thomas, 2000). During solution development, the students work cooperatively and relatively autonomously to reach explicitly defined educational goals (Thomas, 2000). Project-based learning provides students with deeper understanding of key principles and concepts, connecting class topics to real-life practices, improving their learning and metacognitive skills, and increasing their motivation (Blumenfeld et al., 1991) by engaging students as active and reflective learners (Barron et al., 1998).

To assess student learning in engineering courses, instructors often employ traditional assessment methods such as quizzes, homework, and exams as well as team-based projects. Most instructors operate under the assumption that these assessment types can be used as indicators of potential success in engineering work. While active learning strategies such as team-based and project-based learning are beneficial for students, they may assess both similar and different knowledge and skills than traditional assessments. Quizzes, homework, and exams focus on engineering content knowledge and problem solving skills via predominantly closed-ended problems. Projects may focus on modelling or design concepts, communication, and teamwork in addition to students' knowledge and skills in more open-ended situations. From an educational perspective, it is important to understand the relationship between these different types of assessments.

Different assessments occur at different times inside and outside of class. Students take quizzes during class, complete homework outside of class, and work on problem-solving activities and projects in teams. In this paper, we conceptualized learning and their associated assessment activities in three categories: in-class (assessed via quizzes), outside of class (assessed via homework), and engineering skills development (assessed through projects and open-ended problem solving activities). Typically, mid-term and final exams are a significant portion of engineering students' course grades. Thus, exam grades are often used as an overall indication of an individual student's learning in a course.

Research question

In this study, we aim to investigate the relation of different learning activities including in-class learning, out-of-class learning, and the course overall learning (measured by different types of assessments) to the acquisition of important engineering skills. The specific research question is: What types of learning activities, and to what extent, lead to learning engineering skills (measured via open-ended problem solving and project work)?

Methods

Data source and settings

This study uses secondary data collected in Spring 2012 as part of a first-year engineering (FYE) course. Approximately 1650 FYE students at a large mid-western United States university were enrolled in this course in Spring 2012. Approximately 20% of FYE students were female and about 20% were international students. The course was broken into 15 class sections, with each section enrolling as many as 120 students. Each section had one instructor, one graduate teaching assistant (TA), and four undergraduate TAs. The course is a required second semester, 2-credit hour course for all first-year engineering students. In this course, students learn how to use computer tools to solve fundamental engineering problems; how to make evidence-based engineering decisions; develop problem-solving, modeling, and design skills; and develop teaming and communication skills.

During class, students participated in twice weekly quizzes which tested students' class preparation or understanding of low level learning objectives. For the first two-thirds of the semester, the students spent the majority of their class time working on ungraded homework-like problems in teams of four, with occasional classes devoted to working on open-ended problem solving activities. In the last third of the course, the students spent the majority of their time on their team project. Students individually completed weekly homework outside of class focused on the prior week's topic and took three written exams spread evenly across the semester.

Data

Existing FYE course performance data (i.e. course grades book) from Spring 2012 (1650 students) were used in this study. The data include grades for seven assessment types: weekly attendance, quizzes, and homework assignments, as well as a project, open-ended problem solving activities (referred to as Model-Eliciting Activities (MEA)), and exam grades (Table 1). The project and MEAs were team-based collaborative activities and were graded for modeling and design concepts as well as meeting criteria. Since there were three written exams five week apart (two mid-terms and one final), the data were clustered into three main sections: weeks 1-5, weeks 6-10, and weeks 11-15. The number of grades for each assessment type in the three sections of the semester is shown in Table 2. For quizzes and homework assignments, which had more than one grade in each section of the semester, an average score was calculated and used in this study.

Table 1: Distribution of grades for FYE course

Assessment Type	Spring 2012
Class Participation (CP)	5%
In-Class Quizzes	
Team Participation (TP)	5%
Homework (HW)	20%
Project	15%
Open-Ended Problem Solving (MEA)	20%
Written Exams	35%

Table 2: Number of grades for each assessment type in the FYE course

Week	Quiz	Homework	Project	MEA	Exam
1-5	10	5	0	0	1
6-10	10	5	0	1	1
11-15	10	5	1	1	1

Hypothesized models

To answer the research question, the relation of different course assessment types to the project and MEA grades (as the indicators of learning engineering skills) is hypothesized in Model 1 (Figure 1).

Data analysis

Descriptive statistics and correlation between variables were calculated (in SPSS software) to show an overall description of the data and the relations between different variables. Structural equation modeling (SEM) in MPLUS (Muthen & Muthen, 2007) was used to test the relations between different learning activities, overall learning, and engineering skills. As

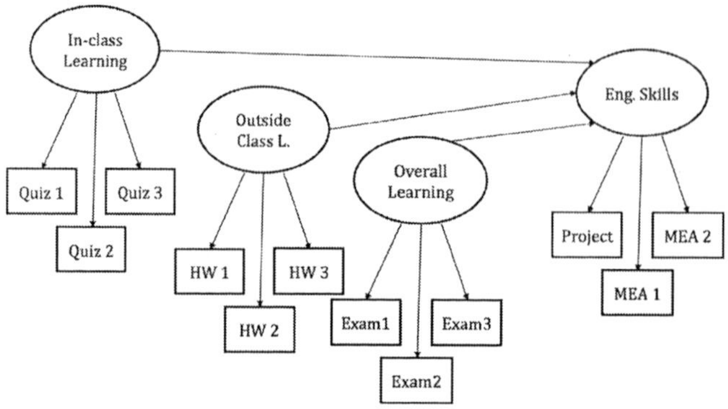

Figure 1: Model 1, the relation of acquiring engineering skills (measured by engineering-based project and MEAs) with other types of learning activities.

shown in Figure 1, Quiz grades were used as a measurement of in-class learning, homework grades were used as a measurement of out-of-class learning, exam grades were used as an indicator of overall learning, and the project and MEA grades were used as a measurement of engineering skills development.

To account for the skewed nature of the data, the Maximum Likelihood Robust (MLR) method was used to fit the models. Missing values were handled using the MPLUS default option. First, the measurement model was evaluated. The variances of latent variables were set to 1 and all factor loadings were freely estimated. Based on the results of the measurement models and because one of the observed variables were loading very highly on the latent variable, one alternative model was tested. In the alternative model (Model 1A in Figure 2), the Exam 3 grade (with factor loading ~0.9) was used instead of the "Learning" latent variable. To evaluate the overall fitness of the model and compare them with their alternative model chi-square, Root Mean Square Error of Approximation (RMSEA), Comparative Fit Index (CFI), Standardized Root Mean Square Residual (SRMR), Akaike Information Criterion (AIC), and Bayesian Information Criterion (BIC) were used. A chi-square goodness of fit test determines if the sample data matches the population. In general, large sample sizes will result in a statistically significant difference between the observed data and model expectations, suggesting misfit, regardless of the true situation. RMSEA is a mechanism for accommodating large sample sizes as a supplementary fit (Steiger and Lind, 1980). The RMSEA is widely used in SEM to provide a mechanism for adjusting for sample size where chi-square statistics are used. CFI compares the model of interest with some alternative such as the null or independence model. SRMR is the square root of the discrepancy between the sample covariance matrix and the model covariance matrix. AIC and BIC are regarded as an information theory goodness of fit measure, applicable when maximum likelihood estimation is used (Burnham & Anderson, 1998). These indices are used to compare different models.

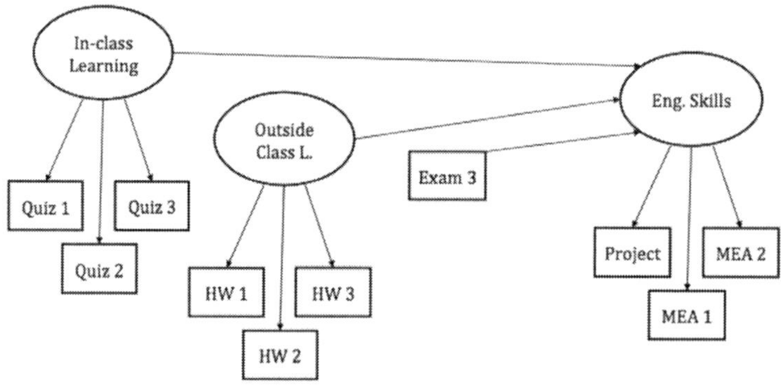

Figure 2 – Model 1A, Alternative for Model 1, replaces Overall Learning with Exam 3.

For RMSEA the recommended value is close to or less than 0.06, for CFI the recommended value is close to or greater than 0.95, and for SRMR the recommended value is close to or less than 0.08 (Hu & Bentler, 1999). Because the model and its alternative was not a nested model, the difference in AIC and BIC (instead of chi-square) was used to compare the models. A nested model is a model that uses the same variables as another model but specifies at least one additional parameter to be estimated.

Results

Descriptive statistics are shown in Table 3. All observed variables had negative skewness and positive kurtosis. Correlations between the observed variables are shown in Table 4. All variables were significantly correlated; however, the correlations were moderate (between 0.3 and 0.7).

Table 3: Descriptive statistics.

Assessment	N	Minimum	Maximum	Mean	Std. Dev.	Skewness	Kurtosis
Quiz1	1650	0.000	10.000	7.642	1.367	-1.228	2.814
Quiz2	1647	0.000	10.500	7.629	1.912	-1.111	1.727
Quiz3	1520	0.000	10.500	7.564	2.571	-1.147	0.832
HW1	1650	0.000	51.900	35.429	8.778	-1.451	2.423
HW2	1646	0.000	50.000	34.487	11.600	-1.178	0.854
HW3	1651	0.000	50.000	38.016	9.999	-1.120	0.789
Project	1647	0.000	100.000	87.469	17.838	-3.572	13.594
MEA1	1651	0.000	102.000	82.185	12.655	-1.228	3.565
MEA2	1642	0.000	102.000	80.932	14.620	-1.479	4.491
Exam1	1642	0.000	99.000	75.054	14.490	-1.578	4.426
Exam2	1620	0.000	100.000	80.870	15.440	-2.627	10.568
Exam3	1619	0.000	100.000	68.135	20.634	-1.015	1.212

Table 4: Correlation table (all correlations are significant at 0.001 level).

Assessment	Q1	Q2	Q3	HW1	HW2	HW3	Proj	MEA1	MEA2	Ex1	Ex2	Ex3
Quiz1	1.000											
Quiz2	0.462	1.000										
Quiz3	0.406	0.451	1.000									
HW1	0.546	0.347	0.378	1.000								
HW2	0.474	0.468	0.435	0.633	1.000							
HW3	0.350	0.334	0.491	0.462	0.567	1.000						
Project	0.483	0.300	0.459	0.404	0.460	0.368	1.000					
MEA1	0.353	0.204	0.240	0.391	0.385	0.293	0.327	1.000				
MEA2	0.338	0.289	0.374	0.377	0.448	0.411	0.468	0.369	1.000			
Exam1	0.578	0.478	0.332	0.482	0.482	0.314	0.401	0.288	0.270	1.000		
Exam2	0.469	0.443	0.369	0.397	0.499	0.409	0.503	0.233	0.334	0.599	1.000	
Exam3	0.550	0.468	0.449	0.474	0.545	0.483	0.472	0.274	0.356	0.638	0.700	1.000

The first step was to confirm the measurement model of the In-class Learning, Outside Class Learning, Overall Learning, and Engineering Skills. Then, the direct effects were added to the models to investigate Model 1 and its alternative, Model 1A. All factor loadings were significant and in acceptable range except Exam 3, which was close to 0.9. Thus, the alternative model was proposed and analyzed.

To evaluate the relationship between acquiring engineering skills and the different types of learning activities, Model 1 and its alternative model were analyzed. Model fit statistics for Model 1 and Model 1A are shown in Table 5. Models 1 fits relatively well, however Model 1A does not fit very well. The chi-square is significant for both models. However, because of the large sample size (~1650) a significant chi-square was expected. The chi-square is smaller for Model 1 than the alternative Model 1A. However, because models are not nested, it is not possible to evaluate the significance of the chi-square difference. For Model 1, the CFI, RMSEA, and SRMR values are close to the recommended values. Thus, Model 1 is an overall good fit. For Model 1A, the CFI, RMSEA, and SRMR values are not close to the recommended values. Thus, alternative Model 1A does not fit well.

Table 5: Fit statistics for Model 1 and Model 1A.

Statistic	Recommended Values	Model 1 (all variables)	Model 1A (Only Exam 3)
Chi-square	---	471.128**	1098.60**
df	---	48	32
RMSEA	<0.06	0.073	0.144
RMSEA 90% CI	---	0.067 0.079	0.136 0.151
CFI	>0.95	0.930	0.759
SRMR	<0.08	0.045	0.172
AIC		129,825.01	89,260.06
BIC		1300,52.01	89,427.05

** *p < .001*

Both factor loadings and standardized factor loadings for Model 1 and the alternative are shown in Table 6. All factor loadings are significant. The standardized factor loading for Exam 3 is 0.895, which indicates that the observed variable Exam 3 can be used instead of the Overall Learning latent variable. Exam 3 was administrated at the end of the semester and was somewhat more cumulative than the previous two exams. In addition, the project content, to some extent, was more related to this exam. Comparison of Model 1 and Model 1A indicates that this replacement does not improve the model.

Table 7 and Figure 3 show the direct relation between In-class Learning, Outside Class Learning, Overall Learning and Engineering Skills. In-class Learning (measured by quiz grades) and Outside Class Learning (measured by homework grades) have a significant relation to Engineering Skills (measured by engineering based design project and MEA grades). Based on Model 1, Exams have no significant relation to Engineering Skills.

Discussion and Conclusion

In this paper, we investigated the relation between students' grades on a number of assessments employed in a first-year engineering course. Specifically, the relation between learning activities (quizzes, homework, and exams) and acquiring engineering skills (projects) was investigated. Among quizzes, homework, and exams, quizzes, followed by homework, had the strongest relation to the engineering skills measured by the project.

Table 6 - Factor loadings for Model 1 and alternative Model 1A.

Factors	Model 1	Model 1A
In-Class Learning		
Quiz 1	1.004**/0.736	0.837**/0.668
Quiz 2	1.215**/0.636	1.064**/0.587
Quiz 3	1.597**/0.617	1.608**/0.643
Out-of-Class Learning		
Homework 1	6.562**/0.746	6.148**/0.740
Homework 2	9.848**/0.846	9.047**/0.811
Homework 3	6.531**/0.653	7.005**/0.714
Overall Learning		
Exam 1	11.249**/0.768	-----
Exam 2	13.475**/0.828	-----
Exam 3	19.816**/0.895	-----
Engineering Skills		
Project	6.529**/0.745	5.331**/0.636
MEA 1	3.143**/0.502	3.392**/0.468
MEA 2	4.952**/0.668	5.047**/0.609

** $p < .001$

Table 7 - Direct Effects for Model 1 and alternative Model 1A.

Direct Effects	Model 1	Model 1A
Engineering Skills		
In-Class Learning	1.329**/0.659	0.468*/0.281
Out-of-Class Learning	0.556**/0.276	0.871**/0.522
Overall Learning	-0.078/-0.039	-----
Exam 3	-----	0.016**/0.200

* $p < 0.01$ ** $p < 0.005$

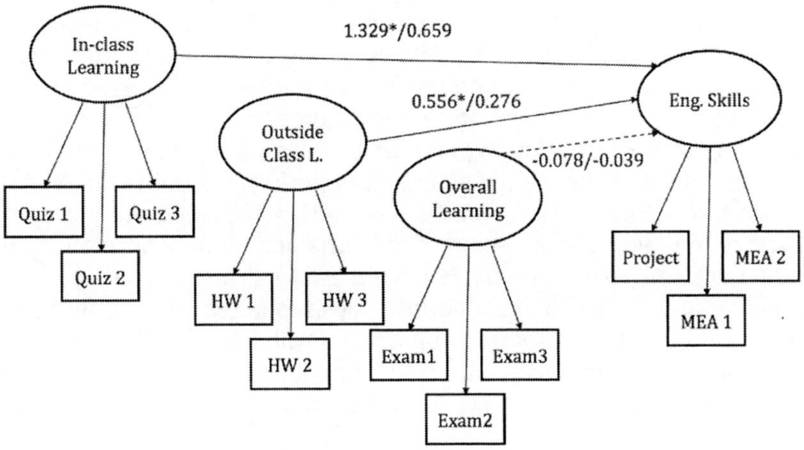

Figure 3: Model 1, the relation of Engineering Skills and different types of learning activities. Non-significant effects are shown with dashed lines.

Exams were not significantly related to the engineering skills. This might be due to the fact that most of the learning objectives associated with the projects and MEAs (e.g., design and communication skills) cannot be evaluated in a quiz or written exam. This makes it surprising that quizzes have a greater significant relation than homework. While skills like design were not be measured via the exams, other learning objectives (i.e. data representation and MATLAB programming skills) were related to the content learning assessed on exams.

Quizzes, were designed by instructors and varied across the different sections from essentially attendance checks to assessments of understanding on the course content. Overall, quizzes were typically easier than other assignments and students received high grades. Students also typically received high grades on their project work, in part from responding to formative feedback and lenient grading by the undergraduate teaching assistants. Thus, it is likely that high quiz and project grades correlated with each other. More analysis is needed to investigate the reasons behind this high relation between quizzes and engineering skills.

Homework assignments were mostly related to the topics of data representation and MATLAB programming. Content-wise they were related to the projects. Homework assignments were the same across all sections. Undergraduate teaching assistants were responsible for grading the majority of homework. Homework grades were likely more variable than quiz grades but, as we have found in our other research, undergraduate teaching assistants do not always follow the grading rubrics and do a poor job of grading often resulting in inflated grades (see Hicks & Diefes-Dux, 2016). This, like quizzes, may have impacted the relation between the homework grades and the engineering skills.

Exams were graded with considerably more fidelity to rubrics than homework as there was more instructor participation in and oversight on grading. There was likely the most variation in the grades, and these grades were likely more representative of actual individual student performance. Because of this variation, exams may not have a significant relation with the engineering skills. Also, the grading of MEAs and projects focused on some different learning objectives than exams (i.e. design and communication). This is particularly evident in the lower correlations between MEA and exam grades (Table 4). The project, with its greater emphasis on MATLAB programming, had a better correlation to exams.

At first blush, is appears that the lack of relation between exams and engineering skills (i.e. project) confirms that not all the assessments are measuring the same skills and knowledge in students. However, there are a number of limitations that make drawing conclusions about the relationships between the learning activities (quizzes, homework, and exams) and acquiring engineering skills (projects) that make drawing this conclusion difficult. These limitations include instructor variation in quiz design, potential issues with grading reliability, and differences in grading strategy across course elements (e.g. points for completion or correctness on quizzes, homework, and exams versus holistic quality assessment of projects), assignment of individual grades (quizzes, homework, exams) versus team grades (projects). Similar limitations exist in any course where a traditional score-based assessment strategies, and a mix of other strategies, are applied across course elements. These limitations could be minimized by adopting standards-based grading a strategy wherein all assessments explicit and consistently assess student performance on clearly stated learning objectives (e.g., Carberry, Siniawski, Atwood & Diefes-Dux, 2016). Such a grading strategy would enable a more meaningful discussion of relationships, or lack of relationships, between grades on different assessments. Such a grading strategy would also make the results of grading and any relationships actionable by instructors as it would be clear how students are performing on specific learning objectives within and across assessments. Actions that would be enabled might include re-design of activities or assignments, re-envisioning instruction on a particular topic, or re-alignment of instruction, activities, and assessments.

References

ABET. (2014). Accreditation Criteria and Supporting Documents. from http://www.abet.org/accreditation-criteria-policiesdocuments/

Barron, B. J., Schwartz, D. L., Vye, N. J., Moore, A., Petrosino, A., Zech, L., & Bransford, J. D. (1998). Doing with understanding: Lessons from research on problem-and project-based learning. *Journal of the Learning Sciences, 7*(3-4), 271-311.

Blumenfeld, P. C., Soloway, E., Marx, R. W., Krajcik, J. S., Guzdial, M., & Palincsar, A. (1991). Motivating project-based learning: Sustaining the doing, supporting the learning. *Educational Psychologist, 26*(3-4), 369-398.

Braunstein, A. W., Lesser, M. H., & Pescatrice, D. R. (2008). The impact of a program for the disadvantaged on student retention. *College Student Journal, 42*(1), 36-40.

Burnham, K, P., and D. R. Anderson (1998). *Model selection and inference: A practical information-theoretic approach.* New York: Springer-Verlag.

Carberry, A., Siniawski, M., Atwood, S., & Diefes-Dux, H. A. (2016). Best practices for using standards-based grading in engineering courses. *Proceedings of the 123rd ASEE Annual Conference and Exposition, New Orleans, LA.*

CCSSM. (2010). Common core state standards for mathematics. from http://www.corestandards.org/assets/CCSSI_Math Standards.pdf

Hicks, N. M., & Diefes-Dux, H. A. (2017). Grader consistency in using standards-based rubrics. *Proceedings of the 124th ASEE Annual Conference and Exposition, Columbus, OH.*

Hu, L. t., & Bentler, P. M. (1999). Cutoff criteria for fit indexes in covariance structure analysis: Conventional criteria versus new alternatives. *Structural Equation Modeling: A Multidisciplinary Journal, 6*(1), 1-55.

Hurtado, S., Eagan, K., Pryor, J. H., Whang, H., & Tran, S. (2012). *Undergraduate Teaching Faculty: The 2010–2011 HERI Faculty Survey.* Los Angeles, CA: Higher Education Research Institute, UCLA.

Leake, M., & Lesik, S. A. (2007). Do remedial english programs impact first-year success in college? An illustration of the regression-discontinuity design. *International Journal of Research & Method in Education, 30*(1), 89-99.

Miller, J. W., Janz, J. C., & Chen, C. (2007). The retention impact of a first-year seminar on students with varying pre-college academic performance. *Journal of The First-Year Experience & Students in Transition, 19*(1), 47-62.

NAE. (2004). *The engineering of 2020: Visions of engineering in the new century.* Washington, DC: The National Academic Press.

NCATE. (2008). Professional standards for the accreditation of teacher preparation institutions. from http://www.ncate.org/LinkClick.aspx?fileticket=nX43fwKc4Ak%3D&tabid=669

Purdie, J. R., II, & Rosser, V. J. (2011). Examining the academic performance and retention of yirst-ear students in living-learning communities and first-year experience courses. *College Student Affairs Journal, 29*(2), 95-112.

Starke, M. C., Harth, M., & Sirianni, F. (2001). Retention, bonding, and academic achievement: Success of a first-year seminar. *Journal of the First-Year Experience & Students in Transition, 13*(2), 7-35.

Steiger, J. H., & Lind, J. C. (1980, May). Statistically based tests for the number of common factors. *In annual meeting of the Psychometric Society*, Iowa City, IA (Vol. 758, pp. 424-453).

Thomas, J. W. (2000). A review of research on project-based learning. from http://www.bobpearlman.org/BestPractices/PBL_Research.pdf

Vinson, T. L. (2008). *The relationship that summer bridge and non-summer bridge participation, demographics, and high school academic performance have on first-year college students: effects of grade point average and retention.* (Doctoral Dissertation), Grambling State University. Retrieved from http://www.eric.ed.gov/ERICWebPortal/detail?accno=ED527512

Copyright statement

AN EDUCATIONAL PROPOSAL FROM THE IDENTIFICATION OF CONSTRUCTIVE BAD PRACTICES IN INFORMAL HOUSING (CASE VILLAS DE KENNEDY)

Camilo Alberto Torres Parra
Universidad Católica de Colombia, Bogotá, Colombia
ctorresaranda@gmail.com

Janneth Arias Hernández
Corporación Universitaria Minuto de Dios, Bogotá, Colombia
jnarias@uniminuto.edu

Abstract: *This paper presents a constructivist and participatory educational proposal led by the Safe and Sustainable Build Cooperation Network to which the Corporación Universitaria minuto de Dios and the Distrital, Católica and Piloto universities belong to from the civil engineering programs; During the process the network identified 45 malpractices linked to habitability in informal housing and translated into instructional objectives through learning units in a course model to communities. This initiative made it possible to bring 25 families from the Kennedy Township, Kennedy Villas neighborhood to good constructive practices in habitability in informal housing, and allowed students and related teachers to socialize their research findings in a vulnerable social context.*

Context

Habitability at the general level can be understood as the capacity of spaces built to meet the objective and subjective needs of individuals and groups, that is to say, the psychosocial relationships established in the housing unit must be taken into account, which allow The healthy physical, biological, psychological and social development of the person, hence their close relationship with the concept of quality of life (Zulaica, Celemín, 2008).

One of the most important social problems of the last decades is the enormous increase in the world's precarious habitability. One-third of the world, 2.240 million people, lack basic shelter needs and 925 million live in environments that are harmful to their health. The United Nations estimates that more than a third of the world's population (some 2 billion people) live in precarious urban settlements, so that communities in these sectors live in vulnerable conditions, staying in temporary housing, with waste materials and located on the outskirts of the capitals (United Nations Human Settlements Program, 2012).

Likewise, in Colombia, given the conditions of inequality in relation to income, education, occupation and access to housing, informal housing alternatives emerged, especially land invasions and urbanizations known as pirates, which allowed families Low income access to a partial housing solution, initially in very precarious conditions, but motivated by an urgency to have a roof that would provide protection and well-being to their families. That is why approximately 3,828,055 households are in deficit at the level National: 1,307,757 have a quantitative deficit of housing, while 2,520,298 have a qualitative deficit (Rueda, 2011).

Bogotá is no stranger to this panorama and within its 20 localities, 13 present a panorama of vulnerability in habitability related to the high prices of formal housing, self-construction and unskilled labor to build homes, poor materials, land Insecure, disorderly occupation of the territory and ignorance of the risks to which it is exposed, due to the subjective perception of the inhabitant, where the conformism is generalized since no other environment is known, and its economic, educational and health limitations, place them at a high risk level in terms of urban habitability (Escallón, 2012).

The study proposed in this document emerges through SWISSCONTACT in its Colombian chapter, where through its professionals and cooperation networks the project "Promotion of Practices of Sustainable Construction in Vulnerable Urban Sectors" has been executed. The aim of this project is to improve the living conditions of the population in the main vulnerable urban areas of Colombia and to reduce poor housing practices through the training of the workforce. The main goal of this foundation is to develop curricula that involve sustainable, safe and healthy construction practices and, from them, to undertake training to improve the skills of construction workers, construction auxiliaries, hardware owners and the owner population in vulnerable areas.

Due to the above, Kennedy Locality was used as the center of action to carry out the educational proposal, which presents various problems regarding habitability. It has a high housing deficit (11,493 units), is affected by the presence of landfills by 13.7%, and tenants and quarters are in the range of 0.5% to 1.2% in relation to apartments (62.4%) and houses (36%). Housing units in 35% do not have adequate ventilation, their materials are in poor condition, structural safety is low and habitability conditions are precarious for their inhabitants (Universidad Javeriana, 2012).

Research questions

Taking into account the housing situation and informality in the dwelling present in the study population, the following research questions emerged: ¿How to design a participatory free course for communities living in informal housing that have poor constructive practices in habitability? ¿How to transmit the knowledge gathered in the field with students of civil engineering in relation to the bad constructive practices of habitability? And ¿What elements must have a constructivist proposal around good construction practices, which contributes to the improvement of habitability conditions in the community housing?. In order to respond to these questions, it was necessary to approach this context from the educational aspect, since it allows people to interpret their experience according to their own knowledge, since any change in the cognitive organization is a personal construction based on the learning experiences And the present reality of each participant, in which people put their capacities into play and expand them (Cubero, 2005).

In this sense, the consolidation of a socially responsible educational proposal from engineering must be seen as the capacity of the individual and the academic community to recognize their reality and that of others, questioning and analyzing the cause-effect relationship and interests It underlies, and from there, to build a position in front of it. Precisely an education for human development is one that maintains a direct relationship with research and social projection, seeking that the student is linked in processes of social transformation that benefit and foster the well-being of populations or groups with some degree of vulnerability, Whether due to their social, economic, age, cultural or political status (Morán, 2010).

Also, the importance of this type of space for engineering education must be understood, because within the training proposal the pedagogical practices are directed to the students to recognize the reality of their environment and build strategies of action through the which participate and influence the positive transformation of the environment, having the capacity to identify the problems that compromise the quality of life of disadvantaged populations and placing at the service of civil society and knowledge the training acquired in the classroom, demonstrating with this pedagogical exercise the importance of vulnerable contexts for engineering and vice versa, and would be contributing to the formation of students who are ethical, responsible and sensitive to the problems of society.

Theoretical framework

Informal housing is one that is not framed in technical regulations in relation to its materials, space distribution, comfort, structural security, location and security of tenure, among others; This type of housing unit, is characterized by its construction by processes of self-construction and empirical labor, affecting this panorama even the so-called developed countries, as it is a scourge that impacts society when it lacks political and social mechanisms to Meet their minimum habitability needs (Programa de las Naciones Unidas para los Asentamientos Humanos, 2012).

In relation to the above, informal housing is directly linked to habitability at the urban level, so it is so important that construction practices are aimed at providing satisfactors related to accessibility, mobility, continuity, location, public space, uniform provision of infrastructures, services, urban furniture, green spaces, equipment, elements of symbolic cultural heritage and the actual availability of public transport, among others. It is also related to housing and its quality in construction, which is why it must be articulated with compliance with minimum standards specified through regulations and technical assistance. (Espinoza, Gómez, 2010).

Consequently, constructive good practices in habitability seek to reduce informality in housing, understanding that the capacity of spaces built to meet the objective and subjective needs of individuals and groups should take into account technical quality standards that promote psychosocial relationships. In this way, the quality of life is defined as the quality of the life of the person, in relation of the physical, biological, psychological and social development of the individual (Zulaica, Celemín, 2008).

So to answer the questions proposed in the study, a response was found in education at the constructivist level, since these types of pedagogical orientations are mostly characterized by understanding that knowledge is built through participation in specific learning communities, within frameworks of values and specific social groups, so social constructivists insist that the creation of knowledge is a shared, social, rather than individual experience (Cubero, 2005). Therefore, the interaction between organism and environment makes possible the emergence of new characters and traits, which implies a reciprocal and complex relationship between the individual and the context; It is by this learning from reality, that populations can link their conditions to theoretical knowledge, but without ignoring their tacit and cultural knowledge (Araya, Alfaro, and Andonegui, 2007).

Therefore, a systematic process was developed to translate the principles of learning and instruction into plans for materials and instructional activities, and this pedagogical design is essential in the educational field, since it becomes the path or guide that every educator should draw when pretending to direct a course, regardless of the modality of this one (Martínez, 2009). In this way, this process involves the development of materials and pedagogic activities, and then the tests and evaluations of the student's activities. Also, the instructional design guides the planning of different moments and concrete educational actions, both in the preparation of materials and in the planning of courses (Méndez, 2010). Besides, the instructional design consists of 5 general phases, which are framed in: 1. Analysis, 2. Design, 3. Development, 4. Implementation and implementation and 5. Evaluation (Lozano, 2007).

Methodology

Regarding the methodology to identify poor constructive practices in habitability in the field and then to carry out an instructional design framed in constructivist education, it was essential to observe in the first instance the bad construction practices that were linked to the habitability in a house. The team proposed to develop a quantitative research approach, since it was necessary to statistically measure the present reality of the population and its direct relation with the living conditions in which they are living. This served to establish the instructional problems and working modules of the proposed course.

The research was also addressed from a descriptive point of view, since phenomena, situations, contexts and events were described, because these studies seek to specify the properties, characteristics and profiles of individuals, groups, communities, processes, objects or any another phenomenon that is submitted to an analysis (Hernández, Fernández and Baptista, 2010, p.102). In this case, the Safe and Sustainable Build Cooperation Network to which the Coorporación Universitaria Minuto de Dios belongs and the Distrital, Católica and Piloto universities from their civil engineering programs, identified 45 bad practices related to habitability in informal housing.

Also the non-experimental design was the approach used in the study, since the research did not manipulate the research variable because only the phenomenon was observed as it was presented in its natural context and later analyzed. For this case, the constructive conditions related to habitability in informal housing were evaluated, in which 25 families participated. Finally, the proposed research was also framed in transectional or transverse research designs, which are characterized by the collection of data in a single moment, in a single time. Its purpose is to describe variables and analyze their incidence and interrelation at a given moment (Hernández, Fernández and Baptista, 2010).

The study was carried out in Bogotá, Colombia, in Kennedy Township, Kennedy Villas neighborhood. This epicenter of action was chosen because there was an approach and previous work by the Network of Cooperation with 25 families in the sector, which recognized the work of the group of professionals in housing issues and also characterized by being leaders who work in order to improve their quality of life and that of their community.

Based on the above, a non-probabilistic sampling was performed, where the choice of the elements does not depend on the probability, but on the causes related to the characteristics of the research and the interest of the researcher (Hernández, Fernández and Baptista, 2010), which is why in these 25 families was found the element of representativeness of the population needed for the proposed research, since this small portion of the community met the necessary characteristics that surround informal housing with low constructive development in habitability, high socioeconomic vulnerability due to their stratification and the initiative of wanting to be trained to improve their living conditions.

With the selected sample, a tool was presented with closed questions, which related categories or response options that were previously delimited, that is to say participants were presented with the possibilities of response, which were limited to these. In addition, teachers and students belonging to the Cooperation Network ensured that the participants to whom the instrument was administered knew and understood the categories of response. Whenever a statistical analysis of the information obtained is sought, it is necessary to codify the participants' answers to the questionnaire questions, for that reason it was necessary to assign them symbols or numerical values (Hernández, Fernández and Baptista, 2010).

As a consequence, the application of a tool such as the survey was proposed for the study, which is used mainly to know the opinion of the people on a variety of subjects, in addition it is the typical modality of collection of the quantitative investigation and the form which is the most common way of obtaining empirical data from quantitative studies, which is why this instrument was used to inquire about the bad housing construction practices present in informal housing (Sylvain, Ginette, 2008).

After completing the process described above, the nine steps were taken according to Lozano (2007) to design a course, where methodological rigor was used to design the educational proposal framed in the 45 constructive bad practices in habitability. This was designed with students and teachers of the Network following the steps listed below:

- Step 1. Identify the instructional problems and specify the objectives for which the instructional program is designed.
- Step 2: Examine those characteristics of the student that should receive attention during the planning.
- Step 3. Clearly identify the contents and analyze the components related to the purposes and objectives of the instruction.
- Step 4. Develop a list of instructional objectives that should be communicated to the student.
- Step 5. Sequencing the contents in didactic units and that respond to a logical learning.
- Step 6. Design didactic strategies that allow the student to achieve the objectives that are proposed.
- Step 7. Plan the didactic strategies.
- Step 8. Design assessment tools to determine if the instructional goals of the student were met.
- Step 9. Select the resources that support the instruction and the proposed didactic activities.

After having the information to transmit to the community framed in a course proposal, the methodology of participatory social formation was applied, which includes four stages, dynamic, flexible and pertinent to the professional and social needs of the participants in this type of socioeconomic scenarios, in addition to bringing the civil engineering student to community work.

Stage 1: Recognition and analysis of the problem scenario. This first stage invites the students, teachers and communities involved, to approach the history and evolution of the scenario on which it will intervene. This recognition of the problem scenario is based on participatory techniques such as the realization of social cartography exercises, focus groups, and visual tools such as the use of graphs, symbols and statistics, among others.

Stage 2: Identification and empowerment of skills and resources. This second stage identifies the skills and potential of people, ie seeks to articulate the knowledge of experiences and scientific and popular knowledge in a way that generate different solutions to the previously identified problem situations, generating leadership profiles and organizing teams with responsibilities and roles.

Stage 3: Putting ideas into practice. The objective of this third stage is to make work hypotheses a reality, this part is crucial in the development of the methodology and in itself of a project, for which the support and the monitoring in the execution are fundamental. Likewise, resources, time and tasks, must generate tools for management and evaluation, prototype development and testing.

Stage 4: Social transformation of the environment. In this last stage it is questioned if the aims and objectives of the course were achieved and if these were agreed, accepted and satisfactory by the participants.

Findings / Conclusions

After the identification of poor constructive practices in the field (Table 1), the information obtained was analyzed, and thus the degree of compliance of these in the community with the objective of obtaining firsthand the instructional problems that they needed special attention during the course proposed.

Table 1. Poor housing practices in informal housing

Poorly install window and door frames.	Do not build devices to take advantage of natural light	Do not build railings on stairs.	Do not install hydraulic points
Do not construct frames in the spaces of windows or doors.	Inadequate location of luminaires.	Do not use glue mortar on vertical joints	Use smaller diameters to evacuate rainwater.
Do not build channels to divert water drops on slab eaves and window frames	Use of luminaires not suitable for space / function.	Separate the top tie beam cover.	Use smaller diameters for evacuation of sewage from the sanitary appliance.
Do not build filters for rainwater management in land containment systems.	Use of glossy paints or coatings inside the house.	Do not build walls to the side or behind the house to separate buildings	Do not veneer or use non-porous materials for hygiene in kitchens and bathrooms.
Do not waterproof the elements of foundations and slabs.	Do not build fixed furniture for storage, in bedrooms, kitchen and bathroom.	Have electrical leads uncovered and in poor condition..	Use rustic or sliding materials on stairs.
Do not waterproof deck slabs	Do not consider basic principles of the functionality of a house.	Install insufficient circuits or bad distribution of electric charge.	Construct light covers with different materials.
Do not install flanges on walls or against upper, lower or lateral walls.	Do not build yard / Do not install ventilation in sanitary networks.	Do not install electrical boxes and conduits for electrical conductors.	Construct lightweight roofs with poor slope and overlap.
Use rainwater collection and management channels with little slope and poor connections	Poorly anchor the stair rails.	Inspection boxes without cover.	Construct poorly leveled or sloped roof slabs.
Do not separate wet areas with hydraulic shut off valves or valves.	Build high divisions in the showers or do not build them.	Connect sewage network to storm water drainage	Making poor shackles of tiles to the support elements.
Do not install siphons at points or sanitary outlets.	Build railings with a lot of distance between elements.	Mix rainwater and sewage in the internal networks.	Do not leave windows or ventilation ducts.
Build edges and vertices dangerous (tips) in bathrooms, kitchens and fixed furniture.	Build stairs with steps (tread and riser) of different dimensions.	Do not build technical pipelines for installations.	Do not install ventilation elements for the evacuation of gases from the burning of fuels.

The previous findings were transmitted to the community through the learning units provided in the constructivist and participatory educational proposal (Table 2). From this model course is intended on the one hand to respond to questions generated in research, and secondly, to the problems encountered in collected during the data collection community, providing populations with precarious habitability an educational space that will explain the importance of good construction practices for health in housing spaces. (Constructivism-contribution in engineering).

Table 2. Course model focused on mitigating constructive bad practices in habitability

LEARNING UNITS	CONCEPTS	GENERAL OBJECTIVE	DURATION
Comfort in the house	Hyrothermicity	To determine the importance of ventilation, lighting, humidity and comfort in the home, linked to design themes in spaces, rainwater management, windows, skylights, bathrooms and kitchens	3 WEEKS
	illumination		
Functionality of housing	Enough paces	Establish the importance of the functionality of the house from the study of electrical installations, the yard, stairs, bedrooms and sewage network	2 WEEKS
	Separate and secure spaces		
Social responsibility and housing ethics	Housing as a vital and social space.	Raise awareness of the social impacts generated by various informal construction processes and how they alter the dynamics of families and communities.	2 WEEKS
Finished housing	Floor, wall and ceiling management	Identify the materials to build the house from the aspects of quality and construction process	2 WEEKS

This experience also contributed to civil engineering students clear working space in which it required commitment and communication skills and techniques to meet the urgent needs of populations living in precarious living conditions. This experience also humanized the engineering student's role in helping him to understand the importance of his professional practice in social contexts, working alongside the communities, without imposing or promoting welfare processes that will limit the transmission of knowledge and the sustainability of this type of processes. This exercise also helped students to apply both their technical and social skills and achieve personal and professional achievements in the framework of bilateral community work.

With this, it was possible for attendees and the academic community to work together and become aware of the importance of education with a social meaning and its close relationship with housing, health and rights; also with this space was able to transmit knowledge and build a bridge of dialogue between theory and practice through each of the learning units, which allowed to socialize the research advanced by the Knowledge Network.

On the other hand, it was possible for the participants to recognize their individual and collective capacities of work and the importance of acquiring previous knowledge in the framework of housing and habitability to improve their personal and community situation. In this sense, it was important to inform and educate the populations about the risks and the social responsibility that implies building empirically, and without any prior knowledge. Only in this way could the community be persuaded to worry about their habitability situation and the improvement of their quality of life, since an educational scenario was created that was pertinent to their needs and abilities.

In this order of results, it is also worth mentioning the various significant learning that the engineering students who participated in the development of the proposal were able to develop (Table 3).

Table 3. Educational learning acquired

SPECIFIC LEARNING	INTERSOCIAL LEARNING	LEARNING INSTRUMENTS
- Situate the problems in the socio-technical and operational contexts. - Identify holistic solutions to various situational schemes for the integral development of individuals. - Raise and solve problems using available resources. - Link the different knowledges and abilities of the different actors with the processes and activities of the communities, respecting the conditions of the environment.	- Critical capacity. - Ability to work collaboratively. - Ability to think creatively and interdisciplinary. - Ability to communicate and lead. - Ability to humanize in the face of diversity and multiculturalism.	- Ability to observe and analyze social problems around housing. - Dexterity to observe and analyze reality, get involved in it and contribute from human and professional experience. - Acquisition of sufficient tools, pedagogical, theoretical and methodological, to interact adequately with the environments in which it is. - Disposition to get involved in the management of training projects from the knowledge itself contributing to the human and social development of vulnerable communities. - Ability to learn focused on experimentation and on the construction and deconstruction of knowledge.

The previous significant learning emerged within the framework of various pedagogical strategies such as: discussion tables, dialogue circles and workshops on social awareness among others. Such strategies of teaching and learning served for both engineering students as people project participants recognized that engineering in a dynamic, responsible profession, and high social commitment it is able to provide answers to the many problems presents in society; Among them the quality of housing and for that reason one of its challenges is to be in continuous renovation.

Finally, it was evidenced by an evaluation that the advance exercise between the community and the students of engineering, had a positive result because the assistants passed from a basic knowledge to a more elaborate one when understanding the bad constructive practices and their relation with quality of life. The community cooperated with civil engineering students, and the students realized that their research had a positive effect on the community under study.

Acknowledge

Given the results and learning obtained during the implementation of the project it is recommended to follow the training methodologies named in this document when considering to approach from the educational and participatory aspect to communities living in homes and environments in which habitability is seriously compromised by the bad constructive practices and the ignorance of the risks and impacts to the physical, mental and social health that it takes to inhabit in dwellings whose degree of informal development generates a detriment in the quality of life of its residents.

In the context of engineering education, it is important that academies, institutions or other social initiatives that are concerned with improving the well-being and quality of life of people from the different areas of knowledge, are concerned with finding and providing students and participants of the projects innovative ways to transmit knowledge, so that these can be translated into a concrete contribution that helping to improving the environments and conditions in which people live and participate. This will promote a permanent attitude of "learning to learn" and application of knowledge to interpret and transform reality.

References

Araya, V. Alfaro, M. y Andonegui, M. (2007). Constructivismo: orígenes y perspectivas, *Revista de educación Laurus, 13*(24), 76-92. Recuperado de http://redalyc.uaemex.mx/pdf/761/76111485004.pdf

Cubero, R. (2005). Elementos básicos para un constructivismo social, *Avances en psicología latinoamericana, 23*, 43-61. Recuperado de http://redalyc.uaemex.mx/src/inicio/ArtPdfRed.jsp?iCve=79902305

Escallón, C. (2012). La vivienda de interés social en Colombia, principios y retos, Revista de Ingeniería Universidad de los Andes (35), 55-60

Espinoza, A. Gómez, G. (2010). Hacia una concepción socio-física de la habitabilidad: espacialidad, sustentabilidad y sociedad, Palapa 5(10), 59-69. Recuperado de http://www.redalyc.org/pdf/948/94820714006.pdf

Méndez, J. (2010, agosto*). http://www.schoolofed.nova.edu*. Recuperado el 11 de marzo de 2016 de http://www.schoolofed.nova.edu/dll/spanish/modulos/diseno/jorge_mendez.pdf

Hernández, R. Fernández, C. Baptista, P. (2006). Metodología de la investigación (5ª ed.). Iztapalapa, México: McGraw Hill Interamericana.

Lozano, A. (2007). Actualidades del diseño instruccional. En Universidad Virtual del Tecnológico de Monterrey, Desarrollo del Nuevo Perfil Docente (pp. 1-19). Moterrey, México.

Morán, A. (2010). Un modelo de formación ciudadana. Soporte de procesos de transformación social. Revista Prospectiva No. 15. Universidad del Valle. Cali – Colombia

Pontificia Universidad Javeriana (Instituto javeriano de vivienda y urbanismo, facultad de arquitectura y diseño e Instituto de promoción de la salud, facultad de medicina). (2012). Identificación de las principales amenazas para la salud física y mental de las poblaciones que habitan en sectores urbanos de origen informal de Bogotá, Bogotá, Colombia.

Programa de las Naciones Unidas para los Asentamientos Humanos, ONU-hábitat Colombia (2012, agosto). Primer Foro Urbano Nacional Colombia: Camino al VI Foro Urbano Mundial, El futuro Urbano. Medellín, Colombia.

Rueda, N. (2011). La integración de la demanda informal a la política de vivienda: algunas opciones iniciales, Revista de Ingeniería Universidad de los Andes (35), 108-115

Sylvain, G.Ginette T.(2008).Metodología de las ciencias humanas. Cap. IV Métodos y técnicas de muestreo. Distrito Federal, México: Fondo del Cultura Económica.

Zulaica, L. Celemín, J. (2008). Análisis territorial de las condiciones de habitabilidad en el periurbano de la ciudad de Mar del Plata (Argentina), a partir de la construcción de un índice y de la aplicación de métodos de asociación espacial, Revista de Geografía Norte Grande (41), 129-146. Recuperado de http://www.scielo.cl/scielo.php?pid=S0718-34022008000300007&script=sci_arttext

Copyright statement

Relationship between *stricto sensu* graduate program structure and patent registrations: A study of Santa Catarina higher education institutions

Caroline Dallacorte; Claudio Alcides Jacoski; Marcelo Costella; Adriana Freitag Migott; Cesar Augusto Seidler; Rafael Medeiros; Fábio Correa Gasparetto

Pos Graduate Program in Technology and Management of Innovation - Chapecó Regional Community University, Chapecó, Brazil

carodl@unochapeco.edu.br; claudio@unochapeco.edu.br; costella@unocahpeco.edu.br, adriana.migott@unochapeco.edu.br, caseidler@unochapeco.edu.br, rafaelmedeiros@unochapeco.edu.br, fabio.gasparetto@unochapeco.edu.br

***Abstract**: This study uses an analysis of patent registrations to evaluate the level of innovative initiatives of Santa Catarina higher education institutions. It also seeks to understand the importance of the structure of graduate programs to success in the advancement of technological production and industrial property registrations. The number of registrations in the National Institute of Industrial Property (Instituto Nacional de Propriedade Industrial - INPI) and United States Patent and Trademark Office (USPTO) databases was established. Given that the patent filing process represents an innovative initiative, the results show that the level of innovation is growing in these institutions.*

Introduction

The definition of a patent is the encoding of a set of information with a degree of innovation. It is considered an advantage offered by the state that rewards an inventor for his effort. Patent registrations presume the description of something new, in which the inventor must describe the content of the invention in detail. This makes patent databases excellent information sources (AMADEI; TORKOMIAN, 2009).

For a long period of time, patents were filed only by companies and contributed to technological advancement through internal research that was oriented only toward each company's interests and needs. Currently, however, universities are playing a key role in regard to technological growth through innovative initiatives. The importance of linking academic research, which is endowed with valuable resources, to industry, with the two working together toward the development of their region, is well known. According to Amadei and Torkomian (2009), universities have made a marked contribution to the growth in innovation in today's developed countries. In the United States, universities are considered strategic elements in the country's economic development and technological advancement.

Therefore, it is important to evaluate the level of higher education institution (HEI) innovative initiatives by analyzing patent registrations and to observe the relationship between such registrations and the structure of graduate programs. For the purposes of this study, the HEIs that are based in Santa Catarina were selected. The number of patent registrations requested by each university was sought from the national database of the National Institute of Industrial Property (Instituto Nacional de Propriedade Industrial - INPI) and international database of the *United States Patent and Trademark Office* (USPTO).

Industrial property registration and patent databases

The INPI defines a patent as a title to an invention or useful model, which has temporary property and is granted by the state to individuals or legal entities, inventors or authors who have rights over its creation (INPI, 2015).

According to Moraes (2014), patent standardization in Brazil began to occur in the eighteenth and nineteenth centuries. The first Brazilian patent was granted in 1822, attributed to Luiz Vestir and Simão Louvain, who created a horse-drawn machine to peel coffee without breaking it.

The first record of a patent being filed by a university occurred in 1979. This registration was by the Federal University of Rio de Janeiro (Universidade Federal do Rio de Janeiro - UFRJS) and was titled "improved process to reduce the molecular weight of elastomers". The letters patent for this filing were obtained in 1985 (PÓVOA, 2006).

A patent is granted by the national public patent administration or regional public patent administration of a group of countries. The rights that are granted by a patent are territorial, with the grant being limited by the borders of the country or region (INSTITUTO NACIONAL DE PROPRIEDADE INDUSTRIAL, 2015). Each country has in its territory an office that is responsible for maintaining national patent filings.

In terms of the relationship between countries, in 1970, the Treaty of Cooperation in Patent Material was signed; it is intended to act as a technology transfer system between countries.

Created in 1970, the INPI is a federal agency associated with the Brazilian Ministry of Development, Industry and Foreign Trade (Ministério do Desenvolvimento, Indústria e Comércio Exterior - MDIC). Its responsibility is to perfect, disseminate and control the management of the Brazilian concession system and to guarantee intellectual property rights for industry, allowing access to patent documents filed in Brazil. The INPI offers information on the progress of each registered case via a search of its database. The database is updated weekly according to information published in the Journal of Industrial Property (Revista da Propriedade Industrial) (INSTITUTO NACIONAL DE PROPRIEDADE INDUSTRIAL, 2015).

The *United States Patent and Trademark Office* (USPTO) is the federal agency that is responsible for granting patents and trademark registrations in the US to promote science and the arts, ensuring authors and inventors exclusive rights to their discoveries for a limited period of time (USPTO, 2015). Searching the patent database system provides an opportunity to access numerous records (since 1791).

Universities and patent registers

According to Guerzoni et al. (2014), scientific discoveries originating from universities can contribute to the emergence of new industries. Naturally, not all research leads to radical changes in the development of existing technology, but the generation of patents is an essential factor in this growth in new technologies. These authors also note that research funded by universities tends to generate more original patents than research funded by industry or non-university institutions.

In the context of innovation, Etzkowitz and Leydesdorff (2001) have coined the term "triple helix", which is an organizational model based on co-participatory interaction between a university, the government and business, in which the university evolves from an educational institute into an educational and research institute, even involving service provision; the government acts as a beneficiary from this alliance, without requiring performance in specific sectors; and the company presents a broader and more sustainable concept, changing the perception of profitability.

This transformation enables a better planning and implementation of innovative action, and it promotes regional development by inserting academic actors into the labor market and assuming the entrepreneurial role of technology incubator through the university. This dynamic model guarantees the interaction of these three societal agents and promotes scientific and technological innovation in a more intense manner (ETZKOWITZ, 2009).

The *Center for World University Rankings* (CWUR, 2014) has published a global ranking of 1,000 universities, considering the quality of education and training of its students as well as the reputation of its faculty and the level of research activity. The classification data include the number of patents registered by the university. This variable has a specific ranking, with *Massachusetts Institute of Technology (MIT)* ranking first, followed by *Harvard University* and *Johns Hopkins University*. The only Santa Catarina university listed in this ranking is UFSC, which ranks 918th. Of the 1,000 institutions presented, 18 are Brazilian, and USP is ranked the highest, at 132nd.

The ranking of the 100 universities with the greatest number of utility patents (US terminology) in 2013 is published by the *National Academy of Inventors* (NAI) and the *Intellectual Property Owners Association* (IPO). The universities in the top three positions are the *University of California* in first, followed by *MIT* and *Tsinghua University*. According to Noonan (2014), American universities are predominant in this ranking (62%), followed by China as the second most represented country (18%).

The Folha University Ranking (Ranking Universitário Folha - RUF) is an annual evaluation of Brazilian higher education institutions performed by the newspaper Folha de São Paulo. It is based on five indicators: research, internationalization, innovation, education and market (FOLHA DE S. PAULO, 2015). A total of 192 Brazilian public and private institutions, are classified in the RUF ranking. In the most recent general ranking, published in 2014, UFSC is the highest ranking Santa Catarina university (7th place), followed by FURB in 53rd, UNIVALI in 58th, UNESC in 65th, and UNISUL and UNIVILLE, which jointly rank 95th.

The changing role of higher education institutions

Currently, HEIs find themselves in an energetic and changing environment that requires constant reorganization, subjecting them to changes in their research lines and thus requiring them to find methods of strengthening the construction of academic skills. This highlights the importance of stimuli to innovation within institutional organizations, given that they enhance the generation and transmission of knowledge and new technologies and stimulate the institution's social role in regard to territorial development .

For Tartaruga (2010), universities are favorable environments for the permanent development of innovation because they facilitate the creation of cooperative networks by interacting with companies and other institutions, contributing to the development of innovation in the region in which they are located. In this context, it is appropriate to highlight the case of Silicon Valley, a cluster of companies located in a region of California in the United States. It is considered to be the land of opportunity in regard to innovation. Normand (2014) emphasizes the fundamental role of Stanford University as a driving force in the development of the Valley because, since 1930, more than 40,000 companies have been founded with the university's help and on account of the entrepreneurial profile of its students. This institutional model is defined as the entrepreneurial university.

Entrepreneurial universities are institutions that aim to generate knowledge to contribute to their region's economic and social development. In this sense, universities seek to act as a model dedicated not only to the transmission of knowledge but also to the incorporation of research to serve as a service provider to the nation.

Currently, this form of activity is growing in institutions, which are increasingly acting as technology incubators and transforming from a purely teaching profile into educational, research and outreach institutions. Therefore, clearly, the adoption of an entrepreneurial university model offers considerable help to regional economic and social development, providing students with an entrepreneurial profile and boosting local innovation.

Santa Catarina State Institutions

Figure 1 shows the public, community and private universities and technological research institutes recognized by the Ministry of Education (Ministério da Educação - MEC) that are incorporated in the state of Santa Catarina.

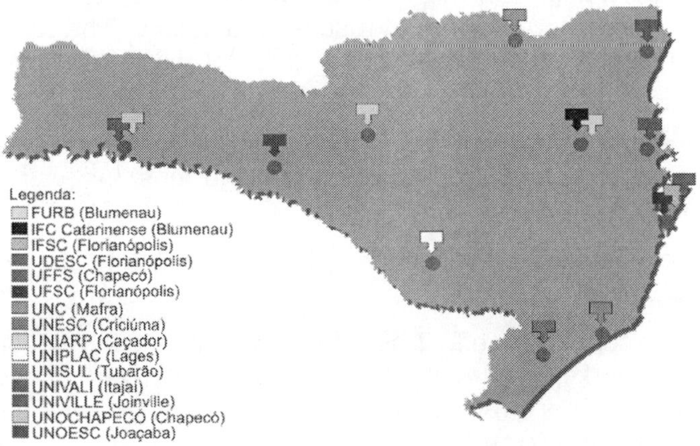

Figure 1 - Santa Catarina Universities recognized by MEC

This research began by defining criteria for the institutions to be evaluated. It was decided that the search should be restricted to Santa Catarina HEIs that are accredited by the MEC and included in the following administrative categories: public municipal, federal or state and community, named universities and technological institutes.

In this case, 15 institutions were identified in the MEC information database.

Table 1 - SC higher education institutions according to the established criteria

Institution (HEI)	Organization	Category
Santa Catarina State University - UDESC	University	Public
Santa Catarina Federal Institute for Education(IF Cat)	Federal Institute	Public
Santa Catarina Federal Institute for Education, - IFSC	Federal Institute	Public
Alto Vale do Rio do Peixe University - UNIARP	University	Community
Chapecó Regional Community University - UNOCHAPECÓ	University	Community
University of Joinville Region - UNIVILLE	University	Community
Contestado University - UNC	University	Community
University of Southernmost Santa Catarina - UNESC	University	Community
University of Western Santa Catarina - UNOESC	University	Community
University of Santa Catarina Plateau - UNIPLAC	University	Community
University of Southern Santa Catarina - UNISUL	University	Community
University of Vale do Itajaí - UNIVALI	University	Community
Federal University of the Southern Border - UFFS	University	Public
Federal University of Santa Catarina - UFSC	University	Public
Regional University of Blumenau - FURB	University	Public

Patent search procedure

The search for patents in the 15 accredited institutions then began. Two databases were used as references, one national – the INPI – and the other international – that of the USTPO. In both cases, an advanced search was performed: for the INPI, keywords were sought in the filer's name field for both patents filed and those already granted, whereas for the USPTO, *Patent Full-Text and Image Database* search links, were used.

The search identified the following 10 institutions with registered patents: UDESC, IF Catarinense, IFSC, Unochapecó, UNC, Unisul, Univali, FURB, UFFS and UFSC. The remaining institutions were therefore discarded from the results evaluation. The data collected for each patent were: the registration number, the date of filing, and the international classification.

To classify the HEIs according to the number of patent registrations, a ranking system was drafted, both overall and by database, to identify which institution had the most innovative initiatives, as determined by the volume of registered patents.

Based on the international classification obtained in the research, it was possible to identify the focus of innovation in Santa Catarina.

Comparative indices were prepared to ascertain the effect of patent registrations in relation to the institution's academic structure, seeking to relate the number of patents to the following factors: the number of master's and doctoral programs, the number of researchers and the volume of scientific, technical and artistic production.

Analysis and results

The analyses were performed according to the described methodological procedures. Each institution's number of registrations was ascertained; the results are shown in Table 2.

Table 2 - Number of patent registrations per institution

Acronym	USPTO	INPI	Total*
UDESC	0	5	5
IF Catarinense	0	1	1
IFSC	0	17	17
UNOCHAPECÓ	0	2	2
UNC	0	2	2
UNISUL	0	5	5
UNIVALI	1	10	10
UFFS	0	1	1
UFSC	9	139	140
FURB	0	12	12

The value refers to the total number of patents requested by the university, assuming that a same patent was sent to both databases.

Figure 2 shows each academic institution's participation as a percentage of the total patent registrations filed in the 1982-2015 period for the academic organizations studied.

It should be noted that UFSC ranks first, with 71.79% of all registrations in the period, exceeding the sum of all of the other academic institutions by more than 20%. It is followed by FURB, with 6.15%, and UNIVALI ranks third, with 5.13% of the total filings.

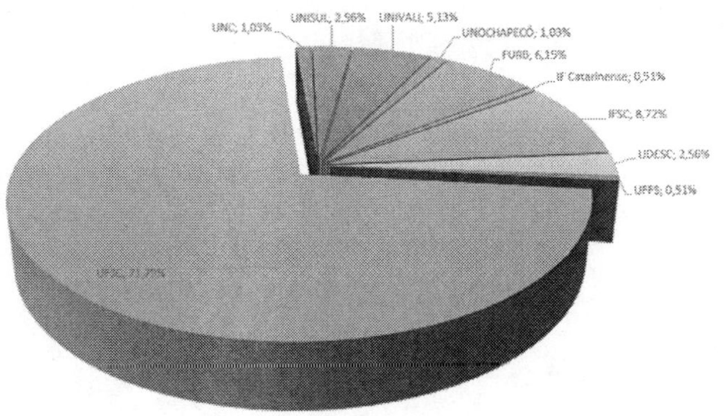

Figure 2 - Percentage of filings by academic institution (overall)

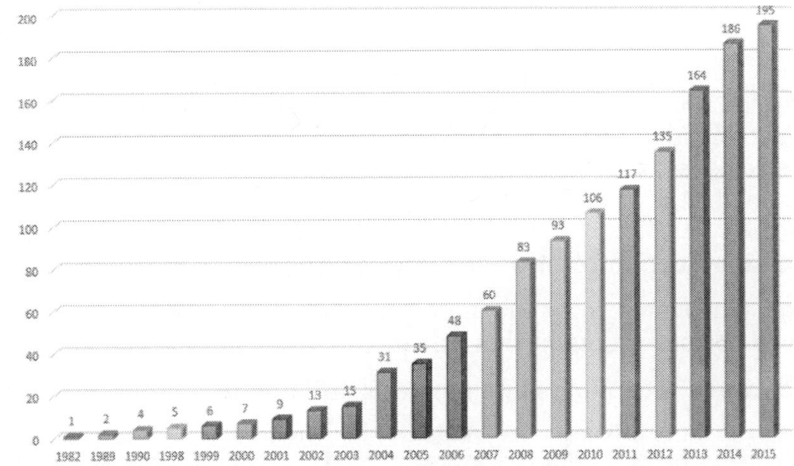

Figure 3 - Accumulated annual patent filings (1982-2015)

The analysis of the accumulated filings shown in Figure 3 reveals a significantly increasing trend, and it can also be noted that since 1982, when the first patent was filed, there has been a considerable annual increase. The most significant difference was observed in the year 2012 to 2013, during which filings increased from 144 to 175. UFSC was the institution that most contributed to this increase, being responsible for more than 70% of all filings.

The section are: (A) Human Needs; (B) Processing and Transport Operation; (C) Chemistry and Metallurgy; (D) Textiles and Paper; (E) Fixed Constructions; (F) Mechanical Engineering and others; (G) Physics; (H) Electricity.

The sections shown in Table 3 are represented by the initial letter of the international classification code. Not all patent filing registrations have an international classification code; thus, those that did not have it were categorized as "not classified". Some registrations have more than one international classification code and were therefore considered to belong to various sections.

Figure 4 reveals that most filings were "not classified" (35.90%), followed by section C, relating to chemistry and metallurgy (15.38%). Section G, relating to physics, ranked third, with 27 filings (13.84%), and had six more filings than section A, relating to human needs (10.77%), and 10 filings more than section B, relating to processing and transport operation (8.72%). Section F, relating to mechanical engineering, lighting, heating, weapons and explosion, recorded 15 filings (7.69%), followed by section H, relating to electricity (6.15%), section D, relating to textiles and paper (1.03%), and section E, relating to fixed construction, with one filing (0.51%).

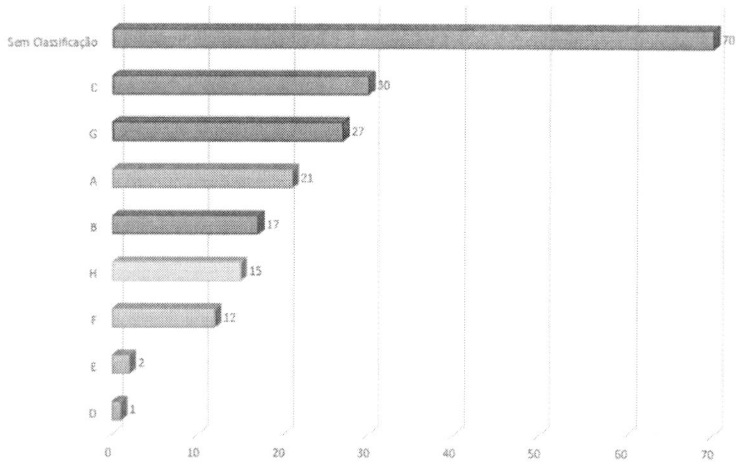

Figure 4 - Ranking by international classification code section

The ranking presented above suggests that most of the developed studies belonged to sections C, G and A, chemistry and metallurgy, physics and human needs, respectively. These represented 40.00% of the total filings, and the sections with the fewest filings do not have a proportional level of studies, suggesting the potential for research in sections B, H, F, E and D, which together represent 34.87% of the total filings.

Analysis of *stricto sensu* structure in relation to patent filings

Table 4 represents the relationship between the total number of filings found for each institution and the master's and doctoral programs that it offers.

Table 4 - Relationship between total filings and master's and doctoral programs

University	Total Filings	Master's p	Index	Doctoral	Index
FURB	12	10	1.20	3	4.00
IF Cat	1	2	0.50	0	-
IFSC	17	1	17.00	0	-
UDESC	5	23	0.22	10	0.50
UFFS	1	8	0.13	1	1.00
UFSC	140	76	1.84	55	2.55
UNC	2	1	2.00	0	-
UNISUL	5	4	1.25	2	2.50
UNIVALI	10	10	1.00	6	1.67
UNOCHAPECÓ	2	7	0.29	0	-
TOTAL	195	142	1.37	77	2.53

Because graduate programs are responsible for research within HEIs, it was deemed important to relate the number of patent filings to the number of master's and doctoral programs. Relating the total filings per master's program revealed that, despite having only one master's program, IFSC had the highest index (17.00), followed by UNC (2.00) and UFSC (1.84). Despite the importance of graduate programs to researcher training and, consequently, to the generation of patents, the values obtained reveal that the number of master's programs does not directly affect the number of patent filings because it does not represent the commitment of the institution to producing patents.

Regarding the indices found for the relationship between the total filings and doctoral programs, FURB had the highest index (4.00), followed by UFSC (2.55) and UNISUL (2.50). The number of doctoral programs did not have a direct relationship with the number of patent filings of an educational institution. For example, IFSC has a total of 17 filings, but it has no doctoral programs and, therefore, has no index. UNOCHAPECÓ has two filings and no doctoral programs, and, therefore, also did not have an index.

This study seeks to show the importance of technical knowledge in the development of new products and processes, which lead to new patents. The profile of Brazilian students contributes positively to the research and development in different institutions, providing greater generation of patent possibilities.

Table 6 - Relationship total filings and number of researchers/Drs./ production

University	Filings	Researchers	Index	Drs.	Index	production	Index
FURB	12	121	10.08	204	17.0	-	-
IF Cat	1	150	150.00	161	161.0	-	-
IFSC	17	425	25.00	280	16.5	2,658	156.35
UDESC	5	346	69.20	587	117.4	-	-
UFFS	1	231	231.00	293	293.0	3,849	3,849.00
UFSC	140	2,714	19.39	2,400	17.14	13,366	95.47
UNC	2	28	14.00	42	21.00	-()	-
UNISUL	5	74	14.80	172	34.40	-	-
UNIVALI	10	130	13.00	224	22.40	-	-
Unochapecó	2	43	21.50	75	37.50	2,752	1,376.00
TOTAL	195	4,262	21.86	4,438	22.76	22,625	141.41

The best index of the number of researchers to the number of filings was obtained by FURB, with only 10.08 researchers per patent filing, followed by UNIVALI (13.00) and UNC (14.00).

Relating the number of doctors to the number of patent filings, IFSC had an index of 16.50 doctors per patent filing, followed by FURB (17.00) and UFSC (17.14). As with the number of researchers, the number of doctors is relevant in relation to the number of patent fillings.

Regarding the scientific production index of the four analyzed institutions, UFSC had the lowest (95.47) and, therefore, most efficient index. IFSC's index was 156.35 and Unochapecó's 1,376.00 publications per filing. UFSC had the highest index because it had the lowest number of publications per filed patent.

In this sense, it is important to encourage research within the institution in order to produce new products and processes. Universities that stimulate researchers to think outside the theoretical environment and produce new studies will consequently have a higher generation of patents.

Final Considerations

The present study aimed to evaluate the level of innovative initiatives of Santa Catarina HEIs by analyzing patent registrations. HEIs were used as references, and the number of registrations requested in the INPI and the USPTO databases was evaluated.

Given that the patent filing process represents an innovative initiative, the results show that the innovative initiative level is growing in these institutions. It was not possible to associate this growth with any of the indices that were calculated in this study because the relationships that lead to the filing of a patent are complex and involve relationships between the study variables and possibly also involve other variables that were not included in the study. However, it can be seen that institutions with the highest generation of patents are mostly institutions with more researchers working on new products and processes, developing studies in the search for new solutions. Thus, the importance of stimulating innovation and technology within the institutions is noted so that the generation of patents is increasingly a factor for the growth and development of the regions.

It is concluded that the emergence of policies that encourage technology and innovation has had a positive and significant effect on research, leading to the filing of patents that may be beneficial to local, regional and national development.

References

AMADEI, J. R. P.; TORKOMIAN, A. L. V. (2009). As patentes nas universidades: análise dos depósitos das univ. paulistas. *Ciência da Informação*, Brasília, v. 38, n. 2, p. 9-18.

CENTER FOR WORLD UNIVESITY RANKINGS (2014). World university ranking. Accessed at: http://cwur.org/2014/ on 26 may 2015.

ETZKOWITZ, H. (2009). **Hélice Tríplice:** Univ.-Indústria-Governo. Porto Alegre: EDIPUCRS.

ETZKOWITZ, H.; LEYDESDORFF, L. (2001). The triple helix of university-industry-government relations and the globalization of national systems of innovation. In: SCIENCE UNDER PRESSURE, Finlandsgade.

FOLHA DE S. PAULO (2015). Ranking Universitário Folha. Accessed at http://ruf.folha.uol.com.br/2014/ on 14 may 2015.

INSTITUTO NACIONAL DE PROPRIEDADE INDUSTRIAL. Guia. 2015. Accessed at http://www.inpi.gov.br/menu-servicos/patente/guia-completo-de-patente on 12 jun. 2015.

GUERZONI, M. et al. (2014). A new industry creation and originality: Insight from the funding sources of university patents. Research Policy, Indiana. Accessed at <http://ac.els-cdn.com/ on 21 may 2015.

MORAES, K. M. et al. (2014). As patentes de Santa Catarina segundo seus depósitos. *Revista ACB: Biblioteconomia em SC,* Florianópolis, v. 19, n. 1, p. 33-51, jun. 2014.

National Academy of Inventors; Intellectual Property Owners Association. (2013) Top 100 worldwide universities granted U.S. utility patents in 2013. Accessed at http://www.academyofinventors.org/pdf/NAI-IPO-Top-100-Universities-2013.pdf on 21 may 2015.

NOONAN, K. E. (2014). IPO - Top 100 Patenting Universities. Accessed at http://www.patentdocs.org/2014/07/ipo-names-top-100-patenting-universities on 21 may 2015.

NORMAND, R. (2014). *Vale do Silício:* entenda como funciona a região mais inovadora do planeta. São Francisco: Spice Agency. 75 p.

PÓVOA, L. M. C.(2006) Depósitos de patentes de universidades brasileiras (1979-2004). In: XII *Seminário Sobre a Economia Mineira*, Belo Horizonte: UFMG/CEDEPLAR

TARTARUGA, I. G. P. (2010). As inovações nos territórios e o papel das universidades: notas preliminares para o desenvolvimento territorial. In: XI *Seminario Internacional Globalización y Territorio*, Mendonza: Universidad Nacional de Cuyo.

UNITED STATES PATENT AND TRADEMARK OFFICE. (2016). About us. Accessed at http://www.uspto.gov/about-us on16 feb. 2016.

Acknowledgements

Acknowledgements to Pos Graduate Program of Technology and Management of Innovation - Unochapecó.

Copyright statement

Design of an educational management system model oriented to reduce student dropout in engineering programs.

Paola Andrea Chinome Becerra
Universidad Pedagógica y Tecnologica de Colombia, Sogamoso, Colombia
paola.chinome@uptc.edu.co

Consuelo Ruiz Cárdenas
Universidad Pedagógica y Tecnológica de Colombia, Sogamoso, Colombia
consuelo.ruiz@uptc.edu.co

Liliana Fernández-Samacá
Universidad Pedagógica y Tecnológica de Colombia, Sogamoso, Colombia
liliana.fernandez@uptc.edu.co

***Abstract**. This work presents a of educational management system model oriented to reduce the student dropout in engineering programs by using a process-based approach and general system theory. To design the model, the authors selected a set of variables from those considered in the studies analyzed during the literature review stage by using the Pareto principle as a prioritization tool. The proposed model is based on the process approach defined in ISO 9001:2015- Quality Management System, taking as a studying process the vocational training, as inputs, the dropout variables and factors and as an output, the status of the student, this means dropout or active student, as sources of inputs, the student, her/his Family and the University and finally, as receivers of outputs, the society. Likewise, the educational management system model comprises an information system and a system of indicators in a continuous improvement process, which takes into account the necessary resources, procedures, and the responsible people.*

INTRODUCTION

Taking into account that education is considered as the main axis of a country's social and economic development (Tas, Bora, Selvitopu & Demirkaya, 2013) and because high dropout rates negatively affect its purpose, the study of student dropout has become an issue of interest not only for higher education institutions, but also for educational government authorities (Castaño, Gallón, Gomez, & Vasquez, 2008). The student dropout brings along social, institutional and personal implications; at the social level, may increase poverty and unemployment; at institutional level negatively affects the quality of education and in the personal level influences in the social insertion of the individuals as consequence of their feelings of failure (Osorio, Bolancé & Castillo Caicedo, 2011).

In the Latin American context, several examples of high student dropout rates can be observed. In 2013 in Chile and Venezuela, for example, the dropout rate by cohort was around 54% and 52%, while in Argentina and Mexico it was 42% and 43%, and in Costa Rica 62% (Ministry of National Education [MEN], 2013). According to (Higuera, 2016), the dropout rate of Higher Education for this year in Colombia was 44.9% and for public universities, 49.03%. In engineering, architecture, urbanism and other related areas, the dropout rate reached 48.82% and it was 50.85% for public high education institutions. In 2015, the dropout by cohort in the area of Engineering increases to 49.57% (Ministry of

National Education [MEN], 2016), indicating that almost half of the students who begin their university studies in programs of this field leave the University.

In view of this problematic issue, the countries have defined mechanisms to analyze the student dropout at the institutional level. In Colombia, for example, the Ministry of National Education has designed different strategies oriented to monitoring the causes of student dropout in higher education, among which is the system of prevention of the dropout of higher education (SPADIES) (Ministry of National Education, 2016).

Given the importance of analyzing the phenomenon of higher education dropout and the formulation of plans and strategies to reduce it, this work presents the design of an educational management system model oriented to reduce student dropout in engineering programs, by using a process-based approach from a systemic perspective. This research aims to answer the research question: ¿How to design an educational management system model to characterize the factors that influence the student dropout in engineering programs? The proposed educational management system model intends to serve as a tool for higher education institutions, and its design takes into account the necessary resources, procedures, and responsible people, considering the university education system as a process of continuous improvement.

This contribution stresses on describing the proposed model, which is based on the process approach defined in ISO 9001:2015- Quality Management System. Results about the validation of the system will be presented in future contributions.

Method:

This research uses as a case study of the Engineering programs at Sogamoso campus of the *Universidad Pedagógica y Tecnológica de Colombia* (UPTC), which is located in the province of Boyacá. This is a National Public University, linked to the Ministry of National Education of Colombia in terms of policies and planning of the education sector. This Campus is an interesting case study since it offers five programs related to the fields of Electronics, Industrial, Geological, Mining and Computing Engineering and with a dropout rate of 41.04%, very close to the national average (Universidad Pedagógica y Tecnológica de Colombia [UPTC], 2016).

The proposed educational management system model has been designed from a literature review identifying different concepts of student dropout, their classifications, analysis models, associated factors and variables by factor. The variables found have been prioritized using the Pareto principle (Cuatrecasas, 2012) in order to establish the most relevant ones as it is presented in Chinome, Ruiz & Fernández-Samacá (2016). The educational management system model is based on a mathematical model in which the variables selected by each of the factors considered are correlated, in order to find its impact on the student's dropout rate of the Engineering programs. The analysis of the behavior of the variables and factors seeks to define strategies that favor academic performance, student retention and, therefore, the reduction of dropout rate.

The content of the paper is organized as follows: Section 1 presents the theoretical background, Section 2 shows the procedure used to prioritize the variables of factors; In Section 3, the authors describe the educational management system model oriented to reduce student dropout; and finally, the last section presents some observations and concluding remarks.

1. Theoretical background

The student dropout phenomenon can be analyzed from different models, in the literature review, authors found psychological, sociological, economic, organizational and interaction models. Among these last ones, it is the model of 'Student Integration' proposed by Tinto (1975) and the 'Student Attrition' model of Bean (1980). Although there is no unified concept of this phenomenon, concepts proposed by authors like Erika Himmel (Himmel 2002), Tinto (cited by Guzmán, 2009), Spady (cited by Montes, Almonacid, Gómez, Zuluaga & Tamayo, 2010) y Frías Ortega, (cited by Osorio & Hernández, 2011) agree that the student dropout is 'the leaving of study program by an individual or the prolongation of the time established to carry it out", due to the interaction of different variables associated with individual, institutional, academic and social factors.

For some authors such as Erika Himmel (Himmel, 2002) and Rodriguez (Rodriguez & Londoño, 2011), student dropout can be classified, from the 'decision' point of view, as voluntary or involuntary. Other works, like those presented in (MEN), 2018) and Castaño (2008), classify it from the point of view of 'space' as a leaving of program, the institution or the education system. Finally, other researchers like Tinto (Rodríguez & Londoño, 2011) and Castaño, Gallón, Gomez, & Vasquez (2008), categorize it according to the 'time' in which this happens, as precocious, early and late.

Moreover, from the literature review, authors observed that studied contributions deal as factors of student dropout the followings: the individual factor related to student's variables and their family environment; the academic factor, which takes into account students' performance aspects and the academic program they are in; the institutional factor in which it´s related the academic regulation, student support strategies and stimuli and, finally the socioeconomic factor that takes into account variables mainly related to the social environment and to the student's economic support.

Among variables related to the individual factor are: age, sex, marital status, participation in extracurricular activities, teenage pregnancy or paternity, attendance at school, coming from another region, time to begin university studies, family size, family structure, level of quality of students, cultural index, relationship with parents, use of psychoactive substances, type of school, commitment, degree of retention –academic credit accumulation–, previous dropout, attitude and behavior. Table 1 shows how individual factor variables have been related in different contributions.

Table 1: The individual factor variables vs studied authors in literature review

Variable	Authors
Age	Tinto (1975), Bean (1980), Pascarella y Terenzini (1980), Pascarella (1985) (1991), ICFES (2002), Smeyers (2006), Pinto et al. (2007), Lopera (2008), De Witte Et al. (2013).
Gender	Spady (1970), Tinto (1975), Bean (1980), Pascarella y Terenzini (1980), Bynum y Thompson (1983), Remerger (1983), Pascarella (1985) (1991), Paragua et al. (2000), Giovagnoli (2002), Business Council of Australia (2002b), ICFES (2002), Aina (2005), Duchesne et al. (2005), CEDE (2006), Ou y Reynolds (2006), Pinto et al. (2007), MEN Spadies (2008), De Witte Et al. (2013).
Academic ability	Rumberger (1983), Ekstrom et al. (1986), Lamb y Rumberger (1998), Herbert y Reis (1999), Ball y Lamb (2001), Teese y Walstab (2002), Entwisle et al. (2004), Allensworth (2004) (2005), Vizcain (2005), Dustmann y Soeast (2008), Dalton et al. (2009), De Witte Et al. (2013).

Source: Chinome et al. (2016)

Regarding to the Academic Factor, studies have included variables such as professional orientation, intellectual development, academic performance, quality of the program, methods of study, results in the entrance examination, satisfaction with the program, number of subjects, and admission processes.

In the case of the institutional factor, studies take into account, among other variables, academic regulations, scholarships and funding, university resources, level of personal interaction with teachers and other students, academic and psychological support, structural characteristics, type of university, size of groups, number of students per teacher, social and academic environment and teaching experience.

Finally, the variables of the socioeconomic factor include: socioeconomic status, labor status, parental status and income, parental education level, parental profession, friends, geographical location of the residence, neighborhood characteristics and economic support.

2. Prioritization of the variables

Due to the high number of variables associated to each factor, for the design of the educational management system model, the authors made a selection of variables applying the Pareto Principle (Cuatrecasas, 2012) as a prioritization tool, which establishes that the 20% of the causes (variables studied by factor) determine 80% of the problem (factor characterization). For the application of the Pareto Principle, the number of authors (frequency) that related each of the variables in the different factors was identified.

Figure 1 shows the number of authors associated with each variable of the Individual Factor, as follows: 1) Gender, 2) Family structure, 3) Academic ability, 4) Relationship with parents, 5) Degree of retention - accumulation of academic credits, 6) Age, 7) Attitude and behavior, 8) previous dropout, 9) Commitment, 10) Proceeding from another region, 11) Family size, 12) Teenage pregnancy or paternity, 13) Family support, 14) Participation in extracurricular activities, 15) Use of psychoactive substances, 16) Cultural index, 17) Attendance to classes, 18) Civil status, 19) Quality level of student, 20) Starting time of university studies, 21) Type of school, (Chinome, Ruiz & Fernández, 2016).

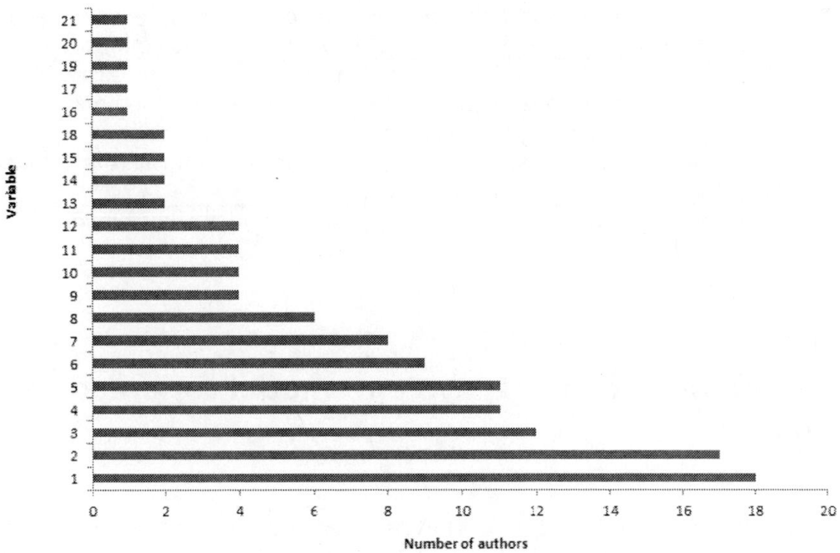

Figure 1. Number of consulted authors vs studied variable for individual factor

Figure 2 shows the prioritization of the variables for the Individual Factor. Applying the Pareto principle, the variables that represent 80% correspond to 1) Gender, 2) Family structure, 3) Academic ability, 4) Relationship with the parents, 5) Degree of retention - accumulation of academic credits, 6) Age, 7) Attitude and behavior, 8) Previous dropout and 9) Commitment. The same procedure was carried out for the rest of the factors, (Chinome, Ruiz & Fernández, 2016). Table 2 presents the most representative variables in each dropout factor.

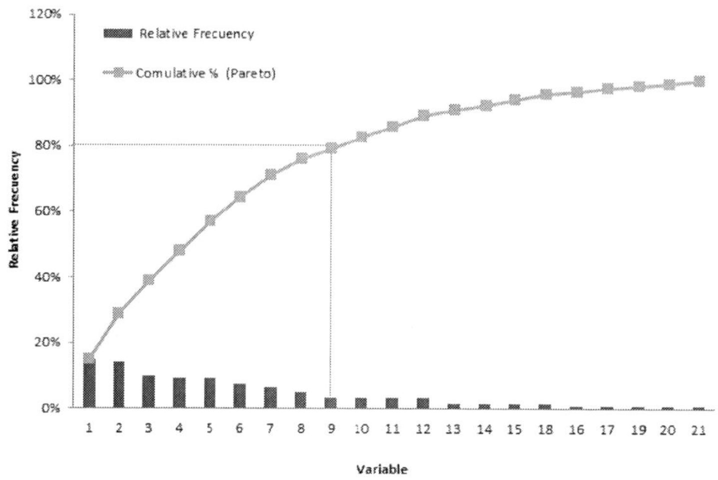

Figure 2. Pareto's diagram for the individual factor.

Table 2: Prioritized variables for each student dropout factor

Factor	Variables
Individual	Gender, family structure (with whom they live, dependents, position), academic ability, relationship with parents, age, degree of retention - accumulation of credits, attitude and behavior, previous dropout and commitment.
Academic	Academic achievement, satisfaction with the program (expectations), program quality, professional orientation and intellectual development.
Institutional	University resources, scholarships and forms of financing, type of university, academic regulations, level of personal interaction with teachers and students and social and academic climate.
Socioeconomic	Level of education of parents, stratum and work situation of parents and income

Taken of: Chinome et al. (2016)

3. Educational management system model oriented to reduce student dropout

According to Thomas Kuhn (cited by Atehortua, Bustamante and Valencia de los Ríos, 2013), science is based on a constant construction of models, as a way of interpreting the universe and systems; thus, when a new model comes up a "paradigm shift" emerges that breaks with some elements of the previous models, but at the same time retains elements that are valid, concluding that a model is understood as 'a theoretical construction' that pretends to represent the reality without to apprehend it completely.

Taking into account the dropout models found in literature review stage and general system theory, researchers proposed an Educational Management System Model to reduce the student dropout in freshmen and junior engineering student, which are described in Figure 3. The proposed model is based on the process approach defined in ISO 9001:2015- Quality Management System. The methodology used to conceptualize has three stages, namely, i) identification of process, ii) characterization of elements (according to ISO 9001:2015), iii) definition of system components. These stages are described as follows:

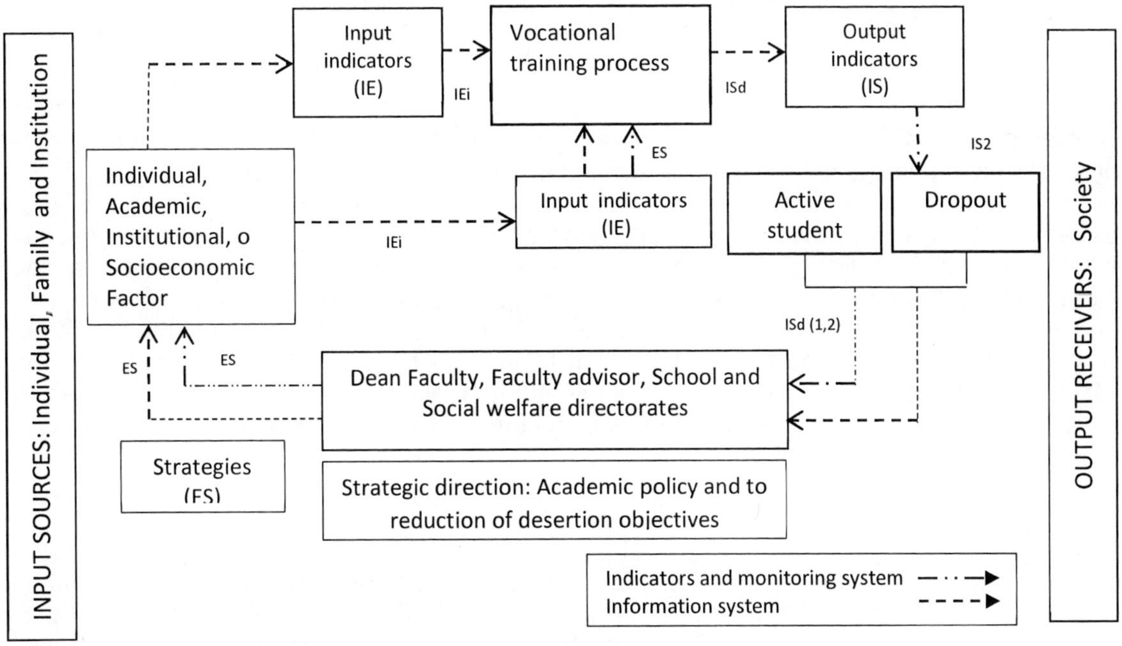

Figure 3. Proposed educational management system model.

i) Identification of the studying process: As a process, researchers defined 'vocational training', highlighting that this process is the 'axis' of the proposed educational management system model. This process begins when a student is enrollment in an academic program of Higher Education and ends when he/she obtains his/her professional degree. In this process, besides the student, there are other actors such as teachers, parents and the support staff. All actors interact with each other in order to achieve the institutional objectives and the desired output of the system.

ii) Characterization of elements. Based on elements of ISO 9001:2015- Quality Management System, the following elements were defined for the chosen studying process, as inputs, the dropout variables and factors; as an output, the status of the student, this means dropout or active student; as sources of inputs, the student, her/his Family and the University and finally, as receivers of outputs, the society. See Figure 4.

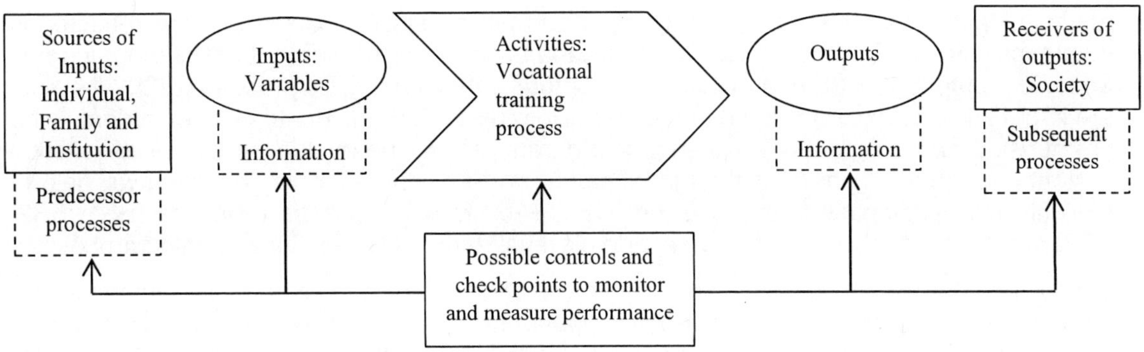

Figure 4. Schematic representation of the elements of vocational training process according to ISO 9001:2015.

iii) Definition of system components. Once the process and elements were established, researchers defined the components of the educational management system model, namely, a) strategic direction, b) information system and c) Indicators and monitoring system.

a) Strategic direction: Regarding strategic direction, the design of management system considers the definition of academic policies and objectives related to reduce the student dropout rate at seeking to meet the institutional mission and vision.

b) Information system: this is one of the most relevant aspects of the design of the proposed educational management system model. For this instance, the system considers that the information of the entry corresponds to the student profile at the time in which he/she was enrolled, by using a survey based on the selected variables and factors. Each Input Indicator (*i-th* indicator, IEi) associated with a dropout factor (*i-th* factor) is defined by equation (1), where each variable (*j-th variable, Vij*) that integrates a factor (*i-th factor*) is characterized by a weight (*Gij*), under the assumption that each factor is a linear combination of its variables.

$$\text{IEi} = \sum_{j=1}^{n} Gij * Vij \quad \text{para i} = 1, 2,3, 4. \qquad (1)$$

This information will arrive to the leader of the process, designated by the Institution (Responsible), who will study and analyze it by using a PDCA cycle (Planning, Do, Check and Act), evaluating the performance of each factor (*i-th* factor) and its influence on the result of the academic process (Dropout student Index, *ISd*) and calculating the student's probability to dropout.

Other input indicators related to student performance factors (*i-th factor*) at the psychological, physical, social and academic level (IEi) and their respective variables (*k-th variable*) are added to the vocational training process. For these factors, researchers also considered a linear behavior, assuming that each factor is a linear combination of its variables, where each *k-th* variable of *i-th* factor is characterized by a weight (*Gik*), see equation (2).

$$\text{IEi} = \sum_{k=1}^{n} Gik * Vik \quad \text{para i} = 1, 2,3, 4. \qquad (2)$$

These indicators will allow recognizing some characteristics of the academic process of professional formation, based on information reported in documents, forms and records of the educational management system model. The information obtained as a result of this analysis will be sent in a document of the system to the Faculty Dean, Faculty advisor, and Social Welfare Director, in order to they can design and implement strategies (ES), which help to reduce the dropout rate.

c) Indicators and monitoring system. Finally, in the indicators and monitoring system, the results obtained from the interaction of input and performance indicators will be reflected in output indicator (ISd), which in turn is subdivided according to the probability of dropping out, undesired output (IS1) (dropout) or desired output (retention) of the process (IS2). Likewise, the information obtained from the output indicators (IS1, IS2) will identify those students who are at risk of dropping out of an academic program and will be evidenced by the information system through a document of the management system (Check). This document will help to the university authorities to improve the model (Act) through corrective actions (ES1, ES2, ES3) at the individual, institutional or academic level during the vocational training process.

Based on ISO 9001: 2015 standard, the Institution must continuously improve the management system, in order to identify opportunities for improvement. In this instance, these opportunities can be seen as mechanisms that contribute to the decrease the dropout in the engineering programs based on the output indicators (ISd) by applying the different strategies or actions (ES) according to the influence of each dropout factor in the vocational process.

Conclusions

The educational management system model oriented to reduce student dropout is formulated using the process approach defined in ISO 9001:2015 and the general system theory. The model uses a set of variables identified in literature review stage, which were selected applying the Pareto's principle. The model takes as a case study the engineering programs of Sogamoso campus at Universidad Pedagógica y Tecnológica de Colombia, Colombia. The designed model proposes to deal the student dropout as an output of a process that corresponds to the vocational training, and in which the risk of dropping out is measured by a system indicator, which will allow planning and developing strategies and actions at individual, family or institutional level, at seeking to improve the students' conditions to guarantee their retention in the Institution.

The educational management system model is based on a mathematical model in which each of the variables that integrate the associated factors is characterized by a weight that determines its contribution in the corresponding factor. The system considers an input diagnosis as a starting point for the analysis of each factor. Moreover, the system considers that leader of the dropout analysis will evaluate the collected information under a cycle PDCA (Plan, Do, Check, Act) observing the influence of each factor on the outcome of the academic process. As a future work, the proposed model will be validated by using documented information of the management system.

References

Castaño, E., Gallón, S., Gómez., K., & Vásquez, J. (2008). Análisis de los factores asociados a la deserción estudiantil en la educación superior: un estudio de caso. Revista de Educación, 345 (1), 255-280.

Cuatrecasas, L. (2012). Organización de la producción y organización de las direcciones. Madrid, España: Ediciones Díaz de Santos.

Chinome Becerra, P., Ruiz Cárdenas, C., & Fernández Samacá, L. (2016). Priorización de variables en el diseño de un sistema de gestión integral de la deserción estudiantil. Revista Educación en Ingeniería, 11(22), [69-77]. Recuperado en http:/www.educacioneningenieria.org/index.php/edi/article/view/683/299.

Donoso, S., & Schiefelbein, E. (2007). Análisis de los modelos explicativos de retención de estudiantes en la Universidad: una visión desde la desigualdad social. Estudios Pedagógicos, 33 (1), [7-27].

García, J. C., González, M. L., & Zanfrillo, A. I. (2011). Desgranamiento Universitario: Perspectiva Estudiantil En Ingeniería. Universidad Federal de Santa Catarina. http://repositorio.ufsc.br/xmlui/handle/123456789/25923.

Guzmán, S. P. (2009). Deserción y retención estudiantil en los programas de pregrado de la Pontificia Universidad Javeriana (Tesis de Maestría). Pontificia Universidad Javeriana, Bogotá, Colombia.

Higuera, O. (2017) Deserción estudiantil en Colombia y los programas de ingeniería de la UPTC, Seccional Sogamoso. 17 (1), [70 - 76].

Himmel, E. (2002). Modelos De Análisis De La Deserción Estudiantil En La Educación Superior. Calidad en Educación, 17 (1), [91 – 108].

International Organization for Standarization (2015). Norma Internacional ISO 9000:2015. Sistemas de gestión de Calidad - Fundamentos y vocabulario. [Traducido al español de Quality management systems — Fundamentals and vocabulary]. Ginebra. Suiza: Secretaría Central de ISO.

Ministerio de Educación Nacional (MEN) República de Colombia. (2008). Análisis De Determinantes De La Deserción En La Educación Superior Colombiana Con Base En El SPADIES. Bogotá, Colombia: Imprenta Nacional de Colombia.

Ministerio de Educación Nacional (MEN) República de Colombia. (2012). Cómo rinde la Educación Superior: Así vamos en acceso y permanencia de la Educación Superior. Diapositivas de PowerPoint. Recuperado de Ministerio de Educación Nacional: http://www.mineducacion.gov.co.

Ministerio de Educación de Colombia. (2013). Destacados históricos, Estadísticas 2013. Recuperado de http://www.colombiaaprende.edu.co/html/micrositios/1752/w3-article-343426.html

Ministerio de Educación de Colombia. (2016). Sistema de Información SPADIES. Recuperado de http://www.mineducacion.gov.co/sistemasde información/1735/w3-propertyname-2895.html

Montes, I. C., Almonacid, P. M., Gómez, S., Zuluaga, F. I., & Tamayo, E. (2010). Análisis de la deserción estudiantil en los programa de pregrado de la universidad EAFIT. Serie Cuadernos de Investigación. Medellín, Colombia: Universidad de EAFIT.

Osorio, A., Bolancé, C., & Castillo Caicedo, M. (2011). Deserción y graduación estudiantil universitaria: una aplicación de los modelos de supervivencia. Revista Iberoamericana de Educación Superior (RIES), 3(6), 31-57.

Osorio, I., & Hernández, M. (2011). Prevalencia de deserción escolar en embarazadas adolecentes de instituciones educativas oficiales del Valle del Cauca, Colombia. Colombia médica, 42(3), 303 - 308.

Rodríguez, L. H., & Londoño, F. J. (2011). Estudio sobre deserción estudiantil en los programas de educación de la Católica del Norte Fundación Universitaria. Revista virtual Universidad Católica del Norte, 33(1), 328-355.

Universidad Pedagógica y Tecnológica de Colombia. (2016). Misión - Visión. Recuperado de http://www.uptc.edu.co/ universidad/acerca _de/inf_institucional/index.html.

Tas, A., Bora, V., Selvitopu, A., & Demirkaya, Y. (2013). Reasons for Dropout for Vocational High School. Educational Sciences: Theory & Practice. DOI: 10.12738/estp.2013.3.1398.

Description of the professional and personal perception of engineering students regarding their academic performance in a first semester course

Darío Esteban Recalde Morillo
Universidad Autónoma de Occidente, Cali – Colombia
derecalde@uao.edu.co

Abstract: *Introduction to Engineering 1 is the first course in nine Engineering programs of the Universidad Autónoma de Occidente. Freshmen start the course taking their expectation into account of being the best professionals in their areas, and teachers support them with learning activities, examples and study cases. Is necessary to know and to work in how much the personal and professional characteristics of the students are associated with their academic results at the end of the semester. This study shows that relation using functions and graphics of descriptive Statistics and correlation coefficients in a base data of 14 groups of classes since 2015 with a self-evaluation instrument with 18 variables of different attributes and skills. This is a powerful way to improve the grades and the profile of the future engineers and, at the same time, to be better human beings for our society.*

Introduction

Introduction to Engineering 1 is currently taught to students of nine programs of the Faculty of Engineering of the Universidad Autónoma de Occidente, UAO. The purpose of the course is to guide the students to understand the professional work through the knowledge, application and articulation of concepts, methods, techniques and conducts, which allow them to provide a methodical, creative and collaborative solution to Engineering problems in a given context. It's a window to see what Engineering studies are coming in next semesters, and a phase for the students to adjust themselves the behavior between high school and the university.

Context

The study starts from the concern of the teacher-author to see the position of its students in front of the duality and apparent distance between their personality and the professionalism that represents the Engineering. This course in first semester is a unique opportunity for students to start their university life by affirming their current characteristics in harmony with their future qualities as a graduate.

In addition, if the students do not identify themselves with the course contents, the probability of low performance is very high, affecting the university dropout variables. Students and teachers contribute the best of each to the construction of an academic scenario, as identical as possible to the reality of work and close to their innate and acquired characteristics. The overall structure of the study is presented in Figure 1.

¿What?

- Reflective activity for students through a quantitative estimation of the learning in variables related with characteristics of the personality and the basic knowledges of Engineering.

¿How?

- Short open questions about students wishes, fears and present and future capabilities.
- Self-evaluation formats with 18 personal and professional characteristics of Engineering.
- Review of academic grades of the students at the end of the course.

¿When?

- In 10 of 16 classes of the semester, taking 10 or 15 minutes in each one at maximum

¿Where?

- In assigned classrooms for Introduction to Engineering 1 course in University campus.

¿Who?

- 14 groups of 28 students (approximately) of different Engineering programs.

¿Why?

- Students need to identify their personal capabilities related with the professional challenge to study an Engineering program.
- Teachers should support the development of personal and professional characteristics of their students during the course.

¿What for?

- Students could strengthen their innate skills and maximize their acquired capabilities in an Engineering first semester course.
- Teachers could have a tool to analyse the impact of motivation in students academic performance.

Figure 1. Structure of the reflective activity to evaluate personal and professional characteristics of the students in an Engineering course of first semester.

Research Question

Students of Introduction to Engineering 1 say at the end of their first semester: "*Teacher, my final grade in this course is not corresponding to how I much I learned in your class*". So, this investigation is moved by that question: What is the gap between the institutional academic evaluation and the personal and professional self-evaluation?

The answer is important because: 1) the students can find a better identity with the Engineering program they are studying (reflective learning); and 2) the teacher can find a better way to explain the topics and to engage the students to the course (active learning).

Theoretical Frameworks

In the course of Introduction to Engineering 1, students know the steps of design process in Engineering defined by Grech (2013), where looking for information is an iterative action all the time. Romero (2008) presents a general planning process for Engineering project, where administration of resources is the key for an effective solution. And Krick (1978) uses the traditional design in five practical and conceptual processes for any Engineering career to respond in a problematic situation. From these references and from his teaching experience of seven years in a course of first semester, the author defines in Table 1 nine formal requirements to solve a hypothetical or real study case for a professional in Engineering.

Table 1. Professional requirements in Engineering.

A	B	C	D	E	F	G	H	I
Identification and description of problems in Engineering	Relationship of causes and effects of a problem	Analysis of variables of a problem.	Approaches and processes for solving problems	Generation, evaluation and choice of solution alternatives	Graphical communication and technical specifications of the solution	Planning of resources and activities for the implementation of the solution	Operation and technical evaluation of the solution	Documentation of the process and the products of Engineering

Bain books talk about what does best university teachers (2007) and what does best university students (2014). In both of them, the role of a teacher is to show great confidence in students through active teaching and encourage them to do explode their talents through reflective learning. At the same time, knowledges, skills and attitudes are demanded for each Engineering programs that Universidad Autónoma de Occidente offers, but the personal characteristics that correspond to the process of solving engineering problems in the first semester are not defined because the course is part of the Engineering Basic Cycle. Thus, the author defines in Table 2 nine personal characteristics that are required to face the study of engineering in the best possible way. If teachers and students focus on them, they will take advantage of teaching and learning experience from the beginning of their careers.

Table 2. Personal characteristics of an Engineering student.

A	B	C	D	E	F	G	H	I
Will: love for learning and joy for living new things in any context and environment	Curiosity: desire to know and try beyond normal by measuring interest and risk	Discipline: judgment in the fulfilment of the instructions and the commitment to the achievement of the objective	Attitude: Positive thoughts and feelings of self and life	Originality: proposition of own valid ideas but different from the traditional ones	Elegance: appreciation of what needs to be done and how it should be done	Organizing: logical and anticipated order of activities to be done and resources to be used over time	Attention: special attention to what is going to be said or done, observing with pause and detail	Communication: sincere and clear contact with people near or far from us

These two perspectives were the motivation to think about how could improve the relation between the nine requirements of a professional (described in Table 1) and the innate or acquired nine characteristics of the Engineering student (proposed in Table 2) in first semester. 18 variables are organized in Table 3, doing a parallel with the attributes proposed by the National Academy of Engineering of USA in The Engineer of 2020 Project (2004). In this publication, NAE attempts to predict the roles that engineers will play in the future. It's clear that the variables are related taking care of the policies of one of the most important association in Engineering

Table 3. Match between professional requirements and personal characteristics.

	Professional requirements in Engineering	Personal characteristics of an engineering student	Relation with the Attributes of The Engineer of 2020 Project.
1	Identification and description of problems in Engineering	Will	"Yesterday, today, and forever, engineering will be synonymous with ingenuity. Using science and practical ingenuity, engineers identify problems and find solutions".
2	Relationship of causes and effects of a problem	Curiosity	"Encompassed in this theme is the imperative for engineers to be lifelong learners. They will need this not only because technology will change quickly but also because the career trajectories of engineers will take on many more directions—directions that include different parts of the world and different types of challenges and that engage different types of people and objectives".
3	Analysis of variables of a problem.	Discipline	"Engineers in 2020, like engineers of yesterday and today, will possess strong analytical skills. Engineering employs principles of science, mathematics, and domains of discovery and design to a particular challenge and for a practical purpose".
4	Approaches and processes for solving problems	Attitude	"Given the uncertain and changing character of the world in which 2020 engineers will work, engineers will need something that cannot be described in a single word. It involves dynamism, agility, resilience, and flexibility".
5	Generation, evaluation and choice of solution alternatives	Originality	"The creativity requisite for engineering will change only in the sense that the problems to be solved may require synthesis of a broader range of interdisciplinary knowledge and a greater focus on systemic constructs and outcomes".
6	Graphical communication and technical specifications of the solution	Elegance	"We envision a world where communication is enabled by an ability to listen effectively as well as to communicate through oral, visual, and written mechanisms".
7	Planning of resources and activities for the implementation of the solution	Organizing	"Complementary to the necessity for strong leadership ability is the need to also possess a working framework upon which high ethical standards and a strong sense of professionalism can be developed".
8	Operation and technical evaluation of the solution	Attention	"Even though the scientific knowledge that defines operating principles is expected to be more fluid and more complex, the core analysis activities of engineering design—establishing structure, planning, evaluating performance, and aligning outcomes to a desired objective—will continue".
9	Documentation of the process and the products of Engineering	Communication	"As always, good engineering will require good communication. Engineering has always engaged multiple stakeholders—government, private industry, and the public".

Methodology

Data collection

During the classes: teacher delivers small sheets of self-adhesive paper in 10 of 16 classes to be pasted in their notebooks. The open question is formulated and the intention is explained with respect to the personal and professional formation. After five minutes, one by one the students read what is written in the paper while the classmates listen. If anyone does not want to make known their thoughts or feelings that day, they can choose not to respond. After that moment of free expression, teacher continues with the class agenda relating some answers with engineering topics. Almost all the students keep the papers in their notebooks with the class notes of the basic concepts.

Last class: teacher gives a format of personal and professional self-assessment to evaluate the appropriation and comprehension of each of the 18 variables. Grades are from 0.0 to 5.0. The most important instruction is to fill the document following the rules of the game proposed by the teacher at the beginning of the course: truth (understood as honesty), respect (understood as esteem) and enjoyment (understood as pleasure). Then, students use the radial-type statistical graph to locate their grades in the format. After seeing their spatial distribution, they decide a grade for their performance they believe they deserve. Additional comments about the course whether positive or negative are welcome as part of this activity. Teacher considers the self-assessment grade with a value of 5% of final grade.

After the classes. At the end of semester, teacher collects all formats of self-assessment and tabulates the values taking into account the engineering program and the student's gender. At the same time, organizes the data to compare the three variables of this study: professional requirements, personal characteristics and final grades of the course.

Data analysis

Descriptive Statistics with graphical and numerical functions are used for the systematization of the data, and the Pearson correlation coefficient, widely used in science to measure the linear dependence between two variables X and Y (García, 2015). It has a value between +1 and −1, where 1 is total positive linear correlation, 0 is no linear correlation, and −1 is total negative linear correlation (Guilford and Fruchter, 1984). Table 4 shows the averages (arithmetic mean) of 353 students of 14 groups since 2015 until 2016 and the final grades:

Table 4. Records of averages grades for study groups.

Period	Group	Professional Learning Estimation	Personal Learning Estimation	Final Grade
2015-01	*G8-15.01*	4,22	4,24	3,86
2015-01	*G14-15-01*	4,00	4,01	3,38
2015-01	*G52-15.01*	3,90	4,16	3,59
2015-03	*G4-15.03*	3,92	4,03	3,47
2015-03	*G6-15.03*	4,02	4,12	3,27
2015-03	*G7-15.03*	4,21	4,20	3,73
2016-01	*G4-16.01*	3,94	3,99	4,08
2016-01	*G6-16.01*	3,99	4,20	3,54
2016-01	*G7-16.01*	3,85	4,14	3,87
2016-01	*G52-16.01*	4,07	4,19	4,02
2016-03	*G5-16.03*	3,83	4,11	3,38
2016-03	*G6-16.03*	4,07	4,20	3,72
2016-03	*G7-16.03*	4,05	4,23	3,69
2016-03	*G55-16.03*	3,89	4,13	3,24

Figure 2 shows how close the estimates of professional self-evaluation and personal self-evaluation of each other are through the time; they are consistent. The value of the final grade is fluctuating with respect to self-evaluation curves:

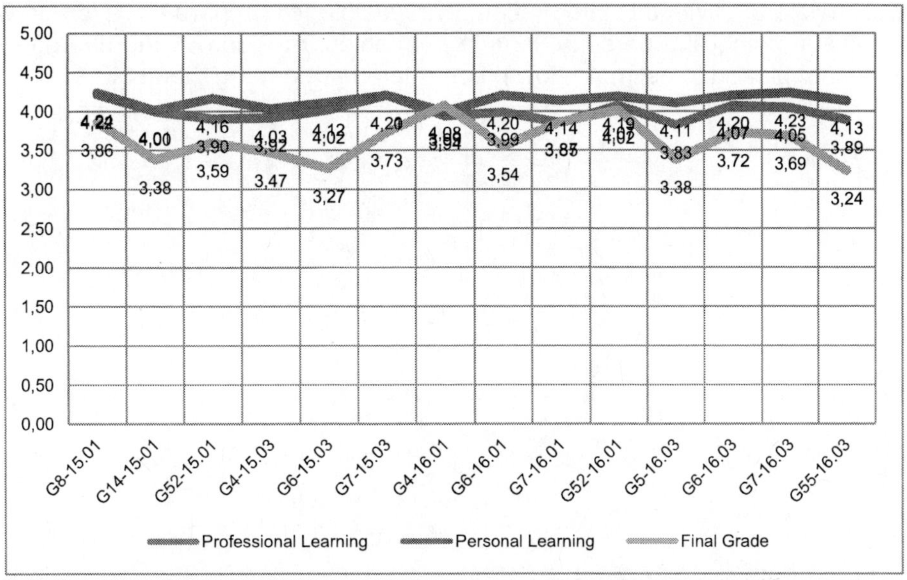

Figure 2. Values of the self-evaluation and academic grades in 14 study groups.

And if we compare in Figure 3 the total average values over time, a constant line is shown, which estimates that what has been learned at a professional and personal level in the classroom is higher than final academic grades.

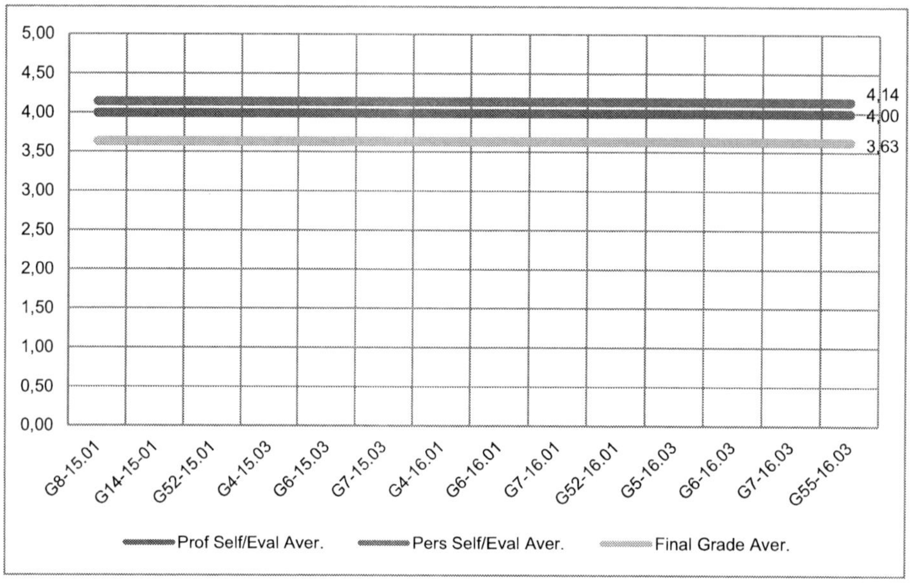

Figure 3. Average of the evaluation in 14 study groups.

For the first analysis, Professional Learning Estimation vs. Final Grades, the coefficient has a value of +0.358, and the correlation is significant at the 0.01 level (bilateral). Given its positive sign, it shows the proportional relationship between the variables. For the second analysis, Personal Learning Estimation vs. Final Grades, the coefficient has a value of +0.288. Its magnitude shows a relation between these two variables and it is significant at the 0.01 level (bilateral). With this information, it is feasible to compare the average values of the professional learning and personal learning variables. In Figure 4 the average value of such variables are plotted to estimate how much related are among themselves:

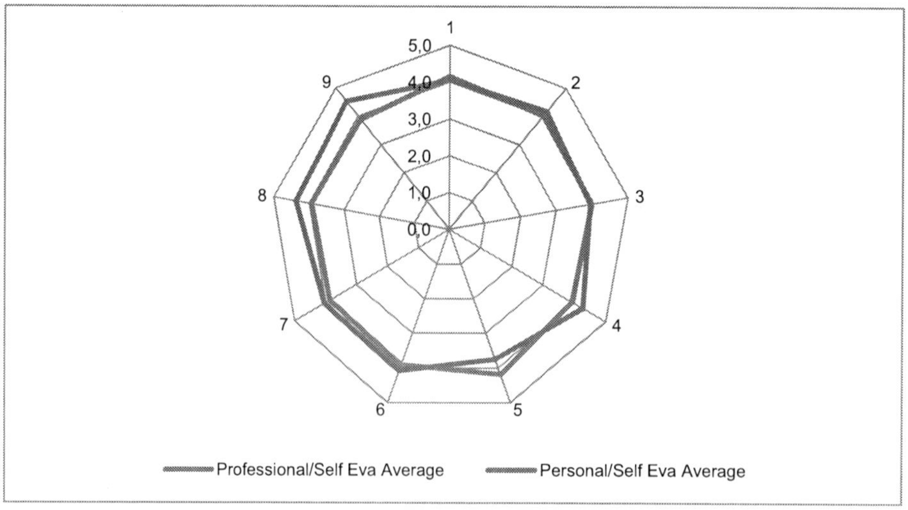

Figure 4. Relation between averages self-evaluation variables.

And finally, in the analysis between Professional Learning and Personal Learning, the value is +0.641 and the correlation are significant at the 0.01 level (bilateral).

Findings

The most important finding is to corroborate a non-obvious premise to study Engineering: If the student believes he knows a lot about the subjects of the course, his final grade does show it to the same extent. And if the student believes he knows much about the subjects inherent to his person, his final grade does not prove it to a greater degree.

In short:

About the reflective experience during classes. Teacher gives examples that motivate students to think that personal training cannot be disconnected from their professional training. If they find a connection of these characteristics with the contents of the course, the meaning for them is greatly enhanced.

About the reflective experience after classes. Students can judge how much they learned in the course, regardless of final grades. In itself, this auto evaluation becomes a learning activity, allowing them to analyze their performance and make decisions based on information, an ideal process for an Engineering professional.

About qualitative analysis of variables. The gap between the curves of self-assessment and final qualification can be reduced with the joint work of students and teachers, based on learning based on both success (active teaching) and error (reflective learning) in class.

About quantitative analysis of variables. The positive and significant correlation is the evidence that the teaching-learning process can be done from a non-academic perspective (doing emphasis on the person above the class topics) and not only in a single academic perspective (emphasis in the course contents).

Recommendations

This classroom experience supports the process of self-identification of the students with respect to the project of life in an Engineering program, and it allows teachers to adjust basic concepts that maybe were not clear in the course development according to student self-perceptions.

What if... Engineering students make decisions on time, but based on the knowledge of a strong relation between their personal characteristics with their professional capabilities? They could continue with the career doing their best effort in specific areas of the programs instead of withdrawing early from the university.

What if... Engineering teachers show more life examples of colleagues or their own professional experiences as an evidence of what students will get through the study? They could change the topics or the practices of the traditional programs in classes to expand the meaning of learning for freshmen far away of numbers, formulas, facts and rules.

With this research is possible in the future: 1) to compare the Engineering programs of the University and classified in specific categories to match with their existing professional profile; 2) to formulate new personal characteristics and professional capabilities for other courses of Engineering Basic Cycle; 3) to evaluate the practices of the teachers who leads at the same time a personal growing in the students and a better academic performance in the groups.

Acknowledgements

To the students of Introduction to Engineering 1, who shared their thoughts and feelings with truth, respect and joy in classes. I shared them too.

References

Bain, K. (2007). *Lo que hacen los Mejores Profesores Universitarios.* España: Universitat De Valencia. Servei De Publicacions.

Bain, K. (2014). *Lo que hacen los Mejores Estudiantes de la Universidad.* España: Universitat De Valencia. Servei De Publicacions.

García P., A. (2015). *La Interpretación de los Datos: Una Introducción a la Estadística Aplicada* [e-book]. Available from: e-Libro Premium, Ipswich, MA. Accessed March 3, 2017.

Guilford, P., Fruchter, B. (1984). *Estadística Aplicada a la Psicología y la Educación* [e-book]. México McGraw-Hill. Available from: Universidad Autonoma de Occidente Catálogo, Ipswich, MA. Accessed March 3, 2017.

Grech, P. (2013). *Introducción a la Ingeniería: un enfoque a través del diseño.* Colombia: Pearson.

Krick, E. (1978). *Introducción a la Ingeniería y al Diseño en la Ingeniería.* México: Limusa.

National Academy of Engineering. 2004. *The Engineer of 2020: Visions of Engineering in the New Century.* Washington, DC: The National Academies Press. doi:https://doi.org/10.17226/10999.

Romero, O. Muñoz, D. Romero, S. (2006). *Introducción a la Ingeniería.* México: Thomson.

Copyright Statement

“It is helpful but having it due a week out might be too long”: How engineering students experience reflection activities

Jennifer Turns, Ahmer Arif, Terri Lovins, Bonnie Chinh, Cynthia J. Atman
University of Washington, Seattle, USA
jturns@uw.edu, ahmer@uw.edu, tlovins@uw.edu, bchinh@uw.edu, atman@uw.edu

Introduction

Reflection can be understood as a form of thinking that involves stepping out of a situation to create knowledge. Reflection is a significant topic for engineering education because of the significance of reflection to professional practice (e.g, Schon, 1983), the connection of reflection to metacognition, and the role of reflection in learning from experience (e.g., Kolb, 2014; Dewey, 2007). In order to support student engagement with reflection, educators create activities that help students reflect.

In this paper, we focus on the reactions that students have when engaging in reflection activities. The quote in the title of our paper offers an example of reactions that students may have. In the quote, the student is suggesting a positive reaction (i.e., “It is helpful”) but also a concern (i.e., “having it due a week out might be too long”). Exploring students’ reactions to reflection activities is a place where we see a gap in our understanding of reflection in engineering education. Core to this paper is showing how reactions can be characterized on both a conceptual and an empirical level. We also discuss implications of our findings on student reactions. Specifically, we identify implications for the design of reflection activities, and we introduce the idea that engaging in student reactions to reflection activities could serve as a site of educator professional learning.

Literature Review

Papers from ASEE 2016 provide a sense of the role that reflection and reflection activities play in the engineering education community. For example, there are papers that describe a specific reflection activity done for a specific purpose: Niño (2016) talks about an autobiographical reflection activity to help students develop as leaders while Goswami and Walia (2016) describe how they have students reflect on mistakes during software inspections in order to improve their software inspection skills. There are also papers that describe collections of reflection activities that can be used toward particular learning goals. For example, Bankhead, Olmstead, & Mannard (2016) discuss a collection of reflection activities used to help entering engineering students prepare to function effectively in engineering education. Scholars also focus on how reflections can be used to gain specific insights into student understandings. For example, Cian, Cook, & Benson (2016) discuss collection of post-hoc problem solving reflections via audio annotations, and how such reflections can be used to explore students’ metacognitive strategies. On a cross-cutting level, Csavina, Nethken, & Carberry (2016) describe an investigation into students’ understandings of reflection in the context of design education--understandings that can be leveraged when designing reflection activities for students and/or introducing reflection activities to students, and Thomas et al. (2016) present a collection of suggestions for educators interested in adding reflection activities to their teaching.

This level of discourse suggests an opportunity to increase foundational, scholarly work on reflection. Scholarship to support the use of reflection in engineering education could proceed via multiple perspectives. For example, an expertise perspective might build on the work of Csavina, Nethken, & Carberry (2016) by looking more closely at the reflection abilities of engineering students. Scholarship leveraging an instrumental perspective might focus on how engaging in reflection leads to changes in distal performances or results in proximal knowledge gains. Finally, scholarship could leverage a phenomenological perspective by exploring students' experiences with and reactions to engaging in reflection. This latter perspective, with its focus on reactions to engaging in reflection, is interesting because it can provide guidance to efforts to design and redesign reflection activities.

Student reactions to reflection activities

We are interested in student reactions to reflection activities and are characterizing student reactions along two dimensions. The first dimension pertains to the valence of the reaction--the extent to which students feel positive and/or negative toward the activity. The second dimension has to do with the basis (or bases) of the reactions.

In terms of valence, it is easy to be drawn to the negative. In fact, a body of work in engineering education involves looking at resistance to active learning (e.g., Shekhar et al., 2015; Recabarren et al., 2015). In bringing the idea of resistance to reflection activities, it is interesting to note that there are understandable and even appropriate reasons for students to resist (e.g., the activity is too personal, the timing of the activity is poor). In the case of reflection and reflection activities, we believe it would be valuable to also look beyond resistance. If focusing on resistance is a form of deficit thinking, there is the opportunity to also look from an asset perspective (Valencia, 2012). Thus, we are interested in seeing the extent to which students are resistant to and/or resonant toward reflection activities.

In order to examine the bases of student reactions, we generated a preliminary reactions framework consisting of a set of possible bases for student reactions. The framework was created through literature review efforts. For example, the power basis (the last one in the list) is included because of works such as Ellsworth (1989) and Shor (1996).

1. **Cognitive basis.** Reflection requires cognitive resources. Students differ in the cognitive resources they bring. Sometimes resources are not readily available (e.g. attention or memory).
2. **Activity basis.** The design of reflection activities (e.g., the different features of an activity, the instructions, the sequences) might work for some and not others.
3. **Self-preservation basis**. Sometimes reflection can destabilize an individual's positions in ways that leave them feeling disoriented or vulnerable; sometimes it can help them preserve their positions.
4. **Cultural basis.** Reflection can run counter to an individual's sense of what is socially valued and reinforced; but can also be experienced as in alignment with social values.
5. **Epistemological basis.** Students can have pre-existing assumptions about the nature of knowledge as discussed in theories of epistemology development.
6. **Mindset basis.** Students can approach reflection activities with a fixed mindset or a growth mindset.
7. **Personal basis.** Students might have privacy concerns when sharing their reflections with others but could also appreciate having the warrant to do so.
8. **Instrumental basis.** Reflection may (or may not) be seen as being pragmatically useful in achieving one's goals.

9. **Situational basis.** Maslow's hierarchy of needs suggests that some people might have unmet needs that are antecedent to reflection. Others might have needs that can be met through reflection.
10. **Time basis.** The fast pace of modern life could mean a reflection activity is seen as a welcome opportunity to slow down or that there is no time for such slowing down.
11. **Power basis.** Students may not see teachers as having the authority to ask them to reflect. Some may welcome having reflection emphasized in a structured environment.

The reactions framework supports multiple insights. First, the framework illustrates the idea that, from a student perspective, there can be multiple reasons to experience resistance to and/or resonance with reflection activities. From an educator perspective, this framework illustrates the challenge of designing reflection activities. Further, this framework foreshadows how thinking about student reactions to reflection activities could function as a site of educator professional learning. If thinking about students' reactions to reflection activities involves thinking about issues such as epistemology, culture, and power, then such thinking might advance the educators' own understanding of such fundamental issues. How though, do we further explore these ideas? How do we collect data that might provide insight on students reactions to reflection activities? Before turning to our specific method, we provide background on the Consortium to Promote Reflection in Engineering Education. Work conducted through the consortium provided data for our analysis and also the basis of our interest in educator professional learning.

CPREE as a lab for studying reflection in engineering education

The Consortium to Promote Reflection in Engineering Education (CPREE) was established in 2014 to help engineering educators bring reflection activities into their teaching. Consortium members are 12 diverse campuses (4 research-extensive, 4 teaching-focused, and 4 associates-degree-granting), with dedicated PI scholars from a variety of engineering disciplines. During the first year, the campus PIs documented over 120 reflection activities.

Interestingly, the process of interviewing some of these educators became an opportunity for educators to talk about teaching and their philosophy of education (Turns et al., 2015). This observation has raised questions about whether, why, when and how a focus on engineering student reflection might function as a site of educator professional learning. For example, might a focus on student reflection help educators to (1) function as a critically reflective teacher (and "hunt for assumptions"), as described by Brookfield (1995), (2) grapple with significant questions of the field (e.g., questions related to engineering epistemologies, learning mechanisms, learning systems, diversity and inclusiveness, and assessment, such as described in Adams et al., 2008), and even (3) think about the multiple concurrent centerings that educational practice requires (i.e., learner-centered, content-centered, assessment-centered, community-centered, Donovan et al., 1999)?

During the second year, the goal was to promote more use of reflection activities. Educators on the campuses were encouraged to have students engage in reflection activities and, where possible, have students share their experiences through surveys. This effort resulted in over 60,000 student reflection experiences and survey data on around 3000 of these reflection experiences (for many reasons, surveys were passed along to only a subset of students).

Approach

In this work, we analyze open ended survey responses where students offered "additional observations" concerning reflection activities they had recently completed. These survey responses represent the perspective of undergraduate students at twelve universities and colleges across the United States. The dataset contains 1082 non-null responses. The analysis focuses on the following questions:

1. What does the data suggest about trends concerning resistance to and/or resonance with reflection activities?
2. What does the data suggest about the bases for resistance to reflection activities? The bases for resonance with reflection activities? How do the different bases manifest in the data? What is the alignment of observed bases for resistance and resonance with the preliminary ideas that sensitized the research?

During the data familiarization phase, a team of five researchers coded random samples of the dataset. This was done to determine whether the data could productively be coded with the reactions framework and to develop an approach for coding the data. This exploratory work confirmed that the survey data did afford exploration of the bases for student resistance to and/or resonance with reflection activities.

For this paper, the two co-authors subsequently coded all of the data (including recoding of subsets of the data that had been coded during the initial work). The responses were coded on two levels. On the first level, each survey response was coded as to whether it indicated resistance and whether it suggested resonance. A single response could be coded for both resistance and resonance. At the second level, each identification of resistance and resonance was then sub-coded using the reactions framework.

Consider the example from the title, "It is helpful but having it due a week out might be too long." At the first level, this student response would be coded as both resonance ("it is helpful") and resistance ("having it due a week out might be too long"). At the second level, the resonance coding would be sub-coded as instrumental (because of reference to the term "helpful"), and the resistance would be sub-coded as activity (because the identified issue has to do with a feature of the activity as designed).

In order to characterize the resulting levels of prevalence that a given code had in our dataset, we first counted the number of responses that had been categorized using that code by one or both of the authors. Using these counts, we then grouped codes into one of three frequency categories: high (>100 instances), medium (>50) or low (<50).

Results

Resonance responses outnumbered resistance responses. From the aggregated results, we found that 67.3% of the responses were positive, and thus evidence of resonance. Two major "bases" dominated our coding: instrumental and activity. *Instrumental* captured student comments about the reflection activity in relation to a goal. For example, instrumental-based resistances included "it didn't do anything" and "I don't feel that I actually learned some specific skills," while instrumental-based resonances included "worth my time," "useful in seeing what progress was made," and "helpful to do in the future." *Activity* captured student comments about the design of the reflection activity itself. For example, activity-based resistances included "at times they were annoying" and "it was worded somewhat vaguely" while activity-based resonances included "the freeform nature of this activity led me to communicate using multiple facets" and "I like that it actually counted for points."

Because a comprehensive description of what we learned for each basis is outside of the scope of this paper, we focus the remainder of the paper on unpacking student reactions relative to three bases. We focus on student reactions that had *situational* bases because such reactions have a practical quality, and concerns can potentially be addressed easily. We focus on student reactions that had *epistemological* bases because of their potentially more challenging significance (e.g., addressing a student who does not appreciate the type of knowledge generated by reflection is more challenging than addressing a situational concern about timing). Finally, we focus on student reactions that had *connectedness* bases because connectedness was something that emerged through our coding effort.

Situational

We observed that 89 student comments described reflection activities positively or negatively in terms of how well these activities aligned with their personal situations. These comments, which we categorized as instances of the situational code, gave voice to student perspectives on how reflection activities can feel like they were 'just in time' or were complete non-starters depending on the situation of the student at the time. 24 of the 89 student comments expressed a positive attitude towards the reflection activity on the basis of the student's situation, while the remaining 65 were expressing a resistance to the activity. This suggests that students may be more likely to notice ill-timed compared to well-timed activities.

Students praised two things when their comments expressed a resonance towards the reflection activity on the basis of situatedness: timing and alignment. The most salient feature was timing with students making comments such as, "It could not have been better timed, I had just finished all of my tests so it was a great time to reflect on all of my midterms" and "The reflection activity was well timed. After a quarter of college, it can be easy for a student to regress to high school habits and lose their motivation to succeed. This activity reminded of why I want to succeed." The second feature was alignment; or when students expressed that the reflection activity gave them the chance to work on exactly what they had been thinking about. For example, one student wrote, "I do a small amount of reflection after each problem already this activity just made me quantify it."

Similarly, the 65 negative situational comments also focused on two issues: timing and existing workload. For instance, one student commented, "The weekend this assignment was assigned was already a busy one and having an assigned, but ungraded assignment, felt like unnecessary added work". Another wrote, "I wish we had done it earlier in the quarter, like maybe after one of the exams". However, occasionally comments also touched on other issues, having to do with the student's unique and situated challenges. For instance, students wrote about how language barriers could make the reflection activity feel more difficult for them: "For the first place, this hard for me to do it because I am not fluent in English but when Ms. [snip] help me, I can work with my essay do well."

Epistemology

A significant number of comments in our dataset described reactions related to what had been learned through the reflection activity from the student perspective. 239 of these comments were coded by us using the conceptual category of epistemology on the basis that they lent insight into the authoring student's assumptions about the nature of knowledge.

Of these 239 comments, 203 were about the student expressing positive feelings towards the reflection activity that they had engaged in because the activity had placed particular value on the kind of knowledge that was involved. One type of knowledge that students resonated with had to do with metacognition or learning about learning. This was succinctly captured in student statements such as: "Good for students to observe their mental states and work habits during assigned programs," "The activity was effective and allowed me to better observe the flaws in my study habits" and, "I think this helps me and the teacher in order to understand my brain and learning style. It also helped me to stayed encouraged that i can do better and i can learn what works in testing for me and what doesn't." Statements such as these highlight the value that some students find in activities that surface the 'muddiest points' in their own learning.

Several student comments also placed emphasis on a more personal kind of self-knowledge. For instance, some students wrote about how reflection helped them with their 'big picture'. Particular examples of such student remarks include: "The activity showed me what are priorities in my life, education and my career", and, "..This activity reminded of why I want to succeed". Some students linked this type of learning to resilience and being critical thinkers. For example, one student wrote, "I liked this activity because it forced us to reach out to our network of people and realize what are true strengths are." While another remarked that, "It was also great to think about the values I've grown up with and those I've takes as my own."

On the other hand, 36 student comments expressed a resistance towards the reflection activity on the basis that the student did not see the knowledge being constructed through the activity as valuable (or even as knowledge at all). For instance, in contrast to previous comments, one student wrote: "Did not seem to be important to anything I see in the future," while another more emphatically expressed, "I DON'T FEEL THAT I ACTUALLY LEARNED SOME SPECIFIC SKILLS". And in a similar vein, another student asked, "Why do we do these, nobody learns anything from them and they are simply taken for busy work and are often incredibly confusing and vague in directions." Collectively these comments foreground the variations in students' epistemological stances.The latter comment also speaks to the blurring between epistemological and cultural bases. The statement not only refers a lack of knowledge resulting from reflection activities, but gives voice to the student's understanding of cultural norms (i.e. "nobody learns" from activities that are generally "taken for busy work").

Connectedness

We observed a type of student comment in the data that did not readily map to any of the bases articulated in the reactions framework. These were comments (84 total, 69 as resonances, 15 as resistances) that cast the reflection based on how well the activity functioned to create *connections* between the student and others.

As a resonance, students viewed the reflection activity as a positive pathway to deeper connections such as through conversations. This sense of finding deeper connection was often expressed in relation to the educator. For instance, one student who felt more connected to the educator wrote: "It was cool. My professors have asked sometimes about how we feel in our own progress, but usually toward the end of quarter. It was nice to have a full term of activity to keep up, and also have so much open communication with the professor." Other student comments address this same idea by further describing the quality of the communication or what those qualities seemed to enable. For instance, one student wrote: "helped me portray my feelings about the project to my professor." This demonstrates an emphasis on emotions related to the project, which can be difficult to elicit or work with in traditional circumstances.

A positive sense of connectedness also involved fellow students. Student comments suggested appreciation of the inclusiveness that their educator's reflection activities created: "I liked how activities like this involved most people in the classroom" and "It went smoothly and was nice listening to the reports from other students going through the same problems!" Similarly, some students remarked that their reflection activities promoted teamwork: "this is something that should be incorporated into every degree plan across the united states to emphasize teamwork in all college graduates".

As a resistance, comments were generally an expression of disappointment for lack of connections and communication. One student captured this with the comment: "It was pretty quiet, I thought we'd talk more with others." Notice that this student might have been given a more contemplative experience, but actually desired collaboration or connection. Touching on the role of feedback, one student remarked: "I wish that we would have discussed the reflection briefly after turning it in."

Discussion

In this paper, we have used survey data to illuminate three bases for reactions students may have to reflection activities. Two of these bases had been identified by our research team prior to the analysis; the remaining basis (connectedness) emerged from the analysis. Our data is likely to underestimate the prevalence of the issues we have identified since students were not required to take the survey and offered only what was on the top of their mind (as opposed to a comprehensive sense of their reaction). Methodological creativity will be needed to further explore student reactions to reflection activities.

In future work, we will explore issues related to coder agreement such as the level of agreement that may be possible given the data source and the interpretive nature of the coding. Additional development of the reactions framework will help resolve coder disagreement. In addition, there is a need to address the overlapping nature of the issues representing the framework, such as the blurred boundary between epistemological and cultural bases for resonating and/or resisting reflection activities. A next step may be to move beyond a list in order to show the relationships among these ideas.

Implications for Practice

Here we touch on two types of implications for practice - the practice of using reflection activities and the practices of educator learning. In other words, we address the questions: (1) what might the overall pattern of resistance and resonance, as well as the observed bases for resistance and resonance, suggest for successfully leveraging reflection activities in engineering education and (2) what do our observations suggest about why "engaging students with reflection" might function as a site of learning for educators themselves? What types of educator learning might arise through work with reflection activities?

Concerning the practice of using reflection activities, the findings about the situatedness of reflection activities serve as a practical reminder to take into account the mundane when asking others to engage in any task. In this case, taking into account practical situational issues by students may be important for helping students engage with reflection. Our findings concerning students reactions on an epistemological level are more difficult to see as immediately actionable, although they are worthy of discussion. Finally, our findings about the role of connectedness in student reactions to reflection is reasonably actionable; the findings suggest that reflection activities that help students feel connected may be more successful.

It is the possibility of such discussion that brings us to the issue of educator professional learning. Indeed, as we (the two authors of this paper) discussed the issue of epistemology as it relates to reflection and engineering student learning, we found ourselves seeing engineering student learning, engineering teaching, and the role of reflection in different ways. In general, as our reactions framework suggests, the different elements of students' reactions to reflection activities are fundamental issues to educator learning. Furthermore, because reflection activities operate on the fringe, the issues may seem more "open" for discussion. For example, while it may arguably be critical to think about the epistemology of engineering, such a conversation may not seem necessary when we "know" that students need to learn theories and do homework problems and take tests. But, when reflection activities come into the picture, there is more room for discussion. Moreover, having such conversations in the context of reflection activities may also be "safer" since reflection activities may feel more experimental. Also, having such conversations in the context of reflection activities may engender more willingness to explore other strategies.

Conclusion

Although the use of reflection activities in engineering education is of interest, the understanding around student reactions to reflection activities remains largely unexamined. The contribution of this work is to empirically describe three bases (two pre-identified and one emergent) for reactions that students may have to reflection activities. Future work can focus on empirical descriptions for other bases as well as extending our analysis of the three bases that we present in this work. Implications of our results for educators include addressing the instrumental value of the reflection activity up front and carefully designing reflection activities to acknowledge situational considerations.

Acknowledgements

This material is based on work supported by The Leona M. and Harry B. Helmsley Charitable Trust through funding of the Consortium to Promote Reflection in Engineering Education (CPREE), a collaboration of twelve educational institutions http://cpree.uw.edu/.

Bibliography

Adams, R., Aldridge, D., Atman, C., Barker, L., Besterfield-Sacre, M., Bjorklund, S., & Young, M. (2006). The research agenda for the new discipline of engineering education. Journal of Engineering Education, 95, 259-261.

Bankhead, R. B., & Olmstead, T. A., & Mannard, J. (2016, June), Changing Student Behavior through the Use of Reflective Teaching Practices in an Introduction to Engineering Course at a Two-Year College Paper presented at 2016 ASEE Annual Conference & Exposition, New Orleans, Louisiana.

Brookfield, S. D. (1995). Becoming a critically reflective teacher. San Francisco: Jossey-Bass.

Cian, H., & Cook, M., & Benson, L. (2016, June), Utility of Post-Hoc Audio Reflection to Expose Metacognition and Strategy Use by First-Year Engineering Students for Different Problem Types Paper presented at 2016 ASEE Annual Conference & Exposition, New Orleans, Louisiana.

Csavina, K. R., & Nethken, C. R., & Carberry, A. R. (2016, June), Assessing Student Understanding of Reflection in Engineering Education Paper presented at 2016 ASEE Annual Conference & Exposition, New Orleans, Louisiana.

Dewey, J. (2007). Experience and education. Simon and Schuster.

Fink, L. D. (2013). Creating significant learning experiences: An integrated approach to designing college courses. John Wiley & Sons.

Donovan, M. S., Bransford, J. D., & Pellegrino, J. W. (1999). How people learn: Bridging research and practice. National Academy Press: Washington, DC.

Ellsworth, E. (1989). Why doesn't this feel empowering? Working through the repressive myths of critical pedagogy. Harvard educational review, 59(3), 297-325.

Goswami, A., & Walia, G. S. (2016, June), Teaching Software Requirements Inspections to Software Engineering Students through Practical Training and Reflection Paper presented at 2016 ASEE Annual Conference & Exposition, New Orleans, Louisiana.

Kolb, D. A. (2014). Experiential learning: Experience as the source of learning and development. FT press.

Niño, D. (2016, June), *Developing Engineering Leaders Using a Reflective Autobiographical Exercise* Paper presented at 2016 ASEE Annual Conference & Exposition, New Orleans.

Radcliffe, D. F. (2006). Shaping the discipline of engineering education. Journal of engineering Education, 95(4), 263-264.

Recabarren, M., Alvarez, C., & Dıaz, M. M. I. (2015). Modifying the student's resistance towards active learning with more active-learning. In Proceedings of the 122nd ASEE Annual Conference & Exposition, June (pp. 14-17).

Schon, D. (1983). The Reflective Practitioner Basic Books. New York.

Shekhar, P., Demonbrun, M., Borrego, M., Finelli, C., Prince, M., Henderson, C., & Waters, C. (2015). Development of an observation protocol to study undergraduate engineering student resistance to active learning. International Journal of Engineering Education, 31(2), 597-609.

Shor, I. (1996). When students have power: Negotiating authority in a critical pedagogy. University of Chicago Press.

Thomas, L. D., & Orand, M., & Shroyer, K. E., & Turns, J. A., & Atman, C. J. (2016, June), Tips & Tricks for Successful Implementation of Reflection Activities in Engineering Education Paper presented at 2016 ASEE Annual Conference & Exposition, New Orleans, Louisiana.

Turns, J. A., & Sattler, B., & Thomas, L. D., & Atman, C. J., & Bankhead, R. B., & Carberry, A. R., & Csavina, K. R., & Cunningham, P., & Faust, D. K., & Harding, T. S., & Yasuhara, K. (2015, June), Reflecting on Reflection: How Educators Experience the Opportunity to Talk About Supporting Student Reflection Paper presented at 2015 ASEE Annual Conference & Exposition, Seattle, Washington. 10.18260/p.24660

Valencia, R. R. (Ed.). (2012). The evolution of deficit thinking: Educational thought and practice. Routledge.

Measurement of the impact of the application of the scientific method on experiments of material sciences

Camila Jaramillo-Monroy
CMUA. Department of Electrical and Electronics Engineering. Universidad de los Andes. Carrera 1E # 19A - 40, Bogota, Colombia.
c.jaramillo56@uniandes.edu.co

Caterine Pardo-Mojica
Parque Cientifico de Innovacion Social. Coorporacion Universitaria Minuto de Dios. Calle 81B # 72B – 70, Bogota, Colombia.
pardoleidy@javeriana.edu.co

Johann F. Osma*
CMUA. Department of Electrical and Electronics Engineering. Universidad de los Andes. Carrera 1E # 19A - 40, Bogota, Colombia.
jf.osma43@uniandes.edu.co

Abstract: *This paper shows the results on the implementation of the scientific and engineering methods on material sciences experiments with 70 secondary students. In order to strengthen the scientific and engineering knowledge of students and make them aware of the resources they have in their territory, three experimental routes to reinforce logics and mathematical analysis with incremental complexity were successfully carried out: granulometric analysis, hydrophobic concrete and DIY water filter. All experiments were based on soil, sand and SiO2 nanoparticles. A baseline and a final line were measured including logics and mathematics, citizenship and reading comprehension competencies. As a result, the experimental routes only impacted on the logics and mathematics skills, passing from 50.0% to 75.0%; while citizenship and reading comprehension, competencies kept constant during the process as expected. In addition, a test was carried in order to measure the regional and local knowledge that the students acquired during the process, where local knowledge (73.6%) was preferred over the regional one (60.8%).*

Introduction

One of the main problems in science and engineering education are the use of traditional teaching methods, the lack of prepared teachers able to develop new teaching dynamics and the difficulty of linking the learnings with the students context (Riedinger et al. 2010; Karsli and Ayas 2014). As an alternative to these problems, in the last decades, in a globalized context, technology has become a necessary tool in teaching and learning processes of the scientific method. The implementation of analysis methods has strengthened the knowledge in youth, as it can be seen in various studies (Cocking, Mestre, and Brown 2000).

Creativity is a key factor, not only because it invigorates the learning process, but because it allows students to understand the phenomena more deeply, allowing the teaching-learning process strengthens the cognitive development. This is reflected in studies such as those conducted by Usta and Akkanat, Zhou et al., Veselinovska, Gudeva and Djokic (Usta and Akkanat 2015; Zhou et al. 2016; Veselinovska, Gudeva, and Djokic 2011), who pointed out that scientific and engineering creativity "can be defined as, depending on previous experiences and knowledge, sensitivity to problems and their solutions, understanding nature of the science and fascinating with it, and development of new, extraordinary and useful scientific knowledge, experiments, theories and products" (Usta and Akkanat 2015). Thus, it can be seen that the development of scientific creativity is important for the development of science education, since it allows students to develop skills for solving problems creatively and sensitively, and enhancing their motivation towards science and engineering.

Currently, there are two major trends converging toward new ways to improve the teaching-learning of science and engineering. On one hand, researchers in cognitive sciences are producing new knowledge about the way the brain works and also the processes involved in the acquisition of new knowledge and understandings; on the other hand, public education policies and educators are focusing on reforming education in science and mathematics, so that learning processes can be improved. The interaction of these two currents has led to the design of science curricula and instructional practices based on rigorous research on how people learn (Cocking, Mestre, and Brown 2000; DeHaan 2005; Hurd 1998).

Whilst it is essential to foster changes in formal education, it is important to emphasize that changes in non-formal education processes can lead to important transformations in the learning and educational processes. Danciu (2014) highlighted that processes involved on non-formal education, integrated to the use of non-conventional methods, are more effective in a cognitive level and allows generating a direct relation between new learning and the student in a short-term (Danciu 2014). The study performed by Riedinger et al., showed that the use of non-conventional methods prepares the students for more complex knowledge in the domain of science, turning into a tool for the production of innovations (Riedinger et al. 2010).

The processes of non-formal education in science and engineering have been characterized for the use of creativity, which has demonstrated to lead to a better learning in youth, as shown in the study done by Johnston et al., who understood "play" as a model for science teaching, integrating not just the student, but also its family and community. In this sense, learning should not be limit to the classroom, but it should go beyond, involving the life of the students (Johnston et al. 2011) and their local context.

Other authors have shown the importance of science education as a strategy for recognition of the local. In the case of Melo et al. (2001), they highlighted the importance of linking science education with the realities of the students; this does not means an exclusion of the global, by the contrary, the authors stated that local knowledge can turn into a broader understanding of the global, as there is no comprehensive division between local and global, but a close relationship that parts from the local to broadly understand the global. In their study, Kumar and Altschuld (1999), highlighted the importance of integrating technology in the teaching-learning process and in all facets of human life (Kumar and Altschuld 1999).

In several countries (MEN 2004), standards have been developed to regulate how education must be implemented in science and engineering, seeking for the development of scientific and design skills to explore phenomena and solve everyday problems. However, a decrease in the development of science skills in youth can be noticed. In response to this situation, local authorities implement non-formal educational programs, which seek to strengthen science knowledge in youth, so that educational quality can be improve.

This paper evaluates the implementation of a non-formal educational project with 70 secondary students, in the town of Soacha, Colombia that included a total of 7 practices, 3 of them related to experimental routes. The three experimental routes based on soil, sand and SiO_2 nanoparticles were carried out, to strengthen the scientific and engineering knowledge of students and make them aware of the mining resources they have in their territory. The first experimental route was a granulometric analysis; the second consisted on the production of hydrophobic concrete using nanoparticles; and the third was the construction and test of a DIY water filter. 30 different competencies were measured during the non-formal educational project to determine the base line and final stage of the 7 practices, including competences in citizenship; environmental awareness; innovation and entrepreneurship; science, technology and design; reading and comprehension skills. However, other specific instruments were designed to only measure the impact of the three experimental routes, including an instrument to measure the regional and local knowledge that the students acquired during the routes. Our objective was to determine the effect of the experimental routes in the competencies of logics and mathematics; however, the specific instruments included measures of citizenship and reading comprehension competencies to evaluate possible unintentional effects of the experimental routes in these areas.

Materials and methods

Groups

The town of Soacha is a well-known mining area located nearby Bogota; however, the added value in the use of the soil components is very low on local inhabitants. Thus, generating knowledge related to biotechnology and nanotechnology based on a better exploitation of soil resources was a major objective of this scientific method experimental pilot. Three experimental routes were designed based on the use of soil components, each of them with a different complexity in terms of experimental design. 70 students between 14 and 19 years old were divided into groups of 6 people to carry on the experimental routes. A group of 14 students, 20% of the total initial population, was monitored as the sample group (SG_1).

Experimental routes

Three experimental routes were created in order to introduce the students in the scientific method. Each experimental route contributed to increase the complexity on the analysis in order to train the students into more precise and complex systems to experiment with. Table 1 summarizes the complexity of each experimental route and details the characteristics of the problem that had to be solved by the students. Although the routes were conceived to promote only competencies of logic and mathematical analysis, the specific tests included parameters to measure citizenship and reading comprehension in an attempt to identify if other competencies were addressed unintentionally.

Table 1: Characteristics of the three different experimental routes.

Characteristics	Experimental route 1 Granulometric Analysis	Experimental route 2 Nano Concrete	Experimental route 3 Water Filter
InputVariables	X= Sand mixture	X= Presence or not of nanoparticles in concrete	g= Waste solid material s= Geometry of the container m= Concentration of suspended solid in waste water
Output Variables	$Y = Y_1 + Y_2 + Y_3 + Y_4$ Y_1= Small particle sand Y_2= Medium particle sand Y_3=Large particle sand Y_4=Very large particle sand	Y= Degree of hydrophobicity of concrete	Y= Water quality
Replicas	1	3 for each case	2
Measurement	Direct	Indirect	Indirect
Infered result	Quantity of sand of each type	Degree of hydrophobicity	Treated water quality
Measured variable	Weight	Contact angle	Color Turbidity Suspended particles
Measuring method	Weighting using a balance	Geometric quantification using a protractor	Microscopic observation
Type of system	Linear	Non-linear	Non-linear
Formula	$x = \sum_{i=1}^{4} Y_i = Y$	$Y = f(x)$	$Y = f(g, s, m)$

Granulometric analysis

The first experimental route was a granulometric analysis and consisted in 7 hours of experimentation. The experiment consisted in weighing an initial sand sample, passing it through 3 different sieves, first through the one with the largest pore size and at the end through the one with the smallest pore size (Fig. 1 Left). Then a qualitative and quantitative analysis of the sample retained in each of the sieves was performed. This experiment was designed as a linear system, with a direct measurement, without losses and with just one replica per group. It had 1 input variable: the sand sample (*X*); and 4 output variables: very large size particle sand, large size particle sand, medium size particle sand and small size particle sand (Y_1, Y_2, Y_3 and Y_4). This experiment gave the students the basis for statistic measurements and they learned how to use equipment such as weighing scales and sieves.

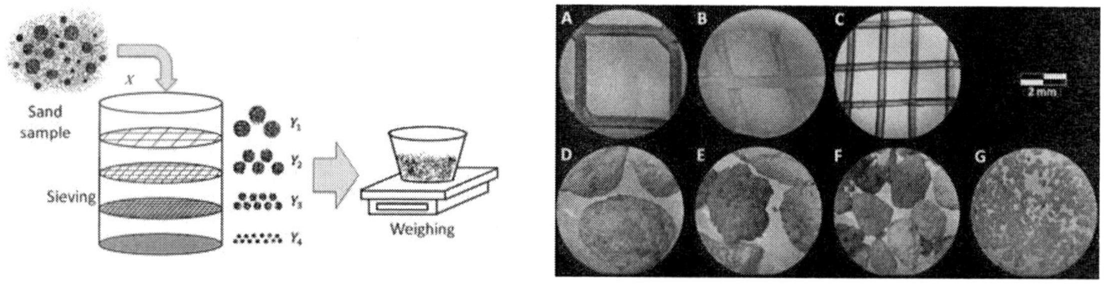

Figure 1: Left: Experimental route 1: Granulometric analysis. Right: Different pore size and particle sand size obtained in Experimental route 1: Granulometric analysis. A) Sieve with pore area of 13.14±0.45 mm². B) Sieve with pore area of 3.90±0.09 mm². C) Sieve with pore area of 1.69±0.3 mm². D) Very large size particle sand. E) Large size particle sand. F) Medium size particle sand. G) Small size particle sand.

Hydrophobic concrete using nanoparticles

The second experimental route consisted in 4 hours of experimentation to produce concrete blocks and evaluate the surface hydrophobicity. The students prepared 6 concrete blocks using local materials of their own territory and applied 85 nm SiO_2 nanoparticles, especially on the surface, to 3 of the 6 concrete blocks made by each team. After letting the blocks dry, they measured the contact angle of a drop of water applied on the surface of the block (Fig. 2 Left). This was done by taking a photograph, and then measuring the angle between the base of the drop and the tangent line to the surface of the drop of water. This was an indirect measurement because an analysis of the result must be performed in order to conclude about the degree of hydrophobicity of the surface of the material. In addition, this experimental route was non-linear, had one input variable: the presence or not of nanoparticles in the concrete (*X*); and 1 output variable: the degree of hydrophobicity of the concrete (*Y*). Three replicates were performed by each group, which implies the use of more advanced statistics. It can be seen that this was a much more complex experiment, compared to experimental route 1.

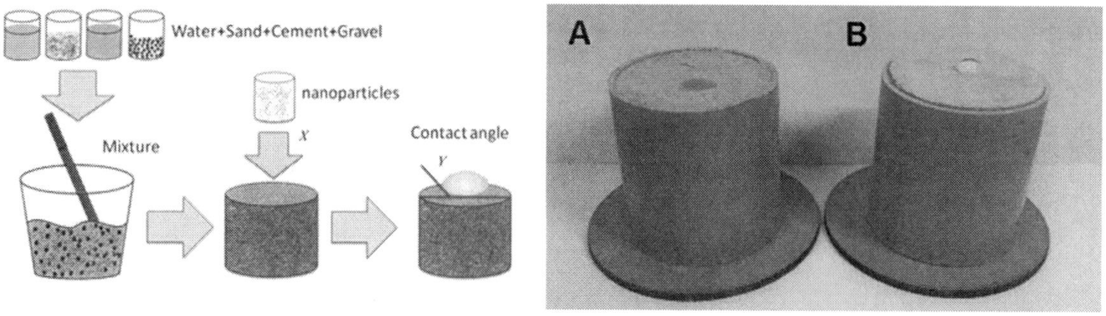

Figure 2: Left: Experimental route 2: Hydrophobic concrete using nanoparticles. Right: Hydrophilic (A) and hydrophobic (B) concrete blocks obtained in Experimental route 2.

Building a DIY water filter

The third experimental route consisted in the construction of a water filter during 4 hours of experimentation. The materials needed included a plastic bottle as container, sand, gravel, activated carbon and cotton; making the composition and distribution of the filter one input of the experiment (*s*). After building the filter, the students tested it, using a sample of artificial wastewater (*m*) composed by water and local waste solid material (*g*) previously collected by the same students. They performed 2 replicas and focused on a qualitative analysis of the water (*Y*) (Fig. 3 Left). In this case there was neither a unique design of the filter nor a unique sample of wastewater. Each group built its own filter and prepared the wastewater sample, therefore it was a multivariable and non-linear experiment with a major complexity than the previous experimental routes. The measurement was indirect, since observable characteristics such as color, suspended particles and turbidity, gave the information necessary for the analysis of quality of the treated wastewater.

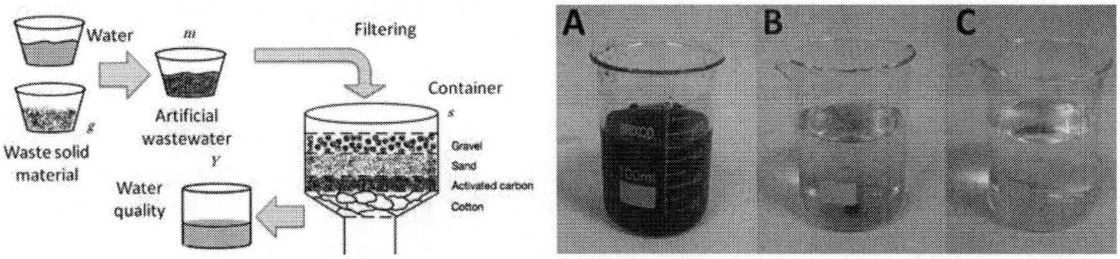

Figure 3: Left: Experimental route 3: Building a DIY water filter. Right: Water filtered in experimental route 3. A) Artificial wastewater before passing through the filter. B) Filtered water. C) Tapwater.

Methodology of regional and local knowledge test

A written test was carried out during the third experimental route in order to measure the regional and local knowledge the students acquired during the experimental sessions. The test consisted on a crossword with 30 different questions: 11 about the town of Soacha (local) and 19 about other towns in the department of Cundinamarca (regional). The results used in the analyses were taken from a random sample group (SG_2) of 20 students. The level of complexity of the test was medium and included questions of popular knowledge.

Methodology of measurement

Due to the heterogeneous population, the baseline was measured through a test after the first two experimental routes were carried out; so that the students already had basic knowledge about the use of equipments and basic statistics. The final line was measured through a test after the last experimental route was carried out, in order to evaluate the knowledge the students acquired during the experimental sessions. Both tests had similar and related questions, so that they could be compared. The baseline test included 11 multiple choice questions with single correct answers: 2 of logics and mathematics, 6 of citizenship competencies and 3 of reading comprehension; and the final line test included 11 multiple choice questions with single correct answers: 2 of logics and mathematics, 5 of citizenship competencies and 4 of reading comprehension. The questions were randomly taken from different Colombian official state exams from 2012 to 2014. The baseline test was taken by 70 students; however, only 20% of the population was randomly selected and tracked as the SG_1, equivalent to 14 students for the final line.

Methodology of social divulgation

Two science fairs were organized in order to show the projects related to the experimental routes, done by the students, to the local inhabitants. The first fair, had a massive attendance of more than 400 people from all ages, and the students presented the results of the experiments they did in the experimental routes 1 and 2 to the community, highlighting the impact of the projects in their municipality. In the second fair, with an attendance of more than 300 people, the students presented the final results of the experimental routes 1, 2 and 3, showing the social transformations that the processes could generate in the community of Soacha. Moreover, the students exposed all the scientific and technical procedures carried out throughout a newspaper mural

Results

Granulometric analysis

The first experimental route was a granulometric analysis to promote data analysis, and experimental design, where students separated and analyzed a sand sample (*X*) into 4 output variables: very large size particle sand (Y_1, Fig. 1D), large size particle sand (Y_2, Fig. 1E), medium size particle sand (Y_3, Fig. 1F) and small size particle sand (Y_4, Fig. 1G). During this experiment, students used three different sieves with pore areas of 13.14±0.45 mm^2 (Fig. 1A), 3.90±0.09 mm^2 (Fig. 1B) and 1.69±0.3 mm^2 (Fig. 1C). After experimentation, students obtained the following results summarized in Table 2. The first sieve retained the 0.09%, the second sieve the 3.06% and the third sieve the 3.91% of the sand sample.

Table 2: Results obtained in Experimental Route 1.

	Pore area (mm^2)	Retained Weigh (g)	Retained percentage (%)	Cumulative retained percentage (%)	Passing through percentage (%)
Sieve A	13.14±0.45	0,90	0,09	0,09	99,91
Sieve B	3.90±0.09	29,60	2,97	3,06	96,94
Sieve C	1.69±0.30	8,50	0,85	3,91	96,09
Bottom	-	958,80	96,09	100,00	-

Hydrophobic concrete using nanoparticles

The second experimental route was an experiment to produce concrete blocks and evaluate the surface hydrophobicity to strength the discussion on the use of engineering solutions to real life problems such as social housing construction. The students prepared 6 concrete blocks using local materials of their own territory and applied 85 nm SiO_2 nanoparticles (*X*), mainly on the surface, to 3 of the 6 concrete blocks. After letting the blocks dry, they measured the contact angle (*Y*) of a drop of water applied on the surface of the block, and found that in all cases, the blocks with SiO_2 nanoparticles (Fig. 2B) presents a contact angle greater than 115°, therefore they can be consider hydrophobic and in some cases even superhydrophobic (~140°); while the blocks without nanoparticles presented contact angles smaller than 15°, therefore hydrophilic (Fig. 2A).

Building a DIY water filter

The third experimental route was the construction of a water filter to strength the use of design processes for the solution of water treatment. The students used plastic bottles of different sizes as containers and different proportions of sand, gravel, activated carbon and cotton; therefore all filters were different. Students tested the filter using a sample of artificial wastewater (*m*), composed by water and local waste solid material (*g*) previously collected by them. After the filtering process of the artificial wastewater, students proved an enhancement in the wastewater quality. Figure 3 shows an example of the original wastewater (Fig. 3A), the filtered water (Fig. 3B) and tap water (Fig. 3C), remarking the visible changes by the action of the DIY water filter.

Results of regional and local knowledge test

A written test was carried out in order to measure the regional and local knowledge the students acquired during the experimental sessions. The test consisted on a crossword with questions about Soacha (local), and about other towns of Cundinamarca (regional). The test was an informal method in the form of a crossword (game) as part of a newspaper given to the students during an exhibition of their projects; therefore, students answered it spontaneously. Accordingly with our attempt to measure the geographical impact of our experimental route, due to the non-formal education characteristics of our model incorporating non-conventional methods to develop scientific methodological thinking in students, other authors have presented similar methodologies. In the study performed by Danciu (2014) in Romania, she used non-conventional methods, such as exhibitions and cultural and artistic manifestations, to enhance the development of new knowledge in adolescents.

Figure 4 shows the results obtained in the test by SG_2. It can be seen that the students answered a larger number of correct questions related to the local knowledge (73.6% ± 17.9) provided by the experimental routes, rather than the questions related to the regional topics (60.8% ± 26.5). These results were expected, since the experimental routes and discussions were related with the local resources and industry, rather than with regional or global aspects. Other studies, like the one developed by Dunbar in Canada (Dunbar 2000), also focused on the local aspects when carrying out the scientific method to develop the scientific thinking by analyzing local problems. On the other hand, Kaptan and Timurlenk, in their study in Turkey (Kaptan and Timurlenk 2012), highlighted the main problems of science education, taking into consideration the context of the students by starting with the identification of local problems, in the same way that we initiated our experimental routes.

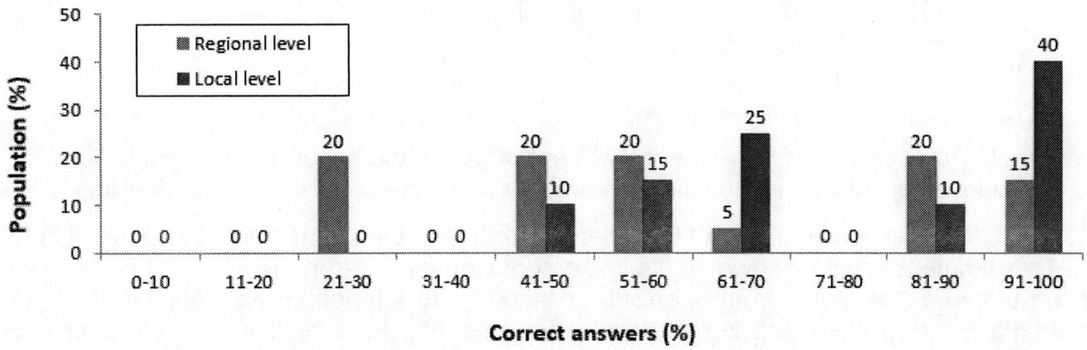

Figure 4: Achievement at the regional and local level by SG_2.

Results of measurement

Due to the heterogeneous initial population, the baseline was measured through a test after the first two experimental routes were concluded in order to reduce heterogeneity in the baseline. The final line was measured through a test, after the last experimental route was carried out. Both tests had similar and related questions of logics and mathematics, citizenship competencies and reading comprehension. Figure 5A shows the results obtained in both, the baseline and the final line by the SG_1. It can be seen that the students (SG_1) improved in the number of correct answers from the baseline to the final line. However, a more detailed analysis revealed that the experimental routes only impacted on the logics and mathematics skills, passing from 50.0% ± 51.9 to 75.0% ± 25.9; while citizenship competencies and reading comprehension skills kept constant during the process (Fig. 5B).

Discussion and conclusions

The development of these experimental routes was carried out during a period of three months. This allowed us to show that in a short period of time it is possible to develop logics and mathematical skills based on local knowledge in the students, through the use of scientific and engineering tools and methods. However, as originally designed and expected, the experimental routes impacted only on the ability to use logics and mathematics in the solving of problems and analysis of data, and for the development of citizenship and reading comprehension skills, other approaches or activities must be taken into consideration.

The three experimental routes carried out by the students of the town of Soacha, while emphasizing in the use of scientific and engineering tools and methods in daily life, sought to strengthen the recognition of local aspects such as resources and industry of their territory, specifically the sand mining industry. According to (González-Labrada and Pérez-Ponce de León 2015), in their studies it has been shown that strengthening the knowledge of the local context allows students, in a long-term, to develop a better understanding of the global context, making possible the creation of a critical thinking at a global level, based on the own thinking.

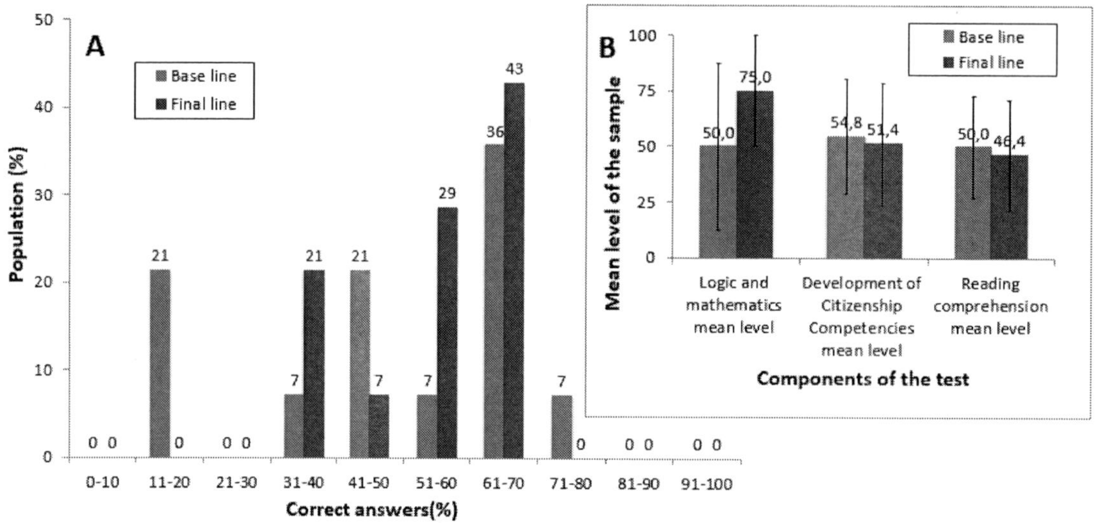

Figure 5: Achievement of SG_1. A) Achievement of SG_1 in the baseline and the final line. B) Achievement of SG_1 in the different components in the baseline and the final line.

In this work, the experimentation of the scientific and engineering methods was present in the three experimental routes; however, its implementation included the increment of complexity of the problem statements from one route to another, to strengthen and develop logic and mathematical new skills and knowledge in the students. In a similar way, other national initiatives to strength the knowledge in science and engineering through the use of materials and resources, has been tested by the implementation of projects and seminars in biotechnology, where young people developed scientific attitudes through the use of materials and resources (González-Labrada and Pérez-Ponce de León 2015). The projects developed in the seminar, like the experimental routes implemented by us, focused on the immediate environment of the students, proving that beyond the generation of new knowledge, the recognition of the local context is crucial.

Other studies, like the one conducted by Gomes Heck et al. (2012) in Rio Grande do Sul, Brazil gave a step further in the initiation of students in science and technology, with the reform of the curricular secondary education for solving everyday problems a formal education strategy. Through the development of real experiments, students understood science and engineering in a practical way. In a similar way, we validated that the designed experimental routes, as part of a greater 7 stages project, impacted only onto the logics and mathematical competencies of the students, as originally designed and intended. At the same time, we included the focus on the use of local resources, that was evidenced in the carried tests. This could empower the youth and make them interested in engineer areas, as they would be able to apply the knowledge to their daily life.

References

Cocking, R. R., Mestre, J. P., & Brown, A. L. (2000) New Developments in the Science of Learning: Using Research to Help Students Learn Science and Mathematics. *Journal of Applied Developmental Psychology*, *21*(1): 1 – 11.

Danciu, E. L. (2014) Specificity and Efficiency In Using Non-Conventional Methods For Adolescent Education. *Procedia - Social and Behavioral Sciences*, *163*: 104 – 109.

DeHaan, R. (2005) The Impending Revolution in Undergraduate Science Education. *Journal of Science Education and Technology*, *14*(2): 253 – 269.

Dunbar, K. (2000) How Scientists Think in the Real World: Implications for Science Education. *Journal of Applied Developmental Psychology*, *21*(1): 49 – 58.

Gomes Heck, T., Maslinkiewicz, A., Gil Sant'Helena, M., Riva, L., Lagranha, D. J., Moreno Senna, S., ... Homem de Bittencourt, P. I. (2012) Scientific initiation in the High School: a model for bringingthe basic school closerto the university through the scientific method. *Ens. de Ciencias E Da Matemática*, *8*(2): 447 – 465.

González -Labrada, R., & Pérez -Ponce de León, N. P. (2015) La actividad investigadora escolar en el proceso de enseñanza-aprendizaje de las Ciencias Naturales. School researching activity in the learning -teaching process of Natural sciences. *Innovación Tecnológica*, *21*(3): 1 – 12.

Hurd, P. D. (1998) Linking Science Education to the Workplace. *Journal of Science Education and Technolog*, *7*(4): 329–335.

Johnston, A., Butler, M. B., Moore Mensah, F., & Williams, B. (2011) Playing with Science: Models for Engaging Communities. *Children, Youth and Environments*, *21*(2): 312–324.

Kaptan, K., & Timurlenk, O. (2012) Challenges for Science Education. *Procedia - Social and Behavioral Sciences*, *51*: 763 – 771.

Karsli, F., & Ayas, A. (2014) Developing a Laboratory Activity by Using 5e Learning Model on Student Learning of Factors Affecting the Reaction Rate and Improving Scientific Process Skills. *Procedia - Social and Behavioral Sciences*, *143*: 663 – 668.

Kumar, D., & Altschuld, J. (1999) Evaluation of Interactive Media in Science Education. *Journal of Science Education and Technolog*, *8*(1): 55 – 65.

Melo S., C., Mondragón, C., Wilches, F., Valbuena, E., & Bolaños, P. (2001) *Desarrollo de proyectos escolares en Biotecnología. Propuesta de trabajo para la enseñanza aprendizaje de las ciencias naturales en el nivel de educación media* (pp. 1 – 6). Colciencias. Retrieved from http://www.colciencias.goc.co/sismbiosis/forovirtuales.htm

MEN. (2004). *Formar en ciencias: ¡el desafío! Lo que necesitamos saber y saber hacer*. Espantapájaros Taller. Retrieved from http://www.mineducacion.gov.co/cvn/1665/articles-81033_archivo_pdf.pdf

Riedinger, K., Marbach-Ad, G., McGinnis, J. R., Hestness, E., & Pease, R. (2010) Transforming Elementary Science Teacher Education by Bridging Formal and Informal Science Education in an Innovative Science Methods Course. *Journal of Science Education and Technology*, *20*: 51–64.

Usta, E., & Akkanat, Çiğdem. (2015) Investigating Scientific Creativity Level of Seventh Grade Students. *Procedia - Social and Behavioral Sciences*, *191*: 1408 – 1415.

Veselinovska, S. S., Gudeva, L. K., & Djokic, M. (2011) The effect of teaching methods on cognitive achievement in biology studying. *Procedia Social and Behavioral Sciences*, *15*: 2521 – 2527.

Zhou, S., Han, J., Koenig, K., Raplinger, A., Pi, Y., Li, D., ... Bao, L. (2016) Assessment of scientific reasoning: The effects of task context, data, and design on student reasoning in control of variables. *Thinking Skills and Creativity*, *19*: 175 – 187.

Acknowledgements

This work was supported by the Science and Technology General System of Royalties through the Government of Cundinamarca, Pontificia Universidad Javeriana, Corporacion Universitaria Minuto de Dios and Universidad de los Andes under the project "Smart Town". The authors want to thank the Government of Cundinamarca and to all the research groups of the participant universities of the related project: Pontificia Universidad Javeriana, Corporacion Universitaria Minuto de Dios and Universidad de los Andes.

Copyright statement

The following copyright statement should be included at the end of your paper. Substitute authors' names in final copy only.

Project and problems based learning at the Facultad de Minas of the Universidad Nacional de Colombia: a teaching strategy

Velez Juan Manuel.
Universidad Nacional de Colombia, Medellín, Colombia.
jmvelez@unal.edu.co

Castro Karem Johanna.
Universidad Nacional de Colombia, Medellín, Colombia.
kjcastr0@unal.edu.co

Fula Manuel Alejandro.
Universidad Nacional de Colombia, Medellín, Colombia.
mafula@unal.edu.co

Abstract: *In 2007 the Universidad Nacional de Colombia (UNAL) began to implement legal agreement 033(2007). As part of the process, the Facultad de Minas (FacMinas) understood the need for the adequate training of engineering students. It focused on providing the students with enough skills and tools that allow them to engage in project planning, and problem solving by means of high complex strategies.*

A new sequence of three courses named Engineering Project Seminar was established in 2009. It was added to the curriculum of all careers belonging to the Faculty of Mines.

It aims to promote the use of active methodologies in teaching processes as well as school learning. With the help and input of experiences reported by teachers, students, and others accumulated throughout the seven years that the course has been active, the FacMinas adopted the methodology of PjBL and PBL as a means to reach the end goal.

KEYWORDS

PBL, teaching

INTRODUCTION

In all country associations, such as the G20, the OECD, and the BRICS, a common denominator is present. All of these seek to improve their economies by means of developing national industries and establishing a standard of both quantity and quality for their products.

The mass production of complex products in a sustainable manner requires the active involvement of diverse fields of work, and specialties that are outside of the Engineering common knowledge bounds. This is understood by the University and the Facultad de Minas, which is why the latter has sought to implement the PBL, and the PjBL methodologies so that the formation of students ensures they are competent enough to enter the workforce under these new conditions.

For several decades, Colombia has sought out economic and social strategies that may aide it in exiting the scenario characterized by technological dependence, the marketing of raw materials lacking in added value, low income or informal employment, and high poverty and inequality rates that it currently faces (ECLAC-SEGIB 2010).

The relevance of technology and innovation in economic growth over the last decades has promoted the use of the terms "knowledge society" and "innovation economy" to describe current socio-economic dynamics (David and Foray 2002).

The concept of knowledge society is based on: 1) An accelerated production of knowledge; 2) The expansion of intangible capital at the macroeconomic level; 3) Innovation, turned into the dominant economic activity 4) The revolution of the instruments of knowledge, embodied in information technologies and the implementation of science (David and Foray 2002).

The economic and social progress of countries, as well as business success depends on the efficiency and effectiveness with which techno-scientific knowledge is produced, transferred, disseminated and incorporated into products and services (Reis 2004). The activities developed in engineering have a potential to generate cumulative scientific and technological capabilities in individuals, companies, and regions (Malpas 2000).

According to (Auyang 2006) these capabilities are based on four fundamental components: intellectual capital, human capital, physical capital and social capital.

It is necessary to recognize which global aspects that may be implemented in education and use them to form graduates capable of working locally with a global vision in mind.

DIAGNOSIS

Engineering plays an important role and has the great responsibility of leading the way towards the technological change necessary for development in Colombia. It bridges the gap between science and technology, as well as technology and innovation (Malpas 2000). However, the important role that engineering plays does not correspond solely to the capabilities of investigation and knowledge transfer. These elements, therefore, generate challenges for the complex curricular, cultural, and organizational structures of engineering education.

In this area, important changes are required: new teaching methods, more flexible and attractive curricula for youth, establishing greater connections between the educational and productive systems, among others. The faculty should make a great emphasis on this type of pedagogy so that all professors are adequately prepared to teach engineering. Just as a university would not accept a teacher who has problems or holes in their theory formation i.e. Math, Physics, etc. a university should not accept a teacher who does not know how to properly broadcast information to his or her students. It is emphatic to implement different pedagogical methodologies so that all his or her students understand said information. Being a well-formed engineer, and being capable of understanding all phenomena associated with each field does not mean that these individuals are sufficiently prepared to be professors. There should be a program implemented by faculties that specifically focuses on the formation, and active capacitation of its professors. These elements are vital for the formation of leaders, teachers and graduates who act as promoters of innovation in their workspace and in their civic coexistence, instead of information manipulating engineers.

"The engineer of the future must be able to act as an innovation manager and entrepreneur, with the ability to run a company, develop business plans, develop marketing strategies and generate economic growth. To achieve this professional profile it is necessary to develop new knowledge and skills within the engineers. This calls for major changes in curricula around the world, which will increasingly focus on sustainable development, internationality and interdisciplinary approaches "(Fuchs 2012).

This was not the case of the graduates of the Faculty. In the successive curricular reforms, the evaluations of the graduates found deficiencies in two important aspects: Skills in resource management in projects or companies, and the ability to perform interdisciplinary and team tasks.

The graduates of the Faculty were very strong in the mathematical approach of problems and solutions involving the technical training of their specialty. However, they were not always able to integrate their fraction of knowledge with that of other fields of work when solving problems, not only in areas of engineering, but also humanities.

During the last fifty years, there have been a considerable number of diagnoses, studies and prospects on engineering education made by expert panels using different methodologies. (Salvat & Navarra, 2009; STREVELER & SMITH, 2006). Some conclusions identified in all these studies are:

- The need to re-engineer the teaching and learning processes used in engineering education.
- Recognize the practice of engineering and research in engineering as fields of action equally important for the graduate.
- Overcome the reduced vision that identifies engineering with a discipline or occupation.

However, professional training based on the engineering sciences and courses of a specialty is a limited model for the training of the engineer of the 21st century. It does not confer to the graduate the attitudes and skills that the current world scenario demands of the practice of engineering. The graduate faces a Workforce characterized by high competitiveness and governed by innovation, in a planet with an exponential growth in population and with great environmental fragility.

Higher education institutions fail to incorporate innovation into their principles and purposes, missionary functions and organizational structure. Specifically, in the processes of training Future engineers and researchers there are three broad fields of action: 1) curricular renewal and pedagogical strategies; 2) research on engineering education and 3) academic organizational structure (Salvat & Navarra, 2009; STREVELER & SMITH, 2006).

FACMINAS STRATEGY

The proposal of innovation of the education in engineering in the Faculty of Mines follows the international and national tendencies. The engineering innovation strategy consists of three main elements, closely related, as illustrated in Figure 1.

In order to respond to these challenges it is necessary to make modifications to the structure used in the training of the engineer (curricula, contents, methods). The training should focus on active learning methodologies, in which the student is the center and main actor of the learning process (Savin-Baden, 2007).

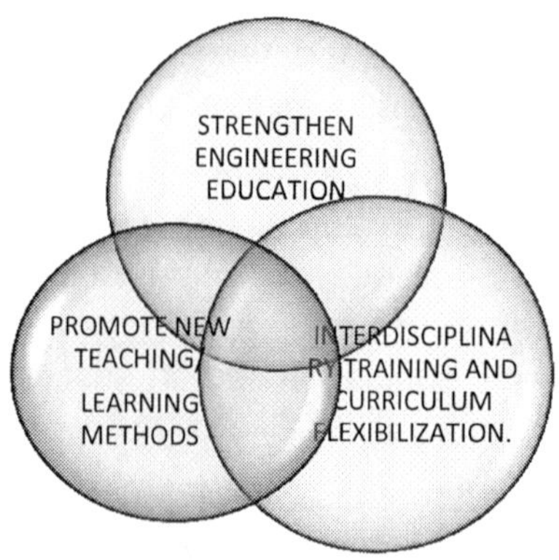

Figure 1: The innovation strategy in engineering education.

New Pedagogies in Facultad de Minas: Problem-Based Learning and Project-Based Learning

Problem-Based Learning (PBL) and Project-Based Learning (PjBL) methodologies aim to organize learning processes in a way that actively engages students to find answers for themselves.

The concept that identifies these methodologies has references in classic and Chinese philosophy. Socrates questioned his apprentices to activate what he called latent knowledge; Confucius, in his part, expressed the importance of active learning in the following lines: "Tell me and I will forget, show me and I will remember, get involved and I will understand." More recent manifestations of these strategies are found in concepts such as 'Learning by Doing', 'Experimental Learning' and 'Student-centred-learning', which show the constant search for effective learning strategies (Savin-Baden 2007).

It is "An apprentice-centered approach that empowers students to research, integrates theory and practice, and applies knowledge and skills to develop a viable solution to a defined problem" (Savery 2006). The use of PBL has spread rapidly in medical education and in other disciplines such as law and engineering. The problem that the student wants to solve is the motor of the PBL method and the starting point of the learning process. In this context, students are able to develop skills in teamwork, problem solving and leadership within a framework where the student is in control of what he or she needs and should learn (De Graaff and Ravesteijn, 2001).

The PjBL method reflects, in general, the proper mode of action of the professional practice of engineering, i.e. when facing a complex problem or opportunities, the individual responds with a project strategy based on real-world situations; and provides learning experiences with which the student identifies, the method also finds and uses the appropriate resources. The work is done in groups, preferably interdisciplinary; learning is active, integrated, cumulative and connected to the real world; communication skills are developed, both inside and outside the working team.

The PjBL method is assumed as a student-centered teaching-learning process, where it applies basic, professional and transversal knowledge to manage the solution of real problems. This gives students the opportunity to present their own solutions in engineering projects and allows them to have confidence in the proposal and realization of work projects within the engineering discipline chosen by them (Chandrasekaran et al., 2012).

According to current international and national scenarios in which the exercise of engineering is developed, it is required to provide graduates with transversal competences, additional to the disciplines of each professional emphasis:

1. Ability to understand the dynamics of the current socio-economic system: economic, political, technological, human, environmental and ethical.
2. Strong communication skills, teamwork, knowledge integration.
3. Competences in the management of physical and human resources.
4. Leadership, ethics and lifelong learning.

The model of learning based on problems and projects that has been implemented in the Engineering Projects Seminars I, II and III at the Faculty of Mines of the National University of Colombia assimilates and uses aspects of various references with international backgrounds. Their principles are described below (Figure 2):

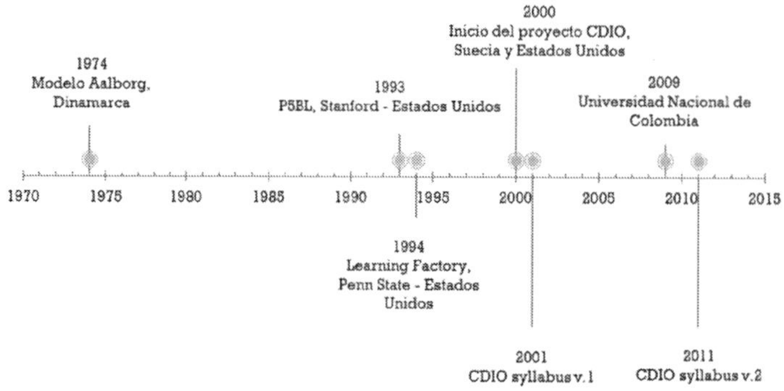

Figure 2: Pedagogical models of reference (Velez, Benjumea, Castro & Ríos (2017))

Background and Current Structure

The general objective of Engineering Project courses was to cover the fields of training related to leadership, effective communication, project management, teamwork, leadership and teacher training; Rather than engineers manipulating information. Training with this vision should improve the profile of the graduate and, certainly, increase the motivation to enter engineering programs (Kolmos, Du, et al., 2008).

This methodology may be applied to a course or a specific part of it, to an area or component of the curriculum or to a curriculum (Kolmes, Holgaard and Dahl 2013, Harmer 2014, Gerhart and Fletcher, 2011).

METHODOLOGY.

The inclusion of this curricular component in the undergraduate curricula of the FacMinas had the general objective of instilling personal values (ethical, social and environmental) and developing professional competences, transversal and essential, in the formation of the engineer of the 21st century and later on. The strategy used to reach the objective was to implement an academic space where the student must face the challenges recognized in the exercise of the engineering profession, among which are:

- Identify and pose problems, needs and opportunities (PNO)
- Formulate projects in the field of engineering and propose the technical and economic feasibility.
- Develop a project start-up exercise, which includes the components: management, financial, economic, social, environmental, regulatory and legal.
- Work in collaborative and interdisciplinary environments.
- Express their arguments and concepts, in a clear manner by oral, written, graphical means.
- Identify and measure the environmental, social and political impacts derived from engineering projects.

For the structuring of the group of subjects, it was taken into account that the Agreement 033 of 2007 of the Higher University Council establishes that the curricular engineering program consists of three major components identified as foundation, disciplinary or professional training and free choice. Under the concept that training in Engineering Projects is an essential part of the disciplinary (professional) component, the Engineering Project Seminars were created, offering three courses distributed in the different stages of the engineer's training, for the twelve (12) undergraduate programs of the Faculty of Mines (FacMinas). Some important elements that support this structure are:

- The search for solutions to problems originating in industries and in communities (or the identification of needs and opportunities), through the formulation of projects leads the student, and the teacher, to the terrain of open questions, which happen to be unstructured, complex and with multiple solutions.
- The development of transversal competences requires long periods in academic life.
- It is desirable that the student has an approximation to the solution of real problems from the beginning of his or her curricular program.
- The ability to work in teams and with other academic disciplines were considered to be the seal of the graduate.

The current grouping comprises three obliged courses, with the objectives indicated in Table 1.

Table 1: Objectives of the current courses of Engineering Projects Seminar - SPI

Signature and credits	Objectives
SPI 1 (3)	1. Identify and raise problems, needs or opportunities (PNO) based on the analysis of the development needs of a community. 2. Use information and concepts developed autonomously in problem solving. 3. Generate teamwork skills in collaborative and interdisciplinary environments. 4. Improve oral and written argumentation skills.
SPI 2 (3)	1. Know the methods of formulation of projects in the field of engineering taking into account the elements of the methodology of projects 2. To develop capabilities Project evaluations in the financial, economic and market, environmental and political areas of knowledge and applicability. 3. Generate teamwork skills in collaborative and interdisciplinary environments.
SPI 3 (4)	1. Develop management and design capabilities of industrial plants. 2. To carry out financial and risk evaluations of projects focused on the production of consumer, and technological products and services. 3. Generate teamwork skills in collaborative and interdisciplinary environments.

New structure of Engineering Project Courses

Based on the study of the literature on the strategy of Problem-Based Learning and Projects-Based Learning in the international engineering training programs, and the analysis of the results of the Engineering Project Seminar courses in the curriculum the 12 undergraduate programs offered by the Facultad de Minas since January of 2009. A reorganization of names, objectives, contents, and methodologies used, as well as operational strategies and evaluation of results within the project group teaching was proposed for the academic periods 03-2017 and 01-2018. It is presented below:

It is hoped that this new stage will consolidate the use of active methodologies in the teaching/learning processes and generate benefits for the institution. Furthermore, students are challenged to develop and apply transversal competences in their engineering training. That means:

1. The decrease in the dropout rate at the beginning of the program, attributed by some authors to the distance between theory and practice in the engineering curriculum structure and the lack of connection between the contents of the basic courses and the real problems facing the practice of Engineering (Froyd & Ohland, 2005).

2. The linkage of undergraduate students to research and extension projects.
3. The identification of topics aimed towards final thesis work.
4. Facilitate access to postgraduate programs.
5. Open doors to entrepreneurship for the undergraduate student and bring the future graduated individual to the reality of working life.

It is proposed that the strategy of Engineering Project courses in the Faculty of Mines of the National University of Colombia has two objectives: the development of skills in project management, and training in engineering design.

The first purpose is to consolidate project management training in each phase of the project life cycle: Identification, Formulation, Budget, Implementation, Evaluation and Closing (report).

The second purpose refers to the recognition of design as a substantial component of engineering; emphasis is placed on conceptual design, decision-making, sustainability criteria and the use of advanced tools in the design process.

The teacher evaluations alongside other surveys carried out by the course participants of the different undergraduate programs of the FacMinas, at the end of the Engineering Project Seminar courses (2009 and later), acknowledge and highlight the work done in them. As well as improve interdisciplinary teamwork and the development and improvement of transversal skills in students.

Basic knowledge and reasoning capacity, considered as the first level, are supported by the set of subjects of the foundation components such as Calculus, Statistics, Physics, among others. On the other hand, Engineering Projects Seminars complement disciplinary or professional training that students of each of the curricular programs attend.

The goals of Engineering Project courses are: strengthen and develop personal, professional and interpersonal skills and engineering design skills.

Table 2: General description of the new structure of Engineering Projects courses (PI)

Current Name	Proposed Name	NC	Prerequisites	OC	Course description
Engineering Projects Seminar I	Fundamentals of Engineering Projects -PIF	3	Introduction to Engineering + 40% Rationale	Yes	• Analysis of problems or opportunities • Project planning methodologies • Analysis and choice of solutions • Identification of costs
Engineering Projects Seminar II	Structuring and Evaluation of Projects - PIEE	3	PIF + 20% Discipline	Yes	• Tools and techniques for project management • Structuring of engineering projects • Market elements • Financial, economic and environmental project evaluation tools
Engineering Projects Seminar III	Integrated Engineering Project - PII	4	PIEE + 100% Rationale + 70% disciplinary	Yes	• Engineering design • Elaboration of prototypes, models and/or simulations of engineering proposals • Leadership • Entrepreneurship

Notes: NC: number of credits; CC: Obligatory Course in FacMinas

RESULTS

In order to update the methodological model it is necessary to redefine the objectives and competences of the courses. This was done by means of follow-up surveys. Statistical analysis is performed with data from the last two years, taking into account a population size of 1200 people in academic periods and about 335 people for summer course periods. In addition, a confidence interval of 99% and a margin of error of 5% are considered in order to measure the development and improvement of transversal competences.Through surveys conducted on the students at the end of the courses it has been found that:

1. More than 91 percent of students consider that the level of improvement of their transversal competences with the subjects is of High level.
2. More than 89 percent of the students identify the importance of interdisciplinary in teamwork.
3. More than 85 percent of students value this set of signatures as signatures that bring them closer to the real conditions of the engineering exercise.
4. More than 84 percent of students consider that the PBL and PjBL methodologies are adequate for signature works.

The evaluations carried out by the certifying peers of the different undergraduate programs of the Faculty of Mines, carried out after the implementation of the Engineering Project Seminar courses (2009 and later), acknowledge and highlight the work done in them, as well as interdisciplinary teamwork and the development and improvement of transversal skills within students.

CONCLUSIONS

In order to carry out the changes required in engineering education, the engineering faculties must understand that this is a complex and multidimensional problem; therefore, it requires direct, not marginal actions. The highlighted activities in this component of the strategy are training of teachers in new methodologies, structuring new courses of transversal character to the engineering training, and new strategies to foster university-industry links.

The FacMinas of the UNAL is adopting the methodology of teaching based on projects and problems as a strategy for teaching and learning, for all its twelve engineering programs. These are intended to cover the fields of training related to leadership, effective communication, project management, and teamwork. Thus, teachers instead of being manipulators of engineering information (or knowledge) become advisors to students' works as well as their substance. Training with this vision seeks to improve the graduate profile and increase motivation to enter engineering programs.

The Engineering Project Seminars, set of signatures is at the level of global academic trends, at the same time engineering students at the FacMinas have generated the capacity to integrate productive dynamics. Interdisciplinary work allows the engineering graduate to know their strengths and weaknesses, recognizing that it requires the mutual help of others to carry out their work. This type of methodologies and strategies applied at the FacMinas is in continuous evolution, in favor of the formation of the students.

REFERENCES

ASME. American Society for Mechanical Engineering. (2008). 2028 VISION FOR MECHANICAL ENGINEERING. A report of the Global Summit on the Future of Mechanical Engineering. ASME, Three Park Avenue, New York, NY 10016, USA. (www.asme.org).

Auyang, S. Y. (2006). Engineering -- an Endless Frontier. New York, USA: Harvard University Press.

BOKOVA, I. (2010). Engineering: Issues, Challenges and Opportunities for Development.

Chandrasekaran, S., Stojcevski, A., Littlefair, G., & Joordens., M. (2012). Learning through projects in engineering education. In SEFI 2012 Engineering Education 2020: Meet The Future:. 40th SEFI Annual Conference 2012. European Society for Engineering Education (SEFI).

Crawley, E. (2001). The CDIO Syllabus. A statement of goals for Undergraduate Engineering Education (2001). The CDIO Syllabus V 1 . Boston, USA.

Crawley, E. F., Malmqvist, J., Lucas, W. A., & Brodeur, D. R. (2011). "The CDIO syllabus v2. 0. An updated statement of goals for engineering education. In Proceedings of 7th International CDIO Conference, Copenhagen, Denmark. 2011.

David, P. A., & Foray, D. (2002). Una introducción a la economía y a la sociedad del saber. Revista Internacional de Ciencias Sociales (171).

Foray, D., & David, P. A. (2002). Una introducción a la economía y a la sociedad del saber. Revista Internacional de Ciencias Sociales (171).

Fuchs, W. (2012). The New Global Responsibilities of Engineers Create Challenges for Engineering Education. Journal of Education for Sustainable Development, 6 (1), 111-113.

Gerhart, A. L., & Fletcher, R. W. (2011). "Project-Based Learning and Design Experiences in Introduction to Engineering Courses-Assessing an Incremental Introduction of Engineering Skills". In American Society for Engineering Education. American Society for Engineering Education.

Harmer, N. (2014). Project-based learning Literature review. PLYMOUTH UNIVERSITY .

Kolmos, A., Du, X., Holgaard, J. E., & Jensen, L. P. (2008). Facilitation in a PBL environment. Aalborg: UCPBL UNESCO Chair in Problem Based Learning.

Kolmos, A., Kuru, S., Hansen, H., Eskil, T., Podesta, L., Fink, F., y otros. (2007). Problem Based Learning. In TREE – Teaching and Research in Engineering in Europe. Recuperado el 04 de 08 de 2016, de https://lirias.kuleuven.be/bitstream/123456789/239445/2/b5.pdf&gathStatIcon=true

Malpas, R. (2000). THE UNIVERSE OF ENGINEERING A UK PERSPECTIVE. (R. A. Council, Ed.) London, UK.

OECD. (1996). THE KNOWLEDGE-BASED ECONOMY (GENERAL DISTRIBUTION OCDE/GD(96)102 ed.). (O. F.-O. DEVELOPMENT, Ed.) Paris: OECD.

Reis, D. R. (2004). Gestao da Innovaçao tecnológica. Barueri, Sao Paulo, Brasil: Manole.

Salvat, B. G., & Navarra, P. L. (2009). ESTRATEGIAS DE INNOVACIÓN EN LA EDUCACIÓN SUPERIOR: EL CASO DE LA UNIVERSITAT OBERTA DE CATALUNYA. REVISTA IBEROAMERICANA DE EDUCACIÓN. , 49, 223-245.

Savin-Baden, M. (2007). Challenging Models and Perspectives of Problem-based Learning. En E. &. de Graaff (Ed.), Management of Change: Implementation of Problem Based and Project Based Learning in Engineering (págs. 9-30). A. Utrecht: Lemma Publishers.

STREVELER, R. A., & SMITH, K. A. (04 de 2006). Conducting Rigorous Research in Engineering Education. Journal of Engineering Education 1 , 103-105.

Velez, J.M.,Benjumea, P.N., Castro, K.J., & Rios, D.C. (2017). Estrategia de Innovación en Educación en Ingeniería.Universidad Nacional de Colombia, Facultad de Minas, Medellin, Colombia. Documento de trabajo.

Acknowledgements

We thank all the authors of the working paper "Estrategia de Innovación en Educación en Ingeniería" of the FacMinas for permitting that we use it as a base for the elaboration of this article.

Copyright statement

Assessing Transformative Learning about Instructional Design: An Example with Colombian Engineering Faculty

Juan D. Ortega-Alvarez
Universidad EAFIT, Medellin, Colombia | Purdue University, West Lafayette, IN, USA
jortega@eafit.edu.co

Camilo Vieira
Purdue University, West Lafayette, IN, USA
cvieira@purdue.edu

Matilde Sanchez-Pena
Purdue University, West Lafayette, IN, USA
sanch105@purdue.edu

Ruth A. Streveler
Purdue University, West Lafayette, IN, USA
streveler@purdue.edu

Abstract: *The authors designed and delivered a workshop on Instructional Design for Colombian engineering faculty that drew from the backwards design model and research in engineering education. The participants were asked to assess the workshop using a survey with pre-post items and open comments. The authors argue that a professional development program can transform faculty's perspectives, particularly when aimed at helping them inform their teaching practice using evidence-based educational research. Therefore, the assessment of such a program must move from an incremental to a transformational notion of learning. The research question driving this work was: How can transformative learning of engineering faculty about instructional design be assessed? Our experience with Colombian faculty suggests that a transformative learning framework can inform the assessment of participants' learning in professional development programs for engineering faculty.*

Introduction

Engineering faculty, like faculty in many other fields, face a particular challenge: they often have not been formally trained to undertake teaching jobs. It is reasonable to expect that they are competent in their disciplinary fields and are, at least, able to teach as they were taught. Nevertheless, this level of teaching performance usually does not meet today's desired standards of engineering education. In response to this situation, the authors designed, delivered, and evaluated a workshop on Instructional Design for Colombian engineering faculty.

The overarching goal of this workshop was to offer an overview of a strategy for course design and different techniques for delivery. In addition, the Scholarship of Teaching and Learning (SoTL) was introduced to participants as a stance they can adopt to inform innovative implementations in their classroom. Similarly, through the adoption of a SoTL perspective, they can contribute to the conversation on educational innovation through the scholarly dissemination of their initiatives (Gayle, Randall, Langley, & Preiss, 2013). Familiarity with SoTL can be relevant in the Latin-American context, where an increasing number of institutions are shifting their focus from teaching to research. Therefore, introducing SoTL to faculty members with an interest in improving their teaching may help them balance their careers as their institutions undertake this transition (Rugarcia, 1991).

Backwards design was the basic tenet for both the development and execution of the workshop. This model was used to develop the workshop's learning objectives, which envisioned that participants would become able to: 1) write appropriate learning objectives; 2) design innovative instructional experiences; 3) evaluate evidence-based course interventions; and 4) document their educational innovations. Various assessments and pedagogies were also introduced. During the workshop, participants were encouraged to consider courses they had an interest in improving. In this way, the workshop provided the opportunity to develop an educational innovation they could easily transfer to their practice.

The twenty-hour workshop was conducted during five consecutive mornings in June 2016 and took place in a private Colombian university with participants from different engineering disciplines. The course design strategy and most delivery techniques were taught through their actual application by the instructors; for instance, teaching about flipped-classrooms through a flipped-classroom session. Most activities in the workshop were hands-on and often involved group discussion. Lecturing was purposely limited to a maximum of 15 minutes.

After this introduction, the following sections of this paper explore: 1) the wider context of professional development programs for faculty and their assessment; 2) the theoretical frameworks considered in this work, including backwards design and transformative learning; 3) the research question driving this study; 4) the methodology used to execute the workshop and the methods for data collection and analysis; and 5) the findings, conclusions, and their implication for the assessment of transformative learning experiences of faculty.

Context

There is abundant information on assessing professional development programs (PDP). Most of the scholarly literature on the assessment of PDP draws upon the work of Guskey. In the early 90s, Guskey and Sparks (1991) advanced a model for staff development that included four factors and their interactions: program content, program quality, organizational climate, and improvement in learning outcomes. Building upon this framework, plus a model for evaluating training programs in business created by Kirkpatrick, Guskey advanced a model for PDP assessment that included four distinctive levels: 1) participants' reactions; 2) new knowledge and skills participants gained; 3) influence of new knowledge in participants' professional practice; and 4) increase in productivity, measured on the basis of students' achievements. Later, Guskey saw the need to add a new level that accounted for organizational support (Kreider & Bouffard, 2005). The refined model, known as the Critical Levels of Professional Development, has, therefore, five levels (figure 1). Although independently observable, each level builds on the previous one. In other words, success at a particular level is arguably required for success at higher levels.

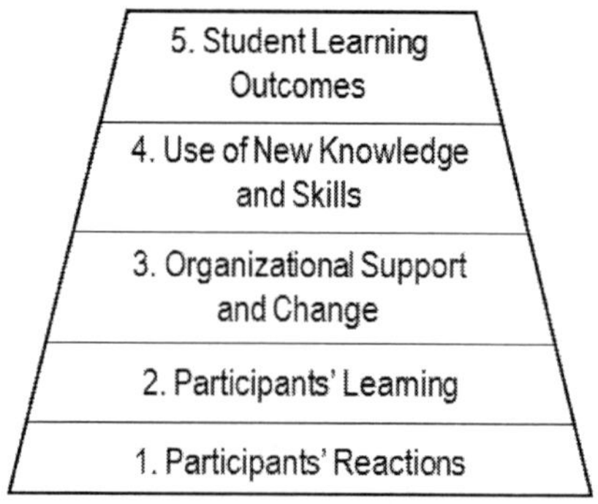

Figure 1. Critical Levels of Professional Development (Adapted from Guskey, 2002)

Instruments intended to assess faculty development workshops and courses usually draw upon the Critical Levels of Professional Development model. Based on this model, the Ohio State University uses a multi-level survey to evaluate their faculty development program, known as the ABLE System (Mullins, Lepicki, & Glandon, 2010). The authors designed an assessment instrument for their workshop based on the ABLE assessment survey, explicitly aimed at levels one and two of Guskey's model. The inclusion of open-ended questions allowed an unintended glimpse into levels three and four.

The assessment of transformative learning experiences has also been scholarly documented (e.g., Snyder, 2008). Most of the work drawing from the transformative learning framework has been applied to teachers and faculty development programs that deal with the design and description of learning experiences. Studies aimed at determining whether perspective transformation occurred usually use the Learning Activities Survey (LAS) developed by King (1997). If transformation occurred, the LAS aims to identify the activities that contributed to this outcome (King, 2009, p. 14). The LAS is a comprehensive instrument that usually requires administering a four-step survey and conducting follow-up interviews, although it has been adapted and simplified for specific research projects (King, 2009, Chapter 13). For instance, Caruana and colleagues adapted the LAS to examine the perspective transformation of participants in two graduate courses for pre-service teachers (Caruana, Woodrow, & Pérez, 2015).

Other scholars have created their own instruments to assess educators' learning experiences based on the ideas of transformative learning. For instance, Kitchenham (2006) used transformative learning to describe the experiences of ten school teachers as they learned to adopt educational technology in their classrooms. His findings suggest that "teachers do experience perspective transformations as they learn to use, adopt, and teach educational technology." (Kitchenham, 2006, p. 222). Similarly, Goulet (2010) described the application of transformative learning in informal instruction contexts. Exploring the intersection of transformative learning and assessment, Goulet draws two main conclusions: 1) formative assessment may lead to perspective transformation, and 2) transformative learning is better fostered and captured by self-assessment. This conclusion seems crucial for the design of assessment instruments for transformative experiences.

Research Questions

The research question driving this work was: How can transformative learning of engineering faculty about instructional design be assessed? This research question could be phrased in a more general sense: What is a reasonable way to assess the impact of professional development programs aimed at helping engineering faculty inform their teaching practice drawing from the findings of educational research? The relevance of this question lies in the need to design PDP that effectively help engineering faculty look at teaching in a scholarly way, and advance their teaching through evidence-based practices. Without proper assessment and reliable feedback, PDP may fail in changing faculty's perspectives of teaching, and the improvement of PDP becomes, at best, a trial-and-error exercise.

Theoretical Framework

Backwards Design

As mentioned earlier, backwards design was both the design strategy and the overarching theme of the workshop. Backwards design is a model for designing instructional experiences, whether they are workshops, short courses, or formal courses within a curriculum. The idea underlying backwards design is that the instructor should first define the outcomes of the instructional experience, and then align the assessment and pedagogical practices to those outcomes (Wiggins & McTighe, 1998). Although such an approach is rather intuitive, it challenges the typical practice of instructors who usually think of a course in terms of the activities they use to deliver some predefined, structured content (Wiggins & McTighe, 1998, p. 8). In other words, backwards design challenges the notion of structuring courses around the idea of covering the content, and advocates considering the logic of learning the content instead (Svinicki, 2004).Backwards design is a sound model for ensuring that a professional development experience responds to deliberate instructional aims and learning goals.

There are three major stages in the backwards design process: 1) identification of the desired results; 2) determination of acceptable evidence of learning; and 3) preparation of the learning experiences. However, far from being linear, backwards design is an iterative approach where the design and definition of each component in and of itself is as important as the alignment of all the pieces together. In other words, there is little value to participants' enjoying class activities and learning from them if what they learn is not relevant to the curricular priorities of the learning experience. Alignment between the different components of instructional design is the key feature of backwards design.

Transformative Learning

Transformative learning is a framework originally devised by John Mezirow to theorize how adults learn when they return to educational settings. After years of application and evolution, this framework has witnessed changes and criticism, but the underlying idea of exploring adult learning remains at its core (Kitchenham, 2008). Transformative learning describes a profound change in assumptions, views, and even values, spurred by a perspective transformation (King, 2009, p. 4). Mezirow's original work identified ten stages of perspective transformation (King, 2009 p. 5). However, the transformation process can be summarized into three major elements: 1) critical reflection on assumptions; 2) use of empirical evidence or critical discourse to explore the assumptions; and 3) take action aligned with the transformed perspective (Mezirow, 2006).

The workshop examined in this study was not deliberately designed to foster perspective transformation. However, the participants are expert learners being presented with thought-provoking, innovative materials. Therefore, it is expected that they will critically examine these materials in light of their experience and knowledge. Moreover, they are being presented with the notion of scholarly teaching as an evidence-based practice; i.e., a perspective supported by empirical evidence. Similarly, the discussion spaces deliberately built into the workshop became the venues for critical discussion and discourse. Transformative learning may have occurred, spurred by critical reflection on teaching practice, exposure to relevant research outcomes, and negotiation through participants' discourse about the values and meaning they associate with teaching.

Methodology

Participants and Procedures

Twenty-three Colombian engineering faculty participated in the twenty-hour workshop. The participants had diverse backgrounds in engineering disciplines and experience as faculty members. The average time working as a faculty member was 11.7 years, ranging from 0.5 to 25 years. Their areas of expertise included mechanical engineering, process engineering, product design engineering, production engineering, and systems engineering. The workshop was held in Spanish and the instructors were native speakers. However, most readings and class materials were provided in English to have participants familiarize with the terminology widely used in engineering education research.

The workshop took place on five consecutive days with two sessions of 110 minutes per day, as shown in figure 2. As mentioned earlier, the workshop was constructed and delivered based on the ideas of backwards design (i.e., alignment between content, assessment, and pedagogy). In addition, the workshop included an introduction to engineering education research on the first day, and participants' presentations of their projects were scheduled on the final day.

Before and after the workshop, the participants completed pre- and posttest surveys in Spanish aimed at identifying changes in their perceived understanding, competence, and interest in instructional design. Participants were presented with five pretest and six posttest Likert-type items with five levels. English translations of these items are presented below:

Please indicate your level of agreement regarding the following statements from one (strongly disagree) to five (strongly agree):

a. *I have the ability to perform my duties as an engineering faculty in this institution.*
b. *I have the required competencies to design courses appropriately.*
c. *I feel motivated to improve my courses.*
d. *I have the required competencies to conduct educational research.*
e. *I feel motivated to conduct educational research.*
f. *This workshop reinforces my abilities as an instructional designer [posttest only].*

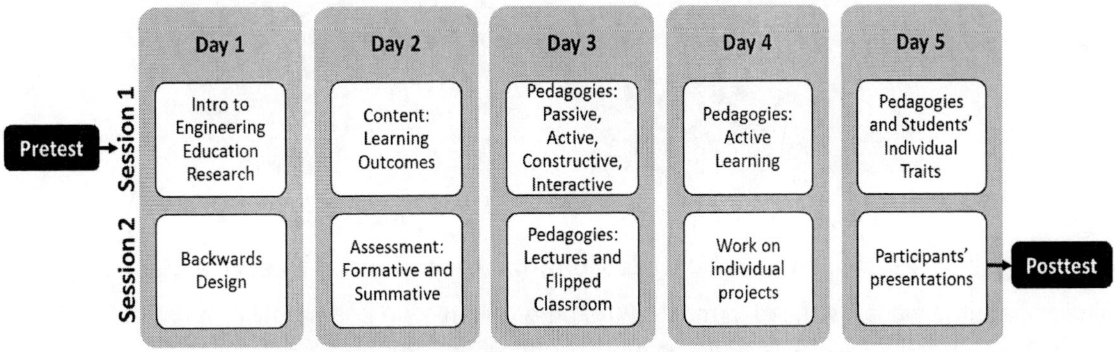

Figure 2. Workshop activities and data collection procedure

The posttest instrument also included three open-ended questions, which traslated into English read as follows: 1) *Please describe the most relevant aspects of this workshop and how you will use them in your future teaching practice*; 2) *Please provide any additional comments you may consider relevant*; and 3) *Please describe what opportunities you see for improving this workshop*. The procedures and instruments for data collection were approved by the Institutional Review Board from Purdue University under the protocol #1605017685.

Data analysis

The research team first analyzed the quantitative Likert-type items looking at the distribution of participants' responses to each of them. Eighteen participants answered the pretest, and fifteen participants completed the posttest. Given the scale nature of the data, the Wilcoxon signed-rank test was used to identify significant differences among the distribution of participants' responses for each question between the pretest and the posttest (McCrum-Gardner, 2008). Participants' responses to the open-ended questions were qualitatively coded independently by three researchers who are native speakers of Spanish. The three researchers were involved in both the design of the workshop and the survey instruments. After completing this open coding process, the three researchers met to compare their resulting codes and themes. The discrepancies that were found among the codes were negotiated until an agreement was reached.

Findings

The analysis of the quantitative data from the pre-post Likert-type items yielded unexpected results about participants' perceptions of their instructional design abilities. Figure 3 depicts the distribution of participants' responses in the pre- and posttest for statement b: *I have the required competencies to design courses appropriately*. This item is particularly important because of its direct connection to the overarching goal of the workshop. The Wilcoxon signed-rank test did not show any significant difference in the distribution of responses between the pretest and posttest. However, looking at the individual responses, it was noticed that only one of the participants moved from level 2 (i.e., disagree) to a higher level, while two of the participants moved from level 5 (i.e., strongly agree) to a lower level and the rest reported no change.

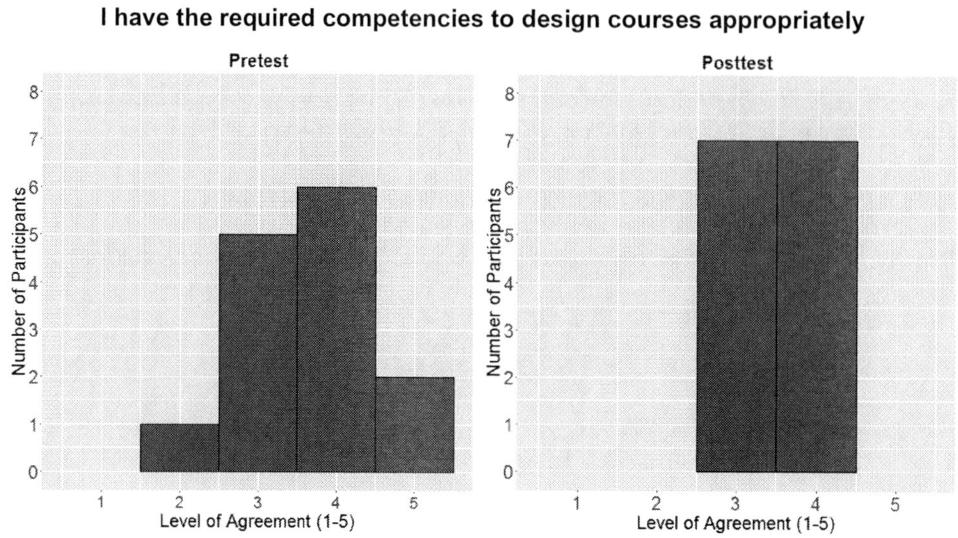

Figure 3. Pretest and posttest distributions of responses to survey item b:

Similar patterns were identified in most of the other Likert-type items between pretest and posttest. On the other hand, responses to the open-ended questions and informal interactions with participants portrayed a more favorable perception of the workshop and its impact. In light of this discrepancy, the research team hypothesized that the workshop was a transformative experience that prompted participants to change their perception of what they needed to design effective courses. In other words, some of the participants might have felt underprepared for instructional design after learning what this task entails. The research team decided to focus on and explore the responses to the open-ended questions of the survey to support this hypothesis. Analysis of these responses provided evidence of changes in participants' conceptions of teaching and learning. This evidence aligned with three major elements of the transformative learning process discussed in the frameworks section. Therefore, each question and the open codes resulting from the initial analysis were examined through this framework.

Question 1: Relevant aspects of the workshop

Three major themes emerged from the participants' comments when they were questioned about the most relevant aspects of the workshop and how they would use them in their future teaching practice. Below, each theme is presented with the number of participants whose comments aligned with it. Exemplary excerpts from the survey in Spanish accompany the themes, along with their free translation into English.

Theme 1: Expansion of understanding about teaching and learning (5 participants)

Quotes aligned with this theme made specific reference to the participants' motivation to reflect on their current practices, and how these practices could be improved based on the evidence presented. One participant expressed:

"Este taller me permitió ver de manera integral cómo es el proceso enseñanza aprendizaje y reflexionar sobre mi proceso de enseñanza y cómo éste puede ser modificado para que un mayor número de estudiantes puedan comprender los temas de la clase de una manera más efectiva" – This workshop allowed me to see the teaching-learning process in an integral way and reflect on my teaching process and how it can be modified so a higher number of students can understand the class topics in a more effective way.

Similarly, another participant indicated:

"Creo que entender el proceso de diseño inverso y lograr mejorar en la definición de objetivos y formas de evaluar el aprendizaje de los estudiantes, así como entender un poco

más sobre las diferentes formas de aprender de los estudiantes, me hace reflexionar" – I believe that understanding backwards design and improving the definition of objectives and ways to evaluate student learning, as well as understanding a little bit more about different learning styles of the students, makes me reflect.

Critical reflection is the key element that triggers transformative learning experiences. As exemplified by the previous quotes, participants explicitly expressed how the workshop content and activities ignited their critical reflection.

Theme 2: Exposure to relevant content of teaching practices (8 participants)

Excerpts aligned with this theme recognized the value of being exposed to research-based teaching practices. This theme relates to the second key element of perspective transformation: gathering evidence through research and discourse. An exemplary quote for this theme versed as follows:

"La aproximación a los temas expuestos, aun solo desde el hacernos conscientes de la diversidad de hallazgos y propuestas que pueden contribuir a la labor docente, lo considero un gran aprendizaje" – The approach to the topics presented, even if just becoming conscious about the diversity of findings and ideas that can contribute to the teaching practice, I consider it to be a great lesson.

Another participant asserted that the most relevant part of the workshop was

"Tener la oportunidad de ver diferentes herramientas pedagógicas en el curso y algunas teorías de enseñanza-aprendizaje. Lo primero me sirve para implementar algunas de ellas en mis cursos (tratando de mejorar el entendimiento de algunos conceptos que son complejos de entender) y los segundo para escribir mejor los artículos sobre prácticas pedagógicas." – Having the opportunity to see different pedagogical tools of the course and some theories of teaching and learning. The former helps me implementing some of them in my courses (trying to improve the understanding of some complex concepts), and the latter to write better articles about pedagogical practices.

Participants also made explicit reference to the relevance of being exposed to educational research. One participant assessed that the workshop was useful to:

"Permitir conocer un panorama de la investigación educativa acompañada de buen material que puede ser aplicada en nuestras aulas de clase." – Allows me to become acquainted with a view of educational research together with good material that could be applied to our classrooms.

This theme shows how participants articulated a new discourse as a result of the exposure to the new paradigm being presented to them. Moreover, they started to provide hints of possible application to their faculty functions.

Theme 3: Applicability of learned techniques to practice (4 participants)

As discussed in the framework, the final key element of transformative learning deals with taking action in accordance with the transformed perspective. This element was illustrated by the third theme identified in the participants' comments. A quote aligned with this theme versed:

"Particularmente el concepto de flipped classroom (en combinación con otras herramientas) me gustó y buscaré aplicarlo." – I particularly liked the concept of flipped classroom (in combination with other tools), and I will seek to apply it.

When referring to specific things they would try, a participant referenced:

"Con base en esto [el taller], actualizaré objetivos de aprendizaje, material educativo, y evaluaciones." – Based on this [workshop], I will update learning objectives, educational material, and assessments.

Although no action has been yet taken, this theme speaks to the planning of a course of action. At this point, the learner has internalized the value of the materials presented and has assessed that their adoption would represent a benefit in their practices (Kitchenham, 2008).

Question 2: General comments

When asked for additional comments they may consider relevant, participants reported reflecting on the relevance of the topics covered in the course. In particular, many participants expressed how the workshop allowed them to reflect on their current teaching practices, making potential space for their improvement. A participant expressed:

"Excelente curso, me parece muy necesario ponernos a pensar sobre cómo hacer mejores nuestras materias." – Excellent workshop, I think it is so necessary to make us think about how to improve our courses.

Through a critical examination of their teaching practice, participants recognized the relevance of the workshop to engineering education and expressed their interest in having similar workshops ingrained in their PDP. Moreover, some participants were explicit about the need for similar follow-up courses.

Question 3: Opportunities for workshop improvement

Participants provided a variety of suggestions to improve the workshop. Most of them perceived that the time was limited and some expressed that they would have preferred to allocate more time, and even suggested a full-time, week-long schedule. Participants also perceived that many topics would have been better supported by a more extensive use of examples and application case studies. Some participants also mentioned their interest in more sophisticated topics related to education, such as neuroscience and learning science.

Conclusion

Transformative learning can be fostered by courses and workshops aimed at helping faculty improve their teaching practice drawing from educational research. This was found to be the case in a workshop the authors offered to a group of twenty-three engineering faculty at a Colombian private university. A survey designed to assess the effectiveness of the workshop yielded contradicting results. On the one hand, the self-reported perception of gain in skills measured through pre-post Likert-type items showed no change, and sometimes a non-significant decrease. On the other hand, stories of positive transformation were present in much of the qualitative data gathered from responses to open-ended questions. Participants' views of what is entailed in instructional design were expanded, and they might have realized that they had overestimated their knowledge about it. Beyond the expected knowledge expansion, the workshop fostered perspective transformation, and the assessment survey was not appropriately designed to account for it. In this situation, the design of an effective and appropriate assessment instrument should draw upon the key elements of the transformative learning process: 1) critical reflection on assumptions, 2) exploration of assumptions, and 3) action aligned with the new perspective.

Implications

The results of this work provide further support for a well-known conception: Instructors must look carefully at what kind of learning is happening in their interventions, and choose the correct assessment measures. The Assessment Triangle (Pellegrino, Chudowsky, & Glaser, 2001) can be used to guide alignment between the learning process taking place and assessment of the observed learning. Interventions designed for faculty development may result in transformative learning. We believe that the use of the assessment triangle, drawing from transformative learning as the theory supporting the cognitive process, could lead to the design of a more appropriate instrument to assess faculty development. Therefore, our future work will be directed toward designing and testing such an instrument.

References

Caruana, V., Woodrow, K., & Pérez, L. (2015). Using the learning activities survey to examine transformative learning experiences in two graduate teacher preparation courses. *InSight: A Journal of Scholarly Teaching, 10*, 25–34.

Gayle, B. M., Randall, N., Langley, L., & Preiss, R. (2013). Faculty learning processes: A model for moving from scholarly teaching to the scholarship of teaching and learning. *Teaching and Learning Inquiry, 1*(1), 81–93.

Goulet, G. (2010). Transformative learning and assessment in non formal education. In *Adult Education Research Conference Proceedings*. Sacramento, CA: New Prairie Press.

Guskey, T. R. (2002). Does it make a difference? *Educational Leadership, 59*(6), 45.

Guskey, T. R., & Sparks, D. (1991). What to consider when evaluating staff development. *Educational Leadership, 49*(3), 73–76.

King, K. P. (1997). Examining learning activities and transformational learning. *International Journal of University Adult Education, 35*(3), 23–37.

King, K. P. (2009). *The handbook of the evolving research of transformative learning, based on the learning activities survey* (10th Anniversary ed.). Charlotte, NC: Information Age Publishing, Inc.

Kitchenham, A. (2006). Teachers and technology: A transformative journey. *Journal of Transformative Education, 4*(3), 202–225.

Kitchenham, A. (2008). The evolution of John Mezirow's Transformative Learning Theory. *Journal of Transformative Education, 6*(2), 104–123.

Kreider, H., & Bouffard, S. (2005). A Conversation with Thomas R. Guskey. *The Evaluation Exchange, 11*(4).

McCrum-Gardner, E. (2008). Which is the correct statistical test to use? *British Journal of Oral and Maxillofacial Surgery, 46*(1), 38–41.

Mezirow, J. (2006). An overview of transformative learning. In P. Sutherland & J. Crowther (Eds.), *Lifelong learning: Concepts and contexts* (pp. 24–38). New York, NY: Routledge.

Mullins, D., Lepicki, T., & Glandon, A. (2010). *A professional development evaluation framework for the Ohio ABLE system* (p. 26). The Ohio State University.

Pellegrino, J. W., Chudowsky, N., & Glaser, R. (2001). *Knowing what students know: The science and design of educational assessment*. Washington, DC: National Academy Press.

Rugarcia, A. (1991). The link between teaching and research: Myth or possibility? *Engineering Education, 81*(1), 20–22.

Snyder, C. (2008). Grabbing hold of a moving target: Identifying and measuring the transformative learning process. *Journal of Transformative Education, 6*(3), 159–181. 3

Svinicki, M. D. (2004). *Learning and motivation in the postsecondary classroom*. Bolton, MA: Anker Publishing.

Wiggins, G. P., & McTighe, J. (1998). *Understanding by design*. Alexandria, VA: ASCD.

Acknowledgements

We want to thank the academic administrators at Universidad EAFIT (Medellin, Colombia) for supporting the realization of the workshop. We are grateful to the engineering faculty who enthusiastically participated in both the workshop and the study.

Copyright statement

Identifying Engineering Students' Design Practices Using Process Data

Camilo Vieira
Purdue University, West Lafayette, USA
cvieira@purdue.edu

Alejandra J. Magana
Purdue University, West Lafayette, USA
admagana@purdue.edu

Senay Purzer
Purdue University, West Lafayette, USA
purzer@purdue.edu

***Abstract**: The assessment of the development and quality of students' engineering design competencies is a challenging task. The open-ended nature of design process, with multiple paths and possible solutions for a given problem, requires creative ways of assessment. Technology-based assessment techniques are a promising approach to achieve this goal. This paper uses clickstream data from an educational CAD tool to identify what elements from student design processes change after instruction. Specifically, we focus on two design strategies: (1) generating ideas and (2) conducting experiments. Colombian engineering students from a Caribean public university worked on a design challenge to create an energy efficient building using an educational CAD tool. This tool logs all student interactions with the software, which the researchers process to identify student design strategies. The results from this study suggest that the process data from an educational CAD tool can be analyzed both visually and using common statistical techniques to identify changes in student design strategies.*

Introduction

Engineering design is an important process to solve real-world complex problems that do not have a unique solution nor a unique path to find the solution. Such complexity and its open-ended nature make the engineering design process difficult to measure. Some of the qualitative approaches that have been used to assess this process include: (1) student descriptions of how they used the design process (Hirsch, Berliner-Heyman, Carpinelli, & Kimmel, 2012); verbal protocol analysis (Atman, Chimka, Bursic, & Nachtman, 1999); and student critiques of a given design process (Bailey, 2008). However, these approaches are often time consuming and therefore, non-scalable.

Technology-based assessments have recently emerged as an opportunity to assess these complex processes. Students' process data can be captured from their interactions with computational tools while working on a design task, and then be analyzed using pattern recognition and statistical analysis tools. For instance, Worsley and Blikstein (2014) used screenshots captured while students worked on a design task to understand their process. Xie and colleagues (2014) studied the sensitivity of CAD logs to instructional interventions related to the science behind the design challenge. Goldstein and colleagues explored screen cast reproductions and students' interactions with the CAD tool to describe their ideation strategies. Vieira, Goldstein, Purzer, and Magana (2016) used students' interactions with an educational CAD tool to characterize student experimentation strategies in design, and Seah and colleagues (2016) expanded this work by exploring the effect of providing an instructional intervention on students' experimentation strategies in design.

This study explores the sensitivity of CAD logs to characterize student design process before and after instruction related to effective design strategies.

Context

The study takes place at a Colombian institution with collaboration among researchers from Colombia and United States. In the United States, engineering design has acquired a great relevance both at the high school and at the college level. The design process is introduced both as a learning outcome and as a pedagogical strategy to support student learning. Conversely, Colombian institutions and Colombian educational researchers have not formally adopted this approach to science and engineering education (Vieira, Aguas, Goldstein, Purzer, & Magana, 2016). To start addressing this gap, engineering educators from the United States conducted a workshop in Colombia during the summer 2015. The workshop was offered to engineering students and focused on completing a design challenge related to energy efficiency using an educational CAD tool called Energy 3D. Students' interactions with the CAD tool were captured into log files, which were analyzed to identify students' design strategies.

Research question

The research question for this study is: *How do engineering students' design strategies change while solving a design challenge as captured by process data from CAD Logs?*

Theoretical Framework

The theoretical framework that guides this study is the informed design teaching and learning matrix (Crismond & Adams, 2012). This matrix describes nine design strategies that beginning and informed designers approach differently: Understand the challenge; Build Knowledge; Generate Ideas; Represent Ideas; Weigh Options and Make Decisions; Conduct Experiments; Troubleshoot; Revise/Iterate; and Reflect on Process. For instance, beginning designers do not spend a lot of time on understanding the challenge and building knowledge because they usually consider the problem to be well-structured, where there is a single solution. Conversely, informed designers wait to make design decisions until the challenge is well understood. Informed designers also identify associated "key issues" to be sure they do not commit to a final design decision in a very early stage.

There are two implications of this framework for the research study. First, students were introduced with these differences and were asked to reflect on where their own design process within each of the design strategies. Second, student interactions with the software will be analyzed to identify whether they depict any changes that represent students moving from beginning to informed designers. For the context of this study, we focused on two of the design strategies that have been previously explored using process data. Table 1 describes these design strategies, how designers with different levels of expertise approach them, and how these can be characterized using process data.

Methodology

Participants and Procedures

During the summer 2015, an eight-hour workshop was delivered for two groups of engineering students from two different programs: Systems Engineering and Industrial Engineering. Sixteen students from Systems Engineering participating in these activities are included in this study. The goals of the workshop were (Vieira, Aguas, Goldstein, Purzer, &

Magana, 2016): (1) to promote student understanding about engineering practice; (2) to foster acquisition of solar science knowledge; and (3) to increase student understanding about engineering design. The workshop was created as an active learning pedagogical strategy guided using Learning by Design™ (Kolodner et al., 2003). The purpose of this design was to provide engineering students with active learning strategies that were more engaging than traditional classroom lectures.

Table 1: Characterization of beginning and informed design strategies with process data.

Design Strategy	Beginning Designers Approach	Informed Designers Approach	Process Data
Generate Ideas	Get fixated into early ideas with little interest on changing it.	Generate multiple ideas for their design, avoiding fixation	Number of designs generated using the CAD tool. Number of actions with a given object.
Conduct Experiments	Perform no experiments or confounded ones.	Perform multiple controlled experiments using relevant variables.	Model: Experimentation a Strategies in Design (Vieira et al., 2016)

Students worked on finding a solution for a design challenge using a computer-aided design (CAD) tool called Energy 3D (Concord, 2017). This is a free open source software that allow students to create energy efficient buildings. The challenge consisted of building a house located in Boston-Massachusetts that generated enough energy for itself in a one-year period. Besides the net-zero energy goal, there were other criteria and constraints involved in this challenge: max cost of $60,000; area of 100-200 m^2; each side of the house must have at least one window; tree trunks must be at least two meters away from the house; and attractive exterior or "curb appeal".

The workshop was divided into four two-hour sessions as illustrated in Figure 1. During the first session, the software was introduced to the students and they were allowed to get familiar with it. The second half of this session, students were introduced with the design challenge and started working on it. During the second session, the differences between novice and informed designers were discussed as described by Crismond and Adams (2012) with the group. In order to avoid fixation of ideas, students were asked to initially create three different designs and make an informed decision to choose their best design during the third session. During the fourth and final session, students finished their selected design and presented it to the group.

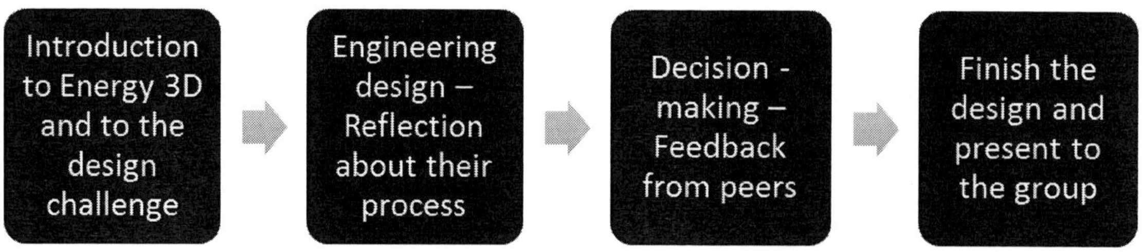

Figure 1: Sequence of the four workshop sessions

Data Analysis

We then analyzed the data on student interactions with Energy 3D (herein called process data) before and after the discussion on session 2, to identify how students' design strategies changed in their own design process. For instance, Figure 2 represents the number of actions of a single student over time during the second session. The top four line plots represent construction actions on four key elements that had an effect on the energy efficiency of the building. The slope of the roof changed both the volume and cost of the building, as well as the efficiency of the solar panels for a given location. The trees could provide shade to the building to keep it cool during the summer months, but inappropriate choices of trees (e.g. deciduous vs. evergreen) or locations for these trees could also increase the need for heating during the winter months. The size and location of windows is also an important element of this design, because they allow sunlight to get into the house, and heat to escape from it. The bottom line plot corresponds to the number of energy consumption analysis actions this student carried out. The blue box in the middle of the figure represents the time when the discussion and reflection about effective design strategies took place.

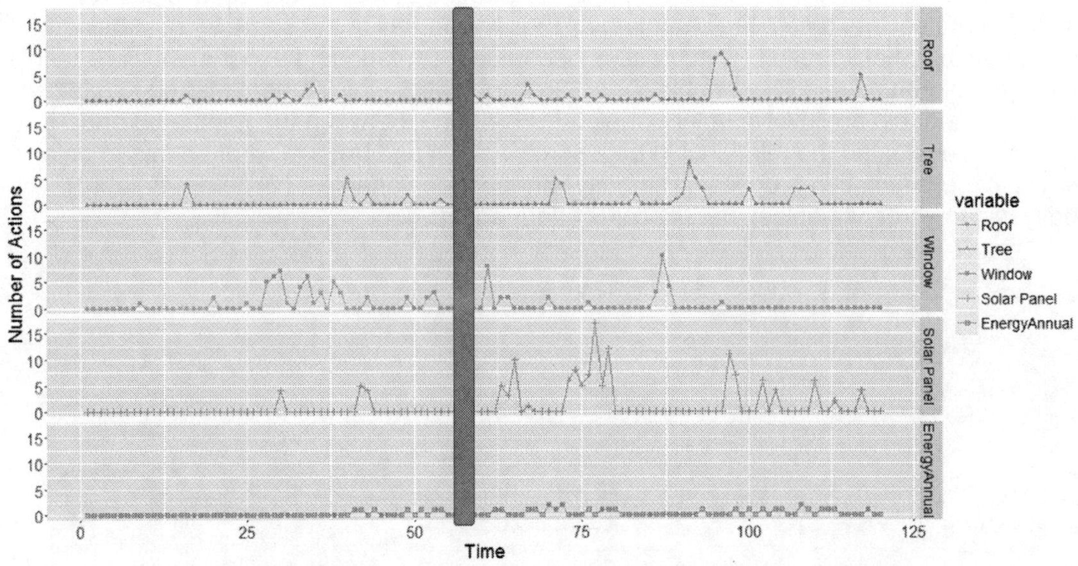

Figure 2: Number of actions from a single student over time

There are a couple of elements that can be visually identified as differences in this student design process before and after the intervention. For instance, this student started to focus on installing and moving solar panels after the intervention, always in parallel with data collection procedures to identify changes in energy efficiency of the building. This suggests that the student was informing her decisions based on data collected from the CAD tool. This exploration seem to be taking place in parallel with changes in the roof, which can potentially have an effect on the efficiency of the solar panels. That connection between solar panels and the roof is not evident before the intervention.

Besides the visual analysis of students' interactions, this study also explores the two design strategies in terms of changes in specific objects. The *generation of ideas* is explored at a building level and at an object level. At the building level, we explore whether students started working on a completely new design after reflecting about their fixation on a single idea. At the object level we identify if they started exploring completely new objects, interactions between objects (e.g., windows and trees), or if they stopped being fixated about a particular object (e.g. the solar panels in Figure 2).

Students' *experimentation strategies* were analyzed using the model experimentation strategies in design (Vieira et al., 2016), which describes a continuum of strategies from beginning to informed designers. We hypothesized that students would start to do more controlled systematic experiments after the intervention.

Findings

We present the study results under three subsections. First, we discuss the visual analysis of student interactions on the second session before and after the instructional intervention. For the visual analysis, we present two representative examples of student interactions. Then, we discuss the changes on two of the design strategies averaging these numbers for all students.

Visual Representation of Students' Interactions with Energy 3D

Figure 3 and 4 depict two representative examples of students' interactions before and after the intervention. For student A in Figure 3, most of the actions were initially between windows, solar panels, and the roof. This student seemed to be modifying the three variables at the same time between the minute 20 and 45. Starting at the minute 25, this student collected data several times, but changing all these variables at the same time, should not give her an informative result. This process becomes a confounding experiment were results cannot be linked to a specific variable. After the intervention, this student started to explore other ideas that did not involve solar panels. For instance, between the minute 60 and 75, this student evaluated the relationship between windows and trees in what seems to be an iterative experimentation process: (1) modify window/tree variables; (2) collect data; (3) modify window/tree variables; and (4) collect data again.

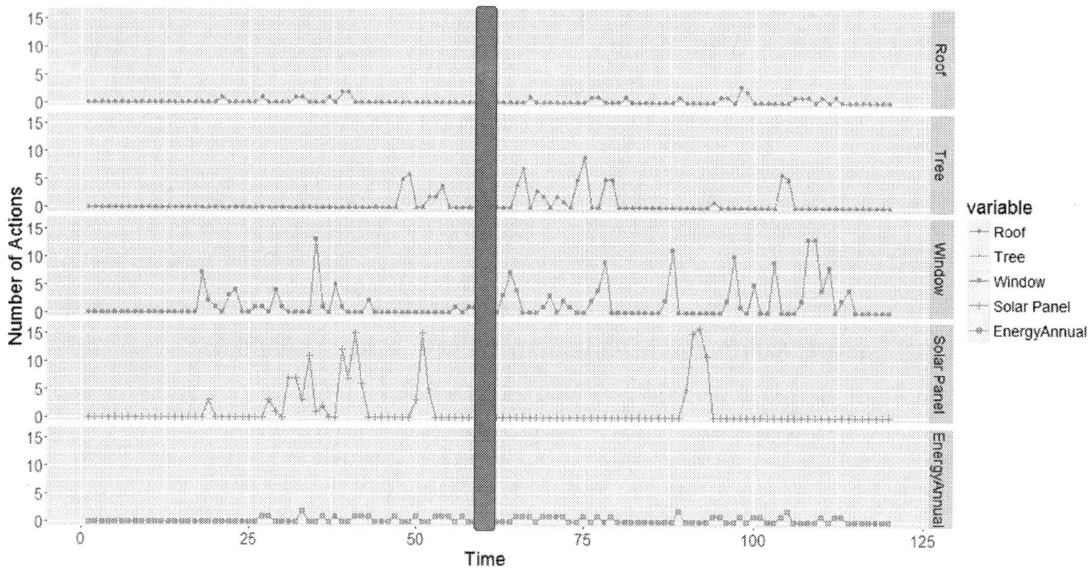

Figure 3: Number of actions from a Student A over time

The student B's interactions with the CAD tool can be visually analyzed in Figure 4. Contrarily to student A, this student continued with very similar actions after the intervention as she was doing before. For instance, the energy annual analysis was only carried out every 10 minutes or so, after several changes were made to the design. This represents confounding experiments. There are no important changes either related to which object this student is focusing on.

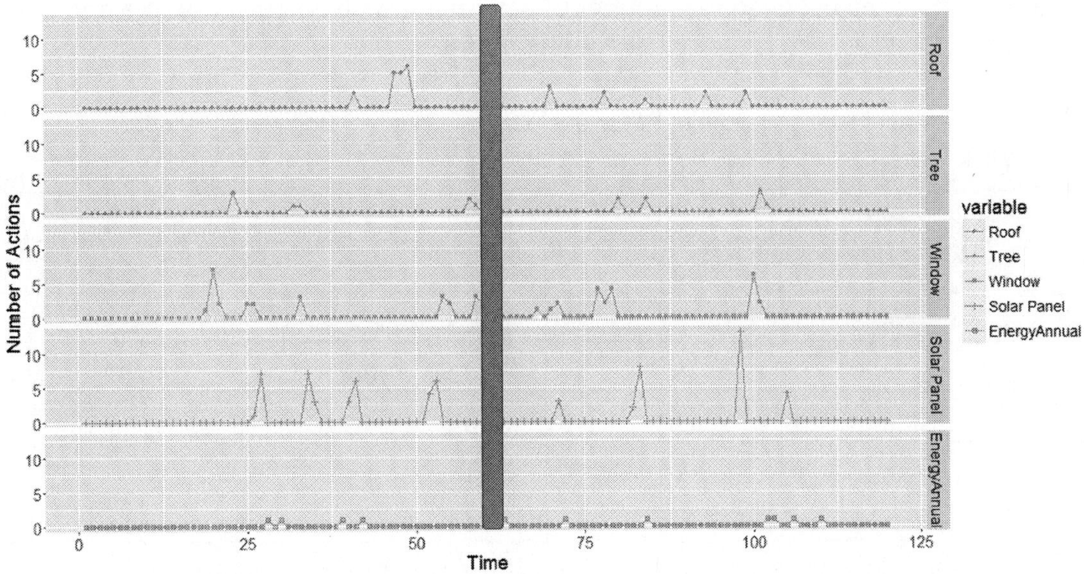

Figure 4: Number of actions from a Student B over time

Generating Ideas

In order to identify how ideation strategies changed after the effective design strategies were discussed in the classroom we considered two different levels. The first level of idea fluency corresponds to students' intention to work on a completely new design, avoiding fixation. Thirteen students out of 16 started to work on a new design after the intervention. Moreover, four of these students actually work on three different designs during this second session of the workshop. The average number of actions for the first, second, and third designs were 209.8, 136.8, and 137.8 correspondingly. This result suggests that students started exploring idea fluency at the design level after the intervention.

The second level of idea fluency corresponds to scenarios in which students would start focusing on different objects after the intervention. Table 2 represents the difference (after-before the intervention) on the number of actions over a given object. A negative number means that the student did less actions on that object after the intervention, while a positive number represents a larger number of actions after the intervention. Since the amount time for students to work on their design is comparable (i.e, approximately 55 minutes), a large positive or a small negative number should suggest the exploration of new ideas.

Students 1, 2, 4, 5, 8, 13, and 15 significantly increased their number of actions with solar panels, some of them (e.g., 4 and 15) reducing their focus on windows. Student 15 instead focused on solar panels (106) and roof actions (30) significantly more after the intervention. Students 8, 11, 12, and 13 started to explore the effect of windows and trees on the energy efficiency of the building. Students 10 and 16 do not seem to have explored many different things after the intervention, as most of their number of actions on different objects remained stable.

Conducting Experiments

The experimentation strategies within student design process were analyzed by identifying both the number of experiments students did before and after the intervention, as well as the number of systematic experiments. The average number of experiments (i.e., data collection) after the intervention (mean= 21.06, SD=12.82) was significantly different (t (15)= -2.44, , p = 0.03) from the number of experiments before the intervention (mean= 12.18, SD= 12.84).

Table 1: Difference between the number of actions for a given object before and after the intervention

Student	Roof			Tree				Window	Solar Panel
	Custom	Hip	Pyramid	Maple	Oak	Pine	Dogwood		
1	0	6	-8	0	0	34	-5	-2	100
2	0	3	12	-2	-2	17	0	1	45
3	13	0	0	2	6	8	4	2	10
4	-6	-5	0	7	-2	-4	1	-31	34
5	4	-5	15	0	0	0	0	32	49
6	0	0	13	0	0	4	6	7	7
7	-14	1	5	0	1	1	-2	-5	-7
8	1	-12	10	4	4	-1	9	56	26
9	0	-15	0	0	0	-2	-9	-23	-22
10	0	0	-4	-1	3	-7	0	5	20
11	0	6	2	4	1	17	14	73	-54
12	-1	0	1	0	0	8	2	31	19
13	0	5	-4	0	0	5	8	20	27
14	0	0	-3	2	0	9	3	2	9
15	30	-2	2	1	21	5	2	-21	106
16	0	4	7	0	0	0	9	-5	21

Likewise, the number of systematic experiments after the intervention (mean= 10.88, SD=8.41) was significantly different (t (15) = -2.565, p = 0.02) from the number of systematic experiments before the intervention (mean= 5.81, SD= 6.75). This result suggests that students started doing more controlled experiments after the intervention, and that the clickstream data was able to capture these changes in students' design behaviors.

Implications and Future Work

This study explored changes in student design strategies as characterized by process data from student interactions with an educational CAD tool. Specifically, we identified changes in the number of ideas generated and the experimentation strategies before and after an instructional intervention about effective design strategies. The results suggest that the process data can be used to identify changes in student design behaviors after a given condition. However, additional studies are required to validate what has been identified from the process data with what the student is actually considering in these changes. A future study can implement a think-aloud protocol that would enable us to compare this process data with students' rationale for their design process.

The implications of this study fall into the use of technology-based assessment techniques to characterize open-ended tasks such as engineering design. The study has demonstrated how this process data can be analyzed both visually, and using common statistical approaches to identify changes un student design strategies.

References

Atman, C. J., Chimka, J. R., Bursic, K. M., & Nachtman, H. L. (1999). A comparison of freshman and senior engineering design processes. *Design Studies, 20*(2), 131–152. http://dx.doi.org/10.1016/S0142-694X(98)00031-3

Bailey, R. (2008). Comparative study of undergraduate and practicing engineer knowledge of the roles of problem definition and idea generation in design. *International Journal of Engineering Education, 24*(2), 226–233.

Concord. (2017). *Energy3D: Learning to build a sustainable future.* http://energy.concord.org/energy3d/

Crismond, D. P., & Adams, R. S. (2012). The informed design teaching and learning matrix. *Journal of Engineering Education, 101*(4), 738-797.

Goldstein, M. H. Purzer, S., Vieira, C., Douglas, A. and Zielinski, M (2015) Assessing Idea Fluency through the Student Design Process Proceedings of the 45th Annual Frontiers in Education (FIE) Conference. El Paso, Texas. October 21-24, 2015.

Hirsch, L. S., Berliner-Heyman, S. L., Carpinelli, J. D., & Kimmel, H. S. (2012). Introducing middle school students to engineering and the engineering design process. Proceedings of the American Society for Engineering Education Annual Conference & Exposition. San Antonio, TX.

Kolodner, J. L., Camp, P. J., Crismond, D., Fasse, B., Gray, J., Holbrook, J., ... & Ryan, M. (2003). Problem-based learning meets case-based reasoning in the middle-school science classroom: Putting learning by design (tm) into practice. *The journal of the learning sciences*, *12*(4), 495-547.

Vieira, C., Aguas, R., Goldstein, M. H., Purzer, S., & Magana, A. J. (2016). Assessing the Impact of an Engineering Design Workshop on Colombian Engineering Undergraduate Students. *INTERNATIONAL JOURNAL OF ENGINEERING EDUCATION*, *32*(5), 1972-1983.

Vieira, C., Goldstein, M. H., Purzer, Ş., & Magana, A. J. (2016). Using Learning Analytics to Characterize Student Experimentation Strategies in the Context of Engineering Design. *Journal of Learning Analytics*, *3*(3), 291-317.

Seah, Y. Y., Vieira, C., Dasgupta, C., & Magana, A. J. (2016). Exploring Students' Experimentation Strategies in Engineering Design using an Educational CAD Tool. Proceedings of the 46th Annual Frontiers in Education (FIE) Conference. Erie, PA. October 12-14, 2016

Worsley, M., & Blikstein, P. (2014). Analyzing engineering design through the lens of computation. *Journal of Learning Analytics, 1*(2), 151–186.

Xie, C., Zhang, Z., Nourian, S., Pallant, A., & Bailey, S. (2014). On the instructional sensitivity of CAD logs. *International Journal of Engineering Education*, 30(4), 760–778.

Acknowledgements

This paper was partially supported by the NSF under the awards DUE 1348547 and DRL 1503436. The research team would also like to thank our colleague Roberto Aguas and the Universidad del Magdalena for inviting us to do these activities at their campus.

Copyright statement

Assessment Formats and Student Learning Performance: What is the Relation?

Khondkar Islam, Pouyan Ahmadi, Salman Yousaf
George Mason University, Fairfax, USA
{kislam2, pahmadi, syousaf3}@gmu.edu

***Abstract**: Although compelling assessments have been examined in recent years, more studies are required to yield a better understanding of the several methods where assessment techniques significantly affect student learning process. Most of the educational research in this area does not consider demographics data, differing methodologies, and notable sample size. To address these drawbacks, the objective of our study is to analyse student learning outcomes of multiple assessment formats for a web-facilitated in-class section with an asynchronous online class of a core data communications course in the Undergraduate IT program of the Information Sciences and Technology (IST) Department at George Mason University (GMU). In this study, students were evaluated based on course assessments such as home and lab assignments, skill-based assessments, and traditional midterm and final exams across all four sections of the course. All sections have equivalent content, assessments, and teaching methodologies. Student demographics such as exam type and location preferences are considered in our study to determine whether they have any impact on their learning approach. Large amount of data from the learning management system (LMS), Blackboard (BB) Learn, had to be examined to compare the results of several assessment outcomes for all students within their respective section and amongst students of other sections. To investigate the effect of dissimilar assessment formats on student performance, we had to correlate individual question formats with the overall course grade. The results show that collective assessment formats allow students to be effective in demonstrating their knowledge.*

Introduction

Assessment and student learning performance are two inseparable elements in higher education. Assessment plays an important role in shaping students' way of thinking towards their learning. The progress of engineering education relies on assessment to a great extent. One of the reasons why assessment has had little impact on the teaching-learning process is due to the absence of studies on the link between assessment type and student performance. A good understanding of this association would lead to mature assessment methods that could evidently enhance student-learning process.

Typically, each course has well-defined learning objectives. Due to several constraints such as class size, restricted time, etc., summative assessments are employed in many universities (Malapati & Murthy, 2013). This type of assessment allows instructors to gauge students' proficiencies in several courses. In summative assessments which is a common method in engineering programs, the instructor generally uses different assessment methods like written exams, papers, projects, presentations and quizzes in order to have a comprehensive understanding of students' learning process. Besides, students have the opportunity of improving their skills throughout the course.

The design of acknowledged practices of instructive assessment have a substantial influence on the advancement of engineering programs and the performance evaluation of students. Imprecise assessments lead to unproductive results, whereas good quality assessments allow instructors to achieve the course outcomes. Also, as it is stated in the research study of

Meyers and Nutly (2008), assessment is the first place that students look at in a distinct curriculum. This requires profound analysis of different assessment methods.

The employment of the alleged Bologna process in European universities has revolutionized the structure of teaching, learning approaches and also assessment methods (Flores & Veiga Simão 2015). This paradigm shift suggests an academic rearrangement focusing on both university instructors, students and new assessment methods. Previous empirical works have revealed the significance of assessment and its crucial impact on students' learning (Black and Wiliam 1998; Knight, 2008; Stassen and Kathryn, 2001; Struyven et al., 2005; Olds, Moskal, & Miller, 2005).

In (Almatrafi, Islam & Johri, 2015), authors consider the rate of accessing the material and the time duration used in assessments. They conclude that there is a relation between students' use of learning management system (LMS), Blackboard (BB) and student performance, but there was no proof of remarkable variance across a variety of demographic characteristics. The aim of Pérez et al. (2017) research study was to establish a set of measures to assess one of the most crucial engineering skill, problem solving capability. Gibbs and Simpson (2005) focused on the way assessment arrangements affect student learning. By referencing to both hypothetical and empirical verifications they propose a framework under which assessment supports learning. This framework can be used by teachers to analyse the efficiency of their own assessment practice.

The performance of students across different assessment methods and courses using correlation analysis, is examined in (Malapati & Murthy, 2013). The results of this study confirm that experimental findings when appropriately understood, can benefit educators in planning and fine-tuning the course to enhance student learning. Also, another assessment methodology, the use of correlation analysis is described in the work of (Gomes, Correia & Abreu, 2016). To locate the best exam question for assessing programming knowledge, they devised a game which is played by students in the first class. According to their findings, it is feasible to evaluate student difficulty levels and programming skills by means of dissimilar forms of questions. Likewise, Sean et al. (2007) examined the effects of conducting an initial test in different formats with regard to the way students received corrective feedback. They concluded that regular quizzes with feedback may be more productive in enhancing student learning than taking a multiple choice (MC) test.

Another study (Shankararaman & Gottipati, 2015) has been conducted to offer an automatic alignment technique in which assessments are associated with competencies to determine the consistency between them. Fletcher et al. (2011) research study is an attempt to answer to a seemingly simple question: whether faculty and students have different conceptions of assessments or not, and if so, what kind of implications this would cause in higher education policies? Authors in (Poza-Lujan, Calafate, Posadas-Yagüe & Cano, 2016) also investigate the relation between instructor workload, different evaluation strategies and student grade. Their findings show that continuous evaluation improves student grades whereas exhaustive continuous evaluation (adding written test at the end of each unit) is likely to impose an unnecessary load to instructors without having a noteworthy influence on student grades.

Although diverse assessment approaches have been scrutinized many times recently, more studies are required to gain a better understanding of the abundant methods in which assessment practice affects student learning process. In addition, most of the works have been done in this area, suffer from scholarly flaws such as insignificant sample size, lack of demographic comparisons, and substantial differences in course materials and assessments.

The aim of this paper is to address these drawbacks by comparing student learning outcomes of in-person and asynchronous online web-assisted sections of a core networking undergraduate course with regard to varied assessment methods. We also factored in demographics data in our analysis for all sections. Furthermore, students' preferred assessment is considered as another important source of information on the essence of connotation between assessment and learning. This study is based on previous research conducted at George Mason University to understand the effectiveness of assessments by discovering students' experiences with several dissimilar evaluation methods in multiple sections of the core course.

Research Study

This empirical study was carried out in the Department of Information Sciences and Technology at George Mason University for three in-class sections (001, 002, 004) and one distance learning (DL) section of a Computer Networking core course (IT 341) that was offered in Fall 2016 semester. In the online section, most of the activities such as lecture materials, pre-recorded lectures, home and lab assignments, etc. were achieved online via BB. However, it was mandatory for students in this class to meet to take their midterm and final exams in a proctored classroom on campus. The other three face-to-face (F2F) sections were all presented in class, but used BB to share course resources with students in similar manner as the DL section. The same instructor taught two sections, and different instructors taught the other sections.

Sessions of each section were administered as follows: half of each session was committed to the theoretical concepts of the course, and the second half was dedicated to practicing skill-based lab assignments conducted with Cisco Packet Tracer network simulator. Throughout this course, other than the classic home/lab assignments and written midterm/final exam, there was one skill-based assessment (SA), which comprehensively gauged students' hands-on skills and one online exam which was taken via Cisco Networking Academy (NetAcad) website.

In order to evaluate students' learning performance, we drastically changed the exam structure in each section. Table 1 summarizes the final exam format for different sections. It's worth noting that, all sections had the same home/lab assignments and similar conceptual materials based on a common syllabus. We considered the same exam structure of 25 multiple-choice questions (MCQs), fill-in-the-blank questions and a numerical problem for sections 001 and 004 because the same instructor taught them. Section 002 had 25 MCQs followed by two essay questions, and the DL section had 45 MCQs followed by a simple numerical problem. It is important to state that for each section, we devised dissimilar essay questions and numerical problems. Students in all sections were given equal amount of time to complete their assessments, and were evaluated by their respective instructor using the same grading rubric across all sections. The course grade was used to evaluate students' learning performance.

Table 1: Final Exam Formats for Different Sections of IT 341

Final Exam Format	Section 001	Section 002	Section DL	Section 004
MCQs		✓	✓	✓
Essay questions	✓	✓	✓	
Fill in the blanks			✓	
Numerical problem			✓	

A questionnaire was also designed to gather student demographics data including their gender, work status, age, proximity to campus, prior networking experience, and marital status whose results were reported in (Ahmadi et al., 2017). Moreover, students were queried about the type of question (MCQ, short answer, problem, etc.) they prefer for their exam, and whether they like in-class or take-home exams. The research questions which this paper pursues to highlight are:

- How do diverse assessment arrangements affect student learning performance in different sections of one course?
- Is there a relationship between students' preference for a particular type of question and their final grades?
- Can we find one specific assessment that influence students' final grades the most?
- Do students show concrete connotations with one single type of question? Or are they leaning towards amalgamate of different types of questions?
- How do students observe assessment regarding efficiency in terms of different components of their final grades?

Participants

In total, 115 undergraduate students participated in the study from 4 different sections. Table 2 summarizes the final grade statistics for all the sections.

Table 2: IT 341 Sections Statistics

Statistics	Section 001	Section 002	Section DL	Section 004
Count	32	27	28	28
Minimum Value	39.85	51.43	15.37	13.37
Maximum Value	92.33	95.20	96.87	91.43
Range	52.48	43.77	81.50	78.06
Average	72.86	77.47	79.61	76.07
Median	72.34	77.78	82.41	79.23
Standard Deviation	11.79	9.31	18.06	18.23
Variance	139.01	86.66	326.00	332.20

Correlation Analysis

The term correlation usually means the relationship between two or more variables, objects, data points etc. In statistics, the term correlation is measured by the correlation coefficient. A correlation coefficient typically shows the linear dependence between two variables. In mathematical notation, correlation coefficient of two data sets/variables equals to their covariance divided by the product of their individual standard deviations. As it is denoted in the below equation, *r* is the correlation coefficient *S.D(x)* is the standard deviation of *variable x,* and *S.D(y)* is the standard deviation of variable *y* (Dalgaard, 2008).

$$r = \frac{Covariance\ (x, y)}{S.D(x)S.D(y)}$$

If the correlation coefficient of two variables is equal to one, then it means that variables are linearly related and if plotted all the data points will lie on the straight line with positive slope. In our analysis, we used Pearson's product-moment correlation method (Dalgaard, 2008) to test if there is a statistically significant relation between two variables of interest. Our null hypothesis is that the correlation between two variables is equal to zero. To further understand the hypothesis testing, the output of one such test is presented in Figure 1. According to this example, the correlation coefficient equals 0.9200831 with p-value less than 0.05. Therefore, we reject the null hypothesis and confirm that, there is a statistically significant relationship between the variables essays score and total final exam score.

```
> cor.test(b$Essays, b$Total.Score)

        Pearson's product-moment correlation

data:  b$Essays and b$Total.Score
t = 16.774, df = 51, p-value < 2.2e-16
alternative hypothesis: true correlation is not equal to 0
95 percent confidence interval:
 0.8648783 0.9532985
sample estimates:
      cor
0.9200831
```

Figure 1: Pearson's Product-moment Correlation

Final Exam Format Study

Based on our analysis in the above section, we can see from Figure 2, final exam has the highest correlation coefficient amongst the other assessments with the value of 0.86. Hence, we can nominate this exam as the best predictor towards the total course score, which can be regarded as a good indicator for the course performance.

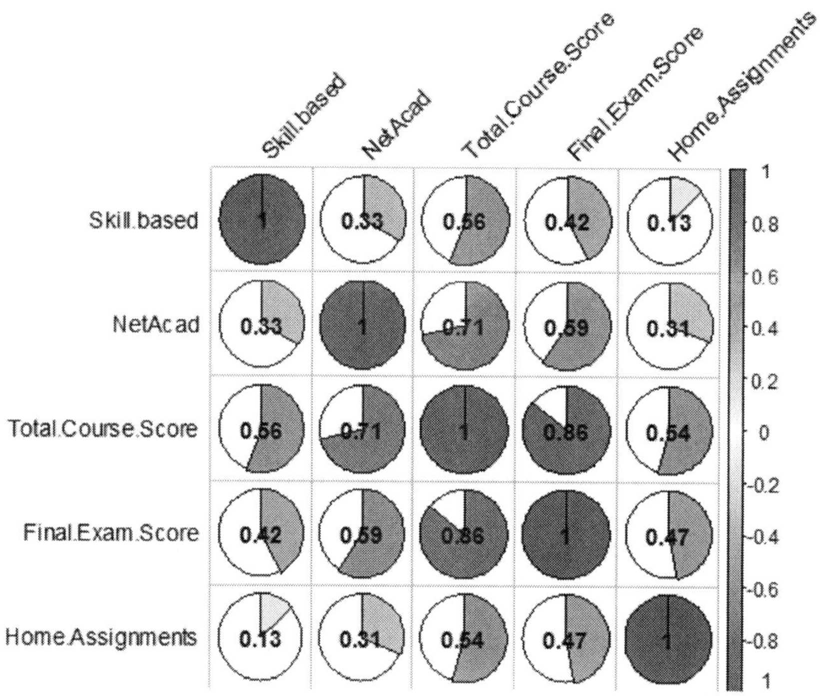

Figure 2: The Best Total Course Score Predictor Chart

Since final exam is the best predictor, we examined the impact of individual final exam question format to determine which one is the best indicator for the overall final exam score. To validate our hypothesis, correlation coefficients of final exam question formats are presented in Table 3 where we see that the highest value is for the essay question format. Thus, we can conclude that essay format is the best predictor for the final exam score.

Table 3: Correlation Coefficients of Final Exam Components

Correlation Variables	Correlation Coefficient	p-value	Hypothesis
(Essays, Final Score)	0.920	2.2e-16	Ha (true correlation is not equal to zero)
(MCQs, Final Score)	0.869	4.143e-09	Ha (true correlation is not equal to zero)
(FIB, Final Score)	0.830	1.612e-07	Ha (true correlation is not equal to zero)
(Numerical Problem, Final Score)	0.430	0.0282	Ha (true correlation is not equal to zero)

The correlation variables column of Table 3 compares the question format with the final exam score to calculate the correlation coefficient and the corresponding p-values. Figure 3 provides a snapshot of final course grade distribution for all sections.

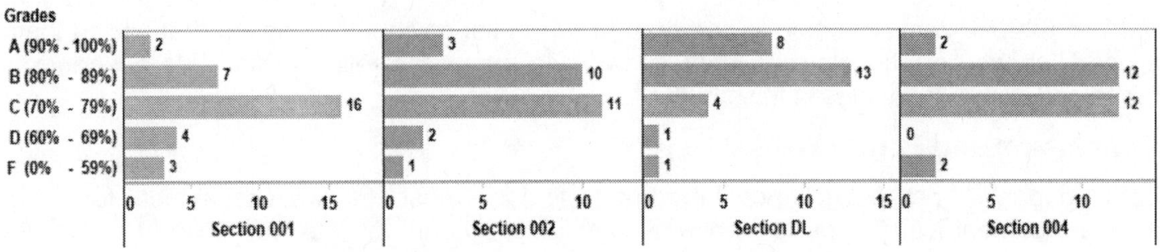

Figure 3: IT 341 Final Grade Distribution

According to Figure 3, section 001 was the only section that had the essay question format in the final exam. Based on the results of Figure 2 and respective discussion above, we now turn to the final grade distribution of Figure 3 in which we find student performance of section 001 to be the worst. This validates our findings that the final exam is the best indicator for the overall course grade because section 001 used the essay questions for its final exam.

Figure 4 illustrates the median value of each section that is represented by the horizontal line in each box. The median is based on the final exam score of respective sections. To recap from Table 1 the assessment formats of the final exam in each section: section 001 used essay format only; section 002 had MCQ and essay formats; section DL included MCQ, essay, FIB and numerical problem formats; and section 004 used MCQ format. Section DL final exam median is the highest and section 001 is the lowest. From our correlation analysis done in this paper, the poor performance of section 001 lives up to our expectation that this section would be the worst. Section DL did the best since it used a combination of all four-assessment formats where the students had the opportunity to demonstrate their knowledge via problem solving and objective questions, in addition to the subjective essay questions.

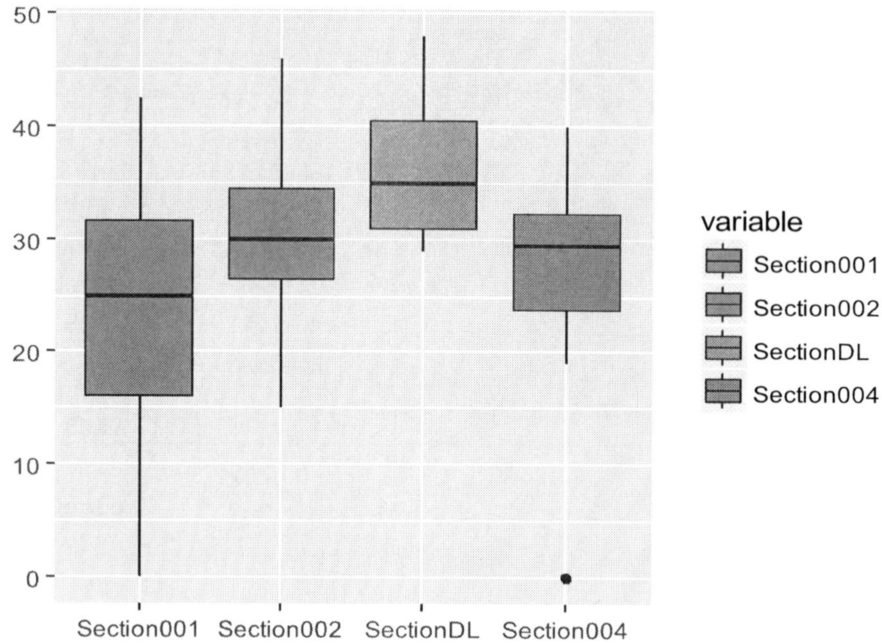

Figure 4: IT 341 Sections Mean Value Comparison for Final Exam Score

Moreover, the demographics analysis of Ahmadi et al. (2017) show students tend to lean toward objective questions over subjective essay formats. This is perhaps because with MCQ format, students are able to see the answer options with the question that allows them a better chance of choosing the right answer via process of elimination. This assertion is reinforced from our findings in this study.

Conclusion and Future Works

In this paper, we presented a novel method to find the most effective assessment format to evaluate student learning performance. A significant amount of data was collected from the learning management system (LMS), Blackboard (BB) Learn that made it possible for us to perform extensive statistical manipulation via correlation analysis. Our findings demonstrate that final exam assessment format is the best predictor to determine the course outcome for a core networking course in our department. Since the performance of section 001 was the worst and it used only essay question format in the final exam, we were able to conclude that this format is the best predictor for the final exam score. Section DL did the best since it used a combination of all four-assessment formats where the students had the opportunity to demonstrate their knowledge via problem solving and objective questions, in addition to the subjective essay questions. The demographics survey that was administered in all sections presented in our previous work also support our findings since students expressed their preference towards combined assessment techniques over a single format.

For our future work, we will investigate the relationship between student navigational behaviour and their learning performance while they access course content in Blackboard LMS.

References

Ahmadi, P., Islam, K., & Yousaf, S. (2017). *An Empirical Study of Teaching Methodologies and Learning Outcomes for Online and in-class Networking Course Sections*. Paper presented at the 2017 ASEE Zone II Conference, Puerto Rico.

Almatrafi, O., Islam, K., & Johri, A. (2015). *An Empirical Study of Face-to-Face and Distance Learning Sections of a Core Telecommunication course*. Paper presented at the 122nd ASEE Annual Conference & Exposition.

Black, P., & Wiliam, D. (2010). Inside the black box: Raising standards through classroom assessment. *Phi Delta Kappan, 92*(1), 81-90.

Dalgaard, P. (2008). *Introductory statistics with R*: Springer Science & Business Media.

Fletcher, R. B., Meyer, L. H., Anderson, H., Johnston, P., & Rees, M. (2012). Faculty and students conceptions of assessment in higher education. *Higher Education, 64*(1), 119-133.

Flores, M. A., Veiga Simão, A. M., Barros, A., & Pereira, D. (2015). Perceptions of effectiveness, fairness and feedback of assessment methods: a study in higher education. *Studies in Higher Education, 40*(9), 1523-1534.

Gibbs, G., & Simpson, C. (2005). Conditions under which assessment supports students' learning. *Learning and teaching in higher education*(1), 3-31.

Gomes, A., Correia, F. B., & Abreu, P. H. (2016). *Types of assessing student-programming knowledge.* Paper presented at the Frontiers in Education Conference (FIE), 2016 IEEE.

Kang, S. H., McDermott, K. B., & Roediger III, H. L. (2007). Test format and corrective feedback modify the effect of testing on long-term retention. *European Journal of Cognitive Psychology, 19*(4-5), 528-558.

Knight, J. (2008). Higher education in turmoil. *The Changing World of Internationalisation. Rotterdam, The Netherlands: Sense Publishers.*

Malapati, A., & Murthy, N. B. (2013). *Performance of students across assessment methods and courses using Correlation analysis.* Paper presented at the Innovation and Technology in Education (MITE), 2013 IEEE International Conference in MOOC.

Meyers, N. M., & Nulty, D. D. (2009). How to use (five) curriculum design principles to align authentic learning environments, assessment, students' approaches to thinking and learning outcomes. *Assessment & Evaluation in Higher Education, 34*(5), 565-577.

Olds, B. M., Moskal, B. M., & Miller, R. L. (2005). Assessment in engineering education: Evolution, approaches and future collaborations. *Journal of Engineering Education, 94*(1), 13-25.

Pérez, J., Vizcarro, C., García, J., Bermúdez, A., & Cobos, R. (2016). Development of Procedures to Assess Problem-Solving Competence in Computing Engineering. *IEEE Transactions on Education.*

Poza-Lujan, J.-L., Calafate, C. T., Posadas-Yagüe, J.-L., & Cano, J.-C. (2016). Assessing the Impact of Continuous Evaluation Strategies: Tradeoff Between Student Performance and Instructor Effort. *IEEE Transactions on Education, 59*(1), 17-23.

Shankararaman, V., & Gottipati, S. (2014). *Aligning assessments with competencies using keyphrase extraction.* Paper presented at the Teaching, Assessment and Learning (TALE), 2014 International Conference on.

Stassen, M. L., Doherty, K., & Poe, M. (2001). *Course-based review and assessment: Methods for understanding student learning*: Office of Academic Planning & Assessment, University of Massachusetts Amherst.

Struyven, K., Dochy, F., & Janssens, S. (2003). Students' perceptions about new modes of assessment in higher education: A review *Optimising new modes of assessment: In search of qualities and standards* (pp. 171-223): Springer.

Copyright statement

Obtaining products in Chemical Engineering through Reproducing an Environment of Innovation in the Classroom

Ronald Márquez, Laura Tolosa, Rubén Gómez, César Izaguirre, Leonardo Rennola,

Johnny Bullón, Beatriz Sandia

Facultad de Ingeniería, Universidad de Los Andes, 5101. Mérida –Venezuela.

marquezronald@ula.ve

***Abstract**: Creativity and capacity for the implementation of products are today part of the bases of university education, aimed at training professionals who possesses skills and competencies to contribute to the development of the society in which they live. The teaching-learning process in traditional university education has used strategies that place the student as a receiver of information which is conceptualized as knowledge. This type of training has been shown to have deficiencies because in present days there is a global access to information, which can be found through different media, mainly electronic. This makes the main goal of education the development of the capacity to generate solutions in obtaining products and services that can meet the needs of people, known as development of skills or know-how. In this paper, various products obtained through the application of "Reproducing an Innovation Environment in the Classroom" (RAIS) in courses in the curriculum of Chemical Engineering at the Universidad de los Andes are presented. These include personal care products such as soaps, sanitizers, cosmetic creams, sun and lip balms; cleaning products; and foods such as liqueurs and beberages. This strategy has enabled students to develop creativity and entrepreneurial skills fundamental in their succesful professional life. It has been shown that this strategy also interconects the course objectives with the development of a product, making available an implementation of product engineering in Chemical Engineering Education.*

Introduction

The current debate on education has generated an important theoretical production and a cumulative experience that has led to intelectual exchanges of all kinds. Reflections on this subject at the international level have allowed a multidimensional approach from different perspectives to what we conceive as education and its purpose (UNESCO's documentation is a good example of this). Not only the conception we have of it is object of study, but also the context in which it is framed and the reality that it must face in the coming years. In this sense, education faces new and changing challenges (Morin, 2002), to which it must respond from its own reality. This leads to a changing enviroment that needs the constant update of knowledge (Morin, 2000) that validates it and makes it effective. Currently education thinkers focus on how to articulate and organize knowledge, in order to be able to recognize and know the problems of the world and to give them adequate and pertinent answers and correctives. This is what UNESCO (1998) has agreed to call "the orientation [of education] to the long-term relevance" that is achieved when education is increasingly at the service of society, with better articulation with their problems and those of entrepreneurial work.

The understanding of the service role that education must render to society is an issue in the teaching of different disciplines, being of particular relevance with regard to engineering education. Not only because the authors are teachers in this area, but also because the socio-economic development of mankind is one of the main pillars of the training and constant professionalization of engineers (UNESCO, 2010). In this sense, engineering, in its different conceptions - research, teaching, application, consulting, technology, techniques -

offers an important factor of innovation, which, when properly empowered during its teaching, which allows the generation of pertinent solutions to new and Increasingly complex challenges facing the world today.

In order to offer an engineering education that responds to these challenges, UNESCO has grouped the different and very varied skills that should characterize an engineering professional in four broad categories: technical, personal, professional and management. Knowing that the current trend of curriculum updating processes is to include a wide variety of important subjects, this identification of the key attributes in the profile of an engineer makes a call to not forget them: they include in addition to the reasonable understanding of the technical aspects of the discipline, the development of skills to analyze, solve problems and generate innovative solutions from each discipline (UNESCO, 2010).

When this new vision of education is adopted, it is possible to understand that educational practices and teaching strategies must be modified in order to overcome the mere cognitive mastery of the disciplines and foster the development of knowledge, skills, communication, creative and critical analysis, teamwork, the ability to undertake projects, and the combined application of traditional or local theoretical and practical knowledge with state-of-the-art science and technology. These new objectives of education arise, motivated mainly by the current world reality: the Information and Communication Society (ICS) in which we live. In the ICS, the democratization of the internet, open content and collaborative environments, give the possibility not only of consuming but of creating information and participating in it, among other aspects, changing the ways of accessing, developing and communicating knowledge. These new ways of relating to knowledge have important implications for education, as they mean new ways of learning, new spaces, greater flexibility and more personalization of the learning process.

As far as higher education is concerned, this reality forces us to transform our conception of teaching and learning and, therefore, to update and rethink the academic processes within university institutions. Reform in educational practices becomes more and more urgent, when the deficiencies and failures of the prevailing model are evident. This reform requires profound curricular transformations, and although such transformations are a long and slow road, the teaching action in the classroom does not need to wait for them to be updated. We would rather say that it is from the classroom, the place from which the process of educational change should begin.

In this sense, the authors, a group of professors of the School of Chemical Engineering of the University of Los Andes of Mérida, Venezuela, have been interested in proposing and carrying out new educational actions, aimed at forming competencies, develop potentialities and promote attitudes committed to the professional and social development of students. In the teaching of different key courses in the Chemical Engineering program, the RAIS strategy "Reproduction of an Innovation Environment in the Classroom" has been implemented (Marquez, et al., 2016). According to this strategy, students learn by developing an engineering product. Knowledge is built around the execution (manufacture) of a final product, whose choice depends on the students themselves, mediated by the experience and curriculum guidelines indicated by the teacher. The students are motivated and oriented to fully handle the course and also general knowledge, that are necessary for the successful development of the project.

In this paper we present some of the most illustrative experiences of the RAIS pedagogical scope in the courses Physicochemistry for Chemical Engineers and Laboratory of Industrial Chemistry of the Chemical Engineering School of the University of Los Andes. Highlighting how the traditional academic objectives of teaching (focused on the teaching of conceptual and procedural contents) with broader objectives through the development of a product are achieved, which aim to discover, awaken, and increase the creative possibilities and the capacity of students to take entrepreneurial initiatives, fundamental conditions for the professional, personal and social development of young people (Delors, 1996).

Methodology

What is RAIS?

RAIS is a teaching strategy that seeks to integrate learning theories by construction and discovery, cooperative and collaborative learning, and project learning, with the reproduction of an innovation environment with the dynamics and characteristics that are presented in an entrepreneurial setting (Sandia, Et al, 2011a). It is a didactic proposal oriented to transform the traditional process of teaching and learning, based on the explanation and transfer of knowledge, in a process based on the application and the integral and pertinent management of knowledge. RAIS shares its fundamental goal of developing professional skills through the learning processes that are similar to those of authentic engineering practice, with models that propose a reform of engineering education, such as Problem / Project Based Learning (PBL) and the proposed CDIO: Conceive-Design-Implement-Operate. For reasons of space and with the purpose of maintaining the attention in RAIS, we invite the consultation of the work published by Edström and Kolmos (2014) where they present a greater explanation and comparison of these last two models for tengineering education.

RAIS, originally called "Reproduction of an Industrial Environment in the Classroom", was created and developed by Sandia et al. (2010) at the Universidad of Los Andes, after a continuous demonstration of a high level of demotivation among the students of the Faculty of Engineering. Subsequently, the word industrial was changed to "Innovation", to account, on the one hand, for a wider universe of applications, and on the other, the authentic creative process that takes place in the classrooms where it is developed. Sandia and coauthors propose the incorporation in the teaching-learning process of activities for the formation of competences related to creativity and entrepreneurship, which contribute to develop a favorable attitude and a disposition to learn, fostering academic success and the achievement of comprehensive vocational training. The RAIS strategy has been used in different courses and subjects of different engineering careers. The published works (Sandia et al., 2011a and b; Gutiérrez et al., 2015; Márquez et al., 2016) are an excellent example of the variety of proposals that can be made and of the flexibility of adequacy and application.

RAIS is based on the productive and effective working environment characteristic of a entrepreneurial setting. The strategy transforms the classroom into a set of start-ups, aimed at generating a final product. The development of the selected product requires the study, understanding, integration and application of a wide range of knowledge and their intrinsic relationships, where both the knowledge inherent in the subject that the students are taking, as well as those that allow the successful execution of the project are taked into account. In this sense, RAIS becomes an effective platform to implement the integration of knowledge and disciplines to develop a product. For a better description of the methodological strategy, its fundamentals, its components and gears, as well as the evaluation strategies, the reader is advised to refer to the previous works of Sandia et al. (2011a and b) and Marquez et al. (2016).

How is RAIS being implemented for teaching in Chemical Engineering?

One of the advantages of the RAIS strategy is that, despite the key and indispensable components necessary for its implementation, and therefore, to ensure the achievement of the objectives, RAIS is a flexible and adaptable strategy, both for the requirements and curricular conditions of the course and the characteristics of the group of students (interests, previous knowledge, expectations, availability of time and resources, ease of access to information, among others). Thus, every RAIS course is based on the following components or steps: establishing clear teaching-learning objectives, organization of the set of companies (work teams), selection of the products to be developed, definition of roles and functions to be fulfilled by both students and the teacher, scheduling activities (product execution plan), meetings and partial reports, and specifying the strategies that will be used for formative and summative evaluation of the process.

The above steps are common to all courses based on RAIS. The particularities, which are given by the nature of the subject and by the group of students, differentiates one RAIS experience from another. In the case of the Chemical Engineering courses where the strategy has been applied, we want to highlight some aspects that have been adapted according to the possibilities and limitations of the courses:

- Different products per company, proposed by the students.

Unlike previous experiences (Sandia et al., 2011b and Gutierrez et al., 2015), for the courses Physicochemistry for Chemical Engineers and Laboratory of Industrial Chemistry, each company proposed several possible products to be developed. With the guidance of the teacher, the students had to select the one that allowed the most optimal fulfillment of the cognitive, procedural and attitudinal objectives proposed. The participation of the students in this stage, favored the diagnosis, a primary educational element in educational planning (Diaz Barriga and Hernández, 2002), which allowed the curricular content to adapt to the interests and needs of the young people. On the other hand, it allowed the activation of previous knowledge, coupled with the search for related and relevant information. From the constructivist approaches (Bruner, 1959) and cognitive theory (Ausubel, 1963, Ausubel et al., 1978), essential for the construction of knowledge and the acquisition of meaning to acomplish an integration between the new Information and pre-existing conceptual structures.

- Goals and schedules adjusted to each company.

When establishing a single product for the course, objectives and work planning is common for all companies. In our case, each company presented varying requirements and levels of complexity (within an average range), which depended on the product and the group of students. This situation demanded that the objectives and schedules should be adjusted to the conditions of each group. In spite of the difficulties that this variety of cases might entail, we believe that it simulates more closely the global and complex reality (Morin, 2002) that students in training will face tomorrow. RAIS, in its conception, seeks to reproduce from the classroom this reality "from the outside", showing that any aspect of this "outside" reality implies a context of its own, which can and should be approached from different perspectives. The contextualization and relevance of the process of study, design and elaboration of the product, and consequently, of the knowledge and the way of approaching them, is another important adaptation to highlight in our work.

- Broader interconnection of knowledge from other disciplines.

As a consequence of the previous point, the variety of products and fields of application, generated a greater wealth in contents and extracurricular knowledge. The interaction between groups during the development of the process, and the final presentation of the projects, allowed all students the access to this wealth of knowledge. We believe that through this way of implementing RAIS, students can form a broader and global vision of the scope and usefulness of the knowledge acquired, and therefore, will have the opportunity to establish greater relations of meaning that promote the positive modification of their behavior and attitude towards professional life.

- Strengthening skillsand competencies for experimental laboratory work.

The particular nature of the knowledge area of Chemical Engineering offers the possibility of developing both cognitive and procedural aspects related to the transformation of raw material into products of greater value, utility and versatility. The experience at the undergraduate level, of developing a product based on technical specifications, in a set time and as a result of a laboratory procedure, is an ideal scenario for learning. In this "controlled" environment, it is possible to promote the specific development of skills and competences, to encourage self-confidence, and to strengthen decision-making capacity in practical laboratory work. This learning will have a positive effect on the future performance of students in all kinds of jobs, especially those involving practical laboratory or production work in a company.

- Promotion of a healthy and proactive digital culture within the framework of the ICS.

As a final part of the work plan for obtaining the product, the publication of the completed project of each of the companies with completely free access on the Web was established. In the Information and Communication Society (ICS) in the that we are living, are generating new ways of accessing, developing and communicating knowledge. We consider that the task of publishing the works produced by the students themselves allows them to become aware of several important aspects: (a) that they are not only information consumers, but also creators and participants; (b) that knowledge must be oriented (being useful) and must be able to be applied in the resolution or satisfaction of the needs of society, otherwise it has no value to communicate it, because (c) the relation of the learner to the world must be authentic, identifying the contributions that as a citizen can offer to that world; (d) that today's society is not restricted to the physical limits of its closest environment, but extends to the planetary community, and consequently, (e) that the solutions and proposals of "here" may also be useful "there".

Results

The RAIS strategy was applied in several courses of the School of Chemical Engineering of the University of Los Andes, during the period 2010 to 2014 (Márquez et al., 2016). This report shows the work done in two of these courses: one theoretical, Physicochemistry for Chemical Engineers (FIQ); and another practical, Laboratory of Industrial Chemistry (LQI). Between the two courses, and in the two opportunities in which RAIS was applied in them, a total of one hundred and twenty-six (126) students were attended. Table 1 shows the particular characteristics of each course.

Table 1 Characteristics of the courses studied

Course	Semestre	Nº de estud.
Physicochemistry for Chemical Engineers *Theoretical course* *6th semester* *6 h / week* *Objective: To introduce the conceptual development of physicochemical phenomena and their importance in Chemical Engineering*	B2010	49
	A2013	37
Laboratory of Industrial Chemistry *Practical course* *7th semester* *4 h / week* *Objective: To handle the practical Industrial Chemistry and its applications in Process Engineering*	B2011	20
	U2014	20

The application of RAIS in both courses was oriented for obtaining an experimental product by company, following the same general objective, structuring of the work in four

(4) phases, strategies and evaluation criteria, for a lapse of sixteen (16) Weeks (Table 2 in Márquez et al., 2016). From each course, the most representative projects were selected, as shown in Table 2 of this paper.

Table 2. Selected products developed with RAIS*

Physicochemistry for Chemical Engineers	
B2010	**A2013**
1. Liqueurs flavored by fermentation of sugar cane syrup 2. SPF 50+ Solar Protector	1. Homemade beer 2. Anesthetic cream based on eugenol
Laboratory of Industrial Chemistry	
B2011 and U 2014	
1. Beeswax products: Healing cream Moisturizing lotion Lip balm Soap	

* Each of the projects are published with free access on the website: http://webdelprofesor.ula.ve/ingenieria/marquezronald/

Physicochemistry for Chemical Engineers:

Physicochemistry for Chemical Engineers is a key course in the basic and applied training of the Chemical Engineer, it integrates the subjects of Solutions, Phase Equilibrium, Reaction Kinetics, Interfacial Phenomena, Electrochemistry, which are the basis for the development of unit operations courses: Unit Operations III and IV and Calculation of Reactors. The objective was to obtain products that would use natural raw materials as much as possible for their formulation. The following products developed in the course are presented:

1. Liqueur by fermentation of sugar cane syrup

1.1 Objective: To produce a creamy liqueur with an alcoholic degree of 40° of three different flavors using the processes of fermentation and distillation to obtain the alcohol.

1.2 Contents:

Conceptual: Solutions, phase equilibrium, chemical equilibrium, reaction kinetics, interfacial phenomena. Other related content: Food science

Procedural: Calculation of colligative properties, determination of reaction rates and suitable operating conditions to obtain higher yields of the product, use of binary and ternary phase diagrams, use of the principles of liquid-vapor equilibrium and distillation for the purification of substances.

1.3 Description of the Procedure: The liquor is made from the fermentation of sugar cane syrup and distillation to obtain high purity ethanol, using brewer's yeast as a fermentation agent, as well as agents such as ammonium sulfate and sodium phosphate to be a suitable medium for the action of yeast. The ethanol produced at 97% (obtained through double distillation) was diluted to 40% (commercial concentration) and then used to produce flavored liquors of chocolate, vanilla and coffee.

2. SPF 50+ Sunscreen

2.1 Objective: To formulate a sunscreen with a sun protection factor SPF greater than 50 that has high stability, appearance and pleasant organoleptic properties.

2.2 Contents:

Conceptual: Solutions, Interfacial Phenomena. Other related contents: Cosmetology, Health sciences.

Procedural: Calculation of colligative properties, use of binary and ternary phase diagrams, use of the principles of physicochemical formulation for the formation of emulsions.

2.3 Description of the procedure: The sunscreen is formulated based on an emulsion of triglyceride oils of vegetable origin (coconut and canola), nonionic ethoxylated and non-ethoxylated sorbitan ester surfactants, chemical sunscreens as active ingredient (Benzophenone, Parsol), preservatives, stabilizers and thickeners. It is obtained through a low energy method.

3. Artisan beer

3.1 Objective: To produce artisan beer based on malted barley and malt extract, which possesses adequate organoleptic properties: color, flavor, aroma.

3.2 Contents:

Conceptual: Solutions, Chemical equilibrium, Reaction kinetics. Other related contents: Food science.

Procedural: Calculation of the composition of the components of a sample, determination of the reaction rates and operating conditions suitable to obtain a higher yield of the product.

3.3 Description of the procedure: Artisan beer is produced through a fermentation process by activating the yeast and preparing the must with the malted barley and the malt extract. The baking is done and the hops are added. The wort is cooled to 35 °C and the yeast is added. The fermentation time is 2 to 3 weeks, an adequate amount of sugar is added and bottled to proceed to the gasification, through the formation of CO2 in 1 or 2 weeks. The reaction kinetics are studied by measuring the concentration of glucose in the system as a function of time.

4. Anesthetic cream based on eugenol

4.1 Objective: To obtain a formulation for the use of eugenol extracted from cloves as a local anesthetic, which has high stability, appearance and pleasant organoleptic properties.

4.2 Contents:

Conceptual: Solutions, Phase Equilibrium, Interfacial Phenomena. Other related contents: Pharmacology, Dentistry, Health Sciences.

Procedural: Calculation of colligative properties, use of binary and ternary phase diagrams, use of the principles of physicochemical formulation for the formation of emulsions.

4.3 Description of the procedure: A base emulsion formed by paraffin oil, non-ethoxylated and non-ethoxylated sorbitan esters of non-ionic surfactants is formulated in which the recommended amount of eugenol is solubilized as an oil phase for application as a topical and oral anesthetic.

Laboratory of Industrial Chemistry

The course Laboratory of Industrial Chemistry is divided into practices of subjects related to each other in that all are organic chemistry processes, including evaluation of properties of petroleum products, soap production, production of pectins, obtaining dispersed systems and evaluation of some of its properties. For this reason the strategy used in the development of RAIS was the creation of companies that were in charge of planning and coordinating the manufacture of the product in the laboratory environment where twenty researchers where available to obtain the products. The products selected in this work are the following:

1. Healing cream, moisturizer and lip balm

1.1 Objectives: Technological use of beeswax to obtain organic synthetic products, non-toxic to humans. Obtain a healing cream, moisturizer and lip balm based on beeswax and extra virgin sesame oil that has adequate wettability and stability.

1.2 Contents:

Conceptual: Natural raw materials, Phase equilibrium, Solutions. Other related contents: Formulation of products for personal care, health sciences.

Procedural: Use of binary phase diagrams, control of the variables of operation: temperature and quantities of the reagents to produce a substance with the required properties, measurement of the viscosity of the creams produced.

1.3 Description of the procedure: The healing cream, moisturizer and lip balm is formulated by heating the beeswax to about 60 °C, then it is added in the appropriate proportion in the sesame oil under agitation and the system is then cooled to form the dispersion of solid particles of beeswax in sesame oil. Creams with a beeswax content of 15 to 30% can be used as a healing and moisturizing agent. Above 30% a solid paste is obtained which is used as a lip protector.

2. Soap based on beeswax

2.1 Objective: To formulate soap from saponification of triglyceride oils of sesame oil and beeswax esters with sodium hydroxide, which has detergency, foamability, consistency, pH and suitable organoleptic properties.

2.2 Contents:

Conceptual: Chemical equilibrium, Phase equilibrium, Interfacial phenomena. Other related contents: Formulation of products for personal care, organic chemistry.

Procedural: Use of binary and ternary phase diagrams, control of the operating variables: temperature and reagent quantities to generate a saponification reaction with a high yield, foamability measurement and stability of solutions with surfactant.

2.3 Description of the procedure: The soap is formulated through the saponification of the mixture of 20-40% of beeswax and sesame oil. Soap is produced by the cold process, which consists in heating the mixture of beeswax and sesame oil to 60 °C until the beeswax melts, adding the sodium hydroxide in an amount calculated with the index of saponification of the mixture of beeswax and oils. The soap produced is then cooled and stored for one month, during which time the cold saponification reaction continues.

Discussion of results

The products presented and developed in each of the courses fulfilled the objective of integrating the greatest amount of programmatic contents to obtain them. The additional contents that could not be part of the formulation of the product were evaluated through other instruments such as short exams and partial evaluations. The objective of the RAIS strategy is that the developed product integrates most of the knowledge and competences of the course, since, as Morin (2002) states, fragmentary knowledge serves no purpose other than technical uses. The methodology, work plan and evaluation method has been described by the authors (Márquez et al., 2016), achieving that competences are developed in all the contents of the course. This strategy motivates the student, developing skills of creativity and innovation, necessary in the professional field of Chemical Engineering and in the training to carry out applied research. In addition, it was observed that the subjects where RAIS is applied present an average increase of 28% of approved students with respect to the courses that develop the contents with master classes, decreasing the number of students dropping out by 17% on average. This is mainly because the RAIS strategy requires continuous monitoring and compliance with the objectives and competences of the course, both by the young person and the teacher (Márquez et al., 2016).

In this sense, surveys were carried out on the 128 students who completed courses through the RAIS strategy. Where 89% considered that education should focus on product and project engineering, and more than 80% indicated that this strategy contributes to educational training to be applied in their professional lives, in the industrial and research fields (Márquez et al., 2016). More than 50% of them have had some experience of entrepreneurship or have started their own private small business, indicating that the strategy has contributed to their development in this area.

Conclusions

The execution of products through the Reproduction of an Innovation Environment in the Classroom is a teaching-learning strategy that promotes the development of the innovation and entrepreneurial skills necessary for the student's personal and professional development, integrating the objectives, contents and competences of the courses in Chemical Engineering. This integration is achieved through the active participation of the young person, achieving that there is commitment in the fulfillment of the educational objectives of the course. In addition, the RAIS strategy leaves an added value, as is the innovation and obtaining products for the social benefit of the community in which the university is integrated. In the future, it is expected to expand this strategy to a greater number of courses in the Chemical Engineering curriculum, which will allow the implementation of interconnected projects and the development of final products that can meet the universal objectives of engineering education.

References

Ausubel, D. (1963). *The Psychology of Meaningful Verbal Learning*. New York: Grune and Stratton.

Ausubel, D., Novak, D. and Hanesian, H. (1978). *Educational Psychology: A Cognitive View* (second edition), Holt, Rinehart and Winston: New York.

Bruner, J.S. (1959). Learning and Thinking. Harvard Education, 29: 184-192.

Bruner, J. S. (1966). Toward a theory of instruction (Vol. 59). Harvard University Press.

Delors, J. (1996). *La Educación encierra un tesoro.* Informe a la UNESCO de la Comisión Internacional sobre la Educación para el Siglo XXI. Madrid: Santillana/UNESCO.

Díaz Barriga, F. and Hernández, G. (2002) *Estrategias docentes para un aprendizaje significativo. Una interpretación constructivista*. McGraw-Hill. México.

Edström, K. y Kolmos, A. (2014) PBL and CDIO: complementary models for engineering education development. *European Journal of Engineering Education*. 39(5), 539-555.

Gutiérrez, D., Sandia, B., Hernández, D., Páez, G. and Lischinsky, P. (2015) Integración de conocimientos a través del desarrollo de un producto RAIS. Caso de estudio: ingeniería de software and bases de datos. *Educere*, 19 (62): 169-179.

Márquez, R., Tolosa, L., Gómez, R., Izaguirre, C., Rennola, L., Bullon, J., Sandia, B. (2016) Reproducción de un ambiente de innovación en el salón de clase. Una estrategia para promover la creatividad en la educación en Ingeniería Química. *Educación Química,* 27, 249-256

Morin, E. (2000) *Los siete saberes necesarios a la educación del futuro*. Ediciones FACES/UCV. Caracas, Venezuela.

Morin, E. (2002) *La cabeza bien puesta. Repensar la reforma. Reformar el pensamiento*. 1st edition. Nueva Visión, Buenos Aires.

Sandia, Beatriz, and col. (2010). RAIS: Una Estrategia para el Manejo Integral de Conocimiento. Experiencias en Ingeniería, III Congreso Venezolano de Enseñanza de la Ingeniería, Mérida, Venezuela. Octubre 2010.

Sandia, B. E., Gutiérrez Pinzone, G. D. and Páez Monzón, G. (2011a). Enseñanza de la Ingeniería reproduciendo el Ambiente Industrial. Un manejo integral de conocimientos. *Educere* 15:(51), 379-388.

Sandia, B., Gutiérrez, D., Hernández, D., Alvarado, J., Puig, J. E. P., Paredes, M. J. V., Monzón, G. P. (2011b). RAIS: una estrategia para el manejo integral de conocimiento. Experiencias en ingeniería. *Ciencia e Ingeniería*, 115-122.

UNESCO. (1998). Declaración Mundial sobre la educación Mundial en el Siglo XXI: Visión and Acción. Available in: http://www.unesco.org/education/educprog/wche/declaration_spa.htm on 5 january 2017.

UNESCO (2010) *Engineering: Issues, Challenges and Opportunities for Development*. UNESCO Publishing. Available in http://unesdoc.unesco.org/images/0018/001897/189753e.pdf on 3 may 2017.

Teaching Mathematics in Engineering Careers: a Permanent Challenge

María Alicia Gemignani

Universidad Tecnológica Nacional, Facultad Regional Paraná, Paraná, Argentina
alicia.gemignani@gmail.com

María Itatí Gandulfo

Universidad Tecnológica Nacional, Facultad Regional Paraná, Paraná, Argentina
mariagandulfo@gmail.com

***Abstract**: This paper reports on the actions implemented by teachers of mathematics in the first course of engineering careers of the Parana Regional Faculty (FRP) belonging to the National Technological University (UTN). The aim is to bring about meaningful learning by making students face learning experiences in the context of their careers. The building of modern engineering should be built on the foundation of physical-mathematical concepts, designed as a process that strengthens interdisciplinary articulations between Mathematical Analysis 1 (MA 1) and Algebra and Analytical Geometry (A&AG), with the specific subjects of the careers: civil, electromechanic and electronic engineering. The FRP of the UTN has implemented an experience promoted by the Chairs of MA 1 and A&AG that aims to ensure that the first year students achieve the necessary skills to perform two Practical Integrative Works (PIW) that involve real problematic situations of each career using mathematical software.*

Context

According to Trejo, Gallardo, and Trejo (2013) the challenges of universities have nowadays changed. Technological development requires that engineers trained in higher education be competitive at the national and international levels to face the globalization process, so it is necessary to rethink the justification for mathematics, its contents and the methodology of teaching, so that the students acquire the capacity to be creative, innovative and be able to reason about the solution of problems of the area of development that competes to them.

In the meetings of Directors of the Department of Basic Subjects of the different Regional Faculties belonging to the UTN, in its 2014-2015 agenda, the need arose to focus training on the future profession and to consider the professional activity, from the beginning of the career, as a significant reference of the formative trajectory.

For Camarena, and Olazábal (2008) Mathematics in the context of science is based on the following educational paradigm: With math courses the student will acquire the elements and tools that he will use in the specific subjects of his career, that is to say, the subjects of the area of mathematics are not a goal in themselves but a tool of support to his career, without leaving aside the fact that mathematics must be formative for the student. These tools should provide the students with the bases that enable them to successfully tackle problems that require their analytical skills and innovation.

In this context, it is fundamental to propose innovations in teaching processes focused on interdisciplinary approaches, on the use of new technologies, on the development of complex cognitive processes, on new relations between theory and practice. Since it is on the foundation of mathematical physical concepts that the modern engineering building must be constructed, thought as a process that begins in the need to strengthen interdisciplinary actions and horizontal and vertical articulation of MA 1 and A&AG, with the specific subjects of the career, planning strategies of action and application.

The new methodologies and teaching techniques should lead the teacher to act as a facilitator in the teaching-learning process, so that students develop skills and abilities that allow them a good social, personal and independent professional development.

So that the present university meets the demands of training required by society, it is essential to adopt new approaches in teaching, especially in what refers to teaching mathematics and problem-based learning, which reflect the nature of engineering itself.

In this sense, our proposal was to implement two Integrative Practical Works which are carried out in the Chairs of MA 1 and A&AG of our faculty, presented as objectives to be achieved by students and which are established as of compulsory approval for the regularization and/or promotion of both subjects.

Framework

The problem of learning mathematics in general and Mathematical Analysis 1 and Algebra and Analytical Geometry in particular is perhaps one of the greatest challenges for didactics, and one of the major concerns for those who are in charge of teaching them. There are investigations that have approached the problem from different dimensions: the psychological, the didactic, and the epistemological and from different theoretical frames of reference.

The insufficient mastery of the basic concepts, the scarce formalization of them and the lack of skills for the analysis and resolution of problems, as well as the mechanistic and strongly operative learning of mathematics in the high school level, show that only a small percentage of the candidates for admission to the engineering courses of the FRP, have the basic knowledge and skills to begin the course of MA 1 and A&AG, as shown by the diagnostic tests of the introductory seminar. All this carries a very powerful negative psychological load in the beginner students and it is the function of first year teachers to try to reverse this situation so that the students develop the necessary attitudes and skills.

Different researches tackle these problems by proposing didactic strategies, with several objectives, some of them, to promote significant learning. According to Ausubel, Hanesian, Ausubel, Novak, and Hanesian (1997) didactic strategies are the teaching strategies which specify a series of learning activities aimed at students and adapted to their characteristics, to the available resources and to the contents of study, with the objective of favouring the understanding of concepts, their classification and relation, the students' reflection, their training of ways of reasoning and the transfer of knowledge.

The students devote very little time to the autonomous activities and support lessons, with little attendance to the tutoring spaces. They reduce their activity, in most cases, to listening to the teacher and taking down notes. This implies high conduction in their work activities and passive processing of information.

The study and gradual implementation of Information and Communication Technologies (ICT), in the development of the different issues of the subjects involved, is an essential factor in the motivation of the student.

Visualization plays a central role in the learning of sciences. Several articles (e.g., Zimmerman, Cunningham 1991, Duval 1999, Hitt 1998) relate experiences using ICT as a mediator in the process of teaching and learning of concepts related to the Calculus. The use of these technological tools in no way invalidates the clear and precise teaching of mathematical concepts. First students must acquire the concepts and then, the technological tools as a strengthening element.

Problem solving proves to be one of the difficulties that in recent times are being approached with great interest and concern by educational research. For Gaulin (2001) to talk about problems involves considering those situations that demand reflection, research, and where to respond we must think about solutions and define a strategy of resolution that does not lead, precisely, to a rapid and immediate response.

The emergence of the problem solving approach as a didactic concern arises as a consequence of considering learning as a social construction that includes conjectures, tests and refutations based on a creative and generative process. Teaching from this perspective aims at emphasizing activities that pose problematic situations whose resolution requires analyzing, discovering, hypothesizing, confronting, reflecting, arguing and communicating ideas.

Thus, problem-based learning helps the student to develop and work on various competences. The main competences highlighted by De MIguel (2005) are:

- Problem resolution
- Decision making
- Teamwork
- Communication skills (argumentation and presentation of information)
- Development of attitudes and values: precision, revision, tolerance.

Similarly, Benito, and Cruz (2005) in addition to the acquisition of the competences already mentioned indicate that problem solving favours the development of effective reasoning and creativity.

It is for this reason that from the first year of the career, the students must perform two Practical Integrative Works (PIW) that the Chairs MA 1 and A&AG jointly develop, in which the aim is the resolution of simple and real problems connected with their profession and that can be solved with the concepts developed within these two chairs.

Objectives

With the raising of problems, the protagonists of learning are the students themselves, who assume the responsibility of being an active part in the process.

Prieto (2006) defending the active learning approach points out that problem-based learning represents an effective and flexible strategy that, based on what students do, can improve the quality of their university learning in very different aspects.

In agreement with the ideas of this author, we propose as objectives, those of the subjects involved through the development of competences in:

- Identification of relevant problems in the professional context
- Awareness of one's own learning
- Planning the strategies to be used
- Critical thinking
- Self-directed learning
- Assessment and self-assessment skills
- Permanent learning

Methodology

As mentioned above, within the requirements of the Chairs to regularize and/or promote the subjects, is the approval of two PIWs that must be solved in groups of two or three students using Mathematica software.

These PIWs consist of exercises that have a greater degree of complexity than those that can be solved manually and application problems according to their future profession. The concepts involved in each PIW are previously worked out in theoretical and practical lessons.

This activity is carried out by about 180 students who attend both subjects. They belong to five different groups of approximately 36 students each. Two of these groups attend electronic engineering, two attend civil engineering and one group attend electromechanical engineering. These PIWs are carried out in class.

Apart from this, students must attend the computer lab every 3 weeks at extra-curricular fixed hours. Here, they solve exercises assigned in the practical classes and others of greater operational complexity using the specific software and analyzing the results obtained. This activity aims at motivating and encouraging learning with the computational tool, tending to optimize the teaching and learning process. These classes are guided by student trainees who elaborate extra guides with commands and applications, under the tutelage of the teachers responsible for the course. This activity constitutes a training work for what the students must later produce in the integrating PIW.

As additional support for this task, special classes are set up in charge of the different professors of both Chairs.

The PIWs are formulated by the directors of each Chair with the collaboration and contribution of all the members. They are settled in articulation meetings prior to the beginning of the school year.

Each course has a teacher responsible for the theory and another one for the practice. They are responsible for carrying out the activities with the groups. They can also introduce modifications that they consider relevant and that may arise from the interactions with their students in the classroom. These interactions serve as a feedback for teachers to obtain information about the process and are useful for making decisions that allow students to be guided towards the proposed learning objectives.

Some of the problems to be solved by the students are presented here, since for reasons of space it is not possible to incorporate also the productions.

Model of Application Problem:

The flooded area of a zone which is shown in the attached chart was determined by satellite photos. Using the method of Simpson, calculate approximately:

1. Area **A** of the interior region (flooded surface).
2. The coordinates of the centroid of the region R (x_c and y_c).

You must submit the sheet with the points considered. In the image corresponding to your group mark the contour of the region in question. It is suggested to divide the interval with a grid of separation of no less than 25 parts. Consider each grid as an area unit.

Figure 1: Group 3 – Rio Salado Overflowing

Model of Application Problem:

Special laminators are used for the lamination of rings, discs and profiles such as those shown in Figure 1.

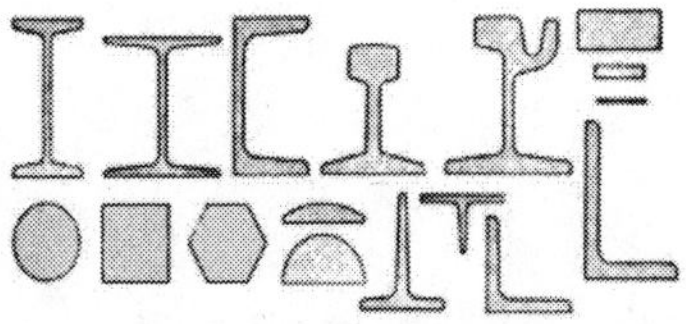

Figura 1

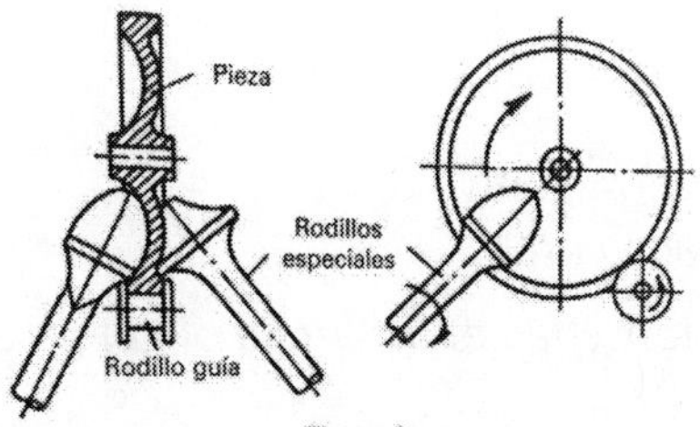

Figura 2

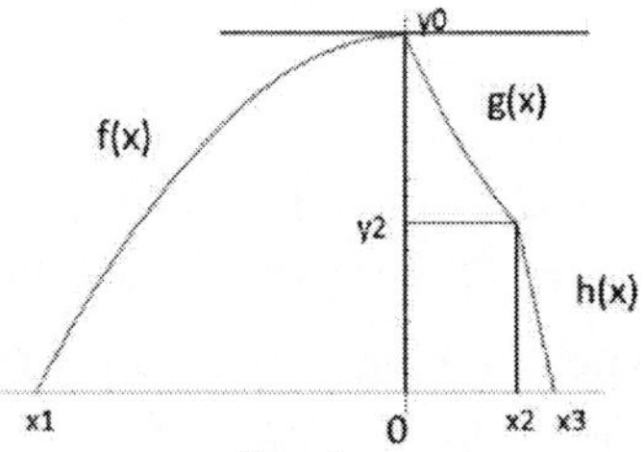

Figura 3

A multilateral ingot cut into pieces is used as raw material. After compressing the part in a hot press, it arrives at a rolling mill in which, with the help of the special rollers, the part acquires the necessary profile. Figure 2 shows the part, the special rollers and the guide roller. It is desired to mathematically model the head of a special roller, in particular the one on the left of the diagram according to the mathematical model of figure 3.

You are asked to solve the following items:

3. Determine the equation of the exponential function g(x).
4. Determine the equation of the linear function.
5. Determine the equation of the parabola f(x) and its characteristics.
6. Calculate the total cross-sectional area of the part of the figure.
7. Calculate the perimeter of that section.
8. Calculate the weight of the piece considering that it is steel with a density of 7850 kg/m³.
9. The piece must be bathed with an anticorrosive coating. The yield of the coating is of 4 lts/m². Calculate the amount of coating required.
10. Calculate the coordinates of the centroid of the piece. Graphic it.
11. Calculate the moments Mx and My and the moments of inertia Ix and Iy of the cross-sectional area with respect to the axis of symmetry.
12. Determine the equation of the surfaces generated by the exponential function, the line and the parabola by rotating around the x-axis. (Note: take into account that it is a surface given by parts). Graphic it.

Note: all measurements must be considered in centimetres.

In the above examples, we can observe the horizontal integration between the subjects, where the concepts of A&AG are used, such as the conic equations, line in the plane and surfaces of revolution, together with contents of MA 1 such as areas, volumes and inertia moments.

The authors will present in the REES some productions of the students.

The issues involved in PIW N° 1 are:

- Approximate graph of functions
- Taylor and Mclaurin Polynomial
- Complete study of functions with different degrees of difficulty
- Application problems of linear equation systems
- Optimization problems
- Application of derivatives
- Practical Applications of Differential Calculus Theorems

The issues involved in PIW N° 2 are:

- Application problems of integrals defined in Cartesian, polar and parametric coordinates (Calculus of: areas, perimeters, lateral areas, centroids, moments of inertia, volumes, forces, works)
- Conics
- Quadric
- Numerical derivation and integration
- Problem of application to each career

The planning of both Chairs establishes an instance of defence of the PIW presented.

Evaluation

Once the PIW is revised, the teachers give feedback to the student group who then makes corrections.

After this, a colloquium is performed with the participation of all the members of the group; however they get an individual final mark. The criteria of evaluation are: the ability to apply the concepts studied, the ability to determine the linkages between concepts, the ability to interrelate theory and practice, the procedures performed, the results obtained, their feasibility and software application. This is shown in the following form:

Tabla1. Evaluation form. Aspects to be assessed in the correction of the PIW

	Evaluation criteria	Outstanding	Very good	Good	Fair	Poor
1	Identification of data and goals					
2	Interpretation of the assignment, relating the theoretical concepts, hypotheses and properties					
3	Selection of the appropriate approach to the solution					
4	Procedures and calculations					
5	Results obtained and their adequate communication					
6	Graphical representations, if applicable					
	Final mark					

To evaluate the learning process, the teachers have devised the following individual form to monitor permanently each student´s performance. Later, this evidence is confronted with the student´s academic achievements in both subjects.

Table 2. Process evaluation form

Surname:... First names:		
Group:.................. Average of partial evaluations:............Final mark PIW:......		
Evaluation Criteria	Indicators	Assessment
Relevance: correspondence between ideas and concepts	• He proposes ideas that fit a situation raised	5 – 4 – 3 – 2 – 1
Creativity: ability to produce and generate new ideas	• He is original • He takes advantage of obstacles • He takes initiative • He incorporates new ideas	5 – 4 – 3 – 2 – 1
Fluency: easiness to express	• Shows clarity and precision in	5 – 4 – 3 – 2 – 1

ideas, technical vocabulary, clarity and coherence	the presentation of ideas • Has the ability to produce and express coherent ideas • Has rich vocabulary • Provides sufficient information	
Coherent articulation: interrelation, sequence and continuity in productions	• Establishes cause-effect relationships • Establishes sequential connection of ideas and concepts	5 – 4 – 3 – 2 – 1
Participation: constructive communicative interaction in carrying out joint activities to solve problems	• Suggests ideas to contribute to solutions • Argues to seek answers to problems • Is active and becomes involved in the situation • Cooperates in the activities	5 – 4 – 3 – 2 – 1
Coexistence: interaction and communication within his group and with other working groups	• Accepts the opinion of others • Enjoys helping others • Collaborates with peers • Assumes fulfillment of tasks • Reaches consensus in constructive communicative interaction	5 – 4 – 3 – 2 – 1
5.- Outstanding - 4.-Very good - 3.-Good - 2.- Fair - 1.- Poor		

To determine the relation between the student's performances in the PIW with the average mark of the partial exams, the following grouping is made:

Average mark of the partial exams:

1. 80≤P≤100
2. 50<P<80
3. 0<P<51

Final mark of the PIW:

a. 80≤TPI≤100
b. 50<TPI<80
c. 0<TPI<51

With these values the following nine combinations were established: 1a; 1b; 1c; 2a; 2b; 2c; 3a; 3b; 3c. (For example, a student with an average of 83 and a PIW grade equal to 95 belongs to class 1a). Out of a total of 174 students in 2016, the result was summarized in Table 2 and Figure 1.

Table 3. Number of students distributed according to partial average and PWI

	1: 80≤P≤100	2: 50<P<80	3: 0<P<51
a: 80≤PIW≤100	51	7	5
b: 50<PIW<80	7	38	11
c: 0<PIW<51	0	17	38

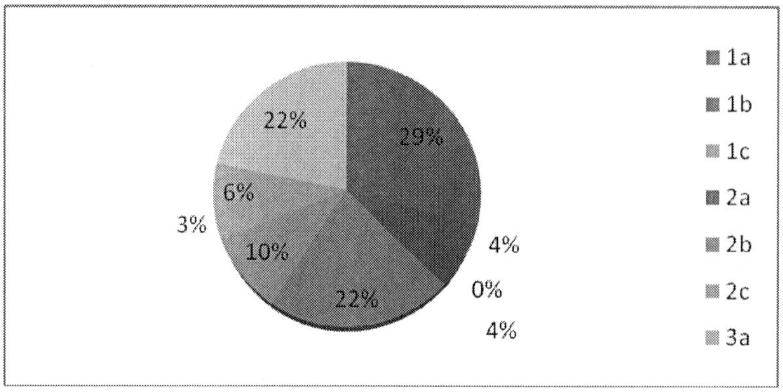

Figure 1. Distribution of students in the different combinations in percentages

The analysis of these results showed 80% of correlation between good performance in the subjects (approval of the partial exams) and approval of the PIWs with a grade above 80 (Of 63 students who scored higher than or equal to 80 in the PIW, 51 (80%) had an average of more than 80 in the partial exams).

At the end of the course, students are given an opinion survey on general aspects of the chair that includes their assessment of the PIWs. This assessment is measured with the following scale: Very Beneficial MB, Beneficial B, Regularly Beneficial RB, Nothing Beneficial NB. In the year 2016, 75% of the students who successfully completed the course of Mathematical Analysis I, (115 respondents), considered as Very Beneficial the realization of the PIWs to relate the concepts of Calculus with the resolution of engineering problems.

Conclusions

Year after year, we have realized that the nature of this approach mainly based on cooperative integral practical works promotes the collaboration among peers, developing interpersonal skills and mutual commitment to the task to be solved.

Students become deeply interested in the real problems because they feel that they are beginning to solve problematic situations, similar to those they may have to face in their future professional life. In many cases the students deepen certain topics that are of their interest and are encouraged by their teachers to continue work within the Chairs in future school years as fellow students.

They also show great enthusiasm for learning the software. The teachers of Mathematical Analysis 2 have informed us that in the second level the students use the software with great solvency to solve problems in R^3.

The visualization capability provided by the software offers a wide environment for exploration, experimentation, creativity and favors the understanding and appropriation of concepts. In many cases, students have exceeded teachers´ expectations in the handling of the software, posing animations and exercises with commands that allow interaction, using the programming language offered by Mathematica software.

Cooperative learning focuses on the student and aims at promoting the learning of certain strategies through the exchange of information that takes place in small group activities. We adhere to Pifarré and Sanuy (2001) who establish that the opportunity for students to help one another in solving a task, to negotiate new meanings, to develop new strategies and to build up new knowledge has a positive effect on their learning.

Recommendations and implications

Although the students realize that the subjects MA 1, and A&AG are fundamental pillars of their integral professional studies as engineers, there exists a certain divorce between contents and applications.

This commits us to go on with research on methodology and didactics on the teaching of mathematics looking for strategies that shorten the breach between theory and practice and between Mathematics and Engineering, encouraging in the classroom the development of assessment skills, individual and collective reflection and, above all, the students' commitment in their academic and professional training. We intend to continue enriching this activity that has been showing great acceptance by the students and has encouraged research and independent learning.

References

Arceo, F. D. B., Rojas, G. H., & González, E. L. G. (2002). *Estrategias docentes para un aprendizaje significativo: una interpretación constructivista* (p. 465). Mcgraw-hill.

Ausubel, D. (1983). Teoría del aprendizaje significativo. *Fascículos de CEIF, 1.*

Ausubel, D. P. N., Hanesian, J. D., Ausubel, H. D. P., Novak, J. D., & Hanesian, H. (1997). *Psicología educativa: un punto de vista cognoscitivo* (No. 370.15). Trillas,

Benito, Á., & Cruz, A. (2005). *Nuevas claves para la docencia universitaria en el Espacio Europeo de Educación Superior: en el espacio europeo de educación superior* (Vol. 10). Narcea Ediciones.

Camarena, G.P., Olazábal, A. (2008). Teoría de la Matemática en el Contexto de las Ciencias. *P). Actas del III Coloquio Internacional sobre Enseñanza de las Matemática* (pp. 83-107).

Costa, V. A.; Di Domenicantonio, R. M.; Vacchino, M. C. (2010). Material educativo digital como recurso didáctico para el aprendizaje del Cálculo Integral y Vectorial. *Revista Iberoamericana de Educación Matemática.* 21, 173-185. Accessed at http://www.fisem.org/descargas/21/Union_021_018.pdf on 20 Oct 2015.

De Miguel, M. (2005). *Metodologías de enseñanza para el desarrollo de competencias. Orientaciones para el profesorado universitario ante el Espacio Europeo de Educación Superior.* Madrid: Alianza.

Duval, R. (1999). Representation, vision and visualization: Cognitive functions in mathematical thinking, *Proceedings of the 21st National Convention of the Association for Educational Communications and Technology* (pp. 3-26). Houston, TX: AECT.

Gaulin, C. (2001). Tendencias actuales de la resolución de problemas. *Sigma: revista de matemáticas= matematika aldizkaria.* 19, 51-63.

Hitt, F. (1998). Visualización matemática, nuevas representaciones, nuevas tecnologías y currículo. *Revista de Educación Matemática*, 10, 23-45.

Pifarré, M. & Sanuy, J. (2001). La enseñanza de estrategias de resolución de problemas matemáticos en la ESO. *Enseñanza de las Ciencias,* 19, 297-308.

Prieto, L. (2006). Aprendizaje activo en el aula universitaria: el caso del aprendizaje basado en problemas. *Miscelánea Comillas. Revista de Ciencias Humanas y Sociales, 124*(64), 173-196.

Trejo, E. T., Gallardo, P. C., & Trejo, N. T. (2013). Las matemáticas en la formación de un ingeniero: una propuesta metodológica. *REDU. Revista de Docencia Universitaria, 11*, 397-424.

Zimmermann, W., Cunningham, S. (1995). Visualization in teaching and learning mathematics. *Mathematical Association of America, 68*(3), 219-232.

Copyright statement

The Enhancing Students' Employability Project: Literature Synthesis of Assessment Instruments.

Melanie Fleming
The University of Queensland, Brisbane, Australia
melanie.fleming@uq.edu.au

Andrea Mazzurco
The University of Queensland, Brisbane, Australia
a.mazzurco@uq.edu.au

Homero Murzi
The University of Queensland, Brisbane, Australia
h.murzi@uq.edu.au

Abstract: *A synthesis of the engineering education literature was conducted in order identify reliable and valid methods to evaluate students development of professional practice skills. The review was conducted with the goal of developing a coherent program of research to explore the development of our students' professional practice skills over the four years of their undergraduate engineering studies. These items came from general academic self-efficacy scales, design and tinkering self-efficacy scales, communication, teamwork, leadership, sustainability, and other instruments relevant to professional practice. Studies with a social cognitive career theory perspective provided a more holistic focus on persistence and retention, transition into study and employment. A smaller pool of qualitative self-efficacy studies were also identified. These studies formed the basis for the development of an interview protocol to explore the development of professional practice competence.*

KEYWORDS

Employability, Professional Skills, Assessment, Literature Synthesis.

CONTEXT

In Australian engineering education, there is a renewed push to enhance the employability and professional competencies developed by engineering students through curriculum design and redevelopment (Male, Bush & Chapman 2009). In keeping with international accrediting agencies such as the Accreditation Board for Engineering and Technology (ABET), Engineers Australia (2011; 2006) have published a set of outcomes based competency standards expected of all graduates entering the profession (EA, 2011). These accreditation standards revolve around a constellation of skills; typically, these include communication, teamwork, problem solving, sustainability, self-management and interdisciplinary skills (Male, Bush & Chapman, 2009). These skills have been identified in the past as important for graduate success in terms of employability and career development. Students have diverse views on how these types of skills can be best incorporated into the curriculum, with industry placements or projects being a preferred learning experience (Edvardsson & Jungert, 2010).

To address this issue, the research team has developed a multi-year research project with the purpose of enhancing the students' experience. This involves understanding their development of competence in professional practice skills. We are also interested exploring how the variety of learning experiences that form the engineering curriculum might contribute to students' feelings of readiness for professional practice. The first step of this project is to evaluate the current educational situation in order to identify best practices and areas in need of improvement. Other authors (Rugarcia et al. 2000; Kumar and Hsiao, 2007) have argued that engineers need to have not only technical or discipline specific knowledge, but also interdisciplinary problem solving skills and a range of professional competencies (McKenna et al, 2016; Mohan et al 2010; Donnell, 2011). This quote from Donnell et al (2011), although about communication skills, is indicative of the attitude in the engineering education literature to the development of professional practice skills in general.

...we do not need yet another study that comes to the final conclusion that communication skills in engineering are important. No one disputes this. What we need is a study that mines down to determine the important things about communication we are teaching well and what we are failing to teach...(n.p.)

The goal of engineering education should be to develop competent graduates ready to participate in the workforce. If students do not feel that they have been well prepared or lack confidence in their own abilities, this can have serious implications for their studies in terms of persistence and for their employability as they enter the workforce. Therefore, we reviewed the engineering education literature to identify best practices for evaluating the development of professional skills, particularly as they related to student perceptions of their career readiness.

Research overview

In this paper, we report on our review of research literature relevant to students' competence in their development of professional practice skills. We focused on research studies in engineering education that had the potential to explain students' development of skills relevant to professional practice from their perspective. There are several lines of effective research in the engineering education literature that link students' self-confidence to performance and which build on the work of Bandura (1977, 1986, 2006) and Lent et al. (1987; 2003**).** Where these skills relate to academic performance (for example expected grades or ability to complete a major in engineering) they are linked to persistence and retention (Lent et al, 2003; Lent et al, 1987). However, where the skills are related to professional practice, they are linked to intentions to pursue or continue in a particular profession. Given that we are interested primarily in how engineering curricula prepare students for their eventual practice as engineers, we chose to focus on research papers with valid survey instruments that measured students' self-efficacy in their development of skills relevant to professional practice. This would serve our long-term goal of developing a longitudinal program of research exploring students preparation for employment, the results of which could serve as an early warning system that educational strategies may need to be adapted if students rated their development of professional practice self-efficacy skills as low in any year of the four year engineering program.

We had two primary aims;

1) To identify a set of measures that from amongst the existing validated instruments that might be useful for evaluating the development of students' self-efficacy in their professional practice skills over time.
2) To assist us to review how our courses prepare students for professional practice across the four years of the undergraduate engineering degree by providing information about how well prepared students feel they are for professional practice based on obtained results.

The literature search was designed to answer the following question:

How has the development of students' self-efficacy in their professional practice skills been previously evaluated?

The following section provides a review of the quantitative and qualitative studies of self-efficacy and social cognitive theory. Our search focused on studies using self-efficacy or social cognitive career theory as theoretical frameworks, and where skills relevant to professional practice also formed part of the research focus.

Quantitative self-efficacy studies

In the context of this research, self-efficacy is defined as "confidence in one's ability to successfully perform career-related tasks" (p. 3 Flores et al, 2014). Self-efficacy theory (Bandura 1977; 1986; 2006) posits that both confidence in one's own ability and the expected or anticipated outcomes of a particular behaviour are affected by four main sources or learning experiences. These learning experiences include prior performance of a skill, vicarious learning (or learning from watching others), verbal persuasion (in the form of feedback from others) and physiological arousal or the emotional state or stress associated with the learning (Bandura, 1977, 1986, 2006). Self-efficacy has been linked to students' intentions to persist, their achievement and their intrinsic motivation (Lent et al, 2003; Lent et al, 1987).

It has also been suggested, that different people will have different levels of self-efficacy based on their exposure to these learning experiences (Hutchison et al., 2006, Hutchison-Green et al, 2008). So different people will have different levels of exposure to learning experiences that involved a sense of mastery or accomplishment, or different levels of stress associated with learning particular material (Bandura, 1986), such as mathematics or thermodynamics.. Additionally, a range of demographic factors may affect access to and interpretation of these learning experiences and these may influence choice of study, intention to persist with study, career choice and goals and performance (Hutchison et al, 2006; Hutchison-Green et al, 2008). Knowing how and which learning experiences contribute positively to self-efficacy would be useful to design curriculum that support students' development of professional practice skills (Carberry, Ohland, & Lee, (2010). This supports the use of self-efficacy theory to understand students professional practice skills and also clarifies its link to learning. An additional theoretical framework, that extends upon and builds upon Bandura's social cognitive theory is social cognitive career theory (SCCT). The role of SCCT in assisting in the development of our study will also be explored.

There are three types broad of self-efficacy study in the engineering education literature as identified by Mamaril et al (2016). These are general measures of academic self-efficacy, adaptations of general self-efficacy items to domain general measure and measures related specifically to engineering tasks (Mamaril et al, 2016). The first includes measures made from general academic self-efficacy items (Lent et al 2005; 2015). General academic self-efficacy items measure students' perceptions of their abilities to perform in a range of academic settings; a typical item might be "I'm certain I can master the skills taught in class this year" (p. 40, Midgley et al., 2000). The second includes adaptations of existing measures (Carberry et al, 2010; Mamaril et al., 2015; Flores et al, 2014; Byars-Winston, Estrada, Howard, Davis & Zalapa, 2010). For example, "I can master the content in the engineering course I am taking this semester" (p. 375, Mamaril et al, 2016) is an example of the second type of measure. The third type of study consists of specific measures of engineering related tasks or skills (Mamaril et al., 2016), for instance, "I can apply technical concepts in engineering" and "I can develop design solutions" are examples of the latter type of self-efficacy.

The remainder of this section of the review will provide an assessment of how relevant the studies are to students' perceptions of development of self-efficacy in skills relevant to professional practice with reference to the three types of self-efficacy studies identified by Mamaril et al. (2016). While all of these studies are useful in exploring students' capacity to pursue a career in engineering at some level and to their development of professional practice skills (Mamaril et al, 2016), some have more utility than other studies. For example, general academic self-efficacy measures are useful in predicting persistence and retention, more specific academic self-efficacy measure have utility in terms of choice of major (Bong 2001) and self-efficacy measures related to particular engineering tasks may have utility in predicting more specific skills required for professional practice and intentions to pursue or remain in an engineering related career.

Lent et al's (2005; 2015) four item academic engineering self-efficacy scale is of more relevance for our purposes, but is mostly concerned with academic self-efficacy in the engineering context and was adapted from a general academic self-efficacy scale. It was not specifically concerned with how students develop self-efficacy of relevance to professional engineering practice. There is however, evidence that academic self-efficacy impacts subsequent career choices (Bandura, Barbaranelli, Caprara, & Pastorelli, 2001). It could be hypothesised that there would be a link between academic engineering self-efficacy and self-efficacy for professional practice, however, there is also evidence that these are separate constructs (Bong 2001; Bong & Skaalvik, 2003). In a longitudinal study, Bong (2001) has shown that certain self-efficacy measures are better predictors of students' course performance, while others are better at predicting degree completion or persistence, which indicates that more precise measures are required in order to explore students' development of self-efficacy skills required for professional practice. MacPhee et al (2013) also explored academic self-efficacy and found that upon graduation and completion of a mentoring program, women and other STEM-minority students had similar academic self-efficacy scores to men. However, although their academic self-efficacy scale had good reliability (Cronbach's alpha .85) their scale was built on a non-standard academic self-efficacy measure and is not directly comparable to other self-efficacy research. It appears that self-efficacy studies more directly related to professional practice are required.

There is also a further smaller group of studies examined self-efficacy in a way that is more strongly related to professional practice. For example, McKenna et al (2016) used the Engineer of 2020 and the Engineering Design Self efficacy (EDSE) survey to explore students' attitudes to engineering. Flores et al (2014) and Byars-Winston, Estrada, Howard, Davis and Zalapa (2010) explored maths and science self-efficacy in order to understand students' goal commitment. Although aptitude for self-efficacy in mathematics and science might be relevant to the pursuit of an engineering degree, such scale items may be of limited utility when exploring students' development of self-efficacy in skills needed for professional practice. Their 18 item scale is based on Lent et al.'s (2001) study on choice of mathematics and science study options. Bandura (2006) states that self-efficacy scales should be constructed from items with a strong connection to the domain of competence, rendering these types of study of marginal utility for our purposes. A subset of studies explore the self-efficacy in a more engineering task specific way. These will now be explored in more depth.

Carberry et al (2010) developed a 9 item engineering design self-efficacy scale based on a model of the engineering design process. Exploratory factor analysis revealed that the survey had high content validity (the factor explained 60% of the variance and had a Cronbach's α of 0.948) criterion and construct validity and reliability. It could be argued that design is a fundamental and key part of an engineer's professional practice, particularly in the early stages of a graduate's career. However, the need for this skill may diminish over time or show different levels of significance depending on the industry in which graduates practice. Mamaril et al (2016) used a more domain specific and task-based set of engineering self-efficacy measures of relevance to professional practice. These included academic engineering self-efficacy, but also items more relevant to professional practice such as experimental, tinkering and design self-efficacy. Explanatory and confirmatory factor analysis revealed two surveys. The unidimensional General Engineering Self-Efficacy Scale and the Engineering Skills Self-Efficacy Scale. The latter survey had three factors, experimental, tinkering and design skills. The surveys had good reliability, validity and predictive utility. One issue for this scale was its lack of ability to explore other skills relevant to professional practice, such as communication or teamwork skills in an engineering context. This scale included 5 items on engineering design self-efficacy adapted from the Carberry et al (2010) items. The Mamaril et al study (2016) is one of the most comprehensive studies of engineering self-efficacy (from the perspective of development and analysis of task or domain specific self-efficacy). Ro and Knight (2016) also specifically explored the development of engineering competence with a survey based on contextual competence, fundamental skills, design skills, leadership skills, teamwork skills, communication skills, and interdisciplinary skills. They defined contextual competence as "an engineer's ability to anticipate and understand the constraints and impacts of social, cultural, environmental, political, and other contexts on engineering solutions" (Ro et al 2015; Ro et al., 2012). These studies provide the basis for further exploration of students' development of skills relevant for professional practice.

MacPhee et al (2013) conducted a longitudinal study of self-efficacy development which provided promising evidence of the benefits of a mentoring program as an educational intervention for students from STEM-minority backgrounds. Their focus was specifically on students who are traditionally under-represented in STEM fields, however it did not systematically examine an entire cohort or cohorts of students from entry until graduation. Longitudinal studies are needed not only to explore variations in self-efficacy development for different groups of students, but also to determine how well an engineering education prepares students for professional practice based upon an agreed set of professional competencies.

An extension of the self-efficacy literature, was a related set of papers that utilised social cognitive career theory (SCCT) to explore issues relating to persistence with regards to intention to study engineering, as well as key variables relating to the development of engineering competence, including skills relating to professional practice and career readiness (Atedero, Rambo–Hernandez & Balgapol, 2015; Casper et al. 2013; Flores et al 2013). These types of studies focus not only on entry characteristics of students' choosing to study particular fields but also include motivational and career goals. They therefore offer the potential for a more complete explanation of students' choice of an engineering degree, their persistence with their studies and their expectations and career values over time since they have the potential to bring the three different types of self-efficacy study identified in the literature together.

Flores et al (2014) used SCCT as part of their longitudinal studies exploring the links between learning experiences, self-efficacy and outcome expectations of engineering students, whilst Lee et al. (2015) used SCCT to examine a model of academic persistence. Flores et al (2014) were particularly interested in persistence of women and minority groups (Latino) in terms of persistence with engineering. They explored the role that prior performance accomplishments, vicarious learning, verbal persuasion, and low levels of emotional arousal had on students' development of self-efficacy in a capstone design course and the development of professional practice skills. Their self-efficacy survey had good reliability for items based on Realistic and Investigative items (.88 and .79 respectively) and good content validity.

In contrast, to the study by Flores et al (2014) the studies by Atedero, et al. (2015) and Casper et al. (2013) provided an exploration of an applied use of SCCT theory to an engineering statics course in a design project context. In these studies, they utilised an intervention in a group design project in order to attempt to positively affect self-efficacy and other variables relevant to SCCT. They determined that there were no significant differences between self-efficacy for the comparison and intervention groups. However, they explained that this may be due to the fact that students received group feedback on their projects, not individualised feedback and that this may have had minimal effect on students' self-efficacy. The Atedero et al (2015) study also used a non-standard measure of self-efficacy and their self-efficacy items had less relevance to professional practice.

The preceding section has indicated that the three broad classes of skills (academic self-efficacy, domain-general self-efficacy and engineering specific self-efficacy) identified by Mamaril et al. (2016) have differing utility when it comes to understanding students' development of self-efficacy in skills relevant to professional practice. It has also explained how they have differential utility in evaluating students' perceptions of their development of professional practice skills. The academic and more engineering specific self-efficacy have utility in identifying students' persistence and choice of major, which influences the pipeline of graduates for different engineering careers. The self-efficacy surveys that have items more specifically related to professional practice are likely to enhance our understanding of students' preparedness for their careers, and therefore their employability. Additionally there is room for more longitudinal studies in this area, particularly those that explore professional practice. What these studies do not provide a clear picture of, is how students develop their self-efficacy through a variety of learning experiences. This review led us to create our own survey based on items from the Carberry et al, (2010), Mamaril et al (2016) and Ro and Knight (2016) studies. This allows to focus our future research on students self-efficacy development in skills relevant to professional practice by focusing on communication, teamwork, design, tinkering, experimentation and contextual skills, instead of a narrower focus on academic self-efficacy.

Qualitative self-efficacy studies

An even smaller group of three studies was identified from the engineering education literature that used a qualitative research methodology to explore the self-efficacy beliefs of students in engineering. The studies used either survey (Berg, 2015) or interview techniques to explore self-efficacy (Hutchison, Follman, and Bodner, 2006; Hutchison-Green et al., 2008).
Of these studies, the self-efficacy scale developed by Berg (2015) was actually quantitative in nature and used a non-standard survey instrument to explore self-efficacy and therefore is of limited utility. Additionally, no validity or reliability criteria were listed. The focus of this study was on examining students' competence in using various classroom learning experiences, in their knowledge of statics and dynamics principles; as well as students' perceptions of their role in the engineering community. The final two studies presented novel, qualitative approaches to exploring skills relevant to engineering education or practice using student interviews of their university (college) experience in education. Hutchison, Follman, and Bodner's (2006) and Hutchison-Green et al.'s (2008) interview based studies using a protocol derived from self-efficacy theory to explore their experiences of each of the four learning experiences that may contribute to students' development of self-efficacy in a given domain. This is an underappreciated variant on self-efficacy theory that could contribute more to our understanding of students' development and maintenance of self-efficacy beliefs through provision "deep appreciation of the variations in the ways they had experienced the engineering learning environment" (Hutchison et al., 2008, p. 178). Zeldin and Pajares (2000) claim that an understanding of students learning experiences from a more in depth and qualitative perspective would assist us to explore

> how efficacy beliefs are developed, how students perceive that these beliefs influence their academic attainments and the academic paths they follow, and how the beliefs influence choices, effort, persistence, perseverance, and resiliency. (p. 221)

This application of self-efficacy theory has the potential to inform curriculum renewal and design efforts as engineering educators seek to make university learning experiences work for a diverse cohort of students, but also paints a rich phenomenographic picture of the learning environment and its connections to previous learning experiences (Hutchison-Green et al, 2008). The study by Hutchison-Green et al. (2008) showed that students' development of self-efficacy is complex and multifaceted and influenced by a range of learning experiences. One of the significant, and as yet unresolved issues of such qualitative studies, is the degree to which male students report mostly positive performance comparisons, whilst females report mostly negative performance comparisons. It would be interesting to ascertain if this is born out in performance either on specific tasks, assessment items, industry placement experiences or on the job performance. If female engineering students continue to negatively experience performance comparisons in the workplace, it could have significant explanatory power for career persistence and advancement.

CONCLUSIONS

Whilst our literature synthesis identified a variety of rigorous measures to assess students' professional development, we also identified a lack of studies rigorously analysing the development of professional skills longitudinally. Longitudinal studies are critical to evaluating current engineering educational practices and to providing support for students to develop the skills required to be successful graduates. Consequently, our next steps are to conduct our own longitudinal study based on a selection and refinement of the instruments identified in this project. This will enable us to explore further engineering educational practices, student motivations and engineering self-efficacy particularly as it relates to professional practice and career readiness. However, another issue with the scales identified, is that they may be more suited to exploring student's development of skills required to complete an engineering degree and have less utility for success in professional practice. This is something that the project team will consider in more detail as the study unfolds. We will also use the small pool qualitative self-efficacy studies in order to develop an interview protocol to explore the learning experiences' that students deem most relevant to their development of skills relevant to professional practice.

There are still significant challenges to the engineering curriculum to provide the range of learning experiences needed, and to ensure that students are able to capitalise on the experiences provided in order to provide the engineer of the future. The challenge for educators is to embed such experiences and to ascertain appropriate tools for measuring students' acquisition of such skills.

In conclusion, the contributions of this paper to engineering education are:

1) An up-to-date review of the available research instruments relevant to evaluating students' development of self-efficacy skills relevant to professional practice.
2) Synthesis of the literature, explication of the relevance of the different types of self-efficacy study to research in engineering education

The next steps for the research team are to pilot the survey constructed from the items identified in our review of qualitative and quantitative self-efficacy and SCCT literature. The surveys and associated items identified to date need to be validated in a broader context to support their utility in understanding students' development of self-efficacy skills in professional practice in Australia.

References

Bandura, A. (1977). Self-efficacy: Toward a unifying theory of behavioral change. Psychological Review, 84, 191-213.

Bandura, A. (1986). Social foundations of thought and action: A social-cognitive theory. Englewood Cliffs, NJ: Prentice-Hall.

Bandura, A. (2006). Guide for creating self-efficacy scales. In F. Pajares & T. Urdan, *Self-Efficacy Beliefs of Adolescents* (307-337). USA: IAP-Information Age Publishing, Inc.

Bandura, A., Barbaranelli, C., Caprara, G. V., & Pastorelli, C. (2001). Self-efficacy beliefs as shapers of children's aspirations and career trajectories. Child Development, 72 (1), 187-206.

Barnard, S., Hassan, T., Bagilhole, B., & Dainty, A. (2012). They're not girly girls': an exploration of quantitative and qualitative data on engineering and gender in higher education. *European Journal of Engineering Education, 37*(2), 193-204.

Bong, M. (2001). Role of self-efficacy and task-value in predicting college students' course performance and future enrollment intentions. *Contemporary Educational Psychology*, 26(4), 553–570. http://dx.doi.org/10.1006/ceps.2000.1048

Bong, M., & Skaalvik, E. M. (2003). Academic self-concept and self-efficacy: How different are they really? Educational Psychology Review, 15(1), 1–40. http://dx.doi.org/10.1023/A:1021302408382.

Carberry, A. R., Lee, H. S., & Ohland, M. W. (2010). Measuring engineering design self-efficacy. *Journal of Engineering Education, 99*, 71–80. http://dx.doi.org/10.1002/j.2168- 9830.2010.tb01043.x

Carter, D. F., Ro, H. K., Alcott, B., & Lattuca, L. R. (2016). Co-curricular connections: The role of undergraduate research experiences in promoting engineering students' communication, teamwork, and leadership skills. *Research in Higher Education, 57,* 363-393.

Casper, A. M., Atadero, R., Balgopal, M. M., Rambo-Hernandez, K. E, & Fontane, D. (2013). The impact of project-based group work on engineering college students' content knowledge and affect. Paper presented at the annual conference of the National Association of Researchers in Science Teaching, Rio Grande, Puerto Rico.

Donnell, J.A., Aller, B.M., Alley, M.A. and Kedrowicz, A.A., (2011) Why industry says that engineering graduates have poor communication skills: what the literature says. In *ASEE Annual Conference* and Exposition, Conference Proceedings.

Engineers Australia (2011). National generic stage 1 competency statements. Sydney, Australia. Engineers Australia.

Flores, L., Y., Navarro, R. L., Lee, H.-S., & Luna, L. L. (2014). Predictors of engineering-related self-efficacy and outcome expectations across gender and racial/ethnic groups. *Journal of Women and Minorities in Science and Engineering, 20*(2), 149-169.

Hill, C., Corbett, C., & St. Rose, A. (2010). Why So Few? Women in Science, Technology, Engineering, and Mathematics. Retrieved from https://www.aauw.org/files/2013/02/Why-So-Few-Women-in-Science-Technology-Engineering-and-Mathematics.pdf

Hutchison, M. A., Follman, D. K., Sumpter, M., & Bodner, G. M. (2006). Factors influencing the self-efficacy beliefs of first-year engineering students. *Journal of Engineering Education.*

Kumar, S. and Hsiao, J. K. "Engineers learn soft skills the hard way: Planting a seed of leadership in engineering classes," Leadership Manage. Eng., vol. 7, pp. 23–28, Jan. 2007.

Lee, H.-S., Flores, L. Y., Navarro, R. L., & Knagui-Munoz, M. (2015). A longitudinal test of social cognitive career theory's academic persistence model among Latino/a and White men and women engineering students. *Journal of Vocational Behavior, 88*, 95-103.

Lent, R.W., Brown, S.D., and Hackett, G. (1994). Toward a unifying social cognitive theory of career and academic interest, choice, and performance. *Journal of Vocational Behavior, 45*(1), 79–122.

Lent, R.W., Brown, S.D., and Larkin, K.C., "Comparison of Three Theoretically Derived Variables in Predicting Career and Academic Behavior: Self-Efficacy, Interest Congruence, and Consequence Thinking," *Journal of Counseling Psychology*, Vol. 34, No. 3, 1987, pp. 293–298.

Lent, R.W., Brown, S.D., and Hackett, G. (2000). Contextual supports and barriers to career choice: A social cognitive analysis. *Journal of Counseling Psychology, 47*(1), 36–49.

Lent, R. W., Brown, S. D., Brenner, B., Chopra, S. B., Davis, T., Talleyrand, R., et al. (2001). The role of contextual supports and barriers in the choice of math/science educational options: A test of social cognitive hypotheses. Journal of Counseling Psychology, 48, 474-483.

Lent, R.W., Brown, S.D., Schmidt, J., Brenner, B., Lyons, H., and Treistman, (2003) D., "Relation of Contextual Supports and Barriers to Choice Behavior in Engineering Majors: Test of Alternative Social Cognitive Models," *Journal of Counseling Psychology*, Vol. 50, No. 4, 2003, pp. 458–465.

Lent, R.W., Brown, S. D., Sheu, H., Schmidt, J., Brenner, B. R., Golster, C. S., et al. (2005). Social cognitive predictors of academic interests and goals in engineering: Utility for women and students at historically black universities. Journal of Counseling Psychology, 52, 84-92.

Lent, R. W., Miller, M. J., Smith, P. E., Watford, B. A., Hui, K., & Lim, R. H. (2015). Social cognitive model of adjustment to engineering majors: Longitudinal test across gender and race/ethnicity. Journal of Vocational Behavior, 86, 77– 85. http://dx.doi.org/10.1016/j.jvb.2014.11 .004

MacPhee, D., Farro, S., & Canetto, S. S. (2013). Academic self-efficacy and performance of underrepresented STEM majors: Gender, ethnic, and social class patterns. *Analyses of Social Issues and Public Policy, 13*(1), 347-369.

Male, S. A, Bush, Mark B and Chapman, E. S. Identification of competencies required by engineers graduating in Australia [online]. In: 20th Annual Conference for the Australasian Association for Engineering Education, 6-9 December 2009: Engineering the Curriculum. Barton, A.C.T.: Engineers Australia, 2009: 882-887.

Mamaril, N. A., Usher, E. L., Li, C. R.., Economy, D. R., & Kennedy, M. S. (2016). Measuring undergraduate students' engineering self-efficacy: A validation study. *Journal of Engineering Education, 105*(2), 366-395.

Marra, R. M., & Bogue, B. (2006). Women engineering students' self-efficacy: A longitudinal multi-institution study. Proceedings of the 2006 Women in Engineering Programs and Advocates Network, Pittsburgh, PA.

Marra, R. M., Bogue, B., & Shen, D. (2008). Engineering classroom environments: Examining differences by gender and departments. *Proceedings of the 2008 American Society for Engineering Education Conference.*

McCormack, J., Beyerkein, S., Brackin, P., Davis, D., Trevisan, M., Davis, H., Lebeau, J., Gerlick, R., Thompson, P., Khan, M. J., Leifer, P., Leifer., P., & Howe, S. (2011). Assessing professional skill development in capstone design courses. *International Journal of Engineering Education, 27*(6), 1308-1323.

McCormick, M., Lawyer, K., Wiggins, J., Swan, C. W., Paterson, K. G., & Bielefeldt, A. R. (2015). Sustainable engineering assessment using rubric-based analysis of challenge question responses. *Journal of Professional Issues in Engineering Education and Practice, 141(2).*

McKenna, A. F. (2007). An investigation of adaptive expertise and transfer of design process knowledge. *Journal of Mechanical Engineering, 129,* 730-734.

McKenna, A. F., Hynes, M. M., Johnson, A. M., & Carberry, A. R. (2016). The use of engineering design scenarios to assess student knowledge of global, societal, economic, and environmental contexts, *European Journal of Engineering Education, 41*(4), 411-425.

McMartin, F., McKenna, A. F., Youssefi, K. (2000). Scenario assignments as assessment tools for undergraduate engineering education. *IEEE Transactions on Education, 43*(2), 111-119.

Melton, R. B., Cardella, M. E., Oakes, W. C., & Zoltowski, C. B. (2012). Development of a design task to assess students' understanding of human-centered design. *Proceedings of the 2012 IEEE Frontiers in Education Conference.*

Midgley, C., Maehr, M. L., Hruda, L. Z., Anderman, E., Anderman, L., Freeman, K. E., … . .Urdan, T. (2000). Manual for the Patterns of Adaptive Learning Scales. Ann Arbor, MI: University of Michigan.

Mohan, A., Merle, D., Jackson, C., Lannin, J., & Nair, S. S. (2010). Professional skills in the engineering curriculum. IEEE Transactions on Education, 53(4), 562–571. doi:10.1109/TE.2009.2033041

Ro, H. K., Lattuca, L. R., Merson, D., & Terenzini, P. T. (2012). Measuring engineering students' contextual competence. Paper presented at the ASEE Annual Conference and Exposition, San Antonio, TX. Retrieved from http://www.asee.org/public/conferences/8/papers/4880/view

Ro, H. K. & Knight, D. B. (2016). Gender differences in learning outcomes from the college experiences of engineering students. *Journal of Engineering Education, 105*(3), 478-507.

Ro, H. K., Merson, D., Lattuca, L. R., & Terenzini, P. T (2015). Validity of the contextual competence scale for engineering students. *Journal of Engineering Education, 104*(1), 35-54.

Copyright statement

The following copyright statement should be included at the end of your paper. Substitute authors' names in final copy only.

The rise of knowledge schools in the Amazon: inspirations from the football school model

Celson Pantoja Lima
Federal University of Western Pará, Santarém, Brazil
Celson.ufopa@gmail.com

Cássio David Borralho Pinheiro
Federal University of Western Pará, Santarém, Brazil
cdbpinheiro@gmail.com

***Abstract**: The principle of developing football skills (or any other high performance skill) is to develop professional players beginning at the earliest possible primary age, continuing into the teens and leading up to adulthood. Skills keep on developing and evolving as individuals grow in proficiency in a continuous process. This is an analogy for the work of educational development reported, with kids challenged to begin with simple play and to remain continuously curious in their learning ("playing" with evolving ideas) from home and kindergarten, through primary school, continuing to senior school, and arriving at university possessing highly developed knowledge learning and absorption skills. This paper covers key aspects of the work, namely theoretical foundations, research methodology, expected results, and future outcomes and the challenges faced.*

Introduction

Education, as a key element in the process of stimulating the creation of knowledge, is one of the most valuable assets of any country nowadays. Indeed, deeper thought would probably indicate that education has been considered that way since the very early days of schools, those meant to produce the workers to keep the industrial revolution moving forward and fearlessly (Goldberg & Sommerville, 2014; Robinson & Aronica, 2015). However, as many influential and relevant thinkers on the topic have emphasised, education on the whole has changed little in approach since those early days (Blikstein, 2013; Goldberg & Sommerville, 2014; Negroponte, Resnick, & Cassell, 2015; Robinson & Aronica, 2015). A quick glance at schools in many parts of the world will certainly confirm that classrooms retain the same format, same chairs, same models, same nearly everything, in a world outside that has radically changed in almost every way. Clearly, the original model of schools may not best fit the revolutionised world we are living in (Ito, 2011).

On the one hand, transforming the approach to education is a very long process requiring the participation of all the main stakeholders – which, in some countries, like Brazil, means involving several ministries, regulatory agencies, state and city level organisations, teacher training institutions, society itself, and more. On the other hand, definitive solutions do not just appear, and small studies and trials can be conceived and conducted on a small scale, yet producing concrete and motivational results. This is the very motto of this work: *think, conceive, create, try, and assess* the results of small trials carrying a powerful transformation message. The good news is that there is a grass root movement happening in many parts of the world, going in the same direction and trialling a variety of theories and technical approaches (Blikstein, 2013; Goldberg & Sommerville, 2014; Ito, 2011; Robinson & Aronica, 2015).

The research question guiding this research is the following: is it possible to connect the University and the Primary school allowing knowledge to be exchanged between them, aiming at increase the knowledge level of the kids? As consequence, the main goal pursued here is to integrate those two (currently) distant but highly interdependent worlds.

Metaphorically, this integration can be viewed as a bridge connecting those two worlds. Structurally, two main pillars support the bridge, namely *constructionism* and *constructivism*. The main deck of the bridge is built using concepts like Making Movement for Education and Design Thinking for Educators. Knowledge is the commodity crossing the bridge supported by Knowledge Management theory, which provides models, concepts and tools enabling knowledge to flow back and forth between universities and primary schools.

This paper is structured as follows. Goals and context of this work follow the introduction. Theoretical and conceptual foundations are then discussed. The research methodology guiding the work is introduced. The expected results of this work are discussed. Future challenges and the main conclusions reached so far close the paper.

Goals and Context of the Work

As already mentioned, this research intends to conceive and then deploy a "bridge" to link University and Primary School, across which knowledge flows back and forth. Such a knowledge flow aims at creating the appropriate environment to reduce the knowledge gap currently existing between the two education worlds, hoping that this will help shape a new generation of *learners* embodying a rich knowledge acquisition background, on the one hand, and, on the other hand, motivated to make a difference by transforming their intrinsic situations socio-economically. Above all, in the long run, this research dreams about improving the education level of the region where it is conducted: the very heart of the Brazilian Amazon area.

Some key topics supporting and guiding our work must be discussed first in order to frame the context in which this research has been conducted. First of all, it has been partially developed within a doctoral work (running its 4th year) in the Federal University of Western Pará (UFOPA), located in Santarém city, Pará state, Brazil. Other aspects are described as follows.

1. Participants targeted in the trials

At first sight, it seems that this research is focused only on the primary school students and indeed, it started like that. However, it was realised that undergraduate students could also be included in the research, since many were found not to be highly motivated to learn nor had a strong background in some core disciplines, such as maths and physics. In other words, they share the same problems that were identified (and tried to solve) with children. This conclusion came from many years of experience teaching disciplines in the computer science program at Federal University of Pará. Both groups of students are called *learners* and the person orchestrating the knowledge flow is called *maestro.*

It is worth mentioning that a third group of *learners* was also included in this research: the teachers from the small villages in the Amazon area. Moreover, we were involved in a national program specifically created to help raise the educational level of those teachers and, not surprisingly, some of the characteristics identified in both children and undergraduates were also found in the teacher training' classes. Later in this paper we will say more about assessing these three types of *learners*.

2. Knowledge Creation: Roles of University and Primary School

It is necessary to define knowledge here. Ranging from philosophy to Knowledge Management theory, knowledge has a myriad of definitions. But instead of presenting a formal definition of knowledge, we describe one example where knowledge is identified for illustrative purposes only. According to Robinson and Aronica (2015), the very essence of education lies on the relationship between "maestro" and learner. The former has something (let's call it "primary knowledge") to transfer to the latter, and this transfer process is completed when the learner has received something (call it "acquired knowledge") as a result of that process.

Nonaka and Takeuchi (1995) work with two types of knowledge, namely Tacit and Explicit. Tacit is the kind transferred between maestro and learner, head knowledge. Explicit is a tangible representation of knowledge, which can take many forms, the most obvious in traditional education terms being a printed textbook. In this work we name Knowledge Sources (KSs) the myriad of tangible representations of explicit knowledge, such as digital documents, videos, and tutorials. We adopted this definition considering that the knowledge embedded in any source will only be captured from it if the user of that source has the skills and background to accomplish that. For instance, only people with an engineering background can properly understand certain technical drawings (which are in principle tangibly available to anyone).

KS play a key role in this work. They are deemed to represent learning objects that can be used in the knowledge transfer process previously described. They may also be used as part of a repertoire of trials to support some of the interactions between maestros and learners. A synonym for KS adopted here is artifact, exemplified by tutorials, videos, web pages, and physical parts (e.g. robots, 3D printed parts, and hand crafted objects).

3. Knowledge School: What is it?

If we take the fact that not all kids are football lovers nor have the natural ability to develop and master the required skills to reach their national teams, the football metaphor used here may not seem to make sense. However, this is not the approach taken to support the use of this metaphor. Rather, it is the joy of learning something (improving skills) through playing and with that to build a natural chain of evolution that will unavoidably lead the children to next steps in their development – being better prepared and eager to learn and progress. Football is merely a topic of interest. Considering this vision, then, children who have attended a primary school that is actively "playful" and with a high emphasis on learning, will most likely arrive at the intermediate school much better prepared to gain more from both the institutions and teachers with a more solid knowledge foundation. The same pattern of behaviour is also expected at the college level and beyond.

A knowledge school is a way to refer to an alternative to a traditional school (a content-oriented school) that has been the norm in the educational system since its inception. A content-oriented school, rooted in the industrial revolution and targeting essentially the preparation of a labor force to satisfy physical industrial needs, seems to be somewhat of a misfit in our present society, commonly called the knowledge society, characterized by the technological innovations that surround us. Moreover, new developments and gadgets are offered at an astonishing pace in terms of both quantity and price (e.g. Apple Store has more then 1 billion apps, and Google launched Virtual Reality (VR) glasses that came free to subscribers of the New York Times). This could represent a sufficiently good justification to propose changes to the educational model in a broad sense. However, we believe that the reasons pressing for change go beyond that.

When considering a knowledge school some fundamental features may be highlighted. Blikstein (2013) emphasizes that knowledge construction is evident when students are able to plan, build and share their knowledge artifacts publicly. Similarly, other authors (Negroponte, Resnick & Cassell, 2015) say that children learn more easily collectively, and the fact that we are an active part of a global network may influence the establishment of a community for construction of knowledge, where children work collaboratively and learn together even though not co-located. An example of such a community is the repository of programs based on the Scratch software platform built by Resnick et al. (2013), where users are sharing their experiences and activities through more than 20 million projects.

Muir-Herzig (2014) points out another feature of the knowledge school, which relates to the fact that teachers must view learning as an active, creative, and socially interactive process, where knowledge is understood by the students not as content to be transferred, but as something to be built by themselves for themselves. The same author argues that the knowledge school may be more efficient than a content-oriented school due to the fact that a student must respond positively to alternative forms of expressing his/her knowledge, which in turn grows self-esteem and improves their relationship with teachers and peer students.

4. The pilot cases: disciplines and organisations

The disciplines used to assess this work are robotics, electronics, and visual/tangible programming. However, the results produced here will be tested with other disciplines and in other scenarios. Regarding the organisations involved, UFOPA is the university used as the reference institution, together with several primary schools (both public and private) from the city of Santarém. It is worth reemphasizing that the cases conceived to assess this work, which were initially targeting primary school students, were extended to include undergraduate students also. The reason is that the current level of motivation shown by undergraduate students to learning is also low and, as such, they became part of the trial as well. The complexity and sophistication of the artifacts used in the trails may differ, but the main goal remains the same.

In terms of conceiving the trials, the bottom line is the application of well-established Knowledge Management cycles (Dalkir, 2005; Nonaka & Takeuchi, 1995). Considering the set of disciplines involved in this research, the main actions are: identifying the knowledge to be transferred; putting it in the appropriate form; creating tangible instances; providing suitable environments and support mechanisms to transfer the knowledge, devising tools and the means to evaluate and assess the results of the transfer process (testing whether the primary knowledge was actually acquired at the other end of the chain).

Needless to say that the last action, the assessment itself, represents a big challenge in our work. On the one hand, there is the assessment of the transfer process *per se*, which can be done fairly straightforwardly. On the other hand, assessing whether the knowledge transferred is actually captured and really "internalised" by the learners, seems to be a rather more subtle task that can be (only partially) accomplished, taking advantage of external measures correlated to this research.

Theoretical and Conceptual Foundations

As previously mentioned, the two theoretical foundations of this work are the Constructivism and the Constructionism. On the conceptual side, we rely on: the Maker Movement for Education; the Design Thinking for Educators; and the concepts from the Knowledge Management area.

1. The Theoretical Foundations

The "hands on" approach has been part of the formal learning process for many years. It is worth recalling some historical references in the field of education, such as Piaget, Papert, and Montessori, the pioneers who worked in the pre-Internet era, and Robinson, Negroponte, Resnick, and Blikstein, who have been dealing with both positive and negative impacts of the technological revolution both inside and outside the classroom. Piaget's constructivist theory (Piaget, 1971) argues that intellectual development is closely related to interactions happening between a person and the environment she/he is part of. That person acquires knowledge as a result of those interactions.

Complementing Piaget's theory, Papert's constructionist theory (Papert, 1983) argues that construction of knowledge is based on the execution of a given action that aims at developing a particular result of interest in the person involved. Papert (1983) defines constructionism as the construction of knowledge through the interaction between the learner and the learning tool (a computer, in this case). Such a tool behaves like a machine to be taught, and the learner (supervised by the educator) can think through and analyse his/her own mistakes. Therefore, experimental activities, group research, and an incentive to question, allow the learner to participate more actively in his/her own learning process by constructing the characteristics of his/her personal learning universe. Additionally, it should be highlighted that both theories emphasise the importance of trial and error (test and see) processes as valid in the quest to learn.

Last but not least, another classical reference in this area is Montessori, who argues that the way to achieve intellectual maturity is in our own hands, since children love to touch the objects they wants to interact with and to know about. Montessori developed a method, famous worldwide, to help build knowledge from both tangible and abstract elements, stating that all kinds of object and means used in the learning process should be reachable (i.e. capable of being understood) by the learners (children, in this case) (Montessori, 1914).

2. The Conceptual Foundations

Both constructivist and constructionist theories argue in favor of intellectual development through: (i) a strong interaction between learner and the learning environment; and (ii) reinforcement and expansion of the learner-computer relationship as a means to maximize the learning process (Ito, 2011). These theories drove several initiatives and movements inspiring also this work, namely:

- *The Maker Movement* (MM): modern and highly technological, it argues that anyone can design, build, and fix/repair a myriad of devices if they are taught how to use the appropriate tools to carry out the task (Hatch, 2013);
- *Design Thinking for Educators* (DTE): a methodology that provides educators with tools to help them face their daily challenges. Such a methodology relies on four key values, namely empathy, collaboration, optimism, and experimentation (Muir-Herzig, 2004);
- *The Maker Movement for Education* (MME): the education-oriented version of the maker movement, giving high importance to practical (hands on) learning in classrooms and labs, where the theory becomes simply added-value in the learning process (Blikstein, 2013).

For the sake of clarity, in this work *learners* are not only children but also the educators working with children. In terms of age, the target range is students between nine and twelve years old, and educators of any age. School selection is based on their community's motivation (principal, educators, and students) to support the work and to fully engage with it.

Electronics is the first discipline selected in the pilot cases. The reasons behind this choice are: (i) the researchers driving this project come from engineering and computer science fields and, as such, they are familiar with this discipline; (ii) there are many relevant international initiatives, including LittleBits (developed by MediaLab at MIT that offers a set of small electronic devices/modules easily interconnected and purposely designed for prototyping and learning of basic electronic concepts) and MaKey MaKey (learners interacting with computers through real world objects such as fruits and paper drawings).

Recently, MM has gained considerable attention around the world, through two specific manifestations: MM for industry (e.g. Techshops and FabLabs across US and other parts of the world, and Artisans Asylum in Cambridge) and MM for education. Making, in this context, is not only about learning a procedure or a sequence of steps to produce something; rather, it focuses on looking for both the tools and the knowledge required to realise some specific product. MM's motto is Do-It-Yourself, and currently MM is a very technology-based approach having no restrictions for a target audience. The single requirement is the ability to learn how to use the tools (and the associated knowledge) necessary to accomplish the task (Papert, 1983). The Maker Movement for Education is more than an "upside down" learning process. It redefines, in a very pragmatic way, the meaning of participation, sharing, democratization, and failure. This is a message published by well known opinion makers, such as: (i) Joi Ito, who wants to unleash the 'rebel' inside learners, and who invites bright talent anywhere, academically qualified or not, to be part of the world's leading "anti-disciplinary" research lab he currently heads, that of MIT Media Lab with a myriad of works leading to some well-recognized products (e.g. Little bits and Scratch) (Ito, 2014); (ii) Sugata

Mitra, who defends the auto-learning process as a viable way for children to learn how to use computers, arguing that education is a self organizing system, where learning is an emergent phenomenon (Mitra, 2005); and (iii) Sir Ken Robinson, the British educator who argues that traditional school is "killing" creativity and to remedy matters, he proposes the creative school as a way to transform education (Robinson & Aronica, 2015).

In the educational context, five basic guidelines are deemed fundamental to actions and projects in this area, namely (Muir-Herzig, 2004): Making together (curiosity and the necessity to exchange ideas, advice, and guidance naturally brings makers closer); Exhibition, not competition (motivation comes from making and sharing our best results); Embracing failure (failure is unavoidable, we must learn from it); Tools keep improving (emerging technologies and infrastructure are getting better and cheaper, and encourage creativity); and Everyone can make (no restrictions or barriers of any kind; everyone can make and learn).

In line with these guidelines, DTE brings a vision that educators can make the difference through innovations capable of creating a positive impact, transforming big challenges into opportunities (IDEO, 2012). DTE has the following features: it is human-centered (deep empathy and understanding of needs and motivations of people), collaborative (a group is stronger than an individual, when solving problems), optimistic (we are all able to contribute to finding solutions no matter how big the challenge is), and experimental (fail and learn from it is part of the learning game). The DTE educational model is founded on the fact that an educator's work never ends since learning does not stop and students will keep learning, no matter what (Muir-Herzig, 2004).

The Research Methodology

The approach adopted in this work targets the reduction of the "knowledge gap" between universities and primary schools through simple and feasible measures. It highlights three fundamental knowledge-related aspects: (i) identification: what are the types of knowledge produced by universities; (ii) classification: what types can be transmitted to primary schools; and (iii) preparation and transformation: knowledge being properly prepared and transformed in order to be (as easily as possible) absorbed by learners from primary schools.

Following a holistic way based on the theories and initiatives previously described, this work adopts principles from the Action Research (AR) scientific methodology. This choice was motivated mainly by the fact that AR is grounded in a strongly empirical approach conceived to link action with solution for a given problem, where researchers and participants of the trials must work cooperatively and collaboratively together to guarantee proper knowledge acquisition and transmission. It is worth mentioning that the adoption of concepts borrowed from the Knowledge Management domain is deemed to improve the quality of the knowledge cycle (i.e. identification, acquaintance, representation, transmission, and acquisition) since those concepts (which involve people, processes, infrastructure and environment, and technologies) are essentially targeting concrete reuse of knowledge.

Design of preliminary Trials & Assessments

The core idea behind the trials is to bring to life events that can be used to assess: (i) Is a bridge between university and primary school really being built? (ii) If yes, are the bridge's pillars providing the necessary foundations to sustain it? If not, what is missing or what are the problems that must be solved in order to make that happen? What about the main deck, is it really supporting the knowledge flow in a proper manner? Regarding the knowledge itself that is being exchanged, what can we say about it in terms of its representation, ease of use, and the tools to promote the knowledge exchange? Concerning the learning process, is it possible to assess it, to detect any kinds of change regarding the previous situation and, even better, any signs of improvement? Therefore, with this long list of questions, we devised and planned a series of trials, some already performed and more on the way to being done.

1. Background and General Guidelines

The first trials inspiring this doctoral work began in 2013 (motivated by weak performance of students in previous years), at Federal University of Western Pará (UFOPA), in two disciplines (electricity and electronics) of the computer science program. It was clearly demonstrated many of the arriving students that: (i) a huge lack of motivation (a lethargy in attitude) to be engaged in learning processes; (ii) a lack of knowing how to learn; (iii) tests, scores and grades was their only concern; and (iv) difficulty in unleashing (freeing up) their creativity (after many years being moulded by traditional schooling). This was the "light bulb" moment where the idea of developing the concept of knowledge schools, mimicking the success of football schools, was born, putting constructive fun and joy in first place to hook them, and to return students to their natural state of creativity and motivation to learn.

In the primary classroom, MME and DTE concepts and guidelines are part of lectures and this introduces a novelty for both educator and students: hands on activities always precede the theoretical part. This seemingly small change has a two-fold impact: it awakens students' curiosity and it eases them into the universe of theory. Students are constantly motivated, during classes, to make suggestions and contribute with new ideas and new tests. Lecturers also often recite an additional, subtle parameter to undergraduate students: that tests and exams are not the target of the interaction between teachers and learners. Rather, knowledge acquisition has jumped to the first place, together with the fulfilling motivation to learn. Of course, evaluations still remain as part of the discipline regime for one (legal) reason: the university formally obliges having tests and grades.

An interactive way to evaluate students has been used, preferably embedded in the learning environment itself. When a trial involves both teachers and learners, both groups are evaluated separately, and teachers also evaluate the performance of their students (as traditionally). Additionally, after each class students are challenged to propose new products (goals) for next classes. They are also encouraged to produce a piece of knowledge from their trials, documenting it for potential reuse by other colleagues.

2. Types of Trials

There are several types of trials that are tailored to the targeted audience, namely: Workshops and Talks (conveying information and practical perspectives, they can be organised and held independently or as part of bigger events); The Metaeletrônica WebSite (https://sites.google.com/site/metaeletronica/fabrica, a static repository used to store sources of knowledge, e.g. tutorials, videos, "how to" manuals, in Portuguese for logical reasons); Interactive fairs (visitors are invited to build new objects using those exhibited, supported by knowledgeable monitors and tutors); and Interactive tutorials (this type of trial tackles the educators' lack of time, aiming at providing a convenient mechanism through which educators can learn).

3. Designing both Trials and Assessment

The outcomes of the trials will be the way to find answers to the list of questions unveiled in the beginning of this section. They will take place at the schools (public and private), either in a single session or as part of an ongoing program. Another important aspect is the methodological support to prepare the events and their respective assessment. Data collection and their analysis are required to be formalised, the respective results produced also need to be supported by a solid conceptual and formal approach.

As a way to gather instructional materials, a good number of events have been conducted since second semester of 2014 to date, involving private and public primary schools, and the university itself. Those events promoted and/or produced a total of 16 prototypes (or devices), all made available (in terms of project, schemas, drawings, videos, tutorials, and the rest) for free for the whole community. The preliminary results so far are exciting and have motivated us to keep pushing this work to its limits. We are currently working on the formalization of the results achieved, by categorizing our results and comparing them with results available in current relevant literature.

Although this work is at a very embryonic stage and needs to be enriched with further criteria, the preliminary not formal results are very encouraging indeed. To illustrate this, consider the following parameters: (i) approval rates in one of the disciplines evaluated in the events were 55%, 80%, and 78% respectively for 2012, 2013, and 2014, which clearly shows that growth is visible; (ii) scientific production of students has increased in the UFOPA events creating a sort of "dynamic" around both disciplines and learning process; and (iii) a Scratch workshop was partially assessed through a questionnaire, whose evaluaion clearly indicated that students were 'hooked' by the experience, potentially suggesting that the hands on experience had helped them to learn much more.

Future Outcomes and Challenges

MM4E concepts and guidelines may have a highly positive impact on the learning process. Moreover, it allows taking the academic research developed on the creation and use of learning objects and putting it into the best place, which is at primary and intermediary school levels. In the development scenario targeted by the work, the Amazon area in Brazil, those school levels particularly lack these types of resources and tools. The core idea is keep developing new *knowledge pieces* (IT-based or not, such as tutorials and step-by-step videos) permanently available on the web (through the *MetaEletrônica* website) together with some smartphone-based tools, in order to facilitate and simplify the learning process.

The possibilities to interact with universities from the south and southeast part of Brazil are very promising, and replication of our events will help enrich the quality of results. Our knowledge sources have been shared freely, together with our technological support and our experience with the trials themselves. Potential changes in the computer studies "universe" at UFOPA, touching both on teaching and learning, are also a highly motivating research exploitation path. We just received a micro-grant from SCiBR, a NGO from Boston that promote and support innovative education-oriented projects. These funds will allow us to built prototypes and take them to students in small villages in the inner Amazon area, where schools open only few months a year due to the fact that during 6 months every year, their lands are flooded. Last but not least, the Distrital University Francisco José de Caldas, from Colombia, is taking our trials and replicating them there, in the social neighborhoods of Bogota.

Conclusions

Education is a lifetime quest, which sometimes flourishes strongly and sometimes is damaged by parties gravitating around and exploiting it in various ways. Governments are commonly obsessed with good numbers in international evaluations, such as PISA evaluations, considering that good participation in such tests implies good education within the country borders. Nothing more wrong than that could be thought, and here we are specifically talking about Brazilian reality, where "fake" schools (associated to real schools) have a small group of students carefully selected and well trained to have top performance in the national evaluation exams. Thus, the fake school gets a top position on the ranking but the real school linked to it, gets a very low position and it is here that the parents, looking for a good school for their children, will in reality enroll them. Despite this, we firmly believe, based on the examples coming from other countries reinforced by our own results, that we can at least provide an alternative way to improve the education level in our region. We want to change what is achievable ourselves, without relying on approvals or recommendations of committees and the like. Hands on is our motto and, in this case, it means exactly that we will keep working and experimenting with everything we can put our hands on, aiming to help change the education reality in the heart of the Amazon jungle. The road ahead is surely not a soft one, but every small success will count and be celebrated.

References

Blikstein, P. (2013) Digital Fabrication and 'Making' in Education: The democratization of Invention. In J. Walter-Herrmann & C. Büching (Eds.), FabLabs: Of Machines, Makers and Inventors. Transcript Publishers.

Dalkir, K. (2005) Knowledge Management in Theory and Practice. Oxford: Elsevier Butterworth–Heinemann, 2005. 345p.

Goldberg, D., & Sommervile, M. (2014) The whole new Engineer. Threejoy Associates. ISBN: 978-0-9860800-2-9.

Hatch, M. (2013) The Maker Movement Manifesto. Accessed at: http://www.techshop.ws/images/0071821139%20Maker%20Movement%20Manifesto%20Sample%20Chapter.pdf on 27 Feb 2017.

IDEO, Design Thinking for Educators Toolkit. 2 ed. (2012).

Ito, J. (2011). The Internet, innovation and learning. Accessed at http://joi.ito.com/weblog/2011/12/05/the-internet-in.html on 27 Feb 2017.

Mitra, S. (2005) Self organising systems for mass computer literacy: Findings from the 'hole in the wall' experiments. International Journal of Development Issues, Vol. 4, No. 1, p 71-81.

Montessori, M. (1914) Dr. Montessori's Own Handobok. Frederick A. Stokes Company.

Muir-Herzig, R. (2004). Technology and Its Impact in the Classroom. In: Computers & Education, n. 42, p. 111–131. PERGAMON.

Negroponte, N., Resnick, M., & Cassell, J. (2015) Creating a Learning Revolution. MIT Media Lab. Accessed at http://www.unesco.org/education/lwf/doc/portfolio/opinion8.htm, 17 March 2015.

Nonaka, I., & Takeuchi, H. (1995) The Knowledge Creating Company: How Japanese Companies Create the Dynamics of Innovation. New York: Oxford University Press.

Sahlberg, P. (2014) , Finnish Lessons 2.0: What Can the World Learn from Educational Change in Finland? Teachers College Press, 2014. 264 pages.

Papert, S. (1993) MINDSTORMS: Children, Computers, and Powerful Ideas. Basic Books, 2nd edition, ISBN 0-465-04629-0.

Piaget, J. (1971) Genetic Epistemology. Columbia University Press. Norton Library.

Resnick, M. (2013) Lifelong Kindergarten. Cultures of Creativity. LEGO Foundation.

Robinson, K. & Aronica, L. (2015) Creative Schools: The grassroots revolution that's transforming education. Viking Penguin.

Acknowledgements

To Dr. Graham Storer, for his priceless contributions improving the readability of the paper.

Copyright statement

Does the project-based-learning (PBL) strategy change first-year students' perceptions of the engineering profession?

Antonio Miguel-Cruz
Biomedical Engineering Program. School of Medicine and Health Sciences. Universidad del Rosario, Bogotá, Colombia
antonio.miguel@urosario.edu.co

William R Rodríguez-Dueñas
Biomedical Engineering Program. School of Medicine and Health Sciences. Universidad del Rosario, Bogotá, Colombia
william.rodriguez@urosario.edu.co

Néstor Flórez-Luna
Biomedical Engineering Program. School of Medicine and Health Sciences. Universidad del Rosario, Bogotá, Colombia
nestor.florez@urosario.edu.co

Daniel A. Quiroga-Torres
Biomedical Engineering Program. School of Medicine and Health Sciences. Universidad del Rosario, Bogotá, Colombia

daniel.quiroga@urosario.edu.co

Abstract: *The purpose of this study is to describe how students' perceptions of the engineering profession change overall as a result of the implementation of the project-based-learning (PBL) strategy in an introductory biomedical engineering course. This is a pre- post-test with no control group study design. Two classes (n = 113) of the course participated in a hybrid lecture+PBL learning strategy. A survey comprising a demographic and 5-point Likert sections measured students' perceptions of the engineering profession. The survey was administered three times; (first week, baseline, T_0), when students were exposed to lectures only (sixth week, T_1), and after the implementation of the PBL strategy, T_2. No statistically significant results in the students' perceptions in the baseline compared with T_1 ($p<0.566$) and statistically significant change in the students' perceptions in T_1 compared with T_2 ($p<0.024$) were found. The PBL educative strategy positively changed first-year students' perceptions of the engineering profession.*

Introduction

The Engineering Education Research Colloquies (EERC, 2006)identified five emerging research areas that serve as the foundation for the new discipline of engineering education. One of these research areas is the Engineering Learning Mechanisms that is the understanding of the factors that affect how engineers learn, build knowledge and develop competencies in real contexts. One factor that influence learning in engineering is students' previous knowledge or perceptions about the engineering profession. This previous knowledge influences not only students' motivation to engage in learning activities but also influences students' retention that is the success in completing the engineering program. Attrition in engineering programs is a concern in university education. Dimitriu & Karimi (2005), found that the attrition in an engineering program was about 51% being one of the main causes of the attrition the fact that the first-year engineering students had not developed a strong affinity with the engineering profession. Likely, in Colombia, after their first year, roughly 50% of engineering students do not continue their engineering programs (MinEducación, 2014). Anderson-Rowland (1997) found that first-year engineering students may drop out of university during or after the first year due to the misunderstanding of what engineering is. When a potential engineering student selects to pursuing an engineering career, it is commonly that his decision is based on uncomplete information because he understands that engineers create things, as a final products, but do not understand what other skills are involved to become a good engineering (Nesbit & Hummel, 2005). If freshmen do not understand during the first year in the career what the engineering program they are enrolled in is about, the likelihood of attrition increases.

The lack of understanding by students about what engineers do is especially a problem in relatively new and emerging engineering programs. This is the case for biomedical engineering programs. One way to mitigate this problem may be providing attractive engineering courses to first-year. These courses may serve as an initial approach to the engineering profession which may improve students' understanding about their engineering career. These courses also should be designed in a way that allows students to develop skills such as creative problem solving and product design and that provides the opportunity for students to "act" like engineers from the very early stages in their career ((Cruz, Rincon, Dueñas, Luna, & Torres, 2016; Wang, Kang, & Lee, 2013). The teaching strategies used in these introductory courses critical. Project Based Learning (PBL) is a teaching strategy which has proven to be highly successful in technical educational fields even in senior courses (Rodriguez et al., 2015). In PBL students gain knowledge and develop self-directed learning skills (Awang & Ramly, 2008) where students have the opportunity to explore real-world problems and the possibility to solve that problems by proposing prototype scale products. In order to know how PBL can modify the first-year student' understanding about the engineering as a profession, this study is guided by the following research question:

How do overall student perceptions of the engineering profession change as a result of their educational experience in an introductory biomedical engineering course?

We believe that responding this research question is important to the biomedical engineering program because we can determine if first-year students' perception about engineering is positively affected with the implementation of PBL strategies.

Theoretical framework

The constructivism and the social cognitive career theory (SCCT) provide the theoretical framework of this study which explores how a hybrid course (lectures+PBL) impacted the students' perceptions about the skills that engineers should have. We used the foundations of constructivism to design an introductory course that used a hybrid method (lectures+PBL). Piaget's learning theory constitutes the theoretical foundations of constructivism which states that children build knowledge organized as cognitive models or schemas through the dialectical relationship between two processes called assimilation and accommodation. During the assimilation, children build knowledge by assimilating new experiences into existing cognitive schemas, while during accommodation, existing cognitive schemes are adjusted to new experiences when these new experiences do not fit into the old schemes (Piaget, 1952). The constructivism and its associated concepts such as active learning, cooperative learning, and active application of knowledge also explain learning in young and adults stating that students are builders of their own knowledge structures (Wadsworth, 2004). Being the constructivism the opposite of a passive mode of learning that is based on lectures, it is a paradigm shift in the teaching-learning process. Under the approach of constructive learning, learning is a natural and dynamic process where the student is active in his learning process; thus, the most effective learning should be meaningful and relevant to the student (Duran, et al., 2007). One of the strategies of the constructive learning is PBL, which integrates self-learning, concept-learning, problem-solving skills, and creativity (Chua, 2014). In PBL, instructors encourage students to have learning-oriented interactions with their peers and independent learning trying to understand some topics by themselves (Visschers-Pleijers, Dolmans, Wolfhagen, & Van Der Vleuten, 2005).

The SCCT (Lent, Brown, & Hackett, 1994) explains the way in which students' careers and academic interests mature, how career choices are developed, and the way in which these choices became in action. The foundations of the SCCT are in Bandura's social cognitive theory (Bandura, 1986), being self-efficacy and outcome expectations critical factor in career selection. Self-efficacy is an individual's beliefs about their capability to succeed in a particular task (or career into the SCCT) while outcome expectations are the anticipated positive consequences of engaging in a particular task or career (e.g. career satisfaction, engagement, performance, and persistence) Thus, self-efficacy and outcome expectation regulate career interests as students develop their interests in activities when they experience self-efficacy and positive outcomes (Lent, Brown, & Hackett, 1994). In education in engineering, the SCCT has been used to explain that those students who have positive self-efficacy beliefs and outcome expectations regarding engineering are more likely to pursue a career in this field. A clear idea about what being an engineer means and what types of professional activities an engineer should be able of doing is necessary for students to have the intention to enroll in an engineering program (Winters, Brunhaver, Chen, Yasuhara, & Sheppard, 2013).

Methods

Study design

This is a pre- post-test with no control group study design.

Setting and participants

This study was conducted into a biomedical engineering program. In 2011 a partnership between the School of Medicine and Health Sciences of School of Medicine and Health Sciences of Universidad del Rosario and the School of Engineering Julio Garavito of Bogota opened a biomedical engineering program. This is a nine semesters and 154 academic credits biomedical engineering program taught by the two universities.

The study involved 113 students from two classes (2015-II and 2016-I) on a first-year biomedical engineering two credit introductory (BEIC) course that used hybrid lecture-PBL learning strategy. This sample size was sufficient, using as parameters an effect size of 0.6-0.7 and a statistical power of 0.8 (statistical significance was set at $p<0.05$) (Portney & Watkins, 2000).

The BEIC hybrid two-credit course is understood as a course in which students are exposed to both PBL and lectures teaching learning experiences. The lectures provide the students with a general vision about the most current themes in biomedical engineering. The design and implementation of the modest complexity project by using PBL strategy is to introduce students to, and encourage the use of, engineering tools and systems to design a system, component, or process to meet desired needs. This course has not prerequisite as this course is taken by students in the first semester of their program. The course grade is based on an essay about biomedical engineering as a specialty (30 percent), a mid-term (30 percent) in which the students have to present a plan of their design project, and a final defense of their designed project (40 percent). The project is assigned to a functional group of 3-4 students, and they have to work as a team throughout all the phases of the project. Also, the project requires the team to hold meetings (two or three) with the finals users (patients or therapists, or both). A mentor is assigned to each group of students to advise them on how to conduct the project successfully. For the defense of project, a multidisciplinary panel grades the proposal, taking into account a multi-angle assessment. The assessment of the project is done by means of a specific rubric designed for this purpose. Examples of modest complexity medical technology projects are: the design and implementation of a resistance-capacitor-inductor (RLC) defibrillator circuit, and the design and implementation of low-cost switches to be adapted in a computer mouse for children with cerebral palsy (Cruz, Rincon, Dueñas, Luna, & Torres, 2016).

Variables

In this study we included demographics variables such as gender (male/female) and socio economical strata (Low/Medium /High) (see result section) in order to characterize our sample.

The outcome variables of this study were the students' perceptions of the engineering profession, measured using a 30-item survey adapted from (Cruz, Rincon, Dueñas, Luna, & Torres, 2016). What we expected is that after students were exposed to the introductory biomedical engineering course, their perceptions about what skills are needed for an engineer would improve.

The instruments

We used a questionnaire proposed by (Cruz, Rincon, Dueñas, Luna, & Torres, 2016) already validated with high levels of internal consistency in a prior research to measure the specific variables examined in this study. In the survey questionnaire the content included 42 items designed to measure the students' perceptions of the engineering profession and to collect demographic variables. The 42 items were grouped into three sections. In the first section (section A, 10 items), we asked for the participants' demographics. In the second section (section B, 2 items) we inquired students' literacy on the competence or skill concept. Section C includes three categories or subscales depicting three types of engineering skills portrayed as certain tasks or things that an engineer would do: technical (items C1-C11, e.g. an ability to design and conduct experiments, as well as to analyze and interpret data), professional (items C12-C24, e.g. an ability to work on multidisciplinary teams, an ability to communicate effectively, behave ethically,), and project management skills (items C25-C20, e.g. an ability to write a business plan; prepare a budget for a project) based on their understanding of the importance of the task for the practice of engineering (Cruz, Rincon, Dueñas, Luna, & Torres, 2016). In section C (30 items) students were asked to rate each item on a Likert scale from 1 to 5 (where "1" was strongly disagree, and "5" was strongly agree).

Variables operationalization

Dichotomous variables were coded as "0" or "1" (e.g., gender). Each polytomous variable was represented by a set of binary design whose values codified each variable category (e.g., Parents' highest educational level). All the items in section B from the questionnaire were scored on a 5-point Likert scale, ranging from strongly disagree "1" to strongly agree "5". This is the most common coding or operationalization technique used in the literature (Cohen, Cohen, West, & Aiken, 2003).

Procedures: Survey administration, data collection, and bias control

We followed a modification the procedure proposed by (Cruz, Rincon, Dueñas, Luna, & Torres, 2016). After receiving the ethical approval by the School of Medicine and Health Research Ethics Board, during the first week of the semester, the research team, informed the all students from BEIC course the aims of the study and asked for volunteers to sign a consent form. In order to guarantee their anonymity and confidentiality, each participant was assigned with a questionnaire code. Before administering the questionnaires we conducted a pilot study with ten students representing of BIEC course, therefore they did not complete final version of the questionnaire. The aim of the pilot study was to obtain feedback from students as to whether the questionnaire items were clear and concise. Minor changes to some items were made as needed.

The survey was administered three times; during the first week (baseline, T_0), during the sixth week, when students were exposed to lectures only (T_1), and during the sixteenth week of the course (after the implementation of the PBL strategy, T_2). The questionnaire was administered in a quiet room and at a time when students did not feel any anxiety, physiological pressure, or fatigue due to any exam. A research assistant who was not involved in the teaching session, was assigned to assist the research team to administer the questionnaires. This mechanism guaranteed that the students did not feel any pressure producing any biased responses in either filling out the questionnaires or talking openly during the focus group sessions.

Statistical analyses

Sum scores for all items was calculated. We used descriptive statistics to summarize students' demographic data. To respond to our research question Wilcoxon signed-rank tests for two dependent samples were performed on the data gathered from the T_0 and T_1, and the T_1 and T_2. The T_0 - T_1 comparison was made in order to control for maturation threats for validity.

The reliability and internal validity measures for each construct of the instrument (questionnaire) was tested using the F-test of the intra class correlation coefficient (ICC) (i.e. average and single measures) and Cronbach's Alpha (CA). Values of Cronbach's alpha higher than 0.70 indicates good internal consistency between the items within each variable (Portney & Watkins, 2008).

The alpha level of significance for all tests were set at $p \leq 0.05$ (two-sided). Missing data was handled in the following procedure: (1) missing data in continuous variables such as students' age were replaced by the average values of these variables; and (2) for categorical and ordinal variables such as the gender, and their socio economical strata, missing data were replaced by the medians of these variables. SPSS® V 22.0 statistics package were used to generate descriptive, univariate and bivariate statistics

Results

Participants' description

The study involved all (113 students) from the biomedical engineering program, being 61.1% female and 38.9%male. More than a half of them from low-middle income families 55.8%. The 19.5% of students obtained their high school degree in public schools, in 21.2% of cases any of students' parents have a higher education degree in engineering, whereas in 7.1% of cases any of students' parents have a higher education degree in health sciences.

Impact of hybrid lecture+PBL course on students' perceptions of the engineering profession.

Table 2 a) and b) shows the descriptive statistics and the results of the paired Wilcoxon signed-rank test statistics for the students' perceptions baseline compared with T_1 and in T_1 compared with T_2 both per group of skills (i.e. technical skills (TS), Professional skills (PS), and Project management skills (PMS)) and all skill together (i.e. TS+PS+PMS). Noticeable, descriptive statistics showed the first year biomedical engineering students have a clear imaginary on what an engineer must be. This assertion can be supported by the fact that, the median/mean and standard deviation were located in the top third of the Likert scale of the questionnaire items (mean > 3.33), even at baseline measurement.

Our statistical analysis of the paired Wilcoxon signed-rank test statistics showed that overall, no statistically significant results in the students' perceptions about engineering profession in the baseline compared with T_1 was found (baseline mean (SD): 4.26 (0.54); T_1 mean (SD): 4.20 (0.36), $p<0.566$). This indicates, that a lecture-only teaching learning strategy does not significantly change students' perception of the engineering profession (Table 2a). On the other hand, the statistically significant change in the students' perceptions in T_1 compared with T_2 (i.e. TS+PS+PMS: T_1 mean (SD): 4.20 (0.36); T_2 mean (SD): 4.30 (0.40), $p<0.024$)

revealed that the students' overall perceptions of the engineering profession had significantly improved by the end of the course, only after the PBL strategy had been implemented (Table 2b). The same is true for the case of individual groups of skills, with the exception of professional skills, i.e. our statistical analysis showed that technical and project management skill dimensions of the engineering profession questionnaire had also changed by the end of the course, only after the PBL strategy had been implemented (i.e. TS: T_1 mean (SD): 4.27 (0.42); T_2 mean (SD): 4.36 (0.42), $p<0.014$, $\Delta TS_{T1\text{-}T2}=9\%$; PMS: T_1 mean (SD): 4.01 (0.52); T_2 mean (SD): 4.23 (0.51), $p<0.023$, $\Delta PMS_{T1\text{-}T2}=22\%$) (Table 2b). Being the higher change the case of project management skills, with a 22% of change in T_2 compared with T_1

Table 2. Summary of results: Impact of hybrid lecture-PBL course on students' perceptions of the engineering profession. (a) Baseline compared with T_1 **(b)** At T_1 compared with T_2

(a)

Group Of Skills	Baseline				T_1				Wilcoxon signed-rank test Statistics
	Mean	**Median**	**SD**	**CI95%**	**Mean**	**Median**	**SD**	**CI95%**	**p-value**
TS	4.21	4.27	0.421	(4.10,4.32)	4.27	4.27	0.42	(4.18,4.35)	0.421
PS	4.17	4.23	0.514	(4.06,4.29)	4.24	4.23	0.39	(4.16,4.32)	0.514
PMS	4.03	4.16	0.353	(3.89,4.17)	4.01	4.00	0.52	(3.91,4.11)	0.353
All	4.16	4.26	0.566	(4.05,4.27)	4.20	4.20	0.36	(4.13,4.28)	0.566

Notes
n: 113; S.D: Standard deviations, C.I: Confidence interval, Technical skills (TS), Professional Skills (PS), Project management Skills (PMS), 1: After exposing students to introductory course with lectures. T2 After students to introductory course with PBL pedagogical approach; All statistical significance was set at $p<0.05$); All: TS+PS+PMS

(b)

Group Of Skills	T_1				T_2				Wilcoxon signed-rank test Statistics
	Mean	**Median**	**SD**	**95% CI**	**Mean**	**Median**	**SD**	**95% CI**	**p-value**
TS	4.27	4.27	0.42	(4.18,4.35)	4.36	4.36	0.40	(4.27,4.44)	0.014
PS	4.24	4.23	0.39	(4.16,4.32)	4.27	4.23	0.49	(4.17, 4.37)	0.066
PMS	4.01	4.00	0.52	(3.91,4.11)	4.23	4.33	0.51	(4.12, 4.33)	0.023
All	4.20	4.20	0.36	(4.13,4.28)	4.29	4.30	0.40	(4.21, 4.38)	0.010

Notes
n: 113; S.D: Standard deviations, C.I: Confidence interval, Technical skills (TS), Professional Skills (PS), Project management Skills (PMS), 1: After exposing students to introductory course with lectures. T2 After students to introductory course with PBL pedagogical approach; All statistical significance was set at $p<0.05$); All: TS+PS+PMS

Reliability and internal validity of the instrument

Table 3 shows the statistics of the reliability and internal validity of the questionnaire used in our study. The statistical significant of the p-values (i.e. $p<0.05$) for the F-test of the intra class correlation coefficient (ICC) (i.e. average and single measures) for all the variables/constructs tells us that the subjects are different from each other, which is a necessary condition for a reliability test. Since Cronbach's Alpha (CA) measures in 100% of cases are above 0.70 and the F-test of the ICC (i.e. average and single measures) for all the constructs were statistically significant ($p<0.05$), one can reach the conclusion that the instrument used has acceptable levels of reliability and internal validity (Portney & Watkins, 2000, p. 595)

Table 3. Reliability and internal validly of the instrument. (a) Baseline compared with T_1 **(b)** At T_1 compared with T_2

(a)

Group Of Skills	Baseline			T_1		
	CA	CI	p-Value	CA	95% CI	p-Value
TS	0.842	(0.258;0.409)	0.000	0.768	(0.169, 0.312)	0.000
PS	0.910	(0.367;0.520)	0.000	0.811	(0.188, 0.324)	0.000
PMS	0.850	(0.401;0.573)	0.000	0.721	(0.219, 0.396)	0.000
All	0.945	(0.299;0.440)	0.000	0.898	(0.176, 0.296)	0.000

Notes
n: 113; C.I: Confidence interval, Technical skills (TS), Professional Skills (PS), Project management Skills (PMS1: After exposing students to introductory course with lectures. T2 After students to introductory course with PBL pedagogical approach; All statistical significance was set at $p<0.05$); CA: Cronbach alpha; All: TS+PS+PMS

(b)

Group Of Skills	T_1			T_2		
	CA	95% CI	p-Value	CA	95% CI	p-Value
TS	0.768	(0.169, 0.312)	0.000	0.832	(0.241,0.395)	0.000
PS	0.811	(0.188, 0.324)	0.000	0.889	(0.307, 0.467)	0.000
PMS	0.721	(0.219, 0.396)	0.000	0.792	(0.303, 0.482)	0.000
All	0.898	(0.176, 0.296)	0.000	0.932	(0.250, 0.391)	0.000

Notes
n: 113; C.I: Confidence interval, Technical skills (TS), Professional Skills (PS), Project management Skills (PMS1: After exposing students to introductory course with lectures. T2 After students to introductory course with PBL pedagogical approach; All statistical significance was set at $p<0.05$); CA: Cronbach alpha; All: TS+PS+PMS

Discussion

The main motivation in conducting this study was to determine how the students on the biomedical engineering program conceptualize the engineering profession. In doing so, we posited the following research question: *How do overall student perceptions of the engineering profession change as a result of their educational experience in an introductory biomedical engineering course?*

The results of this exploratory study revealed that students' overall perceptions of the engineering profession improved as a result of their educational experience on the introductory biomedical engineering. More importantly, the PBL strategy implemented in this hybrid course significantly influenced the change of students' overall perceptions of the engineering profession. This result is consistent with other research, showing that the attitudes and perceptions of engineering students towards the engineering profession have improved in courses that implemented hybrid lecture+PBL teaching strategies (Cruz, Rincon, Dueñas, Luna, & Torres, 2016; Dabbagh & Menascé, 2006; Vinogradov & Plumb, 2008; Wang, Kang, & Lee, 2013;).

Our study found that not only that technical skills dimension of the survey was statistically significant T_1 compared with T_2 but also that technical skills dimension showed the highest scores in the baseline, T_1 and T_2. These result contradicts previous research literature on engineering education, which curiously had found that technical skills dimension of students perception about engineering profession showed the lowest scores (Cruz, Rincon, Dueñas, Luna, & Torres, 2016); but is in the line of other studies which has consistently confirmed that students perceive engineering as a technical field of study (Dabbagh & Menascé, 2006). Although Cruz and colleagues (2016) provided a plausible explanation of why they found this contradictory result (e.g. biomedical engineering is a profession in its early stages and freshmen did not have a clear idea about what the technical skills of a biomedical engineer are) we believe this result was associated to limitations of the study itself such as low sample size, or the the absence of a control group, etc.

Surprisingly, we found that students' perceptions project management skills items did improve. This indicate, contrary what other scholars have found (e.g. Cruz, Rincon, Dueñas, Luna, & Torres, 2016; Sheppard, et al., 2010; Atman, et al., 2010), that our PBL strategy helped to freshmen to understand how project management skills such as communication and teamwork are important outside the classroom in their future practice as engineers. The explanation we have to this results is that although such skills require some preparation and maturity that are expected to be achieved soon after graduation during their work experience (Winters, Brunhaver, Chen, Yasuhara, & Sheppard, 2013), perhaps the regulations imposed during PBL part of the course such as working in group, time management, prepare a budget for the project, and doing some scheduling tasks (e.g. scheduling the project by phases and to do a Gantt chart) to conduct the project had positive effects on the change in the perceptions that management skills are also important to be a good engineer.

Finally, we found that although students' perceptions profesional skills items did improve, it was not statistically significant at T_2 compared to T_1. This indicate that our PBL teaching strategy implemented in this course did not cause a possitive impact on students perceptions that profesional skills are important for engineers. The explanation we have to this results is that for a first-year student, understanding that to abide an engineer's professional code of ethics and transparency, for instance, is a concept that require some degree of maturation, that these students do not have yet at this point in the program. This assertion have been supported by findings in other studies. For example, in a study conducted by VanDeGrift, Dillon, and Camp (2016) showed that altought the first-year engineering students want to earn their degree without cheating, less than a half of the them would report on another cheating student. In another study conducted in a population of 18 universities in the USA and 3914 engineering and computer science students, showed that students consistently self-report high levels (58–82%) of cheating behaviours (Finelli, et al., 2012). These findings suggest that major effort should be done in order to change the student culture about ethics and professional skills in engineering profession.

Conclusions

The first year biomedical engineering students have a clear imaginary on what an engineer must be The PBL educative strategy positively changed first-year students' perceptions of the engineering profession. Professor and academics should implement better learning strategies in order to change the student culture about ethics and professional skills in engineering profession.

References

Anderson-Rowland, M. (1997). Understanding Freshman Engineering Student Retention through a Survey. *Proceedings of the 1997 ASEE Annual Conference* (págs. 1-7). Milwaukee, WI, USA: American Society of Engineering Education.

Atman, C., Sheppard, S., Turns, J., Adams, R., Fleming, L., & Stevens, R. (2010). *Enabling Engineering Student Success: The Final Report for the Center for the Advancement of Engineering Education.* San Rafael, CA: : Morgan & Claypool Publishers.

Awang, H., & Ramly, I. (2008). Creative Thinking Skill Approach through Problem-Based Learning: Pedagogy and Practice in the Engineering Classroom. *International Journal of Social, Behavioral, Educational, Economic, Business and Industrial Engineering, 2*(4), 334:339.

Bandura, A. (1986). *Social foundations of thought and action: A social cognitive theory.* Englewood Cliffs. NJ: Prentice-Hall.

Chua, K. (2014). Performance Differences between First-time Students Undergoing Hybrid and Pure Project-based Learning. *International Journal of Engineering Education, 30*(5), 1200–1212.

Cohen, J., Cohen, P., West, S., & Aiken, L. (2003). *Applied multiple regression/correlation analysis for the behavioral sciences,* (3rd ed.). NY: Taylor and Francis Group.

Cruz, A., Rincon, A., Dueñas, W., Luna, N., & Torres, D. (2016). The Impact of an Introductory Biomedical Engineering Course on Students' Perceptions of the Engineering Profession. *Source of the DocumentInternational Journal of Engineering Education, 32*(1), 136-149.

Dabbagh, N., & Menascé, D. (2006). Student Perceptions of Engineering Entrepreneurship: An Exploratory Study. . *Journal of Engineering Education*, 153-164.

Dimitriu, D., & Karimi, A. (2005). Exploring the engineering profession - A freshman engineering course. *ASEE Annual Conference and Exposition, Conference Proceedings*, (págs. 6283-6292). Portland, OR.

Duran, J., Gallardo, S., Toral, S., Martinez-Torres, R., & Barrero, F. (2007). A learning methodology using Matlab/Simulink forundergraduate electrical engineering courses attending to learner satisfaction outcomes. *Int J Technol Des Educ, 17*, 55-73.

EERC. (2006). The Engineering Education Research Colloquies. *Journal of Engineering Education, 95*(4), 257–258.

Finelli, C. J., Holsapple, M. A., Ra, E. B., Burt, B. A., Carpenter, D. D., Harding, T. S., & Suktus, J. A. (2012). An assessment of engineering students' curricular and co-curricular experiences and their ethical development. *Journal of Engineering Education,, 101*(3), 469–494.

Lent, R., Brown, S., & Hackett, G. (1994). Toward a Unifying Social Cognitive Theory of Career and Academic Interest, Choice, and Performance. *Journal of Vocational Behavior, 45*(1), 79-122.

MinEducación. (16 de 02 de 2014). *Caracterización de estudiantes y diagnóstico de deserción en educación superior. Bogotá.* . Obtenido de Retrieved from http://www.mineducacion.gov.co/sistemasdeinformacion/1735/w3-propertyname-2895.html#

Nesbit, S., & Hummel, S. (2005). A design and assessment-based introductory engineering course. *International Journal of Engineering Education, 21*(3), 434–445.

Piaget, J. (1952). *The origins of intelligence in children. (Margaret Cook, Trans.).* New York: International University Press.

Portney, L., & Watkins, M. (2000). *Foundations of clinical research: Application to practice.* New Jersey: Prentice-Hall. Inc.

Rodriguez, J., Laveron-Simavilla, A., del Cura, J. M., Ezquerro, J. M., Lapuerta, V., & Cordero-Gracia, M. (2015). (2015). Project Based Learning experiences in the space engineering education at Technical University of Madrid. *Advances in Space Research, 56*(7).

Sheppard, S., Gilmartin, S., Chen, H., Donaldson, K., Lichtenstein, G., & Eris, O. (2010). *Exploring the engineering student experience: Findings from the Academic Pathways of People Learning.* Seattle, WA: : Center for the Advancement for Engineering.

VanDeGrift, T., Dillon, H., & Camp, L. (2016). Changing the Engineering Student Culture with Respect to Academic Integrity and Ethics. *Science and Engineering Ethics*, 1-24.

Vinogradov, A., & Plumb, C. (2008). Enhancing Student Understanding of and Interest in Mechanical Engineering. *2008 ASEE Annual Conference and Exposition;* (pág. (p. 12p)). Pittsburg, PA; United States: American Society for Engineering Education.

Visschers-Pleijers, A. S., Dolmans, D. J., Wolfhagen, I. A., & Van Der Vleuten, C. M. (2005). Development and validation of a questionnaire to identify learning-oriented group interactions in PBL. *Medical Teacher, 27*(4), 375–381.

Wadsworth, B. (2004). *Piaget's theory of cognitive and affective development: Foundations of contructivism.* Upper Saddle River, NJ: Allyn & Bacon.

Wang, H. H., Kang, S. C., & Lee, T. K. (2013). Cultivating Imagination in Entry-Level Civil Engineering Students: Exploring the Effects of an Innovative Instructional Model. *International Journal of Engineering Education, 29*(5), 1248–1259.

Winters, K. M., Brunhaver, S., Chen, H., Yasuhara, K., & Sheppard, S. (2013). From freshman engineering students to practicing professionals: Changes in beliefs about important skills over time. *120th ASEE Annual Conference and Exposition* (pág. 18). Atlanta: American Society for Engineering Education.

Acknowledgements

The authors wish to thank the study participants, instructors, and mentors for his invaluable advice in preparing this paper. We also want to thank Oliver Jarvis for his proofreading of this paper. The authors assume overall responsibility for the manuscript.

Copyright statement

The following copyright statement should be included at the end of your paper. Substitute authors' names in final copy only.

Application of the Jigsaw method for solving problems in Ocean Engineering

Gianina Morales
Universidad de Valparaíso, Valparaíso, Chile
gianina.morales@uv.cl

Mauricio Molina
Universidad de Valparaíso, Valparaíso, Chile
mauricio.molina@uv.cl

Recent global changes have encouraged higher education institutions to generate transformations in their teaching processes, so as to enable students to successfully face the current demands of the professional world. In this context, this paper presents the applications of Jigsaw technique *as a concrete way to develop skills and knowledges* in *the classroom, in combination with Cooperative Problem-Based Learning, in an Ocean Engineering undergraduate program's introductory course. From the disciplinary point of view, the learning experience addresses the topic of tsunami evacuation (a threat that poses a significant risk to Chile) and the resolution of typical problems related. Under the action research methodology, here presented preliminary results of two applications that seems indicate that the implemented changes in the way of teaching specific engineering subjects, positively impacts in the appropriation of student's knowledge and their motivation by the learning, beyond the activities request.*

Introduction

In the current global scenario, higher education is an important driver of development for countries (OECD, 2000). Social and political actors agree that universities have a special responsibility in the creation of knowledge to cope with the complex global problems; therefore, countries have made great efforts to improve and guarantee the quality of higher education (UNESCO, 2009).

The teaching of engineering is not exempt from this challenge and since the 80s, the approaches of the new educational paradigm - constructivist or also called socio-constructivist - have begun to be noticed in engineering classrooms (Felder, 2006; 2012). Some research has addressed the role of engineering teacher (Fynewever et al., 2008); others have focused on effective teaching strategies (Felder, Woods, Stice & Rugarcia, 2000) and others have addressed the teaching-learning processes as part of many factors linked to the change in the training of engineers (King, 2008; Jamieson & Lohmann, 2009; Kindelán & Martín, 2008). The results of the referred research determine that successful pedagogical practices incorporate methodologies and strategies of the constructivist paradigm, which supports the idea that the orientation of the processes towards the learner and towards learning (Garret, 1988) allows improvements in educational quality.

In Latin America, and specifically Chile, these methodologies have recently been incorporated and their adoption has been strengthened since the engineering programs innovated his curricula, in the new century. The place where the experience is situated, the Ocean Engineering School at Universidad de Valparaíso, has revamped in 2015 its curricula in a competence-oriented study program, considering the current (and hopefully future) needs of the discipline. In this context, this article describes an experience of active learning in the first year course "Introduction to engineering", a multidisciplinary and motivational course whose main purpose is to discuss the most relevant aspects of the subject from its ethical, technical and scientific dimensions. In addition, the course aims to develop transverse skills in students.

The experience was applied in two opportunities (years 2015 and 2016) and is based on the adaptation of the Jigsaw technique for addressing problems associated with tsunami evacuation-walkout drill in two coastal communes, Viña del Mar and Valparaíso, in Chile. The Jigsaw technique is a cooperative learning tool aimed at improving student relationships, enhancing motivation and increasing meaningful learning of large volumes of knowledge (Smith, 2013). With learning objectives of develop transverse and specific skills associated with the course and particular topic (tsunami evacuation), this method was combined with Cooperative Problem-Based Learning (CPBL), a way of using Problem-Based Learning (PBL) in the engineering curricula.

The two methods considered are recently applicate in Chile and this work presents a preliminary research with the purpose to evaluate their combined effectiveness for the learning objectives achievement. As teachers of the course, we used an action research methodology, an emerging method in engineering education research (Case & Light, 2011) focuses in the systematic professional reflection by teachers about their own practice. By this way, we evaluate not only the learning of the student, but also our pedagogical work, and we used the information gathered in each of the applications of the experience for an improvement of its next implementation. Hence, the inquiry is considered preliminary, because during the current year we will again apply the experience and hope to implement other methods of gathering information to achieve new conclusions.

In this article, the research aims to answer these questions: these methodologies can develop the motivation and the cooperative learning, beyond the requested in the activity? (RQ1), How can the methods strengthen problem-solving skills that allow the appropriation of relevant disciplinary knowledge? (RQ2) and What is the importance of the teaching role for the achievement of the learning objectives with the methods used? (Q3) These questions are answered through the analysis of the results of the experience.

Theoretical framework

Cooperative Learning is an active teaching-learning method where the teacher, for an instructional purpose, forms small groups where students work together to maximize their own learning and that of others (Smith, Sheppard, Johnson & Johnson, 2005). The application of this methodology involves five essentials conditions: positive interdependence, individual responsibilities, direct interaction (face to face), use of interpersonal skills and group work process (Johnson, Johnson & Smith, 1998; Felder & Brent, 2007).

The literature recognizes three general forms of Cooperative Learning, which in turn involve different techniques of classwork (Smith, 2013): Informal cooperative learning groups, Formal cooperative learning groups and Grassroots cooperative groups. The Jigsaw method is considered as a way of working the implementation of formal Cooperative Learning. Is defined as a cooperative learning technique (Aronson & Patnoe, 1997) developed by Elliot Aronson in the early 70's with his students from the universities of Texas and California. Its application has been shown to reduce racial conflict and to develop cooperative and learning skills in the classroom.

Basically, the technique consists of the division of students into Jigsaw groups to deal with a disciplinary topic. Guided by the teacher, each member is in charged of study segment and is responsible for sharing his/her knowledge with the group. The groups are then separated and reorganized into Expert groups formed by the students grouped by study segment of the general theme. In these Expert groups, the students investigate and analyze the segment of the topic that corresponds to them.

Finally, the groups reorganize again and students return to their original Jigsaw group where they teach about their area of expertise and learn from areas which were deeply analyzed by their peers in order to understand the phenomenon as a whole. Another possibility of implementation is to divide the students first into the Expert groups and then organize the Jigsaw groups.

As mentioned earlier, in this paper the Jigsaw Method is combined with the Cooperative Problem-Based Learning (CPBL), a way of using Problem-Based Learning (PBL) in the engineering curricula, based on the work that engineers perform in practice. This concept has been coined by Smith et al. (2005) as a primary part of what they call Pedagogies of Engagement, a set of active strategies and methods that enable students to take charge of their learning and that of others.

Considering that PBL is a proven active teaching methodology that basically consists of the learning which results from the process of work on the comprehension and eventual solution of a problem, his characteristics (Barrows, 1996 cited in Smith et al, 2005) are used in the experience: (i) Learning is student-centered, (ii) Learning occurs in small groups of students, (iii) Teachers are facilitators or guides, (iv) Problems are the focus of organization and stimulation for learning, (v) Problems are the vehicle for developing problem solving skills and (vi) The new information is acquired through self-directed learning.

Methodology

As it mentioned above, in this study has been used the action research methodology. This is defined by Bassey (1998:93) like: "an enquiry which is carried out in order to understand, to evaluate and then to change, in order to improve some educational practice". In higher education, there are studies that indicate its benefits for improving teachers through their application in collaborative projects (McNiff, 1993).

A recognized model on this methodology is formed by 4 elements: Plan, Act, Observe and Reflect (Costello, 2003). According to this model, we designed the Jigsaw activity considering its methodology and evaluation (Plan), then, it was applied in 2015 and 2016 (Act), where the students were evaluated (Observe) and from the evaluation, we raised conclusions to improve the following applications (Reflect).

The instruments used to collect information correspond to a report and an oral presentation, tried to value the achievement of learning objectives of the course, and a survey, to know the motivation of the students. With these procedures, the research questions are answered in section "Findings and Discussion."

The experience

Chile is located in the Pacific Ring of Fire, and their coast is frequently stroke by big earthquakes and tsunamis, as the Valdivia event in 1960, the biggest earthquake and tsunami ever registered. But, ancient registers show a periodicity of this phenomenon, highlighting the 1730 event, produced in the present most populated zone of the country. Population is well prepared in earthquake case, but in tsunami case the country have a dramatic ignorance, that cost many lives in 2010 event. In this context, many institutions have made efforts to prepare population in a tsunami evacuation case.

This issue is a relevant topic in the Ocean Engineering professional performance and training, because this phenomena is the mainly design conditions for coastal works. With the purpose of develop transverse and specific skills of the Ocean Engineer (e.g. effective communication, teamwork, decision making, problem solving, application of knowledge and managing information tools), the course of Introduction to Engineering in 2015 and 2016 involved the application of the Jigsaw technique combined with the CPBL for addressing problems associated with tsunami evacuation in two coastal communities, Viña del Mar and Valparaíso, in Chile.

First, students were organized into four Expert groups who investigated personally and collaboratively on different subtopics associated with the tsunami evacuation processes (figure 1). Each Expert group was guided by monitors from superior courses in two knowledge discussion sessions to be ensured that all the experts of a subject have a similar background.

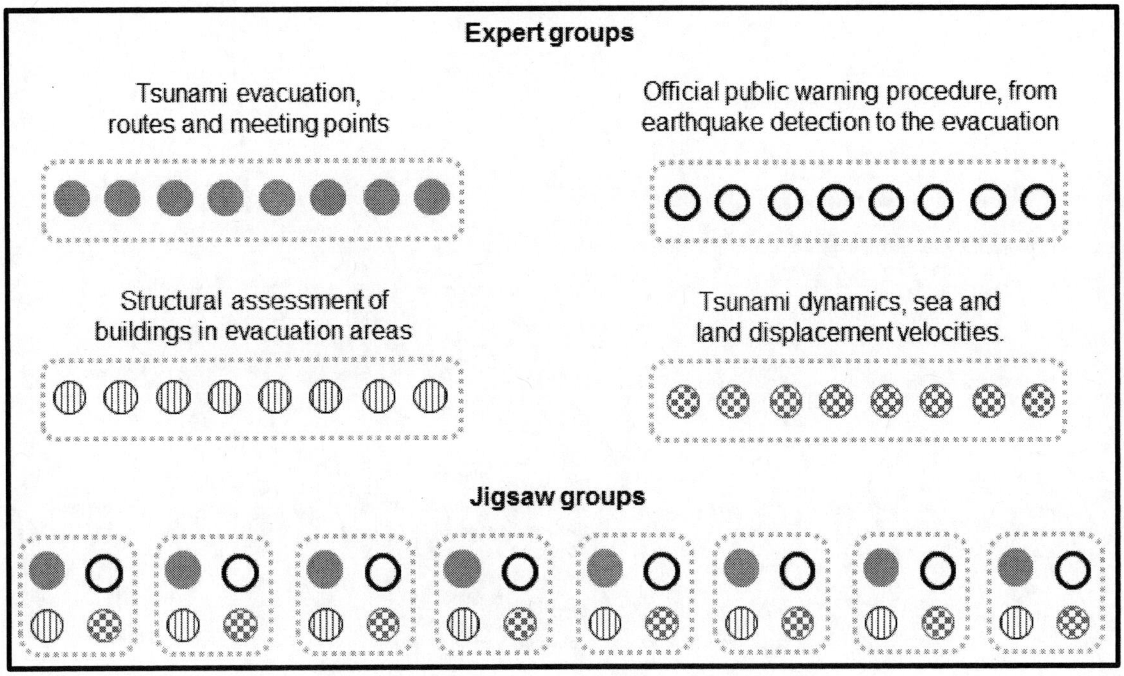

Figure 1: Organization of groups in the Jigsaw experience.

Once the subjects were worked out in depth, the students reorganized themselves into eight Jigsaw groups, who were coordinated to solve the following field:

> *A 9.2 earthquake on the Richter scale struck three regions in the country whose rupture zone extends under the sea in the subduction zone between points 30° 0'S 72° 20'W and 34° 28'S 73° 12'W, approximately 500 kilometers*

(figure 2 right). The earthquake triggered a tsunami with similar characteristics to the 1730 event between both points so it is necessary to generate an evacuation to safe *areas.*

Considering *that you are placed nearest Valparaiso bay coast (one of the cases in figure 2, left), go immediately to the nearest meeting point defined by ONEMI (National Emergency Office of Interior Ministry, 2017), respecting the transit laws (for safety purposes).*

From this experience, answer the following questions:

1. *How long you take to arrive to the meeting point defined by ONEMI?*
2. *Considering the timeline evacuation and the tsunami propagation velocity in sea and land, do you avoid been* reached *by the tsunami?*
3. *What difficulties do you face in the evacuation process?*
4. *What is the feasibility of evacuating by car?*
5. *Is the risk of being hit by the tsunami reduced if you evacuate above the 5th floor of the building closest to your evacuation route? (Vertical evacuation)*
6. *How long would it take to get a tsunami alert from the authorities?*

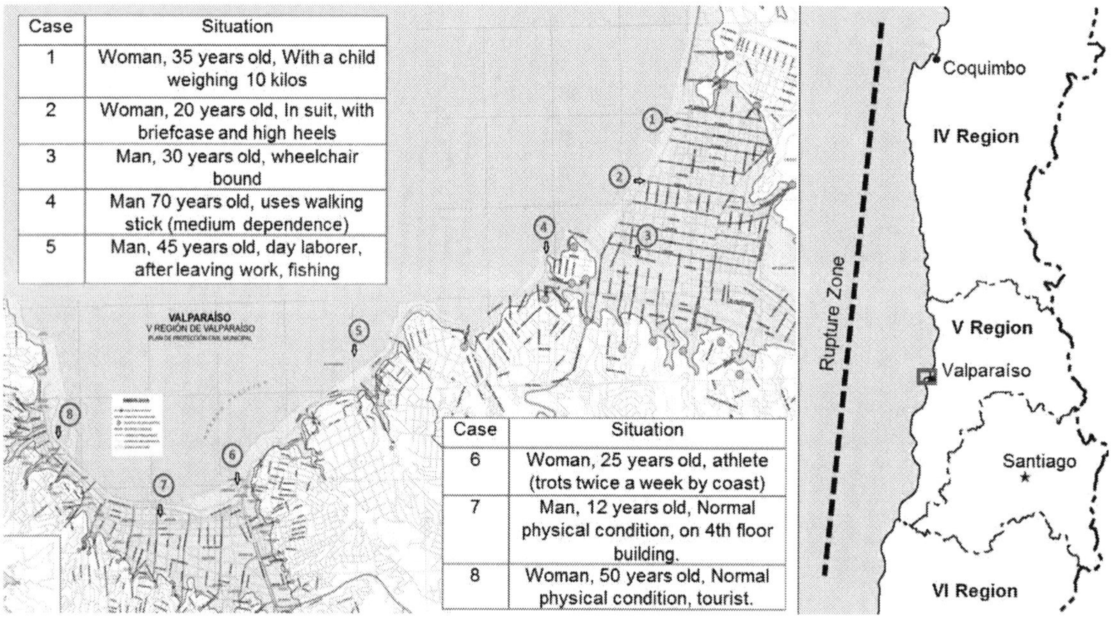

Case	Situation
1	Woman, 35 years old, With a child weighing 10 kilos
2	Woman, 20 years old, In suit, with briefcase and high heels
3	Man, 30 years old, wheelchair bound
4	Man 70 years old, uses walking stick (medium dependence)
5	Man, 45 years old, day laborer, after leaving work, fishing

Case	Situation
6	Woman, 25 years old, athlete (trots twice a week by coast)
7	Man, 12 years old, Normal physical condition, on 4th floor building.
8	Woman, 50 years old, Normal physical condition, tourist.

Figure 2: Case study description.

The development of this activity was coordinated by the members of each group. Once the activity was completed, the students presented a report and their results to the course where teachers and students was consulted on the specific ground experiences.

The evaluation of the activity was carried out by applying an evaluation guideline to the oral presentation and the report. This guideline considered the learning objectives associated with the activity and allowed a qualitative answer to the research questions.

Findings and discussions

At this part, will be presented the principal findings and conclusions generated by the analysis of the experiences. The information is organized according to the research questions mentioned above.

RQ1: *These methodologies can develop the motivation and the cooperative learning, beyond the requested in the activity?*

A high level of motivation was observed during and after the activity implementation, in fact, some groups was presented results in addition to those requested in the evaluation. This was observed especially in the topics that allowed to improve the precision of the experiment, among which are:

1. A member with physical characteristics closer to the case study was chosen to simulate their walking speeds.
2. Implements such as wheelchairs, canes, moving objects with equivalent weight, clothing and footwear, were used.
3. The experiment was repeated to determine time evacuation variability.
4. The experiment was repeated with people with and without knowledge of the evacuation route, to evaluate the importance of this factor.
5. More efficient alternative routes was searching and finding than those defined by the ONEMI.
6. Obstacles maps were made along the evacuation route, analyzing availability of evacuation under different mobility conditions.
7. Comparison between hypothetical and real evacuation conditions was reached, contrasting the evacuation routes defined by ONEMI with the state of way and surrounding buildings.

This support that the case study was very well internalized and analyzed by students, where their performance was evidenced by the various modalities used to finding a solution close to reality.

Among the extra information obtained by student, it's found: i) an interview with police officers regarding the instructions they have to carry out the evacuation and (ii) a survey of pedestrians on what to do in case of evacuation.

A survey realized the next semester from the activity, questioning "Which level of learning motivation produce in you the tsunami evacuation activity" scaled from 1 (no motivation), 2 (low motivation), 3 (medium motivation), 4 (high motivation), and 5 (highest motivation), the results indicate that the 75% of student score answer that the activity produce High (4) or Highest (5) motivation to learning.

RQ2: *How can the methods strengthen problem-solving skills that allow the appropriation of relevant disciplinary knowledge?*

The combination of Jigsaw technique with the CPBL show students doing a high level analysis of the problem to be addressed, which is evident when considering the actions developed by the students to look for proximity to the real situation, described in the previous paragraph. Indeed, this was reflected in the achievement percentage obtained in the activity (figure 3), where all students approve the experience and 50% (2015) and 69% (2016) of student obtain an achievement percentage greater than 75%. The achievement percentage considered the report and the exposition of results.

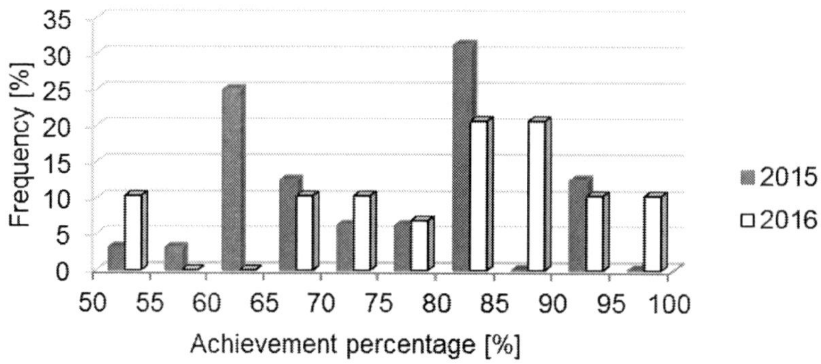

Figure 3: Case study achievement percentage.

The use of the Jigsaw technique led to the development of self-learning and subsequent assessment of knowledge acquired from the discussion with the other members of the Expert group. Assign two discussion session in the Expert groups provided the opportunity to re-direct the investigation in the case of students who did not perform a successful investigation in the first week. This also allowed the students deepen understanding of the scope of the field activity, so that the work planning process in the Jigsaw groups was limited to the definition of strategies for solving the field problem.

In each Jigsaw group oral presentation, in front of all course, a reinforcement of students' argumentation skills was observed as a consequence of the previous discussions about how to proceed in certain situations (depending on the experience they had under similar conditions) and the complementation of experiences and ideas on how to address the problem. We observed that the oral presentation process served as complementary experience of leveling of knowledge.

RQ3: *What is the importance of the teaching role for the achievement of the learning objectives with the methods used?*

The role of the teacher was to guide and monitor the development of each stage, to solve doubts and to synthesize the concepts learned at the end of the process. In fact, the final session of synthesis was developed mainly with the contributions and consultations of the students, thereby covering all the contents provided.

From the activity develop; it is emphasized that with the adoption of these methodologies, the teacher leaves their comfort zone to transform himself in a knowledge mediator and not in the protagonist. This is observed through the following actions: granting instances of self-learning and discussion with peers (both with respect to the information investigated and in relation to the results), not transferring their problem solving models to the student, assess students in a communitarian way and using error as a vehicle of learning, through critical discussion. All these actions were well received by the students.

At the end of the activities in the two years in which they were implemented, feedback instances were generated with the students, which allowed adopting improvements for future applications.

Conclusions and Recommendations

In conclusion, the proper use of the Jigsaw technique in combined with Cooperative Problem Based Learning allowed students to improve problem solving skills in a practical context of study. Through individual and group actions, the students also internalized self-care knowledge, since they considered problems of evacuation before a disaster (tsunami).

Finally, from this experience we conclude that teaching with active learning techniques like the Jigsaw method and CPBL improves student motivation on an individual and collective levels. The activity presented is highly replicable in any disciplinary field of engineering for problem solving, since experience in itself can cover all areas of a specific curricular content.

Regarding the development of our research, future work will address other relevant aspects of learning, such as the impact of the methodologies worked on the development of outcomes. It remains a challenge of the future to investigate the formal short-term and long-term scope of the application of methods such as the treaty in this document, fundamentally along the professional formation and work performance.

References

Aronson, E. & Patnoe, S., (1997). The jigsaw classroom: Building cooperation in the classroom (2nd ed.). New York: Addison Wesley Longman.

Bassey, M. (1998) Action research for improving educational practice, in Halsall, R. (ed.)Teacher Research and School Improvement: Opening Doors from the Inside, Buckingham: Open University Press, pp. 93-108.

Case, J. & Light, G. (2011). Emerging methodologies in engineering education research. Journal of Engineering Education, 100(1), pp. 186–210.

Costello, P. (2003), Action Research, New York: Continuum.

Felder, R. (2006). Teaching Engineering in the 21st Century with a 12th-Century teaching model: How bright is that? Chemical Engineering Education 40(2), pp. 110-113.

Felder, R., Woods, D., Stice, J. & Rugarcia, A. (2000). The future of engineering education II. Teaching methods that work. Chemical Engineering Education 34(1), pp. 26-39.

Felder, R. & Brent, R. (2007). Cooperative Learning. Chapter 4 of P.A. Mabrouk, ed., Active Learning: Models from the Analytical Sciences, ACS Symposium Series 970. Washington, DC: American Chemical Society, pp. 34-53.

Felder, R. (2012). Engineering Education: A Tale of Two Paradigms. In B. McCabe, M. Pantazidou, and D. Phillips, eds., Shaking the Foundations of Geo-Engineering Education, Leiden: CRC Press, pp. 9-14.

Fynewever, H., Henderson, C., Barry, S., Erwin, S., Hanson, S., Lindow, M., Huber, M., & Mirakovits, K. (2008) Honing Teachers' Behind The Scenes Work: Pragmatic Ideas From Best Practices, MSTA Journal, 53 (2), 2-7.

Garret, M.R. (1988). Resolución de problemas y creatividad: implicaciones para el currículo de ciencias. Enseñanza de las Ciencias, 6(3), pp. 224-230.

Jamieson, L. H., & Lohmann, J. R. (Eds.). (2009). Creating a culture for scholarly and systematic innovation in engineering education: Ensuring U.S. engineering has the right people with the right talent for a global society. Washington DC: American Society for

Engineering Education.

Johnson, D. W.; Johnson, R. T.; Smith, K. A.(1998). Active Learning: Cooperation in the College Classroom (2nd ed.). Interaction Book: Edina, MN.

Kindelán, M.P y Martín. A. (2008). Ingenieros del Siglo XXI: importancia de la comunicación y de la formación estratégica en la doble esfera educativa y profesional del ingeniero. ARBOR, Ciencia, Pensamiento y Cultura CLXXXIV 732, julio-agosto (2008), pp. 731-742.

King, R. (2008). Engineers for the future: Addressing the supply and quality of engineering graduates for the 21st century. Sydney: Australian Council of Engineering Deans with support from Australian Learning and Teaching Council.

McKniff, J. (1993). Teaching as learning: an action research approach, London: Routledge.

OECD (2000). Education at a Glance. OECD Indicators 2000: Education and Skills.

ONEMI (2017): Mapas zonas seguras (visitado en octubre de 2015) http://www.onemi.cl/mapas/region/valparaiso/#V

Smith, K.A., Sheppard, S. D., Johnson, D.W., & Johnson, R.T. (2005). Pedagogies of engagement: Classroom-based practices (Cooperative learning and Problem-based learning). Journal of Engineering Education Special Issue on the State of the Art and Practice of Engineering Education Research, 94 (1), 87-102.

Smith, K.A. (2013). Teamwork and Project management. New York: Mc-Graw Hill, 4º Edition.

UNESCO (2009). Conferencia Mundial sobre la Educación Superior - 2009: La nueva dinámica de la educación superior y la investigación para el cambio social y el desarrollo (Sede de la UNESCO, París, 5-8 de julio de 2009).

Copyright statement

Video-related Pedagogical Strategies in Massive Open Online Courses: A Systematic Literature Review

Nathan M. Hicks, Wei Zakharov, Kerrie A. Douglas, Judith M. Nixon, Heidi A. Diefes-Dux, Peter Bermel, and Krishna Madhavan
Purdue University, West Lafayette, Indiana, USA
hicks80@purdue.edu

Abstract: *For engineers who work with rapidly changing technology in multi-disciplinary teams, massive open online courses (MOOCs) offer the unique ability to deliver free, convenient professional development by providing up-to-date information spanning a wide range of disciplines. However, the MOOC boom has not been without its criticisms; many question the effectiveness of MOOCs. In response, many research studies are being conducted across the world to explore the effectiveness of various pedagogical approaches in MOOCs for different stakeholders. As videos constitute one of the most prominent features of MOOCs, it is important to analyse the empirical evidence of best practices for MOOC videos. Through a systematic literature review, we identify a series of important considerations and actions for three groups: instructional teams, video production teams, and platform developers. Considerations include instructor actions, content design and navigation, video style and length, production quality, video annotation tools, viewing options, and embedded assessments.*

Context and Research Questions

The rise of online technologies for delivering educational opportunities in recent years has led to open access to educational resources via massive open online courses (MOOCs). The growing popularity of MOOCs is evident in the number of courses and breadth of subject areas offered, and the number of learners and higher education institutions participating across the globe. Surveys tracking online learning showed that the percentage of higher education institutions offering MOOCs nearly doubled between 2012 and 2013 with nearly twice as many additional institutions actively planning to offer MOOCs in the near future (Allen & Seaman, 2014).

In engineering education, MOOCs offer a particularly promising solution for educators coping with the challenges of the rapid development of new academic fields (Rauf, Daud, & Said, 2016). Even as total knowledge grows at an exponential rate, the amount of instructional time for engineering educators to prepare future engineers remains static. As such, dynamic outlets providing relevant, up-to-date information are necessary for engineers to effectively supplement their education and professional development. MOOCs have the potential to do just that—they can be updated from one run of the course to the next within weeks, making them a potentially superior source of cutting edge engineering-related information compared to traditional textbooks, which suffer from long publication and revision timelines. Further, engineering students and professionals have the ability to select the exact courses or units that cover the specific content they require without being bogged down with other course obligations. The usefulness of MOOCs for engineers is easily highlighted by the large number of engineering courses offered on platforms like edX (https://www.edx.org/course) and Coursera (https://www.coursera.org/courses), each of which had approximately 200 courses categorized as relevant to engineering as of March 1, 2017.

Whether the content of a MOOC is engineering or another subject, MOOCs enable the use of unique instructional strategies, including innovative forms of assessment and socialized learning through discussion forums that connect perspectives of diverse learners across the planet. Capitalizing on this potential, over one-third of the institutions with current or planned MOOCs stated intent to "experiment with innovative pedagogy" and/or "provide more flexible learning opportunities" (Allen & Seaman, 2014, p. 29). While some of these pedagogical strategies vary from platform to platform or course to course, one of the most critical components of nearly every MOOC is the instructional video. Videos excel at promoting acquisition of basic facts and stimulating higher-order thinking while being grounded in more familiar contexts (Bonk, 2008). This strategy of anchoring to previous knowledge and understanding has been shown to improve the meaningfulness of new information (Ausubel, 1978). Further, the on-demand nature of pre-recorded videos, with their ability to be rewound, sped up, slowed down, re-watched, or paused as the learner sees fit, frees learners to personalize their learning experience.

The digital format and massive audiences participating in MOOCs generate enormous quantities of data, contributing to a remarkable increase in MOOC-based research. While MOOC researchers tend to be educationally oriented (Gasevic, Kovanovic, Joksimović, & Siemens, 2014), publications are distributed across a diverse array of journals and conference proceedings (Veletsianos & Shepherdson, 2016). Thus, MOOC stakeholders would benefit from a systematic analysis and synthesis of the literature. A number of studies have synthesized focuses and methodologies of MOOC research (Liyanagunawardena, Adams, & Williams, 2013; Raffaghelli, Cucchiara, & Persico, 2015; Sa'don, Alias, & Ohshima, 2014; Veletsianos & Shepherdson, 2016). However, no studies have specifically investigated empirically-based pedagogical strategies for MOOCs. Out of a general investigation into MOOC pedagogies, we were able to ask the research question: *What are the empirically-based best practices for the development of videos in MOOCs?*

Methodology

We performed a systematic literature review, following the approach outlined by Borrego, Foster, and Froyd (2014) to address the broader question of general pedagogical best practices in MOOCs. Like traditional narrative reviews, systematic literature reviews are intended to synthesize literature covering a given topic, but systematic reviews go further by providing a critical appraisal to direct practice through "transparent, methodical, and reproducible procedures" (Borrego et al., 2014, p. 46). Our results span several major course components, including the use of discussion forums, social media, assessment strategies, general instructional materials, overall course design, motivational strategies, and videos. Given the variability of course aspects identified, we narrow our focus here on just videos, expecting to address others in future papers.

Data Collection

We developed our search strategy through consultation with both education and engineering librarians. Guided by the Population, Intervention, Comparison, and Outcome (PICO) framework (Borrego et al., 2014), and despite our research being driven by a desire to focus on engineering-related MOOCs, we included research involving MOOCs from any subject area to prevent an overly limited pool of articles, as long as the research reported some aspect of pedagogy. On April 15, 2016, we conducted our search within the EBSCO databases *Education Source, Education Full Text, ERIC, Education Administrative Abstracts, PsycINFO*, and *Applied Science & Technology Full Text,* using three search string components: MOOC or "Massive* Open Online" (in the title or subject fields), pedago* or teaching or learning or educat* or instruction, and research or qualitative or quantitative or empirical or "case study" or assessment or analy* or evaluat*. The search was limited to academic journals, journals, reports and conference papers. This produced 1,068 records that the system reduced to 504 records by automatically removing duplicates. An identical search was run in *Compendex* and *Inspec*, limiting the search to English language records and the Classification Codes "Education" and "Computer-aided instruction," which produced 776 records after the duplicates were automatically removed. The records were imported into Endnote (1,280 records, in total) and two passes were made to further manually remove duplicates by matches on author/title/year and then by just by title, leaving 1,170 records.

Data Analysis

We applied four major inclusion criteria in our study: (1) the article had to be written in English; (2) the article had to present data obtained from a MOOC or MOOC participants; (3) the study had to investigate some pedagogical consideration (i.e., some consideration for teaching or supporting student learning); and (4) the article had to be empirical (i.e., it had to have a research purpose, outlined methods, and results). Four researchers independently analyzed a common set of 20 titles and abstracts and conducted entire group discussions to reconcile application of these criteria. This was repeated for two more rounds, filtering through 60 records, in total. Next, we applied the inclusion criteria to the remaining 1,110 articles in pairs, using discussions to reconcile any remaining disagreements. This process led to the rejection of 912 articles. Of these rejections, six were identified as duplicates, twenty were excluded for having non-English abstracts, 432 were not being grounded in a MOOC, 306 were not studies of a pedagogy, and 148 were not being empirical. Note that the documented reason for exclusion was based on the first criterion that was violated in the list (i.e., if it was not grounded in a MOOC and not empirical, the documented reason was the lack of grounding in a MOOC). At this stage, we generally erred on the side of retaining articles when inclusion decisions were uncertain.

The remaining 258 articles passed to the full paper stage of analysis and were each read independently by two researchers, again with any disagreements being resolved through discussion. Through this process, we rejected an additional 185 articles (five had English abstracts but non-English full papers, 30 were not based in a MOOC, 88 were not studies of a pedagogy, and 62 were not empirical). We also calculated a quality score for each article based on clarity of purpose, provision of context, trustworthiness, significance, applicability, and theoretical underpinning. Four of the remaining 73 articles were removed for receiving low quality scores. In a series of group research meetings consisting of five researchers conducted in October and November of 2016 and January of 2017, we sorted the remaining 69 articles into categories based on the pedagogical aspect of MOOCs investigated. Fifteen of these articles were selected for this paper based on their inclusion of content related to videos. The findings of these papers were then separately analyzed and qualitatively synthesized by the first two authors. These authors compared and discussed the themes they separately identified until they reached consensus on the reported themes.

Findings

The set of 15 articles included in this study, and whose content related directly to videos, are summarized in Table 1. These employed a wide range of methodological approaches, including surveys (seven studies), interviews (five studies), learner analytics (five studies), and content analyses (three studies). Five studies employed multiple strategies. Collectively, these studies explored a wide spectrum of aspects related to videos in MOOCs that we identified as relevant to three groups: instructional teams, production teams, and platform developers. As a few aspects of videos may be relevant to multiple groups (e.g., content navigation might be relevant to the instructional and production teams), our inclusion within a given group is based on the following definitions. Instructional staff consists of all individuals developing and delivering course materials, including videos. Production teams consist of those who record, direct, and/or edit videos. Platform developers create the platform software that provides the technical capabilities, such as video, made available to producers and instructors. Of course, it is expected that some individuals may span multiple groups.

Instructional team

There were three broad themes of video aspects identified that relate to the instructional staff: instructor actions, learning content design, and content navigation.

Table 1: Descriptive information of included articles

Author (year)	Purpose of study	MOOC platform	Methods	Video aspect(s)
Adams, Yin, Madriz, & Mullen (2014)	To explore xMOOC completers' learning experiences	Not reported	Qualitative (interviews)	Instructor actions
Evans, Baker, & Dee (2016)	To determine course-, lecture-, and student-level factors that best predict engagement, persistence, and completion	Coursera	Quantitative (Learner analytics)	Navigation, length
Grunewald & Meinel (2015)	To explore effectiveness of personal and collaborative video annotations	openHPI	Quantitative and Qualitative (surveys)	Navigation, annotations
Guo, Kim, & Rubin (2014)	To determine which kinds of videos produce the best learning outcomes	edX	Mixed methods (Learner analytics and interviews)	Instructor actions, content design, style, video
Haavind & Sistek-Chandler (2015)	To explore the role of instructors in MOOCs	Not reported	Qualitative (interviews)	Instructor actions
Hew (2015)	To investigate preferences of MOOC learners	Coursera	Qualitative (Content analysis of discussion boards)	Content design, length, viewing options
Kelder, King, Carew, O'Reilly, Robinson, & Vickers (2013)	To understand student perceptions of structure, usability, accessibility, navigation, and support in a pilot MOOC	Not reported	Mixed methods (Surveys, content analysis)	Content design
Kim, Guo, Seaton, Mitros, Gajos, & Miller (2014)	To understand how students watch and interact with videos	edX	Quantitative (learner analytics)	Content design, style, length, viewing options
Kizilcec, Bailenson, & Gomez (2015)	To study the effect of including instructors' faces in videos	Coursera	Quantitative (Surveys, learner analytics)	Style
Lai, Young, & Huang (2015)	To explore learner perceptions of style and quality of videos	NTHU	Quantitative (surveys)	Style

Table 1, continued

Author (year)	Purpose of study	MOOC platform	Methods	Video aspect(s)
Li, Kidziński, Jermann, & Dillenbourg (2015)	To identify the relationship between behaviour and perception of difficulty	Coursera	Quantitative (learner analytics, surveys)	Content design
Lin, Lin, & Hung (2015)	To investigate learner perspectives of implicit platform attributes and learning consequences	Multiple	Mixed methods (interviews, content analysis)	Length
Mamgain, Sharma, & Goyal (2014)	To compare features offered in different MOOC platforms	Coursera and edX	Quantitative (surveys)	Content design, Style, quality, length, viewing options
Tan, Goh, & Sabastian (2014)	To identify the contribution of various factors to student outcomes in a MOOC	Not reported	Qualitative (interviews)	Style, viewing options
Yousef, Chatti, Schroeder, & Wosnitza (2014)	To identify specific criteria related to the successful design of MOOCs	Not reported	Quantitative (surveys)	Instructor actions, content design, navigation, style, quality, length, annotations, viewing options

Instructor actions. In sum, the four articles related to instructor actions suggested that learners desired instructors who were engaging, understandable, and inclusive. Learners claimed to feel more engaged by instructors who spoke "fairly fast" and enthusiastically (Guo, Kim, & Rubin, 2014) and when videos started with "surprise information" (i.e., unexpected or exciting connections to everyday life) to attract them (Yousef et al., 2014). Learners considered instructors to be more understandable when they used good inflection, diction, and articulation (Haavind & Sistek-Chandler, 2015) and when they spoke in shorter sentences (Yousef et al., 2014). Finally, learners said they preferred when instructors used a warm, inviting tone (Haavind & Sistek-Chandler, 2015) and made relevant interjections or directed, encouraging comments (Adams et al., 2014).

Learning content design. The design of learning content spans individual lectures to the entirety of the course, including lecture structure, purpose, and difficulty. In several studies, learners consistently demonstrated a preference for greater structure (Kelder et al., 2013; Kim et al., 2014; Mamgain et al., 2014; Yousef et al., 2014). In one study, learners suggested structure could be reinforced by an introductory video outlining the layout and important features of the learning environment (Kelder et al., 2013). This structure could also be enhanced by the inclusion of a limited number of clearly defined learning objectives in any given video (Kelder et al., 2013; Yousef et al., 2014). These learning objectives and key points should be summarized and made available without having to watch the video (Kelder et al., 2013; Kim et al., 2014).

The research indicates that learners interact differently with videos depending on the video's purpose (Guo, Kim, & Rubin, 2014; Kim et al., 2014). For instance, instructors should be aware that learners generally watch entire lectures only once, but often return to specific times within tutorial videos (Guo, Kim, & Rubin, 2014; Kim et al., 2014). Also, learners expressed a desire for videos that directly addressed questions or issues that appeared repeatedly or generated considerable interest in the forums (Hew, 2015).

Finally, the learners felt that the level of detail and difficulty of the presented content, as well as the underlying values demonstrated throughout the lecture, should align with the interests, abilities, and values of the course's primary target audience (Yousef et al, 2014). While the objective of delivering an appropriate level of difficulty may not always be easy to achieve upfront, the literature suggests that retroactive analysis of the moments or segments in videos that are replayed, paused, or sped up, can indicate when material was too easy or too challenging (Li et al., 2015). Regardless of difficulty, however, learners expressed an interest in viewing lectures that took cultural diversity and values into consideration (Yousef et al., 2014).

Content navigation. In general, the studies related to content navigation suggested that MOOC learners want to be able to easily find the specific material in which they are interested and want access to additional resources. One study indicated that accompanying each video with a strong set of keywords and tags would help students locate videos related to specific information they want (Yousef et al., 2014). Additionally, learners indicated that videos should have clear, descriptive titles to help learners decide if the videos are of interest (Yousef et al., 2014). Still, care must be taken with the wording of titles, as it was found that learners are less likely to watch videos with titles including the words "optional," "conclusion," and "exercise," suggesting that such phrases should be avoided for the most important content (Evans, Baker, & Dee, 2016). Two additional studies reported that learners wanted help connecting the content presented within specific videos to other topics or content in the same course (Grunewald & Meinel, 2015) or to related information outside of the course (Yousef et al., 2014).

Video production team

The production team is responsible for directing, editing, and overseeing the general production of videos. The related themes include video style, length, and quality.

Video style. Seven of the fifteen studies explored presentation styles including, but not limited to, recorded standard classroom lecture, Khan-style (i.e., videos emulating the style of Khan Academy videos that simulate text or images appearing on a chalkboard accompanied by spoken explanation), slide-view, and slide-view with a "talking head" (Guo, Kim, & Rubin, 2014; Kizilcec, Bailensen, & Gomez, 2015; Lai, Young, & Huang, 2015; Yousef et al., 2014). Each of these studies suggests that recorded standard classroom lectures are the least engaging approach for most learners, though the styles that were most preferred varied across studies and were by no means definitive. Besides the general format, however, it is important to note that learners claim to be distracted by abrupt visual changes (Kim et al., 2014; Yousef et al., 2014), excessive text on the screen, and poor framing, but benefit from the creative use of different colours to highlight important and useful information (Yousef et al., 2014).

Video length. Video length was the single most discussed aspect of videos across the included articles. Learners' preferences for shorter videos were nearly unanimous, though specific maximum recommended lengths varied. Shorter videos were consistently found to be more engaging (Guo, Kim, & Rubin, 2014; Hew, 2015) and facilitated ease of time management (Lin, Lin, & Hung, 2015). Recommended lengths vary from 6 minutes (Guo, Kim, & Rubin, 2014; Kim et al., 2014), to 10 minutes (Mamgain, Sharma, & Goyal, 2014), to 20 minutes (Yousef et al., 2014). However, while one study showed that dropout rate (i.e., "navigating away from a video before its completion") increased with video length (Kim et al., 2014), another indicated that video length had no predictive power for persistence or completion, so instructors should not feel obligated to divide long concepts (Evans, Baker, & Dee, 2016).

Production quality. Learners were less concerned with a minimum resolution of video than the need for clear sound (Yousef et al., 2014). However, learners who had taken courses through both Coursera and edX preferred the better video quality and clarity in Coursera courses, particularly when the lectures included slides with large amounts of text (Mamgain, Sharma, & Goyal, 2014).

Platform developers

Different MOOC platforms allow for a variety of different technical capabilities in their learning environments. Capabilities addressed throughout the included articles relate to video annotation tools, viewing options, and embedded assessment.

Video annotation tools. While learners wanted access to lecture summaries and notes, they also reported wanting the ability to make video annotations and to see the annotations made by other learners (Grunewald & Meinel, 2015; Yousef et al., 2014). Learners suggested that the ability to embed their own notes within the videos and to construct a group-based annotation would be a motivational learning experience (Grunewald & Meinel, 2015).

Viewing options. Based on the five articles related to viewing options, it is clear that learners find having a full set of these options to be important. For instance, learners appreciate the availability of captions or synchronized transcripts (Hew, 2015; Tan, Goh, & Sabastian, 2014; Yousef et al., 2014), which is a required accommodation under the Americans with Disabilities Act. Three different studies also mentioned the need for learners to be able to play videos on full screen, to replay, pause, rewind, fast forward, slow down, speed up, and stop (Hew, 2015; Mamgain, Sharma, & Goyal, 2014; Yousef et al., 2014). Learners also expressed a desire to download or stream videos using multiple connection speeds (Hew, 2015; Yousef et al., 2014). Further, they were interested in having lecture notes that were synchronized with the lecture (Yousef et al., 2014). Finally, given the different use patterns for tutorial videos, learner behaviour suggested the need for a convenient way to be able to navigate to specific points within videos (Kim et al., 2014)

Embedded assessment. Learners found quizzes that were embedded in the middle of video lectures to be less beneficial than quizzes presented after the lectures (Mamgain, Sharma, & Goyal, 2014). One of the primary reasons for this was that embedded quizzes required the learner to be connected to the internet to watch the videos. Alternatively, post-lecture quizzes allowed the learner to download videos when the internet was available, watch at their convenience, and return to the quizzes at a later point in time.

Conclusions, Recommendations, and Future Work

This systematic literature review provides several important considerations when preparing videos in MOOCs in a practical format for three different groups of developers. However, despite learners expressing similar sentiments regarding instructor actions, content design, navigation, and viewing options, learners' preferences toward aspects such as video style varied across the different studies. This variation supports Kizilcec, Bailensen, and Gomez's (2015) claim that there may not be a "one-size-fits-all approach" for every aspect of a MOOC video. If possible, it might be beneficial to allow different users to customize the presentation style. However, such a solution may be more feasible for some styles than others. For instance, it would be reasonable to allow learners to adjust the presence of the instructor's face, but would require considerably more effort to have both traditional lecture style approaches and the previously described Khan-style approaches for a single topic.

A major takeaway is that MOOCs appeal to a wide, diverse spectrum of users who do not represent any single, unifying perspective and often enrol in MOOCs with their own unique goals. As such, the greater the degree of customization available, the more likely the MOOC will be able to meet the needs of its users. Thus, it would be reasonable for platform developers to build their platforms to have as much flexibility as possible while accommodating many learners' specific personal and educational needs. Similarly, production teams should consider ways to film videos that would allow maximum versatility for viewers while meeting at least their minimal expectations for quality. All the while, the production team should work with the instructional team to ensure that videos are presented in digestible lengths with coherent structure and easy navigation. Finally, instructors should remember that even though they may be speaking to a camera, they are still addressing human beings with typical human needs for warmth and encouragement.

This paper only reports on a small segment of the literature collected for our systematic literature review. Many other themes have been identified that relate to best practices in MOOCs. As MOOC offerings continue to reach wider audiences, it is becoming increasingly important to organize and synthesize relevant research findings. These syntheses can help develop meaningful metrics for evaluation of MOOC quality, to help course designers and other stakeholders. As Rauf, Daud, and Said (2016) noted, MOOCs represent a promising approach to update, deepen, and broaden the education of practicing engineers. Thus, it is vitally important that future MOOCs be developed using evidence-based best practices to maximize their benefits to these learners.

References

Adams, C., Yin, Y., Madriz, L. F. V., & Mullen, C. S. (2014). A phenomenology of learning large: The tutorial sphere of xMOOC video lectures. *Distance Education*, *35*(2), 202–216. http://doi.org/10.1080/01587919.2014.917701

Allen, I. E., & Seaman, J. (2014). *Grade change: Tracking online education in the United States.*

Ausubel, D. P. (1978). In defense of advance organizers : A reply to the critics. *Review of Educational Research*, *48*(2), 251—257.

Bonk, C. J. (2008). YouTube anchors and enders : The use of shared online video content as a macrocontext for learning inter-institutional collaboration. In *American Educational Research Association (AERA) 2008 Annual Meeting* (pp. 13–24). New York, NY.

Borrego, M., Foster, M. J., & Froyd, J. E. (2014). Systematic Literature Reviews in Engineering Education and Other Developing Interdisciplinary Fields. *Journal of Engineering Education*, *103*(1), 45–76. http://doi.org/10.1002/jee.20038

Evans, B. J., Baker, R. B., & Dee, T. S. (2016). Persistence patterns in massive open online courses (MOOCs). *The Journal of Higher Education*, *87*(2), 206–242. http://doi.org/10.1353/jhe.2016.0006

Gasevic, D., Kovanovic, V., Joksimović, S., & Siemens, G. (2014). Where is research on massive open online courses headed? A data analysis of the MOOC Research Initiative. *The International Review of Research in Open and Distributed Learning2*, *15*(5). http://doi.org/10.19173/IRRODL.V15I5.1954

Grunewald, F., & Meinel, C. (2015). Implementation and evaluation of digital e-lecture annotation in learning groups to foster active learning. *IEEE Transactions on Learning Technologies*, *8*(3), 286–298. http://doi.org/10.1109/tlt.2015.2396042

Guo, P. J., Kim, J., & Rubin, R. (2014). How video production affects student engagement: An empirical study of MOOC videos. In *1st ACM Conference on Learning at Scale, L@S 2014* (pp. 41–50). MIT CSAIL, United States: Association for Computing Machinery. http://doi.org/10.1145/2556325.2566239

Haavind, S., & Sistek-Chandler, C. (2015). The emergent role of the MOOC instructor: A qualitative study of trends toward improving future practice. *International Journal on E-Learning*, *14*(3), 331–350. Retrieved from http://search.ebscohost.com/login.aspx?direct=true&db=eric&AN=EJ1067944&site=ehost-live

Hew, K. F. (2015). Towards a model of engaging online students: Lessons from MOOCs and four policy documents. *International Journal of Information and Education Technology*, *5*(6), 425–431. http://doi.org/10.7763/IJIET.2015.V5.543

Kelder, J.-A., King, C., Carew, T., O'reilly, J., Robinson, A., & Vickers, J. (2013). Evaluation of a MOOC pilot: Impacts on pedagogical and technical design and dementia education research. In G. M., H. J., & C. H. (Eds.), *30th Annual conference on Australian Society for Computers in Learning in Tertiary Education, ASCILITE 2013* (pp. 456–460). Quality Evaluation Learning and Teaching Unit, University of Tasmania, Australia: ASCILITE. Retrieved from http://www.scopus.com/inward/record.url?eid=2-s2.0-84913554068&partnerID=40&md5=320f6fd66256ba8264bef53ae9621c98

Kim, J., Guo, P. J., Seaton, D. T., Mitros, P., Gajos, K. Z., & Miller, R. C. (2014). Understanding in-video dropouts and interaction peaks in online lecture videos. In *1st ACM Conference on Learning at Scale, L@S 2014* (pp. 31–40). MIT CSAIL, United States: Association for Computing Machinery. http://doi.org/10.1145/2556325.2566239

Kizilcec, R. F., Bailenson, J. N., & Gomez, C. J. (2015). The instructor's face in video instruction: Evidence from two large-scale field studies. *Journal of Educational Psychology*, *107*(3), 724–739. http://doi.org/10.1037/edu0000013

Lai, Y.-C., Young, S. S., & Huang, N.-F. (2015). A preliminary study of producing multimedia online videos for ubiquitous learning on MOOCs. In *2015 8th International Conference on Ubi-Media Computing (UMEDIA)* (pp. 295–297). http://doi.org/10.1109/umedia.2015.7297473

Li, N., Kidzinski, L., Jermann, P., & Dillenbourg, P. (2015). MOOC video interaction patterns: What do they tell us? *Lecture Notes in Computer Science*, 197–210. http://doi.org/10.1007/978-3-319-24258-3

Lin, Y.-L., Lin, H.-W., & Hung, T.-T. (2015). Value hierarchy for Massive Open Online Courses. *Computers in Human Behavior*, *53*, 408–418. http://doi.org/10.1016/j.chb.2015.07.006

Liyanagunawardena, T. R., Adams, A. A., & Williams, S. A. (2013). MOOCs: A systematic study of the published literature 2008–2012. *The International Review of Research in Open and Distributed Learning2*, *14*(3). http://doi.org/10.19173/irrodl.v14i3.1455

Mamgain, N., Sharma, A., & Goyal, P. (2014). Learner's perspective on video-viewing features offered by MOOC providers: Coursera and edX. In *2014 IEEE International Conference on MOOC, Innovation and Technology in Education (MITE)* (pp. 331–336). http://doi.org/10.1109/mite.2014.7020298

Raffaghelli, J. E., Cucchiara, S., & Persico, D. (2015). Methodological approaches in MOOC research: Retracing the myth of Proteus. *British Journal of Educational Technology*, *46*(3), 488–509. Retrieved from https://www.learntechlib.org/p/151031

Rauf, A. S., Daud, M. F., & Said, M. N. H. B. M. (2016). MOOCs in engineering education – A literature review. In *6th Regional Conference on Engineering Education 2016* (pp. 312–315).

Sa'don, N. F., Alias, R. A., & Ohshima, N. (2014). Nascent research trends in MOOCs in higher educational institutions: A systematic literature review. In *2014 International Conference on Web and Open Access Learning (ICWOAL)* (pp. 1–4). IEEE.

Tan, L. F., Goh, K. S., & Sabastian, E. (2014). Motivation and engagement in MOOC - Teachers' perspective. In T. Supnithi, S. C. Kong, Y.-T. Wu, T. Kojiri, C.-C. Liu, H. Ogata, & A. Kashihara (Eds.), *22nd International Conference on Computers in Education, ICCE 2014* (pp. 289–297). Office of Learning Technologies, Temasek Polytechnic, Singapore, Singapore: Asia-Pacific Society for Computers in Education. Retrieved from http://www.scopus.com/inward/record.url?eid=2-s2.0-84924001594&partnerID=40&md5=5c78957758ded7eedb65d0b51254f61b

Veletsianos, G., & Shepherdson, P. (2016). A systematic analysis and synthesis of the empirical MOOC literature published in 2013–2015. *The International Review of Research in Open and Distributed Learning*, *17*(2). http://doi.org/10.19173/irrodl.v17i2.2448

Yousef, A. M. F., Chatti, M. A., Schroeder, U., & Wosnitza, M. (2014). What drives a successful MOOC? An empirical examination of criteria to assure design quality of MOOCs. In *2014 IEEE 14th International Conference on Advanced Learning Technologies* (pp. 44–48). http://doi.org/10.1109/ICALT.2014.23

Acknowledgements

This work was made possible by a grant from the National Science Foundation (DGE 1544259). Any opinions, findings, and conclusions or recommendations expressed in this material are those of the authors and do not necessarily reflect the views of the National Science Foundation.

Concept Inventories for Control Engineering: Past and Future

Ferdinand Kieckhäfer
Hamburg University of Technology, Hamburg, Germany
ferdinand.kieckhaefer@tuhh.de

Julie Direnga
Hamburg University of Technology, Hamburg, Germany
julie.direnga@tuhh.de

Dion Timmermann
Hamburg University of Technology, Hamburg, Germany
dion.timmermann@tuhh.de

Christian Kautz
Hamburg University of Technology, Hamburg, Germany
kautz@tuhh.de

Abstract: *There are many publications about innovations in control engineering education. However, in order to substantially advance control engineering education a deeper discourse about its teaching is needed. This discourse would greatly benefit from having a common instrument to measure the success of the innovations. This can be done by a concept inventory. In physics education, the Force Concept Inventory started such a process. This paper proposes properties of a suitable concept inventory in control engineering, compares them to existing concept inventories in control engineering and concludes that a new concept inventory has to be developed. A research design for the development of this concept inventory is proposed taking into account the control engineering-specific challenges.*

Introduction

Control engineering is an important part in many engineering curricula as can be seen by the importance of automation in the industry. Many students at Hamburg University of Technology (TUHH) seem to have problems learning the basic concepts of control engineering in an introductory course (Eichler et al., 2013). There is no indication that this is significantly different at other universities.

The publication archive of the International Federation of Automatic Control (IFAC) shows many publications tagged 'control education'. This shows interest in improving teaching in control engineering. However, most of these publications are individual projects focusing on improving a certain course and mostly evaluating the success by observations of the instructor and self-reported student impressions. Furthermore, these papers show no coordinated efforts in comparing courses and innovations across different institutions. In summary, it is difficult for instructors to evaluate their own teaching and identify the teaching innovations most promising for improving their courses.

A good solution to this problem would be a concept inventory. A concept inventory for introductory control engineering courses could spark the discourse about teaching in control engineering across institutions like the Force Concept Inventory (Hestenes et al., 1992) did in physics education (Libarkin, 2008). A concept inventory is also an easy-to-use instrument for instructors, especially those new to engineering education research, to get another view on the understanding of their students. This new perspective could give them food for thought concerning their own teaching.

There where two initiatives to create a concept inventory in control engineering. One resulted in the Control Systems Concept Inventory (CSCI) (Bristow et al., 2012) while the other did not yield a final instrument (Lundberg, 2012).

The aim to provide a concept inventory as proposed above raises two questions:

- Q1: What are the properties of a good concept inventory for introductory control engineering courses?
- Q2: Does the existing concept inventory (the CSCI) fulfil these properties?

Each question will be answered in a separate section in this paper. Since the answer to the second question is negative, another section will introduce a research design for developing a suitable concept inventory in control engineering taking into account the domain's special difficulties.

Theoretical framework

Measuring conceptual understanding is a good way to gauge student understanding (Mazur, 1997, pp. 5-7). Concept inventories, like the Force Concept Inventory (FCI, see Hestenes et al., 1992), are widely used instruments to measure conceptual understanding (Libarkin, 2008). As teaching is essentially helping student understanding, concept inventories are one possible measure for the quality of teaching (Evans et al., 2003). Going further, it is possible to compare courses and teaching methods using concept inventory scores (Hake, 1998; Garvin-Doxas et al., 2007).

Q1: Properties of a good concept inventory

General properties

In order to foster engineering education research as described in the introduction of this paper, a concept inventory should

1. measure expert-like thinking
2. be easy to administer and evaluate
3. be applicable to many different courses at different institutions
4. provide comparable results
5. allow valid interpretations of the test results with regard to the intended use

Property 1 is the agreed upon measure of a concept inventory as described in e. g. Adams and Wieman (2011). Properties 2-4 all help spreading the thinking and the discussion about teaching in control and thereby fostering control engineering education research. The concept inventory should not take too much time and effort to use. An easy administration of the test and a simple evaluation of the answers (property 2) lead to more instructors using it. This is especially important for reaching instructors with no prior experience in evaluating their teaching. This property is already fulfilled by using concept inventories with multiple choice questions that do not require any other special arrangements like watching an experiment. Being applicable in many courses and institutions (property 3) also leads to a greater dissemination of the concept inventory. Comparability of test results (property 4) is another factor that sparks the discourse about teaching because it makes comparing teaching methods possible by using the results of concept inventories as indicators of successful teaching. In order to make the results comparable, two points have to be taken into account. First, the result should be a number. This is fulfilled by concept inventories being multiple choice tests. Second, the test results have to be normalized for initial knowledge because the entry level of knowledge can be different from course to course (Adams and Wieman, 2011, p.6). This can be done by using a pretest to measure the understanding of the students entering the course.

The fulfilment of property 5 depends not only on the right development of the test but also on the usage and administration of the test. The developer of the test has to state the usage and administration for which theory and evidence show validity of the interpretations of the test results. If the test is used and administered in a different way conclusions drawn from the results cannot be assumed to be valid (American Educational Research Association., 2014, Ch. 1). In our case, the intended usage of the test is to measure expert-like thinking (property 1) about basic control engineering among groups of university students. In order to give valid interpretations, the test has also to be a reliable instrument for measurement (American Educational Research Association., 2014, Ch. 2). The other properties (1-4) have to be taken into account when thinking about the validity of the conclusions drawn from the test results. Especially the test population matters a lot in these issues as described in the next section.

Control engineering-specific properties

Now that some general properties for concept inventories have been established, one can look at the implications of these properties on a concept inventory in control engineering with regard to the intended usage described above.

Most concept inventories use some kind of example system which illustrates the item at hand. This makes it possible to indirectly ask about a concepts, yielding richer data which is easier to analyse (Cousin, 2009, p.81). Since control engineering plays a role in all engineering disciplines, introductory control engineering courses are often taken by students of different study programs. This has some implications on the choice of example systems. One should choose these systems in a way that they do not favour students from a certain engineering discipline. For example, items using spring-mass-damper-systems favour mechanical engineers whereas electrical engineers could do worse despite having understood the control concept in question. The same is true about cultural differences. Some people understand an example system intuitively because of familiarity with the situation from daily life while others struggle with it. Ice skating would be an example for such a setting in a physics concept inventory.

Another important aspect is the choice of concepts covered in the inventory. In order to measure the understanding of basic control engineering, the concept inventory should cover the concepts related to introductory control engineering courses. Since it should also be applicable at different institutions and in different courses (property 3), it should also be restricted to the most commonly used and agreed upon concepts (Adams and Wieman, 2011, p.11). It should not test concepts that are not covered by most courses.

As mentioned before, the normalization for the initial level of knowledge makes it necessary to administer a test before the instruction, the pretest. The items of a pretest should be understandable for the students prior to instruction in order to deliver reliable results and allow valid interpretations about the initial knowledge. The best case scenario is that the concept inventory used as a posttest is also suitable as the pretest making it easy to analyse the data. Therefore, a new concept inventory in control engineering should be developed with the pretest use in mind. If a concept inventory is not possible to use as a pretest, there is the possibility to use different tests as pre- and posttest (Direnga et al., 2014). If this is the case, one possible pretest in an introductory control course could be the Signals and Systems Concept Inventory (Wage et al., 2005), because knowledge from signals and systems is often needed in introductory control courses.

Q2: The Control Systems Concept Inventory (CSCI)

The existing concept inventory about control engineering, the CSCI, was developed to evaluate an introductory course in control engineering for mechanical engineers at University of Waterloo (Bristow et al., 2012). Because of this target population, only example systems from mechanical engineering are used. Therefore, the CSCI does not fulfil property 3. Since the test was only administered to 147 students, it was possible to add an open ended question to enrich the gathered information with qualitative data about the students' thinking in control engineering. As open ended questions are hard to evaluate as part of an automated process, the CSCI also violates property 2.

The CSCI consists of several segments, each one containing one example systems and sev-

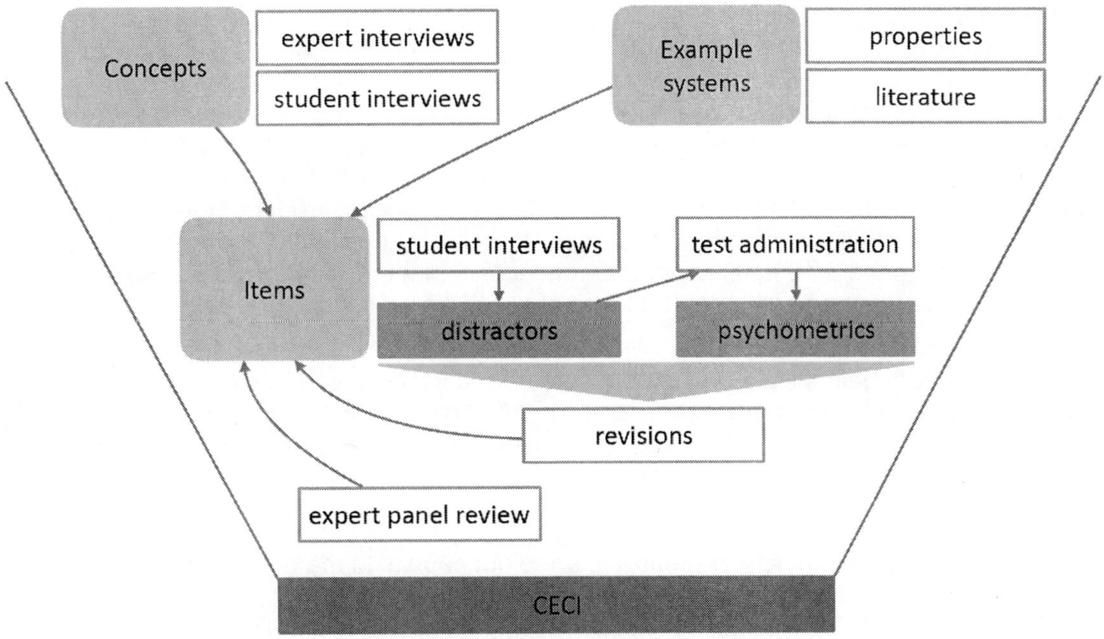

Figure 1: Road map to the CECI

eral items addressing this system. On the one hand, this is nice because the example system does not change for each item allowing for shorter items (Libarkin, 2008, p.7). However, on the other hand, this segmentation is a drawback as it may lead to several wrong answers when a student does not understand the example system. Consequently, all items in a concept inventory should stand for themselves (Engelhardt, 2009, p.11).

Additionally, most segments contain so-called "mini-tutorials", intended to make the pretest use of the CSCI possible (Bristow et al., 2012). The "mini-tutorials" often explain control theory which means that the test is not only assessing but also teaching. Such a test would not only measure the understanding of the test takers but also their ability to learn new content while taking the test. Another aspect of these "mini-tutorials" is that reading them takes a lot of valuable testing time (about 3 of 12 pages are "mini-tutorials"). This is problematic because a concept inventory should not take longer than 30 minutes for all students except the slowest ones (Adams and Wieman, 2011, p.12).

In conclusion, the CSCI is not suitable as a concept inventory for evoking a discourse about teaching in control engineering. A new concept inventory in control engineering is needed. Therefore, we plan to develop the Control Engineering Concept Inventory (CECI). In the following section, the research design for this development is laid out.

Research design for developing the Control Engineering Concept Inventory (CECI)

A graphical representation of the research design for developing the CECI is shown in Figure 1. The following three subsections describe each of the key components, Concepts, Example systems, and Items. The subsections Concepts and Items are heavily based on observations and suggestions by Adams and Wieman (2011), Engelhardt (2009) and Libarkin (2008). The fourth subsection explains the reasoning behind the overall structure of the research design.

Concepts

The first step is to determine the concepts that should be covered by the CECI. For this, a panel of experts is used. Several experts in control engineering (mainly instructors) will be asked in interviews about the concepts taught in introductory control engineering courses. Each interview will yield a list of concepts and ratings on the importance and the difficulty (from the view of the instructor) of the concepts. These lists will then be used to construct a compiled list of concepts ordered by importance which is then shared with the experts so they can make remarks. When all experts agree on a final list, the 5 to 10 most important concepts are the concepts covered in the CECI. This process ensures that the CECI can be used in many different courses and institutions because the concepts covered are agreed upon by instructors of basic control engineering courses from different institutions. A positive aspect of the restriction to the 5-10 most important concepts is that there can be at least 3 items per concept as suggested by Engelhardt (2009) without creating a test that takes to much time.

Furthermore, the final list of concepts will then be discussed in think-aloud-interviews with students in order to gain insight on the concepts from a novice's perspective. This shows the difficulties of the concepts from the test taker's point of view giving starting points for items. It also generates statements about the concepts in student language making it easier to draft items which are understandable for students.

Example Systems

The search for suitable example systems is carried out parallel to the determination of the concepts. This is different from most other concept inventory development processes. In many other fields, the example systems in concept inventories are quite simple and the students have experiences with these systems from daily life observations, i.e. a ball rolling from a table illustrating an item concerning Newton's second law. In control engineering, most example systems commonly used in textbooks are not suitable for the CECI because of two different reasons. The first case is the use of examples from a certain engineering discipline, i. e. spring-mass-damper-systems for mechanical engineers or electric circuits for electrical engineers. The other case is the complexity of some example systems, for example an airplane, which makes it hard to grasp the dynamics of the system while taking the test. Both cases are not desirable, as they lead to an item that not only measures the understanding of the underlying control concept but also the proficiency with the example system. Therefore, a student may give the wrong answer solely because of unfamiliarity with the example system despite understanding the underlying control concept.

The first step of this search is to find out what makes an example system a good example system. This first step has two advantages: first, it helps to better understand example systems and their advantages and problems. Second, a definition of a good example system helps to find a metric for judging the quality of example systems. This metric helps to look for high-quality example systems in textbooks and collections of exercise problems. The insights gained from the process of defining good example systems and the inspiration from the literature research can also spark ideas for good example systems.

It very well could be that this search for example systems does not result in enough example systems or fails to find example systems for a certain concept. In this case, it is possible, but not preferable, to use items that ask about the concept on an abstract level without using an example to illustrate the item and the concept.

Items

After establishing the list of concepts and finding a pool of example systems, the actual items are developed. For each concept at least 3 questions illustrated by example systems are needed for the final concept inventory. An aspect one should keep in mind is a broad range of difficulties for the items overall and preferably also within the items concerning a single concept. It is also better to use items that only ask indirectly about a concept as mentioned above. Asking indirectly means that the question asks for the solution of a problem or for a prediction about a future state of the example system in a way that makes the understanding of the underlying concept necessary in order to give the right answer.

The drafted questions are used as open ended questions in think-aloud-interviews with students. The data collected in these interviews is used to rework the questions in case the question was not understood in the intended way. Often it is useful to rephrase the questions in the words the students use to achieve understanding (Adams and Wieman, 2011; Libarkin, 2008). If it is not possible to rework the question in a way that students show the intended understanding of the question in further interviews the question should be discarded. The students' answers in the think-aloud-interviews are used as answer options. This has the advantage that the right answer and the distractors are in student language. The whole process of interviewing is carried out for smaller batches of drafted questions allowing shorter interviews and giving the flexibility to draft new questions while already interviewing.

When a larger number of items has passed the process of interviewing and finding answer options the items can be administered in written form to larger groups of students. This makes it possible to calculate psychometric measures from classical test theory and item response theory to further evaluate the items(Engelhardt, 2009). A larger number of items is needed because some of these measures, the discrimination index for example, are about the performance of single items with regard to the score on other items. Items that perform not in the acceptable range of these measures have to be reworked or discarded. The reworking can either mean to start again with interviews or to just change the question or the answer options and try it again in a written test.

At the end, the final test will be given to experts in order to see if the test is still compatible with their list of concepts. This procedure also ensures that the items are all correct from an expert's viewpoint.

Structure of the research design

The overall idea is to do as much qualitative work in the beginning in order to need less iterations in the Item segment. With good insight in the important concepts, both from the instructors and from the students, it is more likely to design good items right from the start. This shortens the time spent with revising items which is probably the longest segment of the test development. For example, the student interviews about the concept list give possible topics for questions. Furthermore, the interviews allow to formulate the questions in student language right from the start fostering student understanding of the questions (Adams and Wieman, 2011; Libarkin, 2008).

In control engineering, it seems to be difficult to find example systems suitable for concept inventories. Therefore, it makes sense to analyse and search for example system in an independent process. Solving this problem before the item drafting process allows again for better items right from scratch.

This whole process (see Figure 1) hopefully leads to a set of items that make up a concept inventory allowing for valid assumptions on the proficiency of students in basic control engineering.

Summary and conclusion

In this paper, the properties of a concept inventory that fosters engineering education research have been established. The one existing concept inventory in control engineering does not fulfill these properties. Furthermore, the implications of the general properties for the special case of a concept inventory in control engineering have been analysed. In principle, it seems possible to develop a concept inventory that does not violate these properties. Therefore, we decided to develop a new concept inventory in control engineering, the Control Engineering Concept Inventory (CECI). The research plan for developing the CECI has been mapped out taking into account the challenges described in this paper. Once the CECI has been developed and successfully tested, it can be easily used by instructors all over the world to assess their teaching in control engineering giving them new insights on their students' understanding. Furthermore, we hope this will allow to compare different methods of teaching, sparking a substantial advance in control engineering education.

We invite all education researchers and everyone interested in this topic to make suggestions concerning the research plan. We further welcome instructors of introductory courses in control engineering to contact Ferdinand Kieckhäfer (`Ferdinand.Kieckhaefer@tuhh.de`) if they are interested in taking part as an expert or using the test or selected questions in their courses.

References

Adams, W. K. and Wieman, C. E. (2011). Development and validation of instruments to measure learning of expert-like thinking. *International Journal of Science Education*, 33(9):1289–1312.

American Educational Research Association. (2014). *Standards for educational and psychological testing*.

Bristow, M., Erkorkmaz, K., Huissoon, J. P., Jeon, S., Owen, W. S., Waslander, S. L., and Stubley, G. D. (2012). A control systems concept inventory test design and assessment. *IEEE Transactions on Education*, 55(2):203–212.

Cousin, G. (2009). *Researching learning in higher education : an introduction to contemporary methods and approaches*. The staff and educational development series. Routledge, New York u.a.

Direnga, J., Timmermann, D., Brose, A., and Kautz, C. (2014). A statistical method for assessing teaching effectiveness based on non-identical pre-and post-tests. In *Proceedings of the SEFI 2014 Annual Conference*, Birmingham, UK.

Eichler, A., Hoffmann, C., Kautz, C., and Werner, H. (2013). Design of tutorial activities and homework assignments for a large-enrollment introductory course in control systems. In *Advances in Control Education*, volume 10, pages 43–48.

Engelhardt, P. (2009). An introduction to classical test theory as applied to conceptual multiple-choice tests. In *Getting Started in PER*, volume 2.

Evans, D. L., Gray, P. M. G. L., Krause, S., Martin, J., Midkiff, C., Notaros, B. M., Pavelich, M., Rancour, D., Reed-Rhoads, T., Steif, P., and others (2003). Progress on concept inventory assessment tools. In *Frontiers in Education, 2003. FIE 2003 33rd Annual*, volume 1, pages T4G–1. IEEE.

Garvin-Doxas, K., Doxas, I., and Klymkowsky, M. W. (2007). *Ed's tools: A web-based software toolset for accelerated concept inventory construction.*

Hake, R. R. (1998). Interactive-engagement versus traditional methods: A six-thousand-student survey of mechanics test data for introductory physics courses. *Am. J. Phys.*, 66(1):64.

Hestenes, D., Wells, M., and Swackhamer, G. (1992). Force concept inventory. *The Physics Teacher*, 30(3):141–157.

Libarkin, J. (2008). Concept inventories in higher education science. In *BOSE Conf.*

Lundberg, K. H. (2012). Personal communication.

Mazur, E. (1997). *Peer instruction: A user's manual.* Prentice Hall.

Wage, K., Buck, J., Wright, C., and Welch, T. (2005). The signals and systems concept inventory. *IEEE Transactions on Education*, 48(3):448–461.

How do students meaningfully interpret the role of empathy in engineering? A social phenomenological study

Michael A. Brewer, Nicola W. Sochacka, Joachim Walther, & Shari E. Miller
University of Georgia, Athens, United States
mabrewer@uga.edu

Abstract: *This study uses a social phenomenological approach to examine how students experience and give meaning to empathy in engineering education and practice. Data for the investigation included students' written reflections on a series of four empathy modules, which were integrated into a compulsory, sophomore engineering and society course in a mechanical engineering curriculum at a large state university in the southeastern United States. The findings indicate varying degrees of dissonance between the empathy-based exercises and students' preconceptions of engineering learning and practice. Consequently, students interpreted the role of empathy in engineering in diverse ways, including, for example, a tool with which to handle "difficult" stakeholders and a professional communication skill. More broadly, this study shows how teaching empathy has the potential to expose, and potentially disrupt, dominant cultural norms in engineering education, such as the ideology of meritocracy and the perceived primacy of technical over social competencies.*

Introduction

Leading design companies (Battarbee, Fulton Suri, & Gibbs Howard, 2014; Duhigg, 2016; Hansen, 2010) and cutting-edge university programs (Stanford d.school, 2017) across the United States point to empathy as one of the most crucial ingredients for professional success. A review of the many books on empathy that have been published in the popular literature over the past decade conveys a similar message—empathy is central not only to healthy human relationships (McLaren, 2013; Perry & Szalavitz, 2011) and but the peaceful continuation of civilization (De Waal, 2009; Rifkin, 2009; Trout, 2009).

These discourses have not been lost on engineering education. As reported by Hess and Fila (2016), references to empathy in the American Society for Engineering Education's (ASEE) conference proceedings have increased fourfold over the past five years. In this paper, we share preliminary results from a social phenomenological study (Schütz, 1967; Vagle, 2014) of how undergraduate engineering students meaningfully interpret the role of empathy in engineering.

Engineering schools fail to teach empathy

In the United States, the Code of Ethics for Engineers requires that engineers "[h]old paramount the safety, health, and welfare of the public" (National Society of Professional Engineers, 2007). It stands to reason that in order to fulfil this duty, engineers must be aware of, concerned with, and attuned to, society's needs. Cech (2014) explored this fundamental relationship between engineers and society in a longitudinal study of how undergraduate students' public welfare beliefs change during their engineering education. In other words, Cech sought to investigate if engineering education in the United States is "actually succeeding at nurturing students' sense of professional responsibility to the welfare of the public" (p. 44). To do so, Cech surveyed 326 engineering students at four US colleges "from students' freshman year through their graduation and, for some respondents, into the workforce" (p. 46). The surveys asked students to rate, using a Likert scale, the relative importance of four constructs related to professional/ethical responsibilities, understanding the consequences of technology, understanding how people use machines, and social consciousness. Statistical analysis of the survey results revealed that student perceptions of the importance of all four constructs decreased over the course of their education. In other words, students' interest in public welfare concerns seemed to decline as they progressed through their engineering studies. When Cech's study was picked up by the media, reporters interpreted her findings as a failing on the part of engineering programs to teach students empathy (Marcus, 2013; May, 2014).

Cech attributed her research findings to a "culture of disengagement" in engineering education, which she described as having three ideological pillars:

> . . . the ideology of depoliticization, which frames any "non-technical" concerns such as public welfare as irrelevant to "real" engineering work; the technical/ social dualism, which devalues "social" competencies such as those related to public welfare; and the meritocratic ideology, which frames existing social structures as fair and just. (p. 45)

We hypothesize that these three pillars are particularly salient to efforts to foster empathy in undergraduate engineering programs as they highlight the risk that empathy, as a "non-technical" concern, will be interpreted by students as irrelevant, de- or under-valued, or potentially even deemed wholly unnecessary in light of the presumed fairness of existing social structures.

Research question

In our experience, the introduction of empathy-based activities into an engineering classroom evokes a range of responses from students (Walther, Miller, & Sochacka, 2016). In order to effectively foster empathetic engineers, it is important for engineering educators to understand the nature and origin of these responses, as well as the opportunities they present for transformational learning. With this goal in mind, this study examined the following research question:

> Given the opportunity to engage in and reflect on empathy-focused, active-learning modules in a classroom setting, how do undergraduate engineering students meaningfully interpret the role of empathy in engineering education and professional practice?

Theoretical framework

In our prior work, we developed a theoretical model of empathy in engineering (Walther, Miller, & Sochacka, 2017). This model conceptualizes empathy as a skill, a practice orientation, and a professional way of being. As shown in Figure 1, this model is purposefully composed to illustrate the mutually dependent and supportive nature of each dimension without ascribing a conceptual hierarchy or developmental trajectory.

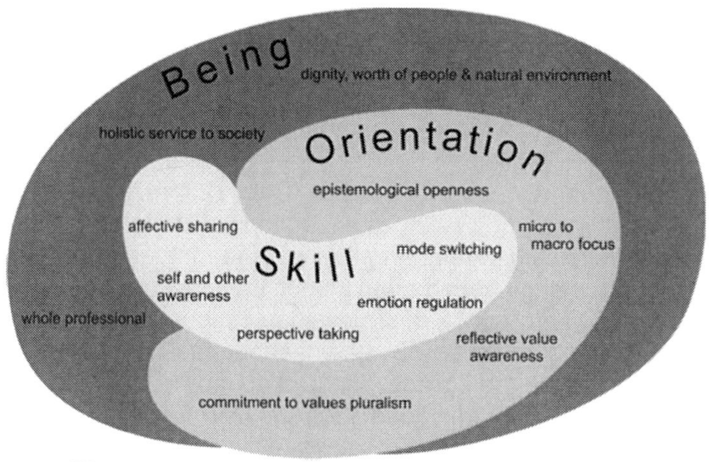

Figure 1: A model of empathy in engineering

The skill dimension comprises five distinct, socio-cognitive processes that interact with each other to form the foundation for empathic communication, relationship building, and decision-making. The orientation dimension captures a range of mental dispositions that influence how engineers or engineering students engage in practice situations. These orientations around epistemology and the role of values influence efforts to empathically engage with others. Finally, the being dimension highlights the need to situate empathic skills, practice orientations, and their development within a contextualizing framework of broader commitments to self, society, and the natural environment. For more details on this model, we refer readers to Walther et al. (2017).

Study setting

Based on the theoretical model in Figure 1, we designed and implemented a series of empathy modules into a sophomore engineering and society course, which is a compulsory component of the design sequence in a mechanical engineering curriculum at a large state university in the southeastern United States.

The course design combines group-based, open-ended design challenges that focus on problem framing in complex, socio-technical contexts, with a series of readings and critical discussions that provide the theoretical and conceptual underpinnings of socio-technical systems. The readings and facilitated group discussions guide students to understand engineering work as inherently socially-situated (e.g., Jonassen, Strobel, & Lee, 2006) and explore aspects of socio-technical complexity such as the relationship of technology to politics and power (e.g., Winner, 1988) or the challenges of sustainable development (e.g., McDonough, Braungart, & Clinton, 2013). Other readings invite students to critically question assumptions around technological determinism (e.g., Pool, 1999) and perceptions of engineering expertise (e.g., Pacey, 2001), and explore the relationship between engineering work and a broader service to society (e.g., Bornstein, 2007).

In this context, students were offered the opportunity to participate in a series of four empathy modules that focused on: i) Encountering Others, ii) Self and Other Awareness, and Affective Sharing, iii) Affective Responding, and iv) Mode Switching. Each module was one hour and 15 minutes long and began with an introduction to the focus of the module. Next, students were given instructions to explore a particular facet of empathy through skill-based exercises. For example, in Module 3, students were shown how to *attend* to their partner (i.e., give their undivided attention), *paraphrase* what they heard their partner say, and *reflect back feelings* they observed or heard. After the skills-based exercises, the students were invited to apply these skills in a real-life scenario that involved engineers, other professionals, and members of the community. These role-plays were designed to give students an opportunity to contextualize and explore empathy skills in situations they might encounter as practicing engineers.

After each module, students were required to answer a series of reflection questions and submit their responses on-line to the eLearning Commons (eLC) site for the course. Students received 1% for submitting each reflection (4 reflections = a total of 4%). The content was not graded. Instructors paid particular attention to reinforcing their appreciation for both positive and critical responses by selecting a range of comments from the students' reflections to share with the class at the beginning of each subsequent module.

The reflection prompts for the first three modules invited students to think back through the module and discuss moments of enjoyment or discomfort, and reflect on the role of empathy in engineering. The reflection prompt for the last module was designed according to Walther, Sochacka, and Kellam's (2011) emotional indicator approach. Examples of both types of prompts are given in Table 1.

Table 1: Examples of reflection prompts

Reflection prompt for Module 2	Reflection prompt for Module 4
Today you participated in exercises about **self and other awareness** and **affective sharing.** 1. Describe your feelings during the body proximity exercises. Why do you think you felt this way? 2. What was your experience as the engineer or stakeholder during the interview exercises? 3. What did you learn from this module? What does it mean for you becoming a professional engineer? You are not limited to just answering these questions in your reflection. Please include any other thoughts and feelings you think are important. Submit your reflection (about two pages) as a PDF file on ELC.	Read through the following statements and think of specific incidents they bring to mind. Write about them. You don't have to directly or specifically answer the statements. • There was a moment during today's module when I started to realize that... • During today's module, I struggled to get my head around... • When my partner responded analytically to my story, I was surprised to feel that... • When my partner responded empathetically to my story, I was surprised to feel that... • As the engineer/stakeholder in the role play, I suddenly understood how... • During the role play, I found it challenging to... Submit your reflection (about two pages) as a PDF file on ELC.

Methodology and methods

Data collection and analysis

Data for the study included the students' empathy module reflections for the Fall 2015 (n=110) and Spring 2016 (n=36) semesters. These two iterations of the course were taught by the third and second authors, respectively. The research team obtained IRB approval to collect and analyze these data.

The data were imported into NVivo 11, a qualitative data analysis software package, and analyzed through the lens of social phenomenology, which is described in more detail in the following sections. Data analysis was conducted primarily by the first author. The research team met weekly to review emergent codes, discuss methodological questions, and collaboratively code selected excerpts from the raw and analyzed data to interrogate and validate the findings (Walther, Sochacka, & Kellam, 2013).

Social phenomenology

We used the social phenomenology of Alfred Schütz (1967) as a methodological framework to examine how students meaningfully interpret the role of empathy in engineering. The following paragraphs discuss specific aspects of this methodology that are relevant for the preliminary research findings we present in this paper.

Building primarily on the work of Edmund Husserl (Husserl & Hardy, 1999; Zahavi, 2003) and Max Weber (Heydebrand, 1994), Schütz's phenomenology of the social world deals with "social actions," which are meaningful experiences that have some reference to other people in order to understand the subjective meaning these experiences have for the actor (Schütz, 1967). Our decision to approach the data from a social phenomenological perspective (rather than some other variant of phenomenology) was driven by two main considerations, first, the uniquely social nature of the experiences presented to the students, and second, the power of engineering culture in shaping those experiences. By emphasizing the intersubjective and contextual nature of the social world (Schütz, 1967), we were better able to investigate how students made sense of the empathy exercises and integrated those interpretations into their pre-existing and developing identities as engineers, as opposed to generally describing empathy (i.e., as in phenomenology) or experiences of it (i.e., as in phenomenography).

The unit of analysis in social phenomenology is the *intentional relationship* between a subject and an object (Schütz, 1967; Vagle, 2014). For example, **Juliet** (the subject) is **in love** (intentional relationship) with **Romeo** (the object)—social phenomenologists study the in-loveness. Intentionality refers to the "aboutness" or "orientation towards" something (Smith &

McIntyre, 2012, p. 336). In our example, Juliet's mental act of love is "about", "oriented towards", or directed at Romeo. It is through such intentional processes that meaning is constructed. The first and most basic intentional act is to give one's attention to an experience, selecting it out of many candidates as the meaningful experience of the moment. A further intentional act might be to take an attitude toward the experience, or to generalize it into an abstract conceptual object (Schütz, 1967).

When trying to understand what subjective meaning an intentional relationship has for another person, we must know something about the *meaning-context* (Schütz, 1967, p. 75) in which that relationship exists. The meaning-context is the synthesis of multiple experiences into a unified object of attention. For an engineering student, experiences such as "applying for scholarships" or "studying calculus" would not make sense in a vacuum; they are only meaningful in the context of "pursuing a degree" or "becoming and engineer." These examples of meaning-contexts have the additional characteristic of being motivated or goal-oriented, and we can call such a meaning-context a *project*(Schütz, 1967). An important feature of meaning-contexts is that they are intersubjective; they are constructed under social influence in reference to the same world, and are therefore shared to some degree.

Social phenomenology distinguishes between intentional relationships with human ("Others") and non-human objects. More specifically, Others are conceived of as other Selves. When "I" recognize that another person is a Self, which has subjective experiences just like I do, I can have attitudes about them, and intentional relationships with them. This is called Other-orientation (Schütz, 1967).

Intentional relationships with Others are categorized by how readily they can be experienced. In face-to-face interaction, the two parties directly experience one another as *consociates*. This is called a We-relationship (Schütz, 1967, p. 164). When people are alive at the same time but do not directly experience one another, they exist as *contemporaries* in a They-relationship (Schütz, 1967, p. 183). In a They-relationship, the existence of other people is known but not directly experienced, and so intentional social relationships of this kind rely on *ideal types* to be made meaningful. For example, there are postal workers and farmers with whom "I" have an indirect social relationship, but they are anonymous to me; I know of them as simplified ideal types who must exist somewhere. Ideal types serve as interpretive schemes for understanding parts of the social world that cannot be readily accessed as in a We-relationship.

Findings

Our analysis of the student reflection data revealed three categories of intentional relationships. These relationships were between the Self (i.e., the students) and Learning, Others, and Content (i.e., the notion of empathy as interpreted by the students), respectively, and are illustrated in Figure 2. The following sections describe the first two of these intentional relationships, that is, between Self and Learning, and Self and Others.

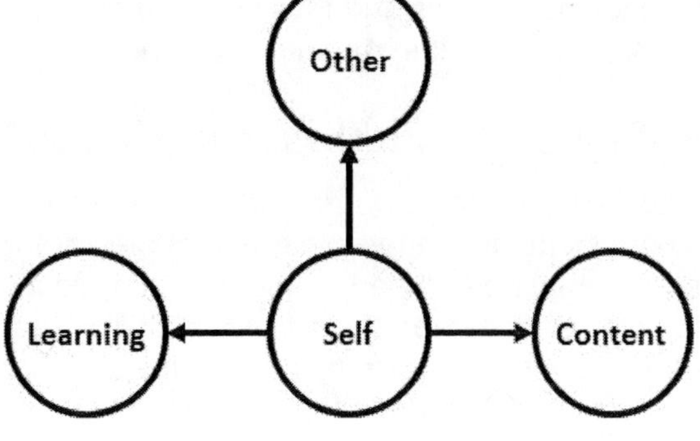

Figure 2: Three types of intentional relationships identified in the data analysis

The act of learning: New experience in context

As discussed above, in Schütz's framework of social phenomenology experiences are interpreted within a meaning-context. A mechanical engineering student attending class is operating within a meaning-context in which attending class is a reasonable thing to do. This meaning-context is also likely to be associated with a project (such as "becoming an engineer"), which motivates the act of attending class.

Schütz (1967) described how projects carry with them an "aura of expectations", which are influenced by "the plan or project we have in mind" (p. 61). When these expectations are apparently fulfilled, it is "the result of the 'feeding' of the project into this particular moment" (p. 61). Interpreting the acts of learning recounted by the students in these terms revealed a range of intentional relationships students had with the empathy-focused exercises.

We begin with a series of quotes from Samir, a sophomore mechanical engineering student (all student names referred to in this paper are pseudonyms).

> *Thursday's in-class modules were interesting in that they were different from what one typically expects in a college classroom, let alone an engineering one. Although I was aware that we would be participating in communication activities, I didn't expect them to be as enjoyable and challenging . . . I was fairly surprised at how much this activity resembled real life scenarios and was a good way to start me thinking about my interactions with people in the professional world. Something as simple as being able to approach and talk to others can be extremely useful (Samir, Module 1 reflection).*

In the first part of the excerpt, we see that Samir held certain expectations within the meaning-context of "a college classroom," particularly "an engineering one." His reflection indicates that the empathy exercises did not fulfill those expectations, and so they did not readily fit into the meaning-context. However, through further consideration of his expectations of "the professional world", Samir was able to make sense of the experience as part of his larger project of becoming a professional. This resulted in a modification of the meaning-context to accommodate the new experience and relate it to his personal motivation.

While many of the students experienced a dissonance between the empathy modules and what they typically expect in an engineering course, not all students were able to reconcile this dissonance, as illustrated in the following excerpt.

> *It seemed like you didn't want us to solve the problem and just make small talk. Next time you do the activity I think it should be an investor, an engineer and a local talking discussing how to solve the food/poverty/job issue in* [city] *(Justin, Module 1 reflection).*

This quote suggests that Justin came to class expecting to solve problems, an expectation influenced by a meaning-context in which problem solving is the motivation for engineering. To Justin, actions that are not problem solving are not engineering, and therefore not appropriate for an engineering classroom. His meaning-context remained unchanged by the new experience, as further indicated by his suggestion that the experience should be modified to fit into what he deems a suitable meaning-context.

Relationship with the Other: Who are They, and what do They want?

The intentional relationships between participants and Others are particularly relevant in the context of the empathy modules because the empathy modules are *about* relationships with others. The relationships with others described in the student reflections were complex and diverse but were generally We-relationships (face-to-face) or They-relationships (distant, ideal typical). Relevant We-relationships were, by definition, limited to interactions that took place in the classroom. The range of They-relationships was comparatively vast—the reflection prompts did not specify any particular people, and so students who talked about Them were obliged to define or assume who They are. A special in-between case existed in the context of the stakeholder role-plays, in which participants effectively had face-to-face relationships with anonymous ideal types, who could be portrayed with varying degrees of seriousness and intimacy. The case of intentional relationships with others in role-play is unusual, but we argue that they can still be studied with methodological rigor considering Schütz's (1967) acknowledgement that We-and-They-relationships are limiting cases on a spectrum. In fact, in the purest They-relationships in the data, Others were not completely anonymous, because the participants expect to someday engage in We-relationships with representatives of these ideal types and are in the process of learning how they will interact with them as real individuals.

The following quotes represent a pattern in the data that concerns how students conceptualize the relationship between engineers and members of the general public ("Stakeholders"), and how empathy can be employed to facilitate these relationships.

In the next excerpt, Blair recounts a role-play experience in which he experienced a We/They-relationship with a certain character—John, a struggling single father who is afraid that his water rates will go up as a result of a city project. Blair played the role of an engineer on the project. His experience led him to ascribe certain characteristics to a particular ideal type:

> *I remained professional, but John, who would experience very real consequences from higher water prices, began to get upset as he told me his opinions about the water crisis. I listened and tried to take his suggestions and opinions into account, but I found it difficult at times to understand everything he was saying because he was getting emotional. From this, I learned that stakeholders may not be clear in their explanations for a few reasons. First of all, most of the time, stakeholders have little technical expertise. Second of all, stakeholders are not always calm when they tell you what they want (Blair, Module 2 reflection).*

To Blair, John was part of the vague ideal type "Stakeholder". He exhibited the traits of Stakeholders, which are lack of expertise and lack of emotional self-control. Blair's experience in the role-play exercise created or reinforced this typification of Stakeholders.

We suspect that Blair would describe his intentional relationship with John as "empathy". He had previously described himself as "empathic and hav[ing] a natural tendency for empathy," and concluded his reflection with a comment on what an empathetic relationship with Stakeholders meant to him:

> *This exercise made me see that sometimes stakeholders are difficult to understand, or sometimes do not know what they want themselves. Additionally, some stakeholders see only one solution to a problem and are unwilling to change their minds or see any other solutions. However, through empathy, an engineer can successfully navigate these situations. The engineer may ask for clarification, try to see the situation from another perspective, or attempt to empathically persuade the other party (Blair, Module 2 reflection).*

From these excerpts, we see how Blair considered his relationship with the Other as an Engineer-Stakeholder relationship. As the engineer, his side of the relationship was to be an objective expert with a cool head and the ability to help and persuade stakeholders who do not know what they want.

Implications for engineering education

These research findings have important implications for engineering education. First, we note how adopting a social phenomenological approach allowed us to examine students' experiences of the empathy modules in relation to broader cultural norms and pre-existing, inter-subjective understandings about what engineering learning and practice are or should be. Viewed in this way, we see that teaching empathy to engineering students does not occur in a vacuum—a fact which, in our view, creates both challenges and opportunities for instructors.

Earlier, we hypothesized that Cech's (2014) three ideological pillars, which she argued support and reinforce a culture of disengagement in engineering education, are particularly salient to efforts to integrate empathy training into undergraduate programs. Here we return to that discussion in light of our study's findings.

Cech's first ideological pillar, "the ideology of depoliticization", concerns the framing of non-technical concerns as irrelevant to "real" engineering work. This cultural understanding means that instructors who teach empathy to engineering students must first recognize and confront and, then, broaden students' expectations of "real" engineering before students can effectively engage with the learning experience. The excerpts from Samir's reflections provide an example of such a process.

This challenge, however, is reinforced by Cech's second ideological pillar, engineering's "technical/social dualism", which devalues social competencies compared to technical expertise. Justin's reference to "small talk" compared to "solv[ing] the problem" demonstrates the ease with which social competencies can be dismissed by students as frivolous actions that do not befit engineers in their capacity as professionals. Once again, it falls to the instructor to acknowledge the presence and power of, and work to deconstruct, such dualistic conceptions of engineering work.

Finally, Blair's conceptualization of the Engineer-Stakeholder relationship demonstrates a tendency for engineering students to ignore and/or discount indications of social injustice, particularly in contexts where engineering work takes place. In this case, rather than interpret the father's distress as a justified grievance and cue to potentially reconsider the proposed solution strategy, Blair describes how empathy can be used to "successfully navigate" such situations.

These findings provide rich, empirical evidence of how Cech's three ideological pillars manifest in classroom settings. Moreover, this study demonstrates the potential for empathy training to break open, expose, and potentially disrupt these dominant cultural norms.

References

Battarbee, K., Fulton Suri, J., & Gibbs Howard, S. (2014). Empathy on the edge: Scaling and sustaining a human-centered approach in the evolving practice of design. *IDEO*, 1-14. http://www.ideo.com/news/archive/2014/01/#pos5142

Bornstein, D. (2007). *How to change the world: Social entrepreneurs and the power of new ideas*: Oxford University Press.

Cech, E. (2014). Culture of disengagement in engineering education? *Science, Technology, & Human Values, 39*(1), 42-72. doi:10.18260/p.2435510.1177/0162243913504305

De Waal, F. (2009). *The age of empathy: Nature's lessons for a kinder society.* New York: Harmony Books.

Duhigg, C. (2016). What google learned from its quest to build the perfect team. *The New York Times Magazine.* https://www.nytimes.com/2016/02/28/magazine/what-google-learned-from-its-quest-to-build-the-perfect-team.html?_r=0

Hansen, M. T. (2010). Ideo ceo tim brown: T-shaped stars: The backbone of ideo's collaborative culture. http://chiefexecutive.net/ideo-ceo-tim-brown-t-shaped-stars-the-backbone-of-ideoae%E2%84%A2s-collaborative-culture/

Hess, J. L., & Fila, N. D. (2016). *The development and growth of empathy among engineering students.* Paper presented at the American Society for Engineering Education Annual Conference & Exposition, New Orleans, LA.

Heydebrand, W. (1994). *Sociological writings: Max weber.* New York, NY: Bloomsbury Academic.

Husserl, E., & Hardy, L. (1999). *The idea of phenomenology.* Boston, MA: Kluwer Academic Publishers.

Jonassen, D., Strobel, J., & Lee, C. B. (2006). Everyday problem solving in engineering: Lessons for engineering educators. *Journal of Engineering Education, 95*(2), 139-151.

Marcus, J. (2013). Engineering schools fail to teach empathy. *PTC.* http://blogs.ptc.com/2013/11/27/engineering-schools-fail-to-teach-empathy/

May, V. (2014). Are women too empathetic to be engineers? *The Huffington Post.* http://www.huffingtonpost.com/vicki-may/are-women-too-empathetic-_b_5522153.html

McDonough, W., Braungart, M., & Clinton, B. (2013). *The upcycle: Beyond sustainability--designing for abundance*: Farrar, Straus and Giroux.

McLaren, K. (2013). *The art of empathy: A complete guide to life's most essential skill*: Sounds True.

National Society of Professional Engineers. (2007). Code of ethics for engineers. https://www.nspe.org/sites/default/files/resources/pdfs/Ethics/CodeofEthics/Code-2007-July.pdf

Pacey, A. (2001). *Meaning in technology*: MIT Press.

Perry, B. D., & Szalavitz, M. (2011). *Born for love: Why empathy is essential and endangered.* New York: HarperCollins.

Pool, R. (1999). *Beyond engineering: How society shapes technology*: Oxford University Press.

Rifkin, J. (2009). *The empathic civilization: The race to global consciousness in a world in crisis.* New York, NY: Penguin.

Schütz, A. (1967). *The phenomenology of the social world* (F. L. George Walsh, Trans.). Evanston, IL: Northwestern University Press.

Smith, D. W., & McIntyre, R. (2012). *Husserl and intentionality: A study of mind, meaning, and language.* Boston, MA: D. Reidel Publishing Company.

Stanford d.school. (2017). Use our methods. http://dschool.stanford.edu/use-our-methods/

Trout, J. D. (2009). *The empathy gap: Building bridges to the good life and the good society.* New York: Penguin Books.

Vagle, M. D. (2014). *Crafting phenomenological research.* New York, NY: Left Coast Press.

Walther, J., Miller, S. E., & Sochacka, N. W. (2016). *Fostering empathy in an undergraduate mechanical engineering course.* Paper presented at the American Society for Engineering Education (ASEE) Annual Conference and Exposition, New Orleans, LA.

Walther, J., Miller, S. E., & Sochacka, N. W. (2017). A model of empathy in engineering as a core skill, practice orientation, and professional way of being. *Journal of Engineering Education, 106*(1), 123-148. doi:10.1002/jee.20159

Walther, J., Sochacka, N. W., & Kellam, N. N. (2011). *Emotional indicators as a way to initiate student reflection in engineering programs.* Paper presented at the American Society for Engineering Education Annual Conference & Exposition, Vancouver, BC.

Walther, J., Sochacka, N. W., & Kellam, N. N. (2013). Quality in interpretive engineering education research: Reflections on an example study. *Journal of Engineering Education, 102*(4), 626-659. doi:10.1002/jee.20029

Winner, L. (1988). *The whale and the reactor: A search for limits in an age of high technology*: University of Chicago Press.

Zahavi, D. (2003). *Husserl's phenomenology.* Stanford, CA: Stanford University Press.

Acknowledgements

This material is based upon work supported by the National Science Foundation under grant 1463829. Any opinions, findings, and conclusions or recommendations expressed in this material are those of the authors and do not necessarily reflect the views of the National Science Foundation.

Copyright statement

A Didactic Sequence for a Meaningful Learning of Linear Functions in Engineering Education

Bruna C. Boff, Ivete A. S. Booth, Laurete Z. Sauer and Valquíria Villas-Boas
University of Caxias do Sul, Caxias do Sul, RS, Brazil
brucboff@gmail.com; iasbooth@gmail.com; lzsauer@ucs.br; vvillasboas@yahoo.com

Abstract: *In many countries, an expressive number of students with difficulties in mathematics chose engineering programs and, for that reason, the number of failures and dropouts are outrageous. This work has as its main purpose the elaboration of a didactic sequence, its application and the evaluation of its contribution to the occurrence of meaningful learning of concepts related to linear functions. The didactic sequence is based on the theory of meaningful learning, that is, it is a potentially meaningful teaching unit. This didactic sequence was applied to some students of the precalculus course, of the engineering programs of the University of Caxias do Sul. Preliminary results showed that the potentially meaningful teaching unit had a relevant contribution, and it was a methodological strategy with the potential to promote meaningful learning, reduce evasion and increase retention rates in the context of engineering education.*

Introduction

In engineering programs, it is possible to identify, on the basis of the performance analysis in the evaluations, the number of failures, dropout and retention rates, and data from the application of diagnostic evaluations, an expressive number of students with difficulties in mathematics. Such difficulties are complicating, sometimes hindering, following studies in mathematics courses and others that use mathematical knowledge. They lead to discouragement, because they generate anguish and a sense of inability to learn, causing low performances of students in engineering programs. These difficulties hamper achievement in other disciplines, which justifies the relevance of research on issues related to the teaching and learning processes of mathematics in the initial stages of engineering courses (Armstrong & Croft, 1999; Soares, Lima, & Sauer, 2004; Araújo et al., 2007; Gill, O'Donoghue, Faulkner, & Hannigan, 2010; Carr, Bowe, & Fhloinn, 2013; Carr et al., 2014; Sonnert, & Sadler, 2014; Nite et al., 2015).

The National Curricular Guidelines for Engineering courses (BRASIL, 2002), as well as INOVA Engenharia (CNI, 2006) and "Rising above the gathering storm" published by the Committee on Prospering in the Global Economy (CPGE, 2007), propose that future engineers be capable of conceiving, designing, analyzing systems, products and processes; planning, supervising, conducting experiments, interpreting results, working in multidisciplinary teams, communicating effectively, evaluating the economic viability of projects and the impact of engineering activities in the social and environmental context.

Engineering instructors, in general, need to consider innovative pedagogical actions in their courses, designing and implementing new proposals for the processes of teaching and learning. According to Booth and Villas-Boas (2014), teaching in many engineering schools is still synonymous with presenting information. In this conception of education, the instructor's action is focused on content exposure, and students are listeners. In this sense, it is important in the current scenario, to provide a groundwork for studies and actions to enable changes of this paradigm.

Given these needs, what strategies and methods have the potential to develop skills required for future engineers? What aspects of teaching mediation need to be present in learning environments when the focus is on the occurrence of meaningful learning?

Teaching and learning processes, consistent with this trend, need to be increasingly focused on students' actions in situations that favor interaction, collaboration, knowledge exchange and the development of meaningful learning (Ausubel, 2012). Learning strategies, using didactic sequences (also known as teaching-learning sequences), understood as planned, tested and analyzed sets of activities, can be a favorable means for knowledge acquisition. Potentially significant teaching units (PMTU) (Moreira, 2011), which are didactic sequences that consist of activities that intend to enable the student to develop new knowledge to deal adequately with reality, seem to be a viable path.

In this context, the PMTU presents a dynamic mechanism of intervention that considers the needs of the students presented during the activities. And one of the instructor's roles is to organize learning situations, with varied materials and activities aimed at achieving learning objectives. To do so, it is necessary to plan the teaching process, organizing and coordinating actions for the instructor and the students, evaluating the process, articulating activities and diagnosing new needs so that it is possible to achieve learning objectives.

In view of these considerations, this paper reports an experiment about the construction, application and analysis of a didactic sequence in the occurrence of meaningful learning of concepts related to linear functions in the daily life of engineering students. The didactic sequence was proposed as a potentially meaningful teaching unit (PMTU), focusing on meaningful learning and based on the contributions of Ausubel (2012) and Moreira (2006, 2011), among others.

The theoretical framework, methodological course, results and discussion are presented below, as well as some final remarks.

Theoretical Framework

Studies by educational researchers such as Fry, Ketteridge and Marshall (2015), Demo (2014), Carvalho (2011), Moreira (2011), Biggs and Tang (2011), Zabala and Arnau (2010), Barr and Tagg (1995) have contributed to the discussion of important issues about the role of education and fundamented ways of conducting the teaching and learning processes. These same researchers point out the need to transform learning environments into privileged spaces of articulation between theory and practice and of a multiplicity of experience. According to ASEE (2012), many engineering schools can improve student retention by changing the way they currently operate and adopting best practices and strategies for retaining students. One solution would be to take a "holistic" approach to improving retention, pursuing multiple strategies rather than focusing on one approach.

Problematic situations are indicated as having the potential to generate meaningful learning in teaching and learning activities to compose didactic sequences. Firme, Ribeiro and Barbosa (2008) emphasize their importance for evoking contextualised teaching, with problematic situations related to real contexts, both socially and environmentally, thus bringing the engineering students closer to the problems that will be faced in the profession. The didactic sequences are based on the theoretical bases of teaching-learning sequences by Méheut and Psillos (2004).

These authors emphasize that didactic sequences aid in the learning process, since it is possible to observe a better performance of students compared to those who had more traditional teaching approaches. Problematic situations allow contextualized questioning and with different dimensions such as: teaching objectives, varied resources, real problems, experimental activities, and technological, environmental and socio-cultural issues, among others. These actions are strategic to develop new skills, with the purpose of making the teaching process an instrument of human construction committed to the needs of the future engineer and the society.

Zabala (1998) and Zabala and Arnau (2010) emphasize that in order to understand the educational value of a didactic sequence, and the reasons that justify it, it is necessary to identify its stages, defining activities and the relationships that are established in this space of design. With the intention of contributing to this new educational scenario, the proposal for the construction of the PMTU appears (Moreira, 2011).

They have guiding principles such as: identification of previous knowledge or subsumers, previous organizers, problematic situations, progressive differentiation, integrative reconciliation and consolidation. The evaluation is understood as the search for evidence and the role of the teacher is that of a mediator, focused on the acquisition of meanings aimed at non-mechanical learning, and, consequently, have a greater potential for success in the occurrence of meaningful learning (Moreira, 2011).

Uhmann and Zanon (2013) emphasize that the instructor can act as a regulating guide to learning, using teaching strategies, such as conceptual maps, until the student develops greater cognitive capacity in learning activities. According to Lopes (2007), learning difficulties seem to relate to each other by the way content is selected, developed, organized and planned. In this sense, it is relevant to rethink the curricular organization of contents of engineering programs, as well as, the pedagogical strategies and methods, in an attempt to create conditions for meaningful learning and to increase retention rates in these courses.

Methodology

In this research work, a didactic sequence was proposed as a PMTU and applied in a group of students of the precalculus course of the engineering programs of the University of Caxias do Sul. The University of Caxias do Sul (UCS) is a community university in the southermost Brazilian state and the city of Caxias do Sul is the second metalurgical center in the country. This combination leads to a population of students that is mostly composed of people employed in the industries of the region and seeking engineering degrees.

In the first phase of the research, a diagnostic questionnaire was delivered on the first day of class to the forty-nine students, to be answered at home and was collected in the next class. The questionnaire contained questions about student personal data and contextualized questions about linear functions. Upon returning the diagnostic questionnaire, an invitation was made for everyone to participate in the second phase of the research.

It was explained to the students that, in the second phase, a PMTU would be applied, with practical activities and would be accompanied by one of the researchers, aiming at learning linear functions. It was also clarified that the activity would be without cost, and it would yield complementary hours and would be held according to an extracurricular schedule, with biweekly meetings on Saturday mornings. Of the forty-nine invited students, only five enrolled to participate in the PMTU. Its development began three weeks after the beginning of the precalculus course.

Three problematic situations, involving the linear functions, in increasing order of difficulty, made up the PMTU. Each problem was worked out in a 3-hour meeting, namely:

(i) First Meeting:

At the beginning of the first meeting, students were asked to construct a conceptual map on linear functions based on what they have learned in high school and in the precalculus course, in order to evaluate their subsumers and select the next material to be used in the meetings. Then, in the physics laboratory, the students conducted an experiment on Ohm's law. With the data obtained, they constructed electric voltage graphs as a function of the electric current and were able to determine the relationship between these physical quantities.

After the experiment, a short lecture was conducted on linear functions, formalizing the concept and its various representations. The mastery of a field of knowledge, as well as the occurrence of meaningful learning, is progressive. Thus, the purpose of the PMTU is the evolution of student learning throughout the process. Aiming at progressive differentiation and integrative reconciliation (Ausubel, 2012), at the end of the first meeting, a list of eight exercises on linear functions was proposed and solved by the participants. Six of these exercises involved contextualized engineering issues.

(ii) Second Meeting:

The second meeting began in the chemistry laboratory, where an experiment was carried out on the solubility of urea in water. With the data obtained, the students constructed graphs of density as a function of the quantity of urea and could determine that the phenomenon was described by a linear function with y-intercept. After the experiment, a short lecture was held on linear function with y-intercept, formalizing the concept and its various representations. A list of exercises, contextualized with the engineering area, was also proposed at the end of the second meeting.

(iii) Third Meeting:

In the third meeting, a uniformly accelerated motion experiment was carried out using the apps: the "Calculation of velocity as a function of time" from Microsoft and "Fiza", an application to simulate motion with constant acceleration (free app available on the Android Play Store system). In this meeting, the concept of speed as a function of time was explored through algebraic, numerical, verbal and graphical approaches, which allowed the students to relate this concept to that of a linear function. Following this, aiming at integrative reconciliation and consolidation of knowledge, one of the researchers recapitulated the main elements of the linear function, followed by a final summative evaluation and the construction of a second conceptual map on linear functions.

Results and Discussion

In this section, the main evidence that the PMTU showed potential for the occurrence of meaningful learning by the students who participated in the activity are presented. The conceptual maps elaborated by the students and the results of the initial diagnostic evaluation and the final summative evaluation are presented in this section.

About the Conceptual Maps:

Many researchers have studied the potential of conceptual maps as a tool for evaluation, organization of knowledge and meaningful learning in higher education and engineering education (Vega-Riveros, Marciales-Vivas, & Martínez-Melo, 1998; Turns, Atman, & Adams, 2000; Darmofal, Soderholm, & Brodeur, 2002; Hay, Kinchin, & Lygo-Baker, 2008; Souza, & Boruchovitch, 2010; Piá, Blasco-Tamarit, & Muñoz-Portero, 2011; Moreira, 2013; Fang, 2015; Fang, 2016).

In a conceptual map, the concepts contextually more important are indicated by arrows, which are related to secondary or specific concepts. The student presents his map and the important thing is not whether the map is correct or not, but rather if it provides evidence that the student is learning the content meaningfully (Moreira, 2013).

In this work, conceptual maps were used as an evaluation tool to highlight the knowledge built by the students throughout the application of the PMTU. The conceptual maps were analyzed according to the topological taxonomy for conceptual maps of Cañas and co-workers (2006).

Comparing the conceptual maps constructed by the students in the first meeting and in the third meeting, it was possible to observe that there was a growing evolution in the ideas represented in them. As an example, in Figure 1, the maps prepared by Student 2 are presented.

It is observed in these maps that Student 2 brings in his cognitive structure some previous knowledge about the concept of linear function. This knowledge is arranged in simple branches, connected directly to the main concept "linear function". In the conceptual map constructed by Student 2 at the end of the third meeting, it is possible to observe the association of new branches connected to the main concept, which denotes an advance in the number of interconnections of new concepts.

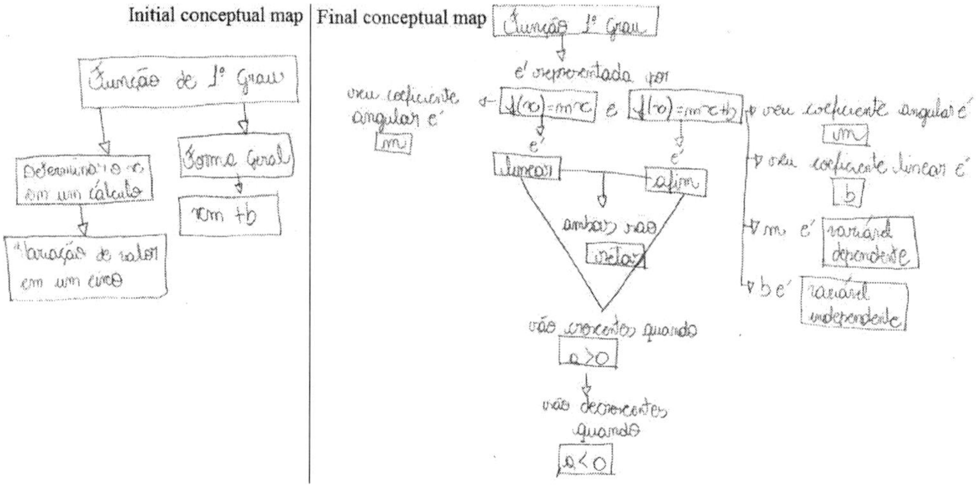

Figure 1 – Initial and final conceptual maps constructed by Student 2.

It is also possible to observe the establishment of explicit relations between the new and existing knowledge. This change in the number of relations between concepts in the construction of these maps denotes a reorganization of the student's cognitive structure, a characteristic of the process of integrative reconciliation that, according to Ausubel (2012), occurs in the course of meaningful learning.

According to the topological taxonomy for conceptual maps of Cañas and co-workers (2006), the initial map obtained level zero, while the final map obtained level 6 (on a scale of 0 to 6), proving that there was an acquisition of meanings about the subject studied in the PMTU. Although the second map still presents some incorrect information such as considering the independent variable "b" and the dependent variable "m", whereas they should be respectively "x" and "y", it is evident that there was meaningful learning.

The initial and final conceptual maps on linear functions produced by the other three students were also evaluated and they also presented evidence of the occurrence of meaningful learning for these students.

Initial elements in the cognitive structure (subsumers), with some degree of differentiation, after the application of the PMTU, are now seen in a related way, acquiring new meanings. Thus, in this movement, concepts that previously belonged to different branches of the main concept begin to relate, providing a reconciliation, a connection between them, that was not previously observable.

This dynamic movement in which concepts and propositions, already existing in the cognitive structure, undergo reorganizations and acquire new meanings and more comprehensive concepts, is characteristic of the occurrence of meaningful learning.

About the initial diagnostic evaluation and the final summative evaluation

The initial diagnostic evaluation and the final summative evaluation were also used to search for evidence of the knowledge constructed by students during the application of the PMTU.

These assessments were carried out individually, in order to obtain a more realistic analysis of the conceptions of each student. The range of application of these instruments was of ten weeks, during which the activities programmed for the PMTU were carried out. The summative evaluation consisted of all questions of the diagnostic evaluation, in addition to some questions worked out in first and second meetings.

One defined criterion was that the issues of diagnostic evaluation were not discussed at subsequent meetings, nor were any comments made on its resolution. The investigation of the occurrence of conceptual evolution was made through an analysis of the students' responses to the set of open questions.

In Figure 2, the results of the diagnostic and summative evaluations of the four students who completed all the activities proposed in the PMTU are presented.

The analysis of the evaluations was qualitative, with descriptive statistics. The distribution of students' answers to each question, both in the initial and final evaluations, was verified in order to make a comparison of the conceptual evolution between the before and after the application of the PMTU.

Analyzing the data presented in Figure 2, it is possible to observe an increase in the percentage of students' correct answers regarding the content of the function studied when compared to the results of the initial diagnostic evaluation. These results show evidence of a meaningful learning in relation to the contents worked during the implementation of the PMTU.

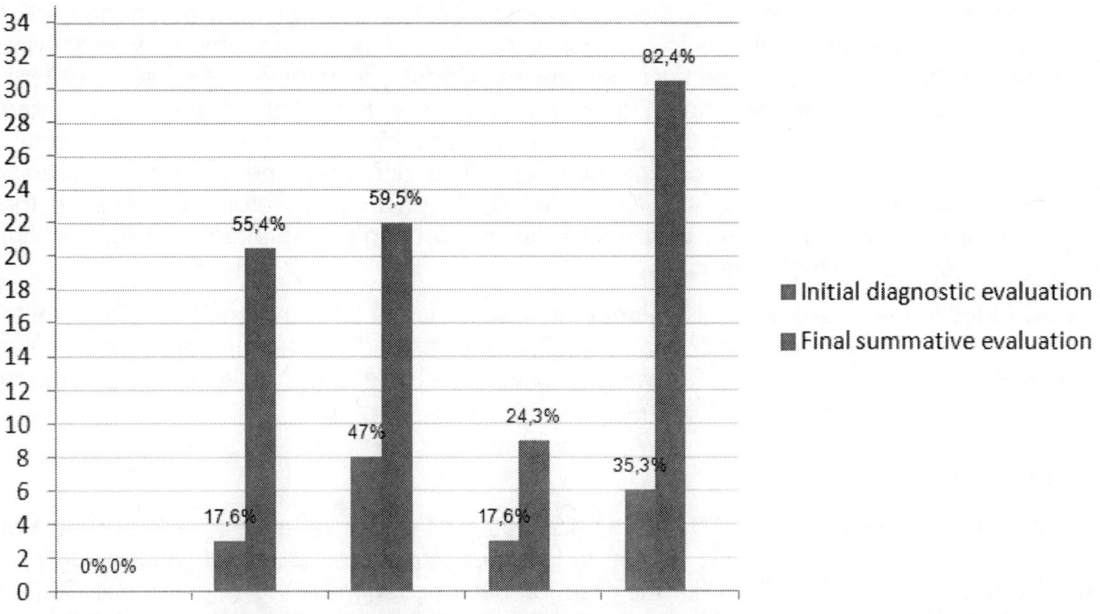

Figure 2 – Results of the initial diagnostic evaluation and the final summative evaluation

In relation to the high comparative indices, between the final and the initial evaluations of students 2 and 5 with a score of 37.8% and 47.1% respectively, it is possible to affirm that there occurred a substantive and non-arbitrary relationship of the students with the new information and the existing subsumers in their respective cognitive structures which favored a process of learning with more meaning.

These results show that the PMTU, based on the Meaningful Learning Theory (Ausubel, 2012; Moreira, 2011), favored students' learning. Thus we can affirm that the students mobilized knowledge acquired in new situations, different from those in which the diagnostic evaluation occurred.

As a consequence of the analysis of the conceptual maps and the results obtained in the initial diagnostic evaluation and the final summative evaluation, it is possible to confirm the occurrence of the reconstruction of students' knowledge, highlighting the conceptual extension and their relations through the contact with diverse materials developed by the researchers, considering the student as the main subject of learning process. Uhmann and Zanon (2014) emphasize that the instructor can act as a guide regulator of the learning process, using evaluative strategies of teaching, until the student develops greater cognitive capacity in the learning activities.

Although this article reports on a small scale study, since the evidence on student learning is limited because only a few students completed the PMTU, it is observed that the work carried out through this PMTU on linear functions was able to assist in the reconstruction of students' knowledge and that the PMTU is a pedagogical strategy that has the potential to attend the students in an integral manner, giving account of their needs and creating the conditions for a meaningful learning. For a comparison, a study carried out by Sonnert and Sadler (2014), where they examined student performance in college calculus, using regression discontinuity to estimate the effects of taking college pre-calculus or not, in a national US sample of 5507 students at 132 institutions, showed that students who take a regular college pre-calculus course do not earn higher calculus grades.

Nite et al. (2015) reported on a personalized precalculus program (PPP) that includes an online study plan with practice problems in common areas of difficulty in algebra and trigonometry and required online tutoring sessions for 30 hours over a six-week period. PPP students taking Engineering Calculus I in their first semester did not perform as well in terms of grades as those taking it in the second semester. In fact, according to Nite et al. (2015), "any pre-college program can contribute to leveling the playing field, if only by allowing students more time to adjust to the differences college life offers and providing them a glimpse of the expectations in college courses". The difference between the proposal reported in this article and that from Nite et al. (2015) is that this proposal relies on the diagnosis of students' subsumers and in the use of problematic situations of Engineering, instead of conventional math exercises.

In view of these considerations, it is evident the importance of teaching proposals that aim to improve the learning process, since teaching is the means and the meaningful learning must be the end (Moreira, 2011). In other words, this proposal offers a promising way to engage students in contrast to the traditional precalculus courses.

Final Remarks

Believing in the potential of a PMTU for the acquisition of meanings and in the occurrence of a meaningful learning of concepts related to the linear function, this investigation sought to show possibilities or limitations in the use of new pedagogical strategies. The PMTU elaborated for this study proved to be a strategy with great potential in the context of the calculus courses in the initial stages of the engineering programs. The experience described and analyzed is still small, but indicated that PMTU is an important pedagogical resource.

Also noteworthy is the ease with which students deal with the technological resources (apps and software) that can quickly be obtained and installed on their own smartphones or notebooks. Allied to this, the use of such resources, available today, to illustrate and contextualize real situations, is one of the factors responsible for the motivation to learn. In the three meetings, the analyses and discussions of the experiments were enriched with the construction and exploitation of graphs, using free software.

The aspects involved in this process of teaching and learning with a PMTU do not allow us a conclusive position on an application of the Meaningful Learning Theory, but a feasible possibility of improvement in the teaching performance in engineering programs

This study is not a complete analysis of the examined phenomenon, but with the results obtained, it was possible to show that PMTU has the potential to promote meaningful learning, reduce evasion and increase retention rates. It is hoped that this work will serve as an inspiration to offer an alternative pedagogical strategy for the occurrence of meaningful learning in engineering education.

References

Araújo, F. J., da Silva, G. C., Silva, R. V., Adissi, P. J., Melo, H. N. D. S., & Araújo, M. A. (2007, September). Dificuldades encontradas por iniciantes nos cursos de engenharia da Escola Politécnica da Universidade de Pernambuco. Proceedings of the *XXXV Congresso Brasileiro de Educação em Engenharia (COBENGE)*, Curitiba, PR, Brazil. Retrieved from http://abenge.org.br/cobenges-anteriores/2007/artigos-2007/artigos-publicados_7

Armstrong, P. K., & Croft, A. C. (1999). Identifying the learning needs in mathematics of entrants to undergraduate engineering programmes in an English university. *European Journal of Engineering Education*, *24*(1), 59-71. doi: 10.1080/03043799908923538

ASEE (2012). Retention Project. *The American Society for Engineering Education*. Retrieved from https://www.asee.org/retention-project

Ausubel, D. P. (2012). *The acquisition and retention of knowledge: A cognitive view*. Springer Science & Business Media.

Barr, R. B., & Tagg, J. (1995). From teaching to learning - A new paradigm for undergraduate education. *Change: The magazine of higher learning*, *27*(6), 12-26. doi: 10.1080/00091383.1995.10544672

Biggs, J. B., & Tang, C. (2011). *Teaching for quality learning at university: What the student does*. Berkshire: Open University Press.

Booth, I. A. S., & Villas-Boas, V. (2014, September). Aprendizagem significativa sobre reações químicas em Engenharia usando uma sequência didática baseada em uma UEPS. Proceedings of the *5º Encontro Nacional de Aprendizagem Significativa (ENAS)* (pp. 321-332). Belém, PA. Retrieved from http://media.wix.com/ugd/75b99d_37a230c8a3b5413aa62a146694374771.pdf.

Brasil. (2002) Ministério da Educação - Conselho Nacional de Educação - Câmara de Ensino Superior. *Resolução CNE/CES, de 11 de março de 2002, que institui Diretrizes Curriculares para os cursos de graduação em Engenharia*. Brasília: Ministério da Educação. Retrieved from http://portal.mec.gov.br/cne/arquivos/pdf/CES112002.pdf

Cañas, A. J., Novak, J. D., Miller, N. L., Collado, C., Rodríguez, M., Concepción, M., Santana, C., & Peña, L. (2006). Confiabilidad de una taxonomía topológica para mapas conceptuales. Proceedings of the *2nd International Conference on Concept Mapping*. San José, Costa Rica. Retrieved from http://cmc.ihmc.us/papers/cmc2004-223.pdf

Carr, M., Bowe, B., & Ní Fhloinn, E. (2013) Core skills assessment to improve mathematical competency. *European Journal of Engineering Education*, *38*(6), 608-619. doi:10.1080/03043797.2012.755500

Carr, M., Fidalgo, C., Bigotte de Almeida, M.E., Branco, J.R., Santos, V., Murphy, E., & Ní Fhloinn, E. (2014): Mathematics diagnostic testing in engineering: an international comparison between Ireland and Portugal. *European Journal of Engineering Education*, *40*(5), 546-556. doi: 10.1080/03043797.2014.967182

Carvalho, J. G. D. (2011). Aula de física: do planejamento à avaliação. São Paulo: Editora Livraria da Física.

CNI (2006). INOVA Engenharia: Propostas para Modernização da Educação em Engenharia no Brasil. Brasília: IEL.NC, SENAI.DN.

Committee on Prospering in the Global Economy of the 21st Century. (2007). *Rising above the gathering storm: Energizing and employing America for a brighter economic future*. Washington, DC: National Academies Press.

Darmofal, D. L., Soderholm, D. H., & Brodeur, D. R. (2002, November). Using concept maps and concept questions to enhance conceptual understanding. Proceedings of the *Frontiers in Education (FIE)*. Boston, MA: IEEE. doi: 10.1109/FIE.2002.1157954

Demo, P. (2014). Educação Científica. *Revista Brasileira de Iniciação Científica*, *1*(1), 02-22. Retrieved from http://itp.ifsp.edu.br/ojs/index.php/IC/article/view/10/421

Fang, N. (2015, June), *Using Students-Generated Concept Maps to Assess Students' Conceptual Understanding in a Foundational Engineering Course* Paper presented at 2015 ASEE Annual Conference & Exposition, Seattle, Washington. doi: 10.18260/p.25020

Fang, N. (2016, June), *Using Concept Maps to Illustrate the Evolution of Key Concepts: Student Learning Experience in a Foundational Undergraduate Engineering Course* Paper presented at 2016 ASEE Annual Conference & Exposition, New Orleans, Louisiana. doi: 10.18260/p.27140

Firme, R. N.; Ribeiro, E. M.; Barbosa, R. M. N. (2008, July) Análise de uma sequência didática sobre pilhas e baterias: uma abordagem CTS em sala de aula de química. Proceedings of the *XIV Encontro Nacional de Ensino de Química.* Curitiba, PR. Retrieved from http://www.quimica.ufpr.br/eduquim/eneq2008/resumos/R0756-2.pdf

Fry, H., Ketteridge, S., & Marshall, S. (2015). A handbook for teaching and learning in higher education: Enhancing academic practice. Abingdon: Routledge.

Gill, O., O'Donoghue, J., Faulkner, F., & Hannigan, A. (2010). Trends in performance of science and technology students (1997–2008) in Ireland. *International Journal of Mathematical Education in Science and Technology, 41*(3), 323-339. doi: 10.1080/00207390903477426

Hay, D., Kinchin, I., & Lygo-Baker, S. (2008). Making learning visible: the role of concept mapping in higher education. *Studies in higher education,* *33*(3), 295-311. http://dx.doi.org/10.1080/03075070802049251

Lopes, A. C. (2007). Currículo e Epistemologia. Ijuí, RS: Editora Unijuí.

Méheut, M., & Psillos, D. (2004). Teaching-learning sequences: aims and tools for science education research. *International Journal of Science Education,* *26*(5), 515-535. doi: 10.1080/09500690310001614762

Moreira, M. A. (2006) A teoria da aprendizagem significativa e sua implementação na sala de aula. Brasília: Editora da UnB.

Moreira, M. A. (2011). Unidades de enseñanza potencialmente significativas. *Aprendizagem Significativa em Revista,* *1*(2), 43-63. Retrieved from http://www.if.ufrgs.br/asr/artigos/Artigo_ID10/v1_n2_a2011.pdf // For a version of this article in English go to: https://www.if.ufrgs.br/~moreira/UEPSing.pdf

Moreira, M. A. (2013) Aprendizagem significativa em mapas conceituais, *Textos de apoio ao professor de física,* *24*(6), 1-49. Retrieved from http://www.if.ufrgs.br/public/tapf/v24_n6_moreira_.pdf

Nite, S. B., Capraro, R. M., Capraro, M. M., Allen, G. D., Pilant, M., & Morgan, J. (2015, October). A bridge to engineering: A personalized precalculus (bridge) program. Proceedings of the *Frontiers in Education Conference (FIE)* (pp. 2053-2058). El Paso, TX: IEEE. Retrieved from http://fie2015.org/sites/fie2015.fie-conference.org/files/FIE-2015_Proceedings_v11.pdf

Piá, A. B., Blasco-Tamarit, E., & Muñoz-Portero, M. J. (2011). Different applications of concept maps in Higher Education. *Journal of industrial engineering and management*, *4*(1), 81-102. doi: 10.3926/jiem.2011.v4n1.p81-102

Sarker, P. K. (2015). Use of concept maps for problem-solving in engineering. *Global Journal of Engineering Education,* *17*(1), 29-33. Retrieved from http://www.wiete.com.au/journals/GJEE/Publish/vol17no1/04-Sarker-P.pdf

Soares, E. M. S., Lima, I. G., & Sauer, L. Z. (2004, March). Discutindo alternativas para ambientes de aprendizagem de matemática para cursos de Engenharia. Paper presented at the *World Congress on Engineering and Technology Education* (pp. 1159-1162). Guarujá, SP, Brazil: COPEC, Retrieved from http://copec.eu/congresses/wcete2004/.

Sonnert, G., & Sadler, P. M. (2014). The impact of taking a college pre-calculus course on students' college calculus performance. *International Journal of Mathematical Education in Science and Technology*, *45*(8), 1188-1207.

Souza, N. A., & Boruchovitch, E. (2010). Mapas conceituais: estratégia de ensino/aprendizagem e ferramenta avaliativa. *Educação em Revista, Belo Horizonte*, *26*(03), 195-218. Retrieved from http://www.scielo.br/pdf/edur/v26n3/v26n3a10.pdf

Turns, J., Atman, C. J., & Adams, R. (2000). Concept maps for engineering education: A cognitively motivated tool supporting varied assessment functions. *IEEE Transactions on Education*, *43*(2), 164-173. doi: 10.1109/13.848069

Uhmann, R. I. M., & Zanon, L. B. (2013, October) O paradigma da avaliação escolar em discussão na docência em ciências/química. Paper presented at the *33º Encontro de Debates sobre o Ensino de Química*, Ijuí – RS. Retrieved from https://www.publicacoeseventos.unijui.edu.br/index.php/edeq/article/view/2750/2326

Vega-Riveros, J. F., Marciales-Vivas, G. P., & Martínez-Melo, M. (1998). Concept maps in engineering education: a case study. *Global J. of Engng. Educ*, *2*(1). 21-28. doi=10.1.1.114.2814

Zabala, A. (1998). A prática educativa: como ensinar. Porto Alegre: Artmed.

Zabala, A., & Arnau, L. (2010). Como aprender e ensinar competências. Porto Alegre: Artmed.

AUTHOR INDEX

AUTHOR INDEX

AUTHOR INDEX

7th Research in Engineering Education Symposium (REES 2017)

Research in Engineering Education

Bogota, Columbia
6-8 July 2017

Volume 2 of 2

ISBN: 978-1-5108-4941-9

Printed from e-media with permission by:

Curran Associates, Inc.
57 Morehouse Lane
Red Hook, NY 12571

Some format issues inherent in the e-media version may also appear in this print version.

Printed by Curran Associates, Inc. (2017)

For permission requests, please contact Research in Engineering Education Network
at the address below.

Research in Engineering Education Network
c/o Bill Williams
Escola Sup. De Tecno Barreiro
Rua Americo Da Silva Marinho 1
2239-001 Lavaradio, Portugal

www.reen.co

bwbillwilliamsbw@gmail.com

Additional copies of this publication are available from:

Curran Associates, Inc.
57 Morehouse Lane
Red Hook, NY 12571 USA
Phone: 845-758-0400
Fax: 845-758-2633
Email: curran@proceedings.com
Web: www.proceedings.com

TABLE OF CONTENTS

VOLUME 1

VOLUME 2

Beyond active learning: Critical factors for learning in labs

Jonte Bernhard
Linköping University, Campus Norrköping, Norrköping, Sweden
jonte.bernhard@liu.se

***Abstract**: Active learning is generally defined as an approach that engages students in the learning process and is supposed to lead to consistently better and deeper understanding. In an earlier study students in mechanics were offered the choice between labs using probe-ware (MBL) [FMCE normalised gain: 48%] and experimental problem-solving labs [18% gain]. Both options were considered to employ active learning, but the difference in gains was remarkable. As this contradicts the conclusions in the literature a follow-up study was performed. Analysis of video recordings from the labs showed that in probe-ware labs students linked observed data to concepts, whereas students in the problem-solving labs made little use of physical concepts in their modeling of phenomena. One implication of this study is that we have to go beyond surface interpretations of "active learning", and in a detailed and nuanced way look into the ways in which students are actually active in a learning environment.*

1. Introduction

Active learning is an approach to education that has received considerable scholarly attention; it is, for example, one of the CDIO standards (Crawley, Malmqvist, Östlund, Brodeur, & Edström, 2014), and has been the topic of two special issues of the *European Journal of Engineering Education* (De Graaff & Christensen, 2004; Lima, Andersson, & Saalman, 2016). Active learning is generally supposed to lead to deeper learning, a better understanding of key concepts, and a greater ability to apply knowledge in new and unfamiliar settings (e.g. Crawley et al., 2014; Freeman et al., 2014; Prince, 2004). However, in an earlier study I found that students in two different options for a lab-course achieved very different gains in conceptual understanding despite the fact that both options were considered to employ active learning principles (Bernhard, 2010). Indeed, the gain in one of the alternatives was equivalent in magnitude to that seen in traditionally designed labs (ibid.). As this apparently contradicted the conclusions in much of the literature about active learning, as well as Trumper's (2003) conclusions regarding learning in labs, these findings prompted a follow-up study to more thoroughly investigate the learning that took place in the labs in the previous study.

The paper is organised as follows: Section 2 describes the background and theoretical framework of the study, i.e. active learning and variation theory, as well as providing a brief review of the results from the earlier quantitative study; Section 3 describes the purpose of the paper and sets out the research questions; Section 4 presents the qualitative methodology used; Section 5 presents the findings of the current study; finally, Section 6 presents the conclusions and implications.

2. Background and theoretical framework

Active learning

Despite the widespread interest in active learning, the term is often only rather vaguely defined; indeed, a surprisingly large number of studies do not define the term at all, the authors relying instead on an intuitive – and therefore somewhat unsatisfactory – understanding of its meaning. The most common definition, based on the classic work of Bonwell and Eison (1991, p. iii), is: "instructional activities involving students in doing things and thinking about what they are doing". In a similar vein, Prince (2004, p. 223) suggested that "active learning requires students to do meaningful learning activities and think about what they are doing". It should be noted that Freeman et al. (2014) in their meta-analysis focused only on "in class" activities and explicitly excluded laboratories. On the other hand, Trumper (2003) focused on the laboratory and concluded that a common attribute of successful activities "is that they are learner-centered. They induce students to become active participants in a scientific process in which they explore the physical world, analyze the data [and] draw conclusions".

The Interactive-Constructive-Active-Passive (ICAP) framework proposed by Chi (2009) offers somewhat more precise definitions and differentiated distinctions. In this framework *overt* learning activities are classified as passive, active, constructive, and interactive. In *active* activities a student does something physically; in *constructive* activities a student also generates an output that extends beyond the presented materials; and finally in *interactive* activities two or more students cooperate on the same topic and have a substantive dialogue that does not ignore and exclude each others' contributions. Based on the literature, Chi suggested that student learning is better in interactive than in constructive activities, which in turn are better than merely active activities, which are better than simple passive learning. For students in a material science course Menekse, Stump, Krause, and Chi (2013) have confirmed Chi's hypothesis.

Variation theory

A further framework that will be employed here is *variation theory* (e.g. Marton, 2015; Marton & Tsui, 2004). To promote students' learning effectively, it is important to ensure that the learning environment enables them to focus on the object of learning and discern its critical features. Central to variation theory is the notion that we discern certain aspects of an environment by experiencing variation, rather than by recognizing similarities. Important concepts are hence *discernment*, *simultaneity*, and *variation*. When one aspect of a phenomenon or an event varies, while one or more aspects remain the same, the aspect that changes is the one that will be discerned. One of the main themes of variation theory is that the pattern of variation inherent in the learning situation is fundamental to the development of certain capabilities. Experiencing variation amounts to experiencing different instances of the object of learning simultaneously. This simultaneity can be either *diachronic* (experiencing, *at the same time*, aspects of something that we have encountered at different points in time), or *synchronic* (experiencing different coexisting aspects of the same thing at the same time).

Marton (2015, p. 27, my italics; cf. Marton, Runesson, & Tsui, 2004) related learning to what students could *possibly* experience in a particular classroom situation, stating that in a learning situation "the critical aspects that it is *possible* [for a student] to discern … make up the *enacted* object of learning". Another important distinction in a learning situation is the difference between the *intended* object of learning (the knowledge, values, and skills the teacher or curriculum designer wants the students to learn) and the *lived* object of learning (the critical aspects that *could be discerned* and that the student *actually discerns*, i.e. what the student learns in the end).

Two active learning labs

As already briefly discussed in the introduction, the current author (Bernhard, 2010) found quite different learning gains for students participating in two alternative sets of labs. In this section a short recapitulation of the results from this earlier study is presented. Engineering students studying mechanics as a part of an introductory physics course were offered the possibility to choose between two options for the lab-course. The regular set of labs within the course was the so-called experimental problem-solving (EPS) labs, but as an alternative the students were offered a set of labs utilizing so-called probe-ware. The students participating in the probe-ware labs were volunteers. For legal reasons, students could not be randomly assigned to groups since the EPS-labs were described as the lab-course in the official curriculum documents. As can be seen in Table 1, the between-group differences in pre-course conceptual understanding of mechanics were very small and not statistically significant.

Both sets consisted of four 4-hour lab sessions, i.e. a total of 16 hours of lab work. All students attended the same 20 hours of lectures (in a lecture hall) in mechanics, and participated in similar sets of 12 hours of problem-solving sessions (with groups of approximately 30 students led by a doctoral student). Thus, the only difference between the groups, in terms of teaching, was the 16 hours of lab work. How the options were organized is summarized in Table 1.

To investigate students' conceptual understanding of mechanics a Swedish translation of *Force and Motion Conceptual Evaluation* (FMCE) (see Thornton & Sokoloff, 1998) was used both pre- and post-test. FMCE is a widely used and well-established instrument that probes student understanding of one-dimensional kinematics and dynamics in depth by utilizing verbal and graphical representations. The average pre- and post-tests, as well as the normalized gain and the effect size, are also presented in Table 1. The *normalized gain* (Hake, 1997) is defined as g = *Actual gain*/[*Gain(max possible)*] where *Actual gain* is the difference between group averages in the pre- and post-test. The *effect size* (Cohen's d) is defined as $d = (M_E - M_C)/S_p$; here, M_E is the experiment group average (in this case the probe-ware labs), M_C is the control group average (in this case the EPS-labs), and S_p is the pooled standard deviation (Cohen, 1988).

Table 1. Organization of the mechanics part of the physics course, and results from the pre- and post-tests using the FMCE (means with standard deviations in parentheses), normalized gains, and calculated effect size from the results in the earlier study (Bernhard, 2010).

<table>
<tr><th>Group</th><th>No.</th><th>Lec-tures</th><th>Problem-solving sessions</th><th>EPS-labs (groups of 2–3 students)</th><th>Probe-ware labs (groups of 2–4 students)</th><th>Pre-test %</th><th>Post-test %</th><th>Norm Gain %</th><th>Effect size</th></tr>
<tr><td>EPS-labs</td><td>86</td><td rowspan="2">20 h (all)</td><td rowspan="2">12 h (groups of approx. 30 students)</td><td>16 h (2×[4+4 h])</td><td></td><td>29.3 (16.4)</td><td>42.3 (22.9)</td><td>18.4</td><td>0 (by def.)</td></tr>
<tr><td>Probe-ware labs</td><td>25</td><td></td><td>16 h (4×4 h)</td><td>34.3 (23.1)</td><td>65.8 (21.8)</td><td>47.9</td><td>1.04</td></tr>
</table>

Although the pre-test differences were insignificant, the differences in post-test results are striking. The students participating in the probe-ware labs achieved a normalized gain of 48%, compared to just 18% for the students participating in the EPS-labs – which corresponds to what is achieved by traditional labs (Hake, 1997). This difference is strongly statistically significant ($t = 2.93$, $p = 0.0003$). The effect size (Cohen's d) calculated from the post-test averages is $d = 1.04$. An effect size $d > 0.8$ is usually regarded as large (Cohen, 1988).

3. Purpose and research questions

The purpose of this follow-up study was to try and explain why, despite the fact that both sets of labs were supposedly using "active learning", considerable differences in conceptual learning had been observed. This led to the following research questions:

Would a closer investigation confirm or reject the hypothesis that both sets of labs could be seen as examples of active learning, and how would the labs be classified according to the ICAP framework (Chi, 2009)?

How did the students act and approach the learning environment in the two sets of labs and how could any observed differences be understood in relation to students' learning, i.e. the enacted object of learning?

It was hoped that answers to these two main questions would provide the basis for additional insights into learning in labs and the general concept of active learning.

4. Methodology

Analysis of the learning that had occurred in the EPS-labs and in the probe-ware labs was performed in several steps: Students' conceptual understanding was investigated through pre- and post-tests using the FMCE test (Thornton & Sokoloff, 1998). The results of this investigation are reported in more detail in Bernhard (2010) and summarized in Table 1. Activities in both sets of labs were recorded using digital camcorders, resulting in 70 hours of video evidence. This data were subsequently used to detect typical interaction patterns and find evidence supporting, or refuting, hypotheses regarding the generality of these patterns (Jordan & Henderson, 1995). I was particularly looking for how the students were actually working together, what they did, what they made relevant, and how they oriented themselves towards the object of learning. After repeated viewings, some episodes were found to contain more interesting and comparable activities in relation to the research questions. Particularly interesting parts of these episodes were transcribed to allow for detailed examination of interactional patterns. In the transcriptions, standard conventions used in conversation analysis have been used (ten Have, 2007) and transcripts included in this paper were ultimately translated from Swedish into English. As a final step the analyses for the two optional lab-courses were contrasted to find similarities as well as "differences that make a difference" (Bateson, 1972; cf. Lindwall & Ivarsson, 2010).

5. Findings

Analysis of the video recordings from both sets of labs showed that the students' actions were framed by the task at hand and encounters with the instructions, the technology and experimental setup, the teacher, and other students. Thus in terms of Chi's (2009) ICAP framework students' activities in both sets of labs could be seen as interactive, which further implies that they were also active and constructive. In addition to this, the criteria for active learning proposed by Bonwell and Eison (1991) as well as by Prince (2004) were fulfilled (although according to Freeman et al. [2014] the course could not be legitimately described as active learning as the lectures and problem-solving sessions were traditional). Yet although the labs fulfilled most of the theoretical criteria to qualify as active learning environments, the way that the students actually worked on their tasks was much more complex and differentiated than might at first be assumed. Some of these complexities will be explored below.

Probe-ware labs

The labs in the probe-ware lab-course were a subset (4×4 h) of labs used in an earlier developed lab-course (7×4 h), utilizing probe-ware technology and instructions in line with active learning (Bernhard, 2010). The approach used in the development of these labs was inspired by the pedagogical techniques outlined in *RealTime Physics* (Sokoloff, Laws, & Thornton, 2007; Sokoloff, Thornton, & Laws, 1998), incorporating "a learning cycle consisting of prediction, observation, comparison, analysis and quantitative experimentation".

Findings from an activity in the first lab (the *motion* lab, where kinematics concepts were introduced using probe-ware) will be presented in detail here. This lab is based on Laws' (1997) contention that a good understanding of kinematics (the laws of motion) is essential for understanding dynamics (the laws of force and motion), and the findings from various researchers (e.g. Bowden et al., 1992; McDermott, 1997; Trowbridge & McDermott, 1980, 1981) that a large proportion of students have problems understanding basic concepts of kinematics. Many students, for example, reportedly have problems distinguishing between velocity and change of velocity. They commonly believe that acceleration is always in the direction of motion and that zero velocity implies that acceleration must be zero.

Hence, one of the tasks in the motion lab is *acceleration with zero velocity*. In this activity students monitor the motion of a cart propelled by a fan that provides almost constant acceleration (see Figure 1a). The students give a cart with a fan an initial push in the opposite direction to that in which the force of the fan is acting, so that the cart will slow down and reverse its direction of motion. Students are first asked to observe the motion of the cart (without measuring it) and then to sketch their predictions of how the motion will be represented by position-time, velocity-time, and acceleration-time graphs. After they have made their predictions the motion of the cart is once more observed and this time the probe-ware equipment is used to measure the motion, and simultaneously display it as a graph (a typical graph is shown in Figure 1b). To make accurate predictions, not only do the differences between position, velocity, and acceleration have to be discerned – the relationships between these concepts must also be perceived and understood. Velocity and position vary, but students have to recognize that the acceleration is constant, and that a zero velocity does not imply that the acceleration is zero – as is commonly believed. In Excerpt 1, below, students discuss what the acceleration should be when the velocity is zero at the cart's turning point, and what the acceleration-time graph should look like around this point.

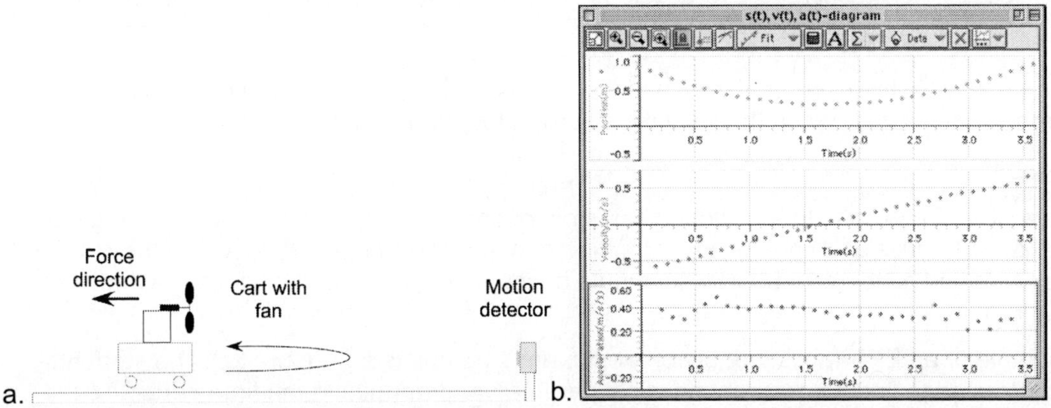

Figure 1. a) A typical setup in a probe-ware experiment. A low friction cart is pushed towards a motion sensor. A fan unit attached to the cart provides an approximately constant force in the opposite direction to the initial movement and, thus, the cart's direction of motion. **b)** The results (which show that acceleration is not zero at the turning point) are presented to the right.

Excerpt 1

```
1.   Beata     here I don't think the acceleration will be constant
2.   Cecilia   no for it will only [increase then
3.   Beata                         [it will
4.   Cecilia   =then stop
               ((a few turns are missing here due to a change of tape))
5.   Beata     something like that increases
               ((makes a sketch))
6.   Cecilia   then it becomes zero
7.   Beata     =for a little while when it turns
```

It will be seen that the students suggest that the acceleration "becomes zero for a little while when it turns". However, after performing the actual experiment Cecilia finds that "the acceleration turns out strange"; contrary to their prediction it is not zero, as shown in Figure 1b. After discovering that their prediction was not correct, and the acceleration was not in fact zero, the students discuss the results for a long time and finally in Excerpt 2 they decide to ask the instructor.

Excerpt 2

```
1.   Beata     it is so [strange
2.   Cecilia            [acceleration in this case
3.   Cecilia   the acceleration can't be constant (.) since it stops and when it
               starts again
4.   Cecilia   can it be constant?
5.   JOHN      yes
6.   Cecilia   because it feels weird
```

After some discussion between the students and the instructor the issue is resolved in Excerpt 3.

Excerpt 3

1.	JOHN	there you have zero (.) but if you look at delta v:: even at this point
2.	Cecilia	=you mean that the velocity doesn't change much
3.	JOHN	no but you [you have
4.	Beata	[no
5.	JOHN	the whole time a constant [change in velocity
6.	Cecilia	[okay
7.	JOHN	=per unit time
8.	Cecilia	yes
9.	Beata	if you have a straight line (.) you will have the same slope on it (.) then you will have the same acceleration the whole way (3.7)
10.	Cecilia	°m::°
11.	Beata	because acceleration is
12.	Cecilia	[it's because
13.	Beata	[the derivative of velocity

As can be seen in the excerpt above, it took several turns before the students realized why acceleration is not zero when the velocity is zero at the turning point.

EPS-labs

The aim of an EPS-lab is to "highlight what is essential for creating good physics; imagination and initiative to develop hypotheses, use of experimental techniques to test hypotheses, and the ability to find concepts that provide simple descriptions" (course description, author's translation).

An example of an EPS-lab is the torsion pendulum lab depicted in Figure 2a. The students were instructed to find the relevant physical parameters determining the time period for the oscillations of the torsion pendulum, and to construct and experimentally verify a mathematical model for the time period. No further instructions were given.

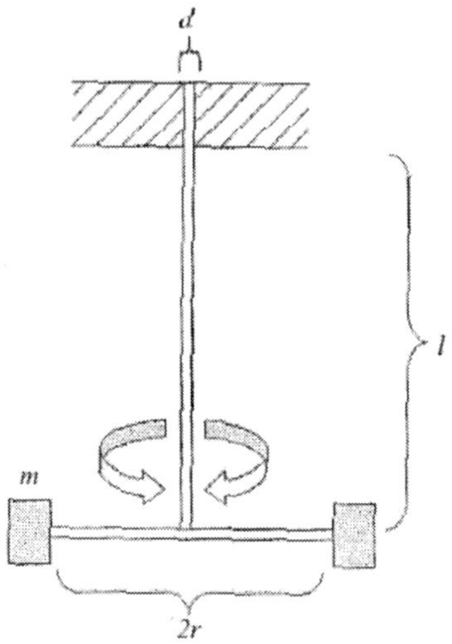

a.

$$T = C_1 \frac{1}{d^2}\sqrt{\frac{l(mr^2 + C_2)}{G}}, \text{ where}$$

T = period time,

C_1 = experimental constant,

C_2 = experimental constant (moment of inertia of the rod),

G = shear modulus,

d = diameter of the wire,

l = length of the wire,

m = mass of the additional weights, and

r = distance of the masses from center.

b.

Figure 2. a) The torsion pendulum setup (drawing from student lab report). **b)** A mathematical model for the oscillations of this torsion pendulum.

At their disposal the students had two lab sessions of four hours each, with one week between the sessions. To solve the problem of finding a suitable mathematical model the students were expected to use a combination of empirical measurements of the physical system and dimension analysis. Besides the torsion pendulum, there were five more EPS-labs, one example being a sinking boat and an oscillating beam. The students worked in the labs in groups of 2–3, and were expected to do two of these six labs; for each lab they had at their disposal eight hours (4 + 4 hours). The EPS-labs were designed after a proposal by Richards (1971).

The initial discussions in the EPS-labs centered largely around which variables were important, how experiments should be planned, and how measurements should be performed. Excerpt 4 is an example from a discussion between Adam and Bengt that took place some 45 minutes into the first session of the first lab about the measurement of period time for the torsion pendulum.

Excerpt 4

1.	Bengt	thus we measure (.) we start here
2.	Adam	yes
3.	Bengt	it goes out one time ((swings the rod with masses by holding it in his hand a full period))
4.		(2 s) ((Bengt swings the rod a full period a second time))
5.	Bengt	two
6.		(2 s) ((Bengt swings the rod a full period a third time))
7.	Bengt	three
8.		(2 s) ((Bengt swings the rod a full period a fourth time))
9.	Bengt	four
10.		(2 s) ((Bengt swings the rod a full period a fifth time))
11.	Bengt	five
12.	Bengt	if we should take (.) if we say before it slows down markedly (.) if we take like three swings or something (.) two swings (.) we can do it a few times and take an average

From this excerpt it could look as if the conversation lacked interactivity, and that Adam was rather passive: after all, his contribution seems to amount to the single word "yes". However, after about a half a minute Adam moves to the pendulum and suggests that they should take the time every time the pendulum stops at the maximum excursion. After some discussion they eventually settled for Adam's suggestion. However, they found that this was quite difficult to do with the available stopwatch. Yet instead of changing their strategy, Bengt recalled that his old mobile phone had a function allowing the recording of several times; one of the students in another lab group had such a phone and they were allowed to borrow it.

In their first series of measurements the position r of the masses on the rod was varied with other variables constant. In the second series the length l of the wire was varied with other variables constant. In the third series the diameter of the wire vas varied.

Much of the second session was spent on analyzing the data from the first session, and testing different hypotheses for a model of the period time for the torsion pendulum. Adam voiced concerns about kinetic energy and wanted to include it in the relationship but struggled to see how this could be done. A hint from the instructor and dimension analysis led the students to the insight that the shear modulus must be included. Near the end of the second session the students had arrived at the mathematical model displayed in Figure 2b. However, their model was lacking C_2 (corresponding to the moment of inertia of the rod). Even when questioned by the instructor, the students failed at first to see that this implied $T = 0$ if there were no masses mounted on the rod and that this was contrary to what was possible to observe. After some Socratic dialogue the students realized that C_2 must be included and arrived at the model in Figure 2b. Some other groups, where the instructor failed to notice this during the session, submitted reports with a model lacking C_2.

Discussion and comparative analysis

The intended object of learning in the two sets of labs was quite different. For the probe-ware labs it was to develop students' conceptual understanding, while in the EPS-labs it was to develop students' abilities to plan experiments and model a physical system. The data show that both labs were successful in what they had set out to do, and the enacted and lived object of learning was in line with the intended object of learning.

Looking at the students' actions in line with Chi's (2009) ICAP framework, it is apparent that their activities were significantly different. The design of the probe-ware labs framed students to construct a conceptual understanding, while the students in the EPS-labs were steered to construct experiments and mathematical models. The content of students' interactions with each other clearly reflected these differences.

In both labs, students displayed misconceptions related to important physical concepts. The probe-ware labs were deliberately designed to develop conceptual understanding and to challenge such misconceptions. In the EPS-labs, on the other hand, misconceptions were not explicitly challenged (as can be seen in the results in Table 1), and so it is apparently possible for students to arrive at an adequate mathematical model of a physical system whilst having an inadequate understanding of the underlying physical concepts. However, an analysis of the students in the EPS-labs suggests that the inadequate conceptual understanding was indeed detrimental for them in their modeling process, and that they might have arrived at an adequate model faster and with a better and deeper understanding.

Variation (Marton, 2015) played an important role in both sets of labs. In the probe-ware labs variation is built into the task structure and supports students' learning of concepts. On the other hand, in the EPS-labs the variation was designed by the students in their experimental planning and helped them to identify the relevant physical parameters. Thus in both cases variation helped students to be aware of central aspects of the intended object of learning.

Finally, it important to emphasize that the instructor played a significant role in both labs by helping students notice and highlight important aspects that might otherwise have gone unnoticed, and by scaffolding their reasoning through Socratic dialogues.

6. Conclusion and implications

Although the probe-ware labs in this study demonstrated an unquestionable superiority in developing conceptual understanding, it would be wrong to conclude that these labs were therefore wholly superior to the EPS-labs in every respect. Much depends on the intended object of learning. Conceptual understanding, as well as skills in experimental planning and mathematical modeling, should *all* be important learning objectives in engineering education. For each lab and each lab-course we have to make conscious and deliberate decisions about our aims and goals. Indeed, Streveler and Menekse (2017, p. 189) have argued very recently for "a more nuanced approach to active learning", and urged the "engineering education community to more precisely describe the active learning strategies they are using" to allow for instructions to be more wisely designed. The findings of this study echo that of Streveler and Menekse in underlining that we have to go beyond surface interpretations of "active learning": In short, when designing learning environments, rather than consider the ways in which, according to the literature, students are *supposed* to learn actively, we should bear in mind the ways in which they are *actually* learning.

References

Bateson, G. (1972). *Steps to an ecology of mind.* New York: Ballantine Books.

Bernhard, J. (2010). Insightful learning in the laboratory: Some experiences from ten years of designing and using conceptual labs. *European Journal of Engineering Education, 35*(3), 271-287.

Bonwell, C. C., & Eison, J. A. (1991). *Active learning: Creating excitement in the classroom. 1991 ASHE-ERIC Higher Education Reports.* Washington, DC: ERIC.

Bowden, J., Dall'Alba, G., Martin, E., Masters, G., Laurillard, D., Marton, F., Stephanou, A. (1992). Displacement, velocity, and frames of reference: Phenomenographic studies of students' understanding and some implications for teaching and assessment. *American Journal of Physics, 60*, 262-268.

Chi, M. T. H. (2009). Active-constructive-interactive: A conceptual framework for differentiating learning activities. *Topics in Cognitive Science, 1*(1), 73-105.

Cohen, J. (1988). *Statistical power analysis for the behavioral sciences* (2nd ed.). Hillsdale: Lawrence Erlbaum.

Crawley, E. F., Malmqvist, J., Östlund, S., Brodeur, D. R., & Edström, K. (2014). *Rethinking engineering education: The CDIO approach* (2nd ed.). New York: Springer.

De Graaff, E., & Christensen, H. P. (2004). Editorial: Theme issue on active learning in engineering education. *European Journal of Engineering Education, 29*(4), 461-463.
Freeman, S., Eddy, S. L., McDonough, M., Smith, M. K., Okoroafor, N., Jordt, H., & Wenderoth, M. P. (2014). Active learning increases student performance in science, engineering, and mathematics. *Proceedings of the National Academy of Sciences, 111*(23), 8410-8415.
Hake, R. R. (1997). Interactive-engagement vs traditional methods: A six-thousand-student survey of mechanics test data for introductory physics courses. *American Journal of Physics, 66*, 64-74.
Jordan, B., & Henderson, A. (1995). Interaction analysis: foundations and practice. *The Journal of the Learning Sciences, 4*(1), 39-103.
Laws, P. (1997). A new order for mechanics. In J. Wilson (Ed.), *Proceedings conference on introductory physics course* (pp. 125-136). New York: Wiley.
Lima, R. M., Andersson, P. H., & Saalman, E. (2016). Active learning in engineering education: a (re)introduction. *European Journal of Engineering Education*. doi: 10.1080/03043797.2016.1254161
Lindwall, O., & Ivarsson, J. (2010). Differences that make a difference: Contrasting the local enactment of two technologies in a kinematics lab. In S. Ludvigsen, A. Lund, I. Rasmussen, & R. Säljö (Eds.), *Learning across sites: New tools, infrastructures and practices* (pp. 364-380). Amsterdam: Elsevier.
Marton, F. (2015). *Necessary conditions of learning*. New York: Routledge.
Marton, F., Runesson, U., & Tsui, A. B. M. (2004). The space of learning. In F. Marton & A. B. M. Tsui (Eds.), *Classroom discourse and the space of learning* (pp. 3-40). Mahwah: Lawrence Erlbaum.
Marton, F., & Tsui, A. B. M. (Eds.). (2004). *Classroom discourse and the space of learning*. Mahwaw: Lawrence Erlbaum.
McDermott, L. (1997). Students' conceptions and problem solving in mechanics. In A. Tiberghien, E. L. Jossem, & J. Borojas (Eds.), Connecting Research in Physics Education with Teacher Education: ICPE. Retrieved from http://www.physics.ohio-state.edu/~jossem/ICPE/C1.html.
Menekse, M., Stump, G. S., Krause, S., & Chi, M. T. H. (2013). Differentiated overt learning activities for effective instruction in engineering classrooms. *Journal of Engineering Education, 102*(3), 346-374.
Prince, M. (2004). Does active learning work? A review of the research. *Journal of Engineering Education, 93*(3), 223-231.
Richards, M. J. (1971). An ABC of dimensional analysis. *Physics Education, 6*(4), 244-249.
Sokoloff, D. R., Laws, P. W., & Thornton, R. K. (2007). RealTime Physics: Active learning labs transforming the introductory laboratory. *European Journal of Physics, 28*(3), S83-S94.
Sokoloff, D. R., Thornton, R. K., & Laws, P. (1998). *RealTime Physics*. New York: Wiley.
Streveler, R. A., & Menekse, M. (2017). Taking a closer look at active learning. *Journal of Engineering Education, 106*(2), 186-190.
ten Have, P. (2007). *Doing conversation analysis: A practical guide* (2nd ed.). Los Angeles: SAGE.
Thornton, R. K., & Sokoloff, D. R. (1998). Assessing student learning of Newton's laws: The Force and motion conceptual evaluation and the evaluation of active learning laboratory and lecture curricula. *American Journal of Physics, 66*(4), 338-352.
Trowbridge, D. E., & McDermott, L. C. (1980). Investigation of student understanding of the concept of velocity in one dimension. *American Journal of Physics, 48*(12), 1020-1028.
Trowbridge, D. E., & McDermott, L. C. (1981). Investigation of student understanding of the concept of acceleration in one dimension. *American Journal of Physics, 49*(3), 242-253.
Trumper, R. (2003). The physics laboratory: Historical overview and future perspectives. *Science & Education, 12*(7), 645-670.

Acknowledgements

I am indebted to Dr Anna-Karin Carstensen for valuable discussions. Support from the Swedish Research Council through grant VR 721-2011-5570 is gratefully acknowledged.

Copyright statement

Ideas about relationship between physics and mathematics in engineering students

Vizcaino. A., Diego. F.
Antonio Nariño University, Bogotá, Colombia
dfvizcaino@udistrital.edu.co

Castiblanco. A., Olga L.
District University, Bogotá, Colombia
olcastiblancoa@udistrital.edu.co

Abstract: *We investigate students' ideas about relationships between physics and mathematics from their discourses and study methods. Our research question was how a group of students in a civil engineering program understands the relationship between physics and mathematics? We developed a qualitative research through a questionnaire and a semi-structured interview. Questions design was based on characterizations about this relationship by the authors in a previous research. We found that students have a restricted perception related to finding and using right equations to solve specifically "pen and paper problems". No express ideas about mathematical processes can help them to solve other kind of problems, nor that physics study can be developed without starting by equations. They are not looking for comprehension about physical phenomena, neither intentions to deepen into meanings of equations beyond of proportionality relations, but all of this seems to be a result the way they are taught.*

Context

In education engineering it is necessary to design teaching methodologies that lead to skill training indispensable to this profession, such as the ability to think in a creative an autonomous way, or linking scientific concepts with real problems. In the teaching and learning processes of physics we can reinforce this ability, because this objective is also part of scientific education. Moreira (2000) mentions how from the physics teaching these skills can be worked with students, he classified three types, (1) representation and communication, (2) research and understanding and (3) sociocultural contextualization. In addition, the author says that it is only possible if educational processes are assumed taking into account the construction of knowledge, modelling of meanings, relationship with the context and meaningful learning.

Several studies in science teaching such as Driver, Newton & Osborne (2000), and Viennot (2002), teach us within its fundamental premises that physics students bring to the class many previous ideas about the topics to be addressed into the classroom and useful for designing teaching processes. This implies, that teaching is only about bringing encyclopaedic knowledge to the student ends up failing, because it does not start from considering the knowledge that students already bring, what can be considered as epistemological obstacles to be solved taking as a reference the conceptual model, instead that perspective assume that scientific conceptual model already contains the teaching methodology itself, which leads to disregard the personal explanatory capacity that students can have and that can be used to collaborate in their learning process.

One of the challenges that today must be solved in physics teaching is the level of autonomy that can be achieved in physics students, considering that, according to authors such as Castiblanco and Nardi (2016), many students can manifest an adequate domain of mathematical algorithms but fails to construct an autonomous discourse about physics he has learned, and instead generates a relationship of dependence on theory. This is evidenced when there is difficulty in speaking of any concept of physics beyond the description of a mathematical algorithm. At this point also authors such as Vizcaíno, Terrazzan (2015) talk us that relates between physics and mathematics in general is diffuse for teachers and students, because many students associate it directly with descriptions of relations of proportionality between variables involved in a physical phenomenon, and ignore everything related to historical, philosophical and epistemological aspects that support the existence of a certain mathematical expression, and even more ignore why the proportionality relation is in this way. Bernhard (2000) reinforces this problem by finding physics students have functional comprehension problems, a fact that was found even in highly selective training centers and postgraduate students. Although this deficiency in conceptual understanding diminishes when the level of students is more advanced, it never reaches zero and even occurs among physics teachers.

It is clear that physics training influences student ideas about the relationship between physics and mathematics, but the problem is to understand how that influence is. Authors such Kragh (2015) given in this regard "while physics students often think of mathematics as a tool, math students are often aware of the role that physics plays in constructing o mathematical concepts and techniques." This is an evidence that relationship between physics and mathematics is very complex, but the difficulty increases when teachers confuse the meaning of this relationship in physics and mathematical constructions with the processes of learning physics and mathematics, according to Vizcaíno and Terrazzan (2015), there is a difference between mathematization in physics history and mathematization in a classroom, also Kanderakis (2016) mentions that students interpret equations as "the commands to calculate" where the equal sign is "an operator symbol- a signal to do something in order to get the answer", which is a barrier to understand physics.

In this way, it is clear that the training of thinking skills of engineering students based on physics teaching requires solid fundamentals that clarify the relationship between physics and mathematics. But changing methodological principles involves exploring new perspectives designs in order to improve learning processes. Therefore, we present in this work a methodology to determine the meanings that engineering students present about the relationship between physics and mathematics and, consequently, some ideas about on what points need a reflection to innovate in the classroom.

Methodology

Data were taken with a group of 15 students of second semester in a course of physics within the civil engineering program of a particular university in the city of Bogota. We seek to identify and classify the meanings they manifest about the relationship between physics and mathematics. To this end, was designed and applied a survey that asked them how they understood the relationship between physics and mathematics from their routine exercise, based on Vizcaino's (2013) proposal, which categorized three different approaches for this relationship, namely, mathematical models, physical-mathematical processes, and conceptual physics. These categories were simultaneously a reference for data analysis. To complement the data obtained in the survey, a semi-structured interview was designed and applied with 6 students from the same group who voluntarily wanted to participate in this activity, investigating their habitual methods of studying physics.

Data collection and analysis was developed on the principles of qualitative research methodology proposed by Flick (2009) from the perspective of content analysis, in three phases, firstly the pre-analysis, then a coding and categorization and finally an interpretation.

The survey contained the following four items:

1) Describe the process you usually use to solve a physics problem.

2) Choose an undulatory phenomenon, name it and represent it.

3) One way of describing Archimedes' principle is: "the upward buoyant force exerted on a body immersed in a fluid, totally or partially submerged, is equal to the weight of the fluid that the body displaces and acts in the upward direction at the center of mass of the displaced fluid." How would you explain this principle to someone else?

4) For you, what is the meaning of the mathematical expression: $F = m.a$?

The interview began with the question about how to study to prepare participation or assessments in physics classes.

Findings

1) On the meanings of the relationship between physics and mathematics

- The first topic of the questionnaire asks about possible sequence of steps used by students to solve a physics problem. Students respond by listing the steps necessary to solve a physics problem, in general, they state as the first step the action of reading the problem statement to be able to decide what equations to use according to data offered, and then proceed to apply equations to solve the problem, which shows that students think only in a "pen and paper problem" in the description of Becerra-Labra, Gras-Martí and Torregrosa (2012), where a theoretical statement is placed and a question is asked whose answer is a value found, based on the application of one or more equations. There is no evidence that the problem can be understood as an experimental type, a type of conceptual analysis or a real problem. This already indicates to us a characteristic of the teaching processes that are applied in its subjects of physics, probably centred in the preparation to solution problems like "pen and paper".

Although not mentioned directly, it is important to note that in an engineering program, students have expectations about how to learn scientific knowledge applications in real situations but it does not occur. Then, we can deduce that when teacher guide this kind of students just for "pen and paper" problems, appears a lack of possibilities to think physics in engineering situations that require mathematization processes and conceptual developments. In this sense we would agree with Castiblanco and Nardi (2015) and Niss (2016) when they mention that if we want students to acquire the ability to work correctly problems from real world situations that involve mathematization, it is necessary to give them that opportunity, and in turn, give them the corresponding formative feedback about the work they do.

Analysing answer details, we find that eight of the fifteen students propose 3 steps for solving a problem, three propose 4 steps, three propose 5 steps and one proposes 6 steps. Even with this variety in steps quantity, the essence of these steps is summarized in (1) reading the problem to understand it that is to identify kind of information, (2) finding the right equations to be used and (3) applying them to find the solution. The variants correspond to criteria to choose equations and identify data offered by the statement. None of the answers includes verification or results validation, showing us that it is not a usual fact in their classes, neither in their studying customs, or their understanding of importance of this step in solving a problem. The whole problem-solving scheme is presented in a linear way, without repetition of steps or without improving understanding of the phenomenon studied, this means that it is granting to the fact of using mathematical schemes (equations) all their confidence in obtaining the correct result, looks like it is no necessary to reflect on it, or they are not educated to do that.

Students who propose 3 steps mention that they firstly analyse the problem, however, this analysis consist in read carefully to identify data, variables and the question, and then, to decide how linking data and equations. An example is the student 10 who says:

> *(S10): "I Identify values in the problem, replace data in the corresponding equation and calculate the result."*

For students who answered 4 steps the difference is they talk about finding the answer just in the final step after 2 steps to find and apply the equation. Students who answered with 5 steps, mentions that final step consist in compare the answer obtained respect to the question, they talk about finding data and choosing equations as different parts of the process. Whoever chose 6 steps mentions as final step to review whether the result is right or not, respect to answers of similar exercises or problems.

These results lead us to infer that, although some students think about finding solutions as a process going beyond the simple use of equations and looking for some essence of what they do, in general, everyone considers that after finding the correct equation, answer must be correct.

This is confirmed in the interview, where students openly state that they privilege mathematical part because they study physics and mathematics in the same way, that is, they solve exercises where they have to find an answer that finally is a unique numerical value. Occasionally some students say:

> *(S1): "In the physics class, the teacher always leaves us some exercises in a book, then it is to do them and let's also say that sometimes I look at tutorials to better understand the subject."*
>
> *(S2): In calculus, the teacher leaves us a certain ... some points of the book that could appear in the Quiz, then with our classmates we go to the library to solve them"*

We can see that, when students solve a certain amount of exercises that teacher assigns, normally from the textbook, not original problems formulated by teachers, students learn a mechanical processes and their study method is to review exercises solved in class. No one talks about taking a time to talk with the teacher to solve doubts or discuss problems, but they talk about going to watch tutorials in Internet when they find a solution models.

- The second item asks them to choose an undulatory phenomenon, to name it and to represent it. With this question we want to see meanings of "represent" in physics. We observed that four students did a drawing without accompanying any explanation, four of them drew a wave form of the sinusoidal type but without specifying values or type of variables represented in the axes, two of them presented only the x axis without data, values or names to the axis. One made a drawing with a small explanation, and two wrote a short phrase about what they consider to be a wave phenomenon. (See example in fig.1). Four of them did not respond.

In cases where a sinusoidal shape was used as a representation, the axes drawn are not intended to represent the relationship between two specific variables, nor are any units of measurement used in them. Therefore, we can say the specific utility of having used this type of graphical representation is like a symbol that itself talk about characteristics of undulatory movement, for example repetitions. In all the answers there was no representation of the phenomenon through equations, it was not accompanied the drawing of some explanation or conceptualization, so we can say that students understand "to represent" as a way of communicating something that happens, trough iconic images and, in this case, it does not include a mathematical or conceptual form as a support for the representation, but rather an image that contains itself a content.

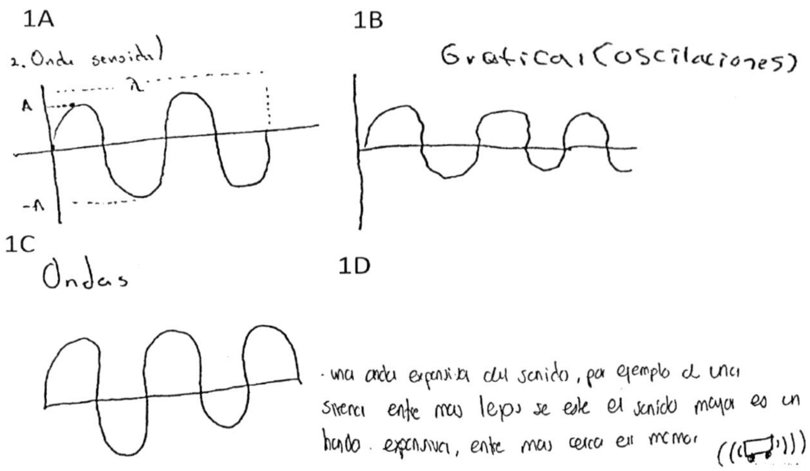

Figure 1. Example of student answers to item 2. (a) Sinusoid wave, (b) Graphic (oscillations), (c) waves, (d) An expansive sound wave, for example a siren, the further away the wave is more expansive and the closer it is the less.

In interviews we verified this perspective, since when asking students if in resolving an exercise they made graphs or drawings, we got answer like:

> *S1-if the exercise requires it*
>
> *S2-well, sometimes, for example, I do them to identify some things and I make a graphic or something like that.*

This shows us that students make little use of different types of representations for a phenomenon. It does not means students can not use it or can not represent a phenomenon with different forms of representation, is just they do not usually do it, probably because teaching and explanation processes in classroom does not include different forms. At least for this group, there is not emphasis in conceptual understanding going beyond a numerical response, for instance, looking for different representation types, such as graphical, pictorial or argumentative to expand phenomena analysis.

- In third item students are asked how would explain Archimedes principle for another person. We found that nine students merely paraphrased the same paragraph that was presented to them as the definition of Archimedes' principle. Of these, six paraphrased with conceptual errors. Two mentioned that they would explain it through Newton's third law, two paraphrased by changing the vessel by a pool of water, and one of them changed the object by a ball. Only one included a drawing but just to talk about volume change and finally, one student did not respond.

We can see that explanation meaning consist in be able to repeat the Archimedes principle using the same or similar words. None of the answers shows indications about explanation in steps or using different representations, or analysing experimentation possibilities, neither expanding conceptual definitions implicit there. Curiously, in spite of importance they give to equations, none makes reference to the fact that some mathematical formalism could be included in the explanation. Although, two mentioned they would explain it with third Newton´s law, they do not write this law, nor mentions what variables are involved in this law, for example, relation between densities of liquid and object, or forces of gravity, weight and buoyancy, or different equilibrium conditions that could be achieved. Hence, explanation is understanding as the action of describing a law statement without assign processes, again, may be because is the way that they are learning physics.

- In the fourth item, we inquire by the meaning that students give to the expression $F = ma$.

We find that thirteen of them responded by describing the equation with the words "Force is equal to the product of mass by acceleration" or something similar. The remaining two students changed some words by making the relationship between the variables wrong. None mention of measurement units, or dimensional analysis.

Taking into account that question inquires about what would mean to the student this mathematical expression and noting that most of them put the emphasis on repeating proportionality relationship between the variables presented there, we can say that physical meaning of the equation is the equation itself. Main function of equation is to obtain a numerical value from two other data, without taking into account the physical meanings of each of the variables neither the conditions that must be fulfilled so that this equation can be used correctly. In general, it is seen that mathematical description does not include qualitative aspects or description of experimental facts; in addition they assume algorithms use like privileged language to represent physical phenomena.

This finding contrasts with results in the second item, where, when asking for phenomena representation, they used iconic images and not equations, so we can interpret a certain degree of incoherence or incomprehension about the meanings they construct in. Why they decide using an iconic image as the best representation for phenomena and no equations? Probably they did not know Archimedes principle equations, it means that students actually use equations when teacher give them equations, because is a formal language that make them feel they are saying something right, no matter if they do not understand it. But, when they must to explain from their own comprehension they go to images that are in their minds.

Conclusions

We conclude that students understand the relationship between physics and mathematics based almost exclusively in the use of equations to find solutions of "pencil and paper" problems, in some cases seem a kind of intention to explain the relationship between variables, but in no cases there are comments on "why" such relationship or meaning is given. When they use equations to describe the phenomenon occurrence, whether expressed in the algorithm or description in words, they do not go to other forms of representation that could expand understanding, for example do not go to graphic representacion, experimental representations or relation with any situation lived in daily context.

This fact talk to us about need to re-think the way that equations are being introduced in the physics teaching, since equations seem to become a barrier that prevents students from generating their own speeches and their own explanations, because when they assume mathematics as the physics language, such language is restricted to the simple presentation of equations and does not refer to mathematization processes, which is the natural form as has been mathematizing physics throughout its history. We are not saying it is not important to work equations in teaching processes, we are questioning the way it is presented and used, based on this results.

No scientist began to postulate his contributions to knowledge, starting to work with an equation like absolute truth. On the contrary, those who manage to run the barriers of knowledge are because they construct enough argumentation to verify the established truths, but also, the equations are representations of ideal physical systems that scientists were organizing in a squence of random, nonlinear and non-predetermined sequence of phases, so, why do we pretend to teach only the final results of scientific research and not enable them to participate in this scientific world?

But even more, our questioning based on these results is about how the processes of mathematization in the classroom should be developed, since they should definitely not be the same ones developed throughout history because that processes occurred in different and diverse socio-cultural contexts, and under particular conditions, besides that it has not been developed by a single person but by many people from different parts of the world, in many moments of history, which is very difficult to recreate in the classroom. Neither can it be to save the student the whole process of building knowledge and delivery only final results to be consumed in the application of theoretical problems. Thus, the design of specific mathematical processes for physics teaching in the context of each professional career is required.

In the case of engineers training, it must start from the reality of the expected profile of this professional, where it requires be educated to understand scientific knowledge in order to make decisions about how and where to apply it, it must also relate this knowledge with problems of the real engineering life. But also, future engineers must be educated for autonomy, creativity, argumentation, construction of explanatory models, observation and, in general all skills involved in processes of physics mathematization.

It is clear that all of this aspects have a starting point from teachers perspective, since it is seen students imitate the way the teacher presents physics, therefore it is necessary for teachers to understand physics, not with the only purpose of making their students learn a particular model, but to design learning environments where students have the opportunity to participate in processes of explanation construction, to invite students to continue investigating outside the school on aspects that allow them to gain understanding and deepen his knowledge, and not where extra class work is limited to training for the best imitation of model exercise solution presented in classroom.

References

Becerra-Labra, C., Gras-Martí, A., & Torregrosa, J. M. (2012). Effects of a problem-based structure of physics contents on conceptual learning and the ability to solve problems. *International Journal of Science Education,* 34(8), 1235-1253.

Bernhard, J. (2000). Improving engineering physics teaching: Learning from physics education research. *Physics Teaching in Engineering Education (PTEE 2000),* 14-17 June 2000, Budapest, Hungary.

Castiblanco, O; Nardi, R. (2013) Un uso de la historia en la enseñanza de la didáctica de la física. *Góndola, Enseñanza y Aprendizaje de las Ciencias* v8n2

Castiblanco, O; Nardi, R. (2016) Percepções e expectativas de estudantes de licenciatura sobre a Aprendizagem para o ensino de física. *Experiências em Ensino de Ciências* V.11, No. 1

Driver, R., Newton, P., & Osborne, J. (2000). Establishing the norms of scientific argumentation in classrooms. *Science education,* 84(3), 287-312.

Flick, U. (2009). An introduction to qualitative research. Sage.

Kanderakis, N. (2016). The Mathematics of High School Physics. *Science & Education*, 25(7-8), 837

Kragh, H. (2015). Mathematics and physics: The idea of a pre-established harmony. *Science & Education,* 24(5-6), 515-527

Moreira, M. A. (2000). Ensino de Física no Brasil: retrospectiva e perspectivas. *Revista brasileira de ensino de física.* São Paulo. Vol. 22, n. 1 p. 94-99.

Niss, M. (2016). Obstacles Related to Structuring for Mathematization Encountered by Students when Solving Physics Problems. *International Journal of Science and Mathematics Education*, 1-22

Viennot, L. (2002). Razonar en física: la contribución al sentido común. Antonio Machado.

Vizcaino, D. (2013). Papel da matematização nas explicações de professores e alunos em disciplinas de física na formação inicial de professores. (Doctoral thesis). Paulista State University. (UNESP)

Vizcaíno, D; Terrazzan, E. (2015) Diferencias trascendentales entre matematización de la física y matematización para la enseñanza de la física. *Tecné, Episteme y Didaxis*, (38), 95-111

Vizcaino, D; Terrazzan, E. (2013) Significados de matematização de professores e estudantes de um curso de licenciatura em física: um estudo de caso. Góndola, Enseñanza y Aprendizaje de las Ciencias. v8n1

Acknowledgements: To the institute of studies and educational investigations (IEIE) of the District University and its support to the project "Relation of the results of investigation in the area of sciences and the teaching practice".

Copyright statement

The Role of Theory in Engineering Education Research

Bruce Kloot
University of Cape Town, Cape Town, South Africa
bruce.kloot@uct.ac.za

***Abstract**: This is a paper in progress that takes as its point of departure the shift towards 'systematic, evidence-based research' (Jesiek et al., 2011, p. 79) in about the 1980s in the USA and thereafter in Europe. Drawing on the work of Merton (1968) and Prediger et al. (2015), it sketches the context of global trends in engineering education and reviews the literature in an attempt to understand how theory is used in the field of engineering education broadly and how this has shifted in recent years. In order to narrow the search, citation indices are used to identify the papers that have had a significant impact on the field. These papers are analysed for evidence of theory (in the broad sense, which includes methodology/paradigm) and the various ways in which theory is used. The findings indicate that less than half the papers analysed use theory in some way.*

Introduction

The use of theory in engineering education research has gained attention in recent years. This paper endeavours to understand the trends that have shaped the use of theory over the years in order to better understand the trajectory of engineering education research and to inform theory use in the future. As such, it focuses on the Discipline-Based Education Research (DBER) field known as engineering education (Singer et al., 2012). Although engineering training obviously draws on the natural sciences, in particular mathematics and physics, the particular disciplinary-based research endeavours in these areas are not directly considered in this paper. Of course, the research trends in science education, and also from the social sciences and other fields, have certainly shaped engineering education, especially in recent years. These influences are important and are part of the investigation that explores the use of theory that follows even though the literature from these sub-fields is not directly considered.

Although *engineering education* may be broadly understood to refer to any formal training in engineering, from technical training at secondary school level to post-tertiary training within the context of an apprenticeship, the focus here is engineering education within the context of tertiary institutions, i.e. in the context of a university or polytechnic or an equivalent. It has been argued that in Britain engineering training was deliberately inserted into the science faculties at universities in the context of the rise to dominance of what Lundgreen (1990) calls the 'school culture'. This model underpins the type of training in the former colonies, such as in the USA and South Africa (Kloot and Rouvrais, 2017). Even though other models of engineering education exist, namely the German and the French models, engineering training globally appears increasingly to be narrowly located within an expanding and internationalising higher education system (Gardelle 2017).

Moreover, while classroom has long been an interest of engineering educators, the specific endeavour of *engineering education research* is of interest in this paper. As will be seen, this is contested terrain since it revolves around what is considered to be legitimate research. Nevertheless, the emergence of formal societies and dedicated journals are taken to delineate the field of engineering education *research* and the substance of the articles published by these dedicated journals are taken to represent the intellectual ideas of the field. Citations are used to indicate the currency and reach of intellectual trends in the field, something that is explained in more detail in the research methods section.

Context

Perhaps because of the technical background of most engineering educators involved in publishing their work (Adams and Cummings-Bond, 2004), engineering education research has traditionally been closely coupled with the contextual knowledge of the discipline. As suggested in the introduction, 'contextual knowledge' in this sense is taken to mean the technical knowledge (and associated skills) that constitute engineering curricula worldwide. Felder and Hadgraft (2013) observe that for 'most of the twentieth century, engineering education research mainly consisted of using student satisfaction surveys and instructors' impressions to assess the effectiveness of teaching methods, courses, and curricula' (p. 339). In his analysis of the articles published in the *JEE*, Wankat (2004) suggests that papers in engineering education are typically of the 'type "we tried this method and it was great" ... that contain no data other than student evaluations' (p. 19).

The 1980s marked an important shift in the USA since, at this time, the National Science Foundation (NSF) made substantial support for scholarship in engineering education available. According to Wankat et al. (2002), this has probably done more to raise awareness of the scholarship of teaching and learning in engineering than any other single factor. In the early 1990s, the flagship journal of the American Society of Engineering Education, *Engineering Education*, was re-named the *Journal of Engineering Education* (*JEE*) and positioned as the society's 'scholarly journal, seeking to serve the needs of the engineering education enterprise' (Ernst 1993, p. 1), a move which drew inspiration from Ernest Boyer's (1990) model of the various facets of scholarship. Wankat et al. (2002) suggest that in this decade, the literature on the scholarship of teaching in engineering education grew rapidly, using surveys and quantitative research methods, approaches with which engineers tend to be relatively comfortable, but also noted that the 'qualitative research methods used widely in the social sciences are gradually percolating into the engineering education literature' (p. 237).

At the start of the new millennium, particular attention was paid to the promotion of *rigour* in engineering education research, an approach that seemingly superseded the scholarship of teaching and learning since it requires that the researcher (i) begins with a research question; (ii) ties the research question to learning, pedagogical or social theory; and (iii) pays careful attention to the study design and methods (Streveler, Borrego and Smith, 2007, p. 143). Although this was driven strongly by an NSF-funded project, there has been some contestation of the assumption that the use of theory or a theoretical framework, possibly derived from the social sciences, necessarily translates into more rigorous research (Felder and Hadgraft, 2013). This was coupled with the concern that an overly strict requirement for theory to ensure acceptance for journal publication would disadvantage researchers wishing to publish their more applied research findings.

Engineering education in Europe was established more recently than in the USA. The European Society of Engineering Education (Société Européenne pour la Formation des Engénieurs – SEFI) was founded in Brussels in 1973 and its official journal, the *European Journal of Engineering Education* (*EJEE*) was launched two years later. The trajectory of this community is similar to that in the USA in that research was traditionally quantitative and close to the contextual knowledge of the discipline. In the late 1990s the SEFI community began to explore the synergies between research and engineering education (Seitzer, 2000) and about a decade later, at the annual SEFI conference in 2008, a working group for EER was established. This demonstrated a shift in the field towards more research-based approaches within engineering education in Europe. It also resulted in a special issue in the *EJEE* entitled 'Educational research impacting engineering education' (Baillie and Bernhard, 2009). The articles in this special issue were considered by the editors to be 'thoroughly anchored in the emerging field of EER' (p. 293) and generally employed qualitative methodologies.

Although engineering education in Europe is smaller than in the USA and the shift towards evidence-based research has occurred more recently, there are a few other important differences between these communities. Firstly, Europe does not benefit from a central fund like the NSF in the USA which drives the field in a particular direction (de Graaff et al., 2012). Secondly, there is significantly more variation within Europe in terms of models of engineering education and thus more attention is paid to national diversity and issues of compatibility (prompted by the Bologna Process) within engineering education. Thirdly, there is a stronger synergy between engineering education and broader research-based support in higher education in Europe within the social sciences. Indeed, the Editor-in-Chief of the *EJEE* in 2010 suggested that the recent development of engineering education into a 'field of scientific research in its own right' appeared to 'unite engineering and social sciences research traditions' (de Graaff, 2010, p. 2).

Although Borrego and Bernhard (2011), colleagues on opposite sides of the Atlantic, compare the USA and (Northern and Central) Europe as 'two examples of the diversity of approaches' (p. 14) it is important to note the *global unevenness* of the endeavour of engineering education as a whole in the context of the shift towards more research-based or theory-rich approaches. In terms of output and sheer number of members involved, the USA is by far the biggest centre of activity. This is followed by the Northern and Central European region which, although smaller than the US, is more international in terms of sources cited and authorship (Williams et al., 2014) in the most important local journal (see below). To give an indication of the relative size of these centres, the most cited paper in the *JEE* has been cited more times than the citations of the top 25 papers in the *EJEE* combined.

Apart from Australia which has a professional organisation and a dedicated journal which serves a vigorous community, the activity in the rest of the world is limited. South Africa is the most active hub within Africa, with the African Engineering Education Association (AEEA) bringing members together from the rest of the continent. There are pockets of activity in Asia (such as in Turkey, Malaysia and India), as well as in the countries of Central and South America but many commentators have questioned the quality of the work in these areas in terms of the parameters of engineering education *research* (Jesiek et al., 2011). This said, there are platforms that bring researchers from different parts of the world together such as the *International Journal of Engineering Education* (*IJEE*) and the Research in Engineering Education Network (REEN), an international organisation that holds a biennial conference.

Overlaying this geographical unevenness are the varying extents to which engineering education communities are involved in the shift towards more 'systematic, evidence-based research'. An analysis of the articles in the most prominent journals in the field at the turn of the century suggests that even in the USA, a substantial proportion of published research is characterised as 'descriptive accounts' of projects, programmes and courses (Osorio and Osorio, 2004). The recognition and prioritization of teaching innovations and technical studies over educational research in Australia is explained by what Jesiek et al. (2009) call the 'historically peripheral nature of EER' which accounts for the 'show and tell' nature of much EER in this region (p. 3). Finally, the journals embedded in the engineering sub-disciplines such as *Chemical Engineering Education*, the *International Journal of Mechanical Engineering Education* and the *IEEE Transactions on Education*, represent a particular subset of the community, the work of which is strongly dependant on the contextual knowledge of their particular sub-disciplines.

Theoretical framework

As noted in the Introduction, this paper takes a broad view on theory. However, before continuing, it is helpful to propose to a working definition of theory as a starting point: a 'theory is a statement of concepts and their interrelationships that shows how and/or why a phenomenon occurs' (Corley and Gioia, 2011, p. 12). Such a definition is consistent with the conception of theory in the natural sciences where mathematical formulae might be used to express interrelationships between variables rather than concepts. In physics the General Theory of Relatively is an obvious example.

In order to analyse the sorts of theories that are used in engineering education, as well as how they are used, this paper draws on the work of Robert Merton (1968). Merton was concerned with theory-building in the field of sociology and described three levels of theory: so-called 'grand' theory, middle range theory and 'micro-theory'. For Merton, grand theories were too distant and abstract to be able to provide important contextual insights into social behaviour which might arise from empirical study. While they could provide descriptive accounts at an elevated level, such as for society as a whole, he did not believe this was of much use. For this reason he preferred to work with middle range theory to bridge the gap between the abstract level of grand theory and context-laden propositions that emerge from empirical study at the level of what might be called 'micro-theory'.

In order to distinguish between grand theories and middle range theories, Trowler (2012) suggests that grand theories are abstracted from the empirical, as noted above, and are also 'reasonably unambiguous [from an] ontological and epistemological standpoint' (p. 274). Examples in sociology include Marxism and postmodernism. These differ from middle range theories which have an empirical base and are 'corrigible but of general validity' (ibid. p. 275). In the field of education, the Communities of Practice learning theory developed by Lave and Wenger (1991) is an example of a middle range theory. What distinguishes middle range theory from micro-theory is generalisability.

In the field of maths education, Prediger et al. (2015) describe something similar in a survey paper about design research, an approach which they note has its roots in engineering. Close to the empirical level, Prediger et al. advocate the development of humble, pragmatic theories: 'humble in the sense of being concerned with topic-specific learning processes, and pragmatic in that they effectively inform prospective design' (p. 879). For Prediger et al. (2015) design research also operates at a level between humble and 'grand' theory but in case of the learning sciences, grand theories include those of Piaget, Vygotsky and Dewey and their intellectual heirs.

In applying these notions to engineering education research, it is important to take note of three things: *firstly*, grand theory is not necessarily better than a middle range or a humble theory. This paper aims to analyse articles that use theory in any way, keeping in mind each level of theory has a particular purpose. This paper interests itself with a device which, at any level, describes interrelationships between some concepts that shows how and/or why a phenomenon occurs. In this way, it takes the broad view that a what might be called a 'paradigm', an 'approach', a 'proposition', a 'methodology', a 'model' an 'hypothesis' or a 'conjecture' but does theoretical work will be categorised at some level as a theory in the analysis that follows.

Secondly, since these levels of theory are derived from ideas about theory building and design research, there is fluidity between these levels. Humble theories can be combined or can contribute in some way to the development of a middle range theory which, in turn, can become a grand theory. Furthermore, apart from the description above about the distinction between the levels, there are no strict criteria for distinguishing one level of theory from another.

Thirdly, while the discussion in this paper is about explicit theories, Trowler (2012) alerts us to the notion of *tacit* theories, implicit and often unexamined assumptions of the researcher which underpin the propositions put forward in a paper that may seem to be atheoretical. Although not the focus of this paper, this is particularly relevant in the content of a DBER area such as engineering education. According to Singer et al., (2012), 'DBER investigates learning and teaching in a discipline using a range of methods with deep grounding in the discipline's priorities, worldview, knowledge, and practices' (p. 9). Further work is required in exploring the deeply grounded worldview and knowledge practices that underpin engineering education (Johnston, Lee and McGregor, 1996).

Research method

Given the dominance of the USA within the field, papers published in the flagship journal of the American Society of Engineering Education, the *JEE*, are analysed in terms of what kind (or level) of theory is used and how it is used. Drawing on the work of Kandlbinder (2013) who argues for the use of citation indices to analyse the field higher education studies, this paper narrows the search of the literature by examining the ten most cited papers in each year. This information was obtained using Harzing's (2007) program 'Publish or Perish', starting in the year 1988. This year was chosen because the most cited paper within the entire engineering education literature was published in this year (Felder and Silverman, 1988). At this time, the *JEE* was simply *Engineering Education*. Also, the late eighties are important with regard to the shifts in the field described in the Context section above so it is necessary to begin in the late 1980s.

At the time of writing, five years worth of the papers from the *JEE* had been analysed but the intention is to analyse the years up until 2012. The reason for this choice is to ensure sufficient opportunity for citations to emerge in the most recently published papers so that the ten most cited works are chosen. Due to unavailability of the hard copy journals in the early years, the years 1991–1992 were not examined. The analysis therefore includes 1988–1990 and 1993–1994. These missing years will hopefully not have an undue effect since the aim is to try to track how the use of theory shifted through the 1990s and into the new millennium, in particular whether the effect of call for more scholarly, rigorous and research-based approaches with engineering education can be noticed in the literature.

It is hoped that paper from the *EJEE* may later be analysed to allow a comparison with what emerges from the *JEE*, particularly given the similarity in trajectory of these two centres of engineering education research activity.

Preliminary findings and discussion

The 50 papers from the five years under consideration were analysed to determine what theories were used (if these were explicitly mentioned), at what level the theory was used and whether there was any form of theory-building in the paper, an act which necessitates the inclusion of empirical work. As noted already, these are preliminary findings and it is hoped that a more complete and recent data set will indicate trends in theory use in later decades.

As Wankat (1999, 2004) found in his analysis of papers published in the *JEE*, the proportion of papers *explicitly* referring to theory in this data set was low at 18% (nine papers). A good example of such a paper is the Felder and Silverman (1988) article already mentioned. This paper can be thought of as introducing learning (and teaching) styles to the engineering education community which is probably one of the reasons that it has been so widely cited. Another reason is probably that the paper draws on principles derived from educational psychology, enabled by Silverman's expertise, but made relevant or pertinent to the field of engineering education by Felder, at the time already a well known figure in the engineering education community.

According to Coffield et al. (2004), who sharply critiqued the vast array of models of learning styles that had appeared by 2004, a 'host of empirical and conceptual problems' lie beneath the apparently unproblematic appeal of learning styles' (p. 2). Nevertheless, the notion of learning styles can be traced back to an empirical base and the idea is certainly generalisable. It is thus clear that learning styles may be classified as a middle range theory. Most of the other theories explicitly mentioned in the papers are also middle range theories and include Astin's student involvement theory, Kolb's Learning Cycle, the Myers-Briggs

Type Indicator (a survey instrument developed from Carl Jung's model of psychological types) and Perry's model of intellectual development.

Of these nine papers, eight do not put the theory to use with empirical data. Examples of such papers are those by Landis (1988), who used Astin's student involvement theory to make a case for minority engineering programmes, and Culver (1990), who used Perry's model of intellectual development to inform a curriculum design process. Koen's (1994) paper draws heavily on behaviourial and cognitive psychology to provide a perspective on teaching design. These may actually be considered grand theories but Koen does not use them to build theory in the sense of Merton (1968). This paper is more a discussion of the merits – and instances of possible misuse – of ideas drawn from psychology in their application in engineering education. As a final example, Harb, Durrant and Terry (1993) combine Kolb's Learning Cycle and the 4MAT learning system to develop 'a simple model which maybe be applied by all faculty, independent of their disciplines and particular teaching styles, to improve and enhance engineering education' (p. 76). So while there appears to be some evidence of theoretical innovation, there is no engagement with empirical data to qualify this as an example of developing a middle range theory as conceived of by Merton.

Perhaps the paper to come closest to doing this is the paper by Felder et al. (1993) entitled 'A longitudinal study of engineering student performance and retention: I. Success and failure in the introductory course'. This paper involves some theory building in that it analyses student profiles in terms of their scores on the Myers-Briggs Type Indicator and the Learning and Study Strategies Inventory and develops a model, actually more like a set of factors, to assist in the prediction of student performance.

Although the remainder of the papers did not explicitly refer to theory, a further 26% (13 papers) contain some form of empirical investigation and conclude with proposition of some sort, however small, regarding a relationship between concepts attempting to explain how/why a phenomenon occurs. For this reason they were categorised as engaging with micro- or humble theory. For example, Barber et al. (1989) used survey questionnaire data to explore why students pursue graduate engineering degrees. They came up with a series of factors including financial reasons, overloaded curricula and family encouragement to pursue PhD studies. Another example is Brickell et al. (1994) who tested five different methods for assigning students to group design projects. They found that groups that are heterogeneous with respect to GPA but homogeneous with respect to curriculum interest appear to perform better.

Together, this means that 44% of the papers surveyed over five years of the *JEE* engaged with theory in some way. The other 56% of the 50 papers analysed did not appear to engage with theory at all and were typically descriptions of curricula interventions or opinion pieces.

Conclusion

This preliminary study has analysed what theories are used and how they are used in the *JEE*, arguably the most prestigious journal in the field of engineering education research. Drawing on the work of Merton (1968) and Prediger et al. (2015), three levels of theory were identified which were used as categories to understand the use of theory in the ten most cited papers over five years for the journal, starting in 1988. These three levels of theory were derived in the context of theory building and this aspect was also examined for the 50 papers under consideration.

While only 18% of the papers explicitly referred to theory, and a smaller proportion brought the theory in contact with empirical data or engaged in some form of theory building, a larger number of papers, 44% overall, used theory in some way, some at the micro-level, so-called humble theory, which typically entailed putting forward some sort of proposition embedded in the context of disciplinary knowledge.

It will be interesting as this study proceeds to discover the trends that develop into the 1990s and 2000s when the discourse of the scholarship of teaching and learning began to have an influence and, later, when the importance of research rigour came to the fore and engineering educators were encouraged to draw theories from the learning sciences. Further, it will be useful to reflect on whether trends from research in the other disciplinary-based areas such as science, physics and mathematics education made a noticeable influence on the use of theory in engineering education. It is worth noting, for example, that the notion of learning theories were enthusiastically taken up in the field of higher education studies before the notion was introduced to the engineering education community by Felder and Silverman in the late 1980s. Perhaps other theoretical trends in neighbouring fields have had a similar effect.

Finally, it will be interesting to see how papers in the *EJEE* compare with the trends that emerge from the *JEE* since this centre of activity followed a similar trajectory although there were also some important differences. Overall, it is hoped that this study will provide a clearer picture of the important trends in the role that theory is playing in engineering education research worldwide and will also inform our how theory is used in shaping future research.

References

Adams, R. S., & Cummings-Bond Jr, T. (2004). Career trajectories in engineering education – Where are they now?. *Proceedings, ASEE Annual Conference and Exposition.*

Astin, A. W. (1984). Student involvement: A developmental theory for higher education. *Journal of college student personnel*, *25*(4), 297–308.

Baillie, C., & Bernhard, J. (2009). Editorial: Educational research impacting engineering education. *European Journal of Engineering Education, 34*(4), 291–294.

Barber, E. G., Morgan, R. P., Darby, W. P., & Sallmensmith, L. (1989). To pursue or not pursue a graduate engineering degree. *Engineering Education*, *79*(5), 550-555.

Borrego, M., & Bernhard, J. (2011). The Emergence of Engineering Education Research as an Internationally Connected Field of Inquiry. *Journal of Engineering Education, 100*(1), 14–47.

Boyer, E. L. (1990). *Scholarship reconsidered: Priorities of the professoriate*, Princeton University Press, 3175 Princeton Pike, Lawrenceville, NJ 08648.

Brickell, L. C., Porter, L. C., Reynolds, L. C., & Cosgrove, C. R. (1994). Assigning students to groups for engineering design projects: A comparison of five methods. *Journal of Engineering Education*, *83*(3), 259-262.

Coffield, F., Moseley, D., Hall, E., & Ecclestone, K. (2004). Should we be using Learning Styles? What research has to say to practice. *London: Learning and Skills Development Agency.*

Corley, K. G., & Gioia, D. A. (2011). Building Theory About Theory Building: What Constitutes a Theoretical Contribution? *Academy of Management Review*, 36(1), 12–32.

Culver, R. S. (1990). Gaining Professional Expertise through Design Activities. *Engineering Education*, *80*(5), 533-36.

de Graaff, E., Alpay, E., Bernhard, J. Ihsen, S., & Kolmos, A. (2012). Editorial. *European Journal of Engineering Education, 37*(1), 1–2.

Ernst, E. W. (1993). The Editor's Page. *Journal of Engineering Education, 81*(1), 1.

Felder, R. M., & Silverman, L. K. (1988). Learning and teaching styles in engineering education. *Engineering education*, *78*(7), 674-681.

Felder, R. M., Forrest, K. D., Baker-Ward, L., Dietz, E. J., & Mohr, P. H. (1993). A longitudinal study of engineering student performance and retention: I. Success and failure in the introductory course. *Journal of Engineering Education*, *82*(1), 15-21.

Felder, R. M., & Hadgraft, R. G. (2013). Educational Practice and Educational Research in Engineering: Partners, Antagonists, or Ships Passing in the Night? *Journal of Engineering Education*, *102*(3), 399–345.

Gardelle, L. (2017) Editorial, *European Journal of Engineering Education*, *42*(2), 127–130,

Harb, J. N., Durrant, S. O., & Terry, R. E. (1993). Use of the Kolb learning cycle and the 4MAT system in engineering education. *Journal of Engineering Education*, *82*(2), 70-77.
Harzing, A.W. (2007) Publish or Perish, available from http://www.harzing.com/pop.htm
Jesiek, B., Beddoes, K. Sangam D., & Borrego, M. (2009). Mapping Local Trajectories of Engineering Education Research to Catalyze Cross-National Collaboration. *Proceedings of the SEFI Annual Conference, Rotterdam,* The Netherlands, 1–4 July.
Jesiek, B. K., Borrego, M., Beddoes, K., Hurtado, M., Rajendran, P., & Sangam, D. (2011). Mapping global trends in engineering education research, 2005–2008. *international Journal of engineering education*, *27*(1), 77-90.
Johnston, S., Lee, A., & McGregor, H. (1996). Engineering as captive discourse. *Techné: Research in Philosophy and Technology*, *1*(3/4), 128-136.
Kandlbinder, P. (2013), Signature concepts of key researchers in higher education teaching and learning. *Teaching in Higher Education, 18*(1), 1–12.
Kloot, B. and Rouvrais, S. (2017). The South African engineering education model with a European perspective: history, analogies, transformations and challenges. *European Journal of Engineering Education 42*(2): 188–202.
Koen, B. V. (1994). Toward a strategy for teaching engineering design. *Journal of Engineering Education*, *83*(3), 193–201.
Landis, R. B. (1988). The Case for Minority Engineering Programs. *Engineering Education*, *78*(8), 756-61.
Lave, J., & Wenger, E. (1991). *Situated learning: Legitimate peripheral participation*. Cambridge University Press.
Lundgreen, Peter. 1990. Engineering Education in Europe and the U.S.A., 1750–1930: The Rise to Dominance of School Culture and the Engineering Professions. *Annals of Science 47*(1), 33–75.
Merton, R.K. (1968). *Social theory and social structure*. New York: The Free Press.
Osorio, N., & Osorio, M. (2002). Engineering education in Europe and the USA: An analysis of two journals. *Science and Technology Libraries 23*(1), 49–70.
Prediger, S., Gravemeijer, K., & Confrey, J. (2015). Design research with a focus on learning processes: An overview on achievement and challenges. *ZDM, 47*(6), 877–891
Seitzer, D. (2000). Editorial: Research and engineering education, *European Journal of Engineering Education, 25* (2), 111–113.
Singer, S. R., Nielsen, N. R., & Schweingruber, H. A. (2012). Discipline based education research. *Washington, DC: The National Academies.*
Streveler, R. A., Borrego, M., & Smith, K. A. (2007). Moving from the 'scholarship of teaching and learning' to 'educational research': An example from engineering. In D.R. Robertson (Ed.), *To Improve the Academy, Volume 25* (pp.139-149). Bolton, MA: Anker.
Trowler, P. (2012). Wicked issues in situating theory in close-up research. *Higher Education Research & Development, 31*(3), 273–284.
Wankat, P. C. (1999). An Analysis of the Articles in the Journal of Engineering Education. *Journal of Engineering Education*, *89*(1), 37–42.
Wankat, P. C. (2004). Analysis of the First Ten Years of the Journal of Engineering Education. *Journal of Engineering Education*, *93*(1), 13–21.
Wankat, P. C., Felder, R. M., Smith, K. A., & Oreovicz, F. S. (2002). The Scholarship of Teaching and Learning in Engineering. In M.T. Huber and S. Morreale (Eds.), *Disciplinary Styles in the Scholarship of Teaching and Learning: Exploring Common Ground*. AAHE/Carnegie Foundation for the Advancement of Teaching, Washington.
Williams, B., Wankat, P., Neto, P. C., & Tiago, C. (2014). *Is Engineering Education Research Global? The Answer May Surprise You.* Paper presented at the ASEE Annual Conference, Indianapolis, 15–18 June.

Acknowledgements

Thanks to Saalih Allie, Kate Le Roux and Dale Taylor for their help in developing the ideas for this paper.

Mathematical vs. Physical Modelling for calculating the dynamic properties of rigid bodies

Giacomo Barbieri
Universidad de Los Andes, Bogotá, Colombia
g.barbieri@uniandes.edu.co

Ibrahim Ignacio Barbery
Universidad Simón Bolívar, Caracas, Venezuela
12-10766@usb.ve

Carlos Francisco Rodriguez
Universidad de Los Andes, Bogotá, Colombia
crodrigu@uniandes.edu.co

***Abstract**: Simulation tools utilized for the synthesis and analysis of physical systems have raised their abstraction through the introduction of physical modelling approaches. In comparison with traditional mathematical modelling, physical approaches allow reducing complexity, and modelling time and efforts. However, when dynamic properties obtainable through both methods are considered, physical approaches constitute a subset of the mathematical ones. On the basis of our experience, this comparison is not provided within the classes of rigid body dynamics. The objective of this paper is to define a new teaching approach which introduces mathematical and physical modelling. The approach will underlie the results that can be achieved through each method in the context of the analysis of rigid body dynamics.*

1. Introduction

Complexity of modern systems is growing: increasing functionality, flexibility and performance are required at reduced costs and within shorter delivery cycles.
The problem of managing system complexity has started within the *embedded domain*. In fact, an embedded system is intrinsically a complex system since must perform predefined tasks satisfying a set of requirements on responsiveness in terms of quality and timeliness. Moreover, implementation requirements must also be fulfilled such as cost, power consumption, and utilization of limited physical resources. Eventually, its interaction with the environment should be continuously available for the entire life of the artefact.
Sangiovanni-Vincentelli (2007) identifies solutions adopted within the embedded domain in order to face complexity:

- Utilize virtual tools for the synthesis and analysis of systems;
- Enhance modularity and reuse;
- Raise the layer of abstraction at which designer express systems.

The same has happened within the *mechanical engineering domain*. Mathematical models are utilized for describing a system in a mathematical form and its resolution (i.e. simulation) shows how system behaves in a particular environment. *Mathematical models* are standard practices both for the synthesis and analysis of physical systems since shorten times and costs as illustrated in Norton (2004).
While maintaining the possibility of "low-level" mathematical modelling, simulation tools have started raising the layer of abstraction in order to speed-up the modelling process. Examples of this tendency are Modelica as illustrated in Fritzson (2004) and Matlab through its toolboxes. Simulink toolbox allows building the mathematical model of a system by interconnecting basic mathematical blocks. Whereas, toolboxes as Simscape are

characterized from blocks implementing "high-level" functionalities. Model is built creating a network of these blocks by imitating the functional elements of the modelled system.

In rigid body dynamics, high-layer of abstraction is achieved through the modelling of the system physical structure: *physical modelling*. Modeller builds a structural representation of the system and the simulation software converts it into an equivalent mathematical model saving time and efforts. Moreover, physical structure modelling can be automated by importing the CAD model into the simulation software. Examples of simulation software allowing this modelling approach are multibody simulation tools as Matlab through the SimMechanics toolbox and MSC Adams.

Academic must reflect this tendency in order to provide an appropriate "up-to-date" education. Traditionally, mathematical formulations have been taught in courses of rigid body dynamics. However, physical modelling has many advantages as shorten modelling time, and can be used for the verification of the correct implementation of the mathematical approach.
While teaching the two methods, a balance must be identified since the utilization of physical modelling without the former knowledge of mathematical approaches may be "dangerous". Since simulation tools automatically convert high-level physical descriptions into mathematical descriptions, the "blind" utilization of physical modelling may lead to the following problems:

- *Black box*: simulation works as a black box since modeller has not the knowledge to predict expected behaviour. Examples are shown in Clariana et al. (2008);
- *Debug*: in the event results are identified to be incorrect, modeller is not able to debug the model since it is not clear what the high-level actors implement;
- *Obtainable results*: physical approaches constitute a subset of mathematical approaches when dynamic properties obtainable through each method are considered. The knowledge of just physical modelling limits the set of obtainable results.

This paper is meant to propose an approach for the introduction of mathematical and physical modelling. The novelty consists in the comparison of the results obtainable through the two methods when same inputs are provided to the same mechanical system. The objectives are to: (i) clarify mathematical equations implemented within physical modelling; (ii) identify application areas for the two modelling methods.
The paper is structured in this way: state of the art is illustrated in section 2, while the proposed teaching approach is introduced in section 3. Section 4 applies the approach in an MSc class of 3D dynamics of rigid bodies. Obtained results are summarized in section 5, while conclusion and future work are discussed in section 6.

2. State of the art

Traditionally, just mathematical approaches have been taught in dynamic courses. There is strong evidence of this in the literature as Meriam and James (2012), Kane and Levinson (1985) and Baruh (1999). During the last years, physical approaches have been introduced due to the widespread of multibody simulation tools as illustrated in Zi et al. (2011).
Different works can be cited about the comparison between mathematical and physical modelling.
Frankovský et al. (2013) show that same results are obtained when dynamic properties of particles are computed with the two methods. However, they do not investigate the methods within the calculation of rigid body dynamics.
Shaoqiang el al. (2008) compare the results obtainable when the same control system and input signal are applied to an inverted pendulum modelled both through mathematical and physical approaches. Similar results are obtained and they conclude that the only drawback of physical modelling is that motion equations are not provided. Therefore, control system theories such as pole configuration, optimal control and so on, cannot be applied.
Schlotter (2003) compares mathematical and physical modelling based on numerical choices

utilized to convert differential algebraic equations of a physical model into ordinary differential equations solvable with numerical methods. However, a comparison based on the obtainable results is not provided.

Authors noted that a complete comparison from a user perspective have not been performed yet. Application areas for mathematical and physical modelling has not been explicitly identified for calculating the dynamic properties of rigid bodies.
On the basis of our experience, this generates confusion on students since they finish these courses without having a clear idea on the differences and implications of using one or other method for the calculation of dynamic properties.
This paper is meant to propose a new teaching approach for rigid body dynamics introducing the comparison between mathematical and physical modelling, and studying the results that can be achieved through each method. This teaching approach will make students conscious about which is the most adequate method for the calculation of the dynamic properties for given situations.

3. Proposed approach

Considering the number of tools / software available today, one may think that engineering is the art of utilizing software. However, the capability of utilizing one or more software makes a person a good technician but not a good engineer. Abraham Maslow said:

"If your only tool is a hammer, you tend to see every problem as a nail"

In fact, a good engineer is a person able to select either the proper tool or a combination of them or to create new tools based on the problem to be solved. This is the reason why authors think that academic must mainly focus not on the utilization of software but on all the concepts behind them. In this way, students will be able to correctly select and utilize them.

Based on these considerations, authors propose the approach illustrated in Figure 1 for a course on analysis of rigid body dynamics. Class and evaluation will be subdivided into:

- *Theory:* 60% of the total grade;
- *Project*: 30% of the total grade;
- *Engineering report:* 10% of the total grade.

Next, the three parts are presented.

Theory: mathematical formulations are introduced in class. "Mathematical modelling" box within Figure 1 illustrates typical topics learnt in dynamic classes. The a criteria of ABET (2016) affirms that students need to acquire the skill to: "Apply the engineering fundamentals, basic science and mathematics for modelling, analysing and building physical systems, components or processes, and for working professionally within the areas of thermal and mechanical systems". In the context of a dynamic class, this means that students need to obtain the ability to solve kinematic and kinetic problems, and to operate on mass properties once inertia matrices have been provided. Therefore, the evaluation of the acquisition of capabilities concerning mathematical formulations must worth 60% of the total grade.

Project: throughout the class, students self-taught a multibody simulation tool and all the activities listed within "Physical modelling" box in Figure 1. Nowadays, self-taught of tools is almost a daily activity for an engineer and self-taught of physical modelling has the advantage to not subtract time to the teaching of mathematical formulations.
Moreover, in order to accustom students to work in group, a group project will be performed. Professor will provide the CAD assembly and the target kinematic trajectory of a dynamic mechanical system. Then, students calculate the dynamic properties (i.e. kinematic, mass and kinetic properties) of the selected case study using both methods: mathematical and physical modelling. Eventually, a comparison among the two methods is performed, and pro and cons for each method are identified.
This activity will worth 30% of the total grade and will teach students to: (i) work in group; (ii)

apply the studied approaches to a realistic case study; (iii) utilize software in a critical way. In fact, through the comparison between mathematical and physical approaches, students must predict the expected results and identify why the approaches may provide different results.

Engineering report: communication of ideas and results is a very important aspect of engineering. Therefore, students will present through an engineering report the performed work and the obtained results. This activity will worth 10% of the total grade.

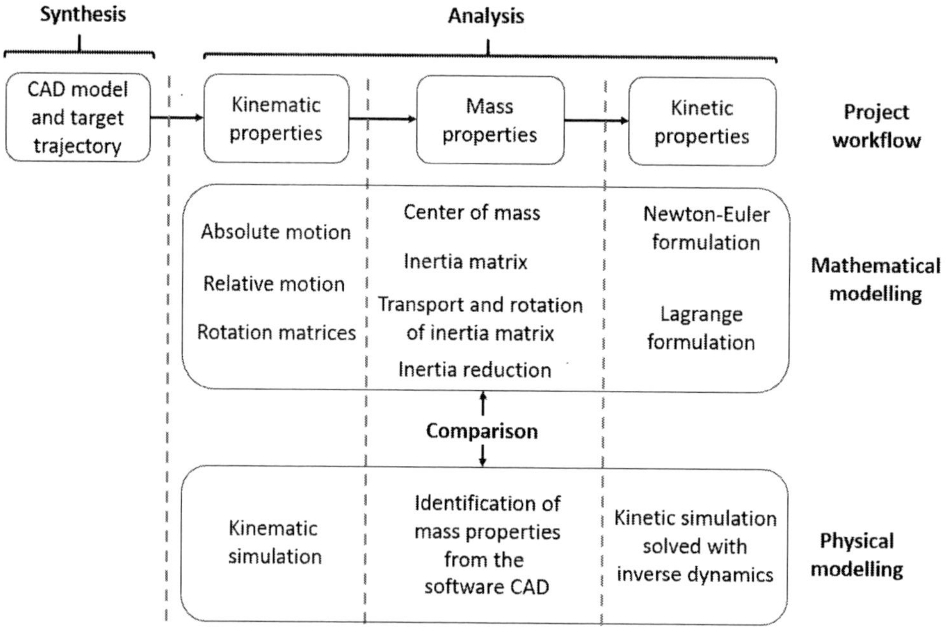

Figure 1: Graphical representation of the proposed approach.

4. Application of the proposed approach

This section summarizes the application of the proposed approach in an MSc class of 3D dynamics of rigid bodies. A robotic leg moving in space was considered as case study. Geometry of the 4DoFs leg was provided as project input, along with the kinematic trajectory of each rotational joint. Then, students applied mathematical and physical modelling for calculating the dynamic properties of the robotic leg while implementing target trajectory. Eventually, a comparison between the two methods was performed.
Figure 2 and 3 respectively show the CAD assembly of the robotic leg and the kinematic trajectory of the joint between body B and C.

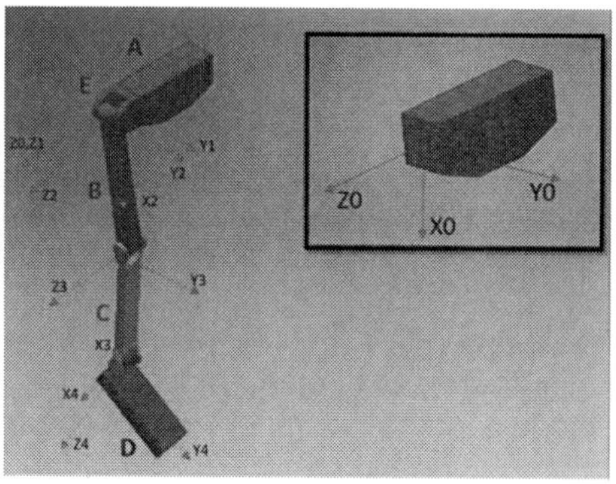

Figure 2: CAD assembly of the leg.

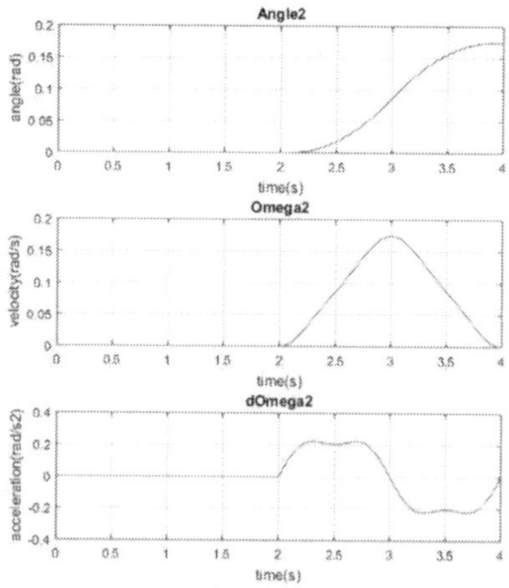

Figure 3: Gutman 1-3 kinematic trajectory.

4.1. Kinematic properties

Students were asked to calculate kinematic properties of each body while implementing the target trajectory. Selected kinematic properties consisted in:

- *Angular properties:* angular velocity and acceleration of each body;
- *Linear properties:* linear velocity and acceleration of each body centre of mass.

In order to mathematically compute selected properties, the two formulations introduced in Kane (1985) were utilized:

- *Absolute motion:* calculation of position vectors with respect to an inertial frame and their time derivation for identifying velocities and accelerations;
- *Relative motion:* joint trajectory identifies relative angular velocity and acceleration between the two connected bodies. This information can be utilized for the application of relative motion formulations as additional theorem and chain rules.

Eventually, each vector can be described with respect to any arbitrary coordinate system. In order to compare mathematical and physical modelling, properties must be expressed with respect to the same coordinate system. The inertial frame was selected and rotation matrices were utilized for expressing each kinematic property with respect to the inertial frame.

Autodesk Inventor Dynamic Simulation was utilized in this phase for physical modelling. Target trajectory was input to each rotational joint and kinematic properties were sensed with respect to the inertial frame.
Figure 4 shows the linear velocity of body B centre of mass obtained with mathematical and physical modelling. It can be appreciated that same results are obtained with the two methods.

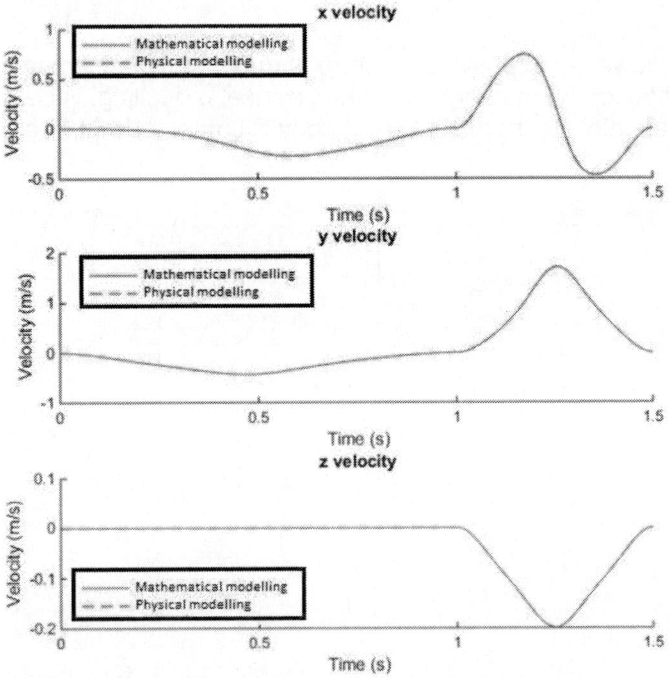

Figure 4: Velocity of body B centre of mass with mathematical and physical modelling.

4.2. Mass properties

As second task, students had to acquire the skill of operating with mass properties. Centre of mass location was obtained from the CAD model, along with the inertia matrix computed with respect to a frame with principal directions and placed on the body centre of mass. Then, students calculated how these properties changed according to the implemented trajectory:

- *Leg centre of mass:* calculation of the overall system centre of mass described with respect to the inertial frame, and identification of its time variation;
- *Leg reduced inertia matrix:* principal inertia matrix of each body was converted into an inertia matrix computed with respect to the inertial frame, and its time variation was identified. Eventually, system overall inertia matrix was calculated.

Mathematical formulation utilized for computing the centre of mass of the whole system consisted in considering each body as a particle with its mass placed in its centre of mass. Then, system centre of mass was computed with the principle of superposition illustrated in Kane (1985).
Whereas, theories for transporting and rotating inertia matrices were applied in order to describe each body inertia matrix with respect to the inertial frame. Parallel axis theorem was utilized for the translation, while rotation matrices for the rotation as shown in Ginsberg (2008).
Eventually, leg reduced matrix was computed by summing the inertia matrix of each body described with respect to the inertial frame.

Autodesk Inventor Dynamic Simulation was utilized for physical modelling.
However, system centre of mass cannot be sensed while implementing a given trajectory. In order to compare the results, time variation of each body centre of mass was sensed and time variation of the whole leg centre of mass was obtained with the superposition principle. Moreover, inertia matrix of a body cannot be sensed in inventor with respect to any arbitrary frame. Since inertia matrix can only be expressed in the principal coordinate system, body inertia matrix remains constant while implementing the trajectory. Reader can notice that all the operations concerning the manipulation of inertia matrices cannot be performed within physical modelling.

Figure 5 plots the time variation of system centre of mass while implementing the trajectory. Same results are obtained for mathematical and physical modelling. Wheareas, Figure 6 shows the time variation of the inertia matrix of body B computed with respect to the inertial frame.

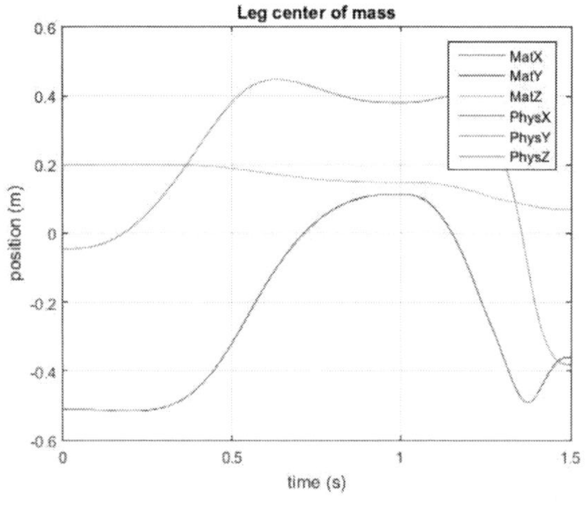

Figure 5: Leg centre of mass trajectory.

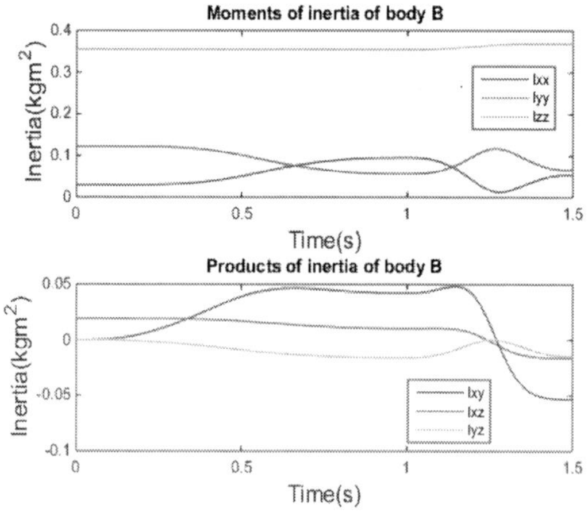

Figure 6: Inertia matrix of body B.

4.3. Kinetic properties

As final task of the project, students computed kinetic properties due to the application of the target trajectory. These properties consisted in:

- *Actuation torque*: torque necessary for implementing each joint target trajectory;

- *Reaction efforts*: reaction forces and bending moments exerted on the supports and actuator shafts.

Mathematical formulations used to compute kinetic properties were:

- *Newton-Euler:* in this formulation, each body is analysed independently. After having drawn the free body diagram, Newton-Euler equations are applied for calculating the inverse dynamics: actuation torque necessary for implementing the target trajectory and consequent reactive efforts on the supports;
- *Lagrange:* in this formulation, the whole system is analysed through an energetic approach, without the need of being separated in its constituting elements. However, this method provides only the actuation torques, while reaction forces and bending moments cannot be computed.

SimMechanics was utilized in this phase for physical modelling. Target trajectory was input to each rotational joint, and actuation torques and constraint efforts were sensed. Actuation torque is described with respect to the joint axis, while constraint efforts with respect to body principal coordinate system.
Whereas, within mathematical modelling, efforts are obtained with respect to the coordinate system in which kinematic properties are expressed. For this reason, rotational matrices were applied for converting kinetic properties in the same coordinate system. In this way, the comparison between the different approaches was possible.

Figure 8 illustrates the actuation torque of the joint between body A and E necessary for implementing its target trajectory. It can be appreciated that Newton-Euler and Lagrange formulations (i.e. mathematical modelling) and the physical modelling approach provide the same results. Whereas, Figure 9 represents the reaction forces exherted on the supports of the motor between body E and B. Again, same results are obtained with mathematical and physical modelling.

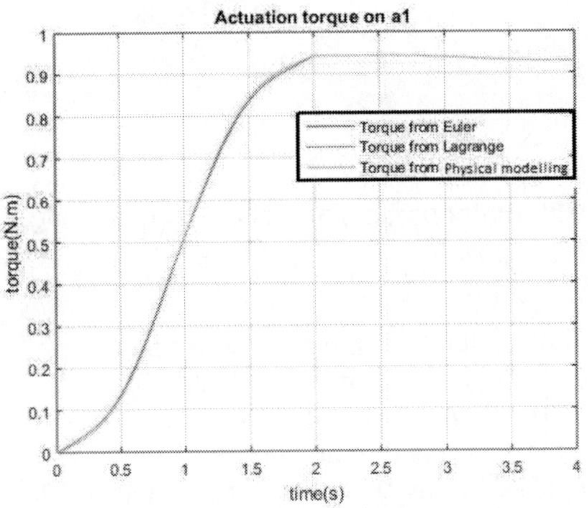

Figure 8: Actuation torque of the first motor.

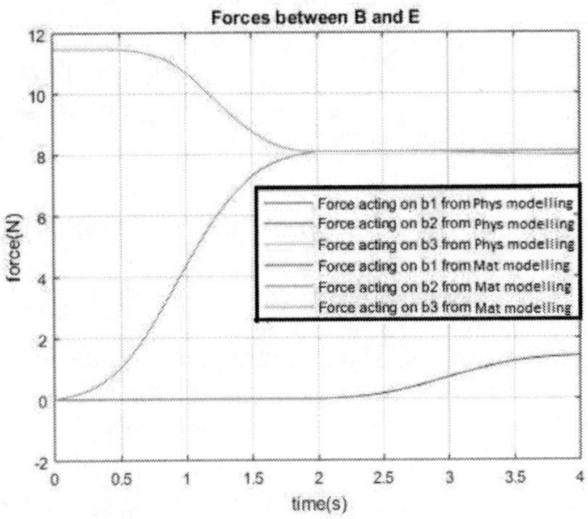

Figure 9: Constraint forces between E and B.

5. Results

Section 4 illustrated how same results were obtained for kinematic and kinetic properties. Concerning mass properties, the two methods calculate same trajectory for the system centre of mass. Whereas, physical modelling does not provide any support for operating on body inertia matrix. Therefore, transport and rotation of inertia matrices, and inertia reductions can only be performed with mathematical modelling.
Figure 10 resumes the results obtained by comparing mathematical and physical modelling for the calculation of dynamic properties.

Within the implemented case study, same results were obtained because same conditions were applied within mathematical and physical modelling. Same conditions imply same input kinematic trajectories, same materials but in particular same geometric properties of the bodies. In fact, real geometries introduce offsets due to the physical joints; within mathematical modelling, joints are generally approximated as points. The inclusion of these geometric offsets within mathematical modelling increases modelling efforts without providing an appreciable contribution to the results.
In conclusion, it was an useful exercise for showing how same results can be obtained with the two methods. However, in a practical case study approximate geometries can be considered since they not provide an appreciable contribution.

During each task, students noticed that mathematical modelling implies larger work than physical modelling. This is due to the identification and implementation of the proper mathematical formulation; with the risk of "human errors". Since same results are provided, physical modelling is more convenient for computing these properties when the knowledge of the underlying mathematical model is not necessary.
However, they concluded that it is necessary to learn both the approaches since the knowledge of the mathematical model is fundamental for solving many engineering problems as optimizations, control system design. In these cases, physical modelling can be used as a verification of the correctness of the implemented mathematical formulations.

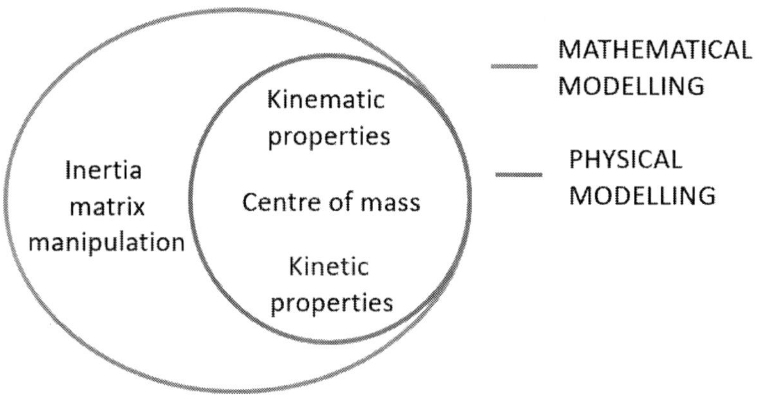

Figure 10: Comparison of the results obtainable with mathematical and physical modelling.

6. Conclusions

Outcomes of this experience were the identification of similarities and adequate application areas for the two methods, along with the acquisition of competences for calculating dynamic properties of rigid bodies, both through mathematical and physical modelling.
Proposed approach was applied to a class of dynamic analysis of rigid bodies moving in space. However, the approach can be potentially utilized to any dynamic class as plane rigid bodies and mechanisms.

Moreover, case study modelling abstraction followed the workflow used in class for introducing mathematical formulations: definition of kinematic properties due to the implementation of a target trajectory, calculation of mass properties and computation of necessary kinetic properties for implementing the required trajectory. Consequently, case study complexity was broken-down and students were able to apply theoretical concepts as soon as they were introduced in class.

Within the class, the ideal hypothesis that system can perfectly implement target trajectory was assumed. An integration with control system classes may be investigated in order to ends up with an implementation of a real device. The following workflow (ideally implemented in a real modelling process) may be applied after the calculation of the mass properties:

- *Mechanical design*: implementation of Lagrange and Newton-Euler formulations, and verification of their correctness through the comparison of the obtained actuation torques. Selection of the required actuator;
- *Control system design*: mathematical model is enriched with the motor inertia and control system is designed. Implementation of a physical model including control system law and verification that same results are obtained with the two methods;
- *Deployment*: dimensioning of the supports, software deployment and device building. Validate the model by tuning physical parameters through the comparison between simulation and experimental results; e.g. friction, damping, stiffness, etc.

References

ABET: "*Criteria for accrediting engineering programs*": http://www.abet.org/accreditation/accreditation-criteria/criteria-for-accrediting-engineering-programs-2016-2017/, 2016

Baruh, Haim. "*Analytical dynamics*". Boston: WCB/McGraw-Hill, 1999.

Clariana, R. & Strober, J. "Modeling technology". Handbook of research educational communication and technologies, 2008.

Frankovský, P., Hroncová, D., Delyová, I., & Virgala, I. "*Modeling of dynamic systems in simulation environment MATLAB/Simulink–SimMechanics*". American Journal of Mechanical Engineering, 2013.

Fritzson, Peter, and Vadim Engelson. *"Modelica: A unified object-oriented language for system modeling and simulation".* European Conference on Object-Oriented Programming. Springer Berlin Heidelberg, 1998.

Ginsberg, J. "*Engineering dynamics*". Cambridge University Press, 2008.

Hroncová, D., Binda, M., Sarga, P., & Kicák, F. "*Kinematical analysis of crank slider mechanism using MSC Adams/View*". Procedia Engineering, 2012.

Kane, Thomas R., and David A. Levinson. "*Dynamics, theory and applications*". McGraw Hill, 1985.

Meriam, James L., and L. Glenn Kraige. "*Engineering mechanics: dynamics*". Vol. 2. John Wiley & Sons, 2012.

Norton, Robert L. *"Design of machinery: an introduction to the synthesis and analysis of mechanisms and machines"*. McGraw-Hill Professional, 2004.

Sangiovanni-Vincentelli, Alberto. "*Quo vadis, SLD? Reasoning about the trends and challenges of system level design.*" Proceedings of the IEEE 95.3 (2007).

Schlotter, M. "*Multibody System Simulation with SimMechanics*". University of Canterbury, 2003.

Shaoqiang, Y., Zhong, L., & Xingshan, L. "*Modeling and simulation of robot based on Matlab/SimMechanics*". In 27th Chinese Control Conference, 2008.

Zi, B., Cao, J. B., & Zhu, Z. C. *"Dynamic simulation of hybrid-driven planar five-bar parallel mechanism based on Simmechanics and tracking control"*. International Journal of Advanced Robotic Systems, 2011.

Acknowledgements

Authors want to thank the 2016-20 MSc class of "Advanced Dynamics" at University of Los Andes for the feedback provided throughout the application of the approach. Eventually, authors want to thank prof. Camilo Enrique Olaya for his ideas about the role of the engineer and for the sentence of Abraham Maslow that have inspired the definition of the proposed approach.

Copyright statement

Introducing a new framework for education in engineering based on the competence loop and Bloom's updated taxonomy: Learnings from a longitudinal study

Alicia Medina
Umeå School of Business and Economics, Umea University, Malmo, Sweden
alicia@medina.se

Daniel Jurburg
Engineering Faculty, Universidad de Montevideo, Montevideo, Uruguay
djurburg1@um.edu.uy

Claudio Ruibal
Engineering Faculty, Universidad de Montevideo, Montevideo, Uruguay
cruibal@um.edu.uy

Abstract – *One of the greatest current challenges for engineering education is to relate the education to what is needed in the fast-changing world of today, while also making it more relevant for students. One way of solving this problem involves reviewing the appropriateness of the existing curricula and finding new ways of teaching and developing competences. This study investigates what graduates from an Engineering school in Uruguay consider to be their acquired competences and to what extent their expectations of the labor market are fulfilled. This is a qualitative study, based on a single case approach and a longitudinal perspective in which a purposive sample of 18 graduates participated in two rounds of interviews with an interval of 12–18 months. The results from this study points towards the introduction of a new educational framework for engineering education based on the merge of a newly published framework called the competence loop and Bloom's updated taxonomy.*

Keywords: Engineering, Bloom's taxonomy, Competences, Higher Education

Introduction

Higher education has been undergoing a paradigm shift since the 1990s. Barr and Tagg (1995) introduced the concept of *learning paradigm* as a new way of seeing the mission of higher education, instead of the traditional way of *instructing* or *teaching* to produce learning.

In line with this paradigm shift, the American Society of Civil Engineers (ASCE) developed in 1998 the Body of Knowledge (BOK) in which the need for change in engineering education is stated, and the definition of the knowledge, skills, and attitudes necessary for engineers is presented. This first BOK was reviewed and updated in 2005 when BOK 2 was created. ASCE based its body of knowledge on Bloom's Taxonomy from 1956 (knowledge, comprehension, application, analysis, synthesis, evaluation) because it provided an appropriate framework for the definition of the levels of achievement in the engineering education for the twenty-first century. In 2002, Krathwohl published a revision of Bloom's taxonomy, developing the following categories: Remember, understand, apply, analyze, evaluate and create. Other authors, such as Male, Bush, and Chapman (2011), Hung, Choi, and Chan (2003), and Passow (2012), further developed this view and call for a change in the educational system.

All these authors expressed the need for new approaches to learning in order to meet future needs. In particular, Beanland and Hadgraft (2014, p.92) recognize that we are living in a time of a paradigm shift which they explained as: "*Significant benefits can result when the new environment is acknowledged and addressed. It is necessary to question everything associated with our current engineering education programs and to rethink how the desired outcomes can be most effectively delivered. A paradigm shift requires action, action requires leadership, and leadership requires courage and conviction."*

In connection with this need to find new approaches in learning in engineering education, the current paper has the main purpose of presenting and analyzing the results of the first two rounds of a longitudinal study in which a group of 18 undergraduate students from an engineering school in Uruguay reflect upon their learning during their studies at the university, the application in the workplace of the competences developed at university, the undergraduates' working ambitions and assessment of whether they feel they can achieve their ambitions in relation to the education they received. The objective is to gain a deeper insight about the educational system applied at this School of Engineering and assess its degree of successfulness, in order to propose new insights about how engineering education should be developed in the future. The results from this study concludes towards the introduction of a new educational framework for engineering education.

The paper is structured as follows. First, a review of education in engineering and development of competences is presented. Second, the research methodology is presented including an outline of the main characteristics of the case university under study, followed by the main results. Finally, a discussion of the results along with the main conclusions from this research are shown. These include the presentation of a framework and both theoretical and practical implications.

Education in Engineering

The world is constantly changing at an exponentially speed rate. Much of this acceleration is caused by technological improvements, led by engineers. Paradoxically enough, engineering education has failed to adapt to these changes at the same pace. One example of this situation was noted in the UNESCO Engineering Report (2010) which highlighted that one of the greatest challenges for engineering education is to relate the education to what is needed in the fast-changing world of today, and make it more interesting and relevant for students. Universities should address these needs by reviewing the appropriateness of the desired outcome and curricula, the effectiveness of learning, and new ways of teaching through a problem-solving approach. The Royal Academy of Engineering in the UK has a similar view and in 2007 published a detailed study, *Educating Engineers for the 21st Century*, which considers many issues about engineering education. Among other findings, it states that universities and industry should work together to ensure that the contents of engineering education programs reflect what the industry requires from a professional engineer; and to provide real world work experiences as part of engineering education. They also argue that the ranking-driven emphasis on publishing and research in many universities may slow down the development of innovative learning and teaching in engineering. The focus should be on the importance of output competence.

In addition, the Washington Accord (2009) states that the main purpose of engineering education is to produce graduates with an ability to continue learning and to be able to develop the competences required for independent practice. Moreover, in a study of engineering education in leading universities in the USA, Sheppard, Macatangay, Colby, and Sullivan (2008) emphasized that engineering education should focus on developing the student's identity as a professional engineer and should be a professional practice in which technical knowledge is combined with skills of practice. Furthermore, Beanland and Hadgraft (2014) argue that there is a critical need for a transformation of education in engineering, so that young people will seek to enter a challenging career that is interesting, essential, satisfying, and important for the future. They emphasize that this change is urgent so that the role of an engineer can be adapted to future needs.

Development of competences

It is not difficult to agree that the goal of an institution for higher education is to develop certain competences, but what does it mean? Different authors will have different answers depending on their standpoint. In this paper, we apply Sanford's (1989) view of competence as being a combination of knowledge, skills, and performance, in which performance is the ability to use the knowledge and skills.

Moreover, based on Crawford (2005) competency model (input, personal and output competences), we conclude that the goal of an institution for higher education is to generate competence by creating knowledge and skills as well as developing personal characteristics and enabling the demonstration and measurement of performance.

In addition to the above definitions and categorization of competence, Medina and Medina (2015) have developed a framework for competence management in knowledge intensive and project intensive organizations. This framework is a loop called "the competence loop" that consists of four mechanisms: assimilation, transformation, utilization and accumulation. Those mechanisms are described as follows. *Assimilation* is the mechanism to assess, interpret, and understand new knowledge. *Transformation* is about knowledge integration as the state in which new and existing competences merge. *Utilization* is about utilizing existing competences. *Accumulation* is a mechanism to create or acquire new competence through practical work. Apart from these four mechanism, they have also identified two learning strategies as well as an interaction between learning and the application of the newly acquired competences.

Method

A qualitative exploratory approach based on a case study methodology was chosen as the strategy, and a longitudinal outlook of 5 years was selected as the time horizon. A longitudinal study is a correlational research study that involves repeated observations of the same variables or/and phenomena over long periods of time (Saunders, Lewis, & Thornhill, 2009). In this case, it means that we will follow a group of engineering graduates for five years after they have completed their studies. By conducting a longitudinal study the differences in the sample will be observed, i.e. we will track changes in the graduates throughout the length of the research project by repeated observations on an individual level (Cooper & Schindler, 2008).

The case selected was the School of Engineering at Universidad de Montevideo (UM). UM offers university programs in engineering and is ranked first among Uruguayan universities in the 2016 University Rankings (QS ranking): Latin America; first among the private universities in Uruguay, and number five among the Latin American young universities. Quacquarelli Symonds is a British organization founded in 1990 and has established itself as the leading global provider of specialist in higher education and careers information and solutions.

This University aims to apply a new learning paradigm in educating the engineers of the future based on a mix of theoretical and practical activities intended to develop certain competences in the undergraduate students. In order to measure the achievement of its objectives, in 2013 the university started an exploratory longitudinal study. The aims of the study are to assess students' perceptions of the knowledge, skills, and competences they have gained at the university, evaluated at the point of graduation and at different time points after graduation. The students' insertion on the labor market and their career opportunities and development is also being followed within the frame of the study. To consider the educational model successful, those perceptions should be positive and some evidences about the interest for starts ups and postgraduate studies must be found.

This paper presents the result of the first two rounds; the rest of this section describes how the study was conducted during these two rounds. Semi-structured interviews were used in order to capture the participants' perceptions, thoughts, and feelings as well as areas and topics beyond the researcher's questions (Miles & Huberman, 1996). The interview guide was designed to focus on the research goals and was piloted with subjects external to the university in order to ensure that the questions and language were clear.

The target population was all the graduates from the School of Engineering who had completed their studies during the six months prior to starting the first round of interviews and also students who were within three months of finishing their studies at the university. An invitation with a presentation of the study was provided to the target population and the secretary of the School of Engineering made the appointments with the 18 candidates who accepted to participate.

The first round of interviews was performed between August and September 2013; the interviews were recorded and lasted between 45 and 60 minutes. The interviewees were guaranteed their anonymity, and codes were used instead of the names. Theoretical saturation was reached after 14 interviews, but because of the longitudinal perspective we conducted the 18 booked interviews. We also interviewed one of the academic mentors, the dean, and the faculty member in charge of the improvements in order to have a complete picture of how the university operates, and have data from different sources, therefore, improving the quality of the findings.

Following Miles and Huberman (1994), the data collected was analyzed by looking for patterns, classifications, and categories. To secure reliability, the codification was conducted by two researchers independently and was compared and reworked until consensus was reached (Miles & Huberman, 1994).

The data collection of the second round started in November 2014 and lasted until August 2015 (the long period of time is because we aimed for a 100% participation rate). The data analysis was conducted in the same way as for the first round and included a comparison between the first and the second rounds.

The results from these two rounds were analyzed and discussed using literature from different domains and not only the higher education literature. Since the respondents regarded as unique and most important the way that the engineering program was constructed with a combination of theory and practical field periods, this led us to consider the domains of knowledge intensity and learning-by-doing perspectives.

Results

In terms of demographics of the sample, eleven participants were male while seven were female. All participants' age ranged from 23 to 27 years old at the time of the first round of interviews.

Following the methodology aforementioned, the following results were obtained. When coding the data from the semi-structured interviews in the first round, we identified the following categories: *Academic content* (perceived usefulness and importance), *Preparation for professional life*, and *Future professional and personal expectations.*

Furthermore, for each of the four categories, subcategories for each one of them were identified from the data. Tables 1 to 3 present the categories and subcategories created by showing the numbers of interviewees who mentioned each aspect.

The first category (Table 1) is about the perceived usefulness and importance of the academic content.

Table 1 – Academic content

Basic skills	Practical skills	General skills	Open & friendly environment

	to form a specific analytical reasoning	Usefulness & importance of technical courses	Usefulness & importance of applied courses to complement the theoretical courses	Usefulness & importance of elective courses in order to define your career profile	Get in touch with reality through interships	Importance of teaching more than only technical skills	Near & kind relationship with faculty staff and advisors
Total	12	7	4	2	9	7	8
Total Category	12	22				7	8

The next category is about which aspects from the engineering education students received at the University are valued the most for the preparation for future professional life (Table 2).

Table 2 – Preparation for the professional life

	Complete and fully-competent professional				Attitude towards work	Interpersonal skills		
	Capacity for Continuous learning	Problem solving capability	Self-confidence to actively participate in engineering related jobs	Holistic education	Responsibility, self-organization & demanding standards for ourselves	Capability of interacting with other colleagues	The importance of teamwork and its virtues	Possibility of knowing & learning from people with different cultural backgrounds
Total	4	3	5	7	2	2	7	2
Total Category	19				2	11		

The third category is about the professional and personal expectations that the participants expressed they wished to achieve after finishing their university degrees. (Table 3).

Table 3 – Professional and personal expectations for the future

	Learning by working		Economy and security		Balance Job-personal life		Continuous development			Entrepreneurs
	Gain capabilities through experience	passion within a specific area	Job security	Economic stability	Find personal growth through work	balance between working and leisure	Achieve post-graduate degrees	Achieve international working experience	Be ambitious, stand out, be above the average	Try having your own business
Sub cat.	6	5	2	2	5	16	12	12	3	8
Cat.	11		4		21		27			8

The comparison of the factors between the first and the second rounds are shown in Table 4. In this table, all the percentages refer to the total number of participants.

Table 4 – Comparison between first and second round

Factor	Description	Level	Year 1	Year 2
Satisfied with the choice to be an engineer	The satisfaction with the choice of both the education as engineer and the same program. About if they would do the same choice today.	Definitely do the same choice	94%	94%
		Some thoughts about doing a different choice	6%	6%
Prepare for the profession	About if they feel that they are prepared to work as engineers	Feel prepared	89%	100%
		Not yet prepared	11%	0%
Consider to go abroad	Having thoughts or aspirations about working abroad, for short or long time.	Willingness	67%	56%
		Whereof already realized	0%	6%
		Willingness	44%	66%

Factor	Description	Level	Year 1	Year 2
Interest to start up own business	Own business that in some cases is together with other people.	Whereof already realized	11%	22%
Desire to conduct post-graduate education	About doing a Master or a specialization. In Uruguay or abroad in a short or long horizon.	Willingness	83%	72%
		Whereof already realized	11%	11%
Meet career expectations	Involves the perception about whether the participants have achieved their working expectations after one year in the job market	Below expectation	n/a	6%
		In line with expectations	n/a	83%
		Exceed expectations	n/a	11%
Factor	**Sub Factor**	**Description**	**Year 1**	**Year 2**
Useful & important knowledge obtained at the university	Engineering specific	Specific and technical engineering courses	39%	50%
	Non-engineering specific	Non-engineering specific courses such as project management, finance, etc. that provides management knowledge	39%	61%
	Practical knowledge & internships	Applied courses such as laboratories and workshops, and internships exemplified by learning by doing, real life experiences, practical work	56%	39%
Applied Skills & capabilities developed during their engineering education	Problem solving skills	Problem solving capabilities such as logical and systematic thinking, good reasoning, efficient decision-making, etc.	17%	28%
	Interpersonal skills	Interpersonal skills such as capability of interacting with other colleagues, team working, communication skills, etc.	50%	50%
	Innovation skills	Innovation skills including entrepreneurship skills	n/a	17%

Discussion

Analyzing results from Round 1 and Round 2

From Table 1 we can see that the students' perceptions about what has been useful content very much emphasized the practical parts of the curriculum. From Table 2 we can see that what graduates value the most about the education received involves learning interpersonal social skills such as the need to work in teams to solve problems, and to engage other students and people in their work, as well as the holistic learning. This is aligned with previous studies that explain learning as a process that in itself is social and collective, with the goal of performing tasks more efficiently by using different methods such as repetition, experimentation, and understanding complex problems (Teece, Pisano, & Shuen, 1997).

From Table 3 we can see that job security and financial compensation have only been mentioned by two of the participants, while having a balance between work and family seems to be a very important future aspiration. This is an interesting finding that could be explained as a cultural characteristic. According to Hofstede (2005), Uruguay scored 36 on the cultural dimension called "Individualism versus Collectivism", meaning that Uruguay is a highly collectivist society which is related to acting as a group instead of putting personal ambitions in the first place. Moreover, in terms of their expectations for future working life, the most valued thing was continuous development, where many participants said that they "aspire to be leaders within different engineering fields of expertise" and also many talked about achieving personal development beyond the average, and being able to innovate within their fields by starting their own companies and not by working as employees all their lives.

When looking at Table 4 which contains the comparison of the two rounds, the following findings can be appreciated. In the first round, some students hesitated about how well the university had prepared them for their profession. In the second round, 100% considered that the university had prepared them accordingly for future jobs. This could be explained as some new undergraduates were feeling that they did not know enough, but after being confronted with situations where they did not have some specific knowledge they found they could manage to acquire the knowledge without difficulties, which led them to realize that they were well prepared. This is in line with the findings of Achcaoucaou et al. (2014) about people's ability to remove constraints by using their potential. Furthermore, the aspirations to work abroad have decreased from 67% to 56%, and just 6% of the total sample were working abroad at the time of the second round. The reason for this decrease lies in their current family and/or personal life situations. One of the participants explained this in the following way: "I joined this large company and because I am getting married next year, I have changed my plans to go abroad." The same explanation applies to the decrease regarding the ambition to pursue postgraduate study, which decreased from 83% to 72%, although it remains quite high in general showing the intention of students to keep their continuous professional and academic development.

The ambition of starting one's own business increased from 34% to 66%, with 22% of all the participants already having their own business, in some cases in parallel with their employment. Regarding the aspirations that they have about the future, 11% report having exceeded them, while 83% are in line with and 6% feel that they are below. Furthermore, regarding what the participants considered useful and important from their engineering studies for future jobs, we see some differences between the first and the second round. The newly graduates tended to give more importance to the practical moments while in the second year more importance is given to management related knowledge. Moreover, in terms of the competences seen as most important for future professional growth, competences such as problem solving, interpersonal skills, and innovativeness have shown an increase in the second year.

In summary, the participants are satisfied with the education they received, the skills and competences they developed, and with entering the labour market. In addition, they have general high ambitions to serve the society and to continuously develop as professionals, and the majority still have the ambition of pursuing postgraduate education. Almost two-thirds are also considering starting their own businesses.

Towards a new framework for education in engineering

Analyzing the interviews performed with the dean, the academic advisor and teachers, as well as the documentation related to the curriculum and the ways of working at the faculty, we see that there is a good mapping with the revised Bloom's taxonomy mentioned in the introduction section (Krathwohl, 2002). The education that the faculty is offering covers all six elements of the taxonomy: *Remember*, which is the recall of knowledge from the memory; *Understand*, which is the interpretation of the knowledge; *Apply*, which is the process of implementing and using the knowledge in a specific situation; *Analyze*, which is about creating structures, classifying and presenting them; *Evaluate, which* is about applying critical thinking and creating recommendations and seeing the pros and con; and *Create,* which is about reorganizing the elements into a new pattern or structure, and creating something new (Krathwohl, 2002). The mapping between the UM way of performing engineering education and Bloom's updated taxonomy is also outlined in the interviews with the students; the plans of study start with basic and factual knowledge, which is later interpreted in specific scenarios. Then, the students are encouraged to use their critical skills to apply, analyze and evaluate.

The students also go through *the competence loop* by having classes that are both theoretical and practical and they undertake compulsory and scheduled internships. The importance of this mix between practical and theoretical approach is highlighted in the results from the students' interviews shown in Table 1. Figure 1 presents an adaptation of Medina and Medinas (2015) competence loop merged with a mapping to Bloom's revised taxonomy for the case of engineering education.

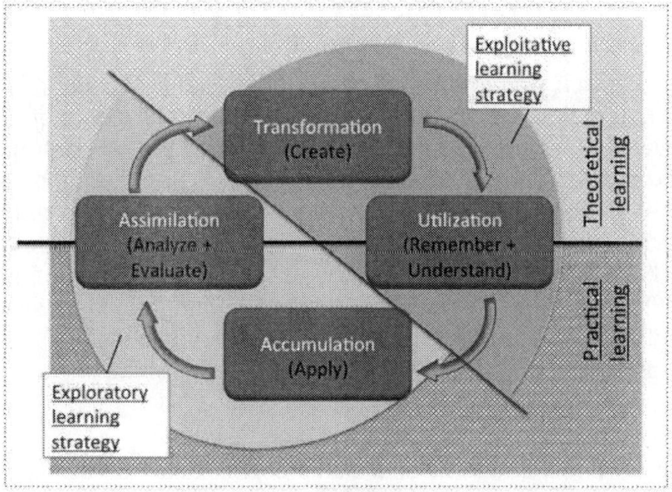

Figure 1 – Mapping the adapted Competence Loop and Bloom's revised taxonomy

Looking at the *utilization* mechanism of the competence loop (Medina & Medina, 2015), we can see that in an educational context there is a theoretical part where the theoretical foundation is presented for the students, and a practical part that is composed of laboratory work, group presentations, cases, and internships. In the *utilization* mechanism, the *Remember* and *Understand* elements from Bloom's revised taxonomy take place. This mechanism takes place both within the presentation of the theoretical foundations and within the practical activities. Moving forward on the loop, the next mechanism, *accumulation*, is about doing and practicing a specific topic. In an educational context, this mechanism takes place when the students are doing the internship, laboratory work, group exercises, thesis, etc. In this mechanism the *Apply* from the updated Bloom's taxonomy take place. The learning from the previous phase of the loop is now being put into action, implemented and applied. The *assimilation* mechanism is the process of interpreting and understanding what the practical actions have generated, which maps very well with *Analyze* and *Evaluate* from the revised Bloom's taxonomy, where the mental process of inspecting, checking, criticizing, following up, and reflection take place. This is what the students do when they write the report about their practical work and when they reflect on the previous "doing" together with their academic advisors. The last mechanism of the loop, *transformation*, is where the learning is integrated with the theoretical base. The last element of the taxonomy, *Create*, takes place in this part of the loop, and it is here that a kind of reorganization of the previous knowledge is transformed into new patterns or structures creating new knowledge.

In addition to the mechanisms of the loop and the elements of the taxonomy, we found the *explorative* and *exploitative learning*. In a higher education context, these work as follows. *Explorative learning* occurs when the students explore new competences by applying previous knowledge in practice and reflect and summarize their findings in reports about the practical work. On the other hand, *exploitative learning* occurs when the students integrate new competence into their theoretical base and new meanings are constructed based on the combination of old and new competence, which later can be used for new practical application such as group exercises, laboratory work and thesis writing.

The high ratings of the students regarding their preparation for the profession, namely 100%, together with the external ranking that placed UM as the first in Uruguay, and the fact that 94% of the participants have been able to meet their high career development ambitions and even in some cases (11%) exceed them as shown in Table 4, implies that the approach followed by the School of Engineering under study and represented in the framework shown in Figure 1, could be seen as an innovative and effective approach to perform engineering education in order to meet the new educational paradigms reviewed in this paper.

Conclusion

The purpose of this research was to analyze the educational system applied at the School of Engineering at UM. The main contribution of this paper is the proposal of a new educational framework for engineering education based on the merge of the competence loop and Bloom's updated taxonomy. We hope this framework can be applied and further developed in other higher education institutions, contributing to meet the changes required to succeed in future engineering education. Although this paper presents the first two rounds of the project, the results produced are clear and could be used as a guide for similar studies. The limitation of the study is that it has been performed in one private education institution. Future research studies should include similar studies in other private and public institutions as well as comparisons within different countries with different cultural habits.

References

Achcaoucaou, F., Guitart-Tarrés, L., Miravitlles-Matamoros, P., Núñez-Carballosa, A., Bernardo, M., & Bikfalvi, A. (2014). Competence assessment in higher education: A dynamic approach. *Human Factors and Ergonomics in Manufacturing & Service Industries, 24*(4), 454–467.

Barr, R. B., & Tagg, J. (1995). From Teaching to Learning: A New Paradigm for Undergraduate Education. *Change 27*(6), 12–25.

Beanland, D., & Hadgraft, R. (2014). *Engineering education: Transformation and innovation.* Melbourne: RMIT University Press.

Bloom, B. S. (1956). *Taxonomy of educational objectives: The classification of educational goals. Handbook 1, cognitive domain.* New York: David McKay.

Cooper, D. R., & Schindler, P. S. (2011). *Business research methods.* New York: McGraw-Hill/Irwin.

Crawford, L. (2005). Senior management perceptions of project management competence. *International Journal of Project Management, 23*(1), 7–16.

Hofstede, G., & Hofstede, G. (2005). *Culture and organizations. Intercultural cooperation and its importance for survival, software of the mind.* New York: McGraw-Hill.

Hung, I. W., Choi, A. C. K., & Chan, J. S. F. (2003). An integrated problem-based learning model for engineering education. *International Journal of Engineering Education, 19*(5), 734–737.

International Engineering Alliance. (2009). Washington Accord. Retrieved from www.ieagreements.org

Krathwohl, D. R. (2002). A revision of Bloom's taxonomy: An overview. *Theory into Practice, 41*(4), 212–218.

Male, S. A., Bush, M. B., & Chapman, E. S. (2011). An Australian study of generic competencies required by engineers. *European Journal of Engineering Education, 36*(2), 151–163.

Medina, R., & Medina, A. (2015). The competence loop: Competence management in knowledge-intensive, project-intensive organizations. *International Journal of Managing Projects in Business, 8*(2), 279–299.

Miles, M. B., & Huberman, A. M. (1994). *Qualitative data analysis: An expanded sourcebook.* Thousand Oaks, CA: SAGE.

Passow, H. J. (2012). Which ABET competencies do engineering graduates find most important in their work? *Journal of Engineering Education, 101*(1), 95–118.

Royal Academy of Engineering. (2007). Educating engineers for the 21st century. Retrieved from http://www.raeng.org.uk/publications/reports/educating-engineers-21st-century/

Sanford, B. (1989). *Strategies for maintaining professional competence: A manual for professional associations and faculties.* Toronto: Canadian Scholars' Press.

Saunders, M., Lewis, P., & Thornhill, A. (2009). *Research methods for business students.* Harlow: Prentice Hall.

Sheppard, S. D., Macatangay, K., Colby, A., & Sullivan, W. M. (2008). *Educating engineers: Designing for the future of the field.* San Francisco: Jossey-Bass.

UNESCO. (2010). *Engineering: Issues, challenges and opportunities for development.* Paris: UNESCO Publishing.

Mapping the motivation of biomedical engineering students: A cross-sectional study

Antonio Miguel-Cruz
Biomedical Engineering Program. School of Medicine and Health Sciences. Universidad del Rosario, Bogotá, Colombia
antonio.miguel@urosario.edu.co

Adriana M. Rios Rincon
Occupational Therapy Program. School of medicine and health sciences. Universidad del Rosario, Bogotá, Colombia
adriana.rios@urosario.edu.co

William R Rodríguez-Dueñas
Biomedical Engineering Program. School of Medicine and Health Sciences. Universidad del Rosario, Bogotá, Colombia
william.rodriguez@urosario.edu.co

Daniel A. Quiroga-Torres
Biomedical Engineering Program. School of Medicine and Health Sciences. Universidad del Rosario, Bogotá, Colombia
daniel.quiroga@urosario.edu.co

***Abstract**: The purposes of this study were: (1) To determine to what extent the basic psychological needs of biomedical engineering program students are satisfied; and (2) how students' motivation is related to both their perceived learning outcomes and program regulations. This is a cross-sectional study in which we used a mixed-method research approach that included a paper-based 5-point Likert questionnaire (from "1"- strongly disagree to "5" - strongly agree) to measure the students' amotivation and intrinsic motivation, as well the different types of extrinsic motivation, and focus groups. The study involved 364 students (1st - 9th semesters) from a biomedical engineering program. Our study showed that the basic psychological needs of biomedical engineering program students are satisfied. The motivations of biomedical engineering students are more inclined toward behaviors that are more self-determined rather than being driven by external regulations. The more motivated students are, the more they perceive they learn.*

Introduction

One of the priority research areas that the Engineering Education Research Colloquies (EERC) presented for the new discipline of Engineering Education (NAE, 2006) was to conduct research in order to understand how engineers learn and develop knowledge and competencies in real contexts (i.e. engineering learning mechanisms). In this research area, studying the learning progressions of learners and their educational experiences and processes during their engineering careers has attracted special interest from scholars. Motivation, in Kleinginna & Kleinginna's (1981) words, is simply an individual's internal process that makes him/her move toward a goal, a critical aspect of learning as motivation regulates the way a student engages in his or her learning activities. Therefore, students' motivation deserves to be studied in-depth in engineering programs because this motivation permeates students' retention, persistence, self-efficacy, achievements, and assimilation of their program's engineering regulations (French, Immekus, & Oakes, 2005; Seymour & Hewitt, 1997). Motivational studies on engineering programs have shown that motivation improves students' engagement in, learning during, and completion of traditionally foundational courses such as math and science (Gasiewski, Eagan, Garcia, Hurtado, & Chang, 2012). Other sudies demonstrated that highly motivated students have better self-efficacy, thus fostering their engineering learning and academic performance (Lent, et al., 2008). Students with high levels of pressure and anxiety about their performance are correlated with low levels of motivation (intrinsic and extrinsic), thus reducing their learning assimilation and performance quality during the educative process (Carberry, Lee, & Ohland, 2010), especially for students in their first year of engineering programs (Jones B. , Paretti, Hein, & Knott, 2010). With regard to learning environments and motivation, studies have demonstrated that course climates have a possitive effect on students' motivation, improving both their engagement, peer collaboration, and student-instructor interaction (Carberry, Lee, & Ohland, 2010).

In spite of the fact that studies on students' motivation in almost all engineering disciplines are increasing in number, research on students' motivation on biomedical engineering programs is scarce. Understanding students' motivation to enroll on and persist with a biomedical engineering program is important because as biomedical engineering is a relatively new interdisciplinary field of study that combines engineering with life sciences and medicine, it is still not well established in the minds of students' parents, candidate students yet to be enrolled, or employers (Linsenmeier, 2003). This lack of understanding about what biomedical engineering is may result in students' motivation to persist with their program and to obtain a degree being affected by a conflict between freshmen's expectations and the real academic demands of engineering learning activities. Therefore, this study is guided by the following research questions:

1. Are the basic psychological needs of biomedical engineering program students satisfied?
2. How is students' motivation related to both their perceived learning outcomes and the program's regulations?

We believe that responding to these research questions is important for our biomedical engineering program because we can determine whether students are more inclined toward behaviors that are more self-determined rather than being driven by external regulations (e.g. being enrolled on a biomedical engineering program as a result of external pressures such as family), thus allowing us to implement interventions for students and their program's learning systems in order to improve their motivation and encouraging them to finish their degree in biomedical engineering.

Theoretical framework

The Self-determination Theory (SDT) was chosen as the underlying theoretical framework for this study (Deci & Ryan, 1985). Although SDT has been used as a single compact theory, SDT integrates previous theories such as Cognitive Evaluation Theory (CET), which helps to explain how factors in social contexts can produce variability in intrinsic motivation, and the Organismic Integration Theory (OIT), which is used to introduce in more detail the different forms of extrinsic motivation and to explain how contextual factors promote or hinder the internalization and integration of the regulation of individuals' behavior (Deci & Ryan, 1985). According to the OIT, internalization is the process of taking in a value or regulation, whereas integration is the process whereby individuals transform a regulation more fully on their own (Ryan & Deci, 2000). As a result, the SDT has been widely used as a theoretical framework to explain how individuals can internalize the responsibility and sense of value of extrinsic goals (Ryan & Deci, 2000). This can be applied to the case of students and educators in different areas of educative environment, as students are expected to internalize the value of learning and their responsibility for their learning process for extrinsic goals such as achieving graduation, becoming a professional, or starting work. In this regard, an education in engineering is not an exception (Jones, Paretti, Hein, & Knott, 2010; Koh, et al., 2010).

In short, under the SDT framework, different types of motivation are distinguished in a continuum. At one extreme of the continuum is amotivation, at the other, intrinsic motivation, and, in the middle, different forms of extrinsic motivation. Amotivation is the state of lacking an intention to act (Ryan & Deci, 2000). On the contrary, intrinsic motivation "*refers to doing something because it is inherently interesting or enjoyable*" or doing it "*by free choice*" (Deci & Ryan, 1985). Intrinsic motivation can be observed in activities such as free play which individuals engage in just because of the pleasure and satisfaction of playing. The SDT states that in order to be inherently intrinsically motivated three psychological needs must be satisfied; autonomy, competence, and relatedness. Autonomy is understood as being the universal urge to be causal agents of one's own life and acting in harmony with one's integrated self. One's competence seeks to control the outcome and experience mastery. Finally, relatedness is the universal need to interact, be connected to, and experience caring for others (Ryan & Deci, 2000). The extrinsic motivation "*refers to doing something because it leads to a separable outcome*" (Deci & Ryan, 1985). The SDT recognizes different types of extrinsic motivations including external regulation, introjected regulation, identification, and integrated regulation. The least autonomous form of extrinsic motivation is external regulation, whereas integrated regulation is the most autonomous (i.e. the one closest to intrinsic motivation) form of extrinsic motivation (Ryan & Deci, 2000). One can say that an individual is extrinsically motivated, or "driven," by an external regulation when that individual shows behaviors that satisfy an external demand, or to obtain an externally imposed reward (i.e. I do my classwork so that the teacher won't yell at me). Also, one can say that an individual is extrinsically motivated by an integrated regulation when the identified regulation has been fully assimilated into the self (i.e. I wear a seat belt because it will save my life in case of an accident and not because I will be fined if I do not).

Since most of the tasks that students perform in their educative environment are not inherently interesting or enjoyable, investigating students' extrinsic motivation has become a very important aspect of education. The key aspect to study is to understand how to foster the internalization and integration of students' values and behavioral regulations in educative environments. The SDT predicts that increasing internalization in students results in greater persistence, more positive self-perceptions, and better quality of engagement (Ryan & Deci, 2000); it is crucial for students to persist with these aspects in their goal of starting their engineering careers.

Methods

Study design

This is a cross-sectional study in which we used a mixed-method research approach that included a paper-based questionnaire and focus groups in order to map students' amotivation and intrinsic motivation, as well as all the types of their extrinsic motivation and their perceptions of achieving the expected learning outcomes.

Setting and participants

This study was conducted in a biomedical engineering program. In 2011 a partnership between the School of Medicine and Health Sciences of Universidad del Rosario and the School of Engineering Julio Garavito of Bogota opened a biomedical engineering program. It had nine semesters and 154 academic credits, and was taught at these two universities.

The study involved 364 students (I-IX semesters) from the biomedical engineering program. This sample size was sufficient, using as its parameters an effect size of 0.6-0.7 and a statistical power of 0.8 (the statistical significance was set at $p<0.05$) (Portney & Watkins, 2000).

Variables

In this study we included demographics variables such as gender (male/female) and socio-economical strata (Low/Medium/High) (see the Results section) in order to characterize our sample. We also included autonomy, competence, and relatedness as measures of basic psychological needs. To map our students' motivation we included amotivation, intrinsic motivation, and all the types of extrinsic motivation, including external regulation (i.e. behaviors that satisfy an external demand or to obtain an externally imposed reward), introjected regulation (i.e. behaviors performed with a feeling of pressure in order to avoid guilt, anxiety, or to increase their contingent self-esteem), identification regulation (i.e. behaviors shown because the person has identified the importance of this behavior, it brings some benefits to him/her. Thus she/he has accepted its regulation as his or her own), and integrated regulation (i.e. behaviors shown because identified regulations have been fully assimilated into the self). In addition, we included other variables in order to have improved mapping of the factors that affect students' motivation to enrol on and study our biomedical engineering program. These variables included perceived choice (i.e. the extent or level of perception that the individual has in terms of their ability or discretion to choose; free choice), usefulness (i.e. the degree to which a person believes that using a particular system/or showing a particular behavior would enhance his or her job performance), anxiety /pressure or tension (i.e. evoking anxious or emotional reactions when showing a particular behavior), and self-efficacy (i.e. the extent or strength of one's belief in one's own ability to complete tasks and reach goals), and the perception of achieving expected learning outcomes (Ryan & Deci, 2000). We also included the Grade Point Average (GPA) as a measure of actual achievement of the expected learning outcomes (Jones, Paretti, Hein, & Knott, 2010)

Instruments

A questionnaire was created by adapting scales and items already validated with high levels of internal consistency in prior research to measure the specific variables examined in this study (McAuley, Duncan, & Tammen, 1987; La Guardia, Ryan, Couchman, & Deci, 2000; Jones, Paretti, Hein, & Knott, 2010). The content of the survey questionnaire included 65 items designed to measure the students' motivation and their perceived learning outcomes. These 65 items were grouped into two sections. In the first section (section A, 15 items), we asked for the participants' demographics. In the second section (section B, 47 items) we enquired about the students' psychological needs, as well as their amotivation, motivation (intrinsic and extrinsic), and perceived achievement of the expected learning outcomes. In section B the students were asked to rate each item on a Likert scale from 1 to 5 (where "1" was "strongly disagree," and "5" was "strongly agree"). We made sure to include the same number of items (i.e. four) per variable in order to have a balanced score for each variable. For example, the components of less autonomous or less self-determined extrinsic motivation such as external regulations were assessed using questionnaire items such as "*I study biomedical engineering because if I withdraw now I will be in trouble (e.g. economic, credit, family, etc.*"), whereas the components of more self-determined extrinsic motivation such as integrated regulation were assessed by questionnaire items such as "*I work on my classwork because I want to learn new and challenging things.*" Section B of the survey also included three open-ended questions in order to explore the three most motivating and non-motivating things about being enrolled on our biomedical engineering program.

The data in the focus group sessions were gathered using a six-item questionnaire in order to explore which aspects of the academic program motivated them and which ones did not.

Variables operationalization

The dichotomous variables were coded as "0" or "1" (e.g., gender). Each polytomous variable was represented by set binary design set whose values codified each variable category (e.g., the parents' highest educational level). All the items in Section B of the questionnaire were scored on the 5-point Likert scale, ranging from "strongly disagree" "1" to "strongly agree" "5". This is the most common coding or operationalization technique used in the literature (Cohen, Cohen, West, & Aiken, 2003).

Procedures: Survey administration, data collection, and bias control

After receiving the ethical approval by the School of Medicine and Health Sciences Research Ethics Board during the first week of the semester, the research team informed all the students from the 1st to 9th semesters of the aims of the study, and asked them to voluntarily sign a consent form. In order to guarantee their anonymity and confidentiality, each participant was assigned a questionnaire code. Before administering the questionnaires we conducted a pilot study with ten students representing every semester of the biomedical engineering program, so they did not complete final version of the questionnaire. The aim of the pilot study was to obtain feedback from the students as to whether the questionnaire items were clear and concise. Minor changes to some items were made as needed. The questionnaire was administered to all the students during the sixteenth week of the semester. The questionnaire was administered in a quiet room and at a time when the students did not feel any anxiety, physiological pressure, or fatigue due to any exam. After the questionnaires had been done two focus groups were formed from a sample of students from the 7thto 9th semesters (n=50) in order to explore which aspects of the academic program motivated them and which ones did not. A research assistant who was not involved in the teaching session was assigned to assist the research team to administer the questionnaires and to conduct the two focus groups. In this way the students were guaranteed not to feel any pressure, or to give any biased responses when filling out the questionnaires or talking openly during the focus group sessions.

Statistical analyses

The total scores for all the items in each variable (e.g. the four items for intrinsic motivation, etc.) were calculated. When mapping the motivation of our biomedical engineering students we performed different statistical analyses. We used descriptive statistics to determine to what extent the students' basic psychological needs were satisfied. In doing so, we calculated the median, mean, and standard deviation of scores, and we also calculated the total of the questionnaire items measured in each variable. In addition, descriptive statistics helped us to summarize the students' demographic data. To interpret the descriptive statistics, the median/mean and standard deviation located in the top third of the Likert scale of the questionnaire items measuring autonomy, competence, and relatedness constructs (mean > 3.33) indicated that the students' basic needs were satisfied. Bivariate correlations (i.e., Spearman's rho correlation) with a total score in each variable were calculated to determine how the students' motivation was related to both their perceived learning outcomes and the biomedical engineering program's regulations. Therefore, a positive and statistically significant correlation between intrinsic motivation and autonomy, between competence and relatedness is also an indication that the students' basic needs are satisfied. Positive and statistically significant correlations between motivation and perception learning outcomes would suggest that students' motivation affects the perception of their achieved learning outcomes. Likewise, positive and statistically significant correlations between the more intrinsic components of extrinsic motivation (i.e. identification and integrated regulation) and intrinsic motivation indicate that the students are more inclined toward behaviors that are more self-determined, rather than being driven by the external regulations of their biomedical engineering program. To interpret the correlation coefficients we followed the recommendations in (Portney & Watkins, 2008), where a low correlation was taken as $r \leq \pm 0.39$; a moderate correlation as $\pm 0.40 \leq r \leq \pm 0.74$; and a high correlation as $\pm 0.75 < r \leq \pm 1.00$.

The reliability and internal validity measures for each construct of the instrument (questionnaire) were tested using the F-test of the intra class correlation coefficient (ICC) (i.e. average and single measures) and Cronbach's Alpha (CA). Cronbach's Alpha values higher than 0.70 indicate good internal consistency between the items within each variable (Portney & Watkins, 2008).

The alpha level of significance for all the tests and correlation coefficients was set at $p \leq 0.05$

(two-sided). Missing data were handled according to the following procedure: (1) Missing data in continuous variables such as the students' age were replaced by the average values of these variables; and (2) for categorical and ordinal variables such as the students' gender, and their socio-economical strata, missing data were replaced by the medians of these variables.

All the focus groups were audio recorded and transcribed verbatim. They were analyzed for common words or phrases that appeared for each of the two focus group questions. An SPSS® V 22.0 statistics package was used to generate descriptive, univariate and bivariate statistics.

Results

Participants' description

This study involved all (364 students, I-IX semesters) from the biomedical engineering program; 58.6% were female and 41.4% male. Half of them came from low-middle income families (50.6%). 49.2% of the students had obtained their high school degrees from public schools. In 40.4% of cases either of the students' parents had a higher education degree in engineering, whereas in 48.6% of cases either of the students' parents had a higher education degree in health science.

Psychological needs and motivation of biomedical engineering program students

The basic psychological need variables means (SD) were located in the top third of the scale (i.e. autonomy, 4.09 (0.99); competence, 3.89 (0.87); relatedness, 3.98 (0.90)). Positive and statistically significant correlations between intrinsic motivation and autonomy ($r_{\text{Intrinsic motivation-Autonomy}}=+0.57$, $p<0.001$), relatedness ($r_{\text{Intrinsic motivation-Relatedness}}=+0.36$, $p<0.001$), and competence ($r_{\text{Intrinsic motivation-Competence}}=+0.37$, $p<0.001$) were found. This indicated that students' basic psychological needs were satisfied during their educational experiences on the biomedical engineering program.

Table 1a shows the Spearman's rho bivariate correlations between intrinsic motivation and the extrinsic motivation components (Table 1a, the row colored in dark gray); and between amotivation and the extrinsic motivation components (Table 1a, the row colored in light gray). Table 1b shows the Spearman's rho bivariate correlations between intrinsic motivation and the other variables measured (e.g. anxiety, etc. Table 1b, the row colored in dark gray) and between amotivation and the other variables measured (e.g. anxiety, etc. Table 1b, the row colored in light gray)

Table 1: Bivariate correlations (*p<0.05, ** p<0.01). **(A) Amotivation, (ER) External regulation, (IR) Introjected regulation, (ID-R)) Identification regulation, (IT-R) Integrated regulation, (ANX) Anxiety, (SE) Self-efficacy, (PCH) Perceived choice, (U) Usefulness, (G) Average student grades, ($ELOP_{1-6}$) Expected learning outcome perceptions (1st - 6th semester, freshman and sophomore), ($ELOP_{7-9}$) Expected learning outcome perceptions (7th - 9th semester, seniors), (IM) Intrinsic motivation**

	Extrinsic motivation components				
A	ER	IR	ID-R	IT-R	IM
	-0.13	+0.22**	+0.51**	+0.60**	
	+0.27**	+0.043	-0.31**	-0.42**	

(a)

	Other variables							
A	ANX	SE	PCH	U	G	$ELOP_{1-6}$	$ELOP_{7-9}$	IM
	-0.22**	+0.45**	+0.67**	+0.48**	-0.10	+0.219**	+0.354**	
	+0,32**	-0.42**	-0.56**	-0,41**	+0.04	+0.08	-0.518**	

(b)

Table 1a shows that the components of extrinsic motivation related to behaviors that are more self-determined such as identification regulation (ID-R) and integrated regulation (IT-R) have the strongest and statistically significant positive correlations with intrinsic motivation (IM). The contrary is also true; the components of extrinsic motivation components related to behaviors that are less self-determined such as external regulations (ERs) and introjected regulation (IR) have weaker statistically significant either positive or negative correlations with intrinsic motivation (IM). Table 1a also show a negative and statistically significant correlation between amotivation and the components of extrinsic motivation related to behaviors that are more self-determined, in other words as amotivation increases the students feel that the regulations are adopted more due to external pressures rewards rather than because they have internalized the regulations themselves (e.g. family or to obtain a reward from the instructor).

Table 1b shows that the students' grade point averages (G) are not correlated either with amotivation or with intrinsic motivation and, on the contrary, the expected learning outcome perceptions ($ELOP_{1-6}$ and $ELOP_{7-9}$) are statistically significantly correlated with intrinsic motivation. This indicates that the more motivated students are, the more they perceive they learn. This result is also supported by the fact that there is a positive and statistically significant correlation between self-efficacy (SE) and intrinsic motivation (IM).

Reliability and internal validity of the instrument

Table 2 shows the statistics of the reliability and internal validity of the questionnaire used in our study. The statistical significance of the p-values (i.e. $p<0.05$) for the F-test of the intra class correlation coefficient (ICC) (i.e. average and single measures) for all the variables/constructs tells us that the subjects are different from each other, which is a necessary condition for a reliability test. Since the Cronbach's Alpha (CA) measurements in most cases (7/10, 70%) are above 0.70 and the F-test of the ICC (i.e. average and single measures) for all the constructs were statistically significant ($p<0.05$), we can reach the conclusion that the instrument used has acceptable levels of reliability and internal validity (Portney & Watkins, 2000, p. 595).

Table 2. Reliability and internal validly of the instrument

Variable name	CA	ICC	95% C.I.	F test p-value
Amotivation	0.704	0.704	(0.63,0.760)	0.000
Intrinsic motivation	0.763	0.764	(0.705,0.812)	0.000
Extrinsic motivation	0.568	0.568	(0.476, 0.650)	0.000
Autonomy	0.710	0.710	(0.639,0.770)	0.000
Competence	0.468	0.468	(0.338,0.577)	0.000
Relatedness	0.747	0.747	(0.686,0.799)	0.000
Anxiety	0.793	0.793	(0.742, 0.836)	0.000
Self-efficacy	0.500	0.500	(0.377, 0.602)	0.000
Perceived choice	0.765	0.765	(0.708,0.813)	0.000
Usefulness	0.742	0.742	(0.674,0.800)	0.000

Discussion

Our study aims to answer two research questions, (1) *Are the basic psychological needs of biomedical engineering program students satisfied?* and (2) *How is students' motivation related to both their perceived learning outcomes and the program's regulations*?

When answering the first research question, we can assert that the basic psychological needs of the biomedical engineering program students are satisfied since the basic psychological need variables means were located in the top third of the scale (i.e. $\bar{x} > 3.33$), and a positive and statistically significant correlation between motivation and psychological need variables was found. Our results are in the line with similar results in other studies in which scholars found that engineering students feel more motivated by the way their psychological needs are satisfied (Koh, et al., 2010; Jones, Paretti, Hein, & Knott, 2010). Likewise, similar results can be found in studies either in high or elementary schools (Leptokaridou, Vlachopoulos, & Papaioannou, 2014), or for higher education students in disciplines different from engineering such as law programs (Sheldon & Krieger, 2007)

In addition, this assertion can be validated by the comments obtained from the focus group, where the participants almost unanimously stated that their decision to be enrolled and to persist with finishing their degree on the biomedical engineering program was driven by autonomous decisions and motivations such as "*I want to study biomedical engineering because I want to help others*"; "*I want to study biomedical engineering because I like to design and innovate [new medical technologies]*"; or "*I want to study biomedical engineering because it is a good project of life.*" As an aside, just 3% of the students' commented that they enrolled on the biomedical engineering program due to "*family pressures*" or parental influence owing to one of their parents having obtained a higher education degree in either engineering or health science. Other comments related to the biomedical engineering program itself seem to have some importance in the decision to be enrolled and persist with obtaining a degree in biomedical engineering, e.g. "*I like the alliance between a medical school and an engineering school,*" "*there are good labs to do research,*" "*there are good professors [faculty members],*" and "*there is a variety of internationalization and internship possibilities [both national and international] to complement our experiences.*" Comments related to the relatedness variable also help to explain why we can assert that our students have their basic psychological need satisfied. For example, we received comments such as "*I like to study this biomedical engineering program because of my friends [friendship and camaraderie on the program].*" Interesting comments such as "*I like that students in more advanced semesters are almost going to graduate [become a biomedical engineer],*" thus indicating a degree of relatedness and fellowship among students.

In answering the second research question we can say that the more motivated students are, the more they perceive they learn, which is also supported by the fact that the students believe they are competent and self-sufficient, thus indicating that the students believe that regardless of how difficult or easy a subject/task will be, they are sure they will be able to understand it. We also found that the students' motivation to be enrolled on and to persist with our biomedical engineering is more inclined toward behaviors that are more self-determined rather than being driven by external regulations. This result is supported by similar studies of engineering programs, in which the authors found that learning orientations such as metacognition and self-regulation had a close association with intrinsic motivation (Koh, et al., 2010; Liu & Chye, 2008). Our results can also be validated by the comments obtained from the focus group participants, who stated that they "*like to be enrolled on our biomedical engineering program because of the high level of demand from the professors, and much effort is required to pass the subjects, thus indicating the quality of the program,*" or "*I study biomedical engineering because I want to understand hard subjects and do challenging things,*" or "*I do additional activities [above and beyond what my professor asks for] because this is important to be a good biomedical engineer.*"

Conclusions

The basic psychological needs of biomedical engineering program students are satisfied. The motivations of biomedical engineering students are more inclined toward behaviors that are more self-determined rather than being driven by external regulations. The more motivated students are, the more they perceive they learn.

References

Carberry, A., Lee, H., & Ohland, M. (2010). Measuring engineering design self-efficacy. *Journal of Engineering Education, 99*(1), 71–79.

Cohen, J., Cohen, P., West, S., & Aiken, L. (2003). *Applied multiple regression/correlation analysis for the behavioral sciences,* (3rd ed.). NY: Taylor and Francis Group.

Deci, E. L., & Ryan, R. M. (1985). *Intrinsic motivation and self-determination in human behavior.* New York: Plenum.

French, B., Immekus, J., & Oakes, W. (2005). An examination of indicators of engineering students' success and persistence. *Journal of Engineering Education, 94*(4), 419–425.

Gasiewski, J., Eagan, M., Garcia, G., Hurtado, S., & Chang, M. (2012). From gatekeeping to engagement: A multicontextual, mixed method study of student academic engagement in introductory STEM courses. *Research in Higher Education, 53*(2), 229–261.

Jones, B., Paretti, M., Hein, S., & Knott, T. (2010). An analysis of motivation constructs with first-year engineering students: Relationships among expectancies, values, achievement, and career plans. *Journal of Engineering Education, 99*(4), 319–336.

Kleinginna, P. J., & Kleinginna, A. (1981). A categorized list of motivation definitions, with suggestions for a consensual definition. *Motivation and Emotion, 5*, 263–291.

Koh, C., Soon-Tan, H., Tan, K.-C., Fang, L., Fong, M., Kan, D., . . . Wee, L. (2010). Investigating the Effect of 3D Simulation-Based Learning on the Motivation and Performance of Engineering Students. *Journal of Engineering Education*, 237–251.

La Guardia, J. G., Ryan, R. M., Couchman, C. E., & Deci, E. L. (2000). Within-person variation in security of attachment: A self-determination theory perspective on attachment, need fulfillment, and well-being. *Journal of Personality and Social Psychology, 79*, 367–384.

Lent, R., Sheu, H., Singley, D., Schmidt, J., Schmidt, L., & Gloster, C. (2008). Longitudinal relations of selfefficacy to outcome expectations, interests, and major choice goals in engineering students. *Journal of Vocational Behavior, 73*(2), 328–335.

Leptokaridou, E., Vlachopoulos, S., & Papaioannou, A. (2014). Experimental longitudinal test of the influence of autonomy-supportive teaching on motivation for participation in elementary school physucal education. *Educational Psychology*, 2–22.

Linsenmeier, R. (2003). What makes a biomedical engineer? *Eng Med Biol Mag, 22*(4), 32–38.

Liu, W., & Chye, S. (2008). The importance of perceived needs satisfaction. En C. Koh, & H. Tan, *In Motivation and practice for the classroom* (pp. 251–264.). Rotterdam: Sense Publishers.

McAuley, E., Duncan, T., & Tammen, V. V. (1987). Psychometric properties of the Intrinsic Motivation Inventory in a competitive sport setting: A confirmatory factor analysis. *Research Quarterly for Exercise and Sport, 60*, 48–58.

NAE. (2006). Special Report on The Research Agenda for the New Discipline of Engineering Education. *Journal of Engineering Education*, 259–261.

Portney, L., & Watkins, M. (2008). *Foundations of clinical research: Application to practice.* New Jersey: Prentice-Hall Inc.

Ryan, R., & Deci, E. (2000). Intrinsic and Extrinsic Motivations: Classic Definitions and New Directions. *Contemporary Educational Psychology, 25*, 54–67.

Seymour, E., & Hewitt, N. (1997). *Talking about leaving: Why undergraduates leave the sciences.* Boulder, CO: Westview Press.

Sheldon, K., & Krieger, L. (2007). Understanding the negative effects of legal education on law students: A longitudinal test of self-determination theory. *Personality and social psychology bulletin*, 883–897.

Acknowledgements

The authors wish to thank the study participants, instructors, and mentors for their invaluable advice in preparing this paper. We also want to thank Oliver Jarvis for his proofreading of this paper. The authors assume overall responsibility for the manuscript.

Copyright statement

Facilitating conceptual understanding and reflective learning ability of undergraduate engineering trainees studying strength of materials via instructional scaffolds.

Eyitayo Olatunde Olakanmi[a] and Eunice Eyitayo Olakanmi[b].
Department of Mechanical, Energy & Industrial Engineering, Botswana International University of Science & Technology, Palapye, Botswana[a].
Department of Science & Technology Education, University of South Africa, Pretoria, South Africa[b].
Corresponding Author Email: olakanmie@biust.ac.bw

Abstract

The difficulties encountered by engineering students in employing their prior knowledge for proffering solutions to situations different from those studied in class could be attributed to their inability to organise their knowledge for facilitation, retrieval and application. This study explores how instructional scaffolds assisted the participants in developing concept maps for problem solving and reflective learning ability in strength of materials module. Data from participants' personal learning journals; groups' concept maps; video recording and natural observation of collaborative discussions and interviews were collected for evaluation of the attainment of the research objective. Analysis of the data outcomes from this study revealed that the development of instructional scaffolds helped the participants to become more aware of the strength of materials concepts.

Keywords
Reflective learning; conceptual understanding; problem solving approaches

Introduction

Most engineering students show very little higher order thinking skills and knowledge construction that enables them to apply what they were being taught in the classroom to real world situation in the engineering field (Streveler, Litzinger, Miller, & Steif, 2008).This is generally due to the students' lack of conceptual understanding and difficulty in applying their mathematical knowledge while solving engineering problems (Streveler *et al.,* 2008). Also, most of the engineering lecturers have not been able to change from the traditional lecturer centred learning which emphasises on transfer of knowledge from lecturers to students into the constructivist learning environment such as student centre learning. Meanwhile, due to the growing pace in the global innovation in terms of products and business models, engineering jobs is likely to be more challenging than before. Engineering trainees are therefore expected to learn to work with ambiguity and diversity of disciplines and with humility regarding their own skills and abilities (Felder, 2008).

Generally, the field of the engineering education has been criticised for lack of characteristics necessary for developing creative problem solving skill which often leads to the development of higher order thinking skills among students. According to Magee, *et al.,* (2003), most engineering students focus on (1) memorisation of knowledge and procedures, rather than higher order thinking skills; (2) a rapid pace of learning that undermines the self-reflection and self-assessment; (3) highly structured learning formats that constrain the expression of ideas; and (4) inadequate balance between building a body of knowledge and creative use of the knowledge. Most of these students are unable to received explicit holistic instruction based on cognitive science findings, which emphasises students' metacognition and self-reflection on the processes and strategies of creative problem solving skills. Although many engineering lecturers have acknowledged that creative problem solving skills are necessary for students in order to develop their higher order thinking skills and knowledge construction, most of them have not been trained as teaching professionals and are not fully aware of how people learn from the cognitive science perspectives as well as how creativity can enhance students' academic performance in engineering courses. As a result of this, most engineering lecturers are often unable to integrate cognitive and metacognitive theoretical frameworks into their instruction in engineering education. Their efforts in helping students to develop higher order thinking skills may consequently be less effective than they might expect.

Based on the experience of teaching African students, the authors believed that one of major causes for low retention among the engineering students in Africa include lack of divergent thinking and motivation in learning engineering. Besides most students find it difficult to regulate their learning processes which might have resulted from the poor preparation in their early schooling. When faced with problems, those students who have lower performance usually fail in establishing the connection between the problems and their prior knowledge and solution procedures that have been discussed in the classes. Those students also lack cognitive and metacognitive strategies for guiding their thinking and finding possible solutions to the given problems. Therefore, it is very necessary for engineering lecturers to adopt effective instructional strategies to inspire and guide students' thinking ability as well as their problem solving skills.

Meanwhile, previous research in the field of the engineering education had established the fact that there is gap between what the undergraduate engineering students are taught and assessed on and what the professional engineers and industry practitioners expect them to be capable of doing (Felder, 2008). The gap will continues to grow wide unless engineering educators decide to introduce various interventions into the heavy content based traditional engineering curriculum. Felder (2006) emphasises that engineering education has been teaching the wrong thing by focusing mainly on equipping students with analytical and problem-solving which are not in alignment with the professional practice. In order to bridge this gap and prepare engineering students for the global work place, students need to develop high order thinking skills as well as the reflective learning ability; instructional scaffolding needs to be embedded into the learning process.

Instructional scaffolding is a process by which a lecturer supports a student in executing a learning task (Putnam, O'Donnell, & Bertozzi, 2010). Instructional scaffolding incorporated into the course material allows students to understand the material better than they could have without the instructional scaffold (Putnam et al, 2010). Conceptually, scaffolding implies providing students with instructions during the early stage of learning before slowly shifting the responsibility to them as they develop their own understanding and skills (Kim & Hannafin, 2011). This temporary strategy for supporting students helps them in developing their higher order thinking skills. It functions in much the same way that scaffolding does when providing safety and access for a window washer or painter. In proving scaffold for students, lecturer should be aware that too much or too little support can surely interfere with the development of higher order thinking skills. According to Crowl, Kaminsky, & Podell, (1997), scaffolding can change the way by which students learn new information and skills. Crowl *et al.,* (1997) gave an example that when teachers give students help even when they do not ask for it, students get the message that they cannot do the task on their own. As students differ in the ways they learn, lecturers should therefore be knowledgeable enough to know when and how to introduce scaffold into the instructional materials. Some educational researchers have suggested some of the strategies that could be used in order to provide the necessary support needed for developing higher order thinking skills. These strategies are using scaffolding during initial learning, and should be used along with a variety of examples to describe the thinking processes involved; using scaffolding only when needed which involves checking the students' understanding first; and using scaffolding to build on student strengths and accommodate weaknesses (Clarke, 1990; .Crowl et al., 1997; Kauchak & Eggen, 1998).

In recent years various researchers across different subject domains have recognised the use of instructional scaffolds as pedagogical tools to assist students in making connection between metacognitive knowledge and metacognitive control (Mayer, Moeller, Kaliwata, Zweber, & Stone , 2012; Schmidt, Loyens, S, Van Gog, & Paas, 2007). Mayer *et al.,* (2012) noted that scaffolding works by reducing the amount of cognitive effort that student must expend to learn the materials; by providing students with concepts structuring their learning in useful ways. Davis & Linn (2000) also found that the introduction of reflective prompts into the learning material help the students in knowledge integration and encouraged reflection at a level that students did not generally consider. Reflective prompts helped students to self-monitor and study strategically. It is expected that reflective prompts may play an important role in helping students to self-monitor their problem-solving processes and consider various perspectives and values regarding their selected solutions. While different studies in the field of learning science has established that effective scaffolding can facilitate conceptual understanding, unfortunately, most research in this comes from science and math education communities. Efforts among engineering education community have been limited in using cognitive findings to facilitate engineering students' learning and problem solving, particularly for African students in engineering.

In order to improve engineering students' higher order thinking ability and knowledge construction that enables them to be able to integrate and apply their classroom knowledge to real world situations, this paper attempts to gain much needed empirical evidence to show a link between instructional scaffolding and the development of concept maps in an introductory strength of material module. The goal of this study to find out if instructional scaffolding and reflective thinking strategies can lead to higher order thinking skills and knowledge construction as measured through the students' concept maps presentations in the module. Meanwhile, this study portrays a range of responses while enlightening our understanding of the pedagogy underlying the use of instructional scaffold in developing concept mapping for meaningful learning in an introductory strength of materials module in addition to helping engineering educators to establish effective teaching strategies to help students improve their understanding of strength of materials concepts, problem solving principles as well as their academic performance.

Constructivism as the Theoretical Framework for this study

As this study is looking at the effect of scaffolding of students' higher order thinking and reflective thinking skills, there is a need for a clear articulation of the basic theoretical principles which ensured its appropriate use in the classroom. The study was considered form the constructivism perspective of learning. Constructivism is a learning theory which posits the phenomenon that learning is the construction of knowledge on the basis of personal understanding (Adam, 2007). Bruner's discovery learning and Vygotsky's zonal of proximal development (ZPD) were the prominent work in the field of constructivism. Bruner's view poses the active engagement of students in the learning process, while Vygotsky revealed that learning is constructed by sharing meaningful information in a social context.

Specifically, this study is based Vygotsky's ZPD work on how people learn; his theory of cognitive development focused on the social aspect of learning and the need for support during the learning process. Vygotsky suggested that in order for a student to learn a concept or skill, the concept or skill had to be within what he called the student's "zone of proximal development (ZPD)" The ZPD is defined as the distance between what a student can do with and without help (Vygotsky, 1978). If a concept or skill is something that a student could do with the help of a "more knowledgeable person," then that concept or skill is something they could perform on their own after learning it with support. Vygotsky called the support that students receive in order to learn "scaffolding." Vygotsky's concept of the ZPD focuses on supporting students to be active learners in order to become self-regulated learners.

In this study, Vygotsky's theory has been very helpful in planning the instructional material for the module. The learning materials were planned in such a way that it helps students to think through the knowledge and skills that they were expected to master and determine the order in which they will learn those things. Using the ideas of scaffolding and ZPD helped in understanding that students are on a journey of cognitive development and they should be supported anytime they are learning things which is outside their ZPD. While we had some level of expectations from the students, we still provided them with some scaffolding that helped the students to master the concepts in a sequential way in order to improve their higher order thinking skill.

The activities used in this module were structured in a way that enables the students have appropriate scaffolding for practicing the concept of engineering module. Once they've had plenty of practice with scaffolding at both individual and group levels, the scaffoldings were slowly remove until the student has mastered the concept on their own. Based on our experiences in this module, Vygotsky's theory of scaffolding can be one of the most beneficial concepts that engineering lecturers can make use of in their lecture rooms. It relies on the way in which students develop cognitively, and it provides a healthy and structured environment for learning. Scaffolding helps students develop higher order thinking skills. However, too much or too little support can hinder development.

Research questions:

(i) How do instructional scaffolds affect students' learning processes of introductory strength of materials module?
(ii) How do students' use the instructional scaffolds to develop concept maps in an introductory strength of materials module?
(iii) What are the participants' views on the instructional scaffolds and concept maps?

Methodology

Qualitative research design dwelling on ethnographic methodology was adopted to achieve the objectives of this study; with a view to obtaining rich observational data via video observation of participants' group collaborative activities, participants' daily learning journals, work artefacts and interview data; as second year engineering students assessed the effects of the usage of instructional scaffolds on their learning processes of an introductory strength of materials module and the development of concept maps. Therefore, a detailed understanding of how the usage of instructional scaffolds influenced the participants' learning processes of the strength of materials as well as the development of concept maps for problem solving purposes is generated.

Participants

Blackboard announcement was sent to 400 students offering strength of materials in the first week of the second semester of 2015-2016 academic session that participants were needed to engage in a study investigating the effect of instructional scaffolds on learning processes of strength of materials module. Students were informed that participants would receive a bonus of 5 marks in addition to their module marks at the end of the semester. Eventually, only forty students agreed to participate in this study. Participants were drawn from ten engineering programmes namely: mechanical & energy engineering; industrial & manufacturing engineering; mechatronics & industrial instrumentation; electrical & electronics engineering; telecommunications and computer engineering; materials & metallurgical engineering; chemical engineering; civil engineering; civil & environmental engineering and mining engineering. The participants consist of twenty-five male and fifteen female students. Students from each programme were requested to form the collaborative group for developing the concept map. Each group consisted of between three to five participants.

Task

Forty second year engineering students who participated in this study were given instructional scaffolds on the following topics in strength of materials module: mechanical properties, axial load, bending and transverse shear over a period ten weeks in the second semester of the 2015-2016 academic session. Instructional scaffolds for the strength of materials module (i) described the nature of the problem required to be solved; and then highlighted on (ii) various concepts required in solving a question and their inter-relationships; (ii) the sequential procedure required in planning how to obtain correct answer to a problem; (iv) reasons for applying a concept and mathematical relationships and (v) provided for questioning the rationale behind the procedure adopted for problem solving in the instructional scaffolds. Figure 1 shows typical instructional scaffolds on bending that was used in this study.

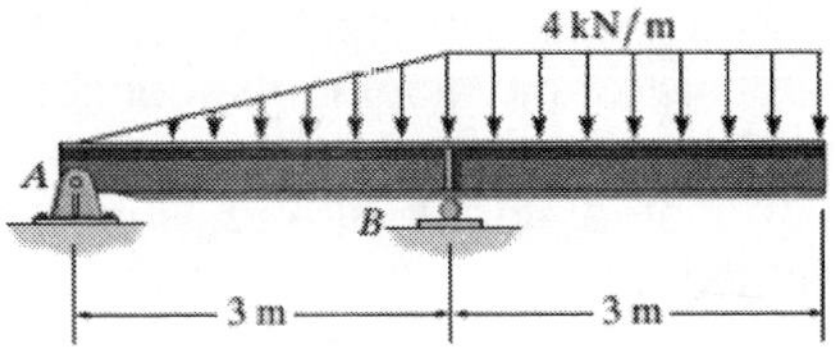

Fig. 1

This problem illustrated in Fig. 1 combines both uniform (to the right side of support B) and non-uniform (to the left side of support B) loadings. Te representative point load for the non-uniform loading would be (1/2 x 4 x 3 kN = 6 kN) and it is placed at a distance of 2 m from support A while the representative point load for the uniform loading would be (4 x 3 kN = 12 kN) and it is also placed at a distance of 1.5 m from support B.

Recall:

- Representative point load for the non-uniform loading would be the area of the triangle as shown on Fig. 1. It acts at 2/3 of the distance AB.
- Representative point load for the uniform loading would be the area of the rectangle as shown on Fig. 1. It acts at the mid-span of the overhang.

(i) Draw the free body diagram with a support reaction acting vertically upward at A, representative non-uniform load of 6 kN acting vertically downward 2 m from support A, a support reaction acting vertically upward at B, and representative uniform load of 12 kN acting vertically downward 1.5 m from support B.

(ii) Write down the force equation and the moment equation by taking moment about point A. Use both force and moment equations to simultaneously determine the magnitudes and directions of the reactions at supports A and B. [Does your support reactions at A = -4 kN and B = 22 kN? What does the result for support A imply?]

(iii) In Fig. 1, the loading system combines both uniform and non-uniform loadings with the sectioning of the beam carried out to the left of support B. Therefore, discontinuous loading (combination of uniform and non-uniform loadings) necessitate that two regions namely $0 \leq x < 3\,m$ and $3\,m < x \leq 6\,m$ shall be considered in establishing the shear force and moment equations.

Figure 1: A typical instructional scaffold for solving a problem on bending.

Participants were requested to use the instructional scaffolds during their personal study time and group collaboration for developing the concept maps on each topic. During the personal study time, participants developed learning journals to record how the use of instructional scaffolds influence their learning processes of strength of materials. They were asked to answer the following questions so that insight may be gained on their learning experiences on each topic when using the instructional scaffolds: (i) What were my learning goals? (ii) Did I achieve my learning goal by using the instructional scaffolds? What are the concepts I learnt about by using the instructional scaffolds? How did the instructional scaffolds affect my learning processes of strength of materials module? How did I use the instructional scaffolds in planning solutions to the concepts learnt?

In the tenth week of this study, each of the ten groups engaged in a collaborative learning activity on a topic of interest for which instructional scaffolds were developed. This entailed working together to identify the knowledge and variables that are necessary to solve multifaceted problems on a topic of interest and how they built upon that knowledge to develop its mastery. Each group also constructed concept map for a topic of interest by using the knowledge and variables already identified by the members during private study. It is expected that this activity will increase both self-efficacy and collaboration since students will realise that their ideas were, in fact, the basis for developing the concept map. The concept map will help students to understand why they were learning and solving each multifaceted problem, how new concepts fit in with the other ideas already learned and where, conceptually speaking, they will be headed in the future.

Data Sources

To assure trustworthiness of findings, triangulation of data which allows for investigation of alternate explanations for study outcomes was engaged as follows: (i) Participants kept track of how the use of instructional scaffolds affected their learning process of strength of materials module in a learning journal during private study time. The learning journal was collected for analysis after the completion of data collection. (ii) Notebook recorded observations by the teaching assistants (TAs) during group collaboration activities for developing concept maps. (iii) Work artefacts such as the concept maps developed by the participants. (iv) Video recording of participants' collaborative sessions when developing concept mapping. (v) Participants' interview which was carried out by the TAs who employed a semi-structured format with open ended and directed questions. The interview served as a check on the interpretation of data from other sources. The video recorded interview took about ten minutes for each participant and was held in a meeting room located at the Engineering Faculty Building on the campus. The video recorded collaborative and interview sessions were transcribed by the first author. During interviews, participants were asked the following questions: How did the instructional scaffold influence your learning process? Give example using any of the topics. What approaches/practices did you adopt in developing the conceptual mapping for a topic of your choice? How would you rate the effectiveness of the introduction of instructional scaffold? Deeper levels of conversation were elicited when interviewing the participants by using follow-up questions. Hence, the outcomes which address each of the research questions are reported.

Data Analysis

Research data including participants' learning journals, work artefacts, observation and interview transcripts obtained from this study were analysed with a view to developing and interpreting how the usage of instructional scaffolds shaped the participants' learning process of strength of materials as well as the development of concept maps. The transcribed video observation and interview data were read with a view to carrying out open coding to allow for emergent themes. For instance, emergent themes describing how students' engagement with the instructional scaffolds influenced their learning processes of strength of materials module included: (i) understanding of the nature of the problem required to be solved; (ii) understanding of the sequential procedure required for obtaining correct answer to a problem; (iii) evaluation of the success or failure of the problem solving and learning processes; (iv) awareness of various strength of materials concepts required in solving a question and their inter-relationships; and (v) providing the rationale for adopting a particular procedure for problem solving. Therefore, the existence, meanings, and relationships of the words or concepts that were related to the influence of students' engagement with instructional scaffolds on their learning processes were explored and noted down during the process of analysis. Furthermore, content analysis was applied to the concept maps made by the participants. Hence analysis of the participants' learning processes provided insights into how it was shaped by the usage of instructional scaffolds. The interview data from the students were also analysed by categorisation. To ensure objectivity in the coding process, validity and reliability aspects were also considered. Two coders (the second author and another expert in education research methods) participated in the categorisation process. The coders separately searched for items in various data which demonstrated each item of the emergent themes and then compared and discussed their selections. After coding, the inter-rater reliability was calculated (Holisti's coefficient) for both coders. After the computation of the inter-rater reliability, the coders discussed any controversial cases until they reached 100% agreement.

Results and findings

Effects of instructional scaffold on learning processes of introductory strength of materials module.

Analysis of the daily learning journals submitted by the participants indicates that all the participants agreed that it affected their learning processes of introductory strength of materials module positively in the following ways:

- ➢ It provided a better understanding of the problems in strength of materials and what it entails by explaining the question in details and its components.
- ➢ It provided the participants with question approach skills because the scaffolds clearly stated the approach process as it is.
- ➢ It aided the participants to confirm and cross-check whether the answers obtained were correct or not. Hence, scaffolds helped the participants to challenge themselves and work out the questions all over again and until a correct answer was obtained.
- ➢ The instructional scaffolds enabled the participants to become aware of the fundamental laws, relevant rules/principles/concepts and assumptions in strength of material as it affect different topics in the module by asking questions related to the answers that needed to be worked.
- ➢ In comparison to the solution manual, instructional scaffolds are much better than the solution manual in that they create room for students to think thoroughly about the problem to be solved as well as the procedure to be adopted.

Findings from the daily learning journals were also corroborated with the interview data. For example, **Amantley** reported that "*The instructional scaffolds enabled me to focus my study of strength of materials module on relevant concepts. Hence, it aided my ability to grasp important concepts and shortened the time I spent studying as I do not have to go through the boring explanations in the text-book.*" Furthermore, **Itumeleng** was also of the opinion that "*the instructional scaffolds helped me to understand the procedure to adopt in solving problems as related to bending, mechanical behaviour, hence, developing skills in thinking through problem solving strategies during his study time*" when asked "*how did the instructional scaffolds influence your learning process?*" In response to the question "*How much would you rate the instructional scaffolds in helping your learning process of strength of materials?*" **Micchelle** stated that she would rate the instructional scaffold 8/10 as it taught her how to think through problem solving by connecting relevant concepts to the problem being solved.

Students' use the instructional scaffolds to develop concept maps in an introductory strength of materials module.

The video data analysis of group collaboration established that the participants referred to the instructional scaffolds in addition to the worked examples in the text-book (Mechanics of

Materials, by Hibbeler, 12th edition) while developing the concept map. For example, in developing the concept map on axial loads, participants from industrial and manufacturing programme used the instructional scaffolds to organise how various concepts such as elastic deformation, relative displacement, force analysis, equilibrium equation and compatibility condition related to each other. They also referred to the instructional scaffolds to identify the various conditions/principle guiding the usage of the concepts. It is also interesting that they opened to chapter of the recommended text-book which dwelt on axial loads to refer to procedures used in solving the worked examples and the arrangement and inter-relationship of various principles in order to develop the concept map. Figure 1 shows the axial loads concept map as developed by participants from industrial and manufacturing programme using the instructional scaffolds and the recommended textbook.

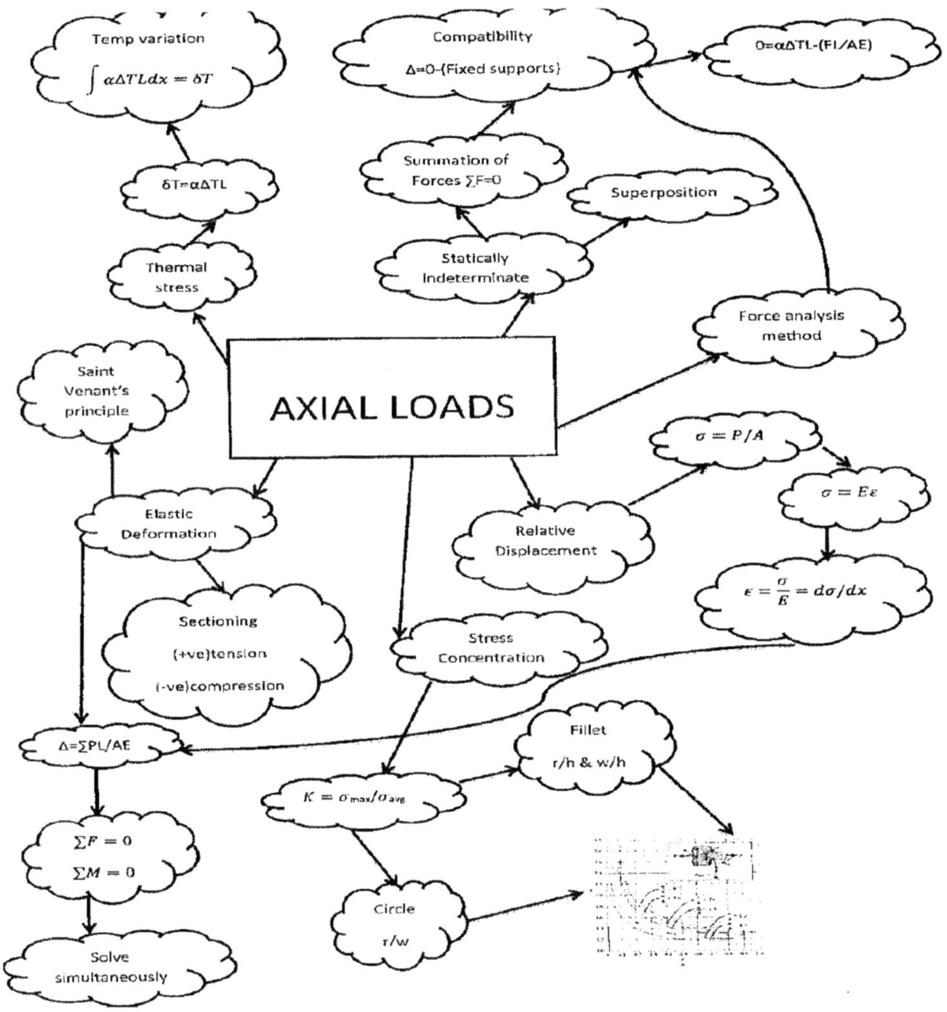

Figure 1: Concept map developed for axial load using the instructional scaffolds and the recommended text-book.

Evidence from participants' daily learning journals revealed the following facts:

- Instructional scaffolds provided insights in summarising various ideas constituted in a topic of interest when developing the concept map.

- Instructional scaffolds provided relevant information on the commonalities among the concepts on a topic and these are included on the concept map so as to solve problems of interest.

Moreover, a participant reported the following, in his daily learning journal, as approaches adopted in using the instructional scaffolds to develop the concept map on bending.

- I took time to study the instructional scaffolds and the chapter on bending (from the recommended text-book) in order to get familiarized with the relevant concepts and their inter-relationships.
- I arranged the concepts and sub-concepts on bending in hierarchical form
- I developed a rough sketch of all the formulae and selected the relevant ones from the rest.
- The rough sketch was developed into concept map using the mind jet software (student version)
- The concept was printed and presented to my group for brainstorming.

Students' views on the instructional scaffolds and concept maps.

Data from the participants' learning journals indicated the following views on the instructional scaffolds and concepts:

- Even though, the topic "axial load" seemed so long and not easy at all, the instructional scaffolds helped me to identify important concepts such as superposition principle and the saint-Venant's principle which I need to grasp. The instructional scaffold also helped me to realise the importance of other concepts (e.g. equilibrium of forces, Hooke's law, mechanical properties) learnt in first semester statics and materials science modules.
- "Instructional scaffolds are a useful facet in improving understanding and academic performance. Ever since I came to this University, I thought any module to do with Strength of Materials was difficult and I always thought answers were obtained by some kind of magic. Even though, I always had Strength of Materials solution manual but never knew where certain steps came from or why certain formulas were brought. However, when I used the instructional scaffolds, I got a larger picture of where does this come from, why do we need this or that formula and how to use what I know to solve a problem........"

Interview data also established the following views of the participants about the instructional scaffolds and concept maps.

- **Itumeleng:** The contents of the instructional scaffolds were too detailed; hence, it should be summarized as it is clearly noticeable that many students do not love reading lengthy paragraphs. Moreover, instructional scaffolds should be sent much earlier before tutorials so that students can be able to prepare for tutorials using them.
- Instructional scaffolds helped me to improve my marks because I understood how to solve the questions much better than before. For example, even though I did not get the average 50% in both my tests, my marks from test 1 to test 2 improved due to my study of the instructional scaffolds.

In summary, participants agreed that the use of instructional scaffolds helped them to grasp relevant concepts in strength of materials and related prior knowledge in statics and materials science to concepts in strength of materials; changed their negative perception that strength of materials was difficult and aided their understanding of how and why certain formulae were used in solving a particular problem in a manner it was done; improved academic performance and that its contents should be concise rather than being too detailed.

Discussion

While finding ways of improving students higher order thinking ability that will make them relevant as practicing engineers, an alternative to the traditional forms of teaching engineering modules were introduced in this study. The results of this study indicate that, all the engineering students who participated in the study were aware of the practice of scaffolding and appreciated its potential use in problem solving. The participants saw the scaffolding that was introduced to their learning material as a useful concept that allows them to move away from the direct instruction of a traditional classroom and search for a richer and more sophisticated educational tool in engineering classroom. Students demonstrated understanding of the concept of strength of materials and they were able to link this concept with other important concepts in the module. They reported improvement in their conceptual understanding of the learning content as they interact with peers in their collaborative group. The instructional scaffolding also enabled the students to study and understand the material within a short time; this was quite encouraging from the students' perspective.

The quality of group work that students presented in their concept maps show that instructional scaffolds enabled them to summarise their ideas about the given topic in just one page. The scaffolds allowed the students to see the inter-relationships among the concepts on a topic of their interest. Students were able to organise their thinking skills by arranging concepts and sub-concepts on bending in hierarchical form. This report gives evidence that students do not only learn the content of bending in the study, they were able to develop other critical skills that the employers are looking for. The students were able to develop their skills in using external resources such as mind jet software which helped them in presenting their thinking in a logical way. Along the line, students were able to develop their interpersonal relationships as they were able to work successfully in their collaborative group.

The students' general views about the use of the instructional scaffolds and concept maps in the strength of material module pointed out to the fact that they were happy with their learning processes. They were even able to revisit their first semester module on statics and materials science concepts with the strength of material concept. This shows that students have improved in their reflective thinking ability. The participating students recounted that instructional scaffolds improve their self-efficacy (confidence) as they can positively contribute to discussion during their group task on concept map. Generally, they demonstrated improvement in their confidence in learning strength of materials which they always think was a very difficult topic to understand. With the instructional scaffolds, students recorded increase in their marks from test 1 to test 2.

The analysis of the results of this study revealed the need for improvement in the way in which instructional scaffoldings were introduced into the learning materials. Some of the students expressed their concerns about the fact that the contents of the instructional scaffolds were too detailed which they do not like. They also felt that the instructional scaffolds should be sent much earlier before tutorials so that they can be able to prepare for tutorials using them. These feedbacks provide the authors with insights into how to improve instructions through scaffolds. For improvements, more efforts will be placed on refining the scaffold questions in such a way that it will not increase the cognitive loads of the student but rather makes learning processes easier and increase students' motivation towards engineering modules.

Conclusion

In this paper, the authors described how instructional scaffolds introduced into the strength of material module help the students in developing their higher order and reflective thinking skills. The students' learning outcomes as measured from the concept maps presented revealed their disposition toward the extra support given to them in the module. Our goal is to communicate the potential impact of scaffolding in supporting engineering students in order to improve their problem solving skills. Meanwhile, we intended to address issues regarding the enhancement of engineering students' creativity, which has becoming high demands in the current data-supported and knowledge-driven economy. Based on our implementation experience, we discussed the details of instructional scaffolding with the group of students and implementation procedures. It is expected that experimental research will be conducted in the future to confirm the hypothesis that students' higher order thinking skills capability significantly improves with the extra support introduced into the module.

References

Adam, N. B. (2007). Towards a Model for Knowledge Development in Virtual Environment: Strategies for Student Ownership. *International Journal of Social Sciences, 2*(2), 71–77.

Clarke, J. H. (1990). *Patterns of thinking: Integrating learning skills in content teaching.* Boston: Allyn and Bacon.

Crowl, T. K., Kaminsky, S., & Podell, D. (1997). *Educational psychology: Windows on teaching.* Madison, WI: Brown and Benchmark.

Davis, E. A., & Linn, M. C. (2000). Scaffolding Students' Knowledge Integration: Prompts for Reflection in KIE. *International Journal of Science Education, 22*(8), 819-37.

Felder, R. (2006). A Whole New Mind for a Flat World. *Chemical Engineering Education, 40*(2), 96-97.

Felder, R. M. (2008). On-the-Job Training,. *Chemical Engineering Education, 42*(2), 96-97.

Kauchak, D. P., & Eggen, P. D. (1998). *Learning and teaching: Research-based methods* (3rd ed.). Boston: Allyn and Bacon.

Kim, M. C., & Hannafin, M. J. (2011). Scaffolding problem solving in technology-enhanced learning environments (TELEs): Bridging research and theory with practice. *Computers & Education*, 56(2), 403-417.

Magee, C. L., Flemings, M. C., Cutcher-Gershenfeld, J., Murphy, W. P., Perkins, D., Petroski, H., . . . Walker, D. F. (2003). *Advancing Inventive Creativity through Education, the report of the Lemelson-MIT Workshop on Advancing Inventive Creativity through Education.* Retrieved January 2017, from http://www.creativetallis.com/uploads/2/2/8/7/2287089/inventive_creativity.pdf

Mayer, R., Moeller, B., Kaliwata, V., Zweber, B., & Stone , R. (2012). Educating engineering undergraduates: Effects of scaffolding in a problembased learning environment. *Proceedings of the Human Factors and Ergonomics Society Annual Meeting*, *56*, pp. 2507-2511.

Putnam, C., O'Donnell, J., & Bertozzi, N. (2010). Scaffolding and fading within and across a six-semester CDIO design sequence. *Proceedings of the 6th International CDIO Conference Ecole Polytechnique, Montreal.*

Schmidt, H. G., Loyens, S, M. M., Van Gog, T., & Paas, F. (2007). Problem-based learning is compatible with human cognitive architecture: Commentary on Kirschner, Sweller, and Clark (2006). *Educational Psychologist, 42*, 91–97.

Streveler, R., Litzinger, T., Miller, R. L., & Steif, P. S. (2008). Learning Conceptual Knowledge in the Engineering Sciences: Overview and Future Research Directions. *Journal of Engineering Education*, 279-294.

Vygotsky, L. S. (1978). *Mind in Society: The Development of Higher Psychological Processes.* (M. Cole, V. John-Steiner, S. Scribner, & E. Souber, Eds.) Cambridge, MA: Harvard University Press.

Significant Learning Assessment Model Using Digital Badges: A case study in the Systems Engineering Program of El Bosque University

Sandra Milena Merchán Rubiano

Universidad El Bosque, Bogotá, Colombia
smerchanr@unbosque.edu.co

Daniel Felipe París Gallardo

Universidad El Bosque, Bogotá, Colombia
dparis@unbosque.edu.co

Julián David Pedraza Acosta

Universidad El Bosque, Bogotá, Colombia
jpedrazaa@unbosque.edu.co

***Abstract**: In the Colombian university context engineering curricula are designed aimed to develop competencies, obeying national and international standards. In the context of the El Bosque University Systems Engineering program, in addition to these competencies, students are expected to develop six kinds of learning, according to the Significant Learning taxonomy. Nonetheless, the majority Faculties´assessment systems are numeric-grades based and do not retrieve information about the achievement level of such competencies or kinds of learning. Based on the background review and interviewing curriculum designers, this work presents the process for constructing a continuous assessment model responding to this need, using Digital Badges. The model is presented through two main results: A conceptual network of relations between each problematic nucleus competencies that are expected to achieve, and the six dimensions of significant learning. The second result is a two-dimensional digital badges network which determines what badges should be awarded and when to award them.*

Introduction and Context

The Colombian universities´ learning assessment system is mainly based on numeric grades and it is discontinuous. Regardless the pedagogical model or practices, the students´ learning outcomes by course determine if one completes the requirements to earn a final diploma or program certificate. It is only enough determining if all the grades obtained for those curriculum courses were above 3.0/5.0.

This system is discontinuous because it lacks some mechanism to retrieve the student´s performance information on a specific course excepting the grades, and those grades themselves do not have the information needed to evaluate his performance in the area which the course belongs to. Besides this, current assessment systems do not provide student´s performance quality information enough to assess the competencies/skills level achieved.

In addition to the above, in the Systems Engineering Program at El Bosque University the assessment system does not support neither the valuation of competencies proposed by the program´s curriculum, nor the institutional pedagogical model. Firstly, although the curriculum for the program is structured in three main problematic nuclei based on international and national competencies standards as ABET (ABET, 2015), ACM/IEEE (Durant et al., 2015) and ACOFI (ACOFI, 2015) and these three nuclei define what engineering competencies the students´ must develop at the end of the stage using what knowledge in what specific scenarios; current assessment system does not provide quality information about students´ performance regarding those competencies.

Secondly, although the courses that compose each problematic nuclei are strictly designed based on Dr. Dee Fink´s course design guide for Significant Learning (Fink, 2003), which determines that the students learn in six kinds of learning, numeric assessment still predominates at the end of the courses and it does not provide information about the student´s performance on each of those six dimensions.

All of the above means that in the studied context, a discontinuous, disintegrated and numeric-grades based assessment system still predominates. From this context it is necessary to design an additional (or maybe the next) qualitative and continuous assessment model that integrates the six dimensions of significant learning and the skills proposed for each problematic nuclei for the Systems Engineering Program, based on national and international standards. This work present a proposal aimed to answer the question: ¿How to construct a continuous assessment model, able to determine which skills the students develop for the six dimensions of the significant learning and for each problematic nuclei of the System Engineering Program, throughout the curriculum?

Theoretical framework

This work is mainly based on the Digital Badges theoretical framework, the Dr. Dee Fink proposal for designing courses for Significant Learning, and the studied context´s curriculum design.

Digital Badges

The proposed model takes into account the principal characteristics determined by outstanding investigators in this topic such as Bixler & Layng (2013), Mozilla Foundation and Peer 2 Peer University, (2012) and (Hickey, Wills, & Quick, 2015)

- **Learning extends across multiple contexts, experiences and interactions, not just schools.** Digital Badges are useful for capturing, promoting, transferring, and making critical skills attractive or accessible across those contexts. Here, the proposed model intends to be the first step to generate a competencies or skills certificate to be granted together with the final diploma.
- **Course completion badges may consist of a series of other nested competency-based badges** (Bixler & Layng 2013). With badges, individuals find new opportunities to learn following digital pathways, in which exist multiples levels of badges. Accordingly with this, badges motivate users to "level up" their achievements and permit to have multiple levels of assessment (Mozilla Foundation and Peer 2 Peer University, 2012). In the proposed model these characteristics are taken to design two interconnected and hierarchical badges taxonomies, giving us a multidimensional assessment model.
- **Competency-Based badges gives students a way to demonstrate competency derived from individual course objectives**(Bixler & Layng 2013). For the proposed model, the assessment system not only informs which skills the students achieved in each course that is part of each problematic nucleus, but what skills achieved in the Significant Learning dimensions and engineering competencies (ABET, 2015).
- **Badges work better where content and technology already exist** (Hickey, Wills, & Quick, 2015). In the institutional context all teachers have to support their courses with the LMS Moodle. Beside this, the institution permanently conducts ICT training programs for teachers.

- **Badges may also be used to explicitly capture and transfer social skills** Mozilla Foundation and Peer 2 Peer University, (2012). The proposed model includes interpersonal kinds of learning as *human dimension* or *caring.*

Having in mind other projects and experiences implementing Digital Badges as Deakin Digital Credentials System (DeakinDigital), Ireland´s All Aboard,University of Notre Damme (Universiy of Notre Dame), School of Business of Saint Lawrence College (Saint Lawrence College), Mahoney (2012) and Tally (2012); the proposed assessment model tries to face the challenges posed by these same authors. It presents an alternative to conventional credentials, providing information that can be shared both inside and outside the university. Other projects and experiences implementing Digital Badges were reviewed (Law, 2015).

Designing Courses for Significant Learning

Dr. Dee Fink course design guide proposes other kinds of learning beyond the cognitive domain stated by the Bloom´s taxonomy (Churches, 2008). These additional types of learning are found in the personal and metacognitive domains, and complement the cognitive domain by structuring a significant learning taxonomy composed by six interacting dimensions: *Foundational Knowledge, Application, Integration, Human Dimension, Caring* and *Learning how to learn.* This is the first principle or basis of this model. From this principle, is stressed that in the *integration* dimension students connect what they learn in their academic work with other areas of their lives, even after finishing their careers. In the *learning how to learn* dimension, the students should learn how to become a better student and how to inquire a construct knowledge.

As a second principle, this pedagogical model states that college courses´ design must be student learning centered. It means that all the design process is oriented by the learning objectives, and the learning and assessment activities are designed to reach those objectives, in a significant fashion.

As third, the course design must be connected with the *Faculty dreams: "Think holistically rather than in compartments, develop key skills in life, such as communication skills, be able to apply and use what they learn in real-life situations, think about problems and issues in integrated ways, rather than in separated and compartmentalized ways, see the connections between themselves and their own beliefs, values, and actions, and those of others".*

The fourth principle refers to how teaching is composed by four elements: *Knowledge of subject matter, teacher-student interactions, course management* and *design of instruction.* The latter is responsible for solving the traditional course integration problem: The learning goals, the teaching/learning activities, and feedback and assessment are frequently disconnected. Nonetheless, this integration problem remains along the curriculum, because the guide is not focused on applying the same principles on it.

The fifth principle is educative evaluation and its expanded version. It states that for better learning, all evaluation processes must have three main components: "FIDeLity" feedback, self-assessment and forward-looking assessment. The first one means the student requires a back-and-forth dialogue between those giving him the feedback. Self-assessment is important for the student to learn how to assess his own work. Finally, forward-looking assessment means that evaluation activities should be designed based on what competencies are expected to be reached and develop for the student in his near and long-term future. Figure 1 shows this model from the author.

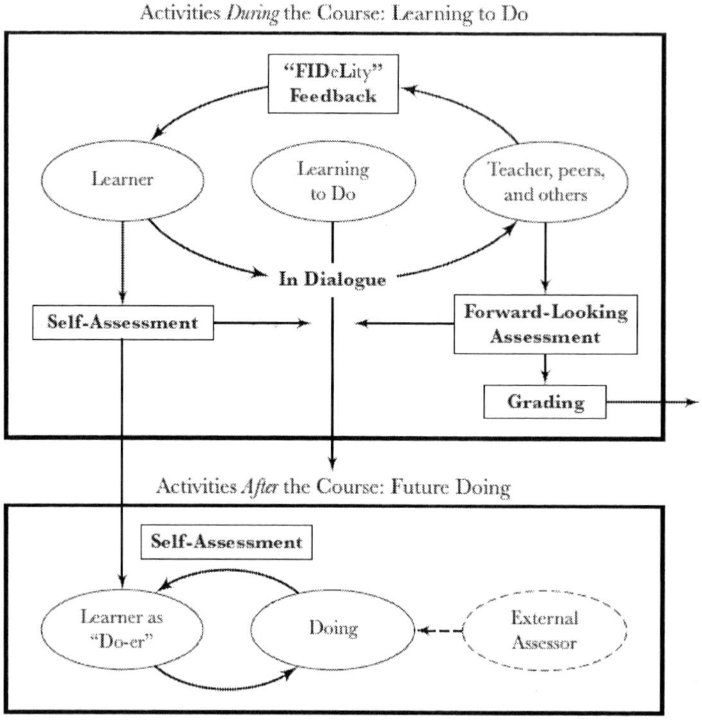

Figure 1: Educative Assessment (Expanded Model)

Systems Engineering Program Curriculum

The program´s curriculum defines the profile of its graduates as follows:

"The systems engineer of El Bosque University trained within the Biopsychosocial and Cultural framework is capable of designing models to propose engineering projects, according to the context and the complex situations of his environment. Constructs information systems for health, education, the productive sector and government, interpreting variables and using updated techniques and tools" (Comite Curricular, 2012)

To achieve previous profile, curriculum is divided into three formative cycles and each of them has its own profile defined as *problematic nucleus*. Those nuclei define what engineering competencies or skills each student must develop when completing the third, sixth and ninth semesters.

Methodological Development

To construct the assessment model, three main stages or phases were conducted. In the first one, named *"Context comprehension and conceptual modeling"*, based on research background and theoretical framework the current program assessment model was studied finding some weaknesses and opportunities for improvement, and identifying the contextual and pedagogic elements that allowed to construct a conceptual model of relations between the skills of each problematic nuclei and the six dimensions of the significant learning for each course that composes those nuclei. At the second stage named Badges Integration, a multidimensional competencies taxonomy and a hierarchical badges awarding system were created. It was determined which badges must be awarded and what should be the best process to award them. At the final stage, the model was evaluated to determine if the objectives were accomplished or not.

First stage: Context comprehension and conceptual modeling

This stage started with a background review, determining what characteristics the model must have. The results of this step helped to construct the theoretical framework and permitted to diagnose current evaluation systems, resulting in the following findings:

- New assessment model must be continuous, valuating engineering competencies and significant learning dimensions.
- New assessment model must solve curriculum´s integration problem. It means the system must permit valuating competencies achievement along the curriculum.
- The principles of the expanded educative evaluation must be applied for the entire curriculum
- Using digital badges in a continuous assessment model throughout the curriculum, permits to determine which skills the students develop for the six dimensions of the significant learning and for each problematic nuclei of the System Engineering Program.

Having determined the principles and characteristics that the new model must have, it proceeded to create a big network of relationships between each courses learning objectives belonging to each nucleus, and the skills or competencies determined for those problematic nuclei. The tasks developed are as follows:

1. It was evaluated if all the learning objectives were correctly classified in the Significant learning six dimensions. To do so, it was analysed if the objective´s main verb corresponded to the dimension in which it was classified
2. The problematic nuclei were conceptually decomposed in competency categories, and the competencies or skills for each category were specified.
3. Achievement indicators were defined for each learning objective, for helping the links establishing.
4. A two-dimensioned relations schema was created, linking the courses´ pedagogical design and the problematic nuclei competencies. Figure 2 shows this schema.

The next step for conceptual modelling was creating a two-dimensional taxonomy for competencies and skills, needed to accomplish the curriculum. These dimensions are created in order to provide the needed student assessment valuation focus. The two dimensions evidently are *Significant Learning* and the *Systems engineering Problematic Nucleus competencies* (hereinafter called variables).

It is important to note that applying these focuses on different levels of this taxonomy is the base for determining what badges and when award them. Having done this, it was possible to go to the next stage of the methodology: *Badges integration.*

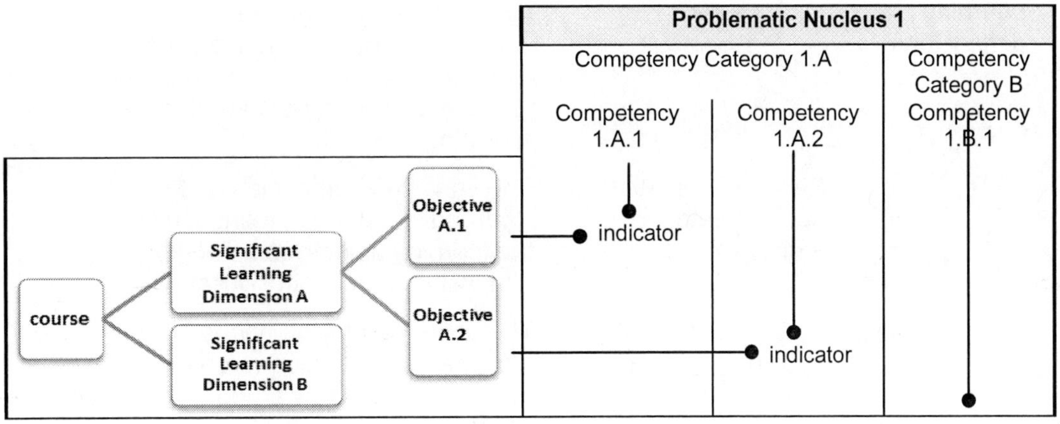

Figure 2. Conceptual Relationships Network

Second stage: Badges integration

The main purpose of this stage was to determine what badges will be awarded in the system, having in mind the dimensions defined previously.

For the first focus (Significant Learning Dimensions) it was determined that it is necessary to assess the students´ significant learning in three contexts: The courses, the problematic nucleus categories competencies, and the all problematic nucleus. For the second focus (Systems engineering Problematic Nucleus competencies) it was determined the need for student´s assessment in three contexts too: The systems engineer competencies for each problematic nucleus, the nucleus categories, and the entire problematic nucleus. Table 1 shows the overlapping two dimensions, shaping the final taxonomy and identifying the types of badges to be awarded.

Table 1. Final bidimensional taxonomy

		Problematic Nucleus 1			
		1.A. Designing and implementing solutions		1.B. Integrating software development, data structures and OOP	...
		1.A.1. Designing models	1.A.2 Designing information systems	1.B.1. Developing information systems	...
Significant learning	Foundational Knowledge	Course1.ObjectiveX, Course2.ObjectiveY	Course3.ObjectiveZ, Course4.ObjectiveW	Course3.ObjectiveA, Course4.ObjectiveB	...
	Application	Course5.ObjectiveX, Course6.ObjectiveY	Course7.ObjectiveZ, Course8.ObjectiveW	Course5.ObjectiveA, Course6.ObjectiveB	...
	Integration	...	...	...	...
	Caring	...	...	...	...
	Human Dimension	...	...	...	...
	Learning how to learn	CourseN.ObjectiveX,	CourseN,ObjectiveZ,	CourseN.ObjectiveA,	...

The last step for this stage was to design the necessary procedures to award the badges. Table 2 shows the defined procedures for each type of badge.

Table 2. Badges types and procedures to award them

Focus	Badge type	Awarding Procedure
Significant Learning dimensions	Inside the courses	Teachers will award a badge for each dimension at the end of the courses, when the student completes all the learning objectives belonging to each dimension. This is the lowest level of awarding.
	Each problematic nucleus category´s variable	The variable´s dimension badge is awarded when the student achieves all the learning objectives belonging to the dimension, only for the observed variable
	Entire problematic nucleus	The entire dimension badge is awarded in a problematic nucleus when the student has already been awarded with all the variables badges, only for the observed

		dimension.
Systems Engineering competencies	Problematic nucleus variables	This badge is awarded by completing all the learning objectives belonging to the variable
	Problematic nucleus category	This badge is awarded by completing all the variables badges belonging to a category
	Entire problematic nucleus	This badge is awarded by completing all the categories belonging to a nucleus

Final stage: Model validation

To ensure that the proposed model is consistent for the studied context, some validation procedures were done using interviews and forms. Firstly, the established relations between learning objectives and each problematic nuclei competencies where validated by learning experts. Those experts helped to validate if the learning objectives were classified in the significant learning dimension correctly. Curriculum designers validated the variables and categories competencies decomposition, for each problematic nucleus. Finally, the awarding procedures were validated by teachers and curriculum designers too.

Results and Analysis

Proposed assessment model is conceptually analyzed, based on the facing challenges and characteristics stated by the theoretical framework and the institutional pedagogical model. It can be said that:

1. The model accomplishes the proposed challenges: It is qualitative and helps to determine the skills or competencies achievement. It is continuous, because students´ performance is not only valuated from each separated course, but giving permanent feedback along the curriculum.
2. This continuous assessment model allows observing the engineering competencies achievement level for each student, based on national and international standards.
3. The proposed model extends and applies the principles of educative evaluation to the curriculum, offering the possibility of transcending from this to the external professional scenarios.
4. The model integrates the main digital badges principles:

- It gives the students a way to demonstrate competency derived from individual course objectives or non-class experiences, such as extra-curricular involvement. In the model, competency achievement could be valuated from the minimum unit: learning objectives for each courses, and the maximum curriculum focus: the systems engineering program´s problematic nuclei.
- It helps students to find new opportunities to learn. Having hierarchical taxonomies the students are motivated to level up their learning to the next level. In the other hand, this kind of hierarchy gives the curriculum designers flexibility to have multiple levels of assessment (learning objectives, significant learning dimensions and engineering competencies)
- Being the six kinds of significant learning and the engineering skills the principal valuation focuses, the model is strongly connected with the competency-based education. This kind of evaluation is much broader than current.

- Being the teacher the actor who grants the lower level badges, assessment is vetted and personalized, not only based on traditional testing practices.
- Identifying the achievement of significant learning kinds of learning allows to observe the students´ personal skills, making them available or accessible for all the interested.

Implications for engineering education and prospectives

One of the most important implications of the present work for engineering education is to meet the need to value specific engineering competencies achieved by the student. Most probably the majority of the engineering curricula are designed and aimed to develop competencies, accomplishing the requirements for the recognition and accreditation of the programs, and the parameters given out by the national and international organizations such as ACOFI, ABET and ACM/IEEE.

The model allows accomplishing some of the parameters established by ACOFI for the system engineers' professional education. Firstly, the association suggests that personal skills are determinant for a system engineer's success. Skills such as being critical, being able to adapt to changes, cooperative learning and interaction with others, are included in the Significant Learning taxonomy, addressed in the model. Observing such skills transversally to the curriculum, the model recognizes that they are not only taught in each course, but that they are also the result of the students´ life experiences. Regarding this, parameters of the undergraduate in computing engineering ACM/IEEE state that the learning outcomes are not only stuck to units of knowledge (the courses) and must be related to the entire degree program.

The proposed model contributes to the improvement of quality for the academic program and the institution, because it becomes a tool of self-evaluation and permanent improvement allowing determining what changes should be made on the curriculum design. This criterion is included as criteria 2 and 3 for the accreditation of engineering programs (ABET, 2015), which indicates that programs must regularly use appropriate and documented process for validating and valuing the accomplishment of the students´ learning outcomes. This revision helps the program to be coherent with its identity and the level of adjustment to the institution's mission and the international standards.

Considering the six dimensional system of significant learning as the evolution of the Bloom's taxonomy, and considering that the systems engineering curriculum in El Bosque University includes the competencies for the ECAES (ACOFI - ICFES, 1996) test at the end of the career; the proposed model meets the needs to integrate the external evaluation into the program.

Thinking about approval and transfer systems based on competencies allows to be separated from the traditional model which depends on the numerical approval of the subjects in a specific scale. So the reference contents are scenarios of competence, not the goal of the programs.

The next step to be carried out is the technological implementation. At present we are advancing in the design of an information architecture to be implemented in the LMS Moodle. Additionally e-portfolios integration is being planned.

The work brings other challenges. One of them is the possibility to reconstruct the same two-assessment dimensional model from the competencies of engineering established by the national and international standards, not only by the ones determined for the specific academic program. Another challenge to accomplish is the introduction to the flexibility of the curriculum, making the competencies certification not be limited to the accomplishment of the learning objectives in each course, but instead that each student can do it in a personalized way.

References

ABET, E. A. C. (2015). *Criteria for accrediting engineering programs.* (). ABET:

ACOFI - ICFES.*Actualización y modernización curricular en ingeniería de sistemas.* ().

ACOFI. (2015). *Marco de fundamentación conceptual especificaciones de prueba ecaes ingeniería de sistemas* ACOFI. Retrieved from http://www.acofi.edu.co/wp-content/uploads/2015/07/Marco-de-Fundamentaci%C3%B3n-Ingenier%C3%ADa-de-Sistemas-ICFES-ACOFI-2005.pdf

All Aboard. All aboard: Digital skills in higher education. Retrieved from http://allaboardhe.org/

Bixler, B., & Layng, K. (2013). *Digital badges in higher education: An overview* Penn State University.

Comite Curricular. (2012). *Universidad el bosque programa de ingeniería de sistemas proyecto educativo del programa*

DeakinDigital. Deakin professional practice credentials for high-performing organisations. Retrieved from https://www.deakindigital.com/how-credentials-work

Durant, E., Impagliazzo, J., Conry, S., Reese, R., Lam, H., Nelson, V., . . . McGettrick, A. (2015). *Computer engineering curricula 2016 - CE2016.* ().ACM/IEEE. doi:10.1145/3025098

Fink, L. D. (2003). *A self-directed guide to designing courses for significant learning*

Garret, J. J. (2000). Un vocabulario visual para describir arquitectura de información y diseño de interacción. Retrieved from http://www.jjg.net/ia/visvocab/spanish.html

Grant, S. (2012). Questions about badges in higher ed. *HASTAC, Changing the Way we Teach+learn,* Retrieved from https://www.hastac.org/blogs/slgrant/2012/10/05/questions-about-badges-higher-ed

Grant, S., & Shawgo, K. E. (2013). Digital badges: An annotated research bibliography v1. Retrieved from https://www.hastac.org/digital-badges-bibliography

Guia para diseño de objetivos

Hickey, D. T., Wills, J. E., & Quick, J. D. (2015). Where badges work better. *Reference Services Review, 43*(1), 31-44. doi:10.1108/RSR-07-2014-0026

La taxonomía de bloom y sus actualizaciones

Law, P. (2015). Recognising informal elearning with digital badging: Evidence for a sustainable business model. *Open Praxis, 7*(4), 299-310. doi://dx.doi.org/10.5944/openpraxis.7.4.247

Law, P. (2015). Recognising informal elearning with digital badging: Evidence for a sustainable business model. *Open Praxis, 7*(4), 299-310. doi://dx.doi.org/10.5944/openpraxis.7.4.247

Mahoney, J. (2012, April 25,). Designing a badge system for universities. Retrieved from http://coursedata.blogs.lincoln.ac.uk/2012/04/25/designing-an-badge-system-for-universities/

Mozilla Foundation, & Peer 2 Peer University. (2012). *Open badges for lifelong learning* Retrieved from https://wiki.mozilla.org/images/5/59/OpenBadges-Working-Paper_012312.pdf

OpenBadges. (2014). Retrieved from https://docs.moodle.org/dev/openbadges

Tally, S. (2012). Digital badges show students' skills along with degree. Retrieved from http://www.purdue.edu/newsroom/releases/2012/Q3/digital-badges-show-students-skills-along-with-degree.html

Evolution of engineering education research as a field of inquiry in the UK

Abel Nyamapfene
University College London, UK
a.nyamapfene@ucl.ac.uk

Bill Williams
Instituto Politécnico de Setubal, and CEG-IST, Universidade de Lisboa Portugal
bwbillwilliamsbw@gmail.com

***Abstract**: We report on a preliminary investigation to determine the current state of Engineering Education Research (EER) within the UK, as well as its likely growth trajectory. The investigation involved the analysis and categorisation of articles published in the period 2000 -2017 by UK authors in the Journal of Engineering Education (JEE), the International Journal of Engineering Education (IJEE) and the European Journal of Engineering Education (EJEE). A similar analysis was also carried out on EER- focussed PhD theses published in the UK in the same period. Our study establishes that the level of engagement in EER in the UK is still relatively low, as evidenced by the small number of published articles. In addition, our study also finds that EER publications from the UK are predominantly single-authored, or from single institutions. This suggests low levels of internal and external collaboration amongst EER researchers in the UK.*

Introduction

Over the last decade there has been increasing interest in mapping the evolution of engineering education research (EER) as a field of inquiry. Lohman and Froyd (2010) looked at the process from a predominantly U.S. perspective, van Hattum-Janssen et al. (2015) studied EER evolution in Portugal while Edstrom et al. (2016) have characterised it in the Nordic context. This work-in-progress study aims to characterize the evolution of EER in the UK with a particular focus on the years from 2000 to the present.

Within the UK and other western countries it is now generally believed that technology is the bedrock for future economies, and that engineering education could contribute to this by ensuring that a sufficient number of young people acquire the necessary skills appropriate for a technology-based 21st century economy, and remain within engineering so that the economy can benefit from their acquired skills (EngineeringUK 2015). However, engineering education is still struggling with issues of underrepresentation, recruitment and retention, and some researchers now believe that these issues can only be resolved by resorting to well researched, and well-funded intervention strategies (Henderson et al. 2011). This therefore calls for well-resourced EER strategies to underpin engineering education.

However, a 2013 survey of EER researchers (Shawcross and Ridgman 2013) suggests that in the UK, EER remains under-resourced, especially when compared to mainstream discipline-specific science and technology research, with most of the EER research being carried out by individual academics without any supporting funding. Because of this, the time allocated by UK-based academics to EER is also small, to the extent that 65% of UK-based EER researchers spend no more than 20% of their time on EER (Shawcross and Ridgman 2013). The objective of the research reported in this paper is to map out the UK EER landscape, with a view to providing a concrete and accurate picture of the state of EER within the UK, and to provide evidence of how it measures up to EER in related Organisation for Economic Co-operation and Development (OECD) countries. We have used the following research question to guide our study:

What does an analysis of data on publications by UK-based researchers tell us about the evolution of EER in the UK?

Organisation and Structure of the Paper

The specific focus of the work reported in this paper is on the period beginning from 2000 to the present. However, EER has been taking place within the UK for a longer period than this. Therefore, to provide the context within which this work is situated, we will provide a brief overview of the history of EER in the UK. We will follow this with a discussion of our approach to mapping the current state of EER in the UK. Jesiek et al. (2011) developed a scheme for categorising EER journal papers which they used in their work on mapping global trends in EER over the period 2005-2008. We have used their categorisation scheme in the study reported in this paper. After this, we discuss the preliminary results that we have obtained to date, and this leads to a discussion of the potential implications of these preliminary findings.

Earlier UK-based EER

In contrast to many other European countries (see for example van Hattum-Janssen et al. (2015)), EER is not a recent development in the UK. Heywood (1970) describes how in the post-war period there was a national focus on the need to increase the qualified technological workforce. There were a number of Ministry of Education sponsored reports in those years: the Percy Report (Ministry of Education 1945), a White Paper(Ministry of Education 1956), the Crowther Report (Education 1959) and the report of the Bosworth Committee (Bosworth 1966).

These national initiatives were paralleled by a number of studies carried out by university faculty, often in collaboration with industry. Examples include the work of Heywood on graduate careers (Heywood 1969) and on engineering practice (Youngman 1978), Hutton and Gerstyl on engineers' attitudes to their education (1964) and Bosworth's study of engineering design (1963). The University of Lancaster made significant contributions to engineering education in this period and introduced "engineering analysis" to its engineering programs, an approach that today would be called problem based learning (PBL) (Heywood 1970). Heywood and his colleagues, firstly at Lancaster and later Salford University were particularly prolific from the 1960s through to the 1980s; see, for example, Heywood (1970), and Heywood and Monk (1977).

Meanwhile in Scotland John Cowan was carrying out research into self-managed learning and in 1982 Heriot-Watt University appointed him to be the first professor of engineering education in the UK.

Method

In this paper we report on preliminary research aimed at identifying and categorising the nature of engineering education research taking place within the UK. Our work follows similar studies that have been undertaken in the United States such as the work by Wankat (2004), Whitin and Sheppard (2004), and the more recent work by Jesiek et al. (2011). Like Wankat (2004), we restrict ourselves to the main EER journals, in our case the International Journal of Engineering Education (IJEE), the Journal for Engineering Education (JEE), and the European Journal for Engineering Education (EJEE). This is because, as Jesiek et al. (2011) observe, such publications have committed to publishing and disseminating high quality EER work.

We have also included in our dataset all the EER related PhD theses published in the UK between 2000 and 2017. Our reason for focussing on PhD theses is that the publication of a PhD thesis follows a sustained period of research on a substantive topic. A PhD thesis also strongly suggests the presence of expertise within an institution and faculty who are committed to researching a particular field of interest. Hence PhD theses are likely to indicate areas deemed to be important and worth researching by academics with an interest in EER.

Jesiek et al. (2011) came up with a list of 38 categories for classifying the nature of EER research based on the keyword used by authors in their papers, as well as the key terms arising out of the paper contents. For the three journals, we identified all the papers authored by UK-based EER researchers and recorded the author keywords. Where keywords were not available, especially in the earlier journal issues, we scanned the abstracts of UK-authored papers to come up with appropriate keywords. Using this method, we identified for the period 2000-2017, 109 EJEE articles, 58 IJEE articles, and 5 JEE articles.

For the PhD theses we accessed the British Library E-Theses Online Service (EThOS), and used Boolean combinations of the terms "engineering", "learning", "education", "practice", "PBL" and "problem-based" to identify theses related to engineering education and practice. We then scanned the theses' abstracts to establish their relevance. Using this method, we identified 93 PhD theses that had been submitted and published in the period 2000 – 2017.

Findings

As pointed out earlier, the research question guiding our study is: What does an analysis of data on publications by UK-based researchers tell us about the evolution of EER in the UK? To find answers to this question we broke it down into three sub-questions:

1. What are the specific categories, as defined by Jesiek et al. (2011), that characterise UK research publications?
2. What is the nature of research collaboration amongst UK EER researchers as evidenced by the categorised research publications?
3. What are the emerging trends in the evolution of UK EER as evidenced by changes in research categories over time?

Categorisation of UK EER publications

Table 1 shows the categorisation of UK EER research by journal ordered by the number of instances of occurrence. The table indicates that 6 categories feature in the top ten of both the EJEE and IJEE journal. These six are *edtech, assessment, design, collab, curriculum* and *profession*. The scope of each of these categories, as specified in Jesiek et al. (2011) is:

Edtech	Educational/Instructional Technology, inc. Online, Distance, Web-based Learning
Assessment	Assessment (of Student Performance), inc. Evaluation, Exams, Grading, Outcomes-Based
Design	Design, inc. Education, Methods, Practices, Processes, Projects, Skills, and Thinking
Collab	Collaborative/Group/Peer/Team Learning, Performance, Roles, and Skills
Curriculum	Design, Reform, and Development of Curricula and Courses (general/unspecified)
Profession	Studies of Alumni, Careers, Continuing Education, Employment, Postgrads, Professionals

Edtech ranked highest in both journals. This suggests that educational technology and its use in higher education is currently an active area of research. Curriculum and assessment also feature highly in EER research. Our findings also suggest that design based learning and collaborative learning are also active areas of investigation. Whilst the category "problem based learning" does not feature highly, the preoccupation of researchers with design and collaborative learning would however seem to indicate that activity-based, collaborative learning, of which problem–based learning is one, features highly in current UK EER.

Table 1: Categorisation of Journal articles

EJEE		IJEE		PhD Theses		JEE	
Category	**Count**	**Category**	**Count**	**Category**	**Count**	**Category**	**Count**
edtech	21	edtech	15	global	25	assessment	2
students	17	PBL	10	profession	15	mathsci	1
assessment	16	design	8	industry	12	curriculum	1
design	16	collab	7	socpolorg	11	learning	1
learning	12	curriculum	7	edtech	9	faculty	1
envsust	11	mathsci	4	learning	9	competencies	1
collab	10	assessment	4	diversity	8	students	1
curriculum	9	industry	4	envsust	7	global	1
progeval	9	labs	3	support	7	teaching	1
profession	7	profession	3	students	6	labs	1
diversity	7	teaching	3	graduate	5		
teaching	6	diversity	3	assessment	5		
industry	6	progeval	2	comm	4		
competencies	6	graduate	2	design	4		
PBL	5	genskills	2	engskills	4		
graduate	5	conceptual	2	faculty	4		
faculty	4	active learning	2	design	4		
conceptual	4	reflection	1	progeval	4		
graphics	3	learning	1	curriculum	4		
recruit retain	3	engskills	1	mathsci	3		
first year	3	students	1	competencies	3		
global	3			interdisciplinary	3		
mathsci	3			compskills	2		
active learning	2			collab	2		
reflection	2			conceptual	2		
labs	2			graphics	1		
compskills	2			PBL	1		
interdisciplinary	2			first year	1		
service learning	2			business	1		
comm	1			learning styles	1		
k12	1						
business	1						
learning styles	1						

There are however some significant differences between the rankings for categories for the two journals. For instance, the *students* category, which was second highest in the EJEE comes in at position 21 in the IJEE. Similarly, the *learning* category, which is 5th in the EJEE is 19th in the IJEE. Likewise, the *PBL* category, which is second in the IJEE ranks 15th in the EJEE; and the *mathsci* category, which is 6th in the IJEE is 24th in the EJEE. This may be due to differences in the focus of the two journals.

Six of the top ten categories in the EJEE are also in the top ten of the PhD Theses, with only 3 top ten IJEE categories making it into the PhD Theses top ten list. This suggests a closer alignment between PhD work and the EJEE compared to the IJEE. Given that EER is aimed at both practitioners and researchers, it may be that the EJEE and IJEE have a different research orientation on the researcher-practitioner continuum. We intend to explore this aspect in more detail as the research progresses.

The top category in the PhD theses is the *global* category. This category is 23rd on the EJEE ranking, and does not feature at all in the IJEE. Its frequency in PhD research may be a result of international PhD candidates coming to UK universities to do research on engineering education in their own national contexts.

Another category, *socpolorg*, is 6th on the PhD Theses list, and does not feature at all on both the EJEE and IJEE categories. Jesiek et al. (2011) define the *global* category as:

> Global Competence and Education, Intercultural Skills, Foreign Language, Mobility

and the *socpolorg* as:

> Social, Organizational, Philosophical, and Political Studies, inc. Ethics, Policy, STS.

It may be that PhD theses are going beyond researching classroom practice to considering global and societal issues related to engineering education. This may suggest that PhD theses are more oriented towards the researcher end of the researcher-practitioner spectrum.

When the individual category instances are summed up to come up with an overall ranking, *edtech* comes highest, with the most popular categories in the IJEE and EJEE generally making it into the top ten. However, the *global* category, which only features in the PhD Theses, comes second highest overall. This suggests a significant level of popularity for the category as an area of PhD research.

Collaboration

Our preliminary findings suggest that EER articles from the UK are predominantly single-authored, and in instances of multiple-authorship, the evidence suggests that typically all the authors belong to the same UK-based institution. For instance, over the period 2000 – 2017, of the 108 EJEE, 79 articles had either been single-authored, or authored by people from the same institution. 20 were the result of co-authorship with another institution. However, 7 of these co-authored articles comprised only UK-based institutions. Moreover when an article had 3 or more co-authors, it was more likely to be the result of collaboration with non-UK-based institutions. Similar trends are evident for the IJEE articles. Of the 58 IJEE articles over the same period, 42 articles were either single-authored or authored by people from the same institution. For those articles with two authors, five were by authors from the UK alone. Again, for the remaining 5 articles that had 3 or more co-authors, only one was by authors from the UK alone.

These data suggest that UK-based EER researchers are more likely to work and publish their work alone, suggesting rather low levels of collaboration with international researchers. Such low levels of collaboration may also point to low levels of engagement with the EER literature by UK engineering academics and researchers, and may suggest that both teaching practitioners and researchers may waste time and resources re-inventing work that has been discovered and evaluated elsewhere. Wankat (2004) suggests that increases in co-authorship may be achieved by the formation of larger research teams, which can be achieved by access to EER funding.

With regard to the JEE, only two of the five identified articles were single authored. Of the three co-authored papers, only one was by UK-based authors alone. The other two were as a result of collaborations with at least a USA based author. This may point to a perception amongst UK-based EER researchers that the JEE is more focussed towards the USA, or it may indicate that the quality of UK-based EER falls short of the requirement of the JEE. Further investigations, using surveys and/or interviews would be needed to determine the correct position.

Category trend analysis

There is no definite trend signifying definite changes in the popularity of categories over the period 2000 – 2017 across all the three journals and the PhD Theses. This suggests that research within EER in the UK is not focussed to any particular institutional or national objective. This is a plausible suggestion given that EER research in the UK is generally not funded, and there is currently no national funding bodies directing EER research in any way (Shawcross and Ridgman 2013). Given such a scenario, individual researchers would probably only focus on EER areas in which they have an intrinsic interest. However, given the relatively low output of EER compared to other countries like Australia and the USA Jesiek et al. 2011), it may simply be that the small number of articles masks the presence of any trend. One possible way of establishing the existence or absence of research trends in EER would be to augment this analysis with survey instruments targeted at EER researchers in the UK. Findings from this preliminary study will be used to develop such a survey instrument. An advantage of such an approach would be the possibility of galvanising interest in EER within the UK.

Concluding Remarks

We have carried out a preliminary investigation into emerging trends in the evolution of EER in the UK. We have done this by categorising articles from the JEE, IJEE, and EJEE, as well as PhD theses in EER from UK–based institutions. Compared to countries like the USA or Australia, the level of EER in the UK is still relatively low. Whilst EER in the UK spans the entire range of available categories, there is no discernible trend to signify the direction of evolution. In addition, UK EER is predominantly single-authored. We suggest that this may be a result of lack of access to research-funding by EER researchers in the UK. Our preliminary conclusion is that UK EER is under-supported and this has led to a stunted development of the field, unlike in other countries where there is access to research funding.

However, our findings are based on a relatively small number of research articles, again a direct consequence of the low level of EER taking place in the UK. Hence more investigation, using surveys and interviews with UK EER researchers, are needed to enable us to reach fully-evidenced conclusions. Use of these instruments will now form the next stage of our investigation.

Copyright statement

References

Bosworth, G. (1963). "Towards creative activity in engineering." *Higher Education Quarterly*, 17(3), 286-297.

Bosworth, G. S. (1966). *The Education and Training requirements for the Electrical and Mechanical Manufacturing Industries Committee on Manpower Resources for Science and Technology*. HMSO, London.

Edström, K., Kolmos, A., Malmi, L., Bernhard, J., and Andersson, P. (2016). "A bottom-up strategy for establishment of EER in three Nordic countries – the role of networks." *European Journal of Engineering Education*, 1-16.

Education, M. o. (1959). *Report of the Central Advisory Council for Education (Crowther Report)*. HMSO, London.

EngineeringUK. (2015). *EngineeringUK, Engineering UK 2015: The state of engineering* London.

Henderson, C., Beach, A., and Finkelstein, N. (2011). "Facilitating change in undergraduate STEM instructional practices: An analytic review of the literature." *Journal of research in science teaching*, 48(8), 952-984.

Heywood, J. (1969). *An Evaluation of Certain Post-War Developments in Higher Technological Education*, Ph. D. thesis, University of Lancaster.

Heywood, J. (1970). "Qualities and their assessment in the education of technologists." *International Bulletin of Mechanical Engineering Education*, 9, 15-29.

Heywood, J., and Monk, J. (1977). "The education and career patterns of professional Mechanical Engineers in design and management." *The Vocational Aspect of Education*, 29(72), 5-16.

Hutton, S., and Gerstl, J. "Engineering education and careers." *Presented at Proceedings of a Symposium on Education and Careers. Proceedings of Institution of Mechanical Engineers*.

Jesiek, B. K., Borrego, M., Beddoes, K., Hurtado, M., Rajendran, P., and Sangam, D. (2011). "Mapping global trends in engineering education research, 2005–2008." *international Journal of engineering education*, 27(1), 77-90.

Lohmann, J., and Froyd, F. "Chronological and ontological development of engineering education as a field of scientific inquiry." *Presented at Second Meeting of the Committee on the Status, Contributions, and Future Directions of Discipline-Based Education Research, Washington, DC. Available: http://www7. nationalacademies. org/bose/DBER_Lohmann_Froyd_October_Paper. pdf.*

Ministry of Education. (1945). *Higher Technological Education (The Percy Report)*. HMSO, London.

Ministry of Education. (1956). *Technical Education (1956) Whitepaper. Cmnd 9703*. HMSO, London.

Shawcross, J. K., and Ridgman, T. W. (2013). *Publishing Engineering Education Research*. HEA Academy Working Paper. Higher Education Academy.

van Hattum-Janssen, N., Williams, B., and Nunes de Oliveira, J. M. (2015). "Engineering Education Research in Portugal, an Emerging Field." *International Journal of Engineering Education*, 31(2), 674-684.

Wankat, P. C. (2004). "Analysis of the first ten years of the Journal of Engineering Education." *Journal of Engineering Education*, 93(1), 13.

Whitin, K., and Sheppard, S. (2004). "Taking stock: An analysis of the publishing record as represented by the Journal of Engineering Education." *Journal of Engineering Education*, 93(1), 5.

Youngman, M., Oxtoby, R., Monk, J. D., Heywood, J. . . (1978). "Analysing Jobs". City: Gower Press: Farnborough, Hampshire, UK.

Integrating creativity in the design of chemical products

Oscar Alvarez.
Chemical Engineering Department, Universidad de Los Andes, Bogotá / Colombia
oalvarez@uniandes.edu.co

Abstract:

In this work, we would like to present a novel alternative in chemical engineering design that focuses on chemical process and product design and that we call the "Integrated chemical process and product design". When designing a chemical process or product all the considerations related to formulation, composition, properties (molecular, microscopic and macroscopic) and the energy required for such process are taken into account as a whole instead of individual elements (multi-scale). Using this multi-scale approach does not guarantee the success of a process or product at the industrial level, but it does guarantee that is adequately designed. To achieve success at the industrial level it is imperative to set the properties of the desired product from an innovation standpoint. With this, an opportunity in the given sector can be correctly identified. Once this step finishes, a working prototype (developed using the multi-scale approach) is analyzed from an entrepreneurship perspective to identify how the product would eventually perform in the open market. This coupled analysis of multi-scale properties, innovation and entrepreneurship is what we call the "Integrated chemical process and product design" or DIPP for its acronym in Spanish.

Introduction:

One could say that within the realm of Chemical Engineering (IQUI) there are three paradigms that could be used as a starting point for explaining the evolution that this area has had throughout many decades. The first paradigm is closely related to the principle or concept that eventually gave birth to IQUI. And That is, the concept of Unit Operations: individual elements which combined give rise to a chemical process. The second paradigm is related to the integration and scale-up of these unit operations to the transport phenomena principles (i.e. momentum, mass and heat). And the third paradigm is of course, the direct result of a chemical process: a chemical product.

The focus of chemical engineering on molecular transformations, processes, products, quantitative analysis, and multi-scale treatment of problems provides an ideal platform for productive interactions with a wide range of other science and engineering disciplines at boundaries that are among the most exciting technology areas of research today. A multi-scale approach should be understood in this context as the process of relating the chemical composition of a material to its properties (molecular, microscopic and macroscopic), which are often determined on larger length and time scales than those pertaining to the atoms and molecules of which it is comprised (Vargas, Alvarez and Gómez 2015; Alvarez and Vargas 2014). This broad applicability of chemical engineering fundamentals has become a defining feature of the discipline, and we are challenged to convey this vast domain to our students. Teaching students more about these particular areas of technology is exciting educationally, but it does tend to have a fragmenting effect on the discipline and represents a challenge for integration into the current curriculum (Armstrong 2006).

During the last decades, Chemical Engineering has been taught following two trends:

1. Classical IQUI: schools that are oriented towards the formation in process design. These schools treat the product as one of the many elements that are to be considered within the design.

2. Schools that have as their core element, the final properties of the product. These schools treat the process as one of the elements to be considered for that end.

We try to develop a new idea that put together analysis of multi-scale properties, innovation and entrepreneurship is what we call the "Integrated chemical process and product design" or DIPP for its acronym in Spanish.

The main challenge using the DIPP concept is to be able to create the space-time required where Chemical Engineers in formation can exploit it. At the Chemical Engineering Department of Universidad de los Andes (Bogotá, Colombia) we have filled such space-time requirement through the creation of a course called "Mid-career project" or PMC for its acronym in Spanish. This concept is based on the methodology developed for the design of chemical products by Cussler and Moggridge (Cussler and Moggridge 2011).

PMC project class are for the most part year 2 chemical engineering students that have only basic knowledge and skills with respect to the practical application of math, sciences and fundamentals of chemical engineering. The introduction to product design and innovation has to be focused on well-defined problems, whereby the students start with a generic requirement for a product or process and trough ideation approaches, such as those, provided by the inside the box method they narrow down the ideas to be implemented. (Vargas, Alvarez and Gómez 2015)

In this course, the main event using the DIPP concept takes place. To include innovation in the designing of a process, creativity is taken as the starting point for that innovation to occur. This means that students taking the course are required to go through something we call "The creativity challenges" (Rey and Feyfant 2012). There are four creativity challenges to be completed during four months. The main feature of these challenges is that students can learn how to use idea-generating tools such as brainstorming and develop skills using Systematic Innovation Thinking (SIT) (Boyd and Goldenberg 2013). Parallel to developing these skills and abilities, students are required to develop a novel chemical product through idea-generating strategies that is needed at a given sector of the industry. This novel product must be presented with a working prototype that must be developed using the multi-scale approach. Once the working prototype is in place, the last part of the PMC course is focused on developing a preliminary business model of the CANVAS-type (Osterwalder, Pigneur, Bernarda, and Smith 2014) through which the added value of the product is established.

Our PMC course not only allows our students to develop the technical aspects related to the design of a product, innovation and entrepreneurship but also it allows them to develop teamwork activities and communication skills using TED talks approach (Donovan 2013). Being these last two elements essential when acting at the industry level.

In this work we present how at the Chemical Engineering Department at the Universidad de los Andes does an insertion of multi-scale approaches to product and process design as well as the incorporation of innovation initiatives both developing of skills for creative thinking and entrepreneurship in the Chemical Engineers formation.

Integrated chemical process and product design:

Creativity

The central and most essential element to develop the learning space-time for the integrated product and process design is creativity. It is the sine qua non condition for innovation, and this it is the reason why the course focuses on three aspects to enhance this skill (Figure 1):

1. Story telling. The professor must attempt an almost elimination of the classical strategy in teaching which consists of making a very good summary of a text or reference to convey a fundamental concept or theory. To achieve this, stories and situations that can be related to the everyday life of a person or to the future life as engineers in the industry are conceived. The application of a fundamental concept is included in such stories to either understand the concept itself or to give a solution to the specific problem. In order for this to be successful, the student must previously explore by himself the fundamentals aspects of the subject, actively interact in the discussion of the story and finally make a decision based on what has been presented.

2. I am not an evaluator; I am a guide. It is clear that one of the main concerns that are part a student's life is the grade. To avoid this imposed strain, I have implemented in all my courses a grading system that consists of both positive and negative points. Negative points are deducted from the positive record when the fundamentals of a concept are not correctly applied to the task at hand. Positive points are related to creativity. In these cases, the student must show novel solutions to the problems stated, for example, in the midterm exams or the development of the challenges proposed for the laboratory sessions.

3. Space-time scenarios. Based on the premise that the students must be actively involved in the courses (i.e. participation is essential), they are exposed, at least every 15 days to communication and dissemination activities and to team work and brainstorming sessions through what I call the "creative challenges". The students must work on these scenarios during class hours as well as outside the classroom.

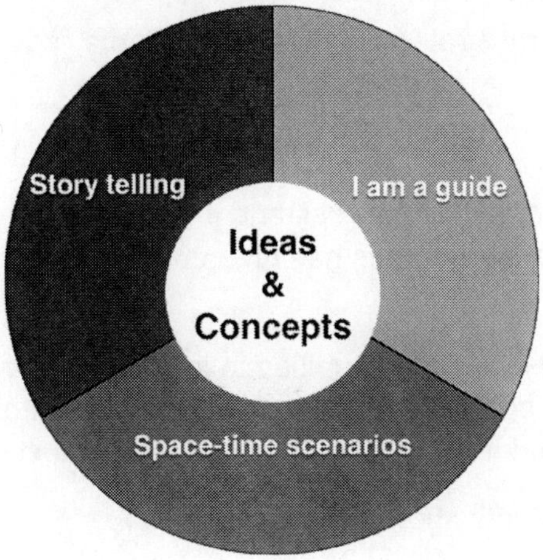

Figure 1. Integrating creativity and product and process design learning.

The concept

The general strategy in order to design a product or process is based on (Figure 2):

1. Identifying opportunities or needs.
2. Producing ideas (processes/products) to take advantage of these opportunities or needs.
3. Selecting the best idea.
4. Specifying the relationships between properties (macroscopic, microscopic and molecular), product variables (formulation and composition) and process conditions to manufacture the product.
5. Integrating a business model (CANVAS).
6. Prototyping.

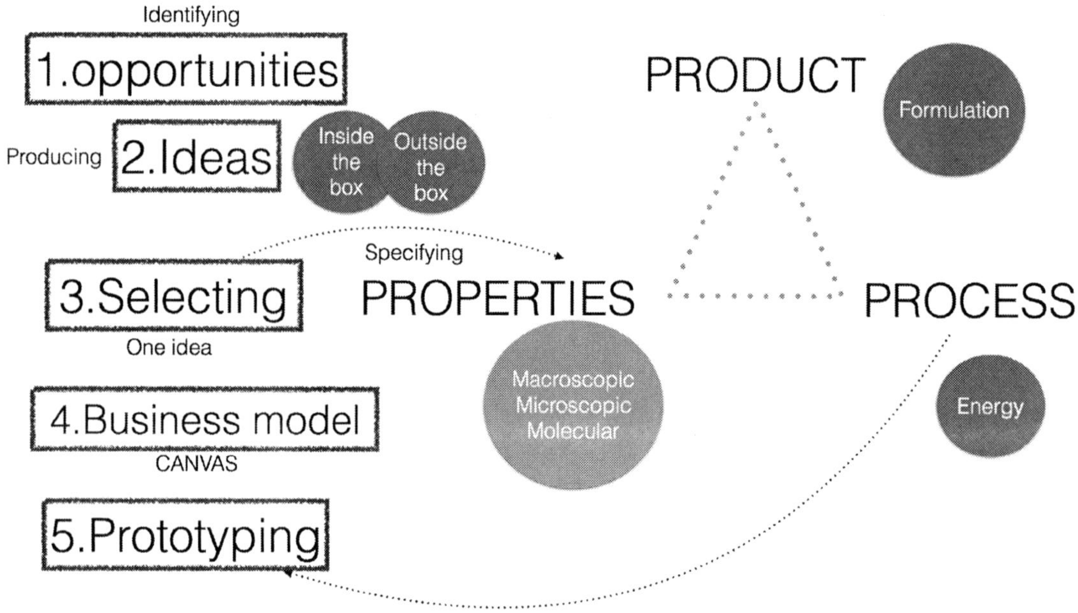

Figure 2. General strategy in order to design a product or process

The evaluation

As a result of our ABET accreditation process and in line with our commitment to continuous improvement, input from different sources has given us the drivers for further revision of our current curriculum.

This course provides students with opportunities to integrate knowledge, skills and basic attitudes in order to correctly complete a task, an action or an intellectual process according to their professional area. Students must act in a defined context of teamwork.

By the end of the course students should be able to:

- Integrate and apply knowledge and competences previously obtained in order to develop a project.
- Develop teamwork skills.
- Work on projects with autonomy, responsibility and taking initiatives
- Use criteria from basic sciences and Engineering to solve problems.

- Use a general strategy in order to design a product or process.
- Organize strategies in order to solve problems based on:

  Effective compilation of information.
  Planning of alternative solutions to solve the problem.
  Planning an appropriate methodology for the possible solution.
  Appropriate implementation of a work schedule.

- Show communication skills (oral and written).
- Recognize topics and contemporary issues.

We use different rubrics to evaluate how the students show a good performance in the following points:

- An ability to design a system component or process to meet desired needs. The idea is to develop topics associated to ideation and multi-scale approach to design chemical products. The evaluation of this point is performed in the final presentation of the prototype II using the rubric A shown in the table 1.
- A broad education to understand the impact of engineering solutions in a global and societal context. The idea is to develop topics related to integrating entrepreneurship and innovation in designing chemical products. The evaluation of this aspect is performed in the final project presenting the business model (CANVAS) using the rubric B shown in the table 2.
- An ability to communicate effectively. The evaluation of this point is performed in the final oral presentation of prototype II using the rubric C shown in the table 3.

Table 1. Rubric A to evaluate an ability to design a system component or process to meet desired needs.

A	1 (poor)	2 (satisfactory)	3 (good)	4 (excellent)
Idea & prototype	The sector where opportunities for innovative development can be identified is not clearly identified	The sector is clearly identified, but the opportunity is not clear for the development of an innovative product or process	The sector and the opportunity are clearly identified, but the prototype proposed from an idea does not take full advantage of the opportunity from an innovative approach	The proposed idea and prototype are innovative and take advantage of a clear opportunity in a given sector

Table 2. Rubric B to evaluate a broad education to understand the impact of engineering solutions in a global and societal context.

B	1 (poor)	2 (satisfactory)	3 (good)	4 (excellent)
CANVAS	It does not indicate the relationship / importance / impact of a given topic in the social, economic or environmental context	Identifies relationships but does not describe any of them	Describes some relationship or impact in the global context but leaves out some important ones	Describes the main relationships or impacts of the research topic in a social, economic or environmental context

Table 3. Rubric C to evaluate an ability to communicate effectively.

C	1 (poor)	2 (satisfactory)	3 (good)	4 (excellent)
Introduction and Organization	Not: Agenda Introduction Context Conclusions	It complies with the agenda and the introduction /context and conclusions, but they are neither clear nor complete nor well-articulated	It complies with the agenda and almost all the fundamental aspects of the work are included in the development	Clear, complete and articulate
Content	Not: Key points	It presents an excess or lack of details given the time and the level of the hearing, several of the important results of the work are omitted	It covers almost all the key elements / results of the work and does so with an adequate level of detail	Clear, complete and articulate
Poster	Lot of text Not figures and tables	Occasional use of appropriate charts and tables	Poster well balanced, supported by graphs and tables, but the narrative is needed to understand it	Poster clear, with short observations that allow the reader to understand much of it even without narration
Voice Quality and Stage Management	Voice very low, very high or very monotonous. Exaggerated gesture, does not look at the audience	It shows in some moments the capacity demanded but in general it fails	In general, it is clear and pleasant to listen but has one or two minor aspects to correct	It is clear and uses his voice and gestures, visual, gesticulation, as a tool for attracting interest

Preparation and Timing	Very long or very short presentation	Read a lot the poster or rely on notes	Improvises sometimes	Proper preparation of the presentation, meets the allotted time.
Oral Communication	Wrong language, poorly structured sentences and colloquial vocabulary	Clear and structured language but colloquial vocabulary	Generally formal but mix styles occasionally	Clear and structured language

Taking as a base line the 84 students (teams of 3 to 4 students) that were part of the PMC course during the second semester of 2016 and using the evaluation methods described in tables 1-3, the results presented in figure 3 were obtained.

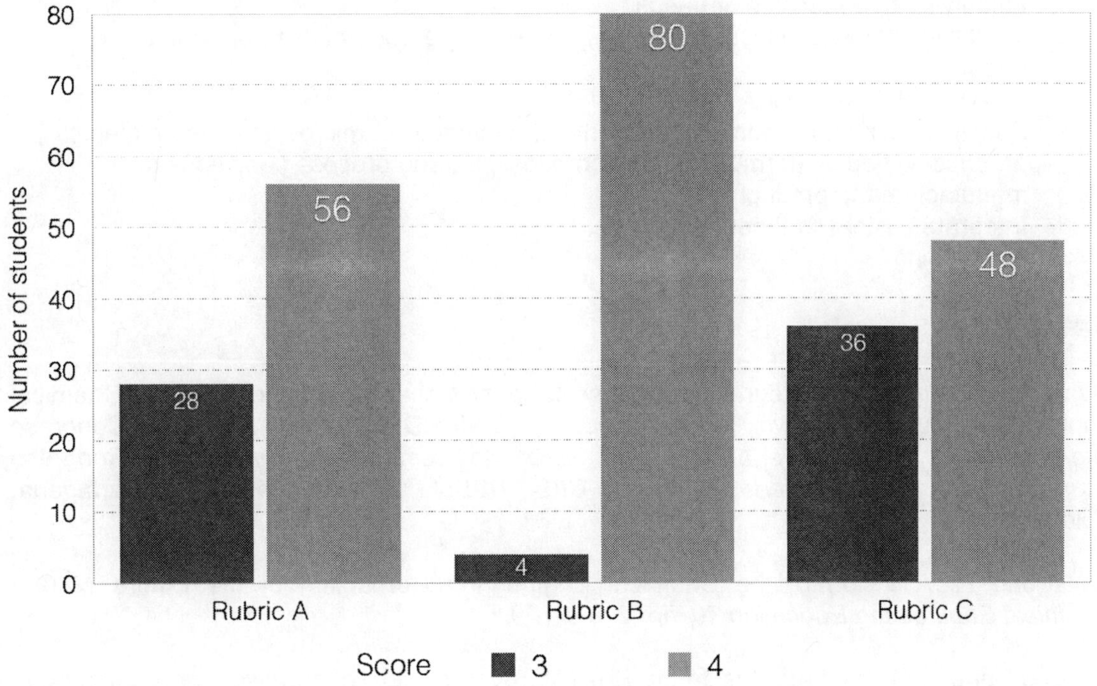

Figure 3. Number of students with score 3 or 4 applying the rubrics A, B and C. For the rubric C is reported de total average.

The results are very clear in the sense that 100% of the students obtained final grades between 3 and 4 in the aspects that were evaluated. It can be said that the integrated product and process design (DIPP) is an adequate tool that allows the professor to introduce and include elements of innovation and entrepreneurship in designing a chemical product (manufacture of a working prototype) where skills related to creativity and oral communication are enhanced.

The implementation of these scenarios started during the first semester of 2014. The first two years were part of the exploring stages in which adjustments were constantly made to have an adequate finished scenario. Since 2016, information is being gathered to establish how this new concept allows the integration of creative aspects in designing chemical products and processes. The next step is to include similar scenarios in the Chemical Engineering curriculum that allow the application of this concept in courses such as "Special project" and "Graduation project".

Conclusions:

The Department of Chemical Engineering at Universidad de los Andes has begun the process of incorporating the integrated process and product design concept in the Mid-career project. The first stage of implementation was introduced in 2014 and will continue through different stages in the upcoming years with the new scenarios in special project and graduation project for undergraduate students. This concept has as its starting point the innovation, which is understood as the prime condition for creativity.

The general strategy, in order to do Integrated chemical process and product design is based on:

1. Identifying opportunities or needs.
2. Identifying ideas (processes/products) to take advantage of this opportunities or needs.
3. Selecting the best idea.
4. Specify relationships between properties (macroscopic, microscopic and molecular), products variables (formulation and composition) and process conditions to manufacture the product.
5. Integrate business model (CANVAS).
6. Prototyping.

References:

Alvarez O, Vargas W. *I*ntegrating multi-scale approaches and innovation in Chemical Engineering curricula: Two experiences at UNIANDES (2014). *XXVII Congreso Interamericano de Ingeniería Química. libro de resúmenes del XXVII Congreso Interamericano de Ingeniería Química CIIQ (ISBN 978-958-58438-0-6)* Cartagena, Colombia.

Armstrong R. A Vision of the Chemical Engineering Curriculum of the Future (2006). *Chemical Engineering Education 40, no. 2 104-109.*

Boyd, D., and J. Goldenberg. Inside the Box: A proven system of creativity for breakthrough results (2013). New York: Simon & Schuster.

Cussler E, Moggridge G. Chemical Product Design (2011), Cambridge, Edition 2.

Donovan J. Método TED para Hablar en Público (2013). Editorial Planeta S.A.

Osterwalder A, Pigneur Y, Bernarda G, Smith A. Value Proposition Design (2014). WILEY.

Rey O, Feyfant A. Vers une Éducation plus Innovante et Créative (2012). Dossier d'Actualité Veille et Analyses (70). Institut Français de l'Education.

Vargas W, Alvarez O, Gómez J. Integrating Multi-Scale Approaches and Innovation into Product and Process Design in Chemical Engineering Curricula (2015). *122nd ASEE Annual Conference & Exposition June 14-17 2015 Seatle, USA.*

Analysis of macroethics teaching practices and perceptions in engineering: results of an international survey

Madeline Polmear, Angela Bielefeldt, Daniel Knight
University of Colorado, Boulder, USA
Madeline.polmear@colorado.edu, Angela.bielefeldt@colrado.edu, Daniel.knight@colorado.edu

Christopher Swan
Tufts University, Medford, USA
Chris.swan@tufts.edu

Nathan Canney
Seattle University, Seattle, USA
canneyn@seattleu.edu

***Abstract**: Students today must be equipped with the skills to understand the nuances of engineering and technology in global and societal contexts. Part of educating students about the broad implications of their work is fostering a sense of ethical awareness and responsibility. This paper explores the results of a survey of faculty who teach engineering and computing students about ethics and broader impacts. The survey probed the ethics-related topics included in instructors' courses and the teaching methods employed in the classroom. The survey, along with follow-up interviews with select respondents, also provides insights on perceptions of the importance of ethics education and its role in the engineering curriculum. The results presented in this paper illustrate how ethics are being taught in the United States and internationally and shed light on global perspectives of the value of ethics education and how it can be improved.*

Context

Ethical awareness and reasoning are imperative skills for students entering the workforce and tackling complicated engineering challenges. Therefore, ethics education in the engineering context must span both microethical and macroethical issues (Herkert, 2003). Microethics instruction is focused on individual responsibility and day-to-day ethical decisions, such as codes of ethics and safety, while macroethics instruction includes the broader ethical considerations of the profession and the societal impacts of engineering, including sustainability and social justice (Herkert, 2001). In engineering education, there is often a disconnect between the two domains of ethics as instruction focuses on microethical topics at the expense of broader issues relating to engineering and technology. Despite the importance of ethics education, there are a number of barriers to the meaningful inclusion of these topics in engineering and computing programs (Haws, 2001).

Accreditation standards, licensing processes, historical practices, and normative contexts in which engineering operates vary between countries thus shaping differences in ethics education across borders. In Germany and the Netherlands, "engineer" is an academic title with two types of degrees based on the institution, in Britain, there is emphasis on learning-on-the –job, and in the United States (US), professional societies and licensing play a large role in engineering education (Brumsen, 2005). Australia also shares a focus on professional societies and codes of ethics, which is reflected in accreditation criteria of professional ethics (Zandvoort, 2000). No accreditation organizations in Latin America and the Caribbean are part of the International Engineering Accreditation Alliance but different countries seek accreditation from agencies in US, Canada, and UK (Larrondo-Petrie, 2015). These differences are manifested in diverse educational practices. Within the European Union, there are agreements to make degrees equivalent and a shared interdisciplinary approach to ethics as an integration of technology, philosophy, and social science (Zandvoort, 2000). However, the emphasis on licensing and professional societies in the US is reflected in ethics education being intertwined with professional codes of ethics. There are many reports in the literature of the integration of ethics into engineering courses and programs within the US and abroad (Haws, 2001;Herkert, 2000). However, the literature is largely descriptive without synthesis of ethics-related topics and pedagogical approaches. Furthermore, the ways in which macroethics are taught in engineering programs and the differences in teaching practices and perceptions of ethics instruction between international and US institutions have not been well studied. The study presented in this paper was guided by the motivation to explore instructors' feedback on ethics education and delineate similarities and differences between faculty within the US and abroad.

Research Questions

The following research questions were examined in this study:

RQ1: What ethics-related topics are incorporated into courses taught to engineering and computing students? Do the topics vary between respondents from international and US institutions?

RQ2: What methods are used to teach ethics and broader impacts to engineering and computing students? Do these methods differ between respondents from international and US institutions?

RQ3: What are educators' perceptions of the importance and sufficiency of the ethics and broader impacts education of engineering and computing students? Do these perceptions differ between respondents from international and US institutions?

The research questions probe at the topics and methods related to ethics education for engineering and computing students. The questions also distinguish between respondents at institutions in the US and internationally to discern similarities and differences in terms of macroethics instruction and perceptions between the two cohorts.

Theoretical Framework

Developing an awareness of broader impacts and sense of ethical responsibility in engineering students necessitates educators to foster effective learning experiences. Vanasupa et al. (2009) proposed the Four-Domain Development Diagram (4DDD) as a guide for holistic engineering education. The model shows that learning is shaped by cognitive, social, psychomotor, and affective domains. The causal loop illustrates the interconnectedness of "engagement active learning", "mastery", "systems thinking", "understanding the broader context", and "moral and ethical development" (p. 74). Motivation, at the core of the model, is driven by "interest, value, and autonomy" (p. 70) and guides students in the learning process. The model also proposes that an understanding of the broader context increases moral development, which in turn reinforces an interest in the broader issues relating to engineering. The 4DDD illustrates the important connection between the broader context and ethical development while also conveying the constructs that should guide the design of learning environments to best motivate engineering students. The model also proposes that multiple instructional settings and methods most effectively facilitate ethical development by tapping into the different domains and the reinforcing loops that link them.

Methodology

All research was conducted in accordance with a protocol approved by the Institutional Review Broad for Human Subjects Research. The data for this study were collected from two online surveys developed to explore how faculty teach ethics; one focused on curricular settings and the other on co-curricular activities (Bielefeldt et al., 2016; Knight et al., 2016). The curricular survey was distributed to lists including the American Society for Engineering Education (ASEE) ethics and community engagement divisions, National Science

Foundation grantees, and authors who have published on engineering ethics (Bielefeldt et al., 2016). The curricular survey reached international educators through their membership in ASEE. Individuals in the international research community who published papers relating to ethics (n=93 with 85 emails successfully delivered) were also invited to complete the survey, generating a 28% response rate. The co-curricular survey was distributed to current and former mentors of professional societies, engineering service groups, engineering design competitions, and undergraduate research experiences; for additional information see Knight et al., 2016. The curricular survey included multiple-choice questions covering what, where, and how ethics issues are taught and assessed in undergraduate and graduate classes. The co-curricular survey included the same questions but started with questions on teaching ethics in informal learning activities and then moved to courses. Responses were received from the US and international groups to both surveys.

The survey was open between February and May 2016 with 1448 survey responses collected. The respondents represented 375 US institutions and all 50 states. The surveys also generated 53 international responses with faculty representing 41 institutions in 24 countries. Table 1 details the countries that are included in the international group and the number of survey respondents from each country.

Table 1: Counts of international respondents

Australia 3	Germany 1	Kuwait 1	Spain 4
Belgium 1	Greece 1	Netherlands 5	Sweden 2
Canada 17	Hong Kong 1	New Zealand 1	Taiwan 1
Chile 1	India 1	Nigeria 1	Turkey 1
Colombia 1	Ireland 2	Pakistan 2	UK 2
Finland 1	Italy 1	South Africa 1	Yemen 1

Table 2 provides a summary of demographic information for survey respondents at US and international institutions. There were no statistically significant differences in the disciplines taught by educators within and outside the US.

Table 2: Survey respondent demographics

Characteristics	*US respondents (n=1359) %*	*International respondents (n=53) %*
Disciplines taught (could select more than one)		
Civil engineering	20.9	17.0
Mechanical engineering	21.2	20.8
Computer	16.6	22.6
Electrical engineering	11.9	20.8
Chemical engineering	10.0	13.2
Environmental engineering	11.3	5.7
Biomedical	9.4	7.5
Institution type		
Public	71.3	84.9
Private	28.5	15.1
Gender		
Male	64.8	67.9
Female	32.6	30.2
Prefer not to say	2.7	1.9

Statistical tests, including chi-square to understand differences between topic selection, pedagogy, and perceptions of macroethics instruction between groups, were conducted in IBM SPSS Statistics 24. Given the small samples for some of the international cases, Fisher's tests were conducted in GraphPad for any cases with p values of 0.05 or under in the chi-square tests. Statistically significant differences were noted when the two-tailed p values from the Fisher's test were 0.05 or lower.

Near the conclusion of the surveys, respondents were asked "Please share your thoughts about the education of engineering students regarding broader impacts and ethical issues." The open-ended question generated 406 responses total, including 25 from international institutions. The responses were analysed using emergent, thematic coding (Creswell, 2007). A codebook with 60 codes was developed and selected subsets of responses were shared with other authors for inter-rater reliability analysis. For more detail of the free-response data analysis, see Canney et al. (2017). The codes and themes were compared across the US and international cohorts.

At the end of the survey, respondents were asked to share their email address if they were interested in being contacted for a follow-up interview. The interview phase was designed to gain in-depth feedback on how macroethics are taught to engineering and computing students in curricular and co-curricular settings. The interviews also provide insight into exemplary teaching practices while probing at the educators' perceptions of their efficacy. Interview selection was based on willingness to participate and representation of a range of pedagogical approaches in various disciplinary and institutional settings. In total, 229 survey respondents volunteered to be interviewed, including 16 individuals from international institutions. Between October 2016 and April 2017, 54 individuals were invited to participate in the interviews including seven individuals from institutions outside the US. As of May 2017, 39 interviews have been completed. The semi-structured interviews are conducted over Skype or phone and are 30-60 minutes in duration. After the interview, 2-page summaries were written and emailed to the interviewee for a member check to increase the validity and accuracy (Koelsch, 2013). The interviews were recorded for transcription and pseudonyms were assigned to the interviewees using a random name generator to protect the confidentiality of the participants (Given, 2008). Thematic analysis of the interviews is on-going but comments from the one international interview are provided to add richness to the results and insights into differences in teaching practices and perceptions.

Findings/Conclusions

RQ1: Ethics-related topics

Ethics spans macro and micro domains and includes a range of topics related to engineering. The survey asked educators to indicate which topics they include in their courses to understand the context in which ethics are being taught to engineering and computing students. Table 3 displays the percentages of educators in both groups who reported teaching the ethics-related topics. The most commonly taught topics for the respondents at institutions in the US (in order of preference) were professional practice issues, societal impacts of technology, engineering decisions under uncertainty, and safety. The most commonly selected topics for the respondents at international institutions were sustainability/sustainable development, societal impacts of technology, professional practice issues, and environmental protection issues. On average, respondents within the US taught 9.8 topics and educators abroad included 7.8 topics.

Chi-square and Fisher's tests indicate statistically significant differences in topic selection between the two groups for sustainability/sustainable development, environmental protection issues, ethical theories, engineering and poverty, and other. In all of these cases, a higher percentage of international respondents taught the topics. The respondents were also asked in which types of classes they teach these topics. The most commonly selected course types for US respondents were capstone design (33.8%), sophomore/junior core engineering

(33.7%) and first year introductory (25.4%). The most commonly selected topics for the international respondents were first year introductory (22.6%), sophomore and junior core engineering (22.6%), and graduate courses (22.6%). There were statistically significant differences between the two groups in terms of teaching ethics for capstone design (18.9% for international) and full engineering ethics courses (5.4% for US and 17% for international).

Table 3: Percentage of educators reporting teaching ethics-related topics in their courses

Ethics-related topics	*Educators at US Institutions (%) (n=1359)*	*Educators at International Institutions (%) (n=53)*
Sustainability/sustainable development	39.4	56.6*
Societal impacts of technology	48.1	54.7
Professional practice issues	53.3	52.8
Environmental protection issues	30.5	45.3*
Engineering code of ethics	40.6	43.4
Engineering decisions under uncertainty	43.6	43.4
Ethical failures/disasters	38.6	39.6
Safety	42.8	37.7
Ethical theories	19.6	35.8*
Responsible conduct of research	28.9	35.8
Ethics in design	34.6	34.0
Risk and liability	31.9	28.3
Engineering and poverty	13.4	28.3*
Social justice	15.5	24.5
Other	7.7	20.8*
Privacy and civil liberties	11.8	15.1
War, peace, military applications of engineering	8.1	13.2
Nanotechnology	4.0	5.7
No topics	11.3	15.1
		* $p<0.05$

RQ2: Methods used to teach ethics and broader impacts

The survey solicited information on how respondents teach ethics and broader impacts to engineering and computing students. The percentages of educators at US and international institutions who reported using the various teaching methods are reported in Table 4. On the survey, the respondents were invited to select methods for up to two classes that they teach, the results shown in Table 4 are for the one course that respondents indicated they believed was most effective. For both groups, lectures, case studies, and in-class discussions were the most commonly selected choices. The methods with the lowest percentage of US respondents were moral exemplars and humanist readings. The least commonly employed methods for the international respondents were problem solving heuristics, think-pair-share, service-learning, and guest lectures. There were statistically significant differences between the cohorts in terms of humanist readings and moral exemplars. International respondents used these pedagogical approaches more commonly than the American educators.

Table 4: Percentage of educators reporting use of particular teaching methods

Types of methods used to teach ethics/societal impacts	*Educators at US Institutions (%)*	*Educators at International Institutions (%)*
Lectures	51.7	58.5
Case studies	52.8	54.7
In-class discussions	54.2	45.3
Examples professional scenarios	48.3	45.3
Engineering design	32.7	43.4
Project based learning	29.9	35.8
Reflections	19.7	30.2
Videos, movie clips	22.0	28.3
In-class debates and/or role plays	17.4	22.6
Humanist readings	7.3	17.0*
Moral exemplars	6.5	15.1*
Guest lectures (e.g., philosophers, social scientists)	23.6	13.2
Service-learning, community engagement, and/or LTS	10.3	13.2
Think-pair-share	10.6	13.2
Problem solving heuristics	11.1	9.4
		*p<0.05

The results indicate that both the US and international respondents rely on lectures, case studies, and in-class discussion most heavily to teach ethics and broader impacts. These findings are consistent with the teaching practices most often reported in the literature for engineering ethics (Herkert, 2000). The cohorts diverge in terms of their incorporation of pedagogy traditionally associated with disciplines outside of engineering. The use of humanist readings and moral exemplars is consistent with the European perspective of ethics as an interdisciplinary study drawing from philosophy and the social sciences (Didier, 2015). An interviewee, who teaches at a technical institute in Ireland, echoed this idea in stating, "American [approach] is way more individualistic than Europe, Europe is more philosophical and sociological in its approach" of educating engineering students about ethics and broader impacts.

RQ3: Perceptions of importance and adequacy

Beyond understanding which ethics-related topics are taught to engineering and computing students and the methods chosen to incorporate them into the classroom, the survey sought feedback on educators' overall perspective on ethics education. After answering questions regarding their courses, the respondents were asked if they believe students in their programs receive sufficient education regarding the broader impacts of engineering and technology, the results are shown in Table 5.

Table 5: Percentage of respondents with ratings of satisfaction with the sufficiency of education regarding ethics and broader impacts

Level of satisfaction	*Educators at US Institutions (%)*		*Educators at International Institutions (%)*	
	Undergraduate (n=1084)	Graduate (n=848)	Undergraduate (n=49)	Graduate (n=39)
Yes, too much or sufficient	32.6	19.8	12.2	7.8
Sufficient microethics, insufficient macroethics	16.1	9.0	14.3	12.8
Sufficient macroethics, insufficient microethics	11.8	9.6	18.4	17.9
No, not enough	39.5	61.6	55.1	61.5

The results show a general lack of satisfaction with the amount of ethics instruction that engineering and computing students receive. Both groups indicated less satisfaction with the ethics and broader impacts instruction of graduate students. When the response options are grouped into "enough" (too much or sufficient) and "not enough" (insufficient microethics, insufficient macroethics, or both), there are statistically significant differences between US and international respondents in terms of undergraduate ethics exposure (p=0.0025). The results also indicate varying perceptions of micro and macroethics between the US and international groups. At the undergraduate level, more respondents within the US believe that microethics instruction is sufficient while macroethics is insufficient than the opposite (16.1% versus 11.8%). Conversely, more international educators believe that macroethics is sufficient and microethics is insufficient (18.4% versus 14.3%). This finding supports the emphasis of professional issues and codes of ethics, microethical topics, in the US and the broader focus on the relationship between science, technology, and society, macroethical concerns, in Europe and Australia (Zandvoort, 2000). The open-ended question near the end of the survey allowed educators to elaborate on these responses.

Responses to the open-ended question regarding respondents' thoughts on education of broader impacts and ethical issues fell into four general categories. The emerging themes included goals/opportunities, topics, current practices, and challenges. The most commonly selected codes for the international group, which appeared three or more times in the 25 responses, are shown in Table 6.

Table 6: Most common codes for the international respondents

Theme	*Code*	*Definition*	*Frequency (n=25)*
Goals and opportunities	Improve	When motivation to improve is noted, or when 'we need more' is stated	5
	Important	When a response says, generically, that ethics and/or societal issues is important	7
	Engineering impact	When the impact on communities by engineering is discussed (negative or positive)	4
	Incorporate into existing classes	When it is recommended or encouraged that ethical topics should be incorporated into existing classes as an effective way to teach ethics	3

Goals/opportunities

The respondents from US and international cohorts noted the importance of ethics education and the desire for it to be improved. Overall, these codes were used 57 and 69 times, respectively. The results suggest widespread value being placed on educating engineering and computing students with a holistic perspective of the profession and understanding of the societal implications of their work while motivating more meaningful inclusion of these topics. One international educator commented: "I strongly believe that this education needs to be integrated throughout the curriculum, not approached in a 'bolt-on' fashion." Another international respondent echoed the importance of ethics across the curriculum and incorporating ethics into existing classes; "I would like to see more of a cross-curriculum integration, with these topics integrated within engineering science and design courses more broadly. I believe this would be a more effective and authentic way of teaching/learning the subject matter." An international educator suggested a paradigm shift to achieve this aim in stating "engineering programmes are conceptualized as mainly technical programmes with some social add ons. This needs to change and engineering needs to be understood better as a social and technical activity." Although embedding ethics into courses was a commonly expressed goal, there was also discussion of the challenges to meaningful inclusion of ethics in engineering and computing programs including, "these issues are regarded by many engineering faculty as 'soft'" and they are "not given sufficient importance by faculty or students." One difference between the two groups was the more frequent mention of barriers in the American responses, such as fitting ethics into a full curriculum and the bridging the gap between learning ethical theory and behaving ethically.

Limitations and Future Work

One limitation of this study is that the data is based on individuals who selected to participate in the survey. The survey invitation and consent information indicated that the study was focused on engineering ethics and thus educators who do not teach ethics might not have participated. As a result, the respondents are not representative of the entire engineering education community. Additionally, although responses were received from Canada, Europe, South American, Australia and Africa, the international group is not large enough to be representative of the global community. Given the relatively small number of international respondents, this 'international' group cannot be further segregated into any continental or regional groupings. The binary analysis also does not account for intra-group variation and does not examine similarities and difference for individuals who were born or educated outside the US and now teach at American institutions and vice versa. Disaggregating the 'international' group would provide even more opportunity to evaluate ethics-related teaching practices and perceptions around the globe.

The interview phase of the study is on-going and future work will delve deeper into exemplary teaching practices, perceptions of macroethics instruction, and comparisons across the American and international groups. The next phase of the study is selecting case studies for in-depth analysis.

Implications

The survey results provide insight into how educators in the US and abroad incorporate ethics and broader impacts of engineering and technology into their courses. The selected topics and teaching methods illuminate the similarities and differences in instructional practices between the two groups. A larger percentage of international educators infuse sustainability, environmental protection issues, and engineering and poverty into their courses. The international respondents also report using moral exemplars and humanist readings more than the US group. Despite a significantly higher percentage of international educators integrating the aforementioned topics and methods, these respondents were less satisfied with the sufficiency of undergraduate ethics education. This disparity could arise from the higher standards that international educators are setting for programs to teach engineering in a broader context while US institutions are more satisfied with teaching ethics as it fulfills accreditation requirements and professional codes. For example, there are sustainability requirements for accreditation in New Zealand (IPENZ, 2009) and the UK (Engineering Council, 2013) and curriculum in Canada must include the impact of technology on society, sustainable development, and environmental stewardship (Engineers Canada, 2015). Engineering is global in its reach with projects spanning national borders and companies drawing employees from diverse programs and cultures so it is important to understand how engineering students are being educated and what similarities and differences define teaching practices in the US and abroad.

References

Bielefeldt, A.R., Canney, N.E., Swan, C., and Knight, D.W. (2016). Efficacy of macroethics education in engineering. ASEE Annual Conference, New Orleans, June.

Brumsen, M. (2005). Ethics in engineering in the Netherlands: The role of professional associations, universities, and law. *International Journal of Engineering Education*, *21*(3), 391-401.

Canney, N.E., Bielefeldt, A.R., Swan, C., Knight, D., Polmear, M., Simon, E. (2017) Challenges and opportunities: Faculty views on the state of macroethics education in engineering. American Society for Engineering Education (ASEE) Annual Conference and Exposition. Columbus, OH.

Creswell, J. W. (2012). *Qualitative inquiry and research design: Choosing among five approaches.* Sage publications.

Didier, C. (2015). Engineering ethics: European perspectives. *Ethics, science, technology, and engineering: A global resource*, *2*, 87-90.

Engineers Canada. (2015). 2015 Accreditation Criteria and Procedures. Engineers Canada Accreditaion Board

Engineering Council. (2013). The Accreditation of Higher Education Programmes: UK Standard for Professional Engineering Competence; Engineering Council: London, UK.

Given, L. M. (Ed.). (2008). *The Sage encyclopedia of qualitative research methods*. Sage Publications.

Haws, D. R. (2001). Ethics instruction in engineering education: A (mini) meta-analysis. *Journal of Engineering Education*, *90*(2), 223.

Herkert, J. R. (2000). Engineering ethics education in the USA: Content, pedagogy and curriculum. *European journal of engineering education*, *25*(4), 303-313.

Herkert, J. R. (2001). Future directions in engineering ethics research: Microethics, macroethics and the role of professional societies. *Science and Engineering Ethics*, *7*(3), 403-414.

Herkert, J. R. (2003). Professional societies, microethics, and macroethics: Product liability as an ethical issue in engineering design. *International Journal of Engineering Education*, *19*(1), 163-7.

IPENZ (Engineers New Zealand). (2009). IPENZ Requirements for Initial Academic Education for Professional Engineers, Version 3.1, Nov. 2009.

Knight, D.W., Bielefeldt, A.R., Canney, N.E., and Swan, C. (2016). Macroethics instruction in co-curricular settings: The development and results of a national survey. Frontiers in Education, Erie.

Koelsch, L.E. (2013). Reconceptualizing the Member Check Interview. *International Journal of Qualitative Methods*, 12: 168-179.

Larrondo-Petrie, M.M. (2015). Engineering Program accreditation in Latin America and the Caribbean, in: American Society of Engineering Education. Presented at the ASEE International Forum, Seattle, WA, pp. 19.15.1-19.15.12.

Vanasura, L., Stolk, J., & Herter, R. J. (2009). The four-domain development diagram: A guide for holistic design of effective learning experiences for the twenty-first century engineer. *Journal of Engineering Education*, *98*(1), 67.

Zandvoort, H., VanDePoel, I., & Brumsen, M. (2000). Ethics in the engineering curricula: Topics, trends and challenges for the future. *European journal of engineering education*, *25*(4), 291-302.

Acknowledgements

This material is based on work supported by the National Science Foundation under Grant Nos. xxx. Any opinions, findings, and conclusions or recommendations expressed in this material are those of the author(s) and do not necessarily reflect the views of the National Science Foundation.

Reflecting on out-of-class involvement of engineering students as a path to retain underrepresented students

Denise R. Simmons, PhD
Virginia Tech, Blacksburg, VA, USA
densimm@vt.edu

Abstract: *While curricular activities have long been the focus of engineering education research and practice, engineering educators have only recently begun to recognize the value of what engineering students do outside of the classroom. As part of a larger project to develop and implement a survey to measure engineering students' engagement in and outcomes of out-of-class activities, information about out-of-class activities of engineering undergraduate students has been collected through a range of methods (including literature reviews, Q-sorts, interviews, focus groups, and surveys). This paper presents a reflection on findings from synthesizing existing literature with data collected during survey development. Recent work in this area suggests that additions to Astin's model of student involvement are necessary. Improved understanding of engineering student involvement in out-of-class activities could help support academic success, persistence, and development of the whole person, and address the shortage of engineers and lack of a diverse engineering workforce.*

Introduction

By 2050, half of the United States (U.S.) population will identify as Black, Hispanic, or American Indian (Vazquez-Akim, 2014). As minorities begin to constitute more of the population, it is essential to break down educational barriers that persist in engineering to address the low retention rate of engineering students and lack of diversity in undergraduate university student bodies. In 2010, underrepresented racial minorities–Blacks, Hispanics, and American Indians–earned just 17% of all engineering bachelor's degrees (National Science Foundation, 2013). Not only does engineering face a racial/ethnic minority deficit, but also a gender deficit. Females are becoming more interested in STEM as a whole; in 2010, women were awarded 59% of all biological science degrees, but only 18.4% of engineering degrees (National Science Foundation, 2011). The U.S. produces 30% of the world's scientist and engineers and is responsible for 40% of the world's development and research spending (Freeman, 2005). If it intends to hold this distinguished position, retention, racial, and gender barriers in engineering must be addressed.

The engineering education community has dedicated attention and research to curricular changes (e.g., Brown, Hershock, Finelli, & O'Neal, 2009) and more recently to out-of-class support for engineering students (e.g., Finelli, et al., 2012; Wilson, et al., 2014; Simmons, et. al., 2017). Involvement in out-of-class activities has long been assumed to be complementary to educational and developmental processes and specifically undergraduate STEM learning (Ahren, 2009; Thiry, Laursen, & Hunter, 2011) and persistence. This assumption has been a driving force behind literature on out-of-class engagement. Space limitations preclude a complete survey of this rapidly growing field of research. Instead, this paper focuses on student's personal traits, college-related environmental factors, motivation to participate, and outcomes from participation in out-of-class activities as found in recent literature.

Context

Throughout this paper, out-of-class activities are defined as activities that occur outside of the formal classroom and curriculum. These activities are connected to the academic curriculum and educational institution and can include “community service groups, student government, fraternities, athletics, honor societies and religious clubs” (Ahren, 2009, p. 4).

Literature shows that findings on whether out-of-class engagement produces positive academic outcomes are mixed, with some studies showing limited significance (Ahren, 2009; Wilson, et al., 2014) or even adverse effects (Suleman, Singh, & Zeeshan, 2014). Still, a number of studies show a positive significant relationship (Jenkins, 2009; Yu & Simmons, 2015; Simmons, Yu, & Creamer, 2017). For example, research shows that college students who score higher on the SAT have higher college GPAs. Other research suggests that students engaged in out-of-class activities, however, have higher college GPAs than might be predicted by their SAT scores (Zehner, 2011). Out-of-class activities have been consistently shown to also have a positive impact on college persistence (Astin, 1984; Pascarella, Terenzini, & Wolfe, 1986; Pascarella & Terenzini, 1991).

Out-of-class activities not only promote academic success and persistence, but also help develop the student as a “whole”. Out-of-class engagement has been shown to support personal development; help students to be productive and civically responsible members of society (Ahren, 2009); decrease racial bias (Denson, 2009); contribute to significantly positive outcomes for ethical development (Finelli, et al., 2012), and assist with mental, physical, moral, and social development of an individual (Suleman, Singh, & Zeeshan, 2014). This research suggests that when other academic indicators are controlled for, out-of-class involvement has a significantly positive relationship with academic success.

Finally, findings also varied on what influences students to participate in out-of-class activities. Studies have shown that structural influences (Pepper, 2009) and personal traits (Bauer, 2003) influence out-of-class engagement. Primary influences that motivate students to participate in out-of-class activities are the social aspect of the activity, having an opportunity to learn something new, and preparation for a future job.

Research questions

This reflection piece revolves around two questions: (1) what are some of the findings from an extensive literature review and survey development process that have focused on engineering student involvement in out-of-class activities and (2) how can the findings from this literature review and data collected during the survey development process contribute to the retention of underrepresented students and to the field of engineering education?

Theoretical framework

This paper is grounded in Astin's Theory of Student Involvement (1984) and more recent extensions of the theory by other scholars (e.g. Terenzini and Reason, 2005) and operates under the assumptions of this theoretical framework. Astin defines student involvement as the amount of physical and psychological energy that a student allocates to the academic experience. A highly involved student is one who expends a considerable amount of energy studying, spends time on campus, is engaged in out-of-class activities, talks to faculty, and participates in an array of other academic activities (Astin, 1984). This theory argues that simply exposing students to course material may not be enough to achieve academic success; rather, to achieve intended goals, students must devote a sufficient amount of effort and energy to their academic experience. Additionally, this theory emphasizes students playing an active role in the learning process (Astin, 1984). Based on this theory of involvement, students who devote a sufficient amount of effort and energy to out-of-class engagement will have higher levels academic success, social development, and moral development than their peers with little to no out-of-class involvement.

Methods

This paper provides scholarly reflection on a number of reviewed and related studies that provided data about engineering student involvement in out-of-class activities. The importance of students reflecting on and learning from their educational experiences is crucial to academic and career successes (Turns, Sattler, Borgford-Parnell & Atman, 2014). Benefits to students include time to consider how past experiences will influence future actions (Turns, Sattler, Borgford-Parnell & Atman, 2014). Similar benefits may be attributed to researchers who reflect on research and also supports checking and auditing all steps of the analysis. Self-reflection also allows researchers the opportunity to think strategically about the coherence of their research agenda.

Findings

The results presented in this paper illustrate the need for future research to further examine the academic and developmental outcomes of out-of-class engagement. For out-of-class activities to aid engineering education in student retention and diversifying university student bodies, there needs to be a clearer idea of what specific components within out-of-class activities that produce positive academic outcomes. This paper concludes by highlighting key implications of this research and offer suggestions for future research to explore the relationship between out-of-class activities and academic achievement, specifically in engineering. The suggestions are not comprehensive in the sense it will solve all research issues, but rather they serve to highlight new ways in which to study out-of-class engagement.

Given the vast array of out-of-class activities that are accessible to students it is implausible to highlight every aspect of out-of-class activities that has produced positive student outcomes. Three specific components of out-of-class involvement have emerged from the literature and data collected as being associated with positive outcomes. These are not the only significantly positive components of out-of-class activities, but rather three primary components that have been identified in an array of scholarly journals, presentations, reports, and primary data collected during the development of a survey of student affective engagement in out-of-class activities. The three primary components of out-of-class activities that produce positive academic, social, and professional outcomes are: (1) interaction with faculty, (2) promotion of campus engagement, and (3) a hierarchal structure of student leadership opportunities.

Interaction with faculty outside of class

Student interaction with faculty has been shown to improve student GPA and insure persistence and degree completion (Umbach & Wawrzynski, 2005; Tierney, Colyar, & Corwin, 2003; Hazeur, 2008). While much of this engagement occurs in conjunction with a class, students report higher levels of engagement and learning at institutions where faculty members engage students in experiences outside of class (Umbach & Wawrzynski, 2005). Faculty play an integral role in socializing students into the college culture because they serve as the primary point of contact between students and the university. To create a dynamic learning environment, faculty members' roles should extend beyond the classroom. Out-of-class activities can serve as a medium in which faculty and student interactions can occur outside of the formal classroom.

Out-of-class activities that give students an opportunity to interact with faculty allow students to develop social capital and cultural capital while simultaneously increasing student involvement. Such interactions have been found to be particularly impactful for first generation college students in civil engineering (Simmons, Kirn, Valdes-Vasquez & Pearson, unpublished) If one of the primary roles of out-of-class activities is to augment the in-class curriculum, it seems rational to have in-class instructors (faculty) play a role in supplementing the formal curriculum. Student interaction with faculty through out-of-class engagement gives students agency to shape their interaction with faculty in a manner they could not in the classroom. Specific high impact examples of how engineering faculty can interact with students in out-of-class activities include as research advisors, principles in living learning communities, leaders of study abroad experiences and advisors of student organizations.

Previous literature shows positive academic and developmental outcomes when engineering out-of-class activities incorporate engineering faculty (Lam, Srivatsan, Doverspike, Vesalo, & Mawasha, 2005; Bruning, Ames, Rover, & Williams, 2011). Finelli (2012) argues positive outcomes from out-of-class support is due to the ability of out-of-class interventions to help integrate engineering students into the existing environment and to adapt to the needs of students in real time. Faculty, serving as instrumental proponents in constructing the engineering curriculum and department climate, may be equipped to efficiently socialize engineering students into the engineering culture.

Studies show that engineering faculty play a momentous role in engineering student's overall experience in the engineering field with high levels of faculty engagement positively contributing to student engagement and academic success. A study by Chen and Hamilton (2008) showed that faculty-student interaction is a predictor of student satisfaction and personal development in engineering and satisfaction with instructors contributes positively to student's levels of confidence. Out-of-class activities can be spaces in which faculty-student interactions can be facilitated. A study by Burt et al. (2011) found that when students were involved in out-of-class activities, faculty members influenced their student involvement. Continuous interactions with engineering faculty can increase student's social and cultural capital, both of which impact degree persistence and attainment.

Promotion of campus engagement

Campus engagement is viewed as the amount of physical and psychological energy a student invests in the university. An out-of-class activity is a form of campus engagement in and of itself. However, students may receive more benefit from that activity if the activity encourages students to expand the manner in which they engage with the campus. Astin's theory argues that the involved student is not a passive agent in the educational system, but plays a role in constructing their own educational experience. Out-of-class activities that allow students to engage in dynamic ways with their university produce positive academic and developmental outcomes. Out-of-class involvement should give students opportunities to interact with individuals from different areas of the university. An example of an out-of-class activity promoting campus engagement is participating in and attending sporting events. This may enhance the students' sense of belonging at the university and allows them the agency to help construct their learning environment.

Sense of belonging, which likely changes over time, is an important component of student persistence and overall success at a university. Involvement in out-of-class activities increases self-efficacy (Wilson et al., 2014) and sense of belonging to a university (Hazeur, 2008). First-year students who earned higher grades than they expected engaged in learning strategies and were more integrated into the campus society (National Survey on Student Engagement, 2014). Out-of-class activities that promote campus engagement allow students the opportunity to be socialized into the campus culture. For underrepresented students, out-of-class activities helped them balance different aspects of well-being and contributed to their remain engagement with their work and studies (Purchase & Simmons, 2017).

Previous literature shows that students who are highly involved persist in completing their college degree are more likely to attend graduate school, and have higher GPAs (Pepper, 2009). Having the opportunity to engage in dynamic ways with different university personnel and becoming comfortable may explain why highly involved students receive such benefits. Pepper (2009, p. 3) suggested that studies have "confirmed Astin's theory that students are not passive recipients of an institution's efforts to educate, but that an individual's effort, engagement, or involvement with the institution is the critical determinant of the impact of college."

Hierarchal structure of leadership opportunities

When out-of-class activities build leadership qualities into students, they receive more positive academic outcomes compared to students with no leadership experience (Pepper, 2009). Students with a motivation to lead are more likely to be engaged students. Numerous studies have cited that participation in out-of-class activities contribute to student's development of leadership qualities (Astin, 1993; Baxter Magolda, 1992; Kuh, 1995; Pascarella & Terenzini, 1991; Burt, et al., 2011). However, these studies do not highlight what specifically in out-of-class activities promoted the development of leadership qualities.

Activities that provide hierarchal structures may promote leadership qualities in students more than out-of-class activities without a hierarchal structure (Coffman, 2012). An example of an out-of-class activity with a hierarchal structure that has been shown to be particularly impactful for underrepresented students is peer led group such as National Society of Black Engineers, which gives students the opportunity to hold different levels of leadership positions.

Hierarchal structures have clear and defined leadership roles. It is not solely about joining a hierarchal structured activity, but rather the process of obtaining and fulfilling the leadership role. Using Astin's theory, hierarchal structures can be conceptualized as a process that creates leadership positions as a form of engagement that requires a great deal of physical and psychological energy to be invested. This invested energy allows students in leadership roles to have the agency to help construct their college experience, which may promote academic achievement.

Discussion

Having diverse and well-qualified engineers is vital to the global economic position of the U.S.. However, the U.S. faces a shortage of engineers due to 40 to 50 percent of engineering students switching majors or dropping out (Matthews, 2012). Additionally, engineering student diversity is a persistent problem given the lack of bachelor's degrees awarded to underrepresented populations (Ohland, Lord, & Layton, 2015). Out-of-class activities may serve as one component of an initiative by universities to reverse these trends.

Three primary components of out-of-class activities that produce positive academic, social, and moral outcomes are: (1) interaction with faculty, (2) promotion of campus engagement, and (3) a hierarchal structure of student leadership opportunities. These three themes were found across a range of literature and primary data about different successful out-of-class activities. Out-of-class activities that incorporate these components are more likely to produce positive academic and developmental outcomes; however, these are not the only positive components within out-of-class activities and more research is needed to identify important elements of out-of-class activities.

Developing effective out-of-class activities is not the only step in increasing student engagement. Students need to be convinced to participate in out-of-class activities and existing research has explored the incentives and barriers of out-of-class activity participation. Students often participate in out-of-class activities for social reasons (Tan & Pope, 2007; VanZile-Tamsen & Bissonnette, 2008). These reasons come in the form of professional networking, meeting new people, and pressures from friends to participate. Factors that reduced out-of-class participation were off campus work, family commitment, and an overly demanding academic life (Tan & Pope, 2007; VanZile-Tamsen & Bissonnette, 2008). Out-of-class activities must be accessible to students who come to college with numerous responsibilities and obligations.

Implications

This paper does not cover the entire scope of how out-of-class involvement can aid engineering education, but it does offer useful and practical recommendations for future research and program development. Out-of-class activities aimed at increasing positive academic and developmental outcomes should incorporate the three components presented in this paper (interaction with faculty, promotion of campus engagement, and hierarchal structure of student leadership opportunities). Out-of-class activities aimed specifically at aiding engineering students persist should have engineering faculty involved, motivate students to be engaged in multiple facets of the university, and provide engineering students opportunities to lead within the activity. Interaction with engineering faculty will allow engineering students to develop social and cultural capital within the engineering field. Encouraging and providing students opportunities to engage with campus gives students a sense of belonging at their particular institution. Students who hold leadership positions can actively create their learning environment.

Student outreach attempts should ensure that the social aspects of the activity are highlighted. Networking and making new friends are primary motivators for participation and essential to academic persistence. Out-of-class activities must be accessible to a variety of students. Family obligations and working due to financial need can undermine participation in out-of-class activities. Students who work and/or have family obligations can be surveyed to understand what types of accommodations these students need. Students who have to work for pay and have family obligations are often non-traditional students, first generation college students, low-income students, and racial and ethnic minorities (Engle & Tinto, 2008). This opens the door for future research to access what types of accommodations need to be given to non-traditional students, first generation college students, low-income students, and racial and ethnic minorities to allow them the opportunity to participate in out-of-class activities. Additionally, to increase out-of-class engagement, social, academic, and professional incentives should be advertised to students. The targeted students may be more willing to participate if the out-of-class activity is relevant to their reality. Finally, studies have shown that too much involvement undermines academic success (Ahren, 2009; Burt, et al., 2011; Suleman, Singh, & Zeeshan, 2014). Future studies should address the question: what is too much participation? After researchers define and have measures for what constitutes too much involvement, students can be advised on the amount of energy that should be expended on out-of-class activities to have a healthy and flourishing college experience.

Future research

More research is needed to examine what components of out-of-class activities motivate participation, produce positive academic and professional outcomes, and promote student development. Higher education literature speaks to students' motivation to participate in out-of-class activities, but the engineering education literature is silent on what motivates underrepresented groups to participate. Research has examined personal characteristics that influences student participation in out-of-class activities, but findings have been mixed.

Out-of-class activities need to be designed to remove barriers to participation. Pilot-scaled interventions can aid the discovery of effective designs that remove such barriers. Astin's Theory of Involvement is not sufficient to explore the design of out-of-class activities and the many aspects of student engagement research. Future research is necessary to answer what influences and discourages out-of-class engagement given that having ample out-of-class opportunities is inadequate if participation is minimum. Such research needs to define out-of-class activities and the array of diverse out-of-class opportunities offered to students, and employ additional theoretical frameworks in doing so. Future research includes answering three questions: (1) What specific components within out-of-class activities produce positive academic outcomes?; (2) What influences students' decisions to become involved in out-of-class activities?; and (3) What, if any, are the connections among the reasons students choose an activity, the activity itself, and the perceived outcomes?

References

Ahren, C. S. (2009). *Disentangling the unique effects of co-curricular engagement on self-reported student learning outcomes.* Indiana University, ProQuest, UMI Dissertations Publishing.

Astin, A. (1984). Student Involvement: A Developmental Theory for Higher Education. *Journal of College Student Personnel 25(4)*, 297-308.

Astin, A. (1993). *What matters in college: Four critical years revisited.* San Francisco: Jossey-Bass.

Bauer, K. (2003). The Effect of Personality and Precollege Characteristics on First-Year Activities and Academic Performance. *Journal of College Student Development, 44(3)*, 277-290.

Baxter Magolda, M. B. (1992). *Knowing and reasoning in college: Gender-related patterns in students' intellectual development.* San Francisco: Jossey-Bass.

Brown, M., Hershock, C., Finelli, C., & O'Neal, C. (2009). *Teaching for retention in science, engineering, and math disciplines: A guide for faculty (Vol. 25, pp. 1-12).* Ann Arbor: The University of Michigan.

Bruning, M., Ames, I., Rover, D., & Williams, A. (2011). Work in progress — Developing engineers for 2020 — An innovative curricular and co-curricular approach. *Frontiers in Education Conference (FIE)* (pp. S4D-1 - S4D-3). Rapid City: IEEE.

Burt, B. A., Carpenter, D. D., Finelli, C. J., Harding, T. S., Sutkus, J., Holsapple, M., Bielby, R. & Ra, E. (2011). Outcomes of engaging engineering undergraduates in co-curricular experiences. *American Society of Engineering*, (pp. 00556 - 00511).

Chen, H., Lattuca, L., & Hamilton, E. (2008). Conceptualizing Engagement Contributions of Faculty to StudentEngagement in Engineering. *Journal of Engineering Education, 97(3)*, 339-353.

Coffman, N. W. (2012). *An exploration of high school co-curricular activities.* University of Pennsylvania, ProQuest, UMI Dissertations Publishing.

Denson, N. (2009). Do Curricular and Cocurricular Diversity Activities Influence Racial Bias? A Meta-Analysis . *Review of Educational Research, 79(2)*, 805-838.

Engle, J., & Tinto, V. (2008). *Moving Beyond Access: College Success For Low-Income,.* Washington, DC: The Pell Institute.

Finelli, C., Matthew, H., Ra, E., Bielby, R., Burt, B., Carpenter, D.D., Harding, T.S., & Sutkus, J. (2012). An Assessment of Engineering Students' Curricular and Co-Curricular Experiences and Their Ethical Development. *Journal of Engineering Education, 101(3)*, 469-494.

Freeman, R. (2005). *Does globalization of the scientific/engineering workforce threaten U.S. economic leadership?* Cambridge: National Bureau of Economic Research.

Hazeur, C. (2008). *Purposeful Co-Curricular Activities Designed to Increase Engagement: A Practice Brief Based on BEAMS Project Outcomes.* Washington, DC: Institute for Higher Education Policy.

Jenkins, C. (2009). *What Factors Contribute to the Achievement Gap: A Case Study of Multicultural/Disadvantaged Student Participation in Co Curricular Activities at a Large Urban High School.* Madison: ProQuest LLC, ProQuest, UMI Dissertations Publishing.

Lam, P. C., Srivatsan, T., Doverspike, D., Vesalo, J., & Mawasha, P. R. (2005). A Ten Year Assessment Of The Pre-Engineering Program For Under-Represented, Low Income And/Or First Generation College Students At The University Of Akron. *Journal Of STEM Education: Innovations & Research, 6(3/4)*, 14-20.

Matthews, M. (2012, August 27). *Keeping students in engineering: A research-to-practice brief.* Retrieved from American Society For Engineering Education: http://www.asee.org/retention-project/keeping-students-in-engineering-a-research-guide-to-improving-retention

National Science Foundation. (2011). *Women, minorities, and persons with disabilities in science and engineering.* Arlington: National Science Foundation.

National Science Foundation. (2013). *Women, Minorities, and Persons with Disabilities in Science and Engineering.* National Science Foundation.

National Survey on Student Engagement (NSSE). (2014). *Bringing The Institutions Into Focus Annual Results 2014.* Bloomington: Indiana University Center for Postsecondary Research.

Ohland, M., Lord, S., & Layton, R. (2015). Student Demographics and Outcomes in Civil Engineering in the United States. *Journal of Professional Issues in Engineering Education and Practice*, 10.1061/(ASCE)EI.1943-5541.0000244.

Pascarella, E., & Terenzini, P. (1991). *How college affects students.* San Francisco: Jossey-Bass.

Pascarella, E., Terenzini, P., & Wolfe, L. (1986). Orientation to college and freshman year persistence/withdrawal decisions. *Journal of Higher Education, 57*, 155-175.

Pepper, R. C. (2009). *The Impact of Motivation to Lead on College Students' Cocurricular Involvement.* Virginia Beach: ProQuest LLC, ProQuest, UMI Dissertations Publishing.

Purchase, J., & Simmons, D. R. (2017). *Exploring the meaning of fun: A missed opportunity to retain underrepresented groups in engineering?* 124th American Society for Engineering Education Annual Conference & Exposition, Columbus, OH.
Simmons, D. R., Kirn, A., Valdes-Vasquez, R and Pearson, N. (under review). Are Constructors Experiencing a Warm Climate? A study of Generation in College, Citizenship Status and Affective Engagement. *Journal of Professional Issues in Engineering Education and Practice.*
Simmons, D. R. & Ye, Y., Hunsu, N. J. & Adesope, O. O. (accepted for publication). Development of a Survey to Explore Out-of-Class Engagement of Engineering Students. *International Journal of Engineering Education.*
Simmons, D.R., Yu, R. & Creamer, E.G. (2017). Involvement in Out-of- Class Activities: A Mixed Research Synthesis Examining Outcomes with a Focus on Engineering Students. *Journal of STEM Education: Innovations and Research.*
Suleman, Q., Singh, T., & Zeeshan. (2014). Effects of Over-Scheduled involvement in Co-Curricular Activities on the Academic Achievement of Secondary School Students in Kohat Division, Pakistan. *International Journal of Learning & Development*, 62-71.
Tan, D., & Pope, M. (2007). Participation in Co-Curricular Activities: Nontraditional Student Perspectives. *College and University, 83(1)*, 2-11.
Terenzini, P. T., and Reason, R. D. (2005, November). Parsing the first year of college: A conceptual framework for studying college impact. Paper presented at the annual meeting of the Association for the Study of Higher Education, Philadelphia, PA.
Thiry, H., Laursen, S., & Hunter, A.-B. (2011). What Experiences Help Students Become Scientists? A Comparative Study of Research and Other Sources of Personal and Professional Gains for STEM Undergraduates. *The Journal of Higher Education, 82(4)*, 357-388.
Tierney, W., Colyar, J., & Corwin, Z. (2003). *Preparing for College: Building Expectations, Changing Realities.* Los Angeles: Center for Higher Education Policy Analysis.
Turns, J. A., Sattler, B., Yasuhara, K., Borgford-Parnell, J., & Atman, C. J. (2014, June). Integrating reflection into engineering education. In *Proceedings of the ASEE Annual Conference and Exposition. ACM* (Vol. 35, p. 64).
Umbach, P., & Wawrzynski, M. (2005). Faculty Do Matter: The Role of College Faculty in Student Learning and Engagement. *Research in Higher Education, 46(2)*, 153-184.
VanZile-Tamsen, C., & Bissonnette, K. (2008). *Encouraging Participation in Enriching Educational Experiences – Co-Curricular Activities and Community Service.* University of Buffalo.
Vazquez-Akim, J. A. (2014). *Crack in the Pipeline: Why Female Underrepresented Racial Minority College Students Leave Engineering.* Los Angeles,: University of California ProQuest, UMI Dissertations Publishing.
Wilson, D., Jones, D., Mee Joo, K., Allendoerfer, C., Bates, R., Crawford, J., Floyd-Smith, T., Plett, M. & Veilleux, N. (2014). The Link between Cocurricular Activities and Academic Engagement in Engineering Education. *Journal of Engineering Education, 103(4)*, 515–651.
Yu, R. & Simmons, D.R. (2015). *Synthesis of Engineering Undergraduate Students' Out of Class Involvement.* Paper presented at 2015 ASEE Annual Conference and Exposition, Seattle, WA. https://peer.asee.org/24787, DOI: 10.18260/p.24787
Zehner, A. (2011). *Co-Curricular Activities & Student Learning Outcomes.* West Lafayette: Purdue University.

Acknowledgements

The author would like to gratefully acknowledge the National Science Foundation for their support of this work under the CAREER grant #EE-1351156. Any opinions, findings, conclusions, or recommendations expressed in this poster are those of the author and do not necessarily reflect the views of the National Science Foundation.

Copyright statement

Service-Learning: An approach to Disaster Risk Reduction in Engineering Education

Paola Moraga
Pontificia Universidad Católica de Valparaíso, Valparaíso, Chile
paola.moraga@pucv.cl

Abstract: *The students entering higher education today are part of a new generation (millennials), for which the traditional teaching strategies fail in delivering the expected learning outcomes. In this light, innovation is needed in the teaching strategies applied, in order to boost their motivation. This paper presents a case study on the incorporation of the Service-Learning methodology (SL) into the Risk Management course at the Construction Engineering School at PUCV in which the Risk Management and Reduction Program for Vulnerable Schools is implemented. Since incorporation of the SL methodology, the results of the teaching-learning process have improved considerably, achieving a high motivation and commitment from the students in the process of formal learning as well as in the service activity performed.*

INTRODUCTION AND MOTIVATION

Today's university student generations (Millenials) are quite different from their teachers' (Baby Boomers and X Generation). This adds a new level of complexity to the teaching-learning process because teaching practices that used to work before no longer work with these new generations. The result is a lack of motivation from the students, poor learning outcomes and low participation.

Millennials have been widely researched and studied by different authors, who agree in the traits that define them and in how the millennials face the world. Some of these established traits are that new generations of students are self-sufficient and autonomous, value participation and collaboration and demand new values such as transparency, sustainability and social commitment. An important differentiation from previous generations is the high motivation they show to carry out social, educational, personal, sports related, activities (Coomes M.D., and De Bard, R. (2011), Howe, N., and Strauss (2000, 2003), Lancaster and Stillman (2002), Martin and Tulgan, (2001)). These authors agree that millennials have a very high potential for work and development, which they can only exploit with the right motivation.

Considering all of the above, as academics we must reflect on how we influence and motivate and how we can achieve a type of teaching that is significant for our students and will be useful in their professional performance and their personal development as active and responsible members of society.

On these days, and in order to achieve motivated students, new teaching strategies have been applied in undergraduate teaching (service learning, flip classroom, problem based learning, among others), each of them having special characteristics that make them more or less appropriate for each of the different course of the program

Within the different available strategies, in the Risk management course, the SL has been selected. This strategy has been widely implemented and studied in several disciplines obtaining differents positive outcomes such as improvements in the learning process, an increase in the motivation and retention of students (Astin AM at al (2000), Eyler, J. (1999 and 2000), Hatcher, JA et al. (1997), Stanton TK, at al (1999) and Zlotkowski, E. (199 and 2002)). However and despite the benefits identified of this teaching technique, in engineering, the strategy has not been generalized as in other areas of study, some positive experiences that encourage the application of this strategy in engineering are the cases presented by Oakes WD (2002) and Ravel et al. (2015)

In this paper, we present implementation of the SL methodology in a course from the Construction Engineering program at the Pontificia Universidad Católica de Valparaiso, Chile.

THEORETICAL FRAMEWORK

Hatcher and Bringle (1997) provide one of them most comprehensive definitions of the SL strategy: "We view service-learning as a credit-bearing educational experience in which students participate in an organized service activity that meets identified community needs and reflect on the service activity in such a way as to gain further understanding of the course content, a broader appreciation of the discipline, and an enhanced sense of civic responsibility".

In the last years, the strategy has been widely studied by different authors (Tapia (2006), Eyler (2009), Oakes (2002 and 2015), Jouannet, C. et al (2013) among others, defining this methodology as a form of experiential education that combines academic studies with community service activities; and in every case all author agrees that the combination derives in mutual benefits for both processes because the students achieve a better result from their learning process by incorporating these experiences while improving the conditions in their own community.

In this research, the Kolb's Learning cycle was adopted (Kolb 1984). According to Kolb the learning process can be separate in 4 different stages (Figure 1), where a metacognitive activity (reflection) becomes a link between the experience and the conceptualization of topics covered in a course.

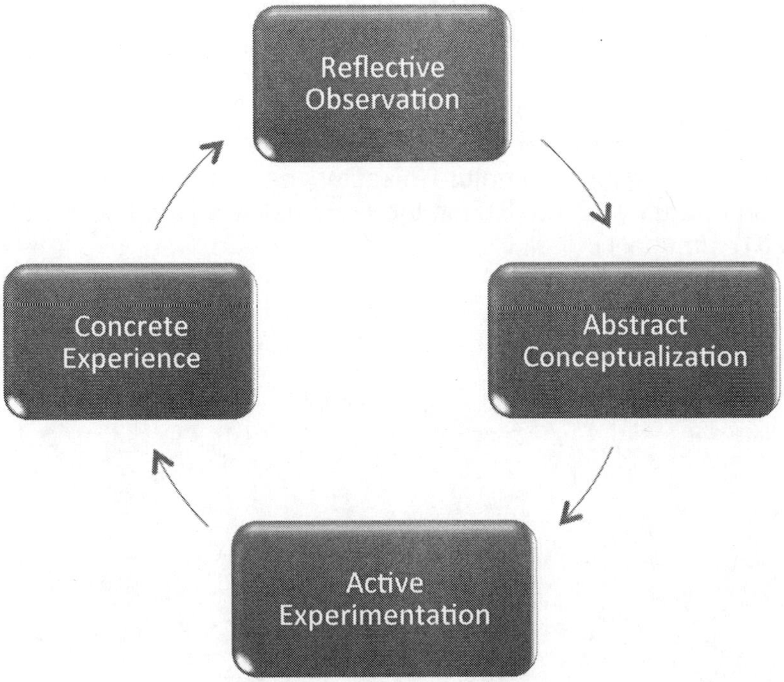

Figure 1. Kolb's learning cycle.

According to Kolb (Kolb, D.A. (1984)), learners gain new knowledge by testing and adapting their existing knowledge through a process that involves an abstract conceptualization, an active experimentation, a concrete experience and a reflective observation (Figure 1).

The SL strategy consider, the combination of classroom instruction and a service activity providing an environment in which students can enhance their knowledge. Since the learning process proposed by Kolb (Kolb, D.A. (1984) is a learning cycle, this allows students to enter the cycle at different places and still complete the full circuit.

Chile is a country that faces a wide variety of natural hazards such as earthquakes, tsunamis, landslides, volcano eruptions, floods, landslides and draughts among others, which will eventually affect part of the territory. Chile is also affected by anthropogenic threats like fires, especially in the urban-rural interfaces, as it has recently been the case, in which fires have caused important loss of human life and property.

Considering these factors and due to the professional nature of the Construction Engineering program, the School of Construction Engineering at the Pontifical Catholic University of Valparaiso took a step forward and created the course *Risk Management and Reduction*, which addresses risk management and preparedness for natural or anthropogenic disasters. Given the opportunity that these contents present for application and their high repercussion in society, an innovation in the classical teaching methods was considered in order to achieve a higher motivation and active participation of the students in their learning process.

METHODOLOGY

The SL methodology aimed at addressing the students' lack of participation and motivation observed in earlier versions of the course. As mentioned above, this strategy envisions an education that combines the academic studies with community service. This technique was chosen considering the profile of the program's students, who are active members of society and have continuously been willing to participate and collaborate when they are needed, especially when coping with natural disasters as it was the case of the earthquake that struck Chile on February 27th,2010 and the fires that engulfed the city of Valparaiso in 2014 and early 2017 (Image 1).

Image 1. Students of the Construction Engineering School collaborating with the community after the colossal fire of Valparaíso in April, 2014 (Source: Paola Moraga).

In this paper, we depict the experience of the implementation of SL strategy in the Risk Management course, in which a plan of Risk Management and Reduction in Vulnerable Schools must be carried out by the students under the direct supervision of their instructor.

The methodology applied has four phases, as the Kolb learning cycle presents, and each one of them is strictly related to it.

Phase 1 is a learning phase (Abstract Conceptualization of Kolb´s learning cycle), which includes the formal training of students, guided by the course instructor at the beginning of the academic term (semester). In this stage, students must fully grasp the theoretical contents delivered by the instructor so they can be implemented afterwards. Once the contents encompassing the different threats and vulnerabilities of the territory are delivered, the students must analyze the educational facility selected and identify all the instances of vulnerability and threat such facility may have.

Phase 2 is the planning stage (Active experimentation of Kolb´s learning cycle), in which students start to design a Risk Management and Reduction activity considering the characteristics of the vulnerable school (location, number of students per class, review of the Integral Safety Plan of the school, etc.). In this phase, students also prepare support material, keeping in mind the educational levels of the institution.

The third phase is the actual SL activity (Concrete experience of Kolb´s learning cycle), which in this case corresponds to the implementation of a Risk Management and Reduction plan. In this activity, the students train the members of the educational community about the school's vulnerabilities and threats as well as how to respond to them and adapt the facilities to safely cope with them (namely determining safety zones, evacuation routes and updating the School Safety Plan) (Image 2).

Finally, in Phase 4 (Reflective Observation of Kolb´s learning cycle) students perform an assessment of the activity through a SWOT matrix (Analysis of Strengths, Opportunities, Weaknesses and Threats). Through this analysis students determine what can be improved in future implementations of the activity as well the students' personal evaluation of the activity.

Image 2. Risk Reduction and Management activity. (Source: Paola Moraga)

RESULTS, DISCUSSIONS AND CONCLUSIONS

The SL methodology provides an opportunity for a three-dimensional teaching activity. The first dimension is academic, in which contents and methodologies are delivered; the second dimension is the real application, in which students observe the problems that affect their community and become active part of the solution; finally, there is a social dimension in which the students commit to the values acquired throughout their educational process, according to the hallmark values promoted by the university.

The Risk Management and Reduction course for the Construction Engineering School of the Pontificia Universidad Católica de Valparaíso is already in its fourth version. The first version was implemented through a traditional teaching process, while in the three subsequent versions an SL approach was used. Ever since, this optional course has incremented the number of enrolled students from 14 to 22 in the last version.

The course perception and evaluation from the students has improved since the incorporation of the SL methodology. The students are willing to devote more hours and dedication to the course, in addition to the regular hours, in order to complete the SL activities.

Image 3. SL students after the end of the activity. Source: Paola Moraga.

Since the SL methodology was installed, the students demonstrate high motivation and active participation. This also reflects in the fact that once the course is completed the students join the research group led by the course instructor, seeking to extend their participation in future SL projects (Image 3).

The motivation generated in the students has been so important that, with the support of their instructor, the Risk Management and Reduction plan has been successfully implemented in five vulnerable schools in the region of Valparaiso.

During the two years that the Risk Management and Reduction course has been given under the SL model, it has been observed that it generates a positive impact on the professional and human development of the students, who, by the end of the course still want to work and participate in future activities of the program or in any other opportunity to help the community.

It is very well known that implement SL strategy involve economic resources needed to develop the activities, been this issue some time difficult to address, however in this particular case, past student are so motivated that they do not just participate in the research group of the professor, they even submit proposals to different available grants (for students) in order to obtain the necessary funding to keep doing SL strategy in the following versions of the course. As a results of this until today former students of the course has applied for two different grants, been awarded with the grant in both cases, also the professor has submitted two other proposal also been selected in both contest. As a result of the implementation, so far 5 vulnerable schools has been implemented, having under preparation the implementation of three more school during this academic year.

One challenge that needs to be acknowledged in the implementation of this teaching strategy is that SL activities demand more time than the stipulated in the curriculum for the course as well as inside and outside the classroom activities.

To conclude, the SL methodology increase the competences of the students and gives them additional abilities, attitudes and values such as the commitment to the society, team working, problem solving, communication skills, responsibility and a positive attitude toward their communities. Therefore, the methodology translates into a tool for the training of competent professionals committed to the society.

REFERENCES

Astin, A. M., Vogelgesang, L. J., Ikeda, E. K., Yee, J. A . (2000) How Service-Learning Affects Students, Higher Education Research Institute, University of California Los Angeles.

Coomes M.D., and De Bard, R. (2011) Serving the Millennial Generation: New Directions for Student Services, Number 106}, isbn={9781118216958}

Eyler, J., (1999). Giles, D. E. Where's the Learning in Service-learning?, San Francisco:Jossey-Bass.

Eyler, J. (2000)"Reflection: Linking Service and Learning-Linking Students and Communities", Journal of Social Issues, vol. 58, no. 3, pp. 517-534.

Eyler, J. (2009) "The power of experiential education", en Liberal Education, 95(4), otoño, pp. 24-31.

Hatcher, J. A., Bringle, R. G. (1997) Reflection: Bridging the gap between service and learning, College Teaching, vol. 45, no. 4, pp. 153-158.

Howe, N., and Strauss, W. Millennials Rising: The Next Great Generation. New York:Vintage Books, 2000.

Howe, N., and Strauss, W. (2003) Millennials Go to College. Great Falls, Va.: American Association of Registrars and Admissions Officers and LifeCourse Associates,

Jouannet, Chantal, Salas, María Helena, & Contreras, María Antonieta. (2013). Modelo de implementación de Aprendizaje Servicio (A+S) en la UC: Una experiencia que impacta positivamente en la formación profesional integral. Calidad en la educación, (39), 197-212. https://dx.doi.org/10.4067/S0718-45652013000200007

Kolb, D.A. (1984). Experiential learning: experience as the source of learning and development. Englewood Cliffs, NJ: Prentice Hall.

Lancaster, L. C., and Stillman, D. (2002). When Generations Collide: Who They Are. Why Clash. How to Solve the Generational Puzzle at Work.New York: HarperBusiness,

Martin, C. A., and Tulgan, B. (2001) Managing Generation Y. New Haven, Conn.: HRD Press.

W. Oakes *et al.*, "Service-learning in engineering," *32nd Annual Frontiers in Education*, 2002, pp. F3A-1-F3A-6 vol.2. doi: 10.1109/FIE.2002.1158178

Ravel, M.K., Linder, B., Oakes, W. C. and Zoltowski, C. B. "Evolving engineering education for social innovation and humanitarian impact" Lessons learned across a range of models," *2015 IEEE Global Humanitarian Technology Conference (GHTC)*, Seattle, WA, 2015, pp. 169-176.

Stanton, T. K., Giles, D. E., Cruz, N. I. (1999). “Service-Learning: A Movement's Pioneers Reflect on Its Origins Practice and Future" in Jossey-Bass Higher and Adult Education Series, 350 Sansome St., San Francisco, CA 94104:Jossey-Bass Inc., Publishers, 1999.

Tapia, M., (2006) La solidaridad como pedagogía. Buenos Aires, Ciudad Nueva.
Zlotkowski, E. (1999) Successful Service-Learning Programs. New Models of Excellence in Higher Education, Ballville Road, PO Box 249, Bolton, MA:Anker Publishing Company, Inc.176, pp. 01740-0249.

Zlotkowski, E. (2002) "Service-learning and the First-year Experience" in Columbia SC: National Resource Center for the First-Year Experience & Students in Transition, University of South Carolina.

Constructing a Measure of Affect Towards Professional Practice: What matters for Engineers?

Anita Patrick
University of Texas at Austin, Austin, USA
apatrick@utexas.edu

Luis Martins
University of Texas at Austin, Austin, USA
luis.martins@mccombs.utexas.edu

Maura Borrego
University of Texas at Austin, Austin, USA
maura.borrego@austin.utexas.edu

Nathan Choe
University of Texas at Austin, Austin, USA
nathanchoe@utexas.edu

Carolyn Seepersad
University of Texas at Austin, Austin, USA
ccseepersad@mail.utexas.edu

Meagan Kendall
University of Texas at El Paso, Austin, USA
mvaughan@utep.edu

Abstract *Identity, or how people choose to define themselves, is emerging as an explanation for who pursues and persists in engineering. Recent developments in the study of engineering identity, including studies of math and science identity, tend to emphasize the academic aspects of engineering without considering aspects of professional practice central to the development of an engineering identity. This paper outlines the methods used to create a new survey measure: affect toward elements of engineering practice. We followed the item generation, refinement, and instrument validation steps required for psychometric validation of a new survey measure. Through this process a final list of 34 items was administered in a survey in the fall of 2016 to engineering undergraduates. We conducted an exploratory factor analysis and established internal consistency using Cronbach's alpha on a subset of the data sample (n=384). The resulting factors reflect key elements of affect towards engineering professional practice.*

Background and Theoretical Framework

Identity is gaining in popularity as a lens for studying persistence in engineering. Much of this work is based on adaptation of identity studies in math and science (e.g., Godwin, 2016; Godwin, Potvin, Hazari, & Lock, 2013; Prybutok, Patrick, Borrego, Seepersad, & Kirisits, 2016). In math and science, identity as a predictor of choice and persistence in STEM fields is composed of recognition, performance/competence, and interest factors (Carlone & Johnson, 2007; Cass, Hazari, Cribbs, Sadler, & Sonnert, 2011; Chemers, Zurbriggen, Syed, Goza, & Bearman, 2011). While engineering has much in common with other STEM fields, engineering is unique because it is a profession defined by common practices and career paths (Downey, 2005). Prior work, which adapts math and science identity measures to engineering identity, has to date largely ignored professional practice aspects of engineering.

The goal of the current work is to develop a measure of affect toward elements of professional engineering practice. We define elements of engineering practice as the skills and tasks that are involved in engineering work. Developing a measure of affect toward elements of engineering practice is important for several reasons. Research finds that when individuals develop affect for certain skills, methods, subject matter, or tools, they are more likely to engage in educational and professional pursuits that are consistent with their attitude towards the knowledge and skills composing that practice (e.g., Lopatto, 2004; Sheppard et al., 2010). Furthermore, affect for elements of engineering practice is likely to drive identification with the engineering profession. As more advanced students are exposed to engineering through formal and informal education, there are more opportunities to study how engineering identity develops over time as students are making decisions about engaging in authentic engineering experiences such as co-ops and internships. However, little emphasis has been placed on defining and measuring professional practice and its potential connection to identity and persistence. This gap in the literature presents a unique opportunity to make an important contribution by explicitly examining how students' affect towards key elements of engineering practice predicts their engineering identities. The first step towards narrowing this gap is establishing a measure of affect towards key elements of engineering practice. In a broader project, we seek to understand how an individual's affect toward elements of engineering practice, i.e., the extent to which one likes these elements, predicts their attraction to and retention in engineering education and the engineering profession, via its effects on identification with the engineering profession.

Two aspects of research are potentially relevant to this work: studies of professional activities and identity, and frameworks for defining engineering professional practice. Work in the area of the former has established that many activities related to the professional aspects of engineering positively impact engineering-related outcomes; those activities include building things, taking things apart, programing, playing computer games, being interested in how things work, or the impact of engineering projects and work experiences on career choice and persistence (e.g., Atman et al., 2010; Pierrakos, Beam, Watson, Thompson, & Anderson, 2010). Yet few studies directly measure engineering identity. Only one prior study focused on the professional aspects of engineering identity development in undergraduate students (Meyers, Ohland, Pawley, Sillman, & Smith, 2012), and found that design, teamwork and professional responsibility were most commonly associated by undergraduates with engineering. This study was inconclusive about how such elements predicted engineering identity.

Secondly, there are few existing frameworks for professional engineering practice that could inform the development of a measurement scale. In "Towards a theoretical framework for engineering practice," Trevelyan (2014) primarily argues that engineering work is more about technical coordination than applying math and science. Williams and Figueiredo (2014) reinforce this finding by listing frequencies of specific tasks completed by practicing engineers in Portugal. A lack of an effective tool to measure affect towards professional engineering practice with the eye towards measuring engineering identity motivated the work in this study.

Methods

Instrument Development

We followed the item generation, refinement, and instrument validation steps required for psychometric validation of a new survey measure (e.g., Hinkin, 1998) to create a new survey measure of individuals' affect toward elements of engineering practice.

Step 1. Item generation. We used an inductive method to generate survey items to assess affect towards elements of engineering practice, and we compared them to deductively derived items generated in parallel by two engineering members of the research team who were blind to the inductive process, as described below. Two members of the research team who were not involved in the inductive item generation (Seepersad and Kendall) created a list of items deductively based on ABET's EC2000 Criterion 3a-k as a theoretical basis for defining elements of engineering practice. There is precedent in other fields, including medical education (Crossley & Vivekananda-Schmidt, 2009) and teaching (Cheung, 2008) for using accreditation standards as a starting point for developing professional identity scales. The ABET EC2000 outcomes for engineering (ABET, 2012) were developed collaboratively between industry and engineering stakeholders through an extended, multi-year process (Prados, Peterson, & Lattuca, 2005).

Concurrently, Borrego conducted inductive interviews with 7 young alumni (3 women and 4 men) who graduated with bachelor's or master's degrees in mechanical, civil or biomedical engineering from LPI within the last 2-5 years, and Patrick and Choe conducted focus groups with 15 undergraduate students 5 graduate students in engineering, across the three disciplines at LPI.

The interview/focus group questions were:

1. Students: Let's form a working definition of an engineer. Alumni: What is your definition of an engineer?
2. What does it mean to think like an engineer?
3. What do engineers do?
4. What are the qualities of engineers?
5. What does it take to be successful in an engineering career?

We audio recorded all interviews and focus groups, and Choe and Patrick later coded them thematically. Prior to establishing semi-structured focus groups for the engineering students, these two authors conducted a pilot of the interview protocol with a group of graduate students with backgrounds in STEM teaching. We wanted to test the process of asking questions and summarizing responses on whiteboards in conjunction with audio recording. The parallel process allowed us to listen to the recordings and see the development of the written artifact through photographs because we used different dry erase marker colors for each question. Patrick and Choe independently coded both the focus group data and the summary notes from the alumni interviews. They then met to create one combined list. Alumni responses were particularly helpful in developing specific item wording. We combined both the inductively and deductively derived item lists to generate a master list of 96 items.

Step 2. Content validity assessment. The initial master list of scale items was content validated using two methods. First, we established face validity by circulating the list of items to a set of experts. Modifications to the list were made based on their feedback. Next, we used the procedure for substantive validity assessment developed by Anderson and Gerbing (1991), whereby we administered a survey to 5 researchers with relevant expertise. Each researcher rated the item and included any written feedback on the extent to which each response assigns an item to the intended construct.

Step 3. Initial item reduction. Using the rule of thumb of an item-to-respondent ratio of 1:10, we administered the final set of 34 items derived from Step 2 in a survey to the target population along with other constructs. We conducted all data analyses using StataCorp. 2015. Stata Statistical Software: Release 14. College Station, TX: StataCorp LP. With a sub-sample, we conducted an exploratory factor analysis (EFA) to determine the sub-dimensions of our new measure of affect toward elements of engineering practice. As we expected the factors to be correlated, we used principal axis factoring, with a promax rotation to extract factors. We retained factors and items based on both theory and quantitative checks. We used a scree plot to determine the number of factors to extract in the exploratory factor analysis. However, this method is subjective, at times producing inconclusive results. Using the scree plot we identified an approximate range of factors to be extracted as 5-8. Subsequently we conducted analyses that forced the extraction of this range of factor solutions to determine the structure that best fit with the data. We examined the final factors for internal consistency using Cronbach's alpha.

Instrument Administration and Participants

The target population for this study was architectural engineering, civil engineering, mechanical engineering (ME), and biomedical engineering (BME) undergraduate majors. We administered surveys to a cross-section of engineering students at two institutions. We secured IRB (ethics) clearance at both study institutions. We refer to these institutions as LPI and HSI, respectively. LPI is a large public institution in the U.S. with high-ranking engineering programs where the students are admitted directly into specific majors (there is no general or freshman engineering program). Specific to LPI, architectural and civil engineering students are in the same department and share many required courses; for this analysis they were grouped together (collectively labeled CE). HSI offered a unique opportunity to survey students from an institution with a much more open enrollment policy and a predominantly Hispanic population (80%), with many students who commute to school daily. Unlike LPI, HSI does have a first year pre-engineering program. Therefore, HSI survey participants included students from the mechanical engineering department as well as pre-engineering students enrolled in first year mechanical engineering courses. At both institutions, freshmen, sophomores, juniors, and seniors were surveyed.

We administered the survey, which took approximately fifteen minutes to complete, in class electronically during the second week of the fall 2016 semester in a total of 22 engineering courses. Students with more than one major were retained in the analysis as long as one major was CE, ME, or BME. A total of 1465 participants consented to the survey; we only examined responses with complete data on affect towards professional practice items. The response rate was approximately 70%. The sample demographics were approximately 66% male and 34% female. Based on first semester of enrollment, 27.6% were freshmen, 27.1% were sophomores, 18.8% were juniors, and 25.8% were seniors across the two institutions. We took a random sample of n=384 for our analytical sample.

Results

It was clear during the initial coding process that the items generated in the focus groups and interviews mirrored many of the ABET criteria. Table 1 lists a sample of preliminary items and themes.

Table 1: Sample of preliminary items and themes from Step 1. Item Generation

Item	**Theme**
Innovation	Professional Skills
Networking	
Curiosity	
Efficiency	Design related activities
Planning	
Designing with constraints	
Building stuff	Attributes of Engineering
Collaboration	
Applying math and science knowledge and skills	
“Getting people to do things”	Teamwork Practices
Working with multidisciplinary teams	
Explaining technical content to “other people without engineering backgrounds”	

Exploratory Factor Analysis

We determined that the 6-factor solution best fit the data and theory. Table 1 presents the results of exploratory factor analysis (EFA) of this solution. We retained nearly all items, 30 of the 34 items hypothesized to characterize affect towards professional practice, in the final EFA. We compared all items to the minimum item communalities of 0.40 and threshold loading of 0.32 to determine future item trimming (Osborne & Costello, 2009). We removed problematic items based on these criteria. In the final analysis, "Communicating visually, for example using drawings or prototypes" continually caused severe cross-loadings of other items in the analysis; removal of this item resulted in a much cleaner factor structure. A set of 5 other items – c, d, h, n, and ee – also had potentially problematic cross-loadings (items at the minimum threshold or items with loadings less than .15 difference from the main loading). Each of these items consistently split on 2 factors in our analyses. However, we retained these items in the analysis as they strengthened the resulting overall factor structure. Three other items, "generating creative solutions to challenging problems", "knowing how to teach myself something if I have to", and "analyzing problems to identify their root causes" did not load onto any factor. We ultimately named the newly constructed factors as follows: framing and solving problems, design, project management, analysis, collaboration, and tinkering. The factor loading and internal consistency for retained factors are listed in Table 2.

Table 2: Retained Factors of Affect Towards Engineering Professional Practice

Latent Construct	**Item**	**Factor Loading**	**Unique Variance**
Framing and Solving Problems (α = 0.83)	(a) Solving problems that allow me to help a lot of people	0.58	0.66
	(b) Learning new things from other people I'm working with	0.76	0.44
	(d) Finding a better way of doing something	0.41	0.42
	(f) Continually learning new things	0.63	0.45
	(o) Using my skills and knowledge to address societal problems	0.39	0.56
	(p) Applying my science knowledge and skills	0.41	0.49
	(r) Being curious	0.42	0.54
Design (α = 0.86)	(k) Keeping up with contemporary issues involving technology	0.40	0.59
	(q) Identifying technical solutions that are as simple as possible	0.40	0.50
	(s) Designing and conducting experiments to test an idea	0.55	0.55
	(y) Searching for innovative ways to do things	0.38	0.50
	(v) Improving a design to make it more efficient (faster, better, cheaper)	0.46	0.47
	(z) Using technology to solve environmental problems	0.41	0.68
	(aa) Creating prototypes to test an idea	0.78	0.36
	(cc) Designing a system, a part/component of a system, or a process based on realistic constraints	0.68	0.45

Project Management (α = 0.72)	(l) Planning a project and staying organized to complete it	0.64	0.51
	(m) Using facts and information, instead of opinions, to make decisions	0.45	0.55
	(x) Tracking various aspects of a project to ensure that it stays on track	0.36	0.73
	(bb) Seeing a project through to its end	0.45	0.54
Analysis (α = 0.76)	(u) Applying my math knowledge and skills	0.56	0.47
	(gg) Using calculations and equations to evaluate things	0.68	0.42
	(hh) Identifying what I need to know to solve a problem or complete a project	0.50	0.46
Collaboration (α = 0.79)	(e) Presenting my work to others	0.46	0.55
	(n) Working with people with different skills and interests	0.40	0.54
	(t) Communicating verbally, for example in discussion with others	0.60	0.45
	(dd) Convincing others to accept my ideas	0.42	0.68
	(ee) Breaking a complicated problem into smaller parts	0.35	0.42
	(ff) Working collaboratively in teams	0.64	0.49
Tinkering (α = 0.75)	(c) Taking something apart to see how it works	0.50	0.41
	(h) Fixing things	0.57	0.49

To examine our assumption of the correlation between factors, we calculated Pearson's correlations between each of the retained factors. This correlation matrix reflects significant ($p \leq 0.001$) moderate to large relationships across nearly all the factors. The weakest correlations were between tinkering and project management (0.23) and tinkering and collaboration (0.29). Design shared correlations of 0.60, 0.63, 0.68 between tinkering, analysis, and framing and solving problems. These high correlations reflect the problems we saw in cross-loadings in the EFA. All other correlations ranged from 0.40 to 0.58.

Discussion

Facilitating the focus groups using white boards as well as recordings proved to be an effective means to conduct the interviews. While most focus group/interview protocols only allow the researcher to see the artifacts of the discussion, our methodology allowed the focus group participants to see artifacts of the discussion as they emerged. This allowed for immediate member checking between the participants and the researchers. Oftentimes students were able to clarify the meaning of their responses in real time and make connections between points of discussion. This technique often drew out more detailed information from the participants, particularly for the focus groups of 5 or more participants.

The goal of this study was to generate a new scale for measuring affect towards engineering professional practice. While not all items were ultimately retained from the item generation process in the factor analysis, the subsequent validity steps yielded a robust set of factors. Careful examination of the factors at the item level reveals characteristics of each latent variable. Surprisingly the factors did not align with either our *a priori* assumptions or with the ABET criteria. Clearly, ABET criteria influence our thinking about engineering. Yet the resulting factor structure reflects a more nuanced interpretation of the ABET criteria based on students' attitudes. While some criteria (e.g., ethics) were not retained in the final set, the overall structure suggests yet another way to consider the revision of these criteria currently underway.

Despite a few cross-loadings, mainly due to ambiguity in the wording of the items, the results of this study are very promising. The key elements that compose affect towards engineering practice – framing and solving problems, design, project management, analysis, collaboration, and tinkering – collectively captured the practices we intended to measure. Tinkering is perhaps the most intriguing. "Taking something apart to see how it works" and "fixing things" are often questions used in the studies of engineering identity. While this was only a two-item factor in our study, it would be worth adding items to the factor to see if tinkering is a distinct professional practice or simply a common activity among those interested in engineering as a whole.

The most cohesive factor in our results was collaboration. Students in this study seem to associate teamwork with communication skills. This is not surprising as teamwork and communication skills are often emphasized together in capstone design, and successful teamwork often entails effective communication. Another notable result was the separation of "applying math knowledge" versus "applying science knowledge" into two factors. Math loaded on the analysis factor whereas science loaded on framing and solving problems. Perhaps the emphasis on a prescribed method in schooling led students to associate science knowledge with framing and solving problems, and math knowledge with analysis. The most indistinguishable factors during the item generation process were problem solving (renamed framing and solving problems) and design. At several points we considered collapsing these items on a theoretical basis. Criterion I, "a recognition of the need for, and an ability to engage in life-long learning," spanned the two factors: framing and solving problems and design. A set of ABET criteria: a. "an ability to design and conduct experiments, as well as to analyze and interpret data," c. "an ability to design a system, component, or process to meet desired needs within realistic constraints such as economic, environmental, social, political, ethical, health and safety, manufacturability, and sustainability" and e. "an ability to identify, formulate, and solve engineering problems" also spanned these two factors. Of this set of criteria, c. is arguably the densest in terms of content. Critical to engineering professional practice, both these factors accounted for nearly half of all the items included in our scale. Clearly the emphasis on these practices has not gone unnoticed among undergraduate students.

To be clear, this study was measuring affect towards engineering professional practice, not professional identity or engineering identity. Affect can best be thought as an affinity towards or liking, whereas identity is seeing oneself as or being recognized as a particular type of person. There are similarities between affect and identity, however it was necessary to distinguish this two constructs to investigate the difference between liking a particular practice versus being a particular type of person.

Conclusion

In sum, this work illustrates a new dimension of measuring attitudes related to professional practices in engineering. The factors derived from this study are intended to be a first step in creating a robust scale of affect towards engineering professional practice. We have described the step-by-step development of items based on qualitative studies, theory, and criteria of the discipline. These steps are essential to the validation process discussed in this paper. By first gaining an understanding of key elements of engineering practice, we can then make more refined changes to our measures of affect towards engineering professional practice, identity, and other engineering related outcomes such as persistence. Richer descriptions and measures of student experience are needed as we as engineering educators seek to further illuminate a path from affect towards engineering professional practice to full participation in the engineering profession and community.

Implications and Future Work

This study explored the key elements of affect towards the engineering profession. In future work, using the remaining data from this sample, we will conduct a confirmatory factor analysis on the reduced set of items resulting from EFA, and assess the final factor structure using various goodness of fit indices. As part of a larger longitudinal study, we will continue to survey students from over 100 high schools, plus university freshmen, sophomores, juniors, seniors, master's and doctoral students. The broader goal of that study is to model engineering identity and persistence based on these and other newly constructed factors from our research. This work has important implications for prospective engineering students, undergraduate students, graduate students, and professionals in engineering settings. In conjunction with qualitative data from a purposeful sample of alumni participants, which provides a richer description of attitudes and experiences, we can further characterize what factors contribute to students' decisions to major and persist in engineering. Addressing the professional aspects of engineering is essential to furthering the work on engineering identity.

Acknowledgements

This research was funded by the U.S. National Science Foundation through grant numbers 1636449 and 1636404. The views presented are those of the authors and not necessarily those of the NSF. The authors wish to thank Catherine Riegle-Crumb for her partnership on this project as well as the student participants, instructors, department chairs and other department liaisons for assisting with data collection.

Copyright statement

References

ABET. (2012). *Accreditation Policy and Procedure Manual (2012-2013)*. Retrieved from Baltimore, MD:

Anderson, J. C., & Gerbing, D. W. (1991). Predicting the performance of measures in a confirmatory factor analysis with a pretest assessment of their substantive validities. *Journal of Applied Psychology, 76*(5), 732.

Atman, C. J., Sheppard, S. D., Turns, J., Adams, R. S., Fleming, L. F., Stevens, R., . . . Lund, D. (2010). *Enabling engineering student success: The final report for the Center for the Advancement of Engineering Education*. San Rafael, CA: Morgan & Claypool Publishers.

Carlone, H. B., & Johnson, A. (2007). Understanding the science experiences of successful women of color: Science identity as an analytic lens. *Journal of Research in Science Teaching, 44*(8), 1187-1218.

Cass, C. A. P., Hazari, Z., Cribbs, J., Sadler, P. M., & Sonnert, G. (2011). Examining the Impact of Mathematics Identity on the Choice of Engineering Careers for Male and Female Students. *2011 Frontiers in Education Conference (Fie)*.

Chemers, M. M., Zurbriggen, E. L., Syed, M., Goza, B. K., & Bearman, S. (2011). The role of efficacy and identity in science career commitment among underrepresented minority students. *Journal of Social Issues, 67*(3), 469-491.

Cheung, H. Y. (2008). Measuring the professional identity of Hong Kong in-service teachers. *Journal of In-service Education, 34*(3), 375-390.

Crossley, J., & Vivekananda-Schmidt, P. (2009). The development and evaluation of a Professional Self Identity Questionnaire to measure evolving professional self-identity in health and social care students. *Medical teacher, 31*(12), e603-e607.

Downey, G. (2005). Are engineers losing control of technology?: From 'problem solving'to 'problem definition and solution'in engineering education. *Chemical Engineering Research and Design, 83*(6), 583-595.

Godwin, A. (2016). *The Development of a Measure of Engineering Identity*. Paper presented at the American Society for Engineering Education Annual Conference, New Orleans, LA.

Godwin, A., Potvin, G., Hazari, Z., & Lock, R. (2013). *Understanding engineering identity through structural equation modeling.* Paper presented at the Frontiers in Education Conference.

Hinkin, T. R. (1998). A brief tutorial on the development of measures for use in survey questionnaires. *Organizational research methods, 1*(1), 104-121.

Lopatto, D. (2004). Survey of undergraduate research experiences (SURE): First findings. *Cell biology education, 3*(4), 270-277.

Meyers, K. L., Ohland, M. W., Pawley, A. L., Sillman, S. E., & Smith, K. A. (2012). Factors relating to engineering identity. *Global Journal of Engineering Education, 14*(1), 119-131.

Osborne, J. W., & Costello, A. B. (2009). Best practices in exploratory factor analysis: Four recommendations for getting the most from your analysis. *Pan-Pacific Management Review, 12*(2), 131-146.

Pierrakos, O., Beam, T., Watson, H., Thompson, E., & Anderson, R. (2010). *Gender differences in freshman engineering students' identification with engineering.* Paper presented at the Frontiers in Education Conference (FIE), 2010 IEEE.

Prados, J. W., Peterson, G. D., & Lattuca, L. R. (2005). Quality assurance of engineering education through accreditation: The impact of Engineering Criteria 2000 and Its global influence. *94*(1), 165-184.

Prybutok, A., Patrick, A., Borrego, M., Seepersad, C. C., & Kirisits, M. J. (2016). *Cross-sectional Survey Study of Undergraduate Engineering Identity*. Paper presented at the American Society for Engineering Education Annual Conference, New Orleans, LA.

Sheppard, S., Gilmartin, S., Chen, H. L., Donaldson, K., Lichtenstein, G., Eris, O., . . . Toye, G. (2010). Exploring the Engineering Student Experience: Findings from the Academic Pathways of People Learning Engineering Survey (APPLES). TR-10-01. *Center for the Advancement of Engineering Education (NJ1)*.

Trevelyan, J. (2014). Towards a theoretical framework for engineering practice. In B. Williams, J. Figueiredo, & J. Trevelyan (Eds.), *Engineering practice in a global context: understanding the technical and the social* (pp. 33-60).

Williams, B., & Figueiredo, J. (2014). Finding workable solutions: Portuguese engineering experience In B. Williams, J. Figueiredo, & J. Trevelyan (Eds.), *Engineering practice in a global context: understanding the technical and the social* (pp. 159-184).

LEARNING AND BECOMING TRANSDISCIPLINARY IN MULTIPLE CONTEXTS: THE CURRENT CHALLENGE FOR ENGINEERING EDUCATION (A CASE STUDY)

Hernán Lopéz Garay
Universidad de Ibagué, Ibagué, Colombia
hernan.lopez@unibague.edu.co

Alfonso Reyes Alvarado
Universidad de los Andes, Bogotá, Colombia
areyes@uniandes.edu.co

***Abstract**: Engineers are confronted with an ever-increasingly complex world. Often, they have to face ill-structured socio-technical problem-situations that require cross and trans-disciplinary approaches, a description from multiple points of view of the situation and serious considerations about the interactions between different but relevant stakeholders involved. To cope with such complexity engineers need to learn a new language that allows them not only to think beyond disciplinary boundaries but also design and implement artefacts and systems able to deal with the variety required to cope with a complex world. We suggest that new language is the systems language. This paper describes the rudiments of a model to teach-learn the systems language to freshmen at the Universidad de Ibagué. We will present some of the theoretical grounds of the model and report the main outcomes of this experience conducted in one introductory course to systems thinking for the past three years. We will conclude with some final reflections for the particular case of research on engineering education.*

INTRODUCTION

We live in a complex, highly interdependent globalized world. In order to educate engineers to perform well in such a complex setting, new models of education and ways of organizing the learning process need to be researched and developed. In this direction, over the last three years an action-research program is underway at Universidad de Ibagué, Colombia. The program is focused on the teaching of a systems-thinking course to first year students (not only to engineering students). The original model of teaching-learning put on scene in this course was primarily based on constructivist and phenomenological theories of education and project-based learning. The model has evolved through action research. The continuous observations of the performance of this model in practice are organized using the principles of qualitative research applied to the construction of grounded theories (Glaser and Strauss, 1967). In the following we will present the main outcomes of this evolving grounded-model and conclude with some final reflections for the particular case of research on engineering education.

1. THE INCREASINGLY COMPLEX WORLD OF THE PRESENT

In 1972 Donella Meadows, Dennis Meadows, Jtsrgen Randers, and William W. Behrens II published their well-known systems study on the *Limits to Growth*. It was the first, large scale, systems study of the global economy that began to reveal the complex global interrelations between population, resources, the environment, patterns of consumption, world politics, distribution of wealth and economic growth. The study raised awareness over the future sustainability of the way of life the West was pursuing. Forty-four years later we are approaching (and perhaps irreversibly surpassing) not merely the limits to growth but of life as a whole in this planet. This is creating a very turbulent and complex world, which requires cross and trans-disciplinary approaches. To cope with such complexity engineers and other professions need a new way of thinking and acting. Equivalently, they need to learn *a new language* that allows them not only to think beyond disciplinary boundaries but also design and implement complex systems to deal with the variety required to adapt to a complex world.

2. THE SYSTEMS *LANGUAGE*

The systems movement has long been concerned with an ever-increasingly complex world. Bertalanffy (1976), one of the founding fathers of this movement, argued in favour of a *general systems language* meant to bridge the gaps between different disciplines. He thought that by providing scientists with a common language we could help to stop the increasing disciplinary fragmentation of science and retake the ideal of developing a unified view of the world. The idea of a systems language was for him something more than metaphorical, because to be able to describe non-linear phenomena a language that allows a description of "circular" causality relationships was required, since a linear language impeded us to focus on circular relationships. In fact the language of science is linear (like most Western languages) because it naturally describes a reductionist view of the world as Bertalanffy stated:

> "It is necessary to study not only parts and processes in isolation, but also to solve the decisive problems found in organization and order unifying them, [problems] resulting from dynamic interaction of parts, [which makes] the behaviour of the parts different when studied in isolation or within the whole." (Von Bertalanffy 1945 p. 49-50).

In sum, Bertalanffy helped us to see systems thinking as a language useful for communicating about complex relationships and interdependencies (for instance negative or positive feedbacks) needed to describe and understand a complex world.

3. TEACHING SYSTEMS THINKING AT THE FRESHMAN LEVEL: AN ACTION-RESEARCH PROGRAM AT UNIBAGUE

Several engineering educators (Craig, 2016; Foster 1999, 2001; Godfrey, et.al 2014; Jain, et.al. 2008) consider engineering education is in dire need of exposing engineering students to a systemic view of the world and train them to become more inter and transdisciplinary in action:

> "Society's problems are getting harder, broader, and deeper and are multidisciplinary in nature. They require a multidisciplinary systems approach to solve them and present-day engineering education is not adequately preparing young engineers for the challenge." (Craig, 2016, p.1).

Accordingly, in order to meet this challenge new models of education and ways of organizing the learning process need to be researched and developed, which is a challenge for higher education in general (Escrigas, 2016). In keeping with this need, a pilot study to teach systems thinking was implemented at the University of Ibagué in the year 2014. To test and evolve some basic ideas of systems thinking and acting in education in engineering in particular and systems education in general the study was conceived as an action-research project (see figure 1. Based on Checkland, 1999),

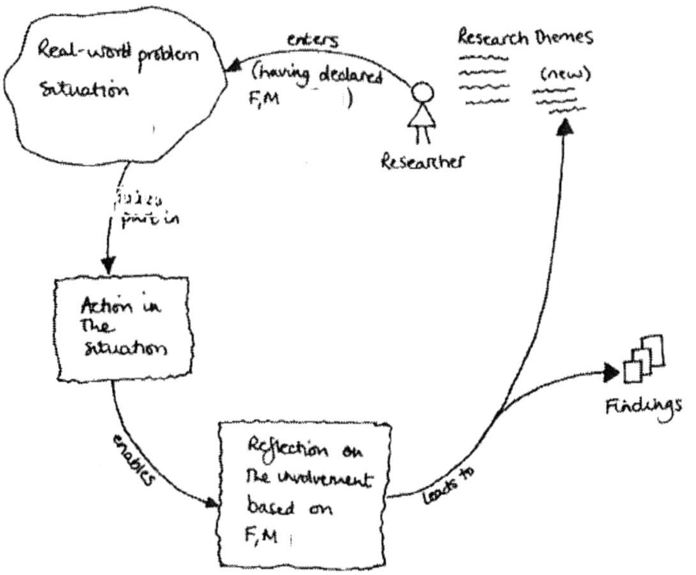

F= Framework of ideas
M= Methodology

Figure 1. Action Research methodology

The pilot study was designed to teach systems thinking to first-year students of all careers, not only engineering (López-Garay, 2013). The idea came out from the realization that often, students of main professions have to face ill-structured socio-technical problem-situations that require cross and trans-disciplinary approaches, and to understand multiple interacting points of view relevant to the situation.

3.1. The initial Framework (F) of ideas and Methodology (M)

Let us see some of the systemic ideas (based on Banathy, 1996) which constituted the initial framework (F) guiding the pilot study. To begin with, learning institutions were seen as instructional systems that have a purpose, a process (i.e. the operations and functions needed to accomplish the purpose) and a content (i.e. the parts that comprise the system). The purpose of an instructional system is to create and maintain a dynamic learning environment. In such an environment, the learner would be on the stage, rather than the teacher. The latter becomes more of a guide on the side than (like the old saying goes) "a sage on the stage". To be more precise, the classroom is considered a stage where different "stage productions" (learning environments) must be easy to assemble and easy to dismantled. This is so because the dynamics of the class is such that at any moment the "classroom" may need to be turned into a design studio. At another moment the teacher needs the classroom to become a debate stage. And at another moment it is the institution as a whole which needs to become a learning stage, a place where the learners and the teacher-guide develop instructional goals together, research together, and guide each other in the process of constituting a dynamic learning community. Hence, whereas in the classic model of education the learner plays basically a reactive role ---in response to the teacher's active role--- and whereas the particular motives of the learner seldom if ever are accepted or encouraged, in the instructional system developed here the learner plays an active role, participating in the design of the learning experiences and the organization of the instructional system as a whole.

As for (M) (see figure 1) we started with a basic model of learning (see figures 2 and 3), based on constructivist theories of education (Vygotsky 1978) and through an action research methodology (figure 1) this model has evolved into a language-game model (to be explained in further sections). Certainly, our pilot study did not start from scratch. There has been much research and knowledge accumulated on this matter (Ausubel (1968,1978), Goleman and Senge (2014), Kolb (1984), Kolb and Fry (1975), Novak (2011). Schön (1983), Vygotsky (1978)). However, this is the type of knowledge that can only be meaningful assimilated once one "makes it alive" by rediscovering it and evolving it according to one´s own situations and idiosyncrasies. Thus, although we may know, theoretically speaking, that learning is not something that takes place in a classroom per se but in a learning situation or that the student-learner is not a *tabula rasa* but a human being with a history behind, in front, and ahead, nevertheless the real meaning of these notions only become "real" in the actual learning situation. Such is the case of a central notion in constructivist theories, namely the "context of experience" of the learner (Dewey, 1963), which for the purpose of this initial model was reduced to the simplistic idea of a network of intuitive concepts and relationships. About the vital importance of the context of experience of the learner, Ausubel (1968, p.vi) says:

> If I had to reduce all of educational psychology to just one principle, I would say this: The most important single factor influencing learning is what the learner already knows. Ascertain this and teach him accordingly.

Therefore the process of learning consists of the construction/development and assimilation of a new concept into the pre-existing knowledge network. The construction is anchored in a network whose nodes represent intuitive notions from which the learner can begin to make sense of the new knowledge being learnt (i.e. assimilated into the network). If there are no anchor points in the experience of the learner, then we say there is a "hole" in the network, i.e., there is no clear point to anchor the process of construction of knowledge. In such a case, the learning process will be impeded until we build bridges to connect the hole with nearby nodes in the network that can play the role of anchor points. Processes of induction and abduction build bridges that go from the intuitive notion to the more abstract concept target of the learning process. For example, teaching the concept of *percentage* may start from the anchor intuitive notion of portion. Then, it may continue from *portion* to the abstract notion of *part of a whole*…etc.

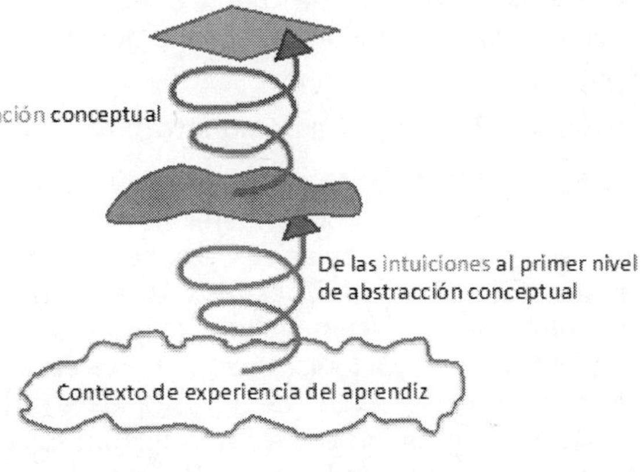

Figure 2. Inducting/Abducting from the context of experience

But learning is not completed till the concept (discovered/constructed) is really assimilated into the learner´s context of experience.

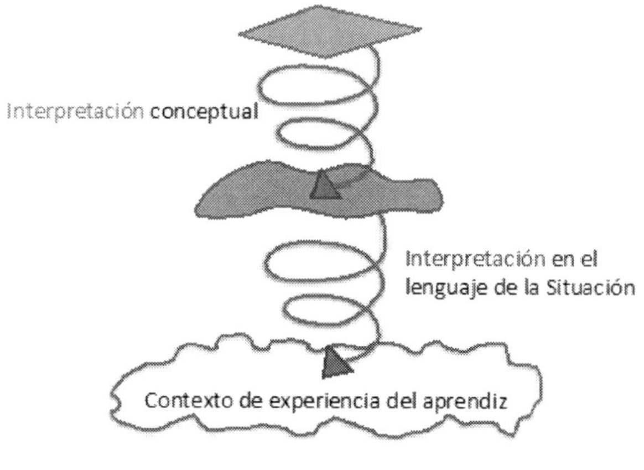

Figure 3. Assimilation into the context of experience

3.2. PBL as the "engine" of the learning process and the guidance of the learner

Now, the question arises as to how to ignite and guide the learning process starting from the context of experience. Project-based learning (Buck Institute for Education, 2015) is one possible answer. It fits well within the constructivist conceptual framework. In fact, as we all know, the learner responds well to an authentic, engaging and complex question, problem, or challenge which is strongly related to his/her own interests and context of experience. Therefore, in the PBL methodology the project must be the learner´s own choice. As a result the learner will be eager to find the knowledge or acquire the skills and tools needed to see his/her objectives reached. In this context, the role of the teacher is to be a guide but also a learner model i.e. a model to others of how one can learn in a project situation (Schön, 1983). Guidance, therefore, is a particularly subtle process. Just as in sailing a boat the sailor must be very attentive and wait for the "right" direction of the wind to show, similarly, the teaching of a concept or practice must wait for the right time to arrive, within the evolution of the learner´s project.

3.3. Teaching systems thinking: a basic constructivist model

To teach the rudiments of systems thinking to freshmen at the University of Ibagué, using project-based learning, and the above-mentioned principles, we completed our initial educational model defining the particular view we have of what is called *systems thinking*. In this task we used the book *Disolver Problemas*, written by one of the authors of this paper (Aldana & Reyes, 2004), as our basic systems thinking text. The idea was to teach a basic set of systems concepts and visual tools (based on *Disolver Problemas*) that would allow the students to conduct the process of structuring complex socio-technological situations proper of the complex world of the present. Armed with this framework of ideas and methodologies (F and M. See figure 1), we launched our first course. Since some of the students that took this course came from the surrounding communities, we selected one of them, Gaviota, as the place to carry out a systems project. This community had a history of community action to better their quality of life, particularly in their relationship with a mining operation close to it. The idea was to anchor their learning in those issues they were aware of. In consequence, the students began to visit the community and elaborate a social map, guided by a teaching team composed of three teachers (two of them were engaged in a two-year training program to become teachers-researchers of systems thinking) and two teacher assistants to support the teachers in the logistics and various other chores related to the course´s projects. Emphasis was made in the type of issues one currently has to face when dealing with problematic situations in social systems. They are situations where the problem is precisely to define the problem. The course was organized into three phases. In the first phase the learner learnt to recognize the systemic nature of non-structured problem situations: multiple and interacting points of view constituting the problem situation, visualizing circular-causality relationships between different variables, and identifying changing, fuzzy, non-intentional outcomes. In the second, they learnt to structure the situation identifying relevant causes and points of leverage for changing the system. In the final phase they learnt to develop strategies to *dissolve* rather than solve the problem situation. By dissolving a situation we meant to reframe it, that is, to develop a new way of looking at the situation which gives it a very different meaning to those aspects that allowed the original problem situation to emerge. And the systems concepts were introduced at each of these stages, according to need.

4. EVOLVING THE FIRST MODEL

The next step in our research emerged as we examined the learning difficulties the students experimented with the project-community-based learning approach. If we recall, they were freshmen students, which means they were experiencing a radical change in their lives: changing high school life for university life. There are many things to say about how this context of experience of the students affects the learning process. But for the purposes of this paper we will focus only on one idea that came out of a comment made by a student in relation to how he experienced the course: "It was like trying to learn a foreign language."

4.1. Teaching Systems Thinking as a *Foreign Language*

The student´s point of view was very interesting to explore: we were not simply teaching a set of systems concepts and tools useful to frame and reframe (dissolve) problem situations. We were teaching a *language* useful for communicating about complexities and interdependencies. Our initial literature survey on the matter revealed that the idea of learning systems thinking as a foreign language had been also proposed by Gittings (2017) and Goodman (2010).

In this regard, Goodman says:

> *Learning systems thinking can be likened to mastering a foreign language*. First one memorizes the essential *vocabulary words and verb conjugations*. Then one starts to put the pieces together into *simple sentences*. In the language of systems thinking, system diagrams such as causal loops can be thought as sentences constructed by linking together key variables and indicating the causal relationships between them. By stringing together several loops, we can create *a paragraph* that tells a coherent story about a particular problem under study. If there were a Berlitz guide to systems thinking, archetypes such as...shifting the burden [or the tragedy of the commons] would be listed as "commonly used phrases". (p. 11. Emphasis added).

As a matter of fact, according to conventional approaches to teaching a second language, to master a foreign language one needs to master not only the vocabulary, the grammar, and the syntax but also the phrases and idioms proper of that language. Accordingly, we began to plan some of our courses keeping in mind three aspects: First, to identify the basic language of the learners. This was part of the task of identifying the context of experience of the learner (the network of basic intuitive notions and concepts about a given subject), a task that we were already carrying out in the learning model introduced in the previous sections. Recall that in the first model we represented the context of experience by means of a network of intuitive concepts (something we organized using the tool of concept *maps*, Novak, 2011, pp.3-4).The idea now was to relate the concept maps and language to structure the basic language of the learners, adapting the method used by (Marriott and Torres, 2016) to teach English as a second language. The second aspect was to structure the basic concepts and tools associated with systems thinking also in the format of a language, a systems language. It is worth to notice before we go on, that to do this structuring we got inspiration among others from Kim and Anderson (2011) and their systematization of the notion of archetypes. We got also guidance from Goleman and Senge (2014) with their ample experience teaching the systems language at High School levels. In this latter case the emotional intelligence is taken into consideration. Accordingly, we defined the systems "words", phrases and idioms, using as the main source the previously cited book *Disolver Problemas* and a pocket guide vocabulary (Lannon, 1991). The language thus structured comprises a set of words, diagrams, and archetypes to speak "systemically" about two basic situations: How to structure problematic situations and how to dissolve them. The third aspect we considered in our planning was related to a main obstacle for learning a foreign language, namely the unrestrained tendency of the learner to transfer phrases and idioms from the "native" (the basic language in our case) to the "foreign language" (the systems language in our case). What we planned to overcome this problem was to identify situations ---in the context of experience of the learner--- similar to the ones a systems thinker finds herself in when tackling complex socio-technical problems. Next we rephrased the learner´s very same situations in the language of systems and invited/challenged the learner to compare and find for herself if there was any advantage in using such a language in their personal context. For instance, we asked the students to give advice to an ill friend, who was also having a hard time in his personal life. In this class exercise we managed to show that the difference between tackling systemically the situation or not was very important: it was the difference between life and death! This exercise was also carried out at the global level, and we managed to show that non-systemic decisions ---even with the best of intentions--- about how to deal with our natural resources were destroying the planet.

In our effort to make the learners more proficient in the use of the "systems" language, we began to realize that we had to create learning situations more akin to those a native "systems" speaker has to face in real life. At this point, two things were important to consider. One was that the use of the systems language is always in context. Words are words-in-use. Which means that the learner must become familiar with the "form of life" proper of the community whose language he/she is learning. By doing so, the learner "learns to play the (language-) games" of that community, as it were. The other thing, derived from the first, is that those guiding the learning process must expose the learner, ideally, to real systems communities. Short of that, the metaphor of flight simulators came to mind. In the next two sections we will describe how these two requirements were met. We will see that the "solution" we found was not that of the flight simulator (although we tried that) but students and teachers embarking together in the creation of a systems community. *The idea was that in the process of building and operating such a community they would learn the language-games of systems thinking and systems practice.*

4.2. Teaching Systems-Thinking as *Language-Games*

We found Wittgenstein (1953) concept of *language-game* useful to think about the educational problem previously stated. Let us see why.

Wittgenstein introduces the concept of 'language games' because of the analogy between using a language and playing a game according to certain rules. Wittgenstein idea was that our whole use of language was similar to game playing. We take part in several language-games, which occur and unfold within a given social context, in a given situation or "form of life" which gives language its meaning: "...speaking a language is part of an activity, or a form of life, " (p. 23). Thus, saying something in a language is analogous to making a move in a language-game, which forms part and defines a form of life. Some language-games may be highly specialized (like for instance the language-game of the systems community), some very common to all (like for instance the formal greeting of an old friend in the street). In all these cases the important thing to notice is the unity formed by language and the actions in which language is interwoven (p.7). Notice also the basic structure of the game and what is expected of each gamer to say/play: In every case we act *as if* we were following certain rules of a game (e.g. the "game" of seduction, the "game" of making friends) although we may not be conscious of it. When we enter a known situation we know what moves to make, that is, what sort of statements are proper to say and respond to in the context of the "game" being played (which is proper of that situation). However, if the situation is not familiar, we are forced to make sense of it and hence learn the rules of the game that go with that "unfamiliar situation". In this connection language-games may be seen as *semiotic practices*: "...semiotic practices, that is, the socially shared ways of using signs, of signifying and of representing... [like for instance in the "game" of] translating from one language into another; constructing an object from a description in a drawing, speculating about an event, reporting about an event, forming and testing a hypothesis... (Xanthos, 2017, p.1). Another example is the interpretation of legislative acts as "...a rule-guided way of attributing meaning. This is the particular language game proper of the world of legislative acts. On the other hand, a particular interpretation of a particular law, composed of a defined set of arguments, would be a series of *moves in the language game* of interpreting legislative acts. Their interpretation as laws presupposes the concepts of rights, duties, obligation, possibility, responsibility, action, etc., which compose *the grammar* of this particular language game. (Ibid. p.1)

The important point here is that one cannot learn a language unless one is exposed to its semiotic practices, and more precisely to what (as mentioned before) Wittgenstein calls "form of life", which is the cultural or community environment in which the language game occurs. In turn, the semiotic practices help to constitute the form of life.

Which brings us back to our systems thinking course and the learning situation seen as *learning systems thinking as a foreign language*. To sum up: Learning a "foreign" language means, then, to learn the *language games, the moves and the grammar* which constitute the *form of life* of a community, be it the German community or, similarly, the community of systems thinkers. In consequence, the design of our course turned to the design of learning environments that allowed the recreation of practices appropriate to the form of life of the community whose language we want to learn. But rather than simulating in class a systems community with special emphasis on its semiotic practices, *we decided to actually build one*. This idea looked interesting because we realized that several themes we discussed above could be tight together. Where else the learner could have a first hand experience of the semiotic practices proper of a systems community than in his/her own systems community? But of course the redundancy of this statement claimed to be cleared: Are we saying that we will learn to speak/play "the systems language-game" by speaking/playing the systems language-game? In keeping with our project-based learning teaching methodology, this was to be the class project.

5. LEARNING TO SPEAK "SYSTEMS" BY BUILDING SYSTEMS THINKING COMMUNITIES

What is a systems-thinking community? It is a community where "systems" is the main "spoken language". And we know that a language is interwoven with semiotic practices. An example is the Society for General Systems Research founded by Bertalanffy and others in 1954. Reading their journals throughout the years, one can see how the systems language-and-practices gradually evolved and constituted that community of theoreticians and practitioners of systems thinking. Hence one option to teach the systems language was to simulate in the classroom a systems community. The idea was further worked out and we decided rather to undertake the class project of building (not simulating) a systems community. The pedagogical principle behind this project was "learning by doing". For instance, in the process of building a small English community I learn to speak English. Analogously, in the process of building *a systems community* one learns to speak the systems language (that is the language-games of the systems community). The logistics of this community-building process were as follows.

Given that every freshman course always starts with a class formed by "social atoms" (since the majority of the students do not know each other), the idea is to take these social atoms and gradually transform them into social molecules, complexes, and finally communities and in so doing develop a learning systems community.

5.1.From "social atoms" to "social molecules"

As mentioned before, we started with a class formed by "social atoms". Our first pedagogical challenge was to get their transformation from "social atoms" to "social molecules". We began by unfolding their contexts of experience, next identifying common interests, and finally inviting them to form clubs. The clubs covered a wide range of interests. For instance, this semester the students formed clubs of anime, gamers, otakus, cars-racing lovers, foreign language clubs, literature, sports, musicians, painters, social networks, feminists, ecologists, animal lovers, entrepreneurs, agnostics, destitute caretakers, "graffiters". When deep-probe their interests most of these youngsters revealed to have ideals of a different type of society to the current one. But due to the lack of real opportunities in our society to let them participate and propel a significant change, students opt in real life for surrogates where their creativity and desire to change society could find a place. To help clubs develop, they were asked to plan a series of activities aimed at gaining new adepts and let the university community know of their existence and ideals. After a presentation of their plans, we asked the clubs to see whether they could form alliances with other clubs (for instance the car-racing lovers with the artists) in order to enhance their mutual development and learning.

5.2. From "social molecules" to Complexes

After a month, clubs (social molecules) organized themselves into complexes (systems of two or three clubs) interacting between them in different ways to reinforce their common interests. The classroom became the learning space where these complexes emerged and evolved. Although the initial thrust of this activity was to seek their mutual development, we wanted them to go one step further. Can we think of the fusion of clubs to create a new complex, like the "fusion" of 2 atoms of hydrogen and one of oxygen creates a totally different emergent substance called water? Each couple or triad of clubs were asked to develop a motto which synthetized the common higher purpose or long-term objective to follow. Let us explain.

5.3. From Complexes (systems) to the formation of new communities

If we recall, at the beginning of the course we discovered that behind every club proposal lurked the ideals of a new society. For instance, members of the otakus club manifested their attraction to certain aspects of the Japanese culture. They admired in the otakus a mixture of some traditional Japanese values, an exquisite sense of beauty, their discipline together with a new and emergent counterculture as depicted in Anime and Manga cartoons. They said they would like to live in a society where space for social innovation was provided, and new ways of life could be freely explored (for instance like the otakus did it in Japan). Lack of such spaces created much frustration in a society, restrained the development of social innovation skills and forced millions of people to live one and only one form of life. The otakus thought they could form complexes with the gamers, the artists, and the foreign language clubs. They could invent a new way of living in society. Inspired by these comments, the teacher's team proposed the classroom to become a laboratory for social innovation and the formation of new types of communities. The next class they brought a video where they enthusiastically gave us a glimpse of how this ideal society could be. Other groups did the same.

5.4. Fertilizing the seeds of systems thinking at each stage

At every stage of this community-building process the teachers-team guide the process and challenge the students with different tasks they have to undertake in order to consolidate their clubs or their alliances or fusions into communities. As the clubs develop these tasks in their own terms, the teachers-team invited them to re-consider in systemic terms what they were doing, thus fertilizing systems thinking and system practices in these groups. Three examples illustrate the point.

First: "Fertilizing" the systemic seeds of Different Perspectives and Interpretive Variety

Take for instance the foreign languages club:

> "...we can say that speaking more than one language opens us up to entire systems of thinking about every aspect of life. When we learn a language well, we discover a whole system of thought, with *ways of seeing the human condition that are often new to us. The process opens our minds to the reality that there are a multiplicity of points of view, a variety of ways to conceptualize the world, many forms of self-expression, and a wide diversity of cultural filters*. (Sadowsky, 2017, p.2. Emphasis added).

Since part of systems thinking is learning to see the world from different perspectives, then it is clear how one could think of different ways of "fertilizing" this seed in the foreign language club. Minati and Collen (1997) inspired us with different classroom activities to engage in systems thinking.

Second: "Fertilizing" the seeds of Synthetic Thinking and multi-causality

In class we challenged the clubs to think of their role in society and how they could contribute to its well-being and at the same time further their own ends. In this particular case we managed to show them that some of their self-centred strategies may not, at the end, served well their own ends. That is to say, their non-systemic strategies could generate counter-intuitive outcomes because they did not consider how their system of activities made part of larger systems or because they disregarded the multi-causal relationships in the particular situation they were involved.

Third: On emergence

When we asked the clubs to form bonds with other clubs, we told them there was a difference between forming a strategic alliance with other clubs and the creation of a new entity altogether. The second case involved the design of a system whose parts were the clubs and the relationships between them were of a particular kind, precisely those that generate a systems property none of its parts has.

5.5. The emergence of a systems community

Each system of clubs began to develop their own project to pursue their own objectives but they were invited to think-systemically their actions: How do they fit within the larger social system of which they are a part? The system formed by the foreign language and sports clubs imagine themselves developing English reading courses for the community surrounding the university campus. A first step in this direction was to develop an English reading methodology for freshman in the University of Ibagué. To do so the group was free to organize and approach the design of the methodology as they saw fit best their needs. However, at each step the teaching-team oversaw the process and, as we said before, fertilized the systemic seeds of their project. In this regard Design Thinking was handy to articulate the design process of the group. The outcome was the prototype of an original methodology which combined body and gaming in the learning process. Other clubs also adopted this methodology in their projects. And they shared their experiences in class using this and other systems thinking methods. *And in so doing, they began to adopt systems practices and "speak" the systems language. The entire class became a systems community in the making.* The rubric to observe this process was based on Senge´s (1999) five disciplines. The advantage of doing so was that we could observe not only if a learning community is emerging but also if a systems community is in the making.

6. SOME REFLECTIONS CONCERNING ENGINEERING EDUCATION AT UNIBAGUE

Although the partial results of the case study presented here pertain to a freshmen course open to students of all careers and not only to engineering, nonetheless in the last three years we have had various courses mainly with engineering students. Their learning behaviour and performance in the course have not been radically different from that of students enrolled in other careers. However, it would be interesting to compare strategies to teach systems thinking courses in other universities of Colombia and other countries to identify some particular traits that would differentiate Engineering freshmen students from others. Certainly, the comparison has to be made with the first model of learning (see Section 3.), since as far as our literature review is concerned, there are little if none similar experiences *teaching systems thinking as a foreign language*. With regard to the first model, there are several similar experiences at the international level. A quick survey on the matter thus revealed it. The REES symposiums for one. Engineering Education Research Centres are another (for instance the EERC´s at Cockrell School of Engineering, Texas University, Swanson School of Engineering, Pittsburgh University, The Northwestern Centre for Engineering Education Research (NCEER), Northwestern University). Engineering education societies such as the American Society for Engineering Education, the European Society for Engineering Education, and the International Technology and Engineering Educators Association, are worth also mention in this regard. In general, the literature on the matter is quite rich. However, we find particularly interesting and close to our own work, the work of Craig (2016). He points out that "Society's problems are getting harder, broader, and deeper and are multidisciplinary in nature. They require a multidisciplinary *systems approach* to solve them and present-day engineering education is not adequately preparing young engineers for the challenge." (p.1, emphasis added). In other words engineering students must learn to work in multiple contexts and multiple disciplines. How Engineering Education is facing this challenge? To begin with, Craig says that Colleges of Engineering must be more than just the sum of the engineering departments, each operating in its own silo. This means, they have to think in ways to integrate their knowledge and teaching strategies and radically change how they are currently organized at all levels, internally as well as externally with their surrounding local and global contexts. The vision that inspires this proposal is that of an integrated engineering curriculum at the under-grad level. Craig also thinks that to build solid foundations the K-12 basic education system must be integrated to the Engineering faculties. In this regard it is important to note that the University of Ibagué is working along similar lines with the K-12 system and it is beginning to integrate the "engineerings" (civil, mechanical, systems, electronics, industrial) through the Common Basic Cycle (common core curriculum at the freshman level), *where the course on systems thinking described in the previous sections of this paper is providing the common language*, and even opening the doors to wider and more drastic changes in engineering faculties and inter-faculty integration. UNIBAGUE´s Fast Lab, a multidisciplinary laboratory to carry out engineering complex systems projects (planned to enter in operation next year) could become an anchor point for the freshman course on systems thinking. In a similar vein we are integrating the course to the program *Paz y Región* (Peace and Regions), where last year engineering students (among others) go to different regions of Tolima and work together with local communities in their development issues.

Another international voice that works with similar ideas for engineering education is Foster´s (1999, 2001). He has recently won an award (year 2016) for his outstanding classroom instruction, and innovative teaching methods, and has worked for many years teaching introductory courses in systems thinking in engineering at the University of Toronto. He points out that because systems thinking is transdisciplinary and synthetic does not fit well within the dominant disciplinary structures of traditional engineering education. Which is why Craig thinks that engineering education, at least at the undergrad level, will need more than a new curriculum to incorporate systems thinking. Now, Foster (2001, pp.3-5) has worked with a general method for teaching systems thinking, inspired in Checkland´s (1999) methodology, which contemplates similar aspects to the ones discussed in our case study (like for instance giving students the opportunity to synthetize *their own definition of systems thinking* through carrying out their own projects). Now, with regard to the contextualization of our study in the Colombian educational context, again we found similar studies to those of the first model. Echeverry and Zuluaga (2011), Quintero and Yepes (2009), Mejía and Zarama (2004). However, the literature survey showed a great deficit of research in engineering education in Colombia and in particular systems thinking education. In this sense, the qualitative research reported in this paper is a minor contribution to a field that awaits further and urgent development.

7. CONCLUSION

Given the complex globalized problems of the present, teaching systems thinking in engineering and other disciplines is not anymore a matter of choice. The question is how to transform higher education to begin to cope with this challenge. At the Universidad de Ibagué we are beginning to find a way which although has similarities with other approaches in Colombia and abroad, nevertheless in many ways is developing its own quest. The introductory courses to systems thinking open a door to face the challenge Craig and others make to engineering education: adequately preparing young engineers for the multidisciplinary and trans-disciplinary problems of our time.

ACKNOWLEDGEMENTS

The authors are very much grateful to the teachers, teachers´ assistants and members of MYSCO research group who have participated in the pilot course described in this paper. Their complete involvement with the project and their support made possible the educational experiences we recount in this paper. Special thanks are due to Daniel Lopera, Reynel Felipe Gómez, Julio Mazorco, and Victoria Ortegón. With them we launched the second phase of this project in the year 2014 (a previous attempt was made in 2013). Without their active participation, their brilliant ideas, and their help to manage this pilot study, nothing of this would have been possible. Last but not least, our thanks to the authorities of the University of Ibague (Rector, Vicerrectora, Consejo Superior) for their total support to the project.

REFERENCES

Aldana, E., Reyes, A. (2004). *Disolver Problemas.* https://isfcolombia.uniandes.edu.co/images/documentos/20112%20aldana%20disolver%20problemas.pdf

Ausubel, D. (1978). *Assimilation Theory of Learning*

Ausubel, D. P. (1968). *Educational Psychology: A Cognitive View*. New York: Holt, Rinehart and Winston.

Banathy, B.H. (1996). *Designing social systems in a changing world.* New York: Plenum Press.

Bertalanffy, L. von, (1945), *Zu einer allgemeinen Systemlehre,* Blätter für deutsche Philosophie, 3/4. (Extracto en: *Biologia Generalis*, 19 (1949), 139-164

Bertalanffy, L. von, (1976). *General System Theory*. Revised Edition. New York: George Braziller, pp. xix-xxi

Buck Institute for Education (2015). Gold Standard PBL:Essential Project Design Elements. http://www.bie.org/object/document/gold_standard_pbl_essential_project_design_elements#

Checkland, P. (1999). *Systems Thinking, Systems Practice: Includes a 30-Year Retrospective*. Wiley.

Craig, K. (2016) Transformational Engineering Education. http://www.ceri.msu.edu/wp-content/uploads/2010/07/Transformational-Engineering-Education.pdf

Dewey, J. (1963). *Experience and Education*, Colliers Books, NY.

Echeverry, E. & Zuluaga, C. (2011). Sistemas dinámicos: Una línea de investigación clave para la formación de los ingenieros del futuro. En J. Arias Ramírez (Presidente), *Jugar; aprender e innovar con la ingeniería*. Simposio llevado a cabo en el VII Encuentro de la Red GEIO, Bogotá, Colombia.

Escrigas, C. (2016). A Higher Calling for Higher Education. Web page: Great Transition Initiative.
http://www.greattransition.org/publication/a-higher-calling-for-higher-education#section-title-one

Foster, J., Kay, J. and Roe, P. (2001) Teaching Complexity and Systems Thinking to Engineers. 4th UICEE *Annual Conference on Engineering Education*, Bangkok, Thailand, 7-10 February 2001.

Godfrey, P., Crick, R.D., Huang, S. (2014). Systems Thinking, Systems Design and Learning Power in Engineering Education. https://www.researchgate.net/publication/268815638_Systems_Thinking_Systems_Design_and_Learning_Power_in_Engineering_Education

Gittings, J. (2017). The Language of Systems Thinking. MediaComUK. http://mediacom.co.uk/en/what-we-think/blink/issues/edition-8-2014/the-language-of-systems-thinking.aspx

Glaser, B. G. and Strauss. A. L. (1967). *The Discovery of Grounded Theory: Strategies for Qualitative Research.* Transaction Publishers.

Goleman, D. and Senge, P. (2014).*The Triple Focus: A New Approach to Education*. More Than Sound. LLC, Florence, MA.

Goodman, D. (2010)."Systems Thinking as a Language," *The Systems Thinker* Vol. 21, No.10.

Jain, R., Sheppard, K., McGrath, E. and Gallois, B. (2008). Promoting Systems Thinking in Engineering and Pre-Engineering Students. https://www.researchgate.net/publication/264839793

Kay, J.J., Foster, J. (1999). About Teaching Systems Thinking in Savage, G. Roe, P. (eds), Proceedings of the HKK conference, 14-16 June, University of Waterloo, Ontario.

Kim, D. H., and Anderson, V. (2011) *Systems Archetype Basics From Story to Structure*. Pegasus Communications. https://thesystemsthinker.com/wp-content/uploads/2016/03/Systems-Archetypes-Basics-WB002E.pdf

Kolb, D.A. (1984). *Experiential learning: experience as the source of learning and development*. Englewood Cliffs.

Kolb. D. A. and Fry, R. (1975) 'Toward an applied theory of experiential learning; in C. Cooper (ed.) *Theories of Group Process*, London: John Wiley.

Lannon, C. (1991) The Vocabulary of Systems Thinking: A Pocket Guide https://thesystemsthinker.com/the-vocabulary-of-systems-thinking-a-pocket-guide/

López-Garay, H. (2013). *Diseño Curso Introductorio al Pensamiento Sistemico*. Internal report, Unidad de Proyectos Especiales, Universidad de Ibagué, Colombia.

Marriott, R., Torres, L.P (2016). Concept maps and language acquisition: An implementation with English Language level 2 students. *Innovating with Concept Mapping, Proc. of the Seventh Int. Conference on Concept Mapping*, Tallinn, Estonia 2016.

Mejía D., Zarama, R. (2004). La Promoción de Pensamiento Crítico en Ingeniería. *Revista de Ingeniería*, (20), pp. 90-104, Universidad de Los Andes, Bogotá,.

Minati, G. and Collen, A. (1997). *Introduction to Systemics*. Walnut Creek, California:

Novak, J.D., (2011). A Theory of Education: Meaningful learning underlies the constructive integration of thinking, feeling, and acting leading to empowerment for commitment and responsibility. *Aprendizagem Significativa em Revista/Meaningful Learning Review* – V1(2), pp. 1-14.

Quintero, J. & Yepes, A. (2009). Diseño de un conjunto de herramientas que permitan la aplicación y enseñanza del pensamiento sistémico en la Facultad de Ingeniería Industrial de la Universidad Tecnológica de Pereira (trabajo de grado). Universidad Tecnológica de Pereira, Facultad de Ingeniería Industrial, Pereira, Colombia.

Sadowsky, J. (2017). *Language learning and systems thinking* - Leadership & Storytelling http://www.johnsadowsky.com/language-learning-and-systems-thinking/

Senge, P. (1999). *The Fifth Discipline*, Doubleday.
Schön, D. (1983). *The Reflective Practitioner*, New York: Basic Book.
Vygotsky, Lev S. (1978). *Mind in society: Development of higher psychological processes.* Edited by Cole, M., John-Steiner, V., Scribner, S., Souberman, E. Cambridge, MA: Harvard University Press.
Xanthos, X. (2017).Wittgenstein's Language-Games, Université du Québec à Chicoutimi http://www.signosemio.com/wittgenstein/language-games.asp
Wittgenstein, L. (1953). *Philosophical Investigations*. Blackwel.

Student learning behaviours around assessments

Teresa Hattingh
School of Mechanical, Industrial and Aeronautical Engineering, University of the Witwatersrand, Johannesburg, South Africa
Teresa.hattingh@wits.ac.za

Laurie Woollacott
School of Chemical and Metallurgical Engineering, University of the Witwatersrand, Johannesburg, South Africa
Lorenzo.woollacott@wits.ac.za

Robert Reid
School of Mechanical, Industrial and Aeronautical Engineering, University of the Witwatersrand, Johannesburg, South Africa
robert.reid@wits.ac.za

Abstract: *An investigation of student perceptions of learning and assessment was conducted in an engineering school at a South African university. Learning in the School is facilitated through lectures, tutorials and assessment in the form of assignments, tests and exams. A survey identified students' current perceptions of assessment and learning. The survey considered distribution of work, student approaches to learning in different assessment contexts, perceptions and use of feedback and the extent to which students assess their performance and learning approaches. It was found that assessments have a significant impact on student approaches to learning and that there are a wide range of approaches and views amongst respondents. The results indicate that a clear assessment strategy that understands student interpretations and use of assessment is critical in driving preferred learning behaviours.*

Introduction

In an influential text on *Re-Thinking Engineering Education*, Crawley, et al. (2007, p.153), suggest that assessment is critical to diagnosing and promoting student learning. The link between assessment and how students learn is therefore an important area of research. In the field of engineering education, this indicates a need to consider how existing assessment practices influence students' learning practices and their approaches to learning. In South African engineering education, this link is not well understood.

Approaches to learning are affected by students' prior learning experiences (Biggs and Tang, 2011) and their perceptions of their current learning context (Prosser and Trigwell, 1999). Ramsden (1979) identified five contextual factors (workload, learning task flexibility, learning goal clarity, teaching quality and assessment type) that strongly influence student learning approaches. Many authors have emphasised the importance of assessment in understanding student learning. Ramsden (1979) states that assessment defines what students perceive as the curriculum, which Sambell and McDowell (1998) refer to as the *hidden curriculum*. According to Biggs and Tang (2011), formal assessment sends very powerful messages to students about what is being valued within their learning context, resulting in what they call, *the backwash effect*.

Rompelman (2000) emphasises the importance of a shift in engineering curriculum and assessment from knowledge and specialist skills to understanding, problem solving and complementary skills. Studies on engineering assessment have considered alternative assessment strategies, moving away from traditional written tests and exams, aiming to promote deeper approaches to learning, (Lundberg, 2010). Green (2005) has explored how accurately exams reflect engineering students' actual understanding of course material and their ability to use it. Musekamp and Pearce (2015) extend these ideas and consider how low stakes assessments, if incorrectly focused can actually distract from the actual purpose of improving teaching and learning.

There is therefore a need to consider more specifically how existing engineering assessment practices influence student learning approaches and encourage deeper levels of understanding.

This study is the first part of a larger triangulated study, which aims to better understand the relationship between assessment systems and practices and student learning behaviours within the School. This first part analyses a student survey, which addresses the following research question,

What are students' current learning practices and, according to the students, how are these influenced by assessment practices in the School of Mechanical, Industrial and Aeronautical Engineering?

The results from this survey will be used to develop the larger study, which will include interviews, document analysis and focus groups.

Context of the study

This study takes place in the School of Mechanical, Industrial and Aeronautical Engineering. In this School, formal in-class lectures are provided for all courses. Some courses also use formal tutorial sessions where students work through problems with tutors. In most courses, assignments, tests and exams are used to provide a summative assessment of student performance. The weighting of assessment types varies but exams usually count at least 50% towards the final mark, tests being the next most heavily weighted. A passing mark for any course is a minimum of 50%. Courses typically only have one test and one exam. Tests are scheduled throughout the semester while exams occur during a dedicated block. Class sizes vary by year and degree speciality. In first and second year, class sizes typically exceed 250 students; reducing in higher years. Students complete two common years before specialising in one of three branches in their third year. The curriculum for the four years of study is set for each degree programme.

Design of the study

The online student survey used a modified Assessment Experience Questionnaire (AEQ) by Gibbs and Simpson (2004).

The AEQ covers three important areas of students' perceptions and experience of assessments. These include volume, focus and quality of effort, feedback and conceptions of learning. A review of literature showed that the aspect of volume, focus and quality of effort is strongly supported by Chalmers and Fuller (1996), feedback is supported by Snyder (1971) and Nicol and MacFarlane-Dick (2006) while conceptions of learning is supported by Carless (2007).

To address other emergent issues, the AEQ was expanded to assess different types of assessment (Sambell, McDowell and Montgomery, 2013), deep and surface learning approaches (Carless, 2007), students' understanding of the gap between expectations and their performance (Sadler, 1989) and engagement and self-evaluation (Snyder, 1971; Sadler, 1989; Andrade and Valtcheva, 2009).

The AEQ specifically considers student learning related to assignments and exams. To adjust for the context of this case, sections relating to tutorials and tests were included.

Many AEQ questions refer to students' behaviour related to "doing well". Based on experience in the School, it was felt that "doing well" may not primarily motivate some students and that these questions could be misinterpreted by students. Some questions from the AEQ were rephrased to focus on "passing" in addition to "doing well". A question was also added to the questionnaire, asking if the main aim of the student was to "pass" or "do well".

The original AEQ is made up of 36 questions. In this study, the questionnaire was expanded to 56 questions, excluding demographic questions. The questionnaire made use of a 4 point Likert scale, using categories of strongly disagree, disagree, agree and strongly agree. Unlike the AEQ, it was decided not to include a neutral position on the scale so as to encourage students to take a position for every answer. A few questions in the questionnaire asked students to rate certain options relative to each other.

There were seven sections in the questionnaire:

1. Amount and distribution of effort
2. Tutorials and learning
3. Assignments and learning
4. Tests, exams and learning
5. Perceptions and experiences of feedback
6. Engagement with feedback
7. General approaches to learning

Online questionnaires were distributed to all students (n= 1011 students) in the School using SurveyMonkey in September 2016 (University terms in South Africa run from February to November). The data were analysed using simple descriptive statistics.

Findings

Respondent profile

Responses were received from 263 students, distributed over the four years of study. Of the 263 students who agreed to take the survey, 240 started answering the actual survey questions. There was further attrition of some 20 students over the balance of the survey. Although not every student answered every question, there was a minimum of 209 responses for every question that was asked.

Completion rates were 19% in first year, 27% in second year, 30% in third year and 24% in fourth year. Students registered for courses from two different academic years are counted as part of the higher year, since they should exhibit learning experiences from this higher year as well as all lower years

Due to high failure rates in the School, many students take more than the standard four years to complete their degrees. This is shown by the number of years a student has added to their degree length. For the respondents, 52% have added 0 years, 22% have added one year, 11% have added two years and 15% have added three or more years.

Introduction to the results

Results for each questionnaire section are introduced by the intent of the questions, followed by a summary of the results in paragraph form and a table of selected results including the distribution of student responses. Each section ends with brief comments on the results. Although the results from most questions are presented, several were omitted where they did not add value to the overall message. When discussed, responses for disagree and strongly disagree and agree and strongly agree are grouped together into disagree and agree, unless otherwise specified.

As a preamble for interpreting the data to follow, it is worth reflecting on students' primary aim when learning. Responses indicated that 51% of students aim to pass while 49% aim to do well. This is important as it sets the scene for interpreting the responses that follow.

Amount and distribution of effort

This section provides insight into student pressures and their reactions to these. This section highlights students' focus, how they plan and distribute work and how they change or adapt their learning behaviours based on their learning context. Workloads are known to significantly affect students' learning approaches (Ramsden,1979).

Table 1 shows that the majority (80%) of respondents believe that they need to study regularly to pass. However, they imply that significantly more work is done in weeks when assignments are due. There is an indication that workloads require students to prioritise certain work, with 91% of respondents indicating that they do this. One question asked respondents, when prioritising work, which activities they give preference to: tests, assignments or general weekly studying. Results show that 68% of respondents prioritise tests as a first priority, 74% prioritise assignments as a second priority and 78% will give general weekly studying the lowest priority. There is a split amongst respondents; 48% state that they can be selective about what they study and still do well while 52% indicate the opposite. This is confirmed by 76% of respondents who indicate that they only study what they know will be covered in an assessment.

Table 1: Amount and distribution of study effort

Questions	Strongly disagree	Disagree	Agree	Strongly agree	Response Count
I do the same amount of study each week, regardless of whether an assignment/test is due or not	20% (46)	58% (133)	19% (43)	3% (8)	230
I can be quite selective about what I study and still do well	8% (19)	44% (101)	43% (99)	4% (10)	229
I have to study regularly, throughout the semester, if I want to pass the course	3% (7)	17% (38)	38% (88)	42% (96)	229
I only study things that are going to be covered in assignments or tests and exams	3% (7)	21% (49)	49% (113)	26% (60)	229
I find that it is possible to do quite well without studying much	53% (122)	30% (68)	12% (28)	5% (11)	229
I often find that I cannot do everything and I end up prioritising some work over other work	3% (6)	6% (14)	53% (121)	39% (89)	230

The results show that the amount of work affects student learning behaviours. Tests and assignments are strong drivers of the weekly amount of work that students do. A significant portion of students are selective about what they study, with the majority of students identifying assessments as a key driver in this selection process. This selective attention does not always lead to success - which could indicate students' mastery of predicting assessment material. These results identify the development of a *hidden curriculum*, very possibly being driven by the workload and the fact that students are forced to prioritise work.

Tutorials and learning

Tutorials are a formal means of providing support to students. This section intends to understand how students perceive and use tutorials. This could indicate whether tutorial sessions are facilitating learning and driving desirable learning approaches.

Table 2 show that the majority of students agree that attending (77%) and doing (94%) tutorials enables them to understand course material better. However, only 54% of students admit to preparing for tutorial sessions. The result from the question, "I find that doing tutorials helps me to understand course material better", is one of the most polarised results in the survey with 44% of respondents strongly agreeing, 50% agreeing. This is a very strong indication that solving example problems is one of the strongest mechanisms that students believe the School currently uses to facilitate deep approaches to learning.

Table 2: Tutorials and learning

Questions	Strongly disagree	Disagree	Agree	Strongly agree	Response Count
I find that attending formal tutorial sessions helps me to understand course material better	5% (11)	18% (41)	45% (100)	32% (72)	224
I find that doing tutorials helps me to understand course material better	1% (2)	5% (11)	50% (113)	44% (98)	224
I make sure that I am always well prepared for tutorial sessions	5% (10)	41% (91)	45% (101)	9% (20)	222

Students' preparation for tutorials could be a function of workload and extent to which students prioritise work. Tutorials usually do not contribute significantly to the overall course mark. Consequently, tutorial preparation falls into the general weekly studying category, which is attended to last when prioritising work. Students may complete tutorials at a later stage, perhaps when preparing for exams. The fact that students believe that tutorials support their learning could be leveraged by incentivising students to prioritise tutorial work and by ensuring that the work that is covered in tutorials encourages a deep learning approach.

Assignments and learning

Assignments constitute work, completed in students' own time, that is handed in for summative assessment, contributing to the final course mark. These include laboratory and design reports or other work, typically solved problems, related to course material. The intent of this section is to understand how students perceive assignments and what learning behaviour assignments encourage.

Table 3 shows that although 88% of students agree that doing assignments makes them think, 48% also agree that they can pass an assignment without actually understanding the course material. There is often uncertainty about the expectations of assignments with only 53% of students agreeing that clear instructions are provided and 77% agreeing that they often do not know what would count as a successful answer.

Table 3: Assignments and learning

Questions	Strongly disagree	Disagree	Agree	Strongly agree	Response Count
Tackling the assignments really makes me think	2% (5)	10% (23)	56% (123)	32% (70)	221
In completing the assignments, I can get away with not understanding and still pass	12% (27)	39% (87)	42% (93)	6% (14)	221
The assignments give very clear instructions about what I am expected to do	12% (26)	35% (78)	49% (107)	4% (9)	220
When I tackle an assignment it is not clear what would count as a successful answer	2% (5)	21% (46)	58% (127)	19% (41)	219

As shown previously, assignments receive the second highest priority when studying indicating that students prioritise work that is "for marks", which they realise may not lead to understanding, over work that helps them to understand but is not for marks (tutorials). There is also an indication that assignments are not designed or communicated to facilitate learning. Chalmers and Fuller (1996) and Sadler (1989) suggest that to support learning, students should have clear expectations of what is required in assessments.

It seems that the current use of assignments is not well aligned to supporting student learning.

Tests, exams and learning

In the School, tests and exams are the highest stakes assessment types - most likely to drive overall student learning. The intent of this section was again to understand how tests and exams affect student learning approaches and behaviours.

Table 4 shows that when approaching tests and examinations, 35% of respondents agree to memorising steps without understanding. A significant portion of respondents, 42% believe that they can pass a course without actually understanding the material. Most respondents (61%), indicate that they target specific material when studying for exams. Although 68% of students agree to having a better understanding of material after a test or exam, the balance (32%), indicate that they do not believe that tests and exams have any learning benefit. Most respondents (66%), indicate that they forget most material after the test or exam.

Table 4: Tests, exams and learning

Questions	Strongly disagree	Disagree	Agree	Strongly agree	Response Count
When studying for tests or exams, I often memorise methods / steps without understanding what I am doing.	18% (38)	46% (98)	29% (62)	6% (13)	211
I understand things better as a result of the tests and exam	3% (7)	29% (62)	52% (109)	16% (33)	211
I usually forget most of what I have learnt after the test / exam	2% (4)	32% (67)	42% (88)	24% (51)	210
In the test or exam you can get away with not understanding and still pass	27% (57)	31% (65)	34% (71)	8% (17)	210
I usually spot (target) which sections to cover for the exams	11% (23)	28% (58)	46% (96)	15% (31)	208

The results suggest that a significant portion of students adopt a procedural, surface approach to learning as defined by Case and Marshall (2004). This is shown by the fact that many students believe that they can pass without understanding, admit to memorising steps without understanding and indicate that they forget material after the assessment has passed. This could also suggest that the design of questions in exams is such that only surface approaches are being tested and encouraged.

The results again suggest the development of a *hidden curriculum* with a significant portion of students considering what they believe will be included in an exam when studying. Students' learning behaviours are being driven by a high stakes assessment.

Not all students see tests and exams as facilitating learning, a key finding considering the importance placed on tests and exams in the School.

Feedback – perceptions and experiences

Feedback (including quality, timing and the use thereof) is critical in the learning process (Angelo and Cross, 1993). This section intends to understand how students experience feedback practices. The section following this will discuss the extent to which students engage with this feedback.

Table 6 shows that only 25% of respondents believe they are getting "plenty" of feedback. When asked where feedback was currently received, respondents indicated that they receive feedback for: tests (79%), assignments (64%), tutorials (41%) and exams (17%). When asked how feedback is typically received, respondents indicated: written comments (55%), class discussions (53%), verbal feedback (44%) and handouts (class notes) (26%).

Students clearly value feedback with 76% of respondents indicating that feedback facilitates understanding. Only 8% of respondents indicated that they did not "bother" to look at feedback. When asked if students would learn more if they received more feedback, 95% of students indicated that they would (58% strongly agreeing).

Table 6: Feedback perceptions and experiences

Questions	Strongly disagree	Disagree	Agree	Strongly agree	Response Count
I get plenty of feedback on how I am doing	18% (38)	57% (123)	24% (51)	1% (3)	215
I don't usually bother to look at feedback	32% (68)	59% (127)	7% (16)	1% (3)	214
The feedback helps me to understand things better	3% (6)	21% (45)	60% (128)	16% (34)	213
Once I have read the feedback I understand why I got the mark I did	3% (7)	21% (44)	62% (132)	14% (29)	212
When I get things wrong or misunderstand them I don't receive much guidance in what to do about it	4% (9)	30% (66)	45% (97)	21% (45)	217
I would learn more if I received more feedback	0% (0)	5% (11)	37% (80)	58% (125)	216

According to this survey, feedback is the single, biggest feature that respondents believe would improve their learning. Feedback is clearly valued and used by students. There are however questions around the current quality and adequacy of the feedback.

Feedback – engagement

Feedback is an important aspect of learning, but it is students' engagement with and use of feedback, where learning occurs (Snyder, 1971).

Table 6 shows that 88% of respondents indicate that they read through feedback carefully to understand what the feedback is saying. In addition, 86% of respondents indicate that they use feedback to go over what they did and 69% indicate that the feedback also prompts them to revise course material.

Table 6: Engagement with feedback

Questions	Strongly disagree	Disagree	Agree	Strongly agree	Response Count
I read the feedback carefully and try to understand what the feedback is saying	1% (3)	11% (22)	65% (135)	23% (49)	209
I use the feedback to go back over what I did	1% (2)	13% (27)	70% (147)	16% (33)	209
The feedback prompts me to go back over material covered earlier in the course	3% (6)	28% (58)	57% (120)	12% (25)	209

It appears that most students are using received feedback to facilitate understanding.

General approaches to learning

This section explores general approaches to learning including study groups, whom students approach for help, what they do when they do poorly in a course and whether students consider and adapt their learning method based on their past experiences.

Table 7 shows that students appear to be actively engaged in understanding why they fail and attempt to adapt to do better. When respondents do poorly, 89% reported trying to understand why. Respondents indicated that they sought help from classmates (91%) and lecturers (60%). Concerning group work, 70% of respondents agree that studying in a group helps them to better understand concepts and 48% of respondents indicate that they actually do work in groups.

In trying to understand the extent to which students think about and adapt their learning practices, 87% of respondents indicate that they have changed the way in which they study over time. The way in which studying occurs could refer to the method used or effort applied

- 74% have considered how to change their study methods while 48% are merely trying harder using the same methods.

The last two questions were asked to understand whether students were encouraged to think about their learning as a result of completing this questionnaire. Of all respondents, 81% indicated that answering the questionnaire made them think about the way in which they learn, with 66% indicating that they may consider changing the way in which they approach learning and studying in future.

Table 7: General approaches to learning

Questions	Strongly disagree	Disagree	Agree	Strongly agree	Response Count
I often study with other students in my class	12% (25)	40% (84)	38% (81)	10% (21)	211
I find that studying together with other students helps me to understand concepts better	9% (19)	21% (44)	51% (107)	19% (41)	211
When I am struggling with a course, I seek help from other students in my class	1% (2)	9% (18)	60% (125)	31% (65)	210
When I am struggling with a course, I seek help from the course lecturers	8% (16)	33% (68)	51% (107)	9% (18)	209
When I fail a test or exam, I try to understand why	1% (3)	9% (19)	67% (139)	22% (46)	207
The way in which I study for tests and exams has changed as I have realised what works and what doesn't	3% (7)	10% (21)	57% (119)	30% (63)	210
When I fail or do poorly in a test, I evaluate my study methods and consider what I need to change	2% (5)	24% (50)	57% (119)	17% (35)	209
When I fail or do poorly in a test, I try harder next time using the same methods as before	2% (4)	50% (104)	38% (80)	10% (20)	208
I use my mark for assignments and tests to give me an indication of how well I am doing in the course	2% (5)	7% (14)	55% (116)	36% (75)	210
Answering this survey has made me think about how I study and learn	3% (7)	16% (33)	56% (117)	25% (53)	210
After answering this survey, I may consider how to change the way I approach learning and studying	6% (13)	27% (57)	49% (103)	17% (36)	209

Results indicate an eagerness in students to do better. It is unclear whether students have the necessary knowledge and skills to adapt and change their learning approaches in isolation from some intervention or supporting structures which encourage more self evaluation and deeper approaches to learning.

Conclusions

The results from this survey provide valuable insight into the effect that assessment practices in the School have on student learning behaviours and approaches. Assessments are a driver of the amount of effort and types of learning approaches that students adopt.

Most students actively try to improve their performance by engaging with feedback and adjusting learning approaches. Students consider feedback to be one of the most significant contributors to their learning. However, student behaviours are currently more concerned with predicting what will be asked in an assessment with assessment types that are for marks appearing to have a greater impact on student learning behaviours.

The findings reported in this paper will inform the next parts of this triangulated study, using interviews, focus groups and document analysis. The results from the larger study will provide a more holistic and in depth understanding of how student learning in the School is influenced by its assessment practices. These findings will also speak more generally to the design and implementation of engineering assessment practice in the country.

References

Andrade, H., & Valtcheva, A. (2009). Promoting Learning and Achievement Through Self-Assessment. *Theory Into Practice*, *48*(1), 12–19.

Angelo, T. & Cross, P. (1993) *Classroom Assessment Techniques. A Handbook for College Teachers*. Jossey-Bass.

Biggs, J. B. & Tang, C (2011) *Teaching for quality learning at university, what the student does*. Society of Research into Higher Education. Open University Press. 4th Edition.

Carless, D. (2007). Learning‐oriented assessment: conceptual bases and practical implications. *Innovations in Education and Teaching International*, 44(1), 57-66.

Case, J., & Marshall, D. (2004). Between deep and surface: procedural approaches to learning in engineering education contexts. *Studies in Higher Education*, 29(5), 605–615.

Chalmers, D., & Fuller, R. (1996). Teaching for learning at university. Kogan Page.

Crawley, E., Malmqvist, J., Ostlund, S., Brodeur, D. & Gray, P.J. (2007). *Rethinking engineering education. The CDIO Approach*. Springer.

Gibbs, G. & Simpson, C. (2004) Conditions under which assessment supports students' learning. *Learning and Teaching in Higher Education*, 1, 3-31.

Green, S. I. (2005). Student Assessment Precision in Mechanical Engineering Courses. *Journal of Engineering Education*, *94*(2), 273–278.

Lundberg, A. (2004). Student and teacher experiences of assessing different levels of understanding. *Assessment & Evaluation in Higher Education*, *29*(3), 323-333.

Musekamp, F., & Pearce, J. (2015). Assessing engineering competencies: the conditions for educational improvement. *Studies in Higher Education*, *40*(3), 505–524.

Nicol, D. J., & Macfarlane-Dick, D. (2006). Formative assessment and self-regulated learning: A model and seven principles of good feedback practice. *Studies in Higher Education*, 31(2),199–218.

Prosser, M. & Trigwell, K. (1999). *Understanding learning and teaching: the experience in higher education*. Open University Press.

Ramsden, P. (1979) Student learning and perceptions of the academic environment. *Higher education*, 8, 411-428.

Rompelman, O. (2000). Assessment of student learning: evolution of objectives in engineering education and the consequences for assessment. *European Journal of Engineering Education*, *25*(4), 339-350.

Sadler, D. R. (1989). Formative assessment and the design of instructional systems. *Instructional science*, 18(2), 119-144.

Sambell, K. & McDowell, L. (1998) The construction of the hidden curriculum: Messages and meanings in the assessment of student learning, *Assessment and Evaluation in Higher Education*, 23 (4). 349-371.

Sambell, K., McDowell, L., & Montgomery, C. (2013). *Assessment for learning in higher education*. Routledge.

Snyder, B.R. (1971). *The hidden curriculum*. New York: Knopf.

Wicked problems and assessment in engineering education: Developing and evaluating an analytic rubric

Johanna Lönngren
Chalmers University of Technology, Gothenburg, Sweden
johanna.lonngren@chalmers.se

Tom Adawi
Chalmers University of Technology, Gothenburg, Sweden
tom.adawi@chalmers.se

Magdalena Svanström
Chalmers University of Technology, Gothenburg, Sweden
magdalena.svanstrom@chalmers.se

Abstract: *Previous research indicates that engineering education does not adequately prepare students to address complex, ill-structured, real-world problems, such as wicked problems (WPs), and that one reason for this may be a lack of robust assessment instruments. In recent years, assessment rubrics have been developed and evaluated for a variety of learning outcomes, but no rigorously tested rubric has yet been developed for assessing engineering students' ability to integratively address WPs. The aim of this paper is to fill this gap by introducing an analytic rubric for assessing engineering students' written responses to WPs and evaluating its reliability, validity, and utility. The results suggest that the rubric can support reliable and valid assessment if raters are carefully trained. The utility of the rubric for formative assessment and teacher professional development was most prominent.*

Introduction

In recent years, there has been an increasing interest in the engineering education research community in how to prepare students to address complex, ill-structured problems that are located in a real-world context and concern technical as well as societal issues (e.g. Jonassen, Strobel, & Beng Lee, 2006; Pierrakos et al. 2010; Riis et al., 2016). Such problems are commonly described as *wicked problems* (WPs) (Rittel & Webber, 1973). WPs are characterized by powerful conflicts of interest and differences in norms and values between stakeholders.

Unfortunately, most engineering education today does not adequately prepare students to address WPs (Jonassen, Strobel, & Beng Lee, 2006). Engineering educators report that one reason for not teaching students to address WPs is a lack of robust assessment instruments (Lönngren & Svanström, 2015), such as *assessment rubrics*. Typically, rubrics provide "criteria for rating important dimensions of performance, as well as standards of attainment for those criteria" (Jonsson & Svingby, 2007, p. 131). Previous research suggests that well-designed, analytic rubrics can improve reliability in scoring student performance, promote learning, and improve teaching (Brookhart & Chen, 2015; Jonsson & Svingby, 2007; Reddy & Andrade, 2010). Rigorously tested rubrics can also provide powerful analytic tools for comparing the effectiveness of different pedagogies or educational materials (Brookhart & Chen, 2015).

Rubrics have been developed and evaluated for a variety of learning outcomes that are closely related to the ability to address WPs, such as sustainable engineering (McCormick et al., 2015), critical thinking (Schamber & Mahoney, 2006), reflective writing (Wald et al., 2012), and argumentative reasoning (Reznitskaya et al., 2009). In this paper, we describe the development of an analytic rubric for assessing engineering students' ability to reason about ten important aspects of WPs, and we report on the results of an analysis of the reliability, validity, and utility of the rubric.

Rubric development

We iteratively developed the rubric in the context of a design-based research project in which we also developed and evaluated an educational intervention. To inform the development of the rubric and the intervention, we primarily used the results of a phenomenographic study of engineering students' approaches to a WP (Lönngren, Ingerman & Svanström, 2016). In that study, we identified four approaches that differed in terms of how much of the complexity of the WP the students considered. In the most complex of these approaches, problem parts and improvement measures were viewed as an integrated, complex system. Students using this approach addressed the WP in an *integrative* way. Rather than trying to "solve" isolated problem parts with isolated improvement measures, they considered the problem holistically as an unsatisfactory situation, and they tried to iteratively improve this situation while constantly adapting to changing circumstances and monitoring for unintended consequences. Based on the description of this approach, we developed a *framework for integratively addressing WPs*. The framework outlines ten *aspects* that are important to consider when addressing WPs (Appendix A).

The rubric (Appendix B) covers ten assessment criteria, which reflect the ten aspects in the above-mentioned framework. For each of the ten criteria, the rubric provides descriptions of three levels of performance, where grade 3 is the highest achievable level. Higher performance levels include lower performance levels and "fail" is assigned for a criterion if a text does not meet the requirements for grade 1 for that criterion. The total score for an assessed text is calculated as the mean value of the grades for all ten criteria. In this calculation, failed criteria score 0 points. The rubric only assesses *what is present* in a given text and does not punish instances of criterion violation, i.e. if a certain criterion is fulfilled in one part of the text but violated in another part of the text, the text should be scored as fulfilling that criterion. Importantly, the rubric does not assess the *viability* of proposed improvement measures; it only assesses whether a text includes a description of improvement measures and whether it includes students' own judgments regarding these measures.

We iteratively developed and evaluated the rubric in the context of an educational intervention, which aimed to train students to reason about the ten aspects in the framework for integratively addressing WPs (i.e. the ten criteria in the rubric) and thus ultimately develop the students' ability to integratively address WPs. The intervention was embedded in an undergraduate course on sustainability for students in information technology engineering. The intervention consisted of a written exam, then two interactive workshops, and finally another written exam. During each of the written exams, the students were asked to reflect on how they would address a wicked problem (*Literacy in Afghanistan* in the first exam and *Dengue fever in sub-Saharan Africa* in the second exam, see Appendix C). The first exam was voluntary and the rubric was used for formative assessment only. The second exam was mandatory and the rubric was then used for summative assessment. The workshops were voluntary and consisted of interactive group exercises that aimed to familiarize the students first with the ten aspects in the framework and then with the three grade levels in the rubric.

We also used the intervention to gather data for evaluating and further developing the rubric in a naturalistic context. Elsewhere, we have provided detailed information about the intervention and reported on an analysis of the students' performance, their approaches to WPs, and affordances for learning to integratively address WPs in different parts of the intervention (Lönngren, Adawi, & Svanström, forthcoming). Here, we report on the reliability, validity, and utility of the rubric.

Reliability

Methods

To evaluate the reliability of the rubric for assessment of students' written responses to WPs, we performed an inter-rater analysis with seven raters. Three of the raters were involved in the development of the rubric and thus trained to interpret and apply the rubric. The other four raters did not receive any training apart from general, written instructions. The trained raters each scored five texts and the untrained raters each scored a subset of three texts.

According to Hallgren (2012), the *intraclass correlation coefficient* (ICC) is an appropriate measure for assessing inter-rater reliability for ordinal variables and multiple raters. Reporting ICC values is also relatively common in studies on assessment rubrics (Brookhart & Chen, 2015). ICC values typically range between 0 and 1, where 0 indicates absence of agreement and 1 indicates perfect agreement. Negative ICC values are also possible and indicate systematic disagreement. Cicchetti (1994) suggests that ICC values of 0.00-0.39 indicate poor reliability, 0.40-0.59 fair reliability, 0.60-0.74 good reliability, and 0.75-1.00 excellent reliability.

To calculate ICC values, different combinations of parameters are possible depending on the purpose of the analysis. For this paper, we calculated ICC values for absolute agreement among raters, which, according to Brookhart and Chen (2015), is appropriate for evaluating the reliability of rubrics for assessing student learning in relation to standards, objectives, or course goals. We performed two-way ICC tests using SPSS software, assuming that raters were randomly selected from a larger pool of possible raters (also called "random effects"), i.e. assuming that any teacher who teaches engineering students to address WPs should be able to use the rubric, not only the raters who were involved in the present study. We performed two analyses: First, we used a *single measure* approach to analyze reliability of a single teacher's assessment. Second, we used an *average measures* approach to analyze whether averaging scores from several teachers could improve reliability (c.f. Hallgren, 2012).

Results and discussion

Table 1 summarizes the results for single and average measures analyses for all raters, untrained raters, and trained raters (values indicating fair to excellent reliability (≥0.4) are bolded; values indicating systematic disagreement (<0) are italicized).

Table 1. Two-way ICC scores

Criterion in rubric	Single measures analysis			Average measures analysis		
	All raters	Untrained raters	Trained raters	All raters	Untrained raters	Trained raters
I	**0.49**	0.35	**0.60**	**0.87**	**0.68**	**0.82**
II	*-0.03*	*-0.29*	**0.54**	*-0.14*	*-2.00*	**0.78**
III	**0.64**	**0.69**	0.31	**0.92**	**0.87**	**0.58**
IV	*-0.13*	*-0.16*	0.33	*-1.33*	*-1.27*	**0.60**
V	**0.58**	**0.57**	**0.79**	**0.87**	**0.80**	**0.92**
VI	0.06	0.00	0.29	0.25	0.00	**0.55**
VII	**0.52**	0.27	**0.52**	**0.87**	**0.52**	**0.77**
VIII	0.39	0.11	0.30	**0.82**	0.33	**0.57**
IX	0.11	0.06	0.16	**0.46**	0.21	0.37
X	*-0.10*	*-0.67*	**0.70**	*-0.57*	*-4.00*	**0.88**
Total score	0.33	0.25	0.26	**0.71**	**0.50**	**0.51**

The results for all raters indicate a large degree of variation in ICC scores for the different criteria in the rubric. On the one hand, the results indicate fair to excellent reliability for criteria I, III, V, and VII for both single and average measures analysis and for criteria VIII, IX, and total score for average measures analysis. On the other hand, the results for criteria II, IV, and X indicate systematic disagreement for both single and average measures analysis. These results suggest that the criteria in the rubric are not equally clear and well-formulated, and that especially criteria II, IV, and X should be clarified. For trained raters, variation between criteria was less than for untrained raters, especially for average measures. Most notably, no instances of systematic disagreement were found for trained raters and, in contrast to results from untrained raters, average measures analysis for trained raters showed good to excellent reliability for criteria II, IV, and X. These results indicate that rater training to some degree could compensate for unclear criteria descriptions.

The results clearly show a difference in reliability scores depending on rater training. For both single and average measures analyses, the results for trained raters were on a similar or higher level than those for untrained raters for most criteria. This result reiterates conclusions in the literature about the importance of rater training to achieve acceptable levels of reliability (Brookhart & Chen, 2015; Hallgren, 2012; Jonsson & Svingby, 2007).

We observed an interesting anomaly for criterion III, for which the results for trained raters were lower than those for untrained raters. A possible explanation for this unexpected result may be that as raters became familiar with the complex rubric, their understanding of the items measured with the rubric evolved, which in turn may have evoked complex questions about the meaning of the items. In fact, the trained raters repeatedly experienced that the more they used the rubric, the more questions arose that they had to address when they tried to apply the rubric. For example, one of the trained raters included the following comment when submitting the scores for two of the texts: ”I think it [my scoring] can vary quite a lot this time; it feels like I've overanalyzed both the texts and the rubric”.

In summary, the results for untrained raters suggest that summative assessment with the rubric may not provide reliable results for all criteria. For trained raters, the results indicate fair to excellent reliability for five criteria for single measures analysis, and all but one criterion for average measures analysis. Thus, the results suggest that assessment with the rubric can provide reasonably reliable results if teachers are carefully trained and if an average of several teachers' scores is used.

While we would have hoped for higher ICC scores for some of the criteria, the results of our analysis are in line with previously reported studies on assessment rubrics for complex competences. For example, Jonsson and Svingby (2007, pp. 139f) found that ”the studies reporting on rubric reliability, reviewed in this article, generally present low reliability coefficients as compared to traditional psychometric requirements, indicating that the use of rubrics might not in itself be enough to produce sufficient reliability for summative assessments”. They also found that assessment of essay writing generally produced low reliability scores. However, they also argued that even when reliability is low for rubric assessment, it is likely to be higher than in assessment without using rubrics.

Validity

Methods

To analyze the validity of our rubric, we used an argument-based approach (Kane, 1992). According to this approach, evaluating the validity of an assessment instrument includes specifying how test scores should be interpreted and providing a convincing argument for that interpretation. For our rubric, we suggest the following interpretation: *The rubric assesses engineering students' ability to reason about ten important aspects of WPs, which in turn is an important part of engineering students' ability to integratively address WPs.* This interpretation is based on four assumptions, which are described below.

To evaluate the assumptions, we used empirical data obtained in expert reviews of the rubric during a series of presentations, seminars, and workshops; interviews with raters who had used the rubric to score student texts; and a professional language review. We also used teachers' observations from and reflections on using the rubric in the educational intervention, and student feedback during and after the intervention. Finally, we used data from the written exams that were part of the intervention, such as the distribution of grades in each of the exams and improvement in student performance between the exams.

Results and discussion

Assumption 1: The ability to reason about important aspects of WPs is an important part of the ability to integratively address WPs. Support for this assumption is provided by the fact that the development of the rubric was heavily based on previous empirical and theoretical research as described above. In addition, educators and researchers from different disciplines confirmed this assumption in expert reviews.

Assumption 2: The rubric measures students' ability to reason about ten important aspects of WPs. Expert reviews provided strong support for this assumption. The experts unanimously considered the rubric to be very comprehensive for assessing whether a text addressed some of the most important aspects of WPs. However, they also pointed out that the rubric does not assess the *viability* of proposed improvement measures or the *adequacy* with which students describe the local problem context. This reflection relates back to the first assumption and indicates that assessment with the rubric focuses on a *part* of students' ability to integratively and adequately address WPs, rather than covering it completely.

To further strengthen the argument for the second assumption, we address two possible challenges to the assumption: (A) Assessment with the rubric could measure other processes than what is intended, and (B) Raters use other criteria than those specified in the rubric. Challenge (A) concerns the possibility that the task of writing a response to a WP could create what Messick (1995) calls *construct-irrelevant difficulty*, i.e. difficulties that are external to the process(es) one wants to measure. We identified three processes that students need to engage in, in addition to the process of reasoning about important aspects of WPs: (a) Reading and understanding the WP description, (b) reading and understanding the rubric, and (c) identifying and articulating a connection between the WP and a specific domain of engineering. Evidence from teachers' observations and student feedback during and after the intervention suggests that processes (a) and (b) did not provide significant difficulties for the students. Students indicated that process (c) provided some initial difficulty as they started to work on the task of responding to a WP. However, we would argue that process (c) is an important part of what the rubric aims to measure and that it therefore is unnecessary to discriminate between process (c) and the task of writing a response to a WP.

Further evidence against challenge (A) is provided by grade distribution and improvement in our intervention. Grading student exams with the rubric resulted in a reasonable grade distribution for most criteria and for the total score, and the rubric was sensitive enough to detect differences in students' performance before and after the intervention (Lönngren, Adawi, & Svanström, forthcoming). Thus, the rubric provided the information that we had expected, which provides evidence that assessment with the rubric measures the intended processes rather than other processes.

Challenge (B) is related to the raters. Most importantly, raters need to understand the criteria in the rubric and use them consistently in assessment. The evidence from the reliability analysis (see above) suggests that this assumption is only warranted for trained raters. However, even if raters fully understand the criteria, they may include additional, construct-irrelevant criteria in their assessment (Hay & Macdonald, 2008). In our study, several teachers used the rubric to assess student texts. The teachers were explicitly instructed to follow the rubric very closely and only assess the criteria given in the rubric. Feedback from these teachers suggests that they made a conscious effort to avoid including construct-irrelevant criteria, such as the viability of proposed improvement measures and the quality of writing (e.g. spelling and grammar). We suggest that the explicit instructions helped minimize the inclusion of construct-irrelevant criteria.

Assumption 3: Assessment with the rubric is generalizable across different domains of engineering (i.e. different engineering majors), different specific WPs, and different raters. Support for *generalizability across different engineering domains* is provided by the fact that teachers from different engineering disciplines were able to use the rubric to score student texts. In addition, expert reviews indicated that the rubric could be used across engineering majors. Support for *generalizability across WPs* is provided by the fact that we have used and evaluated the rubric in relation to two different WPs. Finally, as discussed above, *generalizability across different raters* can be achieved if raters are adequately trained.

Assumption 4: There are no sources of systematic error. One source of systematic error could be that students do not fulfill criteria in a rubric because they do not know that they are expected to do so (Kane, 1992). In our study, we found such an effect when students did not have access to the rubric; the effect disappeared when students were given access to the rubric (Lönngren, Adawi, & Svanström, forthcoming). Another source of systematic error could be that students are unmotivated to perform a task. In our study, we found such an effect when students were asked to respond to a WP in the first exam for which they did not receive any grade credit. In that case, students did not take the task seriously and addressed fewer aspects of the WP than they could have. In contrast, when students were asked to respond to a WP in the context of the second (graded) exam, students were highly motivated to deliver high-quality responses (Lönngren, Adawi, & Svanström, forthcoming). We suggest that eliminating systematic error requires carefully considering the specific context and student cohort in which the rubric is to be used.

We conclude that the interpretation that *the rubric assesses engineering students' ability to reason about ten important aspects of WPs, which in turn is an important part of engineering students' ability to integratively and adequately address WPs* is warranted given five conditions: (1) The specific task that is to be assessed does not introduce construct-irrelevant difficulties, (2) raters are adequately trained such that they fully understand the criteria in the rubric and are able to resist including construct-irrelevant criteria, (3) the rubric is used in an engineering education context, (4) students have access to the rubric when performing the task of responding to a WP, and (5) students are motivated to perform the task.

Rubric utility

Methods

To analyze the utility of our rubric, we relied on the same empirical material that we also used for the validity analysis: Expert reviews of the rubric, interviews with raters who had used the rubric, teachers' observations from and reflections on using the rubric in the educational intervention, student feedback during and after the intervention, and data from the written exams that were part of the intervention. We focused on three aspects of utility: Potential to support student learning, feasibility of use, and potential to support teacher professional development.

Results and discussion

Potential to support student learning. We first used the rubric for formative assessment in the first exam in the intervention. Assessing these texts with the rubric provided detailed information about the students' strengths and learning needs, and the teachers could successfully use this information to design contingent scaffolding throughout the remainder of the intervention. With contingent scaffolding, students' performance during workshop tasks and in the second written exam improved considerably compared to the first (unscaffolded) written exam (Lönngren, Adawi, & Svanström, forthcoming). These results support conclusions in the literature about the value of rubrics for formative assessment (Jonsson & Svingby, 2007).

The performance data from the second written exam also suggest that the rubric could be used for summative assessment: The rubric could be used to differentiate between different levels of student performance with reasonable levels of reliability, and it could be used to indicate performance improvement between the first and second written exam.

Feasibility of use. The raters who had used the rubric to assess student texts indicated that the assessment was very time-consuming and that they therefore were reluctant to use the rubric in their own teaching. While the high level of detail in the rubric may have contributed to achieving reasonable levels of reliability, it may also have contributed to reducing the utility of the rubric for summative assessment (Popham, 1997). This is particularly problematic since our results indicate that reliable grading may require that *several* teachers grade each text and that teachers also need to spend time in rater training. Thus, the results suggest that there may be a trade-off between rubric reliability and feasibility of use for summative assessment.

When we used the rubric for formative assessment, we graded texts that were considerably shorter than the texts used for summative assessment. While the reduced length of the texts may have limited the richness of the information obtained from the grading, it also considerably reduced the amount of time needed for grading and thus rendered the use of the rubric more feasible. An alternative approach to reducing the time spent on grading in formative assessment could be to combine teacher assessment with peer grading. Peer grading could be used for all texts, and the teacher could choose to grade a subset of responses to get deeper insights into the strengths and learning needs of the students.

Potential to support teacher professional development. We found that the rubric provided valuable opportunities for professional development. All practitioners in our study commented that applying the rubric to example texts helped them develop a better understanding of the characteristics of WPs and of what students need to learn to be able to address WPs. Above, we suggested that the high level of detail in the rubric may have reduced the utility of the rubric for summative assessment. In contrast to that conclusion, we found that the high level of detail in our rubric may have contributed to the high utility of the rubric for teacher professional development. The high level of detail in the rubric seemed to facilitate engaged and critical discussions among the participants, discussions that might not have been possible with a less detailed rubric.

Limitations

In the reliability analysis, we achieved reasonable levels of reliability for the rubric under certain conditions (trained raters, average measures analysis). However, the confidence intervals for the reliability scores were quite large, which reduced the strength of conclusions that could be drawn from the reliability data. To achieve smaller confidence intervals, each rater would have needed to assess a considerably larger number of student texts (probably around 20 texts each instead of three to five as was the case in our study). Unfortunately, this was beyond the scope of our study.

Contributions and implications

In this paper, we introduced and evaluated an analytic rubric for assessing engineering students' ability to reason about ten important aspects of WPs. The reported results suggest that assessment with the rubric can provide reasonably reliable and valid results if teachers are carefully trained and if an average of several teachers' scores is used. Thus, the research contributes directly to practice by providing an assessment tool for which educators have expressed a need in previous research (Lönngren & Svanström, 2015). The research also contributes with a framework for integratively addressing WPs that can be used as a tool for teaching and research.

We found that engaging engineering educators in the development and evaluation of the rubric contributed to teacher professional development. We suggest that this may be one of the most important contributions of the rubric and that this aspect of rubric utility therefore deserves more attention in research on assessment rubrics. We also suggest that educational institutions should allocate more time and resources for professional development to support engineering educators in developing competencies in teaching and assessing students' ability to integratively address WPs. Preferably, educators should be given an opportunity to develop these competencies collaboratively through both intensive workshops and continuous opportunities for collegial exchange and support. Our research suggests that our rubric can facilitate discussions about what students need to learn to be able to address WPs. Other important aspects to consider may be student motivation, specific difficulties that specific groups of students may face in learning to address WPs, and the form and content of exams. Based on these discussions, educators can also work together to develop new examples of WPs descriptions, intended learning outcomes, assessment approaches, and contingent scaffolding. Thus, the activities can serve dual functions of teacher professional development and the development of a resource bank that can be used to facilitate other educators' attempts to include WPs in their teaching.

Theoretically, the research contributes to developing a more nuanced understanding of what could be considered adequate levels of detail in assessment rubrics. While the high level of detail in the rubric seems to have reduced the utility of the rubric for summative assessment, it also seems to have increased the utility of the rubric for teacher professional development, and possibly for formative assessment. Thus, different levels of detail may be appropriate for different uses of assessment rubrics.

The research makes a methodological contribution by adding to the empirical base of reliability data for assessment rubrics for complex performance tasks; thus, the research contributes to developing reliability standards for these types of assessment. To improve comparability and eventually establish accepted standards for reliability, we suggest that the research community should agree on one or two kinds of reliability scores and use these consistently to report results of reliability analyses.

References

Brookhart, S. M., & Chen, F. (2015). The quality and effectiveness of descriptive rubrics. *Educational Review, 67* (3), pp. 343–368.

Cicchetti, D. V. (1994). Guidelines, criteria, and rules of thumb for evaluating normed and standardized assessment instruments in psychology. *Psychological Assessment, 6* (4), pp. 284-290.

Hallgren, K. A. (2012). Computing Inter-Rater Reliability for Observational Data: An Overview and Tutorial. *Tutorials in Quantitative Methods for Psychology, 8* (1), pp. 23-34.

Hay, P. J., & Macdonald, D. (2008). (Mis)Appropriations of Criteria and Standards-referenced Assessment in a Performance-based Subject. *Assessment in Education: Principles, Policy & Practice, 15* (2), pp. 153–168.

Jonassen, D., Strobel, J., & Beng Lee, C. (2006). Everyday Problem Solving in Engineering: Lessons for Engineering Educators. *Journal of Engineering Education, 92* (2), pp. 139-151.

Jonsson, A., & Svingby, G. (2007). The use of scoring rubrics: Reliability, validity and educational consequences. *Educational Research Review, 2*, pp. 130-144.

Kane, M. T. (1992). An Argument-Based Approach to Validity. *Psychological Bulletin, 112* (3), pp. 527-535.

Lönngren, J. (forthcoming). Wicked Problems in Engineering Education: Preparing Future Engineers to Work for Sustainability. *Doctoral thesis*. Gothenburg, Sweden: Chalmers University of Technology.

Lönngren, J., & Svanström, M. (2015). Assessing "Wicked Sustainability Problem"-Literacy in Engineering Education. *Proceedings of the 122nd ASEE Annual Conference & Exhibition.* Seattle: American Society for Engineering Education.

Lönngren, J., Adawi, T., & Svanström, M. (forthcoming). Scaffolding strategies and affordances for learning in a rubric-based intervention to promote engineering students' ability to address wicked problems.

Lönngren, J., Ingerman, Å., & Svanström, M. (2016). Avoid, Control, Succumb, or Balance: Engineering Students' Approaches to a Wicked Sustainability Problem. *Research in Science Education.*

McCormick, M., Lawyer, K., Wiggins, J., Swan, C. W., Paterson, K. G., & Bielefeldt, A. R. (2015). Sustainable Engineering Assessment Using Rubric-Based Analysis of Challenge Question Responses. *Journal of Professional Issues in Engineering Education Practice, 141* (2), pp. 1-10.

Messick, S. (1995). Validity of Psychological Assessment: Validation of Inferences From Persons' Responses and Performances as Scientific Inquiry Into Score Meaning. *American Psychologist, 50* (9), pp. 741-749.

Pierrakos, O., Watson, H., Kander, R., Russell, J., & Anderson, R. (2010). Special Session - Not All Problems Are Created Equal: From Problem-based Learning Theory to Research on Complex Problem Solving and Implications for the Engineering Classroom. *Proceedings of the 40th ASEE/IEEE Frontiers in Education Conference.* Washington, DC: IEEE.

Popham, W. J. (1997). What's wrong--and what's right--with rubrics. *Educational Leadership, 55* (2), pp. 72-75.

Reddy, Y., & Andrade, H. (2010). A review of rubric use in higher education. *Assessment & Evaluation in Higher Education, 35* (4), pp. 435–448.

Reznitskaya, A., Kuo, L., Glina, M., & Anderson, R. C. (2009). Measuring Argumentative Reasoning: What's behind the Numbers? *Learning and Individual Differences, 19* (2), pp. 219-224.

Riis, J. O., Achenbach, M., Israelsen, P., Kyvsgaard Hansen, P., Johansen, J., & Deuse J. (2016). Dealing with complex and ill-structured problems: results of a Plan-Do-Check-Act experiment in a business engineering semester. *European Journal of Engineering Education.*

Rittel, H. W., & Webber, M. W. (1973). Dilemmas in a General Theory of Planning. *Policy Sciences, 4*, pp. 155-169.

Schamber, J. F., & Mahoney, S. L. (2006). Assessing and Improving the Quality of Group Critical Thinking Exhibited in the Final Projects of Collaborative Learning Groups. *The Journal of General Education, 55* (2), pp. 103-137.

Wald, H. D., Borkan, J. M., Scott Taylor, J., Anthony, D., & Reis, S. P. (2012). Fostering and Evaluating Reflective Capacity in Medical Education: Developing the REFLECT Rubric for Assessing Reflective Writing. *Academic Medicine: Journal of the Association of American Medical Colleges, 87* (1), pp. 41-50.

Copyright statement

Appendix A. Framework for integratively addressing WPs

To facilitate an integrative approach to a wicked problem, the following ten aspects are important to consider:

I. Problem parts (problem parts)
II. Improvement measures (improvement measures)
III. Interaction between problem parts (problem part interaction)
IV. Interaction between improvement measures (improvement measure interaction)
V. Secondary problems caused by improvement measures (secondary problems)
VI. Stakeholders and their interests in relation to the problem and/or improvement measures (stakeholders)
VII. Actors' spheres of influence (spheres of influence)
VIII. Lack of accessible information (lack of information)
IX. The importance of incomplete control and predictability for the outcome of improvement measures (uncertainty)
X. The importance of the local context for the outcome of improvement measures (local context)

Appendix B: Rubric for assessing written responses to WP

The student demonstrates an ability to consider	Grade 1	Grade 2	Grade 3
I. problem parts	Indirectly indicate at least two problem parts and indirectly indicate how they contribute to the overall problem, e.g. by indicating something that can be improved with specific improvement measures	In general terms, describe at least two problem parts, and in general terms describe how they contribute to the overall problem	In concrete terms, describe at least two problem parts, and in concrete terms describe how they contribute to the overall problem
II. improvement measures	In general terms, describe at least two improvement measures, and indicate how they could contribute to address the overall problem and/or one/several specific problem part(s) that contribute to the overall problem	In concrete terms, describe at least two improvement measures, and in general terms describe how they could contribute to address the overall problem and/or one/several specific problem part(s) that contribute to the overall problem	In concrete terms, describe at least two improvement measures, and in concrete terms describe how they could contribute to address the overall problem and/or one/several specific problem part(s) that contribute to the overall problem
III. interaction between problem parts	Indirectly indicate at least one example of possible interactions between problem parts, e.g. by indicating that one problem parts cannot be addressed without also addressing another problem part	In general terms, describe at least one example of possible interactions between problem parts, e.g. by describing, in general terms, how one problem part can increase or decrease the impact of another problem part on the overall problem	In concrete terms, describe at least one example of possible interactions between problem parts, e.g. by describing, in concrete terms, how one problem part can increase or decrease the impact of another problem part on the overall problem, and, on the basis of the described interaction(s), prioritize which problem part(s) should be addressed primarily
IV. interaction between improvement measures	Indirectly indicate at least one example of possible interactions between improvement measures, e.g. by indicating that one improvement measure can create the conditions required for implementing another improvement measure, or that one improvement measure can increase or decrease the impact of another improvement measure	In general terms, describe at least one example of possible interactions between improvement measures, e.g. by describing, in general terms, that one improvement measure can create the conditions required for implementing another improvement measure, or that one improvement measure can increase or decrease the impact of another improvement measure	In concrete terms, describe at least one example of possible interactions between improvement measures, e.g. by describing, in concrete terms, that one improvement measure can create the conditions required for implementing another improvement measure, or that one improvement measure can increase or decrease the impact of another improvement measure, and, on the basis of the described interaction(s), evaluate the appropriateness of the described improvement measures
V. secondary problems caused by improvement measures	Indirectly indicate at least one example of secondary problems that may be caused by a specific improvement measure, where the secondary problem(s) is/are relevant for the overall problem	In general terms, describe at least one example of secondary problems that may be caused by a specific improvement measure, where the secondary problem(s) is/are relevant for the overall problem	In concrete terms, describe at least one example of secondary problems that may be caused by a specific improvement measure, where the secondary problem(s) is/are relevant for the overall problem, and, on the basis of the described secondary problem(s), evaluate the appropriateness of the described improvement measure(s)

The student demonstrates an ability to consider	**Grade 1**	**Grade 2**	**Grade 3**
VI. stakeholders and their interests in relation to the problem and/or improvement measures	Indirectly indicate at least two relevant stakeholders and indirectly indicate their interests in relation to the problem and/or possible improvement measures	Identify at least two relevant stakeholders, and describe, in general terms, their interests in relation to the problem and/or possible improvement measures	Identify at least two relevant stakeholders, and describe, in concrete terms, their interests in relation to the problem and/or possible improvement measures, and describe, in concrete terms, at least one example of how differences between their interests could be addressed to improve the outcome of at least one specific improvement measure
VII. actors' spheres of influence	For an actor in the given professional role, delineate, in general terms, a reasonable sphere of influence in relation to the problem and/or possible improvement measures, e.g. by describing, in general terms, s.th. within and/or s.th. beyond the actor's sphere of influence	For an actor in the given professional role, delineate, in concrete terms, a reasonable sphere of influence in relation to the problem and/or possible improvement measures, e.g. by describing, in concrete terms, both s.th. within and s.th. beyond the actor's sphere of influence	For an actor in the given professional role, delineate, in concrete terms, a reasonable sphere of influence in relation to the problem and/or possible improvement measures, e.g. by describing, in concrete terms, both s.th. within and s.th. beyond the actor's sphere of influence, and describe, in concrete terms, how an actor in the given professional role could cooperate with another actor who has a complementary sphere of influence
VIII. lack of accessible information	Identify at least one piece of information that is not currently available for an actor in the given professional role, but that could be gathered and then provide valuable input for addressing the problem	In concrete terms, describe at least one piece of information that is not currently available for an actor in the given professional role, but that could be gathered and then provide valuable input for addressing the problem	In concrete terms, describe at least one piece of information that is not currently available for an actor in the given professional role, but that could be gathered and then provide valuable input for addressing the problem, and evaluate the appropriateness of realizing at least one specific improvement measure as long as this piece of information is not available
IX. the importance of incomplete control and predictability for the outcome of improvement measures	Identify at least one example of incomplete control over, or predictability of, the outcome of at least one specific improvement measure, e.g. by identifying one example of random processes that can influence the outcome of an improvement measure	In concrete terms, describe at least one example of incomplete control over, or predictability of, the outcome of at least one specific improvement measure, e.g. by describing, in concrete terms, one example of random processes that can influence the outcome of an improvement measure	In concrete terms, describe at least one example of incomplete control over, or predictability of, the outcome of at least one specific improvement measure, e.g. by describing, in concrete terms, one example of random processes that can influence the outcome of an improvement measure, and, on the basis of the described example of incomplete control and/or predictability, evaluate the appropriateness of the improvement measure(s)
X. the importance of the local context for the outcome of improvement measures	Identify at least one example of how s.th. in the local problem context can influence the outcome of at least one specific improvement measure	In concrete terms, describe at least one example of how s.th. in the local problem context can influence the outcome of at least one specific improvement measure	In concrete terms, describe at least one example of how s.th. in the local problem context can influence the outcome of at least one specific improvement measure, and, on the basis of the described influence, evaluate the appropriateness of the mentioned improvement measure(s)

Appendix C. Descriptions of WPs used in the intervention

Literacy in Afghanistan

Assume the following professional role: You are employed at a large, international IT company. The company has a broad portfolio of services and products, and it wants to be perceived as a company that actively works to support all three dimensions of sustainable development.

The company's sustainability department continuously monitors developments in society to identify possible issues in which the company could get involved. The department has collected the following information about the educational situation in Afghanistan from the Internet:

On 25 September 2015, the UN international conference adopted 17 global Sustainable Development Goals. Goal 4 is to "ensure inclusive and quality education for all and promote lifelong learning" by the year 2030." This goal includes an ambition to ensure that all people should have access to inclusive and high quality primary and secondary education. Current population statistics suggest that, worldwide, 250 million children in 4th grade and 774 million people overall lack basic reading and writing skills. In the latter group, two thirds are women.

An example of a country that faces major problems in relation to the above stated goal is Afghanistan. Current estimates suggest that only 38% of Afghanistan's population is able to read and write; for women, the percentage is only 24%. Reading and writing literacy is generally lower in remote areas of the country than in urban areas.

Afghanistan has a population of 32 million, and 76% of the population lives in rural areas. The capital, Kabul, has 3.3 million inhabitants. All other cities in the country have fewer than 500 000 inhabitants. According to the Human Development Index, Afghanistan was the 15th least developed country in the world in 2011. One of the factors that hinder development in the country is deficient energy supply, especially in rural areas. Another factor is many years of violent conflict. For example, Afghanistan is one of the world's most affected countries with regard to violence against schools. Between 2006 and 2009, 439 teachers and other school personnel have been killed. The country faces a severe lack of qualified teachers.

The sustainability department suggests that the above issue could be a suitable focus for your company. You are asked to investigate how the company could contribute to improving the educational situation in Afghanistan in a way that would support sustainable development.

Describe how you would approach the task of identifying suggestions for how your company could contribute to improving the situation. Describe also which factors of the situation you would take into account, and which improvement measures could be included in your suggestion. Motivate your answer.

Dengue fever in Sub-Saharan Africa

Dengue fever is a viral disease that is spread by mosquitoes in areas with tropical and subtropical climate. In recent decades, the number of dengue fever infections has increased dramatically. Recent research estimates that 390 million people are infected with dengue fever every year.

The World Health Organization (WHO) describes dengue fever as one of 17 neglected tropical diseases (NTDs) because it primarily affects people in poor parts of the world, and because it has not received as much attention as e.g. HIV, malaria, and tuberculosis. WHO has established a specific division that aims to combat NTDs. Also, the United Nations' Sustainable Development Goals contain a sub-target which includes an ambition to eradicate all NTDs by 2030.

Assume the following professional role: You are employed at WHO as an IT specialist. As one part of the organization's ambitions to eradicate all NTDs, WHO has created a number of interdisciplinary expert committees that are asked to focus on specific diseases and geographic areas. You have been assigned to work with a group that aims to combat dengue fever in sub-Saharan Africa. As a starting-point for the group's first meeting, you have been asked to investigate how IT could contribute to combat dengue fever in that area. Write a short summary of the problem situation and what you see as possible IT-related improvement measures. The text shall not exceed a maximum of 1000 words plus references.

Evaluation of the process of teaching learning through the Flipped Classroom Model

Rogerio Tondato
UTFPR, Londrina, Brasil
rogeriotondato@utfpr.edu.br

Silvana Rodrigues Quintilhano
UTFPR, Londrina, Brasil
squintilhano@utfpr.edu.br

Abstract: *The development of Information and Communication Technology in recent years has transformed the way people have access to information. Thus, the traditional process of transmitting knowledge in the teaching-learning system, both in basic education and in higher education, has become obsolete, since the medium that this digital society acquired knowledge, generating knowledge, is the technological support. It is common to check in the classrooms students with no motivation to participate in classes or attentive exclusively to their technological devices, generating a problem for the teacher's performance in the classroom in the process of teaching learning. Thus, new forms of teaching learning should be considered by educators, such as the inclusion of information technology and the realization of classes that are much more practical than theoretical, with real learning experiences and interactive possibilities, as this new network generation requires. Thus, the objective of this article will be to evaluate the differences between the traditional model of knowledge transmission and the flipped classroom model that will be applied to the students of the Production Engineering course at the Universidade Tecnológica Federal do Paraná - Campus Londrina.*

Introduction

The changes that occurred between the 1970s and 1990s, especially in the technological field, as Pereira and Silva (2010) point out, have given revolutionary directions to the world on social and development issues. Through technology jumps, maintained by the constant and immediate use of what is generated, the world is now connected mainly through the Internet and also through the intermediary of Information Technologies. In this way, we find ourselves living in a new social environment, the Digital Society.

It is a society that is defined by means and instruments in constant evolution and, with this, we are faced with a new composition of aspects of greater social relevance: fast obtaining of information, greater acquisition of material for the construction of knowledge, Social, economic and cultural interactivity, the union of leisure, pleasure, work and research in the same environment - the Internet (KOHN & MORAES, 2007).

According to Pereira e Silva (2010), this information and knowledge society has a frequent capacity for innovation, which is renewed on the basis of information - the transmission of messages of meaning, goal-oriented and conscious purpose - reaching knowledge. The medium that this new society, the Digital Society, uses to acquire information, which generates knowledge, is the technological support. Therefore, technology becomes the generator of knowledge.

As Marcuschi (2010) reiterates, the internet has become a "kind of prototype of new forms of communicative behavior", so technological development "reconfigured the way people act, relate and exist" (KOHN & MORAES, 2007).). The Internet has opened up space for the personal, professional and educational administration of citizens, developing large-scale distance education and / or networking formats (PEREIRA AND SILVA, 2010) for the ease of accessing data and information that influence and contribute to the Academic scope and professional, in short, a new projection eminently virtual.

Thus, the new social configuration that we are facing contemplates, above all, the application of the knowledge that we have in constant action between the use and the innovation, making primordial the sectors: technological, communicational, informational, digital and mediatic (KOHN & MORAES, 2007). With this, " we proceed to process knowledge in a multimedia way, joining together pieces of texts from several languages simultaneously " (SCHIAVONI, 2008). It is the technological resources that are available that allow such practice to complement the exchange of information and the construct of knowledge, in constant elaboration and dialogue, in a virtual interpersonal relationship; Establishing a link between technology and knowledge.

In this sense, the information ceases to belong to a fixed location, it is present in the global scope, resenting space and time, extinguishing distances and streamlining practices. It is through the interactivity that is fully present in the network established by the Internet that the individual ceases to act as a viewer and assumes the role of user (SCHIAVONI, 2008). Thus, the user is able to participate in the message, from the materialized content, changing its route.

The new information and communication technologies also innovate the educational scope, reorganize the interaction process between teacher and student, as well as reconfigure the face-to-face space of teaching and learning. It is very common for the student to bring to the classroom the most varied technological devices, which often generates a misalignment of attention in relation to the class proposed by the teacher.

In this sense, it is up to the teacher to adapt these new technologies in favor of the teaching-learning process; Creating and making available new horizons, new approaches, information and interactive possibilities, in a practical and fast way. In higher education, the dynamics are no different, the training process requires methodologies that support the new technological supports, which coexist classroom and classroom environments, with new virtual learning environments, in an interactive and dynamic process.

Therefore, the objective of this article will be to evaluate the methodological application of the traditional model and the flipped classroom model, verifying students' perception of learning in the topic of simple linear regression in the discipline of Data Analysis of the Production Engineering course of the University Federal Technological University of Paraná - Campus Londrina.

Theoretical Framework

The flipped classroom is a methodological strategy that attempts to change the paradigms of the teaching-learning process, changing the logic of traditional organization. Learning does not register only in the context of the classroom, but in activities outside it, using technological inputs.

To paraphrase Trevelin et.al (2013) the first studies on the Flipped Classroom are from the 1990s, conducted by Eric Masur at Harvard University. Similarly, in 2000, Professors Lage, Platt and Treglia of the University of Miami in the United States applied the method, called the "Flipped Classroom" in subjects of introduction to economics. In 2007, there was an account of the experience of using this method in higher education courses at Midwestern Christian Liberal Arts University, Strayer University.

In 2008, Woodland Park High School chemistry teachers Aaron Sams and Jonathan Bergmann developed a method of recording all of their video lessons, as Aaron asked: "What if we recorded all the classes, and if students watched the video As 'homework' and then used all the time in the classroom to help them with the concepts they did not understand? "(2016). Effectively, the first teachers to use screencast videos in the classroom ostensibly. From there they instigated effective research on this new teaching-learning performance: the flipped classroom.

According to Bergmann and Sams (2016) what he basically reverses is: "what is traditionally done in the classroom is now run at home, and what is traditionally done as work from home is now done in the classroom ".

The structure of this new methodological model is characterized according to table 1:

Table 1: Structure of the Flipped Classroom

BEFORE	DURING	AFTER
Preliminary presentation of content	Practice and Feedback	Deepening knowledge

According to Bergmann & Sams (2016) in the classroom model flipped time is restructured. The first moments of the lesson are to ask questions about the video assisted at home, then developed independent and guided practices and / or laboratory activities, which the authors call a "practical problem package." In this way, the student applies the theoretical content assisted in the simulated practice in the classroom, returning the learning to the solution of problems.

The student participates actively in the teaching-learning process, while the teacher acts as a mediator. Students watch the videos, perform practical activities in the classroom, create situations-problems, look for possibilities of solutions. To the extent that the teacher promotes feedback of specialized content, it facilitates the development of practices and motivates, in a challenging way, the learning. As Bergamann & Sams (2016) points out, "the teacher's role in the classroom is to support students, not to convey information."

Another relevant factor of this model is the use of technological resources. The digital society reconfigures itself from the internet and other supports of information technology and communication. Students have grown up in this digital world and easily handle any digital device. In this sense, the flipped classroom creates new digital environments for the promotion of knowledge.

This model understands that information and communication technologies are allied in the process, facilitating the use of time in the classroom and stimulating the participation and use of students. Bergamann & Sams (2016) explain that content transmitted through online videos helps students to respect their own pace of study and learning, and to organize study time in a flexible way.

On the other hand, the type of material or activities that the student performs online and in the classroom vary according to the proposal being implanted, creating different possibilities for this pedagogical approach.

Thus, the basic rules for reversing the classroom are: (1) classroom activities involve a significant amount of questioning, problem solving, and other active learning activities, forcing the student to recover, Apply and extend the material learned online; 2) Students receive feedback immediately upon completion of face-to-face activities; 3) Students are encouraged to participate in on-line and face-to-face activities, which are computed in the student's formal assessment, ie, they are worth note; 4) Both the material being used online and the classroom learning environments are highly structured and well-planned.

According to Bergamann & Sams (2016), to implement the flipped classroom, one must analyze the best technology to be used to pass the content. To do so, one should seek colleagues, professionals and even students to validate if the model to be used is recommended. In the context of the authors, the most used tool is the video lessons, which can be produced by the author or can be used from other classes already recorded. However, the important thing in this case is to produce content that holds the attention of the student, so that the student is interested in watching and learning with the content.

The stages for the production of the video should be as follows (BERGAMANN & SAMS, 2016):

- - Plan the class in order to present the educational objectives of the class;
- - Record the video through appropriate tools, using presentation software, including the recording of the teacher's words;
- - The video should be edited in order to be clear, concise and short, in order to encourage students to attend.
- - Finally, you have to divuklgar the video so that the same can be seen by the students, preferably online.

The next step is in the classroom, applying the contents seized in video, and the role of the teacher is to mitigate the flaws and / or deficiencies presented by the students, either through the resolution of exercises or application projects.

According to Valente (2014), as an example of applying the flipped classroom, Harvard University introduced the Peer Instruction (PI) method, developed by Proff. Eric Mazur. The PI consists of providing background material so that the student can study the content before attending the classroom. Based on the material studied, the student responds to a set of questions via a Learning Management System (LMS). The teacher, before giving the lesson, checks the most problematic questions that must be worked out in the classroom. During class, the discussions are interspersed with Concept Tests, designed to expose the difficulties that students encounter. These tests are answered via an interactive response system, clicker type, so that the class and the teacher can follow the level of understanding about the concepts under discussion. Depending on the answer, if a large percentage of students in the class (usually between 30 and 65%) respond incorrectly, then students should discuss this in small groups, forcing students to think about the arguments to be developed, and allows them (As well as the teacher) can assess the level of understanding about concepts before they even leave the classroom.

After this discussion the students answer the conceptual question again. The instructor provides feedback, explaining the correct answer and completing with related questions if necessary. The cycle is then repeated with another theme, each cycle typically being 13-15 minutes. Using this strategy, it was verified that students present significant gains in conceptual comprehension, evaluated with standardized tests, as well as gain skills to solve problems comparable to those acquired in traditional classes (CROUCH; MAZUR, 2001, apud VALENTE, 2014).

Methodology

The method used in this research is Research-Action, a qualitative basis of approach to action-oriented problems. According to Turrioni and Mello (2012), "it is a type of empirical social research that is conceived and carried out in close association with an action." In this case, the action of the applicability of the Methodology in the course of Production Engineering, seeks the possibility of resolution in the teaching-learning process, for which researchers (teachers) and participants (students) are involved in a participatory manner.

As Tussioni and Mello (2012) describe: "In research action, the term research refers to the production of knowledge and the term action refers to an intentional modification of a given reality." Therefore, methodologically, for the analytical comparison of the traditional classroom and the flipped classroom wing, a theoretical reference was made about the Flipped Classroom, then a traditional lesson plan was elaborated that had been applied in the first semester of 2016 on the content of Advanced Statistics; And an flipped classroom plan was developed that had been applied in the second half of 2016 to the Data Analysis Discipline. Both classes of the Production Engineering course of the Federal Technological University - Campus Londrina.

Application of the Flipped Classroom in the Production Engineering Course

Methodologically, the application was carried out in two semesters: 1 semester / 2016 and 2 semester / 2016, in the discipline of Advanced Statistics.

In the first half of 2016 the traditional process of knowledge transmission was carried out in the initial content of simple linear regression. In the second half of 2016 the flipped classroom was used for the same simple linear regression concept.

Initially, both for the conventional classroom and for the flipped classroom, discipline planning divided the simple linear regression content is divided into the following concepts:

- Class 1: Linear Regression Linear Equation, Construction and Application;
- Class 2: Coefficient of Determination;
- Class 3: Hypothesis Testing.

For the traditional classroom, which was held in the first semester of 2016, 3 meetings of 2 hours each were used to review all the contents on simple linear regression, with the presentation of examples and calculation methods. Subsequently, a list of exercises was passed on to the students to solve at home and deliver this list resolved as part of the evaluation. It is worth mentioning that despite the weekly availability of the teacher in an hour additional class to ask questions of the students, few use this moment.

For the second semester of 2016, where the flipped classroom was applied, video lessons were chosen as a way of transmitting the anticipated knowledge and their inclusion on the internet page.

To elaborate the videos of the flipped classroom were used the planning contents that were divided in 3 topics and that allowed the use of videos of a maximum of 15 minutes. Thus, for each of the three meetings, the videos were made available, two days in advance, so that the students could attend and participate in the class.

To elaborate the videos of the flipped classroom were used the planning contents that were divided in 3 topics and that allowed the use of videos of a maximum of 15 minutes.

After the inclusion of the videos on the Internet, for each of the three meetings, the meeting was held in the classroom, where learning exercises were applied on the content assisted and the consequent attendance to the deficiencies presented by the students. As a work from home, you can include more complex exercises and even application activities in real situations. Table 2 presents the summary of the methodology used.

Table 2: Comparative Table of Methodological Applications

Traditional		Sala de Aula Invertida
There is no initiative on the part of the students for any type of previous research on the content that will be worked in the classroom.	**BEFORE THE CLASS**	Students watch the interactive learning module / video on "Simple Linear Regression" of approximately 15 minutes duration.
The professor presented an expositive lecture on "Simple Linear Regression", using as a didactic resource the blackboard.	**DURING THE CLASS**	Laboratory-oriented practice: Case study developed in groups.
Carrying out exercise at home.	**AFTER THE CLASS**	More complex task: application of the concept in a real situation.

For both cases, the evaluation of learning at the end of the class was carried out to indicate the gain that each of the models raised in the students' knowledge. This evaluation included the performance of individual exercises for the student to survey the learning process.

Results

Learning assessments show an approximately 35% gain in knowledge acquisition through learning assessment (Table 3).

Table 3: Student Revenue Results

	Traditional Classroom		Flipped Classroom	
	1° Sem/2016	**%**	**2° Sem/2016**	**%**
Total Students	30	100	35	100
Average greater than 6.0	11	36,7	18	51,4
Average less than 6.0	19	63,3	17	48,6

In table 3, it can be observed that of the thirty students who performed the activities in the traditional classroom (1st semester), 36.7% were able to effect the knowledge and were above the average grade. When compared to the flipped classroom, it managed to snatch just over half of the students, 51.4%, who managed to get the grade above the average of the discipline.

It can be observed that the flipped classroom model is a good way to attract the participation of this new student that is inside the classroom, since at the time of the lesson, the content is applied in a practical way, staying under The responsibility of the student to contact the theoretical content through digital media. As this student is familiarized with the digital media, a greater participation of the same ones in the realization of the knowledge is perceived.

In spite of the significant gain, the teacher of the discipline emphasizes as main difficulty of the application of the method, is the understanding on the part of the students of how this new modality of class works. The student is not yet accustomed to this type of inversion, that is, for him, to the teacher to pass a video with contents that should be taught by the teacher, it is just a way for the teacher not to have the work to carry out the activity.

On the other hand, the student, not yet accustomed to such a form of teaching teaching process, ends up performing inappropriately his extra class class activities flipped, often not watching the video, or leaving to watch it "in in time". What it indicates, in the same way of the traditional class, that the student is not yet prepared to be auto didáta in small concepts.

Another difficulty was the disparity of previous knowledge among the students, that is, some have a higher degree of knowledge in the basic concepts related to the discipline, which facilitates learning and hinders the student with low achievement in previous disciplines.

On the other hand, since the content is limited to the subject of the video, the biggest difficulty is the doubts that the student encounters when watching the video, which in many cases ends up causing the student to not continue with the subject. It can be verified in some cases that the student did not have the capacity nor to research the subject in which he did not know (but should already have the knowledge).

In general, the flipped classroom presents new horizons to students, who instead of having their time in the classroom spent on basic learning activities, end up in turn understanding more fully the content and even applying it in real situations the content learned. However, this new form of learning is still very incipient, making it difficult to apply, since students are not yet familiar with it, since their basic education and literacy are still in the traditional method.

Recommendations and Implication for Education in Production Engineering

The experience with new practices leads the teacher to reflect on his role as mediator of the teaching-learning process. The classroom flipped behind a new lens in this process of teaching learning. It makes the learner part of the process, facilitates learning, and makes the learner largely dedicated to adjusting the deficiencies of the students.

However, these concepts should be applied more forcefully in the course of Production Engineering, in view of their greater possibility of practical action within the contents. That is, the student would be much closer to practical application and reality, which causes the egress already out of college with much greater application knowledge.

References

Bergmann, J. & Sams (2016). A. Sala de aula invertida: uma metodologia ativa de aprendizagem. Trad. Afonso Celso da Cunha Serra. 1ed. Rio de Janeiro: LTC.

Kohn, K.; Moraes, C. H. (2007). O impacto das novas tecnologias na sociedade: conceitos e características da Sociedade da Informação e da Sociedade Digital. *XXX Congresso Brasileiro de Ciências da Comunicação*. Accessed at: http://www.intercom.org.br/papers/nacionais/2007/resumos/R1533-1.pdf. on: 12/04/2016.

Pereira, D. M.; Silva, G. S (2010). As Tecnologias de Informação e Comunicação (TICs) como aliadas para o desenvolvimento. *Cadernos de Ciência*. Accessed at: http://periodicos.uesb.br/index.php/cadernosdeciencias/article/viewFile/884/891. on 12/04/2016.

Schiavoni, J. E. (2007). Mídia: o papel das novas tecnologias na sociedade do conhecimento. In Cardoso, C. M. (Org.) Diversidade e igualdade na comunicação - coletânea de textos do Fórum da Diversidade e Igualdade: cultura, educação e mídia. Bauru: Unesp. Accessed at: http://diversitas.fflch.usp.br/files/active/0/jaqueline_schiavoni.pdf. on: 12/04/2016.

Trevelin, A. T.; Pereira, M. A. A.; Oliveira Neto, J. D. (2013). A utilização da "sala de aula invertida" em cursos superiores de tecnologia: comparação entre o modelo tradicional e o modelo invertido "flipped classroom" adaptado aos estilos de aprendizagem. *Revista de Estilos de Aprendizagem*, n.11, vol 12. Outubro, 2013. Accessed at: http://www2.uned.es/revistaestilosdeaprendizaje/numero_12/articulos/articulo_8.pdf. on 20/02/2017.

Turrioni, J.B & Mello, C. H. P.(2012). Pesquisa-ação na Engenharia de Produção. In: Metodologia de Pesquisa em Engenharia de Produção e Gestão de Operações. 2.ed. Rio de Janeiro: Elsevier: ABEPRO,

Valente, J. A. (2014). Blended Learning and Changes in Higher Education: the flipped classroom proposal. *Educar em Revista*, Curitiba, Brasil, Edição Especial n. 4/2014, p. 79-97. Editora UFPR

Copyright statement

 This article was supported by the PROPPG of the UTFPR Londrina campus.

Student Outcomes mapping via Professional Internship for Undergraduate Programs

Angélica Burbano Collazos
Universidad Icesi, Cali, Colombia
aburbano@icesi.edu.co

Juliana Jaramillo Ospina
Universidad Icesi, Cali, Colombia
jjaramillo@icesi.edu.co

Maria Isabel Velasco Arango
Universidad Icesi, Cali, Colombia
mvelasco@icesi.edu.co

***Abstract**: The Center for Professional Development (CPD) at Universidad Icesi coordinates the Professional Internship experience for all undergraduates. This experience is used to gather evidence associated with the development of personal and professional skills. Motivated by the ABET (Accreditation Board for Engineering and Technology) accreditation process this work aims to consolidate the information already gathered by the CPD in the form of the final evaluation report done by the employers. This consolidation will help to understand the relationship between the behaviors evidenced at work and the skills developed throughout the curriculum.*

Background

Since the creation of the IE undergraduate program in 1998 the professional internship experience has been offered and coordinated by the Center for Professional Development (CPD). This internship experience has an explicit place in the curriculum which means it has a course code (05198) and a syllabus.

The professional internship experience is a six month student´s exposure to the real world, usually takes place during the last semester of the program. During this semester the student works under the employer´s supervision accompanied by a tutor from the CPD. The information gathered, provides feedback and useful information on how to improve the professional internship experience itself and also to the Industrial Engineering program continuous improvement process.

Research Questions

The research question that motivated this paper is the following:

How could the professional internship experience be mapped to the curriculum and what evidence can be gathered that explicitly illustrate the level of achievement of a set of skills?

Theoretical Framework

This is an important research question for engineering education since evidence of student´s performance in the real world would represent the realization of the final intent of the CDP and the curriculum which is to prepare students for the workplace.

The context in which the education is finally proven represents a challenge and it should be taken in to consideration when an improvement action is going to take place (Enrique et al., 2007). In this case the professional internship experience provides useful information to improve both the professional internship itself and the curriculum.

The proposed set of skills defined by ABET (Crawley & Lucas, 2011) are considered as an international agreement regarding what a student should be able to after finishing an engineering educational program. In this work student outcomes a – k are defined as the set of skills used for assessment purposes. The assessment, evaluation and improvement of these student outcomes are an important component of the quality assurance process at the institution and as defined by (Meek, 2000) the quality assurance (QA) concept is defined as: "*systematic management and assessment procedures adopted by a higher educational institution or system to monitor performance and to ensure achievement of quality outputs program assessment and course assessment.*" This definition could be seen from the institutional culture perspective, the program perspective, or the course perspective. In this paper we are going to develop the program perspective.

Methodology

At the end of the semester the employer evaluates the student´s performance by filling the attached form (Appendix 1). This form has four sections. Sections I-III are a questionnaire with closed questions related to the way the employer evaluates the student´s performance. Section IV includes a set of open questions related to student's contributions to the company and also on how to improve the internship experience from the employer´s perspective.

The first part of the survey (Sections I-III) was mapped to the ABET a – k student outcomes for program assessment purposes during the 2014- 2016 first ABET accreditation cycle at the institution. The second part of the survey (Section IV) is part of an ongoing research project which aims to categorize the information related to the student's contributions. This is performed by categorizing the information, then processing the information using Atlas.ti software, results will be reviewed by the curriculum committee and will be an input for the IE program continuous improvement process. This project is expected to be finished by the end of 2017.

In this paper the mapping and analysis of the first part (Sections I-III) of the employer´s survey is presented. This was done via curricular mapping of the 05198 course. Using CDIO syllabus at the microcurricular level and ABET student outcomes the course objectives were mapped to the student outcomes, the student outcomes associated with the 05198 course are presented in Table 1. Then the assessment results of the student outcomes associated with the course were analyzed for course and program assessment purposes.

The assessment of a given student outcome is performed via an Assessment plan and ends with an Assessment report that includes improvement actions in the short medium and long term. When the report is presented in the curriculum committee plans are validated and integrated the program's continuous improvement process. This process includes the improvement plans that come via course and program assessment and evaluation as well as results of the curriculum committee meetings with different groups in order to validate improvement actions.

For the CPD the mapping of the employer´s survey to the student outcomes provides useful feedback to the CPD core mission which is to prepare students for the workplace and to provide evidence of student's performance. The CPD is interested on continuing sharing information with undergraduate programs as well to adjust and improve its current evaluation methods in order to align efforts to better measure student's performance following international standards.

The professional Internship course provides assessment data for student outcomes D (Ability to function on multidisciplinary teams), E (Ability to identify, formulate and solve engineering problems), F (Understanding of professional and ethical responsibility), G (Ability to communicate effectively) and I (Recognition of the need for and an ability to engage in life-long learning). Table 1 presents the association of student outcomes with their respective performance indicator, a performance indicator is a measure defined by the assessment committee as a valid indicator of what the student outcome looks like as a student´s observable behavior. These performance indicators were mapped to specific sections and questions of the employer´s survey, these mapping can be shared upon request to the contact author.

Student Outcome DESCRIBE AS IN ABET	Performance Indicator
Outcome D – Teamwork	D-PI4 - Multidisciplinary teaming: To interact effectively in multidisciplinary teams
Outcome E – Problem Solving	E-PI3 – Solution and recommendation: Solve the problem, by proposing a strategy compatible with its formulation
Outcome F – Ethics	F-PI2 To demonstrate an appropriate professional behavior during the professional internship
Outcome G – Communications	G-PI3 (Argumentation). Build and defend arguments evidencing coherence between conclusions and their supporting premises.
Outcome I – Life-long learning:	I-PI3 -To demonstrate the ability to engage in lifelong learning by effectively conducting self-learning experiences to achieve the expected results.

Table 1. Student Outcomes mapped to professional internship (Course code 05198)

Findings

For the 2014-2016 ABET accreditation cycle data related to the described student outcomes and the specific performance indicators was collected and analysed, 48 students were evaluated. In table 1 only the performance indicators directly mapped to one section of employer´s survey are shown.

Results for the student outcome F and I are presented. In Figure 1 the performance indicator 2 for student outcome F defined as F-PI2 is decomposed in 14 factors, these factors were evaluated by employers considering the observable behaviour during the professional internship experience. The defined threshold of 70% is exceeded in all evaluated factors. Among the highest are work quality (F- PI2- F3), resource usage (F-PI2-F5), following institutional rules and procedures (F-PI2-F8) and adaptability to change (F-PI2-F11).

These results were presented at the assessment and evaluation committee level for review and decision making. No improvement plans are associated with this performance indicator since the performance was above the threshold.

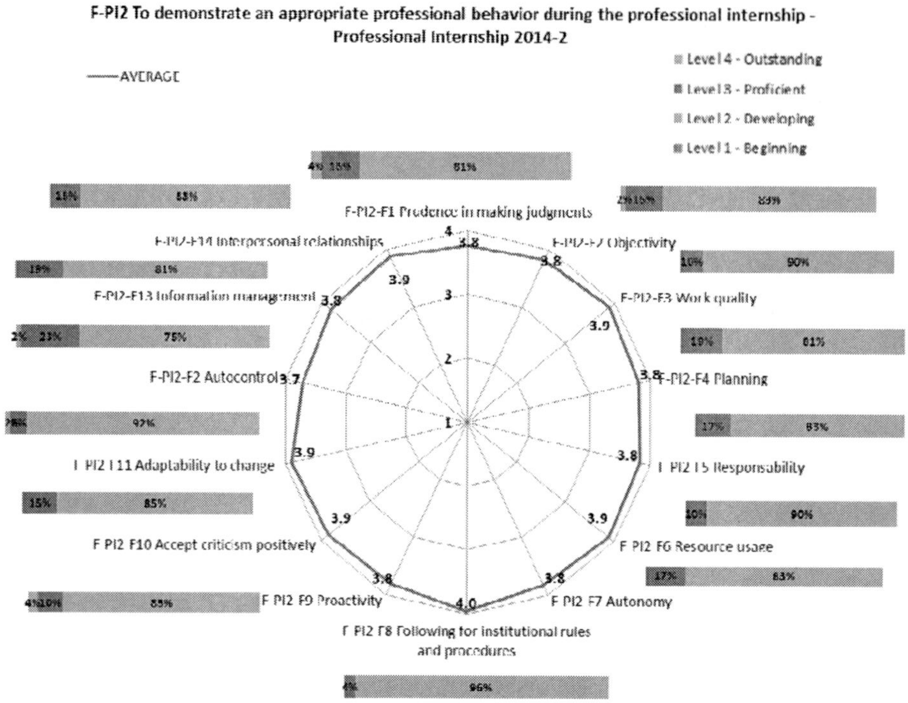

Figure 1. F-PI2

In Figure 2 the performance indicator 3 for student outcome I is shown. This performance indicator is defined as I-PI3 and it is decomposed in 4 factors according to the information evidenced in the employer survey. The defined threshold of 70% is exceeded in all evaluated factors.

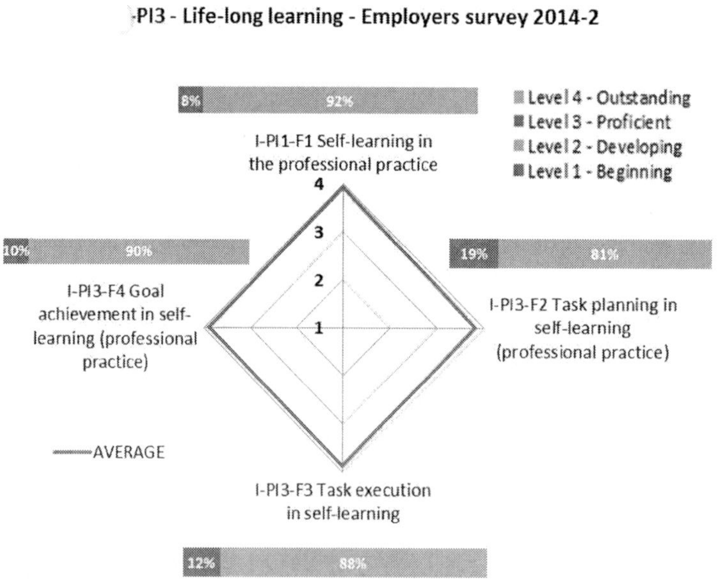

Figure 2. I-PI3

Employers rated 81% of the students at an outstanding level in factor 2 (Q2. Task planning); rated 87.5% of the students at an outstanding level in factor 3 (Q3. Task execution), and 90% of the students at an outstanding level in factor 4 (Q4. Goal achievement in self-learning). Since it was accomplished it is recognized the need of having results consistent over time. It is also important and included in the improvement plans for this student outcome, to motivate students to participate in initiatives like the Solar Decathlon 2018. To develop a multidisciplinary project within the school of engineering.

The answer to the original question: How could the professional internship experience be mapped to the curriculum and what evidence can be gathered that explicitly illustrate the level of achievement of a set of skills? Is provided in two parts.

First, the mapping was done at the syllabus level and at the student outcome level by mapping of performance indicators to the sections and questions of the employer´s survey.

Second, there is evidence of the level of achievement of a set of skills. In this paper evidence of student´s performance on student outcomes F and I is presented in a graphical way via radar figures that illustrate the results of the factors associated with the student outcomes used as an example. Student outcomes assessment plans a reports and be provided upon request.

The preliminary results are a useful input for the IE program continuous improvement process since it provides direct evidence of skills development at the summative level and it can be linked to the formative assessment process as proposed by (Reeves, 2006).

Recommendations and Future research

This paper reports on the findings associated with linking the information already gathered by the CPD mapped with ABET. This could be useful for other institutions with similar centers or new programs considering following the proposed steps to design a professional internship experience mapped or aligned with international standards.

It is important to align to the curriculum the internship program in order to use the information gathered at the work place to feed the IE program continuous improvement process via the curriculum quality assurance process.

This will revert in benefits not only for the IE program since it with have information of students performance at the workplace but also will benefit the CPD because it could provide useful information to improve the program itself and the centers´ relationship with industry.

Next steps in the research include the analysis and mapping of the open ended questions to the student outcomes in particular the student outcome K (Ability to use the techniques, skills and modern engineering tools necessary for engineering practice).

References

Crawley, E. F., & Lucas, W. a. (2011). The CDIO Syllabus v2 . 0 An Updated Statement of Goals for Engineering EducationB. *Engineering Education*, *24*, 1–4. Retrieved from http://files.conferencemanager.dk/medialibrary/59856d54-6d1c-4deb-ac21-0dd98ccd0470/images/CrawleyEtAlCDIOSyllabus2.0Paper_17June2011.pdf

Enrique, R., Caballero, E., Vlademir, A., Moreno, G., Jesús, H. De, & Algarín, H. (2007). L a educación superior frente a las tendencias sociales del contexto. *Educación Y Educadores*, *10*(1), 63–77.

Meek, V. L. (2000). Repositioning Quality Assurance and Accreditation in Australian Higher Education.

Reeves, T. C. (2006). How do you know they are learning?: the importance of alignment in higher education. *Int. J. Learning Technology J. Learning Technology*, *2*(4), 294–309. https://doi.org/10.1504/IJLT.2006.011336

Copyright statement

First Results from Building and Guiding Student Teams for Project-Based Learning in a First-Semester STEM Class

Leonhard E. Bernold
Universidad Tecnica Federico Santa Maria, Valparaiso, Chile
Leonhard.Bernold@usm.cl

Paulina Leddihn
Centro de Innovacion para la Calidad Educativa, Valparaiso, Chile
pleddihn@gmail.com

Arie Aizman
Universidad Tecnica Federico Santa Maria, Valparaiso, Chile
Arie.Aizman@usm.cl

Brayan Díaz
Universidad Tecnica Federico Santa Maria, Valparaiso, Chile
Brayan.Díaz @usm.cl

Hugo Alarcon
Universidad Tecnica Federico Santa Maria, Valparaiso, Chile
hugo.alarcon@usm.cl

***Abstract**: In the future workplace for engineers the effective operation within an interdisciplinary work team is paramount. This paper reports about the design and implementation of a freshmen STEM class, Introduction to Engineering, with a strong emphasis on fostering teamwork. Framed by an authentic project, namely planning a robotic mission inside the nuclear reactor at Fukushima to measure radioactivity, teams of students were charged to develop a path plan for a robot. Only limited training was provided about teamwork but the designed tasks promoted working in teams. Various instruments were used to assess the students' perception and readiness to collaboration. It was interesting to observe how the tools showed poor correlations between each other and especially the final grades in Physics and Math. Qualitative data from weekly journals and focus group meetings clearly showed that students understand the value of teamwork but subsequent classes need to provide opportunities strengthen their initial skills.*

Introduction

The concept of teaching science, technology, engineering, and math (STEM) in an integrated manner, rather than in separate classes, has received increased interest in the past 25 years (Beichner et al., 1999). Wieman (et al., 2010) emphasized: "There are countless reports stressing the economic and societal benefits to be gained from improved STEM) education for all students. But also there is extensive research on alternative teaching methods that increase student learning and are practical to implement."

At the same time, the ability to work in multi-disciplinary teams moved to the second rank of importance for graduating students, just behind oral communication. (Hughes and Jones, 2011). In fact, being able to work in a team commonly appears as a top priority on surveys that involve hiring companies and teachers who apply student-centered teaching. Albertine (2011) reported on a survey by the Association of American Colleges and Universities (AAC&U) that focused on identifying the essential outcomes of college learning in the 21 century. While content knowledge for the different STEM fields are essential, it stresses the increased emphasis on practical and soft skills. Some of those are:

- Inquiry and analysis
- Critical and creative thinking
- Teamwork and problem solving *(project-based)*
- Written and oral communication
- Quantitative literacy
- Information literacy
- Civic and intercultural competences
- Ethical reasoning and action
- Foundations and skills for lifelong learning

Not surprisingly, being able to work effectively in teams also is incorporate into the list of skills required by accreditation boards such as the Accreditation Board for Engineering and Technology (ABET 2009) requires in criteria (d) that every graduating engineers has "an ability to function on multi-disciplinary teams".

Nevertheless a clear definition of the functional capabilities necessary for being an excellent team member is still emerging. What is commonly accepted is the fact that the focus of teamwork training should be on the process rather than on the outcome. The following presents a recommended five item list of characteristics that members of team should exhibit. According to Hughes and Jones (2011) they should:

(1) have a shared collective identity,
(2) have common goals,
(3) are interdependent in terms of their assigned tasks or outcomes,
(4) have distinctive roles within the team, and
(5) are part of a larger organizational context that influences their work and that they in turn can influence

Naturally, these individual characteristics require soft and hard abilities. Lingard (2010) offered an extensive list of skills needed by students:

• Attend meetings and arrives promptly	• Introduce new ideas
• Complete individual tasks promptly	• Openly express opinions
• Gather information as appropriate	• Share opinions and knowledge
• Perform research when necessary	• Listen to views and opinions of others
• Complete tasks with high quality	• Consider the suggestions of others
• Accomplish a fair share of the work	• Adopt suggestions of others when appropriate
• Express him/herself clearly	• Provide help to others

The assessment that have emerged are thus of two types, predictive measures that evaluate person's readiness to work collaborative, for example the tool by Marín-Suárez and Alarcon (2013). Their validated survey intends to measure seven factor such as commitment, assertiveness or argumentation. Other tools focus on measuring the quality of performance of teams moving away from paper based surveys towards observing team in a simulated environment which has shown great promises in medicine (Kiesewetter and Fischer, 2015)

Value of constructive work in teams

Hampden-Thompson and Bennett (2013) confirmed that student engagement increases by engaging students in "constructivist" activities: "Greater levels of student motivation, enjoyment, and future orientation towards science were found in classrooms where students reported that various measures of interaction, hands-on activities, and applications in science took place frequently." In fact, at the end of a 10-yr study, Luckie (et al., 2012) assert that providing students with more time to student-driven inquiries and less material: "...leads to deeper learning in fewer topic areas (less teaching) and can reap gains in scientific thinking and fundamental understanding applicable to a broader range of topic areas..."

Beichner (et al., 1999) demonstrated that inquiry and context-based learning does not always depend on the availability of a laboratory. Using simple devices (e.g., coil spring) measuring tools can be made and used in the classroom:" Of course, the use of technology as a means for students to gather data and analyse it meant that the instructor did not have to be "up front and talking" for most of the class time. He was free to move about the classroom, making inquiries and offering guidance where and when it was needed. The focus was kept on the phenomenon being studied rather than on an authority talk." Based on the summarized principles, a team of USM professors and assistants developed and tested a Project-Based Learning program for semester 1 in 2016 and will be described next.

Introduction to engineering - A Multi-Dimensional course

As the title of this paper indicates, the objective of this course was to expose the new students to a wide variety of aspects related to engineering applying Project-Based Learning and STEM in a holistic manner. A particular goal was the integration with other parallel courses such as Mathematics and Physic to provide authentic engineering problems for the theories learned in those classes.

The pilot class was prepared for the first semester of 2016. Because of the nature of the class, a generic project was selected to allow various engineering disciplines to be included. With this in mind, the 2011 Fukushima disaster was selected as the project framework. On March 11, at 14:46 a large earthquake occurred off the coast of Japan creating a large tsunami that hit Fukushima overwhelming the barriers protecting the nuclear reactors built on the coast. The multi-disciplinary problem was to plan a robotic investigation of the damaged nuclear reactor 1 via the measurement of radioactive levels at various locations inside. Figure 1 presents the architecture of the learning skills, knowledge and experiential activities that were developed around the project.

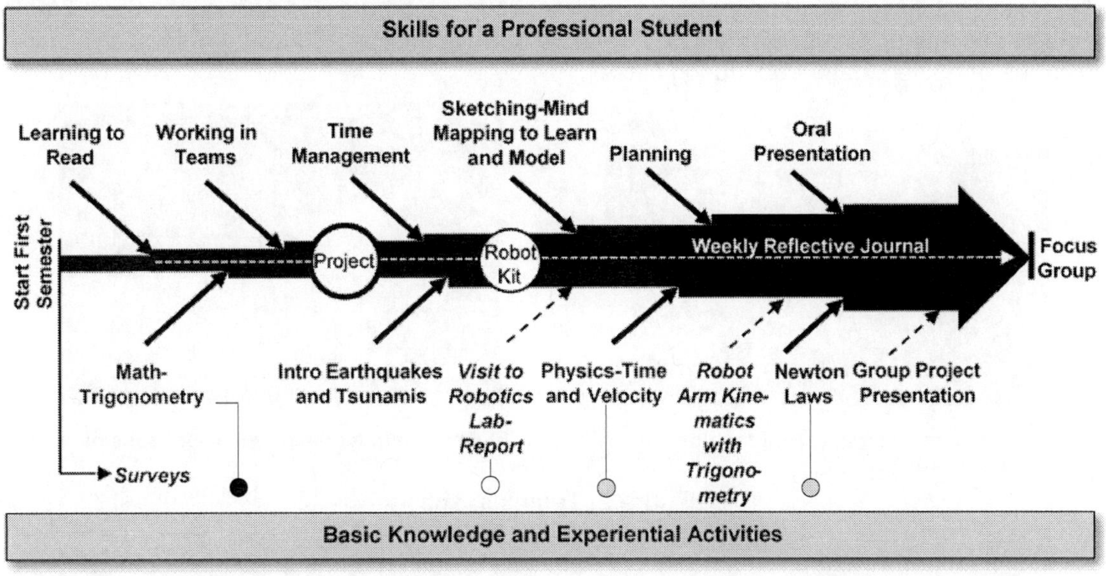

Figure 1: Architecture of holistic and formative introduction to engineering course

As indicated, the class emphasized the generation of both, learning skills and knowledge, with apparent linkages to Mathematics, Physics and the Robotics laboratory. The assessment of students focused on standard surveys, such as Lawson's Scientific Reasoning, Pittsburgh Freshman Engineering Attitudes Survey (PFEAS) (Besterfield et al., 1998) and Cuestionario Habilidades de Aprendizaje Colaborativo (CHAC) (Marín-Suárez and Alarcon, 2013) at the outset of the class. Finally, a selected set Spanish translations of some Learning Ability Analysis Instrument (LASSI) factors were used (Bernold et al., 2007). Parallel to the progression of the class and after its completion qualitative data were collected through weekly journal questions and a Focus group discussion. Finally, the grades in both classes were attained and used for a correlation analysis.

Efforts to guide students to engage effectively in teamwork

The initial question was, how much training the students should receive to be most effective. The research team was led by research reported by Zarske (et al., 2013) who found that although training resulted in: "...an improvement in professional and technical skills over the course of the semester for all students, there was no significant improvement of these skills based on the amount of team dynamics training the participants had received or students' demographics of gender, ethnicity, or first-generation college-bound status." As a result, the training for teamwork was minimal but the project itself was used as a promoter of teamwork. This path is further supported by the observations by Marín-Suárez & Alarcón (2013) who found that Chilean students don't value collaborative work.

Not surprisingly, most students have never learned how to work in a team because the present lecture type teaching actually discourages teamwork. The design of this first pilot class had to consider many concepts new not only to the students but also to the teachers (e.g. authors). In order not to overload the class it was decided to provide a paper-based and a video introduction to teamwork with a signed agreement to commit and to fulfil the duties of predefined roles. Following the recommendations by Vance (et al., 2015), the team designed: ... intriguing group assignments and emphasized the importance of teamwork in the engineering discipline..." To this effect, the project was founded on a letter by Alumnus who had founded a company, requesting assistance in developing a plan that would meet their requirements. These efforts were accompanied by designs that promoted teamwork. For example, the classroom had round tables with online computers built in fostering in-class exercises in teams. The project required the students to work together in order to complete the tasks that invited different learning preferences to contribute and gain confidence. Finally, written and oral project presentations had to be done in groups. Figure 2 presents two pictures of guided team-work during class.

a) Team members present their design

b) Sub-group works on the robot assembly

Figure 2: Teams in action

As indicated earlier extensive amounts of data was collected at the beginning of the course with the goal to establish predictive capabilities in the future. Specifically interesting was to compare PFEAS and CHAC and their relations to the final grades in Math and Physics.

Correlations between teamwork skills and related factors

PFEAS is a Likert type survey with 50 questions can be grouped into three main areas: 1) Perception of the value of engineering, 2) Ability to study and learn, and 3) Abilities to work in teams. CHAC on the other hand, focus specifically to assess students' skills needed to be an effective team member. The 13 factors include: 1) Assertiveness, 2) Value of Working in Teams, 3) Argumentation and 4) Commitment. Figure 3 presents the result of pairwise correlations of some of the most interesting variables.

The most striking initial observation are the very low correlation coefficient r, many times even negative, between the final grades in Physics and Math and any variable related to teamwork and study skills ($-0.1 < r < 0.1$ = extremely weak relationships). It's important to consider, however, that the assessments were done at the start of the class and the grades accumulated over time with a very important final exam providing a major percentage. Can it be that the class' emphasis on developing learning skills made a difference? Further studies have to be made to find the answer. On the other hand, we observe the significant r of 0.603 between those two variables indicating a strong association. These fact might be the result of the common method of using large tests to calculate the final grade. In fact, the r coefficient related to Test taking is the only one that is with 0.042 positive albeit very weak. Astounding is the -0.200 between the Application of Learning Methods and the Physics grade. Of course, it's still a weak correlation but its negative, telling us that the less a student knows about learning to higher the final grade.

The Teamwork component of PFEAS shows moderate to strong correlations with the other PFEAS elements (0.524 and 0.573) but its r-coefficients with CHAC and especially the final grades in Math and Physics are very weak (0.006 and 0.024).

		Final Grade Math	Pittsburgh Freshmen Eng. Attitude		
			Value of Engineering	Abilities to Study	Teamwork
	Final Grade Math	1.000			
PFEAS	Value of Engineering	-0.099	1.000		
PFEAS	Abilities to Study	-0.106	0.315	1.000	
PFEAS	Teamwork	0.006	*0.524*	*0.573*	1.000
CHAC	Assertiveness	0.009	0.113	0.367	0.240
CHAC	Value of Working in Teams	-0.166	0.037	0.049	0.050
CHAC	Argumentation	-0.016	0.139	0.219	0.237
CHAC	Commitment	-0.084	0.140	0.280	0.231
CHAC	Openness to Dialoge	-0.111	-0.152	-0.019	-0.134
	Final Grade Physics	0.603	-0.057	-0.100	0.024
4 Factors Learning	Time Management	-0.191	0.176	0.363	0.202
4 Factors Learning	Taking Tests	0.042	0.038	0.317	0.163
4 Factors Learning	Use of Learning Methods	-0.200	0.159	0.386	0.180
4 Factors Learning	Metacognition	-0.149	0.148	0.405	0.240

Figure 3: EXCEL correlation coefficients for selected factors related to teamwork

The most consistent but moderate r-values result from PFEAS Study component and the 4 dimensions of Learning with Metacognition reaching a positive 0.405.

In summary, the test of how three different assessment instruments related to teamwork correlate showed only weak to moderate r-values. Most interestingly, all factors and dimensions are only poor predictors of the final course grade in both Math and Physics.

The remainder of the paper is dedicated to present the highlights the outcome of the qualitative assessments during and after the semester.

Analysis of qualitative data from journals and focus groups

One important method to capture the attitudes and growth of the semester was the weekly on-line journal where students were asked to respond to specially designed questions. They were queried about their experiences, expectations, perception of own abilities, presence of study and time management skills, motivation for action, reflexion and awareness of the objectives of the activities that or the previous weeks. In addition, a total of three focus groups interviews were performed, with 8 students each time, to obtain information about reflective thoughts about the entire course and the value of acquired skills and their experience working in teams. Table 1 shows the distribution of students providing input.

Table 1: Distribution of students providing qualitative input

Journal Entries (37)		Focus Groups					
		First (8)		Second (8)		Third (8)	
Female	Male	Female	Male	Female	Male	Female	Male
14	23	2	6	2	6	5	3

Content analysis of journal entries

A content analysis of all journals and of the focus provided some interesting insights associated with the most valued aspects of the pilot class, teamwork and the collaborative learning.

Both journals and focus groups, clearly show that one of the most valued aspects from the class was the possibility to work together with others and the possibility to develop professional skills, like teamwork, between others. Male students describe more what specific things they liked than females. They clearly emphasised much stronger their positive experiences to work in teams, the ability to listen to or learn from peers. They felt that the collaboration within the team to prepare reports and oral presentation of their project felt like in the real world. They highlighted as very important to their professional performance one of the elements described in the focus of the STEM module, the emphasis on communication and work in teams with previously unknown people.

The students specifically felt that the pilot class was useful to develop their ability to work in teams. However, when they were asked to specify about de mode they did work together, the old method of operandi was prevalent. The actually worked as a group not as a team where they only divided the assignments but did not synergistic work.

73% of the students provided positive attitudes and benefits of teamwork. The main perceptions mentioned were: a) Learning how to organize, b) improve communication, c) gaining trust to express what they think, c) be able to discuss openly the different points of view, d) approach questions together, e) listen to others, f) complement ideas to reach the solution of problems, g) achieve agreements and solve difficulties together, h) understand clearly the different roles in a team, i) create clarity in what needs to be done, k) maintaining good relations between students, that favours a pleasant work environment and motivates to continue working, l) support between students, m) good time management, n) cooperation to achieve a good performance and o) motiving effect of the others.

37.5% of females report having performed an operative role versus 57% described by male. 37.5% of female report having performed a leadership role versus 43% described by male.

As expected, there were also negative voices. In fact, 27% of the students presented negative feelings about working teams. The most prevalent negative items that were mentions included: a) disorganization in the team, b) lack of interest or motivation to work by others, c) lack of understanding regarding the task to be performed and autonomous work, d) poor division of labour, e) lack of good planning, f) leave work for the last hour, procrastination, g) not everybody contributes to the work, leaving their work to others to do, and h) poor performance of peers.

Different to expectations, there was no difference between male and female students describing negative elements of teamwork. On the other hand 85% of the females clearly emphasized their positive perceptions compared to only 66% of males. 20% of males do not mention positive or negative feelings. The main causes of malfunction of a team identified by females corresponded to bad division of labour and lack of organization. The males noted the lack of planning, individual responsibilities not met and lack of communication.

Content analysis of focus group data

In the focus groups, the following factors were mentioned that hampered teamwork: a) unequal length of contributions, b) no feedback on progress by the teachers, c) lack of coordination within the new team, d) lack of consideration by teachers, who do not consider other responsibilities of students, e) extensive time needed for journals and team work, f) different degrees of commitment among students.

In relation to the role represented in the work teams, in 58% of the appointments, it is described to have played rather leadership roles, such us: planner, organizer of ideas or work, coordinator, brainstorming, teaching how to solve the problem, supervisor, problem solving guide. In a 42% of the quotes the roles are operational, such as: energizer or give moral support, calculator performer, recorder, information collector, problem-solving support, secretariat, facilitator, builder, sketch maker, reach conclusions and results, improver.

Compared to the journals, the focus group highlighted the positive aspects of being able to resolve doubts among the peers, to study much better and learn more, to share points of view, to increase the confidence between the members, overcome difficulties, develop skills, complement gaps and help peers. One student stated succinctly: "It seems interesting to me, because I think that in group we achieve better things than individually, because different points of view are expressed and when we share them, we see from another perspective the ideas". From the focus groups, "you learn to work in teams and to form groups to study in other programs too".

For the participating females, the collaborative work offers the possibility of clarifying and solving doubts together and with this, one learns to trust others, allows a group of work and study with friends.

From the focus groups, we learned that collaborative learning allows to learn from others, complement ideas, which can be applied to other programs and learn how to solve problems, and to work as a team.

It is important to note that among the expectations expressed in relation to one's own performance, students expect to continue practicing teamwork and exchanging knowledge. In relation to expectations towards the nature of the learning process carried out, do activities of the specialty of each student, with multidisciplinary teams to practice teamwork.

Discussion and conclusions

This paper presents the architecture of a pilot freshmen course, Introduction to Engineering, that applied PBL integrated with Math and Physics courses. The learning outcomes stressed two areas: a) Learning skills to become a professional students, b) appreciation and deep understanding of key engineering areas and principles. One of the central capabilities to be developed was working in teams. The course provide minimal training but promoted teamwork through the design of the project. The framework of the project, an investigation inside the Nuclear Reactor damaged by a tsunami in Fukushima, allowed the use of a robot as a key learning tool.

Various quantitative assessment instrument were used but showed only minimal correlations on teamwork. Extremely interesting is a total lack of collaboration between the cooperative skill factors by CHA, teamwork measured by the PFEAS and the final grades in Math and Physics.

Although the central focus of the design of this program was the integration of mathematics and physics subjects through the holistic problems, the students clearly learned about the benefits of teamwork to solve real problems. However, the traditional classes of Physics and Mathematics do not rely on teamwork skills to gain high grades. This might explain the low correlations between teamwork skills with the final grades in Math and Physics.

As long as students do not transfer work methodology to other subjects, the risk is that teamwork is conceived as a tool that is only used one area of learning and is not applicable to other fields. Some comments made in the focus groups make one believe that some students recognise the possibility to carry out this transfer. Some students actually started to make teamwork a cross-cutting tool. More work is needed to increase the awareness of these possibilities.

The positive aspects of a functioning team prevailed. More work is needed to foster synergism and moving them further than just dividing the work. A strategy of a brief introduction about teamwork and the continued engagement of students with different roles, could be a part from the basis of positive assessment and perception of good results. Still, difficulties expressed suggest that students need also guidance on team building and maintenance. Developing skills for the work of this nature, such as decision making and conflict resolution are critically important as well. There were no significant differences on the number of males and females mentioned negative factors of the teamwork, but females had a more positive view than males. This observation was surprising given our cultural characteristics, that there is a higher percentage of male students referring to having performed operative work and that in the case of women. Clearly, more work is needed to create formative developments of teamwork capabilities that will span across multiple courses as our students were barely able to practice and experience the basic potentials of working together.

References

ABET (2009). Criteria for Accrediting Engineering Programs, Accreditation Board for Engineering and Technology, Inc., www.abet.org/ (last accessed March 2017)

Albertine, S. (2011). Systemic Change for Student Success: Goals and Lessons of the LEAP States Initiative, Peer Review, 13(2) https://www.aacu.org/publications-research/periodicals (Feb. 2017)

Beichner, R., Bernold, L., Burniston, E., Dail, P., Felder, R., Gastineau, J., Gjertsen, M., and Risley, J. (1999). Case study of the physics component of an integrated curriculum," *Phys. Ed. Res. Supplement to Am. J. of Phys*, 67, pp. 16-24.

Bernold, L.E., Spurlin J. and Anson Ch. M. (2007). "Understanding our Students: A Longitudinal Study of Success and Failure in Engineering," *J. of Engineering Education, ASEE*, Vol. 96(3), July, pp. 263-274.

Besterfield-Sacre, M., Atman, C.J. and Shuman L.J. (1998). Engineering Student Attitudes Assessment," *J. Engineering Ed*., 87(2), pp. 133–141.

Hampden-Thompson, G. and Bennett, J. (2013). Science Teaching and Learning Activities and Students' Engagement in Science, *Int. J. Sci. Edu*., 35:8, 1325-1343.

Hughes, R. L. and Jones, S. K. (2011). Developing and Assessing College Student Teamwork Skills, *New Directions for Institutional Research*, 149, Wiley Periodicals, Inc., pg. 53, (wileyonlinelibrary.com) Last Accessed March 15 2017.

Denning, P. J. (1992). Educating the New Engineer, *Communications of the ACM,* 35 (12), pp. 83-97, ACM.

Kiesewetter, J. and Fischer M.R. (2013). The Teamwork Assessment Scale: A Novel Instrument to Assess Quality of Undergraduate Medical Students' Teamwork, *GMS, Zeit. Med. Ausbildung*, 32(2), (https://www.ncbi.nlm.nih.gov/pmc/articles/PMC4446650/)

Lingard, R. W. (2010). Teaching and Assessing Teamwork Skills in Engineering and Computer Science, *Systemics, Cybernetics and Informatics*, 8(1), pp 34-37.

Luckie, D.B., Aubry, J.R., Marengo, B.J., Rivkin, A.M. , Foos, L.A. and Maleszewski, J.J. (2012). Less teaching, more learning: 10-yr study supports increasing student learning through less coverage and more inquiry, *Advanced Physio Edu*, 36, 325–335.

Marín-Suárez, T. and Alarcon, H. (2013). Diseño e implementación de cuestionario para medir las habilidades de aprendizaje colaborativo en estudiantes de ingeniería, Proc. Congreso Chileno de Educacion en Ingenieria, October, Vina del Mar, Chile.

Vance, K., Konak, A., Kulturel-Konak, S., Okudan Kremer, G., Esparragoza, I. (2015). Teamwork Efficacy, Attitudes and Interest: Insights on Their Relationships, *Proc. Mid-Atlantic ASEE Conference,* April, Villanova University, online at <https://www.researchgate.net/publication/274709737_Teamwork_Efficacy_Attitudes_and_Interest_Insights_on_Their_Relationships>.

Wieman, C., Perkins, K. and Gilbert, S. (2010). Transforming Science Education at Large Research Universities: A Case Study in Progress, *Change*, July, online at <http://bit.ly/oy2Vgs>.

Zarske, M., Yowell, J., Maierhofer, S., Reamon, D. (2013), Teamwork in First-Year Engineering Projects Courses: Does Training Students in Team Dynamics Improve Course Outcomes and Student Experiences?, *Proc. 120th ASEE Annual Conference & Exposition,* June, online at <https://www.asee.org/public/conferences/20/papers/7069/view>.

Acknowledgements

The authors would like to acknowledge the support of the program Engineering 2030 and the UTFSM Center for Teaching and Learning.

Copyright statement

Introduction to Engineering: a STEM-PBL approach

Arie Aizman
Universidad Técnica Federico Santa María, Valparaíso, Chile
arie.aizman@usm.cl

Leonhard Bernold
Universidad Técnica Federico Santa María, Valparaíso, Chile
leonhard.bernold@usm.cl

Brayan Díaz
Universidad Técnica Federico Santa María, Valparaíso, Chile
brayan.diaz@usm.cl

Hugo Alarcón
Universidad Técnica Federico Santa María, Valparaíso, Chile
hugo.alarcon@usm.cl

Ángela Olivares
Universidad Técnica Federico Santa María, Valparaíso, Chile
Angela.olivares@alumnos.usm.cl

Celin Mora
Universidad Técnica Federico Santa María, Valparaíso, Chile
celin.mora@usm.cl

***Abstract**: The course Introduction to Engineering (ITE) at The Universidad Técnica Federico Santa Maria (USM) was redesigned to provide an integrated and formative learning experience where students needed to apply contents from other first year courses to solve small but authentic engineering problems, all related to the nuclear disaster in Fukushima. Meshed into the problem were exercises to practice study skills such as time-management. The results show that the approach not only helped the students' learning of science, technology, engineering and mathematics (STEM) topics, for example trigonometry, but also improved their confidence in solutions that they provided. Weekly student journals and a focus group organized after the end of the semester, show that this type of integrated course is a viable avenue to implement a STEM-Problem Based learning (PBL) approach, with measurable impact on learning skills and confidence of understanding Math and Science topics by freshmen engineering students.*

Introduction and background

The USM is a leading technical University in Chile. Technologically oriented, it educates students in the fields of engineering, architecture and sciences. Each year, around 1800 freshmen start their engineering studies with standard courses of math, physics and chemistry. While the entering students belong to the top 10% of the country, their high school background in math, science and their study skills are not at a level necessary to fulfill the university requirements. As a result, the students experience a lot of frustration, which negatively influences their satisfaction and retention within the different engineering programs at USM, a problem found in many national and international institutions (Centro de Estudios Mineduc, 2012).

Researchers have long focused on better understanding how to solve this serious problem. For example, Froyd and Ohland (2005) found that engineering programs with an integrative approach clearly increase retention. Early interventions applying student-centered teaching methods have also been shown to be an effective strategy in reducing dropouts in engineering (Jones, 2010). Furthermore, interventions that foster student´s motivation are the most productive in terms of improving learning and retention-satisfaction in engineering programs (Stump et al., 2011, Olson 1997).

In this spirit, the first semester ITE course at USM was redesigned to explore the effectiveness of a STEM-PBL approach. The centerpiece was an authentic problem that served as a framework that allowed the students to work on a real, but simplified engineering problem, practicing experiential and formative learning skills. The project of the pilot class was linked in parallel to the content of the first semester courses in Math and Physics. At the same time, the students learned and practiced study skills such as critical reading, time management, metacognition, and professional writing.

The research question was related to testing if STEM-PBL based Introduction to Engineering course could be designed in a manner that offered freshmen the opportunities to experience actively some of the key topics of engineering, while practicing basic competencies of a professional student. The hypothesis was that such a course would improve student learning and general satisfaction with their first year experience at USM.

Theoretical Framework

Introduction to engineering (ITE) courses are commonly offered in the first semester and are designed to expose the new students to some of the practices of engineering and the roles of engineers in society (Crawley et. al., 2007, Núñez et al, 2013, Dini, Musiak, & Grabiec, 1992; Gunn, 1996; Hoit & Ohland, 1998; Knickle, 1996; McDonald, 1995; Milano, Parker, & Pincus, 1996; R. B. Landis, 2007; Woolston, Shook, & Wilson, 1996). It was recognized that an ITE course at USM offers then an excellent opportunity to use a PBL approach to help students to not only get a deep understanding about engineering, but also reinforce their learning of Math and Physics.

Learning by doing is the central focus of a PBL approach, in accord to John Dewey's dictum "learning will be effective when students put theory into practice". Naturally, designing PBL based classes requires new sets of tools. Early adopters emphasized that a very critical success factor is the design of an authentic project or problem relevant to the students' interests (Kahn and O'Rocke, 2004). A real-world problem is fundamental to the utilization of PBL (Thomas 2000, Bell 2010, Hanney and Savina Baden, 2012).

PBL requires the students to take on a new role, changing from "learning by listening to learning by doing" (Stauffacher 2006). This necessity represents a challenge for freshmen Chilean students, since a traditional passive method of memorization is used in the K12 education curricula (Centro de Estudios Mineduc, 2012).

Research studies published by Goldin (2003), Seymour and Hewitt (2000), and Singh et al. (2002) revealed that the subjects of science, technology, engineering and mathematics (STEM) are closely related to each other. In fact, the objective of a STEM approach is to generate meaningful learning through integrating knowledge, concepts and skills, that are traditionally developed separately (Tseng et at., 2012).

Currently STEM principles are mainly used in K12 education. Only few universities have a full STEM programs such as in Olin College and Griffith University. Still, STEM activities are mainly used during summer camps, short courses, etc. On the other hand, curriculum integration is not new. An example in the engineering field is the Foundation Coalition, where several universities, financed by NSF, developed curricula for engineering programs (Froyd 2005, Borrego 2007, Borrego et al., 2007, Quinn, 1993). In these experiences, the focus was to coordinate the content of regular math, physics and chemistry courses in first and second year curricula. Some of them also included small projects to give the students hand on experience. A successful example of integrating Math, Physics and Engineering was reported by Beichner et.al. (1999). The PBL design of the course resulted in drastic improvements of the students' understanding, confidence and professional skills. The benefits of the integrated curriculum that have been reported are: Higher retention rate, improved academic performance and faster graduation (Picket-May & Avery, 2001, Borrego, 2010, Razzag, 2003). Despite the proven benefits, few of the NSF funded programs survived after the NSF's initiative ended. One wonders why these successes were discontinued. Is it because of the extra effort needed for sustained integration that requires a change from the traditional practices and administrative procedures? Of course, it requires arduous work to integrate multiple courses of the first year engineering curricula (Holtzapple et al., 2014).

This work is framed under the same integration spirit, but with a different focus. The goal of the presented project is to use an authentic engineering problem as the basis for allowing students to experience hands on the basics of engineering while not being overpowered by its complexity. In fact, raising the students curiosity to ask HOW and WHY systems perform the way they do was a focal point of the work.

Methodology

The investigation represents an experimental project using a mixed method approach. Data and evidence were collected using various methods: a) week journal, b) pre-post tests on general skills and attitudes toward engineering, c) pre-post tests on specific math topics, d) time management tasks that students were required to fulfill, and a focus group analysis.

An active learning method was adopted requiring students to work in teams to solve a series of tasks described below. In order to achieve the integration with Math and Physics, engineering problems were designed that matched the progress in those courses. The adopted model is shown in Figure 1.

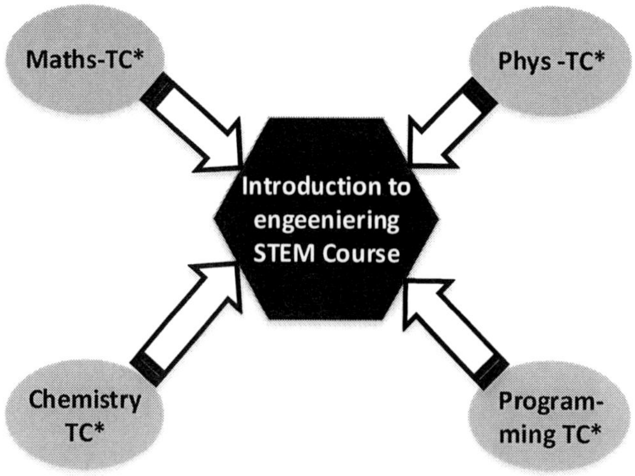

Figure 1: Architecture of the STEM-PBL approach; TC*= Traditional curriculum

The nuclear disaster at the Fukushima I Nuclear Power Plant in March 2011 was selected as the overarching project as it provided many opportunities to design simplified engineering problems. The students were given an introduction to earthquake engineering, of particular interest in Chile, followed by a discussion of the mechanics of tsunamis. Naturally, both attracted the interest of the students. Youtube videos and written reports provided the important sources of information for students. The "framing" project problem given to the students was a "hypothetical" request for creative help from a group of former USM students who had created a robotics company. The company asked the class to design a plan for a robot to enter the damaged nuclear reactor I to take radioactive measurements at a series of points inside the reactor and climb a staircase within a timeframe set by the maximum battery life. Figure 2 introduces two key elements provided to the students in design phase, the scaled annotated floor plan and the dimensions of the robot to be available to each team.

a) Fukushima reactor I floor plan **b) Actual robot used at Fukushima** **c) Robot used in class**

Figure 2: Partial information provided to the students about the reactor and robot

The desire to integrate some of the activities with math (used as an example) requires careful planning, to ensure that the project selected is an integral part of the first year curriculum. Table 1 presents an example of how desired learning outcomes in Math-101 were linked to activities required to develop a plan for the robot to provide data from inside the damaged reactor.

LINKAGE	DESIRED OUTCOMES	LEARNING ACTIVITIES BASED ON PROBLEM
	Utilize trigonometry functions correctly	Calculate the required angles of the robot arm to reach required positions in front of the robot chassis to measure radioactivity. Calculate the angles of the stairs to be climbed by the robot.
	Understand how to represent movements of points using polar as well as Cartesian coordinates.	Sketch the movements of the three sections of the robot arm and describe in Cartesian and Polar coordinates.
MATH-101	Transform Cartesian into Polar coordinates.	Change the Cartesian coordinates of desired points to be reached by the robot gripper into polar coordinates needed to control the different robot arm segments.
	Define a vector in 2-D	Calculate the angle of the camera mounted on the robot gripper to provide a visual image of a desired object at great heights.
	Being able to establish actual distances from a scaled floor plan	Measure distances and turns from the floor plan to create a path that the robot has to travel to reach the various locations for radiation measurements

Table 1: Linkages of PBL activities to learning outcomes in Math-101

As mentioned earlier, this introductory course was designed to provide the students the opportunity to learn about and practice the most critical skills of a "professional" student. Table 2 provides a matrix used to integrate various activities during the entire semester.

Activities	1	2	3	4	5	6	7
Class 2							
• Activities: "Read for Learn".	X		X				
• Explanation of weekly journals (importance of handling a personal log)	X	X					X
• Presentation of the concept and example of weekly and monthly time / study plans for a professional student	X				X		
• Results of the surveys (Feedback). (Study Skills, Learning Preferences)							
Active work done in class							
• "How much time do I have in my day?" Form to fill in (Form)	X	X					X
• Make a weekly time management plan.		X					
• Creative activity related to time management, mathematics, drawing a robot.... Activity using Sketching				X		X	

1.- Time management/Planning 2.- Metacognition 3.- Effective Reading 4.- Creative thinking
5.- Creative Communication 6.- Modeling reality 7.- Reflective Thinking

Table 2: Class activities linked to the practice of learning skills

During the first part of the course, a significant amount of time was set aside to work with the students in building time management skills. It is well known that correct planning and good time management is an important predictor of success in further courses (Bernold, L., 2007).

Results and Discussion

To assess the experience, data were collected using a combination of instruments and methods. General skills and attitudes instruments like the Pittsburgh Freshman Engineering Attitudes Survey (Besterfield- Sacre, M (1989), scientific reasoning (Lawson, 1978), among others, were used in combination with specific content tests, all in pre-post format. Qualitative data was collected via a weekly journal where the students had to answer open-ended questions and three focus group discussions. Surprisingly, 70% of the students of the pilot class participated.

The questions in the weekly journals had a reflective character designed to promote their awareness of the knowledge and skills that they acquired each week. A content analysis was made to study which aspects the students valued. Table 3 shows a summary results of most important "skills" mentioned in the journals, indicated by the number of citations separated by male and female students.

Skills	**Adaptive Skills {29} M = 12 = 83% ; W = 8 = 36%**
	Skill {36} W= 13 = 54% ; W = 10 = 71%
	Motivation {27} M= 11 = 48%; W= 5 = 36%
	Reflection {30} M= 18 = 78%; W= 8 = 57%
	Metacognition skills {27} M = 9 = 38%; W = 6 =43%
	Study Skills {43} M= 15 = 63% ; W= 9 = 64%
	Time management {38} M= 14 = 61%; W = 11 = 79%
	Preparation for class {27} M= 15 = 65%; W= 9 = 64%
	Confidence {25} M= 13 = 57%; W= 7 = 50%

Table 3: Summary of result of skill category in the journals. M= Men, W=Women

The results show the great variety of professional and university skills that the students value. As shown, the highest values are: a) Study skills (43) and b) Time management (36).

The focus group discussions after the end of the semester surfaced other values that sum up the students´ complete experience with comments such as "*...the course has made me think a lot. The design of a mechanical tool for the robot was cool*"; "*I really liked the course because it was not required to be sitting still looking at the blackboard all the time*"; "*looking to something in a paper and the applying it immediately was great*"; "*...the course gave me a feeling of what the work in engineering is all about; I thought that it was only Math and I was wrong*", "*...the project taught me about working together with different kinds of people trying to be creative*". The results confirmed the project's hypothesis that working on an authentic and appropriate STEM problem will lead to high satisfaction and deep understanding, which is in accord with much of the literature in engineering education.

Content specific pre-post instruments and comparisons with a control group were also used. One example was the assessment of improvements in understanding trigonometry. The proper use of trigonometric functions is a learning goal of the Math 101 course that all freshmen take in their first semester. Activities were designed in the module to complement the content taught in regular Math 101. Since basic trigonometric functions are part of the high school curriculum, a test was given to all students before and after completion of both the robot module and trigonometry classes in Math101. The results showed that students who worked with the robot had significantly better performance (> 50%,) than the students in the control group. The trigonometry test additionally included questions of student confidence about their answers. Hands on experience not only improved learning in the specific content, but also improved the student confidence in the solutions that they provided. Results are summarized in Table 4.

	Pre-Confidence Test	Post-Confidence Test [a]	Post-Confidence Test [b]	Pre-Trigonometry Test	Post-Trigonometry Test [a]	Post-Trigonometry Test [b]
Number of students	35	35	36	35	35	36
Response P-1	12	29	16	13	28	36
Response P-2	3	29	27	9	20	2
Total Score	49	193	166	3	16	6
Average points P-1	3.67	3.62	1.62	-	-	-
Average Points P-2	1.67	3.03	2.86	-	-	-

[a] Pilot Group [b] Control Group

Table 4: Summary of confidence test results and results of the trigonometry tests

Predictably, the journal entries reflected the linkages that were intentionally put in place to help the students, with comments such as "*being able to calculate the angles required for the robot arm to use the gripper was very helpful with the trigonometry calculations in Math*".

Conclusions

The paper presented the design and the outcome of a STEM-PBL based, Introduction to Engineering course at USM. The Fukushima nuclear disaster served as an authentic background to design various small problems that challenged the students to apply some of what they studied in parallel courses. As Chile is part of the ring of earthquake zones around the Pacific, the earthquake and tsunami that caused the disaster in Japan were of particular interest to the Chilean students. The presented data clearly show that providing engineering students the opportunity to apply math and science to solve engineering problems improves not only depth of understanding but also their confidence. While not helping them to gain better grades, they recognized, in their weekly journals as well in the focus groups, the value of learning about the different roles of work teams, practicing basic time management skills throughout the semester and to write reflectively about their learning experiences. The journal content analysis and the complementary focus groups confirm the hypothesis that a STEM-PBL course improves student´s learning and satisfaction during their first year experience at USM.

Future efforts will be dedicated to improve the support and guidance of students working in in-effective teams. Furthermore, a tighter integration with Math and Science will be pursued.

References

Beichner, R., Bernold, L., Burniston, E., Dail, P., Felder, R., Gastineau, J., Gjertsen, M., & Risley, J. (1999). Case study of the physics component of an integrated curriculum, *Phys. Ed. Res. Supplement to Am. J. of Phys*, 67, pp. 16-24

Bell, S. (2010). Project-based learning for the 21st century: Skills for the future. The Clearing House, 83(2), 39-43.

Bernold, L.E. (2007). Early Warning System to Identify Poor Time Management Habits. Int. J. of Eng. Ed., 23(6), 1182-1191.

Besterfield-Sacre, M., Atman, C. J., & Shuman, L. J. (1998). Engineering student attitudes assessment. Journal of Engineering Education, 87(2), 133.

Borrego,M. J. E. Froyd, T. S. Hall, (2010). Diffusion of engineering education innovations: A survey of awareness and adoption rates in U.S. engineering departments, *J. Eng. Educ.*, 99(3), 185-207.

Borrego,J. (2007). Development of engineering education as a rigorous discipline: A study of the publication patterns of four coalitions. *J. Eng. Educ.*, 96(1), 5-18.

Borrego, M., Adams, R. S., Froyd, J., Lattuca, L. R., Terenzini, P. T., & Harper, B. (2007, October). Panel-emerging results: Were the engineering education coalitions an effective intervention?. In Frontiers In Education Conference-Global Engineering: Knowledge Without Borders, Opportunities Without Passports, 2007. FIE'07. 37th Annual (pp. F4F-1). IEEE.

Centro de Estudios Mineduc (2012) Serie Evidencias: Deserción en la educación superior en Chile. Accessed at http://portales.mineduc.cl/usuarios/bmineduc/doc/201209281737360.EVIDENCIASCEM9.pdf on 30 September 2012.

Crawley, E., Malmqvist, J., Ostlund, S., & Brodeur, D. (2007). Rethinking engineering education. The CDIO Approach, 302, 60-62.

Froyd, J. E. (2005). The engineering education coalitions program. Educating the engineer of 2020: Adapting engineering education to the new century, 82-97.

Dini, S., Musiak, R., & Grabiec, R. (1992). Helping Today's Student Succeed in Engineering: A Freshman Orientation Course. In Proceedings, 1992 ASEE Annual Conference (p. 90)..

Goldin, G. A. (2003). Developing complex understandings: On the relation of mathematics education research to mathematics. Educational Studies in Mathematics, 54(2), 171-202.

Gunn, C. J. (1996, June). Addressing the Freshman Need for an Engineering Experience. In Proceedings, 1996 ASEE Annual Conference, ASEE.

Hoit, M., & Ohland, M. (1998). The Impact of a Discipline-Based Introduction to Engineering Course on Improving Retention. Journal of engineering education, 87(1), 79-85.

Holtzapple, M., Toback, K., & Holtzapple, C. (2015). Foundation coalition: Impact on chemical engineering education at Texas A&M University 20 years later. QScience Proceedings, 76.

Kahn, P., & O'Rourke, K. (2004). Guide to curriculum design: Enquiry-based learning. Higher Education Academy, 30-3.

Knickle, H. (1996). Foundations of engineering-a first year course. page, 1, 1.

Landis, R. B. (1995). *Studying engineering: A road map to a rewarding career*. Discovery Press (Los Angeles, CA).

Lawson, A. E. (1978). The development and validation of a classroom test of formal reasoning. Journal of Research in Science Teaching, 15(1), 11-24.

McDonald, D. (1995, November). Enhancing the recruitment and retention of first-year students with improved career selection opportunities. In Frontiers in Education Conference, 1995. Proceedings. 1995 (Vol. 1, pp. 3a3-6). IEEE.

Milano, G. B., Parker, R., & Pincus, G. (1997). A freshman design experience: retention and motivation. age, 1, 1.
Núñez, J., Lascano, S., & Esparragoza, I.(2013) E. A project-based learning approach for a first-year engineering course. 11th Latin American and Caribbean Conference for Engineering and Technology. Cancun, Mexico
Olson, G. (1997). Motivation, Motivation, Motivation-Secondary School Educators. Retrieved from sysiwyg.
Piket-May, M., & Avery, J. (2001). Service learning first year design retention results. In Frontiers in Education Conference, 2001. 31st Annual (Vol. 2, pp. F3C-19). IEEE..
Ponce, O. A., & Pagán-Maldonado, N. (2015). Mixed methods research in education: Capturing the complexity of the profession. International Journal of Educational Excellence, 1(1), 111-135.
Quinn, R. G. (1993). Drexel's E4 program: A different professional experience for engineering students and faculty. Journal of Engineering Education, 82(4), 196-202.
Razzaq, Z. (2003). An effective teaching strategy for motivation and retention of engineering and technology freshmen. In Proceedings of the ASEE Annual Conference and Exposition. Nashville.
Seymour, E. (2000). Talking about leaving: Why undergraduates leave the sciences. Westview Press.
Singh, K., Granville, M., & Dika, S. (2002). Mathematics and science achievement: Effects of motivation, interest, and academic engagement. The Journal of Educational Research, 95(6), 323-332.
Stauffacher, M., Walter, A. I., Lang, D. J., Wiek, A., & Scholz, R. W. (2006). Learning to research environmental problems from a functional socio-cultural constructivism perspective: the transdisciplinary case study approach. International Journal of Sustainability in Higher Education, 7(3), 252-275.
Stump, G. S., Hilpert, J. C., Husman, J., Chung, W. T., & Kim, W. (2011). Collaborative learning in engineering students: Gender and achievement. Journal of Engineering Education, 100(3), 475-497
Thomas, J. W. (2000). A review of research on project-based learning.
Tseng, K. H., Chang, C. C., Lou, S. J., Tan, Y., & Chiu, C. J. (2012). How Concept-mapping Perception Navigates Student Knowledge Transfer Performance. Educational Technology & Society, 15(1), 102-115.
Woolston, D. G., Shook, K., & Wilson, J. (1995, November). Same problem, different solutions: attempts at improving retention in engineering at a research vs. a teaching university. In Frontiers in Education Conference, 1995. Proceedings. 1995 (Vol. 1, pp. 3a3-1). IEEE.

Acknowledgements

The authors would like to acknowledge funding from project MECESUP 1408 and program Engineering 2030. Infrastructure and logistics provided by the Center of Learning Basic Sciences at USM is also acknowledged.

Copyright statement

Atarraya STEM: A STEM Hub for Universidad de los Andes

Juan Camilo Rincón
Universidad de los Andes, Bogotá, Colombia
jc.rincon29@uniandes.edu.co

Giovanna Danies
Universidad de los Andes, Bogotá, Colombia
g-danies@uniandes.edu.co

Michael Canü
Universidad de los Andes, Bogotá, Colombia
m.canu134@uniandes.edu.co

Mariana Tafur-Arciniegas
Universidad de los Andes, Bogotá, Colombia
m-tafur@uniandes.edu.co

Abstract: *Universidad de los Andes (Uniandes) is a private higher education institution located in Bogotá-Colombia. Uniandes promotes interdisciplinary collaboration between schools, academic programs, and research groups. This research shows the current state of courses, laboratory facilities, and research groups related with Science, Technology, Engineering, and Mathematics (STEM), as well as the collaboration that currently exists between these disciplines at Uniandes. Interdisciplinary opportunities that emerge from these collaborations are in many cases concealed preventing other people, particularly students and recently hired professors, to perceive the different relationships that have been built in the University.*

Atarraya STEM is a hub initiative for voicing faculty's current STEM-related work and aiming to broadcast new collaborations within the academic community. With this hub, we seek to position Uniandes as a STEM-integrated referent both nationally and internationally.

Introduction

Nowadays the pervasiveness of technology has led to a necessity of closing the gap of technology literacy among citizens. More and more countries are getting involved in promoting educational strategies that engage students into science, technology, engineering, and mathematics - STEM (e.g. The Royal Society, 2014; White House Office of Science and Technology Policy's "Preparing Americans with 21st Century Skills", 2014; Margison *et al.*, 2013). Even more, some developed countries claim that STEM education is required for innovation and competitiveness (e.g. IAP, 2014). Thus, national policies are leading towards increasing STEM workforce (e.g. National Academies of Sciences, Engineering, and Medicine, 2016).

Two main strategies that seek to prepare citizens for the challenges of the XXI century have recently emerged (BHEF, 2010). The first strategy focuses on each discipline (science, technology, engineering, and mathematics) as a unique field of study, with its own needs and challenges. This aims to improve competitiveness by attracting and retaining more people into STEM careers. From this point of view, STEM is the construction of four distinct fields that interact intrinsically. The second approach focuses on the interdisciplinary nature of the STEM fields as a whole, concentrating on the integration of the four disciplines, rather than interactions between separated domains (e.g. Kärkkäinen and Vincent-Lancrin, 2013).

This paper reflects how Universidad de los Andes (Uniandes) has addressed the needs of improving STEM education and how these efforts have been integrated and visualized by and for the community. This paper also suggests a strategy for promoting and voicing STEM education efforts within the community which may be useful for other institutions and STEM groups.

Context

Uniandes is a private higher education institution located in Bogotá-Colombia. This institution is committed with the country's development and with internationalization programs through an integral education and high-quality research. A way to address these goals has been by fostering interdisciplinary collaboration between schools, academic programs, and research groups. In fact, there are several STEM initiatives within the schools of education, engineering, and science that already integrate STEM areas. Some initiatives include curriculum development, interdisciplinary research grants and labs, and outreach programs such as the science clubs and fairs. However, many of the efforts have not been voiced within the Uniandes STEM community, lessening the consolidation of STEM development and competitiveness in addressing today's transversal and interdisciplinary needs. Atarraya STEM is a hub initiative that has been implemented for voicing faculty's current efforts and for promoting new collaborations within the academic community. This hub seeks for Uniandes to be recognized as a STEM-integration referent, both nationally and internationally.

Research Questions

- How is STEM education currently approached at Universidad de Los Andes?
- How can STEM education initiatives be voiced?

To address these questions, we initially investigated: How do the laboratories offer opportunities for disciplinary and interdisciplinary learning in STEM fields? How are the curricula in science, engineering, mathematics, and education programs offering interdisciplinary education for undergraduate students? How do the research groups promote STEM integration and collaboration among schools?

STEM Framework

The main goal of this research was to identify how the STEM disciplines are being represented at different activities in Uniandes. To do so, it is necessary to first explain each of the STEM components (Science, Technology, Engineering & Mathematics).

- Science includes both a body of knowledge by itself (related with the understanding of the nature) and a process to create new knowledge (National Academy of Engineering, 2014). It is related with disciplines like Physics, Chemistry, and Biology. Scientists apply the scientific method to explore, inquire, and discover.
- Technology includes both crafting and operation of technological artifacts and so the respective knowledge associated (National Academy of Engineering, 2014). Thus, it is related with the transformation of the natural world to meet human needs (ITEA/ITEEA, 2000/2002/2007).
- Engineering is understood as both a body of knowledge (related with the design and creation of products) and the process for problem solving (National Academy of

Engineering, 2014). Engineers optimize the use of resources and natural forces for the benefit of mankind (ABET, 2002).

- Mathematics provides an exact language for the other STEM disciplines. It is defined as the science of patterns and relationships (AAAS, 1993).

Methods

A content analysis was performed to investigate how the integration of STEM fields and interdisciplinary collaboration occur within the academic community. Three different types of data were collected: interviews of laboratory facilitators, curricula of undergraduate programs, and an analysis of projects and products of the STEM-related research groups.

Laboratories: All STEM-related laboratory facilities at the Uniandes main campus were studied. The information was collected from the official web page of each lab within the School of Science and the School of Engineering. The School of Science in Uniandes is comprised by the departments of Biological Sciences, Chemistry, Geosciences, Mathematics, and Physics. The School of Engineering, is comprised by the departments of Biomedical Engineering, Chemical Engineering, Civil and Environmental Engineering, Electrical and Electronical Engineering, Industrial Engineering, and Mechanical Engineering. The School of Education, was not included in this section given that their practical sessions are carried out in classrooms.

Given that information regarding STEM integration within laboratory facilities was not publicly available, we interviewed laboratory coordinators and staff. The objective of the interviews was to identify existing collaborations between STEM-related laboratories. Most of the interviews were made in person and only a few of them were conducted via e-mail. The following questions were asked during the interviews:

1. What kind of activities are performed in these laboratories?
2. Which other schools or departments have used this facility?
3. What are the mechanisms that facilitate collaboration among laboratories in different schools or departments?

Curricula: A revision of the curriculum structure was done for each major within the Schools of Science, Engineering, and Education. All required courses within these schools and elective courses offered to all students on campus that are related to STEM disciplines were considered. The main objective was to determine the interdisciplinary content of each major. To do so, the online catalog of undergraduate programs was consulted using the STEM framework mentioned above. Furthermore, information on the pre-required and co-required courses was considered to identify the connection between STEM disciplines.

The information analyzed corresponds to the description of each course of interest. This data was obtained through the Uniandes Courses Catalog Web page (http://catalogo.uniandes.edu.co/es-ES/2016/Catalogo), which is publicly available for both the internal and external community. For courses that did not have a description we used information from its pre-required and co-required courses to identify the possible STEM components related to that specific class (https://registro.uniandes.edu.co/index.php/nuevos/prerrequisitos-cursos and https://registro.uniandes.edu.co/index.php/nuevos/correquisitos-cursos, respectively).

Research Groups: All research groups at Uniandes, which are registered in Colciencias (Colombian Public Administrative Department of Sciences, Technology, and Innovation), were analyzed. For the content analysis, public information released by Colciencias was collected, to identify how their academic and research activities were related with STEM disciplines. Once the research groups working on STEM disciplines were identified, their productivity was assessed using the Colciencias research group classification (A1, A, B, C, and D, being A1 the highest and D the lowest).

First, a list of the research groups from the university was elaborated according to the institutional information available online (https://investigaciones.uniandes.edu.co/es/informacion/grupos-de-investigacion/). Only research groups, whose principal investigator belonged to the Schools of Science, Engineering, and/or Education were considered. The Colciencias research groups database (GrupLac) was considered to identify the STEM discipline (Science, Technology, and/or Engineering) associated to each research group.

Trustworthiness

The data related to the laboratory facilities at the main campus was drawn directly from the Uniandes web page. Here the user can explicitly find the distribution of laboratory facilities per School. Due to the lack of publicly available information on the active collaborations between laboratories laboratory coordinators and some facilitators were interviewed.

The curricula information about the majors within the STEM disciplines was all codified by a single person. The single researcher consulted the information corresponding to the description of the courses based on the STEM framework provided. Only the undergraduate level courses in the Schools of Science, Engineering, and Education were considered.

Data regarding the research groups at Uniandes was also collected and codified by a single researcher.

Findings

Laboratories: Laboratories at Uniandes are essential spaces for the daily operation of academic and research activities. There are three types of laboratories: *i)* teaching labs, *ii)* research labs, and *iii)* core facilities. All of these facilities are administered independently by different faculty members or staff within the institution. We found many ways of interdisciplinary cooperation among laboratories, allowing the interaction between students, professors, and administrative staff.

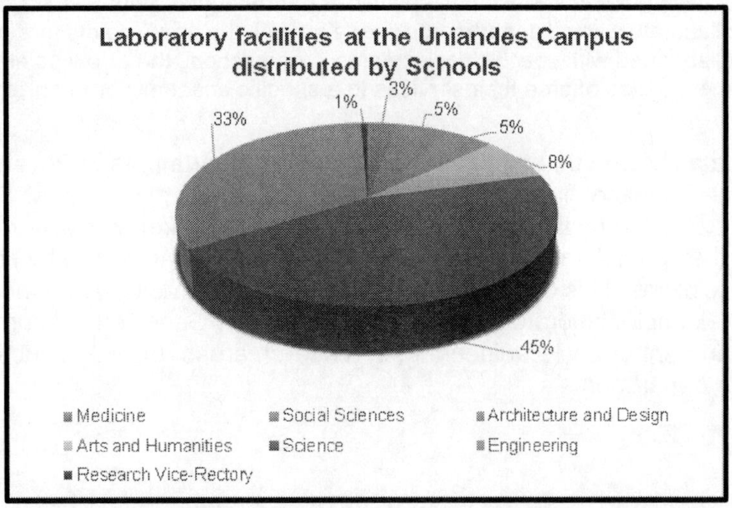

Figure 1: Laboratories within each School at the Uniandes main campus.

The University has a total of 169 laboratory facilities at the main campus. The School of Science and the School of Engineering have a considerable number of laboratory facilities within the institution, as is illustrated in Figure 1, in comparison with other Schools. In fact, 82% of the laboratories are related to STEM areas (45% belong to the School of Science, 33% to the School of Engineering, 5% to Medicine, and 1% to the Vice Presidency for Research).

In addition to the large proportion of STEM related laboratories within each School, there is evidence of collaboration between these facilities and the different STEM stakeholders in Uniandes. The collaborations between laboratories and stakeholders are reflected in Figures 2 and 3. The distinct types of partnerships, both within and outside the Uniandes community, consisted but was not limited to: independent research projects, undergraduate and graduate research theses and dissertations, faculty research, educational workshops, and consulting. Our data suggests that the department of Biological Sciences (Figure 2) and the Department of Mechanical Engineering (Figure 3) have the greatest number of interactions as can be reflected by the fact that both collaborate with all units in the School of Sciences and the School of Engineering.

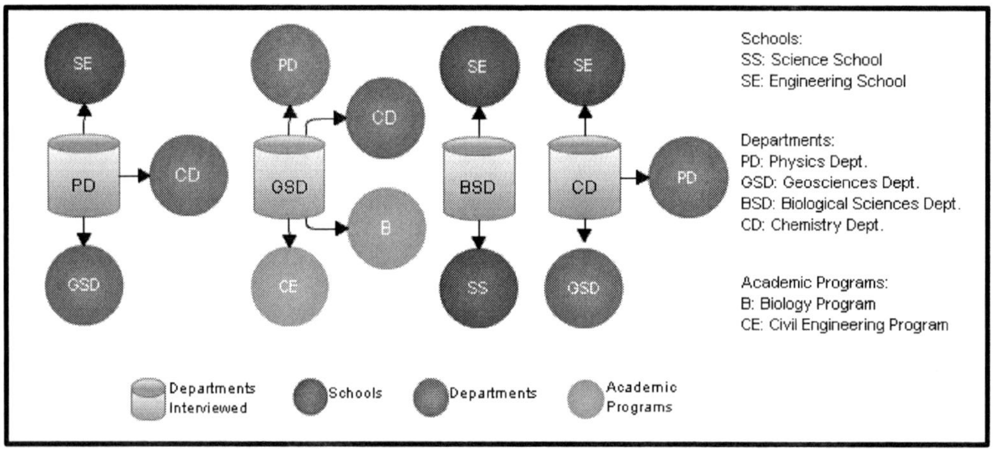

Figure 2. Existing collaborations found in laboratories within the School of Science. Blue cylinders correspond to the Department that was interviewed. Purple, blue, and green circles represent the School, Department, and STEM Academic Programs, respectively. Arrows indicate the relationships between facilities within one specific department and other STEM units. In cases where the facilities collaborate with all departments in one school, the arrow points to a purple circle. When the facilities collaborated with specific departments in one school, the arrow pointed to a blue circle. Finally, when the facilities offered their services to a specific academic program, the arrow pointed to a green circle.

Our data showed that the laboratories in the Departments of Physics, Biological Sciences, and Chemistry had a higher degree of collaboration with other STEM-related units. This may be explained by the fact that these departments offer first-year required core courses and their complementary practical sessions, which are taken by students in a wide number of disciplines. This does not occur in other Departments within the School of Science. For example, laboratories in the Department of Geosciences commonly interact with the Department of Civil Engineering, in research areas related to subsoil and soil, but rarely to other departments.

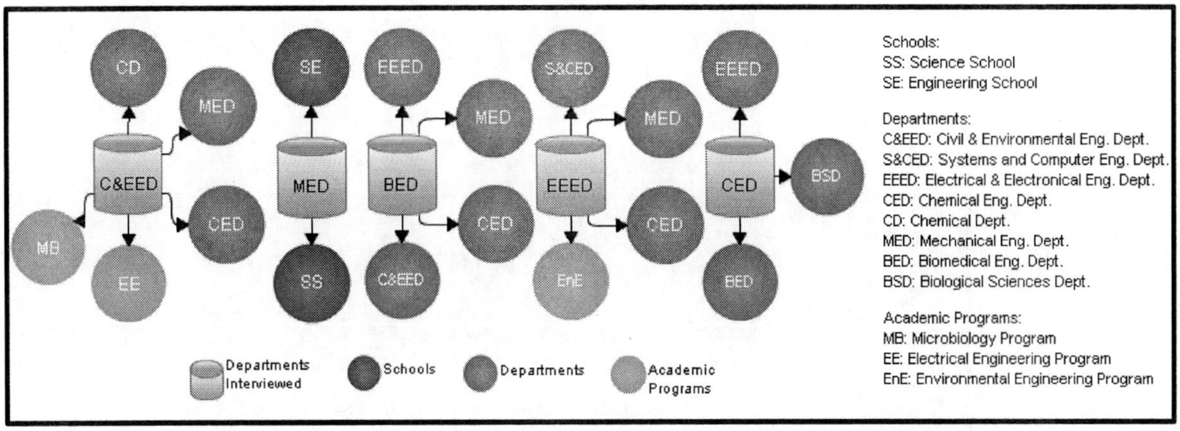

Figure 3: Existing collaborations in laboratories within the School of Engineering. Blue cylinders correspond to the Department that was interviewed. Purple, blue, and green circles represent the School, Department, and STEM Academic Programs, respectively. Arrows indicate the relationships between facilities within one specific department and other STEM units. In cases where the facilities collaborate with all departments in one school, the arrow points to a purple circle. When the facilities collaborated with specific departments in one school, the arrow pointed to a blue circle. Finally, when the facilities offered their services to a specific academic program, the arrow pointed to a green circle.

Within the School of Engineering, interdisciplinary efforts are facilitated by strategies like the laboratory coordination committee, which provides information about resources, needs, capabilities, and the staff associated to each of the laboratories. In addition, there are common facilities known as co-laboratories, where faculty work together in interdisciplinary spaces. One of the most representative examples is the "Colivri" co-laboratory that is managed by Mechanical Engineering and collaborates with the Departments of Electronic Engineering and Systems Engineering in several projects. Other interdisciplinary spaces at the School of Engineering are the Active Learning Plaza and the Prototype Development Room, both administered by the Electrical and Electronical Engineering Department, where interactions of students across the School of Engineering occur.

One interesting aspect about the laboratories within the School of Science, is the clear distinction between teaching and research laboratories. The main goal of the former is to provide experimental support to the courses, while the main goal for the latter is to provide tools for specific projects conducted by research groups. In contrast, at the School of Engineering, the same facility provides the space for both teaching and research activities.

Curricula: According to the University's mission, the institution aims to provide an interdisciplinary and holistic educational environment for students, who are considered the main agents of their own learning process. This is reflected in the way curricula in the STEM-related fields are intertwined. On one hand, a course offered by a STEM department may be a required course in the curriculum of other departments. For example, a calculus course offered by the Department of Mathematics, is a required course for students in other fields in Science and Engineering. On the other hand, some courses themselves, have interdisciplinary STEM objectives. For instance, the learning objectives of an elective course called "Science and Education" address generic skills that are transversal to all STEM disciplines such as scientific thinking and problem solving.

For this study, we investigated all required courses in STEM programs and elective courses related to STEM disciplines. Among the required courses there are some, such as mathematics, physics, and chemistry, that are core to different disciplines. Other required courses are specific to the field of study, and its level of interdisciplinarity may be limited.

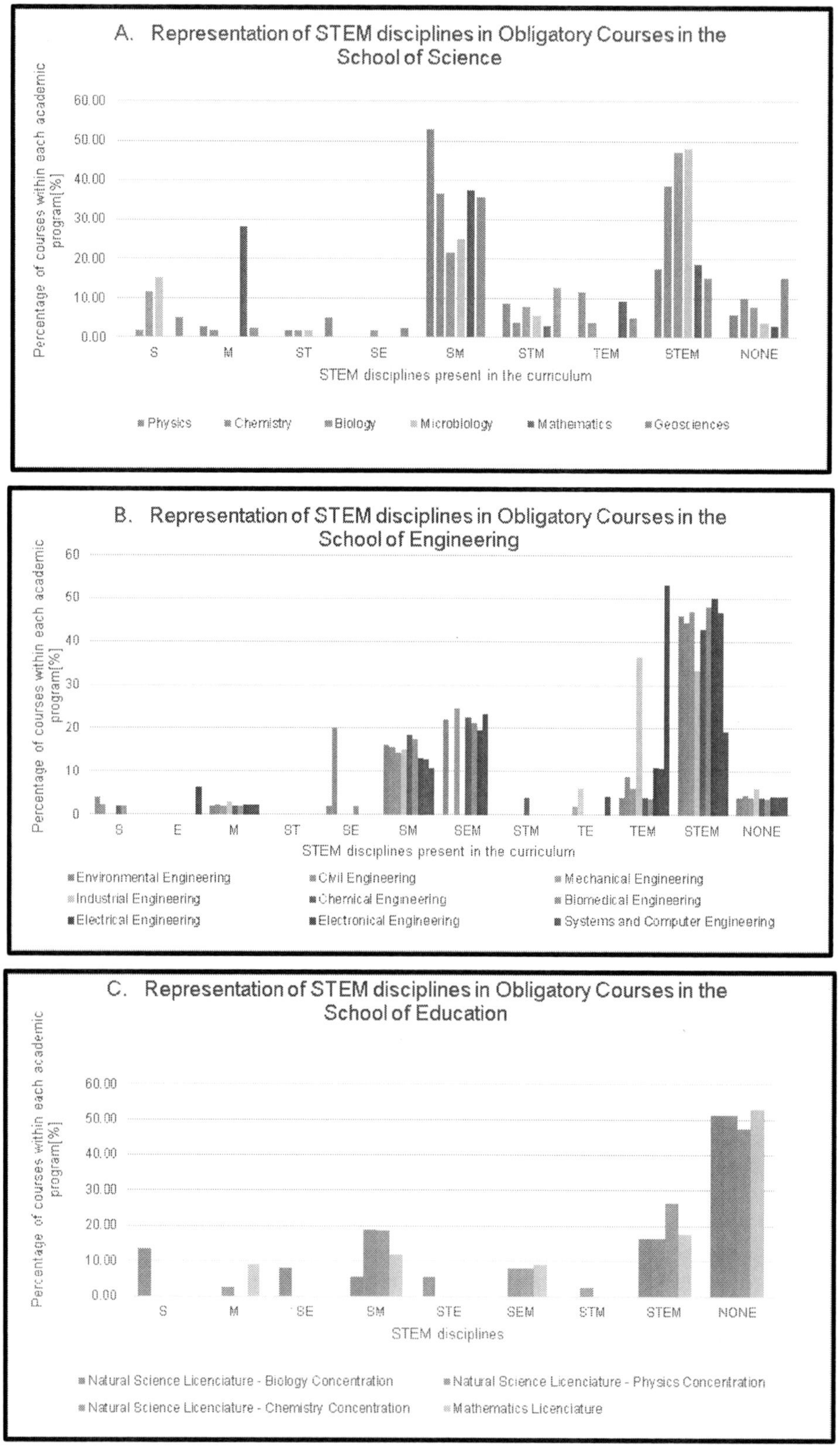

Figure 5. Representation of STEM disciplines in courses within the distinct academic programs in the Schools of **A.** Science, **B.** Engineering, and **C.** Education.

Figure 5 illustrates the percentage of required courses in the Schools of Science, Engineering, and Education that address learning objectives in each STEM discipline. The courses are categorized according to the number of STEM disciplines involved in the course summary shown in the University's course catalogue. For instance, those courses specifically teaching mathematics are categorized under 'M', those specifically teaching science are categorized under 'S', and therefore show no level of interdisciplinarity. In contrast, interdisciplinary courses are categorized under two, three, and four STEM letters. For instance, a course integrating topics of science and mathematics is categorized under 'SM' and a course integrating topics of mathematics, technology, and engineering is categorized under 'TEM'.

Every student at Uniandes must take six courses among the portfolio of elective courses. These courses are known as CBU ('Uniandes Basic Courses' in Spanish). The purpose of this courses is to address the University's mission of offering students an environment for interdisciplinary and holistic education. These courses are focused on three main areas of knowledge: Arts and Humanities, Social Sciences, and Science and Technology. Among these courses, 24.74% are content-related with STEM disciplines. Figure 6 shows the percentage from total amount of CBU courses that address learning objectives in each STEM discipline. The courses are categorized according to the number of STEM disciplines as done for the required courses in Figure 5.

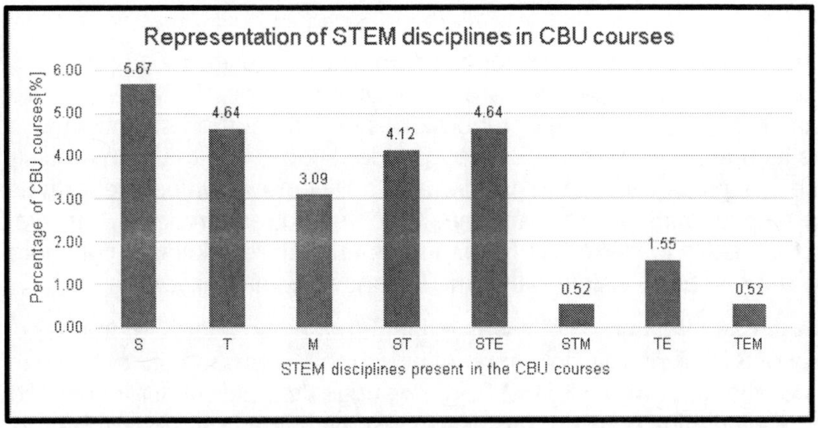

Figure 6: Representation of STEM disciplines in the CBU courses.

<u>Research Groups</u>: Research and knowledge production is fundamental for Uniandes. This is mainly conducted by the research groups at the university. Aside from the creation of new knowledge, these groups are learning spaces for students contributing to the development of their critical thinking, problem solving, and teamwork skills among other generic skills.

Currently, Uniandes has a total of 136 registered research groups in Colciencias. These groups are distributed into five categories according to their level of scientific production, being A1 the highest and D the lowest category recognized by Colciencias. Among the registered research groups, 63 are associated to STEM-related Schools at Uniandes. The categorization was based on the research area reported by each group on the Colciencias platform and was further refined based on the STEM framework mentioned above. The complete distribution of STEM disciplines being represented within each research group in the STEM-related Schools at Uniandes are shown in Figure 7.

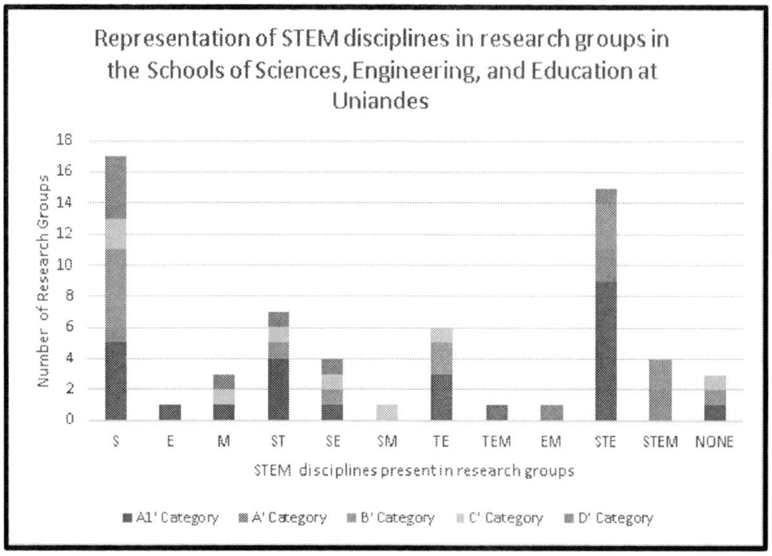

Figure 7. Representation of STEM disciplines in the Uniandes research groups.

Conclusions

Like most research universities, laboratories at Uniandes, are led by a principal investigator. Despite the fact that these laboratories are associated to a specific academic program, we found that many collaborate with other laboratories outside of their specific discipline. Collaborations lead to research questions of grander scope, due to the interdisciplinary nature of these groups as well as the greater availability of resources. Nevertheless, those connections are not completely visible to the students and researchers. Atarraya STEM-hub is an initiative that seeks to make STEM resources and interdisciplinary collaborations, in both teaching and research, visible within the Uniandes community.

With regards to the curricula, it was evident that a considerable number of courses had some level of STEM interdisciplinarity. Although this approach did not give information about faculty collaboration within STEM fields, it suggests a potential for consolidating and strengthening relationships between professors. Finally, data from research groups showed that approximately half are related with STEM disciplines. This offers an opportunity to centralize information on resources and research activities in a platform that would allow the community to become aware of the existing and prospective collaborations at the University. The Atarraya STEM-hub, will do this, allowing not only interactions between professors, but also between students working on different STEM-related fields. Furthermore, it will allow stakeholders from outside Uniandes to identify the distinct researchers working in their area of interest.

Future research should present further information on the existing relationships between the Uniandes STEM community. With a better understanding of these links, interdisciplinary collaborations between students and researchers in Uniandes may be fostered. This will not only benefit the quality of the research, but will also help optimize resources including equipment, materials, reagents, facilities, etc. at Uniandes. In addition, the level of STEM integration on current teaching/learning methodologies should be investigated with the aim of improving the curricula by identifying the best STEM practices.

References

American Association for the Advancement of Science (AAAS). (1993). Benchmarks for science literacy. New York, NY: Oxford University Press.

Accreditation Board for Engineering and Technology. (2007-2008). Engineering accreditation criteria. Baltimore, MD.

BHEM (2010). Increasing the Number of STEM Graduates: Insights from the U.S. STEM Education & Modeling Project, Business-Higher Education Forum ed., Washington (www.stemnetwork.org)

IAP (2014). Beijing Declaration on Science Education and Science Literacy. Science Education Programme (SEP) Biennial International Conference 28-30 October 2014, Beijing, China.

International Technology Education Association (ITEA/ITEEA). (200/2002/2007). Standards for Technological Literacy: Content for the Study of Technology. Reston, VA.

Kärkkäinen, K. and S. Vincent-Lancrin (2013), "Sparking Innovation in STEM Education with Technology and Collaboration: A Case Study of the HP Catalyst Initiative", OECD Education Working Papers, No. 91, OECD Publishing. http://dx.doi.org/10.1787/5k480sj9k442-en.

Marginson, S, Tytler, R, Freeman, B and Roberts, K (2013). STEM: Country comparisons. Report for the Australian Council of Learned Academies, www.acola.org.au. ISBN 978-0-9875798-0-5.

National Academies of Sciences, Engineering, and Medicine (2014). Developing a National STEM Workforce Strategy: A Workshop Summary, National Academies Press. ISBN 978-0-309-39158-0 | DOI: 10.17226/21900

National Academy of Engineering., National Research Council (U.S.), Honey, M., Pearson, G., Schweingruber, H., & Heidi Schweingruber., et al. (2014). STEM Integration in K-12 Education: Status, Prospects, and an Agenda for Research (1st ed.). National Academies Press.

The Royal Society (2014). Vision for science and mathematics education The Royal Society Science Policy Centre, London. ISBN: 978-1-78252-081-8

The effects of curriculum overhaul: investigating the students' experience.

M.E.D. van den Bogaard
Leiden University, Leiden, The Netherlands
m.e.d.van.den.bogaard@iclon.leidenuniv.nl

R.G. Klaassen
Delft University of Technology, Delft, The Netherlands
r.g.klaassen@tudelft.nl

***Abstract**: In this study, the final part of an overarching project on curriculum design and implementation executed at Delft University of Technology is presented. The last step in this research is to establish how the students perceive their new curricula. To that end a survey was developed using standardized questions on self-determination, motivation, self-efficacy, engineering education attributes, doability and study load of the programme. In addition, measures of student progress were collected in a programme for Architecture and the Built Environment and electrical Engineering. Relationships between the variables were established, but the relationship with progress was not strong. There were significant differences between the programmes, some of which may have a relationship with the way the curriculum was designed.*

Introduction

The Dutch Ministry of Education, Culture and Science signalled in 2010 a low student success rate particularly at the technical Universities in the Netherlands. To improve this situation the Ministry and the Universities amongst which Delft University of Technology made performance agreements on a number of KPI's on study duration and student retention. Failure to achieve these KPI's had major implication for the universities funding.

A task force identified a number of essential changes, that had to be made in each curriculum (Werkgroep Didactiek, 2011). The bachelor programme boards of each faculty were responsible for any programme improvements alongside the proposed changes and the implementation of improvements and changes in the programme.

The 4TU (previously 3TU) Centre for Engineering Education (CEE) evaluated these processes using the taxonomy of Van den Akker (2003) focusing on the curricula as designed, as implemented and as experienced. In this model it is implied that the extent to which curriculum goals have been achieved, can be measured by looking at student experiences within the curriculum, after it has been implemented. Fullan (2007) points out sustainable curricular change can only be realised if, amongst others, the intrinsic motivation of students and teachers is stimulated by the proposed reforms. This means that the reform contributes to goals that are important to students and teachers.

In previous work, the intended programme changes of the three Universities were studied to identify their possible success and fail factors (Gommer, Klaassen, & Brans, 2015). In the implementation phase, two programmes were studied in the three universities of technology that form the 4TU Centre for Engineering Education: Architecture and the Built Environment and Electrical Engineering (Verkroost, Van den Bogaard, Oude Alink & Schellen, 2015). The two programmes pursued different goals and used very different approaches to the design and implementation of their curriculum. In this paper we briefly describe the main findings of the previous study and discuss the third part of the evaluation in which we try to find out (1) to what extend these efforts have led to the achievement of the goals the programmes had set for themselves. (2) We will explore how student perceptions of (a) motivating curriculum elements in Architecture and the Built Environment (ABE) and Electrical Engineering (EEng) correlate with (b)student motivation and (c) success. These outcomes will be related to the goals the programme boards had set for the curriculum overhaul. In this paper we present preliminary results of this study.

The curriculum overhaul (implementation) at Delft University of Technology

At EEng student enrolment had declined to about 50 students in 2005. The programme management decided to create a broad programme in which all the disciplines constituting EEng were clearly represented and to create projects aimed at integrating knowledge representing the various disciplines of EEng. In this process it was decided early on that the programme should continue to focus on the technical and analytical side of EEng. Regular courses already in place were left unchanged for two reasons: (1) the programme management wanted to continue to use the existing high-quality learning materials that were tried, tested and complete, and (2) it was of importance to maintain the visibility of the constituent elements which together cover the field of EEng. These courses offered knowledge that had to be be applied in the projects that would run in the second half of each semester. The committee had put a lot of effort into looking at the order in which topics needed to be offered and at the increasing levels of difficulty and complexity of courses and projects throughout the programme. The new curriculum was implemented in 2010/2011 and improved in 2012: it proved to be successful in attracting more students and retention rates went up with approximately 10 per cent.

By 2005, the Bachelor programme in ABE came in last in terms of retention. The programme was fragmented: there were many courses with a small number of EC taught in parallel. As a result of the fragmentation, there were many exams and resits scheduled in the same timeframe. Students tended to procrastinate their studies, large numbers of students were failing their exams and students prioritized the design projects over their foundation courses. Senior management identified three focal points: professionalisation of the staff, assessment policy, and improving the doability of the Bachelor programme by removing barriers within the programme. Besides improving the programme's doability the faculty had a number of content related goals: [1] to update the programme, such as addressing current and future design challenges in the area of redevelopment and transformation; [2] to increase the academic level of the programme; [3] to remove overlap between courses, such as the overlap between foundations of architecture, foundations of urban development and history of architecture. The rationale for the new programme were six content-based 'learning lines' that comprised related and integrated courses over the three years of the Bachelor programme. The core competency in this curriculum is 1) Design. A good design requires a strong background in 2) Building Technology, 3) Societal Issues and 4) Foundations of Architecture and the Built Environment. 5) Academic and Research Competence comprise skills required for designing, and a building engineer also needs to be able to 6) Communicate his or her designs to others. A number of starting points were defined: students need to work on integral assignments that combine knowledge from courses (from the old curriculum) that feed into the design project, there would be fewer partial grades awarded, and all relevant topics should be addressed during the terms in which a module is offered. Every semester contains a design project and the topic of the design project leads the choice of modules in that semester.

In this project three key attributes specific to engineering were identified: the inclusion of design activities; the use of authentic engineering problems, and cross-disciplinarity within and across courses. Although in both programmes a different approach to the curriculum design was chosen, both curricula clearly comply with these three attributes of engineering education (Godfrey & Parker, 2010; Van den Bogaard, 2015).

Theoretical framework of the study

Student success is among the most widely studied topics in higher education. In those studies, student success is often operationalized with units such as the number of credits students obtain, the average grade, or passing or failing some kind of progress requirement. Studies on student success would commonly include some student attributes, attributes of motivation, behaviour and perceptions of the learning environment. In Figure 1 we present a general model for student success, which is explained in detail below.

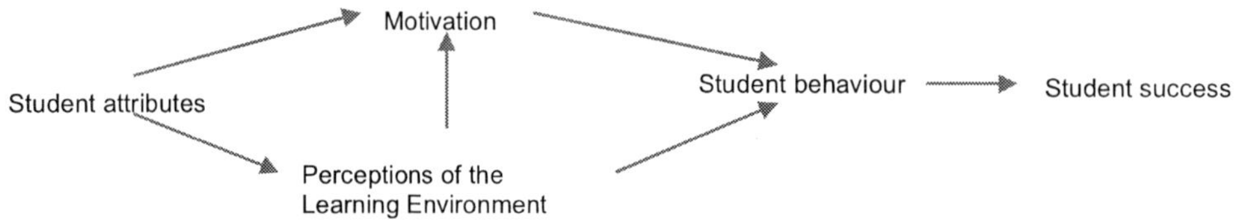

Figure 1: General model of student success (adapted from Jansen & Bruinsma, 2005; Van den Bogaard, 2015).

In this study we have used, in addition to the engineering attributes, this general model for student success that is grounded in research. The variables will be explained briefly below.

Student dispositions: motivation and self-efficacy

Ryan en Deci (2000) have established a well-known theory on motivation and postulate that there are three intrinsic needs that contribute to kindling and sustaining positive behaviour. These three needs are:

1. The need for competence (the extent to which student scan control the outcomes of their learning or experiencing success),
2. The need for autonomy
3. The need of relatedness (the need to interact with one's environment).

In an effective curriculum these needs are met. This theory is known as 'self-determination theory'. The amount of 'self-determination' student experience, can be supported by challenging students, by giving ample high quality feedback, by providing a supportive learning environment and to create opportunities for teachers and students to build effective relationships. As a result, students' interest and creativity can be increased and this would increase the students intrinsic motivation for a programme. (Ryan & Deci, 2000; "www.selfdeterminationtheory.org"). Students with little 'self-determination' tend to attribute their success or failure to external factor, which leads them to experience a lack of control over their situation.

Pintrich and De Groot (1990) found that intrinsic value and self-efficacy are related to student performance. Intrinsic value concerns students' reasons for doing a task: students who attach more value to a task will engage in more strategic study behaviours. According to Bandura (1997) the term 'self-efficacy' indicates the confidence people have in their capacity to achieve goals. Persons with a high level of self-efficacy often view difficult tasks as things that can be learnt, rather than seeing them as things that should be avoided.

Perceptions of the learning environment

A third element in the general model for student success pertains to students' perception of their learning environment. An environment can be supportive to students in objective terms, but if students' do not perceive it as such, the environment becomes a burden (Ginns, Prosser, & Barrie, 2007; Prosser & Trigwell, 1999). Student perceptions of the learning environment are therefore of importance when studying student success.

The term 'learning environment' is quite broad. Van den Bogaard (2015) established in a study at Delft University of Technology that the most important elements of the learning environment pertained to the quality of the teachers, the facilities including student support services, assessment and the educational organisation. The perceived quality of the teachers is important for the amount of relatedness (see self-determination theory), the quality of the educational organisation and assessment are important because organisation and the transparency of assessment procedures are an important element of the extent to which students perceive their curriculum as 'doable' (Werkgroep Didactiek, 2011).

In this study we consider a number of curriculum attributes that are related to the subject of the programme. E.g. at ABE one of the goals of the overhaul was to contribute to the training of better engineers through incorporating more challenging design assignments by combining authentic engineering cases of high societal relevance with topics from the theoretical foundation courses, that require students to collaborate (Goldberg & Sommerville, 2014; Sheppard, Macatangay, Colby, & Sullivan, 2009).

Student attributes

A number of relevant student attributes have been included in this study. Ability is the most consistent predictor of success in higher education and is usually operationalized as achievements in prior education (Pascarella & Terenzini, 2005). Gender is also a relevant variable, as women consistently report different experiences in the learning environment as do men (Seymour & Hewitt, 1997; Tonso, 2007). Besides these student attributes, a number of other attributes are included in the study presented in this paper: programme, student cohort, first year of enrolment in the programme, enrolment status and prior education.

Research questions and methodology

The goal of this study is to find out to what extent the goals of the curriculum overhaul have been achieved. The research questions that are leading in this study:

1. How do the students perceive the elements of engineering that have been embedded in the programme?
2. In what ways do the curricula at Architecture and the Built Environment and Electrical Engineering contribute to student motivation and student success?
3. What knowledge and skills do students think they are still missing in the new curriculum?

Methods

The researchers developed an online questionnaire fort his study. The questionnaire was based on standard scales that have been validated in similar contexts and complemented with questions on other relevant topics. The questionnaire was available in English and Dutch.

All bachelor students in these two programmes were sent a unique link to the questionnaire. Two reminders were sent out. The questionnaire was available for about 18 days.

At EEng there were 99 respondents and at ABE 37 students participated. These students came from the three years of the bachelor programme. The data was cleaned, cases with 50 percent or more data missing were removed. In the end 136 cases were included in the analysis: in ABE: 20 females and 15 males were included (2 respondents did not indicate their gender), at Electrical Engineering 5 females and 85 males were included (9 respondents did not indicate their gender). The response rate for EEng was 21 per cent, for ABE the response rate was 8 per cent. Data was explored visually and analysed using linear techniques.

Analysis

Research question 1 deals with the question to what extent the attributes of engineering education have been embedded in the programmes successfully. The questionnaire contained a scaled question on engineering education attributes, of which the items are included in the table below. Questions were answered on a 5-point Likert scale.

Table 1 Engineering Education (EE) scale items

	Engineering Education (EE) scale items	Alpha
EE1	There are plenty of proejcts, design and/or other project in the programme	
EE2	Project assignments are relevant for and related to the theoretical courses in this programme.	
EE3	The project assignments are appropriate for the level of this programme.	.735
EE4	Project are representative for issues from real life practical engineering problems.	
EE5	The project assignments are challenging and motivating.	

The alpha-score of this scale is higher than the lower cut off score of .7 that is commonly used (Field, 2013). The mean for the EEng students on this scale is 4.15 (SD: .46) and for the ABE students is 3.73 (SD: .55). The means for these groups are significantly different (F=21.202, p=.000). Architecture scores significantly lower on EE2 (F=17,569, p=.000), EE3 (F=11.640, p=.001), EE4 (F=10.410, p=.002) and EE5 (F=10.534, p=.001). There have not been any measurements prior to the overhaul, so it is not possible to compare these outcomes, but based on the means and the standard deviations it is clear that at Architecture there is room for improvement of the elements of EE in the programme. The response rate at ABE was very low and could explain the low scores on this scale.

Research question 2 deals with the issues of motivation and student success. To measure motivation we drew from various theories: Self-determination theory and theories on self-efficacy and intrinsic motivation. The items of the scales are presented below.

Table 2 Self Determination Theory scale items

	Self Determination Theory scale items	Alpha
Competence 1	I have difficulty making decisions during my learning process, in deisgn or project oriented courses.	
Competence 2	I have experienced a lot of pressure in my studies.	.303
Competence 3	I find it difficult to oversee all my options during the learning process in deisgn or project oriented courses.	
Competence 4	I realise in time when I am stuck in the learning process.	
Autonomy 1	The programme offers a number of predefined choice/electives that support my learning process, such as master track choices or minor programmes.	
Autonomy 2	Within courses I can make independent choices ith respect to my personal interests.	.586
Autonomy 3	I have the liberty to make my own decisions regarding the course and course assignments.	
Relatedness 1	In my courses I have plenty of opportunities to ask questions about the topic/assignment.	
Relatedness 2	I feel I am taken seriously as a student in this programme.	
Relatedness 3	My teachers are very approachable.	.787
Relatedness 4	The communication between the teachers and students is clear.	
Relatedness 5	Ther is enough support available in the programme.	

The alpha scores for the Competence and Autonomy scales are too low to be considered any further, as dropping items from the scale do not contribute to a better alpha. The alpha for the relatedness scale meets the criteria. ABE again scores lower on this scale than EEng (F=10.623, p=.001). We cannot establish if students feel motivated by their programmes.

The second element of this research question pertains to student success. To measure student success, we included a scale on self efficacy and one on intrinsic motivation in the questionnaire. These scales were taken from the MSLQ. Again, questions were answered using a 5-point Likert scale.

Table 3 Intrinsic Motivation (IM) and Self Efficacy (SE) scale items

	Intrinsic Motivation (IM) and Self Efficacy (SE) scale items	Alpha
IM1	I like the content of the courses in my programme.	
IM2	I often choose topics of which I will learn something, even if they require more work.	
IM3	It is important to me to understand the quint-essence of the courses.	
IM4	I am very much engaged by the programme's topics.	.652
IM5	I prefer teaching materials that are truly challenging so I can learn new things.	
IM6	I think I will be able to apply what I learn in my courses in the rest of my study programme.	
IM7	Being able to understand and apply the subject matter in diverse settings is important to me.	
SE1	I am certain I will master the skills taught in class during my courses this year.	
SE2	I can master almost all the work in my courses if I do not give up.	
SE3	I am certain I will figure out how to do the most difficult course work.	.825
SE4	Even if the work is hard, I can learn it.	
SE5	I can complete even the hardest assignments in my courses if I try.	

The alpha for the intrinsic motivation scale does not meet the low cut off point. Removal from any items will lower the score further. The alpha for self-efficacy does meet the requirements for scales. For both these scales there are no significant differences for ABE (IM: mean= 3.97, SD=.38. SE mean= 3.94, SD= .58) and EEng (IM: mean=4.01, SD=.41. SE mean= 3.98, SD= .53). An important element of student success pertains to the extent that a programmes is deemed doable by students.

Table 4 Doability and Studyload (DS) scale items

	Doability and Studyload (DS) scale items	Alpha
DS1	The course load is well spread over the year.	
DS2	It is clear how different course elements in my programme are related. (e.g. design, knowledge oriented courses, research, projects etc).	
DS3	The exams and tests are well spread out over the year.	.727
DS4	Assessment procedures are transparent and clear.	
DS5	Teachers and supervisors engage me during lectures and other educational activities.	
DS6	The various courses in this programme are clearly related.	

The ABE mean on Doability and Study Load is 3.43 with a SD of .51. The EEng mean on this scale is 3.84 with a SD of .52. ABE scores significantly lower on all items but DS5: DS1 F=19.107, p=.000, DS2 F=14.098, p=.000m DS3 F=5.015, p=.027, DS4 F=6.100, p=.015, DS6 F=8.733, p=.004. In the ABE programme the board chose to integrate the curriculum fully, which may be a reason why the programme scores low in comparison to EEng where courses and disciplines are more clearly visible. Another explanation could be that overload has always been an issue at ABE. The fact that if the SD is subtracted from the mean the score is close to 3 indicates that the doability and study load of the programme can be improved, but that this is not an urgent issue.

In Table 5 below we present the Spearman correlations of the means of the scales discussed and student progress. Progress is taken here as self-reported number of credits obtained in the first semester of the academic year.

Table 5 Spearman correlations of the means of the scales and student progress in ECTS

	ECTS	EE mean	Rel mean	IM mean	SE mean	DS mean	FS mean
ECTS	1.0	.049	.136	.007	.181*	.189*	.101
EE mean		1.0	.453**	.398**	.298**	.545**	.212*
Rel mean			1.0	.349**	.377**	.659**	.146
IM mean				1.0	.508**	.342**	.314**
SE mean					1.0	.356**	.145
DS mean						1.0	.259**

Spearman correlations. *=p<.05, **p=<.001

What stands out in this table is that ECTS hardly correlates with any of the scales. There are two small correlations with self-efficacy and doability and study load. These relationships are expected, but they are very weak. 'Future skills' shows mediocre correlations with EE and DS. The first correlation makes sense: in project work students are likely to get more exposure to new technologies. The most interesting correlation are between DS and EE and DS and Relatedness. Students who report higher scores on their perceptions of elements of EE in their programme and on their experience with relatedness, also report higher scores on how doable to the programme is. The correlation between intrinsic motivation and self-efficacy is to be expected based on the literature. When we consider the question on how the curricula influence student motivation, we note that EE correlates with Relatedness, IM and SE.

Research question 3 concerns the question what knowledge and skills students feel is lacking in their new curriculum. We asked the students to indicate to what extent future skills have been incorporated in their programmes and to indicate what is still missing in an open question. The questions on future skills are based on work by Kamp (2016). They are presented below.

Table 6 Future Skills scale items

	Future Skills scale items	Alpha
FS1	New innovative technologies are integrated in the lectures and project assignments.	
FS2	The newest insights from science and technology are discussed in class during a course.	.785
FS3	I have plenty of opportunities to experiment with new technologies in my programme.	

The alpha meets the criteria and no differences were found between the two programmes, however, the means for this scale are relatively low: Future Skills ABE mean= 3.21, SD= .85 and EEng mean=3.03, SD= .73. This means that the students feel they are not adequately prepared for the future. This also shows in the answers students gave on the open questions on what knowledge and skills were lacking. Students could give multiple answers. All the remarks were categorized and counted. At ABE a total of 47 remarks were made, at EEng 21 remarks were made.

In terms of skills that are lacking in the curriculum at ABE, students mention design software skills, such as working with Indesign, Revit and Autocad most often (10 times). Nine students mention technical knowledge and skills, such as material sciences, construction, etc. Reflection and critical thinking are mentioned 7 times, presentation skills are mentioned 8 times. This ranges from skills to present in 2D and 3D, but also to give presentations. Four students mention that skills and knowledge regarding design are lacking: they feel there is not enough focus on social implications of design and the design itself, rather than the presentation of design. Three students state they miss practical and entrepreneurial skills. The EEng students mention topics relating to circuit design 6 times. Five students mention presentation and writing skills. Two students mention the soft skills and the link to professional practice is also mentioned twice. A topic that is mentioned only a few times by both groups of students is collaboration. In ABE it is clear that students feel they are not able to experiment with new technologies, specifically in software. The fact that relatively many students state that there is not enough focus on building technology would also explain why the programme scores low on the Future Skills questions. The low score on the questionnaire questions at EEng is harder to explain based on the open question.

Conclusions, limitations and discussion

We set out to finish the overarching research project into the curriculum overhauls. We have no data to compare with the situation prior to the curriculum overhaul. Overall the scores on relevant scales are all 3 and higher on a 5-point scale. The students report that there are plenty of EE activities in their curricula, although ABE scores significantly lower than EEng.

We established some relations between curriculum attributes, motivation, self-efficacy and student success. The correlations between student progress and the other variables however were weak and only significant on a 95 per cent probability level: correlations were only found for SE and DS. The relations between the other variables were mediocre to strong. A few relationships stand out: there is a correlation of .659 between Relatedness and Doability and Studyload. Students who report that they feel supported by their learning environment also report that the programme is doable for them. DS is also strongly related to EE: students who report that the projects are meeting standards that are expected from this level of education also report higher scores on DS. A continuous effort to maintain and, where possible, improve the doability of the programmes should translate in higher scores on relevant variables and progress. Additionally, it will be worth wile to try to understand how the scores on DS emerge from student attributes and students' experiences in the educational environment.

In terms of future skills and knowledge and skills that are lacking, the students report that they feel they are not well prepared for their future. Students in ABE report they want to have more exposure to the use of software, they would like ot have more grounding in technology and construction and they feel that they are not trained well enough to deal with societal issues. Students from EEng report they would like more circuit design. Students from both programmes report they would like to learn more about presentation skills: oral, written, 2D and 3D, with a focus on critical thinking and reflection. Again, it is not possible to say whether or not the current curricula are an improvement compared to the previous curricula. Students do not report values under 3 out of 5 on the scale means and students do not consistently report clear omissions in their programmes. As overload is a serious concern in EE is might prove difficult to add more to the programme. The challenge for the programme boards will be to see where improvements can be made

A limitation of this study is that not all scales that had been selected as part of this research could be verified in this study: two out of three scales representing self-determination theory could not be validated. For future studies it is important to revise the scale items and test them. A more fundamental limitation of this study is the question to what extent student experiences and perceptions of different versions of a curriculum can be compared: student have nothing to compare the curriculum they are enrolled in with. This issue should be further explored to see if it is possible to create benchmarks that would allow such comparisons. When we compare the two programmes, we see that EEng seems to do better in certain aspects than ABE. The question here is if the differences can be attributed to the to the curriculum design or that they are due to other factors.

To further this study we intend to collect more data at the ABE programme and run some more analyses. Additionally, we intend to do a number of focus groups at both programmes to find out how the students look at the data collected in this study and to find out if they come up with more specific issues that need to be addressed. We would like to improve the questionnaire and include more programmes in this study to find out if we can establish benchmarks that will help us understand the curriculum as perceived by the students better. That way we can close the circle and provide better input for future endeavours of curriculum design and implementation.

References

Bandura, A. (1997). *Self-efficacy: the exercise of control.* New York City NY: Worth Publishers.

Field, A. (2013). *Discovering statistics using SPSS.* London UK: Sage Publications.

Fullan, M. (2007). *The new meaning of educational change.* (Fourth). London: Routledge.

Ginns, P., Prosser, M., & Barrie, S. (2007). Students' perceptions of teaching quality in higher education: the perspective of currently enrolled students. *Studies in Higher Education, 32*(5), 603–615. http://doi.org/10.1080/03075070701573773

Godfrey, E., & Parker, L. (2010). Mapping the cultural landscape in engineering. *Journal of Engineering Education, 99*(1), 5–22. http://doi.org/10.1002/j.2168-9830.2010.tb01038.x

Goldberg, D. E., & Sommerville, M. (2014). *A whole new engineer. The coming revolution in Engineering Education.* Douglas MI: Threejoy.

Gommer, E. M., Klaassen, R. G., & Brans, C. H. T. A. (2015). *Comparing Bachelor Curriculum Innovations at the Three Technical Universities - an exploratory study.*

Jansen, E. P. W. A., & Bruinsma, M. (2005). Explaining achievement in higher education. *Educational Research and Evaluation, 11*(3), 235–252. http://doi.org/10.1080/13803610500101173

Kamp, A. (2016). *Engineering education in a rapidly changing world. Rethinking the vision for higher engineering education* (2nd ed.). Delft: The Netherlands: 4TU Centre for Engineering Education.

Pascarella, E. T., & Terenzini, P. T. (2005). *How college affects students. A third decade of research.* (2nd ed.). San Fransisco: Jossey-Bass San Francisco.

Pintrich, P. R., & de Groot, E. V. (1990). Motivational and self-regulated learning components of classroom academic performance. *Journal of Educational Psychology, 82*(1), 33–40. http://doi.org/10.1037//0022-0663.82.1.33

Prosser, M., & Trigwell, K. (1999). Students' perceptions of their learning situation. In M. Prosser & K. Trigwell (Eds.), *Understanding learning and teaching. The experience in higher education* (pp. 58–82). Buckingham UK: Society for Research into Higher Education and Open University Press.

Ryan, R., & Deci, E. (2000). Intrinsic and Extrinsic Motivations: Classic Definitions and New Directions. *Contemporary Educational Psychology, 25*(1), 54–67. http://doi.org/10.1006/ceps.1999.1020

Seymour, E. H., & Hewitt, N. M. (1997). *Talking about leaving. Why undergraduates leave the sciences.* Boulder CO: Westview Press.

Sheppard, S. D., Macatangay, K., Colby, A., & Sullivan, W. M. (2009). *Educating Engineers: Designing for the Future of the Field.* San Fransisco CA: Jossey-Bass.

Tonso, K. L. (2007). *On the Outskirts of Engineering: Learning Identity, Gender, and Power Via Engineering Practice.* (R. M. Wolff, Ed.) (New Direct). Rotterdam The Netherlands: Sense Publishers.

Van den Akker, J. (2003). Curriculum perspectives: an introduction. In J. Van den Akker, W. Kuiper, & U. Hameyer (Eds.), *Curriculum Landscapes and Trends* (pp. 1–10). Dordrecht The Netherlands: Kluwer Academic Publishers.

Van den Bogaard, M. E. D. (2015). *Towards an action-oriented model for first year engineering student success. A mixed methods approach.* Delft University of Technology.

Verkroost, M. J., Van den Bogaard, M. E. D., Oude Alink, C., & Schellen, H. L. (2015). *Comparing Bachelor Curriculum Innovations at Three Universities of Technology : Implementing a new curriculum.* Enschede The Netherlands: 4TU Centre for Engineering Education.

Werkgroep Didactiek. (2011). Koersen op Studiesucces [Mapping the course to student success]. Delft, The Netherlands: TU Delft.

www.selfdeterminationtheory.org. (n.d.).

Copyright statement

The following copyright statement should be included at the end of your paper. Substitute authors' names in final copy only.

Development of a video coding structure to record Active, Blended, and Collaborative pedagogical practice

David Evenhouse[a], Claudio Cesar Silva de Freitas[a], Nimit Patel[a], Rohit Kandakatla[a], Austin Zadoks[a], Taylor Prebel[a], Nick Stites[a], Edward Berger[a,b], Jennifer DeBoer[a,b], Jeffrey Rhoads[b], Charles Krousgrill[b]

[a]School of Engineering Education
[b]School of Mechanical Engineering
Purdue University
West Lafayette, Indiana, USA

freeform@purdue.edu

***Abstract**: Classroom observation protocols (COPs) are valuable tools for collecting information about and characterizing classroom instruction. The purpose of this work is to describe the development and final structure of a specialized COP for use in coding video recordings of an active, blended, and collaborative (ABC) learning environment known as Freeform. Coding these ABC elements in the context of the instructor's actions allows the research team to empirically describe what occurs in the Freeform classroom, generating a unique and rich dataset containing information about the duration and sequencing of these ABC instructional elements. This data informs ABC instruction, facilitates further implementation of Freeform, and can inform further innovation based on reliable and valid observation data. Additionally, we believe documentation about the development process itself will empower other researchers to explore the use of COPs in analyzing and characterizing their own pedagogical practices in undergraduate STEM education.*

Introduction

Recent calls in the Engineering Education community for greater use of research based, innovative pedagogical practices (Jamieson & Lohmann, 2012; Lattuca, Bergom, & Knight, 2014) have increased the demand for tools that can be used to measure the effects, and support the implementation, of such innovative instruction. At times this requires collecting data during class meetings, observing how the various elements of the innovation, the instruction, and the institutional setting interact. Classroom Observation Protocols (COPs) can be used to generate reliable and detailed descriptions of what happens during an observed class period. On their own, or combined with other data such as student academic performance metrics, COPs can act as valuable tools to inform our understanding of pedagogical innovations.

Freeform is an innovative learning environment developed at Purdue University – West Lafayette in the School of Mechanical Engineering. It intentionally combines best practices in active, blended, and collaborative (ABC) learning to improve student acquisition of conceptual knowledge and problem solving skills (Rhoads, Nauman, Holloway, & Krousgrill, 2014). Since Freeform's inception in the Mechanical Engineering undergraduate Dynamics course, the number of students who earn a D, fail (F) or withdraw (W) from their course (the so-called DFW rate) each semester has decreased significantly (DeBoer et al., 2016). To better understand what Freeform looks like in practice, and to empower Freeform developers to bring these benefits to other instructors, courses, and institutions, our team has been using classroom observation techniques to identify characteristics of these Freeform classes through highly structured video analysis.

In this paper, we present our approach to crafting a COP for use in the analysis of video recordings of Freeform class meetings. In so doing, we address the research question:

What are the necessary components of a COP to capture ABC practices, and how do we adapt or develop such a COP for the unique needs of our research context?

We have developed and are currently using this new COP to evaluate instruction within the Freeform environment. Its structure, and method of creation, can inform the work of others as they examine and assess innovative teaching practices in undergraduate STEM education.

Background: Foundations of a classroom observation protocol

Instructional context: Freeform

Freeform combines ABC practices to enhance student learning in conceptual knowledge and problem solving skills. Any COP employed in Freeform research must therefore be able to capture each of these ABC elements in the form of valid and reliable data. The validity of a COP is demonstrated if the instrument being used successfully measures the variables that it is intended to measure (DeMonbrun, Finelli, & Shekhar, 2015). The first step in our process of selecting or developing a COP is to rigorously define those aspects of Freeform that we intend to measure. In this paper, we demonstrate the validity of this instrument through the intentional alignment of our COP with the nature of Freeform and its broader institutional context.

Active learning in Freeform is defined as learning that requires students to engage cognitively and meaningfully with the materials (Bonwell & Eison, 1991). It requires action (Chi & Wylie, 2014) on the part of the students, and cannot include traditional lecturing on the part of the instructor (Freeman et al., 2014). Blended learning is defined as the combination of in-class learning with web-based online learning elements, and does not necessarily require a fully flipped classroom (Oliver & Trigwell, 2010) [Note: we do not view Freeform as a flipped classroom]. Finally, collaborative learning occurs when two or more students work interactively as they progress toward specified learning outcomes (Berkeley, 2014), in this case within the setting of the classroom environment. Collaborative learning is also considered to be a form of active learning (Chi & Wylie, 2014).

In her study on curricular innovation, Hjalmarson (2008) defines a variety of different curricular systems based upon what innovations are being emphasized. Freeform focuses on the pedagogical strategies and learning resources being employed to present information to students, framing it as a pedagogically-based curricular system (Hjalmarson, 2008). Though the resulting environment may be considered student driven, the core of Freeform's impact on education lies in the alignment of instructors' in-class pedagogical choices with both research-based instructional innovations, and the Freeform learning resources made available to students online. Observations of the instructors' actions in their classrooms are therefore essential to understanding Freeform and informing its future implementations.

Institutional context: Resources, data, and limitations

In our work to characterize Freeform, the team decided to video record class meetings, aiming for more accurate, efficient, and less intrusive analysis than can be provided by in-class, real-time observation approaches. Several unique limitations and opportunities arose from the video capture and analysis methods employed. Because Freeform is pedagogically-based, our videos intentionally track instructor actions rather than the student responses to those actions. Coding actions and interactions of the students, for example, would be impossible given the nature of our video samples. However, tracking the actions of the instructor and recording their pedagogical choices provides us with the clearest picture of how Freeform has been applied to the classroom we are observing. This focus also mitigates ethical and professional concerns brought up by the University's Institutional Review Board related to recording students.

We chose to use the StudioCode software package to assist us with our video analysis, as it affords us a variety of unique opportunities. In StudioCode, coding structures such as COPs may be intentionally crafted with internal inclusions and exclusions, building reliability (consistency in coding across users and datasets) into the instrument and allowing for coding multiple instances, or events, simultaneously. Video speed can also be increased or decreased to facilitate time efficiency or coding precision as needed. This all allows researchers to easily extract qualitative and quantitative data from videos for subsequent analysis.

Establishing the need for a specialized COP

Previously-documented COPs have not been designed to capture all the features of interest to us within the Freeform environment. For example, Smith et al. developed an observation protocol to characterize university STEM classroom practices (Smith, Jones, Gilbert, & Wieman, 2013). This Classroom Observation Protocol for Undergraduate Stem (COPUS), which was heavily influenced by the pre-existing Reformed Teaching Observations Protocol (RTOP), was developed to provide a more targeted and context-appropriate tool for classroom observation in collegiate STEM courses. However, COPUS does not explicitly address both conceptual and problem solving knowledge, does not distinguish among ABC learning practices, incorporates observations of the students in addition to the instructor, and employs real-time coding at two minute intervals rather than continuous, offsite coding of video recordings (Smith et al., 2013). Likewise, Finelli et al. developed a coding scheme that observed the degree of student engagement in active learning, a point of interest to the Freeform team (Finelli et al., 2014). However, the innovations of the Freeform environment focus on the pedagogical choices and subsequent actions of the instructor. Thus, it would be more beneficial to us for the purpose of our project to observe the actions of the instructor and record the amount of student engagement the instructor intends to facilitate, a slight but important variation from the work of Finelli et al. Finally, the COP developed by Finelli et al. does not capture blended learning, nor several types of instructor actions, which prevents us from simply taking and adapting this work to our research.

The instrument we developed improves upon existing COPs for application within the Freeform framework by: (i) using a video-based offline procedure (enabling continuous and flexible coding), (ii) differentiating of ABC pedagogies, and (iii) discerning both conceptual and problem solving foci of instructor actions. To do this we integrated Finelli's description of active learning with robust coding definitions from the COPUS instrument, supplemented these with blended and collaborative codes, and finally created a new COP calibrated for observations within an ABC environment.

Developing a Freeform video coding protocol

Overall structure of the COP

We conceptualized the new COP to include instructor Events, Characterizations of those events, and the Degree of Engagement of students in learning activities. Events form the backbone of our coding protocol by denoting mutually exclusive, in-class applications of distinct pedagogical practices or behaviors. This is much like what may be seen in other prominent COPs that demonstrate a focus on instructor actions. For example, the COPUS instrument incorporates twelve distinct codes to record what the instructor is doing during a given classroom observation (Smith et al., 2013). Events are coded continuously in StudioCode (rather than at specified intervals), an affordance of using offline video analysis rather than real-time, in-class observation.

To capture the presence or absence of ABC learning practices, we include a set of Characterization codes, activated in tandem with the Event codes. Characterization codes indicate that the instructor is explicitly incorporating active, blended, or collaborative practices in their teaching. For example, an Event code denoting that the instructor in conducting an assessment could include Characterization codes indicating the assessment is both active (students are solving a problem) and collaborative (in the case of a group quiz) in nature.

Using Characterization codes alongside Event codes informs our understanding of an instructor's in-class intentions in an efficient and manageable way.

Broader literature indicates that the degree to which students are actively engaged is significant to their eventual learning outcomes (Chi & Wylie, 2014; Finelli et al., 2014). Several Degree of Engagement codes capture the extent of student involvement in the learning activities, expressed as 'All', 'Some', or 'None'. These codes capture instructor intentions, not actual student actions, because our data collection and analysis is entirely focused on pedagogical decisions of the instructor. For example, an in-class quiz would be tagged with a Degree of Engagement of 'All', because all students in the class would be involved in completing the quiz.

Lastly, two catch-all terms were included in the coding scheme. This was done later in the development process to facilitate data retrieval, synthesizing existing information rather than generating new data. The 'Assessment' term combines all data from Event codes that involve assessment of students, while 'Direct Instruction' combines all codes related to direct, passive (i.e. lecture) instruction.

StudioCode also enables forced inclusions or exclusions, linking activated codes to other, related codes in an automated way. For example, if an in-class quiz were to be coded as a 'Graded Assessment' Event, the Characterization code 'Active' and the Degree of Engagement code 'All' would automatically be applied to the event by the StudioCode environment itself. These conceptual linkages between Events, Characterizations, and Degrees of Engagement facilitate consistent and efficient coding of instructional events.

Codes and definitions

Multiple rounds of testing and feedback were required to establish the final coding scheme and to determine the finer points of the coding definitions. This process included meetings with Freeform professors and researchers, application of the coding scheme, and reliability analysis using Cohen's Kappa statistic (Spitzer, Cohen, Fleiss, & Endicott, 1967). This progression can be seen in Figure 1.

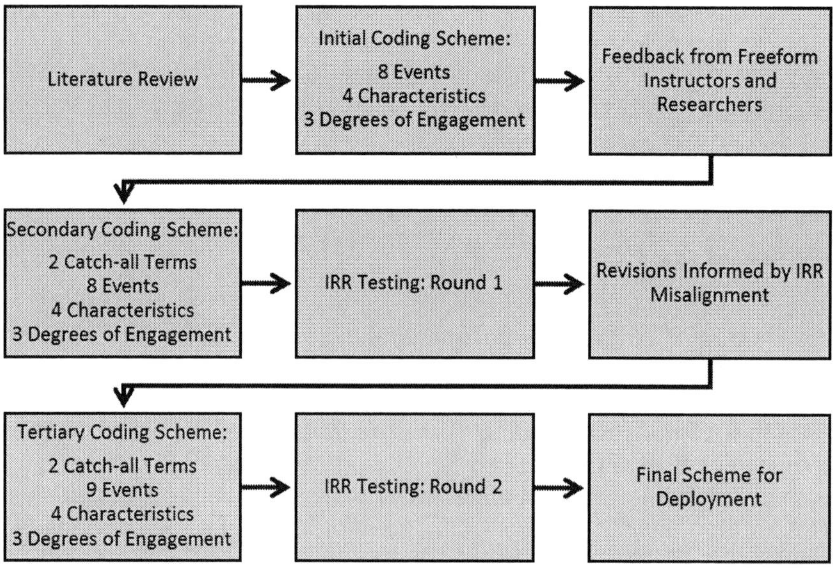

Figure 1: A visual summary of our coding scheme development process.

Events: Our final structure includes nine Event codes. These events are largely informed by the COPUS and RTOP instruments, with slight adjustments having been made to better align them with the purpose and interests of the Freeform project. For example, Talking/Lecturing (T/L) and Real Time Writing (RTW) codes directly reflect the COPUS framework, which distinguishes between the two as in Freeman et al. (2014). Likewise, our Demonstration code is derived from the COPUS D/V (Demonstration/Video) coding, and our Other/Admin code combines both the W (Waiting) and Adm (Administration) codes from COPUS (Smith et al., 2013). Opportunities for student and instructor feedback, assessment, and interactive learning are then captured with the Questions and Assessment codes. The full list of nine Events is below:

- *Conceptual T/L*: The instructor is talking directly to the students; a monologue or purely didactic form of instruction. The content is purely conceptual; theoretical knowledge is delivered to the students
- *Problem Solving T/L*: The instructor is talking directly to the students; a monologue or purely didactic form of instruction. The content is generally a verbal discussion of a problem-solving activity, the reading out of a problem statement, etc.
- *Conceptual RTW:* The instructor is explaining some concept (e.g., free body diagrams, equation derivations, etc.) by writing on the board.
- *Problem Solving RTW:* The instructor is solving some example problem on the board, demonstrating the application of equations, or enumerating a problem solving process.
- *Questions:* This categorization includes both when the students ask a question of the instructor, and when the instructor asks a question of the students. In the second case, this specifically refers to instances where the instructor is not expecting, nor requiring, every student to respond.
- *Graded Assessment:* This categorization includes instances where the instructor asks a question or series of questions of the students that all of the students are expected to answer for a grade; for example, quizzes, exams, extra credit in-class problems, etc.
- *Ungraded Assessment:* This categorization includes where the instructor asks a question or series of questions of the students that all of the students are expected to answer but their responses are not graded; for example, feedback forms, problems or examples given to solve in class.
- *Demonstration:* This categorization includes any kind of demonstration that uses some accessory, digital resource, or real-world object, and is intended to ease the understanding or visualization of a phenomenon or a concept. This includes the use of videos and simulations.
- *Other/Admin:* This categorization includes any other events which may transpire that do not fit the above categories. For example, administrative work, logistics, waiting when the instructor is late to class, etc.

Characterization: Characterization codes classify the events above to verify the presence, and record the duration, of ABC learning practices. Here, we see the influence of the essential nature of Freeform as an ABC learning environment (Rhoads et al., 2014) as well as literature on each of the ABC practices. Events may be classified using four characteristic codes:

- *Active (A):* Includes all interactive, constructive, and active events in the classroom (Chi & Wylie, 2014; Freeman et al., 2014).
- *Blended (B):* Includes any use or mention of online resources (Bowen, Chingos, Lack, & Nygren, 2012; Means, Toyama, Murphy, Bakia, & Jones, 2009)
- *Collaborative (C):* Includes any peer-to-peer interaction within the classroom (Dillenbourg, 1999; Jeong & Chi, 2007).
- *Passive (P):* A null instance representing the lack of active learning, does not expressly promote any activity or interaction on the part of the students.

'Passive' and 'Active' are, by our definitions, mutually exclusive, a nuance programmed directly into our coding scheme. Likewise, collaborative events are assumed to be active, thus classifying an event as 'Collaborative' forces it to be 'Active' as well. Our final scheme is laid out in StudioCode such that most event codes are automatically characterized to facilitate greater coding reliability. For example, selecting a T/L or RTW event code will automatically characterize the event as 'Passive', as demonstrated in Figure 2.

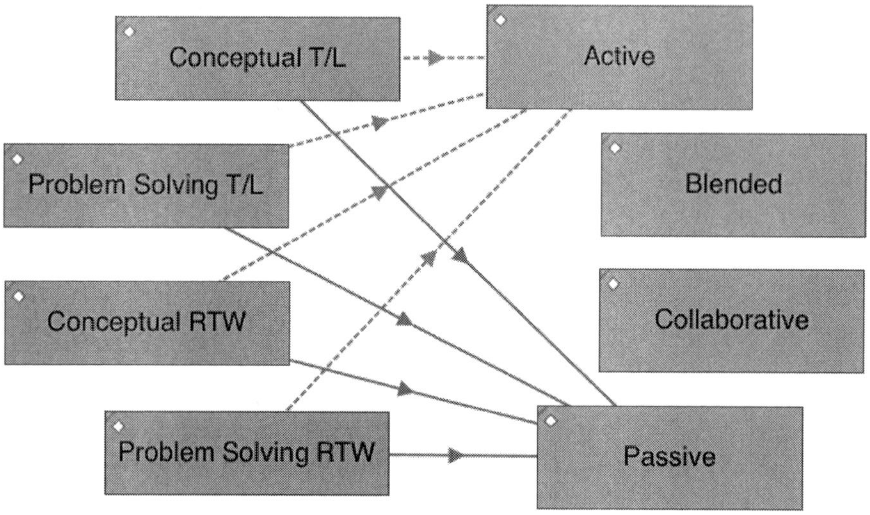

Figure 2: The relationships between T/L and RTW Events and their Characterization codes. Solid lines indicate a forced activation, while dotted lines indicate a deactivation.

Degree of Engagement: Degree of Engagement codes draw heavily on the work of Finelli et al. (2014), and were adjusted to reflect our focus on the actions of the instructor. Like Finelli et al. (2014), we opted to use a three-tiered coding scheme. However, rather than recording student engagement, the 'None', 'Some', and 'All' codes capture instructors' intentions regarding student engagement through a given instructional behavior. Engagement codes again take advantage of built-in dependencies in StudioCode. For example, Passive events are automatically coded as having a Degree of Engagement of 'None'.

Whenever codes are selected, instances of the video being coded are saved to a timeline as is shown in Figure 4. These timelines may then be queried to generate numerical data such as the length of instances, the number of instances, or the amount of overlap between the various codes. Each of these instances also represents an individual video clip, which may be exported, watched, and even transcribed to generate qualitative data for analysis.

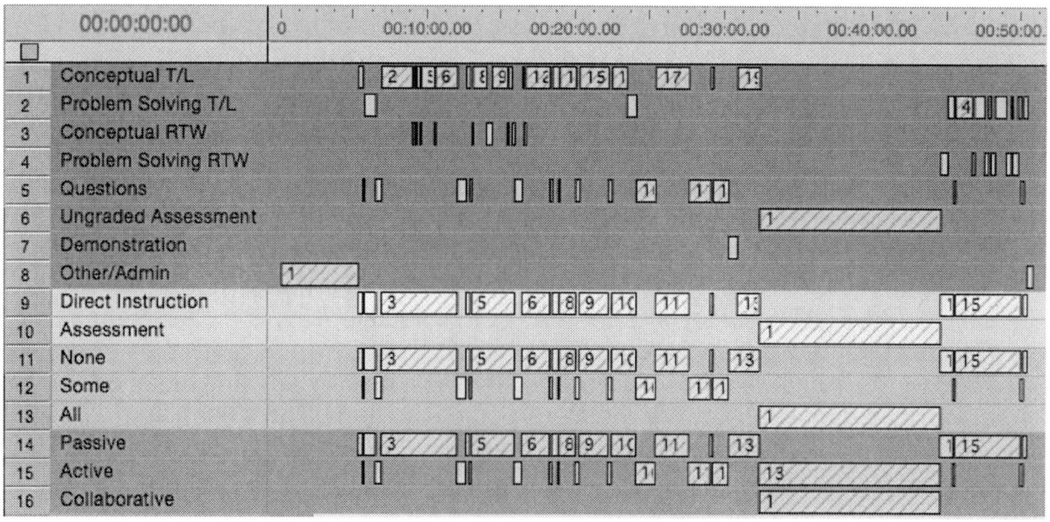

Figure 3. A representative StudioCode timeline. Code names are located in the leftmost column, with numbered instances arranged along the timeline to the right.

Validity, reliability, and further application

This COP has been intentionally designed to measure the pedagogical variables of interest to us within the Freeform environment, making it valid for use in this research context (DeMonbrun et al., 2015). Like in the work of DeMonbrun et al., we foster concurrent validity by drawing from the designs of existing, well-vetted observation protocols, and foster face validity by firmly rooting our code definitions in the theory behind our topics and phenomena of interest. Yet validity concerns do not begin and end with the development of our COP. To ensure that this validity is preserved moving forward, we will have to take care during our sampling process to ensure that the class recordings we take are truly representative of what occurs in the Freeform classroom. Well defined coding procedures can only prove useful if we are collecting the proper data for analysis (DeMonbrun et al., 2015).

In determining reliability, we employed Cohen's Kappa (Hallgren, 2012; Spitzer et al., 1967) as a means of calculating Inter-Rater Reliability (IRR) as seen elsewhere in STEM literature (El Emam, 1999; Kilgore, Atman, Yasuhara, Barker, & Morozov, 2007). During development, we had two coders analyze four videos for the purpose of IRR testing. Both coders were involved in the development of this coding scheme. After coding two videos, an initial Kappa was calculated in order to determine where there may be ambiguities in the coding structure or definitions. For example, an initial Kappa of 0.695 was calculated comparing the coding of Active and Passive events. This number is considered sufficiently reliable for the purpose of research (El Emam, 1999; Kilgore et al., 2007) but certainly suggests room for improvement. After a review of the coding structure and several theory-based revisions to the code definitions, another test was conducted which demonstrated a Kappa value of 0.798. Across all the components of the coding structure (Events, Characterization, and Degree of Engagement), and after adding additional coders, Kappa continues to be at or above 0.6, indicating moderate to substantial agreement among the coders who have now participated (El Emam, 1999).

Discussion

The development process and definitions presented here have resulted in a custom COP that targets key variables of interest within Freeform's ABC learning environment. Its components have been designed to align these target variables, the abilities of the software package available to us, and the insight gained from the literature at large, enabling us to produce a unique COP and resultant dataset detailing what exactly occurs within this innovative classroom environment.

This COP has already been put to use in research, and has demonstrated promising preliminary results. A publication by our team at the 124th ASEE Annual Conference and Exposition (Evenhouse et al., 2017) presents some of these preliminary findings, as well as their implications for Freeform instructors. For example, we are presently working to benchmark and characterize instruction within the Freeform environment across instructors and semesters. Doing so will inform future implementations of Freeform at our institution and others, as well as contribute to the broader body of literature on ABC instructional practices. Locally, this empirically generated dataset can facilitate empirically based professional development, fostering reflection and encouraging Freeform instructors to continuously improve on their own pedagogical practices (Creemers & Reezigt, 1996).

This work also expands on existing classroom observation literature by combining the coding of instructor behaviors with the coding of variables designed to evaluate the use of research-based, innovative instructional practices. In so doing, it demonstrates the potential of video coding to inform and facilitate the implementation of research-based instructional practices, a topic of interest within the broader engineering education community (Dancy & Henderson, 2008; Henderson, Beach, & Finkelstein, 2011; Litzinger & Lattuca, 2014).

Conclusion

This paper details the development of a customized COP to help us understand classroom practices in an ABC environment. Built upon prior COP instruments, our framework and its implementation extend the literature on classroom observation and evaluation, and provide an outline for other researchers to systematically build COPs for their specific needs. Although we do not recommend this tool for use in non-ABC environments, the methods described here can be used to guide the development of tools designed for different learning environments, where other teaching practices could be examined and observed. Though this coding scheme was developed to support the needs of the Freeform research group, we expect that the methods described in this paper can be useful in future applications in a variety of educational contexts.

References

Barkley, E. F. (2014). *Collaborative learning techniques: A handbook for college faculty* (2nd ed.). San Francisco, California: Jossey-Bass.

Bonwell, C. C. (1991). *Active learning: Creating excitement in the classroom*. Washington, D.C.: School of Education and Human Development, George Washington University.

Bowen, W. G., Chingos, M. M., Lack, K. A., & Nygren, T. I. (2012). *Interactive learning online at public universities: Evidence from randomized trials*. Ithaka S+R.

Chi, M. T. H., & Wylie, R. (2014). The ICAP Framework: Linking cognitive engagement to active learning outcomes. *Educational Psychologist*, *49*(4), 219–243.

Creemers, B. P. M., & Reezigt, G. J. (1996). School level conditions affecting the effectiveness of instruction. *School Effectiveness and School Improvement*, *7*(3), 197–228.

Dancy, M. H., & Henderson, C. (2008). Barriers and promises in STEM reform. In *National Academies of Science Promising Practices Workshop*.

DeBoer, J., Stites, N., Berger, E. J., Rhoads, J. F., Krousgrill, C. M., Nelson, D. B., Zywicki, C., Evenhouse, D. (2016). Work in Progress: Rigorously assessing the anecdotal evidence of increased student persistence in an active, blended, and collaborative Mechanical Engineering environment. In *123rd ASEE Annual Conference and Exposition*.

DeMonbrun, R. M., Finelli, C. J., & Shekhar, P. (2015). Methods for establishing validity and reliability of observation protocols. In *122nd ASEE Annual Conference and Exposition*.

Dillenbourg, P. (1999). What do you mean by “collaborative learning”? *Collaborative-Learning: Cognitive and Computational Approaches.* (pp. 1–19). Elsevier.

El Emam, K. (1999). Benchmarking Kappa: Interrater agreement in software process assessments. *Empirical Software Engineering*, *4*(2), 113–133.

Evenhouse, D., Kandakatla, R., Stites, N., Patel, N., Zadoks, A., Prebel, T., Freitas, C., Krousgrill, C., Rhoads, J., DeBoer, J. (2017). What does an in-class meeting entail? A characterization and assessment of instructor actions in an active, blended, and collaborative classroom. In *124th ASEE Annual Conference and Exposition*.

Finelli, C. J., DeMonbron, M., Shekhar, P., Borrego, M., Henderson, C., Prince, M., & Waters, C. K. (2014). A classroom observation instrument to assess student response to active learning. In *2014 IEEE Frontiers in Education Conference (FIE) Proceedings*.

Freeman, S., Eddy, S. L., McDonough, M., Smith, M. K., Okoroafor, N., Jordt, H., & Wenderoth, M. P. (2014). Active learning increases student performance in science, engineering, and mathematics. *Proceedings of the National Academy of Sciences*, *111*(23), 8410–8415.

Hallgren, K. A. (2012). Computing inter-rater reliability for observational data: An overview and tutorial. *Tutorials in Quantitative Methods for Psychology*, *8*(1), 23–34.

Henderson, C., Beach, A., & Finkelstein, N. (2011). Facilitating change in undergraduate STEM instructional practices: An analytic review of the literature. *Journal of Research in Science Teaching*, *48*(8), 952–984.

Hjalmarson, M. (2008). Mathematics curriculum systems: Models for analysis of curricular innovation and developement. *Peabody Journal of Education*, *83*(4), 592–610.

Jamieson, L. H., & Lohmann, J. R. (2012). Innovation with impact. *American Society for Engineering Education*. Washington, DC.

Jeong, H., & Chi, M. T. H. (2007). Knowledge convergence and collaborative learning. *Instructional Science*, *35*(4), 287–315.

Kilgore, D., Atman, C. J., Yasuhara, K., Barker, T. J., & Morozov, A. (2007). Considering context: A study of first-year engineering students. *Journal of Engineering Education*, *96*(4), 321–334.

Lattuca, L. R., Bergom, I., & Knight, D. B. (2014). Professional development, departmental contexts, and use of instructional strategies. *Journal of Engineering Education*, *103*(4), 549–572.

Litzinger, T. A., & Lattuca, L. R. (2014). Translating research to widespread practice in engineering education. In A. Johri & B. M. Olds (Eds.), *Cambridge Handbook of Engineering Education Research*.

Means, B., Toyama, Y., Murphy, R., Bakia, M., & Jones, K. (2009). *Evaluation of evidence-based practices in online learning: A meta-analysis and review of online learning studies*. Washington, D.C: US Department of Education.

Oliver, M., & Trigwell, K. (2010). Can "blended learning" be redeemed? *E-Learning and Digital Media*, *2*(1), 17–26.

Rhoads, J. F., Nauman, E., Holloway, B., & Krousgrill, C. M. (2014). The Purdue Mechanics Freeform Classroom: A new approach to engineering mechanics education. In *121st ASEE Annual Conference and Exposition*.

Smith, M. K., Jones, F. H. M., Gilbert, S. L., & Wieman, C. E. (2013). The classroom observation protocol for undergraduate stem (COPUS): A new instrument to characterize university STEM classroom practices. *CBE Life Sciences Education*, *12*(4), 618–627.

Spitzer, R. L., Cohen, J., Fleiss, J. L., & Endicott, J. (1967). Quantification of agreement in psychiatric diagnosis. *Archives of General Psychiatry*, *17*(1), 83–87.

Tileston, D. W. (2011). *Ten best teaching practices: how brain research and learning styles define teaching competencies* (3rd ed.). Thousand Oaks, CA: Corwin Press.

Acknowledgements

We would like to acknowledge our full Freeform research team for their help in this instrument's development. Likewise, we would like to thank Dr. Tara Kaczoroski, author of the blog "StudioCode for Educators", whose advice was invaluable to our coding and reliability testing. This material is based on work supported by the National Science Foundation (NSF), Division of Undergraduate Education under Grant No. 1525671. Any opinions, findings, conclusions, or recommendations expressed here are those of the authors and do not necessarily reflect the views of the NSF.

Copyright statement

Work in progress: An example of teamwork in engineering education. Case: Management of Information Technologies course.

Amparo Camacho
Universidad del Norte, Barranquilla, Colombia
acamacho@uninorte.edu.co

***Abstract**: Teamwork is one of the most valuable skills in the workforce of contemporary professionals. In engineering schools, we are developing learning environments that promote the development of these skills. In particular, in the course of Management of Information Technologies, the author has been developing a methodology aimed at fostering the development of teamwork culture with the goal of preparing the students for their insertion in the workforce. The study presented has been developed in two consecutive semesters, first and second semester of 2016. During the first semester of 2017, the author has continued the study, including some adjustments to improve the methodology used in previous semesters. This paper shows the context of the study, the methodology used to answer the research question, the major findings, preliminary conclusions, recommendations, and future research plans.*

Introduction

In the formation of contemporary engineers, it is essential to develop soft skills such as communication, comprehension of ethical issues, leadership, lifelong learning, and teamwork in addition to the technical skills they develop throughout their major courses. In particular, teamwork integrates and strengthens other soft skills that are currently necessary for the daily work of engineers such as the participation in projects and other activities that require highly efficient and competent teams for the production of goods and services.

As Oladiran, Uziak, Eisenberg and Scheffer (2011) explain, in the engineering field, engineers are exposed to situations that require the use of soft as well as hard skills in areas such as designing and manufacturing. In the same way, accreditation bodies stipulate the necessity of developing these skills. ABET (2017), for example, explicitly states in student outcome (d) the necessity to develop the ability to work in multidisciplinary teams.

Given the current panorama of globalization and the requirements of the industry, colleges of engineering must implement pedagogical strategies that favor teamwork as a key skill for an engineering graduate. This paper presents one example of how a certain teaching methodology has been implemented in one course of an engineering major program to promote the development of teamwork.

Background

In our academic culture and context, it is common to implement teamwork as part of the teaching-learning process. However, frequently, “teamwork” is confused with “group work.” The distinction between the aforementioned terms is made clear by Katzenbach and Smith (1993) who explain that the difference lies in students’ performance results; the performance of group work is based on what each member of the group does as an individual while the performance of a team involves not only individual work, but also what the authors call “collective work-products,” which are those products produced by a team of two or more members working together. In the same way, group work emphasizes the achievement of individual objectives, with group members only taking responsibility for their own work, not for the results of the group itself. Teamwork, on the other hand, focuses on the team taking responsibility for both their individual work as well as the team's collaborative work. When explicitly defining teamwork, Katzenbach and Smith (1993) define a team as, "A team is a small number of people with complementary skills who are committed to a common purpose, set of performance goals, and approach for which they hold themselves mutually accountable." (p.112)

The above definition highlights the most important aspects of teamwork such as: having a common purpose, shared responsibility, mutual accountability, and the establishment of specific objectives. These abilities are rarely taught in such an integrated way in engineering schools, whose teaching focus is typically more centered on the technical aspects of the engineering profession. Learning environments conducive to the practice and development of these aspects of teamwork, such as in the case presented by Zou and Ko (2012) where explicit instruction in teamwork and opportunities for practice and formative feedback for three years in an undergraduate engineering program, are rare within in the engineering educational field.

There have been training programs like those described by Natishan, Schmidt and Mead (2000) called Building Engineering Student Team Effectiveness and Management Systems (BESTEAMS) that aim to develop and implement standardized training to improve teamwork among engineering students. There have also been courses focused on communication designed within engineering programs in which, as Missingham and Matthews (2014) explain, students are taught about teamwork and at the same time develop abilities for critical text analysis and oral and written presentations. Likewise, courses have been designed to respond to the needs of the engineering industry, specifically the need for functioning in multidisciplinary teams, such as those mentioned by Bhavnani and Aldridge (2000), that are interdisciplinary and are offered to Engineering, Business, and Industrial Design students.

These experiences show the need for theoretical and practical preparation for teamwork, which should be complemented and integrated with project management. Together with communication, these two abilities are those currently required by employees in the engineering field. Heldman (2017) defines a project as "a temporary endeavor to create a unique product or service" (p. 22). In order to achieve the creation of unique products and services, it is necessary to have a coordinated effort among members of a team that share a common goal, shared responsibility, and shared commitment. As Smith (2008) explains, an effective way to develop this teamwork is through cooperative learning. This implies positive interdependence, where members of a team work and support each other mutually, individual responsibility, and group responsibility, where each member is responsible for the final product. In the development of software projects, for example, cooperative teamwork determines the final results of the project, as Rizwan, Qureshi, Alshamat, and Sabir (2014) explain, given that this type of project requires permanent interaction and cooperation among team members. This is just one example that demonstrates the importance of cooperative teamwork for engineering students.

The research presented in this paper describes the application of teamwork methodology implemented in an information and communication technology (ICT) course within an engineering program. The purpose of this implementation was to foment the practice and understanding of cooperative teamwork through creating a solution for a real world problem.

Methodology

The methodology developed aims to answer the following research question:

Can teamwork help students to improve their academic results in a particular subject ?

The classroom research was developed over two consecutive semesters: 201610 (first semester of the year) and 201630 (second semester of the year), during both of which the same methodology was implemented with more support, guidance and feedback were provided in the classroom to the activities developed by the students in the period 201630.

The methodology is as follows:

1. First Phase

Since the main purpose of classroom research was to explore how students work as a team, several teams were formed. The teams was formed and created by the students themselves. Each team was required to prepare a course project consisting of the formulation of a project in the area of ICT (Computer and Communications Technology), following a specific methodology for project formulation and management.

Students were assigned roles within the team, the most important of which was the role of project leader; to select the project leader, each team set their own criteria and made the selection. This was done to promote the development of autonomous decision-making. In the second semester of 2016, each team was asked to define its own "rules of the game" to be implemented during the semester, whereby each team was responsible for the management of interpersonal relationships.

Once the team was structured and the roles assigned, each team independently selected the topic for the project. This was done to foment motivation, commitment and responsibility in regards to the achievement of the project objectives.

2. Second Phase

Each team completed the project following the selected methodology. To assess the teamwork implemented during the project, a rubric was designed and applied in both the 201610 and 201630 semester. In the second semester, an additionally category was added to the rubric. See (Appendix 1).

The categories defined in the rubric were:

1. Establishment and distribution of roles in each team
2. Individual work of each student, in preparation for quizzes
3. Participation in weekly meetings
4. Communication management
5. Efficiency in the project management methodology
6. Class work (defined in the second semester)
7. Quality of the final product

With the above categories, the author tried to account for the assessment of teamwork through the development of a project with real engineering features in the area of ICTs, within a collaborative work scheme.

Throughout the two semesters, the class took place two days a week, as follows:

- In the first class, an hour long, the author explained the theory of methodology to be used in the second class of the same week; the students had to send via the online course platform, between the first and second class, the agenda that they were going to follow in the second class of the same week.
- In the second class, two hours long, in teams, the students carried out the scheduled activities. At the end of the class, the team leader and his/her classmates evaluated how well the planned agenda was followed and turned in this evaluation along with the activities carried out in the class.

In line with the above description, the categories that the author assessed in the second classes, through observation and interaction with the students, were: 3, 4 and 5. The assessment of category 5 was complemented by the review and evaluation of the documentation that each team submitted in the second class. Category 2 was evaluated based on several short quizzes that the author gave occasionally during the first class of the week.

The purpose of these quizzes was to measure how prepared students were regarding the topics and activities to be covered and carried out in the second class of the week.

It is important to add that students also turned in elements of the project throughout the semester, so the author could provide students with feedback on their progress and the quality of their work.

3. Third Phase

At the end of the semester, in the first week of final exams, students formally submitted the project carried out, through the submission of a written report and an oral presentation in front of 2 jurors: an expert professor in effective oral communication and an expert professor in methodology formulation and development of projects in the area of ICT, the teacher of the course and the rest of the students from the course. The judges used a rubric to evaluate the ICT projects, which complemented the author's assessment of the final product quality, category 7 of the rubric a used during the semester.

To assess category 1, the author applied a survey at the end of the semester in order to investigate the assignment of roles and the effort devoted to them within the teams. In the second semester, two additional surveys were designed: one regarding perception of the performance of the team leader and another for each student to self-assess their learning during the semester. Two figures of the methodology are presented below.

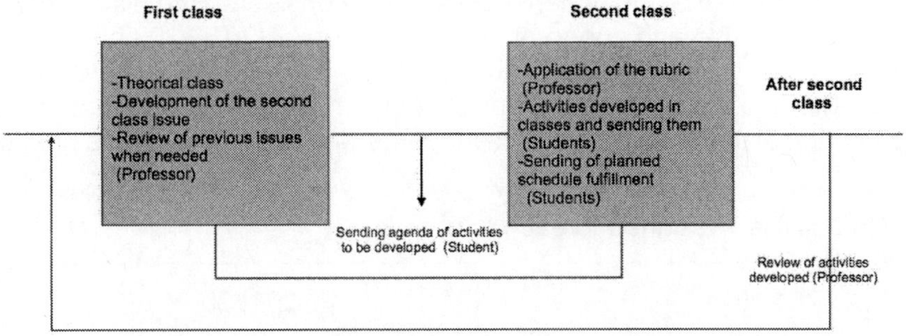

Figure 1: Classroom research developed through the semester

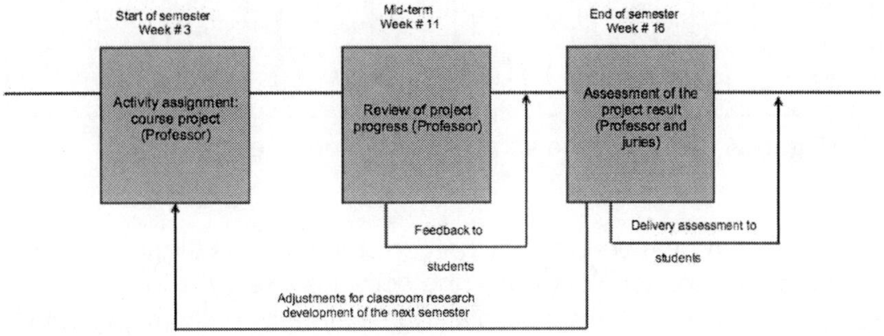

Figure 2: Assessment and administration of classroom research throughout the semester

Results

In this section, the results of the most significant categories assessed in the rubric applied in the 2016-10 and 2016-30 semesters will be presented.

For the 2016-10 period, 6 teams were formed among students, and in the 2016-30 period, 4 teams were formed. Below, is a comparison of the results obtained in the 2016-10 semester with the results for the 2016-30 semester.

Category 1: Establishment and distribution of roles in each team

A survey was designed and applied during both semesters to evaluate this category. The survey assessed the role and performance of each student in the aspects mentioned below, and a 4-level Likert scale was used to record students' answers: Strongly Agree (4), Agree (3), Disagree (2) and Strongly Disagree (1).

- Equitable and coherent distribution of roles:

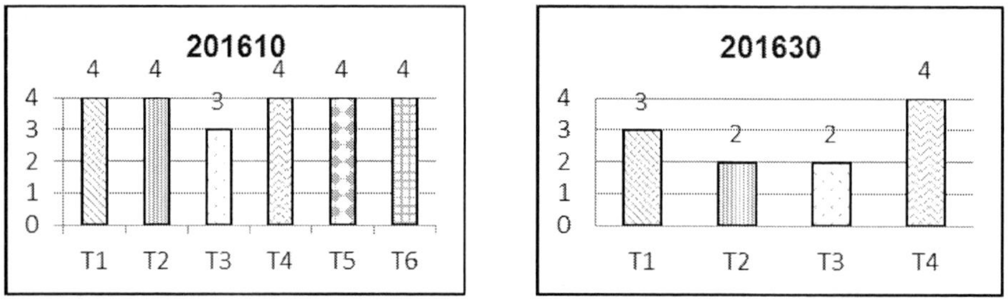

Figure 3: Equitable and coherent distribution of roles

In 2016-10, it can be observed that the majority of students considered the distribution of roles within their team to be equitable and coherent. In the case of the 2016-30 semester, there is variability in the responses, with a tendency towards a low-level in the upper limit (disagreement over the equitable and coherent distribution of the roles). That is, students mostly perceived that the distribution of the roles in teams was neither equitable nor coherent.

- Clarity in the activities assigned to each role:

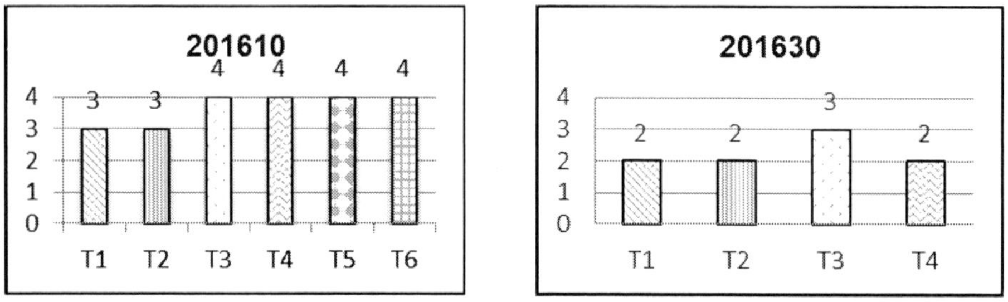

Figure 4: Clarity in the activities assigned to each role

In 2016-10, most of the team members clearly perceived their roles in the team. In 2016-30, most of the team members were not clear about the activities assigned to their role.

- Self-assessment of the performance of team members (except the role of project leader)

For this aspect, a Likert scale of 4-levels was used: Very high (4), high (3), medium (2) and low (1).

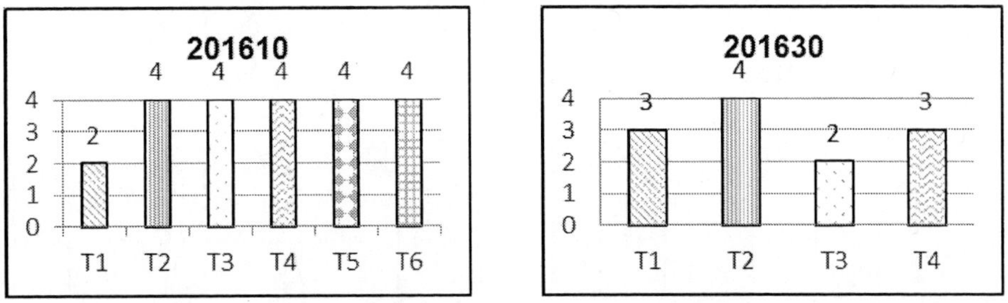

Figure 5: Self-assessment of the performance of team members (except the role of project leader)

In the 2016-10 period, most students self-assessed their performance as "very high." In 2016-30, students' self-assessment of their performance was variable, with a trend towards a high-middle level of performance. In this sense, a higher level of criticality was observed in student self-assessment, meaning students were more critical of their work within the team.

Category 2: Individual work of each student (in preparation for quizzes)

For this category, 5 short quizzes were given in the 2016-10 semester and 3 short quizzes in the 2016-30 semester. All quizzes centered on the subject areas to be developed in the course by means of teamwork activities in the second class of the week. In both semesters, the average grades per team were obtained and correlated to the work they completed in class.

By 2016-10, only one of the quizzes corresponded to the level of performance of the work done in class. In 2016-30, of the 3 quizzes applied, only one corresponded to the level of performance of work done in class. For both 2016-10 and 2016-30, these results were not taken into account because there was no alignment between the topics evaluated in the quizzes and the subject areas developed by the students in the activities of the second class of the week.

Category 4: Communication management

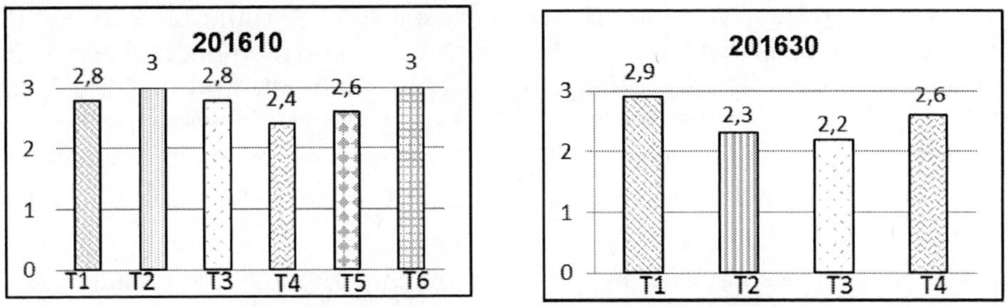

Figure 6: Communication management

Figure 6 shows that in the 2016-10 semester two of the teams obtained the maximum score of the rubric (3), and the other teams had scores close to the maximum score. In the second semester, the maximum score was 2.9, and the trend was towards a medium-high score. One possible cause of lower team performance in 2016-30 could be related to the lack of clarity in team roles and the size of teams, as the number of members in each team was greater compared to the teams formed in 2016-10.

Category 5: Efficiency in the project management methodology

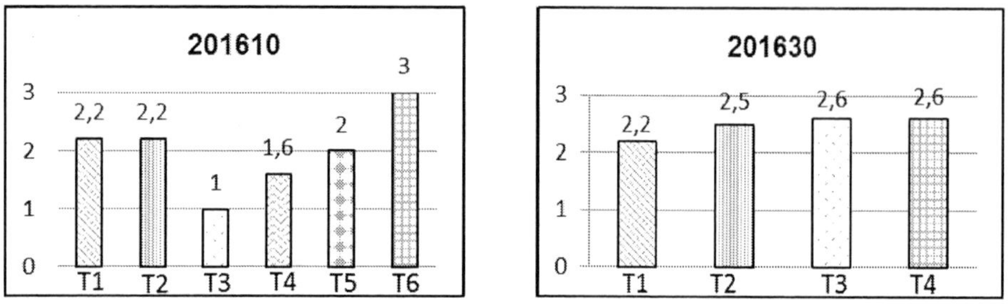

Figure 7: Efficiency in the project management methodology

In this category, the performance of teams in the second semester is better than in the first, given that the average of the first semester was a 2, and the scores obtained by the teams during the second semester were greater than this despite no team obtaining the maximum score (3). One possible cause of the increase in scores in the second semester could be the greater guidance and feedback provided to the teams during the classes which was absent in the first semester of implementation.

Category 7: Quality of the final product

This assessment was carried out in both semesters in an integrated way considering the assessment carried out by the two judges as well as the assessment done by the course instructor, based on the final written report and supporting evidence submitted. For this category, an overall assessment will be presented rather than quantitative results.

According to the overall assessment for the 2016-10 period, the team that presented the final product of highest quality in terms of the final report and the oral presentation of the work, tended to have better participation in weekly team meetings, communication management, and efficiency in the project management methodology. The team that presented the lowest level of quality in the final product stood out only in the handling of communication. In 2016-30, the team that presented the final product of highest quality was the one that presented satisfactory results for participation in weekly team meetings, effective communication, class work, and efficiency in the project management methodology, which shows a high level of correlation between the quality of the final product delivered and the other areas of project development. Regarding this team, it was evidenced in the team member role survey that there was enough clarity in the roles assigned to each team member, and that the team leader performed very well. The team who turned in the final product of lowest quality, lower than those submitted in 2016-10, stood out only in the efficiency in the project management methodology. In terms of the distribution of roles, the self-assessment showed students reported a medium-high score.

It can be observed from the previous analysis that the category that had a satisfactory overall performance was the efficiency in the project management methodology, and, in terms of role management, the best score was obtained in an adequate distribution of roles and sufficient clarity regarding these roles. As for the role of the leader, the most outstanding aspects were: management of peer motivation and effective communication.

If the results of the two semesters are compared, it can be observed that in the 2016-30 semester, better results were obtained in efficiency in the project management methodology, and communication decreased compared to the 2016-10 semester. Regarding the management of team member roles, it is worth noting that in the 2016-30 semester, some students were more self-critical about their performance.

Preliminary conclusions

In 2016-30, more information was collected than in the previous semester, but it is still not conclusive. One of the highlights of this semester is related to class work, through which the course instructor assessed each team's completion of the work done in class by evaluating whether the team followed the required method appropriately. The results were satisfactory. Another outstanding aspect was the efficiency of the project management methodology. In this aspect, the measurement focused on determining how far the agenda of activities established by the same teams to fulfill the required activities was fulfilled. The results of these two aspects are coherent. In this semester (2016-30), the author provided more guidance and was more observant of the activities carried out in the class. She also made a more permanent follow-up on the development of activities, and she assessed and adjusted teams' working agendas when necessary. Although the communication decreased compared to the previous semester, the instructor considers there was an adequate level of communication in the teams, bearing in mind the differences between the members of each team.

The measurement of individual preparation to carry out the activities in class remains one of the greatest challenges of the course, one that the author should explore through the implementation of further activities in the following semester. Although the evaluation of this preparation was carried out, there is still not enough information or data to give any preliminary conclusions.

In general, the author considers that the results obtained from the methodology implemented are consistent with the perceptions and expectations regarding the impact of teamwork to obtain better academic results observed through the work carried out by students.

Recommendations

According to the results obtained from the implementation, the author, considers that the main adjustments that should be made in order to continue this classroom research include: the creation of a strategy for creating teams among students that are balanced and take into account their soft and hard skills and learning styles, aligning individual student preparation with the teamwork activities carried out in class, rotating the role of project leader at least two times during the semester, creating a strategy for better assignment of roles among team members, and including activities that involve teamwork previous to project commencement to train students in assertive communication and efficient behavior within teams. Additionally, one aspect the author plans to investigate in upcoming semesters is the influence of individual effort in the overall effort of the team. With these adjustments, the author plans to include the use of online self and peer review tools such as WebPA or CATME.

In any future research, the author aims to increase the sample size by involving a larger number of students in order to have more valid results. Even so, the number of students involved depends on how many students register for the specific course in which the investigation is taking place.

References

ABET. (1997). *Criteria for accrediting engineering programs, 2017-2018.* Accessed at http://www.abet.org/accreditation/accreditation-criteria/criteria-for-accrediting-engineering-programs-2017-2018/

Bhavnani, S., Aldridge, D. (2000). Teamwork across disciplinary borders: a bridge between college and the work place. *Journal of Engineering Education,* 13-15.

Heldman, K. (2017). CompTIA Project+ Study Guide. Exam PK0-004. John Wile & Sons, Jan, 2017.

Katzenbach, L., & Smith, D. K. (1993). The discipline of teams. *Harvard Business Review*, 71,111-120.

Missingham, D., & Matthews, R. (2014). A democratic and student-centred approach to facilitating teamwork learning among first-year engineering students: a learning and teaching case study. *European Journal of Engineering Education*, 39 (4),412-423.

Natishan, M.E., Schmidt, L.C., & Mead, P. (2000). Student focus group results on student team performance issues. *Journal of Engineering Education*, 269-272.

Rizwman,M, Qureshi ,J., Alshamat,S., & Sabir, F. (2014). Significance of the teamwork in agile software engineering. *Sci.Int(Lahore)*,26(1), 117-120.

Smith, K.A. (2008). *Strategies for developing engineering student's teamwork and project management skills.* Accesed at http://faculty.kfupm.edu.sa/COE/elrabaa/rich%20text/general%20presentations/20565.pdf on 24 Feb 2017

Zou T., Ko, E. (2012). Teamwork development across the curriculum for chemical engineering students in Hong Kong: Processes, outcomes and lessons learned. *Education for Chemical Engineers 7* ,e105-e117.

Copyright statement

Appendix 1

MANAGEMENT OF INFORMATION TECHNOLOGIES RUBRIC. SEMESTER 201630						
		Scale	T1	T2	T3	T4
	Number of students					
Establishment and distribution of roles, in each team	Clarity of role distribution					
	Balanced and coherent distribution of roles					
Individual work of each student in preparation for quizzes	Student does not prepare for quiz (0 to 2 on quiz)					
	Student partially prepares for quiz (2 a 3.5 on quiz)					
	Student prepares for quiz (3.5 a 5 on quiz)					
Participation in weekly meetings	One student does not participate	1				
	Some students do not participate	2				
	All students participate	3				
Communication management	No student listens to nor respects the opinions of his/her classmates	1				
	Some students listen to and respect the opinions of their classmates	2				
	All students listen to and respect the opinions of their classmates	3				
Efficiency in the project management methodology	The agenda is not followed or completed at all	1				
	The agenda is partially followed and completed	2				
	The entire agenda is followed and completed	3				
Class work (defined in the second semester)	Complete and in accordance with the approved methodology	2				
	Incomplete, although approved methodology was used	1				
	Incomplete with no use of approved methodology	3				
Quality of the final product	The project does not meet expectations	1				
	The project partially meets the expectations	2				
	The project meets all expectations	3				

Assessment of Significant Learning from students and professors perspective focused on learning cycles promotion

Nubia Isolina Patarroyo Duran
Universidad El Bosque, Bogotá, Colombia
patarroyonubia@unbosque.edu.co

Constanza Beatriz Pérez Martelo
Universidad El Bosque, Bogotá, Colombia
cbperez@unbosque.edu.co

Beatriz Nathalia Serrato Panqueba
Universidad El Bosque, Bogotá, Colombia
bserratop@unbosque.edu.co

Abstract: *The purpose of this work is research how to assess the dimensions of significant learning from the perspective of students and professors. According to the objective, the work is conducted through a methodological approach which integrates both Fink's learning taxonomy with the OADI learning cycles. This method is applied in the learning evaluation processes, within the frame of a study case in a Colombian university where the Fink approach was adopted as its pedagogical model. The results reflect that the interaction between the learning cycles of teachers and students encourages the creation of shared mental models reinforcing the incorporation of active learning spaces. This study makes it possible to broaden previous research on learning assessment, and provides elements on the articulation of the participant roles of the process and its interactions, to increase the capacity for action in significant learning. This is a contribution to organizations that incorporate, in their engineering programs, active learning approaches.*

Introduction

Engineering education has recognized the impact of promoting spaces that strengthen students' ability to adapt to the changing dynamics of the environment, interact and communicate with others, and learn throughout their own lives. In this context, the incorporation of active learning approaches has increased. In these perspectives, "students are encouraged to take responsibility for their own learning" (de Graaff, Saunders-Smits and Nieweg, 2005, p.7). This paper analyzes and studies the processes of learning among students and teachers in implementing one of the active learning approaches: Significant learning (Fink, 2003). The taxonomy of significant learning proposed by Fink (2003) includes six dimensions of learning: fundamental knowledge, application, integration, human dimension, care dimension, and learning to learn. Some of these studies make clear that this method contributes to the design of courses focused on a learning process that makes sense for the student (Barnes and Caprino, 2016; Krueger, Russell and Bischoff, 2011, Simms and George, 2014; Takeda and Ortiz, 2005). Notwithstanding the foregoing, it has paid less attention to learning processes in organizations incorporating such approaches. This work contributes to the evaluation of significant learning, when this approach is adopted in an educational organization. For this purpose we propose a methodology of assessment which integrates aspects of the active learning approach with the learning cycles in the participant roles of the process.

Through this work it is also intended to contribute to the evaluation of significant learning in the context of a Colombian university that has adopted the Fink approach as its pedagogical model. As part of the implementation phase of this approach, the university has provided training for teaching staff as well as guidelines for course design. In this process, teachers have changed their pedagogical practices, hence the importance of evaluating the results obtained in the generation of learning cycles (Kim, 1993). We have focused on an engineering program, where we have analyzed the interactions between teachers and students in the framework of the evaluation of significant learning. The article is organized into five sections. The theoretical framework is presented in the second section by integrating approaches of significant learning with perspectives of individual and organizational learning. In the next section, the case study methodology is presented and the results are given as well. The final section describes the findings from the study, and some implications of the research project.

Theoretical framework

This case study integrates two theoretical approaches: the Taxonomy of Significant Learning (Fink, 2003) and the OADI learning cycle, (Observe, Assess, Design and Implement) (Kim, 1993).

Significant Learning is a methodology focused on students and their learning process, which was presented by Dr. Dee Fink (2003, 2005, and 2007) who agrees with the methodology of Boom. Fink argues that effective teaching should focus on student learning rather than teaching; however, his approach demonstrated that three perspectives need to be clear: learning-centered teaching, how students learn and the importance of having a good course design. Fink recognizes that learning involves implementing change in order to achieve the proposed goal. He created a model with six dimensions that identify six types of learning that should be given to students to complete the course.

Although Bloom's Taxonomy proposes six main categories of learning, significant learning is not hierarchical but interactive, that is, that students have to achieve any kind of learning that will help them achieve the other five types of learning, Taxonomy is not only focused on cognitive learning but includes the development of human competences both personal and in relation to others.

Fink's taxonomy comprises six dimensions (Fink, 2007) which will be described below. In the dimension of fundamental knowledge students will understand and remember the most important course information. In the dimension of applied learning, students will use the course content effectively. The third dimension is the integration which will identify the inter-relationship between the content of the current course and previous courses. In the human dimension of learning, students learn who they are, what they can do and interact effectively with others. In the dimension of care, they will develop new feelings, interests and values in relation to the subject being studied, the environment and social responsibility. The last dimension is learning to learn, in which the student uses his skills to be a good student and researcher to perform autonomous learning. This approach seeks to make students become co-creators of their own learning (see Figure 1).

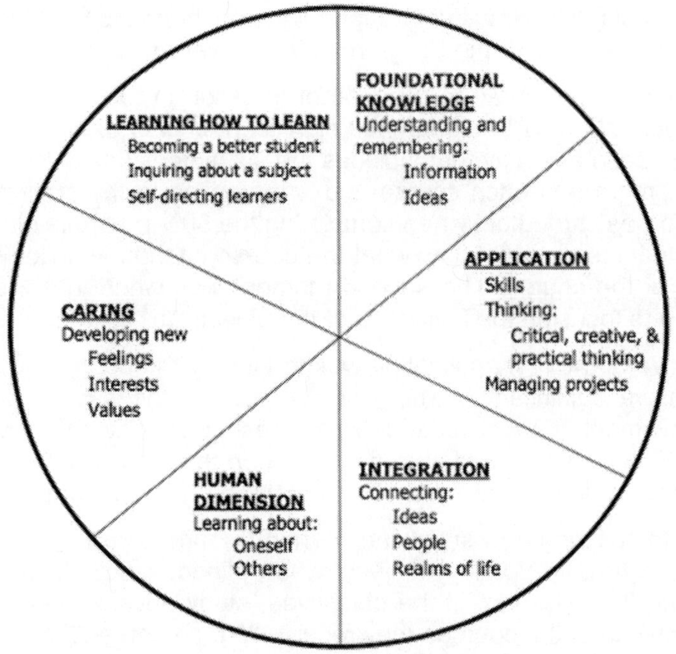

Figure 1. Taxonomy of Significant Learning. Source: Fink, 2005, p.9

In the context of engineering education, course design by taxonomy Fink, shows the enrichment of learning processes and increased students engagement with their training (Apul and Philpott, 2011, Hunsu, Abdul, Van Wie, Adesope and Brown, 2015). Beside this, this approach offers the pedagogical support to minimize the project development time (Kusumawardani et al., 2014). The studies have been conducted mainly towards the resulting learning from the Fink approach-designed courses, but not in how the roles of teachers and students interact themselves in the implementation of this pedagogical model.

In order to evaluate the learning processes related to the adoption of Fink's taxonomy as a pedagogical model, we use the approach proposed by Kim (1993). According to this author, learning is to increase "*a person's ability to take effective action*" (1993, p.38). Learning involves four components: observe, evaluate, design and implement (OADI cycle). People have concrete experiences and observe the situation. "*They assess (consciously or subconsciously) their experience by reflecting on their observations and then design or construct an abstract concept that seems to be an appropriate response to the assessment*" (Kim, 1993, p. 39). They implement actions related to this design and restart a new cycle. In these learning loops, mental models also influence, that "*represent a person's view of the world*" (p. 39). When individual mental models are shared, organizational action is affected.

The OADI model has proved to be useful for evaluating individual learning processes in the organizational context. This allows identifying the learning links and loops of each participant in a project or initiative and supporting the management of the complexity of an organizational task (Reyes, 2012).

Methodology

The question proposed in this research project is "How to assess the dimensions of significant learning from the perspective of students and professors?". Fink (2007) proposes to assess significant learning in two stages during the development of the subjects; For each of them (Fink, 2007) proposes the questions that lead to a reflection on learning and in turn constitutes a possibility to improve the curriculum and the design of the course. Starting from these questions, we developed the first pilot project including some of the subjects contained in ten curriculum of basic engineering training throughout the entire academic process including first, fifth, eighth and ninth semester of the Program under surveillance and research process.

We took a Case Study appropriate to the type of question proposed in this research paper (Yin, 2014) that would allow us to identify the significant impact of learning on students. We designed for this purpose a survey that includes the six dimensions of the Fink taxonomy, as established in the program of each course and was applied to the students in two different moments; first one was carried out when completing the 50% progress of the course since at that moment the students already know what the course intends and likewise what they can expect at the end of the course; The second moment was when 100% of the course was completed, allowing to focus on the fulfillment of the objectives.

The primary objective of the first application was to identify the degree of progress in course information to achieve significant learning objectives. On the second occasion, students analyzed the achievement of goals in the course. Subsequently, we conducted a focus group with the teachers of the program under study, based on the results of the survey. Participant observation was also carried out by the teachers of the program under study.

With the aim of understanding how significant learning is incorporated in the courses, a focus group of teachers of the participating subjects was defined, requiring them to present their perception regarding the fulfillment of the objectives, study development and the evaluation the students currently do of the courses they receive. We presented the results obtained, the categories established in this study and the reflection around the exercise, from the perspective of the students, thus allowing teachers to validate and complement the categorization.

The results of the surveys showed that it was necessary to define emerging categories (Bernard, Wutich & Ryan, 2016; Coffey & Atkinson, 1996) according to the answers, to associate them with each dimension, which allowed the integration of different perspectives as well as analyzing them in a general way; therefrom be able to identify common categories that exist among the same dimensions which are established on the model of significant learning.

We analyze the information collected in the OADI cycle for domains of action related to significant learning (see figure 2).

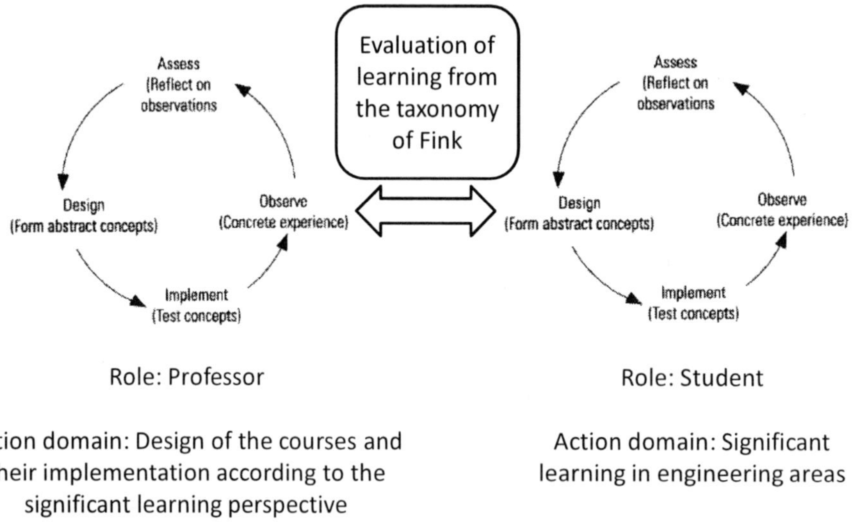

Figure 2: Interaction of learning cycles of students and professors in the domain of action of "significant learning". Source: Prepared from Fink (2005) and Kim (1993)

This paper contents the results of this initial analysis.

Results: Articulation of learning cycles of professors and students

In this section, we present the results associated to the learning evaluation of the analyzed roles, and their interactions.

In the action domain of Design of the Courses and their implementation according to the significant learning perspective, professors begin their process from an e-learning training on the Fink approach. The result of this exercise is the design of one of its courses under this perspective. Lectures observe differences between their previous practices and the new model, and make the evaluation from Fink's taxonomy. Subsequently, they design their course according to the guidelines of this pedagogical approach.

The articulation between the learning cycles of professors and students is achieved through the design of the course and its implementation. In the action domain of Significant learning in engineering areas, the students generate learning cycles based on the development of planned activities in the course, the evaluation of their lecturer and their self-assessment. The actions developed according to the objectives of "learning to learn" allow a design of strategies to develop autonomy in their learning process.

Assessment of learning by the role of the student

The first survey to evaluate student learning, contained 162 answers. The results are explained as follows: the first question is "What are you learning now?" all observations coincided with the information related in course's program. The second question is "Do you feel satisfied with the information you are learning?" 84% think the information is enough, 8% think it isn't enough, 5% is extend and 3% is so short. Another question was "What is the course teaching like?" 47% think the methodology is good, 29% this is a dynamic class and active learning tools, 15% practical activities, 7% good professor's treatment and 2% does not response about the course.

When we ask them about what they would like change about course, 29% say nothing, 13% think could be better if the feedback course is reinforced, decrease course introduction and extend other themes of interest, 9% ask include more activities for active learning, 13% increase practical activities and 3% think in changing professor's relationship. Nevertheless, 33% stated different factors not depending of professor but of the University and mentioned the locations where the course is conducted, schedule, further tuition time, decrease number of students in the course and change the grading system.

In the second learning survey, the question and answers are as follows: first question was "Have you developed basic knowledge in the course?" all answers were "yes". Second question was "Have developed ability to use that knowledge to solve typical complex problems?" 96.3% answered "yes", and the remaining students said "not". About achievement goals, 77.2% think is "high" and 21.1% think is "intermediate". In the table 1, we show all dimension which have a similar behavior, however, there seems to be a difficulty in integration dimension.

Table 1: Level of achievement learning objectives according to students' perspective

	Foundational knowledge (%)	Application (%)	Integration (%)	Human dimension (%)	Caring (%)	Learn to learn (%)	Totally (%)
High	86	86,2	59,4	78,8	87,6	65,4	77,2
Intermediate	13,4	12,6	38,2	18,8	10,4	33	21,1
Low	0	0,4	1	1,6	1,2	0,8	0,8
Deficient	0,6	0,8	1,4	0,8	0,8	0,8	0,9
Total	100	100	100	100	100	100	100

The significant learning methodology generates good results satisfying the learning in engineering students, because they respond to the requirements of active learning. Other important facts like learning to learn and taking care of the dimension, are based on the humanistic aspect. The student's reflection on each dimension of Fink's taxonomy, showing the interrelationship between them, is relevant and must be taken into account to make the design of the courses, making class dynamic to correctly construct the knowledge and a good space to reflect over the information learned around the experience itself. In the application dimension focus attention on the implementation process, but not in bio-psychological and social conditions. As the comments made in the student survey on the assessment of learning was possible to identify a major category in which the dimension was evaluated. The results are shown in Figure 3.

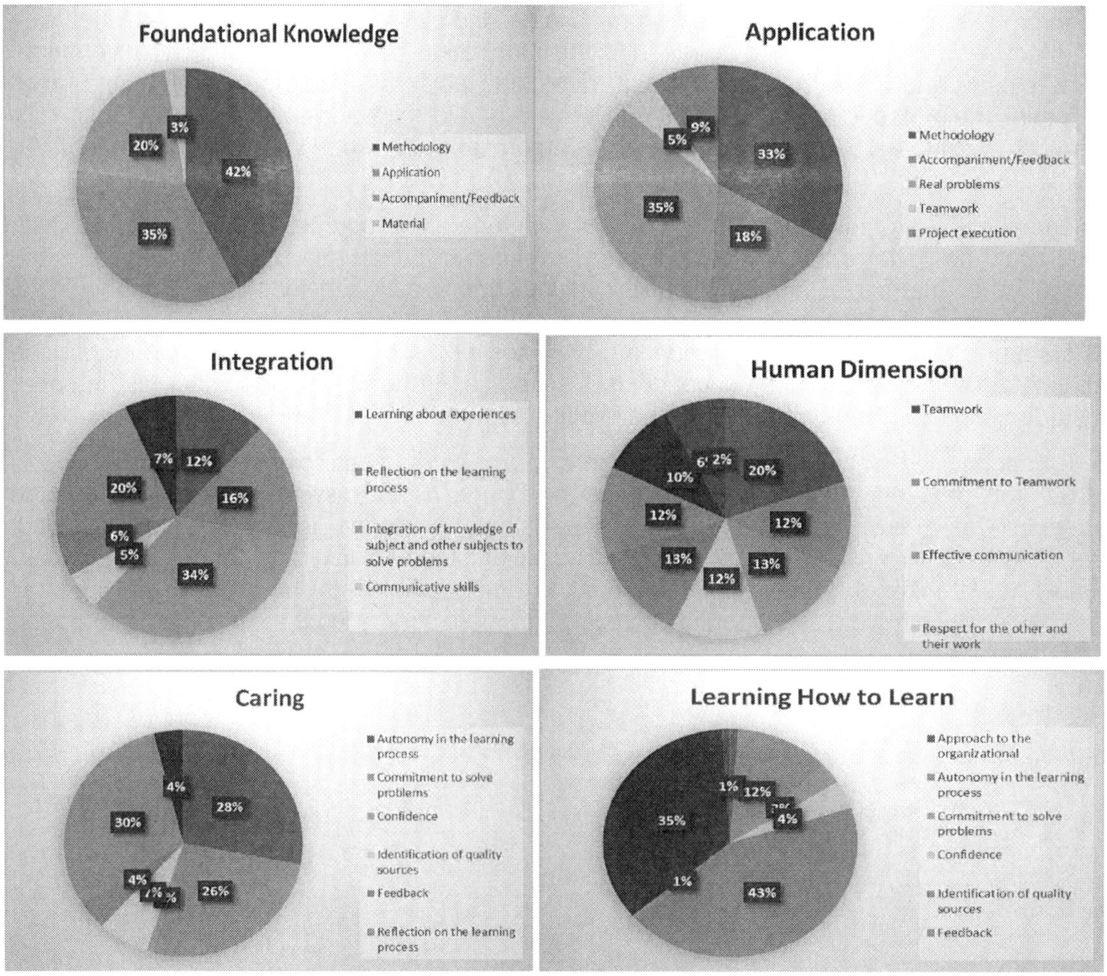

Figure 3: Emerging categories for each of the dimensions of significant learning

In the analysis of the students on the evaluation of the dimensions, it was detected that for the cognitive dimensions, fundamental knowledge, application and integration, the categories of methodology and practice are fundamental. For the first two, the connection is direct, but integration is not evident because the students assume that the methodology of the course must integrate a practical part. In addition, students consider that the accompaniment and feedback actions in the courses are relevant and that despite being part of the methodology, they are visualized as independent.

In terms of the humanistic dimension, we identify common aspects that are related to personal development, integration with others and environmental, community and organizational impact.

Evaluation of the teacher's role

After the analysis of the students' responses, we developed a focus group with teachers, whose feedback allowed them to identify better opportunities as well as to evaluate other points of view.

The first question asked the teachers: "What is surprising for you, of the categories that students consider to evaluate each dimension?" In general, all teachers agree with the comments of the students, and in addition they think that students demonstrate good skills to analyze their learning process and to develop their human capacities, not just cognitive skills. It is good to know that students value active learning, the contextualization of knowledge and practice in real context.

Another question posed to teachers is which category would you include in the student's analysis of each dimension and if they consider that fundamental knowledge, contextualization, construction of deductive logic and critical thinking are very important.

Another question that teachers ask is what category you would include as a teacher in the student's analysis of each dimension and if they consider that fundamental knowledge, contextualization, construction of deductive logic and critical thinking are very important.

In the dimension of the application, the future utility and the transversal organization; and in the integration dimension the interdisciplinary work and the integrity of the propagation can be interesting.

In the human dimension, leadership, form and respect are very important categories. In the commitment dimension, social responsibility and time control should be considered. At the same time, in the dimension of learning to learn, adequate information and quality must be sought to generate abstraction and generalize the information learned and adopt the method of study necessary to include in those categories.

Other feedback realized by professors was what they thought about learning evaluation made by their students for each of the learning dimensions. They agree upon fundamental knowledge and highlighted the methodology the fundamental basis to learn and apply concepts in real life. It is relevant to obtain equilibrium between goals, methodology and practice.

In the application dimension it is important to have the opportunity to solve real problems by applying concepts obtained in the course and analyze the results to make decisions. It is satisfying to know that the message in the course was understood. On the other hand, in the integration dimension identify the motivation of life, work and the University, and the importance of the methodology used. In this sense, it is directly related to the identified problem. The accompaniment of the teacher is key to integrate knowledge.

Identify roles and their importance for work group corresponds to the human dimension, likewise the possibility of being multipliers of knowledge and learn to defend their ideas. In the commitment dimension, the bio-psychological and social approach of the university is detected and the student's identification is a result of his / her learning. Teachers are pleased that their students deliver mature and integrated answers. Finally, in the dimension of learning to learn, they recognize the importance of obtaining information from sources of quality, communication capacity and self-evaluation. Students understand the importance of their behavior for success.

When we ask the teacher about what other strategies could be implemented after this evaluation, they respond that it is important to determine situations and use that information to adjust course design, including periods of relaxation in class and use tools to streamline course activities; better use of resources offered by the University and strengthen the support of the teacher in the integration of knowledge.

Conclusions

This work provides methodological elements to study the articulation of student roles and lectures in the implementation of significant learning approaches. In the case studied, the assessment of learning cycles provides a tool for reflection on student-centered learning in the program. The cycles of learning of teachers and students are in permanent interaction.

The incorporation of Fink's taxonomic categories into evaluation processes allows a dialogue between these two roles and allows the teacher to design new actions to promote significant learning in the courses he orients.

The evaluation of the dimensions of Significant learning from the perspective of the students, favors the change of pedagogical practices and allows the promotion of learning cycles. The differences between the objectives of the changes of practices and the results, allow to promote the component of self-learning in all the actors involved in the process. The dimension of learning to learn and the dimension of caring are very similar because they correspond to the humanistic aspect of significant learning. On the other hand, the evaluation of the learning corroborates the coherence and coincidence between the course design and what the student perceives from the process. Significant learning produces satisfactory results for engineering students because they respond to their active learning requirements.

The student's analysis of each dimension in Fink's taxonomy shows the interrelation between them, but the teacher is not really aware of it when planning and designing the course. Therefore, learning requires a good course design accompanied by a dynamic classroom space to promote an adequate environment to build knowledge and a good analysis of learning in their experience, but for this it is necessary for teachers to take time to plan and design, which becomes critical to achieving the goals of significant learning. By changing their traditional teaching practices towards significant learning, teachers realize that more than content, what counts is the interaction space generated.

In Significant Learning methodology, the students showed integrity under the experience that must transform their cognition and elements for human relationship and personal growth. In the application dimension, students focus on attending the execution process, but not on bio-social, bioethical, technical, scientific, social and humanistic conditions, which must be shown in an integral person.

Self-confidence is relevant to the dimension of commitment in significant learning. It is because each person must know who he/she is, and what he/she must do to respond to impacts on the community and the environment.

The interaction of assessment processes of significant learning by professors and students generates the construction of shared mental models that promote the strengthening of this pedagogical perspective within the institution. The movement from mental models associated to traditional teaching approaches towards active learning perspective, is promoted and achieved through an assessment focused on learning objectives and professor-student interaction.

The results obtained from the application of the tool make it possible to offer the teacher more information for the planning and design of the course and its work itself, considering that the institution currently carries out a generalized evaluation of which there are no available data that allow to improve the Course design.

From this first pilot, adjustments will be made to the tool and it is expected to be applied to all the subjects of the program under study, to complement the research study.

References

Apul, D. S., & Philpott, S. M. (2011). Use of outdoor living spaces and fink's taxonomy of significant learning in sustainability engineering education. *Journal of Professional Issues in Engineering Education and Practice,* 137(2), 69-77.

Barnes, M. E., & Caprino, K. (2016). Analyzing service-learning reflections through Fink's taxonomy. *Teaching in Higher Education,* 21(5), 557-575.

Bernard, H.B, Wutich, A., & Ryan, G.W. (2016). Analyzing Qualitative Data: Systematic Approaches. Second edition. Newbury Park, CA: Sage Publications.

Coffey, A., & Atkinson, P. (1996). Making sense of qualitative data: Complementary research strategies. Thousand Oaks, California, EE.UU: Sage Publications.

de Graaff, E, Saunders-Smits, G.N. and Nieweg, M.R. (2005). Research and Practice of Active Learning in Engineering Education. Amsterdam: Amsterdam University Press. Accessed at http://ebookcentral.proquest.com/lib/bibliobosque-ebooks/detail.action?docID=420126

Fink, L.D. (2003). Creating Significant Learning Experiences. San Francisco: Jossey-Bass.

Fink, L.D. (2005). A Self-Directed Guide to Designing Courses for Significant Learning. Accessed at https://www.deefinkandassociates.com/GuidetoCourseDesignAug05.pdf on November 10 2016.

Fink, L.D. (2007). The Power of Course Design to Increase Student Engagement and Learning. *Peer Review*, 9 (1), 13-17.

Hunsu, N. J., Abdul, B., Van Wie, B. J., Adesope, O., & Brown, G. R. (2015). Exploring students' perceptions of an innovative active learning paradigm in a fluid mechanics and heat transfer course. *International Journal of Engineering Education*, 31(5), 1200 - 1213.

Kim, D. (1993). The Link between Individual and Organizational Learning. *Sloan Management Review*. Fall, 37-50.

Krueger, K. P., Russell, M. A., and Bischoff, J. (2011). A Health Policy Course Based on Fink's Taxonomy of Significant Learning. *American Journal of Pharmaceutical Education*, 75(1), 1-7.

Kusumawardani, S.S. et al. (2014). Reducing student's learning duration on engineering final project by implementing Fink's taxonomy on e-learning. *Journal of Theoretical and Applied Information Technology, 68 (3), 375 – 380.s)*

Reyes, A. (2012). Organizational learning and the effective management of complexity. Kybernetes. 41 (3/4), 318 – 326.

Simms, M., and George, B. (2014). Approaching assessment from a learning perspective: Elevating assessment beyond technique. *Educational Assessment, Evaluation and Accountability*, 26(1), 95–104.

Takeda, N., & Ortiz, C. (2005). Morphologic Design Based on Active Learning. In E. de Graaff, G.N, Saunders-Smits, & M.R. Nieweg (Eds.), *Research and Practice of Active Learning in Engineering Education.* Amsterdam: Amsterdam University Press. Accessed at http://ebookcentral.proquest.com/lib/bibliobosque-ebooks/detail.action?docID=420126

Yin, R. (2014). Case Study Research: Design and Methods (Quinta ed.). Thousand Oaks, California: SAGE Publications.

Implementation of Blended Learning in a Structural Analysis Course

Juan C. Reyes
Universidad de los Andes, Bogota, Colombia
jureyes@uniandes.edu.co

Francisco A. Galvis
Universidad de los Andes, Bogota, Colombia
fa.galvis10@uniandes.edu.co

Diego M. Valencia
Universidad de los Andes, Bogota, Colombia
dm.valencia40@uniandes.edu.co

Mariana Tafur
Universidad de los Andes, Bogota, Colombia
m-tafur@uniandes.edu.co

Alvaro Galvis
Universidad de los Andes, Bogota, Colombia
a.galvis73 @uniandes.edu.co

***Abstract**: The Structural Analysis course is the first in a series in the Civil Engineering program at the Universidad de Los Andes (Bogotá – Colombia) in which the fundamental concepts of this discipline are learned. After the invitation of the Dean of Engineering, the professors of Structural Analysis, who were accompanied by pedagogical, technological and assessment advisers, spent two years redesigning, developing, testing this course to incorporate active pedagogy. The incorporated technology and a blended learning modality was also implemented. This paper presents the course design and answers the research question: What are the advantages provided by the implementation of pedagogical innovations with the support of digital technologies in a professional civil engineering course?*

Introduction

Engineering professionals are facing challenges that go beyond the concepts they learn in their academic program. Namely, learning to learn, and doing so with technology, becomes a necessity. Despite this reality, Galvis (2017, 2) states that most traditional engineering faculty members teach content in a traditional manner, that is, they teach as they were taught. This approach parts from the notion that their role is to transfer relevant mental models to help students understand problems and transform reality with engineering methods and tools, though not necessarily to help students become autonomous learners and problem solvers. According to Mills and Treaagust (2003), this traditional approach also does not prepare students for today's engineering field where professionals deal with incomplete data and competing (often conflicting) demands from clients, governments, environmental groups and the general public.

Recognizing the importance of modifying the educational processes to meet students' new needs, the Faculty of Engineering at the Universidad de Los Andes invited professors to explore the use of innovative teaching methodologies, specifically transformative, active and are digital technology (DT) supported pedagogies. Core courses in the engineering curriculum were chosen as the contexts for the redesign, and highly qualified instructors in this discipline were named as leading professors. One of the courses selected for this innovation was the Structural Analysis class from the Department of Civil and Environmental Engineering.

Literature review

Today, and increasing number of courses are being redesigned to integrate technology as a key tool for enhancing students' learning. As mentioned by Kirkwood and Price (2014), the inclusion of technology can improve the learning environment given its greater flexibility; it complements conventional teaching and provides an increased number of learning opportunities for the students.

A course becomes more flexible using a technology-enhanced learning design (TEL) (Kirkwood & Price, 2014) as it includes spaces in the virtual environment for dialogue, teamwork, and the use of interactive tools. Students can choose which tools to use according to their interests and needs (Garrison & Kanuka, 2004; Kolb & Kolb, 2005). Additionally, a learning environment that provides virtual tools promotes both written and oral communication skills. In student-centered, face-to-face settings, oral interaction predominates, and participants must engage with their peers and teachers synchronously (Garrison & Kanuka, 2004; Smith, Sheppard, Johnson, & Johnson, 2005). On the contrary, asynchronous virtual spaces mainly promote communication in written form through forums and group interactions that make all students' participation visible (Garrison & Kanuka, 2004; Nguyen, et al, 2015). In fact, having flexible environments can provide learning spaces for students with different learning styles (Kolb & Kolb, 2005). Therefore, innovative courses can make the most of both learning environments: face-to-face and virtual spaces.

TEL strategies allow for the creation of student-centered learning environments that complement more traditional learning spaces. Strategies such as the flipped-classroom (Observatory of Educational Innovation, 2014; Nguyen et al., 2015) enable the repurposing of learning environments in which virtual spaces are dedicated to more traditional instructional procedures (transmission of content) by using videos and readings; class time is employed to promote student interaction. Furthermore, in student-centered environments, virtual tools open up a universe of possibilities for learners to take charge of their process through the development of technological resources that promote learning. An example of this is the free-form classroom designed at Purdue University for teaching a mechanical engineering course (Rhoads, Nauman, Holloway, & Krousgrill, 2014).

Similarly, the flexibility and diversity of teaching strategies in a blended approach make it possible to adequately address different student learning styles. This modality offers environments that facilitate the balance between spaces of reflection, action, emotion and metacognition. As mentioned by Kolb and Kolb (2005), it is necessary to design the curriculum holistically to achieve an environment that effectively takes different learning styles into account. The design process should also consider elements such as the students' background knowledge, striking a balance between an environment that is challenging but also supportive, the promotion of deep learning, dialogue and constant feedback, and the promotion of self-directed learning (Kolb & Kolb, 2005). A holistically-designed learning environment promotes students' responsibility for their own learning process since it allows them to flexibly and individually select their own learning tools (Gros, 2016).

Critics of this type of methodology argue that if a traditional design perspective guides the program, this will be reflected in the curricular design regardless of whether or not technology is included (Observatory of Educational Innovation, 2014). However, it should be noted that strategies such as the flipped classroom emerged from the need to use different learning environments, including both collaborative and individual spaces, more effectively. Through this optimization of learning environments, curriculum designers seek to prioritize interaction and collective construction of knowledge in the classroom and individual learning in virtual environments. In fact, studies have shown that this flexibility and diversity of learning opportunities in blended modalities positively affects dropout and passing rates, even when the assessment difficulty is increased (Martin, 2013; Rhoads et al., 2014).

The process of including technology in the learning environment goes beyond the integration of virtual or computational tools; it is a process of conceptual, pedagogical change in the community that requires shifting the locus of control away from the instructor toward the student. This paradigmatic change requires consciousness in the design process to ensure that key elements such as incorporating challenges, collaboration, interaction, experience, care, action, and reflection are incorporated (Kolb & Kolb, 2005; Martin, 2013; Smith et al., 2005). The implications of this kind of effort are reflected in the effectiveness of short-term innovation. However, studies point out that, in the long run, the balance between investment and quality of learning is positive (Garrison & Kanuka, 2004) and it is facilitated when institutional support is provided (Raucent et al., 2005).

Context of the study

Structural Analysis courses are part of Civil Engineering undergraduate programs across the world. At the Universidad de los Andes, this course is taken by students typically in their fifth semester of undergraduate study, and it aims to prepare them to materialize real structures into models and determine the internal forces, reactions, and displacements of these structures. In order to acquire these abilities, students need to understand concepts that require a significant amount of abstraction and that are usually hard to learn with the traditional face-to-face approach. For this reason, the Structural Analysis course at the Universidad de los Andes was redesigned using a Blended Learning methodology accompanied with a series of laboratory practices in small scale structure models. The focus of the innovation was to give the students a more active role in their own learning process by making them accountable for studying the fundamental concepts before class. This allowed the instructor to focus in class on real world engineering problems that include not only the mathematical component, but also the need to interpret the results of those problems.

Blended learning and flipped classroom ideas were put into practice as strategies of active pedagogy, and these were complemented with Digital Technology (DT). This approach necessitated the creation of multimedia materials (video clips with study guides which reduced class time). Face-to-face software laboratories were replaced by virtual sessions and the complementary classes were replaced by optional office hours for students. This led to a 30% reduction of face-to-face time when compared to traditional courses. The redesign was implemented as the development of the materials and activities progressed.

Tensions identified at the onset of the redesign were: (1) large thematic content, which limited class time for the student to analyze and solve problems under the teacher's guidance; these tasks, which are fundamental to the course, must be developed in external scenarios. (2) Experience shows that the student addresses problems from a mathematical operations perspective, without focusing on the analysis of the results obtained. With this approach, mathematics is not the means to solve a real problem, but an end in and of themselves. (3) Many students do not master the prerequisite subjects of the course, which makes it necessary to add thematic content.

In response to these tensions, the areas for redesign were: (1) To provide contextualized content in digital format to complement the printed course materials and encourage the students to explore, understand and apply concepts in real contexts before going to class. The aim is also for previous concepts to be reviewed or reinforced through this medium. (2) To ensure that students analyze the problem and its mathematical solution, as well as the interpretation of the answer. (3) To create alternatives to approach the theoretical concepts of the course, so that the student can confront the application contexts of civil engineering; this includes the systematization of exercises administered by the computer and the teacher's support in the analysis and discussion of solutions.

Over the two-year period of the course redesign, the professors were accompanied by consultants in the creation and development of virtual learning objects. Table 1 summarizes these developments.

The course was redesigned to have four learning moments a week: 1) individual preparation for class supported by video clips; 2) completion of practical problems during class; 3) comparison between theoretical solutions for the analysis of scaled structures with real measurements done at the academic lab; and 4) completion of individual/couples assignments with higher-level thinking problems situated in professional contexts.

The course is divided in five modules. The first module deals with concepts of structural types and loads. The second module addresses the materialization of structures into structural models that are suitable for analysis. In the third module, students learn to apply traditional analysis methods such as direct integration and energy methods. The fourth module focuses on the Direct Stiffness Method of analysis or Matrix Stiffness Method. This is the most robust module. In the final module, students compare precise calculation of structural forces and displacements with results obtained with approximate methods of analysis such as Wilbur for displacements of space frames, MacLeod for displacements and internal forces on dual or combined structural systems, ACI coefficients and Points of Inflexion for internal forces due to vertical loads in beams or frames, and the Portal method for internal forces of frames with lateral loads only. The last couple of classes address the analysis of moving loads on structures.

Table 1: Synthesis of activities by course and resources produced

Activities	Resources produced
Course redesign using active pedagogy	Global structure, with five modules
Material design for class preparation and development	Twenty-six videos plus workshops and quizzes Video clip example: https://youtu.be/PTUWj4Nznjk

Research questions

Since this is the first civil engineering course in the country to use a blended learning methodology, it was deemed appropriate to try to answer this question: What are the advantages of implementing pedagogical innovations with the support of digital technologies in an undergraduate civil engineering course?

To answer this question, the following sub questions were formulated:

From the student's perspective:

1. What are the students' perceptions of the contribution of active methodology and the academic laboratory to their learning process?

2. What are the students' perceptions about the increased interaction with the professor fostered by the active methodology?

Questions regarding the teacher's point of view and curriculum development include:

3. What changes in student passing rate can be attributed to the course's transformation?

4. How does the incorporation of active pedagogy supported by DT impact the level of difficulty of the course, and how is this reflected in the other courses structural engineering courses that follow Structural Analysis in the curricular sequence?

Research Design

To answer the research question, the intervention had several stages. In the first two semesters of implementation (2015-2 and 2016-1) the intervention was partial. In 2015-2, only two of the five modules of the course were innovated in, shifting them from the traditional method to a flipped classroom model. Therefore, students had the opportunity to contrast the two methodologies used for independent modules of the same course in the same semester. In 2016-1, the entire course adopted the blended learning methodology with a 30% reduction of face-to-face time as mentioned before.

In order to evaluate the students' perceptions regarding the innovation, a total of four surveys were carried out in the two semesters mentioned. In addition, a focus group was conducted at the end of 2016-1. The surveys inquired about the quality of the resources developed (video clips), the feasibility of the proposed activities, and the improvement of the interaction between peers and professors encouraged by the methodology. Table 2 summarizes the characteristics of the four surveys, including the semester and week in which they were applied, the modules being tested, the teaching methodology reviewed by the students, and the total number of participants who answered the survey.

The academic results were obtained by reviewing the grade history of the course in its traditional format and comparing passing rates before, during and after the innovation.

The course's level of difficulty was considered by comparing the contents that were addressed before and after the innovation in conjunction with the difficulty of the exams. This was supported by an analysis of the influence of the redesign process in the development of more advanced courses.

Table 2: Survey characteristics

Survey	Semester	Week	Modules of the course	Methodology evaluated	Total students in survey
1	2015-2	9	2 and 3	Traditional	97/113 (83%)
2	2015-2	12	4	Flipped classroom	81/113 (72%)
3	2016-1	7	1 and 2	Blended	93/114 (82%)
4	2016-1	12	3 and 4	Blended	63/114 (56%)

Findings

Students' perception

In all the surveys conducted, students were asked what they thought about their learning process. This question allowed us to compare the students' level of satisfaction with their own learning process for the course in traditional methodology, flipped classroom and blended learning. Figure 1 shows that 84.5% of the students who participated in the first survey of 2015-2 believed that their learning process was good or excellent when using the traditional methodology. This is a baseline indicator that shows that the course was successful from the beginning. However, for the modules that used flipped classroom, the percentage of students that rated their learning process good or excellent increased to 95% of the same population (second survey of 2015-2).

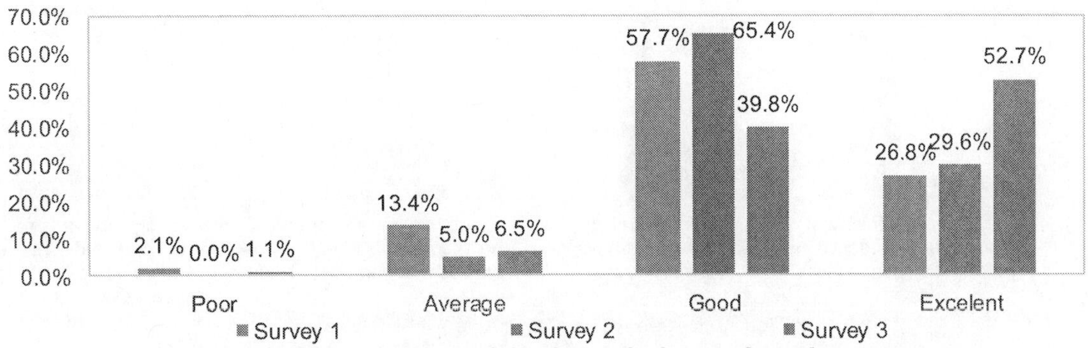

Figure 1: Students' perception about their own learning process.

In the following semester (2016-1), the course had a reduction in face-to-face time and moved from flipped classroom to blended learning. This transformation improved the students' perception in the learning process to the point that more than 50% of the participants declared that their learning process was excellent with this methodology (Survey 3).

This favourable perception by the majority of students is due to the fact that this methodology allowed them to understand the contents of the course. The videos, solving proposed exercises and interaction with instructors during lectures were contributing factors as can be seen in the following testimonies:

> *"The class preparation methodology allows for more clarity in the explanation of topics during lectures" (Student participant in focus group).*

> *"The videos and exercises in class significantly facilitated the learning process. Also, homework assignments—I think they were fundamental, because I could solve problems by myself after class and they allow me to really see how good my understanding is and if I'm really doing things right" (Student participant in focus group).*

Regarding class preparation videos and laboratory sessions, students who participated in the survey have a favorable view, which is consistent with the general opinion they have of their learning process. As indicated in Figure 2, 96.5% of students agreed or strongly agreed that class preparation activities were important to understanding the topics; 90% reported that laboratory sessions were valuable to their learning process.

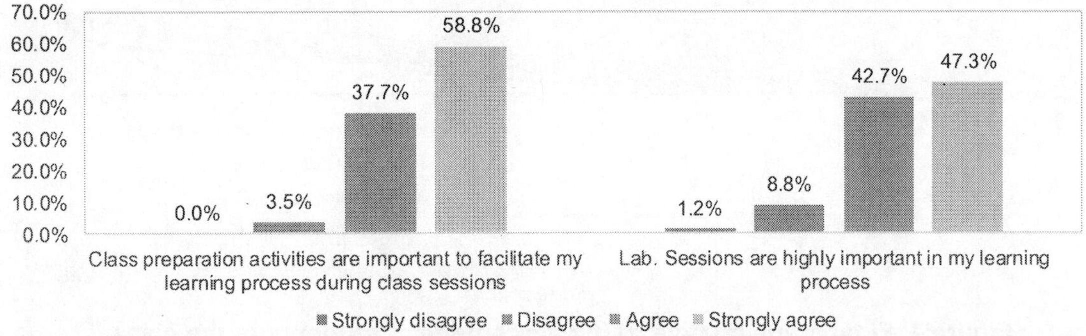

Figure 2: Students' perception of class preparation activities and laboratory activities (average result of surveys 2, 3, and 4)

Finally, students also declare that the methodology accounts for efficient interactions with peers and professors during the lectures, as is shown in Figure 3.

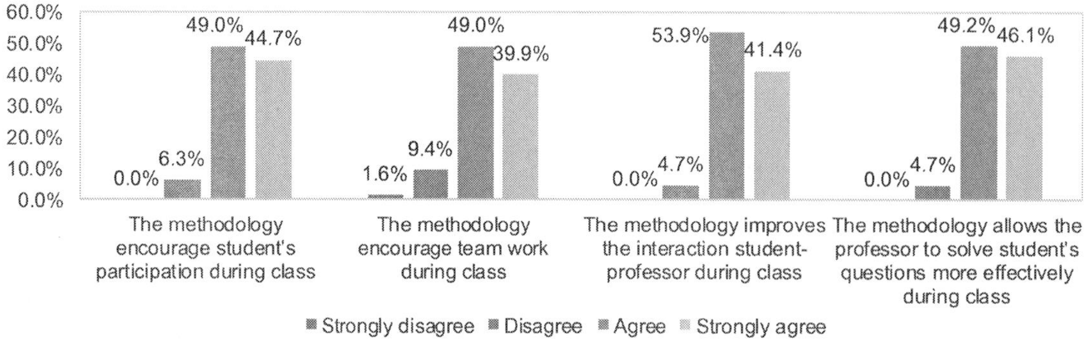

Figure 3: Students' perception about the use of the methodology in face-to-face sessions (survey 4).

The students' active engagement in the lectures is also related with their preparation of course content before class. In their opinion, class preparation activities give them more time to ask questions and solve exercises in class with the teachers than a traditional class gives them. The following testimonies show this:

> *"It seems to me that preparing the theory previously gives a space for one to process this information and discover if any questions exist in class ... and in class one can then take better advantage of the professors" (Student participant in focus group).*
>
> *"In class there is room to actually do exercises, so you do not waste class time just explaining the theory. You already come with that and you can use class time to answer questions and use the theory to solve real problems" (Student participant in focus group).*

Students' passing rates

In the three semesters that the innovation has taken place (2015-2, 2016-1 and 2016-2), student grades have not changed significantly compared with those obtained by students who took the traditional course, as is shown in Figure 4. This indicates that the methodology has not improved the student passing rates. In fact, the final grades are at the lower limit of what has historically been obtained in this course. The behavior of the final grade is partially explained by an increase in the level of difficulty of the course identified by the professors, which we will be address in the following section.

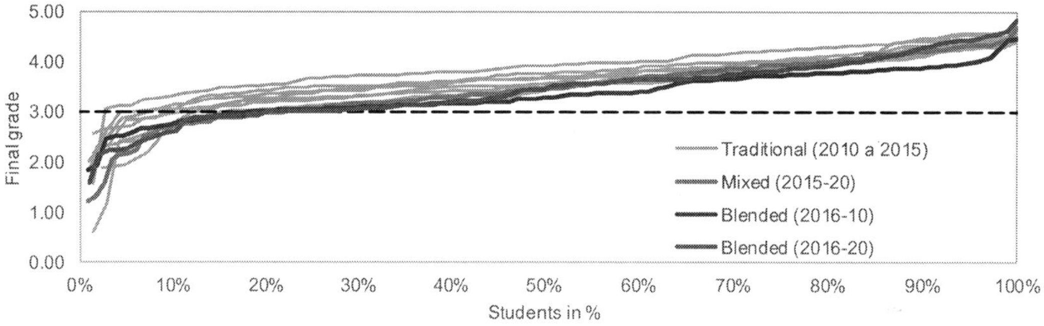

Figure 4: Final grade relative to the percentage of students in the class

Figure 5 shows the percentage of students who failed the course in the last 12 semesters (note that the semesters 2014-2 and 2015-1 are not shown in the statistics because the professors were not the same). These statistics show that in the first semester the innovation was applied, the percentage of failing students was the historical maximum. This seems to be due to the high academic demands of the course in its first implementation. To address this problem, the size of the homework assignments was considerably reduced. The result has been a 10% reduction of failing students.

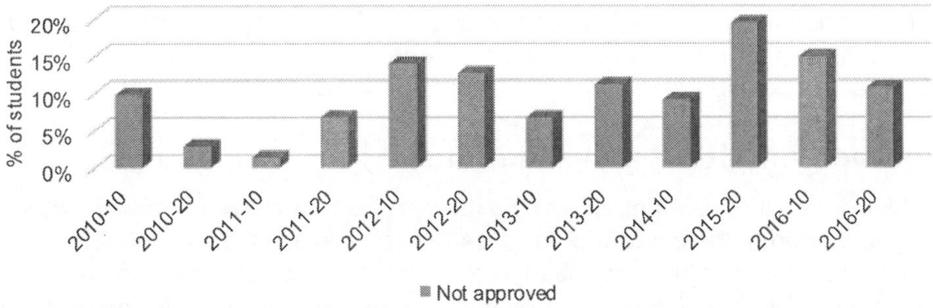

Figure 5: Percentage of students who did not pass the course

Course difficulty level

The course contents remained the same before and after the innovation. However, this innovation allowed for a reorganization of subjects and the implementation of a teaching approach which focused on the "lasting understandings" essential for a civil engineer in this course rather than merely focusing on memorizing contents and procedures.

Because the teaching of the theory has been transferred to a virtual environment through video clips, explanations are no longer limited to the classroom space. Due to this approach, the course concepts can now be explained through real situations more readily. Concepts can be abstracted more readily than they can on whiteboard with drawings or slide presentations. That is why current videos include virtual field visits to existing structures that allow the student to see the applicability of the course concepts to a professional context. Thanks to 3D animation, it is also possible to make difficult abstractions—that the student would normally have to imagine or infer—easier to understand.

During the class, the problems are more complex compared to the examples used in the traditional version of the course. The students are able to solve them in class by collaborating with peers and with the professor's guidance. The reformatting of these activities leads professors to include more complex application problems in at-home assignments and exams, which results in increased difficulty. For example, in traditional versions of the course, the simplified analysis of buildings with shear walls and moving loads was not evaluated; however, now these are recurrent subjects in the evaluation instruments.

Influence on subsequent courses

The Structural Analysis course is the third of four mandatory courses of the structural branch in the Civil Engineering program of the Universidad de los Andes. The last course in the sequence is called Structural Design. This course focuses on applying all the tools students learned in the three previous courses to solving common problems in concrete structure design. Thus, students must have sufficient mastery of the concepts learned in Structural Analysis to pass the Design course.

Even though the Structural Analysis course in its earlier version tried to cover a broad range of topics to prepare students for the Structural Design course, the students did not seem to acquire sufficient domain of the analysis methods. Consequently, the Structural design course spent up to 25% of the lectures reviewing structural analysis contents, with the cost of letting out important content related to design. After the innovation, however, that time was reduced to 10% because students could use the video-clips to revisit structural analysis topics. The time saved allowed the instructor to increase the scope of the Structural Design course.

The influence of the innovation has reached graduate level courses in the Structural engineering master's program as well. In this program, the university receives students from different undergraduate institutions and age groups (typically between 20 and 35 years), so the audiovisual material developed in Structural Analysis has been an important tool in this program to assure all students have learned the fundamental concepts of Structural Analysis before they take advanced courses.

Lessons learned and Recommendations

One of the problems with the new methodology is that professors are wary of how much the students have understood of the content of the videos before a class. This meant that in the first versions of the course after the innovation, more than half of class time was used to review the contents of videos. This practice resulted in a loss of motivation for many students shown in levels of class attendance that were as low as 50%. To overcome this problem, a system of virtual quizzes is being implemented in all lectures. Quizzes seek to evaluate key concepts from the videos, and thus, increase the attendance to the lecture sessions and minimize class time spent summarizing contents.

Another challenge that must be carefully examined in the transition from the traditional course to the blended course methodology is the increase in the work load for students. This variable must be constantly monitored during implementation and must become a priority during course design.

In conclusion, implementing pedagogical innovations with the support of digital technologies in a civil engineering course increases students' satisfaction because it gives them greater autonomy in their learning process and enhances their interaction with their teacher. In addition, innovation allows students to reach a greater depth of understanding of the contents while keeping the average grade constant and student passing rate similar to that of the traditional method of teaching. Additional benefits include the impact of the audiovisual material in more advanced courses, including graduate level courses.

References

Forté, E., and Wentland, M. (1998). The ARIADNE project: knowledge pools for computer-based and telematics supported classical, open and distance education. *AAUC Ariadne Academic Users Group Conference* (pp. 1-23). Lucerna, Swtizerland: AAUC.

Galvis, A.H. (2017). Rethinking engineering education practices with active learning technologies: Pilot experiences at University of los Andes. Paper 184 accepted for presentation in REES 2017. Research in Engineeriing Education Conference (Bogota, July 6-8, 2017).

Garrison, R. and Kanuka, H. (2004). Blended learning: Uncovering its transformative potential in higher education. Internet and Higher Education 7, 95–105.

Gros, B. (2016). The design of smart educational environments. Open Access 15 (3).

Kirkwood, A. and Price, L. (2014) Technology-enhanced learning and teaching in higher education: what is 'enhanced' and how do we know? A critical literature review. Learning, Media and Technology, 39(1) pp. 6–36.

Kolb and Kolb (2005) Learning Styles and Learning Spaces: Enhancing Experiential Learning in Higher Education. Academy of Management Learning & Education, 2005, Vol. 4, No. 2, 193–212.

Martin (2013) Implementing Active Learning Principles in an Engineering Technology Fluid Mechanics Course. American Society for Engineering Education Proceedings

Mills, J. E., and Treaagust, D. F. (2003). Engineering education - is problem-based or project-based learning the answer? Australian Journal of Engineering Education, 1-16.

Nguyen, B., Yu, X., Japutra, A., y Chen, C. (2015) Reverse teaching: Exploring student perceptions of "flip teaching" Active Learning in Higher Education 11(1).

Observatorio de la Innovación Educativa. (2015). Aprendizaje basado en retos. Técnológico de Monterrey. Edu Trends.

Observatorio de la Innovación Educativa (2014) Aprendizaje invertido. Técnico de Monterrey. Edu Trends
Raucent, B., Galand, B., Frenay, M., Laloux, A., Milgrom, E., Vander, B., and Wouters, P. (2005) Is Active Learning More Efficient Than Traditional Learning?
Rhoads, J., Nauman, E., Holloway, B., and Krousgrill, C. (2014) The Purdue Mechanics Freeform Classroom: A New Approach to Engineering Mechanics Education. *American Society for Engineering Education Proceedings*
Smith, K., Sheppard, S., Johnson, D., and Johnson, R., (2005) Pedagogies of Engagement: Classroom-Based Practices. *Journal of Engineering Education*, 94(1), 87-101.

Acknowledgements

This work is the result of collaboration between the Faculty of Engineering at the University of the Andes and the Center for Innovation in Technology and Education (ConectaTE) from the Faculty of Education. The Dean of Engineering, Eduardo Behrenz, led the process in 2015 and 2016, and his successors, Carlos Francisco Rodríguez and Alfonso Reyes, continued. Special thanks to Luz Adriana Osorio, director of ConectaTE, Mónica Patiño, coordinator and technological advisor, and Angélica Avalos, pedagogical advisor. Video production and creation of digital leaning environments was done by the team led by Mariela Rivero. Special thanks to Nicole Bruskewits for her huge help in the review of language style of this manuscript. Many thanks to everyone involved in the process.

Copyright statement

Hands-On vs Instructor Demonstration of Inquiry-Based Learning Activities: A Comparison to Traditional Instruction in Engineering Dynamics

Brian P. Self
California Polytechnic State University, San Luis Obispo, CA, USA
bself@calpoly.edu

James Widmann
California Polytechnic State University, San Luis Obispo, CA, USA
jwidmann@calpoly.edu

Abstract: *Students can often choose a correct equation, substitute in values, and "solve" a dynamics problem without ever truly comprehending the physics concepts involved. Our Inquiry-Based Learning Activities follow a predict-observe-explain cycle, where a physical artefact is the authority rather than the word of the instructor. Students are presented with a physical scenario, and work several cases a learning cycle: (a) make an individual prediction, (b) discuss predictions with teammates, (c) observe the scenario/experiment, and (d) explain the results. Several scenarios are presented, and the instructor provides guided instruction to help repair any misconceptions. In this work, we compare results from a hands-on versus instructor demonstration implementation, and compare these results to traditionally taught classrooms. Specifically, we discuss the analysis of our Spool IBLA, which explores the relationships between applied forces, moments and direction of acceleration, and of the Pulley IBLA, which investigates the relationships between net force, system mass, and acceleration.*

Introduction

The physics educational community has long recognized that students struggle with basic mechanics principles, and pioneering work involving concept inventories (Hake, 1998) and Peer Instruction (Mazur (Mazur, 1997), among others, helped initiate a number of active learning based pedagogies to address these deficiencies. Development of similar pedagogies in engineering has lagged behind that in the sciences, but is becoming more and more prevalent. Our work in Inquiry-Based Learning Activities (IBLAs) builds upon the successful work involving interactive lecture demonstrations, microcomputer-based laboratory tools (Sokoloff & Thornton, 1997), Workshop Physics, and inquiry activities (Laws, Sokoloff, & Thornton, 1999) in physics, as well as recent work in thermodynamics and heat transfer (Prince, Vigeant, & Nottis, 2015).

Our IBLAs are formulated using the elements introduced by Laws and summarized by Prince, as shown in Table 1. We follow a predict-observe-explain cycle, similar to that implemented by Bernhard's group (Bernhard, Lindwall, Engkvist, Zhu, & Stadig Degerman, 2007) during their use of interactive lecture demonstrations. In the IBLAs, students are confronted by a series of physical scenarios. For each scenario, students are first required to make individual predictions about the physical phenomena of interest, then discuss their predictions with a group of 3-4 students. They observe the system experimentally, and finally discuss and explain the experimental results on a team worksheet. The worksheet provides specific prompts to guide the students in their thinking, and asks them to practice engineering skills such as drawing free-body diagrams. At specific intervals, direct instruction is incorporated to make sure students are applying appropriate scientific principles (Figure 1).

Table 1: Elements of Inquiry-Based Learning Activities.

(a) Use peer instruction and collaborative work
(b) Use activity-based guided-inquiry curricular materials
(c) Use a learning cycle beginning with predictions
(d) Emphasize conceptual understanding
(e) Let the physical world be the authority
(f) Evaluate student understanding
(g) Make appropriate use of technology
(h) Begin with the specific and move to the general

In IBLAs, the focus is on conceptual understanding through the integration of hands-on activities in a cycle of predictions, observations, and explanations. Teams of students interactively construct new knowledge, following recommendations in Chi's Active-Constructive-Interactive framework (Chi, 2009). Her definition of active varies from the traditional one; any type of physical manipulation (including taking notes) would be characterized as active. The constructive activities involve "producing outputs that contain ideas that go beyond the presented information." Finally, interactive activities involve dialoguing on a topic, building on the contributions of colleagues. The dialogue may involve collaborative learning and peer instruction, as well as instructional dialogues or "guided construction" involving the instructor. The IBLAs incorporate both of these, allowing construction of new knowledge as well as scaffolding through guided worksheets and discussion.

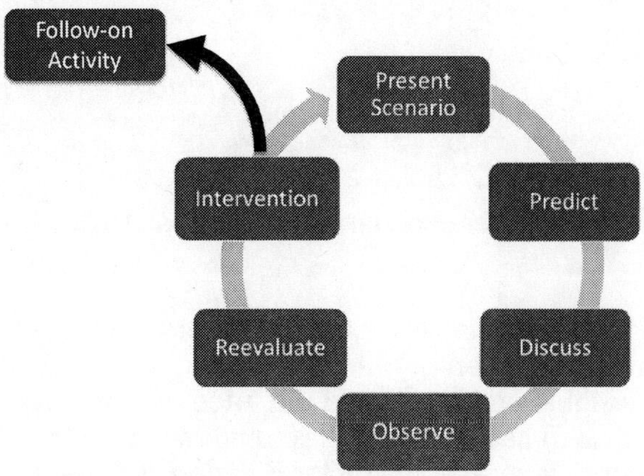

Figure 1. IBLA Learning cycle.

Undergraduate Dynamics

Undergraduate dynamics is taken by most engineering students at our institution, including mechanical, aerospace, biomedical, civil, bioresource and agricultural, architectural, and general engineers. Undergraduate physics is a pre-requisite for the course, so students enter with some prior knowledge of mechanics. Our dynamics classes meet for 50 minutes, approximately 28 times during a ten-week quarter. Section sizes average 35 students each, and at the end of the quarter all sections take a common final exam that is written by a team of instructors. During the time period studied, the final exam consisted of 4 workout problems and a set of short answer and conceptual problems.

Two instructors with similar teaching styles taught two sections each at the same time of day. They developed the same syllabus, used the same lesson notes, slides, homework, quizzes, and tests, and the only difference between the courses was in how four different IBLAs were implemented. During the quarter, four IBLAs were conducted during a regular 50 minute classroom session. A cross-over study approach was used, where Instructor A's sections did the first and third IBLA in a hands-on mode, while Instructor B's sections did the second and fourth IBLA in hands-on mode. Here we present results for the Spool and Pulley IBLAs.

Pulley IBLA and Spool IBLA Implementation

As reported at the last REES conference Self (Self, Widmann, Adam, & Coburn, 2015), our IBLAs are sequenced according to variation theory (Lo, 2012). We attempt to vary one characteristic at a time, forcing students to discern the critical features of the different scenarios. Additional scenarios are then presented where a different characteristic is varied. For example, in Scenario 1 for the Pulley IBLA shown in Figure 2, the net force is the same (50 grams*g downward on the left pulley) while the total system mass is varied.

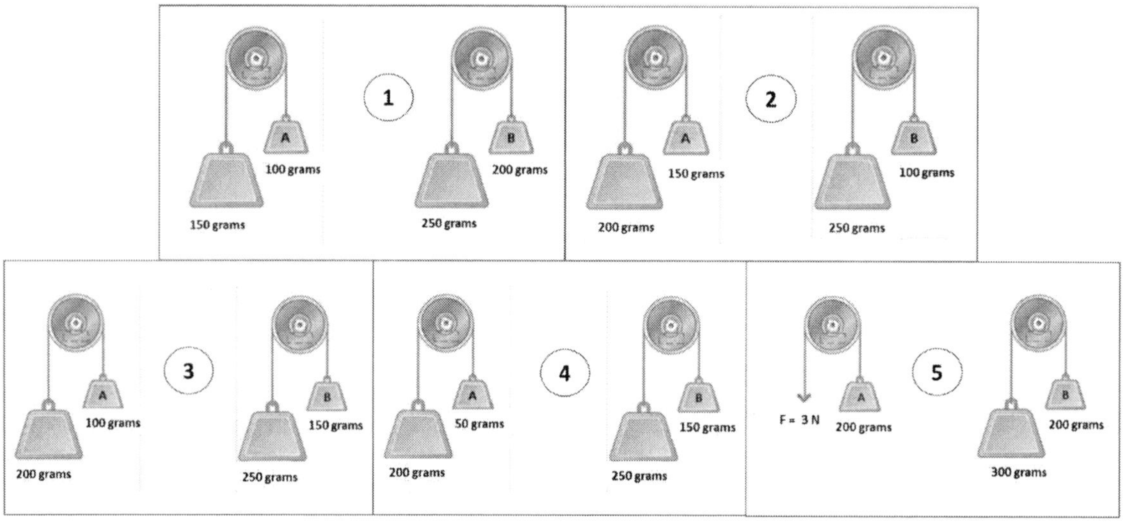

Figure 2. Five different scenarios in the Mass Pulley IBLA.

Students go through the same basic steps whether they were in the hands-on or the instructor demonstration implementation. For Instructor A's two sections, all students were given a set of pulleys and weights and conducted the "race" on their own. They had to switch out the different weights between cases, and figured out how to release the masses so that there would be a fair and consistent race. Instructor B conducted the race at the front of the room as a demonstration. The steps of the IBLA are as follows:

1. Individually predict which mass will accelerate more quickly (we do this using our online course management system quiz) [~ 1 min]
2. Discuss their predictions in teams of 3-4, then record new predictions [2-3 min]
3. Observe the experiment (either as an "experiment" or as demonstrated by the instructor at the front of the room) [2-3 min]
4. Explain their observations and fill out the guiding worksheet [3-4 min]
5. Participate in instructor-led discussion [~ 3 min]

6. Conduct another scenario
7. After the IBLA, a follow-on workout problem is assigned where students calculate the accelerations and rope tensions for different pulley systems.

The Spool IBLA tests the relationships between net force, net moment, linear accelerations, and angular accelerations. The four different scenarios are depicted in Figure 3. The students were told to pull gently on the string – we chose not to use the phrase “rolling without slip” since we often do this activity as an introduction to rolling and students may not be familiar with this exact terminology.

1. Looking at the figure in Scenario # X, if you pull on the string gently, which way do you predict the spool will move?
   Right   Left   Won’t move
2. When pulling, in which direction is the friction force?
   Right   Left   there is no friction force
3. If there is friction, what is the value of the friction force?
   $f = \mu_k N$ $f = \mu_s N$ $f \leq \mu_s N$

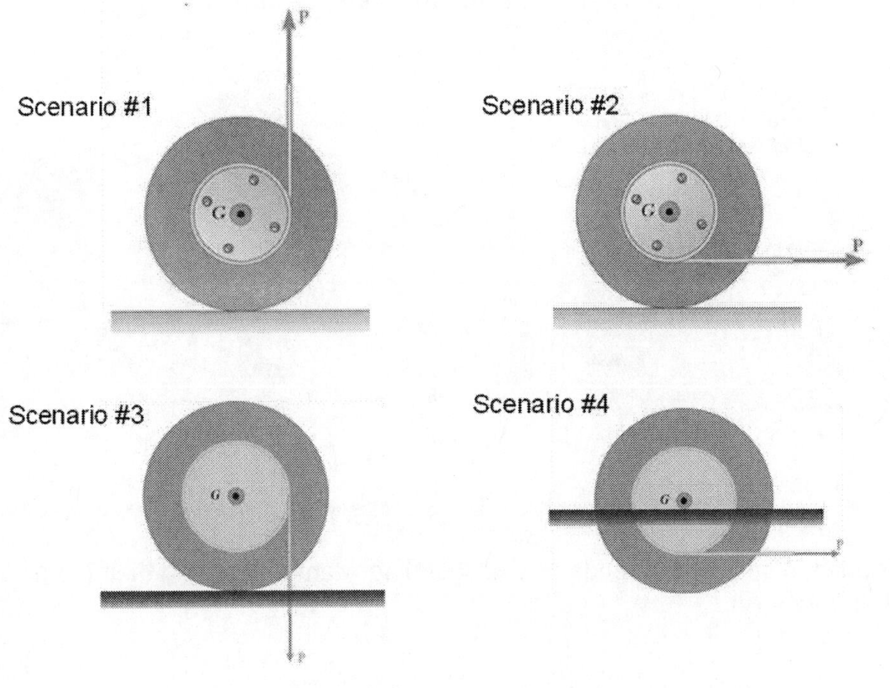

Figure 3. Spool IBLA scenarios.

As with the Pulley IBLA, the primary difference between the two groups is that the instructor performs the demonstration at the front of the room versus having the students interact directly with the artefact. In one of the worksheet prompts, students are instructed to pull the string at different angles, see if they can make the spool slide without spinning, and explain what happens. In the Instructor Demonstration implementation, the instructor simply walks the students through this exercise. One difficulty with this IBLA is that although students can see the acceleration direction of the spool, they cannot see the friction force. This has to be determined by going through the equations of motion to deduce this direction. After the full sequence of scenarios, students are given a homework problem where they calculate the friction force and accelerations of one of the scenarios.

Research Methods

Students from the hands-on, instructor demonstration, and the traditionally-taught classes all took the same final examination. Additionally, scores on targeted questions on the Dynamics Concept Inventory (Gray et al., 2005) were obtained for all IBLA sections and a number of the traditionally-taught sections. Finally, a subjective survey was conducted to obtain student feedback on the IBLAS. A Likert scale question asked if students agreed that "This activity helped me learn dynamics," and a second question asked if the students preferred hands-on or instructor led demonstrations. These questions were asked immediately after the students participated in the IBLA. A final open-ended question asked if they had any comments or suggestions for improvement.

Results

Student responses to the prompt "this activity helped me learn dynamics" were in general positive, as shown in Figure 4. We saw very little difference between the hands-on and demonstration implementations, or between the two different IBLAs.

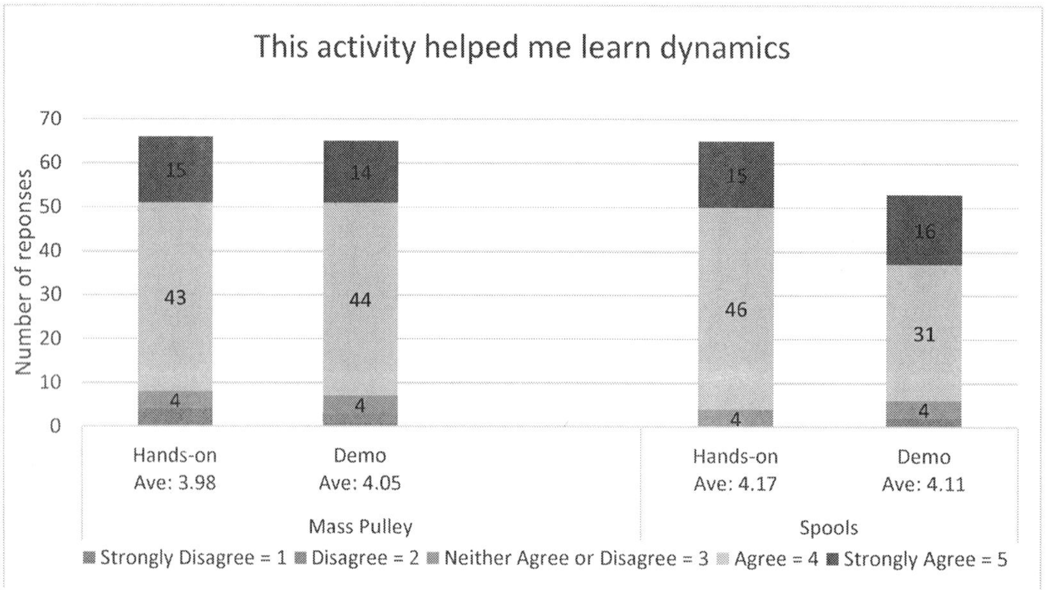

Figure 4. Student response for hands-on and instructor-demo regarding their learning for the Pulley and Spool IBLAs.

Student responses to their preferences on hands-on versus instructor demonstrations are shown in Figure 5. Responses to the open-ended question on how to improve the activity for the Pulley IBLA were somewhat mixed. Some students felt a bit rushed and commented that more discussion time would be helpful, while others thought that the concept was a fairly easy one and did not warrant so much class time. The Spool IBLA involves more complex topics, so several student comments asked for additional instructor explanation. Several students in the instructor demo group did suggest having a spool for each group so that they could actually perform the experiment, and one suggested having more time at the end of class to "play" with the spool.

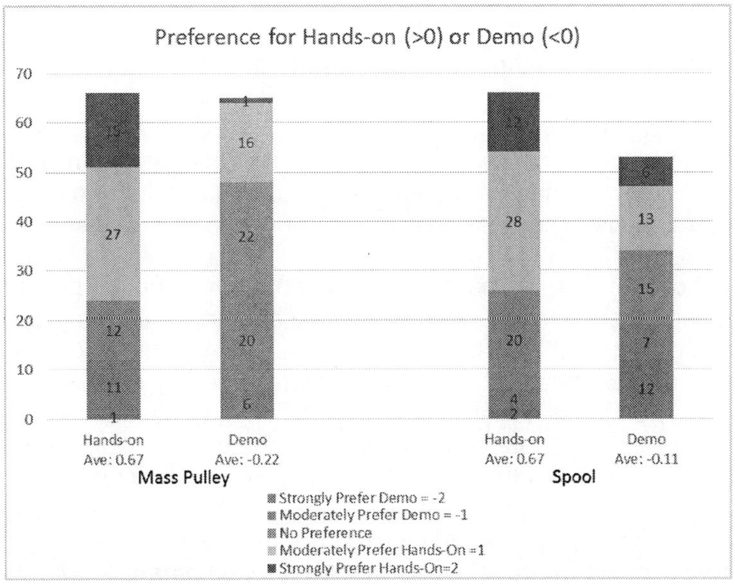

Figure 5. Student preference for hands-on vs instructor demonstration.

Cognitive performance was measured by choosing targeted question on the Dynamics Concept Inventory and scores on the common final exam (see Fig. 6). Question 13 is essentially Case 5 from the Pulley IBLA (see Fig. 2), and asks which mass will accelerate more quickly. Question 14 is more of a transfer question, where now the pulley has mass and has either a weight pulling on one side or a mass on a rope on one side. This tests how the mass moment of inertia affects the acceleration of the spool, and surprisingly there was no difference between any of the groups on this problem. Finally, question 27 is a transfer question related to the spool. It asks the direction of friction on the drive wheel of a car accelerating forward. Final exam scores for the active learning sections that included IBLAs and the more traditionally taught sections are shown in Figure 6b.

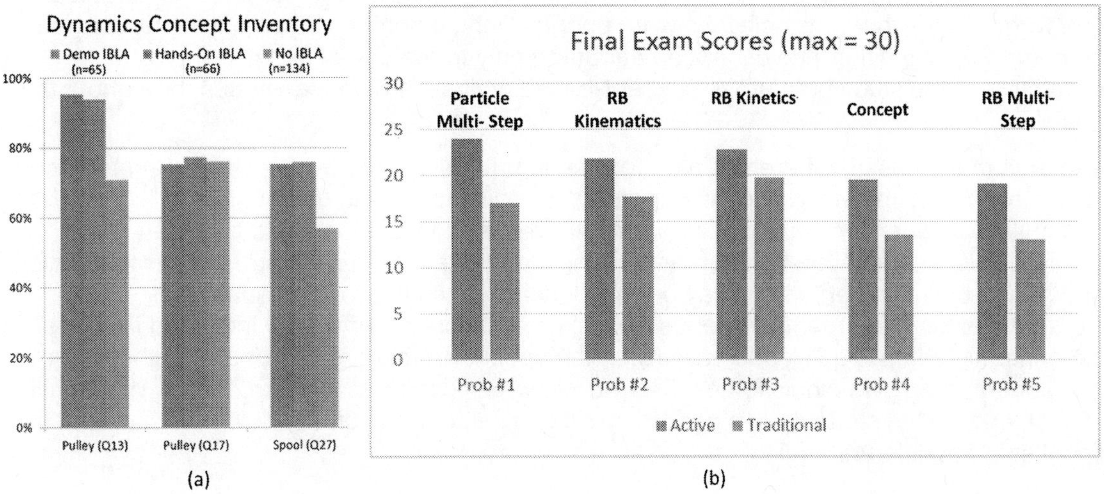

Figure 6. Cognitive assessment when using IBLAs. RB indicates a Rigid Body problem.

Discussion

We found no difference in cognitive performance between using a hands-on IBLA versus an instructor-led demonstration using our IBLAs as measured by concept questions on the DCI. This replicates the findings of Glasson (Glasson, 1989), who found no difference in declarative knowledge between hands-on and teacher demonstrations in laboratories. Glasson's team did, however, find that the hands-on students performed better on procedural knowledge tests than the instructor demonstration group. We did not test this in our study.

More compelling is how any form of the IBLA implementation improved performance on cognitive tests. As can be seen in Figure 6, students who participated in the IBLA course performed significantly better on DCI and final exam questions. These results are similar to those found by Bernhard (Bernhard et al., 2007), who modified work by Sokoloff, Thornton, and co-workers to use computer-assisted data acquisition during lecture demonstrations. A similar predict-observe-explain sequence was utilized, and guiding worksheets were provided to the students. As the authors point out, the use of a data acquisition system can be somewhat expensive and may preclude the use of multiple setups in a lecture-based setting. Bernhard's group also reports finding even greater learning gains when using the DAQs in a laboratory session where students utilized the equipment – we did not see these differential gains in our investigation.

Interestingly, our students did not seem to have a preference for the hands-on versus the instructor demonstration IBLA implementation. In fact, students seemed to prefer whatever implementation method was used in their class. A final survey on student preference in addition to the surveys that were performed immediately after the activities may have provided additional information on student preferences and the factors that affect these preferences.

Conclusions and Recommendations

Our results indicated no differences between a hands-on versus an instructor demonstration when performing IBLAs in undergraduate dynamics. Students in both groups performed higher on targeted DCI questions than traditionally-taught students, but did not perform differently on a pulley transfer question. Students in both IBLA groups performed significantly better on a common final examination than traditionally-taught courses, which is particularly noteworthy since four fewer classes were spent solving algorithmic-type questions. It should be noted that this difference cannot be attributed only to the use of IBLAs – the two instructors in the study both utilize concept-based active learning techniques throughout the quarter.

Although there was no difference in performance on the DCI or obvious student preference for the hands-on implementation in our study, we still support the use of physical manipulatives where warranted. Our manipulatives are low-cost, easy to transport, and highly visible, so can even be adapted to large class sizes. A mix of instructor demonstration and hands-on implementations may be the ideal approach to provide variation of experiences for the students. A hands-on implementation is particularly indicated if (a) the experimental setup is quick and easy (e.g., the spool), (b) there is a strong kinesthetic component (e.g., gyroscopic forces on a bicycle wheel (Bohn, Georgette, Self, & Widmann, 2014)), and (c) the students can perform additional exploration on their own (e.g., pulling the spool string at different angles).

References

Bernhard, J., Lindwall, O., Engkvist, J., Zhu, X., & Stadig Degerman, M. (2007). Making physics visible and learnable through interactive lecture demonstrations. *Physics Teaching in Engineering Education PTEE.*

Bohn, K., Georgette, J., Self, B. P., & Widmann, J. (2014). *The Study of Gyroscopic Motion through Inquiry Based Learning Activities.* Paper presented at the Proceedings of the American Society for Engineering Education Zone IV Conference, Long Beach, CA.

Chi, M. T. H. (2009). Active-Constructive-Interactive: A Conceptual Framework for Differentiating Learning Activities. *Topics in Cognitive Science, 1*(1), 73-105. doi:10.1111/j.1756-8765.2008.01005.x

Glasson, G. E. (1989). The effects of hands-on and teacher demonstration laboratory methods on science achievement in relation to reasoning ability and prior knowledge. *Journal of Research in Science Teaching, 26*(2), 121-131.

Gray, G., Costanzo, F., Evans, D., Cornwell, P., Self, B., & Lane, J. (2005). *The dynamics concept inventory assessment test: A progress report and some results.* Paper presented at the Proceedings of the American Society for Engineering Education Annual Conference and Exposition.

Hake, R. R. (1998). Interactive-engagement versus traditional methods: A six-thousand-student survey of mechanics test data for introductory physics courses. *American Journal of Physics, 66*(1), 64-74. doi:10.1119/1.18809

Laws, P., Sokoloff, D., & Thornton, R. (1999). Promoting active learning using the results of physics education research. *UniServe Science News, 13.*

Lo, M. L. (2012). *Variation Theory and the Improvement of Teaching and Learning* (Vol. 323): Acta Universitatis Gotehoburgensis.

Mazur, E. (1997). *Peer Instruction*: Prentice Hall.

Prince, M., Vigeant, M., & Nottis, K. (2015). The Impact of Inquiry-Based Learning Activities on the Retention and Transfer of Conceptual Learning in Heat Transfer. *Chemical Engineering Education, 49*(1).

Self, B., Widmann, J., Adam, G., & Coburn, A. (2015). *Developing an inquiry-based learning activity using variation theory.* Paper presented at the Research in Engineering Education Symposium, Dublin, Ireland.

Sokoloff, D. R., & Thornton, R. K. (1997). Using interactive lecture demonstrations to create an active learning environment. *The Physics Teacher, 35*(6), 340-347.

Acknowledgements

Support for this work was generously provided by the National Science Foundation, NSF #1044282, "Using Inquiry-Based Activities to Repair Student Misconceptions in Engineering Dynamics." Any opinions, findings, and conclusions or recommendations expressed in this material are those of the author(s) and do not necessarily reflect the views of the National Science Foundation.

Copyright statement

An Ethnographic Investigation into the Culture of an Engineers Without Borders USA Team During Preparation for a Design Project in Peru

Daniel W. Knight
University of Colorado Boulder, USA
knightdw@colorado.edu

Sarah A. Savage
University of Colorado Boulder, USA
Sarah.Savage@colorado.edu

Madeline R. Polmear
University of Colorado Boulder, USA
Madeline.Polmear@colorado.edu

***Abstract**: This study is focused on the engineering co-curricular organization Engineers Without Borders (EWB) USA. This club is growing rapidly with more than 300 chapters across the United States formed since it was founded in 2003. With a membership of ~50% women and a focus on service learning projects in developing countries, EWB is a unique type of organization in undergraduate engineering education which is traditionally less focused on study abroad opportunities and more focused on meeting the stringent requirements for graduation. Although EWB USA is growing rapidly, research into the organization is developing more slowly. The present ethnographic investigation expands EWB research by delving deeply into the culture of one EWB team as the students prepare to revamp the infrastructure of a water source in a small rural Peruvian mountain community.*

Introduction

This study focuses on the club Engineers without Borders (EWB). This club is growing rapidly with more than 300 chapters across the United States formed since it was founded in 2003. With the growth of the club, it has attracted some attention as being a very unique engineering club. EWB is nearly fifty percent women and is centered around service work and exploring other cultures (Wittig, 2013). Both of these things are rare in engineering colleges. The main goal of this study was to explore what makes EWB so different from its fellow engineering clubs. The types of people who join the club, the motivation behind joining the club, and the benefits of joining were explored throughout the study in order to understand what distinguishes EWB and its members.

Context

This study was conducted solely at the University Of Colorado Boulder (CU) which has a very well established club. This club is focused around giving engineering students the opportunity to use the skills they are learning in class and apply them to a project in another country. The projects that the students take on are aimed at helping to improve a community in a developing nation and giving them the ability to meet some of their basic human needs, while also giving the students leadership and problem solving skills. EWB is also unique because after the students finish designing their project they travel to their partner community and implement it themselves. This study worked mainly with the Peru chapter. The Peru chapter was the founding chapter at CU and therefore one of the largest and most developed, with approximately 25 people coming to each of the weekly meetings. This chapter is partnering with a rural village in western Peru to help them improve their water system. This study took place as the team developed plans for building a new water catchment system and distribution line for the community.

Theoretical Framework

One of the large motivations for this study was an investigation of student style. A goal of this study was to look at the different styles of the students who were involved with EWB and see if they differed from engineering students who are not involved with EWB or a similar club. In order to test this the Keirsey Temperament Sorter was used (Keirsey, 2016). This style questionnaire places the user into one of four main temperaments: guardian, idealist, rational, or artisan. The guardian temperament is focused on being dependable and hardworking, making up about 40% of the entire population and 76% of engineering students. The idealist temperament is centered around personal growth and development and enjoying working with others, making up 15% of the population. Rationals are the problem solving temperaments, who also pride themselves on being independent and strong willed. Rationals make up about 5 % of the population. Finally, the artisan temperament has the natural ability to excel in any of the arts including athletic and industrial arts while being unconventional and spontaneous. The artisans make up about 30% of the population (Alperin, 2014). Overall this temperament sorter was used in order to get a better idea of what type of students are attracted to EWB.

Methodology

This was a mixed methods study using both quantitative and qualitative methods. The main focuses of the qualitative methods were observation (approximately 25 meetings were observed) and interview. The weekly EWB Peru chapter meetings were observed on a regular basis along with the technical team meetings and the planning, monitoring, evaluation and learning (PMEL) team meetings. The occasional fundraisers and workshops that were sponsored by EWB were also attended in order to get a more complete understanding of the EWB culture. Field notes were taken at all of the key events to document observations of the meetings, members, and overall club culture. Eight interviews were also conducted with members of the team who had varying levels of involvement. These interviews were focused around what the students believed to be the strengths and weaknesses of the club, as well as their reasons for joining the club and reasons for continuing to stay involved or not continuing with their involvement.

The quantitative piece of this study was a questionnaire that was distributed during one of the weekly EWB Peru chapter meetings in the spring semester. This survey asked a wide range of questions in order to get a better understanding of the club. The questionnaire had demographic questions such as grade level and gender in order to see exactly who was involved. The questionnaire also had questions about the students' involvement in EWB such as how involved, committed and active was the student. Another main purpose of the questionnaire was to identify the style types of the students involved with EWB. The 18 multiple-choice questions Keirsey temperament sorter was added to this questionnaire, which assigned each of the students to one of the four temperaments described above. This questionnaire gave a lot of data about how the students felt about the club as well the distribution of style types among the club.

Findings

Once all of the data was collected, four main themes were identified as being the most important parts of EWB. These themes were labeled as follows:

- Styles,
- Travel,
- Getting involved, and
- Motivations.

Styles

The styles of the students involved with EWB are very different from the style of the students in an engineering college in general. A study by Alperin (2014) found investigated style types of engineering students, which are presented below in Table 1.

Table 1: Style Types in Engineering

Style	Number of students
Guardian	76%
Rational	24%

Alperin found 76% of the students to be the Guardian style type. Guardians pride themselves on being dependable and hard working. Guardians are also dutiful and tend to be focused on credentials and tradition. These students fit in well in an engineering college. However, the questionnaire of EWB members indicated that 11 of the 21 students were the idealist personality type and only one of the 21 students were the guardian personality type (which can be since in Table 2).

Table 2: Style Types in EWB

Style	Number of students
Idealist	11 (52%)
Rational	6 (29%)
Artisan	3 (14%)
Guardian	1 (5%)

The idealist personality types love to work with others and are very concerned about personal growth and development. The idealist personality types also pride themselves on being loving and kindhearted. This personality fits well within a club like EWB, which is focused around working in a group to help less fortunate people in another country. Many interviewees commented that the true value of EWB could be considered personal growth. One student said, "I do this because I want to make myself a better person while also you know maybe making some small impact." This quote aligns with the idealist personality type since the students said their reason for doing EWB was to better themselves. Idealists also tend to be more people-orientated, which could help them get involved when they first join since the club's structure selects for more outgoing students who are not as intimidated to ask older members for tasks. Overall, the large percentage of idealists in the club leads to an outgoing and welcoming culture with many passionate individuals.

Travel

Many club activities are centered around the idea of travel. Travel is a large reason why new members join EWB. During an interview one member said, "They [new members] want to have that international experience, and that was a big motivator for me just wanting to see the world." This shows how important the opportunity for a sponsored trip to Peru can be for new members. Travel is also a reason why members continue to stay motivated and involved throughout the semester. The opportunity to be selected for the travel team gives members a goal to work towards, "seeing your boots hit the ground is incredibly rewarding, it's the realization that this is worth something." This excerpt from an interview shows how much the team values the fact that they are the ones traveling to Peru and implementing the system that they designed themselves. This is very unique to EWB; most other clubs such as Public Health Brigades which is also popular at CU, do not see a project all the way through like this club does.

The selection for the travel team is an important event for members of EWB. The questionnaire results indicated that only 40% of the team has traveled while almost everyone wants to travel. The questionnaire question that asked if the members were interested in travel had an average score of 4.44 out of 5.00 (as seen in Table 3), which shows how many people would like to travel. Each style type had a high rating for travel interest showing that it was something that brings the team together.

In order to be selected to travel you must fill out an application; then some members of the leadership team for the chapter go through the application and select people who have been active members on the team and that they believe will handle the travel well. However, this path to be selected to travel was not clearly presented at the beginning of the semester. The only thing that was said to the team was that members who are very involved have a better chance of being selected for travel. This could deter some members who enjoy structure or were motivated to join because of the opportunity to travel.

Table 3: Travel Interest (5-point scale)

Style Type	How interested are you in traveling with EWB?
Idealist (n=11)	4.55
Rational (n=6)	4.17
Artisan (n=3)	5.00
Guardian (n=1)	5.00
Average	4.44

Getting Involved

Getting involved when first joining the team may be the hardest part of the club; this can be seen through the low retention rate. The first meetings of the semester usually have almost 50 students while only about 25 students consistently come to the weekly meetings. This low retention rate shows that many students are interested in the club but only a few are interested in sticking around and become permanent members of the team. This was reflected in many of the interviews, "a lot of people will just kind of flake after the first month or so." This low retention rate could have multiple causes. The first meetings could be intimidating for new members due to the amount of information that is usually presented. At the first meeting of the semester the team usually discusses the travel team that has recently returned and the work that they did in Peru. However, it could be hard to understand what the team was talking about without background knowledge.

One student who only attended the first meeting of the semester described why she did not continue to attend the meetings, "as a freshman I was like, 'oh wow I don't even know what engineering is, or how to do engineering so I don't know if I could contribute to this at all.' So it kind of intimidated me." The EWB team designs a full water catchment system for a community in Peru. This level of technical difficulty could intimidate younger students who cannot see themselves in a position to help due to lack of experience.

However, there were varying opinions surrounding getting involved. Some new members found it easy get involved right away, "it is exciting how I am actually doing stuff and it is what like week four." This comment shows that while the retention rate is low, some people are able to jump in right away and get involved easily. Another student commented, "It was just about how motivated you are to get involved." This could favor more outgoing students who are less intimidated by the lack of structure and willing to take the extra step and ask older members if they need help with anything. Supporting this idea, one student involved with EWB said, "You had to be an advocate for yourself if you wanted to do things." This comment furthers the idea that more outgoing students would find it easier to get involved. This could also account for why the culture of the club is characterized as being outgoing and social.

Motivations

There are many different things that motivate students to become committed to EWB, such as the opportunity to travel and learn about another culture. In the engineering curriculum there is often not enough room to study abroad or take many language classes. This means that EWB could fill gaps for students who are passionate about travel or other cultures. This idea can be seen throughout the interviews when students were asked why they joined EWB, "I love Spanish… don't have time to take Spanish classes." This shows that EWB could help to fulfill a desire of students who would like to learn or practice another language. Since EWB is such a unique opportunity for engineering students it can motivate them to continue to study engineering. One student said, "I don't think I would still be in engineering without it [EWB]." EWB allows students to see how what they are learning in class could apply to real problems. This makes students excited about what they are learning and want to continue to study engineering.

Most people who are involved in the club tend to either be very involved and committed or not very committed or involved. There are fewer members in the middle. The students who are very committed are also very passionate about EWB. One student even said, "EWB was actually the reason I choose to come to CU." This shows how passionate the very involved members of EWB can be. On the other side there are a few members that just attend weekly meetings without a large amount of participation. However, there are more members who are extremely involved making the average commitment rating of the club fairly high with a score of 4.12 out of 5.00 as seen in Table 4. Every personality type, except guardian, has high scores for commitment, showing that the club does a good job letting each style type feel included and become a committed member. The sample size for the guardian only included one student, which may contribute to the low rating.

Table 4: EWB Participation (5-point scale)

Personality Type	How would you rate your commitment to EWB with 5 being most committed?	How active are you with EWB, with 5 being most active?	How included do you feel in the EWB community, with 5 being feeling the most included?
Idealist (n=11)	3.91	3.45	4.54
Rational (n=6)	4.17	3.83	4.20
Artisan (n=3)	4.67	4.67	5.00
Guardian (n=1)	3.00	2.00	5.00
Average	4.12	3.67	4.54

Correlations

The correlations between questionnaire questions are significant when the values in table 5 are less than .05 (highlighted in grey in table 5). The correlation results indicate many significant relationships between questions. Time involved is correlated to both commitment and how active of a member the student is. This means that the longer a member is involved the more active and committed they tend to be. However, time involved is not correlated with inclusion. This shows that a strength of the club is that everyone can feel included regardless of how long the member had been involved. Another interesting correlation is that travel is correlated with time involved, commitment, and how active the member is. This shows that these are all important factors for being selected on the travel team.

Table 5: Correlations Between Questions (p< .05)

	Time Involved	Commitment	How active	Inclusion	Have travelled	Interested in Travel
Time Involved		0.000	0.001	0.054	0.000	0.048
Commitment	0.000		0.000	0.005	0.000	0.895
How Active	0.001	0.000		0.001	0.001	0.501
Inclusion	0.054	0.005	0.001		0.012	0.491
Have Travelled	0.000	0.000	0.001	0.012		0.359
Interested in Travel	0.048	0.895	0.501	0.491	0.359	

These correlations also allow the links between themes to become clearer. Travel, getting involved and commitment are all heavily related. In order to travel you must be very committed and involved. However, after traveling you are part of the leadership team so travel leads to high levels of involvement and commitment as well. Through observation it was seen that getting involved could also be linked to style because outgoing personality types may find it easier to get involved. This was seen when students would go up to the leadership team members and ask for tasks after a meeting. An outgoing student may find this to be less intimidating than a more reserved student.

Conclusions

This study verified that EWB does attract a unique group of students who differ from other engineering students. The group of students who are involved with EWB have a different distribution of style types than a general engineering college with a higher percent of idealists. The students who are involved with EWB also tend to be extremely passionate and committed to the organization, sometimes using the club to enrich the engineering curriculum with opportunities for travel and learning more about another culture. These findings aligned with the results of other studies. In a study by Knight, Litchfield, and Javernick-Will (2014), they also found engineering students who were involved with EWB to be different than their peers. Furthering that idea, Litchfield and Javernick-Will (2015) found that students who were involved with EWB had a wider range of interests than students who were not involved with EWB or a similar club. Overall, through this study and others supporting studies one can say that EWB students are different from students who are not involved with EWB. One main limitation to this study is the small sample size. Only one EWB chapter was observed from one school. In addition, only 25 students filled out the questionnaire. This leads to a contextual perspective on EWB that can be expanded in future studies.

Implications

The purpose of this study was to learn more about the culture of the club. However, given the large amount of observation, there is some feedback for the club. Overall, the EWB Peru chapter at CU is a strong chapter. Through observation it was seen that this EWB chapter has a strong history of multiple travel trips a year with successful projects. Observations also showed that this club is very well organized with a supportive leadership team. The questionnaire had very high rating for inclusion, showing that the club has an inclusive environment.

However, there are a few suggestions for improvement based on observations and interviews. One suggestion is to change the structure of the club to recruit a more diverse group of students. Guardians make up almost 80% of the typical engineering students and guardians enjoy structure. This could be addressed by changing the first few meetings of the semester. The club could present pathways to get involved and pathways to travel and to membership on the leadership team clearly at the beginning of the semester so students who enjoy structure feel more comfortable. The club could also talk about others things you can do besides travel, as guardians often like to stay at home and not travel. In order to attract more rationals the club could present more ways to get involved on the technical team. Finally, to attract more artisans the club could present additional building or hands on activities such as constructing a prototype of the system to be implemented. Attracting these different personality types could help to achieve a more inclusive environment and resolve the lower retention rate.

Summary

This study focuses on the club Engineers without Borders (EWB) USA. This study was conducted solely at the University of Colorado Boulder which has a very well established club. This club is focused around giving engineering students the opportunity to use the skills they are learning in class and apply them to a project in another country. One of the large motivations for this study was an investigation of student style. A goal of this study was to look at the different styles of the students who were involved with EWB and see if they differed from engineering students who are not involved with EWB or a similar club. This was a mixed methods study using both quantitative and qualitative methods. The main qualitative methods were observation (approximately 25 meetings were observed) and interview. The quantitative piece of this study was a questionnaire that was distributed during one of the weekly EWB Peru chapter meetings in the spring semester. This questionnaire asked a wide range of questions in order to get a better understanding of the club.

Once all of the data were collected four main themes were identified as being the most important parts of EWB investigated in this study. These themes were styles, travel, getting involved, and motivations. The styles of the students involved with EWB are very different from the style of the students in the engineering college as a whole. A study by Alperin (2014) found almost 80% of engineering students were guardians, while EWB has 52% idealists and only 5% guardians. The club is centered around the ideal of travel. Travel is a large reason why new members join EWB, and members continue to stay involved. Getting involved when first joining the team may be the hardest part of the club; this can be seen through the lower retention rate. The first meetings of the semester usually have almost 50 students while only about 25 students consistently come to the weekly meetings. There are many different things that motivate students to become committed to EWB, such as the opportunity to travel and learn about another culture. In the engineering curriculum there is often not enough room to study abroad or take many language classes. This means that EWB could fill gaps for students who are passionate about travel or other cultures. This study verified that EWB does attract a unique group of students who differ from other engineering students and that the students who are involved with EWB also tend to be extremely passionate and committed to the organization.

References

Alperin, Kenneth. "Personality Types." Electrical and Computer Engineering Design Handbook. Web. 09 Apr. 2016.

Keirsey Temperament Website - Overview of the Four Temperaments. "Keirsey Temperament Website - Overview of the Four Temperaments. Web. 10 Apr. 2016.

Knight, D., Litchfield, K., & Javernick-Will, A. (2014, October). Engineers without borders: An empirical investigation of member's defining characteristics. In Frontiers in Education Conference (FIE), 2014 IEEE (pp. 1-5). IEEE.

Litchfield, K. & Javernick-Will, A (2015). "“I Am an Engineer AND”: A Mixed Methods Study of Socially Engaged Engineers." Journal of Engineering Education, 104(4), 393-416.

Wittig, A. (2013). Implementing problem based learning through engineers without borders student projects. Advances in Engineering Education, 3(4), 1-20.

Copyright statement

Pedagogic experience focused on crowdsourcing for text mining teaching in engineering

Author 1 Yaneth Patricia Caviativa Castro
Universidad Manuela Beltrán, Bogotá, Colombia
Janeth.caviativa@umb.edu.co

Author 2 Adan Beltran Gomez
Universidad Manuela Beltrán, Bogotá, Colombia
Adan.beltran@docentes.umb.edu.co

Author 3 Lizeth Vanessa Jaramillo
Universidad Distrital Francisco José de Caldas, Bogotá, Colombia
lvjaramilloc@correo.udistrial.edu.co

Author 4 Valentino Jaramillo Guzmán
Secretaria de Educación del Distrito Capital, Bogotá, Colombia
Valentino jaramillog@yahoo.es

Author 5 María Inés Mantilla
Universidad Manuela Beltrán, Bogotá, Colombia
Maria.mantilla@docentes.umb.edu.co

Author 6 Raúl Ernesto Menéndez-Mora
Universidad Católica de Colombia, Bogotá, Colombia
remenendez@ucatolica.edu.co

Abstract: *In systems engineering education there are pedagogic challenges due to continuous apparition of emerging technologies that do not arrive on time for the curricular design of academic programs in universities. These emerging technologies require that the students correctly appropriate the new concepts and models for their proper labor performance in the near future. One way of facing this challenge is the continuous education through virtual graduated courses or certifications which seek to keep the students updated within new technologies. On this context it is presented an experience with the design, development and implementation of a virtual degree on text mining using techniques of collaborative work or crowdsourcing and five virtual learning objects in systems engineering students. This qualitative descriptive study is based on a case study as the research model. Positive results in building meaningful knowledge were obtained with this learning tool. A characterization of collaborative learning was also generated.*

Keywords: *collaborative work, crowdsourcing, text mining, virtual learning objects*, collective interdependence.

Introduction

This article presents the results of a research project which pretends to strengthen integration strategies of Information and Communication Technologies (ICT) in higher education through five virtual learning objects (VLO). Learning objects of dynamic design and easy application can helps information to be clear and simpler. VLO allows the student to access these virtual pages for building meaningful knowledge and so it can be used as a strategy by systems engineering students characterizing the crowdsourcing or collaborative work.

Background

Law 1341 of July 30th, 2009, establishes the priority to the access and use of Communication and Information Technologies (ICT) in no discriminative conditions for all Colombian society, prioritizing the access in production of goods and services (MINTIC, 2009). Likewise, guide number 21 highlights the importance of training competent students to perform, in effective ways, an ability or a skill in a specific job. The general and specific professional competences, as well as the basic and civic, which must be developed in youth, are described as an integral part of the Colombian educative system. In chapter 1, of the decennial plan of education 2006-2016 (MEN, 2006), a third subject named as pedagogic renovation and the usage of ICT in education in Colombia with the following macro objectives:

Strengthening pedagogic processes through ICT
Pedagogic innovation and interaction of educative actors
Strengthening of educative projects and follow up mechanisms

According to previous studies, it has been evidenced that one of the lines of thought that orientates educational change is the one linked with collaborative. Collaborative systems attempt to find a model that includes the different participants, the tasks to carry out, and collaborative modes. The role of technology is to ease communications, but also it can play the role of promoter and organizer giving support to the work of a group of people in learning tasks (Barros and Verdejo, 2000).

Therefore, the formation activities performed with systems engineering students have evidenced new problems, which make necessary to advance in conceptualization of collaborative learning through educational research. One of the first elements to take into account as base of every learning, and especially collaborative learning, is communicative interaction.

When analyzing the subject of collaborative learning it is necessary to make some precisions. According to Dillenbourg, et al. (1995), it is needed to differentiate problem resolution from learning and collaboration from cooperation. From a psychological approach, it is considered that learning and problems resolution share similar processes, but for computational sciences these are different. In accordance with the above, Wessner and Pfister (2001) describe this subject similarly.

Likewise, Levy (2004) describes that collaborative learning involves participation of "two or more people with the common objective of acquiring knowledge, that are disposed to share their knowledge and experiences, in a framework of interaction and communication actions directed to achieve these goals" (p.33). In this way, collaborative learning is offered as an alternative for learning and improvement of competition and individualism.

Similarly, Arteaga and Fabregat (2003) indicate that "a collaborative team should have positive interdependence, individual responsibility, promoted interaction, and proper use of social skills; these are elements which generate conditions for an effective collaboration". As it can be observed in the described elements, collaborative learning is characterized by its autonomy, leadership, task assignation, time management, responsibility in activities and others; while collaborative work has the purpose of constructing new ideas, the teacher predetermines the groups of work and gives work guidelines.

Research question

What are the characteristics of collaborative learning or crowdsourcing when implemented as a strategy in a diplomate in text mining aimed at systems engineering students?

General objective

Characterize collaborative learning through the development of a didactic strategy that strengthens cognitive abilities in text mining concepts in systems engineering students?

Specific objectives

- Design a didactic strategy to strengthen cognitive abilities for the collaborative learning with text mining concepts
- Implement the didactic strategy in engineering students to strengthen the social appropriation of meaningful knowledge
- Analyse collaborative learning through the application of the didactic strategy

Theoretical framework

ICT can be defined in two ways: traditional technologies of communication, constituted mainly by radio, television, and conventional telephony; and modern technologies such as informatics and telematics. ICT can contribute to universal access to education, instruction equality, teaching experience and learning of high quality, and to professional development. ICT is a tool for the new education (Vargas and Ricardo, 2015).

From a pedagogical approach this proposal is framed in a constructivism model considering that learning virtual objects are a set of digital resources, like in (MEN, 2007) states through its portal web page ColombiaAprende, and they can be used in different contexts, with an educative purpose.

In the approaches of the social constructivism there are open paths for the development of collaborative learning. The psychologist Vygotsky (1978) developed theories that, more than the psychological contribution, make a social contribution that serves as a basis for what collaborative learning implies. Constructivism states that each individual, as his name implies, is able to construct his own learning by making use of his abilities and, from Vygotsky's perspective, through social. Constructivism seeks to help students internalize, rearrange, or transform new information. This transformation occurs through the creation of new learning and this comes from the emergence of new cognitive structures (Grennon and Brooks, 1999). The present work of research from the socio-constructivist position requires a flexible design, which facilitates the construction of meaningful knowledge, through progressive inquiry, participation, organization and structuring of activities, addressed and accompanied by the tutor.

In the case of the implementation of collaborative learning in OVA design, the author Gros (2007) mentions that it is the same partners who make up the social environment. They carry out different intervention processes, contributions, corrections, revisions, suggestions, among others, thus building a network of support and development of collective learning. Group processing, according to Johnson and Johnson (1999), can be defined as the reflection on a group activity to see what actions of its members are useful and which were not useful and to make decisions on what actions to keep and which ones must change. The purpose of group processing is to clarify and improve the individual effectiveness of each of the group members and therefore of the group as a whole.

Dillenbourg (1999) suggests a relation between the conception of knowledge with the new technologies and the area that they cover for the execution of projects and group work that need the constant interaction between the subjects asynchronously. According to Dreves (2000), the main characteristics of collaborative learning are positive interdependence, interaction, individual contribution and personal and group skills. Palloff and Pratt (2005) say that collaborative learning develops critical thinking, knowledge building, ongoing reflection, transformation of initial learning and use of ICT.

Collaborative learning is characterized by three families of analysis: collective interdependence, social habits and evaluation processes. For this project we have focused on the analysis family: collective interdependence. The sub-categories and descriptors within this family are described in Table 1.

Table 1. Family of analysis of collective interdependence's category. Source: Johnson and Johnson (1999).

Collective Interdependence		
Group responsibility	Group promotion	Increased self-esteem
• Each student is concerned about the development of the group. • Passive, dominant, or competitive postures are reduced through peer interaction.	• They learn to complement the individual skills and abilities with those of the group in a way that optimizes the possibilities to obtain the proposed achievement. • The group gives feedback to each individual in the group so that their development corresponds to the level of achievement reached. • Competitiveness is changed by assuring the learning of each member of the group.	• All members are held accountable. • The skills and characteristics of the elements that demand the activity are distributed heterogeneously within the group. • The group recognizes its effectiveness by comparing itself with other groups and to feedback its own progress.

There are several models of instructional design such as the Assure Model and the Instructional Design model of Dick and Carey (Martínez, R. and Azucena, C., 2009), among others. An instructional model containing a structured methodologáy is ADDIE, composed of five basic activities: analysis, design, development, implementation and evaluation. For the development of the present work we follow the model ADDIE (Williams, Schrum and Sangrá, 2004) which we show in Figure 1.

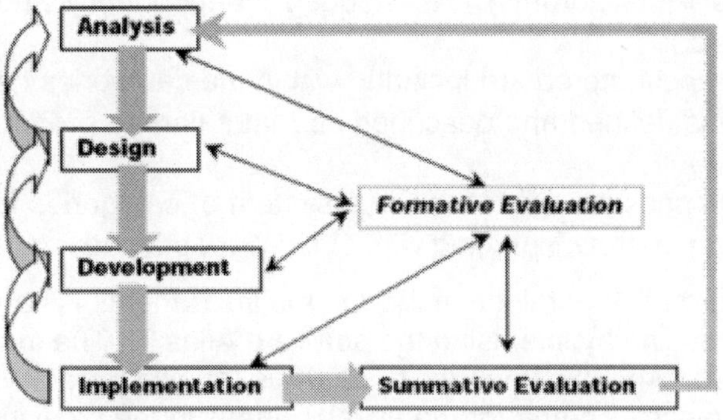

Figure 1. ADDIE model (Williams, Schrum and Sangrá, 2004). Source: https://www.bhten.com/planning-instructional-design

Methodology

The methodology that is implemented in this study is qualitative, being a descriptive and interpretative study, considering the methodological rigor proposed by Castillo and Vásquez (2003). The pilotage was carried out with 20 students of last semester of systems engineering programs of different universities of the country. These students, adults, who count with a variety of electronic mobile devices and with wireless connectivity, enrolled and met the requirements to pass the diplomate in "Text mining and its applications in organizations". They interact through activities defined in the LVOs, which contain readings, case studies and problem solving.

For the construction of the VLOs two main aspects were considered: pedagogical and content design.

Content design: The thematic structure of the VLO was organized by sub themes which were presented in tab format without transition. Each tab represented a subject of the diploma in text mining. The content of the modules was provided with reference material from various sources and formats. In total, open digital educational resources were used, as well as Creative Commons licensed materials: such as PDF documents, slides in ppt, flash, embedded HTML video with embedded code, images and links to specialized web pages. In the evaluative component, in addition to the disciplinary component oriented from the theoretical and practical, the student must have basic knowledge in the management of the platform and its automated protocols such as chats, forums, wikis, concept maps and online simulators.

Pedagogical design: The type of learning that the LVOs support is focused on the need to guide the student toward collaborative learning. The role of the Tutor is: design and develop the VLOs, accompany the student, guide the process of development of the thematic contents, encourage collaborative work through the discussion and analysis of the phenomena in the forums.

The analysis of the automated protocol "forum" was developed as follows:

- The category "Collective Interdependence" and the subcategory "Increase Self-Esteem" of the Johnson and Johnson category system (1999) are developed (see Table 1)
- Based on the category system of Johnson and Johnson (1999) descriptors are created (see Table 2)
- 3 Forums are selected for the application of the system of categories-subcategories-descriptors
- Each forum is codified independently
- The chosen coding unit is the complete message because some messages can be encoded with more than one descriptor
- The selected instruments (forums) are encoded independently and compared with the encodings performed
- The new situations encountered are included within the categories generated
- Each new category is defined and described so that it can be properly integrated into the initial classification
- At the end of this process the quotations of a system of categories for the analysis of the collaborative learning or crowdsourcing with LVOs are obtained

Triangulation of the information obtained in the forums instruments is carried out from its transcription and analysis line by line using the software Atlas TI. The information obtained in the forums is intersected with the categories of analysis (shown in Table 1). The initial codes are acquired from relevance's criteria for the assertiveness in the crossing. The initial codes grouped into categories are the object of the realization of networks of analysis from which cross relations are established that allow initiating the interpretation of the data.

Results

The diploma in text mining was supported by five VLO. A snapshot of one of the VLO is shown in Figure 2.

The characterization of collaborative learning was obtained by crossing the three automated protocols (forums) with the narrative analysis of the messages (line by line). These activities yielded data by line of text as follows: forum 1 (392 textual lines), forum 2 (193 lines) and forum 3 (214 lines). The characterization of collaborative learning was supported through the elaboration of a filter categories system.

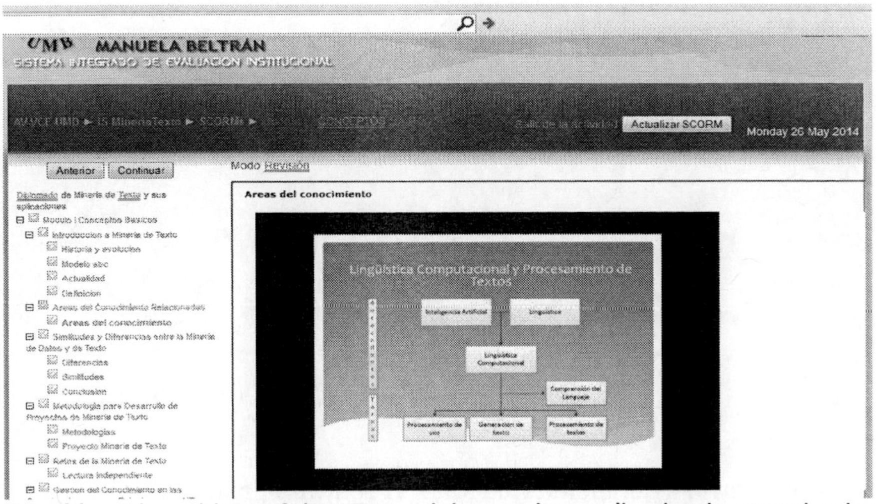

Figure 2. Virtual learning object of the "Text mining an its application in organizations" virtual degree

Table 2. Categories, subcategories and analysis descriptors

Category	Subcategory	Descriptor
Collective interdependence	Group responsibility 149	Shared interests 77
		Ethical group responsibility 72
	Group promotion 132	Contribution to the motivational dialogue 38
		Recognition of the other 64
		Complementation of contributions 10
		Communication within the group 20
	Increase of self-esteem 163	Recognition of one's work 151
		Distribution of responsibilities 12

These filter categories were the descriptors identified by the investigators as shown in Table 2. The number that accompanies the subcategories and descriptors represent the relative frequency of the element once processed in Atlas TI software. The qualitative study performed with Atlas IT generated the semantic network shown in Figure 3.

This research was able to identify narrative support categories that allow for Collective Interdependence from Johnson and Johnson (1999) approach as shown in Table 2. These categories describe criteria such as: group responsibility (shared interests, ethical group responsibility); group promotion (contribution to motivational dialogue, recognition of the other, complementation of contributions, communication within the group); and increase of self-esteem (recognition of own work, distribution of responsibilities). As well as, the relationship of social skills from individual responsibility (contribution to knowledge, autonomy, direct communication as an emerging element, reflective knowledge); respect for the work of the other (ethical commitment); norms of coexistence (communication by etiquette); use of development space (use of ICTs in collective construction), as sources of validity in learning.

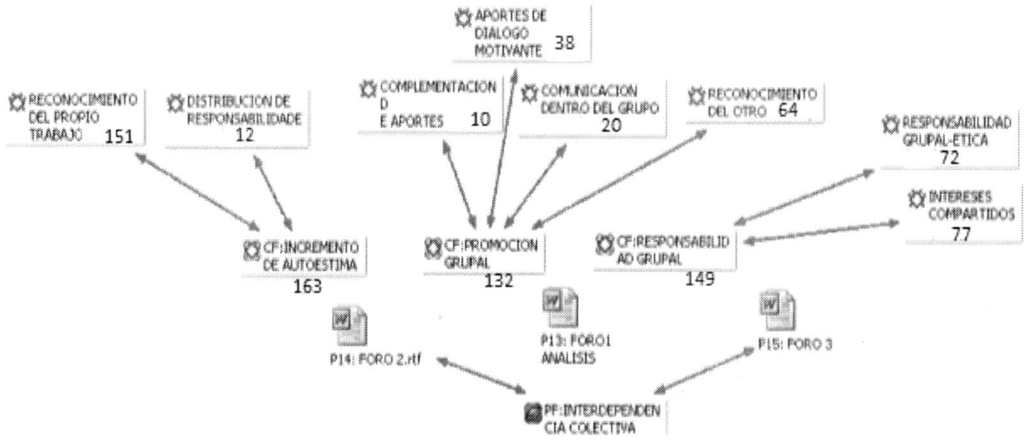

Figure 3. Semantic network generated with Atlas TI software (labels in Spanish)

The form of collaborative contribution depends on the autonomy and leadership considered as characteristics elements in the analysis within the activities, as well as the relation of individual responsibility processes which include an element of individual reflection.

Within the categories of collaborative learning it was found concordance with theory, which established that individual abilities and competences complement with the group ones so that the following possibilities are maximized: achieving the proposed goal, feedback to each individual of the group in ways that their development corresponds with the level of the achievement obtained, and the lack of competence for the assurance of each member learning. In this category etiquette communication was highlighted, as well as the cordial, respectful and inclusive dialogue, which allowed the strengthening of partnership.

This confirms the importance of collaboration as an imperative element in human communication systems. These categories are in convergence with the descriptors of Johnson and Johnson (1999), alluding to a connection with individual processes, which autonomously induce to improve communication skills. These are refined to the extent that communication mechanisms and strategies relate to each member of the group.

Conclusions

The following research was able to characterize the collaborative work in students of systems engineering participating in a diploma in text mining. Being able to design VLO in engineering areas becomes an excellent virtual pedagogic resource that can be used no matter where the person is. These learning objects fulfill the organization scheme that is drawn from a pedagogic planning allowing an improvement and enrichment of educative learning.

The use of the VLO in the text mining degree evidenced the benefits that can be obtained, which consist in having the capacity of organizing, classifying and retrieving texts automatically, as well as the management of time in independent way allowing the accessibility to learning contents, creating scenarios for the co-evaluation and facilitation of synchronic and asynchrony communication, breaking the barriers of time-space. These objects can be focused towards collaborative learning, and it can be identified as an exercise in determining the association scheme. This tool was positive since it reinforces communication, reflection, teamwork and helps strengthening the educative role expected to give nowadays.

The collective interdependence was observed in the construction of learning supported by collaborative instruments in which there was no evidence of postures of competition but of group interaction. Interests were shared within professional training in face of their daily practice and work experience.

From the collaborative learning and the area of the computer science was propitiated the construction of meaningful knowledge in the subject of text mining.

The process of analysis with the Atlas TI tool contributed to the characterization of collaborative learning, evidenced through the semantic network.

From the analysis process it can be said that it was a good strategy. The analysis in Atlas TI contributed to the characterization of collaborative learning, and with the description of the categories the analysis of learning was achieved.

Recommendations

As a continuation of this work we plan to analyze the categories social habits and evaluation processes to fully characterize collaborative learning or crowdsourcing.

References

Arteaga, C., and Fabregat, R. (2003). *Integración del aprendizaje individual y del colaborativo en un sistema hipermedia adaptativo*. Institut d'Informàtica i Aplicacions (IliA), Italia.

Barros, B., and Verdejo, M. (2000). Analysing students interactions process for improving collaboration. The DEGREE approach. *International Journal of Artificial Intelligence in Education, 11*, 221-241.

Castillo, E., and Vásquez, M. (2003). El rigor metodológico en la investigación cualitativa. *Colombia Médica, 34*(3), 164-167.

Dillenbourg, P., Baker, M., Blaye, A., and O'Malley, C. (1995). The evolution of research on collaborative learning. In E. Spada and P. Reiman (Eds.), *Learning in Humans and Machine: Towards an interdisciplinary learning science* (pp.189-211). Oxford, England: Elsevier.

Dreves, C. (2000). The relationship between mother's beliefs about ability an effort and student's motivation to learn. Chicago: Poster presented at the Biennal Meeting of the Society for Research on Adolescence.

Grennon, B., and Brooks, G. (1999). In search of understanding, the case for constructivist classrooms. Virginia, USA: Association for Supervision and Curriculum Development.

Gros, B. (2007). El aprendizaje colaborativo a través de la red: límites y posibilidades. Aula de innovación educativa. Barcelona: Universidad de Barcelona.

Johnson, D. and Johnson, R. (1999). Aprender juntos y solos. Buenos Aires: Aique.

Johnson, D., Johnson, R., and Holubec, E. (1999). El aprendizaje cooperativo en el aula. Buenos Aires, Argentina: Paidós.

Levy, P. (2004). *Inteligencia colectiva: por una antropología del ciberespacio*. Accessed at: http://inteligenciacolectiva/bvsalud.org on 12 Oct 2015.

Martínez, R., and Azucena, C. (2009). El diseño instruccional en la educación a distancia. Un acercamiento a los Modelos Apertura. México: Universidad de Guadalajara. Accessed at: http://www.redalyc.org/pdf/688/68812679010.pdf on 12 Oct 2015.

Ministerio de Educación Nacional, MEN. (2006). Plan decenal de educación 2006 – 2016. Accessed at: http://www.mineducacion.gov.co on 12 Oct 2015.

Ministerio de Educación Nacional, MEN. (2007). Objetos virtuales de aprendizaje e informativos. Accessed at: http://www.colombiaaprende.edu.co/html/directivos/1598 on 12 Oct 2015.

Ministerio de Tecnologías de la Información y las Comunicaciones, MINTIC. (2009). Ley 1341 de 2009. Accessed at: http://www.mintic.gov.co/portal/604/w3-article-3707.html on 12 Oct 2015.

Palloff, R., and Pratt, K. (2005). Online learning communities revisited. 21st annual conference on Distance Teaching and Learning

Vargas, M., and Ricardo, M. (2015). *Capacitación para el conocimiento sobre las TIC's.* Cundinamarca, Colombia.

Vygotski, L. (1978). Mind in society. Cambridge, USA: Harvard UniversityPress.

Wessner, M., and Pfister, H. (2001). *Group formation in computer-supported collaborative learning.* Proceedings of the International ACM SIGGROUP Conference on Supporting Group Work, New York, United States: AMC.

Williams P., Schrum L., Sangrá A., and Guardia L. (2004). *Modelos de diseño instruccional*. España: Ed. Eureka media.

Communication skills training effect on academic performance: the case of ABET student outcomes f and g

Jose D. Soto
Universidad del Norte, Barranquilla, Colombia
jsoto@uninorte.edu.co

Luis Torres
Universidad del Norte, Barranquilla, Colombia
ltorres@uninorte.edu.co

María Calle
Universidad del Norte, Barranquilla, Colombia
mcalle@uninorte.edu.co

Adela De Castro
Universidad del Norte, Barranquilla, Colombia
decastro@uninorte.edu.co

Lucy García
Universidad del Norte, Barranquilla, Colombia
lucyr@uninorte.edu.co

Norelli Schettini
Universidad del Norte, Barranquilla, Colombia
nschettini@uninorte.edu.co

***Abstract**: Engineering students and professors traditionally do not pay enough attention to soft skills such as communications and professional ethics, since their training is focused mainly in technical knowledge. Understanding this difficulty, a multidisciplinary group of professors started working on developing these soft skills in a student population. Professors designed an intervention for different academic programs, developing tools for helping students to improve their skills and assessing ABET student outcomes f and g, while performing a classroom research experience. Students wrote an argumentative essay regarding one particular case in Engineering Ethics. Professors evaluated all data using rubrics developed by other professors at the Engineering College at Universidad del Norte. Results of student outcome f show that students understand the concept of professional ethics and acknowledge the legal framework for ethical responsibility. Assessment of student outcome g shows that students improved their performance in communication skills.*

Introduction

In the engineering curricula of universities there is at least one course on ethics and its location in the program varies in each Institution (Haws, 2001). Universidad del Norte in Barranquilla, Colombia, has a group of elective courses on ethics, but students can only take one. Faced with this situation, which is common in engineering programs, professors should determine the degree of theoretical knowledge and practice of ethics of the future graduate. This is also one of the concerns of ABET (Accreditation Board for Engineering and Technology) (ABET, 2016).

Two of the challenges of higher education in the 21st century will be to train ethical professionals and to prepare them to communicate more effectively, in accordance with the interests of new global markets. This idea is relevant especially for engineers, who are meant to be the leaders of technological and social transformations of this century.

Professors of the Electrical, Electronics and Systems engineering programs at Universidad del Norte have been working for several years with the group of students of the Capstone Design course (Calle Torres, 2013) regarding communication skills. Therefore, professors decided to structure a strategy that allowed them to evaluate professional ethics skills and at the same time, argumentative writing skills.

2. Background and research question

Since 2012, a multidisciplinary group of professors from Universidad del Norte joined forces to work on classroom research. Their goal is to improve the communication skills of the students by performing interventions and creating different tools to evaluate these skills. At the same time, the Engineering College at Universidad del Norte formed a group of professors to develop standardized rubrics for the evaluation of ABET student outcomes. Several professors worked in both groups.

The rubrics team also created a complete guide explaining the methodology and how to evaluate each one of the 11 ABET student outcomes (from a to k) (Camacho, et al. 2014). This was an opportunity for professors of the formed group to evaluate communication skills, because the student outcome f, corresponding to an understanding of professional and ethical responsibility, could be evaluated using argumentative communication methodologies.

Therefore, the research question addressed is: is it possible to improve communication skills and professional ethics of students using the same methodology?

3. Theoretical framework

There are 11 basic criteria (a – k) that ABET establishes for the accreditation of engineering programs (ABET, 2016). Institutions must demonstrate that students have developed the skills and that there is a continuous improvement process in the evaluated programs.

The assessment of some ABET student outcomes related to soft skills is controversial as professors and students seem not to pay enough attention to them. It has been hypothesized that both, professors and students do not consider soft skills as important as the knowledge of the concepts of the engineering field. However, students in an initial survey acknowledged the importance of communication skills during their academic training (De Castro et al, 2014). For this reason, professors have developed diverse and creative ways for measuring student performance for these outcomes. However, the subject of how to adequately measure them with tasks that trigger student motivations remains open to research.

Two of the soft skills evaluated by ABET are our subject of study: students must communicate effectively (student outcome g) and they should also have an appropriate ethical behaviour (student outcome f) (ABET, 2016). Regarding outcome g, some studies explain the need for adequate development of written and oral skills in students to present their reports (Robinson, Sparrow, Clegg, and Birdi, 2005; Leydens, and Schneider, 2009). Nevertheless, many engineering students at the last semesters of their academic programs have deficiencies in their communication skills, both written and oral. One consequence of the situation is poor performance of engineers when writing and presenting reports at the beginning of their professional practice.

As for the problems generated by not understanding their professional ethical responsibilities, faculty will only see the impact of this skill when students start working.

However, faculty has a vision of the time devoted to this subject in the classroom, which may not be enough. It is clear that a great job must be done in this topic, especially considering that every day the news present different cases of corruption in the engineering field. This is one example of the result of poor performance in this important student outcome.

Several studies proposed methods to foster ethical skills in students. One study developed a specific rubric for evaluating ethical skills in a discussion of ethical dilemmas (Schmeckpeper, Kelley and Beyerlein, 2014). Specifically, Yalvac, Smith, Troy, and Hirsch (2007) applied a method to recreate real ethical dilemmas to engage students in their resolution through written text. We followed the same method in this research.

On the other hand, many projects in engineering must be presented in written reports with a clear argumentative component. The situation represents a challenge in the engineer training process, especially in those cases in which secondary education has emphasized the transmission of highly structured and formalized knowledge, while in many university courses students are expected to test and defend their views instead of simply presenting a series of concepts. Additionally, standardized tests that evaluate the quality of secondary education indicate that in Colombia a high percentage of students do not reach the acceptable level of performance in science, mathematics and reading (OECD, 2014). These students will therefore have a clear disadvantage when developing the expertise required in the knowledge society, including communication and ethical skills.

Already in 2007, Holvikivi established that one of the obstacles faced by engineers is not precisely the mathematical reasoning for their discipline's problem solving, but rather the mastery of language structures to express a formal reasoning about the solution of a disciplinary problem. The author concludes that a greater emphasis on the preparation of students in conceptualization and argumentation would have a greater impact on the learning outcomes of the sciences.

4. Methodology

The group of professors has been working in classroom research from 2012 to 2017. Professors discussed the research topics with advisors, external peers and trainers, both in communication skills and the understanding of professional ethics responsibility.

Table 1 shows the different types of classroom research regarding communication skills and the number of professors working in each project.

Table 1 - Classroom research projects over time

Type of research	Subject	# professors	years
Research 1: Quasi-experimental study of descriptive scope (De Castro et al, 2013)	Communication skills in Capstone Design students in Electrical, Electronics and Systems engineering. Topic covered: oral and written communication skills.	4	2012 - 2013
Research 2: Quasi-experimental study of explanatory scope (De Castro et al, 2015)	Communication skills in argumentation in Electrical, Electronics and Systems engineering. Research included the argument structure and the writing of argumentative essays in engineering.	7	2014 - 2015
Research 3: Quasi-experimental study of explanatory scope	Coherence and cohesion in argumentation in engineering in Electrical, Electronics and Systems engineering.	7	2016 - 2017

Professors proposed to use a methodology including a diagnostic test, an intervention, one work session with the students and the application of a final test. To evaluate the achievement of student outcomes, professors employed the standardized rubrics created by the Engineering College. These rubrics were reviewed by external peers and adjusted after pilot tests performed in different years.

For the diagnostic test, professors asked students to write an argumentative essay regarding a technical topic. The intervention included professor training and presenting a clear example of a good argumentative essay to the students. After the intervention, students must read the Code of Ethics in Engineering, and professors assign readings of news and articles where engineers face ethical situations in their professional practice. For the final test, professors created a case study presenting different ethical issues that may happen during their professional lives. Professors asked them to write an argumentative essay answering this question: "what do you think are the most important violations to the Ethics Code presented in the case study?" Students had one hour to write the essay. Essays were evaluated with two rubrics: one for student outcome g and one for student outcome f.

The project is of quantitative type, with an explanatory scope and a quasi-experimental design.

5. Results

In the communication skills area, the initial research determined that morphosyntax and spelling of students improved by 72.88%, and the macrostructure of written texts by 42%, but there were no specific changes in coherence and cohesion of written texts (Calle et al, 2015).

As an example of student outcome evaluation, Figure 1 shows results obtained in one of the programs by evaluating the essay with the rubric for ethics in engineering. The rubric has 5 levels of achievement: below average, satisfactory, very good, excellent and outstanding. Figure 1 shows the percentage of students that achieved each level in 2015, for 5 different criteria evaluated in the rubric. Most students achieved the three higher categories and mainly in the "outstanding" level. Therefore, by writing the argumentative essay, students demonstrated their proficiency regarding the Code of Ethics in Engineering and general ethics issues.

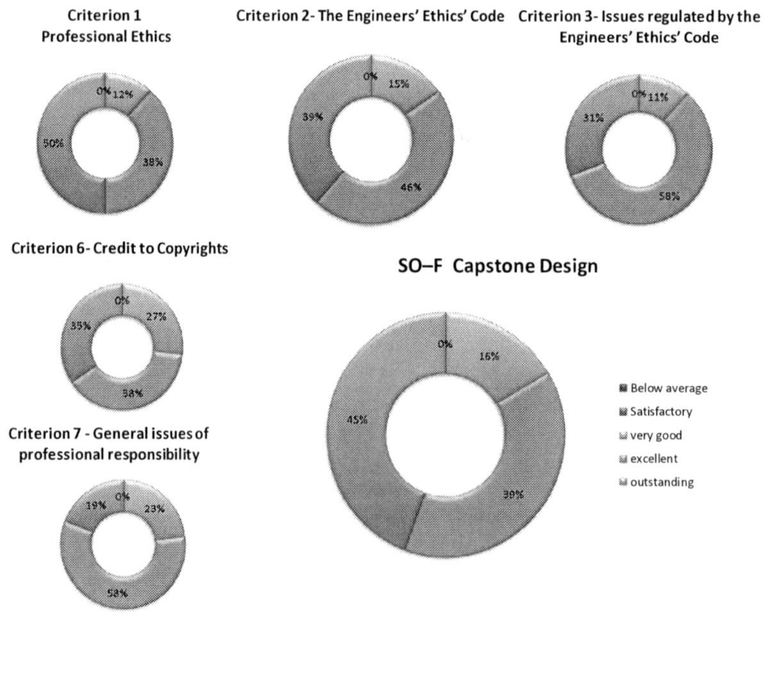

Figure 1: Results of Student Outcome f: An understanding of professional and ethical responsibility.

Figure 2 shows example results from evaluating student outcome g in one engineering program with the specific rubric, in year 2014. Overall, 61% of students performed in the "excellent" category in outcome g. Therefore, students demonstrated their achievements in communication skills.

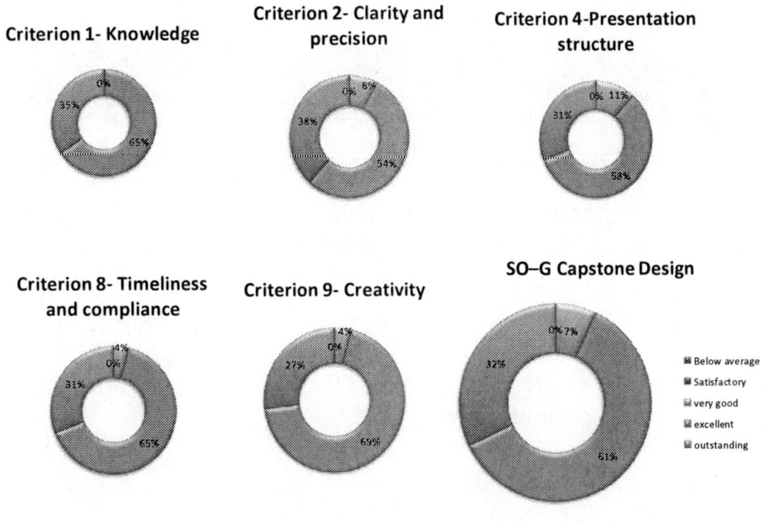

Figure 2: Results of Student Outcome G: The ability to communicate effectively

Figures 3 and 4 show a comparison scenario by academic program and year of measurement, with the average evaluation for each student outcome. Note year 2012 does not show results for Systems Engineering. In that year, the program did not use the same methodology employed in the other two programs. Figure 4 shows a decreasing trend in most cases. The situation may be explained because professors mastered techniques in evaluating outcome g, therefore they may have been more demanding when evaluating the work of the students.

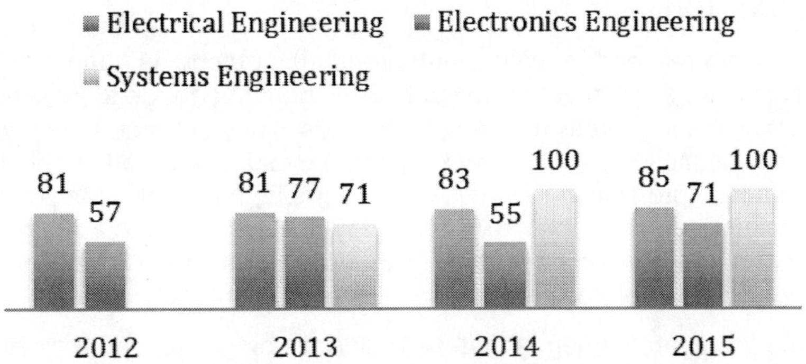

Figure 3: Student performance in different years of evaluation of student outcome f (Understanding of professional ethics).

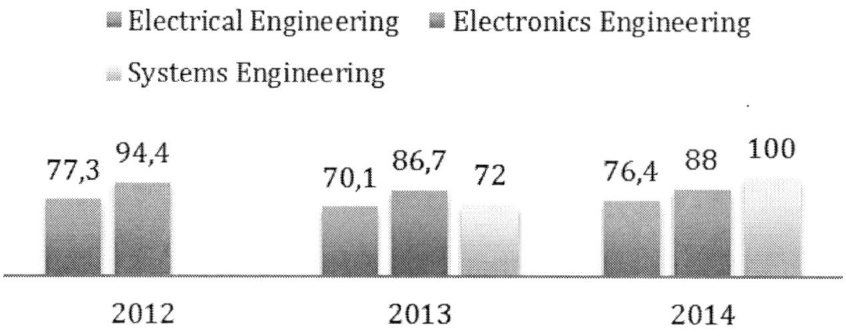

Figure 4: Student Performance in different years of evaluation of student outcome g (Using communication aids/resources).

These results indicate that the intervention favoured the reinforcement of argumentative skills and motivated a better writing from the initial essay to the final essay (De Castro et al, 2015). From the point of view of the ability to understand their professional and ethical responsibilities, an overall improvement was also observed.

5. Conclusions

The research confirmed that the inclusion of transversal communication skills in the curriculum of all subjects is no longer a fact but a necessity. The study showed the engineering case, but it becomes essential to verify with more studies of this same kind in other disciplines.

At the beginning of the study, professors thought that the direct beneficiaries would only be the students. However, the research demonstrated that professors were also benefited by all training and practice acquired while helping the students.

The work done by this group since 2012 has important results. The methodology used allowed interventions to be carried out, without affecting the fulfilment of the objectives of each course.

6. Recommendations

Students and professors had good results when measuring outcomes f and g. However, professors think that more adjustments and strategies should be made to assess the understanding of ethical and professional responsibilities. They propose a review of the rubric currently used, because they consider that some evaluation criteria could be replaced by others closer to actual professional practice. In Colombia, the issue of ethics in the area of engineering is very important, and the news every day confirms major corruption cases linked to infrastructure or other engineering projects. As a consequence, colleges must emphasize the importance of professional ethics to their students.

For this reason, the group recommends that the academic programs and their managers implement this methodology in different courses (not just engineering-related), so there is a deeper impact throughout the curriculum. This year, the group of professors is working on the concepts of coherence and cohesion in the Spanish language, and its vision is to consolidate this work to be implemented by all professors of the Electrical, Electronics and Systems engineering programs

7. References

ABET Engineering Accreditation Commission. (2016). Criteria for accrediting engineering programs 2016-2017. Available at: http://www.abet.org/accreditation/accreditation-criteria/criteria-for-accrediting-engineering-programs-2016-2017/#outcomes

Calle Torres, M. G. (2013). Un modelo de asignatura colectiva en ingeniería: Proyecto Final en Ingeniería Electrónica. *World Engineering Education Forum (WEEF 2013.* Asociación Colombiana de Facultades de Ingeniería (ACOFI), International Federation of Engineering Education Societies (IFEES). Cartagena. 24-27 September 2013.

Calle, M., García, L., Soto, J., Torres, L., De Castro, A., Candelo, J., González, R., & Moreno, M. (2015). Competencias comunicativas en ingeniería: caso estudiantes de último semestre. *Memorias del II Congreso Internacional: Lectura y Escritura en la Sociedad Global*, Barranquilla. 10-12 June 2015.

Camacho, A., Calle Torres, M.G., Coronell, G., Domínguez, M., Durango, N., Jimeno, M., Torres, L. (2014). Manual de rúbricas, División de ingenierías. Universidad del Norte.

De Castro, A., Soto, J.D., Calle Torres, M.G., García, L., Guerra, D., & Hernández, F. (2013). Competencias comunicativas en proyecto final de ingeniería: informe preliminar. *Foro Internacional de Innovación Pedagógica Innova Cesal,* Bogotá. 8-10 October, 2013. Available at: http://www.innovacesal.org/innova_public/archivos/publica/area06_tema04/268/archivos/redIC_ing_interdisc_03_2013.pdf

De Castro, A., Soto, J.D., Calle, M.G., García, L., Candelo, J., Torres, L. & Gonzalez, R. (2015) Developing Argumentative Documents In Engineering: a Case of Study through the Curriculum. In *11th International Scientific Conference: eLearning and Software for Education ELSE 2015*, Bucharest, 23-24 April 2015. DOI 10.12753/2066-026X-15-003.

De Castro, A., Soto, J.D., Calle, M.G., García, L., & Guerra, D. (2014). Communication skills in senior engineering students. In *10th International Scientific Conference: eLearning and Software for education ELSE2014*, Bucharest, 24-25 April 2014. DOI 10.12753/2066-026X-14-000.

Haws, D.R. (2001). Ethics Instruction in Engineering Education: A (Mini) Meta-Analysis. *Journal of Engineering Education, 90*(2), 223-229.

Holvikivi, J. (2007). Logical reasoning ability in engineering students: a case study. *IEEE Transactions on Education*, *50*(4), 367-372.

Leydens, J. A., & Schneider, J. (2009). Innovations in composition programs that educate engineers: Drivers, opportunities, and challenges. *Journal of Engineering Education, (98)*3, 255-271.

OECD. (2014). PISA 2012 Results: What Students Know and Can Do (Volume I). Student Performance in Mathematics, Reading and Science (Vol. 1). BOOK, OECD Publishing. http://doi.org/10.1787/9789264201118-en

Robinson, M. A., Sparrow, P. R.,Clegg, C., and Birdi, K . (2005). Design engineering competencies: future requirements and predicted changes in the forthcoming decade. *Design Studies, 26* (2), 123–53.

Schmeckpeper, E., Kelley, M., and Beyerlein, S. (2014). Using the EPSA Rubric to Evaluate Student Work on Ethics Case Studies in a Professional Issues Course. *2014 Zone 1 Conference of the American Society for Engineering Education (ASEE Zone 1).* Bridgeport. 3-5 April 2014.

Yalvac, B., Smith, H.D., Troy, J.B., & Hirsch, P. (2007). Promoting advanced writing skills in an upperlevel engineering class. *Journal of Engineering Education, 96*(2), 117–128.

8. Acknowledgements

Thanks to CEDU at Universidad del Norte for its academic and financial support in this classroom research.

Copyright statement

Social & Critical Consciousness Across the Engineering Curriculum

Renata A. Revelo
University of Illinois at Chicago, Chicago, USA
revelo@uic.edu

Joel Alejandro Mejia
University of San Diego, San Diego, USA
jmejia@sandiego.edu

***Abstract**: Driven by scholarly work that promotes incorporating the study of social justice issues in the engineering curriculum, this work-in-progress paper focuses on teaching social and critical consciousness in engineering. The purpose of this study is to discuss and compare the implementation of social and critical consciousness in a first-year general education engineering course and a senior design engineering course.*

Introduction

Driven by scholarly work that promotes incorporating the study of social justice issues in the engineering curriculum (Riley, 2008), this work focuses on teaching social and critical consciousness in engineering as a way to nurture development of engineers who are meaningfully engaged with these issues (Cech, 2014). Through such focus, we aim to re-define what it means to be an engineer. The longer-term goal of this work is to re-define the ways that students understand the field of engineering, moving away from a dehumanized, technocratic understanding of the field.

As discussed in Riley (2008), teaching social justice courses in engineering asks students to think about their role as engineers and consider the impact of engineering on communities while considering a diversity of perspectives. Courses that incorporate lessons on social justice in engineering are not new. There have been courses developed to integrate lessons or learning modules of social justice in bioengineering (Catalano et al., 2010), control systems (Johnson, Leydens, Moskal, Fantasky, & Silva, 2015), and signals and systems (Huff, 2015). Leydens, Lucena, and Nieusma (2014) pose criteria for design for social justice wherein when applied to design courses, these criteria incite students to consider structural conditions that are present in the design process and lead to inequitable designs. Furthermore, a course aligned with design for social justice criteria would challenge students to consider the relationship between privilege, disadvantage and affected communities.

Institutional Contexts

There were two institutions where social and critical consciousness were incorporated into engineering courses. At institution A, a first-year, general studies engineering course was developed and offered for the first time, and at institution B, a senior-year electrical and computer engineering design course was revised from an existing capstone course.

Institution A is a public, minority serving institution. The institution is located in the Southwest and primarily awards baccalaureate and master's degrees. In the Fall of 2016, approximately 9,500 students were enrolled and 154 students were part of the engineering program. The first-year engineering course was a general studies course open to all majors. However, 90% of the 52 students enrolled in the course were housed in the engineering department. This course accounted for 1 credit hour and ran for 8 weeks during the Fall semester.

Institution B is a public, research-intensive, urban, minority serving institution in the Midwest. The senior-year course was offered in the electrical and computer engineering department in the Spring 2017 semester. During the 2016-2017 academic year, the enrolment in the department of electrical and computer engineering was ~560 students. There were 53 students registered in the course. Sixteen of those students were registered in the pilot section of the course, where the curriculum was revised to include social and critical consciousness lessons. This course is required for graduation and accounts for 2 credit hours.

Background

The first-year course was offered during the Fall 2016 semester. This course was part of a set of university-wide courses aimed at providing freshman students with the resources necessary to succeed in college. The goal for these courses is to engage students in shared inquiry and the practice of reasoning. Moreover, these courses are structured around writing, oral communication, and information literacy. The engineering and social justice course was designed in a way that provided students with an introduction to engineering and the social, cultural, economical, and environmental impacts of engineering. The idea behind this course was to show engineering from an interdisciplinary perspective in order to foster exploration and scholarly exchange among students with the instructor. The course became a perfect venue to expose students to issues involving engineering and social critical consciousness. The students were presented with topics related to engineering (in the form of case studies or engineering news in social media) every week and asked to gather, evaluate, analyze, and synthesize information relevant to the issues presented in class. Subsequently, students were asked to evaluate choices, actions, and consequences as related to ethical decision-making.

The senior-year engineering design course taught at institution B is part of the core curriculum in the department and meets ABET criteria. In this course, students learn about the engineering design process and are asked to work on a problem of their own choosing. In the spring of 2017, a new section was added to this course. In the new section, students worked with local community organizations on their senior design project. The instructor for the class worked with two students to recruit community organizations for this course. Community organizations who were in the vicinity of the university, had a social justice aligned mission, and were non-profit organizations were approached. Four community organizations were recruited and were asked to provide problem statements. The students in this section of senior design were presented with all of the problem statements and were matched to a community organization based on interest and expertise.

The new section of senior design was developed to meet three goals. First, students work with non-profit, social justice oriented community organizations on a problem statement presented by the organization. Second, social and critical consciousness lessons are implemented in the senior design curriculum. Third, aspects of serving learning are implemented throughout the design process.

While the first-year and senior-year courses were developed independent of one another, the authors noted parallels between their work and used that as an opportunity to share their process with the engineering education community. As a result, the authors focus only on one of the courses' goals, namely, to incorporate social and critical consciousness in the curriculum.

Theoretical Frameworks

Freire (2000) describes critical consciousness as the awareness of privilege, power, and oppression and the human relationships that surround these. Social consciousness allows for incorporation of social identities that are relevant in learning such as race, ethnicity, and gender. In this paper, these identities are specific to the social and cultural context in the United States where the courses were taught. By teaching with a Freirean understanding of oppressor and oppression, students become aware of inequities within oppressive systems and the systematic ways in which inequities persist and disproportionally affect minoritized social groups.

There are teaching frameworks such as culturally responsive pedagogy and culturally responsive teaching that incorporate social and critical consciousness in the curriculum. In culturally responsive pedagogy, there are three criteria that culturally relevant teaching should follow; namely, "an ability to develop students academically, a willingness to nurture and support cultural competence, and the development of a sociopolitical and critical consciousness" (Ladson-Billings, 1995, p. 483).

The intent of culturally responsive teaching is to connect students' everyday lives and classroom instruction. This model recognizes that students learn in different ways, and focuses on the students' learning strengths while fostering their linguistic, literate and cultural practices (Paris, 2012). According to Geneva Gay (2010), culturally responsive education uses "the cultural knowledge, prior experiences, frames of reference, and performance styles of ethnically diverse students to make learning encounters more relevant and effective for them" (p. 31). One of the goals of culturally responsive education is to replace individualistic perspectives with cooperative and active roles among students and between students and instructor. Moreover, culturally responsive teaching recognizes the differences in values, beliefs, ways of being, doing, knowing, and the importance of racial and cultural diversity in learning. Thus, culturally responsive teaching has the potential to validate students' lived experiences because it acknowledges culturally and historically accumulated knowledge, skills, and practices while creating bridges to the classroom experience (Gay, 2010). This framework is also comprehensive because academic achievement is approached via helping students maintain their identity, connect to their communities, develop a sense of shared responsibility and develop critical social consciousness. Through culturally responsive teaching, students see themselves and their communities in a curriculum that is empowering, transformative, and liberating.

As indicated by Freire (2000), we must ensure that our classroom instruction supports our students' own quests. "Culturally sensitive remedies to educational problems of oppressed minority students that ignore the political aspects of schooling are doomed to failure" (Villegas, 1988, p. 263). Therefore, culturally responsive education must address the different criteria mentioned above to be effective. Freire's conceptualization of critical consciousness and culturally responsive pedagogy were both used to guide the development of the relevant lessons for the engineering courses discussed in this study. While the focus of the courses is on engineering design, the main difference between these courses and other similar design courses is the incorporation of lessons that address social and critical consciousness.

Process of Incorporating Social and Critical Consciousness in the Curriculum

In this section, we discuss our independent processes for incorporating social and critical consciousness into the curriculum.

First-year General Studies Engineering Course

The curriculum for the freshman course was delivered over a period of 8 weeks. Students met twice every week for one hour. This course addressed some of the social and ethical implications of engineering and technology. Students were encouraged to think about how different engineering disasters have shaped the national priorities of security, poverty, sustainability, and energy, and how engineering has defined the assumptions, structures, values, and practices within the profession. The discussions were primarily guided by Riley's (2008) *Engineering and Social Justice*.

Three main objectives were addressed in this course: (1) to demonstrate critical thinking skills, (2) to demonstrate communication skills, and (3) to demonstrate personal responsibility. The specific learning outcomes related to the main objectives of the course were: (a) to gather, analyze, evaluate, and synthesize information relevant to a question or issue; (b) to develop, interpret, and express ideas through effective written communication; (c) to develop, interpret, and express ideas through effective oral communication; and (d) to evaluate choices, actions, and consequences as related to ethical decision-making.

The goals of the course were also two-fold: to help students learn about resources available for their academic success and help students develop a critical consciousness. For instance, students learned about using library resources and other electronic databases to gather information. Activities included choosing and locating potentially useful information sources, identifying keywords and terminology, developing searches, retrieving results, refining searches, and writing a bibliography. Other assignments included two written summaries and a reflective essay related to a specific engineering disaster affecting their community. For their main project, students selected three different sources of information and wrote a paper on an engineering disaster. The topic for that year's engineering disaster was the explosion of the West Texas Fertilizer Plant.

Every week, students were required to search for articles that involved "engineering in the news and social media." Students presented their findings to the class in 60 seconds (60-second engineering stories). Within those 60 seconds, students were asked to summarize their engineering study and use it as a prompt to discuss how engineering is portrayed in the media and the ethical and social issues involved within that engineering context. Each week different topics were covered: (1) how engineering influences your life; (2) How engineering influences society, or how society influences engineering; (3) Engineering and the military; (4) Engineering and corporate America; (5) Engineering: for who and by whom? (6) Engineering and injustice; (7) Pro-bono Engineering (engineers without borders or service learning engineering); and (8) Engineering for change (how is engineering used as a vehicle to accomplish social justice). The objective was to expose students to different sources of information, the socio-political construction of engineering work, and to describe the contextual origins of engineering. This type of equity-oriented activities and instruction intentionally engage students and allowed them to use their everyday practices (e.g., media literacy and technology enhanced learning) in ways that encourage group and community transformation (Gutierrez, 2008).

Senior Design Course

The concept of praxis applied to engineering design (as discussed in Riley, 2008) guided the revision of the senior design course wherein students were asked to work with the community on a problem that emerged from the community.

Four community organizations in the Chicago area were approached for this course; namely, Enlace Chicago, Lakeview Food Pantry, Little Village Environmental Justice Organization, and People for Community Recovery. These community organizations were approached based on two criteria: that they have a social justice mission, and that they be located in the communities around the university. Upon agreement to work together, the community organizations were asked to provide a problem statement that they thought could have a potential engineering solution. Via phone conversations, the instructor discussed with the community partners about the types of problems that could be addressed via this project by providing the following categories: everyday tasks, facilities, execution of programs, management, and projects. Examples from previous senior design projects, not completed with community organizations, were provided to guide the community partners in developing their problem statement. A draft of the problem statement was provided to the instructor and after phone conversation with the community partner, the problem statement was finalized and presented to the students during the first week of classes. The process of finalizing the problem statement with the community partner included providing specifics where required and making the problems statements open-ended so that the students are not constrained to a single type of solution. See Figure 1 for an example problem statement.

> *Little Village Environmental Justice Organization (LVEJO)*
>
> LVEJO's community garden is a small and intimate outdoor space that is heavily utilized throughout the gardening season for many workshops and acts as a safe space for community members.
>
> Unfortunately, due to the lack of lighting, the amount of time that can be spent there is limited, thus bringing an abrupt end to events.

The curriculum for the senior design course is divided into about twelve weeks of content delivery and three weeks of student presentations. In the first twelve weeks of the semester, lectures are structured around the engineering design process with a focus on electrical and computer engineering designs (Ford & Coulston, 2007). Bi-weekly assignments emphasize problem scoping, user needs, technical specifications, and design alternatives. The end of semester presentation provides the students with an opportunity to discuss their own design process and receive feedback from their peers and instructor.

In-class discussions around social and critical consciousness were primarily concentrated in the first four weeks of classes, during the problem scoping part of the design process. Similar to the curriculum for the first-year course, these discussions were primarily guided by readings from Riley's (2008) *Engineering and Social Justice* book. In-class activities included summarizing readings and providing applicable examples via group discussions. Reporting out was done in a big group set up where individual groups reported on their discussions to the whole class. Outside of class, students were asked to reflect upon readings from this book in two ways. First, they were asked to reflect upon their own engineering identity development with weekly reflection questions such as "What is my role as an engineer in working with the community partners?" and "What values are salient to me as an engineer in this project?" In response to the reflection, where appropriate, the instructor provided follow up questions or commentary to encourage further thinking on the weekly reflection question.

Another aspect of reflection was on the field and culture of engineering. Students discussed how they view the culture of engineering and how they envision it being different. These discussions were in part facilitated by book readings and also facilitated by news stories that discuss the negative impact of engineering design on humans. While part of the discussion around the culture of engineering was mediated via these news stories, the goal was for students to iterate back to reflecting upon their own engineering education. This aspect of incorporating social and critical consciousness is aligned with Culturally Relevant Pedagogy criteria (Ladson-Billings, 1995) for the development of a socio-political consciousness applied to teaching and education wherein students are welcomed to challenge and critique the educational system to which they are part.

Discussion

While most of the overarching goals for the first-year and senior-year courses were different, these courses shared one common goal to integrate social and critical consciousness into the curriculum.

To achieve this goal, the curricula were developed so that students reflect deeply about engineering and their role within it. The lessons on inciting social and critical consciousness challenged the students to analyse their roles as engineers. The students had to grapple with questions about how they view their current engineering identity development and how they envision their professional engineering life. What incited them to go into engineering? What motivated them to continue in the field? What kind of jobs would they look for when they graduate? These types of reflective questions lead students to look within themselves for answers about engineering for them. The students also had to reflect on the field, separate from themselves, with consideration on the history of the field and current foci of engineering. How was engineering "born"? How is engineering done currently? What kind of impacts has engineering had on U.S. society? These types of reflective questions asked students to evaluate the engineering field as an object. To bridge these reflections, students were asked to analyze their thinking around how they, as engineers, "fit" within their perception of engineering itself. The purpose of this exercise was to have students identify how their engineering values might match or not the subfield within engineering that they envision occupying after graduation. The students faced questions about the meaning of a mismatch in values, and whether the kind of values they hold as engineers are (or not) reflected in the within the field itself. Are there sectors of engineering where the students' values were more or less reflected? This deep reflective process engaged thinking about social, political, and cultural dimensions of engineering and identity.

Aligned with culturally responsive pedagogy, these courses challenged students to think about their engineering education through a critical lens. Students were not only asked to analyze the impact of engineering disasters or engineering design, but also to understand the impact of their education on learning. What structural and inequitable learning conditions are present in engineering education? What attention do these pay to the ways that the students learn engineering? These questions encourage students to activate their agency and consider policies and practices that may impact the way they are taught engineering. Ladson-Billings (2014) argues that this component, of inciting social and critical consciousness, of culturally responsive pedagogy is often missing from courses that claim to use this pedagogy. In engineering and other traditionally positivistic fields, the students' role in shaping the way they learn may have to be even more accentuated.

Future and On-Going Work

In the near term, we plan on completing the assessment part of this work and report on the student outcomes around social and critical consciousness due to the students experiences in these courses. Such work will address the research question "What gains in social and critical consciousness do students have after a one semester exposure of these topics?"

Future work will focus on further developing modular ways to incorporate lessons of social and critical consciousness in engineering courses throughout the curriculum.

References

Catalano, G., Baillie, C., Riley, D., Nieusma, D., Byrne, C., Bailey, M., & Haralampides, K. (2010). Integrating social justice ideas into a numerical methods course in bioengineering. In *American Society for Engineering Education*. American Society for Engineering Education.

Cech, E. A. (2014). Culture of disengagement in engineering education? *Science, Technology & Human Values*, *39*(1), 42–72.

Ford, R., & Coulston, C. (20*07). Design for Electrical and Computer Engineers*. McGraw-Hill, Inc.

Freire, P. (2000). Pedagogy of the oppressed. New York: Bloomsbury Publishing.

Gay, G. (2010). *Culturally responsive teaching: Theory, research, and practice*. Teachers College Press.

Gutiérrez, K. D. (2008). Developing a sociocritical literacy in the third space. *Reading research quarterly*, *43*(2), 148-164.

Huff, J. (2015). Humanizing Signals and Systems: A Reflective Account. In *Proceedings of the ASEE Annual Conference*.

Johnson, K., Leydens, J., Moskal, B., Fantasky, J., & Silva, D. (2015). Social Justice in Control Systems Engineering. *NREL (National Renewable Energy Laboratory (NREL), NREL/CP-5000-65111.*

Ladson-Billings, G. (1995). Toward a theory of culturally relevant pedagogy. *American Educational Research Journal*, *32*(3), 465–491.

Leydens, J. A., Lucena, J. C., & Nieusma, D. (2014). What is design for social justice. In *ASEE Annual Conference and Exposition.*

Paris, D. (2012). Culturally sustaining pedagogy: A needed change in stance, terminology, and practice. *Educational Researcher*, *41*(3), 93–97.

Riley, D. (2008). Engineering and social justice. *Synthesis Lectures on Engineers, Technology, and Society*, *3*(1), 1–152.

Villegas, A. M. (1988). School failure and cultural mismatch: Another view. *The Urban Review, 20*(4), 253-265.

Acknowledgements

Dr. Revelo would like to thank the University of Illinois at Chicago College of Engineering for supporting the senior design pilot section and the students, Ramchandra Sri Harsha Achanta and Karen Canales Salas, who worked on the development of the course.

Copyright statement

The innovation challenge in Probability and Statistics teaching, using ICTs and flipped classroom methodologies

Mario Castillo Hernández
Universidad de los Andes, Bogotá, Colombia
mcastill@uniandes.edu.co

Astrid Johanna Bernal Rueda
Universidad de los Andes, Bogotá, Colombia
aj.bernal218@uniandes.edu.co

Abstract: *In this paper we summarize the experience of the use of virtual and interactive tools, and the method of flipped classroom in the design of an integral methodology oriented to the improvement of the teaching process in engineering in the topic of probability and statistics. The principal expected results are a better level of students understanding, compared with other methodologies, and a change in the dynamic of class with a higher level of participation and autonomy of the students during the learning process.*

Until now we have observed that the proposed methodology is more aligned with the use of the technology of information currently available and the dynamics in the classroom more appropriate. The level of participation, communication and activity in class with students is greater than with the traditional methodology. Statistical analysis conducted about this experience show some evidence of improvement in academic performance results.

Context

The University of Los Andes has declared its intention to maintain the Academic Leadership through its Institutional Development Program. Under this statement and having as their goal to "count with academic programs based on the development of competencies evaluated under an integral and systematic form, with international standards. Under the principles of quality, innovation, flexibility, interdisciplinarity, internationalization and sustainability" (Uniandes, 2015), the School of Engineering has developed the intention to support processes that allow to explore the use of active pedagogies supported by information technologies. Therefore, the Engineering Deanship has supported the redesign of the Probability and Statistics I course (P&E I), as well as the process of redesigning the Rigid Bodies Dynamics and Analysis of Structural Systems courses.

The redesign of the P&E I course is made under the project of REDINGE1 — Redesign of Engineering courses 1, an initiative led by the School of Engineering, expecting it to be a pilot in order to learn i) about a paradigm shift, to be focused on fostering learning instead of just teaching (Bates, 2015); ii) about changes in the means of learning (Forté & Wentland, 1998), to complement the lectures, books, articles and labs that are normally employed, through virtual learning environments with digital learning objects that flexibilize the knowledge attaining process; and iii) about the changes in the role of the teacher, from being the center of the process to be instead a sideline facilitator (Collison, Elbaum, Haavind, & Tinker, 2002). The teachings and learnings from these pilots are presented in the article (in process) REES 2017 accepted paper 184, titled "Rethinking engineering education practices with active learning and digital technologies: Pilot experiences at University of Los Andes, Bogotá" (Galvis et al, 2017).

Regarding the context and the importance of the P&E I course, it has very relevant structural characteristics, which made its intervention regarded as one of high impact for the Engineering School.

P&E I is a mandatory course inside the curriculum of all the careers from the School of Engineering. An average of 550 students takes the course every semester and they can choose to take it in their fourth, fifth or sixth semester, depending on the engineering program in which they are registered. Every semester offers seven to ten lecture sections, each of them with 50 to 80 students and with 19 supplementary or reinforcement sections, each with an average of 35 students.

The program of the course is covered in two weekly sessions of one hour and twenty minutes each that are in charge of a professor, and a weekly supplementary session of equal intensity given by a graduate assistant. The sessions in charge of the teacher are dedicated to the detailed presentation of the main topics of the course and the complementary sessions to the realization of practical exercises. This course has an intensity of 3 credits, which implies an estimated extra-class work of 6 hours per week.

The course has a practical emphasis in the context of engineering; for this reason it has been structured conceptually in matrix form, so that in one dimension there are the thematic contents of the course and in the other a set of cross-cutting activities among which are data manipulation, simulation and experimentation, software use, decision making under risk and case study analysis.

Taking into account the importance of the course in the foundation of basic engineering concepts in the subject of probability and statistics, it is taught in a unified manner in terms of content and academic objectives, seeking to develop the same competencies and skills in every student who takes the course. The evaluation activities are also unified in order to ensure that the same concepts are evaluated independently of the lecture section in which the students are enrolled.

The course is coordinated by a full professor from the Industrial Engineering Department and has a teaching team of about five faculty members and six graduate assistants.

Research questions

One of the main problems of teaching basic Probability and Statistics in engineering is the students' lack of motivation regarding these subjects, and their misunderstanding of the great opportunities of applying probability and statistics in problem solving in the field of engineering. The active and constant participation of students during the class sessions, as well as their permanent work in the review of the theoretical concepts before and after the class, are essential elements for the successful development of the course.

The innovation and pedagogical redesign project of the P&E I course was aimed: i) to increase the student motivation through their active participation during their learning process (before, during and after the lectures); ii) integrate the use of technology as a facilitator of the learning and feedback processes; iii) to improve the understanding of the application of concepts in the solution of real life engineering problems, through the solution of exercises and problems in engineering context; iv) to improve the learning and understanding level of students evidenced through academic achievement in the final exam of the course.

Based on the above, we must ask ourselves if the introduction of structured methodological improvements in the teaching process aimed to promote the active participation and autonomous work of the students, involving the appropriate use of ITCs, leads to an improvement in the level of understanding and performance of the students in the course.

Theoretical framework

Taking into account the established objectives and the characteristics of the P&E I course, we identified the theoretical guidelines that would support the structural innovation initiatives and the process of redesigning pedagogical activities of the course.

Studies about engineering education (Crawley, Malmqvist, Östlund, Brodeur, & Edström, 2014) state that "there is a seemingly irreconcilable tension between two positions in engineering education. On one hand, there is the need to convey ever-increasing body of technical knowledge that graduating students must master. On the other hand, there is growing acknowledgement that engineers must possess a wide array of personal and interpersonal skills, as well as product, process, and system building knowledge and skills required to function on real teams to produce real products and systems".

This situation is present in basic engineering courses, such as P&E I, in which the developed concepts will be useful throughout all the career and professional life, but do not have specific anchoring in situations of a specific discipline.

According to Badia (2006), Information and Communication Technologies (ICT) allow the creation of new conditions in educational contexts for seeking, obtaining, accessing, organizing, transmitting and using the information. According to this author, these new possibilities can change learning processes or the mental functioning of students. The ICT can offer different types of educational help. For example, they may facilitate the temporal planning of learning activities; provide access to the learning contents; provide resources that support the construction of knowledge; support the comprehension of the activity organization and give meaning to its objectives; provide assistance to the joint organization of teachers and students participation; and also promote a regular monitoring of the learning progress (Badia, 2006).

For Ossa (2011) there are three different phases: planning, control-monitoring and evaluating. On the first phase the purpose is that the student finds meaning in the learning process and keeps in mind where his efforts are orientated to; that is, to establish goals, objectives and action plans. On the second phase, control-monitoring, the necessary tasks and activities are performed to achieve the learning goals, so that strengths and weaknesses can be identified. On the third phase, evaluating, a review is made of the achieved goals and the process, as well as a comparison of self-evaluation with external evaluation. Part of this process is to make constructive contributions to the learning process (Ossa, 2011) which includes feedback to redefine the self-regulated learning cycle.

The flipped classroom teaching allows for increased classroom time for students to engage in active learning that fosters analysis, and the solution of problems in context of engineering. Addittionally, the rationale behind the flipped classroom methodology is to increase student engagement with content, increase and improve faculty contact time with students, and enhance learning (Lage, 2000).

Methodology

The redesign process of the P&E I course considered two axes. The first, a curricular axe, was made with the support of ConectaTe, in which the content and the subjects of the program where reorganized around four learning goals: a) to recognize and understand events and phenomena of the probabilistic nature; b) to acquire the abilities to construct probabilistic models to quantify risk; c) to manage and analyze data to identify patterns and make inferences; and d) to build basic statistical models.

The second axis had a methodological nature. In it the material developed allowed: 1) to guide the student into the goals and development of each session through instructions or learning guides; 2) to have activities feedback opportunely to prepare the class and evaluate the learning level of each session; 3) to present and debate the applicability of the course concepts in different real engineering contexts before, during and after each session.

The course redesigned is organized in 15 weekly sessions of three hours. Each session is designed with an active pedagogy approach, using flipped classroom strategy, supported by information technologies such as Blackboard (SicuaPlus) and Learning Catalytics, for the immediate feedback process.

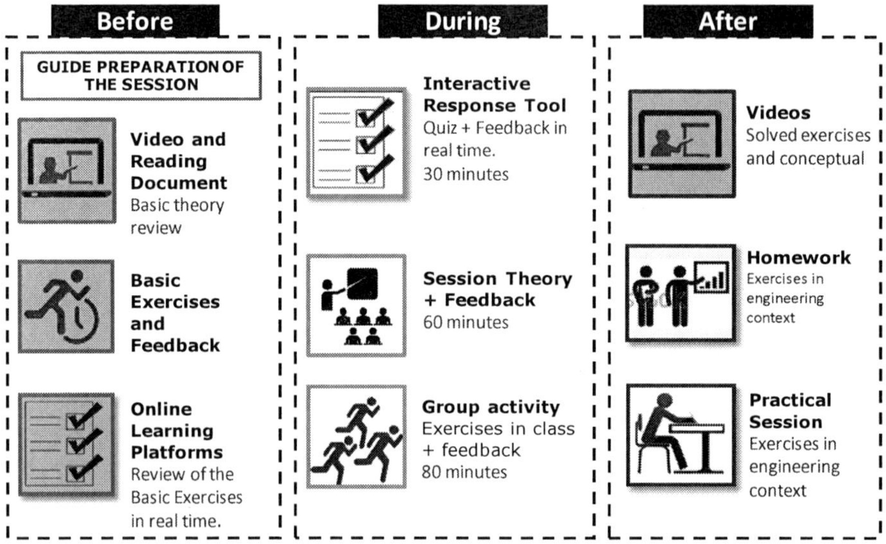

Figure 1. Structure of the new P&E I course session

Figure 1 presents an outline of the structure of each class session from the new P&E I course, where the lecture presentation methodology used by the professor is changed to a dynamic class that involves the active participation of students during the learning process, before, during and after each session.

Taking into account this structure, is possible to observe four moments in which the student has to face different challenges and objectives; 1) before the class, preparing the basic level available material; 2) during class, to be evaluated about the virtual material and to have the professors' feedback; 3) during class, reviewing the theory and carrying out basic level exercises in engineering contexts; and 4) after the class, attending to the practical session and solving the homework.

Clearly established roles for the students and the professor, according to the three moments in each session, focused on the detailed development of the activities planned for the session and on reaching the learning goals.

This methodology propounds: a) that the student has a first approach to the concepts and faces situations that demand the application of knowledge before the classroom session; b) that the class sessions be complemented with videos, lectures and previous exercises, promoting the debate among peers about possible solutions to the problems presented, as well as debates in class after the initial evaluation about the findings made from the achievement evidences, collected with immediate interactive response tool; and (c) that students carry out after-session activities in which the process of contextualization in different areas of engineering is strengthened.

To evaluate the achievement of the defined objectives with the redesign process of the P&E I course, different instruments have been used, among them: 1) a statistical evaluation of the students' performance in their level of ability to answer the final exam of the P&E I course; 2) evaluation of soft skills as an effect of pedagogical interventions, related to variables such as academic self-efficiency, methodological flexibility, autonomy enhancement, and support of the ICT to the teaching-learning process; and the 3) semi-quantitative analyses of the students perceptions about the resources quality, the viability of the activities, and the contribution of the methodology to the interaction in the lecture sessions, as well as to the learning.

This paper is centered on the first instrument described, since it is the one that presents the greatest level of progress on the analyses made so far.

Students performance evaluation

This study considered 1363 undergraduate students of the University of the Andes School of Engineering, enrolled in 30 sections of the Probability and Statistics I course (P&E I) during three semesters: 2015-II, 2016-I and 2016-II. Out of the 1363 students, there are 1456 registrations on the P&E I course (a student repeating the course would have two or more registrations). All students had already approved the Linear Algebra I and Integral Calculus courses.

The final exam of the course which is a multiple-choice exam with between 20 and 25 questions, and in which the principal subjects studied are evaluated, was used as an instrument for a good approximation of knowledge retention measurement. This instrument was adjusted since the first semester of 2015 in order to use anchor questions during the following semesters and validate the reliability and internal consistency level of the test through the Cronbach's alpha, ranging from 0 to 1 (Brown, 2010). The exams used has a reliability (Cronbach's alpha) of the 0.71. In order to evaluate the concepts learned during the course during the second semester of 2015 and throughout 2016, the test was applied to students from both types of sections: intervened (redesigned course) and not intervened.

According to the theory of the response to the item proposed by Muñiz (2010), it is calculated a score for each student that measures the ability to respond to the exam. this score corresponds to the depend variable (Y) of the model presented in Table 1.

For the student's qualification process a score or skill level of probability and statistical knowledge is assigned to all students who took the final exam (both, from the intervened and not intervened sections). In order to facilitate the interpretation of the results, the students scored were standardized, which means that they are expressed in standard deviation units.

The regression model presented in Table 1 is: $Y = \beta_0 + \beta_1 X_1 + \beta_2 X_2 + \beta_3 X_3 + \beta_4 X_4 + \varepsilon$
X_1: The student belongs to a section of intervention class
X_2: The student belongs to the section of the semester 2016-I
X_3: The student belongs to the section of the semester 2016-II
X_4: The student is repeating the course.

Variable	Model		
	β	Standard Error	P-value
Intercept	0,007	0,049	**0,884**
Intervened	0,216	0,075	**0,004**
Sem. 2016-I	0,145	0,065	**0,026**
Sem. 2016-II	0,140	0,064	**0,029**
Repeating	-0,461	0,077	**0,000**
Obs.		1456	

Table 1. Model of intervention and learning 1

The Regression Model state that by controlling each semester and the repeating students, the intervention still has a significant effect on the test results ($\beta = 0{,}216$; $p<0{,}05$).

It must be noted that the differences between the results of the exam for the three semesters analyzed cannot be attributed to the differences between the exams, but to actual differences between the results, as the equalizing process carried out during the development and validation of the exam guarantees the comparability between the applications.

Soft skills evaluation

This study used a quasi-experimental design to compare the control groups with the intervention groups, using pre-test and post-test measures. During the first academic weeks was taken the first measure (pre-test) of the self-regulated learning variables and academic self-efficiency of the students. In addition, three variables related to pedagogical intervention were explored: enhancement of autonomy, methodological flexibility and support of the ICTs to the teaching-learning process. The results of this study do not yet present conclusive evidence regarding the development of the mentioned skills.

Semi-quantitative evaluation of the active methodology/flipped classroom contribution

During the second semester of 2015 and the first semester of 2016, the students' perceptions regarding the contribution of the active inverted class or flipped classroom methodology to their learning process was analyzed through surveys and focus groups.

The results show that around 90% of the students consider that the activities such as videos, basic exercises and lectures to prepare the class sessions were important for their learning process, which was verified during the focus groups. In terms of activity and interaction between teacher-student during classroom sessions, it must be noted that: 1) 62% of students fully agreed with the statement that this methodology encourages students participation in class; 2) 82% agreed that this methodology eases the interaction among students during the lecture sessions; 3) 79% agreed that this methodology eases interaction between teacher and students during the lecture sessions making them more profitable; 4) 89% of the students agreed that this methodology allows the teacher to solve all doubts students might have during a lecture (Valencia, 2017).

Findings and/or conclusions

The implementation of the P&E I course redesign carried out for three consecutive semesters (since 2015 second semester) has evidenced:

- Sustainability of the course and its redesign.
- Potential of the produced material for supporting other courses related to probability and statistics. The process of redesign allowed to undertake the review of the available material for the teaching of Probability and Statistics I, as well as to produce educational and technological materials and resources that can be used in multiple contexts. Among other materials, 18 videos covering the principal and more relevant subjects of the course were produced by the professors with the support the University of Los Andes.
- Considering the statistical analyses for academic performance evaluation undertaken, it can be stated that the intervention has an effect of approximately a quarter of standard deviation (0,22) in the final exam score. Regarding the other variables considered in the models, it was concluded that there is also an effect for the semester, in favor of the two semesters of 2016, in comparison with 2015-II.

- The professors teaching the course have noticed that the interaction with the students increases considerably with respect to class without intervention.
- The experience of the professors regarding the course is improved due to the enhanced communication and interaction with the students, which can be confirmed by the results of the semi-quantitative study on the students' perception and assessment of their learning process.
- The innovation processes discussed in this paper, encouraged other professors of the engineering school to undertake similar innovation processes in the methodology of their courses; additionally, to see the complete experience of this intervention is going to facilitate the development of their innovation projects.

Recommendations and future research

This experience allows to recommend to the department and to the School of Engineering that these type of teaching developments continue to be supported and sponsored, as well as that they continue to accompany the design, the formative evaluation and the effects measurement processes, the last two being relevant evidences for supporting pedagogical innovation initiatives. Additionally, we consider that not undertaking these types of methodological innovations in the engineering courses could cause, in the long run, the engineering teaching to be left behind considerably.

Therefore, we consider important to continue undertaking formal studies that allow: 1) to measure the impact of students' motivation; 2) to analyze the development of soft skills such as autonomy, by effect of the implemented methodology; and 3) to carry out analyses that permit to establish if there is a bigger appropriation of the knowledges of the P&E I course, based on the performance of the students in latter courses that require these concepts.

References

Badia, A. (2006). Ayuda al aprendizaje con tecnología en la educación superior. *Revista de Universidad y Sociedad del Conocimiento.*

Bates, A. W. (2015). *Teaching in a digital age. Guidelines for designing teaching and learning for a digital age.* Tony Bates Associates Ltd. Recuperado el 03 de 11 de 2016, de https://opentextbc.ca/teachinginadigitalage/

Brown, F. (2010). Principios de la Medición en Psicología y Educación. México D.F.: Editorial El Manual Moderno.

Collison, G., Elbaum, B., Haavind, S., & Tinker, R. F. (2002). *Facilitating online learning: effective strategies for online facilitators.* Madison, WI: Atwood Publishing.

Crawley, E. F., Malmqvist, J., Östlund, S., Brodeur, D. R., & Edström, K. (2014). *Rethinking Engineering Education: The CDIO Approach.* Springer Science & Business Media.

Forté, E., & Wentland, M. (1998). The ARIADNE project: knowledge pools for computer-based and telematics supported classical, open and distance education. *AAUC Ariadne Academic Users Group Conference* (págs. 1-23). Lucerna, Swtizerland: AAUC.

Galvis et al. (2017). Rethinking engineering education practices with active learning and digital technologies: Pilot experiences at University of Los Andes, Bogotá. *REES.*

Lage M, P. G. (2000). Inverting the classroom: a gateway to creating an inclusive learning environment. Journal of Economic Educational, 30-43.

Muñiz, J. (2010). Las teorías de los test:teoría clásica y teoría de respuesta a los ítems. En *Papeles del psicólogo* (págs. 57-66).

Ossa, M. (2011). Aprender a aprender: el desarrollo de competencias de aprendizaje en la educación universitaria. En J. Montoya, A. Truscott, & A. Mejía. Bogotá: Universidad de Los Andes.

Uniandes. (9 de 12 de 2015). *PDI 2016-2020 Versión final.* Obtenido de Universidad de los Andes, Colombia: https://goo.gl/biCHcW

Valencia, D. (2017). *Borrador de informe sobre opiniones de estudiantes y docentes acerca de cursos de REDINGE1.* Bogotá: ConectaTE - documento de trabajo.

Acknowledgement

This work is the product of the collaboration between the School of Engineering of the University of Los Andes and the Center of Innovation in Technology and Education – ConectaTE- of the School of Education. The Dean's Office, headed by Eduardo Behrenz, led the process in 2015 and 2016, it was given continuity during the deanship of Carlos Francisco Rodríguez, and it is being enhanced in the deanship of Alfonso Reyes. Andrés Medaglia, director of the Industrial Engineering department for his support during the process. Luz Adriana Osorio, director of ConectaTE, carried out the negotiation of the resources and the follow-up of the process. Alvaro Galvis, of ConectaTE, was the academic director of the process as well as its techno-pedagogical conductor. On the part of ConectaTE, the participants were Monica Patiño, coordinator and technological advisor; Alexa Tatiana Jiménez and Angélica Avalos, pedagogical advisors; Diego Valencia, formative evaluation advisor; and the video and digital surroundings production was in charge of the production and support team led by Mariela Rivero. And the Evaluation Center of the School of Education made the summary evaluation with contributions of Samir Cure, Elvia Sánchez y Julián Mariño. We deeply thank them all.

STE(A)M: ART, DESIGN AND ARCHITECTURE IN THE FORMATION OF ENGINEERS AT UTFSM

Brayan Díaz
Universidad Técnica Federico Santa María, Valparaíso, Chile
brayan.diaz@usm.cl

Leonhard Bernold
Universidad Técnica Federico Santa María, Valparaíso, Chile
Leonhard.bernol@usm.cl

Luis Felipe Gonzalez
Universidad Técnica Federico Santa María, Valparaíso, Chile
luisfelipe.gonzalez@usm.cl

Arie Aizman
Universidad Técnica Federico Santa María, Valparaíso, Chile
arie.aizman@usm.cl

Abstract: *The universal means of communication for engineers across the world are pencil sketches drawn on "napkins". However, with few exceptions, the power of sketching to model reality is not used in engineering education. This paper presents the results of a pilot freshmen course "Introduction to Engineering" with modules that focused on developing this creative skill. Students had to practice these abilities to demonstrate how mathematical models support the engineering problem solving process. The course was built applying the principles of Problem-Based-Learning (PBL) with links to Math and Physics. A variety of quantitative and qualitative assessment tools were used. The paper presents the experiences with using mapping and sketching to enhance understanding. It will be shown that a large percentage of students value the artistic skills as tools to solve different problems that that they may face in the future.*

Introduction and Background

The modern skills and knowledge needed by professionals in Science, Technology,, Engineering and Mathematics (STEM) disciplines today, provide new challenges for the education of Chilean engineers and architects. The employers have recognized that the strength of graduates of the Technical University Federico Santa Maria is their ability to solve problems. However, the same employers point out the increasingly complex and multidisciplinary nature of the problems that modern engineering must solve. These demands require versatility, autonomous learning ability and teamwork (Hurtado, Reyes, Saavedra, Zúñiga & Madrid, 2012). The content integration proposal in STEM for the early development of professional competencies of the engineer and architect at UTFSM, financed by the Program Improvement Plan of MECESUP (FSM1408-PM), seeks to address these challenges. The project makes a contribution in the classroom, through a strategy that generates curricular initiatives with multidisciplinary components of content integration in STEM throughout the different cycles of student training. The aim is to improve the learning experience of students and to promote the early development of professional competences consistent with current requirements. The research presented here focuses on studying the impact of design and conceptualization skills as tools when solving different problems that may be faced by the freshmen students in the future.

The Technical University Federico Santa María in Chile offers undecided engineering students the possibility of taking the so called "Basic Plan" for the first year. The program allows these students to become familiar with the different specialties taught at UTFSM and will earn credits towards the specialty that is selected in the second year. The first two semesters of the program includes Physics, Mathematics, Chemistry and Programming. In addition, it includes in the first semester its own version of the Introduction to Engineering. This course provides an insertion to the training and experience of the UTFSM engineering program that seeks to stimulate students' interest and motivation to become a professional engineer while developing basic skills to become a successful engineering student. Within this course, STEM integrator modules were designed and implemented parallel to the subjects of the basic engineering plan. Particular effort was directed towards developing the ability to perform collaborative work (Bernold et al., 2007), reflective and creative thinking skills, effective reading and time management. These modules were implemented in a pilot program with 30 students which represents 20% of all students entering to study this basic engineering plan.

By definition, to design is to plan, and in the arts to design is to fashion artistically or skillfully within a medium (Eisner, 2002). In this context, it has been proposed that art is able to strengthen STEM skills as well as increase the arsenal of skills and visions that can have an engineer defining what is known as a STEAM education (Science, Technology, Art and Mathematics) the considers the arts as critical addition to STEM learning (Piro 2010). However, Eisner (2002), Hetland (et al., 2007) and others reject the justification for study in the arts alone as a way to improve student performance in other disciplines. "The two fields can be of assistance to one another. Art education need not be teleological in order to be of value Creativity, aesthetic sensibilities and appreciation, higher spatial reasoning skills, sensory awareness, and many other benefits of art have cultural value that are not easily measured" (Bruce, 2010). It is in this context that in this paper defines STEAM as STE+Art+M.

Putting the A into STEM

Architecture and engineering share design as a core competency, even if they understand it differently. For example, Gross and Do (2009) argue that architectural designers are traditionally taught to embrace ambiguity in contrast to engineers who are taught to stay within their field of expertise and reduce or eliminate ambiguity. According to Cross (1996) and Lawson (2004), ambiguity in communication between designers proves beneficial because it does provide non-prescriptive approaches to a design problem. Ambiguity, in this case, supports creative thinking. It also encourages the iterative construction of alternative, not necessarily causally or logically related, models for the same problem. De Bono (1967) describes such processes as lateral thinking, as opposed to what he calls vertical thinking (i.e. logical thinking) and deems it critical to innovation. A similar mode of thinking is described by Cross (2007) as appositional thinking, which in combination with what he describes as abductive thinking helps designers to learn about a problem through a series of attempts to create a solution to that problem. He describes such solutions as conjectural solutions. Architects' solution-focused problem-solving strategy became clear to Lawson (1979) when he compared final year architecture and post-graduate science students. Lateral, appositional, abductive and solution-focused thinking are all closely related to creative thinking. According to Villa and Poblete (2007), creative thinking is a key instrumental-cognitive skill which is characterized by not following a logical process like most modes of thinking. Creative thinking is the mental behavior that drives processes of search and discovery of new and unusual solutions that make sense in different areas of life. Creative thinking needs a flexible mindset that allows perceiving things from multiple perspectives. Analogical thinking, on the other hand, also plays a key role in the creative process. Cross (2007), citing Gordon (1961), argues that analogical thinking is the basis for creative design. Villa and Poblete (2007) define analogical thinking as the mental behavior that manages to establish similarities between different things. Analogical thinking allows associating and transforming ideas, perceptions and concepts to find different ways of structuring a problem. Lawson (2006) sustains that "it is one of the most flexible and powerful tools for conducting the conversation of negotiation between what is desired and what can be realized." Creative and analogical thinking skills are integral parts of architectural thinking. To quote Gross and Do (2009): "It is easier for design students to acquire the technical skills (programming, electronics) they need to carry out projects with a computational component than for engineering and computer science students to acquire the mindset needed to work in ill-defined situations."

The fact that architects and engineers also learn to design in different ways offers an extraordinary opportunity not only to exchange learning methods, but also to develop different modes of thinking in engineering students. Traditional architectural education encourages students to deal with ambiguity and to communicate mostly through nonverbal language. It is worth noting that architectural communication not only involves conversations between people but also between the architect and his or her own sketches, as can be read in the work of Schön (1983), Goldschmidt (2003), or Lawson (2004). Creative and analogical thinking skills are integral parts of architectural thinking. The purpose of introducing some of the learning methods used to develop architectural thinking is not simply to enlarge the engineering students' toolbox, but to develop in them the ability to connect data, vary the approach and combine methods to solve real world problems, often with incomplete information and haste. In other words, we try to train engineers prepared for change, capable of adapting fast to circumstances, and to alternate their ways of thinking in the face of complex, unstructured problems that require planning to be solved and that may have not one but many solutions.

Experiential learning depends on active students

The problem based learning (PBL) approach to teach holistically has evolved over the last 20 years (Lehmann, 2008, Kolmos, 2009). Learning guided by an authentic problem by actively doing is the central focus of the PBL, which coincides with John Dewey's conviction that learning will only be effective when students put theory into practice." Miettinen 2000). This methodology entails a great change in the traditional methodology undergone by students in Chile, because in PBL the role of the students changes from "learning by listening to learning by doing" (Stauffacher 2006). Among the key factors of success in applying this methodology is the generation of an authentic project or problem with a clear relevance to its interests (Kahn and O'Rocke, 2004, de Graaf and Kolmos, 2007). Considering a real-world problem is fundamental to the practice of all PBL disciplines (Thomas 2000, Bell 2010). Froyd and Ohland (2005) found that when engineering programs take an integrative approach they increase retention in the early years. Motivational interventions in engineering students are the most productive in decreasing school dropout in Engineering and increasing the learning quality of their content (Stump et al., 2011, Olson 1997, Helmlinger 1997). Experimenting with the introduction of a real problem, Botha (2010) found that enjoying work on such a project correlates directly and positively with the acquisition of skills. When students are subjected to a large number of motivational experiences during the class, the learning process is optimized (Palmer, 2007, Debnath, 2005, D'Souza and Maheshwari, 2010).

Designing an authentic project with links to science and math

The nuclear disaster after an earthquake and tsunami at the Fukushima I Nuclear Power Plant in March 2011 was selected as a frame that allowed the inclusion of a STE(A)M experience, since, among other factors, earthquakes and tsunami are authentic issues for Chile's students. The central theme of the module was the planning of a robotic inspection within the post-Fukushima disaster reactor in 2011. Self-selected student teams were given a formal contract from a fictitious company that was hired to do an inspection at the plant. The goal was to develop the path and a time schedule for the robot explorer to travel through the damaged building. Teams were given robot kits at a 1:4 scale to be assembled and operated (partially). It was made of 4 fundamental parts (robot arm with gripper, chassis, and geared drive motors with remote controls). The students started with drawing and building 2-D models understand the size and shape of the robot's work envelope (Figure 1a). The robot itself provided an excellent learning tool as the students needed to calculate the angles for the robot arm to reach different heights, study friction and motor gearing, as shown in Figure 1b). Figure 1 c) depicts that the need of a new gripper also served as an excellent tool to practice creative sketching and modeling.

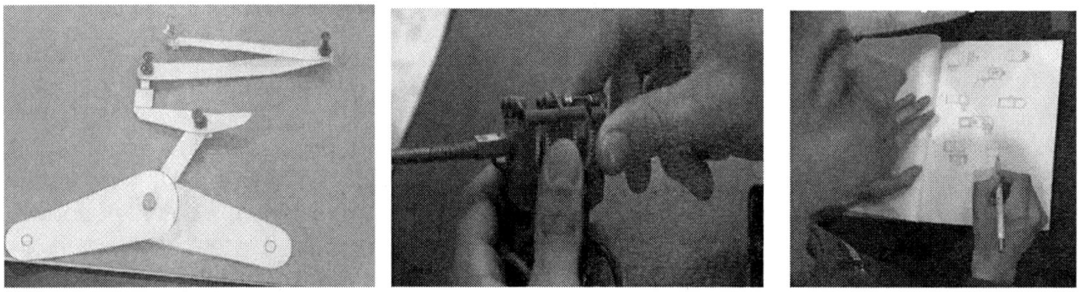

a) Model 2-D of the robot **b) Drive motor gearing** **c) New gripper design**

Figure 1: Application of design tools for problem solving

Figure 2 presents a model of the different components and objectives of the overall learning experience. As shown, the bullet Authentic Problem serves as a link to a list of important components and aspects such as Math and Physics but also to motivation and relevance. It depicts that the problem was designed so that it asked the students to apply content from the Math and Physics regular courses to solve the problem. The required problem solving skills were also developed during the course. As expected, simple failures of the robot mechanism were found to be a challenge that the students appreciated.

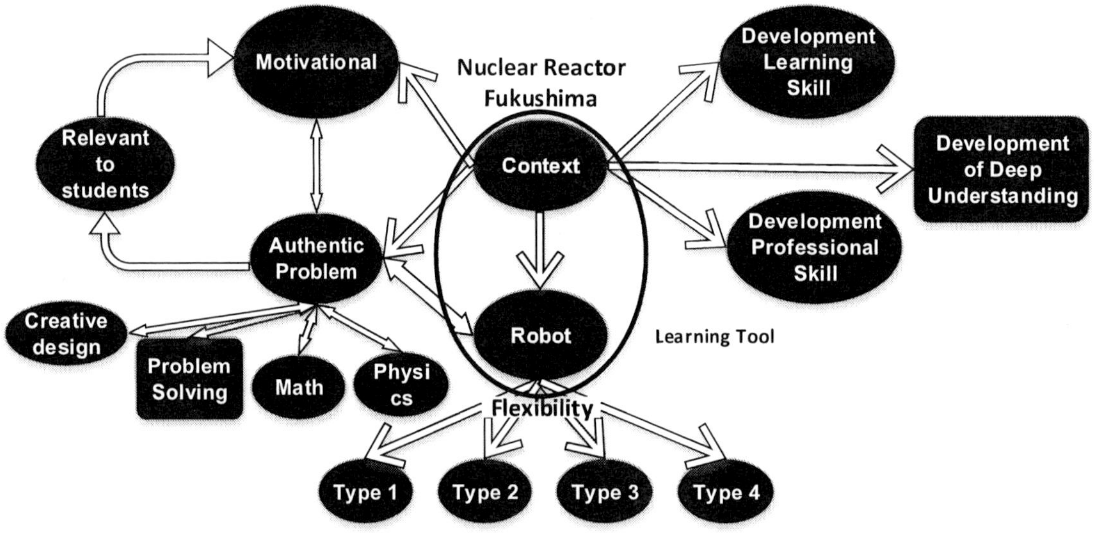

Figure 2: Principle linkages of desired outcomes, motivation and problem based learning strategies

Methods of student assessment

Different instruments were used to assess the whole experience. Standard instruments used in a pre-post setting included: a) The Pittsburgh Freshman Engineering Attitudes Survey (Besterfield-Sacre, 1998) b) Lawson test (Lawson, 1978) and c) skills to learning (based on Bernold L, 2007). In addition, two qualitative methods provided valuable insights into the students' development and experiences. Once a week students were asked to respond to reflective questions using the Moodle platform of the course to answer. These journals were used to monitor problems and progress of each student week by week. Finally, a focus group study was carried out, breaking the volunteering students into groups with 8 students allowing to collect over 70% of their thoughts about their experiences during the course. The assessment process is shown in figure 3.

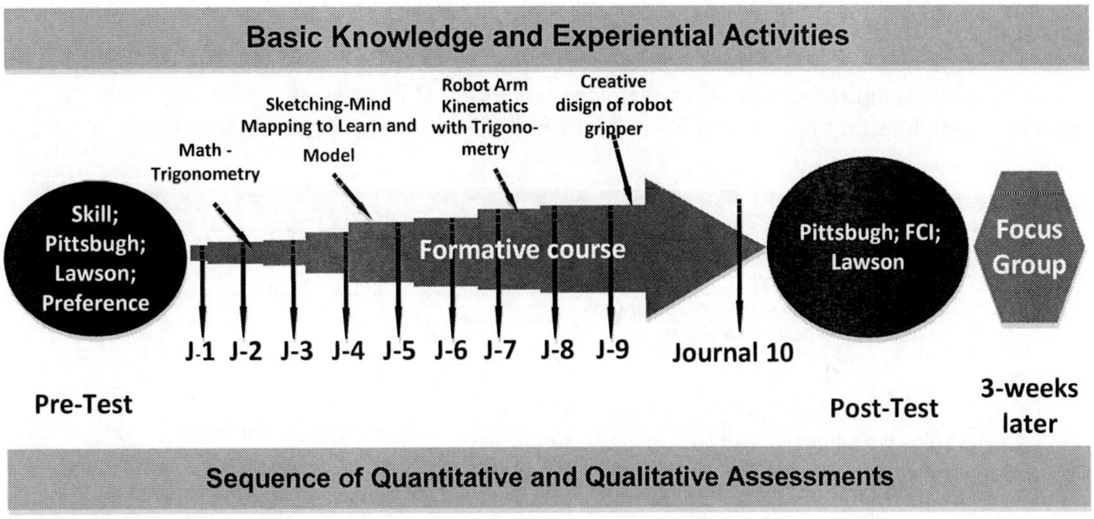

Figure 3. Diagram of administered assessments. J=Journal.

Evaluation of collected data

Qualitative Data

Due to space limitation, only journal questions from week 4 and week 10 will be briefly discussed.

> *Journal Question Week 4:*
>
> *"Looking back over the past two weeks, what are the different elements of knowledge or skills you have seen and find very interesting and useful? Why? (Do not hesitate to draw a mental map!) "*

While still early in the 15-week semester, 48% of the students expressed that mental mapping and sketching were particularly useful as a tool to discuss their ideas and solving problems. As expected, the fostering of good time management skills did not have a large impact although data that showed its importance was presented as a motivation. (Bernold 2007)

The journal in week 10 asked the students to evaluate their experience with the team project:

> *Journal Question Week 10:*
>
> *"On which component of the project would you have liked to work more? Why?"*

As was expected, a majority of 60% affirmed that they would have liked to have worked more in areas of robot design, assembly and sketching. Students' opinions were highly indicative that the robot was very effective as a learning tool to demonstrate many different aspects of engineering, a major objective of the course,.

Finally, focus group studies were carried out in order to gain insights into reflections that students had after three weeks had passed. Three sessions with 8 students covered over 70% of the students' opinions. A main area of discussion was the difficulty of implementing effective teamwork methods. It came as no surprise that math and physics do not encourage teamwork and most students did not recognise the possibility to transfer teamwork skills into those areas. More work is needed to increase the awareness of these possibilities.

Analysis of quantitative data on problem solving and scientific reasoning

Bernold et al. (2015) argued that sketching promotes problem-solving skills in engineering. To further explore this point, Table 1 presents a group of 8 questions from the Pittsburgh Engineering Attitudes used to assess problem-solving ability (Besterfield-Sacre, 1997), where students used a Likert scale from 1 (strongly disagree) to 5 (strongly agree).

Q#	Statement
Q1	I enjoy taking liberal arts courses more than math and science courses
Q2	Engineers are creative.
Q3	Creative thinking is one of my strengths.
Q4	I have strong problem solving skills.
Q5	I feel confident in my ability to succeed in engineering
Q6	I am good at designing things
Q7	I enjoy solving open-ended problems.
Q8	I enjoy problems that can be solved in different ways.

Table1: Selected questions from the Pittsburgh Problem Solving Skills Survey

Three instruments were used with all questionnaires were relocated to all the freshman students the engineering common. We received the response the 111 engineering students using the Moodle platform. Four instruments were used: a) Study skills, b) Pittsburgh Freshman Engineering Attitudes, c) learning preferences. Table 1 presents a group of question from Pittsburgh survey about to assess problem solving ability. The students used a Likert scale from 1 (Strongly disagree) to a 5 (Strongly agree)..-

Regression Statistics				
Multiple R	0.799			
R Square	0.6392			
Adjusted R2	0.158			
Standard	1.935			
Observations	15			
	Coefficients	*Standard*	*t Stat*	*P-value*
Intercept	11.847	5.786	2.047	0.087
X Q1	-1.154	0.589	-1.958	0.098
X Q2	-0.243	1.091	-0.222	**0.831**
X Q3	0,513	1.200	0.427	**0.684**
X Q4	1.504	1.013	1.485	0.188
X Q5	-0.802	0.838	-0.956	0.376
X Q6	-0.059	0.540	-0.109	**0.917**
X Q7	-2.086	2.177	-0.958	0.375
X Q8	0.593	1.553	0.382	**0.716**

Table 2: Multivariable regression analysis for Q1-Q8 on Metacognition (Study skills Bernold 2007)

The regression resulted in a satisfactory Multiple R value of 79.9%.The variables that are significant in this study are: Q2 (p=0.831), Q3 (p=0.6845), Q6 (p=0.917) and Q8 (p=0.716).The significant variables that have a positive coefficient is Q3 Q8, both represent their perception about problem solving and designing things. As expected, the by far best correlation has question Q3 (Creative thinking is one of my strengths) and Q8 (I enjoy problems that can be solved in different ways) which indicates a strong link between the students' perception of their own design skills with their scientific reasoning.

Discussion and conclusions

The theory of experiential learning is based on constructivism which emphasizes the knowledge needs to be constructed and cannot be poured into the students head by a teachers. The paper presents an approach to create a rich and holistically integrated introductory course to engineering. Each student of a cohort of 30 students participated in the first pilot in the first semester in 2016.

The analysis of the reflective weekly journals shows that student a high satisfaction with the STE (A) M. Most students valued the skills of diagramming, mental-mapping, EXCEL and creative design as extremely useful tools when solving different problems that they may face in the future. Throughout the course, students were working in teams which led to a great change in their collaborative skills. At the end they understood that engineering is teamwork. Clearly, the challenges of the 21st century engineering will require creative professionals and innovators who can tackle complex problems needing teamwork, modeling and sketching in addition to their theoretical knowledge of Math or Physics. Multivariable analysis show how design influences scientific reasoning levels. The presented work promises to be an exciting way to motivate students to be curious and active learners. The focus group study highlighted that over 60% of the students understand the importance of the arts and of working together to solve problems. Future pilot studies will focus on improving the introduction of time management as well as the design activities and to provide better support of the teamwork component.

References

Bell, S. (2010) 'Project-Based Learning for the 21st Century: Skills for the Future'. The Clearing House: A Journal of Educational Strategies, Issues and Ideas, 83 (2). pp 39-43.

Bernold, L. E. (2007). Early warning system to identify poor time management habits. International Journal of Engineering Education, 23(6), 1182-1191.

Bernold, L. E., González, L. F. & Maino, S. (2015) Reductive sketching to synergize authentic problem solving in engineering. In: BOWE, B., ed. Proceedings of the 6th REES 2015 Symposium, 13-15 July 2015 Dublin, Ireland.

Bernold, L. E., Spurlin, J. E., & Anson, C. M. (2007). Understanding our students: A longitudinal-study of success and failure in engineering with implications for increased retention. Journal of Engineering Education, 96(3), 263-274.

Besterfield-Sacre, M., Atman, C. J., & Shuman, L. J. (1998). Engineering student attitudes assessment. Journal of Engineering Education, 87(2), 133.

Besterfield-Sacre, M., Atman, C. J., & Shuman, L. J. (1997). Characteristics of freshman engineering students: Models for determing student attrition in engineering. Journal of engineering education-Washington, 86 ,139-150.

Botha, M. (2010). A project-based learning approach as a method of teaching entrepreneurship to a large group of undergraduate students in South Africa. Education As Change, 14(2), 213-232.

Bruce, C. (2010). Blog: People dreaming of androids, May 4, 2010. Retrieved July 11, 2011 from http:// it442-bruce.blogspot.com/2010/05/ going-from-stem-to-steam-artshave-role.html

Cross, N. (1996). Creativity in design: not leaping but bridging. In Creativity and Cognition 1996: Proceedings of the Second International Symposium. Loughborough, UK.

Cross, N. (2007). Designerly ways of knowing. Basel, Switzerland: Birkhäuser Verlag.

D'Souza, K. A. and Maheshwari, S. K. (2010). Factors Influencing Student Performance in the Introductory Management Science Course. Academy of Educational Leadership Journal, 14(3), 99-120.

Debnath, S. C. (2005). College Student Motivation: An Interdisciplinary Approach to an Integrated Learning Systems Model. Journal of Behavioral and Applied Management,6(3), 168-189.

De Bono, E. (1967). *The use of lateral thinking*: Intl Center for Creative Thinking.

De Graaff, E., & Kolmos, A. (2007). History of problem-based and project-based learning. Management of change: Implementation of problem-based and project-based learning in engineering, 1-8.

Eisner, E. (2002). Arts and the creation of mind. New Haven, CT: Yale University Press

Froyd, J. E., & Ohland, M. W. (2005). Integrated engineering curricula. Journal of Engineering Education, 94(1), 147-164.
Goldschmidt, G. (2003). The backtalk of self-generated sketches. Design Issues, 19(1), 72-88.
Gordon, W. J. (1961). Synectics: The development of creative capacity. New York, USA: Harper and Brothers.
Gross, M. D., & Do, E. Y.-L. (2009). Educating the new makers: Cross-disciplinary creativity. Leonardo, 42(3), 210-215.
Hanney, R., & Savin-Baden, M. (2013). The problem of projects: understanding the theoretical underpinnings of project-led PBL. London Review of Education, 11(1), 7-19.
Helmlinger, W. (1997). Motivation. Retrieved on December 15 from http://www.cratus.com/resources/willAug97. html.
Hetland, L., Winner, E. Veenema, S., & Sheridan, K. (2007). Studio thinking: The real benefits of visual arts education. New York: Teachers College Press
Hurtado, M., Reyes, C., Saavedra, O., Zúñiga, M. & Madrid, R. (2012) Informe técnico "Fortalecimiento del desarrollo de Competencias Transversales en la formación del profesional UTFSM", Plan de Fortalecimiento FSM 1199.
Kahn, P. & O'Rourke, K. (2004). Guide to Curriculum Design: Enquiry-based learning. [Online]. Available at: http://www.ceebl.manchester.ac.uk/resources/guides/kahn_2004.pdf.
Kolmos, A. (2009). Problem-Based and Project-Based Learning. University Science and Mathematics Education in Transition. O. Skovsmose, P. Valero and O. Christensen, Springer US: 261-280.
Miettinen, R. (2000). The concept of experiential learning and John Dewey's theory of reflective thought and action. International Journal of Lifelong Education, 19, 54-72.
Lawson, A. E. (1978). The development and validation of a classroom test of formal reasoning. Journal of Research in Science Teaching, 15(1), 11-24.
Lawson, B. R. (1979). Cognitive strategies in architectural design. Ergonomics, 22(1), 59-68.
Lawson, B. (2004). What designers know. London, UK: Architectural Press.
Lawson, B. (2006). How designers think: The design process demystified. Oxford, UK: Architectural Press (4th Edition) - Elsevier.
Lehmann, M., Christensen, P., Du, X., & Thrane, M. (2008). Problem-oriented and project-based learning (POPBL) as an innovative learning strategy for sustainable development in engineering education. European Journal of Engineering Education, 33(3), 283-295.
Olson, G. (1997). Motivation, Motivation, Motivation - Secondary School Educators. Retrieved from sysiwyg://934/http://7-12educators.about... 12educators/library/weekly/aa071897.htm.
Palmer, D. (2007). What Is the Best Way to Motivate Students in Science? Teaching Science-The Journal of the Australian Science Teachers Association, 53(1), 38-42.
Piro, J. (2010). Going from STEM to STEAM: The arts have a role in America's future, too. Education Week 29(24), 28-30.
Schön, D. (1983). The reflective practitioner: How professionals think in action. London, UK: Basic Books.
Stauffacher, M., Walter, A. I., Lang, D. J., Wiek, A., & Scholz, R. W. (2006). Learning to research environmental problems from a functional socio-cultural constructivism perspective: the transdisciplinary case study approach. International Journal of Sustainability in Higher Education, 7(3), 252-275.
Stump, G. S., Hilpert, J. C., Husman, J., Chung, W. T., & Kim, W. (2011). Collaborative learning in engineering students: Gender and achievement. Journal of Engineering Education, 100(3), 475-497.
Thomas, J. W. (2000) 'A review of research on project-based learning'. [Online]. Available at: http://w.newtechnetwork.org/sites/default/files/news/pbl_research2.pdf (Accessed: 18.6.2014).
Villa, A., & Poblete, M. (2007). Aprendizaje basado en competencias: Una propuesta para la evaluación de las competencias genéricas. Bilbao, Spain: Universidad de Deusto.

Acknowledgements

The authors would like to acknowledge the support of project MECESUP 1408, the program Engineering 2030 and Programa de Incentivos a la Iniciación Científica (PIIC) de la USM.

Trends in Engineering Education Research: A systematic database analysis

Fernando José Rodriguez-Mesa

Universidad Nacional de Colombia, Bogotá, Colombia

fjrodriguezm@unal.edu.co

Aida Guerra

Aalborg Univeristy, Aalborg, Denmark

ag@plan.aau.dk

Erik de Graaff

Aalborg Univeristy, Aalborg, Denmark

edg@plan.aau.dk

Jose Ismael Peña-Reyes)

Universidad Nacional de Colombia, Bogotá, Colombia

jipenar@unal.edu.co

***Abstract**: The establishment of professional societies, departments, research centres, academic degrees, international conferences, journals, etc. as Education Engineering Research (EER) infrastructures, constitute a cornerstone to EER further advancement. These infrastructures play a central role in building a community of practice, disseminating knowledge, developing research collaborations, providing research training. Consequently, a systematic analysis of these infrastructures provides a comprehensive and holistic understanding of status quo of EER, its research topics, approaches and underlying assumptions and what kind of innovation to take for further development. The study provides an overview of the main research topics, how they have changed in the last seven years and their relations with different infrastructures. We start by reviewing previous studies with a similar purpose as a point of departure, followed by a brief literature review where we built upon our theoretical and empirical framework. We then present the methods used followed by the main findings.*

Introduction

In the last decades, engineering education research (EER) came a long away to develop itself as a field of inquiry. EER tries to understand how the engineering student learns, how is the nature of engineering knowledge, topics to teach, how and what are the organisations to teach and tools to learning to engineer in general (Borrego & Bernhard, 2011; Jesiek, Newslander, & Borrego, 2009). Even although EER achieved much, some ambiguity remains when it comes to conceptualising its identity, research standards, questions, methods and evidence. It is up to practitioners to agree upon the kind of problems, their order of importance, or priority, means to solve them and what constitute a solution.

There are studies and reports in databases which examined EER infrastructures like journals and conferences proceedings and identify relevant topics, type of papers and the methodologies applied. We refer particularly to the studies of Osorio & Osorio (2002), Wankat (2004), Osorio (2005), Jesiek et al. (2011), Malmi et al. (2013), Borrego, Foster & Froyd (2014)

Osorio and Osorio (2002) analysed a total of 331 articles of the two leading journals within the field of EER; the Journal of Engineering Education (JEE) and the European Journal of

Engineering Education (EJEE), between 1998 and 2000. The authors categorised a list of terms into subjects and defined the main type of articles. For example, most JEE articles are a description of projects, research reports and description of courses, while in the EJEE the most are a review about some topic, case study and project description. According to the authors' definitions, description of programmes, courses and projects refer to a scholarly approach to curriculum development. Regarding the topics covered, the results show a similarity between both journals, i.e. courses, programmes and assessment. According to the kind of articles published, reform approaches dominated the publications in both journals.

Wankat (2004) reports an analysis of JEE publications from a period of 10 years (1993-2002). The study follows up the journal's revision of its mission, where the aim is to constitute "an archival record of scholarly research in engineering education". At this stage, JEE starts to emphasise the need for more rigorous research with the aim to identify and answer relevant questions within engineering education. The results show that the most common keywords remain teaching, computers and design; however, it is a significant increasing of other topics such as assessment, groups/ teams, distance education. The author argues that partly this is a reflection of changes in EER approaches and priorities whereas JEE increases it selectivity to research focus articles and the change of requirements of National Science Foundation (NSF) and ABET criteria. In conclusion, EER topics reflect on one hand the accreditation and societal changes by taking a scholarly approach to curriculum reform but also receives funding for research. In the conclusions, the author also argues for the relation between the analyses of journals content and the quality of the research carried out within the field, where JEE has been improving considerably by being selective and rigours in quality standards for publication.

In another study, Osorio (2005) presents the status of the literature of engineering education and discusses the deficiencies found in the existent databases due to the lack of information indexed. The author claims that extensive analysis of literature and existing databases provide the knowledge to make decisions regarding the relevant topics to investigate, link them to the existent body of knowledge and related theories and select possible avenues for publication.

Jesiek et al. (2011) map the global trends in engineering education by (1) quantifying and identifying the type of research being done, both regionally and globally; (2) identifying collaborative patterns in the field, including size and multi-nationality of the teams. Authors used both quantitative and qualitative approaches to analyse 2173 journal articles published between 2005 and 2008. Using basic criteria, the authors identify 885 empirical research papers followed by the analysis of authors' affiliation, co-authorships, the prevalence of research teams, and identify 38 categories representing different research topics and contexts. The study provides a holistic overview of EER activity by (1) describing research topics priorities through 38 categories identified, (2) link them with relevant theory through a description of the research area. The results show learning, assessment, education technologies and design as the most published. According to the authors, these results are similar with Wankat (2004) outcomes. The authors confirm and prove an increasing of a more systematic and evidence-based form of EER. Furthermore, this study also makes recommendations for EER building capacity whereas journals and conferences should increase the standards and request more detailed, high-quality contributions.

Malmi et al. (2013) investigate the theoretical frameworks applied to EER papers, the general research setting (i.e. data collection, analysis methods and procedures), link with previous research studies and type of articles. It provides a taxonomy to apply in analysing the EER activity and defines six key dimensions: nature of paper, explanatory framework, research strategy, data source, data analysis and reporting. For example, the nature dimension refers to empirical, case reports, theory papers or position papers, while research strategy dimension refers to research design and methodologies. The results show most of the articles position themselves in a serious and rigorous research by, for example, reporting empirical studies, linking them with previous studies, identifying a clear research strategy, including data sources and analysis.

Borrego et al. (Borrego et al., 2014) established framework to conduct systematic literature reviews in engineering education as an emergent interdisciplinary field. According to the authors, the systematic literature review is critical to any field of study mainly because they disseminate and summarise research to inform policy and practice.

According to the above, the studies focus on the systematic analysis of EER literature and its importance for development on three primary levels: (1) establishment of EER as a field of inquiry, where evolution and shift from curriculum reform approach towards more rigorous research; (2) the research activity in itself by identifying, discussing and synthesising relevant research topics, appropriate methodologies, quality standards; (3) building multi-disciplinary collaborations across countries and involving different engineering education contexts and cultures.

Purpose and research question

The above stresses the relevance of systemic analysis of bibliographic databases for the EER development by providing an overview of research priorities (report curriculum reform vs. rigorous research), relevant topics in alignment with different contexts and professional demands; relate research with relevant theories and methods and solutions. Furthermore, a systematic and continuous analysis of the EER *status quo* through its publications and databases contributes to its growth, capacity building and establishment as a field of inquiry.

The study investigates the following research question: What trends can we observe in engineering education research from 2010 to 2016? Moreover, it aims to provide an overview of engineering education research trends regarding challenging themes such as sustainability, globalisation, interdisciplinary and transdisciplinary on educational journals.

Theoretical framework

There are numerous examples of a systematic review of the literature (e.g. Bassir, Sadr-Eshkevari, Amirikhorheh, & Karimbux, 2014; Borrego et al., 2014; Keathley-Herring et al., 2016) in various areas of research. The common denominator of these studies is that they all use a search equation and apply it to a database. What makes them different is the analysis method and the criteria for selection and grouping of the terms. Some of the authors use the citation frequency in which two documents are cited together or co-citation. Visual maps represented the frequency (Small, 1973, 1999). Others use co-occurrence indices to determine common research topics.

There are also examples of authors proposing descriptive taxonomies by thematic areas. They use the findings on the revision in some sources (e.g. Finelli, Borrego, & Rasoulifar, 2016). The problem with this type of procedure is that there is a connection between the topics result and the thematic focus in the sources.

The premise of categorization in this paper takes two parallel procedures. The first seeks to search by the association between keywords, topics of highest recurrence and therefore trends. The second procedure aims to identify critical issues that are recurring in each category, to regroup. Therefore they could consider as a trend.

In the first procedure, for example, being the number of terms A> B> C, the following situations may occur:

- If B always appears with A and never with C, it means that subject B could be unique to A.
- If B appears with A and C in a relatively significant proportion (i.e. percentages of occurrence in each of them), B is a cross-sectional theme and therefore could interpret independently.
- If the co-occurrence of B with A is comparable to the co-occurrence of B with C, both A and C tend to form a category. If the co-occurrence of B with A is greater than C with A, category C tends to let out in the trend.

In the second procedure, which is not the subject of this article, we analyse the words that have cross-sectional co-occurrence. We selected them and grouped in thematic areas. For example, the recurrent appearance of the terms skills, competencies, teamwork, oral communication, could be associated with a group called competencies.

This work of co-occurrence analysis can be done manually or using existing bibliometric tools, such as VOSViewer. It is doing by using the association strength between terms along with all data (Leiden University, 2017; Van Eck & Waltman, 2010, 2015). Then, the categorization or grouping could support by some existing taxonomy (Borrego et al., 2014; Jesiek et al., 2011; Malmi et al., 2013; Osorio, 2005; Osorio & Osorio, 2002; Wankat, 2004). The categories, or keywords, used in the content analysis are also the EER trends ranked after the analysis. Table 1 summarises some that information, where the examples provided are the firsts ranked in the study.

Table 1 Summary of studies and analysis of EER publications.

Study	No. of papers analysed	No. of categories	Examples of categories	Journals
Osorio and Osorio (2002)	331	18	Course Program Assessment Students	Journals: JEE and EJEE
Wankat (2004)	367	20	Computer, group, ABET, gender/woman, distances education	JEE
Jesiek et al. (2011)	885	38	Learning Assessment Educ. Technologies Design Collaboration	Journals: IJEE, EJEE, AJEE, JEE. Proceedings of SEFI, ASEE Global Colloquium, AAEE.

The examples in Table 1 summarise the search done in a given period, and consider the topic in all articles published over that period. The categories can be keywords in content analysis of articles. They also reflect the research or editor's preferences in general but not necessarily the preferences of EER audience (i.e. educators and researchers, academic and non-academic).

Methodology

The study uses content analysis as methodology where we first identify sources of information (i.e. literature databases) followed by definition of categories of analysis (i.e. main research topics). The find trend in the content of the paper we used the procedure shows in **Figure 1**, what is coherent with some methods (e.g. Tranfield, Denyer, & Smart, 2003; Wendler, 2012). To comply with paper length (i.e. maximum ten pages), we only include the mark steps. The following two sub-sections explain in more detail the procedures carried out.

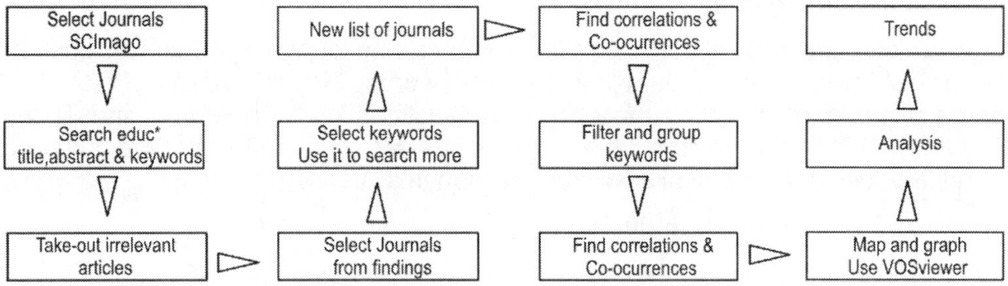

Figure 1 Steps to find educational trends

Identifying sources of information

We started by searching infrastructures among engineering and education using SCIMago (Scimago, 2014). It takes into account the number of citations and journal impact. The results show 914 journals. To illustrates, we show in Table 2 the first 20 ranked.

Table 2 Ranking SCImago for education journals (retrieved February 2016)

1	Journal of Research in Science Te	11	International Journal of Computer-Supported
2	Review of Educational Research	12	User Modelling and User-Adapted Interaction
3	Science Education	13	Journal of the Learning Sciences
4	Handbook of the Economics of Edu	14	Reading Research Quarterly
5	Studies in Science Education	15	Computers and Education
6	Child Development	16	Internet and Higher Education
7	American Educational Research J	17	Journal of Second Language Writing
8	Journal of Educational Psychology	18	Educational Evaluation and Policy Analysis
9	Learning and Instruction	19	Developmental Review
10	Educational Researcher	20	Scientific Studies of Reading

Search for author keywords

The second step is a search for articles regarding with education. It was done using the SCImago list and the word "educ*" on the search engine. Thus, we expect to find publications related to "education", "educación", "educative", "educational", "educacional" or "educando". The results of this search were 6,253 publications. After review, manually we took articles without educational topics. The list was reduced to 6167 articles in only 41 journals. The list of journals and quantity of articles found in Scopus is on Table 3; we use the abbreviated journal title.

Table 3 Select journals to search educational trends and quantity articles between 2010-2017 in Scopus.

	Source title	Can		Source title	Cant
1	Int. J. Environ. Sci. Educ.	686	22	Tech. Technol. Educ. Manage.	96
2	Int. J. Eng. Educ.	644	23	Int. J. Sci. Math. Technol. Learn.	50
3	Educational Technology and Society	585	24	Int. J. Constr. Educ. Res.	49
4	Comput Appl Eng Educ	383	25	Eng. Stud.	46
5	World Trans. Eng. Technol. Edu.	370	26	Int J Hum Comput Stud	37
6	Int. J. Emerg. Technol. Learn.	360	27	J. Inf. Syst. Educ.	37
7	IEEE Trans Educ	308	28	J. Technol. Educ.	33
8	J. Sci. Educ. Technol.	270	29	Eng. Educ.	32
9	Eur. J. Eng. Educ.	267	30	Australasian Journal of Engineering	25
10	Adv. Sci. Lett.	238	31	Int. J. Fash. Des. Technol. Eudc.	24
11	Int J Technol Des Educ	200	32	J. Appl. Secur. Res.	24
12	J Prof Issues Eng Educ Pract	188	33	J. Sch. Violence	20
13	IEEE Trans. Learn. Technol.	150	34	J. Archit. Educ.	14
14	Int J Electr Eng Educ	150	35	J. Mater.	14
15	J. Eng. Educ.	133	36	J. Pre-Coll. Eng. Educ. Res.	12
16	Rev. Iberoam. Technol. Aprendizaje	131	37	Eng. Des. Graph. J.	10
17	Int J Contin Eng Educ Life Long Learn	122	38	J. Sch. Publ.	9
18	Learn. Media Technol.	122	39	Synth. Lect. Eng. Technol. Soc.	7
19	IEEE Potentials	116	40	Eng. Econ.	1
20	Adv. Eng. Educ.	107	41	Int. J. Aviat. Psychol.	1
21	Int. J. Mech. Eng. Educ.	96			

Keywords count and synonym reduction

The previous step focuses attention on title, abstract, and keywords. Now we focus only on the keywords. There were 25,709 words in the keywords. Only the 10.6% of the articles did not have any keyword and, just the World Transactions on Engineering and Technology Education with 370 articles had the 6.6%. The list was running in VOSviewer to find prominent words and in Excel to reduce similar words or synonyms.

Categories and occurrences analysis

Considering only author keywords, and after reducing synonyms and plurals, VOSviewer recognised 13,441 different keywords, 591 with a least five co-occurrences. Figure 2 shows the results after did not consider directly in VOSviewer the keywords "engineering education", "education", "engineering", "engineers", and "higher education" because of the word "educ*" was part of the search equation in Scopus in the first step. The graph is ready to do visual analysis.

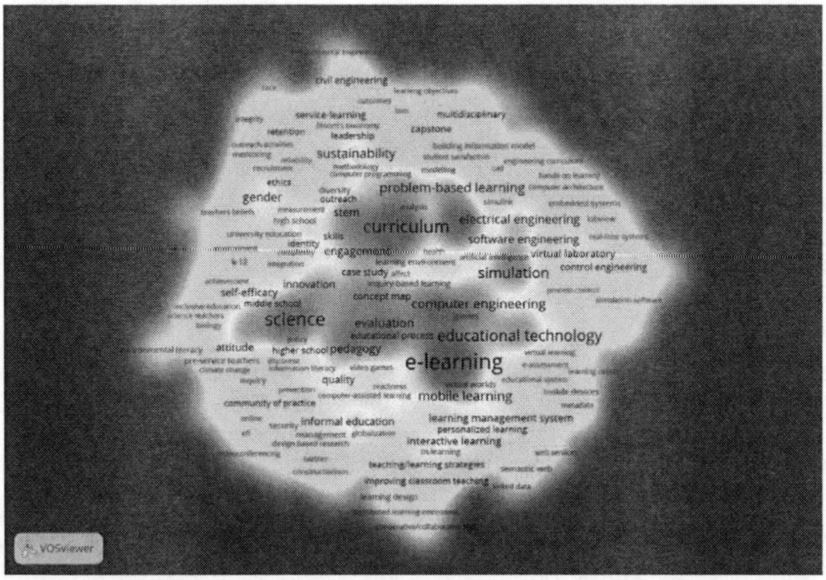

Figure 2 Author keywords density for 41 journals (2010-2016) Process in VOSviewer)

General trends overview

The cluster was organised by an association between its occurrences of keywords in the same article. They could represent topics into thematic areas. With this method, were identified 12 clusters. The cluster was organised according to the number of connections between items. There is a direct and proportional relationship between the number of occurrences and connections. The thematic areas were determined if they were a clear field of study with the following criteria: 1) they are recognised as a topic of study in engineering; 2) they have a strong number of occurrences comparing with the others, i.e. a considerable weight of occurrences and, 3) they were previously classified in other taxonomy, i.e., Borrego, Jesiek and Osorio. The **Error! Reference source not found.** shows the categories established.

Findings

There are 15 main categories summarised in Table 4. They reflect the use of "educ*" key in SCOPUS search of educational articles on selected journal. Therefore, a further study should choose each category to try to find more documents with the main areas and also searching in title and abstract. In this paper, we only report the first five categories found through this approach. In this paper, we only describe the first five categories found through this approach.

In the EER, the category that is most studied is e-learning, which from 2010 to 2016 shows a positive growth. The inspection of the articles indicates that this subject has been the object of study in areas of electronic engineering, computer science and control. It includes topics of distance education, tools for simulation and learning and virtual laboratories. They also relate to active learning methods in the classroom even though with less quantity.

Table 4 Thematic areas published from 2010-2016

id	Category	Links weight	Occurrences
1	E-learning	288	192
2	Science/STEM	214	132
3	Curriculum	202	123
4	Assessment	195	115

5	Project-based learning	171	112
6	Educational technology	160	98
7	Distance learning	159	114
8	Motivation	147	82
9	Simulation/Computer engineering	124	85
10	Computer engineering	123	70
11	Collaborative learning	122	71
12	Design	110	60
13	Sustainability	108	71
14	Active learning	106	80
15	Blended learning	105	58

The following is science education. It also shows a growth trend. The main related publications concern training of teachers and active teaching methods in the classroom. It has a high co-occurrence with mathematics, indicating that this is the main focus of study. On a smaller scale, contents and visual tools have been worked on in the classroom.

Curriculum studies correlate with several topics but mainly student motivation. Also, creativity appears with design cases and assessment. The attitude and curriculum studies towards science education, gender and informal education. Despite, the authors' references to the curriculum have decreased since 2013. According to Streveler & Smith (2006), the need to shift the EER focus from teaching and curriculum development to fundamental research on how students learn to engineer. In this sense, it is important to define the type of questions EER, and how those questions addressed methodologically, what kind of learning, psychologic and developmental theories link to, and how results are generalised and in which ways contribute to the development of educational and learning theories.

The assessment shows a positive trend in recent years. There is a relationship with various studies, from science education, lifelong learning, sustainability to teamwork, and competencies. However, the greatest correlation is in e-learning and STEM work.

Project-based learning is one of the fields of study that showed growth until 2015, but in 2016 it declined. The disciplines with the highest correlation are engineering design, mechanics, mechatronics and robotics. Not surprisingly, problem-based learning, active learning, and collaborative learning have a relation with project-based learning. We found studies focused on implementation and curriculum. In Wankat's (1993-2002) (Wankat, 2004) and Jesiek's studies, team and collaborative topics are on the top side. Nevertheless, collaboration is used in many papers regarding ICT and PBL. The same happens with group and teamwork, and skills. That features are characteristics of PBL (De Graaff & Kolmos, 2003). Therefore, many papers regarding PBL consider these characteristics.

Figure 3 shows the growth in the number of articles published for the whole five categories considered. The progressive growing is consistent with the call of various organisations to improve engineering education (Hoidn & Kärkkäinen, 2014; Union, 2013).

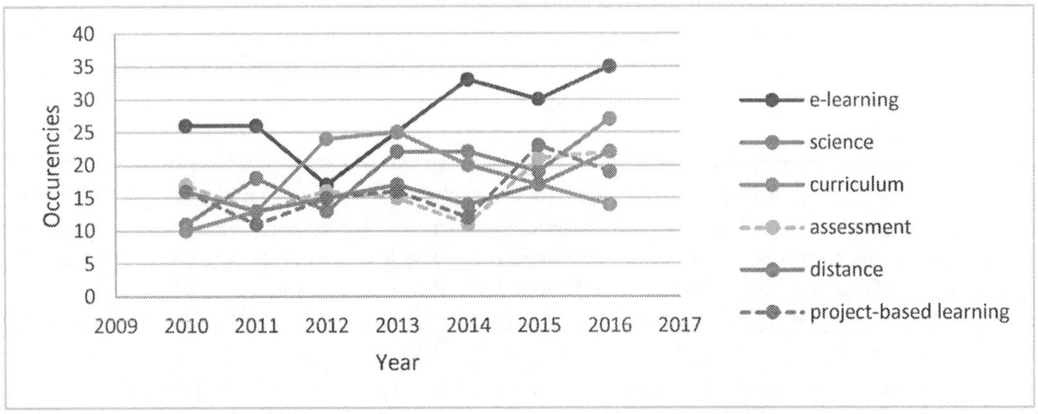

Figure 3 Annual variation in occurrence of categories

However, in contrast to the authors(e.g. Jesiek et al., 2011; Osorio & Osorio, 2002; Wankat, 2004), in our finding the competences have a relationship with undergraduates but does not have a significant number of occurrences about other categories. Nevertheless, it has strong relations with the assessment topic. Figure 4 shows the small relations of competencies with others.

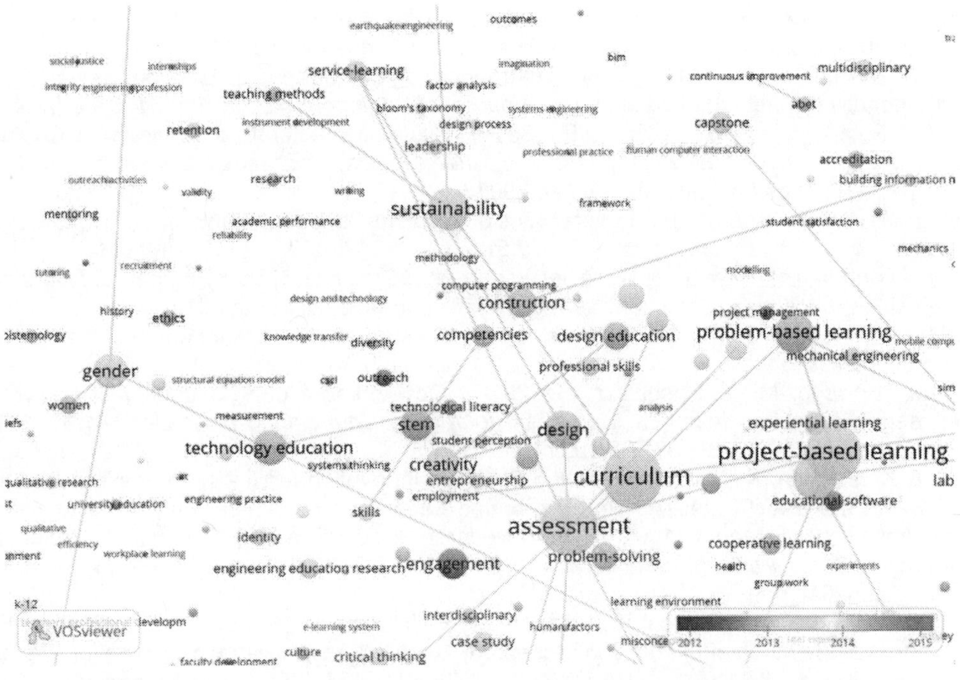

Figure 4 Co-occurrence of some topic in educational articles.

Conclusions

In the last decade, several studies have been conducted with the aim to identify and analyse the engineering education research (EER) main topics and the infrastructures (i.e. journals and other publication databases). In this study, we discuss the need for a continuous and systematic analysis of EER publications and databases to understand the EER growth and evolution as a field of inquiry. We have built our analysis in previous works and their methodological frameworks. Therefore, we have defined fifteen categories that cluster more than 25,000 keywords related to engineering education. These reflect the diversity of research topics within EER. In overall, the results show that an increase of publications in EER in the last years, showing a growth of the community of practice and interest in the scholarly approach to engineering and how engineers learn. The study also shows an increase of publications in topics such as e-learning and technologies used in education. It is not a surprise if we take into consideration that ICT a field in expansion and which adds value to educational practices and how it affects students learning. Other categories like assessment and learning remain as trends as Wankat (2004), and Jesiek et al. (2011) have shown. However, it is important to refine the term learning and have a clear picture of its scope and what specific areas of research covers. Note that learning connects to several categories including the relation between engineering education research practice and relevant theories that explain, for example, how we learn. Finally, yet importantly, the results also show that has been publishing different topics in selected journals, which indicates that even though EER is a multidisciplinary field, it remains anchored to a particular discipline and with small communities of practice within the EER community. Supporting this claim is, for example, the working groups within SEFI, the European Society for Engineering Education.

Here we present the first part of a study focus on EER topics and journals. Following studies involve for example: (1) redefine the learning as a category within EER and how it links with fundamental educational and learning theories, and (2) relate the research topics with different world regions and how they relate to context social and economic challenges.

References

Bassir, S. H., Sadr-Eshkevari, P., Amirikhorheh, S., & Karimbux, N. Y. (2014). Problem-based learning in dental education: A systematic review of the literature. *Journal of Dental Education*, *78*(1), 98–109. Retrieved from http://www.scopus.com/inward/record.url?eid=2-s2.0-84891877027&partnerID=40&md5=c5fb335a8c5be9027d6e4562b032c471

Borrego, M., & Bernhard, J. (2011). The Emergence of Engineering Education Research as an Internationally Connected Field of Inquiry. *Journal of Engineering Education*, *100*(1), 14–47.

Borrego, M., Foster, M. J., & Froyd, J. E. (2014). Systematic Literature Reviews in Engineering Education and Other Developing Interdisciplinary Fields. *Journal of Engineering Education*, *103*(1), 45–76. https://doi.org/10.1002/jee.20038

De Graaff, E., & Kolmos, A. (2003). Characteristics of Problem-Based Learning. *International Journal of Engineering Education*, *19*(5), 657–662. Retrieved from http://sfx.aub.aau.dk/sfxaub?sid=Elsevier:Scopus&%5C_service%5C_type=getFullTxt&issn=0949149X&isbn=&volume=19&issue=5&spage=657&epage=662&pages=657-662&artnum=&date=2003&id=doi:&title=International+Journal+of+Engineering+Education&atitle=Characteristics+o

Finelli, C. J., Borrego, M., & Rasoulifar, G. (2016). Development of a taxonomy of keywords for engineering education research. *European Journal of Engineering Education*, *41*(3), 231–251. https://doi.org/10.1080/03043797.2016.1153045

Hoidn, S., & Kärkkäinen, K. (2014). Promoting Skills for Innovation in Higher Education: A Literature Review on the Effectiveness of Problem-based Learning and of Teaching Behaviours. *OECD Education Working Papers*, (100), 64. Retrieved from /Users/Fernando/Documents/Project Based learning/MPBL Aalborg/Semester 3/Semestre 3 course 1/S3_C1_Ses4_Monna Dams/Hoidn2014.pdf

Jesiek, B. K., Borrego, M., Beddoes, K., Hurtado, M., Rajendran, P., & Sangam, D. (2011). Mapping global trends in engineering education research, 2005-2008. *International Journal of Engineering Education*, *27*(1 PART 2), 77–90. Retrieved from http://sfx.aub.aau.dk/sfxaub?sid=Elsevier:Scopus&_service_type=getFullTxt&issn=0949149X&isbn=&volume=27&issue=1+PART+2&spage=77&epage=90&pages=77-90&artnum=&date=2011&id=doi:&title=International+Journal+of+Engineering+Education&atitle=Mapping+global+trend

Jesiek, B. K., Newslander, L. K., & Borrego, M. (2009). Engineering Education Research: Discipline, Community, or Field? *Journal of Engineering Education*, 39–52. Retrieved from /Users/Fernando/Documents/Project Based learning/PBL Articles/jesiek2009.pdf

Keathley-Herring, H., Van Aken, E., Gonzalez-Aleu, F., Deschamps, F., Letens, G., & Orlandini, P. C. (2016). Assessing the maturity of a research area: bibliometric review and proposed framework. *Scientometrics*, *109*(2), 927–951. https://doi.org/10.1007/s11192-016-2096-x

Leiden University. (2017). Welcome to VOSviewer. Retrieved March 6, 2017, from http://www.vosviewer.com/publications

Malmi, L., Adawi, T., Curmi, R., De Graaff, E., Duffy, G., Kautz, C., ... Williams, B. (2013). Methodological analysis of SEFI EER papers. European Society for Engineering Education (SEFI). Retrieved from http://www.scopus.com/inward/record.url?eid=2-s2.0-84938828710&partnerID=40&md5=9c21746611ebe801a728e5dde5388609

Osorio, N. L. (2005). What every engineer should know about engineering education. In *2005 IL/IN Sectional Conference*. Northern Illinois University, DeKalb, Illinois: American Society for Engineering Education.

Osorio, N. L., & Osorio, M. A. (2002). Engineering Education in Europe and the USA. *Science & Technology Libraries*, *23*(1), 49–70. https://doi.org/10.1300/J122v23n01_05

Scimago. (2014). scimago Institution Rankings . Retrieved from http://www.scimagoir.com/institution.php?idp=1953

Small, H. (1973). Co-citation in the scientific literature: A new measure of the relationship between two documents. *Journal of the American Society for Information Science*, *24*(4), 265–269. https://doi.org/10.1002/asi.4630240406

Small, H. (1999). Visualizing science by citation mapping. *Journal of the American Society for Information Science*, *50*(9), 799–813. https://doi.org/10.1002/(SICI)1097-4571(1999)50:9<799::AID-ASI9>3.0.CO;2-G

Streveler, R. A., & Smith, K. A. (2006). Guest Editorial: Conducting Rigorous Research in Engineering Education. *Journal of Engineering Education*, (April), 103–105.

Tranfield, D., Denyer, D., & Smart, P. (2003). Towards a Methodology for Developing Evidence-Informed Management Knowledge by Means of Systematic Review. *British Journal of Management*, *14*(3), 207–222. https://doi.org/10.1111/1467-8551.00375

Union, E. (2013). Council Conclusions on investing in education and training - a response to “Rethinking Education: Investing in skills for better socio-economic outcomes” and the “2013 Annual Growth Survey.” *Official Journal of the European Union*, *56*(2013/C 64/06). Retrieved from /Users/Fernando/Documents/Project Based learning/PBL Articles/pbl_education/OJ-C-2013-064-Comunidad Europea.pdf

van Eck, N. J., & Waltman, L. (2010). Software survey: VOSviewer, a computer program for bibliometric mapping. *Scientometrics*, *84*(2), 523–538. https://doi.org/10.1007/s11192-009-0146-3

van Eck, N. J., & Waltman, L. (2015). Systematic retrieval of scientific literature based on citation relations: Introducing the CitNetExplorer tool. Retrieved from http://ceur-ws.org/Vol-1143/paper2.pdf

Wankat, P. C. (2004). Analysis of the First Ten Years of the Journal of Engineering Education. *Journal of Engineering Education*, *93*(1), 13–21. https://doi.org/10.1002/j.2168-9830.2004.tb00784.x

Wendler, R. (2012). The maturity of maturity model research: A systematic mapping study. *Information and Software Technology*, *54*(12), 1317–1339. https://doi.org/http://dx.doi.org/10.1016/j.infsof.2012.07.007

An active learning activity to design a system using modern engineering tools: the case of a production line using invention systems kits

Barros-Castro Ricardo Abad
Industrial Engineering Department, Pontificia Universidad Javeriana, Bogota, Colombia
ricardo-barros@javeriana.edu.co

Cruz Joan Paola
Industrial Engineering Department, Escuela Colombiana de Ingeniería Julio Garavito, Bogota, Colombia
joan.cruz@escuelaing.edu.co

Castiblanco Ivonne Angélica
Industrial Engineering Department, Escuela Colombiana de Ingeniería Julio Garavito, Bogota, Colombia
ivonne.castiblanco@escuelaing.edu.co

***Abstract**: Nowadays, there is still a need for establishing pedagogical strategies that can include abstract knowledge as well as practical situations with different technological tools. This need can be supported through the ideas of an active learning approach. Taking these ideas into account, we explore how can we design an active learning activity to combine theory, practice, and technology, circumscribed in a "real" industrial context. In particular, the purpose of this study is to present and reflect upon an active learning activity that seeks for enhancing students' ability to design a production line, using invention systems kits that recreate what typically happens in a real industrial context. Results of applying a series of measures to evaluate the activity in a pilot group of students show that the activity has a potential to enhance learning about the design and development of a production system.*

Introduction

Context

Nowadays, there is still a need for establishing pedagogical strategies that can include abstract knowledge as well as practical situations. These practical situations should help to relate theoretical frameworks typically learned in engineering learning environments to problems in a real context. This mixture of theory and practice involves the promotion of strategic knowledge (cognitive and technical capabilities) as well as the tools associated to solve engineering problems (International Society for Technology in Education – ISTE; International Technology and Engineering Educators Association – ITEEA). Hence, technology is understood as a means and a set of capacities (knowledge, methods, tools, and skills) (Lonsdale and McCurry, 2004). Therefore, as a means to solve engineering problems, technology plays and important role in the engineering student learning process. This implies the challenge of enhancing several skills in engineering' students bearing in mind theory and practice, with the support of technology.

Theoretical background

In order to achieve this challenge, active learning approach has been acknowledged as an umbrella term to different models of teaching (e.g. collaborative learning, experiential learning, problem-based learning, game-based learning, etc.) that helps building learning environments with the idea of generating deep learning (Barkley, 2010). It implies the design of pedagogical strategies to work a concept into participants' personal knowledge and experience to learn, while students reflect and monitor over their learning process and results. One of the core elements of active learning is the transformation of experience as creator of knowledge (Kolb, 1984). This transformation occurs while students participate in hands-on activities, make questions about the phenomenon, discuss about concepts to explain the phenomenon, and experience different levels of understanding in new situations. In this context, active learning activities play a fundamental role in pedagogy. For Chickering and Gamson (1987), the students do not only learn by listen to their lecturer, on the contrary, they analyze, discuss, reflect, and diffuse the knowledge acquired into their day to day in cooperation with others. Therefore, knowledge emerges when the student is questioned about a particular problem being fundamental in their self-taught process.

Within this learning process, social interactions are highlighted. These interactions help individuals to build understanding, in a learning environment that promotes autonomy and meaningful experiences. In addition, the cultural and social meaning that learning activities might have (Salkind, 2014) and the supporting artefacts (e.g. different technology – tools, method, etc.) that can be used to enhance learning (Barros-Castro, Córdoba-Pachón and Pinzón-Salcedo, 2013) should be taken into account for engagement (active involvement) to emerge.

One of the active pedagogical approach to help students to engage involves the use of games. Educational games can provide students with a motivating and stimulating environment while providing them with feedback to promote learning (Bodnar et al., 2016). Research has shown that both, student learning and positive attitudes are improved by game-based activities (gamification). Compared with a transmission classroom model, a game-based classroom can incorporate many of the ABET student outcomes and can generate a context as a "real" situation (Bodnar et al., 2016). Among the different studies in engineering education research some cases are highlighted in industrial engineering programs (e.g.: Ozelkan and Galambosi, 2007; Park, 2012; Pourabdollahian et al., 2012; Chesler et al., 2013). These studies focus on the game to enhance learning regarding product design and manufacturing issues. Nevertheless, games within these studies are primarily digital (or classroom activities) and do not involve challenge, rapid feedback, and changes, characteristics that should be present in games to better engagement (Bodnar et al., 2016). In other words, the majority of the games focus their attention on a single activity (mostly digital, such as a simulation or a computer game) instead of focus on a process. In addition, there is a need to propose more games related to the design and develop of a production system taking into account the product design and the manufacturing constraints (little studies are focused on industrial engineering curriculum). Moreover, although research has shown student learning and positive attitudes by using game-based activities, there is still a challenge in the measures designed to evaluate the impact of games on learning (Alanne, 2016), taking into account that only 35% to 45% of the studies measure learning outcomes (Bodnar et al., 2016).

Research question

Taking these ideas into account, in the context of an industrial engineering curriculum, *we want to explore how can we design an active learning activity (game-based activity) to combine theory, practice and technology, circumscribed in a "real" industrial context.* In particular, the purpose of this study is to present and reflect upon a game (challenge) that seeks for enhancing students' ability to design a production line, using invention systems kits that recreate what typically happens in a real industrial context.

We explain, in the methodology, the case study of a pilot group that was challenged to build a production system with some constraints. We also present the results of three different measures that can contribute to the evaluation of a learning process based on games. These measures are related to a binnacle (to evaluate the process), a survey (to evaluate perceptions), and a checklist of the requirements accomplished and a conceptual test (to evaluate learning and the application of learning in a "real" context). In doing so, this paper contributes to the debate of the effectiveness of game-based activities in engineering education environments.

Methodology

To achieve the purpose, we design an active learning activity that has the goal of promoting students' ability to design a system to meet desired needs and to use the techniques, skills, and modern engineering tools necessary for engineering practice.

In doing so, we present to the students a challenge where they have to design any functional production system, therefore, they should decide a product and the best way to assemble it (as will be seen later, students selected an aircraft production line). In addition, the system must be fully automated, hence using sensors and actuators. The technological resources they may use are four (4) Lego Mindstorms Education EV3 sets and Lego Mindstorms' software licence.

We select a pilot group of three (3) students who must complete the challenge within an academic period (typically four (4) months). The group is composed of students from last year of an industrial engineering program. They completed the product development course successfully and show interest on education research. During the process, students may consult their lecturers in case of doubts. In order to contextualize students about the real industrial environment, we request them to consider guidelines such as consistency of product form, assembly system steadiness, resources optimization, system automation, and absence of reworks.

To reflect about the effectiveness of this activity in students' learning we design three types of measures: first, a binnacle to evaluate students' process and reflections; second, a survey to measure their perceptions about the activity and its utility to enhance learning; and third, the accomplishment of the challenge with all the conditions that were given to the students.

The binnacle

The binnacle is a workbook, used in a personal way by students during the development of the activities needed to complete the challenge such as planning, design, and construction of the production system. It allows collecting experiences about the tests, experiments and results in a chronological way, which let them to understand the use of the technological tools and fulfil all the requirements. On the other hand, the registration of this information along the way allows them to reconsider and justify decisions and provide alerts on priority issues. It is also a traceable mean for consultation in order to achieve the objective successfully. In this regard, with the support of the binnacle, students perform interpretation processes, look for explanations through practice, and find solutions to problems they face during the construction of the system.

Table 1 shows the binnacle template given to the students. It records experiences during the main phases of product development process.

Table 1: Binnacle Template

Student Name		
Weeks		Date:
Challenge		Design of a production system
Activities Developed by the Student		Methodology and Description
Appropriation of basic concepts related to the use of Invention System Kits, specifically Lego Mindstorms EV3		
Idea selection		
Compliance with the finite resource constraint: strategy and planning		
Assembly line and product planning (benchmarking and concepts)		
Development		
Absence of reprocessing or operations repetitions per station: strategy and planning		
Line automation		

Table 2: Survey Template

Taking into account your perception of the activity, please answer the following questions:

1. Mention the concepts and skills you have learned and point out how important are to you, being 5: very important and 1: not important.

Concept	Importance (Mark with an "X")				
	1	2	3	4	5

2. According to your academic background, please indicate the three (3) most important previous concepts you learned before that were useful during the development of the activity.

Concept	Mark with an "X" only 3 concepts
Product Development	
Industrial Processes	
Human Resources	
Management	
Electrical Concepts	
Manufacturing	
Programming	

Taking into account your perception of the activity, please answer the following questions, being 5 completely agree and 1 completely disagree:

		1	2	3	4	5
Cognitive ABET Student Outcome	3. The activity allows me to know elements in how to design and model a real situation.					

Socio-Cultural Context	4. The activity allows me to recognize the complexity of a real problem.					
Cognitive ABET Student Outcome	5. The activity allows me to use techniques, skills and modern engineering tools necessary to design a system.					
Cognitive ABET Student Outcome	6. The techniques, skills and modern engineering tools where useful to accomplish the challenge.					
Meaningful Experience	7. The activity generates more benefits (added value) versus a traditional class.					

Taking into account your perception of the activity, please answer the following question:

		1	2	3	4	5
Whole Activity	8. Please evaluate the whole activity being 5: I liked all of it and 1: I did not like it at all.					

Please also comment about your evaluation:

____

____

The survey

The survey is structured with open and closed questions developed based on Velasco and Barros-Castro (2015). It evaluates students' perceptions about the activity and its utility to enhance learning according to aspects such as cognitive outcomes, socio-cultural context, and meaningful experience. Open questions allow the students to answer freely and extend their reflections. Table 2 shows the survey template given to the students.

The accomplishment of the challenge

In order to measure the accomplishment of the challenge with all the conditions that were given to students, we design an accomplishment challenge checklist to evaluate students' results and a product development test to evaluate the concepts acquired by students during the process. Table 3 shows the checklist template. The qualification is based on the qualitative criteria of the lecturer, with five years of experience in product design and development field, who in this case is one of the authors. Table 4 shows the product development test. These last resources allow us to evaluate students' ability to design a production system twofold. On the one hand, the test is aim to verify the theoretical knowledge acquired by the student in order to achieve the challenge of designing and developing a production systems taking into account the product chosen. In the order hand, the checklist aims to verify the applied knowledge taking into account the guidelines requested.

Results

Results are fourfold: first, students consider that this activity help them to gain interest and motivation about real engineering problems, while they are learning conceptual and technical concepts.

Table 3: Checklist Template

ACCOMPLISHMENT CHALLENGE CHECKLIST		
CHALLENGE: DESIGN AND DEVELOPMENT OF A PRODUCTION SYSTEM		
FINAL GRADE: ___________		
Students Names		
SYSTEM DESIGN AND DEVELOPMENT	CHECK	COMMENTS
Consistency of product form (functional prototype, aesthetic and related with the original idea)		
Assembly line and product steadiness		
Compliance with the finite resource constraint (parts, bricks, motors and sensors of 4 sets Lego Mindstorms EV3)		
Optimizing the use of motors and sensors (Bricks: $500USD Max. 4, Large Motors: $200USD Max. 8, Medium Motors: $100USD Max. 4, Sensors: $50 USD Max. 20. Maximum Amount: $5000USD)		
Line automation		
Absence of reprocessing or operations repetitions per station		

Table 4: Test template

Conceptual test – designing and developing a production system
1. Please, describe the process of designing and developing a production system and how the process is related to the process of achieving the challenge.
2. Please, describe the process of choosing the production system based on considerations about the product selected.

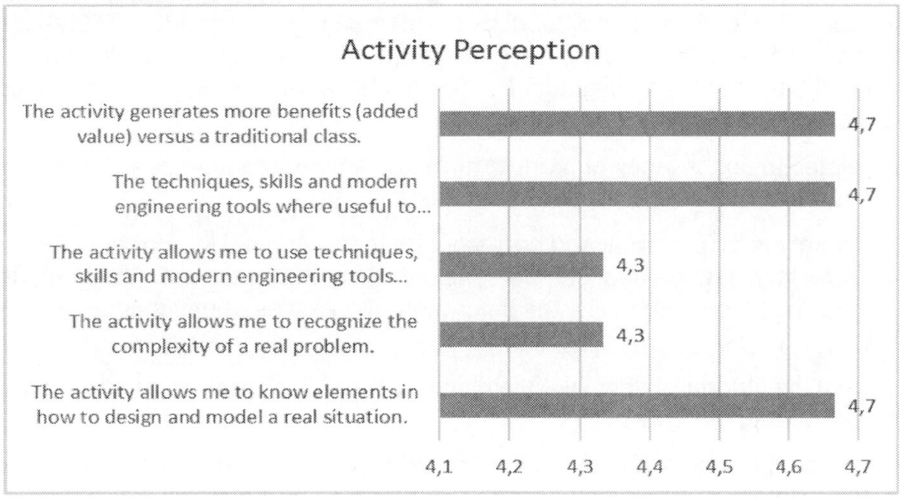

Figure 1: Survey (students' perceptions) about the activity

The interest and motivation about real engineering problems is showed in the answers they gave in the perception survey as shown in Figure 1. According to the survey results, it can be highlighted that the activity allows them to know elements in how to design and model a real situation. It also generated more benefits versus a traditional class. Moreover, the activity allows them to use techniques, skills, and modern engineering tools, which were useful to accomplish the challenge.

Regarding learning conceptual and technical concepts the students identified the concepts shown in Figure 2. Teamwork and communication, product development, programming and automation were expected in this research, while planning and time management and problem solving were not expected. These results show additional skills students perceived they learned, which are important on their learning process even if they were not anticipated as a goal since the beginning.

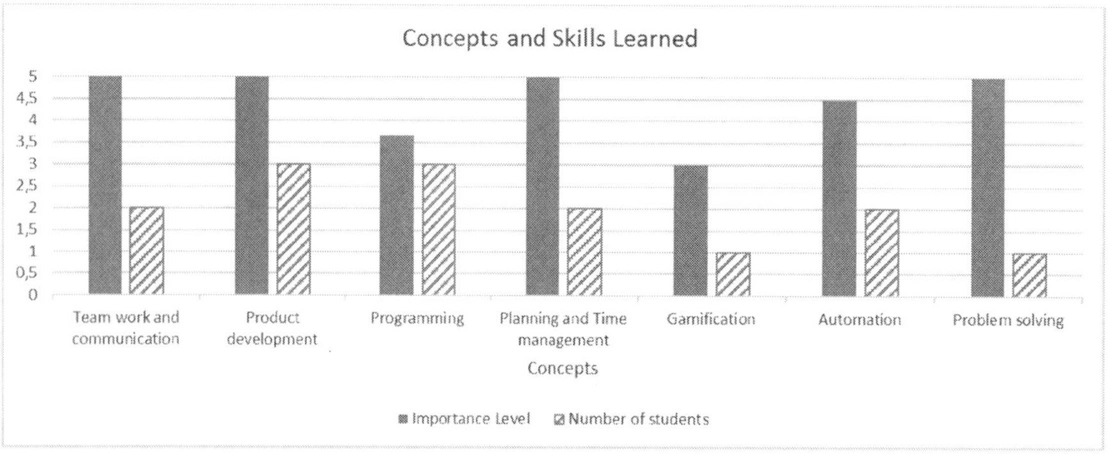

Figure 2: Concepts and Skills Learned by the Students

Second, the challenge helps them enhancing the way they design a system taking into account several constraints and conditions outside the production line. The qualitative criteria and the qualification given by the expert in product development field reflect students accomplished the challenge successfully taking into account the constraints and guidelines they received. Some of the feedback comments given by the lecturer are:

> "Compliance with the finite resource constraint and optimizing the use of motors and sensors: the constraint of using maximum 4 sets of Lego Mindstorms EV3 was fulfilled. A total of 3 bricks, 5 medium motors, 3 large motors and 5 sensors were used for a total estimated of $2850USD. The goal of not exceeding $5000USD was achieved. The fact of using 1 station to execute 2 similar operations which saves 1 brick in the process is highlighted. For future challenges, I suggest to continue trying this kind of cost reduction".

In addition, related to consistency of product form and line automation conditions in general the development process was appropriate as follows:

> "The prototype is functional and corresponds to the original idea of the students, an aircraft factory. The product can be considered aesthetic as well. Although the first part must be placed manually, the line automation assessment is considered very positive."

In addition, from the student's viewpoint we highlight the reflections and strategies they used during the development process:

> "For the production system development, several constraints were found such as creating a totally new line without any programming and building procedure. Furthermore, the kits were limited and a key factor was the time that was destined in

order to carry out the realization of the system. Hence, the strategy was to build a system of similar size and try to use and adapt similar structures to those made previously during the practice process. Having the activities established, it was possible to organize the time and avoid interferences between daily responsibilities and the accomplishment of the challenge".

Third, they acknowledged the use of technology as a mean to accomplish the conditions needed for a design as it can be seen in the reflections of one of the students:

> "Through the study of different sources such as books, guides, tutorials, internet and articles, I used this theoretical basis to build different products and systems, to program and modifiy them to be able to perform different functions. In my experience, the appropriation of concepts related to the use of Invention System Kits particularly Lego Mindstorms EV3 is observed through the different activities that were elaborated step by step both for the building and programming of the system defined for the challenge".

Fourth, according to the test to evaluate conceptual learning and their application in the challenge, students could describe proficiently the different phases to consider when the objective is to design and develop a production systems taking into account the product chosen. Moreover, they could relate every part of the process to their challenge and explain how each step was applied in it and to make decisions regarding the production system appropriately.

Finally, the students liked the whole activity as they scored in question number 8 on the survey and they gave meaningful feedback about the relation between the theoretical frameworks typically learned in engineering learning environments to problems in a real context. They emphasized that although the active learning activity recreates what typically happens in a real industrial context, it does not include the dimensions, specifications and volumes that can be found in the real industry, as it can be seen in the following comment:

> "The activity allows me to know in a playful and dynamic way concepts related to engineering that have been learned throughout the industrial engineering program and gave me additional concepts that have not been taught before but represent a great contribution. I consider it is very important to add them permanently to the curriculum, such as automation. This kind of projects allows me to approach and understand real production systems, however, being an academic exercise, its dimensions, specifications and volumes are different from those present in the real environment. This implies a difficult to know, evidence, and evaluate all the variables that could be involved in a real problem or situation".

As engineers and educators, we are aware of the limitations of this kind of classroom activities and understand that it is difficult to reach a truly real level of application. In addition, we consider that the pilot size equals 3 should be acknowledged as a limitation. However, it is a mid-point between the theoretical class and the actual context faced by a graduate who has not had the opportunity to experience industrial situations. It is not the purpose of this study to generalize these findings but we consider these types of reflections very positive because they show the student's ability to analyze and highlight the usefulness of the active learning activities (including the scope of this type of activities). This is an open field with possibilities for future research.

Final reflections (recommendations, implications future research)

This paper aims to present and reflect about an active learning activity from the game-based learning approach to enhance students' abilities to design and develop a production system taking into account product design and different constraints. In doing so, the paper studies a case of a pilot group and the impact of this game (challenge) on learning effectiveness using several measures. The results obtained from the different measures that evaluated different dimensions of the learning process such as conceptual learning, attitudes, learning process, and operational learning (application), show a potential of this kind of activities to promote engagement and deep learning. However, there is still the need to consider the scope of this kind of games to represent a "real context situation". Furthermore, studies where these games and measures are used in more groups are needed.

References

Alanne, K. (2016). An overview of game-based learning in building services engineering education. *European Journal of Engineering Education 41(2),* 204-219.

Barkley, E.F. 2010. Students engagement techniques: A handbook for a college university. San Francisco, CA.: Jossey-Bass A Wiley Imprint.

Barros-Castro, R.A., Córdoba, J.R., Pinzón, L. (2013). "A Systemic framework for evaluating Computer-Supported Collaborative Learning – Mathematical Problem-Solving (CSCL-MPS) Initiatives: insights from a Colombian case". *Systemic Practice and Action Research 27(3),* 265–285.

Bodnar, C.A., Anastasio, D., Enszer, J.A., Burkeyd, D.D. (2016). Engineers at Play: Games as Teaching Tools for Undergraduate Engineering Students. *Journal of Engineering Education, 105(1),* 147-200.

Chesler, N., Arastoopour, G., D'Angelo, C., Bagley, E., & Shaffer, D. W. (2013). Design of a professional practice simulator for educating and motivating first-year engineering students. *Advances in Engineering Education, 3(3),* 1–29.

Chickering, A., and Gamson, Z. (1987). Seven Principles for Good Practice. AAHE Bulletin 39(7): 3-7.

Kolb, D. (1984). Experiential learning: experience as the source of learning and development. Upper Saddle River, NJ: Prentice-Hall.

Lonsdale, M., McCurry, D. (2004). Literacy in the new millennium. Adelaide: Australian Council for Educational Research, NCVER. Accessed at: http://www.ncver.edu.au/publications/1490.html on 2 October, 2005

Ozelkan, E., & Galambosi, A. (2007). Lampshade game for teaching lean manufacturing. *ASEE 2007 Annual Conference, June 24–27, 2007.* Honolulu, Hawaii.

Park, J. (2012). An application of a manufacturing activity simulation board for teaching product costing. *International Journal of Engineering Education. 28(3),* 734–743.

Pourabdollahian, B., Taisch, M., & Kerga, E. (2012). Serious games in manufacturing education: evaluation of learners' engagement. *Procedia Computer Science. 15,* 256–265.

Salkind, N.J. 2004. An Introduction to Theories of Human Development. Thousand Oaks (Calif, the US): Sage Publications.

Acknowledgements

We acknowledge the active participation of Juan Sebastián Cadena, Mauricio Urrego, and Miguel Ángel Barragán in the design and develop of the production system using invention systems kits.

Copyright statement

Rethinking engineering education practices with active learning and digital technologies: Pilot experiences at the University of the Andes, Bogotá

Alvaro H. Galvis
University of los Andes, Bogotá, Colombia
a.galvis73@uniandes.edu.co

Francisco A. Galvis, Juan P. Casas
University of los Andes, Bogotá, Colombia
fa.galvis10@uniandes.edu.co jcasas@uniandes.edu.co

Diego M. Valencia, Angélica Avalo, Alexa T. Jiménez
University of los Andes, Bogotá, Colombia
dm.valencia40@uniandes.edu.co a.avalo@uniandes.edu.co at.jimenez10@uniandes.edu.co

Abstract. *The Dean of the Faculty of Engineering invited professors to explore the use of pedagogies aimed at transforming their practice by making active use of Information and Communication Technologies (ICT)-supported learning strategies. Three courses and their respective professors were chosen to participate in the innovation process. With the support of pedagogical and technological advisors, the professors redesigned, tested, implemented and researched the effects of this change on the actors. Studies were conducted to answer questions related to possible lessons from this pilot experience and the lessons learned will be used to create competencies-based academic programs led by principles of quality, innovation, flexibility, internationalization and sustainability. This paper presents the rationale and design of the pilot experiences, the studies that were conducted about the intervention, findings, conclusions, and recommendations.*

Introduction

Engineering Education is a field of practice where using digital technologies is a must. Engineers must master both their domain of knowledge as well as the use of up-to-date technologies –most of them digital—to be competitive in their professional lives. In spite of this, teaching engineering is not an endeavour that is always enriched by technology-based pedagogies (Forté & Wentland, 1998). Most traditional engineering faculty members teach as they were taught, reinforcing the idea that the professor's role is to transfer relevant mental models to help students understand and solve problems with engineering methods and tools. This approach does not prepare them for today's engineering field where professionals deal with incomplete data and competing (often conflicting) demands from clients, government, environmental groups and the general public (Mills & Treaagust, 2003). In spite of this, non-traditional teaching ideas have had relatively little impact on mainstream engineering education so far (Bates, 2015).

To break with traditional teaching practices in engineering at our University of the Andes (or UniAndes), we undertook a two-year pilot project in which three engineering courses belonging to different engineering programs became laboratories for the integration of new technology-based innovations and active pedagogy. A multidisciplinary team from the Center for Innovation in Technology and Education (ConectaTE) coached faculty members as they evaluated their teaching practices and reflected on the impact of different aspects of the intervention. This paper presents initial results and lessons from the pilot experiences.

Context of the study

The Engineering portal of the University of the Andes (UniAndes) proudly states that since its foundation in 1948, it has been a national leader in training engineers and researchers at the forefront of the discipline. It also states that its undergraduate, masters and doctoral programs are based on the best international practices of the discipline, which focus not only on the acquisition of technological competencies, but also on the development of soft skills such as communication, leadership, management and interaction with interdisciplinary teams (UniAndes, Faculty of Engineering, 2017). This characterization describes the level of academic leadership that the university intends to maintain. As is stated in objective 1.4 of the institution's 2016-2020 Development Plan, the university will continue "to have academic programs based on the development of competencies that are evaluated comprehensively and systematically against international standards and criteria such as quality, innovation, flexibility, interdisciplinarity, internationalization and sustainability" (UniAndes, 2015).

Within this inspiring framework, the Dean of the Faculty of Engineering made a call for explorations in the use of transformative pedagogies, namely active teaching strategies supported by digital technologies. The targets of these interventions were very representative courses of those in engineering curriculum, and the participants were professors who were highly distinguished for teaching in their discipline. Three courses, Probability and Statistics (PS), Structural Systems Analysis (SSA), and Rigid Solid Mechanics (RSM), and their respective professors, were chosen to participate in a two-year pilot experience. The courses are described below.

Probability and Statistics (PS)

Probability and Statistics is an undergraduate course in the area of the fundamentals of engineering offered by the Department of Industrial Engineering to all students of the faculty. While the course is taught in fourth semester for Industrial Engineering majors, in the rest of the engineering programs, it is taught either in the third or fifth semester. It is a course with high demand every semester, which is why it has at least ten sections with an enrollment of 40 to 100 students per section. The course has one coordinator and the sections are taught by a variety of professors. PS is a three-credit course traditionally taught in two hour and twenty-minute lecture sessions, complemented by independent assignments equivalent to six hours of work per week. The course director, Mario Castillo, and one of the professors, Astrid Bernal, took part in the pilot experience. They were each in charge of one section that met for three-hours a week.

The Probability and Statistics course had thirteen topics. Now it consists of two core modules that integrate the two themes: probability and statistics. The intervention was piloted and the flipped learning strategies were implemented. Before class, students were asked to study the subject and do preparatory exercises; during class, their understanding of digital material seen before class was evaluated, and feedback and more practice was given. After class, it was suggested that students attend optional supplementary classes where additional practice of concepts and feedback was provided. The remaining sections did not undergo any intervention.

Tensions at the beginning of the redesign:

(1) *Curricular tensions*: Because the course integrates two thematic areas, and each one is broad, this affects how deep one can go into the content.
(2) *Instructional tensions*: Lecture classes with many students do not necessarily favor active participation.
(3) *Contextual tensions*: Each section has students from different engineering programs, with differences in interests, previous training, and maturity. Given these conditions, the examples and the exercises do not necessarily address the needs of each student.
(4) *Reorientation tensions*: Homework exercises are long, so by the time students receive feedback, the class has often moved on to another topic. This does not allow students to reflect on their mistakes nor to try new comprehension / study strategies.

Areas of redesign:

(1) *Curricular area*: The course structure was rearranged around four fundamental concepts that meet the requirements of basic engineering training, instead of the original fifteen topics. Redefined learning goals include: (a) recognize and understand events and phenomena of a probabilistic nature; b) acquire skills to construct probabilistic models to quantify risk; c) manage and analyze data to identify patterns and make inferences; and d) build basic statistical models.
(2) *Instructional, contextual and reorientation areas*: Active pedagogy, through the use of the flipped classroom and a variety of short and focused exercises, was used. Operationally, this approach requires three complementary efforts: (a) Students must study topics and test what they learned before the face-to-face session. This required the creation of video clips to complement the printed material of each learning unit, for example https://youtu.be/A3PC1E4nZoY ; (b) During classroom sessions, the idea was to complement what was learned in video clips, books and previous exercises, through peer evaluation of answers to practice problems. Additionally, the classroom sessions were used for professor-facilitated group discussion and feedback on exercises that students completed on their phones or computers within an immediate response and feedback tool; (c) In after-class activities, the student had the chance to practice with exercises situated in the different areas of engineering and receive feedback in breakout sessions.

Structural Systems Analysis (SSA)

SSA is a compulsory undergraduate course in Civil Engineering taken predominantly by students of this program. It is part of the general fundamentals of engineering curriculum and is taught in the fifth semester. It is the first course that offers the student a real context in which to apply skills in the field. The main objective of SSA is for students to interpret reality from the perspective of civil engineering. Before the intervention process started, the SSA course had six thematic units and twenty-nine theoretical sessions. The course is now divided into two main modules that correspond to the general concepts about civil structures and methods of analysis. The director of the course, Juan Carlos Reyes, and a professor of the course, Francisco Galvis, took part in the pilot study, where blended learning, a flipped classroom approach, and active pedagogy strategies complemented with ICT were put into practice. Multimedia materials (video clips with study guides) were created in keeping with this approach. Total class time of some learning units was reduced, and virtual laboratories replaced the sap2000 face-to-face laboratories. The complementary classes were exchanged for optional office hours. This represented a total reduction of 30% in face-to-face sessions compared to the traditional course. The redesign was implemented as materials and activities were developed.

Tensions at the beginning of the redesign:

(1) Since the course has broad thematic content, it is not possible to allocate class time for the students to analyze and solve problems with the guidance of the teacher. These tasks, which are central to the course, should be done outside of class.
(2) Experience shows that students approach the problems from mathematical operations perspective, disregarding the analysis of the results they obtain in their calculations, which makes mathematics an ends, not a means, to solving real world problems.
(3) Many students do not master the prerequisite subjects of the course making it necessary to add such thematic content.

Areas of redesign:

(1) To provide digital, contextualized content to complement course materials and enable the student to explore, understand and apply concepts in a real context before the class sessions, and to review and reinforce previous knowledge about concepts through this channel as well (Example: https://youtu.be/PTUWj4Nznjk).
(2) To focus not only on the analysis of the problem and the mathematical solution to it, but to concentrate on the interpretation of the results of that solution through rubrics and peer evaluation.
(3) To create alternatives for how to approach the theoretical concepts of the course so that the student can apply them in the context of civil engineering. This includes the systematization of exercises administered by the computer and accompaniment of the professor in the analysis and discussion of solutions.

Rigid Solids Mechanics (RSM)

RSM is the first of a series of three compulsory courses in the mechanical systems area of the undergraduate Mechanical Engineering program; however, the concepts taught are transversal to all areas of the Engineering program. The course focuses on the basic concepts of applied mechanics in engineering and is oriented towards the achievement of a central objective, which is the interpretation of reality in the frame of mechanical and civil engineering (Casas & Leon, 2016). The RSM course was composed of ten units which were distributed into sixteen weeks of class. Currently, the course has four main modules that coherently articulate the course contents. It includes two mandatory lectures per week and an optional supplementary session. The former is by given by professor Juan Pablo Casas and the latter by graduate assistants.

Tensions at the beginning of the redesign:

(1) The enrollment and performance statistics of the course were lower than the rest of the Mechanical Engineering Department's courses. In the four years prior to the redesign, the percentage of withdrawals was higher than those of the department's other courses, and the average grades were also lower (Casas & Leon, 2016, p. 6).

(2) The lecture class sessions used conventional pedagogy and were centered on the teacher's explanations and demonstrations; theory predominated over practice.
(3) Feedback on what was learned was deferred over time as it depended on grading, so the focus of the classroom sessions was not on the learning problems that arose during the process; rather, feedback came after the fact.
(4) The learning resources were closely linked to what the texts said and to the teachers' presentations in class.

Areas of redesign:

(1) To use active pedagogy through a flipped classroom approach making sure students prepare for and participate in the class by using exercises that have the student face real situations that transcend the theoretical context.
(2) To promote practice by using exercises with increasing levels of difficulty (three levels: textbook-related situations; simplified, real situations; complex real situations) which receive timely feedback about students' degree of understanding. The information serves as a basis for developing new strategies to reinforce the proposed learning goals. In order for this to happen, weekly solutions to a set of problems that represent the level of difficulty that the student must be able to handle are delivered and revised.
(3) To create multimedia resources (Video clips with study guides such as https://youtu.be/ZpBFBMbeZPk) that the student can access repeatedly to see how the thematic expert interprets the situation and proposes ways of addressing it using what was learned in each learning unit.

Literature review

In the literature review that follows, we seek to clarify concepts that guide the design of each of the pilot experiences.

Engineering Education

Studies about engineering education indicate that "the task of higher education is to educate students to become effective modern engineers—able to participate and eventually to lead in aspects of conceiving, designing, implementing, and operating systems, products, processes, and projects. To do this, students must be technically experts, socially responsible, and inclined to innovate" (Crawley, Malmqvist, Östlund, Brodeur, & Edström, 2014, p. 2). The same authors also point out that "there is a seemingly irreconcilable tension between two positions in engineering education. On one hand, there is the need to convey the ever-increasing body of technical knowledge that graduating students must master. On the other hand, there is growing acknowledgement that engineers must possess a wide array of personal and interpersonal skills, as well as product, process, and system building knowledge and skills required to function on real teams to produce real products and systems" (2014, p.3).

The previous tension is present in each of the engineering courses. In some cases, PS for example, concepts that will be useful throughout one's career and life may not be clearly connected to a specific discipline. This means that achieving competency in this course demands going beyond understanding each concept and method to being able to apply these in diverse contexts. In cases such as the RSM or SSA courses, concepts have empirical/concrete references that make it possible to grapple with real-life problems associated with them and whose solutions require higher level thinking (i.e. analysis, evaluation and synthesis in Bloom's taxonomy) and the use of interpersonal skills to search for, test and socialize valid solutions (Churches, 2009).

Engineering Education Course Design

In the search for anchoring principles and methods for the design of engineering courses, we discovered the following frames of reference: 1) design courses considering the students' needs; 2) use curricular objectives suggested in the study guides; 3) base the design on thematic frameworks proposed by disciplinary authorities; 4) follow processes or sequences for solving problems; 5) use disciplinary norms or standards as a guide (Ziegenfuss, 2007). The aforementioned practices are not exclusive, as is the case with curricular tensions. That said, we decided to begin designing from a multidimensional needs analysis that combines the Engineering Education frameworks with the students' perceptions; this marriage of perspectives provides the framework for the course redesign (Galvis & Pedraza, 2013).

It was decided to use non-traditional approach for the design. According to Hansen (2011, pp. 17-18), course design usually focuses on teaching, aligning the content of the course with the needs it must satisfy. Afterwards, the evaluation criteria and tests are defined and designed. Finally, the learning objectives that guide the treatment of the topics that the instructor will present are defined in a linear and logical sequence. In contrast, in the design of courses focused on learning, after clarifying the needs that the course must attend, everything that the student should learn in order to achieve the objectives is detailed, and a plan to promote deep understanding of concepts in a non-linear way is designed. The students are guided by essential questions whose answers must be found in groups through experimentation with content, rather than by simply memorizing concepts. This distinction between the logic of teaching and the logic of learning is the basis for design for understanding (Wiggins & McTighe, 2005) and can be implemented with the help of what is known as the "backward design model" (Erickson, 2007).

Following this systematic model and proposals for understanding big ideas (Wiggins & McTighe, 2005 and Hansen, 2011 adapted by Galvis & Pedraza, 2013 for active learning supported with digital technologies), the first step is to determine learning outcomes that students should be able to demonstrate, in terms of big ideas, also called fundamental concepts, they should understand in depth, that is, not only outcomes based in curricular content. A detailed "what to learn" design is made, taking into account the concepts that students are supposed to have mastered and, in particular, the misconceptions they usually have. Guiding questions are used as bridges for inquiry and collaboration and aim at filling conceptual gaps, which are the difference between what should be and what is in the student's mind. Only then is the assessment system defined, including both authentic performance tests for measuring observable comprehensive results and ongoing formative assessments during the process to measure strengthen the understanding of complementary concepts. Learning units are organized by big idea and give the learner control over how to approach the themes. This follows the principles of active pedagogy and is supported by the flexibility that virtual or blended learning environments provide. The student can also participate in learning experiences that only virtual learning and collaborative learning environments offer.

Conditions for Efficient and Effective Courses Redesign

Professors from research universities such as UniAndes are usually focused on their teaching and doing disciplinary research, either by themselves, or with support from graduate assistants. Thus, several questions arise: Why should they spend time redesigning courses that are well-structured? Why should they reform courses that, according to the typical retention and passing rates, are successful in developing the expected competencies? Why do such a thing when the courses are well-evaluated by students? These questions were present when the dean opened the call to participate in REDINGE1-Engineering Courses Redesign 1 project. The objective of the pilot was to learn about a paradigm focused on fostering learning and not just teaching (Bates, 2015), to learn about the different means to learn (Forté & Wentland, 1998), to complement the lecture class, books, articles and laboratories with virtual learning environments implemented with digital learning tools that provide flexibility in accessing knowledge, and to foster changes in the professor's traditional role, as a facilitator of the process (Collison, Elbaum, Haavind, & Tinker, 2002).

This promise of value would not have been possible had there not been conditions for change that favored its materialization, such as those studied by Galvis (2016). The dean of the Faculty of Engineering was the champion of the REDINGE1 process. He made resources to the departments available such as reducing the course load of participating. He also provided funds to sponsor the support of the multidisciplinary team from the Center for Innovation in Technology and Education (ConectaTE) of the Faculty of Education during the innovation. This made the process viable, but it was not enough. The university issued a new faculty handbook which recognized the importance of innovation in teaching as an academic outcome related to progress in one's professional trajectory (UniAndes, General Secretariat, 2015, p. 19). This is precisely what the pilot aims to recognize, and this change in the handbook made this pilot initiative acquire academic value for the professors, which is something that motivates them and generates the possibility for them to create learning resources making the value of their teaching experience evident. The innovation initiative was also made possible by the experience of ConectaTE in providing accompaniment to innovative teachers, in creating Virtual Learning Environments (VLEs) and Virtual Learning Objects (VLOs), developing software, videos and websites in particular (UniAndes, ConectaTE, 2016).

Research Questions

Considering that these pilot experiences aim at transforming engineering teaching, it was considered appropriate to try to answer this question: *What can the Faculty of Engineering learn about the development and implementation of pedagogical innovations supported by digital technologies (DT) to achieve what is proposed in Objective 1.4 of the Academic Leadership Plan of the university?*

To answer this question, the following sub questions were formulated:

From the students' perspective:

1. What effects on the retention and results of the students of the redesigned courses is attributable to the transformations of educational conceptions, media and practices in the courses?
2. What changes in students' perceptions are attributable to the processes and products of the innovation process?

From the professors' perspective:

3. What changes in the conceptions, means and educational practices of innovating teachers can be attributed to their participation in the redesigns and its implementation? What lessons does the implementation of active pedagogy leave for the innovative professors?

Considering the interactions in the redesign:

4. What are the key success factors in these innovation processes from the perspective of the professors and sponsors of the transformation process?

Research Design

To answer the questions related to learning outcomes, the research possibilities in each of the courses with interventions were taken into account considering the extent to which the innovation was implemented in each semester of the redesign. In 2015-1 and 2015-2, interventions were implemented in one or more modules of each course according to the stage of the redesign of activities or materials. In 2016-1, all modules from each course were involved. In all three courses, attention was paid to content validity in the performance tests, ensuring coherence between the learning objectives that were proposed and taught and the grades reflecting achievement levels.

In the PS course, the final exam was standardized, so it was possible to collect data for two semesters. A total of 882 undergraduate students from the different majors registered in the ten sections participating during 2015-2 and 2016-1, 127 of whom were enrolled in the sections with the intervention. Results were categorized by major, entry level of the student, and whether or not the student was repeating the course. The students' cumulative average and grades in pre-requisite math courses were also considered (UniAndes, Faculty of Education Evaluation Center, 2016). We controlled for variability in the results by tracking record of students enrolled and withdrawn from the three courses over time.

Regarding students' perceptions of changes in their learning autonomy due to the innovation, soft skills were measured. We applied a survey that measures academic self-efficacy, methodological flexibility, support of autonomy, support of ICT in the teaching-learning process, and self-regulated learning whose scales show good internal validity. See Table 1

Table 1. Internal validity

Instrument	No. items	Alfa
Academic self-efficacy	4	0.83
Methodological flexibility	4	0.89
Support de autonomy	4	0.85
Support of ICT in the T-L process	4	0.87
Self-regulated learning (SRL)	19	0.88
Planning	5	0.75
Self-monitoring	5	0.76
Assessment	4	0.68
Reflection	5	0.83

Source: (Uniandes, Centro de Evaluación de la Facultad de Educación, 2016, pág. 9)

The constructs in this survey were chosen because they build on research that shows that when the learning environment is autonomous, students' motivation and self-regulation are strengthened. This translates into greater interest and confidence, and thus, increased performance, persistence, creativity, and general well-being of students (Deci & Ryan, 2000).

A quasi-experimental design was used in which control and experimental groups were compared. A pre-test and post-test measured students' ability to self-regulate their learning and their academic self-efficacy. Students in five sections of the three courses (two PS, one RSM and two SSA sections) and twelve non-intervened sections (eight of the PS course and four of the RSM and SSA related courses) participated. In addition, three variables related to the pedagogical intervention were explored: a) promotion of autonomy; b) methodological flexibility; c) support of ICT in the teaching-learning process (UniAndes, School of Education Evaluation Center, 2017, p. 6).

To understand students' perceptions of the innovations in their courses, in 2016-1, 110 individuals from the intervention sections (14 of the MSR course, 65 from the SSA course, and 31 from the PS course) took an online survey. The survey asked participants to rate the quality of the resources, the feasibility of the activities, the contribution of the methodology to promoting interaction in the lecture classes, and the overall impact on learning. Information from the surveys was complemented with information from three focus groups, two of which had RSM students and one with SSA students with eight, nine and seven students respectively.

Interviews were carried out with the professors implementing the intervention in their courses. The objective here was to understand: i) the changes generated in their teaching by the accompaniment they received; ii) the effects the accompaniment had on the redesigned courses; iii) and finally, the impact of the support in the other courses they are teaching. These interviews were transcribed, analyzed, categorized and the responses were compared. The perceptions described complemented those written by some of the teachers in their reports at the end of the pilot.

Online surveys were also applied to administrators in the Faculty of Engineering and to the professors innovating in their courses to find answers to the fourth sub-question.

Findings

Cognitive outcomes for students

The findings will be discussed for each course separately.

In the **PS course** show that active learning and flipped pedagogy supported by digital technologies had a positive effect on performance in the final exam of the class. Scores were approximately a quarter of a standard deviation higher when other academic variables of interest (e.g. semester, repetition, GPA, and grades of two previous courses) were controlled for (UniAndes, Faculty of Education Evaluation Center, 2016, p. 4).

In the **SSA course**, it was not possible to have a control group since both of the two sections received the intervention. However, a historical review of the course grades in iterations taught by the same professor before and after the innovation was conducted. The percentage of failures in the course during the last twelve semesters was also analyzed. Before the innovation, the failure rates of the SSA course where around 10%. In the first semester of implementation of the innovation, that is, when the course transitioned from conventional to active pedagogy, it showed the highest historical failure rate (20%). It was detected that the workload was very high during this period. Subsequently, the size of the tasks was reduced, and support material was continuously reduced; this led the percentage of failures to decrease by half in the most recent cohort (Reyes & Galvis, 2017, p.4). Reyes & Galvis (2017, p. 5) comment that one of the problems identified in the implementation of the methodology is the professor's misunderstanding of the students' comprehension of the video content before class. This essentially meant that in the first version of the course after the innovation, more than half of the class time was spent reviewing the content of the video. This practice resulted in a loss of motivation for many students, which was evident due to low attendance (close to only 50%). To overcome this problem, virtual quizzes which evaluated students' understanding of the videos were given in each class. This assessment practice allowed the professor to minimize the time spent reviewing and at once improve attendance.

In the **RSM course**, it was also impossible to have a control group since there was only one course per semester. Thus, students' opinions and achievement scores from 2015-2 were gathered; this was during the semester of the first implementation in which materials were still in production. In 2016-1, the semester of the full implementation, the status of these variables was verified. Per a report by Casas and León (2016), there was a significant improvement between the two implementations. This was due to the fact that the second implementation corrected defects in the first and that students were more comfortable with the active, flipped pedagogy in the second iteration, is shown in the following figure:

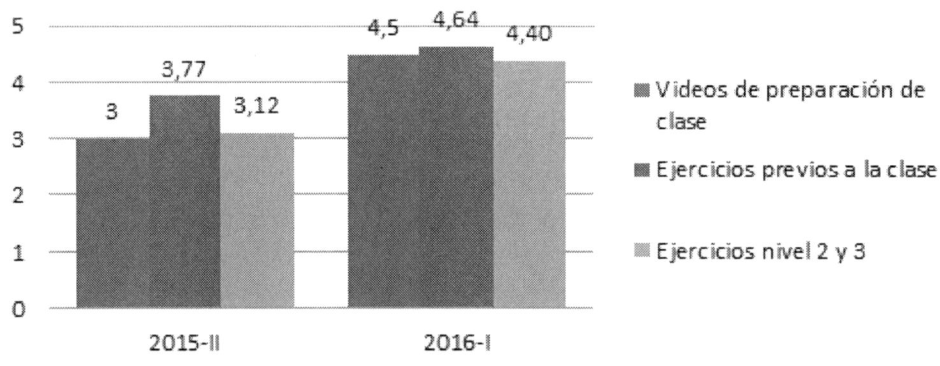

Figure 1. Student grades for some RSM activities on a 0.0 to 5.0 grading scale

When contrasting the achievement indicators obtained in the RSM course with the other mechanical engineering courses over time, Casas and León (2016, pp. 20-23) found that:

- Although the percentage of failures in RSM was still higher than those of mechanical engineering courses, there was a tendency for retention to improve as the innovation progressed.
- Average RSM scores were still below the average grades of the Mechanical Engineering Department, but there was a tendency to improve as the innovation progressed.
- The percentage of withdrawals also continued to have a decreasing trend.

- The final grade of the course depended on the class work grade: in 2015-2, 77% of the people who passed class work passed the course, and the same happened with failing grades. In 2016-1 this happened for 80% of the cases.

Casas and León (2016, p.23) concluded that the active learning methodology used before and during the class allowed the student to take greater responsibility for their learning process and also to organize their study time more effectively. It permitted the professor to have an idea of the problems that the students have before each class, which helped them to orient the class in such a way that they dealt only with the issues that caused the students the most difficulty. On the other hand, students received constant feedback that informed them of their strengths and weaknesses.

Non-cognitive student outcomes

The results of the soft skills study show the following (UniAndes, Faculty of Education Evaluation Center, 2017, pp. 19-20):

- There were no clear significant differences in self-regulation of learning in the sections with the intervention.
- Significant differences in academic self-efficacy cannot be attributed to intervention.
- There were no significant differences in the promotion of autonomy in the sections with the intervention.
- The greater methodological flexibility in the sections with the intervention compared to the non-intervened sections is statistically significant.
- Students' positive perceptions of the support of the ICTs to the teaching and learning process were significantly more positive in the group with the intervention.
- Of the three courses, the students perceived that the one that was the most flexible and most supported by the ICTs was the Structural Systems Analysis course.

To interpret these results, it is important to bear in mind that the development of soft skills can be affected not only by the intervention, but also by other factors related to the learner. As the university puts it, the institution "has students who, within an holistic, interdisciplinary and flexible learning environment, are the main agents in their educational processes" (UniAndes, 2015).

When inquiring in 2016-1 about students' perceptions of the active methodology (blended learning in SSA, flipped classroom in RSM and PS, all supported by the use of video clips and study guides), the following was found (Valencia, 2017):

- In RSM, all respondents considered the learning process to be good or excellent. In SSA, the great majority (87%) shared this opinion, while the remainder rated the process as regular or poor. In PS a good majority (64%) considered the methodology to be good or excellent. In the focus groups, it was determined that the length of the class session affected the differences in perception.
- In terms of class preparation activities, in the three courses the vast majority agreed that videos and pre-class exercises were important (100% in RSM, 93% in SSA, 90% in PS). The focus groups confirmed the positive rating that video clips received as well as rating of the importance of doing the preparatory exercises.

- As for the interaction in the lecture classes, the following table shows the very positive evaluation of the active pedagogy approach in class in general, though with a lower score for the PS group, likely due to the length of the sessions in this course.

Table 2. Student's opinions about active methodology in class

Survey item	Course	Response options			
		Completely disagree	Disagree	Agree	Completely agree
The methodology encourages student participation in the master lectures	Rigid Solids Mechanics	0.0%	0.0%	57.1%	42.9%
	Statistics and probability	6.9%	31.0%	44.8%	17.2%
	Structural Systems Analysis	0.0%	6.3%	49.2%	44.4%
The methodology facilitates interaction among students during the master lectures	Rigid Solids Mechanics	0.0%	0.0%	42.9%	57.1%
	Statistics and probability	10.3%	6.9%	55.2%	27.6%
	Structural Systems Analysis	1.6%	9.5%	49.2%	39.7%
The methodology allows for more helpful and efficient interaction between the students and the teacher in the master lectures	Rigid Solids Mechanics	0.0%	0.0%	57.1%	42.9%
	Statistics and probability	13.8%	6.9%	48.3%	31.0%
	Structural Systems Analysis	0.0%	4.8%	54.0%	41.3%
The methodology allows the teacher to solve the students' questions efficiently in the master lectures	Rigid Solids Mechanics	0.0%	7.1%	28.6%	64.3%
	Statistics and probability	6.9%	3.4%	48.3%	41.4%
	Structural Systems Analysis	0.0%	4.8%	49.2%	46.0%

Changes for professors

Professors were interviewed about the changes and lesson that came from their participation in course redesigns. Specifically, they reflected on the changes in conceptions, in media they use to teach, and in the educational practice transformations attributable to their participation. Below are some testimonies, which help to understand certain effects of the innovation piloted.

Changes in professor´s conceptions

> *I do not think that I have changed the concept that I have of what pedagogy is or how people learn, but what I see is that the process occurs with a different dynamic. The same steps, but with a slightly faster dynamic and I think better, because [the student] does not wait until the last day to try to study everything and sit down to do a test (Reyes, 2016).*
>
> *In my case I do not think it is a change in me as a teacher as such, or in my teaching style, but in the relationship of the teacher with the student, especially in class, during class ... I have always emphasized application in my courses. That is not new for me, and I have also insisted a lot on bringing the abstract concepts down to earth, but now I communicate them better (Castillo, 2017).*
>
> *In my regular classes I try to do things throughout the class (...) so that [the students] are connected. So, I ask questions (...) In the innovation course, this definitely happens because students must work during the whole class (Bernal, 2017).*
>
> *I have just become a teacher. This is my second semester of full-time teaching. I did not really have a previous experience of traditional classes. Rather, what I had was the*

opportunity to do this course blended and another course not blended. That's where I have been able to contrast things (Galvis, 2016).

Changes in media for teaching

Without the resources of this project, it was impossible to think that the student could come to class with content knowledge, because there is no textbook or guide [for SSA]. One had to write the book first. In this case, what has been done is to write the book in video form, so the student watches the video, which is a little more pleasant than reading the book (Reyes, 2016).

Both the opportunity and the need to make the videos themselves is a new experience, and it is very interesting because we thought it was not so easy to reach the students with a probability video taught by a university professor. We have found that, in general, they really appreciate the videos, which they think are of good quality (Castillo, 2017).

The real situations used as examples during the course motivated the students and increased their interest in the course. These examples allow the student to understand the usefulness of the concepts he/she is learning and also to create a valuable link between theory and practice (Casas & Leon, 2016, p. 23).

Changes in educational practices

I feel that the role of the teacher in the three moments of the flipped class [before, during, after] is more valuable, makes more sense... When I'm reviewing a quiz in class and asking why they did something, or elaborating on some content they have already looked at ... one feels that that time is more useful than if one were lecturing for two and a half hours, which is what often happens. One really does see students are more focused on the activity one is doing (Castillo, 2017).

The redesigned version of the PS course makes the process of guiding them on the topic easier because the methodology allows them to participate actively in an hour and twenty-minute-long workshop, and because they have a quiz in which they have to think and I, after, I can tell them what they got wrong and they can raise their hands and say, I still don't understand (Bernal, 2017).

The new methodology creates a challenge for both the student and the teacher. For the student, it creates the challenge of assuming a great responsibility for his or her preparation, and for the teacher it creates the challenge of taking a different role in which he is not the owner of knowledge, but rather a guide during the process (Casas & Leon, 2016, p.23)

The experience of the instructors involved in the innovations is broad. The youngest instructor was in charge of one section of the SSA course; however, academic results from the students of his sections was equivalent to the performance of the students of the other section, suggesting that the skill of the instructor was not a factor in the outcomes of the redesign.

Outcomes of interactions in the redesign process

We surveyed administrators in the Faculty of Engineering and the professors who took part in the pilot about what must be done well to succeed in an initiative like this. We also asked about the things that should also not be done, or the key success factors in an educational innovation supported by ICT. The administration was very generous in sharing their ideas which are presented in the following synthesis.

Table 3. Critical success factors stated by administrators and professors

Category	Administrators' Opinions	Professors' Opinions
Institutional Commitment	Explicit political will is required. High priority must be given to the innovation, and budget should be available to cover costs of lightened work load for professors, accompaniment, production of resources, and evaluation of the pilot.	It is important to lighten the professor's course load, especially during the semester of the innovation design and planning, as is assigning assistants to support research into / creation of digital resources.
Professor's Commitment	The professor must be intrinsically motivated to innovate in his/her practice and participate voluntarily.	Innovation requires commitment from the teacher and the program director as well as dedication throughout the project.
Focus	Active and reflective students are the focus of innovation. The fact that students are digital natives is an exploitable resource. The innovation is a good opportunity to reflect on teaching in the field of engineering.	We must always keep the academic workload of the student in mind and consider the strategy for introducing the change in teaching style. The professor should have the autonomy to define the changes that he makes in the courses.
Scope and target	The pedagogical innovation with ICT must be part of a broader, long-term institutional goal. Giving the opportunity to innovate in course units/modules, courses themselves, and at the program level.	Leadership by the innovation leader helps to define and refine the scope and strategy of innovation. Starting by thinking about the desired lessons for the students' professional training rather than planning beginning with content makes a positive difference.
Accompaniment	The pedagogical and technological accompaniment must be ensured throughout the innovation process.	The accompaniment of educational consultants from ConectaTE throughout the innovation process makes a positive difference. Workshops to improve the fluency and quality of the presentations in the videos should be included.
Formative assessment	It is important to assess innovation set up and to learn from mistakes	Generating information from the beginning permits one to show advances clearly and with scientific support, which encourages replication of this type of experiences.
Assessment of outcomes	It is interesting to know if the amount of face-to-face instruction can be reduced in the blended modality.	From the outset, it is important to define how the impact of the innovation on students and teachers will be evaluated. This includes the entry and exit profiles, as well as information about the inputs and the process.
Socialization	Socialization via publications and presentations helps to motivate other professors to innovate.	It is important to publicize cases of success and to reflect on them in order to invigorate the process of change.

Outcomes in the Faculty of Engineering

Here we return to the central question of this research: What can the Faculty of Engineering learn from this pilot experience with the aim of assuring competency-based academic programs, quality, innovation, flexibility, internationalization and sustainability? Below we present what we can confirm so far from our data:

1. Reflection on teaching practice and the pedagogical, as well as on technological innovation in engineering courses, serves as the basis for designing and implementing innovations that seek to cause changes in the professors and students.
2. The three pilots carried out in REDINGE1 make the collaboration between the innovating teachers and the consultants that accompanied them evident. The objective of this collaboration was for students to gain competencies that go beyond the disciplinary and to the development of competencies necessary to participate in the knowledge society.
3. In the professors' opinion, the accompaniment by ConectaTE throughout the innovation process enriches their ideas about how to promote learning. It impacts the media they use in their teaching. Because of the process, they shift from being the center of their classes to having the role of facilitator.
4. The cognitive impact of the innovation was significant as was shown by the significant differences between the control and experimental groups. In the cases where an experimental design was not possible, there was improved student retention and passing rates. This was largely due to changes in participation before, during and after class because of the use of easily accessible media to transmit knowledge.
5. The evaluation of the non-cognitive effects does not show significant differences for soft skills though there is not a definitive explanation of why. One could imagine that this result is not only the effect of participating in a course, but of the educational ecosystem that leads students to study in a university like UniAndes. The feedback about the learning processes shows that students were satisfied with the media and methodologies used. There were variations in these perceptions depending on how the innovation was carried out.
6. What has been learned in relation to the key success factors for pedagogical transformation with ICT support provides guidelines for the expansion and sustainability of initiatives such as those studied here.

Recommendations

The pilot experiences in engineering education carried out at the University of the Andes, Bogotá, can serve as inspiration for other initiatives within and beyond this university. Internally, only a few departments of the Engineering Faculty at UniAndes have benefited from initiatives such as this, which means that it is necessary to capitalize on this experience to create innovation guidelines to expand the initiative. This experience can also be a reference for exploring such opportunities in other engineering faculties.

Keeping this in mind, we make the following recommendations to the academic community:

1. Encourage teachers from different engineering programs to learn about innovation experiences such as these by reading publications about innovations, talking with other professors innovating in their courses, and exploring the digital resources produced. For these purposes, the Faculty of Engineering has made the video clips produced for each of the redesigned courses open access.
2. Analyze the critical success factors identified by the different actors in the process to define or refine an institutional strategy to foster pedagogical innovation with ICT.
3. Explore interest in participating in experiences such as these, both among faculty members and the directors of their departments; this information can be the starting point for defining the scope of the innovations and determining the human and financial resources required to carry out what is proposed.
4. Ensure that the accompaniment is directed not only at the design and implementation of the innovation, but also toward formative assessment and evaluation of its impacts.

Acknowledgments

This work is the result of collaboration between the Faculty of Engineering at the University of the Andes and the Center for Innovation in Technology and Education (ConectaTE) from the Faculty of Education. The Dean of Engineering, Eduardo Behrenz, led the process in 2015 and 2016, and his successors, Carlos Francisco Rodríguez and Alfonso Reyes, continued. Professors Mario Castillo and Astrid Bernal (Industrial Engineering), Juan Pablo Casas (Mechanical Engineering), and Juan Carlos Reyes and Francisco Galvis (Civil and Environmental Engineering), were the professors who implemented the innovations in the redesign their courses. Luz Adriana Osorio, director of ConectaTE, coordinated resources and monitored the process. Alvaro Galvis (ConectaTE) was the academic director of the process and techno-pedagogical facilitator as well. Mónica Patiño, coordinator and technological advisor, Alexa Tatiana Jiménez and Angélica Avalos, pedagogical advisors, Diego Valencia, consultant in formative assessment, also participated from ConectaTE. The production of educational software was led by Ricardo Calle's team. Video production and creation of digital leaning environments was done by the team led by Mariela Rivero. The Faculty of Education's Assessment Center performed the summative assessment with contributions from Samir Cure and Julian Mariño. Many thanks to everyone involved in the process.

References

Bates, A. W. (2015). *Teaching in a digital age. Guidelines for designing teaching and learning for a digital age.* Tony Bates Associates Ltd. Retrieved 11 03, 2016, from https://opentextbc.ca/teachinginadigitalage/

Bernal, A. (2017, 2 10). Reflexiones acerca de mi docencia con pedagogía activa y TIC. (D. Valencia, Interviewer)

Casas, J. P., & Leon, J. A. (2016). *Pedagogía activa con TIC para aprender Mecánica de Sólidos Rígidos.* Bogotá: Uniandes - Ingeniería Mecánica (Borrador de informe).

Castillo, M. (2017, 02 10). Mis reflexiones sobre el uso de pedagogía activa en Probabilidad y Estadística. (D. Valencia, Interviewer)

Churches, A. (2009, 10 01). *Taxonomía de Bloom para la era digital.* Retrieved from Eduteka: http://eduteka.icesi.edu.co/articulos/TaxonomiaBloomDigital

Collison, G., Elbaum, B., Haavind, S., & Tinker, R. F. (2002). *Facilitating online learning: effective strategies for online facilitators.* Madison, WI: Atwood Publishing.

Crawley, E. F., Malmqvist, J., Östlund, S., Brodeur, D. R., & Edström, K. (2014). *Rethinking Engineering Education: The CDIO Approach.* Springer Science & Business Media.

Deci, E., & Ryan, R. (2000). The "What" and "Why" of goal pursuits: Human needs and the self-determination of behavior. *Psychological Inquiry, 11*(1), 227-268.

Erickson, H. L. (2007). *Concept-based curriculum and instruction for the thinking classroom.* Thousand Oaks, CA: Sage Publications Ltd.

Forté, E., & Wentland, M. (1998). The ARIADNE project: knowledge pools for computer-based and telematics supported classical, open and distance education. *AAUC Ariadne Academic Users Group Conference* (pp. 1-23). Lucerna, Swtizerland: AAUC.

Galvis, A. H. (2016). *Ambientes de aprendiaje híbridos, o mixtos, para educación superior y a lo largo de la vida.* Bogotá: Uniandes - Facultad de Educación - Conecta-TE.

Galvis, A. H., & Pedraza, L. C. (2013). Desafíos del bLearning y del eLearning en educación superior. In N. Arboleda Toro, & C. Rama Vitale (Eds.), *La educación superior a distancia y virtual en Colombia: nuevas realidades* (pp. 113-154). Bogotá, DC: Virtual Educa y ACESAD. Retrieved from https://goo.gl/jBNMpN

Galvis, F. A. (2016, 10 13). Reflexiones acerca de mi docencia con pedagogía activa y TIC. (D. Valencia, Interviewer)

Hansen, E. J. (2011). *Idea-based learning: A course design process to promote conceptual understanding.* Sterling, VA: Stylus Publishing, LLC.

Mills, J. E., & Treaagust, D. F. (2003). Engineering education - is problem-based or project-based learning the answer? *Australian Journal of Engineering Education*, 1-16.

Reyes, J. C. (2016, 10 21). Reflexiones acerca de mi docencia antes y después de REDINGE1. (D. Valencia, Interviewer)

Reyes, J. C., & Galvis, F. A. (2017). *Implementation of blended learning in a Structural Analysis course: Paper 163 accepted for presentation in REES 2017.* Bogotá: Research in Engineering Education Conference Proceedings (julio 6 a 8, 2017).

Uniandes. (2015, 12 9). *PDI 2016-2020 Versión final.* Retrieved from Universidad de los Andes, Colombia: https://goo.gl/biCHcW

Uniandes, Centro de Evaluación de la Facultad de Educación. (2016). *Evaluaciòn del efecto de una innovación tecnológica y pedagógica sobre el aprendizaje en un curso de Probabilidad y Estadística.* Bogotá: Facultad de Educación - Centro de Evaluaciòn.

Uniandes, Centro de Evaluaciòn de la Facultad de Educación. (2017). *Evaluación de innovaciones educativas de ConectaTE sobre habilidades blandas en la Facultad de Ingeniería.* Bogotá: Centro de Evaluaciòn, Facultad de Educación.

Uniandes, ConectaTE. (2016). *Portal del Centro de Innovación en Tecnología y Educación.* Retrieved from ConectaTE : https://conectate.uniandes.edu.co/

Uniandes, Facultad de Ingeniería. (2017). *Facultad de Ingeniería, Información General.* Retrieved 02 16, 2017, from Universidad de los Andes: https://goo.gl/vM44a6

Uniandes, Secretaría General. (2015, 04 15). Faculty Handbook - Approved by the Steering Commitee session 105-15, April 15, 2015. Bogotá, DC, Colombia. Retrieved from https://goo.gl/1kTxnW

Valencia, D. (2017). *Borrador de informe sobre opiniones de estudiantes y docentes acerca de cursos de REDINGE1.* Bogotá: ConectaTE - documento de trabajo.

Wiggins, G., & McTighe, J. (2005). *Understanding by Design.* Alexandria, VA: Association for Supervision and Curriculum Development.

Ziegenfuss, D. (2007). A phenomenographic analysis of course design in the academy. *Journal of Ethnographic and Qualitative Research*(2), 70-79.

Copyright statement

Understanding Engineering Educators' Pedagogical Transformations through the Hero's Journey

Audrey Boklage
Arizona State University, Mesa, Arizona
aboklage@asu.edu

Brooke Coley
Arizona State University, Mesa, Arizona
bccoley@asu.edu

Nadia Kellam
Arizona State University, Mesa, Arizona
nkellam@asu.edu

***Abstract**: Many faculty wish to create better learning environments for their students. Particularly, in Engineering Education, where traditional approaches to learning have been associated with limited engagement for many students. This study investigates the journeys taken by Engineering faculty at various United States higher education institutions to adopt student-centred pedagogies in the classroom using Dewey's Theory of Experience as a theoretical framework and Joseph Campbell's Hero's Journey as a structural way to construct and analyse narratives. Two narratives are highlighted in this paper to elucidate a better understanding of the process to successfully adopting pedagogies, including, for example, the different ways faculty experience their calls to adventure, their road of trials, and their ultimate boons. These journeys shed light on the stages involved in the process, how individuals interact with the environment to evolve over time, and the ways in which the vision for change impacts the way the journey unfolds.*

Introduction

Research recognizes the importance of universally incorporating student-centred teaching in all classrooms, ranging from pre-kindergarten to post-secondary institutions (Bransford, Brown, & Cocking, 2000; McCombs & Miller, 2007; Weimer, 2013). However, incorporating these pedagogies is challenging for many reasons. Professors must identify students' needs while advocating for their own professional development to understand and incorporate such strategies, which can be difficult against the priorities of research and service (Fairweather, 2002). Such challenges are clearly articulated in the literature (Council, 2012; Fairweather, 2008); however, avenues to support and encourage success in adopting student-centred pedagogies are lacking.

Engineering Education researchers have been interested in the process by which educators shift to adopting student-centred strategies in the classroom (Borrego, Froyd, & Hall, 2010; Council, 2012; Fairweather, 2008; Henderson, C., Beach, A., 2011; Seymour, 2002). These student-centred strategies are not being implemented on a wide scale and researchers are still observing a heavy reliance on lectures by engineering educators (Borrego et al., 2010; Council, 2012; Fairweather, 2008; Seymour, 2002). Traditionally, researchers have studied this discrepancy by focusing on why engineering faculty do not succeed in transitioning to student-centred strategies (Council, 2012; Fairweather, 2008). Factors cited that contribute to this gap include faculty rewards, the perceived importance of teaching, and the institutional focus on research funding (Fairweather, 2008). Additional factors cited in the Discipline-Based Education Research (DBER) report include the department, institutional priorities, reward systems, students, and faculty members' beliefs (Council, 2012).

More recently, researchers have begun to try to understand how to encourage faculty adoption of research-based, student-centred instructional strategies through studying faculty motivation (Matusovich, Paretti, McNair, & Hixson, 2014), theories of change (Borrego & Henderson, 2014; Finelli, Daly, & Richardson, 2014), and professional development programs (Lattuca, Bergom, & Knight, 2014). Moreover, a recent push in the literature and from funding agencies to focus on change strategies (Borrego & Henderson, 2014; Finelli et al., 2014) align with our current study. In this work, we are attempting to understand successful change stories that promise to illuminate effective theories of change through constructing narratives to impact other faculty and administrators as they begin to develop their shared vision for innovation in engineering education (Borrego et al., 2010). Besterfield-Sacre et al. surveyed faculty, department chairs and deans in Engineering Education across 156 institutions to find that a shared vision for innovation was critical to achieving change (Besterfield-Sacre, Cox, Borrego, Beddoes, & Zhu, 2014).

This qualitative research project focuses on the journeys of exemplary engineering faculty who have experienced successful implementation of student-centred teaching strategies. From their narratives we are able to begin identifying contextual and individual factors that contributed to their ability to implement student-centred pedagogies while also gaining understanding of the barriers that made the process challenging.

Purpose and Research Questions

The purpose of this study is to better understand the successful transition of engineering faculty from teacher-centred to student-centred strategies. The following research question guided our project: "How do engineering faculty experience their personal journey of changing their teaching practices and pedagogies?"

Theoretical Framework

The theoretical framework guiding this work is Dewey's Theory of Experience (Dewey, 1938) which situates experience as consisting of the principle of continuity (past, present, and future) and the principle of interaction (between an individual and their environment). Dewey explains that an experience has the potential to act as a "moving force" and that through having an experience that "arouses curiosity, strengthens initiative, and sets up desires and purposes" the experience can be an impetus for growth. An experience can only be fully understood if one considers not only the individual, but also the environment in which the individual interacts. As it applies to this study, the faculty members' experience is represented in their journey as they interact with their environment, i.e., the institution, their classroom, in the process of executing change. The faculty member's process, from the impetus of their desire for change to its actualization, is the growth that Dewey describes. This theoretical framework also guided the choices for our methodology.

Methodology

We conducted narrative interviews with eight engineering faculty specifically chosen from diverse disciplinary backgrounds within the engineering community. Next, based on the transcripts of these interviews, we constructed participant narratives using Joseph Campbell's Hero's Journey as a coding and structuralizing scheme, method is described in detail in Cruz and Kellam's paper (2017). Lastly, we selected two narratives from the group to demonstrate the range, in terms of the associated experiences, in the faculty journeys toward transforming their pedagogies.

Participants

Participant selection consisted of a combination sampling strategy. First, we identified participants using a criterion sampling strategy. This involved reviewing abstracts from two United States conferences to identify authors who are at the cutting edge of pedagogical innovations and who report using a student-centred teaching strategy. Snowball sampling was then used to identify additional faculty members who have transitioned to incorporating student-centred teaching strategies. The snowball sampling strategy led us to have many participants from construction engineering so we are now using a maximum variation sampling strategy to identify more participants and to ensure that faculty used in our sample are diverse in their gender, race/ ethnicity, and institution type.

Data Collection

Narrative interviews were conducted in person and via Skype ® with participants that lasted between 29 and 84 minutes. The purpose of these semi-structured interviews was to elicit the narratives of participants. The interview questions aligned with the theoretical framework (Dewey, 1938) in that they encouraged respondents to describe experiences from their past and present and to discuss how they are situated to move into the future (principle of continuity) and encouraged them to discuss their personal experiences and broader social experiences (principle of interaction). The first question, "I understand you are an [job title of discipline], can you walk me through how you have become the engineering educator that you are today," was developed to encourage participants to enter the narration phase of the interview (Kim, 2016) where they described their story starting with their first realization of a need to change their teaching style from the way that they typically taught or from the way they were taught as a student. The interviewer then transitioned into the conversation phase (Kim, 2016) of the interview and asked probing questions to encourage the participants to further elaborate on aspects of their story; these probing questions included questions such as, "You mentioned _______, tell me more about that," "You mentioned ___________, can you give me an example of that," or "You mentioned _________, what was that like for you." Depending on when the participant started or ended their story additional probing questions were asked to encourage them to describe early experiences teaching, recent experiences teaching, and/ or how they improved their teaching (e.g., from formal workshops, mentors, students). Additionally, the interviewer asked the participant to describe catalysts and barriers to changing their teaching practices. These questions, such as "How have others in your unit responded to changes in your teaching?" helped uncover social and cultural aspects that may have contributed or discouraged changes in teaching. In keeping in alignment with Dewey's principle of continuum, the final question looked to their future and focused on future plans for development as a teacher, "How would you like to further develop as a teacher? How do you think you'll go about doing that?"

Data Analysis

The data analysis followed an approach outlined by Cruz & Kellam (2017) that takes a structural approach to constructing and analysing participant narratives. This approach involves using Joseph Campbell's monomyth as a way of uncovering patterns that are common across narratives. The first phase of this approach involves becoming familiar with the transcripts through multiple passes of reading and identifying events within the transcripts. In the next step, the events were coded according to Campbell's 16 monomythic codes (2004) modified by Cruz & Kellam (2017). Each of these monomythic codes and our interpretation or description of that event as it relates to our data set is described below:

- The *call to adventure* is the opening of the narrative and involves the faculty member describing their motivation to change their teaching practice. The refusal of the call occurs when the faculty member considers not answering the call to adventure, and, in turn, not changing their teaching practice.
- *Supernatural aid* occurs when the faculty member receives unexpected aid from a person (e.g., mentor or colleague) or from media (e.g., a book or website) that helps prepare them for the journey on which they are about to embark.

- *The first threshold* is the first trial that the faculty member faces in their journey. This is a challenge that is generally expected by the faculty member and sets them on their journey into the unknown, from the ordinary world into the special world.
- During the *belly of the whale* portion of the journey, the faculty member is "swallowed into the unknown, and would have appeared to have died" (Campbell, 2004). This is the lowest point in the faculty members' journey and becomes a transformative event.
- During the *road of trials,* the faculty member encounters many challenging and difficult events as they experiment with different pedagogies.
- The *meeting with the all-knower* event involves the faculty member meeting with a mentor or colleague who transforms their journey and is essential to the faculty member becoming a teacher who uses active-learning strategies in their classroom. This relationship is crucial for the faculty member to reach a resolution in their journey.
- Moreover, the *meeting with temptations* portion of the journey involves the faculty member being tempted by something that would encourage them to go back to the way that they taught previously. The temptation is generally positive and could be a faculty member moving out of teaching due to new administrative responsibilities.
- In the *atonement with parent* stage of the journey, the parent, here possibly represented by an administrator or a more senior faculty, has influenced the faculty member until their realization of their own power. In this stage of the journey, the faculty harnesses their power and becomes more autonomous and confident in following their own path.
- In the *apotheosis* stage, the faculty member reaches a stage of understanding and begins to take ownership of their new way of teaching becoming comfortable in his or her new role as a teacher who has transformed their pedagogies.
- The *ultimate boon* occurs when the faculty member reaches resolution and feels as if they have mastered their new art of teaching and have reached a resolution of their journey.
- The *refusal of the return* phase occurs after the journey has reached a conclusion when the faculty member consciously refuses to return to 'the ordinary world' and, for example, defers teaching through focusing on research.
- The *magical flight* stage involves a chase or pursuit that helps the faculty member return to the starting point of their journey or to the next journey, for example pursuing tenure or enrolling in a higher level course to incorporate student-centred learning.
- The *rescue from without* phase in the journey occurs if the faculty member is refusing the return. In this stage, there is some type of event helping bring the faculty member back to the ordinary world, thus resolving the journey.
- The *return threshold* occurs when the faculty member returns to the ordinary world and shares with colleagues and administrators about what happened during their journey, thus reconciling their identity that was developed in the special world with their 'ordinary' faculty identity.
- The final phase, *master of both worlds and freedom to live*, enables the faculty member to move back and forth effortlessly between the ordinary world and the special world. The faculty member has reached a point of clarity where they can be an exemplary teacher embracing active learning strategies while also being an exemplary faculty member in the other aspects of their job—research and service.

Findings

We chose two faculty's narratives to illustrate two different pathways of faculty incorporating student-centred pedagogy in their classrooms. These findings will first focus on Dewey's principle of continuity and conclude with a discussion of the principle of interaction. The findings will also highlight the structural analysis that was conducted using Joseph Campbell's monomyth. Each participant's story was unique and did not necessarily use all 16 codes from Campbell's monomythic journey. For the purpose of this paper, we will share excerpts from each interview and constructed narrative with their subsequent codes. Additionally, these findings will highlight the contextual and individual factors that contributed to the successful transitioning of faculty to student-centred teaching strategies. We are using the names of the participants in this section as these are their stories and the participants gave consent to use their names in dissemination efforts.

Case 1: Donna Riley

Donna Riley is a Professor of Engineering Education and Science and Technology in Society at Virginia Tech. Her journey to adopting student-centred pedagogies evolved over a "10-year period of experimentation" with her thermodynamics course.

As an undergraduate, Donna described her professor norm in engineering as the "Cambridge and Oxford educated" that lectured and presented notes fostering very little engagement for the students. Their pedagogies forced students to have to work together to gain an understanding of the material beyond that created in lecture. The outlier in this group consisted of a professor that used transparencies projected onto a screen rather than a chalkboard.

However, in her other courses, those in the humanities and social science, Donna was having a very different experience. Particularly, the professor of her Religious Studies course would engage them in critical conversations related to readings and reflections that would create a space where everyone felt connected. Donna recalled, "What always surprised me was that I felt like I belonged in the room."

This level of engagement stimulated her interest in the subject despite it being a challenge for her. For example, she did not speak Greek, one of the suggested prerequisites for the course, but she still felt a sense of belonging in the class based on the professor's delivery alone. This triggered her to start thinking critically about the way engineering was taught and why similar approaches had not been adopted in such environments. She described,

> I became curious about that time about why engineering couldn't be taught in the same way that my religion classes were being taught. I didn't really get to pursue that question, [it] just rested in the back of my mind for a while and when I got in the grad school, I started...

Donna's experiences prompted her to enter the Ph.D. program in the Center for Teaching and Learning at Carnegie Mellon where she could focus on teaching as a part of a community in Engineering also interested in such topics. Having had such engaging learning experiences in other studies compared to engineering served to be Donna's *Call to Adventure*. Early on in her graduate education, she established a desire to challenge the existing approaches to teaching engineering in hopes of developing pedagogical innovations. The first test of her journey occurred when she received her first faculty appointment at Smith College.

Smith College is a women's, liberal arts college. After teaching a team-taught Introduction to Engineering class in the fall, Donna's first solo taught course was Thermodynamics the following Spring. It was the first time everything was up to her—the syllabus, curriculum, and deliverables. As a new faculty member and first time independently teaching a course, she was noticing familiar outcomes of her teaching despite being more proactive and engaging. This was Donna's *First Threshold* and she recalled it in this way,

> I noticed that I was repeating some of the very same problematic relationships that existed in my prior experience with Thermodynamics. This was true even though I wasn't doing this passive lecture thing. I was doing active learning. I was doing the stuff I was taught to do, but I could tell there were students in front of the class that were engaged, and the students in the back of the class weren't engaged. I could just see it all unfolding in those same ways that I had been taught."

In efforts to understand how she could further improve her teaching, she reached out to a trusted colleague for advice. Her colleague served to provide one of what would become several sources of *Supernatural Aid* along her journey. This colleague recommended a particular text, *Teaching to Transgress* by bell hooks, that enlightened Donna's understanding of classroom dynamics and would ultimately prove to be a critical input into the unfolding of her journey. Upon reading the book, it completely shifted how Donna perceived the dynamics of her classroom. It gave her a deeper understanding of the knowledge that her Religious Studies Professor had held regarding power relationships in the classroom and viewing students in a holistic way. The text provided knowledge regarding valuing the authority of experience as well as what students bring into the classroom. These elements had been integrated into her Religious Studies course, but had been lacking in her Engineering related courses.

Donna continued to innovate the Thermodynamics course, but she wanted to secure a sustainable source of funding to support her research. She entered the *Belly of the Whale* phase of her journey when she found there was no home for her interdisciplinary research at the National Science Foundation (NSF). Her Ph.D. had been received in Engineering and Public Policy which made her research interdisciplinary. Her specific work was in the areas of risk-assessment and risk-communication. As a proactive faculty member, she went to NSF to meet with Program Officers to discuss funding opportunities, one of happened to be a colleague from her research group. The Program Officers informed Donna that there was no fit for her research in either of their Directorates and she was encouraged not to waste her time writing a CAREER award around her research interests.

This upset Donna because it left her uncertain of how to proceed. She was expected to apply for a CAREER award as a part of the tenure process, but here she was being told that her research had no place in the competition. As *Supernatural Aid* would have it, Donna serendipitously encountered a colleague that was known to be a mentor to everyone while in the office of another colleague, just shortly after receiving the discouraging news. When asked how she was doing she shared her frustration with the news she had just received. Unbeknownst to her, the mentor described a unique opportunity for her to apply her research for a CAREER award in Engineering Education, which was a new venue for similar work at the time. Donna remembered the crunch and fulfilment of that exchange, "I had two weeks to write the thing. I wrote it. I submitted it. It wasn't the best grant proposal ever written, but they funded it." Although she described funding it as a potential risk for the NSF Program Officer, through this award, Donna now had a five year funded research program to explore what it would mean to adopt bell hooks-like pedagogies in Engineering Education.

This was a turning point in Donna's journey. She faced several challenges in the *road of trials*, such as receiving push back from the students and other faculty or challenging critical assumptions in the classroom. Nonetheless, she continued to make strides in her teaching to create a critical and reflective way of thinking amongst her students. Donna exceeded a desire to simply engage students in the classroom; she was attempting to create a transformational change in the way things were done and the degree of development that it fostered for the students. She experienced *Apotheosis* after enduring the 10-year experiment with Thermodynamics at Smith College, as she described it. She describes her sentiment,

> Some of it certainly was that I did grow frustrated with the amount of push back I was getting at Smith having gone there to be a change-maker and getting to a place where people didn't really understand what I was doing. The students didn't, my

colleagues didn't. While I had a great cadre of folks that I could start to step up with, it became, I started thinking more about the bigger picture of what was I doing in the field of Engineering Education and how I could have an impact outside of my institution. As the Smith experiment wore on, people paid less and less attention to what was going on there, and kept saying "Well, you can do that because you're at Smith. You're a special case." Doing something at Virginia Tech would obviously directly impact a large number of engineers right away, and have more influence in the rest of the enterprise of Engineering Education, so that made the move make a lot of sense.

After serving in the policy domain at the National Science Foundation as Program Officer in Engineering Education, Donna transitioned to Virginia Tech to have a more widespread influence on the enterprise of Engineering Education. She continues to innovate new pedagogies for transforming the realm of Engineering Education from several vantage points. Her current work involves iterative development and delivery of a course for non-Engineering majors on Citizen Engineering. This course explores the meanings of engineering in society and how members of the public can influence large engineering projects as well as undertake their own engineering projects in their communities, redefining what engineering can and should be.

Case 2: Charlie

Dr. Charlie Pierce is an associate professor of civil and environmental engineering at the University of South Carolina. Charlie's father was a civil engineer and he credits his father with an early exposure to the professional practice of engineering "I had a pretty good idea I wanted to go into engineering from high school into college…I was well aware of what my dad did and I seemed to have those interests." It was not until after his undergraduate experience that Charlie experienced the first step of his journey, *the call to adventure*, earning a PHD with the purpose of eventually teaching engineering. At this point in his life he had a "better sense" of what a PhD entailed and recognized that although he was a good researcher he felt his true strength was in teaching.

After teaching a variety of civil engineering courses during graduate school and eventually as a professor, Charlie's *call to adventure* evolved from teaching to a more student-cantered approach to teaching, mentoring and purposefully empowering his students to use their engineering education to accomplish their dreams.

As a result of embarking on the journey of purposeful, student-centred teaching, Charlie, like all heroes, encountered *trials* along the way. Many of his trials were related to his determination to teach in a way that engaged students with the materials and forced them to think critically. In his interview he reflected on his experiences as an engineering student and the challenges he faced with regard to engaging in his courses. He expressed fond remembrance for instructors and courses that were "engaging and entertaining". Ultimately this served as a motivation for him to make his classes engaging. He described a specific tension regarding engaging his students that he experienced early in his career between his preparation for the class as a professor and the need to formatively assess student learning. Charlie shared an anecdote about his efforts to compile notes for his lecture, something he spent a lot of time on; however, once he began teaching he realized that good teaching was not about covering all the content, but rather about students truly understanding the concepts of the course. This was his first step towards student-centred teaching.

The importance of conceptual understanding as part of his student-centred pedagogy was Charlie's *ultimate boone*. Throughout his interview he referenced the importance of "students understand[ing] the most basic principles" and "critical thinking". He credits teaching students in their first year of engineering courses with forcing him to focus on fundamental understanding of principles. "Trying to introduce to them what civil engineering is" was of utmost importance to Charlie.

As a result of his trials and realization of his *ultimate boone*, Charlie decided to permanently change his teaching towards an active-learning student approach, his *apotheosis*. He realized that worksheets and grading answers as either right or wrong did not contribute towards his vision of student learning. Charlie became more focused on the process of how students learned and grasped a concept. This encouragement of students to work towards finding the correct answer and explain their thinking became a core tenet of his classes a process he described as "active learning, worksheets to document, [and] in-class problem solving."

As Charlie's journey culminated, he experienced the sense of being a *master of two worlds* and *freedom to live* as a professor who is still focused on prioritizing and encouraging student's conceptual understanding of engineering in his classroom. He credits earning tenure with allowing him this freedom to write grants and collaborate with other faculty members to implement innovative pedagogies, such as a flipped classroom, in engineering courses at his University.

> That makes a huge difference for me personally to know that there's a decent size group of faculty that I brought in to writing these kinds of proposals, to doing this kind of work, to knowing that it has meaning and potential impact.

Discussion and Implications

This study applied Dewey's Theory of Experience to understand the journeys of faculty as they made transitions to incorporating student-centred pedagogies in their teaching using the structure of Campbell's Hero's journey. As explained in Dewey's framework, both Donna and Charlie experienced change that was largely impacted by the interaction between the individual and their environment as they navigated a path well aligned with the progressive stages of the Hero's journey. Interestingly, although both individuals were striving to attain a student-centred approach to teaching, their overarching pedagogical goals were different. Specifically, Donna wanted to transform the realm of Engineering Education compared to Charlie who wanted to promote conceptual understanding for his students. Consequently, the Belly of the Whale for Donna and Charlie differed significantly. Donna's involved external forces that held influence over her ability to develop a research program to learn how to create and implement innovative pedagogies where Charlie's involved internal tensions between his perception and practice of teaching. Although the paths of these faculty were distinct, they each demonstrated continuity and growth in their journeys while structurally aligning to the Hero's journey. Even though each participant did not experience every stage of the Hero's journey, it served well as a structural model and in it's applicability across a range of faculty journeys.

This study yields insight into the larger social systems that should be considered as faculty take the initiative to adopt alternative pedagogies. Specifically, the barriers (i.e., trials in the hero's journey) experienced on the journey seem to be correlated with the degree of disruption of the innovation. Of the two cases presented, Donna's vision was to transform a way of doing by incorporating the value of students' experiences into the classroom and viewing them in a holistic way, an atypical approach for most Engineering courses. This effort was met with greater push back from the surrounding community of other faculty and colleagues because it challenged an accepted norm. Charlie had a more local vision, involving optimization of students' conceptual understanding of the material through fostered engagement in the course. The barriers in achieving this vision rested internally to the actual faculty member and his classroom. This work suggests a correlation between the degree of change envisioned through the student-centred approach and the resulting resistance of the institutional context. For faculty wishing to make changes in their teaching, having a clear vision of the impact they want their innovations to have can better prepare them for the types and scales of barriers they will face on their journeys.

Another implication of this work is the importance of having a community of support in moving toward student-centred pedagogies. Both faculty cases in this work shared the sentiment that having people to share and encourage their ideas made all of the difference in their decision to pursue their *ultimate boone*. True impact came from a collective of people with the mindset to create change rather than just one person in a silo trying to make a difference.

References

Besterfield-Sacre, M., Cox, M., Borrego, M., Beddoes, K., & Zhu, J. (2014). Changing engineering education: Views of US faculty, chairs and deans. *Journal of Engineering Education*, *103*(2), 193–219.

Borrego, M., Froyd, J. E., & Hall, T. S. (2010). Diffusion of engineering education innovations: A survey of awareness and adoption rates in U.S. engineering departments. *Journal of Engineering Education*, *99*(3), 185–207.

Borrego, M., & Henderson, C. (2014). Increasing the use of evidence-based teaching in STEM higher education: A comparison of eight change strategies. *Journal of Engineering Education*, *103*(2), 220–252.

Bransford, J. D., Brown, A. L., & Cocking, R. R. (2000). *How people learn: Brain, mind, experience, and school*. Washington, D. C.: National Academy Press.

Campbell, J. (2004). *The hero with a thousand faces*. Princeton, New Jersey: Princeton University Press.

Council, N. R. (2012). *Discipline-based education research: Understanding and improving learning in undergraduate science and engineering.* Washington, D.C.: The National Academies Press. Retrieved from internal-pdf://dber-report-1246205190/DBER-report.pdf

Cruz, J. M., & Kellam, N. N. (2017). Restructuring structural narrative analysis using Campbell's monomyth to understand participant narratives. *Narrative Inquiry*, *27*(1), 172–189.

Dewey, J. (1938). *Experience and Education*. Kappa Delta Pi.

Fairweather, J. (2002). The mythologies of faculty productivity: Implications for institutional policy and decision making. *Journal of Higher Education*, *73*(1), 26–48. http://doi.org/10.1353/jhe.2002.0006

Fairweather, J. (2008). *Linking Evidence and Promising Practices in Science, Technology, Engineering, and Mathematics (STEM) Undergraduate Education: A Status Report*. Washington, D.C.: National Research Council.

Finelli, C., Daly, S., & Richardson, K. (2014). Bridging the Research-to-Practice Gap: Designing an Institutional Change Plan Using Local Evidence. *Journal of Engineering Education*, *103*(2), 331–361.

Henderson, C., Beach, A. (2011). Facilitating change in undergraduate STEM instructional practices: An analytic review of the literature. *Journal of Research in Science Teaching*, *48*(8), 952–984.

Kim, J.-H. (2016). *Understanding Narrative Inquiry*. Thousand Oaks: SAGE Publications Inc.

Lattuca, L., Bergom, I., & Knight, D. (2014). Professional development, departmental contexts, and use of instructional strategies. *Journal of Engineering Education*, *103*(4), 549–572.

Matusovich, H., Paretti, M., McNair, L., & Hixson, C. (2014). Faculty motivation: A gateway to transforming engineering education. *Journal of Engineering Education*, *103*(2), 302–330.

McCombs, B. L., & Miller, L. (2007). *Learner-Centered Classroom Practices and Assessments: Maximizing Student Motivation, Learning, and Achievement*. Thousand Oaks: Corwin Press.

Seymour, E. (2002). Tracking the processes of change in US undergraduate education in science, mathematics, engineering, and technology. *Science Education*, *86*(1), 79–105. http://doi.org/10.1002/sce.1044

Weimer, M. (2013). *Learner-centered teaching: Five key changes to practice* (2nd ed.). San Francisco: Jossey-Bass.

Copyright statement

The following copyright statement should be included at the end of your paper. Substitute authors' names in final copy only.

Exploring how knowledge is enacted within engineering practice

Grace Panther
Oregon State University, Corvallis, United States
pantherg@oregonstate.edu

Dr. Devlin Montfort
Oregon State University, Corvallis, United States
Devlin.Montfort@oregonstate.edu

Abstract: *Describing how knowledge is utilized differs between cognitive scientists and engineering education practitioners but little is known about how this relates to actual use of knowledge in engineering practice. This study aims to explore how overlapping realms of knowledge are enacted by practicing engineers in the context of an interdisciplinary engineering project through the use of phenomenological interviews with practicing engineers. Findings suggest that engineers classify knowledge based on requirements and how these shape their responsibilities in the context of their teams. The emphasis on requirements and responsibilities differs from the emphasis in academic settings and has implications for how students are taught in preparation for entering the workforce.*

Context

Interdisciplinary engineering projects require a vast amount of knowledge spanning multiple disciplines and fields. Previous research suggests that knowledge used within engineering practice is connected by social interactions (Trevelyan, 2010). Furthermore, describing the knowledge used on projects differs between cognitive scientists, who tend to use the construct "domain," and engineering curriculum, which divides knowledge into disciplines. But little is known about the impact on students in regards to how knowledge is studied (domains) and taught (disciplines) and how social interactions shape the knowledge used in engineering practice. Therefore, this work focuses on understanding how knowledge is enacted in engineering practice as a way to identify challenges students might face when transitioning to engineering practice.

Cognitive science research suggest that human knowledge is specialized and tied more directly to specific domains (Hirschfeld and Gelman, 1994). For this research, we define a domain as a cluster of facts, processes, and concepts that are grouped together based on a common theme to form high level and abstract thinking and are shaped by the social context (Alexander and Judy, 1988; Paris, Lipson, and Wixson 1983). Though it is important to note that there is no consensus within the field on what constitutes a domain, how to identify them, or how many there are (Hofer and Pintrich, 1997). Domains then become a way to describe the order and grouping of knowledge. A few examples of how domains have been divided include the sciences versus the humanities and social sciences, hard versus soft, applied sciences versus pure sciences, and school world versus real world. (ex. Jehng, Johnson, and Anderson, 1993; Palmer and Marra, 2004; Paulsen and Wells, 1998).

When discussing engineering curriculum, however, we usually see disciplines as conceptually and administratively different units with some connectedness and interrelationships (Palmer and Marra, 2004). For example, Chemical and Mechanical engineers are two different disciplines but both take courses in thermodynamics. While these courses both have the same name, they have a difference in emphasis as Mechanical engineers learn about machinery and Chemical engineers learn about reactions. The knowledge is presented to students as being essentially different along lines of discipline.

Purpose

The purpose of this study is to explore how overlapping realms of knowledge are enacted by practicing engineers in the context of two interdisciplinary engineering projects.

Methodology

We approach this work from a situative perspective by interviewing engineers about real-world interdisciplinary projects that they are currently involved in as we believe knowing and doing are interlocked and inseparable (Brown, Collins, and Duguid, 1989).

Phenomenological interviews were conducted with 19 engineers from two different interdisciplinary projects each lasting approximately 50 minutes. The interviewees represented civil, mechanical, chemical, electrical, computer, biological, industrial and environmental engineering, as well as backgrounds and expertise in chemistry and computer science. Participants' work experience ranged from recently graduated from a doctoral program to over 35 years of industry experience. The first project focused on designing an autonomous, remotely operated machine to make measurements in extreme environments. The second project involved the design of a water treatment system to remove an industrial solvent contaminating an area of 550-acres. Each project required significant and ongoing interaction between the engineers in an effort to meet all of the project requirements in complex and uncertain environments.

Interviews were semi-structured and followed a pre-designed interview guide. Following an interview guide enabled ease of comparison across interviewees while the semi-structured nature allowed for further exploration of ideas as they organically arose in conversation (Patton, 2002). The interview guide was designed with the assistance of an external research advisory board whom had expertise in epistemology and cross-disciplinary engineering design teams. The questions were intended to explore how the engineers thought about knowledge and how their knowledge related to the knowledge of others on the project. The interview guide was first piloted with engineers not from the projects studied here to ensure the questions were clear and that the terminology used was appropriate for the context. Modifications were then made to the interview guide before further interviews were conducted. Questions spanned a range of topics including having the interviewees summarize what they were responsible for designing, what they did when they got stuck, and what constraints they were working within. The audio recordings of the interviews were then transcribed before being uploaded into a cloud-based qualitative platform for analysis (Dedoose, 2017).

Results reported here are supported by preliminary analysis of four of the interviews. At this stage, broad themes have been identified through a process of identifying relative text and deductive and inductive coding (Auerbach and Silverstein, 2003). Analysis was guided by exploring how engineers talked about knowledge in relation to other engineers on the project. These events were chosen to study as they often signalled a discussion centred on overlapping realms of knowledge.

Findings and Discussion

When looking to see how engineers talk about their work in comparison to other project components, we found that engineers primarily discussed their designs or analyses in terms of external requirements and how these requirements shaped their responsibilities.

Yeah it is [first name of the mechanical engineer]'s job to make sure the thing fits into the structure appropriately. He is the one that said 'I want you to put screw holes there if you can because that is where it is going to attach to the overall spacecraft.' ~Propulsion Engineer

In the above quote, we see the propulsion engineer relating a feature of his design that was determined by the structural engineer. As is characteristic of many such discussions in the interviews, the propulsion engineer discussed this interaction as an exchange of information intended to help both parties achieve their goals: the propulsion system component needed to be attached to a structural component so the two engineers worked together to ensure their overlapping areas of responsibility were both considered. We see here how the design responsibilities of one subsystem (ensure everything fits securely inside the container) instigated an interaction with another subsystem in terms of a necessary requirement (placement of screw holes). It is important to note that the engineer here made no mention of *how* he was going to fulfill this requirement, only that the requirement was placed on him by another system. This again shows how the practicing engineers conceptualized their interactions as a negotiation of "requirements" across subsystems. A similar notion was echoed by a mechanical engineer.

So there are lots of input from other team members on kind of non-physical requirements and stuff and I have to somehow turn that into a physical manifestation that fits within a volume constraints, and mass constraints of the mission. ~Mechanical Engineer

Here the mechanical engineer was referring to how interactions with other subsystems, including non-physical requirements, must be translated into a "physical manifestation" in order to fit within his given set of constraints. These outside requirements partially drove the design and influenced how the engineer worked within his own constraints. Though the engineer never mentioned *how* he would do this, he simply focused on the idea that it was his responsibility to figure it out. He does not see his role in terms of his knowledge of his subsystem (although that is probably assumed), but rather as a kind of spokes-person for the requirements resulting from the constraints and demands on his particular area of responsibility. This idea is clearly evident in the following quote by the same engineer.

...how are we going to verify the fused manifold process [i.e. how can we ensure that using a new manufacturing process for some of our components won't interfere with the function of the system]? Does it have an impact on biocompatibility? Every design that I produce for the bio team, has a review with the scientists where they really pick at the materials, the surface finishes, the processing, all that sort of stuff that is related to that. ~Mechanical Engineer

The mechanical engineer had to work within several different layers of constraints to meet the design requirements. First, they had to work within their own subsystem constraints to verify the fused manifold process (can it withstand the forces it will be subjected to). They were also regularly interacting with other subsystems (biology in this example) in an effort to meet set requirements, in this case, the requirement for the fused manifold to be biocompatible. These requirements, while not always explicit from the beginning of the project, helped shape what the engineer was ultimately responsible for designing and creating.

Engineers only vaguely and indirectly referred to the content knowledge necessary to fulfill their project responsibilities.

So I have had to go back, I mean I took those classes eons ago, so I had to go back and refresh my memory about how they all worked. ~Propulsion Engineer

Subsystem requirements drove the propulsion engineer to use a system he was unfamiliar with, even though familiarity or "track record" was often the most important consideration in choosing design elements. Overall, the engineer exhibited little concern about his ability to understand and re-learn the material but instead placed a greater emphasis on how his system interacted and affected other systems.

Engineers on the other engineering project also described similar interactions with the other engineers on the project.

> *We have a sanitary sewer connection here. For gravity, sanitary sewer is all gravity flow. You have to have the right slope and you need to have the right elevations. You have to have enough size of your pipe and slope of your pipe to make it all work. It takes a lot of coordination to make sure that you can get the flows you want to the existing sewers. ~Mechanical Engineer*

Here we see the engineer talk about the coordination of getting a functional flow of treated groundwater into the existing sewer. Although the physical processes being analyzed are relatively well understood, the true challenge of the design is finding a way to achieve those standards while simultaneously meeting the requirements of all the connecting subsystems. Other engineers determined the elevations at the beginning and end of the gravity system so the mechanical engineer's task was more about coordinating than calculating. Again, we see the emphasis placed not on the content knowledge necessary to achieve the correct flow but instead on the shared responsibility of ensuring correct flow is achieved without impinging on any other subsystems' requirements.

We also see how a change in design from one system can impact the requirements of another system which requires communication.

> *We have weekly design meetings where we sit there with all the discipline leads and say, "Okay, this is what we're working on, these are our challenges, we're changing this over here, we're adding a pump here, and then electrical needs to know, "Oh, you're adding a pump, I need to route power to that and make sure we have enough power for it and so and so forth." There's a lot of interdisciplinary coordination that goes on... ~Environmental Engineer*

The above quote demonstrates how changes in one system are a focal point of interaction with other systems. We see in this quote how the addition of a pump spurs an interaction between the engineer responsible for the pumps and the electrical engineers because the engineer responsible for the pumps have set a new requirement for the electrical engineers (the need for additional power and routing of power). This demonstrates how changes in one system translates to a change in requirements and responsibilities of another system, and how the project practices themselves are structured in terms of this division of knowledge.

Conclusion

Overall, we see through the above examples how the engineers in our study defined their knowledge realms in terms of responsibilities (see also Panther, Montfort & Pirtle, 2017), and then interact across those realms of responsibility through the mechanisms of requirements. The electrical engineers in the last example are in a meeting hearing about a civil engineering design precisely so that they can impose the perspective of their responsibilities. When an overlap is detected – in the example the addition of a powered pump system – the engineers coordinate in terms of requirements. What exactly does the pump need to do, quantitatively, and what does the power system need to meet those demands? They interact in terms of system requirements and how these requirements then shape the engineering design. The engineers may have taken for granted their discipline-specific technical competency throughout the interview for a variety of reasons but by doing so, the engineers revealed a critical aspect of their work – understanding the interconnectedness of disciplines and the associated responsibilities and design requirements.

The findings here suggests that when students choose an engineering discipline, it is less constraining and concrete than originally thought. Instead, the space between disciplines, in terms of the ability to fulfil responsibilities, is more plastic and transferable (with adjustments) to other engineering disciplines and fields. Certainly, engineers need to be able to complete the calculations and quantifications necessary to understand, explain and predict the behaviour of their subsystems, but we find that in the projects studied here the performance of those calculations takes such a relatively short amount of time in comparison to other project responsibilities that they are easily overlooked.

Most engineering education curricula make efforts to emphasize discipline-specific knowledge over system interactions. In school a student might spend two hours performing the calculations necessary to determine a specification given a straight-forward set of requirements. Our findings suggest that practicing engineers may spend months negotiating or researching those requirements. The findings presented here are further evidence that organizing people based on their academic discipline does not accurately reflect how engineering practice is organized. As engineers continue to work on larger and more detailed interdisciplinary projects, it is important to understand how they utilize knowledge on these projects. This is especially important from an academic standpoint as many people believe that engineering education is geared towards preparing students to work in industry. Therefore, to ignore how engineers actually work and organize in practice, could hinder the transition for students to the workplace.

Recommendations/Implications

In many ways engineering education is seen as a way to prepare students to work in industry. But the way engineering education categorizes disciplines compared to where engineers place emphasis is disconnected. Therefore, we suggest research should work on identifying ways to better align these two practices. Additionally, the findings here can be used within industry as a way to better understand and prepare for new hires who have recently graduated.

Alternative Explanation

Our research design and analysis procedure could have directly influenced our finding. The engineers' focus on coordination over calculation could have been inspired or influenced by our line of questions. We argue that our interview questions were sufficiently open-ended to avoid such prompting. As shown in the Methods section, we asked engineers' about what they were responsible for designing or what they did when they got stuck. Additionally, our findings are supported by consistent and repeated themes in the interviews, and it is unlikely that participants would be motivated or able to spontaneously generate detail-rich responses grounded in their own experiences that also refer consistently to a construct they perceived as being promoted by the interviewer. Instead, we assume that these patterns reflect the ways they normally talk and organize their thoughts.

There is a possibility that the projects we studied were unique or unusual, and our findings are not meaningful to other engineering work. This concern will be best addressed through future work, but we would note that the findings were based on themes that were present in both projects, and that the projects were chosen for their contrast to each other as well as their interdisciplinarity. Additionally, the projects came from two different sectors of industry – one had a strong research and innovation focus while the other utilized well-established best-practices – so our findings could be an artefact of the organization but it is unlikely as we purposely sampled from two very different organizations.

Future Research Plans

Future research plans aim to identify the knowledge domain of practicing engineers and engineering students. Defining the knowledge domain of these two groups will assist researchers in understanding in what ways engineers and students differ in their categorization of knowledge. Implications for this include how students are taught engineering and refocusing on linking concepts that are most widely utilized in practice.

References

Alexander, P. A., and Judy, J. E. (1988). The interaction of domain-specific and strategic knowledge in academic performance. *Review of Educational research*, *58*(4), 375-404.

Auerbach, C., and Silverstein, L. B. (2003). *Qualitative data: An introduction to coding and analysis.* NYU press.

Brown, J. S., Collins, A., and Duguid, P. (1989). Situated cognition and the culture of learning. *Educational researcher*, 18(1), 32-42.

Dedoose Version 7.5.9, web application for managing, analyzing, and presenting qualitative and mixed method research data (2017). *Los Angeles, CA: SocioCultural Research Consultants, LLC* (www.dedoose.com).

Estes, D., Chandler, M., Horvath, K. J., and Backus, D. W. (2003). American and British college students' epistemological beliefs about research on psychological and biological development. *Journal of Applied Developmental Psychology*, 23(6), 625-642.

Hirschfeld, L. A., and Gelman, S. A. (1994). Mapping the mind: Domain specificity in cognition and culture. Cambridge University Press.

Hofer, B. K., and Pintrich, P. R. (1997). The development of epistemological theories: Beliefs about knowledge and knowing and their relation to learning. *Review of educational research*, 67(1), 88-140.

Jehng, J. C. J., Johnson, S. D., and Anderson, R. C. (1993). Schooling and students' epistemological beliefs about learning. *Contemporary educational psychology*, *18*(1), 23-35.

Palmer, B., and Marra, R. M. (2004). College student epistemological perspectives across knowledge domains: A proposed grounded theory. *Higher Education*, 47(3), 311-335.

Panther, G., Montfort, D., and Pirtle, Z. (2017). How Engineers Negotiate Domain Boundaries in a Complex, Interdisciplinary Engineering Project. Paper presented at the 124th ASEE Annual Conference and Exposition, Columbus, Ohio.

Paris, S. G., Lipson, M. Y., and Wixson, K. K. (1983). Becoming a strategic reader. *Contemporary educational psychology*, *8*(3), 293-316.

Patton, M. Q. (2002). Qualitative research and evaluation methods (3rd ed.). California EU: Sage Publications Inc.

Paulsen, M. B., and Wells, C. T. (1998). Domain differences in the epistemological beliefs of college students. *Research in higher education*, *39*(4), 365-384.

Trevelyan, J. (2010). Reconstructing engineering from practice. *Engineering Studies*, 2(3), 175-195.

Acknowledgements

This material is based upon work supported by the National Science Foundation under Grant No. 1361107 and No. 1642022. Any opinions, findings and conclusions, or recommendations expressed in this material are those of the authors and do not necessarily reflect the views of the National Science Foundation.

Story of change—Using experience-based critical event narrative analysis to understand an engineering program's culture

Nadia Kellam, Brooke Coley, and Audrey Boklage
Arizona State University, Mesa, United States
Nadia.Kellam@asu.edu, Brooke.Coley@asu.edu, and Audrey.Boklage@asu.edu

***Abstract**: The purpose of this research was to gain a baseline understanding of an engineering program's current culture to inform the progress towards the goal of transforming the culture. This study explored the evolution of an engineering program at a very high research activity university over the course of 15 years. Participants included 13 faculty members involved with the program over the time period of interest. Using Schein's organizational cultural framework and an experienced-centered narrative approach, we focused on the interplay of the experiences of individuals within the program and the larger sociocultural forces. Multiple themes emerged from this qualitative analysis including leadership, process of acculturation, and evolving culture. Implications of this work include the importance of leadership and strategies for acculturation of faculty into a department or program, especially at critical points in an organization's development.*

Context

Despite repeated calls to revamp our engineering education departments to be better aligned with current needs of society and industry (National Academy of Engineering, 2004; National Science Foundation, 2010) and to change the way that we teach engineering to increase student learning (Eiseman & Fairweather, 1996; Fairweather, 2008; Seymour, DeWelde, & Fry, 2010), little widespread change to engineering education has taken hold. According to Henderson, Beach, and Finkelstein, these changes have not taken hold because of the implementation of ineffective change strategies, including testing and sharing of best practices and implementing top-down policies from administration (2011). In order to make widespread changes to engineering education systems we must focus on effective change strategies of changing the beliefs of individuals within the system, implementing long-term interventions, and taking a systems approach to change (Henderson et al., 2011). This work points to a need to develop a systemic understanding of the cultures of engineering programs so that we can begin to understand the underlying reasons of why people behave in certain ways (and subsequently how to make changes to their beliefs) so that we can make strides to develop a systems understanding of engineering programs and the larger institutions that they are embedded within so that we can design strategies that align with these systems.

This research is part of a broader National Science Foundation funded effort in which we are attempting to revolutionize our engineering program. In this research project at Arizona State University (ASU), we are interested in developing a baseline understanding of our current culture and a methodology for doing this. We anticipate this information will help us as we attempt an ambitious goal—changing our engineering program culture to one that embraces concepts of risk taking and additive innovation. In particular, we are looking at an engineering program that started in 2005 at ASU, a very high research activity university. The engineering program is similar to a department and offers undergraduate Bachelor's degrees in General Engineering and Manufacturing Engineering; Master's degree in Systems Engineering; and two Doctoral degrees in Systems Engineering and Engineering Education, Systems, and Design.

Purpose and Research Questions

To better understand the culture of change within our program, we explored the history of change and the current culture captured in faculty experiences as described through their narratives. Specifically, the following research question guided our research: What do experiences of faculty during critical events in our program's history reveal about the program's culture? The long-term goal of understanding our program's culture is to explore how we can best position ourselves to seed a revolution in and through our program.

Theoretical Framework

Schein's organizational cultural framework is used to theoretically frame this study which views culture as abstract, but also as something that can be understood through examining how a group has adapted and evolved over time as it has tried to "survive, grow, deal with its external environment and organize itself" (Schein, 2016, p. 14). Schein further explains, "If we can understand where culture comes from and how it evolves, we can grasp something that is abstract, that exists in a group's unconscious, yet that has a powerful influence on a group's behavior (2016, pp. 14–15). In our study we will develop an understanding of our engineering program's culture, as others have done (Godfrey & Parker, 2010; Merton, Froyd, Clark, & Richardson, 2009), through examining major events in our program's history and how faculty experienced those major events—thus getting to the espoused, underlying, and underappreciated beliefs and values of the program.

Methodology

In this project, we took an experience-centered narrative approach that involved studying narratives as stories of experiences rather than events (Squire, 2008) and using these stories to generate an understanding of our culture in engineering. The experience-centered approach assumes that narratives are meaningful, representative of experiences through a recollection of past events, and demonstrative of transformation or change (Squire, 2008).

This experience-centered narrative approach aligned well with our theoretical framework as it uncovered the rich, complex, diverse, and multifaceted ways that faculty, with varying appointments and time vested into the institution, have experienced the organization over time. Using this approach enabled us to focus on the interplay between the experiences of individuals within our program and the larger sociocultural forces.

Data Sources

The primary data used for this paper were narrative interview transcripts; these interviews were conducted with 13 engineering faculty in compliance with our institutional review board process. Interviews were conducted with founders of the program, mid-career faculty, recent tenure-track hires, lecturers, and adjunct faculty. In this project, we were interested in collecting and studying stories about specific experiences—thus our interview protocol involved questions such as "What are major shifts that have happened during your time at ASU? Could you elaborate on how those changes have influenced your teaching, research, or service?" Moreover, we were interested in the relationship between the individual faculty member and the broader social system. These questions in our interview protocol included, "How do you see your role as a faculty member? How has that changed over time? How has the leadership (dean, director) influenced you and your development as a faculty member? What kind of institutional level support do you need to develop professionally?" We also included probing questions to encourage the participant to describe specific experiences. The authors of this project conducted nine of these interviews and a member of the broader project's evaluation team, Zora Ziazi, conducted the remaining four interviews.

Secondary data included American Society for Engineering Education (ASEE) papers; news articles; university, college, and program vision and mission statements; and ABET documentation. These data were found through database searches, documents provided by the program, and website searches. These secondary data were used to help develop a broader and more holistic understanding of faculty experiences as well as to triangulate these experiences with those expressed through the different forms.

Participants

The participants in this research project were purposively sampled from our engineering program faculty with varying years of experience in the program—ranging founders of the engineering program to new hires. At the time of the interviews there were 42 faculty members in the engineering program—six part-time faculty associates, four lecturers, one clinical assistant professor, one professor of practice, 11 assistant professors, 14 associate professors, and five full professors. We interviewed faculty from each of these faculty positions with a focus on interviewing faculty who teach in the mezzanine level (years two and three) of the curriculum. We conducted 13 interviews with faculty with the following demographics: one faculty associate, three lecturers/clinical assistant professors/professors of practice, four assistant professors, three associate professors, and two full professors.

Analysis

For this project, we were interested in the following events that have happened in our history as an engineering program: Founding program, merging with other programs, and merging with Fulton (Ira A. Fulton Schools of Engineering). When interview transcripts included any of these critical events, the data was coded so that we could develop an understanding of how faculty experienced these critical events.

After this initial pass of coding, we developed plot outlines for each participant or data source—these plot outlines included an abbreviated description of the experience and divided these experiences according to the three critical events. The plot outline enabled the researchers to easily see the experiences of each participant in a cohesive way, enabling us to keep the narratives intact. Secondary data sources were also coded for these critical events so that they could bring additional perspectives to the faculty interview perspectives and capture some of the events closer to the time that the events occurred. In many of the secondary data sources, only one of the critical events was discussed.

Findings

When teasing out the experiences of faculty in our program during the selected critical events in our history, aspects of the culture begin to be revealed. The following sections describe how the faculty experienced each of the three selected critical events—founding program, merger with other programs, and merger with Fulton (see Figure 1 for a timeline of critical events).

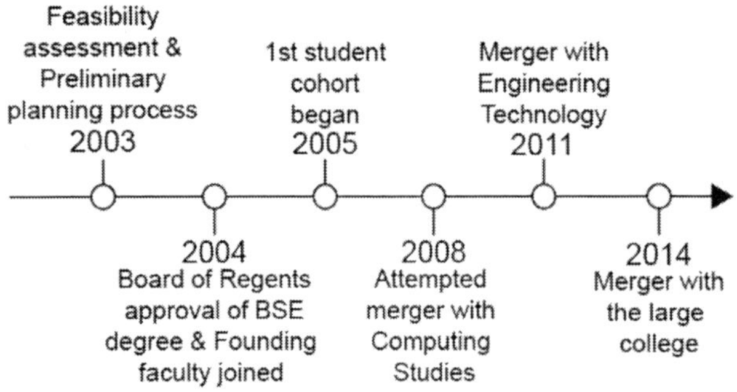

Figure 1: Timeline showing critical events in the program's history

Founding program

Of the 13 faculty interviewed, only three were members of the founding group of faculty. In early discussions about the idea for a new engineering program there was a cohesive vision that many of the founding faculty found compelling and intriguing. This vision was that a new engineering program would be started at a different campus (called East Campus at the time, now the Polytechnic Campus) that was focused primarily on undergraduate education and would be in a new college. The choices to start this engineering program on a different campus from existing engineering programs and outside of Fulton was deliberate. This separation was meant to "provide autonomy from many constraints that would hinder the development of an engineering program" (Roberts & McHenry, 2004, p. 3). Moreover, faculty expressed how they felt that being a part of the team designing this new engineering program was an opportunity that is rarely seen by faculty. Art explains, "The world was our oyster, we could do anything we wanted."

Most faculty were excited about the opportunity to have "free reign" although one faculty member explained that the enthusiasm did not permeate everyone in the new program as there was some trepidation about leaving the established program to develop a startup program without guarantee of success. Art explained that the founding faculty had an "escape hatch" in their contracts—they would still have jobs at ASU if the program did not succeed. "It was not as cut and dried ra ra enthusiasm as one might think." According to Art, many initially maintained their connections with Fulton through teaching courses, but all eventually moved to the new program full time.

Even in spite of the concerns about moving positions from an established program to a new program on another campus, the faculty expressed much excitement and clarity concerning the vision during the first year of planning the curriculum. The faculty engaged deeply with each other and shared a deep concern for education—a value that is in tension with many very high research activity universities where faculty value research over teaching and service. Clarke explains, "We really did, by the time we were done with the year, have a very clear view of what we thought our value proposition was. We had boiled it down to engaged learning, focus on individual, and flexibility, originally."

This culture that was emerging was radical and one that, in many ways, was anti-establishment. This was a different culture than these faculty members had ever experienced before in their careers. Clarke describes this as,

> We saw ourselves as fairly radical. Occasionally, we felt like we were the hippies. I mean, if you've seen College culture, which is very straight-laced, we were the hippies. We're talking about peace and love and learning as opposed to the stuff you get major necessarily, by publication, dollars or publications.

The early cohorts of students that joined this new program were also taking risks with a new engineering program. There was a sense of passion, having fun, and doing things differently that emerged in many of the interviews. There was also a camaraderie that developed with this initial group of faculty, one illustration of this was that their offices were cubicles within a room. Art described this as, "We were all in one room and so we basically were in a perpetual faculty meeting. You could just shout over the top. I thought I was going to hate it, I loved it. I loved it."

This program and its development were different in that there was a strong undercurrent of developing this program based on educational research, not the common practice of developing the curriculum based on intuition and experiences alone. The faculty and program were scholarly in their approach to developing a new program and in disseminating their experiences within the community (Grondin & Roberts, 2010; Morrell et al., 2005, 2006; Purzer, Henderson, McKay, Roberts, & De Pennington, 2008; Roberts et al., 2007; Roberts, Morrell, Grondin, Kuo, & Hinks, 2005). The leader, Charles, provided educational research that faculty were expected to read and to help guide their curricular decisions. The faculty perceived this as positive and exciting as they were building their program in a scholarly way. Sheila described an example of this influence from Charles.

> I started doing this competency-based thing and in my first year annual review he's like, "Well, you're doing this, but can you tell me why?" and I couldn't and he's like, "Next year, when we have this review, I want you to tell me theoretically why you think this is a good approach to use in a classroom." He forced me to read about it and then it really improved my understanding of why I was doing that and made me believe in it even more.

Many of the results in this section have focused on the founding faculty, but there was also a recognition that this program would not have been possible without higher level support within the university (Roberts et al., 2005, p. 12). The group of faculty were radical in their approach to this new program, which was, according to faculty, possible because of the support that they got from the president's office. This aligns with research that shows that perceived organizational support enables workers to take more risks in their jobs (Neves & Eisenberger, 2014). Without that level of support the innovative new engineering program may not have had an opportunity to thrive. In having "the blessing from the president" (Art) many faculty felt safe in taking on this risky challenge.

As the program began to grow, the perceived culture was different for newcomers—especially from the perspective of the 'outsiders' as they became faculty in this program. One interviewee described a "lack of feeling like you're in the club." This experience from someone new to the faculty demonstrated how a small, tight knit group of faculty can make it difficult for new members of the group to gain acceptance. One of the early hires described some experiences that alluded to limited diversity of the program faculty and the existence of some preconceived ideas of who an engineering faculty member should be. This faculty member described a few experiences that were particularly "bothersome" and went on to describe how much they valued some of the more recent hires, as the faculty were becoming more diverse in backgrounds, age, and gender. This growth reflected an opportunity for the program's culture to begin to grow and adapt.

Mergers with other programs

In addition to hiring new faculty into the program, ASU began to restructure many programs which involved mergers with other programs. The first time this occurred was during an attempted merger between Engineering and Computing Studies in 2008. This merger had the potential to completely disrupt the prior culture of the engineering program and to upend the cohesive vision that was guiding faculty actions. Faculty described this merger as being difficult due to incompatible visions of the Engineering and Computing Studies faculty. Clarke explains that they were expected to merge not only in the administration of the program, but also in the curriculum. They worked for two years to identify a set of common first-year courses to satisfy both programs, however, there were different priorities between the two programs. Engineering was focused on professional skills, while Computing Studies was focused on programming skills. As a result of the differences in the faculty's vision and educational foci, the merger was unsuccessful.

The next critical event was a merger of the Engineering Technology Department into the Engineering Program in 2011, which was controversial among faculty and the industrial advisory board. The stated reason for the merger was due to engineering technology graduates having difficulties finding jobs and reaching their career goals. Julia explains, "They [engineering technology graduates] were coming out of degree programs with the wrong perception of where they could go in their career and it was heart-breaking actually for me, when I'm talking to students who are graduating and can't get into Master's programs. It was enough." As the programs began to merge, there was an explicit focus from the leadership on developing relationships between the Engineering and Engineering Technology faculty through meetings and retreats. This facilitated the building of rapport and eventually trust among these colleagues and moved them towards a more cohesive faculty.

Other faculty shared how this merger influenced the vision of Engineering and the overall program culture. The Engineering Technology faculty were described as having a very traditional perspective of education that did not share value for the way the Engineering program taught their students. The project courses were an example that emerged a few times because they were focused on professional skills rather than technical skill acquisition.

When discussing this merger, there was a vulnerability that appeared as faculty from the Engineering Technology department discussed the reality in industry that a "technology degree is not good enough for engineering," (Jason) and that they themselves, and now their students, could not get a job as an engineer because of their major being Engineering Technology. Another vulnerability that emerged was the concern that faculty would be expected to teach the underlying math and science concepts to their students. Jason explained that he had a complex because he did not feel comfortable getting into the underlying equations and theory within his subject area.

These emergent differences across the faculty influenced the culture of the program and forced the faculty to confront their differences in values and visions of undergraduate education. Clarke explains that "there was really no questioning of, 'does this fit our culture? Do these people fit our culture?' It was, 'We're all one big, happy family now.'" It appears that this merger, while initially perceived as non-threatening to the cohesive vision of the faculty, did impact the program's evolving vision and culture.

Merger with Fulton

In 2014 faculty learned that the university was restructuring again but this time it was not a combination of two programs, but rather the College of Technology and Innovation was to be folded back into the Ira A. Fulton Schools of Engineering. The engineering program that had started with a clean slate on a different campus was being reabsorbed into the mainstream engineering program. Of course, the announcement of this merger brought about ample anxiety—anxiety surrounding faculty identity, the collective vision and culture, and whether those could be sustained after being merged back into Fulton. "It was a big deal because it made people afraid and it made people fear that we were going to lose our identity" (Sheila).

A fear that was articulated by many faculty was that the focus on teaching, especially undergraduate teaching, would be replaced by a focus on research. If research were to become the top concern of faculty, there was a potential threat to the quality of the undergraduate engineering program. Many faculty described being sad during the transition as the conversation quickly changed to discussions of graduate programs and PhD production instead of previous conversations focused on undergraduate teaching. These faculty felt that they could see the vision of undergraduate education starting to slip away.

Not all faculty shared this concern. Julia, whom has always kept an active research program at ASU, discussed how she felt that research and teaching could be mutually beneficial. She explained that she did not see a dichotomy between research and teaching. "I always try to explain that those two can come together in really innovative and interesting ways and support each other." Moreover, many of the newer faculty also did not express a concern for valuing research over teaching. When they discussed research, they discussed looking forward to having a PhD program and beginning to advise PhD students.

Many faculty discuss being stressed as a faculty member and a few discussed a concern that time spent doing research is taking away time for teaching. There was a discussion of a concern about the upcoming PhD program and how that would impact their teaching and time to do everything. Deborah, an Assistant Professor, explained that she feels stretched thin—"just trying to do it all, all by myself, all the time." There was a hope expressed that a PhD program would help ease some of these stresses as graduate students can help with some of the work load.

Clarke also commented on the time constraints and how that impacted the likelihood of developing a shared, collective vision, "I'm sort of pessimistic about that ... well, in the sense that I don't know that we have ... everybody's busy teaching, everybody's busy—now, we've got research, we've got PhD programs starting up. I just don't see people having the time to devote to it." Another Associate Professor, Sheila, discussed the difficulties of establishing a collective vision with a faculty that has grown from eight to 80,

> If you have seven people, or eight people, or nine people, which we had in those first years...a shared vision and a shared culture is easier to maintain than it is with the 80 faculty we have now across so many disciplines and so many different positions and everything.

Discussion and Implications

Many of the themes that emerged in the experience-centered narrative analysis align with Schein's cultural framework. These themes include leadership, process of acculturation, and evolving culture.

Leadership

Schein places leadership in a prominent role of forming, establishing, and maintaining cultures (2016). Schein describes the leader as using his or her "personal power to demand some new behaviour directed toward achieving some purpose" (2016, p. 14).

In the founding program experience, the leader, Charles, was described as being instrumental to the development of the new program. The faculty that he brought to help realize his vision were excited about the undertaking of designing a new engineering program from scratch. But, at least a few, describe being reluctant, especially initially, to make the switch from the known (their original department in a traditional college of engineering) to the unknown (a new, general engineering program in a separate college). In addition, the faculty that were interviewed (and some of the conference proceedings), also discussed the critical role that upper administration (the president's office in particular) had in supporting these efforts to establish a new engineering program. During this phase in the program, a strong and cohesive vision was developed that was focused on undergraduate education and the culture was described as being radical.

Moreover, as the program was merged back into Fulton, one of their leaders (who has been in various leadership roles since 2011) really helped make that transition a positive one. This transition could have been detrimental to the culture that had emerged in the program and many faculty attributed the somewhat seamless transition to the leader of our program, Julia. Sheila explains, "I really, really value Julia as a leader. I have a hard time imagining someone being able to more effectively navigate us through a lot of the changes."

Process of Acculturation

Schein also describes a process of acculturation (or socialization) as being critical to sustaining the culture as the group evolves and grows. He explains that a group member must gain permanent status and entry into the "inner circles" before learning the true culture of the group. When Charles first formed the program, he recruited established faculty from Fulton to join the initial, founding group of faculty in College of Technology and Innovation. Later, when early-career junior faculty were hired into the program, there was mention of difficulties gaining acceptance into the community. This could point to an issue with junior faculty members not being able to gain entrance into the "inner circles" and thus not immediately learning about, and eventually adopting the cultures, values, and beliefs of the larger group.

When the program was merged with other programs, the faculty that were brought into the program may not have been socialized into the norms of the group. Many faculty discussed subgroups that formed (and were continuations of previous subgroups) and these subcultures had tensions with the broader program's culture as well as with other subcultures. An example of this was discussed in the findings where Engineering Technology faculty feared that students would lose their coveted hands-on skill development in the curriculum through a shifted focus on professional skill development.

Finally, a not well socialized subgroup of faculty were identified—the adjunct professors. The experiences of these faculty did not appear in the findings section of this paper because they are separated from the program and thus did not talk about the founding of the program, its merger with other programs, or the merger with Fulton as it didn't have much of an impact on their role. Alan, an adjunct faculty member explains, "well, being part-time, one of the really good things here, being part-time, is I'm out of the main flow of everything that happens here... I don't get caught up in a lot of things. My contact with leadership is pretty small." As we bring more adjunct faculty to teach courses without socializing them into our culture, our culture may continue to be weakened.

Evolving culture

Schein explains that a useful way to understand culture is to "think in dynamic evolutionary terms" (2016, p. 14). Early in the program, the faculty established a strong vision for undergraduate education. The faculty described the culture at this time as being anti-establishment and risk taking—they were really trying to change the way that faculty at a very high research activity university envisioned and valued undergraduate education.

During the attempted merger with the Computing Studies program in 2008, faculty described a strain on the culture that was established in the first years of the program. This event highlights the evolving culture of the program—one that values undergraduate education in general, and professional skill development, in particular.

In the next merger with Engineering Technology in 2011, there was a similar strain on the culture as their faculty valued technical skill acquisition and hands-on skill development. Although this merger was successful, it resulted in a subgroup within the broader program with a different culture from the engineering program, which continues to cause tensions.

When the engineering program was then merged back into Fulton in 2014, some faculty viewed this as the largest threat to our culture. The concern was that our program would slip back into the culture from which the founding faculty came—one that values research, publications, and external funding over teaching. Not all faculty shared this concern, with some faculty explaining that they found research and teaching to be mutually beneficial. Whether our culture continues to align more closely with the culture in Fulton remains to be seen, but early signs show that we still maintain an essence of that early culture established with the development of the new program.

Conclusion

The experience-centered narratives indicated a subculture with beliefs and values that are distinct from the dominant Fulton culture. As we move forward and begin to seed our revolution an important question to consider is: How do our plans to seed a revolution in this distinct subculture align with the dominant Fulton culture? And, if we align with this broader culture, will faculty still feel empowered to take risks and make changes to our engineering undergraduate program that could lead to more lasting and systemic cultural change? Or, would we be more likely to be successful if we intentionally develop this risk-taking subculture. As engineering educators, understanding the relationship between the environment, culture, and subcultures can be powerful in successfully starting a revolution.

References

Eiseman, J., & Fairweather, J. (1996). *Evaluation of the National Science Foundation Undergraduate Course and Curriculum Development Program: Final Report*. Washington, D.C.: SRI International.

Fairweather, J. (2008). *Linking Evidence and Promising Practices in Science, Technology, Engineering, and Mathematics (STEM) Undergraduate Education: A Status Report*. Washington, D.C.: National Research Council.

Godfrey, E., & Parker, L. (2010). Mapping the cultural landscape in engineering education. *Journal of Engineering Education*, *99*(1), 5–22.

Grondin, R., & Roberts, C. (2010). Development of perspective in a general engineering degree. In *ASEE Annual Conference and Exposition, Conference Proceedings*. Louisville, KY, United states.
Henderson, C., Beach, A., & Finkelstein, N. (2011). Facilitating change in undergraduate STEM instructional practices: An analytic review of the literature. *Journal of Research in Science Teaching*, *48*(8), 952–984. http://doi.org/10.1002/tea.20439
Merton, P., Froyd, J. E., Clark, M. C., & Richardson, J. (2009). A case study of relationship between organization culture and curricular change in engineering education. *Innovations in Higher Education*, *34*, 219–233.
Morrell, D., Roberts, C., Grondin, R., Kuo, C.-Y., Hinks, R., Danielson, S., & Henderson, M. (2005). A flexible curriculum for a multi-disciplinary undergraduate engineering degree. In *Proceedings - Frontiers in Education Conference, FIE* (Vol. 2005, p. F1D–18–F1D–23). Indianapolis, IN, United states.
Morrell, D., Roberts, C., Grondin, R., Kuo, C.-Y., Hinks, R., Henderson, M., ... Danielson, S. (2006). Beyond measurement: Designing engineering outcomes to foster student achievement. In *ASEE Annual Conference and Exposition, Conference Proceedings* (p. Dassault Systemes; HP; Lockheed Martin; IBM; Micro). Chicago, IL, United states.
National Academy of Engineering. (2004). *The Engineer of 2020: Visions of Engineering in the New Century*. Washington DC: National Academies Press.
National Science Foundation. (2010). *Transforming Undergraduate Education in Science, Technology, Engineering and Mathematics (TUES) Program Solicitation*. nsf.gov. Retrieved from http://www.nsf.gov/pubs/2010/nsf10544/nsf10544.pdf
Neves, P., & Eisenberger, R. (2014). Perceived organizational support and risk taking. *Journal of Managerial Psychology*, *29*(2), 187–205.
Purzer, S. Y., Henderson, M., McKay, A., Roberts, C., & De Pennington, A. (2008). Comparing the design problem solving processes of product design and engineering student teams in the US and UK. In *ASEE Annual Conference and Exposition, Conference Proceedings*. Pittsburg, PA, United states.
Roberts, C., & McHenry, A. (2004). Developing A Multidisciplinary Engineering Program. In *American Society for Engineering Education* (pp. 1–11). Salt Lake City, Utah: ASEE Conferences. Retrieved from https://peer.asee.org/13444
Roberts, C., Morrell, D., Grondin, R., Kuo, C., & Hinks, R. (2005). Developing a Multidisciplinary Engineering Program at Arizona State University ' s East Campus. In *ASEE Annual Conference*. Portland, OR.
Roberts, C., Yasar, S., Morrell, D., Henderson, M., Danielson, S., & Cooke, N. (2007). A pilot study of engineering design teams using protocol analysis. In *ASEE Annual Conference and Exposition, Conference Proceedings* (p. Dassault Systemes; HP; Lockheed Martin; IBM; DuPon). Honolulu, HI, United states.
Schein, E. H. (2016). *Organizational Culture and Leadership* (5th ed.). Hoboken: Wiley.
Seymour, E., DeWelde, K., & Fry, C. (2010). *Determining Progress in Improving Undergraduate STEM Education: The Reformers' Tale*. Washington, D.C.: National Academy of Engineering.
Squire, C. (2008). Experience-centered and culturally-oriented approaches to narrative. In M. Andrews, C. Squire, & Tamboukou (Eds.), *Doing Narrative Research* (pp. 42–64). Sage Publications Ltd. Retrieved from http://dx.doi.org/10.4135/9780857024992

Acknowledgements

We wish to thank Zora Ziazi and Alison Cook-Davis at Arizona State University, who assisted in conducting interviews and are serving as evaluators on the broader project. We also wish to thank members of the broader RED project for their input throughout this project: Ann McKenna, Jennifer Bekki, Samantha Brunhaver, Adam Carberry, Shawn Jordan, Micah Lande, and Jeremi London at Arizona State University. This material is based upon work supported by the National Science Foundation under Grant No. 1519339.

Copyright statement

Assignment: How effective can be lessons developed by students themselves

José Nicolás Pacheco Garavito
Universidad de los Andes, Bogotá, Colombia
jn.pacheco2761@uniandes.edu.co

María Catalina Ramírez Cajiao
Universidad de los Andes, Bogotá, Colombia
mariaram@uniandes.edu.co

José Tiberio Hernández Peñaloza
Universidad de los Andes, Bogotá, Colombia
jhernand@uniandes.edu.co

***Abstract**: Despite the fact that Learning-by-Teaching advantages have been studied extensively in recent decades, there are gaps regarding the research on students' learning levels when they are not in the teacher's role. The purpose of the following paper is to present an empirical analysis of the learning students' perception when the class is taught and directed by another classmate. The target population is restricted to undergraduate students enrolled in the elective course "Science, technology, and innovation" during 2016. Students in groups of five should prepare, design, research and be responsible for the complete development of a session. In accordance with students' self-assessments, their average learning level for all sessions was mid-high. Results indicate that the interest aroused by the session topic is the most closely related factor to students' learning levels.*

Introduction

One of the major tasks of engineering teachers is to provide students with learning environments. In addition, today the information and communications technology (ICT) takes on a more important role in teaching processes, to the extent that such tools can substitute a traditional class for total virtual environments (Piccoli, Ahmad, & Ives, 2001). The above highlights the importance of reflecting on the activities and use of a class in a physical/temporal space which is shared by both the students and the teacher. In recent decades, there have been trends and innovations in engineering education instructor-centered (e.g. active learning and collaborative learning) and student centered (e.g. cooperative learning and Project-based Learning (Smith, 2011)). Even though depending on the methodology the teacher's role is changed to facilitator, tutor, guide, advisor, etc., teachers are still responsible for the learning methodology and students are responsible for the learning outcomes (Rivero et al., 2016).

In the traditional higher education approach the role of the teacher is limited to spreading their wisdom and the students must absorb it (Felder, Woods, Stice, & Rugarcia, 2000). An alternative to the previous paradigm is Learning-by-Teaching, or LDL by its acronym in German. This methodology allows students to prepare and to teach partial or complete lessons to their classmates. LDL was mainly driven in the 1980s by Jean-Pol Martin when he used this methodology to teach French as a foreign language, and he began to systematically develop a theoretical base for this approach (Martin, 1985). Other authors like psychiatrist William Glasser reinforce the advantages of the LDL methodology, stating that "We learn 10% of what we read, 30% of what we see, 70% of what we discuss, 95% of what we teach others" William Glasser (b.1925).

The purpose of this paper is to perform an empirical analysis of students' perception about their learning outcomes (those who are not in a teaching role) when the class is taught and directed by another student. To measure the Learning-by-Teaching methodology outcomes, the role of student as teacher and developer of a full lesson was designed and implemented in a unit of an engineering elective course. In addition, in the course design, we considered the preliminary conditions to implement LDL proposed by Grzega (2005). Subsequently, the learning outcome was analyzed using a mixed method (quantitative - qualitative), like the methodologies proposed by Shekhar et al. (2015) to study resistance to Active Learning.

Context of the study

In recent years, the group of teaches of the elective course "Science, Technology and Innovation" has introduced different methodologies and roles for its students within its active innovative practices in the engineering education process (Ramírez, Avila, Duque, & Hernandez, 2015). This course is offered to undergraduate students from different academic programs of the university at an early stage of their careers, promoting interdisciplinary participation (Duque, Ramírez, Ávila, & Hernández, 2013).The course is conducted by four facilitators with a background in industrial engineering, systems and computing engineering, electricity/electronics engineering and a local entrepreneur. In the course design, active learning strategies are used with a definition of roles, with one role worked through PBL (entrepreneur), another role through active learning (investor) and the last role based on "Learning-by-Teaching." The last student's role we will be called the "developer of a lesson" from now on. This role consists of classes in which the teacher previously assigned a general theme of the course to groups of five students. The students in charge of the class must prepare, investigate, develop and direct a full session (one and a half hours) without the intervention of the teacher. This totally transforms the social system of the class and makes the students in the teacher role aware of their learning (Roma, 2009). Findings from the aspects mentioned above, allows students to have several perspectives beyond a traditional masterclass led by a single teacher. Thus, the course contributes to an increase in the motivation of the students to study the course contents' or in its academic performance (Rodriguez, Maya, & Jaén, 2012).

Research questions

When physical/temporal spaces and resources are used correctly during a session, students have a unique opportunity to develop both social and cognitive skills (Gartner, Kohler, & Riessman, 1971). To analyze the effect of the lessons given by their peers on students, our research considers the following questions:

1. What is the students' perception of learning outcomes of the classes taught by their peers?
2. Which are the main methodologies and activities used in lessons taught by students?
3. What factors are correlated with the learning level when the session is developed by students?

Theoretical framework and literature review

The basic concept of Learning-by-Teaching is that students take the role of the teacher for a limited or indefinite amount of time. This methodology should not be confused with presentations or homework done by students, because they not only communicate specific content but also choose their own methods and didactic approaches to the teaching of other classmates (Hanke, 2011).

As Hatano & Inagaki (1991) explained, there are two types of interaction in this methodology: Horizontal interaction, whereby roles are changed only during an activity or specific moment of the class, and vertical interaction where the roles are completely changed in the class. Therefore, the first interaction can be used as a validation mechanism of what is explained by the teacher being used as a peer tutoring, while the second interaction totally holds the student accountable for the teaching process of their peers and requires more preparation.

Thus, a very important distinction must be made regarding the two types of students in this methodology. On the one hand, there are the students in the role of teachers (tutors) and on the other hand are the passive students (students-tutees). This is relevant since researchers usually generalize the benefits of this methodology for both types of students.

Some of the main advantages for students in the role of teachers are:

-They learn to take responsibility for the learning process of their peers. Their self-concept is reinforced because they feel they are being treated with respect (Goodland, 1989).

-The role encourages and requires creativity, independence, and self-confidence (Grzega, 2005).

-It can be a great tool if the students know in advance and have experience on the subject that they are going to teach (Fisher, 2004).

-The social interaction induced by the process of teaching can produce valuable cognitive benefits and personal development that is not seen through other forms of learning (Hartman, 1990).

Per cognitive psychology, these advantages can be derived from four main factors: the expectation of having to teach, the requirement to explain something, the requirement to explain something to others, and the requirement to answer questions from others (Renkl, 1997). In other words, these advantages arise because of the obligation to perform the activities of the teaching process of a normal teacher. These activities can be preparation (revision, reformulation, and organization of the session), presentation of the session, and evaluation (observation and reflection of the session) (Okita & Schwartz, 509-515) (Martin, 1985) (Bargh & Schul, 1980).

On the other hand, some of the advantages for passive students during the session are:

-The students are more willing to learn from their peers because they have a similar social level, facilitating communication (Goodland, 1989).

-There may be a more pleasant and cooperative class atmosphere (Hedin, 1987).

The list of advantages for the passive students of the lesson is very small because the effectiveness of the learning of the class depends mainly on the methodology and format adopted by their peers in the role of the teacher. Although several studies and experiments related to the effectiveness of this methodology have been carried out, most of them make comparative measurements between the two types of students in the same session (some as the teacher and the remaining group as normal students). In accordance with Bargh & Schul (1980) one of the most referenced studies in the literature, there is quantitative evidence with the help of a post-test that students in the role of teacher ($M = 0.649$) have a higher level of learning than students watching the class ($M = 0.569$). This can be supported by the advantages mentioned in the previous paragraphs and by other researchers who argue the effectiveness of Learning-by-Teaching (Herweg, 2009) (Seidel et al., 2006).

Generally, through case studies, there are important research and information related to teaching practices attempting to change teachers (Bransford, Brown, & Cocking, 2000); however, there are very few studies of the application of Learning-by-Teaching in a university environment. The main contributions to the application of LDL in higher education come from cases such as (Biswas, Leelawong, K., & K., 2004; Matsuda et al., 2012; Grzega, 2003), as well as studies and reviews that investigate what happens around Learning-by-Teaching in the engineering education context (Carberry & Ohland, 2012; Carberry, 2008).

Nevertheless, none of the previous studies show the possible learning outcomes of passive students due to the different factors that may occur during a session directed and imparted by the other students. Contrary to the large amount of research and meta-studies (I.E.: (Hattie, 2009; Hattie & Yates, 2014)), it shows that not necessarily the strategies and didactics considered innovative by the teachers are the most effective.

Methodology

As mentioned in the context of the study, the elective course "Science, technology and Innovation" was used to apply active learning practices with role definition. For this, the teaching group in the role of "developer of a lesson" will use the Learning-by-Teaching methodology in 23% of the course. This practical case study seeks to investigate a contemporary phenomenon within its real context, thus being an appropriate tool to obtain preliminary inquiries (Yin, 2009). Likewise, as mentioned in Grzega (2005) courses under this dynamic should be prepared before starting with certain considerations and special conditions in order for them to be successful.

The pre-session process is outlined below.

The instructor:

- On the first day of the course, the teacher presents the dynamics of the different classes and they give the students a brief introduction to each role.
- Groups of five people are formed at random from different careers, genders and semesters to lead a class.
- Each group is randomly assigned to one of the following topics: Artificial intelligence, materials, global warming, reduction of malnutrition, medicine, agriculture and the internet of things (IoT).
- A specific date is assigned for each of the groups with a minimum of one month's notice.
- A class is given in which the objectives and the expected results will be presented by the students. In addition, a first approach and investigation of the assigned topics is carried out.
- A class is taught to explain the different practices and methodologies of active learning that they could use. Additionally, teachers give advice and tools that allow for better assimilation and interest of the transmitted information.

The students:

- They must submit a research paper related to their previously assigned theme and should mention the specific topics that they will address during their session. In addition, they should explain the activities, methodologies, and resources that they will use in their sessions.

Note: If the topics are general as in this case study, we recommend that teachers be very attentive to the previous deliverables to realign the specific topics that the students are going to give in their session if necessary.

However, during the sessions, teachers are just participants and observers of the dynamics raised by those responsible for the session. At the end of the session, the group of teachers grade students for the teacher role under the following criteria:

-Format of the Event

-Fulfillment of the Agenda

-Quality of Interventions

-Interest in the Promoted Issues

Data collection

This study was conducted during the second semester of 2016 with thirty-five students in their first year in different undergraduate bachelor degree programs. There were seven sessions during the semester, and each of them was totally directed and designed by groups of five students.

The collected data is given below.

Quantitative:

-For the students in a teacher role, the course group of teachers had to evaluate them based on the criteria mentioned above for each session on a scale of 0-100, where 0 is deficient and 100 is excellent.

-For the students who took the session, they had to answer an anonymous questionnaire at the end of each class. The main objective of this instrument was to find out the perceived learning level outcome related to the explained topic. This method served to indirectly measure learning outcomes and value added (Breslow, 2007). This survey inquired about both the students' attitudes towards the new pedagogy and their reflection on the learning achieved at the end of the session. These key questions were used in the questionnaire:

-Do you feel that the introduction of the session motivated you to learn the subject of the class? (yes, no, partially)

-Did you acquire new knowledge through the class today? (Y, N, P)

-Do you feel that the lesson's topic was explained clearly and correctly? (Y, N, P)

-Do you feel that class exercises promoted reflection and learning of new knowledge? (Y, N, P)

-Do you feel you can share the knowledge you learned today with others? (Y, N, P)

-Your interest in the topic of today's class was (low, mid, high).

-Your level of learning achievement related to the topic of today's class was (low, mid, high).

At the end of the course, all students have to answer an anonymous survey related to their perception of the development of competencies, such as:

-Transmit and teach new knowledge to other people.

-Interact and research issues related to science, technology and innovation.

Qualitative:

-For the students in a teacher role, the group of teachers gave feedback about the format and performance of each session (direct measurement method).

Results

The questionnaire results of each session (G1-5) were normalized, grouped and averaged. For example, if all students agreed with the affirmative question type, the maximum score would be 1 (corresponds to yes), and the worst score would be -1 (corresponds to no). For scale-type questions, the maximum score would be 1 (corresponds to high), and the worst score would be -1 (corresponds to low).

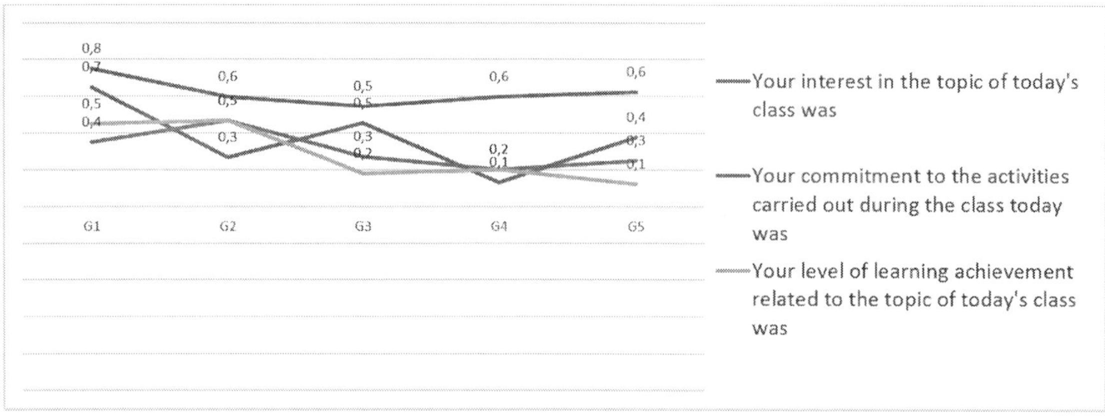

Figure 1. Perception of student learning outcomes

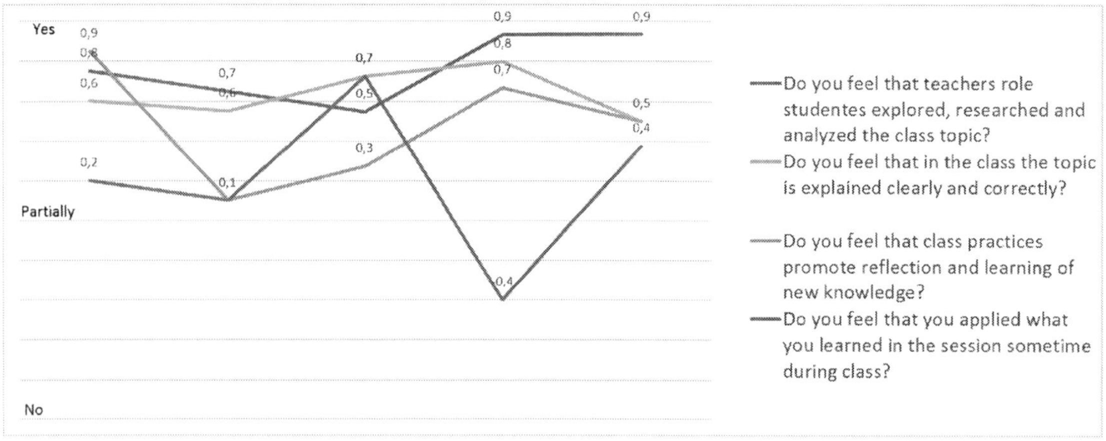

Figure 2. Perception of session format by students

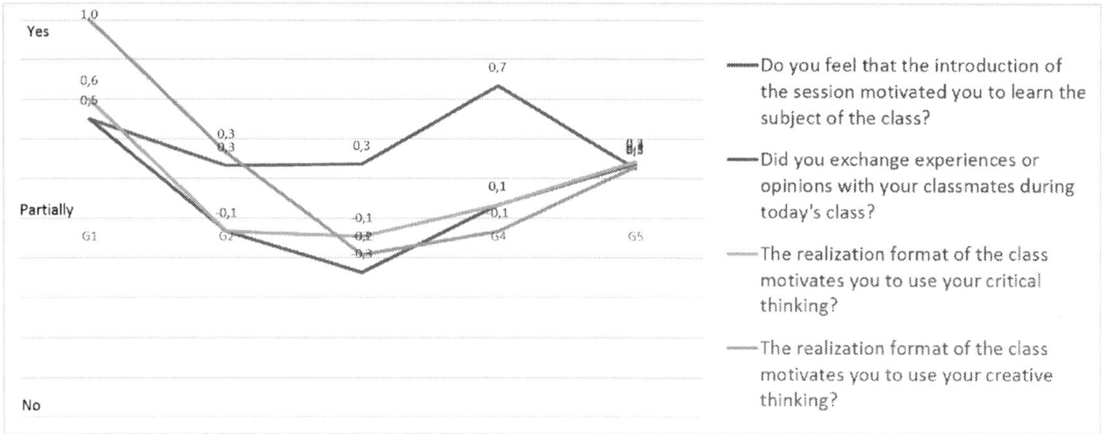

Figure 3. Perception of indirect learning effects on students

From the data presented in Figure 1, the students' perception of their learning outcome level was homogeneous except for the first two groups (M = 0.3). However, in all sessions the level of learning was at a medium-high level. Additionally, the same figure shows a strong correlation (Spearman = 0.86) between the learning level and the interest in the topic of the session.

Secondly, students normally consider that their peers explored, researched and analyzed the class topic (M = 0.8). This is a very important factor, not only because of the quality of the content presented but also because the credibility of the teacher is one of the larger effects (0.9) related to student achievement (Hattie, 2014). Likewise, many students believe the subjects were explained in a clear and correct way (M = 0.6). Nevertheless, in most sessions, the students were not sure that the practices of the session promoted reflection and learning of new knowledge. Furthermore, there is no uniformity regarding if students used their new learned knowledge during some time in the session (see Figure 2). The students are not sure that they applied what they were taught at some point in the session, being that this is the question with the greater standard deviation between sessions (0.4). This may be related to the fact that in some sessions, most of the time was used as a master class instead of an active strategy centered on the students.

As can be seen in Figure 3, students did not believe that they exchanged experiences or opinions during the different sessions. In other words, they did not have the opportunity to employ a kind of collaborative learning. Similarly, apart from the first group, students were not sure that the format of the sessions motivated them to use their creative and critical thinking.

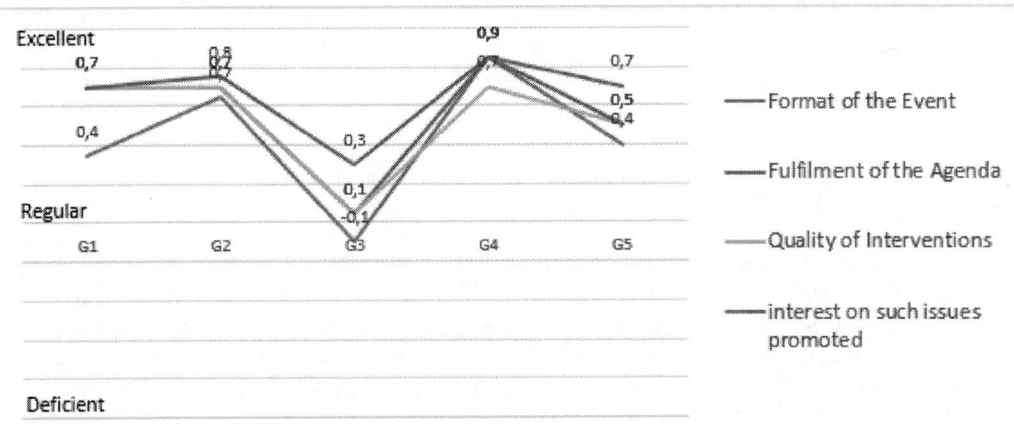

Figure 4. Perception of session format by instructors (grades)

Regarding the grades of the instructors towards the students responsible for each session (teacher role), homogeneous grades (regular-excellent) can be observed in most sessions (see Figure 4). The criteria that had the best overall performance are fulfilment of the agenda and interest in the promoted issues (M = 0.6). If the latter criterion "interest" is compared with the students' responses, teachers overestimated the general interest aroused by the topics of the sessions regarding the opinions per student survey. Moreover, the criterion that had the worst overall performance was the format of the event (M = 0.4). This criterion also had the largest standard deviation (0.4) due to the teachers' perception of having had classes with two excellent formats, two good formats and one regular format.

Some of the main comments that the instructors gave regarding the sessions were:

G1	Too many videos and little intervention of the group. They brought in an interesting guest, although the group could further take advantage of the guest and asked more questions to understand his investigation.
G2	They could have focused more on innovation and not so much on theory.
G3	Lacked preparation and focus on the session.
G4	They could have sent the readings before the session. The reason for the final activity was not understood (Last video without specific relation with the session)

G5 There is research, but they failed to communicate it the right way. They showed a list of everything they found. Encyclopedic presentation. They did not show a focus.

The methodology used by the groups to teach the assigned lesson was in general one of these:

-Completely masterclass by an external expert guest.

-Theoretical introduction to the subject of the lesson using PowerPoint and videos.

-Theoretical introduction by PowerPoint, and presentation and complementary talk from an external expert guest.

P.S. Just a few groups used an active learning activity during the session.

Finally, 66% of the students indicated that they had developed competencies to transmit and teach new knowledge to other people at the end of the course. Nevertheless, 33% responded that they achieved this by working in the role of teacher and 59% responded that they achieved this through other workshops in class, inter alia. In the same way, when asked if they developed competencies, interacted, and investigated issues related to science, technology, and innovation, 58% of the students said yes. But only 4% argued that they reached this conclusion through some of the sessions developed by themselves.

Conclusions

In conclusion, preliminary findings suggest that the interest aroused by the subject is the factor with main incidence in the student's learning level. Most sessions involved a semi-master format supported by audiovisual material (i.e. explanatory videos) and some cases had a presentation by external guests. The session activities developed by the students were not intended to meet specific objectives but to explain a general subject.

The results of the self-evaluation surveys indicated that 71% of the students did acquire new knowledge through the sessions developed by their peers, although their average learning level was 0.3 (medium-high). At the end of the course, 66% of students believed that they could transmit and teach new knowledge to other people. Although, the previous competence is the main expected result due to the teaching process related to role of teacher. Nevertheless, many students indicated that they developed that competence through the workshops in class and not during the teacher role session.

Recommendations for engineering education

The Learning-by-Teaching methodology could be a valuable tool for introducing students to the context of engineering education. It is necessary to change the paradigm of traditional podium presentations as the main teaching tool in engineering students' minds. But if this methodology is not accompanied by preparation, guidance and prior support of students in a teacher role, it could be counterproductive to the learning level of their peers

As can be seen in Figures 1-3, the learning outcomes of each session are directly related to the execution format designed by their classmates. However, the responsibility of the course and the learning outcomes fall on the teacher, so it is very important to know the students' plans before each session to support them with feedback and the necessary corrections.

It is important to push students to think outside the box about a lesson. If the instructor gives a lecture explaining different teaching strategies or active learning practices, it can be of great help to the future assigned class approach. Otherwise, the sessions presented later may only be master classes and fail to attract students' interest and attention.

Future research could include mandatory formative assessments on the subject assigned at the end of each class. This not only could be used as a method of direct measured learning outcomes, but it could also be used to support and guide teacher-students on specific issues that should be taught during the assigned sessions.

References

Bargh, J., & Schul, Y. (1980). On the cognitive benefits of teaching. *Journal of Educational Psychology*(72), 593-604.

Biswas, G., Leelawong, K., K., B., & K., V. (2004). Developing Learning by Teaching Environments That Support Self-Regulated Learning. 7th International Conference Intelligent Tutoring Systems.

Bransford, J. D., Brown, A. L., & Cocking, R. R. (2000). *How people learn.* Washington, D.C.: National Academy.

Breslow, L. (2007). Methods of Measuring Learning Outcomes and Value Added. Massachusetts Institute of Technology.

Carberry, A. (2008). Learning-by-teaching as a pedagogical approach and its implications on engineering education. *TUFTS University* .

Carberry, A., & Ohland, M. (2012). A Review of Learning-by-Teaching for Engineering. *Advances in Engineering Education.*

Duque, M., Ramírez, M. C., Ávila, A., & Hernández, J. T. (2013). Innovar en el aula en ciecnia y tecnologia con ingeneria. *World Engineering Education Forum.* Cartagena.

Felder, R., Woods, D., Stice, J., & Rugarcia, A. (2000). The Future of Engineering Education: II. Teaching Methods That Work. *Chemical Engineering Education*, pp. 26-39.

Fisher, R. (2004). *Learning by Teaching in Engineering Technology Education.* American Society for Engineering Education Annual Conference.

Gartner, A., Kohler, M. C., & Riessman, F. (1971). *Children teach children: Learning by teaching.* New York: Harper & Row.

Goodland, S. (1989). *Peer tutoring: A guide to learning by teaching.* London: Kogan Page.

Grzega, J. (2003). LdL in universitären Kursen: Ein hochschuldidaktischer Weg zur Vorbereitung auf die Wissensgesellschaft. Retrieved from http://www.ldl.de/material/berichte/uni/uni.html

Grzega, J. (2005). Learning By Teaching The Didactic Model LdL in University Classes. Germany.

Hanke, U. (2011). Learning by Teaching. In R. M. Seel, *Encyclopedia of the Sciences of Learning* (pp. 1830-1832). Springer US.

Hartman, H. J. (1990). Factors affecting the tutoring process. *Journal of Educational Development,*, 2-6.

Hatano, G., & Inagaki, K. (1991). Sharing cognition through collective comprehension activity. In L. B. Resuick, J. Levine, & S. Teasley, *Perspectives on socially shared cognition* (pp. 331-348). Washington, DC: American Psychological Association.

Hattie, J. (2009). *Visible learning: A synthesis of over 800 meta-analyses relating to achievement.* London: Routledge.

Hattie, J., & Yates, G. (2014). *Visible learnin for teachers and the science of how we learn.* London: Routledge.

Hedin, D. (1987). Students as Teachers: A Tool for Improving School. 42-47. Social Policy.

Herweg, C. (2009). Zielorientierung im deutschen und schweizerischen Physikunterricht. Leibniz-Institut für die Pädagogik der Naturwissenschaften (IPN) an der Universität Kiel.

Martin, J. P. (1985). Zum Aufbau didaktischer Teilkompetenzen beim Schüler. Tübingen: Fremdsprachenunterricht auf der lerntheoretischen Base des Informationsverarbeitungsansatzes.

Matsuda, N., Yarzebinski, E., Keiser R, V., Raizada, R., Stylianides, G., & K., K. (2012). Motivational Factors for Learning by Teaching: The Effect of a Competitive Game Show in a Virtual peer-Learning Environment. In N. Matsuda, E. Yarzebinski, V. Keiser R, R. Raizada, G. Stylianides, & K. K., *Intelligent Tutoring Systems* (pp. 101-111). Springer Berlin Heidelberg.

Okita, S. Y., & Schwartz, D. L. (509-515). When observation beats doing: Learning by teaching. Bloomington, IN: Proceedings of the 7th international conference on Learning sciences.

Piccoli, G., Ahmad, R., & Ives, B. (2001). Web-Based Virtual Learning Environments: A Research Framework and a Preliminary. *MIS Quarterly*, 401-426.

Ramírez, C., Avila, A., Duque, M., & Hernandez, T. (2015). Active innovative practice in engineering education process. Encuentro internacional de educación en ingeniería ACOFI.

Renkl, A. (1997). *Lernen durch Lehren. Zentrale Wirkmechanismen.* Wiesbaden: Deutscher Universita¨ts-Verlag.

Rivero, M., Gonzalez, M., Muñiz, P., Ena, J., Ortega, M., & Cavia, M. (2016). Student participation in the learning process: Responsibility. *ICERI2016 Proceedings*, 1093-1099.

Rodriguez, K. P., Maya, M. A., & Jaén, J. S. (2012). Educación en Ingenierías: de las clases magistrales a la pedagogía del aprendizaje activo. *Ingeniería Y Desarrollo*, 30(1), 125-142.

Roma, M. (2009). Learning by Teaching in Engineering: a Step beyond Learning by Doing. In A. Lazinica, & C. Calafate, *Technology Education and Development* (pp. 377-394). Madrid: Intech. Retrieved from http://www.intechopen.com/books/technology-education-and-development/learning-by-teaching-in-engineering-a-step-beyond-learning-by-doing

Seidel, T., Prenzel, M., Rimmele, R., Dalehefte, I. M., Herweg, C., Kobarg, M., & Schwindt, K. (2006). Views on instruction in physics – Findings of the IPN video study. *Zeitschrift für Pädagogik*(52), 799-821.

Shekhar, P., Borrego, M., Finelli, C., Prince, M., DeMonbrun, M., Waters, C., & Henderson, C. (2015). A mixed methods Study to Understand Student Resistance to Active Learning. *Research in Engineering Education Symposium 2015.*

Smith, K. A. (2011). Advancing the Practice of Engineering Education. South African Society for South African Society for Engineering Education.

Yin, R. (2009). Case study research: Design and methods. *Thousand Oaks, CA: Sage.*

Zabaleta, M. A. (2004). Innovación en la enseñanza universitaria. *Contextos educativos*, 113-136.

Copyright statement

Student engagement in the engineering design process in design-based learning: A mixed methods approach to video data analysis

Miancheng Guo
University of Massachusetts Amherst, Amherst, MA, USA
mianchengguo@educ.umass.edu

Elizabeth McEneaney
University of Massachusetts Amherst, Amherst, MA, USA
emcenean@educ.umass.edu

Martina Nieswandt
University of Massachusetts Amherst, Amherst, MA, USA
mnieswan@educ.umass.edu

***Abstract**: Recently, video recording has become a common means for researchers to capture interactions in classrooms. However, there has been little research literature on analyzing video data using mixed methods approaches. This paper addresses this shortage by introducing a mixed methods video data analysis case in which high school students' academic engagement in their engineering design process in a design-based science (DBS) activity is examined through the combination of a quantitative data analysis method named sequence analysis and a qualitative data analysis technique named elaborated running records (ERRs). Also, a set of recommendations for using our approach is provided.*

Introduction

Recently, video recording has become a common means for researchers to capture interactions in classrooms (Derry 2007; Shrum and Scott, 2017). This is mainly because such data collection allows the recording of both verbal and nonverbal behavior, which can be analyzed both quantitatively and qualitatively (DeCuir-Gunby, Marshall & McCulloch, 2012). Because of this important characteristic of video data, Jacobs, Kawanaka and Stigler (1999) argue that video data easily lends itself to mixed methods analysis. However, there has been little research literature on analyzing video data using mixed methods approaches (DeCuir-Gunby et al., 2012). Thus, the purpose of this paper is to address this gap by introducing a mixed methods video data analysis case in which high school students' academic engagement in a design-based science (DBS) activity. We examined this through the combination of sequence analysis, a quantitative approach, and a qualitative data analysis technique named elaborated running records (ERRs), with the assistance of qualitative analysis of student interview data.

Context and Research Question

This study was part of NSF-funded project that explores students' collective construction of three-dimensional (cognitive, social, and affective) problem-solving space in small group work in science and engineering activities. The present study focuses on gender and engagement. Specifically, we investigate whether gender grouping in small group work influences girls' behavioral, emotional, and cognitive engagement (Fredricks, Blumenfeld & Paris, 2004) in a DBS activity to design an artificial heart valve. DBS is a pedagogy where students learn science and real-world problem-solving skills through engineering design (Fortus, Krajcik, Dershimer, Marx & Mamlok-Naaman, 2005). Along with the release of the Next Generation of Science Standards (NGSS) in the USA, which advocates integrating engineering design into science education (NGSS Lead States, 2013), DBS has gained unprecedented attention and popularity. However, ensuring that girls are engaged and learning may depend on including peers of the same gender in the small group or team (Dasgupta, Sciricle, & Hunsinger, 2015).

Theoretical Framework

Figure 1 shows the overall logic of this paper. The DBS activity students participated in was basically an engineering design task. As the design process proceeds through various steps, the designer may engage behaviorally/emotionally/cognitively at different levels of intensity. Thus, the videos recording the students' behavioral/emotional/cognitive engagement during their design processes are actually temporal data, and such data can be analyzed by sequence analysis (Abbott, 1995; Maclndoe & Abbott, 2009). As we describe in more detail later, sequence analysis quantizes raw qualitative video data into numerical information so that the changes of levels of students' behavioral/emotional/cognitive engagement across the engineering design process can be visualized and assessed. However, such quantitative information cannot tell why such changes occurred, and so we used a qualitative data analysis method – elaborated running records (Rogat & Linnenbrink-Garcia, 2011; Rogat & Adams-Wiggins, 2014), together with qualitative analysis of student interviews, to capture interactions among students, so as to reveal clues for understanding such changes.

Thus, such data conversion and combination of methods were necessary for answering the research question and made this study a mixed methods study (Teddlie & Tashakkori, 2009; Creswell & Plano Clark, 2011).

Below, we briefly introduce sequence analysis and ERRs.

Sequence analysis originated as a set of bioinformatics methods designed to analyze DNA, RNA, and peptide sequences. When applied in social science, it becomes "a body of questions about social processes and a collection of techniques available to answer them" (Abbott, 1995, p.93). So, it is not a particular technique. Broadly, there are two types of sequence analysis: step-by-step methods and whole sequence methods (Abbott, 1995). In this study, we used a step-by-step method.

ERRs are running records that summarize video data to go beyond capturing the dialogue within the learning group by also describing actions/behaviors relevant to the variables of interest (e.g., facial expressions, gestures, tone used in speaking); importantly, these records are meant to be descriptions without viewer interpretation (Rogat & Linnenbrink-Garcia, 2011).

Methods

Participants and Data Collection

For the purpose of this study, we selected four student groups with different gender constellations:

- Group 4: four girls (100% female)
- Group 3: three girls and one boy (75% female)
- Group 2: two girls and two boys (50% female)
- Group 1: one girl and two boys (33% female)

Students were enrolled in two different biology classes in a high school in Massachusetts, had volunteered to participate, and had received parental/guardian consent for participating. Qualitative data were collected by videotaping each student group's whole design process and through follow-up focus group interviews, lasting approximately 20 minutes, regarding their perceptions of their group work process.

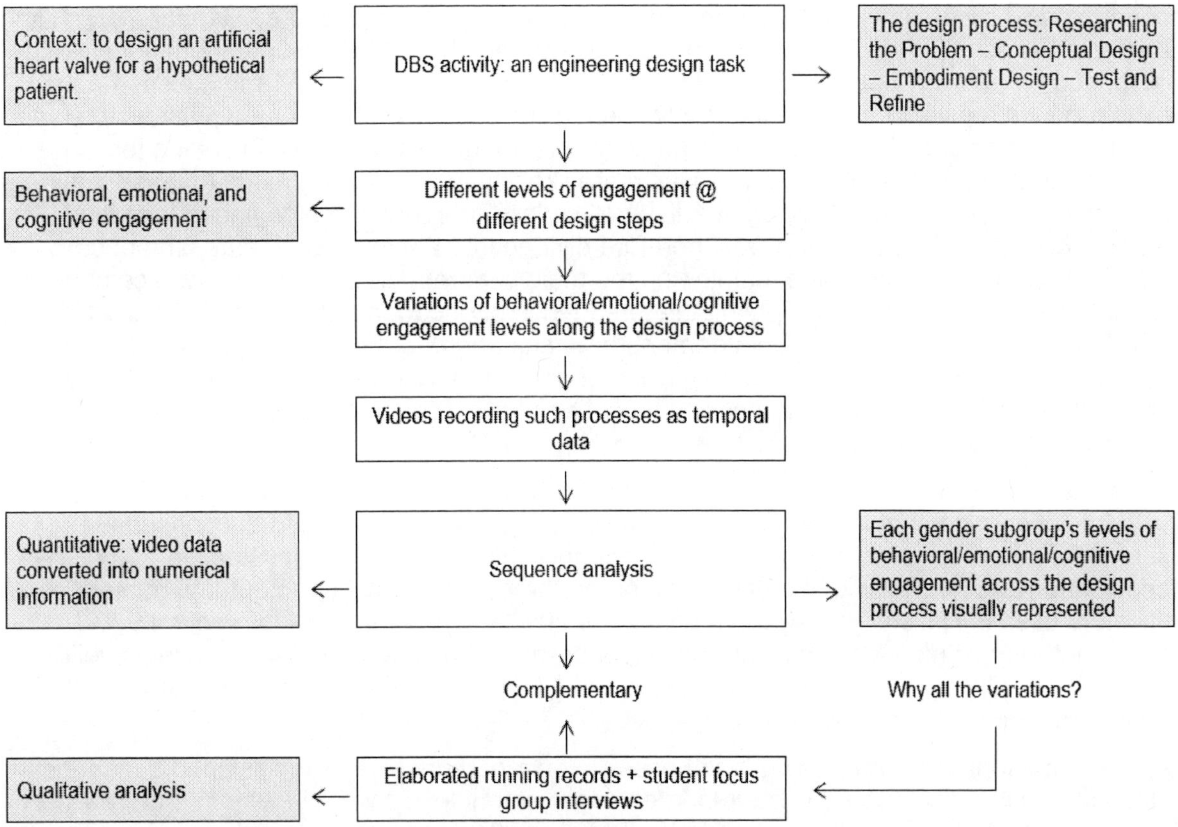

Figure 1: The overall logic of this paper

Data Analysis

We compared the engagement of female and male students in each group separately in order to address our research question. Thus, the unit of analysis is the gender-specific subgroup – all boys in a group as the male subgroup and all the girls in a group constituting the female subgroup. To identify students' behavioral, emotional, and cognitive engagement, we developed a system of indicators for these three types of engagement based on Fredricks, et al.'s (2004) definition of student engagement.

Student Engagement. Fredricks, et al. (2004) conceptualize student engagement as having three interwoven but conceptually distinct dimensions: behavioral, emotional, and cognitive. Behavioral engagement refers to various kinds of learning-related and academic-oriented behaviors, actions and involvements that students engage in. Emotional engagement pertains to students' affective reactions to their teachers, classmates, academic contents/activities, and school. Cognitive engagement involves two important aspects: (1) A psychological investment that incorporates thoughtfulness and willingness to make efforts to comprehend complex ideas and master difficult skills and (2) self-regulated learning (e.g., the use of strategies such as planning learning and self-evaluation).

Sequence analysis. We used this method to analyze and visually represent each female and male subgroup's level of behavioral, emotional, and cognitive engagement. Here, the level of engagement is defined as the proportion of a subgroup that is engaged in any given 2-minute period (i.e., a segment). There were five possible levels of engagement: all engaged, majority engaged, half engaged, minority engaged, and none engaged. Each sequence is composed of temporally arranged and separately coded segments. No prior published research has used sequence analysis to analyze classroom video data, so there is no accepted standard length for segments to code. Our decision to divide the activity sequence into segment of 2 minutes each was based on piloting coding in different ways. We found that: (1) setting 1-minute segments would produce too much and unnecessarily finely-grained data, making the sequences too long (given that each group finished their design activity in about an hour); and (2) setting the segment length to more than 2 minutes was more likely than shorter segments to mask within-segment changes in levels of engagement, while coders' memory of the beginning of a long segment often faded. Each video was therefore divided into 2-minute segments, and each segment was viewed to code the female and the male subgroup separately and was compared with predetermined indicators reflecting the three types of engagement, which enabled us to record the levels of each type of engagement for each subgroup. For complete information of these indicators, see Appendix 1. Here, it is important to report that they were developed by the second and third authors of this paper and have been validated for a very similar purpose in a previous study (Guo, Nieswandt, McEneaney & Howe, 2016).

During the sequence analysis process, we paid careful attention to inter-rater reliability. Since the levels of engagement in this study constituted ordinal data, we adopted the consensus estimate of interrater reliability and used percent agreement to calculate it (Stemler, 2004). Two coders who were familiar with the definitions of the three types of engagement and related indicators used in this study analyzed a sample of 20% of all segments, and reached 92% agreement. As this number was significantly higher than 70%, the generally accepted level of interrater reliability based on consensus estimates (Stemler, 2004), one of the coders proceeded to finish the rest of the coding work.

For later data input and processing, each possible engagement level was assigned a numerical value and a color (see Table 1). The recorded engagement levels were input into the TraMineR package in R (Gabadinho, Ritschard, Studer & Müller, 2010), which then produced visual representations of each sub-group's three types of engagement.

Table 1. Different engagement levels and their numerical values and color codes

Engagement level	Numerical Value	Color Code
All engaged	5	Dark green
Majority engaged	4	Light green
Half engaged	3	Yellow
Minority engaged	2	Purple
None engaged	1	Red

ERRs. We used this method to qualitatively capture the interactions among group members and other salient phenomena that were related to their behavioral/emotional/cognitive engagement. First, we developed an ERR template that contained three columns: Time, Running Record, and Reviewer Comments. Then, the research team watched each video throughout, identified each event/phenomenon that was a manifestation of or related to students' behavioral/emotional/cognitive engagement based on predetermined indicators, and described it in the Running Record column. Each event/phenomenon was recorded by narrative description with master indicators inserted wherever necessary in the passage. Here, a master indicator is a single letter put in a square bracket denoting a certain type of engagement. For example, cognitive engagement is denoted by "[C]" and behavioral engagement by "[B]". In a passage describing an event/phenomenon, a master indicator signifies that the preceding passage is a manifestation of the type of engagement denoted by this master indicator. In the Time column, the period in which an event/phenomenon happened was recorded. In the Reviewer Comments column, ERR coders put in any materials that thought to be helpful for understanding the contents in the Running Record column, such as notes and screenshots from the video. For an example from a completed ERR table, see Figure 2.

Time	Running Record	Reviewer Comments
17. 39:39-40:35	They test the model. During the first tilt, there's sound indicating marbles moving; during the second tilt, there's no sound. Then Malik and Jenna smile, and Jenna does the intimate action to Malik again (see picture on the right). [B] [E]	 **Test and Redesign** begins. The first test.
18. 40:36-43:09	Malik folds a small board in half and then tapes it in the box. Jenna helps. [B] Then, Jenna goes to another group.	A revision.

Figure 2. Part of the elaborated running records table

Results and Discussion

Since the full set of sequence analyses of all the subgroups produced a substantial number of charts, and since the purpose of this paper is to demonstrate how classroom video data can be analyzed by mixed methods, here in the limited space of this paper we only report one case in which sequence analysis and ERRs functioned complementarily. Visual representations of sequence analysis results are represented below. Figure 3 shows that in Group 1 the two boys and one girl had very similar levels of behavioral engagement – they all maintained high behavioral engagement levels throughout their entire design process. However, this is not the case for emotional or cognitive engagement. From Figure 4, one can see that the girl's and the boys' emotional engagement seems to be "complementary" – when the girl's engagement levels are high (the step of Researching the Problem), the boys' are low; where the boys' engagement levels are high (the steps of Embodiment Design and Test and Redesign), the girl's are low. The only step where the boys' and the girl's engagement levels are both high is Conceptual Design. For cognitive engagement (Figure 5), a roughly similar pattern exists, as girl's engagement is the inverse of the boys'.

Why is this? Was it that something happened between the girl and the boys or an external factor that contributed to the significant decreases in the levels of emotional and cognitive engagement of the girl in the design steps after Conceptual Design? The sequence analysis results cannot address this question, and the student focus group interview of this group only provided some vague clues (e.g., the girl mentioned the boys' "leadership") which are not specific to the design steps. However, the ERRs provided additional insight. Since the transitions of the girl's emotional and cognitive engagement occurred at around segment 11, we located the corresponding time period in the ERRs table for this group, and read through the whole table with more attention to the group members' interactions adjacent to this period. We noticed two salient phenomena from these records: when the group transitioned from Conceptual Design (where the main group activity was brainstorming/discussing ideas) to Embodiment Design (where the main group activity was manual construction), the boys became highly active and took the lead. In this phase, the girl was ignored four times by the boys four times when she wanted to contribute ideas or participate in the boys' construction work – once before segment 11, once right after it and two other times following closely thereafter.

This information, together with the clue provided by the interview data (as briefly mentioned above), suggests that the girls' decrease in emotional and cognitive engagement was closely related to some of the boys' behaviors. This is consistent with findings that girls sometimes feel intimidated by the stereotypical male-pattern behaviors of boys in their classes or learning groups and consequently feel restricted in participation (e.g., Campbell & Evans, 1997).

Implications and Recommendations

Since the U.S. NGSS stress that engineering design has the same status as scientific inquiry in science education, DBS has become an important science pedagogy in the U.S. and is also gaining popularity in more and more other areas in the world. How can researchers analyze students' engagement in DBS activities in all its complexity? This study introduces a temporal perspective, that is, we viewed student engagement as sequences of phenomena/events across the different steps of the engineering design process and showed how sequence analysis – an approach rarely used in educational research – could be adopted to analyze such video data. However, as reported above, we needed to use two qualitative methods – ERRs and student focus group interview to find clues for interpreting the quantitative sequence analysis results.

Based on our experience, we now offer several recommendations for using mixed methods to analyze classroom video data. First, to analyze small group video data, one must have recorded such data with high quality. We and our project colleagues have accumulated some experience in this aspect. An important tip is that the researcher must make sure that all the students in a group are recorded. Our team used two video recording devices – one was a digital camera mounted on a stand and the other one was an iPad which could be held by hand so that the researcher could follow and record students when they physically moved away from the group, often times to retrieve needed materials. Another important tip is that the researcher shouldn't assume that with two video recording devices he/she will record sufficiently high-quality audio for research purposes. For the authors' research team, many times this was not the case – in order to record all group members, the camera had to be put a few feet away from their table and the iPad had to be put near the edge of their table, making it difficult to clearly record what everyone was saying. To solve this problem, we added a voice recording device at the center of the group's table, and this proved to be an effective solution. Accordingly, one should understand that such a project cannot be done by one person, and that research teams should include at least one person who is familiar with video production technologies, such as Final Cut Pro X to produce composite videos that show all members of a student group by showing the two camera angles side by side and synchronized. In addition, such editing can draw on the best quality audio at any moment from the various recording devices used. Third, this study's video analysis method (or an adapted but similar method) can be used in any settings where researchers are interested in sequences of events/phenomena (e.g., the levels of students' conceptual design skills within each iteration of the engineering design process for a semester-long design task). Fourth, we recommend that in using mixed methods to analyze video data, researchers should collect data for the same variable(s) from at least three perspectives/sources that could potentially complement each other in terms of analyzing or understanding the variable(s) of interest. Such triangulation may confirm, complement or add complexity to our understanding (Denzin, 2012). Finally, one should realize that such a mixed methods approach to video data analysis is a time- and labor- intensive process in which the issue of inter-rater reliability should receive adequate attention.

Acknowledgements

This study is part of NSF-funded project "Managing Small Groups to Meet the Social and Psychological Demands of Scientific and Engineering Practices in High School Science" (DRL-1252339).

Copyright statement

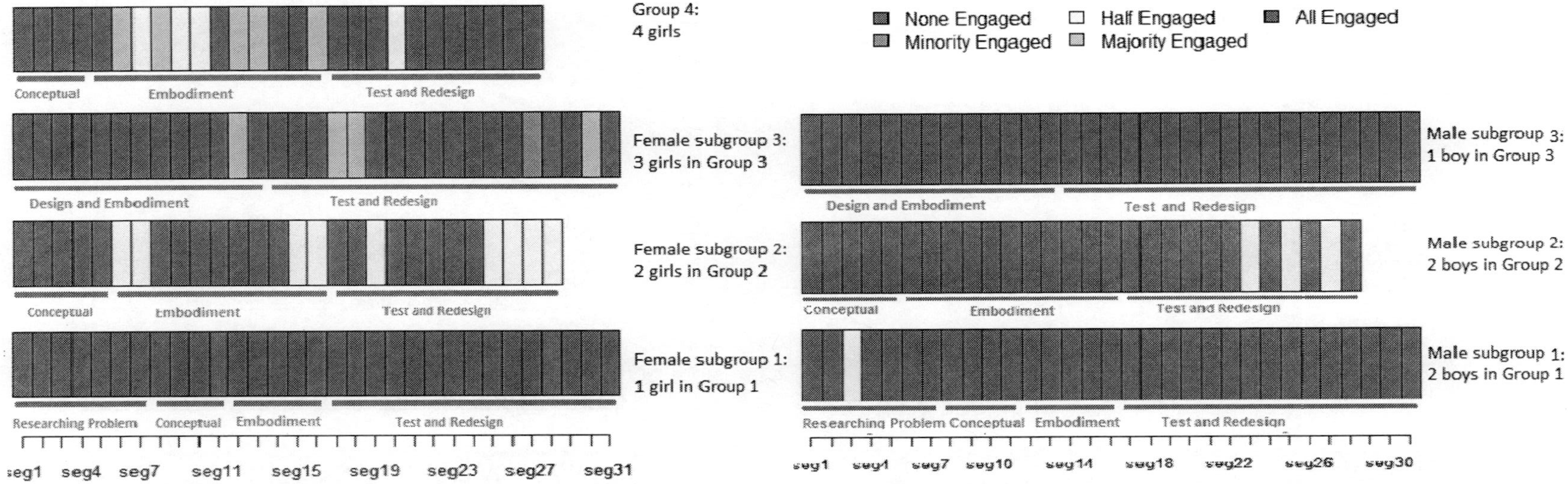

Figure3 Female and male subgroups' behavioral engagement

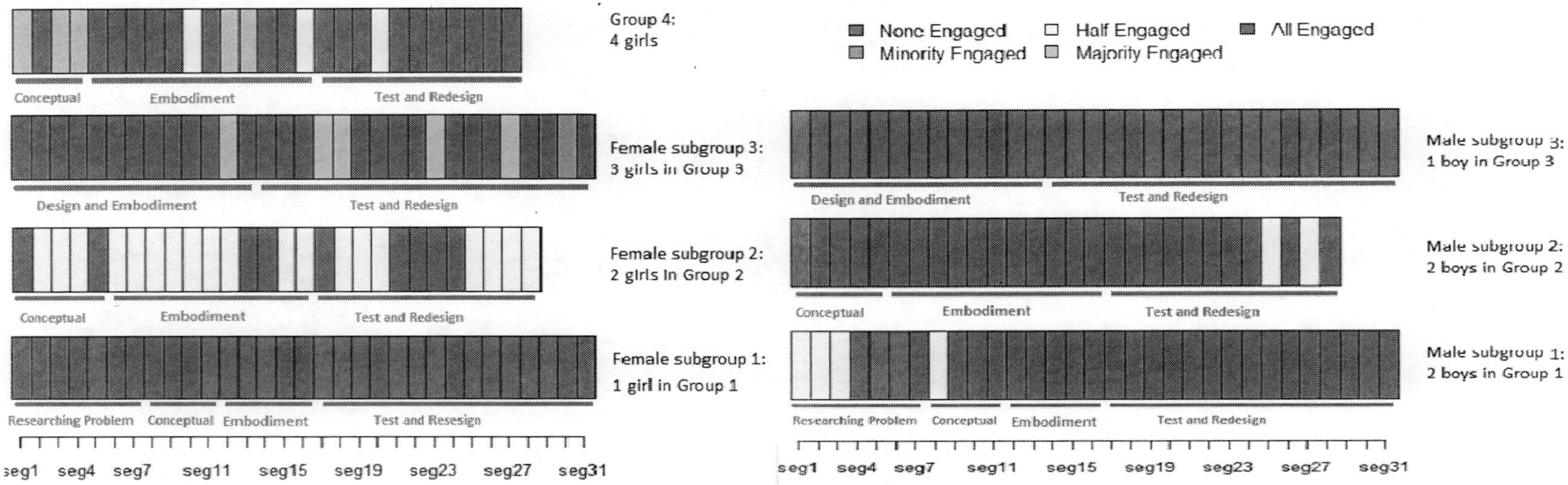

Figure 4 Female and male subgroups' emotional engagement

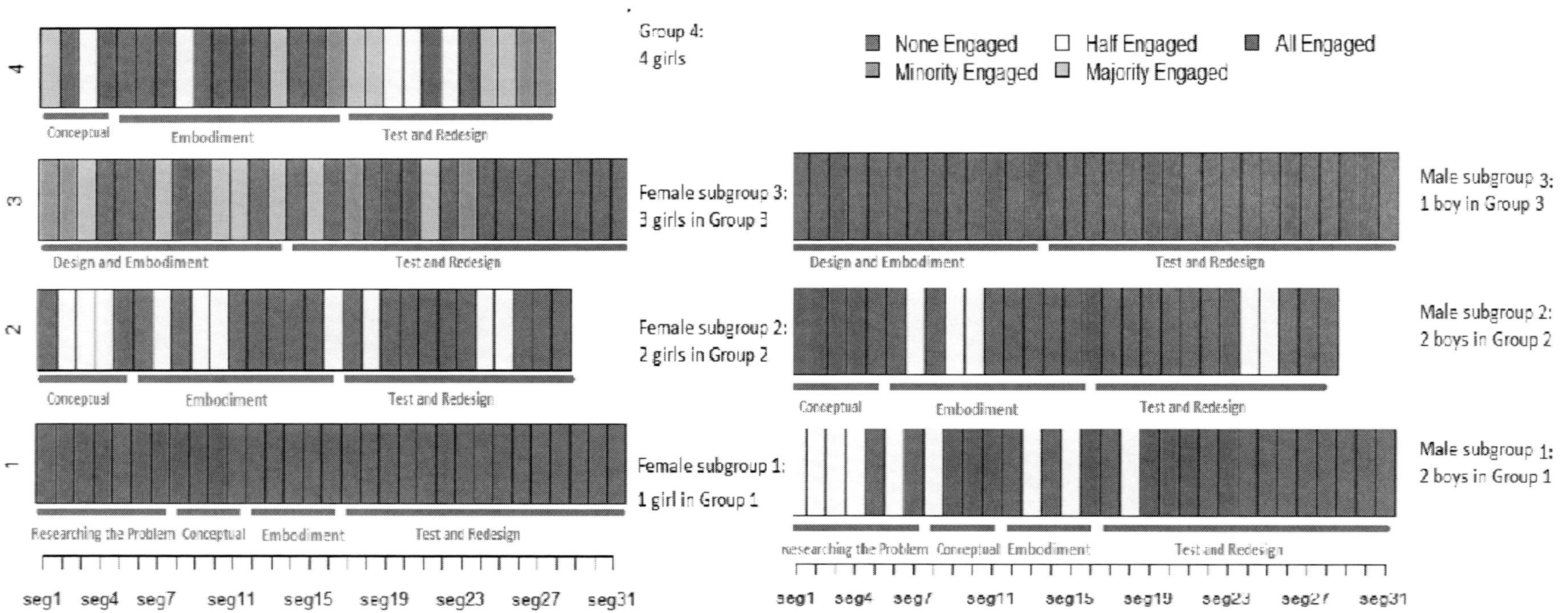

Figure 5 Female and male subgroups' cognitive engagement

References:

Abbott, A. (1995). Sequence Analysis: New Methods for Old Ideas. *Annual Review of Sociology*. http://doi.org/10.1146/annurev.so.21.080195.000521

Campbell, K., & Evans, C. (1997). Gender issues in the classroom: A comparison of mathematics anxiety. Education, 117 (3), 332-339.

Connell, J. P. (1990). Context, self, and action: A motivational analysis of self-system processes across the life-span. In D. Cicchetti (Ed.), The self in transition: Infancy to childhood (pp. 61–97). Chicago: University of Chicago Press.

Creswell, J. W., & Plano Clark, V. L. (2011). Designing and conducting mixed methods research. SAGE Publications. http://doi.org/10.1111/j.1753-6405.2007.00096.x

Dasgupta, N., Scircle, M.M., & Hunsinger, M (2015). Female peers in small work groups enhance women's motivation, verbal participation, and career aspirations in engineering. Proceedings of the National Academy of Sciences 112 (16): 4988–4993. doi:10.1073/pnas.1422822112

DeCuir-Gunby, J. T., Marshall, P. L., & McCulloch, A. W. (2012). Using mixed methods to analyze video data : A mathematics teacher professional development example. *Journal of Mixed Methods Research*, *6*(3), 199–216. http://doi.org/10.1177/1558689811421174

Denzin, N. K. (2012). Triangulation 2.0. *Journal of Mixed Methods Research*, *6*(2), 80-88.

Derry, S. J. (2007). Video research in classroom and teacher learning (Standardize that!). In R. Goldman (Ed.), *Video research in the learning sciences* (pp. 305–320). Mahwah, NJ: Routledge.

Fortus, D., Dershimer, R. C., Krajcik, J., Marx, R. W., & Mamlok-Naaman, R. (2004). Design-based science and student learning. *Journal of Research in Science Teaching*, *41*, 1081–1110. doi:10.1002/tea.20040

Fredricks, J. A., Blumenfeld, P. C., & Paris, A. H. (2004). School Engagement: Potential of the Concept, State of the Evidence. Review of Educational Research. doi:10.3102/00346543074001059

Gabadinho, A., G. Ritschard, M. Studer and N. S. Müller (2010). Mining sequence data in R with the TraMineR package: A user's guide. University of Geneva (http://mephisto.unige.ch/traminer).

Guo, M., Nieswandt, M., McEneaney, E. H. & Howe, S.V. (2016, April). Students' engagement in different steps of the engineering design process in a design-based biology activity. Paper presented at the Annual International Conference of the National Association for Research in Science Teaching, Baltimore, MD.

Jacobs, J. K., Kawanaka, T., & Stigler, J. W. (1999). Integrating qualitative and quantitative approaches to the analysis of video data on classroom teaching. International Journal of Educational Research, 31, 717-724.

MacIndoe, H. and Abbott, A. (2009) Sequence Analysis and Optimal Matching Techniques for Social Science Data. In Hardy, M. A., & Bryman, A. (Eds.), *Handbook of data analysis* (pp. 387-409). Sage.

NGSS Lead States. (2013). Next Generation Science Standards. Achieve, Inc.

Rogat, T. K., & Adams-Wiggins, K. R. (2014). Other-regulation in collaborative groups: Implications for regulation quality. *Instructional Science*, *42*(6), 879-904.

Rogat, T. K., & Linnenbrink-Garcia, L. (2011). Socially Shared Regulation in Collaborative Groups: An Analysis of the Interplay Between Quality of Social Regulation and Group Processes. *Cognition and Instruction*, *29*(4), 375–415. http://doi.org/10.1080/07370008.2011.607930

Shrum, W., & Scott, G. (2017). *Video ethnography in practice: Planning, shooting, and editing for social analysis*.

Stemler, S. E. (2004). A comparison of consensus, consistency, and measurement approaches to estimating interrater reliability. Practical Assessment Research Evaluation, 9(4), 1–19. Retrieved from http://pareonline.net/getvn.asp?v=9&n=4

Teddlie, C. & Tashakkori, A. (2009). Foundations of mixed methods research. Los Angeles: Sage.

Appendix 1. Indicators of behavioral, emotional and cognitive engagement

Type of Engagement	Definition of Engagement	Indicators
Behavioral	Various kinds of learning-related and academic-oriented behaviors, actions and involvements that students engage in. For example following school rules, attending classes, concentrating on academic tasks, asking questions, contributing to class discussion, completing assignments, etc. (Fredricks et al., 2004).	Main Indicator: Paying Attention Sub-indicators: • Listening to group members • Talking to group members • Watching group members' on-task behavior • Listening to teacher • Talking to teacher Main Indicator: Doing Hands-on Work Sub-indicators: • Fetching, organizing, cleaning necessary materials • Manipulating and investigating objects • Constructing prototype • Estimating • Calculating, measuring • Sketching, drawing • Making notes
Emotional	Students' affective reactions to their teachers, classmates, academic contents/activities, and school. For example, students' self-efficacy in academics, sense of belonging in the school context, interest in academic content, perception of connectedness to teachers and other students, as well as their emotions such as happiness, sadness, anxiety, boredom, etc. (Fredricks et al., 2004).	Main Indicator: Interest Sub-indicators: • Curiosity - e.g., asking curiosity questions • Enjoyment - e.g., verbal expression of having fun or liking tasks/activity • Frustration - e.g., verbal expression of frustration or negative feelings • Eager to work - e.g., verbal expression of starting to work or impatience with slower group members • Endurance working on problem /task - e.g. verbal expression of doing more than asked for, of continuing working on problem/task at home or in free time, etc. • Attending to content - e.g., asking questions related to task/activity • Help seeking from others (other groups, teacher) • Focused on problem/task • Recognizing each other's contributions • Actively seeking feedback Main Indicator: Positive Group Interaction Sub-indicators: • Group enjoys working together • Work well together • Caring about what each person thinks Main Indicator: Social Cohesion Sub-indicators: • Group members are friends • Feeling of belonging • Getting along with members of group • Liking group

Type of Engagement	Definition of Engagement	Indicators
Cognitive	Involves two important aspects (Fredricks et al., 2004): (1) A psychological investment that incorporates thoughtfulness and willingness to make efforts to comprehend complex ideas and master difficult skills (e.g., a desire to go beyond requirements, a preference for challenges, etc.) and (2) self-regulated learning (e.g., the use of strategies such as planning and monitoring learning and self-evaluation).	Main Indicator: Asking Exploratory Questions Sub-indicators: • Learning about content • Articulating the specific need/problem for the design task • Examining the current state of the issue and current solutions • Understanding group member's views • Confirming group members' explanations • Alternative explanations Main Indicator: Creating/Designing • Brainstorming possible solutions • Drawing on math and biology knowledge • Articulating the possible solutions • Refining the possible solutions • Identifying the best solution Main Indicator: Cumulative Reasoning Sub-indicators: • Elaborating on each other's arguments. • Motivating arguments • Explanations of group members completed with explanations of other group members • Drawing conclusions from information discussed in group • Making connections from information discussed in group Main Indicator: Handling Cognitive Conflicts Sub-indicators: • Presenting of contradictory beliefs on information concerning the learning content • Group member(s) was/were contradicted by others • Contradicted group member stated a counter-argument Main Indicator: Information Processing Sub-indicators: • Synthesizing information • Summarizing results, conclusions, observations etc. • Evaluating results, observations, conclusions, etc.

Learning outcomes and educational quality assessment for industrial engineering undergraduate programs in Brazil

Felipe Guilherme Melo
Universidade Federal da Bahia/ Graduate Program in Industrial Engineering, Salvador, Brazil
felipeguilherme1@gmail.com

Ricardo Kalid
Universidade Federal do Sul da Bahia/ Graduate Program in Industrial Engineering, Salvador, Brazil
ricardo.kalid@gmail.com

Marcelo Embiruçu
Universidade Federal da Bahia / Graduate Program in Industrial Engineering, Salvador, Brazil
embirucu@ufba.br

Abstract: *The relevance of higher education in the economic and social development of all countries is unquestionable, therefore quality management in higher education institutions has been the subject of many scientific studies worldwide. This study was based on statistical analyses of the 2014 Brazilian exam of student performance (ENADE) database from 11024 Industrial Engineering (IE) students in 347 undergraduate programs, and 278 higher education institutions in Brazil, aiming to identify strategies that foster the excellence of these programs. We identified a set of five educational strategies found in the best IE undergraduate programs in Brazil. Among those strategies, the opportunity to participate in study abroad programs and internship experiences inside and/ or outside the country are the main factors that differentiate the best programs. In addition, considering the results from ENADE 2014, our analysis indicates that the quality of Brazilian IE undergraduate programs has not been proportional to their quantitative growth.*

Introduction

The internationalization of companies has driven the globalization of engineers' professional performance. Prados, Peterson, and Lattuca (2005) emphasize that to work in international projects and within multinational teams, engineering students must have access to high quality education.

Several researchers have been studying the quality of higher education (Dias, Marchelli, & Horiguela, 2006; Cunha, 2014; Diedericks, Klerk, & Bevan-Dye, 2015; Teeroovengadum, Kamalanabhan, & Seebaluck, 2016; Zabadi, 2013). Concerns about higher education quality in Brazil began to be discussed in 1990 (Pinto, Mello, & Melo, 2016). In 1995, the Brazilian Ministry of Education (MEC, *Ministério da Educação*) started the Brazilian System of Higher Education Evaluation (SINAES, *Sistema Nacional de Avaliação da Educação Superior*), which aims to guarantee the quality of Brazilian higher education through indicators that evaluate Higher Education Institutions (HEI), undergraduate programs, and students learning outcomes.

Among the undergraduate engineering programs in Brazil, IE programs are the second most abundant. Between 2000 and 2017, these programs reached a growth rate of 1150%, increasing from 72 to 900 programs in March 2017. Regarding quality, SINAES data on student learning outcomes has shown that 81.8% of these programs have poor educational quality or have never completed the evaluation process. The quality management of undergraduate programs works as an evaluation and accreditation system that guarantees quality professional education to the students.

This study aims to: (1) outline the quality of the Industrial Engineering (IE) undergraduate programs in Brazil; and (2) identify educational strategies adopted by the best IE undergraduate programs in Brazil in order to contribute to the excellence of these programs. In this context, the questions of this study further investigate the quality of IE undergraduate programs, and, consequently, contribute to the professional education of future Brazilian industrial engineers.

Theoretical Framework

Educational Quality Management in HEI in Brazil

The relevance of higher education in the economic and social development of all countries is undeniable. Therefore, quality management in higher education institutions has been the subject of many scientific studies worldwide. Đonlagić and Fazlić (2015) point out that in the current knowledge based economy, higher education is a key factor to reach social, political, and economic development.

Matos, Cirino, Brown, and Leite (2013) emphasize that the Brazilian higher education system has undergone changes and adaptations that involve the formulation of new legislation, new evaluative practices, and new formats of programs curricula, aiming, among others, to meet the new requirements of the marketplace. Quality management in the Brazilian higher education system has been discussed, evaluated, and improved by MEC and by the Brazilian Institute for Educational and Pedagogical Research (INEP, *Instituto Nacional de Estudos e Pesquisas Educacionais Anísio Teixeira*), which is the national institution responsible for developing and applying the Brazilian Exam of Student Performance (ENADE, *Exame Nacional de Desempenho de Estudantes*) (Ribeiro & Belhot, 2002).

The first steps to evaluate the quality of Brazilian HEI started in the 1990's through a gradual process of developing a higher education evaluation system (Pinto, Mello, & Melo, 2016). In contrast to the United States and European countries, Pedrosa, Amaral, and Knobel (2013) claim that Brazil was one of the first countries to develop a system that evaluates HEI in a comprehensive way, which encompasses an evaluation of students learning outcomes, for both public and private HEI. The process of creating the Brazilian higher education evaluation system began in 1995 with the Federal Law 9.131/1995 (Brasil, 1995), which established the National Exam of Programs (ENC, *Exame Nacional dos Cursos*, also known in Brazil as "Provão"). In 2003, this exam was applied to more than 470 thousand undergraduate students of 6.5 thousand programs distributed in 26 fields of study. This was the last time ENC was applied (INEP, 2003).

In 2004, SINAES was established by the Federal Law 10.861/2004 (Brasil, 2004). This national system aims to ensure the national process of evaluating HEI, undergraduate programs and the academic performance of their students (Brasil, 2004, p. 1). Likewise, Pinto, Mello, and Melo (2016) emphasize that SINAES represents an evaluation of education policy, focused on the effective construction of participation, ethics and quality in higher education, regarding the differences among HEI and regional identities (p. 92).

Pedrosa, Amaral, and Knobel (2013) point out that one of the major differences between SINAES and ENC is the establishment of a test to evaluate student learning outcomes (ENADE). Since then, SINAES has worked as a measure for undergraduate programs accreditation in public and private HEI. The elements evaluated by SINAES are represented by three main indicators: (1) ENADE, an evaluation that focuses on student learning outcomes; (2) a general index for the programs (IGC, *Índice Geral de Cursos*), which evaluates the HEI; and (3) a preliminary score for the programs (CPC, *Conceito Preliminar de Curso*), which evaluates the structure of the program curriculum. These indicators are evaluated on a discrete scale from 1 to 5, where 5 is the best score (MEC, 2007). SINAES evaluates the undergraduate programs triennially. Consequently, IE programs were evaluated in 2005, 2008, 2011, and 2014. The next evaluation will take place in 2017.

Focusing on assessing student learning outcomes, ENADE is more than just a quantitative indicator, it is a tool to assess the skills and abilities of undergraduates over a three-year cycle (Verhine, Dantas, & Soares, 2006). According to Tagg (2010), the evaluation of student learning outcomes is a key factor to improve quality management in HEI. In addition, it should be noted that ENADE is one of the most important indicators of SINAES, as it supports CPC and IGC scores (Griboski, 2012; Sousa, Seiffert, & Fernandes, 2016). ENADE has become a required curriculum component, which is now registered in each student's transcript. INEP (2014, p. 5) defines the main goal of ENADE as:

To assess students' performance regarding to the content provided in the curriculum guidelines of the undergraduate programs, their abilities to meet the arising demands of the evolution of knowledge, and their competency to understand subjects outside the specific scope of their profession, including other areas of knowledge and contemporary Brazilian and global issues.

Industrial Engineering Programs in Brazil: an Overview

The pillars of the IE were built during the Industrial Revolution. According to Cunha (2002), since the Industrial Revolution the models of the organizations have changed with the formulation of new techniques and operational methods. In this context, IE programs were maturing throughout the 20th century with the aim of developing methods and techniques that would allow for the optimization of production processes. Therefore, the organizations were able to follow the technological and market evolution characterized by the Industrial Revolution (INEP, 2010). In this way, the industrial engineers' areas of expertise have expanded over time and required increasing levels of skill and ability. Bittencourt, Viali, and Beltrame (2010) state that the competency of the 21st century industrial engineer is in line with the primary needs of these organizations, which depend on technical, scientific, and management knowledge.

The first Brazilian IE undergraduate program started in 1958 at São Paulo University (USP *Universidade de São Paulo*) (Fleury, 2008). Regarding the quantitative growth of these programs, currently MEC recognizes 900 programs in its database, according to E-MEC system on Mach 7, 2017. Figure 1 shows the quantitative evolution of the Brazilian IE undergraduate programs. Between 2000 and 2017, the number of program grew from 72 to 900, which represents a growth rate of 1150%.

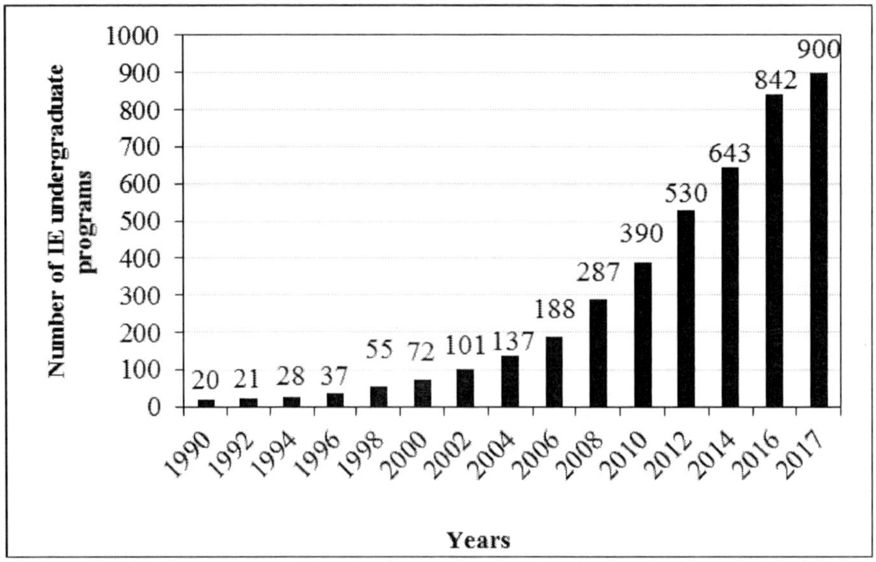

Figure 1: Quantitative growth of the Brazilian IE undergraduate programs (1990-2017).

Considering the increase in the number of Brazilian IE undergraduate programs, Faé and Ribeiro (2005) highlight that the industrial engineer has become a crucial profession for the most diverse branches of industry, commerce, and services, due to competitiveness and integration between global markets, which resulted in a need for high quality products and increasingly efficient organizations.

Regarding the regulations of Brazilian IE undergraduate programs, in 2002 the Brazilian education council (CNE, *Conselho Nacional de Educação*) established the Resolution CNE/CES 11/2002 (Brasil, 2002), which defines national curriculum guidelines for engineering programs. For IE programs in particular, the Brazilian IE association (ABEPRO, *Associação Brasileira de Engenharia de Produção*) published the "Industrial Engineering: major and curriculum guidelines", which established basic guidelines for the pedagogical practices of Brazilian IE undergraduate programs (ABEPRO, 1998).

According to INEP (2010, p. 47), until 2002 IE undergraduate programs were evaluated based on the parameters established for mechanical engineering undergraduate programs. After 2002, INEP launched a new manual to evaluate undergraduate programs, which initiated the first SINAES proposals that are the basis of ENADE, considered one of the main Brazilian educational quality indicators. Table 1 presents a summary of the ENADE scores obtained by the IE undergraduate programs. The programs with the indication "-" have no score or have yet to be evaluated by SINAES.

Table 1: Summary of the results from ENADE 2014 to IE undergraduate programs

Quality scale	ENADE Scores	Number of programs	% of programs	% total
-	-	538	59.8	59.8
Poor	1	52	5.8	22.0
	2	146	16.2	
Fair	3	102	11.3	11.3
Good	4	39	4.3	6.9
	5	23	2.6	
Total		900	100.0	100.0

In contrast to the quantitative growth rate, Table 1 reveals that IE high quality education in Brazil is restricted to only 6.9% of the existing undergraduate programs. Meanwhile, 81.8% had a poor quality score or were not yet evaluated by SINAES, and therefore only 18.2% of the programs were certified as good or satisfactory.

Research Methods

This study was based on the statistical analyzes of the 2014 ENADE database (INEP, 2017). We analyzed the answers of the students' surveys applied by INEP to evaluate the students' perception on the conditions of the educational process within IE undergraduate programs in Brazil. This survey was distributed 15 days before the exam that evaluates the students learning outcomes, and it includes 42 items to assess the didactic and pedagogical structures of the programs, the infrastructure and physical facilities, and the opportunities offered by the programs in order to expand academic and professional training. These items are evaluated on a 6-point Likert scale of agreement.

Here, the students' responses were divided into three categories (quality scale) based on the ENADE score of their respective undergraduate program. The first category has the students' responses from programs with ENADE 2014 scores 1 or 2, which represents programs with poor educational quality. The second category represents programs with fair educational quality and has the students' responses from programs with ENADE 2014 score 3. Finally, the third category represents programs that has a good educational quality and has the student's responses from programs with ENADE 2014 scores 4 or 5. Table 2 shows the distribution of the students' responses based on these three groups of ENADE scores.

Table 2: Summary of the number of students and number of IE undergraduate programs grouped by its ENADE score

Quality scale	ENADE Scores	Number of students		Number of IE programs	
		N	%	N	%
Poor	1 or 2	6495	58.9	190	54.8
Fair	3	3108	28.1	98	28.2
Good	4 or 5	1421	13.0	59	17.0
Total		11024	100.0	347	100.0

In Table 2 we notice that most of the students who answered the survey were within IE undergraduate programs with ENADE score 1 or 2. A total of 11024 surveys were analyzed, which represents 63.7% of the IE students who were chosen to take the exam in 2014, distributed among 347 IE undergraduate programs and 278 HEI. The 36.3% of the students left out of this study (6282 students) either missed the exam (2225 students, 12.8%) or did not answer the student survey (4058 students, 23.5%).

Descriptive statistics and graphics (histograms and box plots) were generated and tests of normality (Kolmogorov-Smirnov test) were conducted on all data sets to determine whether parametric or non-parametric statistical analyses were appropriate. Considering that the data were not strictly normally distributed, we used nonparametric tests for the evaluation of the results. Comparisons of the 42 survey items were done by Kruskal-Wallis test with post-hoc Dunn's multiple comparison tests between all pairs among the three groups. The significance level was set at $\alpha=0.05$ in all tests. All statistical analyses were performed using the software R version 3.3.3.

Results and Discussions

The results from Kruskal-Wallis test showed that only 39 survey items had statistically significant differences by comparing the three groups. We verified that 4 items showed a correlation between higher responses (on a Likert scale) and "good" programs, 1 item showed a relationship between higher responses and "fair" and "good" quality IE undergraduate programs, 26 survey items exclusively corresponded to programs with "poor" quality, and 8 items exclusively corresponded to programs with "fair" quality.

The following two sub-sections present and discuss the educational strategies we found to be associated exclusively with "good" quality IE programs and that associated with both "fair" and "good" quality IE programs. In order to focus on the primary goal of this paper, the other 34 items of the survey will not be analyzed further.

Evaluation 1: Items related to "good" quality programs

This section aims to analyze the items related only to the best IE undergraduate programs in Brazil (ENADE scores greater than 3). Table 3 presents the four items that showed an exclusive correlation with the educational structures of the best IE undergraduate programs in Brazil.

Table 3: Items related to the best IE undergraduate programs in Brazil

	Evaluated items	Average of the student's responses			Kruskal-Wallis chi-squared*
		Poor	Fair	Good	
		Mean (SD)	Mean (SD)	Mean (SD)	
i53	The program offered opportunities for students to conduct exchanges and/or internships abroad.	3.894 (1.952)	4.493 (1.635)	5.030 (1.305)	463.326
i52	The program provided opportunities for students to conduct exchanges and/or internships within Brazil.	4.220 (1.806)	4.609 (1.547)	4.750 (1.447)	126.060
i43	The program provided opportunities for students to participate in programs, projects, or activities related to the community.	4.783 (1.421)	4.933 (1.271)	5.055 (1.174)	35.316
i44	The program provided undergraduate research opportunities or projects and activities that stimulated academic research.	4.723 (1.469)	4.858 (1.326)	4.984 (1.206)	25.912

Notes: SD = Standard Deviation. *All p-values were < 0.05 and Dunn's test p-values were < 0.008.

The four items shown in Table 3 are related to the opportunities offered by the programs in order to expand academic and professional training. Table 3 indicates that the main factors that differentiate the best programs are the opportunities for participation in study abroad programs and internships inside and outside the country. These factors are related to opportunities outside the programs and to whether the HEI provides students with new learning experiences. In 2014, as stated by Pozzo and Nihei (2016), the Brazilian government encouraged academic mobility through international programs such as the Brazilian Scientific Mobility Program (*Programa Ciência sem Fronteiras*) and the establishment of universities with an integration profile. Currently, there is also a program named Paulo Freire's Mobility Program (*Programa Paulo Freire de Mobilidade Acadêmica*). These programs aim to train students to work in different countries, promote local development, and learn from cultural, scientific, professional, and educational exchange experiences. These authors add that these opportunities act as a catalyst for international cooperation and professional training, which has potential benefits for the economic and social aspects of all countries involved, and aims to improve the quality of HEI and to identify and strengthen common international interests (Pozzo & Nihei, 2016, p. 55).

In addition, opportunities to participate in academic programs, projects and research were indicated by the students' responses as important factors for educational quality. These opportunities are directly related to the students' performance in activities and academic research that employ the content of the programs' curriculum and motivate academic research. Massi and Queiroz (2010, p. 174) consider undergraduate research as a process in which essential knowledge is provided to introduce students to techniques and traditions of science. Furthermore, they emphasize the value of participating in extracurricular opportunities to encourage personal development, creating an alternative perspective of science, and cultivating professional socialization.

Evaluation 2: Items related to both "fair" and "good" quality programs

Comparing the survey answers from "fair" and "good" programs, we found only one item with correlated responses (Table 4). We believe that this item may be part of a process of transition and educational reorganization by the "fair" quality programs aiming to achieve a better educational performance, although each program has its own characteristics and peculiarities. Furthermore, due to the comparative nature of these results, adopting the best programs strategies does not necessarily imply in obtaining a higher ENADE score.

Table 4: Item related to both "fair" and "good" quality IE undergraduate programs

Evaluated items		Average of the student's responses			Kruskal-Wallis chi-squared*
		Poor	Fair	Good	
		Mean (SD)	Mean (SD)	Mean (SD)	
i45	The program offered support for students to participate in internal and/or external scientific events.	4.859 (1.354)	4.980 (1.217)	4.928 (1.198)	8.077

Notes: SD = Standard Deviation. *p-value = 0.0017. Dunn's test p-value = 0.029.

The item displayed in Table 4 is related to the scientific opportunities to expand academic and professional training that are offered by the programs. Marchiori, Adami, Ferreira, and Cristofoli (2006, p. 8) emphasize that the main functions of scientific events are to: create opportunities to exchange experiences among researchers; update on recent progress of the study areas; systematize the latest advances in the field; disseminate new knowledge; and outline guidelines and goals for future ventures in a particular study area. In summary, attending scientific events can contribute to the quality of undergraduate programs because it allows knowledge exchange and comparisons about what has been approached by researchers from different HEI.

Conclusions

This study identified a set of five educational strategies found in the best Industrial Engineering (IE) undergraduate program in Brazil (Table 5). All these strategies are related to the opportunities offered by the programs in order to expand academic and professional training. Specifically, we verified that the opportunities to participate in exchange programs both inside and outside the country, the students' involvement in programs and activities related to academic development, and the participation in scientific events are factors that positively influence the student learning outcomes and the quality of IE undergraduate programs.

Table 5: Didactic and pedagogical structures found in the best IE undergraduate program in Brazil

	Strategies
1	The program offered opportunities for students to conduct exchanges and/or internships abroad.
2	The program provided opportunities for students to conduct exchanges and/or internships within Brazil.
3	The program provided opportunities for students to participate in programs, projects, or activities related to the community.
4	The program provided undergraduate research opportunities or projects and activities that stimulated academic research.
5	The program offered support for students to participate in internal and/or external scientific events.

We point out that no strategy in Table 5 is related to the didactic and pedagogical structures of the programs or the infrastructure and physical facilities offered by them. In addition, we note that the best IE undergraduate programs have mostly offered opportunities to expand academic and professional training of its students.

The quality management in Brazilian IE undergraduate programs is a current concern due to the rapid increase in the quantity of programs and the large number of programs that have never been evaluated by SINAES (*Sistema Nacional de Avaliação da Educação Superior*, Brazilian system of higher education evaluation) or have a very low ENADE (*Exame Nacional de Desempenho de Estudantes*, Brazilian exam of student performance) score. The data indicate that the quality of Brazilian IE undergraduate programs has not been maintained in light of its quantitative growth. Therefore, the question of quality of these programs is undeniably prudent.

We reiterate that each program has its own characteristics and peculiarities. Consequently, adopting all these strategies does not necessarily imply the acquisition of a higher ENADE score by the program. We also note that the educational structure of the program is just one of the items evaluated by INEP to compute CPC score. The coordinators of the IE undergraduate programs in Brazil are invited and encouraged to rethink the educational structure of their programs and implement actions related to the five strategies identified in this study. In doing so, this study points to the likely improvement of student learning outcomes and consequently the potential for a higher CPC score during the next program evaluation process, in 2017.

References

ABEPRO (Brazilian Association of Production Engineering). (1998). *Engenharia de Produção: grande área e diretrizes curriculares.* Accessed at http://www.abepro.org.br/arquivos/websites/1/DiretrCurr19981.pdf on 25 Jan 2017.

Bittencourt, H. R., Viali, L., & Beltrame, E. (2010). A engenharia de Produção no Brasil: Uma Panorama dos cursos de Graduação e Pós-graduação. *Revista de Ensino de Engenharia*, 29(1), 11-19.

Brasil. (2002). *Resolução CNE/CES nº 11*, de 11 de março de 2002. Brasília, DF: Diário Oficial [da] República Federativa do Brasil.

Brasil. (2004). *Lei nº 10.861*, de 14 de abril de 2004. Brasília, DF: Diário Oficial [da] República Federativa do Brasil.

Brasil. (1995) Lei nº 9.131, de 24 de novembro de 1995. Brasília, DF: Diário Oficial [da] República Federativa do Brasil.

Cunha, G. (2002). *Um panorama atual da Engenharia de Produção.* Brazilian Association of Production Engineering (ABEPRO). Accessed at www.abepro.org.br on 20 may 2016.

Cunha, M. I. (2014). A qualidade e ensino de graduação e o complexo exercício de propor indicadores: é possível obter avanços? *Avaliação*, 19(2), 453-462.

Dias, C. L., Marchelli, P. S., Horiguela, & M. L. M. (2006). Policies for the assessment of higher education in Brazil: a critical review. *Educação e Pesquisa*, 32, 1-29.

Diedericks, R., Klerk, N., & Bevan-Dye, A. L. (2015). Students' Perceptions of Service Quality at a South African Traditional University and a University of Technology. *J Soc Sci*, 43(2), 161-171.

Đonlagić, S, & Fazlić, S. (2015). Quality assessment in higher education using the SERVQUAL model. *Management*, 20(1), 39-57.

Faé, C. S., & Ribeiro, J. L. D. (2005). Um Retrato da Engenharia de Produção no Brasil. *Revista Gestão Industrial (Online)*, 1, 315-324.

Griboski, C. M. (2012). O ENADE como indutor da qualidade da educação superior. *Est. Aval. Educ.*, 23(53), 178-195.

Instituto Nacional de Estudos e Pesquisas Educacionais Anísio Pereira (INEP). (2017). Indicadores de qualidade: ENADE 2014 [Data file]. Retrieved from http://portal.inep.gov.br/web/guest/conceito-enade.

Instituto Nacional de Estudos e Pesquisas Educacionais Anísio Pereira (INEP); Conselho Federal de Engenharia, Arquitetura e Agronomia (CONFEA). (2010). *Trajetória e estado da arte da formação em engenharia, arquitetura e agronomia - Volume VII: Engenharia de Produção.* Brasília, DF: INEP.

Instituto Nacional de Estudos e Pesquisas Educacionais Anísio Pereira (INEP). (2014). Manual ENADE 2014. Accessed at http://download.inep.gov.br/educacao_superior/enade/manuais/manual_enade_2014.pdf on 26 jan. 2017.

Instituto Nacional de Pesquisas Educacional Anísio Teixeira - INEP. (2003). NC - Provão: Exame será aplicado no dia 8 de junho para mais de 470 mil inscritos. Retrieved from http://provabrasil.inep.gov.br/artigo/-/asset_publisher/B4AQV9zFY7Bv/content/enc-provao-exame-sera-aplicado-no-dia-8-de-junho-para-mais-de-470-mil-inscritos/21206 on 04 mar. 2017.

Marchiori, P. Z.; Adami, A.; Ferreira, S. M. P.; Cristofoli, F. (2006). Fatores motivacionais da comunidade científica para publicação e divulgação de sua produção em revistas científicas. In: Seminário Nacional de Bibliotecas Universitárias, 2006, Salvador – BA.

Massi, L. & Queiroz, S. L. (2010). Estudos Sobre Iniciação Científica No Brasil: Uma Revisão. *Cadernos de Pesquisa*, 40(139), 173-197.

Matos, D. A. S., Cirino, S. D., Brown, G. T. L., & Leite, W. L. (2013). Avaliação no ensino superior: concepções múltiplas de estudantes brasileiros. *Estudos em Avaliação Educacional (Online)*, 24, 172-193.

Ministério da Educação (MEC). (2007). *Portaria normativa nº 40*, de 12 de dezembro de 2007. Brasília, DF: Diário Oficial [da] República Federativa do Brasil.
Pedrosa, R. H.L., Amaral, E., & Knobel, M. (2013). Assessing higher education learning outcomes in Brazil. *Higher Education Management and Policy*, 24(2), 55-71.
Pinto, R. S., Mello, S. P. T., & Melo, P. A. (2016). Meta-avaliação: uma década do Processo de Avaliação Institucional do SINAES. *Avaliação*, 21(1), 89-107.
Pozzo, E. D.; Nihei, O. K. (2016). A Internacionalização da Educação Superior e a UNILA sob a Ótica da Integração. *Pleiade*, 10(20), 52-60.
Prados, J. W., Peterson, G. D., & Lattuca, L. R. (2005). Quality Assurance of Engineering Education through Accreditation: The Impact of Engineering Criteria 2000 and Its Global Influence. *Journal of Engineering Education*, 94(1), 165-184.
Ribeiro, L. R. C., & Belhot, R. V. (2002). A gestão da qualidade e o ensino de engenharia. *Revista de Ensino de Engenharia,* 21(2), 17-25.
Sousa, J. V., Seiffert, O. M. L. B., Fernandes, I. R. (2016). Acesso e expansão de cursos de graduação de alta qualidade no Brasil: outros indicadores de qualidade para a educação superior. *Educação em Revista*, 32(4), 19-47.
Tagg, J. (2010). The learning-paradigm campus: From single- to double-loop learning. New Directions for Teaching and Learning, *123*, 51-61.
Teeroovengadum, V., Kamalanabhan, T.J., & Seebaluck, A. K. (2016). Measuring service quality in higher education. *Quality Assurance in Education*, 24(2), 244-258.
Verhine, R. E., Dantas, L. V., & Soares, J. F. (2006). Do provão ao ENADE: uma análise comparativa dos exames nacionais utilizados no ensino superior brasileiro. *Ensaio: Avaliação e Políticas Públicas em Educação*, 14, 291-309.
Zabadi, A. M. A. (2013). Implementing Total Quality Management (TQM) on the Higher Education Institutions – A Conceptual Model. *Journal of Finance & Economics*, 1(1), 42-60, 2013.

Acknowledgements

The authors acknowledge CAPES (Coordenação de Aperfeiçoamento de Pessoal de Nível Superior) and CNPq (Conselho Nacional de Desenvolvimento Científico e Tecnológico) for their financial support.

Copyright statement

The influence of situated didactics in connectivism for improving the competences through problem-based learning (PBL)

Jairo Gutiérrez de Piñeres Rocha

Fundación Universitaria del Área Andina - FUAA, Valledupar, Colombia
jgutierrez52@areandina.edu.co

Abstract: At this research, you could see as *Problem-Based Learning (PBL) becomes a learning strategy in Statics' Virtual Course (SVC). PBL was used as Learning Virtual Object (LVO) and Interactive Virtual Object (IVO) and in assessment activities. For this reason, all contents, all environments are designed with categories and processes based on PBL. This SVC had enrolled 320 students for this period. Additionally, the situated didactic was used in a website designed with es.wix.com, at which so many different resources, tools, IVO, LVO, contents and assessment activities were focused and based on situated didactic as a treatment environment for an experimental group of 30 SVC's students already approved. Another group of students, from the same SVC, was used as a control group.*

Situated didactic will try of improving the competences in Civil Engineering, shifting the focus from the individual to the sociocultural setting as a cognition that is assumed to be social and situated activity (Kirshner & Whitson, 1997); one learns a subject matter by doing what experts in that subject matter do (Lave, 1990/1997) arguing that knowledge remains inert and unused if taught in contexts that separate knowing from doing, and the activities of the students within that virtual setting called website under parameters of connectivism, in which, learning and knowledge depend on diversity of individual concepts continually evaluating how each new generation learns with regard to instruction and curriculum serves to hold education to high standards.

Introduction – overall layout

Static's Virtual Course (SVC) emphasizes in the integral formation of the human person through communicative action, the pedagogical dialogue and training research has been the constant task of it. For this reason, this course has sought different paradigms to be increasingly relevant to social demands through the nature of situated cognition. Consequently, the theory of situated cognition claims that every human thought is adapted to the environment, that is, situated, because what people perceive, how they conceive of their activity, and what they physically do develop together according to Clancey (1997).

Moreover, SVC is a social context making students perceive, think and do develops learning comprising a set of constructs linking changes in performance or the capability to perform with what is thought to bring about those changes. That is, the results of learning are to be explained through inputs based on connectivism parameters: largely external to the student and means largely internal to them following some elements of PBL as a different perspective that makes about processes and knowledge occurring within the student.

The aim of this paper is to demonstrate that PBL can use the situated didactic supported in some principles of connectivism to improve competences measuring Wenger's learning trajectories: peripheral, inbound, insider, boundary and outbound. PBL can foster sensitivity and effectiveness in handling the practical problem and involve the knowledge as a matter of meaningful competence to experience the world and our engagement with it inside SVC. In addition, SVC was a community of learning in which was used PBL as a means to strengthen the competences in Civil Engineering through assessment activities.

Next, a random control group was formed with 30 students who had enrolled and passed the SVC; and another experimental group, under the same approval and randomness criterion, and with an equal number of students of the same SVC. That is, both groups had been trained in the same learning community with the same content and equal assessment activities based on the PBL and under the same approval scale.

Taking into account the above, situated didactics approach was applied to the experimental group for two months to strengthen the generic and disciplinary competences of the students that passed in the SVC. Consequently, both groups were exposed to pretest and posttest designed with questions focused on the parameters of the situated didactics to verify the influence of situated didactics in the strengthening of the generic and disciplinary competences of Civil Engineering.

SVC's assessment activities based on PBL

Cunningham, Duffy and Knuth (1993) Likewise, problem-based learning (PBL) becomes a lived practice to improve students' competences measuring their learning trajectories on diagnosis, summary statistics and portfolios.

1.Provide experience with the knowledge construction process: Every subject has previous learning experiences in certain field of knowledge or in activities of another order. The didactic function is to create contexts, conditions and environments so that the student can objectify the meanings of their previous experiences reached in their vital world and endow it with methods, techniques and tools to facilitate this process. This is the design of situations and activities that allow the student to move from the unthinking of their experiences to the systematization of the same or the prenotions to the notions.

2.Encourage the use of multiple modes of representation: via video and computer. The use of software resources and tools Interactive character is essential in order to effectively training of these people. Another way to respond to the requirements it raises this style of mediated knowledge is the design of virtual learning objects, whether auditory, fixed or moving; The use of simulators, remote laboratories, searches and Navigation for access to decanted contents and stored in the repositories of specialized information; the use of mobile communication devices for the reception and permanent issuance of information in different mediations and for access to learning devices, etc.

3.Provide experience in appreciation for multiple perspectives: multiple way to solve problems: It refers to the set of situations and activities previously designed in a didactic way, leading to the appropriation of concepts, categories, theories, thought models, or processes, procedures and methodologies of different orders, according to the purposes, objectives, competencies and goals of Established in the academic year.

4.Embed learning in realistic and relevant contexts: to keep the authentic context of the learning task: problem statement : truss is part of a bridge. It is about activating metacognitive structures that allow the student to displace their cognitive schemas and combine their previous knowledge with the fields of knowledge involved in the respective academic course. It adds the generation of learning processes from understanding and not from mere repetition

5.Encourage ownership and voice in the learning process: Tutors help students frame their learning objectives: to choose a joint and make a force addition according to X,Y axis.

All knowledge, ability, skill or competence may Allow the transfer of known situations to unknown situations. The Situations and activities designed in the didactic guide should add values of recontextualization and productivity to the knowledge that is learned and to the derived competences. It is also a question of proving the social utility of knowledge, establish productive relationships with the context and seek the activation of transversal or paracognitive competences.

6.Embed learning in social experience: teacher-student / student-student: teamwork

7.Encourage self-awareness of the knowledge construction process: Analysis is the right scenario for feedback and synchronic and asynchronous interaction of the students organized in small groups and the tutor; of the academic work of students in small groups through forums, chats and social networks and the affirmation of learning using Web 2.0, conference, radio, IP television, etc., for large groups.

Since the definition of Manuel Castells (2006) of The Red Society, it is determined that the frontiers of knowledge have widened their horizons, but not only because they are access to more information in less time, but mainly because has sprouted a new optimal social paradigm for learning. In other words, we participate in a new educational paradigm: the learning networks that help learn to learn, to do, to live and to coexist in a more interesting way. This also indicates that the student of the 21st century seeks, analyzes, integrates and uses information from Interdependent manner and with high co-responsibility and ethics, at best.

Consequently, the collaborative learning environment is an educational context Where the student achieves positive social interdependence and pedagogical interactivity, Through greater interaction in volume and intensity, the exchange of information with ease of access, responsibility, individual commitment and processing in group. According to Fainlhoc (1999), this type of learning allows perceptions and cognitions are modified "as part of the presence and conduct of the other" (p 61).

Consequently, students' ability to explain why or how they solved a problem in a certain way from their so many different disciplines: There are also activities for the transfer of one phase to another with the purpose of consolidating or leveling the domain of the competences acquired. At the end of the process, closing activities or learning balances are carried out consistent in. The transfer of the learning outcomes obtained in an academic year through the development of include feedback from the tutor and from the students themselves and levels of objectivation of determined competences in learning objectives and goals.

SVC has all its contents and evaluation activities based on processes of situated didactic, from which were understood as defining ways of belonging to a community of learning as is shown in Figure 1:

SVC as a Community of Learning

Figure 1. Structure of SVC as a community of learning

This SVC was based on situated didactic and therefore was changing forms of participation and identity of students who participate in a community of learning over a long time. They logged into as newcomers, progress to the point of seeming to be oldtimers with respect to new newcomers, and eventually became the oldtimers themselves. With respect to teaching and learning, this pointed to a richly diverse field of essential actors and, with it, other forms of relationships of participation as Wenger (1998) suggested that learning as participation be conceived on three broad levels:

1. For students
2. For community of learning
3. For process

For this reason, the three categories defined above, student learning means participating in the practices of the communities of which the students is a member. Community learning is a matter of refining practice of the community and ensuring new members, so that new newcomers enter as the original newcomers become oldtimers. Likewise, learning at the level of process involving students and community is an issue of sustaining the interconnected these categories through which a process knows and thus becomes effective and valuable as a community (Wenger, 1998).

Moreover, students had strong goals and motivation, and through engagement in learning, they develop a view of what the common outcomes is all about. In apprenticeship learning there also tends to be communication among peers and near-peers to the extent that relevant information is spread rapidly and effectively, as you often tell newly admitted students at SVC, if you want to know what really goes on in the community, ask for another student, not a tutor member.

While apprenticeship learning serves as an exemplary case of situated didactic, it is not the only means for learning in practice to occur. Students belong to many different communities of learning, and their participation in those communities can take many forms, including nonparticipation as a consequence of being excluded from participation. The concept of a learning trajectory based on connectivism helps to describe a student's participation over time:

1. Peripheral: students on a peripheral trajectory never, for one reason or another, engaged in full participation at SVC's course
2. Inbound: students were invested in the course and headed toward full participation
3. Insider: students were involved with a means for continued evolution of practices within the course through OVI and OVA according to Free-Body Diagrams about rigid bodies
4. Boundary: students sustained membership and participation in related topics to moments of forces and, in essence, "broker" interactions between them
5. Outbound: students were in the process of leaving this course

Table 1 shows these results:

Table 1: Students in learning trajectory

Learning Trajectory	Number of students involved
Peripheral	25
Inbound	193

Insider	**89**
Boundary	**18**
Outbound	**5**

Those strong goals and motivation are bigger if students learn through new technology, this is, connectivism. For designing the SVC, connectivism was used with its Gagne's nine events (Gagne & Medsker, 1996), at which was developed a sequence of requirements that codified all instructional design of course:

1. Gaining attention: importance of Statics in Engineering in diagnosis

2. Informing learners of the objective: acquisition of competences as an Engineer

3. Stimulating recall of prior learning: Static's importance for solving many problems

4. Presenting the content: 8 units in 2 months

5. Providing learning guidance: learning activity guidance

6. Eliciting performance: interactive activities, OVI and OVA.

7. Providing feedback: at collaborative forums

8. Assessing performance: portfolio and summary statistics

9. Enhancing retention and transfer: from portfolios there is an improvement plan for all of SVC's students.

According to above, Gagne's nine events were designed basing on all characteristics of problem-based learning (PBL), used at SVC course as an instructional methodology; that was an instructional solution to learning problems related to a basic mechanics develop in the engineering student the ability to analyze a given problem in a simple and logical manner and to apply to its solution a few fundamental and well-understood principles. This SCV' course was designed with fundamental topics of mechanics of materials offered to engineering students, and the PBL helped students achieve this goal in this particular course in the same way that their other texts might have helped them in dynamics or another engineering's topic.

First of all, the primary goal of PBL at SCV's course was to enhance learning by requiring students to solve problems. It was a methodology with the following characteristics citing to Hung (2006):

1. They were problems focused on the understanding of a few basic concepts and on the use of simplified models, such that students began learning by addressing simulations of an authentic, ill-structured problem.

2. The SCV's content and skills to be learned were organized around problems, rather than as a hierarchical list of topics, so a reciprocal relationship exists between knowledge and the problem. Knowledge building was stimulated by the problem and applied back to the problem: This approach made it possible to develop all the necessary formulas in a rational and logical manner, and to clearly indicate the conditions under which they could be safely applied to the analysis and design of actual engineering structures and machine components derived from Statics.

3. It was self-directed, such that students individually and collaboratively assumed responsibility for generating learning issues and processes through self-assessment and peer assessment and access their own learning materials: throughout the course free-body diagrams were used to determine external or internal forces. The use of "picture equations" also helped the students understand the superposition of loadings and the resulting forces and components.

4. Tutor was a facilitator who supported and modelled reasoning processes, facilitated group processes and interpersonal dynamics, probe students' knowledge deeply and never interjected content or provided direct answers to questions related to SI metric units and U.S. customary units, half the examples, sample problems, and problems to be assigned had been stated in SI units and half in U.S. customary units. Since a large number of problems were available, tutor could assign problems using each system of units in whatever proportion he found most desirable for his activity.

PBL and Acquisition of Competences

SCV used PBL for acquiring competences as a basic assumption that knowledge was an internal representation of a kind of external reality. Information was out there, to be received and stored by the human processing system, whether as memory networks, cognitive structure, or schemata. Fundamental Static's contents; however, was the assumption that learning involves social participation, such as cognition "took place within the course and not in 'minds' construed as somehow separate from or outside the world" (Whitson, 1997), and put another way, if students inside course didn't consist of stored descriptions, what then was the relation of what I said to what I did? Solving force addition must be seen not as bringing out what was already inside, but as a way of changing what was inside. Solving Static's problems was not restating what had already been posted subconsciously inside their brains, but was itself an activity of representing. The Static's foundations and what they mean, its theories, and conceptions in engineering developed in their behavior as they interacted in SVC and re-perceived what they and others had previously said and done. (Clancey, 1997).

Acquisition of competences using PBL, was a matter of sign activity inside SVC, or competence (Cunningham, 1987, 1992; Whitson, 1997). According to the Peircean tradition of competence, is the continuously dynamic and productive activity of signs (Peirce, 1955); and SVC is designed based on signs for acquisition of competences solving Static's problem. A sign is anything that stands for something else, such as the words 'rigid body' or a drawing of a beam or column standing for the topic we experience as a beam or column. Competence theory makes the claim that all knowledge of the Static was mediated through signs, which were jointly determined by the rigid bodies creating and using Static's signs that went beyond their immediate experience.

SVC's signs for acquisition of competences were representing in three ways using PBL:

1. As an index (e.g. the components in axes was a sign of addition of forces in two dimensions)
2. As an icon (e.g., the free-body diagram was a force schema or concurrent forces)
3. As a symbol (e.g. the vector representing any force or a plane truss)

Figure 2 shows the Peirce's triadic sign relations:

Perice's triadic signs as an acquisition of competences

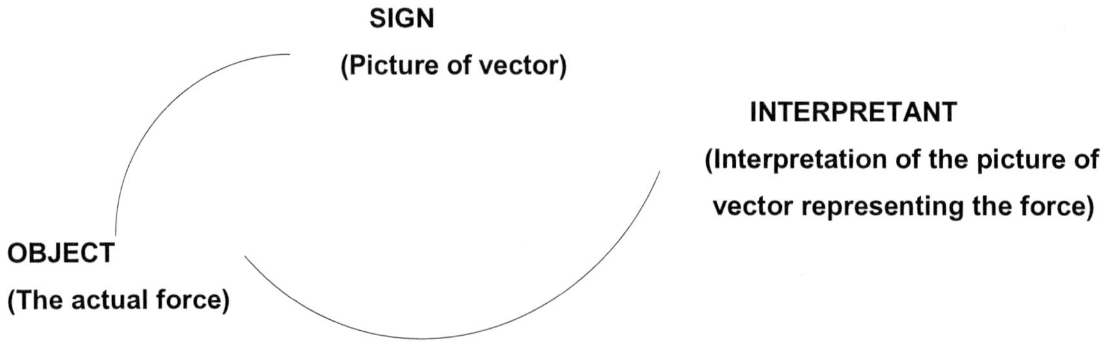

Consequently, results of assessment activities taking into account Peirce's triadic using PBL as a learning strategy in SVC are shown in Table 2:

Table 2: Relationship of students that approved SVC's course

	Value	Gl	Sig. asymptotic (bilateral)
Pearson's Ji-Square	6.371	4	**0.035**
Reason for likelihood	4.957	4	0.354
Association linear by linear	417	1	0.123
Number of valid cases	320		

For this Ji square test, a significance level of 5% was established, which described that the test results were significant if the p value was less than 0.05. For such a reason, Table 2 shows that the value of p in the Ji-square test was 0.035, less than 0.05. Therefore it indicates that there was a significant relationship between the number of students who approved SVC course and acquisition of competences solving problems and learning concept on situated didactic in website .

Table 3 shows the number of students that solved problems based on PBL involving Peirce's triadic signs:

Table 3: number of students that solved PBL assessment activities with Peirce's triadic signs inside SVC

Students	Individuals			Teamgroup		
	Diagnosis	Summary Statistics	Portfolio	Situations PBL	Analysis PBL	Forums
Approved	307	297	305	311	302	307
Didn't approve	13	23	15	9	18	13

Source: Autor

Methodology

Processes of Situated Didactics as a treatment for fostering competences

The situated didactics, as a learning strategy was applied in 30 students of SVC in a treatment group. For such a reason, a website was designed with contents, hypermedia, PDF, IVO, LVO, and assessment activities based and focused on situated didactic. Another group of students was the control, but just came out from the same SVC of the same academic period.

Keep in mind, situated didactic could shift the focus on the individual to the sociocultural setting, and the activities of the students within that setting called connectivism because of students could interact both with contents, IVO, LVO and assessment activities, at which, learning and knowledge depend on diversity of individual concepts; in this case, students were exposed to website 2.0 for fostering their competences through concepts and training and assessment activities focused on local setting: bridges, roads, buildings, pavements and so many different works made in region.

Besides, situated cognition was applied in a website 2.0 taking into account knowledge was accrued through the lived practices that were meaningful actions that had relations of meaning to one another in terms of some cultural system according to Lemke (1997). SVC's all activities were designed assuming the learning viewed as situated setting having as its central defining characteristic a process called 'legitimate peripheral participation' (Lave & Wenger, 1991) as a space accounts for the way a student to a community of learning develops into full participant, or from the viewpoint of a learner, how a learner engages in the activity of a sociocultural practice and becomes increasingly competent in this practice.

It was a cross sectional study under quantitative approach, where the dependant variable was competences measured by website es.wix.com using principles of situated didactic and connectivism. It was applied Ji Square. All information was processed with SPSS, v.19.

Results

Through PBL in Statics' virtual course, 320 students were measured by learning trajectories for one year, factors in diagnosis, summary statistics and portfolios. Therefore, so many different competences were improved on their own, individual learning, the learning of their collaborative group at forums, the learning of the class as a community and suggestions for improvement at Pierce's triadic sign relations to design new PBL models. Then, 30 students that approved SVC were extracted as a treatment group at website applying situated didactic for two months.

The same pretest (30 questions) was applied both control and treatment group putting values between 0 and 2, of according to the rating scale of the test. 2, was used to express the total pretest scores in values from 1 to 20, according to with the official grading scale of the university where the research was conducted. Postest only was applied to treatment group seeingTable 4:

Table 4.

Scores for Questions in Pretest and Postest

	Pretest		Postest
Question	Control Group	Treatment Group	Treatment Group
I	0,90	1,03	1,09
II	0,19	0,06	0,18
III	0,14	0,42	0,27
IV	0,43	0,27	0,39
V	0,43	0,61	0,67
VI	0,95	0,79	1,00
VII	0,40	0,47	0,49
VIII	1,00	0,97	1,39
IX	0,24	0,09	0,21
X	0,05	0,03	0,06

Test t for question scores. The t-test for the Question results in a difference of means of the pretest of the control group less The pretest of the experimental group of -0.034329 with t = -1.343 for p = .05 and 29 degrees of freedom. This value of t falls outside the confidence interval and therefore the difference of the average is not significant and the equivalence between groups can be ensured.

Pretest Versus Postest of the Experimental Group

Experimentally differ somewhat from the results obtained in the pretest by the same group. The total score reached was 13.75 in the pretest and 15.61 in the posttest, these Show a small difference between the results of the final observation with with respect to the initial observation, with the results in the postest score being higher than those of the pretest.

Test t for question scores. Taking as an alternative hypothesis research: students in the experimental group will obtain an average of qualifications superior in the posttest than in the pretest, in symbols HA: X Postest> X Pretest is assumed as null hypothesis H0: X Postest = X Pretest. The test gives t = - 0.0620 as result for the difference between the means of the pretest minus the postest of the treatment group being t = -2.757 for 29 degrees of freedom and p = .05. The resulting value of t falls outside the confidence interval and is greater than the critical value: 1.699, in consequently, the null hypothesis is rejected. It is assumed that there is a 5% probability of that the means of the scores by questions obtained in the pretest and postest by the group are the same. The difference between the scores obtained by the treatment group is significant and therefore the hypothesis is confirmed.

Conclusions

PBL will make virtual mid-course adjustments that will help to meet the emergent needs of students. Sometimes, students request that their confidential reports be shared with the rest of the learning community by websites or learning trajectories, and in these cases, they may have a question or suggestion that they would like the learning community to discuss and resolve as a group with situated didactic. In addition, the situated didactic and connectivist factors help to easily categorize guidelines outlined by the tutor because students can select works that illustrate their progress and achievements over time.

According to the results obtained, it is concluded:

1. Hypothesis 1 is confirmed: the average postest scores of the students in the treatment group is higher than the pretest grades average:

X Postest> X Pretest. The average postest score of 15.61 was higher than the score in the pretest of 13.75. This situation is also evident, in the scale of 1 to 20, the average score in the postest was 10 while in the pretest was 09.

2. Hypothesis 2 is confirmed: the average student treatment group that are approved is greater than 11.84 points: X Approved> 11.84.

On the rating scale from 1 to 20, the average rating on the postest was 12 points.

3. Hypothesis 3 is confirmed: the average grade of students in the treatment group that fail is greater than 2.84 points: X Failed> 2.84.

On the rating scale from 1 to 20, the average rating on the postest was 07 points.

Website as an instructional delivery tool, it could be included in next SVC as a further tool and resource for fostering competences in Civil Engineer. It will be possible to determine the effects on grades, dropout. Moreover, to use e-mail, websites based and focused on situated didactic for next students enrolled in SVC would influence in a meaningful learning

Recommendations and Implications for Engineering Education

1. Students in groups of five could encounter and reason through the problem in learning virtual environments either asynchronous and synchronous resources: collaborative forums, at which they could attempt to define and bound the problem and set learning goals by identifying what they know already, what hypotheses or conjectures they could think of, what they need to learn to better understand the dimensions of the problem, and what learning activities are required and who will perform them through situated didactic.

2. Using PBL in virtual environments fosters autonomous learning, individual students could complete their learning assignments based on PBL collecting and studying resources and preparing reports to the group.

3. Students could share their learning with the group and revisit the PBL in diagnosis, portfolios and summary statistics generating additional hypotheses and rejecting others based on their new acquisition of competences.

4. At the end of the leaning period, usually one week, students could summarize and integrate their learning to experiential settings.

5. Situated didactic might use as a learning strategy in virtual learning communities for fostering competences in Civil Engineer.

References

Bloom, B. S. (1956).Taxonomy of Educational Objectives, Handbook I: The Cognitive Domain. New York: David McKay Co Inc.

Bobbitt, J. F. (1918).The Curriculum. Houghton Mifflin Company. Retrieved from from a hypertext history of Instructional Design Web site

Clancey, W.J. (1997). Situated Cognition: On human knowledge and computer Representations. New York: Cambridge University Press.

Cunningham, D.J. (1987). Outline of an education semiotic. American Journal of Semiotic, 5, 201-216.

Cunningham, D., Duffy, T.M., & Knuth, R. (1993). Textbook of the future. In C. McKnight (Ed.), Hypertext: A psychological perspective. London: Ellis Horwood Publishing.

David Kirshner & James A. Whitson (1997) Situated Cognition – Social, Semiotic, and Psychological Perspectives. (ed.) Lawrence Erlbaum Associates Publishers. 323 sider.

Driscoll, M. (2005). Psychology of learning for instruction. Third edition. U.S.A.

Gagne, R.M. (1983). Some issues in the psychology of mathematics instructions. Journal Of Research in Mathematics Education, 14, 7-18.

Hung, W. (2006a). The 3C3R model: a conceptual framework for designing problems in PBL. Interdiscip. J. Problem-Based Learn., 1(1), 55–77.

Hung, W. (2006b). A 9-Step PBL Problems Designing Process: Application of the 3C3R Model. Paper presented at the 2006 AERA Annual Meeting, April 8–12, San Francisco, CA.

Hung, W., Bailey, J. H., and Jonassen, D. H. (2003). Exploring the tensions of problem-based learning: insights from research. In Problem-Based Learning in the Information Age, edited by D. Knowlton and D. Sharp, pp. 13–23. San Francisco, CA: Jossey Bass.

Kirshner, D. & Whitson, J.A. (1997). Editor's introduction. In D. Kirshner & J.A. Whitson (Eds.), Situated cognition: Social, semiotic, and psychological perspectives.

Mahwah,NJ: Erlbaum.

Lave, J. (1990/1997). The culture of acquisition and the practice of understanding. Reprinted In D. Kirshner & J.A. Whitson (Eds.), Situated cognition: Social, semiotic, and Psychological perspectives. Mahwah, NJ: Erlbaum.

Lave, Jean and Chaiklin, Seth (eds.) (1993) Understanding Practice: Perspectives on Activity and Context, Cambridge: University of Cambridge Press.

Lemke, J. L. (1997). Cognition, Context and Learning: A Social semiotic process. In D. Kirshner & J.A. Whitson (Eds.), Situated cognition: Social, semiotic, and Psychological perspectives. Mahwah, NJ: Erlbaum.

Peirce (1955). "Some Consequences of Four Incapacities", Journal of Speculative Philosophy v. 2, n. 3, pp. 140-57. Reprinted CP 5.264-317, W 2:211-42, EP 1:28-55.Arisbe Eprint.

Wenger, E. (1998). Communities of practice: learning, meaning, and identity. New York: Cambridge University Press.

Art and technology: course designed on Competency - Based Learning to form the professional profile of the actual engineer

González Lara, Gerardo Salvador; Ferchow, Ewelina;
Eguia, Lander; Loera, Valeria; Velázquez, León.
Tecnológico de Monterrey – Campus Monterrey, México
Corresponding Author Email: gsgonzal@itesm.mx

Abstract:

Report of educational innovation of the course for engineering Art and Technology, with Learning Based on Competences. Designed and imparted from the Department of Interdisciplinary Innovation from January to July 2016 by five specialists: in Humanities and Fine Arts, Industrial Design, Mechatronics, Industrial Engineering, and Innovation and Sustainability. For the first time in August 2016, 38 students from five engineering schools were divided into two groups. The challenge was: To create an artistic work that fuses art and technology by tackling 21st century global problems based on solid aesthetic discourse. With a Competency-Based Learning (CBL), this course stimulated in the students the interdisciplinary, the human sense and the development of three competences of egress1. Knowledge of processes to transfer technology, and 2. Knowledge of the processes to apply technology. Results: 55 individual and / or collective artistic works with art and technology, explained with academic essays for its aesthetic foundation. This course impacts on the desired professional profile for the current engineer.

KEYWORDS
Art, technology, learning, skills, competences.

CONTEXT

Art and technology for the museum of life. At the global level, the teaching of engineering is undergoing a process of didactic transformation. On the one hand, the traditional teaching / learning process based on an Objective-Based Learning (OBL) educational model is reoriented towards a Competency-Based Learning (CBL) approach; On the other hand interdisciplinary and human sense charge relevant importance to the relevant professional profile in the development of real projects in the professional world.
The present article presents a project of educational innovation in its first stage of implementation - that is to say, for the time being it is not possible to carry out a

research report - entitled "Art and Technology: Course with Learning Based on Competences to form the professional profile of the current engineer ". Is result of the proposals of the Center for Educational Development of Educational Innovation (CEDDIE) of the Monterrey Campus of Tecnológico de Monterrey. It was designed during the semester of January - May and Summer 2016. The outstanding thing is that this initiative was consolidated in an academic course designed by five experts: in Humanities and Fine Arts, Industrial Design, Mechatronics, Industrial Engineering and Innovation and Sustainability; Today it is taught by four of them. The competencies to be developed are two:
1. Knowledge of processes to transfer technology.
2. Knowledge of the processes to apply technology.
The course promotes interdisciplinary and human sense. Through a challenge that is reported from the beginning of the course, the deployment of skills that impacted on the personal and professional profile of future engineers is detonated.
Because Art and Technology as an innovative course is in its first phase of implementation, so far the results or findings have no precedents or previous studies that have influenced this work.

PURPOSE

The main question of exploration or research for this proposal of educational innovation is: Can art, technology and innovation combine in an academic course to satisfy the current demands of engineers with a comprehensive education?
It was hypothesized that:
"At Tecnológico de Monterrey, Monterrey Campus, Mexico, the course of Art and Technology with Learning Based on Competences will achieve in the first stage of teaching - August 2016 to May 2017 - the development of skills: 1. Intellectual curiosity. 2. Innovation, and 3. Use of technology, through the challenge: To create an artistic work that fuses art and technology in tackling global problems of the 21st century based on a solid aesthetic discourse.
For validation in this hypothesis this course was implemented in the period August - December 2016, and is currently taught in January - May 2017. Both semesters are considered the first stage of the application of the course for the first time.
The topics from which students tackle this challenge are:
1. The self in search of meaning.
2. The encounter with the other.
3. Societies and power.
Thus, this experience of CBL in the course of Art and Technology is significant for the professional profile of a current engineer for the conceptual, procedural and interdisciplinary implications
The purpose of this article is:
Demonstrate the design, methodology with CBL of the course Art and Technology and its results.

FOCUS

The Art and Technology course is the result of the proposals of the Center for Educational Development of Educational Innovation (CEDDIE) of Tecnológico de Monterrey. It was designed during the January - May semester and summer 2016. The outstanding thing is that this initiative was consolidated in an academic course from the Department of Interdisciplinary Innovation and imparted in "team teaching" by four specialists. For the first time in August - December 2016 of 38 students from five careers were taught in two groups as optative subjects in the various engineering curricula at Tecnológico de Monterrey. Actually, in this period January - May 2017 is in the program for the second time.

Theoretical Framework

The didactic foundation of the course Art and Technology is the Educational Model of Learning Based on Competences (CBL); the main elements of this educational model describes below:
In the CBL, the development of predefined competences is sought. Being integrated in practice by knowledge, skills, attitudes and values, these are developed through a challenge applied to concrete situations within real contexts.
After investigating various authors, the competence of the Monterrey Tec Educational Model "... is the conscious integration of knowledge, skills, attitudes and values that allow to successfully deal with both structured and uncertain situations and which may involve mental processes of higher order. The competences integrate both the knowledge and the procedures proper to the discipline, as well as the attitudes and values that allow professionals to be participative and committed to society "(9)
It is assumed that the disciplinary competences are those declared by the educational institution in question as belonging to a specific area of knowledge that will be applied in the professional field; and transversal competences are those that the student develops throughout the academic life through its curriculum. For Tecnológico de Monterrey, the transversal competences are: Innovation, passion for self-learning, leadership, ethics, collaborative work, problem solving, citizenship and payment of social mortgage, communication in Spanish, critical thinking, global perspective, Entrepreneurship, intellectual curiosity and management of information and communication technologies.
For the development of competences, a course is presented in advance a challenge: that is, "... an experiential experience designed to expose the student to an attractive and challenging situation of the environment. The challenges contribute to the development of students' disciplinary and transversal competences, since they apply their knowledge, skills, attitudes and values in an individual and collaborative way "(12). For all of the above, competency-based education also implies a change in the current role of teacher and student. Here's how these changes can be dealt with.
Changes in the current role of the teacher: The teacher should have more approach with the student and more direct and immediate observation and advice, since at the time of the development of competition should be feedback. It also provides a minor approach to information, but must teach in the process of finding information relevant to the solution of the challenge so that in a future, the student in other contexts apply that process of solution and search for information, It is impossible to know everything that has been written of a discipline or area of knowledge.
According to Villarruel Fuentes (2009), the teacher's educational practice in LBC is to be a mediator between the context and the student. Hence, the design of the challenge should be clear, relevant: "The central idea underlying such definitions starts with assuming that the new teacher can not reduce his or her work to the strict transmission of information, not even to be a simple facilitator of learning, but has To mediate the encounter of their students with knowlede, in the sense of guiding and orienting their activity, assuming the role of constructive and reflective teacher "(1).
Changes in the student's role: The student is required to be more willing to be guided by the teacher, but above all, to tolerate failures or when it is evident that a skill is not fully developed or adequately. It must be aware that for that it is in a learning process, and does not come with the perfection of the development of a particular competence, but in the process implies knowledge, attitudes, skills and values; discuss the mistakes of those who learn consciously. In most cases it is also willing to interact with other disciplines and even with people without university studies, if the challenge requires it. Finally, the student must internalize that all involved are equally important in meeting the challenge in the conditions marked by the design, projecting this environment to what will be presented in the world of work.
Valderde Berro *et.al* affirm that in CBL the student notes more of his learning and his responsibilities as a student because "each activity offers students the opportunity to know what their learning results are in terms of "what" they have learned and "how" any

evaluation process should be designed with this basic principle in mind" (52). This becomes more relevant because it is a characteristic to achieve the challenge through teamwork.
As for the academic knowledge required for all of the above, simultaneously the appropriate conceptual framework for the challenge and the respective activities that involve it must be prepared. This element is called Information Module: Timely and relevant information and knowledge for the realization of the challenge.
The way students access the content points to the need to modularize the courses as much as possible. Modularization enables "just in time" instruction, which facilitates the implementation of project-based learning for students on campus and in other parts of the world. (Working Group on the Future of Education in the Institute, MIT, 2014, quoted in Program Models ... 16).
CBL also requires a different approach to evaluation. The assessment of competences involves learning by action and feedback to the student almost immediately to the application of knowledge, attitudes, skills and values, to consciously guarantee the degree or level of development of certain competencies.
In the Model of Vocational Training Programs of the Tecnológico de Monterrey they demand that the feedback be carried out continuously in a process of evaluation, feedback and reflection. The evaluation of the students is based on the evidence of the level of performance of the disciplinary and transversal competences (18). It must demonstrate the application of what has been learned in real and new contexts, and demonstrate the attitudes and values that have been acted upon. That is why the evaluation should be individual and with face-to-face monitoring for efficient feedback.
Finally, on the role of the professor as an advisor in the teaching / learning process Ayala (2000) agrees with the statement that it is important to identify in this type of innovations: "In this human relationship, the advisor has confidence and respect for the possibilities of development that each student has, and in this perspective of respect, intervenes seeking the best development for the student"(48). In this way, the student will develop his full potential for a graduate in integral engineering.

Methodology

The Art and Technology course has eight units per week, ie three hours inside and five outside the classroom. Thus, the course was constituted by 16 sessions of three hours, once a week. The electronic learning platform was Blackboard. The first session was at the Museum of Contemporary Art of Monterrey (MARCO) to visit the relevant exhibition and with an introduction to the course and presentation of the educational model of the ABP course and the teachers. Subsequently, from sessions two through fourteen, the participants focused on the exhibition of the discipline of each expert with information modules, guests and activities focused on the progress, follow-up, evaluation and construction of the course challenge. It should be noted that all teachers are present in all sessions and meet face-to-face and virtual meetings throughout the semester to integrate the disciplines and content of the sessions towards the same academic challenge.
Sessions 15 and 16 are for the presentation of the works together with the essay that justifies the aesthetic discourse, before galleries, artists and relevant guests of museums and galleries. In the period August - December 2016, the guests represented the Museum of Contemporary Art of Monterrey (MARCO: http://www.marco.org.mx/) and the Gallery of Contemporary Art HeartEg: http://heartego.com/. This experience is realized to emphasize more the LBC with a feedback and evaluation by specialists incorporated to the real world of art and technology.

Sequence of sections of the Art and Technology course:

1. Introduction to the course in a exhibition of contemporary art			

2. Creative process I: Introduction	• Form I: Introduction	• Technology as a medium I: Introduction	• Aesthetic speech: Introduction
• Creative Process II	• Form II	• Technology as a medium II	• Aesthetic speech II
10 Creative Process III	11. Form III	12. Technology as a medium III	13. Aesthetic speech IV
14. Production of final works I	15. Production of final works II	16. Exposición de obras. Conclusión del reto: retroalimentación	17. Exhibition of works. Conclusion of the challenge: feedback

Each teacher has 25% participation in the request for activities as well as in evaluation by competences; the sum of the four adds up to 100 percent of the final grade.

Results

During the semester August - December 2016, the Art and Technology course was imparted to two groups of 20 students each, and five different engineering. Throughout the semester, more than 55 individual works of art of individual authorship and 14 pieces of equipment were added to the main collective challenge: Create a work that merges Art and Technology based on a solid aesthetic discourse as aesthetic response to the Problems of the 21st century.

The themes from which they addressed their proposals were:

1. The self in search of meaning.
2. The encounter with the other.
3. Societies and power.

Their respective essay with the aesthetic discourse that underlies them supported all the works, individual and collective.

A finding of this innovative didactic experience was that the students of this course effectively merged the Art and Technology through the information modules of the four experts to achieve the challenge of creating an artistic work whose aesthetic discourse was based on an essay with academic rigor.

The five best-evaluated papers were:

Title	Authors	Theme / Summary of aesthetic speech	Brief description of the work	Image

Barcelona	Arturo Martínez, Valeria Flores, Sofía Garcia	How to travel to another city by memories?	Sculpture connected to internet	
Lovebot	Daniel Silva	Do we care about the likes if they come from a robot?	Twitterbot	
Follow	Ernesto Lobo	How do we move through the city?	Web page, Open source code	
Monterrey	Sofía Romero, Eduardo Casas	How does it feels a polluted city?	Internet connected installation	

The work “Monterrey”, by the student authors Sofía Romero and Eduardo Casas participated in the competition for the Internet Art of Things "Adaptation" (Berlin, Germany) sponsored by the European Commission.
The work “Follow” by the student author Ernesto Lobo will be replicated as a co-curricular didactic project in Week i 2017 at Tecnológico de Monterrey.
Other results were:

It established relationships with contemporary artists who use technology as a medium and served as mentors and example for students. Among the artists involved are Ernesto Walker (http://www.ernestowalker.com/), David Martínez (bio-artist), Alberto Viloria (https://www.youtube.com/watch?v=D-ka_rC-IV0) And Gerardo García de la Garza (http://ggdelag.info/)

A close relationship was established with the Museum of Contemporary Art of Monterrey (MARCO) [http://www.marco.org.mx] and with the Museum of Art, Architecture and Technology (MAAT) of Lisbon, Portugal, recently inaugurated in October 2016 [https://www.maat.pt/]; it was contemplated to make a trip of studies of LBC with students of this course or related to the Art and Technology.

It also detonated the creation of an innovation laboratory related to Art and Technology called ARTEC Medialab to develop more ideas for the same purpose that originated the course.

The objective of this article was achieved:

Demonstrate the design, methodology with LBC of the course Art and Technology and its results.

With the experience of the previous period and what goes so far (March 2017), we can consider as valid the hypothesis with which this course was designed and taught: "At the Tecnológico de Monterrey, Campus Monterrey, Mexico, the Art and Technology Course With Competence Based Learning will achieve in the first stage of delivery - from August 2016 to May 2017 - the development of skills: 1. Knowledge of processes to transfer technology and 2. Knowledge of processes to apply technology. To create an artistic work that fuses art and technology in tackling global issues of the 21st century, and backed by a solid aesthetic discourse.

Conclusions

The Art and Technology are two disciplines of knowledge that in this course are fused by five different approaches to complement a professional performance of the graduates by the competences that they develop. Likewise, through the aesthetic enjoyment and its reflection, the promotion of the value of human sense as part of his personal and professional profile in the performance of profession and in the contribution of the engineering to the world because the course Art and Technology detonates the aesthetic enjoyment of students.

Supported by the theoretical framework of the CBL and according to the Educational Model Tec 21, the students developed interdisciplinary with their colleagues and their teachers the following transversal professional competences: 1. Intellectual curiosity, 2. Innovation and 3. Use of technology.

In a second moment of academic implementation and research, the development of the mentioned competences will be validated more in detail and will identify areas of opportunity in the challenge and in the design of the whole course in general, inviting to replicate to all the Technological Campuses Of Monterrey in all Mexico, and the interested universities.

Finally, it is concluded that the interdisciplinary academic activity must respond to the demands of the solution of problems and search of answers to the problems of the current society that the graduates of the engineering of the XXI century face, is for that reason the importance of designing courses with an CBL to develop the disciplinary and cross-disciplinary skills required by today's world.

References

Ayala, Francisco. (2002). *El profesor como asesor*. México, Trillas.

Modelo de programas formativos de profesional – Documento de trabajo – Tecnológico de Monterrey. (Julio 2015). México.

Valverde Berrocoso, Jesús; Revuelta Domínguez, Francisco Ignacio: Fernández Sánchez, María Rosa. (2012) "Modelos de evaluación por competencias a través de un sistema de gestión de aprendizaje. Experiencias en la formación inicial del profesorado". *Revista Iberoamericana de Educación*. N.º 60, pp. 51-62 (1022-6508) - OEI/CAEU.

Villarruel Fuentes, Manuel. SEP-DGEST. (2009). Instituto Tecnológico Úrsulo Galván, México La práctica educativa del maestro mediador. Revista Iberoamericana de Educación. SEP-DGEST. Instituto Tecnológico Úrsulo Galván, México 2009. *Revista Iberoamericana de Educación*. ISSN: 1681-5653. n.º 50/3 – 10 de septiembre de 2009.

Attachments:

Promotional poster of the course Art and Technology. Tecnológico de Monterrey. Summer 2016.

Humanitarian Engineering: a proposal to articulate Engineering Education with social problems. Study Case: Universidad Sergio Arboleda

María Paula Flórez Jiménez
Universidad Sergio Arboleda
maria.florez@usa.edu.co

Diana María Duarte
Universidad Sergio Arboleda
diana.duarte29@correo.usa.edu.co

Luis Alejandro Angel Acosta
Universidad Sergio Arboleda
luis.angel@usa.edu.co

***Abstract**: Academy has focused on theoretical discussions that move away from problems and social and economic realities of the communities. With the objective of create synergy between engineering education and communities with low income the Engineering Faculty of the Sergio Arboleda University have created the Humanitarian Engineering center. This paper presents the model used in the design of the first academic offer of the center: an elective course where students have to face communities challenges in order to design artefacts, systems or engineering process, under social, environmental, technical and economic constraints. First, is presented a framework around Humanitarian Engineering and second the study case where is analysed the model proposal after it first implementation and the result perceived by the students. As a conclusion, there is presented then the changes proposed to the model for the second version of the course.*

Introduction – overall layout

In the discussion of the main function of universities a gap is perceived between the theoretical and academic components in relation to the problems and social and economic realities of those communities that should be directly impacted (Arango, 2004). According to the aforementioned, shouldn´t the fundamental task of education be considered as the generation of professionals that a capable of assuming an active role in the construction of more equalitarian societies with a better quality of life?

Engineering is by definition, an activity that transforms knowledge into something practical (El periódico, 2013). This practicality must be aligned with community interests because it would not be useful to know how to apply knowledge solely on the resolution of problems in books, and not have the capacity of solving real problems that are contained within economic, social and environmental restrictions (Lucena, 2010). Therefore, the goal of engineering must be oriented towards the design of solutions within real environments that may allow the students to identify restrictions that they may encounter within the context itself. That is why, the true challenge of engineering schools and faculties must aim to create an auspicious environment and an active methodology to improve the abilities of the students in attending the needs of a society. Many engineers graduate from the university without being prepared to design under real, true conditions; which means, under economic, social and environmental restrictions (Lucena, 2010). The academy is educating engineers to design, without really considering the importance that learning has in identifying the diverse problems and evaluating them from diverse areas of knowledge, his causes a lack of humanity in engineering and hinders the creation of a true society. Due to this reason, education in engineering must rethink changes to ensure that these professionals are able to deal with the true needs of the society.

Emphasizing on the need of inculcating these points of focus within engineering education; Julio Esteban Colmenares & Jorge Celis hold that, "in spite of the considerable amount of studies that have been advanced in relation to changes in engineering in the last decade, there is scarce investigation in relation to strategies about accomplishing successful change. Likewise, there is few evidence to evaluate the impact of curricular changes made to different engineering programs." (2016). With the objective of cultivating change in engineering education, a model for a subject is proposed that favors the inclusion of points of focus on community work in a traditional engineering program. Therefore, with this objective, this article presents a design of a subject on Humanitarian Engineering from the Universidad Sergio Arboleda through the implementation of a pilot subject and the analysis of its results. The subject aims that through interdisciplinary groups symbiosis may be generated between the university and the community. By these means, guaranteeing that the students in engineering graduate with a preparation that allows them to attend the diverse needs of society through the design of artifacts, systems or sustainable processes.

In this way, the intention is of answering the question: How to design a subject of Humanitarian Engineering in the context of a traditional program? To accomplish this objective, the structure of the following article is presented as follows: (1) A theoretical framework is introduced where various propositions for change suggested by theoretists for education in engineering are exposed. As well as, a definition on Humanitarian Engineering is provided; (2) Information is synthesized and analyzed regarding Humanitarian Engineering; (3) The methodology used is explained to create a proposal for the design of a subject in engineering on Humanitarian Engineering; (4) The results obtained on the investigation are demonstrated; (5) The investigation is concluded; (6) Some recommendations and implications in regards to the proposal are mentioned.

Theoretical framework

Proposal for changes in Engineering Education

To achieve the goal that engineering education be oriented in the construction of better societies, there are authors that propose different changes in the curriculum of the engineering programs, within which we find the cultivation of an interdisciplinary, sustainable and practical focus.

The change suggested for the cultivation of an interdisciplinary focus is founded on the ability of designing in a participative manner. For this, Verharen, C., Tharakan, J., Middendorf, G., Castro-sitiriche, M., & Kadoda, G. (2013) suggest courses of engineering in which working with interdisciplinary groups is advised in the moment of designing so that the contributions that individuals from the diverse disciplines are taken into consideration. In this way, it holds that positive change may be accomplished in the technology that is designed. By incorporating this interdisciplinary focus in engineering education, as Shields, D., Verga, F., & Andrea Blengini, G. (2014) suggest, it would favor sustainability to the engineering curriculum, because by working with interdisciplinary groups, learning is done under real restrictions (social, economic and environmental). This is of utmost importance, because by taking into consideration these kinds of restrictions in the design of solutions to the diverse problems, one can assure the efficacy of these and the reduction of the probability of them having counterintuitive consequences in the future. Nazzal, D., & Zabinski, J. (2014) contribute that in the -how to-, one must introduce a principle of sustainability in the different subjects of the engineering curriculum and after; create modules or specific subjects where the extent of the relation between engineering and sustainability is addressed.

On the other hand, Colmenares & Ceils (2016) hold in their document: -Hacia una formación más fundamentada y flexible en ingeniería civil- (Towards a more founded and flexible formation on Civil engineering.): "Something that international experience does show is that the changes suggested for the formation on engineering not only aim for it to me more founded and articulate towards mastership, but also that the applicants to a license have a previous professional experience with views on being truly qualified for obtaining the such license." and continue saying: "the professional license must not be assumed solely as a matter of theoretical knowledge but more as a certification of professional competences and necessary experience to have the ability of developing with feasibility projects that impact and have social responsibility". This means that an engineer must take all his theoretical knowledge and put it into practice, in the process of solving real problems.

A tool that allows the search of solutions to social problems, is the poCDIO methodolgy proposed by Arias, Ramírez, Duarte, Flórez and Sanabria (2016), which promotes the active participation of all the stakeholders in the stages of the construction of the theoretical framework. OCDIO meaning observing, conceiving, designing, implementing and operating.

To accomplish these changes in the curriculum that were previously mentioned it is important; as stated by Besterfield-Sacre M., Cox, M. F., Borrego, M., Beddoes, K., & Zhu, J. (2014), to create a shared vision about such innovation, which supposes the support from the faculties to their professor in their own academic teaching and in the implementation of politics that may allow the reward of teacher innovation.

Humanitarian Engineering

One of the other changes that has been developed at the level of engineering education is the cultivation of projects with social and environmental impact. Humanitarian engineering is a focus that has been gaining strength in various programs at a worldwide level as is recognized as a specialty that cultivates the development of communities through the design of technologies, systems and processes. Kevin Passino (2015) in his book on Humanitarian Engineering defines this concept as: "Humanitarian engineering is the creation of technologies that help people." On the other hand, Juan Lucena (2010) in his book, Engineering and sustainable community development; mentions that Humanitarian engineering refers to the development of communities.

Humanitarian engineering and similar programs have been welcomed by different universities in the United States, such as: Colorado School of Mines, University of Canterbury, Arizona State University, Villanova University, Rochester Institute of Technology, University of Wisconsin-Madison, Carleton University, University of Michigan and Purdue University, The Ohio State University; in Mexico: Tecnológico de Monterrey, Universidad del Valle in México; in Latin America: Universidad Nacional de la Matanza and Universidad Católica in Argentina, Universidad de Chile, Pontificia Universidad Católica, and Universidad del Desarrollo in Chile; Universidad del Valle, Universidad de La Salle, Universidad de los Andes and Universidad Minuto de Dios in Colombia.

References on Humanitarian engineering

The abilities that a student acquires depend on the design and the methodology of the subject. With the purpose of understanding a few of the basic characteristics of a class on Humanitarian engineering, the following Table 1 analyses three references, considering aspects such as: the audience to whom the subject is addressed, the type of subject, the weekly hours that the subject has assigned, the theory component, the practice component and how it is evaluated.

Table 1: Examples of references on Humanitarian engineering.

Subject / Aspect	**Ingenieros sin Fronteras Colombia (Engineers without borders Colombia) (ISF-CO)** ***Universidad de los Andes*** ***(Universidad de los Andes, 2017)***	**Community Based Research** ***Colorado School of Mines*** ***(Colorado School of Mines, 2017)***	***Humanitarian Engineering*** ***The Ohio State University*** ***(The Ohio State University, 2017)***
Audience to whom it is addressed	Is a subject for engineers	Is a subject for engineers	Is not an exclusive subject for engineers
Type of subject	Elective	Elective	*Mandatory for a Minor degree in Humanitarian Engineering, and is elective for other degrees.*
Weekly hours that the subject has assigned	4 weekly hours	3 weekly hours	3 weekly hours
Theory component	The topics are focused on the development of the concepts related to the criteria of the projects of ISF-CO: socially inclusive, feasible, environmentally responsible, related to engineering, innovative, technically possible, and of high impact. Lecturers are invited to the class sessions so that they can give talks oriented to the criteria of the ISF-CO projects.	The class sessions are oriented to the development of topics such as qualitative investigation, ethnography, participative observation, collaborative design, community-based mapping, interviewing, and the importance of listening. Lecturers are invited to some of the class sessions to enrich some of the topics.	The class sessions are oriented towards the following topics: poverty, underdevelopment, sustainability, culture, social justice, and development strategies. As well as engineering for community development. Analytical methods and participative humanitarian technology development. During the class sessions discussions in relation to the different topics are held.
Practice component	Through the duration of the course the students must find solutions for a challenge that the community raises in a participative manner. This community may be vulnerable of have development potential. The challenge that the students must solve is chosen but the teachers of the subject. The OCDIO content is used to approach the challenge.	Through the duration of the course the students must answer an investigation question related to a group that is a part of the university community. The group and the question that will be investigated is suggested by the students themselves. Students are taught different tools that are useful in working with communities,	The students work on a final project where they must select and develop an appropriate technology to solve a problem around the world; as well as evaluating its environmental and economic impact. Students are taught different tools for the participative design of technologies. Students are taught some ways to model and simulate

	Field laboratories are performed with the students so that they can get to know the area and interact with the actors involved in the diverse problems. The practice component is developed in groups.	specially tools for participative observation. Students must perform activities outside of the classroom; participative investigation activities with the actors involved in the investigation question. The practice component is developed in groups.	the impact of the given solutions. The practice component is developed in groups.
How is it evaluated?	Quizzes are performed to evaluate the comprehension of the readings assigned for the class sessions. Workshops are graded in which the students can apply some of the tools learned in class. Students present through a portfolio their advances in relation to the challenge. Each portfolio synthesizes and documents the information gathered by at least one of the OCDIO stages. At the end of the term the students have a presentation where they sustain their proposal to experts.	Assistance and participation in class is graded. The application of the tools learned in class is graded; such as: the guide of the interview, the interview, participative observation, and the activity of participative investigation. The students present an analysis of the assigned readings. The students present a final project where they provide a conclusion with regards to their findings during the investigation.	Quizzes are performed about the lectures assigned for the class session and comments on the class are given by the teacher through videos. The students should turn in homework that consists on resolving problems from the main textbook of the subject. *(Humanitarian Engineering: Advancing Technology for Sustainable Development, 3rd Edition) Some homework must be done on the Matlab programming software.* At the end of the duration of the course the students turn in a report on a final project and they sustain it.

Based on what is presented in the Table 1 of references, it is possible to view that there are many shared aspects in the three Humanitarian Engineering subjects that were analyzed. At first, all the courses have a theory and practice component. The theory component provides the student with tools that can be developed in the practice component that will consist in the execution of a project or of the solution of a challenge. It is common in all cases that this component be performed in groups.

It must be clarified that not in all cases the practice component is focused on finding solutions to local problems. The social system that wishes to be analyzed depends on the context and possibilities of each place.

On the other hand, it is made evident that a humanitarian engineering subject does not necessarily need to be conformed by engineers, and thus; is important that all students learn techniques and are able to apply participative methodologies in working with multidisciplinary teams.

Also, one may see that the three reference courses are elective. This may be explained by considering that, even though these subjects deal with pertinent topics for the formation of any professional, and particularly, for the formation of any engineer; that not necessarily all the topics that are dealt with are of interest to all people. It is hopeful that, behind the diverse problems addressed in class the human being is considered as a central actor and the commitment of the students be total.

As far as the manner on how the students are evaluated, each subject has its own different pedagogical strategies. Nonetheless, it is common that in the three courses the final project is evaluated. Additionally, in the three references, apart from the grade obtained from the project, other activities are performed to evaluate theoretical knowledge. In none of the cases midterm evaluations are done as a method of evaluation.

Study Case: Methodology

Following is presented the methodology used to define a course model on Humanitarian Engineering. The proposal was tasted in the Industrial Engineering program of Sergio Arboleda University as a strategy to include new perspectives to its traditional approach:

1. Initial proposal of a course model on Humanitarian Engineering in Sergio Arboleda University
2. Model Implementation
3. Students evaluation by an anonymous survey and professors perceptions of the model
4. Model re-design based on the results

Initial proposal of a course model on Humanitarian Engineering in Sergio Arboleda University

The Humanitarian Engineering course was designed in order to achieve the following objectives:

- To create synergy between engineering and other programs of the university and set up multidisciplinary groups to design artifacts, systems or engineering processes, under social, environmental, technical and economic restrictions.
- To create synergy between low income communities (living in and around Bogota) and the university through the co-creation of engineering solutions.
- To train the students in skills and competencies based on the humanist-scientific approach.

With that scope, the course was defined as obligatory so all the students from industrial engineering program (in its pilot phase) could get an approach to the role of the engineer in the context of communities projects. These types of experiences help the development of design skills with economic, environmental and ecological constrains which are needed to have a complete professional profile in any economic sector in the times of the 2030 agenda.

About the methodology, according to the objectives established, the course required having a practical approach. With that intention, the course model was aligned with an innovative proposal made by Distancia Cero, an social entrepreneurship with the mission of decrease the distance between the reality and the academia by promoting an education strategy named Challenge Based Learning. Distancia Cero provided to the course a challenges portfolio posted by two social organizations:

- ***Casa Taller Armando Sueños:*** A woman collective initiative which looks for the empowerment of the community of Guasca, a town located at 50 km from Bogota. It challenge was: ¿How to define a communication strategy in order to make financial sustainable Casa Taller?
- **Fundación Poca Lana:** An ONG with more than 10 years of presence in Ciudad Bolivar, one of the poorest areas from Bogota. It challenge was ¿How to generate a business model with paper or plastic recycled with a group of mother in condition of vulnerability?

The course was design around the development of these challenges where the students had to propose solutions in teams using the methodological frame po+CDIO, proposed by the organization Ingenieros Sin Fronteras Colombia as an adaptation of the frame CDIO created by MIT.

Following is presented the model course that was implemented for it first time in the second semester of 2016. These aspects are related to the general characteristics that engineering courses with a communitarian emphasis usually focus on (as was shown in the section "References on Humanitarian engineering")

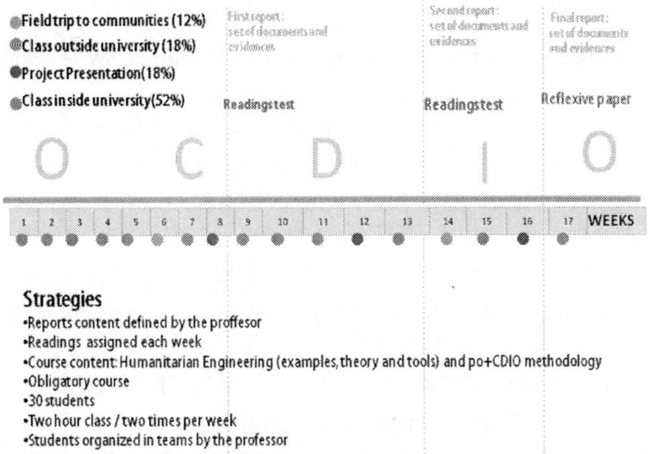

Figure 1: Initial proposal of a course model on Humanitarian Engineering in Sergio Arboleda University

Results

In order to evaluate the model, at the end of the semester the students completed a survey (33 answers) where they were asked about their perception of the course. The main priority was to understand from the point of view of the students, if there was a perception that the course was effectively adding value to their educational experience. Here are the principal results:

General perception

20% of the students indicated that the course was very relevant to their training in engineering; 60% indicated that it was relevant; 20% indicated it was indifferent or little relevant.

About the assessments

- 30% of the students said they dedicated more than the minimum time needed in the assessments; the 50% just the time needed
- The students indicated that the course should be more flexible about the design process they have to face. Not being too methodic
- Although the readings were related to the course, they perceived it didn't have much to do with the challenges that are the main objective of the class.
- There was a quantity of assessments that didn't have a real connection with the challenged and it was perceived as a waste of time.

About the methodology

- 55% of the students indicated they perceived the methodology as very innovative (they hadn't had a class similar); 33% indicated is innovative.

- 88% of the students indicated that the experience of working with real challenges was very valuable for their learning.
- Almost the 95% of the students said that it was clear how to apply the po+OCDIO methodology
- 55% of the students indicated they were interested on continuing the challenge because working with the communities was a truly important experience for them

Findings

After the assessment process and according to the results it was defined that the most valuable aspects of the model was the implementation of the challenge based learning methodology. In the context of the Engineering Faculty of Universidad Sergio Arboleda this approach showed innovation in the student educational experience. Nevertheless it is important to emphasize that there were some aspects that did not function correctly and that should be corrected in order to respond effectively to educational offer. First, the development of the challenge requires an important amount of time so it is necessary to minimize the assessments that are not directly related to it. Minimizing the content of the class is related to have classes more focused on having experiences outside the classroom.
Also, Humanitarian Engineering is an open and complex field so it is necessary to bring experiences and examples that are related to the challenge so the students can connect this information in a easy way to their practice experience along the course. Respect to the assessments, the reading and its tests along the semester did not contribute to the learning process but the reflexive paper (and the methodology given) did help to understand the principal conclusions from a personal perspective. In this way, Humanitarian Engineering should connect both theoretical and practical component.

Conclusions and recommendations

After an analysis of the results, there was included some changes to the course model on Humanitarian Engineering that are now being implemented in the first academic semester of 2017, considering that both theoretical and practical component are important.

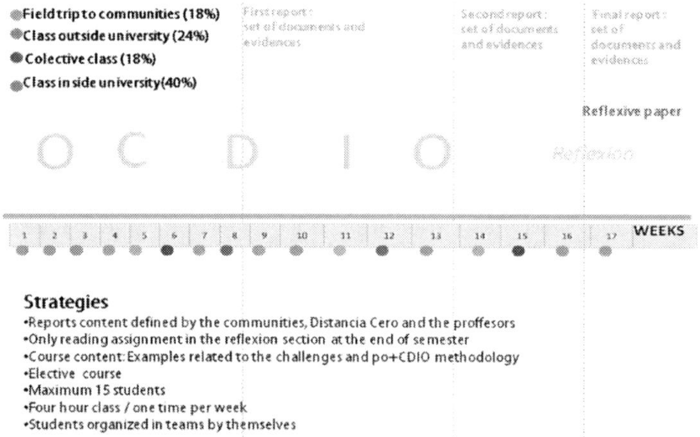

Figure 2: Proposal of a course model on Humanitarian Engineering in Sergio Arboleda University

This focus will ensure that the students will have to make some decisions about a design process in social, economic and environmental constrains. That also means that all the class assessments should be focused on the challenges because the amount of time the students are going to dedicate on facing them is enough to the learning objectives. It is valuable to have a reflexive exercise at the end of the semester where they can analyze their experience using readings related to humanitarian engineering so they can generate their own conclusions. About the impact of the results, the assessments should be defined before the semester begins in a negotiation between the communities and the professors (not only by the professors). In these process is really valuable the role of organizations as Distancia Cero. With that strategy, the students will have goals but they will have the liberty in deciding how to do it. In this process the professors have a roll of facilitator. These also will ensure the immediately impact on communities because it will be measurable in terms of the accomplishment of the assessments.

The course should be elective and no obligatory. Although the professors think the sense that Humanitarian Engineering should be a clue for the development of design skills of the students, there are a common perceptions or paradigm that working with communities is not transversal and is not for everyone. Said that and Thinking on the compromise of the students it should be elective. The po+CDIO frame should be used as a frame and no as a methodology. The students can defined how they embraced it taking into a count that those are the basics steps in a design process

References

Arango, G. M. (2004). La educación superior en Colombia. *Universidad Nacional de Colombia.* Retrieved from http://www.colombiaaprende.edu.co/html/investigadores/1609/articles-73081_archivo.pdf

Arias, L., Rámirez, C., Duarte, D., Flórez, M. & Sanabria, J. (2016)., poCDIO: A Methological Proposal for promoting Active Participation in Social Engineering Projects. Systemic Practice and Action Research. p 1-25.

Besterfield-Sacre, M., Cox, M. F., Borrego, M., Beddoes, K., & Zhu, J. (2014). Changing engineering education: Views of U.S. faculty, chairs, and deans. Journal of Engineering Education, 103(2), 193-219. Retrieved from http://search.proquest.com/docview/1636192778?accountid=48417

Colmenares, J. S., & Celis, J. (2016). Hacia una formación más fundamentada y flexible en ingeniería civil. *Revista Educación en Ingeniería, 11*(21), 4-8.

Colorado School of Mines. (2017). Humanitarian Engineering Program. Retrieved from http://inside.mines.edu/HE-Program

El periódico. (2013). *Una actividad que transforma el conocimiento en algo práctico.* Retrieved from http://www.zetaestaticos.com/aragon/documentos/especiales/2013/ingenieria/001.pdf

Passino, K. M (2015). Engineers' Role in Humanitarianism. Humanitarian Engineering. p 51-68

Lucena, J., et al.(2010) Engineering and Sustainable Community Development. Morgan & Claypool Publishers series.

Nazzal, D., & Zabinski, J. (2014). Introduction of sustainability concepts into industrial engineering education: A modular approach. *IIE Annual Conference.Proceedings,* , 948-954. Retrieved from http://search.proquest.com/docview/1622299468?accountid=48417.

Shields, D., Verga, F., & Andrea Blengini, G. (2014). Incorporating sustainability in engineering education. International Journal of Sustainability in Higher Education, 15(4), 390. Retrieved from http://search.proquest.com/docview/1660703254?accountid=48417.

The Ohio State University. (2017). Courses on Humanitarian Engineering. Recuperado el 2017, de https://hecourse.engineering.osu.edu/humanitarian-engineering/spring-2017.

Universidad de los Andes. (2017). Ingenieros Sin Fronteras Colombia. Retrieved from https://isfcolombia.uniandes.edu.co/index.php/proyecto-intermedio-2017-10/presentacion-2017-10

Verharen, C., Tharakan, J., Middendorf, G., Castro-sitiriche, M., & Kadoda, G. (2013). Introducing survival ethics into engineering education and practice. *Science and Engineering Ethics, 19*(2), 599-623. doi:http://dx.doi.org/10.1007/s11948-011-9332-9

Mechatronic Engineering and its role in the Productive Matrix of Ecuador

Lara-Padilla Hernan
Universidad de las Fuerzas Armadas ESPE, Sangolqui, Ecuador
hvlara@espe.edu.ec

Terneus Páez Carlos F.
Universidad de las Fuerzas Armadas ESPE, Sangolqui, Ecuador
cfterneus@espe.edu.ec

Rea Minango Sylvia N.
Universidad de las Fuerzas Armadas ESPE, Sangolqui, Ecuador
snrea@espe.edu.ec

Paucar T. Annabel
American Society of Engineering Education (ASEE) Member, Columbus, USA
annabel_paucar@hotmail.com

Parra Moreno Vicente
Universidad Central del Ecuador, Quito, Ecuador
mat.vicenteparra@gmail.com

Abstract: *The policy of Transformation of the Productive Matrix (TPM) created by the Ecuadorian government in 2013 focuses on stimulating the local production, reducing imports in raw materials and finished products. The main aim of this research is to study the status of the convergence among industry, university and government policy in Ecuador through the analysis of the mechatronic engineering curriculum of the Universidad de las Fuerzas Armadas – ESPE (ESPE) and its contribution to the TPM in Ecuador. A mixed method research was used in the present study based on the database of the senior theses, graduates' surveys and employers' interviews. The results show a weak convergence between the industry and university in the framework of the TPM. The mechatronic engineering curriculum of the ESPE does not consider either the participation of the Ecuadorian industry by advisory committees, or the feedback of mechatronic engineering graduates from the ESPE.*

Keywords: *Mechatronic Engineering, Mechatronics, Transformation of the Productive Matrix, Curriculum*

1. Introduction

The concept of mechatronics was introduced nearly 50 years ago, in order to express and reflect the increasing use of computers and micro-electronic systems to control a range of mechanical process and systems (Bradley et al., 2015). Some mechatronic products and services, which are going to dominate the technology world, are smart sensors, machine vision, flexible robotic manipulators, and automated guided vehicles (Dixit, Hazarika, and Davim, 2017).

The adoption of mechatronic engineering in Ecuador as an undergraduate program has prompted many debates around the academia and industry about their primary occupational field and their contribution in the Ecuadorian industry. To date, in Ecuador, six universities offer the undergraduate program of mechatronic engineering.

Mechatronic engineering started in Ecuador from 2006 at the Escuela Politécnica del Ejercito (ESPE) (today Universidad de las Fuerzas Armadas – ESPE) in the campuses of Sangolquí and Latacunga. The Departments of Mechanical Engineering and Electronics Engineering provides a 90% of the support to the mechatronic engineering program. The formation is complemented by other Departments as International Language, Humanities, and Social Sciences.

In 2008, the new Constitution of Ecuador promulgated a country policy to expand its sources of income and to reduce the oil exportation dependence (Constitución del Ecuador, 2008). Inside of this framework, one of the most important national objectives is named "Transformation of the Productive Matrix" (TPM) (Villafuerte and Intriago, 2016). Companies operating and the other ones that want to invest in Ecuador have the challenge to follow the TPM plan to be competitive, stimulating the local production and decreasing imports in raw materials as much as finished products (Poveda, 2014).

The Council of Higher Education, Evaluation, Accreditation and Quality Assurance of Ecuador (CEAACES) has established the quality standards for universities in Ecuador and their careers programs (CEAACES 2013). The criteria of CEAACES used for the evaluation of universities are faculty, graduates percentage, research, organization, and facilities. In the case of the undergraduate programs, CEAACES establishes the next parameters: pertinence, curriculum planning, faculty, institutional environment, and students. However, this model is still in progress, and it appears as a generic process, which does not provide a real tool for cooperation among the three components of a knowledge-base economy: government, industry, and university (Leydesdorff, 2011).

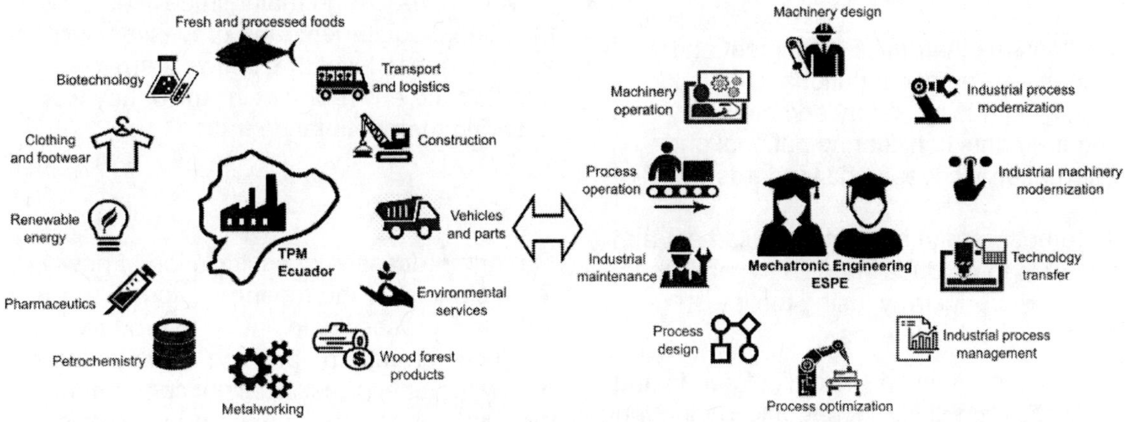

Figure 1. Framework of Transformation of the Productive Matrix (TPM) and Mechatronic Engineering of the ESPE

To the best of our knowledge, there are no results in the literature regarding the influence of mechatronic engineering in the TPM. To the date, there is not a well-documented plan for programs evaluation and the feedback that improve the interaction between the involved actors.

The aim of this work is to present a model for the evaluation of mechatronic engineering program of the ESPE in the context of TPM and generate feedback for future improvements in the curriculum planning. This process constitutes a reference for other undergraduate engineering programs in Ecuador. This process began with the analysis of the mechatronic engineering outcomes. Next, those outcomes were contrasted with the priority industries established by the TPM policies. Finally, graduates, experts, and entrepreneurs were interviewed to create a global outlook of the situation of mechatronic engineers from the ESPE in Ecuador.

The results of the proposed model show a weak convergence between the industry and university in the framework of the TPM, especially because the ESPE mechatronic engineers show a low participation in the sectors considered as primaries by the Ecuadorian government.

2. Institutional background and context

The history of mechatronic engineering at the ESPE began on February 6 - 2006 in both the main campus (Sangolquí) and its extension (Latacunga) ("ESPE – Ingeniería Mecatrónica," n.d.). As stated in the passed document for the creation of mechatronic engineering, this engineering is framed in the Ecuadorian policies related to the function of a higher education institution. In this document, the pertinency of mechatronic engineering was studied also and the conclusions show that the new engineering fulfills the requirement of the Ecuadorian society. Table 1 shows the professional skills and learning objectives of ESPE mechatronic engineering. Up until today, the mechatronic engineering program has about 200 graduates.

Table 1. Formation outcomes of the mechatronic engineers of the ESPE

PROFESSIONAL COMPETENCES	LEARNING OBJECTIVES OF MECHATRONIC ENGINEERING
Applies the basic sciences in solving mechatronic engineering problems, using scientific and technological principles. Designs mechatronics equipment and systems using computational tools and satisfying quality, safety and economy requirements considering national and international codes and standards.	Design, operate and maintain devices, machines, equipment, and processes, with an intelligence level that allows them to adapt to the environment in which they work, ensuring their optimal operation.
Automatizes and innovates equipment and systems to produce goods and services to improve productivity, using the latest technological advances. Provides production systems of goods and services related to mechatronic engineering, so as to improve the productivity and competitiveness of industrial companies according to international standards.	Manage processes of assimilation of new technologies for the modernization of industrial processes intended to produce products through the physical transformation of raw material or assembly of components for power generators, paper industry, wood industry, food process, agro-industrial sector, automotive sector, etc.

Industrialization has been the emblem of national development in the twentieth century. This process is a global phenomenon, and it is made up by the dynamics of the world-system. The Ecuadorian government attempts to generate more wealth, achieve more sustainable development and strengthen human talent, technology and knowledge through the strategy named Transformation of the Productive Matrix. The recession in the international market the last years, due to the reduction of oil prices and other products, has hit the Ecuadorian economy. Unfortunately, to Ecuador, the cost of these goods depends on external factors. The Ecuadorian government plans to rise the production of clean electricity to increase the local and regional development, to export processed products with added value, to reduce imports through increasing the domestic production, and to invest in education and technology. In numbers, (i) the change of the production model means $5,200 billion a year to Gross Domestic Product (GDP) in primary industries such as petrochemicals, steel, aluminum, shipbuilding, copper refining, among others; (ii) $6,100 million more to GDP as cacao agro-business, biofuels, farming, aquaculture, and forestry; (iii) $800 million through the reduction of non-oil imports (e.g. canning and footwear) through the strengthening of the domestic industry; (iv) 9,000 direct jobs in the operation of basic industries; and (v) 485,000 direct jobs in agro-industrial sector ("What is the change in the production matrix in Ecuador?," 2014).

The CEACCES policies highlight the university programs should be evaluated by their level of articulation with strategic areas and sectors of regional and national development. The programs with interdisciplinary and transdisciplinary approaches presented by the universities to solve regional and national problems are considered of great value. As reported by CEACCES, Mechatronics is part of the "Strategic Knowledge Field" in the field of engineering, industry, and construction.

3. Methods

The aim of this study is to answer the question: what is the influence of the ESPE mechatronic engineers in the Ecuadorian productive matrix? A mixed method research, a combination of quantitative and qualitative methods, was used in the present study. In the first part, we used a quantitative research method in order to carry out a comparative study of the formation outcomes of the ESPE mechatronic engineering and the policies of the Ecuadorian government related with the TPM. We used the repository of graduation projects of the University (http://repositorio.espe.edu.ec/handle/21000/5655) to select randomly 100 projects to evaluate their relation with the professional outcomes and the Ecuadorian policies of production.

The second part of the research contained personal interviews to a group of 252 graduates and 20 employers (stakeholders) of different industrial sectors. The survey for graduate included questions about their current employment situation and their criteria about the reached performance in the labor field as a function of the education received during their formation as a mechatronic engineer. Interviews with stakeholders led to search if they know the contribution of mechatronic engineering in the Ecuadorian industry. We did not consider the students and faculty of mechatronic engineering as stakeholders by reasoning that the students use the curriculum but often a lack of information regarding the competences needed in their vocations, and the lack of industrial experience of faculty, in which case their priorities and ranking would differ from the real industrial environment.

4. Results and Discussion

It is noted from Figure 2, less than 60% of the graduated projects have a significate contribution to the priority industries considered in the TPM. 70% of the projects are related to the approach of machinery design and construction. However, there is a considerable number of projects related to other industries, especially with educational projects. 43% of the projects are not associated with the TPM. Additionally, these educational projects are focused on either specific facilities/laboratories requirements (i.e. modernization/automation of lab equipment, old machinery) or faculty research projects. Most of the projects unrelated with the TPM are focused on home-made developments in areas as robotics systems, 3D printers, and computer vision technology. The lack of collaboration between academia and industry is a crucial factor to reach a high-level performance of mechatronic engineers.

Unfortunately, Ecuador as other developing countries has even greater restrictions to such alliances, fomented for the weak government policies to promoting university-industry collaboration.

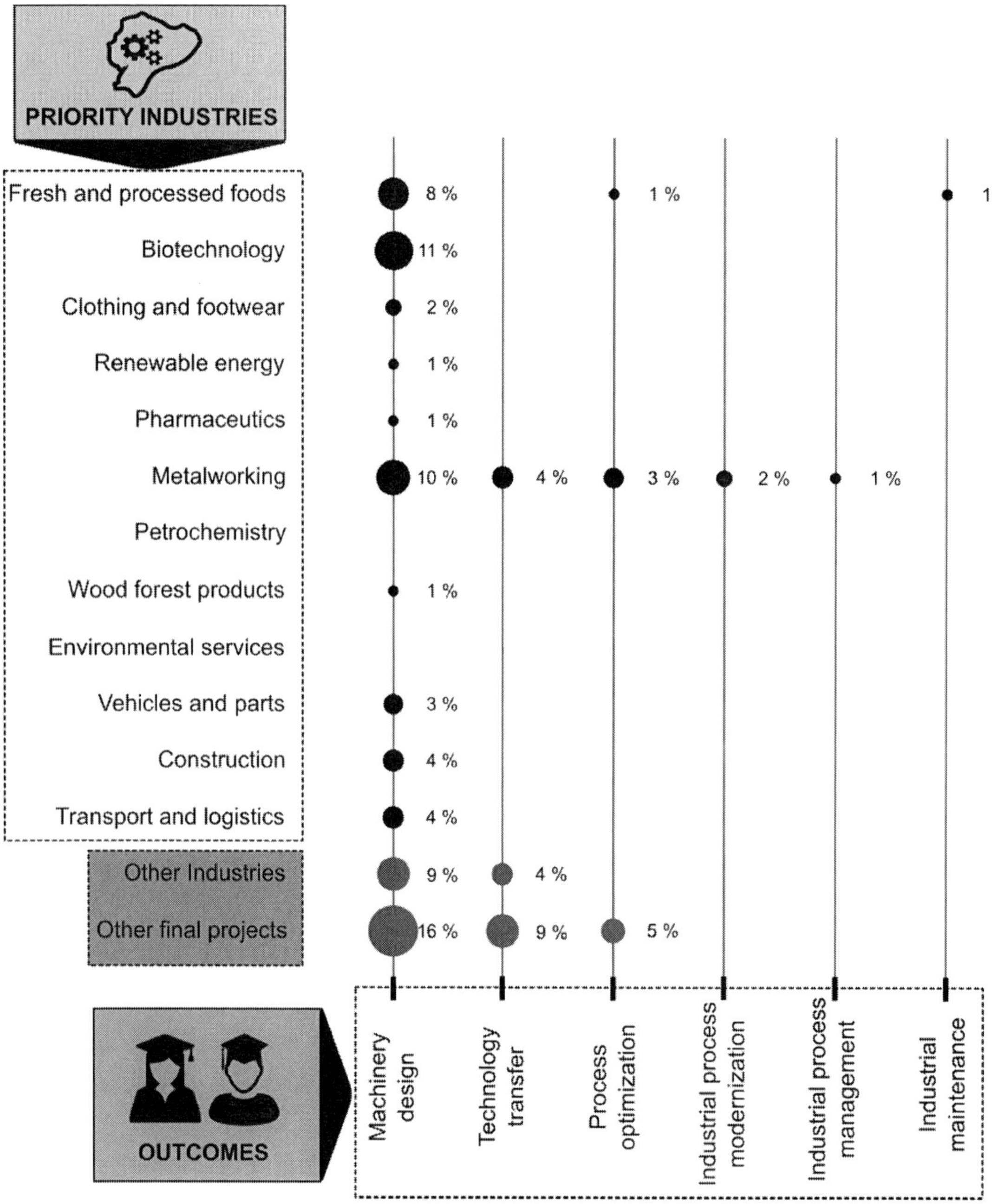

Figure 2. Connection between the TPM and the outcomes of the ESPE mecatronic engineers

Advanced topics as optimization, modernization, management of industrial process, and technology transfer end up being difficult tasks to implement in the curriculum of mechatronic engineering. The poor linkage between the ESPE mechatronic engineering and the Ecuadorian industry has not permitted to create cooperation plans as coordinating research and development (R&D) agendas, stimulating additional private R&D investment and utilizing synergies of scientific and technological capabilities.

Public policies may influence on the companies to collaborate with universities. The scope of such collaborations can cover different aspects, having a direct role in contributing funds to universities and R&D projects. Another purpose of public policies must be to implement the necessary facilities and partnership between institutions such as technology transfer offices, research parks, and business incubators. Furthermore, governments can stimulate cooperation through secure actions, for instance providing specialized support services to companies and universities, searching associates, and conducting predominate projects to encourage networking and raising awareness of the importance of collaboration.

Figure 3 shows the current employment situation of the ESPE mechatronic engineers. The unemployment rate rises about 55%, and just a 30% of the graduates have gotten a job related to their formation in the university. The high percentage of estimated unemployment could be due to some students surveyed were recent graduates. Adding the impact of the economic recession of the last years in Ecuador, the job opportunities have been reduced significantly. Other important factor is the limited employer engagement. In regions like the United Kingdom, higher education institutions have answered to this challenge by creating business units to address responsibility for employer links. Usually, local partnerships are formed with business links and chambers of commerce, as well as outstanding local and regional boards (Altbach, & Knight, 2007). In Ecuador, most of companies recruits skilled graduates following their specific rules. For instance, when company managers or chief engineers are talking about mechatronics, the skills and competences of a mechatronic engineer are not clear for them. Many industries in Ecuador unknown the skills of a mechatronic engineer and they still prefer either mechanical engineers for tasks as manufacturing and machinery design or electronic engineers for industrial automation jobs.

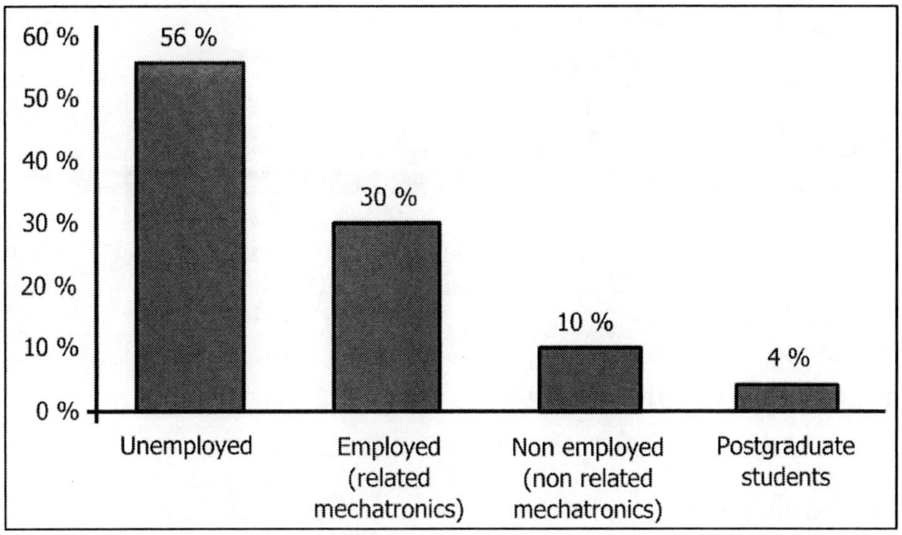

Figure 3. Employment situation of the mecatronics engineers of the ESPE

The voice of the employers is vital in the formation of a new professional, even more in new careers as mechatronic engineering. The interviews show that only a 20% of the employers knows the concept of mechatronics and they talked about the possibility to hire a mechatronic engineer to improve processes or machinery in their companies. The ESPE mechatronic engineer is well-known to the oil companies, especially in the fields of industrial automation and instrumentation. Some employers from the priority industries in Ecuador have an imprecise idea of the real application of mechatronics. Further, the possible workplace for mechatronic engineers are being occupied in its majority by mechanical and electronic engineers, who have several decades of presence in the Ecuadorian industry with a well-earned reputation. Countries such as Germany, United States, and the United Kingdom have a huge demand for mechatronic engineers. In these countries, after completing their program, mechatronic engineers are enrolled easily in automotive, manufacturing, electronics, mining, transport, and biomedical companies (Larsen, Bandara, Esham, & Unantenne, 2016).

Finally, on consideration of the last report of The Atlas of Economic Complexity ("The Atlas of Economic Complexity", 2017), it shows that 90% of the export products from Ecuador is composed of four categories: minral products, vegetal products, animal and animal products, and foodstuffs (Figure 4). Many of the products inside these categories are raw material as oil and minerals. Then the new generation of mechatronic engineers should be formed with competences and skills to incorporate added value in these industries throughout the development and improvement of technology.

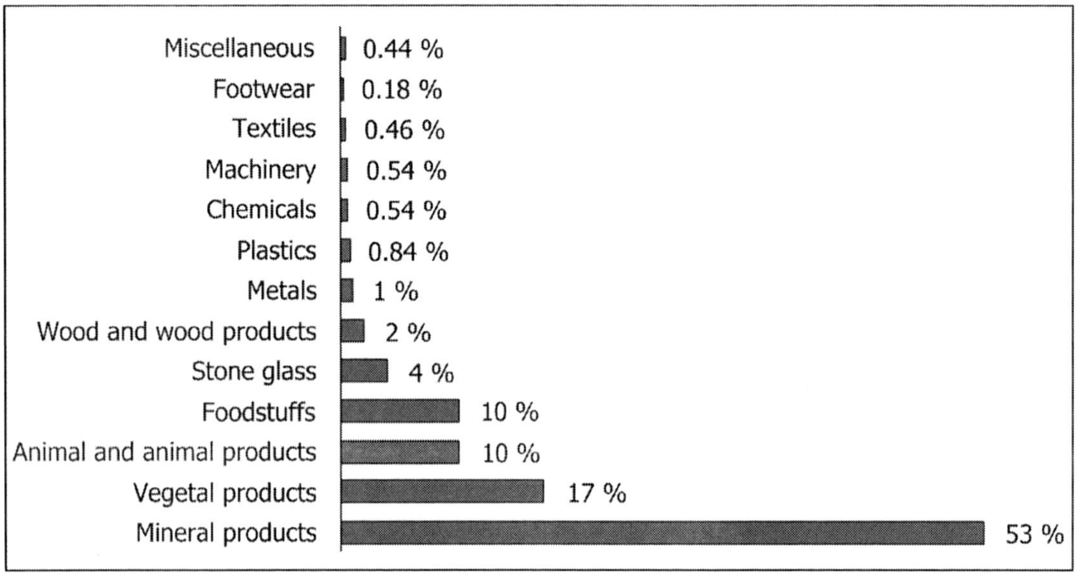

Figure 4. Percentage of exports from Ecuador in 2014. Adapted from ("The Atlas of Economic Complexity", 2017)

5. Conclusions and future work

An evaluation of the Ecuadorian Transformation of Productive Matrix versus the outcomes and the participation of ESPE mechatronic engineers on it, has been carried out. Employers expectations and the voice of graduates show the need for a curriculum revision.

There is an imperative need to create an assessment process with accurate metrics and documented results for the program of mechatronic engineering at ESPE. Additionally, the deans and faculty must have sufficient qualifications and must ensure the proper guidance of the study program, its evaluation, and updates.

The concept of mechatronics is still fuzzy for most of the Ecuadorian industries. Ecuadorian government may seek the way to improve the quality of the ESPE (and universities at large) graduates by fostering a stronger collaboration between universities and industry as part of a comprehensive science, technology, and innovation policy plan.

References

Altbach, P., & Knight, J. (2007). The Internationalization of Higher Education: Motivations and Realities. Journal Of Studies In International Education, 11(3-4), 290-305. http://dx.doi.org/10.1177/1028315307303542

Bradley, D., Russell, D., Ferguson, I., Isaacs, J., MacLeod, A., & White, R. (2015). The Internet of things – the future or the end of mechatronics. *Mechatronics*, *27*, 57–74. doi:10.1016/j.mechatronics.2015.02.005

CEAACES. (2013). *Consejo de Evaluación, Acreditación y Aseguramiento de la Calidad de la Educación superior Comisión de Evaluación y Acreditación de Carreras y Programas MODELO PARA LA EVALUACIO N DE LAS CARRERAS PRESENCIALES Y SEMI-PRESENCIALES DE LAS UNIVERSIDADES Y ESC*. Retrieved from http://www.ceaaces.gob.ec/sitio/wp-content/uploads/2013/10/1.-Modelo_generico_carreras-presentacion.pdf

CONSTITUCIÓN DEL ECUADOR. (2008). Retrieved from http://www.asambleanacional.gov.ec/documentos/constitucion_de_bolsillo.pdf

Dixit, U. S., Hazarika, M., & Davim, J. P. (2017). *A brief history of mechanical engineering*. doi:10.1007/978-3-319-42916-8

ESPE – Ingeniería Mecatrónica. Retrieved January 6, 2017, from http://mecatronica.espe.edu.ec/

Larsen, K., Bandara, D. C., Esham, M., & Unantenne, R. (2016). *Promoting university-industry collaboration in Sri Lanka: Status, case studies, and policy options*. doi:10.1596/978-1-4648-0922-4

Leydesdorff, L. (2011). The triple helix, quadruple helix, …, and an N-Tuple of Helices: Explanatory models for analyzing the knowledge-based economy? *Journal of the Knowledge Economy*, *3*(1), 25–35. doi:10.1007/s13132-011-0049-4

Perfil Profesional. Retrieved February 2, 2017, from http://mecatronica.espe.edu.ec/perfil-profesional/

Poveda, V. (2014). *Ecuador: The change of the productive matrix and the challenge of the import substitution*. Retrieved from http://www.desarrollando-ideas.com/publico/140224_dmasi_Ecuador_challenge_of_the_import_substitution_ENG.pdf

The Atlas of Economic Complexity |. (2017). Atlas.cid.harvard.edu. Retrieved 18 May 2017, from http://atlas.cid.harvard.edu/

Villafuerte, J., & Intriago, E. (2016). Productive matrix change in Ecuador and the petroleum crisis. Case study: Entrepreneurs and productive associations. *Journal of Business*, *1*(1), 1–11. doi:10.18533/job.v1i2.18

What is the change in the production matrix in Ecuador? (2014, November 21). Retrieved March 6, 2017, from ANDES, http://www.andes.info.ec/en/news/what-change-production-matrix-ecuador.html

Copyright statement

Engineering Students: Their Perception of Satisfaction with Traditional and Non-Traditional Course, the Use of POL in Database Course Case

Juan Raúl Esparza Martínez

Tecnológico de Monterrey, Monterrey, México
juan.raul.esparza@itesm.mx

Author 2 (Dolores Guadalupe Lankenau Caballero

Tecnológico de Monterrey, Monterrey, México
dlankenau@itesm.mx

***Abstract**: Universities are facing the challenge of the new generations of students: Millennials and Digital natives who are deeply immerse in the use of new technologies. The purpose of the study is to determine the perception of satisfaction with a course imparted in a non – traditional way, expecting to have better outcomes than in a traditional course. To achieve it we used a simple experimental design, the group "control" was taught in the traditional way with lecture system, the "experimental" group was taught by enouraging students to generate their own knowledge using POL and the teacher acting as facilitator of the process. Students expressed their perception of aspects of the course and their satisfaction with IT. The results were analyzed by Student's "t" , Chi Square and ANOVA tests and show that apparently there are not statistically significant differences between the two groups under study. Discussion and implications are presented*

Introduction

In search of excellence, universities follow a process of strategic planning in order to set the goals to fulfil, at this time they are facing the challenge of the new generations of students: Millennials and Digital Natives who have characteristics different that the ones of generations before and are deeply immerse in the use of new technologies. Professors are updated and trained and listen words like "change", "innovation", "non-traditional teaching" "professors new role", "technological tools" among others and they are encouraged to apply all these theories in planning their courses, at the end professors have the final decision and the results have showed that sometimes are successful but sometimes fail, and impact in the satisfaction of the students with the course so therefore action must be taken to prevent failures and achieve institutional goals. Our institution establish a set of guidelines to be used when implementing this non –traditional courses and promotes de use of Problem based learning (PBL), Case method, Collaborative learning and Project oriented learning (POL).

Research questions

If we change our role in the classroom to become a non-traditional professor that support the students to build their own knowledge and the use of tools that fit with their characteristics, Will students satisfaction with the course be the same than in a traditional course? Moreover, in the implementation, Will be enough to follow the established guidelines for the technique to be successful? The purpose of the study is to determine the perception of satisfaction with the course imparted with the new approach, expecting to have better results than in a traditional course or to get insight in order to improve the teaching process.

Theoretical Framework

García, F., Portillo, J., Romo, J. and Benito, M., (sf) present in their article "Digital Natives and Learning Models" a series of characteristics that outline the learning environment both traditional and non – traditional and establish that it would be the most convenient for the new students who now enter the universities. The authors manage that in the environment of the Classic Model: Knowledge and learning are structured and controlled, using as learning theories behaviorism and cognitivism, where communication is from one to many, learning is linear, memory teaching, centered and managed by the professor where he/she is the transmitter and organizes the course in classes and subjects that promote individualism and competition. The online technology used is Blackboard. In the new model, knowledge and learning is adaptable and dynamic, based on the learning theories of social constructivism and collectivism, where communication is many-to-many, learning takes place in new environments, teaching is built on Knowledge centered on the development of the student and managed by him, where the teacher is the mediator and organizes the course in activities and experiences and that fosters participation and collaboration.

At present, students entering professional studies belong to the generation known as "Millennials" which according to Northern Illinois University, Faculty Development and Instructional Design Center, states that millennials: "Born 1982 through 2000, Millennials are the most diverse generation we have had to teach, so our approaches to teaching must be diverse. In the same reference it is presented: that Millennials expect to be engaged in their learning, they do not act as well being passive learners. If as professors / university do not have technology that will be part of their learning, they will go somewhere else where they can be engaged with, and interactive with, technology. Very important, millennials perceive the sharp contrast between their comfort level of technology and the technology comfort level of their professors. Millenials are also known as "digital natives" for their skills in the management of new technologies, for purposes of this study will be used as synonyms.

Kjermar, F. and Enemark, S. (1996) established that "The design-oriented project work will normally be used for training the necessary knowledge and skills within the disciplines presented. The process and the content of the project work will therefore be organized as fixed standards in advance. The supervisor will teach the students what to do, and the capability of doing it will be trained through the project work.", they establish the steps to follow in order to run the project work process and give guidelines for evaluation and the role of the professor that they called supervisor. Moesby, E. (2006) presents the results obtained of the comparison between 2 universities, Aalborg University that is generally regarded as a POPBL (Project oriented problem based learning) institution and the Technical University of Denmark that is generally regarded as a conventional taught university showed that the personal competences develop in the POPBL approach are significantly higher when compared with conventional techniques, some of the personal competences developed are project and people management, quality of engineering and technical skills, contact and working relations to industry, and innovative and creative skills among others, this research started with his article Moesby (2005) .

There are studies that address the implementation of POL and/or POPBL, Amhed andJabbar (2007) implemented POPBL for students in their first year of Electrical Engineering, they found that the implementation was successful, they compared the final grades of their students against the ones of a course taught in a traditional way, they applied a questionnaire but only to the group under study and present histograms of the answer they obtained to support their affirmation that the learning outcomes obtained improved, but they do not use any statistical tool to test if the differences against the traditional taught group were significant. Ibrahim and Abd.Halim (2013) implemented POPBL in an Introduction of programming course, they established that the results showed a high achievement from the students and consider the POPBL implemented as project-based is suitable for the Engineering course and the students were highly motivated and satisfied with the approach, the authors established that implementation will require a lot of effort from the instructor of the group and maybe the need of an assistant to track individual performance of each student in order to ensure the success of the implementation.

Finally Coronella (2006) implemented POL in the first year of a Chemical engineering course, he considered that the course was successful based on the fact that all the students completed the project assigned and present the comments/complains they expressed at the end of the course. These can be used to improve the implementation of future courses, the author established that the students improve their knowledge of Chemical Engineering but they were somewhat frustrated.

These was very useful in our decision of use POL as the technique for our research and very helpful in the planning of the course.

Methodology

To fulfil the objective of the study we use a single experimental design, the research unit was the two groups of a Database course taught to Engineering students opened in the semester January - May 2016 at a private university in the northeast of the country, the groups were randomly assigned one as "control" and the other "experimental", the control group was taught using the "traditional" approach and the other with the non-traditional, the course was planned mainly including activities that promote lifelong learning, focused in the application of knowledge into a project, and the professor becoming an expert consultant and guide, everything supported in the use of new technologies and based on the POL strategies.

The control group was taught in the traditional way focused on learning the contents of the course through the expository technique and applying this knowledge in a project. Students enrolled in this subject at random, at their convenience, they did not know that the course would be different in each one. In order to be able to consider that the groups are equivalent and therefore comparable, the control group was randomly selected and the other became the experimental group, once defined the control group was assigned in a traditional classroom with support of computer, projector and screen, and the experimental group to a classroom with furniture that facilitates work in groups, computer, projector and the ease of sharing the individual screen content of the student with the whole group. At the beginning of the course the content information, evaluation form, course rules and application project guidelines were delivered to both groups that were taught by the same instructor.

The dependent variable was the satisfaction of students with the course, the independent variables were completely related to the changes implemented to the course that made it different to the "traditional" course. A questionnaire was developed, tested an applied to both groups, the questionnaire consisted of 6 closed end questions with Likert scale from 1 to 5, with 1 being the best answer and 5 being the worst, where they expressed what they considered their perception of their contribution to build their learning (CONTRIBUTION), their understanding of the contents of the course (COMPREHENSION), they also express how much they have applied what they learned to the Project (APPLICATION), their experience working as a team (TEAMWORK), their vision about how traditional and / or innovative consider the course (INNOVATION) and their satisfaction with it (SATISFACTION). An open end question about actions they recommended to improve the course and a set of demographic questions like gender, age, semester, career and whether they currently work were included. The questionnaire was validated by two experts who reviewed whether the questions were pertinent, as well as the scales and wording, based on their comments the questionnaire was modified and the Cronbach Alpha test was used to measure their internal consistency.

The questionnaire was applied twice, the first time almost at the end of the semester and the second at the end of the semester. Data were analyzed using the statistical package SPSS 21, using descriptive statistical measures such as mean, mode, median, frequencies, variance and standard deviation and inferential statistics to determine if there were significant differences in all the questions between the groups, using techniques to compare means such as Chi Square, Oneway ANOVA and Student's "t" the grades of the mid term evaluation, final project, grade and exam of the students were compared too. A file was created to record all the comments that students made about the course throughout the whole semester and the comments were revised and categorized, we took the same procedure for the actions student recommended.

Findings

To meet the research objectives we worked with the 2 groups of the Database course of the January-May 2016 semester. The groups were randomly assigned one as "control" / Traditional and the other as "experimental"/ non-traditional. Below are tables 1 to 3 with statistics for demographic variables by group.

The groups turned out to be uniform with respect to the variables career, semester and work, also they were not uniform in the variable gender, we verified the above using the Chi square test

Regarding age, the mean of the control and experimental groups were the same, 20.21 years meaning they are Millennials/Digital natives, nevertheless it was verified if they were similar through a test of "t" of student and it turned out that there was no difference between them.

Table 1. Gender

		Group		Total
		Control	Experimental	
Gender	Male	23	17	40
	Female	1	6	7
Total		24	23	47

Table 2. Semester

		Group		Total
		Control	Experimental	
Semester	3.00	0	1	1
	4.00	18	16	34
	5.00	3	1	4
	6.00	3	3	6
	8.00	0	2	2
Total		24	23	47

Table 3. Work

		Group		Total
		Control	Experimental	
Work	Yes	9	8	17
	No	15	15	30
Total		24	23	47

Regarding the survey, the results show that the mean in each question is very similar for both groups, based on the scales used the best mean is the closest to 1, in all cases the mean was better for the control group, on the contrary the highest standard deviations are presented by the experimental group. The variables used were: perception of their contribution to their learning (Contribution), their understanding of the contents of the course, (Comprehension), they also consider that they have applied what they learned to the Project (Application), their experience working as a team (Team), their vision about how traditional and / or innovative consider the course (Innovation) and their satisfaction with it (Satisfaction). The internal consistency of the instrument was tested with Cronbach's Alpha which resulted in a value of 0.741 which gives an acceptable reliability. A Oneway ANOVA test with a 95% confidence interval was used to find statistically significant differences between the responses of both groups in the variables and no differences were found in any case. Descriptive statistics for each group are presented in Tables 5 and 6.

Table 4. Statistics perceptions Control

	N	Minimum	Maximum	Mean	Standard deviation	Variance
Age	24	19.00	25.00	20.2083	1.35066	1.824
Contribution	24	1.00	3.00	2.1250	.67967	.462
Comprehension	24	1.00	3.00	2.2083	.65801	.433
Application	24	50.00	100.00	80.2083	19.47569	379.303
Teamwork	24	1.00	3.00	1.7083	.62409	.389
Innovation	24	1.00	4.00	2.4167	.97431	.949
Satisfaction	24	1.00	4.00	2.0000	.83406	.696
Valid N (listwise)	24					

Table 5. Statistics perceptions Experimental group

	N	Minimum	Maximum	Mean	Standard Deviation	Variance
Age	23	18.00	23.00	20.2174	1.04257	1.087
Contribution	23	1.00	3.00	2.2609	.54082	.292
Comprehension	23	1.00	4.00	2.5652	.72777	.530
Project	23	.00	100.00	70.6522	25.73044	662.055
Team	23	1.00	4.00	2.1304	1.09977	1.209
Innovation	23	1.00	4.00	2.2083	2.03623	4.146
Satisfaction	23	1.00	4.00	2.0870	1.04067	1.083
Valid N (listwise)	23					

As a first approach to the objective variables we compared the mean of the grades of the first partial for each group, for the control group the mean was 89.16 and for the experimental 87.83, the Student's "t" test showed that there was no significant difference between them. At the end of the semester the grades of the final project, final grades, and final exam were compared, it should be clarified that this examination is institutional applied by the department where the course belongs. For the comparison we used Oneway ANOVA and the results show that there are no statistically significant differences for any of the variables. The averages are presented in Tables 6 and 7.

Table 6. Descriptive statistics Control group

	N	Media	Standard Deviation	Variance
Mid term grade	24	89.1667	4.08248	16.667
Final exam	24	69.6250	13.59288	184.766
Final project	24	91.0417	8.46722	71.694
Final grade	24	85.0833	6.90253	47.645
Valid N (listwise)	24			

Regarding the final surveys, the response rate was low, only 29.2% and of the received only a lower percentage of them were from the experimental group, this does not allow a comparison of the final perceptions between both groups.

Table 7. Descriptive statistics Experimental group

	N	Media	Standard Deviation	Variance
Mid term grade	24	87.8333	3.66733	13.449
Final exam	24	63.2083	20.33840	413.650
Final project	24	93.3333	6.37022	40.580
Final grade	23	85.5217	5.49272	30.170
Valid N (listwise)	23			

Groups have a very similar distribution in size, age, semester, career and gender. The mean of each question was compare between the groups, the results of perceptions of the students about satisfaction, team work, comprehension of contents of the course, contribution to their own learning and application of knowledge to the project were tested using ANOVA, it seems that there are not significant differences, final grades, final exam and project were compared too and the results were the same, no differences. Even though there were no differences, comparing the means of each item, the dispersion was bigger for the "experimental" group, and the means tend to be bigger for the "control" group. With respect to the comments about the course, only the students of the experimental group mainly expressed not being satisfied with their grades, because even though they worked hard it was not reflected in their grades. There were no recommendations to improve the course.

Conclusions

As we already established the purpose of the study is to determine the perception of satisfaction with the course imparted with the new approach, expecting to have better results than in a traditional course or to get insight in order to improve the teaching process. The findings presented in this first approach showed that the perception of satisfaction of students in the non-traditional course does not have significant difference with the perception of students in the traditional course, this led us to consider that a possible explanation was directly related to resistance to change, the control group was calm and accustomed to the methodology of the course, whereas the experimental group faced the change, moreover it seems that follow the guidelines established by both the POL technique itself and our institution is not enough to plan a successful course.

Based on these results a set of remarks was generated trying to put all the findings together and that will be used to improve the planning and implementation of future Database courses as well as the teaching process:

1. It seems that the students in the experimental group perceive the course as "partially innovative" that is good because that is what it was expected, but the result is the same for the control group, maybe is not clear for them the meaning of innovative or traditional course.
2. The strategies of POL requires more effort of the student in order to build their long life learning, students of the "experimental" group complained throughout the semester about the amount of work required for the course was too high, they are not accustomed to work in this way, but rather to receive the content of the course basically of the explanations of the professor, students of the "control" group have no reason to complain.
3. It seems that both groups consider that their participation in building their knowledge is high, and the students of both groups perceived that their comprehension of the content of the course is between high and medium in addition there is no significant difference in the media of the grades obtained in the project, in the final exam and the final grades of both groups, it seems that the "control" group obtained knowledge enough to thrive in the course even though they do work as hard as the "experimental" group, but they perceive they did it.. For the "experimental" group, they really work hard but it seems that the comprehension of the content was enough but harder to get but at the end similar than the comprehension of the "control" group.
4. According to the results about the application of the content of the course to the project, once again there is not significant differences between the groups, but the percentage of application was rated higher for the "control" group, the expectation was that the perception of the "experimental" group will be higher than for the "control" group. A possible explanation should be related to the fact that the project was made just in paper, they do not have a real client to present their proposal. A very important fact emerged during the semester, students have never taken a course in Project Management.
5. The fact that there were no significant differences in the final grades, the project and the final exam lead us to indicate if the level of comprehension of the contents of the course achieved is the same for the two groups or the form of evaluation of the course is not the most adequate.
6. The satisfaction of students with the course is "high" in both groups but the dispersion in the "experimental" group is much higher than in the "control" group so it seems that there is internal disagreement, there are students very satisfied with the course and others not so much, the students of the experimental group complained throughout the semester that even they work too hard, the grades did not reflect their efforts, probably this have an impact in their evaluation of satisfaction with the course, and once again take us to consider alternative ways of evaluation.

Recommendations and implications

As a result of all the remarks presented, a series of recommendations should be taken into account when planning and implementing a non - traditional course based on POL that uses innovative strategies and seeks lifelong learning and a high level of satisfaction for the student.

It seems that the students in a "traditional" course are accustomed and can thrive easily in it, they know what they have to do, the load of work required, and fulfil due dates, attendance and team work, when these students face a non-traditional course, is a challenge because even though these courses are oriented to fit with the characteristics of their generation of Millennials/Digital natives, they have to face the series of changes involved in these courses and they come in general from taking traditional courses, maybe not the best for them, but know how to manage themselves in them.

The following recommendations will be so useful to improve non – traditional courses and must be seen as a key points to be taken into account:

1. Planning should include time for project management learning, a handbook with the guidelines to manage the Project must be created and will be a good idea to invite an expert to briefly explain the principles of Project Management and its importance.
2. The planning should explicitly include activities that will enable the student to know what he/she is doing and why he/she is doing it. The professor is responsible for keeping the student aware of what he/she is doing throughout the semester.
3. The project must solve a real situation of a formal company, which should give feedback to the student. The responsibility of getting real clients should be shared by the teacher and the students, they should know from the first day of classes that they should look for a company with which to work during the semester, at the same time the teacher should look for possible options to offer to the students.
4. It is necessary to generate a new proposal of evaluation of the course and its activities in a way that reflects in the best way the work done by the student.
5. The teams must be integrated by the professor and have representation of the 3 engineering careers in which they are commonly enrolled and that are: innovation and development, management informatics and computer science.

We met the objectives of the research and at first glance and after reviewing the results obtained, it would seem pointless to devote time and resources to generating a non-traditional course if it does not seem to achieve appreciable improvement for students, this may be the reason why those who have previously attempted to implement non-traditional courses with negative or not appreciable results even though they followed the guidelines marked by the innovative tools used, prefer to continue teaching with the traditional way, but on the contrary, the experiment revealed and gave an understanding of the phenomenon under study as well as a series of insights that open the panorama to new investigations. We must remember that these tools are necessary to achieve our first objective as educators and that is to achieve lifelong learning in the student for that is worth to work on the implementation of a non-traditional course that is success.

We must have remember that we are facing the challenge of teach students of the "digital natives" generation, who study engineering that gives them different characteristics in comparison with students of Marketing or Graphic Design and they are Mexican. This opens a wide field of research, in this first approach, a series of recommendations were generated that were already implemented in the creation of a new non-traditional course and which is expected to generate great improvements in students' perceptions about their learning and to positively impact the satisfaction of students. It will be interesting to determine a model that present the weight of the different variables and their impact in satisfaction, in order to know the key points where to allocate the available resources in an optimal way, and finally determine if the use of this innovative strategies that promote lifelong learning really achieve their objective and the knowledge acquired lasts longer than the one achieved through the methods we call traditional.

We consider very important to express a comment from the instructor in charge of the course, he states that in his perception the traditional group presented the attitudes and behaviors that at the last dates have become recurrent in the students and that are cause of complaints on the part of the professors. We refer to lack of attention, continuous distractions, lack of motivation and commitment and even to have the physical presence of the student but with his/her mind elsewhere. These attitudes and behaviors were greatly diminished in the non-traditional group that showed an improvement in concentration and involvement, generating an environment that allowed an open and close interaction with the students. Therefore, we consider relevant to share it and attribute this improvement to the changes implemented in the form of teaching. Based on the above we recommend that professors implementing these type of courses must be aware that this change possibly happen and seek to foster these new attitudes and behaviors.

Finally, for the professors, it is very important that they put all their efforts in counseling students. but above all, accepting their new role. At the moment the course is based on POL, their role changes radically, from being the center of the process where students are passive, to putting students at the center of the process, becoming facilitaors to make each student take the active role of being part of a team, to work collaboratively, to contribute to the construction of their own knowledge and work hard to deliver a project developed in a way that truly reflects the knowledge achieved. We must be the first to assimilate change and be aware that our actions rmust be consistent with this new role to endure the achievement of better outcomes in students' performance and support the fulfillment of institutional goals.

References

Coronella, C.. (2006). Project-based learning in a first-year chemical engineering course: Evaporative Cooling. Febrero 23, 2017, de American Society for Engineering Education. Sitio web: https://www.engr.uky.edu/~aseeched/papers/2006/1454.pdf

García, F., Portillo, J., Romo, J., & Benito, M.. Nativos digitales y modelos de aprendizaje. Febrero 23, 2017. , de Universidad de País Vasco. Sitio web: htt p://spdece07.ehu.es/actas/Garcia.pdf

Kjersdam, F., & Enemark, S.. (1997.). The Aalborg Experiment. Project innovation in University Education. Denmark: Aalborg University Press. p. 23.

Moesby, E.. (2005). Curriculum Development for Project-Oriented and Problem-Based Learning (POPBL) with Emphasis on Personal Skills and Abilities*. Global J. of Engng. Educ., 9, 121- 128. 2017, Febrero 23, De School of Basic Studies in Science and Engineering, Aalborg University Base de datos. Sitio web: http://www.wiete.com.au/journals/GJEE/Publish/vol9no2/Moesby.pdf.

Moesby, E.. (2005). Personal skills and abilities in curriculum development planning for Project Oriented and Problem-Based Learning (POPBL). UNESCO International Centre for Engineering Education (UICEE), 155 - 158. 2017, febrero 23, De Esbjerg Institute of Technology, Aalborg University Base de datos.

Rouse, M.& Haughn,M.. (2015). Millennials (Millennial generation). Febrero 23, 2017, de TechTarget. Sitio web: http://whatis.techtarget.com/definition/millennials-millennial-g eneration

Evaluation of Methods and Algorithms of Educational Data Mining

Oswaldo Moscoso-Zea
Universidad Tecnológica Equinoccial, Facultad de Ingeniería, Quito, Ecuador
omoscoso@ute.edu.ec

Mayra Vizcaino
Universidad Tecnológica Equinoccial, Facultad de Ingeniería, Quito, Ecuador
vrme94810@ute.edu.ec

Sergio Luján-Mora
University of Alicante, Department of Software and Computing Systems, Alicante, Spain.
sergio.lujan@ua.es

Abstract: *Educational data mining (EDM) is an evolving discipline that allows the creation and exploration of knowledge from academic environments by means of developing and applying data mining (DM) methods and algorithms to information stored in data repositories of higher education institutions. The results of the application of these methods and algorithms allows these institutions to better understand the way the lecturers teach, the way the students learn and the activities of organizational processes to improve decision making. This paper describes DM, EDM and the existing methods and algorithms of the discipline. Furthermore, it presents the experiments carried out for the evaluation of methods and algorithms applied to two key performance indicators in a private university: student dropout and graduation rate. Finally, it compares these methods and algorithms and suggests which has better precision in certain scenarios.*

Introduction

Higher education institutions (HEI) are generating large amounts of data from their organizational systems and applications. These data are not always processed and analyzed accordingly to produce information and knowledge. The production and dissemination of organizational knowledge is a strategic objective that supports HEI in the roadmap for planning, modernization and improvement of academic and research indicators (Baepler & Murdoch, 2010).

The setting of a technological infrastructure is necessary for a sound process of data analysis. One of the core technologies of this infrastructure is a data warehouse (DW). A DW is a data repository with a multidimensional design and used specifically for analysis (Moscoso-Zea, Sampedro, & Luján- Mora, 2016). The information dispersed in different operational data bases is migrated to the DW by means of an extraction, transformation and loading (ETL) process. An approach that guides this knowledge creation process is called knowledge discovery in databases (KDD). KDD uses data mining (DM) as the core element for knowledge creation. DM has been applied to different industries and fields of study in the last years with promising results. Some of the fields of analysis of DM are marketing, health, finances and insurance, among others. The application of DM in educational contexts is known as educational data mining (EDM). EDM is a discipline in evolution that focuses in the design of models to improve learning experiences and organizational efficiency (Huebner, 2013). EDM uses software tools to discover trends and patterns on educational data to improve decision making in HEI.

There are many institutions that are applying EDM and are gaining insights to improve students, lecturers and staff performance. Some examples of what institutions are doing in this field of study are the following. Paul Smith's College uses learning analytics to increment the graduation rate of their students. Washington University is associated with "Persistence Plus" company to improve the evaluation of online courses by using data analytics. Open University uses its historic data to improve the retention rates of its students (Bichsel, 2012). Georgia University carry out experiments using analytic techniques to predict graduation rates and student dropout in online courses (Morris, Wu, & Finnegan, 2015). In Purdue University researchers have been using DM to determine that frequent evaluations in early stages can change the habits of students with low grades; the research team of this university has developed an academic alert system to track the performance of students (Baepler & Murdoch, 2010).

The main motivation for writing this paper is to continue exploiting EDM to discover hidden knowledge in academic data. Therefore, this paper presents the different methods and algorithms of EDM that have been developed by researchers. Furthermore, it describes the experiments carried out to analyze two key performance indicators (KPI) in a private university: student dropout and graduation rate. Classification methods and algorithms of EDM are applied in the analysis. In addition, this study also compares these methods and algorithms and suggests which has better precision in certain scenarios. The conclusions of this paper can support data scientists to choose the right methods and algorithms in further studies of these KPI.

Therefore, the research question of this work is:

"Which are the best methods and algorithms of EDM that can be used to analyze student dropout and graduation rate?"

The answer to this question can be fundamental for researchers in education to reduce the time of experimentation and analysis of student dropout and graduation rate. This paper hopefully can be used to have a clear vision in which methods and algorithms are more adequate to study these academic indicators.

Background

In this section, the state of the art of the EDM topic along with the existing methods and algorithms of EDM that are needed for the design of the experiments are described.

Data Mining

The term DM was coined in the 1990s by researchers in the information systems and databases field. Academics in the literature also use terms as "Data Archeology", "Data Collection", "Knowledge Extraction" and "Data Analysis" when referring to DM. DM is a fundamental part in the process of KDD. DM is an approach that uses different information technologies (IT) systems and tools to analyze and extract knowledge from information contained in data repositories of organizations. The most commonly used framework for understanding the life cycle of a DM project is the Cross Industry Standard Process for Data Mining (CRISP-DM) (Chapman et al., 2000). The main phases, as shown in Figure 1, are: understanding the business, understanding data, preparing data, creating models, evaluation and deployment.

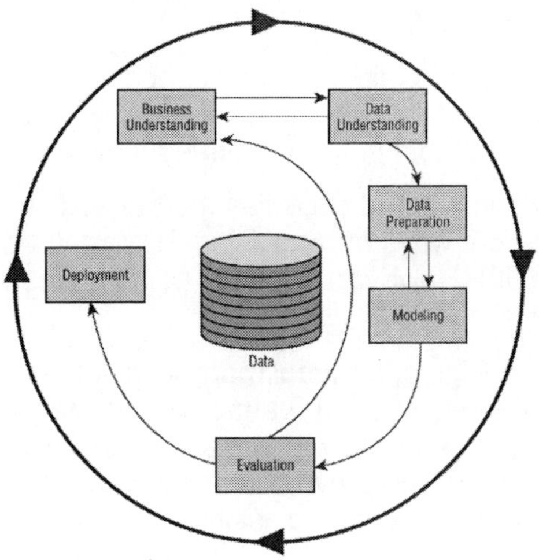

Figure 1: Phases of CRISP-DM. Source: (Chapman et al., 2000)

Educational Data Mining

EDM uses methods, tools and algorithms of DM to investigate data from students, teachers and administrative staff, collaboration between students, administrative data and demographic data of HEI. Hence, EDM is an evolving discipline, there is not a widely accepted definition yet. However, a definition that fits the objectives of our research is provided by the International Society of Educational Data Mining: "EDM is an emerging discipline, concerned with developing methods for exploring the unique and increasingly large-scale data that come from educational settings, and using those methods to better understand students, and the settings which they learn in" (International Society Educational Data Mining, 2015).

The knowledge discovery with EDM process is shown in Figure 2. This figure depicts in the first step the preprocessing of raw data obtained from an educational environment. These raw data is then modified (a new data set is created) and used with EDM methods or algorithms. As a following step a model is defined and the experiments are carried out. The results of experimentation allows the evaluation and refinement of the process with the results of the analysis (Ventura & Romero, 2013).

EDM explores the data in the organizational context of a HEI to transform these data into useful knowledge. This is one of the reasons that it is key in the study and improvement of academic processes as desertion rate, graduation rate, restructuration processes and organizational management (Bienkowski, Feng, & Means, 2012).

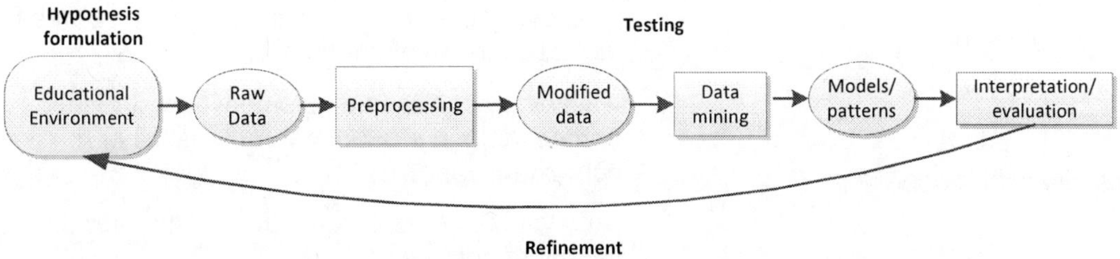

Figure 2: Knowledge Discovery with EDM. Source: (Ventura & Romero, 2013)

Methods and Algorithms of EDM

The field of EDM integrates methods, algorithms and techniques which can be used to perform different experiments and to design models. The output of the model implementation allows researchers to predict or obtain patterns from educational data.

This section presents the algorithms and classifies the methods used by researchers in this scientific field. There are different ways of classifying EDM methods (Kumar & Vijayalakshmi, 2011) (Mitra, Pal, & Mitra, 2002). However, the most popular classification is presented by Siemens & Baker (2012):

- Prediction: Is used in the design of a model that allows inferring in some aspects of the data based on the combinations of other features of the data, for example, information of student dropouts can be collected and with the analysis of this information, predictions can be made to take corrective and preventive actions targeted to new students. There are three types of methods in this group: classification, regression and latent knowledge estimation.
- Structure Discovery: Is a question of finding a structure for the data without a prior idea of the solution. The researcher main task is to identify the natural structure of the data. Within this classification are: clustering, factor analysis, social network analysis and discovery of domain structures.
- Relationship Mining: Is useful to discover relationships between certain variables in a data set. Within this category we can mention: association, correlation, sequential pattern mining and causal mining of data.
- Model Discovery: The results of the mining analysis are used as inputs for further analysis. Normally a model is obtained through prediction methods.

DM methods and the key areas of application in education are presented in Table 1 (Kumar & Vijayalakshmi, 2011).

Table 1: Educational Data Mining Algorithms and Applications

Method	Applications in Education
Prediction	-Detecting student behaviors -Developing domain models -Predicting and understanding student educational outcomes
Clustering	-Discovery of new student behavior patterns -Investigating similarities and differences between schools
Relationship Mining	-Discovery of curricular associations in courses and sequences of courses
Discovery with models	-Discovery of relationships between student behaviors, and student characteristics or contextual variables -Analysis of research question across a wide variety of contexts
Distillation of Data for Human Judgment	-Human identification of patterns in student learning, behavior, or collaboration -Labeling data for use in later development of prediction models

In the international conference on DM of 2006 sponsored by the Institute of Electrical and Electronics Engineers (IEEE), a ranking of the eight best DM algorithms was presented. These algorithms are:

- Decision Trees: Organize data forming branches of influences for decision making. The tree trunk represents the initial decision. This decision starts with a yes and no question for example if the student will graduate or not. The next divergent branches are graduation and no graduation and each further election should have their own divergent branches that conduct to an end point.
- K-Means Algorithm: Based in group analysis. Divides data collected in clusters separated by common characteristics.
- Apriori Algorithm: Controls transactional data. For example the algorithm could predict which products might be bought together in a store.
- EM Algorithm: Define parameters by analysing data and predicts the possibility of a future output or a random event within the parameters of data. For example: EM algorithm could intent to predict the time of the next eruption of a volcano analysing the data of previous eruptions.
- Page Rank Algorithm: Created for search engines. Ranks and estimates the relevance of a piece of data within a big set of data. An example is the ranking of a web site within a big set of all the web sites in the internet.
- Adaboost Algorithm: Anticipates the behavior using observed data in order to be sensitive to statistical extremes.
- Nearest Neighbor Algorithm: Recognizes patterns in the location of data and associates the data with a bigger identifier.
- Naive Bayes Algorithm: Predicts the output of an identity based in the data of known observations. For example if the height of a person is 1.90 meters and the size of the shoes is 14, the algorithm could predict with a determined probability that the person is a man.

Methodology

The methods used in this research are described in this section: KDD for the knowledge discovery process and CRISP-DM for the DM process.

Knowledge Discovery in Databases

One of the methods used in this research is KDD. KDD is an approach to discover useful knowledge from a group of data. This process is shown in Figure 3. It is composed of the following phases:

1. Data selection: Selection of source data from operational databases and the migration to a target data repository, in this case a DW.
2. Data preprocessing: Cleansing and preprocessing of data by deciding strategies to put the data in the right format, removing duplicates and handling missing fields.
3. Data transformation: Creating data sets with the needed variables for reducing complexity of analysis.
4. Data mining: Application of methods and algorithms to the data set in order to predict trends and discover patterns in data.
5. Interpretation and evaluation: Understanding the results and the creation of explicit knowledge by means of visualization of data in reports and dashboards.

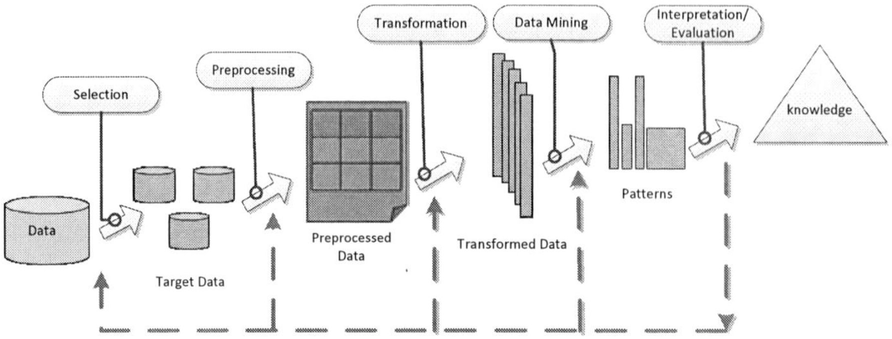

Figure 3: Knowledge Discovery in Databases (KDD)

CRISP- DM

The implementation of DM within the KDD method in this study was performed following the CRISP-DM phases. These phases are:

1. Understanding the business: Comprehension of the mission, vision and goals of the business and how the DM project will benefit the organization. In this phase, it is important to have clearly identified the requirements of the project.
2. Understanding data: Identifying the data tables and fields that will be subject of analysis.
3. Preparing data: Migrating the required data to a data set that will be used in the analysis. A data cleansing process must be performed during this transformation.
4. Creating models: Designing and planning the model that will be used for analysis.
5. Evaluation: Experimenting and selection of algorithms and tools. The output of this phase is the knowledge created with the existing model.
6. Deployment: Presentation of the results. If the results are not relevant to the requirements, a new model should be proposed.

Approach and Findings

This section shows the development of the experiments using the KDD and the CRISP-DM method. It analyzes the academic information of HEI. The primary analysis is realized to discover and predict trends in two indicators: Student dropout and graduation rate. The data analyzed was limited to records of computer science students. The period of analysis was from the year 2002 until the year 2015. There were two groups for data analysis: The students who entered the university since the first semester and the students who came from other universities and entered the university validating courses and subjects.

The data source contains information collected when entering the institution from the academic system of the university (personal and educational data) and data that are collected during the duration of the studies. Data selection and preprocessing are performed using different criteria for the representation and application of classification algorithms such as decision trees, bayesian networks and decision rules. Two influential classes are defined for the analysis: Graduation and desertion of students. These indicators of students are evaluated according to data of enrollment, for example: If the student does not enroll for more than two consecutive semesters, it is considered in desertion as indicated in the regulation of the boards of academic evaluation, accreditation and quality assurance of higher education in the country of the university (Ceaaces, 2015). This evaluation board is a public entity that carries out continuous evaluation and accreditation processes. The detailed process of this research is shown in the following KDD steps.

Data Selection

The data used for this analysis was previously collected in a DW of the HEI. DM could be done directly from this data repository, however, the process of structuring the views for analysis of the data was increasingly complex. The decision to make the process efficiently was to create an external table to the DW with all the fields necessary for the analysis of student dropout and graduation rate. In this step the data from the DW from the year 2002 to the year 2015 was analyzed. The focus group for the analysis was the students of the computer science faculty. The total number of students in this group was 441 students. From these students, 330 entered the university from the first semester, and the remaining 111 students came from other universities by means of validating courses and subjects.

Data Preprocessing

After having the information of the focus group of analysis, the data was cleaned, verifying formats and checking that information was correct. This process was performed using Microsoft SQL Server Integration Services and Microsoft Analysis Services.

Data Transformation

In order to create the dataset, an ETL process was performed. Data was extracted from the DW. Different dimensions and fact tables were the source of these data. The transformation process was a less complicated activity due to the fact that a transformation process was previously performed for the creation of the DW. However, different SQL operations (aggregation and normalization) were executed in the information in order to structure the data set fields with yes, no or 1, 0 in order to facilitate further analysis. Once the data were transformed, they were loaded into the new created table (dataset). The data set was the main input of DM in the different analysis tools used in this investigation, as shown in Figure 4. This data set contained information of names, gender, desertion, date of admission, date of graduation, among others.

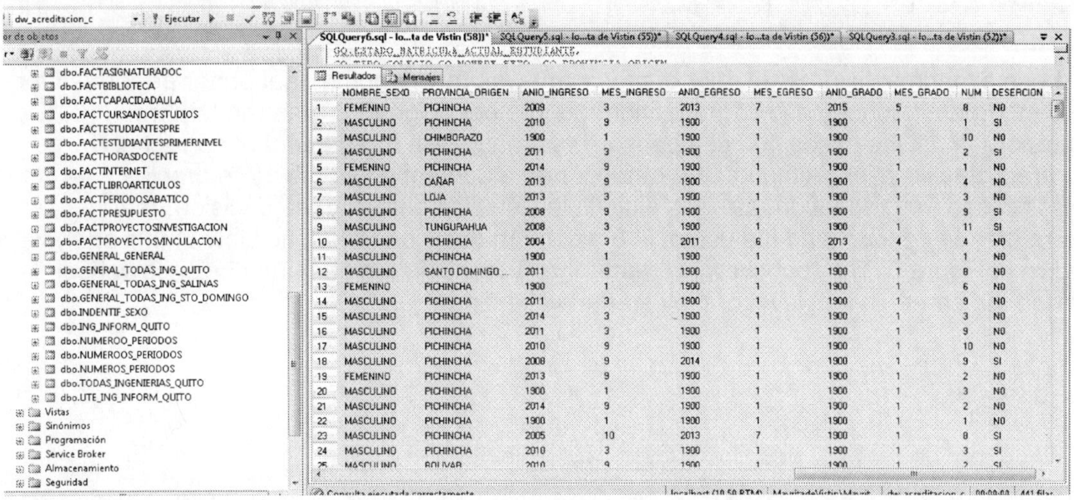

Figure 4: Dataset for analysis

Data Mining

The DM process starts with the creation of a comma-separated file from the resulting dataset. This file is the input for the three tools used in our experiments: WEKA, Orange 3 and Rapid Miner. After performing a feasibility analysis realized to these tools, WEKA was selected as the most adequate for this experiment.

Interpretation and Evaluation

As explained previously, in the IEEE conference of 2006 it was stated that one of the best methods for EDM is classification (decision trees, Naïve Bayes, meta classifiers), therefore, this method is implemented in this analysis. Many algorithms were tested using supervised and non-supervised filters applied to the data set. WEKA works with different classifiers that can be chosen in the tool. In this study the following classifiers were used: Naïve Bayes, Stacking, One R, J48, and Random Tree. The precision of the classifiers applied to student dropout is shown in Table 2. The kappa index is a measure between the categories predicted by the classifier and the categories observed, the absolute error indicates the percentage of error that may exist in the classification

Table 2: Comparison of different algorithms within student dropout

Algorithm	Classifier	Test Mode	Well ranked instances (%)	Badly-ranked instances	Kappa Index	(%)Absolute Error
Naïve Bayes	With 8 data set attributes	Percentage division	71.33%.	28.66%	0.28	0.31%
Stacking	With 8 data set attributes	Percentage division	72.00%	28.00%	0.00	0.40%
One R	With all data set attributes	Cross-validation	72.00%	28.00%	0.02	0.28%
J48	With 8 data set attributes	Percentage division	99.09%	0.90%	0.97	0.01%
Random Tree	With all data set attributes	Cross-validation	74.00%	26.00%	0.33	0.27%

In these experiments, J48 (99.09% of precision) was the algorithm that better performed for student dropout. In the case of graduation rate the best algorithm was random tree (100% of precision). In all the cases, the comparison was made with the percent of correct and incorrect classification. Although the random tree algorithm gives the more precise classification for graduation rate, it is difficult to visualize correctly the decision tree. Therefore, we recommend the use of J48 algorithm decision tree. One view of this analysis is shown in Figure 5. The root shows whether the students in the computer science school started the career in first level or they started validating courses.

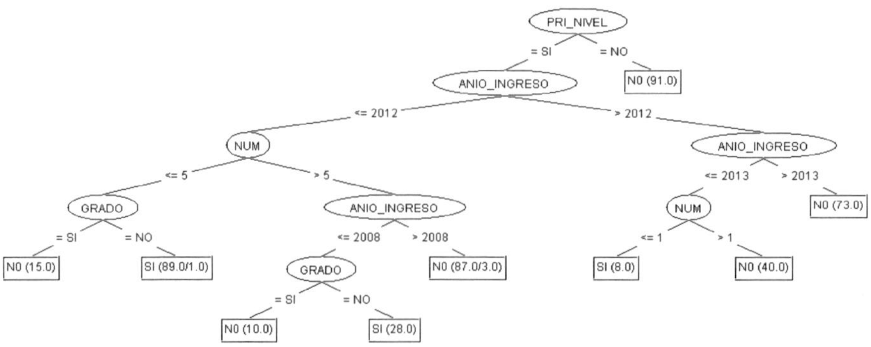

Figure 5: Tree view of J48 analysis for dropout

The discoveries of knowledge from the experiments with the EDM are: Graduation rate is higher for students born and living in the same city and for students that have entered the faculty validating courses. Students that studied high school in public institutions, married students and those that lose a grant have higher risk to dropout

CONCLUSIONS

This paper presents a view of EDM and hopefully it can be a good source for researchers wishing to experiment with EDM in areas such as evaluation, enrollment, planning, student welfare, marketing, etc. EDM is projected as an essential discipline for university management that gives visibility to managers to improve decision making. By using DM tools as WEKA, an evaluation of EDM methods and algorithms was performed. This analysis shows that decision trees algorithms J48 and random trees have better precision in classifying students in the analysis of student dropout and graduation rate. Some potential limitations of the work are that the analysis is done only to classification algorithms.

The knowledge created with these experiments allows the institution to identify students in risk of dropout in early stages to implement corrective actions. It also gives information on the groups that are more likely to graduate and the groups that are not. With the output of the experiments carried out, new strategies can be implemented to improve academic indicators.

References

Baepler, P., & Murdoch, C. J. (2010). Academic Analytics and Data Mining in Higher Education. International Journal for the Scholarship of Teaching and Learning, 4(2), 1–9.

Bichsel, J. (2012). Analytics in Higher Education Benefits, Barriers, Progress and Recommendations. Retrieved May 3, 2016, from https://net.educause.edu/ir/library/pdf/ERS1207/ers1207.pdf

Bienkowski, M., Feng, M., & Means, B. (2012). Enhancing teaching and learning through educational data mining and learning analytics: An issue brief. Department of Education's (ED) Office of Educational Technology, 1–57.

Ceaaces. (2015). Modelo genérico de evaluación del entorno de aprendizaje de carreras presenciales y semipresenciales de las universidades y escuelas politécnicas del ecuador. Quito.

Chapman, P., Clinton, J., Kerber, R., Khabaza, T., Reinartz, T., Shearer, C., & Rudiger, W. (2000). CRISP-DM 1.0. CRISP-DM Consortium, 76.

Huebner, R. A. (2013). A Survey of Educational Data-Mining Research. Research in Higher Education Journal, 19, 1–13.

International Society Educational Data Mining. (2015). Educational Data Mining. Retrieved June 29, 2016, from http://www.educationaldatamining.org/

Kumar, S. A., & Vijayalakshmi, M. N. (2011). A Novel Approach in Data Mining Techniques for Educational Data. In 3rd International Conference on Machine Learning and Computing (ICMLC 2011) (pp. 152–154).

Mitra, S., Pal, S. K., & Mitra, P. (2002). Data mining in soft computing framework: A survey. IEEE Transactions on Neural Networks, 13(1), 3–14.

Morris, L., Wu, S.-S., & Finnegan, C. (2015). Predicting retention in online general education courses. American Journal of Distance Education, 19(1), 23–36.

Moscoso-Zea, O., Sampedro, A., & Luján- Mora, S. (2016). Datawarehouse design for Educational Datamining. In 15th Information Technology Based Higher Education and Training (ITHET 2016) (pp. 1–6).

Siemens, G., & Baker, R. S. J. D. (2012). Learning analytics and educational data mining. In 2nd International Conference on Learning Analytics and Knowledge(LAK 2012) (pp. 252).

Ventura, S., & Romero, C. (2013). Data mining in education. WIREs Data Mining Knowledge Discovery, 3(1), 12–27.

An Inductive-Deductive-Inductive Pedagogical Learning Approach for Mechanics of Materials

Jorge O. Montoya
Universidad De Ibagué, Ibague, Colombia
Email address

Heidi A. Diefes-Dux
Purdue University, West Lafayette, IN, USA
hdiefes@purdue.edu

***Abstract**: An inductive-deductive-inductive pedagogical approach was taken in a mechanics of materials course. The first inductive phase consisted of visualization and experimentation with a simple physical model. The second inductive phase consisted of problem solving and physical model development. The two inductive phases were bridged with a more deductive development of the constitutive equations. The implementation of this approach in a course that previously only used lecture resulted in a significant increase in the student passing rate and decrease in the number of withdrawals. The importance of the three phases is discussed.*

Introduction

The development of conceptual understanding of and practical application of phenomena associated with mechanics of materials are fundamental to students' academic training and their subsequent professional performance in a number of engineering fields, including structures in the civil engineering. Students' acquisition of the basic knowledge of stress and strain relationships in a Mechanics of Materials course presents a challenge for engineering professors and students due to the highly analytic and theoretical content that often appears to students to be in opposition to their lived experiences with objects (Brown, 2013). It is well documented in the literature that students have misconceptions about these phenomena (e.g., Montfort, Brown, & Pollock, 2009). Students' struggles to understand and demonstrate proficiency with the content may be a factor in high failure and dropout rates from mechanics of materials courses. Low academic performance in and a high dropout rate from a mechanics of material course were the drivers for this study.

Mechanics of materials must be understood as an integrated collection of its parts. All sub-topics must relate to the global connection of stress and strain. Observations and studies have demonstrated that students have difficulty observing this global connection and even the brightest students have difficulty remembering some concepts shortly after taking a course (Burns & Egelhoff, 2011). Most students have little to no practical experiences with the theoretical-mathematical content. Theoretical and mathematical representations require comparison, as well as physical and visual proof, to guarantee and favor adequate knowledge appropriation (Roylance, Jenkins, & Khanna, 2001).

Mechanics of Materials courses in most engineering programs, including civil engineering, employ the traditional method of classroom lectures. Yet, evidence points to the classroom lecture approach in engineering courses being ineffective - not leading to the development of advanced problem solving skills, not generating creative or critical thought, and not preparing students for the types of problems they will confront during professional life (Johnson, 1999). Further, theory, presented on its own, has led to lower levels of comprehension and motivation and, consequently, higher rates of failure and withdrawal (Mahendran, 1995). In moving towards more active approaches to instruction, a balance must be found between students' active participation in their knowledge acquisition and the professor's role in and out of the classroom. In finding this balance, it is necessary that the professor consider that each student has a preferred way of learning, particularly visualization and construction of models (Felder & Silverman, 1988). A balance must be struck in instructional methods between engaging students in real-life situations, theories, and mathematical models, thus permitting students to gain and demonstrate comprehension within and across the different representations. That is, it is recommended there be a balance between concrete and abstract information conveyed in a course (Felder, Woods, Stice, & Rugarcia, 2000).

The intervention used in this study had two parts: the inductive-deductive-inductive active learning pedagogical approach used in the classroom instruction and a team project that continued the inductive pedagogical approach. An inductive approach to teaching and learning starts with students making observations and experimenting; the instructor then guides students to general principles. Whereas, a deductive teaching and learning approach starts with instructors presenting the general principles and then moving on to applications of those principles. Engineering instruction has a long tradition of being deductive in nature, with the instructor at the center of learning (teacher-centered). Inductive instruction, in contrast, puts students at the center, requiring them to fit new information into their cognitive structures. Inductive instructional methods all involve active learning (e.g., discussion and problem solving) and collaborative or cooperative learning (working in groups) (Prince and Felder, 2006). Such instructional methods are grounded in cognitive constructivism (originating with Piaget, 1972), which says learning is the result of processing ones experiences, and social constructivism (e.g., Vygotsky, 1978), which focuses on language and interactions with others as a means of making sense of one's experiences. Evidence that inductive teaching and learning approaches have positive impacts on students' education are convincing. For example, Freeman et al. (2014), in a meta-analysis of 225 studies of traditional lecturing versus active learning in undergraduate science, technology, engineering, and mathematics (STEM) courses, found that average examination scores significantly increased and failure rates decreased.

In the context of this study, it was evident that a new pedagogical approach was needed in the Mechanics of Materials course to promote deeper learning of the course content. The overarching goals of adopting a new pedagogical approach were to reduce the number of students failing and dropping the course, increase the students' conceptual and analytical capacity, and improve academic performance in advanced courses in the area of structures in civil engineering. In keeping with an inductive approach, it was desired that students take a greater lead in their learning process and the professor more often assume the role of facilitator and guide, without giving up conceptual and mathematical rigor.

Research Questions

In this study, we sought to quantitatively determine whether an inductive-deductive-inductive learning pedagogical approach used in a Mechanics of Materials course resulted in a change in the percent of students passing and dropping the course and the academic performance of the students completing the course. In this study we also sought to understand how much time the professor and students were active when the inductive-deductive-inductive active learning pedagogical approach was used in the classroom.

Methods

Setting and Participants

This study was conducted in a typical mechanics of materials course in a 5-year civil engineering program offered at a small private university in Colombia. Students enrolled in this course were in their fifth semester of the program. The two course sections offered in 2011B and 2012A, where A and B represent first and second semester of the academic year respectively, served as the Control for this study. Treatment 1 (described below) for this study occurred in the three sections offered in 2012B, 2013A, and 2013B. Treatment 2 occurred in the five sections offered in 2014A (two sections), 2014B, (one section), and 2015A (two sections). Section size ranged from 29 to 45 students. The total enrolment across the treatment sections was 297 students. There were 82 students in the Control group, 114 in Treatment 1, and 183 in Treatment 2.

This course met three times a week for 120 minutes each. There was no formal lab activity for this course. For the Control, the course was taught in a teacher-centered fashion; class time was spent in lecture and in application exercises, mainly done by the instructor with students taking notes and asking for clarification. For the Treatments, the course was student-centered, with the role of the teacher being primarily facilitator. Lectures were given at the beginning of a new topic. For the Control, the final course grade was in three parts: 60% for written exams (one for each third of the semester), 10% for independent problem solving (during the whole semester), and 30% for a final theoretical group project, done mainly in the last 4 weeks of the semester. For the Treatments, the final course grade was divided into three parts: 50% for written exams (one for each third of the semester), 20% for class participation and team problem solving, and 30% for the final team project. The written exams were similar across the Control and Treatment groups, typically 1 or 2 problems. The Treatment team project began in the second week and was submitted in the last week of the semester.

Three main topics (stress, strain and axial load; torsion; and flection stress and strain) linked to seven phenomena or material properties were selected to be taught using the intervention described in detail below. Other course topics (e.g. flexural moment, shear force diagrams, and stress transformation) were taught in a traditional fashion, as they were deemed more difficult to physically represent.

Inductive-Deductive-Inductive Teaching and Learning Intervention

An inductive-deductive-inductive (I-D-I) pedagogical approach was employed to engage students in the construction of knowledge related to select mechanics of materials concepts. The elements of this approach are shown in Figure 1. To support inductive learning, the professor brought simple physical models to the class sessions to demonstrate and allow students to visualize and experiment with a concept being studied. The physical models were used to create a space for discussion of the physical manifestation of the concept. Once the students demonstrated understanding of the concept, the instructor, through a mix of inductive and deductive approaches, led the students through the development of the constitutive equations. The learning strategy then switched to a more deductive, though still active, approach. The instructor led the solving of example problems. Switching back to a more inductive approach, students working in groups solved additional application problems in class. Finally, continuing with an inductive approach, the students, working in groups on the course project, designed and constructed a more complex physical model to deepen their understanding of one concept.

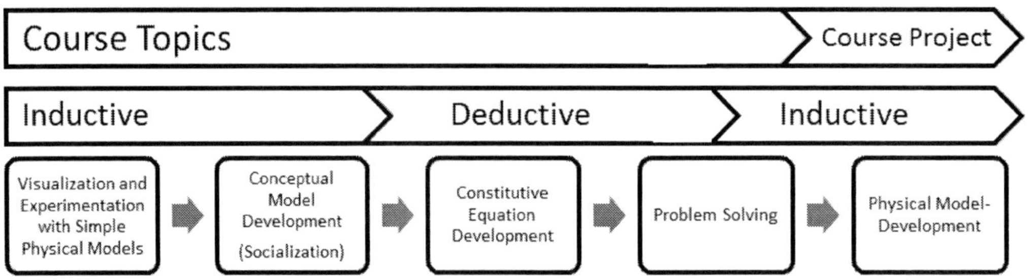

Figure 1: Inductive-deductive-inductive pedagogical approach elements

Visualization of and experimentation with each concept (phenomena or property) was made possible through the use of a simple physical model (Figures 2a-b). Each model was made of a visibly deformable material so that the concept presented and discussed was visible to the students' naked eyes. For example, cylindrical polyisobutylene rubber (Figure 2a) of various diameters between 8 and 16 mm and lengths between 300 and 600 mm was selected to demonstrate what happens when normal forces are applied. For this example, the instructor applied a force to the model in front of the class; then the students, organized in groups of 3 and 4, took the model and subjected it to forces. Similarly, a rigid polyurethane foam cylinder was used to demonstrate the application of torque (Figure 2b). Following visualization and experimentation, student groups reported their findings to the class. The instructor's role was to encourage individual participation and guide elicitation of a full description of the physical manifestation of the phenomena or property.

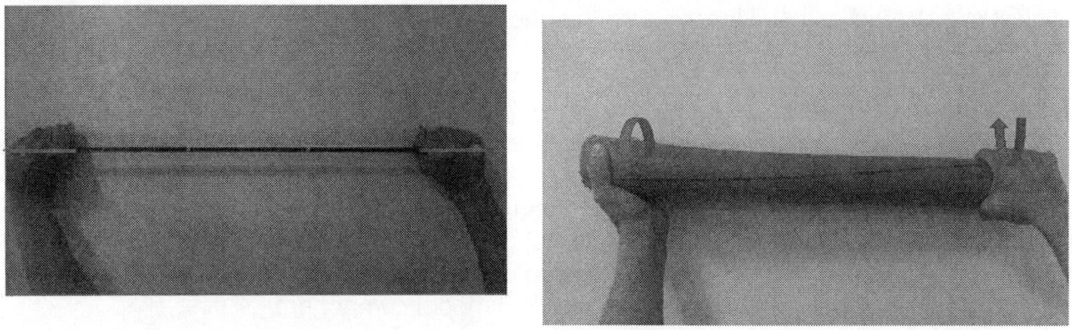

(a) normal force (concept: axial elongation; material: cylindrical polyisobutylene rubber)

(b) torque (concept: torsion stress and strain; material: rigid polyurethane foam)

Figure 2: Simple physical models used for visualization and experimentation

Using the students' experiences with the simple physical model, the professor guided inductive conceptual development followed by core equation development. That is, the practical use equations were developed with active participation from the students, with the professor constantly reinforcing the relationships among the students' conclusions from visualization and experimentation, the concept, and the equations.

Upon completing the development of the constitutive equations, the professor solved one or two carefully selected application exercises with student participation. The objective of these exercises was to relate the constitutive equations to the physical models presented and reinforce the physical model's theoretical and mathematical foundation. The professor formulated loose questions to verify students' understanding and to guide the solution development. After this, the students, working groups, assumed the challenge of solving more complex application exercises. The professor highlighted the importance of group discussion around the concepts and mathematics, as well as of the results obtained. For outside of class, the professor assigned application exercises for individual work. Exercises assigned for individual work, were real-world applications, and related to the profession, whenever possible.

In the final element of the intervention, student groups constructed their own physical models for the final project. Each group selected one course topic or phenomena (i.e., normal stress, shear stress, Poison ratio, elasticity modulus, torsion/rotation angle, or torsion/shear strain, or flection/deflection). For Treatment 1 sections, the objective was solely to have the students construct a model to represent the topic or phenomena. For Treatment 2 sections, the objective was to have the students not only construct a physical model but also construct the model in a way that measurements could be taken using the models. For example, the model for axial tension force assigned as a potential project after period 2014A, had two objectives: (1) to represent the phenomena associated with normal stress and (2) to analyse the behaviour of the materials under axial tension load and, hence, characterize said materials by calculating properties like elasticity modulus and Poisson's ratio. For these purposes, the students constructed physical models to not only represented bending, torsion, shear, or tension but also enabled measurement of the respective strains for each case.

Project selection occurred in the second week of class and counselling was personalized for each group. Besides the six regular class hours per week, six additional hours per week were devoted by the professor to counselling the groups in a lab facility. Some groups attended this counselling regularly; some others did not seem to need it. Given that some of the project topics had not yet been studied in class at the time of selection, the students had to do independent research and request clarification from the professor. At the end of the semester, students' projects were evaluated by a group of professors and research assistants. The projects were evaluated on aspects such as phenomena portrayal and the physical characteristics used to measure the phenomena.

Data Collection and Analysis

To summarize the student groups for this study, the Control group (2011B, 2012A) did not experience the inductive-deductive-inductive active learning pedagogical approach; nor did this group complete a project. Treatment 1 (2012B, 2013A, 2013B) experienced the inductive-deductive-inductive active learning pedagogical approach and completed a physical model only for the project. Treatment 2 (2014A, 2014B, 2015A) also experienced the inductive-deductive active learning pedagogical approach, but the project required not only the physical model but also the mathematical model component.

For each study period, whether or not a student passed (e.g., had a course grade higher than 3.0 out of 5.0) or dropped the course (i.e., officially withdrew or stopped attending after the withdrawal period - the first third of the semester) was recorded. The percentage of students passing the course was computed on the basis of those completing the course. The percentage of students dropping the course was computed on the basis of those initially enrolled in the course. The following null hypotheses were tested using a paired Chi-squared test, with a significance level at $p = 0.01$:

H_o: The percent of students passing the course is the same for the control and treatments

H_o: The percent of students dropping the course is the same for the control and treatments

In addition, students' final course grades were recorded. Those students that stopped attending the class were issued a final course grade; those grades were included in the calculation of the academic average. The following null hypothesis was tested using a paired *t*-test, with a significance level at $p = 0.01$:

H_o: The academic average of the students completing the course is the same for the control and treatments

During each class session of Treatments 1 and 2, a student teaching assistant measured, with a stopwatch, the time spent in class on each inductive-deductive-inductive active learning element disaggregated by the active participation of the professor or the students. That is, the time that the professor was talking and the time that the students were talking was measured. The average duration of a class in which a single topic was introduced was 115 minutes and was divided into 6 elements: attendance check, model observation, group discussion and experimentation, socialization (professor and student discussion of the concept based on the experimentation), equation deduction, and problem solving.

Findings

As can be seen in Figure 3, the percentage of students passing the course increased from 36% (average of the Control sections) to 61% for the last section (2015A). The passing rate for Treatment 1 (55%) was not significantly different than the Control (39%) ($X^2 = 5.0354$, $p = 0.0248$), though certainly it is a meaningful improvement. The passing rate for Treatment 2 (58%) was also significantly different than the Control ($X^2 = 8.1042$, $p = 0.0044$). The treatment groups were not significantly different from each other.

The academic average of those students completing the course increased from 2.50 to 3.21 on a scale of 0 to 5 (Figure 3). The academic average for Treatment 1 (M=3.16, SD=0.99) was significantly different than the Control (M=2.52, SD=1.81) ($t = 3.803$, $p < 0.0001$). The average passing rate for Treatment 2 (M=3.20, SD=1.19) was also significantly different than the Control ($t = 4.330$, $p < 0.0001$). The treatment groups were not significantly different from each other.

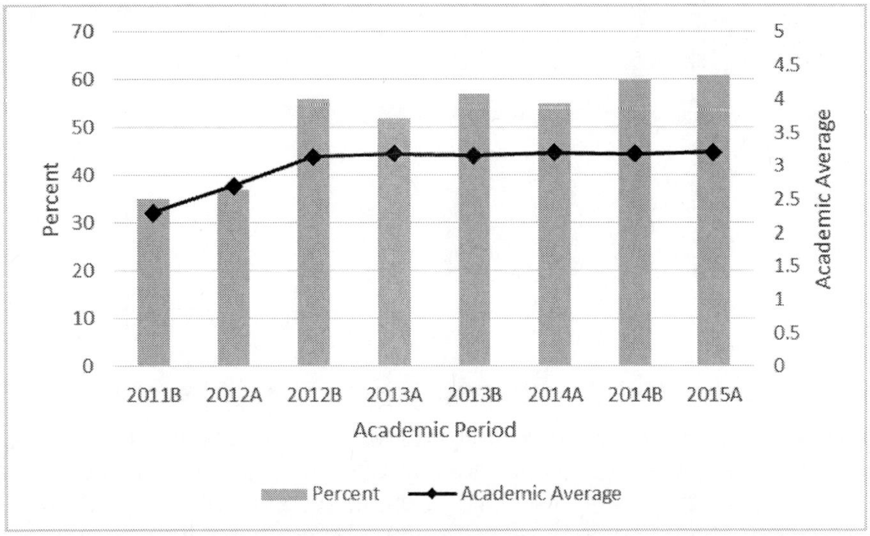

Figure 3: Percent of students passing and their academic average

The course dropout rate diminished from 34% to 8% during the study (Figure 4). The number of official withdrawals from the course remained relatively steady at 1 to 3 per course offering, while the number of students that stopped attending after the withdrawal period dropped from 13 to zero by the end of the study (Figure 5). The dropout rate for Treatment 1 (15%) was significantly different than the Control (32%) ($X^2 = 9.0937$, $p = 0.0026$). The dropout rate for Treatment 2 (7.9%) was also significantly different than the Control ($X^2 = 51.2284$, $p < 0.0001$). The dropout rate for Treatments 1 and 2 were also significantly different from each other ($X^2 = 16.5359$, $p < 0.0001$).

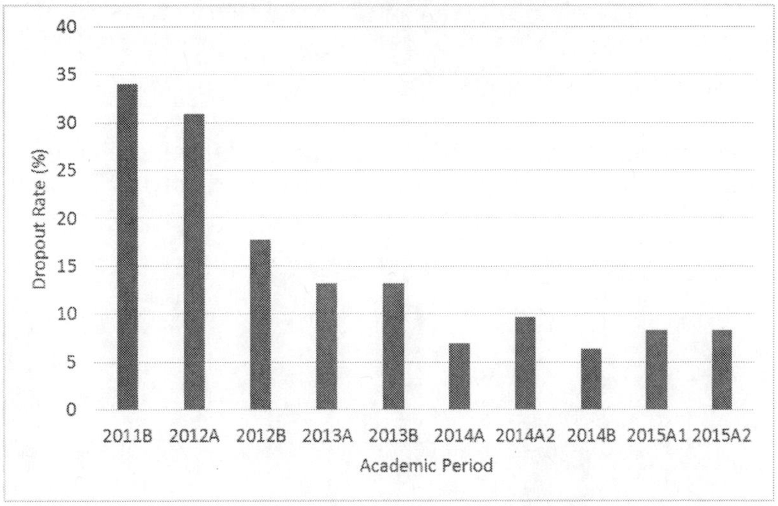

Figure 4: Percent of students that dropped the course

There were a total of 56 class sessions taught across Treatments 1 and 2 using the I-D-I pedagogy. The time registration was carried out in 35 of the total 56 I-D-I class sessions across Treatments 1 and 2. Time registration was not carried out in all sessions due to lack of availability of the teaching assistants. Figure 6 shows a sample, and typical, average breakdown of class time for a single topic. Figure 7 presents the averages time spent on

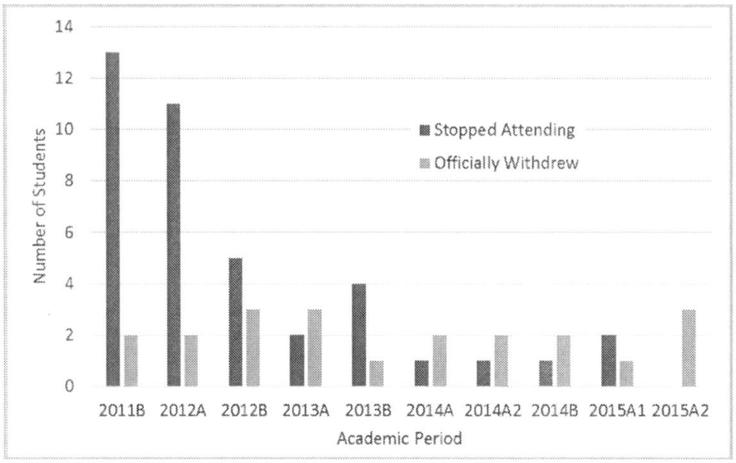

Figure 5: Students officially withdrawing and stopping attending the course

each I-D-I element disaggregated by the professor and students. Students played a leading role in solving application exercises independently, after the foundation exercises were introduced (82%). Socialization and discussion were also course elements wherein students were the main players (88% and 80%, respectively). The professor took the lead when developing the constitutive equations (83%). The professor and students shared the leading roles in observing the phenomena (54% and 46%, respectively).

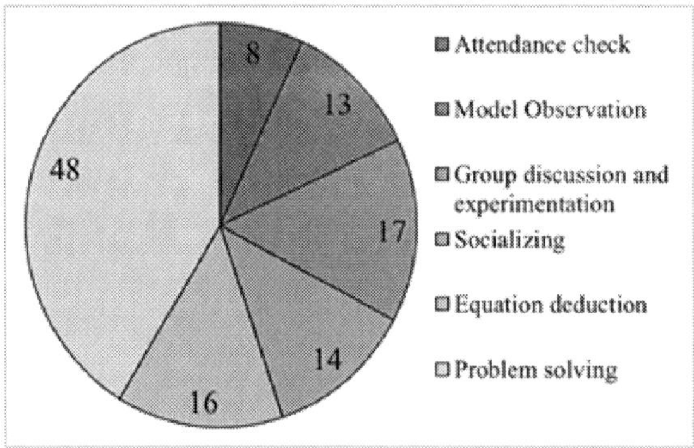

Figure 6: Average time spent on each of the I-D-I active learning elements in five class sessions on the topic of axial elongation and shortening across Treatments 1 and 2

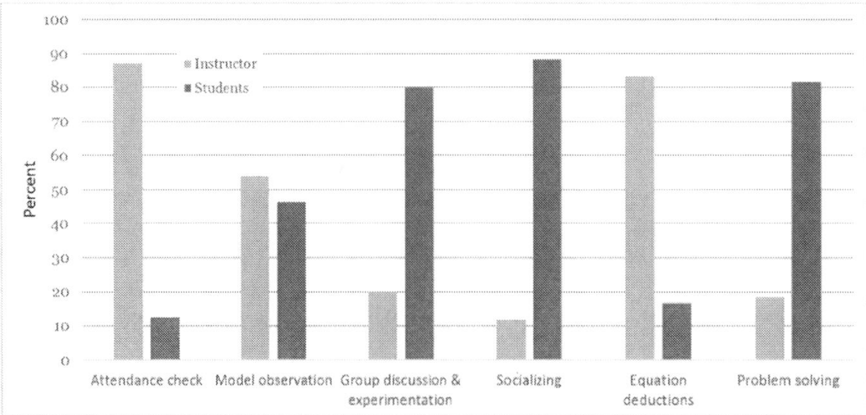

Figure 7: Division of professor versus student participation in each I-D-I class element

Discussion

The I-D-I pedagogical approach taken in this study to teach Mechanics of Materials in the area of structures to civil engineering students did result in a significant increase in academic performance and decrease in course dropout rate. These results are likely attributed to the considerable shift in roles of students from passive learners to active learners and professor from "sage on the stage" (King, 1993) to facilitator. The time being spent in a deductive mode in the Treatments was essentially reduced to approximately one-fifth that of the Control. The increasing level of comfort of the professor with this new role may also be evidenced in the reduction of students that stopped attending class.

The first inductive phase of the I-D-I pedagogical approach followed principles of effective instruction (Prince & Felder, 2006). The professor presented each concept through a concrete example (i.e., simple physical model) that was somewhat familiar to students, enabling the students to relate the concept to their current knowledge structures. But through experimentation, the concrete example challenged students to think more deeply about the concept. The second inductive phase (i.e., problem solving and model building) required students to go beyond the material presented. Both inductive phases capitalized on group interactions, enabling social construction of knowledge. The more deductive phase connecting the two inductive phases appropriately used lecture to transmit knowledge (Bligh, 2000), in this case, to ensure the constitutive equations were developed correctly.

It is not possible to tease apart the impact of the changes in the classroom instruction from the addition of the physical model project as they were both implemented together. The literature would suggest that each plays a critical role in student learning as well as in the increase in student passing rates and reduction in dropouts from the course. The extermination and project combination used in this study may be framed as both project-based learning and problem based learning. Problem-based learning, on the one hand, is related to knowledge acquisition, while project-based formation is directed at the application of knowledge, beyond both being founded on principles of collaboration, multidisciplinary orientation, and self-direction (Perrenet, Bouhuijs, & Smits, 2000). In truth, the I-D-I pedagogical approach developed here did both. Students controlled parts of the content and interacted in groups, while the professor played a role in which he favoured students' active and collaborative participation. The visualization and experimentation that lead to the development of the constitutive equations may be said to be a critical activity, a necessary condition for subsequent learning through the project. (Hadgraft & Young, 1998). However, the creation of external representations is important as that act transforms concepts and processes into symbolic and visual forms required to develop ideas, objects, and relations (Nathan, Srisurichan, Walkington, Wolfgram, Williams, & Alibali, 2013). Confirming the findings of Crone (2002), it may be possible to say that the use of experiments with and the design of physical models to demonstrate mechanics concepts are two strategies that can improve undergraduate mechanics courses.

Conclusion

An inductive-deductive-inductive pedagogical approach to teaching mechanics of materials resulted in a significant increase in the student pass rate and overall course performance and decrease in student withdrawals from the course. The approach taken in this study coupled visualization and experimentation with mechanics of materials concepts (inductive learning) with problem solving and physical model construction (inductive learning) via professor-led, but student co-generation, of the constitutive equations (deductive learning). The considerable impact of the approach may be attributes to the high level of student active learning in the course which included both the changes in the classroom instructional practice and the addition of a physical model development project. Adding mathematical modelling to the project did not significant change the student pass rate, overall course performance, or withdrawal rate, though this addition may have had other impacts not accessed in this study.

References

Bligh, D.A. (2000). *What's the Use of Lectures?* Jossey-Bass, San Francisco.

Brown, S. (2003). Conceptual change in mechanics of materials. *Proceedings of the 120th American Society for Engineering Education Conference*, Atlanta, GA: ASEE.

Crone, W. C. (2002). Using an advanced mechanics of materials design project to enhance learning in an introductory mechanics of materials course. *Proceedings of the 109th American Society for Engineering Education Conference*, Montreal, Canada: ASEE.

Egelhoff, C. J., & Burns, K. L. (2011). A heuristic to aid teaching, learning, and problem-solving for mechanics of materials. *Proceedings of the 118th American Society for Engineering Education Conference*, Vancouver, BC: ASEE.

Felder, R. M., & Silverman, L. K. (1988). Learning and teaching styles in engineering education. *Journal of Engineering Education*, *78*(7), 674–681.

Felder, R. M., Woods, D. R., Stice, J. E., & Rugarcia, A. (2000). The future of engineering education II. Teaching methods that work. *Chemical Engineering Education*, *34*(1), 26–39.

Freeman, S., Eddy, S. L., McDonough, M., Smith, M. K., Okoroafor, N., Jordt, H., & Wenderoth, M. P. (2014). Active learning increases student performance in science, engineering, and mathematics. *Proceedings of the National Academy of Sciences*, *111*(3), 8410–8415.

Hadgraft, R. G., & Young, W. (1998). *Teaching Strategy*. Department of Civil Engineering, Monash University.

Johnson, P. A. (1999). Problem-based, cooperative learning in the engineering classroom. *Journal of Professional Issues in Engineering Education and Practice*, *125*(1), 8–11.

King, A. (1993). From sage on the stage to guide on the side. *College Teaching*, *41*(1), 30-35.

Mahendran, M. (1995). Project-based civil engineering courses. *Journal of Engineering Education*, 84(1), 75–79.

Montfort, D., Shane Brown and David Pollock. (2009). An investigation of students' conceptual understanding in related sophomore to graduate-level engineering and mechanics courses. *Journal of Engineering Education*, *98*(2) 111–129.

Nathan, M. J., Srisurichan, R., Walkington, C., Wolfgram, M., Williams, C., & Alibali, M. W. (2013). Building cohesion across representations: A mechanism for STEM integration: Cohesion across representations. *Journal of Engineering Education*, *102*(1), 77–116.

Perrenet, J. C., Bouhuijs, P. A. J., & Smits, J. (2000). The suitability of problem-based learning for engineering education: Theory and practice. *Teaching in Higher Education*, *5*(3), 345–358.

Piaget, J. (1972). *The Psychology of the Child*, New York: Basic Books.

Prince, M J., & Felder, Richard M. (2006). Inductive teaching and learning methods: definitions, comparisons, and research bases. *Journal of Engineering Education*, *95*(2), 123-138.

Roylance, D., Jenkins, C., & Khanna, S. (2001). Innovations in teaching mechanics of materials in materials science and engineering departments. *Proceedings of the 108th American Society for Engineering Education Conference*, Albuquerque, NM: ASEE.

Vygotsky, L.S. (1978). *Mind in Society*, Cambridge, Massachusetts: Harvard University Press.

Lima Accord: A Multilateral Agreement for Mutual Recognition of Engineering Education in Latin America and the Caribbean

Daniel Sanchez Ruiz
ICACIT, Lima, Peru
daniel.sanchez@icacit.org.pe

***Abstract**: Mutual recognition agreement allow the mobility of graduates in the world. However, this reality remains distant in Latin America and the Caribbean. This paper focus on engineering education and seeks to address if the mutual recognition of accredited engineering programs in Latin America is possible. To this end, the Lima Accord is proposed as a feasible mid-term alternative and describes in detail the motivations and actions undertaken that led up to signing the Lima Accord, as well describe the key role of the IEEE as a promoter.*

Introduction

Globalization of economic activity is a reality that impacts on engineering professionals. Provision of commodities, goods, services and infrastructure increasingly involves engineering professionals and enterprises working across national boundaries. The quality of engineering professional practice and education is an essential facilitator of global practice. (Hanrahan, 2013).

Mechanisms for easing the mutual recognition of engineering qualifications accredited in one jurisdictions by authorities in another have therefore developed, either on a regional basis, for example the European Accredited Engineer (EUR-ACE) accreditation system (Augusti, 2009), or without regional constraints, for example the International Engineering Alliance (IEA) educational accords, namely the Washington, Sydney and Dublin Accords (Hanrahan, 2009). Besides this mechanisms allow accredited programs be widely recognized as equivalents for recognition for academic mobility, graduate studies and by governments for migratory issues.

The formal objective of mutual recognition is to enable graduates of programs accredited by a signatory of an agreement to obtain recognition of academic requirements (e.g. baccalaureate degree) for entry in to the practice of the profession.

In contrast, in Latin America this kind of recognition remains a distant reality. While there are experiences of supranational accreditation and other agreements signed between different types of organizations, to date none of these initiatives has successfully achieved mutual recognition.

Based on previous experiences, this paper describes in detail the development of the Lima Accord, a multilateral agreement that seeks to achieve mutual recognition of engineering education in Latin America and the Caribbean (LAC).

In the first part, review different agreements and attempts of mutual recognition in LAC. The second part, describe the role of the IEEE on accreditation activities in the United States and globally. The third part, describe the three stages of this initiative promoted by the IEEE that came to fruition with the signing of the Lima Agreement. Finally, future actions of the Accord examined in order to put into effect the mutual recognition in LAC.

The table lists the names, abbreviations and jurisdiction of accreditation agencies involved with the Lima Accord in any stage of its development.

Table 1. Accreditation Agencies involved with the Lima Accord

Name	Abbreviation	Jurisdiction
Accrediting Agency Engineers Association of Chile	Acredita CI SA	Chile
Accreditation Council for Engineering Education	CACEI	Mexico
Agency for Accreditation of Engineering and Architecture Programs	AAPIA	Costa Rica
Caribbean Accreditation Council of Engineering and Technology	CACET	Caribbean
Central Agency for Accreditation of Architecture and Engineering	ACAAI	Central America
National Accreditation Council	CNA	Colombia
National Commission for University Evaluation and Accreditation	CONEAU	Argentina
National System of Evaluation, Accreditation and Certification of Educational Quality	SINEACE	Peru
The Institute of Quality and Accreditation of Programmes in Computing, Engineering and Technology Education	ICACIT	Peru

Purpose and Scope

This paper seeks to address the following research question: ***Is it possible to achieve a mutual recognition agreement of accredited engineering programs in Latin America and the Caribbean?***

The purposes of this analysis are (1) to promote the Lima Accord within the accreditation agencies of LAC as a mid-term alternative to achieve mutual recognition of accredited engineering programs; and (2) to involve Engineering Registering Bodies in order to facilitate the mobility at all stages of the engineer's life cycle and ensure the quality of them.

The scope of this paper focus on the mutual recognition in engineering accreditation; however, some accreditations agencies on the Lima Accord also make evaluations of other kind or category of programs (i.e. engineering technology). In this paper, the author describes the progress made by the Lima Accord so far, in search of the mutual recognition.

Methodology

This paper is descriptive and includes details of the Summits carried out and the results of each of these between representatives of accreditation agencies, engineering associations and volunteers and staff of IEEE.

The development of the Accord is explained in three stages, which reflect the level of maturity of the initiative and the extent of commitment of each accreditation agency.

The author address this research question through the following methodology:

(1) Examine the relevant factors of previous agreements of mutual recognition around the world.

(2) Describe the process toward the Lima Accord. Its origins, the annual summits and the several meeting that led to sign of the agreement, identifying any substantive insight.

(3) Make a qualitative judgment about how insights from other agreements were included over the process toward the Lima Accord.

Role of IEEE in Accreditation

The IEEE is the largest engineering association in the world and has more than 420,000 members in more than 160 countries. IEEE contribute with nearly a third of the world's technical literature in electrical engineering, computer science, and electronics; contributing to the growth of technical and engineering knowledge.

Over the decades since its foundation, the roles of the technologies served by the IEEE continued to spread across the world and reach into more and more areas of people's lives. Now the IEEE's field of interest has expanded well beyond electrical and electronics engineering and computing into areas such as micro- and nanotechnologies, ultrasonic's, bioengineering, robotics, electronic materials, and many others like social Implications of technology or engineering education.

IEEE carries out accreditation activities that can be distinguished in two: (1) Accreditation activities in the United States through ABET and (2) Global Accreditation Activities (excluding ABET activities).

With respect to activities within the United States, IEEE as a founding member and the largest of the 30 professional societies comprising ABET, Inc (formerly Accreditation Board for Education and Technology) contributes with more than 300 members who year after year serve as Program Evaluators responsible for applying accreditation criteria in the evaluation of educational institutions.

Outside the United States, IEEE coordinates global accreditation activities including support for existing accreditation agencies, development of new accreditation agencies, and accreditation education in Engineering, Computing and Technology. IEEE state (IEEE 2007) that:

- The IEEE favors the formation of national and regional accrediting bodies and foresees the future formation of multi-national accrediting bodies where appropriate. At the same time, IEEE strongly disfavors permanent arrangements that have the accreditation of programs in one country performed and approved by an accrediting body from another country or region.
- The IEEE favors the formation of local, regional, and multi-national bodies where constituents from the countries where the programs are accredited are fully represented in voice and in vote.
- The IEEE seeks to cooperate with other associations both within and without these fields in creating, maintaining, supporting, and expanding accrediting bodies for programs in engineering, computing, and technology.

1st IEEE Accreditation Summit

On the occasion of the 1st Symposium on Engineering Accreditation organized by ICACIT in 2014 and at the proposal of the IEEE, it was organized the 1st Summit of Accreditation Agencies in Engineering.

After having participated as an observer in initiatives such as *Engineering for the Americas - EftA* (2005) or the *Latin American Agreement on Accreditation in Engineering - ALAI* (2010) that failed to materialize successfully, IEEE proposed bring together all Latin American accreditation agencies of engineering programs to discuss the feasibility of reaching a mutual recognition agreement.

The objective of this first summit was essentially exploratory, it was hoped to know about the expectations of accreditation agencies - at that time 15 at country level and 3 at regional level - to learn about related experiences and to discuss how to achieve mutual recognition of accreditations that will lead to the mobility of graduates. For the latter, representatives of the Regional Accreditation System for University Programs of MERCOSUR (ARCU-SUR) and the Washington Accord were invited.

The First Summit of Accreditation Agencies was held on September 3, 2014, with the participation of 15 representatives of accreditation agencies and engineering associations.

From the presentation given by Daniel Morano, advisor to the Secretariat of University Policies of Argentina, on the ARCU-SUR System the following points were clear:

- ARCU-SUR is a supranational accreditation system in which MERCOSUR member countries (Argentina, Brazil, Paraguay and Uruguay) and partners (Bolivia, Chile, Colombia, and Venezuela) participate;
- The ARCU-SUR accreditation is distinct and independent of national accreditation. However, this is conducted by the national accreditation agency;
- It is restricted to agronomy, architecture, nursing, engineering, medicine, dentistry and veterinary programs;
- Its scope is limited to recognizing the quality of qualifications, without entailing any mutual recognition of qualifications;
- The inclusion of any other state as an associate is through the Ministry of Education.

Professor Hu Hanrahan, president of the Washington Accord, emphasized the following on this agreement:

- The Washington Accord establishes mutual recognition of accreditation. In this way, the accreditations granted by a signatory of the agreement are recognized by the other signatories as equivalent;
- Members are only non-governmental agencies;
- Mutual recognition is based on the substantial equivalence of the signatories' program outcomes, known as graduate attributes, and accreditation processes, verified for each signatory by peer review at the time of admission as a signatory and periodically thereafter;
- The scope of the Washington Accord reach recognition of the educational requirements. This in order that qualifications are recognized by the register/licenses' body that regulate the engineering practice in a jurisdiction;
- The IEA Graduate Attribute Profile and best practices are the result of a long-term process started in 2001, adopted for first time in June 2005, and the most recent update was published in June 2016. The current IEA Graduate Attribute Profile are based on extensive scholarly analysis and coordination.

In addition, ICACIT – that was admitted as provisional member of the Washington Accord and became the first one from Latin America - presented a case of Study base on its experience to demonstrate that Latin American accreditation agencies have the capacity to integrate into global agreements.

The discussion was raised under the question: ***The Path to Take: A New Agreement or Join an Existing Agreement?*** The answer to this question was not easy because of several factors; such as the nature of accreditation agencies or the spectrum of accrediting programs. Added to a lack of knowledge regarding the accreditation processes and the educational context of each country of Latin America.

Although no concrete outcome was achieved at this first summit, important information was revealed on already established agreements and, above all, a matter related to accreditation agencies: the mistrust regarding the quality of the accreditation system conducted by neighbors essentially based by lack of knowledge.

2nd IEEE Accreditation Summit

For the second summit, seven accreditation agencies attended. Based on what was identified in the previous edition, it was necessary to increase the trust between the agencies regarding their accreditation systems. Therefore, each participating agency spoke about its educational context and the operation of its accreditation system in its jurisdiction.

In addition to the accreditation agency attendees, were invited representatives from the Washington Accord, the Iberoamerican Network for Quality Assurance in Higher Education (RIACES), Institutions of Engineering Education (ASIBEI) and International Federation of Engineering Education Societies (IFEES) to help to develop a roadmap to establish a mutual recognition.

Among the previous coordinations, Prof. Hanrahan was consulted - given his deep knowledge and experience in the Washington Accord - on what recommendation he had in order Latin America achieve a mutual recognition of accreditation. Obtaining as a response that accreditation agencies in Latin America can constitute a regional agreement for mutual recognition and in turn each aspiring to be admitted to the Washington Accord as ICACIT already achieved in 2014 upon admission as provisional member.

Consequently, the objectives set for this second summit were: set goals, define processes, organization and actions to implement a Mutual Recognition Agreement in Latin America, in turn that accreditation agencies prepares self and then integrate into agreements such as the Washington Accord.

After two days of presentations and working sessions, the accreditation agencies agreed to go ahead with the initiative, signed the LIMA DECLARATION, which in essence establishes: the search for mechanisms that allow the mutual recognition of the engineering programs of higher education accredited by the signatories of the document, with the support and experience of the IEEE.

The seven Accrediting Agencies signatories of the Lima Declaration are: CACET (Caribbean), Acredita CI (Chile), CNA (Colombia), CACEI (Mexico), ACAAI (Central America), ICACIT (Peru) and SINEACE (Peru).

IEEE, ASIBEI and IFEES as witnesses also signed the Lima Declaration.

In addition to the signing of the Lima Declaration, accreditation agencies established a work agenda with activities between November and March 2016 to submit the first draft of the Lima Accord to the accreditation agencies from other Latin American countries - that for various reasons could not be presented in the second IEEE Accreditation Summit - for consultation. During the process CACEI would act as a leader, ICACIT would assume the role of the secretariat and IEEE would participate in the Planning Committee.

Signing the Lima Accord

The planning committee, composed of the representatives of the Lima Declaration signatories and IEEE, decided to take the Washington Accord as a model for the Lima Accord. Motivated by relying on a "substantial equivalence" for mutual recognition without the need to use the same procedures or criteria among the signatories (IEA 2016), for have started - in 1989 - with a minimum of formality and have being improving over time (IEA 2015) , and mainly for having demonstrated its success by providing international mobility to graduates of engineering programs (Hou 2012).

For a year, the planning committee did a hard work of drafting the accord, soliciting feedback from all accreditation agencies and organizations linked to accreditation in Latin America and the Caribbean, and finalizing the official agreement.

On September 6, 2016, representatives of accreditation agencies and the IEEE met in Lima to sign the Lima Agreement, an international agreement among organizations responsible for accreditation of engineering programs in Latin America and the Caribbean; in a ceremony organized by IEEE and ICACIT.

The signatories of the Lima Accord are: Acredita CI (Chile), AAPIA (Costa Rica), CACEI (Mexico), CACET (Caribbean), ACAAI (Central America), SINEACE (Peru), ICACIT (Peru)

An additional agency, ACAAI (Central America) was not able to attend, but signed the Accord at the end of September in a separate ceremony.

Following the signing ceremony, the first working sessions of the Accord between the representatives of the signatories and IEEE volunteers and staff were held for two days. The purpose of these sessions was to develop a road map for the implementation and operation of the Lima Accord. Priorities included governance and communications. Likewise, the representatives pledged that the Lima Agreement be fully implemented and operational before October 2017.

Finally, the first president and vice-president of the Lima Accord was elected. Enrique Alvarez Rodrich, president of ICACIT, and María Elena Barrera Bustillos, Executive Director of CACEI, were elected by majority vote of all the signatories, as president and vice-president of the Accord respectively.

Future Actions

As part of the conclusions of the working sessions of September 2016, the following deadlines were established:

- 6 months to define the Accord's policies and procedures manual;
- 3 years to put into effect the mutual recognition of accreditations, after signatories passing successfully through the review and verification visits; and
- 5 years for signatories to demonstrate compliance against the standards and graduate profile established by the Agreement. Agencies that fail to demonstrate compliance with the standards and graduate profile established by the Agreement will lose their status as signatory to the Agreement.

In addition, during the *International Forum: Accreditation and Professional Certification of Engineering in Latin America and the Caribbean* held on Friday, December 2, 2016 and organized by the College of Engineers of Peru (CIP), ICACIT and SINEACE; the Declaration of Miraflores was signed. This statement, prepared by representatives of accreditation agencies and professional associations in Latin America and the Caribbean, recommends:

- To promote the participation of accreditation agencies of engineering programs within the Lima Accord.
- Share the requirements that each country establishes to obtain the license or register that enables the engineering practice in the jurisdiction.
- To learn about international experiences of mutual recognition and professional mobility in engineering that serve as reference to establish an agreement adapted to the social, economic and cultural environment of Latin America and the Caribbean.

To this end, the Pan-American Union of Engineers Associations (UPADI) was entrusted with the task of facilitating the consolidation of this space for te exchange of experiences in Latin America and the Caribbean, in order to achieve the mobility of professionals in engineering within the region.

Finally, the Council of Accreditation of Engineering Education (CACEI) of Mexico offered to develop in 2017 the second edition of this forum. Bringing together the meeting of the Lima Accord and the participating institutions of the Miraflores Declaration in order to generate a synergy that accelerates the consolidation of an agreement for the professional mobility of engineers in Latin America and the Caribbean.

Conclusions

Based on explained process of the Lima Accord, the author draw the following conclusions:

- Accreditation agencies in LAC initially mistrust regarding the quality of the accreditation system conducted by neighbours. However, during the process they have gained greater knowledge and trust between them.
- Despite this, verification and quality assurance of accreditation systems in LAC can only be achieved through peer review visits among Accord signatories.
- The participation of chairs of the Washington Agreement allowed a greater understanding of this agreement. This led the Lima Accord took the Washington Accord as model.
- Many accreditation agencies in LAC involved in the Lima Accord aspire to be admitted to the Washington Accord and some of them are already in provisional status.
- Unlike the Washington Accord, the Lima Accord allows the participation of government agencies. Given the nature of the accreditation agencies in this part of the world, this would allow the inclusion of all accreditation agencies in LAC.
- The role played by IEEE as facilitating has proven to be a great catalyst for achieving a mutual recognition agreement.

Based on these conclusions, is it possible to achieve a mutual recognition agreement of accredited engineering programs in LAC? From the author's perspective, the answer to this question is an unequivocal yes. The present paper not only shows the achievement of a mutual recognition agreement in LAC, but also the factors that allowed its achievement.

However, it is necessary to maintain the pace of work among the Lima Accord signatories to establish the rules and procedures of the Accord; as well as the standards on which the substantial equivalence of accredited programs in LAC will be demonstrated.

Recommendations

- Extrapolating the above conclusion, UPADI acting as facilitator of the Miraflores Declaration could be the promoter of a future mobility agreement for engineering professionals in LAC.
- To keep the Accord active, the rules and procedures, and standards of the Accord; as well as the standards should be established as soon as possible.
- Meanwhile, other follow-up activities can be implemented, for example, allowing the observation of accreditation visits among Accord signatories.
- It is important that registration bodies also be involved in this agreement as observers. In a monitoring role, that the standards required for engineering practice in its jurisdiction are being considered in the Agreement standards.

References

Augusti, G. (2009) EUR-ACE: The European Accreditation System of Engineering Education and its Global Context, in Patil, A.S. and Gray, P.J. (eds), *Engineering Education Quality Assurance: A Global Perspective* (41-49). Springer.

Hanrahan, H. (2009) Toward Consensus Global Standards for Quality Assurance of Engineering Programmes, in Patil, A.S. and Gray, P.J. (eds), *Engineering Education Quality Assurance: A Global Perspective* (51-71). Springer.

Hanrahan, H. (2013) Toward Global Recognition of Engineering Qualifications Accredited in Different Systems, *ENAEE Conference. Belgium*. Accessed at http://www.enaee.eu/wp-assets-enaee/uploads/2012/01/HANRAHAN-Paper-130820.pdf on 4 march 2017

Hou, A. (2012). Mutual recognition of quality assurance decisions on higher education institutions in three regions: A lesson for Asia. Higher Education, 64(6), 911–926. Accessed at http://dx.doi.org/10.1007/s10734012-9536-1 1 march 2017

IEEE (2007), IEEE Position Paper on Accreditation of Academic Programs in Engineering, Computing and Technology. Accessed at https://www.ieee.org/education_careers/education/eab/position_statements.html on 4 march 2017.

International Engineering Alliance – IEA (2015), *A History of the International Engineering Alliance and its Constituent Agreements: Toward Global Engineering Education and Professional Competence Standards.* Accessed at http://www.ieagreements.org/about-us/history/ on 2 march 2017.

International Engineering Alliance – IEA (2016), *Educational Accords: Section B:Rules and Procedures*. Accessed at http://www.ieagreements.org/about-us/key-documents/ on 2 march 2017.

Acknowledgements

The Author gratefully acknowledge the support of IEEE, which promoted and provided permanent support to the Lima Accord.

Copyright statement

Student perceptions of group learning quality and learning outcomes in design-based science

Elizabeth H. McEneaney
University of Massachusetts, Amherst, MA USA
emcenean@umass.edu

Miancheng Guo
University of Massachusetts, Amherst, MA USA
mianchengguo@educ.umass.edu

Martina Nieswandt
University of Massachusetts, Amherst, MA USA
mnieswan@educ.umass.edu

Abstract: *This paper analyses how student perception of the quality of the dynamics of small group work on three engineering design tasks in high school biology classes relates to their learning of engineering design process and content. First, we ask how groups may have varied along different dimension of group quality (or whether groups simply tend to be either low on all relevant dimensions or high on all dimensions). We then consider whether an individual student's perception of her or his group's functioning along particular dimensions is correlated to subsequent learning about engineering design, both specifically in terms of engineering design practice and overall. Survey data on various dimensions of group dynamics were collected at two time points during the study, and engineering design process and content learning outcomes were measured after each task. Data from 185 students in 51 small groups were analyzed.*

Context

In the USA, the Framework for K-12 Science Education (NRC, 2011) lists eight scientific and engineering practices that inquiry and engineering design tasks in the classroom should reinforce, including one key practice that is especially challenging for students: Engaging in arguments from evidence. Figure 1 highlights the similarities and differences in this practice between science and engineering as stated in the Framework (NRC, 2011, p. 3-31). Although some of the processes in the two domains are similar, the goals of these processes are different. While scientists search for the best explanation for a natural phenomenon, engineers search for the best possible solution to a problem. In addition to engaging with evidence in order to defend explanations and conclusions, the need to effectively collaborate with peers is essential to both scientific inquiry as well as engineering design. The USA's Next Generation Science Standards (2013) reiterate the importance of use of evidence and collaboration as aspects of practice in both science inquiry and engineering design.

[Figure 1 about here]

Research Questions

This paper analyzes how student perception of the quality of the dynamics of small group work on engineering design tasks in high school biology classes relates to their learning of engineering design process and content. First, we ask how groups may have varied along different dimensions of group quality (or whether groups simply tend to be either low on all relevant dimensions or high on all dimensions). We then consider whether an individual student's perception of her or his group's functioning along particular dimensions is correlated to subsequent learning about engineering design, both specifically in terms of engineering design practice and overall.

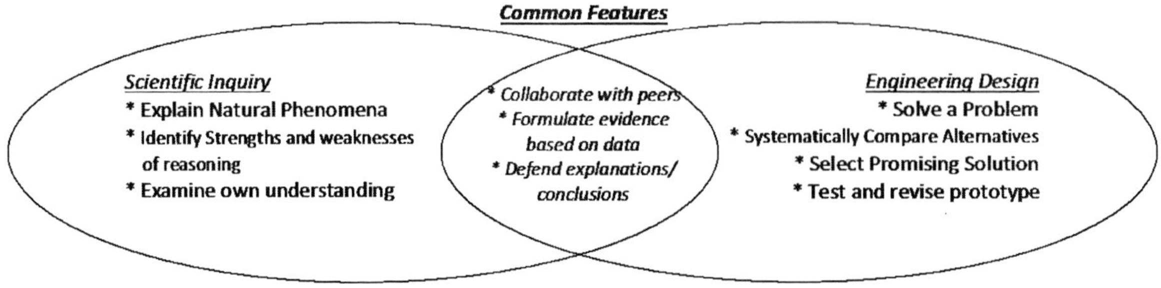

Figure 1: Comparing scientific inquiry and engineering design practices with respect to engaging in arguments from evidence; adapted from National Resource Council, 2011, pp. 3-31.

Theoretical Framework

Classroom activities in K-12 settings have often featured student work in small groups to reduce teacher-centeredness and reliance on pre-packaged material and to maximize the autonomy of students (NRC, 1996). What is the evidence that "group work" works? In an overview of the research on small groups and learning in science, Bennett, Hogarth, Lubben, Campbell and Robinson (2010, p. 86) found that published studies reported positive links between small group discussion and subsequent student understanding, but they caution that authors of these studies were often "advocates" of the small group approach. These authors also note "[t]here is a growing body of evidence … that teachers lack skills and do not feel confident in small group discussion". (2010, p. 71), echoing early concerns about managing small groups by others (Good, McCaslin and Reyes 1992). At the same time, Johnson and Johnson (2009, p. 375) state that one approach to small group work, cooperative learning, is now "a standard and widespread teaching procedure" across the world and "an educational psychology success story." Cooperative learning is based on a clear theoretical foundation (social interdependence theory), and their review of the research validates its positive effects on student achievement, interpersonal relationships and on psychological health across subjects, grades and educational settings.

Proponents claim that collaboration within a group leads to shared goals and values and develops collective and individual responsibility, stronger engagement, interest and motivation. Well-structured and managed group work allows students to develop communication skills by defending their work based on evidence, to learn from other groups, and to engage in problem solving that mirrors future work and life experiences. Asking questions, formulating hypotheses, articulating arguments, using models or analogies to explain concepts, conducting investigations, analyzing and evaluating data, proposing solutions, and creating various ways of communicating results are all aspects of cooperative learning. These processes are key features of engineering practices. Small group activities may therefore support learning of such practices.

One way to view the state of research regarding the challenges of inquiry and engineering design teaching and learning, at least as it pertains to collaboration in small groups, is to consider what group members must *collectively construct* as they work on a task. Most of the research has focused on the need for groups to construct a "joint problem-solving space" (Roschelle, 1992; Teasley & Roschelle, 1993) that involves coming to a collective cognitive understanding of the task. Barron (2003) usefully demonstrates that the nature of the cognitive challenges faced by sixth grade math groups does not explain, on its own, the variability of group outcomes. To address this gap, Barron proposes a "dual-space" model of collaboration in which groups must attend to and develop the "content space" (the problem to be solved) as well as the "relational space" (the challenges based on social interactions within the group). Both content and relational spaces may be deeply shaped by the language abilities of group members, and for English language learners, science and engineering design may well represent a second foreign language (Rosebery & Warren, 2008).

A few studies support Barron's dual space model by highlighting the social dimension of small group work during inquiry by focusing on intra-group interaction and discourse patterns (Engle & Conant, 2002; Keys 1997; Jiménez-Aleixandre, Rodriguez, & Duschl, 2000). Such relational space work may benefit from the interpersonal social competency of group members (Ten Dam &Volman 2005), such as respect for others and insight into others' perspectives. Nonetheless, most of the research focuses only on interactions pertaining to the task. To fully understand group work, all interactions, task-related and non-task-related, must be considered.
In our study, we posit that the role of the affective domain in group work is under-theorized, and so we investigate how the demands of engineering design tasks necessitate collective construction of a "triple problem-solving space" in which content, social/relational, and affective components are developed on a moment-by-moment basis. Affect, whether positive or negative, activated or not, has been shown to impact small group interaction in upper-elementary math tasks (Linnenbrink-Garcia, Rogat, & Koskey, 2011). Moreover, Cartney and Rouse (1996, p. 85), emphasize the importance of the "emotional life of the group" which is distinct from the affective states of individual members, thereby suggesting that the group itself must be understood as co-constructing affect, just as Barron (2003) takes the group as the unit of analysis from a social/relational standpoint. Consequently, Nieswandt, Affolter and McEneaney (2014) argue that academically successful small groups must co-construct the "triple problem-solving space" in which content, social/relational, and affective components are developed and maintained. Since engineering design practices are part of the U.S. Next Generation Science Standards and growing in emphasis worldwide, it is critical to understand how small groups in secondary school can navigate the challenges of design problems that are open-ended and often ill-defined, demanding cycles of divergent and convergent thinking, design development, refinement and evaluation (Dym et al., 2005).

If groups fail to manage any of these three aspects of the collective problem-solving space, group collaboration will be unsuccessful, and learning – even from a well-designed task – is likely to be minimal. While our focus is on collective construction of the group's "triple space," we recognize that students bring various kinds of individual resources to the demands of cooperative inquiry and engineering design work, such as prior knowledge and skills, social competencies, and interest. Tracing the way these kinds of individual student resources influence group construction of the triple problem-solving space will allow us to take an analytical perspective to identify dynamics driving group work rather than the more typical descriptive approach of most research on small group work in an inquiry context (Bennett, et al. 2010, p. 91).

Methodology

Fifty-one small groups of 3 or 4 students completed 3 engineering design tasks (oil spill clean up, designing pill coating, developing an artificial heart valve) aligned to U.S. state and national curricular standards for secondary biology. These groups were formed in introductory biology classes in four high schools in the Northeast of the U.S.. In these schools, biology is generally taken by 9th and/or 10th graders, and in three of the four schools, the biology course is completed in a time-intensive semester (half school year). In the fourth school, the biology course spans the entire school year, with classes meeting in shorter periods of 45 minutes compared to the longer block schedules at the other schools. Five teachers participated in the study, leading 8 different introductory biology sections in total. Three of the teachers had over ten years of experience teaching science, while one of the teachers was in her second year of full-time teaching. The participating schools varied in their sociodemographic profiles. All schools served a predominantly white student population, although one school served a sizeable Hispanic population, primarily Puerto Rican. In two of the communities, less than 20% of adults possessed a bachelor's degree or higher, while in another community, that figure was over 50% (U.S. Census Bureau, 2014). Median household incomes likewise varied, from about $30,000 per year to about $60,000.

Instruments to measure student perceptions of group dynamics were administered after the first and last task while students completed closed- and open-ended learning outcome items after each task, either as a stand-alone assessment or as part of a unit test. Two coders scored open-ended learning outcome items with high inter-rater reliability. Student perceptions of the *cognitive space* were measured with the Group Interaction Questionnaire (Visschers-Pleijers, et al., 2005), which resulted in two distinct factors. The first was "Knowledge Building" (e.g., We posed questions to each other in order to understand the science content, We built on each other's ideas and suggestions; 6 items, α=0.84) and we labelled the second "Handling Cognitive Conflict" (e.g., People shared conflicting ideas; 3 items, α=0.62). We assessed the *social/relational space* of the group using Sargent and Sue-Chang's (2001) Social Cohesion scale (e.g., I feel a sense of belongingness to my group, I like my group; 4 items, α=0.85). In addition, this dimension of group quality was measured by Absence of Social Loafing (e.g., I stopped listening to my group; I tried to get other students in my group to do the hard parts; 6 items, α=0.81) and Positive Group Interaction scales (e.g., We all worked well together; 5 items, α=0.87) (Linnenbrink-Garcia, et al. 2011). The *affective* component of the problem-solving space was measured using Edmondson's (1999) Psychological Safety scale (e.g., It is safe to take a risk in my group; 4 items, α=0.80). The research team developed learning outcome measures of relevant task content and engineering practices in conjunction with participating teachers. A summary measure of perceived group quality summed the average of the two cognitive scales, the average of the three social/relational scales, and the psychological safety scale (affective). Thus, high group quality scores reflected groups that had successfully addressed all three dimensions of the triple problem solving space. Approximately 8 percent of data was missing. The expectation-maximization algorithm (Cox, McIntosh, Reason, and Terenzini 2014) in SPSS was used to address missing cases, yielding a data set of 185 students in 51 groups..

Findings

Before analyzing the relationship between a student's perception of group quality on their learning in engineering design, we present an overview of the distribution of groups in terms of the three dimensions – cognitive, social/relational and affective. Table 1 below shows where the 51 small groups are located along these dimensions based on whether the mean of the small group members' reports (usually 4 members) on any particular dimension was above or below the overall mean for scales under that dimension.

			Social/Relational		
Affective			Low	High	Total
Low	Cognitive	Low	16 (31%)	1 (2%)	17 (33%)
		High	4 (8%)	2(4%)	6 (12%)
High		Low	1 (2%)	4 (8%)	5 (10%)
		High	4 (8%)	19 (37%)	23 (45%)
	Total		25 (49%)	26 (51%)	51 (100%)

Table 1: Distribution of Small Groups by Three Dimensions of Functioning

Two findings are immediately evident in table 1. First, cognitive, social/relational and affective dynamics in the small group work on these tasks do tend to be positively correlated in that 16 groups (31%) score relatively low on all three dimensions, while 19 (37%) score relatively high on all three. Secondly, however, 16 groups do not follow this pattern; they are a mixture of high and low on the three dimensions, and in fact, there are no empty cells among the eight possible combinations across the dimensions. Only 1 group scored high on social/relational (positive group interaction, social cohesion, and absence of social loafing) while scoring low on affective (psychological safety) and cognitive (knowledge building and handling cognitive conflict). Similarly, high affective scores with low cognitive and social/relational was also rare, with only 1 group showing this pattern. More common “mixed” patterns were high affective and social/relational and low cognitive and high affective and cognitive but low social/relational (4 groups each). Thus, about a third of groups reported low functioning across the dimensions, a third high functioning on all three, and a third were mixed. A separate analysis by school (not shown) indicated that all four schools tended to show a similar distribution of groups across the dimensions. These results suggest that while the dimensions are not independent of each other by any means, groups do vary empirically in their location in the triple problem solving space, and so this framework may be a useful lens for understanding small group processes going forward.

Turning to the question of whether an individual student’s perceptions of small group functioning was related to engineering design learning, the results of the 185 individual student cases are mixed. Preliminary bivariate analysis found no significant correlation between an individual student’s perceptions about her group’s knowledge building process and overall learning in engineering design. Perceptions of a group’s capacity to handle cognitive conflict also was not significantly correlated with overall learning. This pattern held when the learning outcomes measure was narrowed to learning about the engineering design process (rather than engineering design content items.) There were also no statistically significant correlations between either the overall learning or process learning measures when we considered key measures of the social/relational dynamics of the group, the positive group interaction and social cohesion scores. The reported absence of social loafing in the group was significantly related to student learning however, with a correlation of +.157 ($p=.033$) with overall learning and $r = +.224$ ($p=.002$) for engineering design process learning in particular. Perceived psychological safety was not significantly related to overall learning, but approached significance as related to learning engineering design process ($r=.130$, $p=.079$). Finally, the summary measure of perceived group quality (summing scores reported by a student across all 3 dimensions) was not significantly correlated with overall learning, but it approached significance when correlated with process learning in particular ($r= +.142$, $p=.093$).

Recommendations and Implications

Results appear to highlight the importance of all three dimensions of small group dynamics to support student learning in design-based science, and small groups in high school biology classes showed variation in their location along these three dimensions. Somewhat surprisingly, it is rare for groups to be strong on the social/relational dimension while weaker on cognitive and affective dimensions. On the other hand, it is fairly common for groups to score either low or high on all aspects of the triple problem-solving space. The impact of students’ perceptions of their small group’s functioning on learning is modest, based on the preliminary analysis. Students appear to learn more during engineering design activities when there is minimal social loafing, even if they themselves are not engaging in it. Perceived group quality seems to have a greater effect on learning about engineering design process than specific content (i.e., knowledge of terms). Feelings of psychological safety seem to support this aspect of learning, as does, especially, the absence of social loafing. Notable here is that measure of the perceived cognitive function of the groups (the capacity to build on knowledge and ideas offered by others in the group, the ability to handle cognitive conflicts and work through disagreements about the material) did not appear to have an effect on learning about engineering design. Future analysis should take into account the individual resources, such as general interest in biology, particular interest in this biology class, self-concept and self-efficacy in biology, as well as social competencies students brought to the group prior to engaging with the three tasks. Incorporating both individual- and group-level measures, hierarchical linear modelling should draw on group-level reports about collective construction of the triple problem solving space, while also controlling at the individual student level for rudimentary pretest measures of engineering design knowledge. For the time being, our analysis suggests that perceived small group functioning most likely influences students' learning about engineering design process more than content. It could be that students' learned content, such as key vocabulary, through other means such as doing reading for homework, completing worksheets, or through teacher lecture, rather than actively using academic language as part of their group process.

References

Barron, B. (2003). When smart groups fail. *Journal of the Learning Sciences*, 12, 307–359.

Bennett, J., Hogarth, S., Lubben, F., Campbell, B., & Robinson, A. (2010). Talking Science: The research evidence on the use of small group discussions in science teaching. *International Journal of Science Education*, 32(1), 69–95.

Cartney, P., & Rouse, A. (2006). The emotional impact of learning in small groups: Highlighting the impact on student progression and retention. *Teaching in Higher Education,* 11(1), 79-91.

Cox, B. E., McIntosh, K., Reason, R. D., & Terenzini, P. T. (2014). Working with missing data in higher education research: A primer and real-world example. *The Review of Higher Education*, *37*(3), 377-402.

Dym, C., Agogino, A., Eris, O., Frey, D., & Leifer, L. (2005). Engineering design thinking, teaching, and learning, *Journal of Engineering Education*, 94(1),103-120.

Edmondson, A. (1999). Psychological safety and learning behavior in work teams. *Administrative Science Quarterly*, 44, 350-383.

Engle, R. A., & Conant, F. R. (2002). Guiding principles for fostering productive disciplinary engagement: Explaining an emergent argument in a community of learners classroom. *Cognition and Instruction, 20,* 399–483.

Good, T., McCaslin, M., & Reys, B. J. (1992). Investigating work groups to promote problem solving in mathematics. *Advances in research on teaching*, *3*, 115-160.

Jiménez-Aleixandre, M. P., Rodriguez, A. B., & Duschl, R. A. (2000). 'Doing the lesson' or 'doing science': Argument in high school genetics. *Science and Education, 84,* 757–792.

Johnson, D.W., & Johnson, R.T. (2009). An educational psychology success story: Social interdependence theory and cooperative learning, *Educational Researcher*, 38(5), 365-379.

Keys, C. W. (1997). An investigation of the relationship between scientific reasoning, conceptual knowledge and model formulation in a naturalistic setting. *International Journal of Science Education, 19,* 957–970.

Linnenbrink-Garcia, L., Rogat, T.K., & Koskey, K.L.K. (2011). Affect and engagement during small group instruction. *Contemporary Educational Psychology*, 36, 13–24.

National Research Council (1996). *National Science Education Standards*. Washington, DC: The National Academy Press.

National Research Council. (2011). *A Framework for K-12 Science Education: Practices, Crosscutting Concepts, and Core Ideas.* Committee on Conceptual Framework for the New K-12 Science Education Standards. Board on Science Education, Division of Behavioral and Social Sciences and Education. Washington DC: The National Academic Press.

NGSS Lead States. 2013. *Next Generation Science Standards: For States, By States.* Washington, DC: The National Academies Press.

Nieswandt, M., Affolter, R., & McEneaney, E.H. (2014). Interest, instructional strategies, and the creation of group space. *International Journal of Education and Psychological Research*, *3*(3), 1-5.

Roschelle, J. (1992). Learning by collaborating: Convergent conceptual change. *The Journal of the Learning Sciences, 2,* 235–276.

Rosebery, A. & Warren, B., (2008) *Teaching Science to English Language Learners*. Washington DC: NSTA Press.

Sargent, L.D. & Sue-Chang, C. (2001). Does diversity affect group efficacy? : The intervening role of cohesion and task interdependence. *Small Groups Research*, 32(4), 426-450.

Teasley, S. D., & Roschelle, J. (1993). Constructing a joint problem space: The computer as a tool for sharing knowledge. In S. P. Lajoie & S. J. Derry (Eds.), *Computers as cognitive tools* (pp. 229–258). Mahwah, NJ: Lawrence Erlbaum Associates, Inc.

ten Dam, G., & Volman, M. (2005). *Educating for adulthood or for citizenship: social competency as an educational goal* (Amsterdam, Graduate School of Teaching and Learning).

U.S. Census Bureau. (2014). 2010-2014 American Community Survey 5-year estimates. Accessed at https://www.census.gov/programs-surveys/acs/ on 1 Mar 2017.

Visschers-Pleijers, A., Dolmans, D., Wolfhagen, I., & van der Vleuten, C. (2005). Development and validation of a questionnaire to identify learning-oriented group interactions in PBL. *Medical Teacher*, 27(4), 375-381.

Acknowledgements

This study is part of NSF-funded project "Managing Small Groups to Meet the Social and Psychological Demands of Scientific and Engineering Practices in High School Science" (DRL-1252339).

Copyright statement

Training engineers in social innovation contexts: The case of the Social Innovation Science Park

Miguel Angel Gonzălez Palacios
Social Innovation Science Park, UNIMINUTO, Bogotá, Colombia
mgonzalezp@uniminuto.edu

Carmen Leonor Avella
Social Innovation Science Park, UNIMINUTO, Bogotá, Colombia
cavella@uniminuto.edu

Juan Fernando Pacheco Duarte
Social Innovation Science Park, UNIMINUTO, Bogotá, Colombia
jpacheco@uniminuto.edu

Abstract:

In Colombia, the presence of several social challenges requires for R&D capacities to focus in generating innovative alternatives to such problems. Since 2010, UNIMINUTO has designed and implemented the Social Innovation Science Park MD - SISP, whose objective, among others, is to articulate researchers with the needs and opportunities in communities, and to transfer to teachers and students the knowledge generated in the development of social innovations carried out by the projects and entrepreneurial initiatives incubated in the Science Park. In this context, this paper aims to present the current work on designing and implementing the 'Social Innovation School' inside the Science Park, and particularly, the projects and programs related to the engineers formation inside the university; special attention is given to some lessons learned throughout this process, regarding how universities' science and technology parks may impact the pertinence of the education strategies for engineers.

Introduction

The Social Innovation Science Park is an initiative carried out by the 'Corporación Universitaria Minuto de Dios – UNIMINUTO', a Colombian university which has been recognized as an 'Inclusive Business Innovation Model' in education by the G20. Although Social Innovation is a quite recent concept, the SISP team has proposed a definition, based on researchers such as Howaldt et. al (2012) , Phills et. al (2008) , and organizations like ECLAC (2010), thus proposing a conceptual approach, that define this concept as follows:

Social innovation refers to the pursuing, finding and implementing solutions which are novel, efficient, participatory and sustainable, to face problems that limit the achievement of better life conditions in a community. These problems include, but are not limited to, the usual problems related to poverty, and the new challenges related to issues such as climate change, water management, food security, the energy security, among others, on the most vulnerable populations.

Based on this definition, and through alliances with other stakeholders, including the Government of Cundinamarca and other research universities, the park has develop and implement a conceptual, strategic and management model, in order to support the development of social innovations through concrete projects.

In 2012, UNIMINUTO in alliance with the National Administrative Department of Science, Technology and Innovation of Colombia, develop a feasibility study for the SIPS, which produced the following conceptual model, based on five components:

- Observatory for Social Innovation.
- Knowledge Network.
- R&D Support Platform.
- Platform for Projects and Social Entrepreneurship.
- Social Appropriation of Knowledge.

The five components configure a systemic solution to strengthen social innovation capabilities, in an innovative scheme inside science parks context. Thus, the five components are not independent areas, but act coordinately to generate greater value to the initiatives involved in the SISP. This model has been operating since 2013, with very important results, such as incubating more than 50 innovation and entrepreneurship projects, with more than USD12 million invested by several stakeholders.

The 'Social appropriation of knowledge' has the objective to articulate the knowledge generated inside the social innovations developed by projects and entrepreneurial initiatives incubated in the Science Park to teachers and students. In this context, this paper aims to present the current work on designing and implementing the 'Social Innovation School', and particularly, the projects and programs related to the engineers formation inside the university; special attention is given to some lessons learned throughout this process, regarding how universities' science and technology parks may impact the pertinence of the education strategies for engineers.

Research questions

Science Parks always face the challenge of finding innovative strategies to articulate students and researchers with the problems and needs of the local industry. In the case of the SISP, this idea has been translated in how to train students with abilities in social innovation, to be able to contribute in the solution of environmental and social problems. Even though the Social Innovation School strategies are not exclusively related to engineering students, their participation is a key aspect, since future engineers would have the purpose to design and implement systems with a potential great impact on these problems.

Thus, the research question for this paper is: How to design and implement strategies to educate engineers in social innovation contexts? The following sections explain the theoretical background that supports the work currently being developed, and then one concrete strategy that has been implemented to pilot this framework.

Theoretical Background:

Social appropriation of knowledge - Definition

Knowledge has become a category that interacts in a mandatory and irreplaceable way with different environments of growth, development and learning of the modern society. According to Chaparro (2001), individuals, as well as territories, begin to configure their own action capacity from the creation, development and transformation of knowledge, based in their own abilities. In this logic, knowledge does not just imply an educative formal process, but also constitutes a scenery of social construction, inside a permanent dialogue between reality - analysis of the reality - development of new realities - transformation in daily praxis, or, as proposes the praxeology approach (Juliao, 2011), see - judge - take action - creative return. Thus, individuals, communities, organizations and the state are actors with "capacity to generate knowledge about their reality and environment, and with capacity to use this knowledge to design, forge and build their own future" (Chaparro, 2001, p.19)

This new perspective has led to a knowledge society, where the scientific and popular knowledge interact permanently framing a new dynamic to build and communicate the knowledge; in this approach, new knowledge emerges from and transforms the same context where it is inserted. Here, the experiences and tacit knowledge are combined into narratives that, with time, consolidate information that has been instituted on the collective knowledge to generate practice and learning that can lead to new ideas, which, once appropriated, consolidate a new form of understanding the knowledge to transform territories. This process of interaction between science and the society has been called Social Appropriation of Knowledge. "The Social Appropriation of Knowledge is understood as an intervention and comprehension process of relationships between Techno-science and society, developed through the active participation of the diverse social groups that generate knowledge" (COLCIENCIAS, 2010, p:22).

The Social Appropriation of knowledge is built, interacted and managed in territories through information that becomes data, in a process of self-reflection, auto-management and auto-analysis that gives new practice and learning, delivers knowledge that can generate paradigm changes and produces ideas to solve problems, challenges or opportunities in different fields of life. In terms of Social Innovation, this signifies redefining the knowledge which is co-constructed at the territory, from the technical and social knowledge, making use of the technologies for increasing the collective knowledge. In this context, the reflection and dialogue become an assertive medium for understanding new forms of seeing the territory.

In accordance with COLCIENCIAS (2010), the production and social management of knowledge, that is to say its creation, transformation and use is not found outside social context; conversely, it is sustained and developed around the interests, codes, knowledge systems and narratives, which empowered actors use in a pertinent manner to address their needs, and not only as simple receptors of a given information. The knowledge does not arrive and is not appropriated at the territory by default or by imposition; it emerges through the interaction, accompaniment and co-construction of confidence, knowledge and results between academics - researchers - territory.

Therefore, the Social Appropriation of Knowledge is a social construction exercise in which the territories' implicit knowledge enters into a dialogue with the technical and specialized knowledge of the facilitator teacher, and from there arise new knowledge and contents to be established and to be interiorized in the collective narratives that are being built; giving as a result a new paradigm to watch, feel and do the things. Reaffirming this view, the Social Appropriation of Knowledge is born and consolidated under the wing of communities through the communication process and participation in their ways of integrating the territory, because in this way narratives and languages come together and, with time, are established, producing own information and data of their daily experiences, which allow them to understand their reality. In this sense, the knowledge building and appropriation by the territories don't depend of others, but of "ourselves", changing thus the community paradigm from "wait for others do things" to "do with the others".

Here the facilitator teacher enters into the territory in a co-construction dialogue, in where he isn't the exclusive source of knowledge, but rather someone who puts into the daily knowledge his technical and specific knowledge in order to produce new analogies, knowledge and interpretations applicable for all; promoting the generation of diverse and new ideas that can produce innovations and development in the given context. This logic of knowledge co-construction, goes from the lineally understood world to a systematic and complex self-understanding one, in which all the different views provide applicable contributions for the territory comprehension, as well as for the definition of solutions and challenges which lead to sustainable environments for all. Considering the Social Appropriation of Knowledge from a formative perspective, the academic contents should be based in experiential cases which contextualize knowledge through an exercise of construction and reconstruction of technics and practice, in order to generate diverse and new knowledge. In this context the classroom becomes a laboratory of reflection on problems, challenges, opportunities and needs of the given context, which encourages the understanding of the science and possible solutions; the participant of the formative process engages in the comprehension and knowledge of his or her environment, assuming a posture of active learning, mediated by the action - the reflection - the theorization; whereas the facilitator teacher acts as a follower-mediator in the appropriation of knowledge process.

During the knowledge process, the student develops abilities and capacities for the critic analysis of their own reality, finding analogies, resources, and new reality interpretation forms, which allows them to generate new perceptions of it.

Conceptual foundations to understanding the Social Appropriation of Knowledge at the Territory

In order to consolidate processes of Social Appropriation of Knowledge inside academic activities working together with territories, three conceptual premises have been defined:

a) Social Construction of Knowledge: From this perspective, language and narrative underlay as a constitutive aspect of the social construction of knowledge, and of the legitimization of the process of social appropriation of this knowledge, as well as of the related relationships and interactions. The above implies understanding the inherent inter-subjectivity of everyday life, recognizing that actions and their meaning are a social construction, that emerges constantly from interaction and it's circumscribed in the context that it develops. In this discursive process, the language - thoughts - reality, interact simultaneously giving content and sense to the stories - concepts - experiences, that are narrated and transformed thanks to their internalization thus, establishing new codes, concepts and categories which nurture the epistemological view of the reality, contributing in this form to achieve higher objectivity levels, critic analysis, evaluative analysis, and of the cognition inside the ' know - thought - take action' exercise; in other words, it is producing a knowledge transformation, based on the distinctions and forms that have been established in this way.

As affirmed by Berger and Luckman, inter-subjectivity and narratives constitute meeting scenarios where individuals build the world collectively, from their own approach and own knowledge. "The everyday life arises as a reality interpreted by humans, and for them have a subjective meaning of a coherent world" (Berger and Luckman, 2001: 36) going beyond the face to face meeting, into the one between shared meanings, which put in manifest the social construction of reality and, therefore, of the knowledge which society generates.

b) Systems Thinking: Thinking in a complex and systemic way, does not mean to complex - hinder - lock processes, knowledge or activities; conversely, the systemic thought constitutes a knowledge model oriented to understand the world and its categories in a multidimensional form, which contemplates the inter-connection between the whole and its parts, in order to solve the challenges which the different social, economic, political and cultural systems propose, "the borders of the map do not exist at the territory, fate on the territory, with barbed wire and customs officers" (Morin, s.f: 37), in other words, the categories and territories which we generate are the result of our mental limitations and of the linearity, with which we watch the reality, if we overcome the mental walls which have been created, multiples changes to challenges, that currently do not have a solution, could be conceived.

In accordance with the praxeological approach, this category constitutes a particular form of understanding the process, which leads to considering the observation (see), auto reflection-analysis (judge), staging (take action) of the flow between the process and the world where it is developed, going through an integral multidisciplinary comprehension exercise by the involved actors (creative return). In this dynamic the process and its parts interact permanently between the academic investigator and the territory, producing new interactions which transform the thinking forms and the "action taking" of the actors involved.

c) Sustainable Local Development: "The human progress always had depended of our inventive technical and of our action capacity... often this capacity has been used of constructive manner to achieve the progress of develop" (Report-Brundtland-Aug-1987: 51). Every territory is endowed with a particular logic, that underlays its trajectories, cultural resources, social resources, economics and politics resources, and provides it with specific learning competences and abilities, which, watched and tapped, produce skills and active knowledge proper to its region. Sustainable development, is the individuals and collective capacity from a territory to empower their resources and produce structural changes in a systematic manner, making use of collective learning, in order to develop and increase the relational, social, business and institutional competences and skills, hence the territories count with an own and invaluable potential for producing its own development and for meeting the challenges of innovation in a competitive and sustainable way, becoming thus regions with competitive advantages faced to other territories.

Learning logic for the Social Appropriation of Knowledge at Territory

In this framework, a learning logic mediated by the praxeological pedagogy is required, where the territories and students assume an active posture in their learning process, and the facilitator-teacher acts as a follower-mediator in the social appropriation of knowledge. In this context, the participants make a self-reflection about their experiences and everyday practices making a reality analysis, and proposing new postures that emerge from the communication exercise and shared narratives, towards the construction of collective knowledge.

The praxeological pedagogy, is an action one, in which the subjects (territories - facilitator teacher - students) interact permanently, turning the classroom into a self-reflection scenario, in order to unveil new inherent postures to their practices and formal and not formal knowledge, attempting to transform narratives and paradigms about their own reality; thus, the praxeological pedagogy "proposed an action, which reflects critically, and a critic reflection molded for the practice and value on this" (Juliao 2016, p. 97). In this context, the teacher becomes a facilitator that guides the co-construction process of knowledge and its appropriation.

In the exercise of everyday praxis, Juliao (2014) proposes four components, which, in the process of social innovation and appropriation of knowledge, will be the transversal axis used by the facilitator-teacher to intentionality guide the pedagogy towards the exercise of reflection - formulation, ideation of action and transformation proposals. These components are:

- See: Involves the exercise of the imagination and observation, to question and see ideas around the knowledge object.
- Judge: Requires to abstract and interpret practices experienced under the knowledge light; connecting the implicit knowledge with the scientific knowledge in order to get new inferences, conceptions and explications of the reality argued and to produce ideas and data concerning the challenges defined.
- Take action: Consolidating a new reality representation, establishing connections that produce changes in the ways of being - thinking - regarding the object of knowledge.
- Creative return: The pedagogic experience is represented and arranged, in this particular case through the knowledge village, allowing to glimpse at the reflections about the pedagogic practices lived, and the changes and innovations generated when the individuals takes control of their territory knowledge.

In this process, the facilitator-teacher introduces territories and students into an interaction exercise between divergent thinking (creative - adductive - analogical) and convergent thinking (logic - analytic - systematic), facilitating for the participants the development of abilities to be innovators, while they acquire tools to generate, manage and make ideas viable, in dynamic and complex contexts.

a) Creative thinking: "the creative talent of humans has been constituted into the basic core of develop processes" (Cerda Gutierrez: 2006, P. 7). Creative thinking is thus a learning strategy, which allows the participants in the formative process to build capacities for transforming their environments, through the imagination, experimentation, conceptualization, invention and theorization of their reality. This way of thinking sets up a different way of conceiving and addressing the problems posed by the world. To activate the creative act in the learning process it is important to change from a lineal and traditional convergent thinking to a divergent and complex thinking, in which the problems, dilemmas, opportunities and needs, which the world faces us with, are seen under a systematic magnifying glass, which enhances the interpretations spectra and possible solutions; exploring different connections and pathways in order to obtain the most appropriate solution for the context and available resources.

b) Analogical Thinking: It constitutes a natural form of thinking, in which it is conducted a search of similarities between objects - situations, establishing possible connections, relationships, representations of the observation object. During the analogical process, the participant extends the capacities to transcend from the particular to the symbolic, using what is being seen - felt - observed as an inspiration source to generate different and new scenarios to establish viable hypothesis. This type of reasoning aims to find the fundamentals of what is known, and, from there, to find new fundamentals which increase the knowledge spectra and the capacity to solve dilemmas, problems, opportunities and needs.

c) Abductive reasoning: It is based in the employment of explanatory inferences, that seek to find conclusions and consequences, as well as the explanation of fact - phenomenon; and to find possible connections between the cognitive process and analogies, from the creation experience of new ideas. "The abduction proposals give a primordial role to the experience, and to being open to the outside" (Barrena: 2007, P. 85), in this sense, and in accordance with the founder of abduction Charles Pierce, knowledge emerges insofar the world is being internalized; the 'subjectivation' that has made of this, inferring, postures and possible ways, in the reality construction process.

d) Logical thinking: While creative thinking activates the brain to think differently and develop multiple ideas and diverse pathways, opposite the needs, dilemmas, problems and opportunities, which the world proposes; logical thinking categorizes, orders and establishes the route to undertake. In other words, it directs the ideas, yielding structure, argumentation and judgment which emerge from knowledge and past experiences. In this process of converting the creativity into concrete solutions, it is possible to develop the understanding capacity to categorize and assimilate ideas, yielding a directional and rational sense, which is applied to find the relationships between effect - cause, allowing the development of an elaborate discourse, which can go through testing and verification.

These four elements are integrated with a systems thinking approach, which has been described above.

Methodology: Pilots design and implementation

In order to implement this theoretical framework into concrete strategies, the innovation method of design and implement pilots was also used as research methodology, in order to figure out the best possible design for academic courses in the training of innovators, particularly engineers.

To achieve this objective, and based on its experience working with communities and territories, the team of the Social Innovation Science Park looked for partners with experience in innovation and with a similar approach. The i.school of Tokyo University was identified as a key partner for the process.

i.School was founded in 2009 as a program developed by the faculty of Engineering professors, and has been focused in the education to develop abilities that drive students to generate human centered innovations, using participative methodologies in practical areas and deep discussions. The structure of the program is based on three fundamental steps: understand + create + realize. This means, carrying out field observations and case studies to define opportunity areas and new ideas, which may further generate reachable prototypes that answer human wellbeing needs. i.school yearly realizes an open summer workshop, with participants (university students and professors) coming from universities from all over the world.

The SISP began the joint work with i.school in 2014 through some academic exchanges and the participation of some staff in the summer workshop, from where the idea to develop the 'IDEAS PCIS' course emerged. Afterward, during 2015, a team from both institutions began to design the curriculum of the course, mainly based on the work developed by professor Shunsaku Komatsuzaki, integrating it with the experience of social appropriation of knowledge of the SIPS developed with more than 30.000 people from communities in Cundinamarca.

Generally speaking, the IDEAS PCIS course was designed to involve the four moments mentioned below, in an agenda described further:

- Preparation of the creative thinking.
- Individual thinking and self reflection.
- Group thinking for idea generation.
- Socialization.

Table 1: Agenda of IDEAS PCIS Course

Day 1	Analysis of Colombian and Japanese regional innovation cases
Day 2	Kamaishi case analysis
Day 3	Field observation
Day 4	Analysis of field observation
	Discussion with entrepreneurs
	Regional Innovation Project Design
Day 5	Regional Innovation Project Design II
	Projects socialization

During the development of the course, several activities were developed, and these were oriented to develop abilities related to creative thinking, analogical thinking, adductive reasoning, logical thinking and systemic thinking.

Main Results

The design of the course 'IDEAS PCIS' presented above has been implemented twice, the first time during October 2015 in Bogotá, and the second one in October 2016. The participants, locations, professors and dates are presented below:

Table 2: IDEAS PCIS Course description

Year	Location	Participants	Main Professor
2015	Bogotá	30	Shunsaku Komatsuzaki i-School Vice-director and PhD in Engineering from Tokyo University.
2016	Bogotá	40	Shunsaku Komatsuzaki
	Ibagué	40	Daniel del Barrio PhD in Engineering and Professor at the same University

According to the student experience survey applied in IDEAS PCIS 2015, more than the 73% of participants want to keep working on social innovation aspects with the SISP, in order to achieve objective such as:

- Change mindsets in finding solutions to social problems.
- Develop social innovation methodologies.
- Formulate and implement social innovation projects.
- Increase conscience of innovation inside our social context.
- Implement the process in their own courses and activities.
- Develop abilities and competences for social innovation.

Thus, the IDEAS PCIS course has proved to be a successful strategy to implement the social appropriation of knowledge framework and to develop social innovation abilities in students, at least according to the qualitative data collected during the courses. Future plans include the introduction of a social innovator competences assessment instrument, which will provide quantitative data to measure the impact of these strategies

Below are some of the results of the course in the years 2015 and 2016 in the city of Bogotá:

- The distribution by gender of the 40 participants was homogeneous, with the women predominating in the experts level and the men in the students level.

Figure 1. Distribution by gender

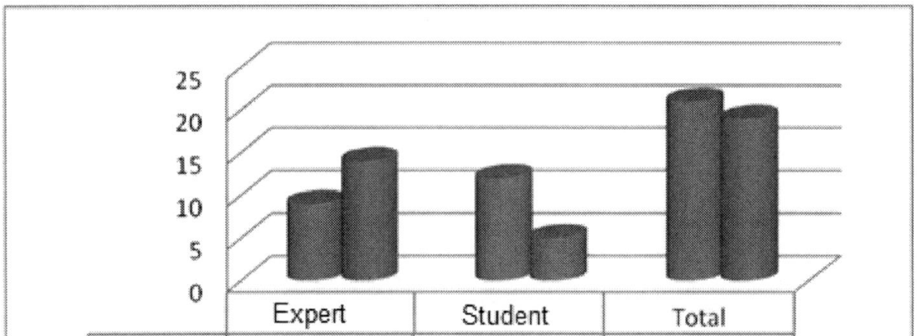

	Expert	Student	Total
■ Men	9	12	21
■ Women	14	5	19

- The age range of the participants was in accordance with the participation requirements which established a maximum age of 40 years in order to participate. Regarding the age range of the two levels of the participants, the students one had mainly persons between 20 and 24 years with three persons over 30 years old, whereas in the experts level the main ranges were 25-29 and 35-40, with only one participant over 40 years old.

Figure 2. Distribution by age range

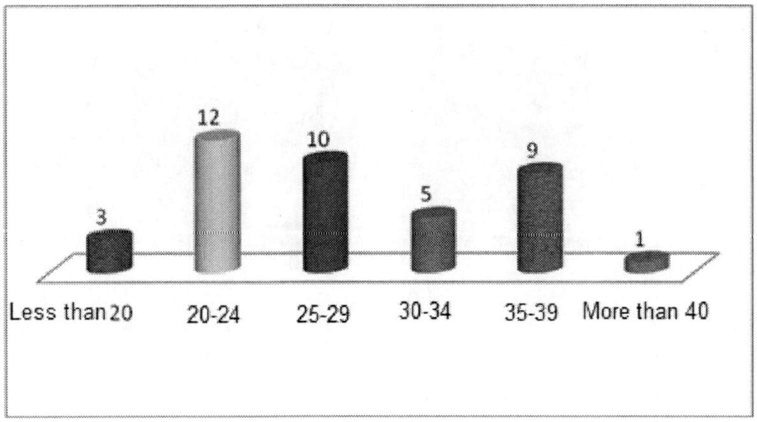

- The main professions of the workshop participants were: business administration (9), public accountant (4), agroecology (3), social worker (3) Presently, 44% (18) are undergraduate students, going through the mid part of their studies, that is to say semesters IV-V-VI.

Figure 3. Participants' professions

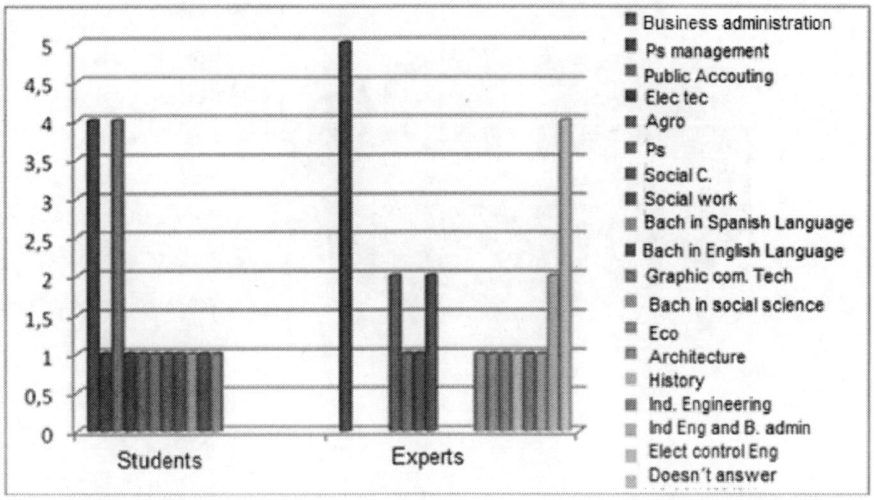

- 52% (12) of the experts level participants have not undertaken post graduate studies; 26 % are carrying out masters' studies; 18% (4) are studying a specialization and 4% (1) is developing doctoral studies.

Figure 4. Post graduate studies level of the experts level participants

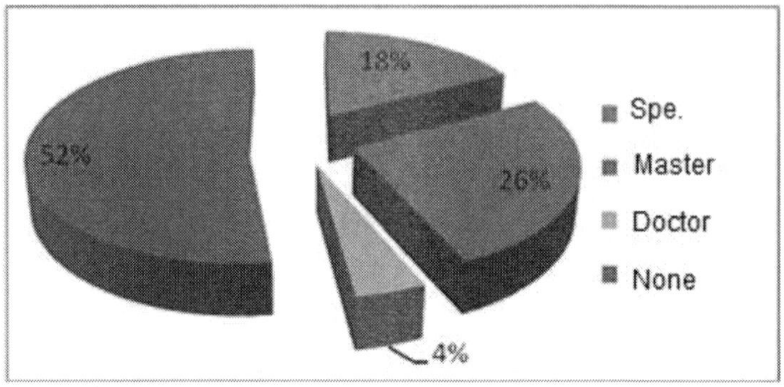

- From the 44% (18) of the participants that are developing their undergraduate studies, 32 % (13) dedicate all their time to their studies, and 12% study and work at UNIMINUTO as teachers assistants and administrative assistants.
  Regarding the faculty there is a group of professors 17% (7) and a group of coordinators in research and project areas, CED, 12% (

Figure 5. Participants' occupations

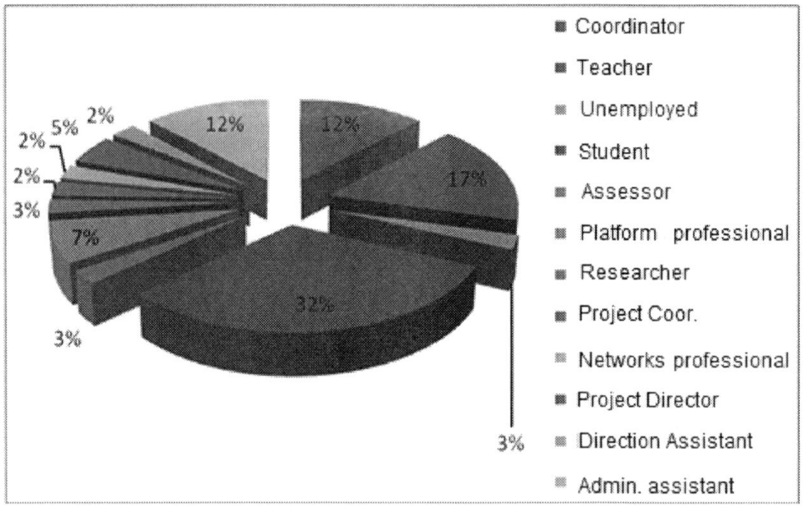

- In broad terms it could be stated that the participants of the course did not have an innovation background, as only 30% said to have carried out or participated in innovation-related activities.

Figure 6. Education and/or activities of the participants in the innovation field

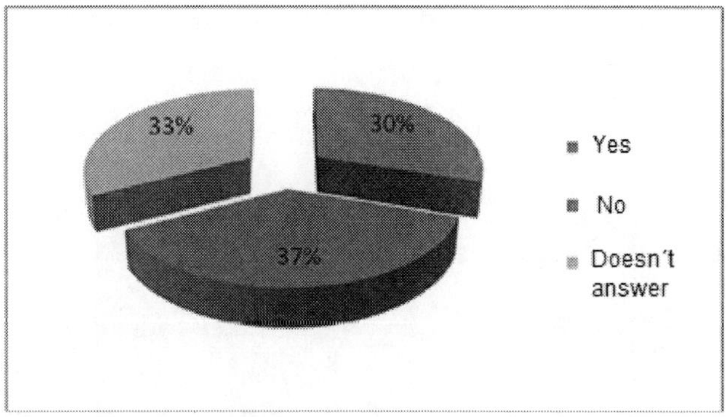

Below there are some results of the course in the year 2016 in the city of Bogotá

In the city of Bogotá the course was directed by the professor Daniel del Barrio, it had 41 participants, 25 professionals and 19 students

Figure 7. Participants´ Gender

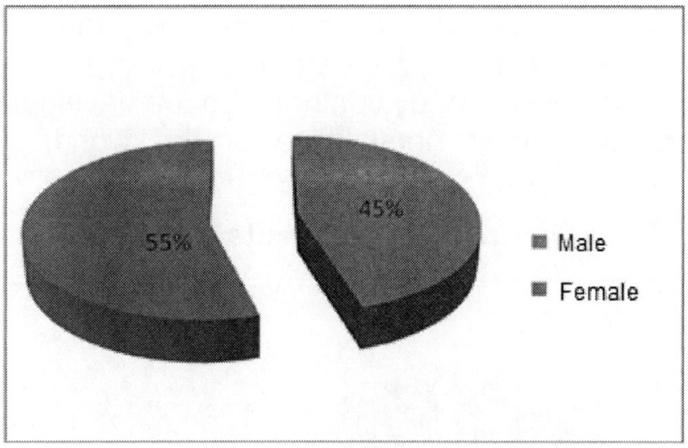

From the total of participants 55% (24) are women and 45% (20) are men, presenting an equitable distribution by gender, which means that social innovation issues are not gender exclusive issues, conversely both men and women show interest in strengthening their skill and competences in this field

Figure 8. Participants´ Age

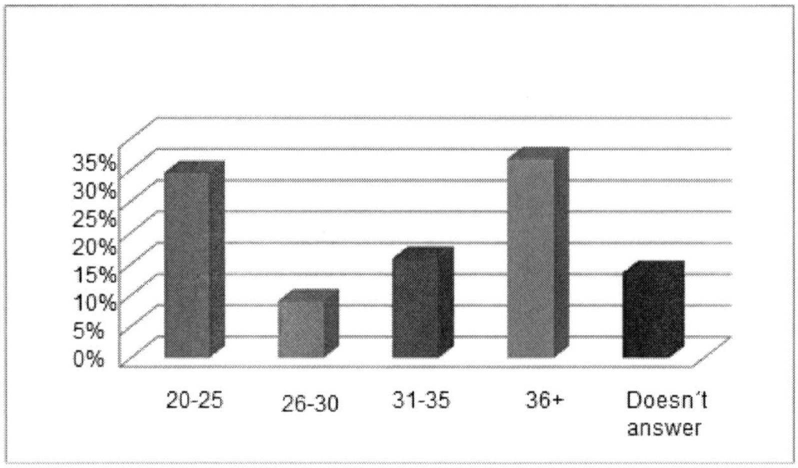

The participation age starts at 20 years old, this is an important factor as the course seeks to create a dialogue between the tacit, formal and experiential knowledge of the participants, and, therefore, it is intended that different generations are included, and that they contribute to the development of viable and sustainable solutions from their perspective of the world, its problems and challenges

Figure 9. Participants´ Age

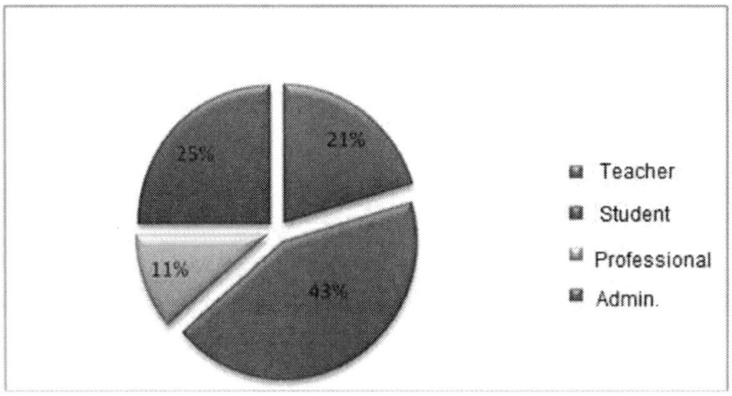

The group of participants includes undergraduate students (43%, 19), and persons who already have their bachelor's degree, some of which have also master's and doctoral degrees (57% -25), the latter group is constituted by professors, professionals, and administrative and managerial staff; this means that the social innovation subject is relevant to every strategic and operational level of the organization

Figure 10. What students study

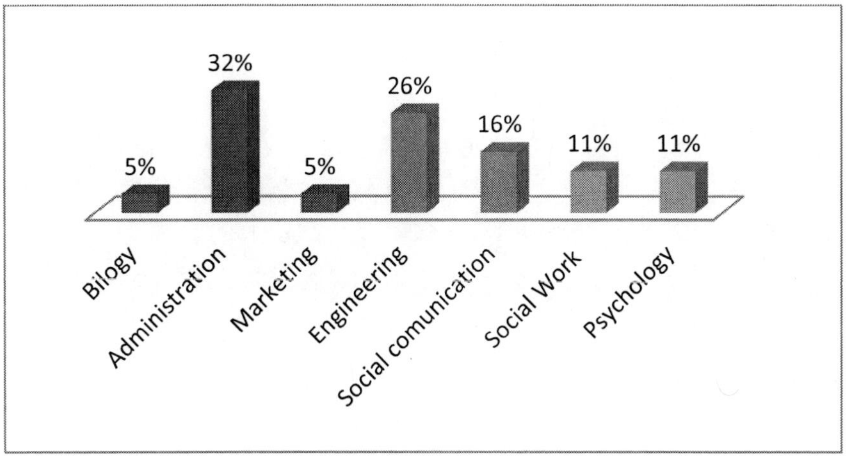

The undergraduate students' majors include Business Administration 32%, Engineering 26%, Social Communication 16%, and Social Work and Psychology in a smaller percentage. Social Innovation is not exclusive to a career and for students and developing tools and methodologies that help to understand problems and to devise solutions focused in the human being is enriching for the professional development of all students, regardless of their specific field of studies.

Figure 11. Current semester of the students

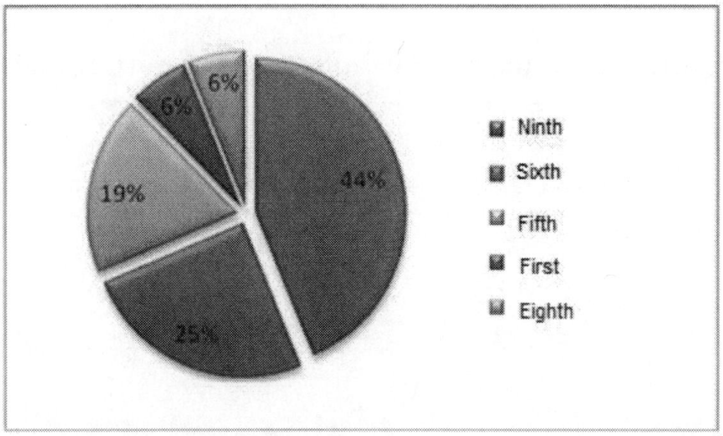

In broad terms most of the students are in the latter semesters of their studies (44% ninth semester, 25% sixth semester)

Figure 11. Participants´ professions

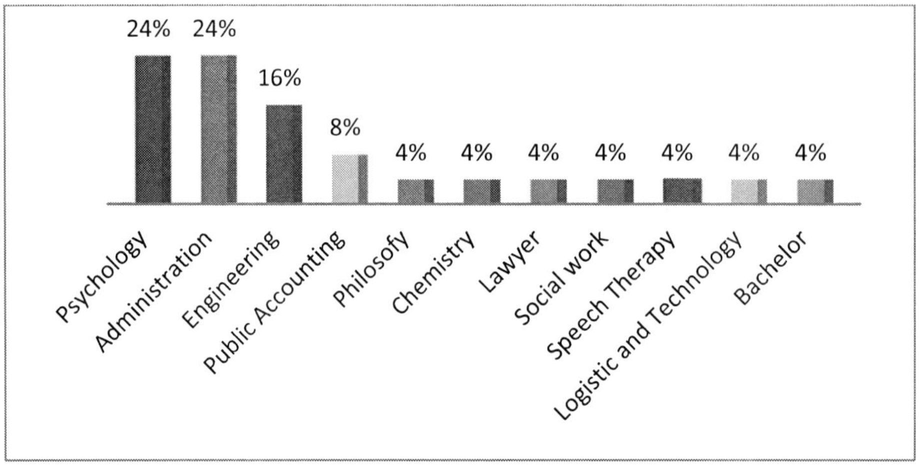

From the group of people who are already in the working world, whether it be as a teacher, administrative or managerial staff, or as a professional, their main work fields are Psychology, Business Administration, Engineering, and, to a lesser degree, social fields such as Philosophy, Social Work, Education.

Figure 12. Course contribution to the participants

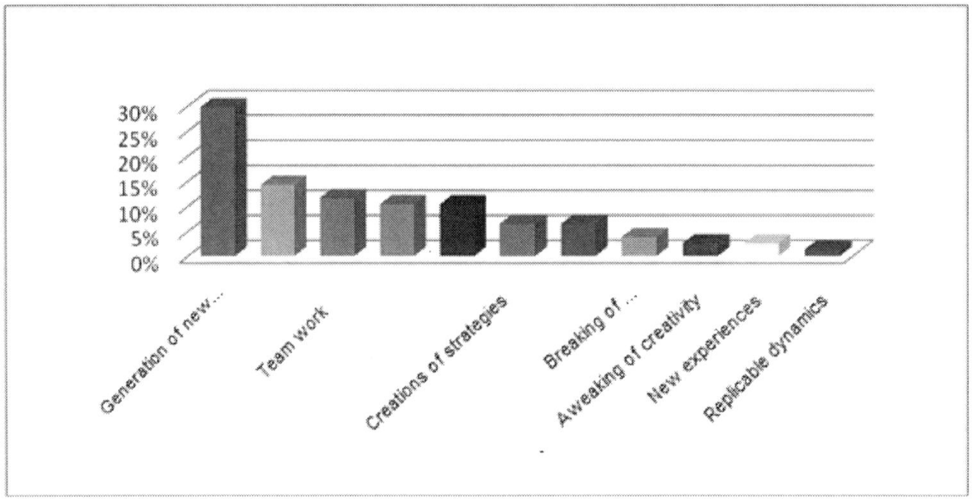

The course provides to its participants the generation of new ideas through a defined methodology that allows them to understand the territory and the people that inhabit it, develops teamwork competences, and facilitates problems segmentation in order to generate solutions using strategies suitable to each situation presented.

In the personal level, it strengthens knowledge, awakes the creativity, breaks paradigms, and generates new experiences that can be replicated in their professional contexts.

Recommendations and Future Research

As has been described above, the Social Appropriation of Knowledge framework has great potential in academic scenarios, particularly in engineering education. Since this an approach currently being developed, further theoretical work would be needed to connect it with similar approaches, such as service engineering, project based learning, humanitarian engineering and others. This would provide interesting insights for both conceptual and practical work.

This paper has focused in a particular strategy being carried out by the Social Innovation School, however, in a parallel experimentation work, other strategies are currently being developed:

- Calculus course. One of the mandatory courses at engineering programs is being redesigned to involve an Action Based Learning approach, based on the results of the STEM Robotics project developed with more than 2,000 students from public high schools.
- Social Innovation courses, specialized in Green Business Management, Innovative Education and other areas of expertise, delivered by the teams working in the SISP, are being designed and will be implemented starting in 2017-2.

Results from the implementation of new strategies will be crucial in the development of this framework. They will be also the scenario to implement and social innovator competences assessment instrument, that is currently being designed.

References

Amartya Sen. (2004) Desarrollo y libertad. Bogotá, Colombia: Planeta, séptima reimpresión.

Barrena Sara (2007) La razón creativa, crecimiento y finalidad del ser humano según Charles Peirce. Madrid, España: RIALP.

Berger Peter, Luckmann Thomas (2001) La construcción social de la realidad. Buenos Aires, Argentina: Amorrortu Editores decimo séptima reimpresión.

Brown Tim (2008) Design Thinking. Harvard Business Riview, América Latina.

Beuchot Mauricio (2004) Introducción a la lógica, México: Universidad Autónoma.

Castro Ramírez Maria Eugenia (1998) Globalización y Desarrollo Sostenible v/s Autodesarrollo y Sustentabilidad.

Centro de Investigación y Desarrollo de la Educación (1997) Técnicas Participativas para la Educación Popular. 3ª Ed. Santiago de Chile

Cerda Gutiérrez Hugo (2006) La creatividad en la ciencia y en la educación. 2ª Ed. Bogotá, Colombia: Aula Abierta Magisterio.

Chamorro Fernando (2001) Conocimiento, aprendizaje y capital social como motor de desarrollo. Brasilia, Brasil: Revista Ciencia de la información V30 n. 1.

COLCIENCIAS (2010) Estrategia nacional de apropiación social de la ciencia, la tecnología y la innovación, Bogotá

Elder Linda, Paul Richard (2003) Pensamiento analítico.

Gergen Kenneth (1996) Realidades y relaciones, aproximaciones al construccionismo social. México.

Hacking Ian (1998) La Construcción social de qué?

Juliao Vargas Carlos G (2011) El enfoque Praxeológico. Bogotá, Colombia: UNIMINUTO

Juliao Vargas Carlos G (2016) Una pedagogía praxeológica. Bogotá Colombia UNIMINUTO.

Hideyuki Horii (2015) Designing Innovatión (documento de trabajo) i. school UTokio.

Komatsuzaki Shunsaku (2015). Introduction to i.School. Curso IDEA PCIS 2015 Bogotá, Colombia

Marchioni Marco (1999) Comunidad, Participación y Desarrollo. Teoría y Metodología de la intervención comunitaria. Madrid

Max Neef Manfred, Antonio Elizalde, Martín Hopenhayn (2006) Desarrollo a Escala Humana. 3a Ed. Barcelona, España: Romanya/valls, S.A.

Morin Edgar (s.f.) Introducción al pensamiento complejo

Oliva Jose Maria (2004) El pensamiento analógico desde la investigación educativa y desde la perspectiva del profesor de ciencias. Cadiz, España. Revista Electrónica de Enseñanza de las Ciencias, Vol. 3, Nº 3, 363-384.

Palacios Gil Juan (2005) El crucigrama: retos e ideas para desaprender a pensar. Madrid: Diaz Santos.

Parque Científico de Innovación Social. Documentos internos de trabajo (2010 – 2015) UNIMINUTO

Raff Carmen (2004) Desarrollo Autónomo. San Jose, Costa Rica: EUNA

Soler Toscano F (2012) Razonamiento abductivo en lógica clásica.

Acknowledgements

Acknowledgements to UNIMINUTO and to the Government of Cundinamarca, which have been the main sponsors of the innovation and research processes developed by the Social Innovation Science Park.

DEVELOPMENT OF STRATEGIES TO EASE EARTHQUAKE ENGINEERING LEARNING

Diana Marcela Soler Rivera
University of Ibagué, Ibagué, Colombia
dianasoler1995@gmail.com

Yvonne Merino Peña
University of Ibagué, Ibagué, Colombia
yvonne.merino@unibague.edu.co

Oscar Javier Araque de los Ríos
University of Ibagué, Ibagué, Colombia
oscar.araque@unibague.edu.co

***Abstract**: The principal goal for structural dynamics courses in civil engineering programs usually focuses on the representations of earthquakes' effects in structures through the formulation of analytical models. In countries in with high levels of seismic risk some universities have concentrated their efforts in developing teaching methodologies to help students understand the dynamic phenomenon. However, in countries with limited resources to acquire technology for education, it is common that these strategies are restricted to a simple numerical analysis in structural dynamics. This research presents some low-cost technological and educational tools developed by engineering students in University of Ibagué in order to support teaching-learning processes in structural dynamics courses at the undergraduate level. The instruments and strategies used to reach the objectives in this project are conceived from technological advance ideas implemented by other institutions with background in engineering education research and also from particular academic needs identified from students conditions in our local context. Finally, results are presented as the design of experiments for structures models tested on a bench-scale shaking table.*

Introduction

Mechanics field has shown to be a challenging science area for engineering students. Different authors report difficulties in learning basic mechanics concepts and in the development of skills for solving typical engineering problems related to basic mechanics (Dwight & Carew, 2006). It is possible to demonstrate these difficulties by observing the students behavior in the face of problems' analysis and solution recorded in their courses assessments, as shown in data below.

Understanding the effects of seismic movements on structures is of great importance in the civil engineering community; by means of this perception, it is possible to develop new criteria for design and minimize the potentially life-threatening consequences of having a vulnerable infrastructure exposed to such natural events. At the undergraduate level of engineering programmes, students have few or no practical and realistic experiences that complement the theoretical material of the books concerning the behavior of structures subjected to earthquake loading (Dyke et al., 2000). Due to this fact, the understanding of some fundamental concepts and principles of dynamics become a problem for the student since they are abstract in nature (Karim, 2011).

Some advantages of learning based on active acquisition of knowledge through real experiences in engineering field against learning from instruction could be summarized as:

- Complete learning, authentic and realistic, it allows for the student to be capable of using all his abilities (Briede, 2013).
- Allows for interaction between the learning process participant and the surrounding world (Briede, 2013).
- This theory is the most efficient in development of professional competences (McHenry et al., 2005).
- Learning is extremely connected with the individual. It makes possible to build his own mental models, rules and meanings in order to make sense with the lived experiences (McHenry et al., 2005).

Constructivist learning theory asserts that for learning to take place, knowledge must be actively constructed or reconstructed by the learner's mind through experiences, this means that learning is not what is simply transmitted from the teacher to the student (Avitabile, 2007). This is why in countries with high levels of seismic risk some universities have developed learning techniques focused on the acquisition of knowledge, skills and self-constructed competences that challenge and promote innovative thinking of students. These techniques are based on the introduction of a series of “hands-on” experiments and classroom demonstrations throughout the courses of structural dynamics.

These type of “hands-on” experiments are based on the use of bench-scale shaking tables, and are thought to be capable of reproducing the movement of the ground during an earthquake, and to allow a control of structures subjected to earthquakes (Dyke et al., 2000). Further, shaking tables provide students access to measure dynamic response of structures models and also the design and construction of scaled models in order to be “seismically” tested. Even so, these strategies are applied in Colombia by few universities, because of the high cost of acquiring sufficient mechanical technology for educational purposes.

Dynamic structural analysis is usually looking at the representation of earthquakes’ effects on structures, through the formulation of analytical models that demand several assumptions. However, undergraduate students in general Civil Engineering programs frequently express difficulties to understand contents in dynamics from visualizations of static representations with numerical models. In recent years, several students taking Structural Dynamics course in University of Ibagué have revealed difficulties in learning processes related with the integration of analytical concepts and practical experiences. Indeed, the need for improve results by strengthen teaching-learning processes has brought to formulate this potential strategies with special emphasis on learning from practice in order to promote critical thinking, by using limited resources.

The focus of this paper is to present the low-cost technological and educational tools that have been developed at the University of Ibagué to support the teaching-learning processes in Structural Dynamics course at undergraduate level in Civil Engineering program. It describes the characteristics of the bench-scale shake tables selected for the development of this project, considering that they contribute to the measurement of the effects of torsion in the plant, since they were built to be operated in two directions; and presents as a result experimental tests in scaled structures subjected to dynamic loading in the shake table, which allow students be capable of making right decisions using sensible judgment, have the capacity of critical thinking, and be able to develop a structural engineering intuition.

Context

Although dynamic analysis in structural engineering represents the principal tool to describe, analyze and understand mechanical behavior of structural systems subjected to time-varying loading, undergraduate students show difficulties for understanding the primary characteristics of motion due to limitations for generating experiences where they experiment with real-life situations. For this reason some universities have focussed their effort in the development of strategies to formulate experiments in Structural Dynamics, making necessary the building or acquisition of technological tools that reproduce dynamics loads in structures, such as the bench-scales shake tables presented.

In this research we looked into Civil Engineering's curriculum for different universities to identify the dynamics' courses scope. Ten institutions were examined; five of these are prominent according to QS world ranking, especially in engineering subject, and five more universities with high quality in Civil Engineering programs in Colombia.

The principal objectives to make these comparisons include:

- Identify similarities in Structural Dynamics' curriculum between University of Ibagué and the rest of institutions examined.
- Identify differences for the time locations for students to take this course in each university.
- Identify the existence of a practical component considered in the course contents.

Table 1: Contents similarities for some universities

Common Concepts/University	U. of Ibagué	A	B	C	D	E	F	G	H	I
1	X	X	X	X	X	X	X	X	X	X
2			X	X			X			
3	X	X	X	X	X	X	X	X	X	X
4	X	X		X	X		X	X	X	X
5	X					X				
6	X	X	X	X	X	X			X	
Practical component		X		X						X

Common concepts are:

- Concept 1: Introduction to dynamics.
- Concept 2: Universal gravitation and conservation laws.
- Concept 3: Single-degree of freedom systems.
- Concept 4: Multi-degree of freedom systems.
- Concept 5: Basic concepts of seismic movements.
- Concept 6: Analysis of the dynamic response of structures.

According to the preliminary review, the basic concepts in dynamics of SDOF systems are similar for most of the subject contents, but they show some variations in MDOF systems scope and in the insertion of fundamentals of design earthquakes, which probably will be studied at postgraduate level. Additionally, only some of the universities involve a practical component.

In many universities around the world, the subjects where students get closer to fundamentals of structural dynamics and earthquake engineering consider a practical component in their curriculum design. In some countries with limitations for education investment, teachers must work by the hand of undergraduate students to build their own labs and equipment to make demonstrations of dynamics systems.

As a fundamental aspect of constructivist teaching, students should be assessed formatively (Briede, 2013). However, for this research, summative assessment has been useful to identify difficulties in the student's' performance. As proposed in (Garrison & Ehringhaus, 2007), we analysed results from summative assessment for dynamics courses in the past 11 semesters, in order to guide efforts and activities for formative assessment in actual courses.

Figure 1 shows the students' performance along 11 different semesters in subject dynamics course, evaluated in three partial reports. Comparing curves shows poor performance in the first report, in which, for usual engineering programs, students should study fundamentals of *single degree of freedom systems*.

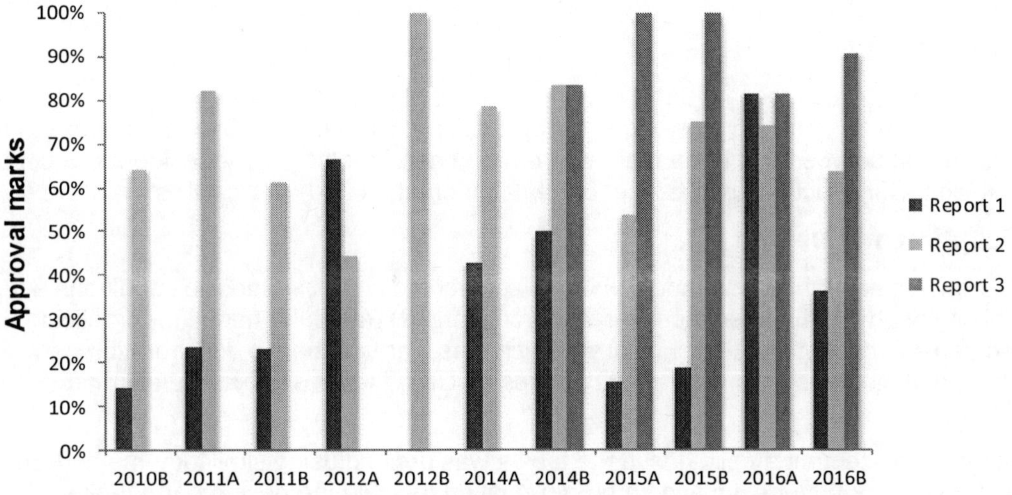

Figure 1. Students that reached the minimum passing score for each report (values in percentage terms).

Results for last 6 years data suggests that the first report of scores is challenging for undergraduates; in this period they must be have clear basics concepts of mechanics regarding to simple harmonic movements and dynamic properties in SDOF systems.

Due to the difficulty to describe dynamic phenomenon using statics images, and that should be presented with waves and spectra, other research groups in Japan (Fukuwa et al., 2008) have developed simple experiment models including motion simulators with inverted pendulum, two-story building models with struts, seismic shear walls, base isolation mechanism, damping mechanism, and soil liquefaction models. Results from the application of these devices have led to implement new technologies to automate the operation for the experiments, and have showed progress in techniques for education in society consciousness about seismic risk starting from undergraduate students helping community to understand importance of seismic adaptation.

In local context, shaking tables have been used in real-life scale to analyze dynamic response of structures, and some of them are used in engineering education. However, experiments with these simulators require significant invest of time and high cost to produce real-life scale models; most of them are better founded by research projects in earthquake engineering. Some institutions have produced bench scale simulators in order to describe dynamic behavior of physical systems that help students observe how mechanical properties define dynamic response of simple models representing common structures in civil engineering. Despite the good educational potential of using mechanical simulators for earthquake engineering courses, economic resources for acquiring materials and equipment to elaborate mechanical devices are limited in most universities. Few institutions have developed simulators that reproduce motion in one direction to formulate experiments within structural dynamic courses.

Civil Engineering Program from University of Ibagué has focussed its effort in constitute a student's research group in collaboration with Mechanical Engineering and Electronic Engineering Programs, in order to integrate three different disciplines to produce two low-cost shaking tables able to reproduce bi-directional motion, controlled by variable displacements and vibration frequencies. These technological tools are designed especially for complement structural dynamics courses with practical hands-on experiments that lead to involve earthquake engineering fundamental concepts in these courses, and generate social consciousness about seismic risk from students to the rest of the community.

Methodology

As a work-in-progress research, this document describes the methodology used to formulate proper mechanical simulators as a tool for reproducing dynamic effects of earthquakes in structures. Objectives have been reached from a general method that considers Conception and Design, Construction and Testing, and Implementation and Applications.

Conception and Design

For design, scaled tables and models were sized according to: 1) Specific conditions and availability of physical and economical resources, and 2) Particular needs for practical experiments in dynamics. All technical specifications were defined in collaboration between teachers and students from different disciplines as Civil, Mechanical, and Electronic Engineering.

Range of displacement amplitudes in the tables was defined using similitude method Eq. 1 (Muhlenkamp, 1997) for estimating a proper relation to scale the design earthquake displacement. In the same way, range of frequencies was determined so that structures models shaking on the tables, allow observing effects of dynamic loading in buildings. For this purpose, analytical models were built to estimate possible frequencies of different types of systems to be tested in the experiments. For the choice of the maximum frequency that must reproduce the shaking tables was taken into account that the strongest earthquake recorded in Colombia had a maximum frequency of approximately 15 Hz (Carrillo et al., 2013).

$$\left(\frac{1}{\lambda_M}\right) M\left[\left(\frac{{\lambda_t}^2}{\lambda_V}\right)\ddot{x} + 1\left(\frac{1}{\lambda_{Ug}}\right)Ug\right] + \left(\frac{1}{\lambda_c}\right)C\left(\frac{\lambda_t}{\lambda_V}\right)\dot{x} + \left(\frac{1}{\lambda_K}\right)K\left(\frac{1}{\lambda_V}\right)x = 0 \qquad (1)$$

Construction and Testing

Prototypes were assembled for the first time by different groups of students, who gathered ideas for mechanisms and materials depending on the needs and resources. This exercise was developed within activities of problem based-learning Design courses of Mechanical Engineering Program (Figure 2), and funded by the state government. During a second construction process, prototypes were improved (Figure 3) in terms of motion mechanism, finishes, and automated control of vibration frequencies; funded by Civil Engineering

Program, as an initiative to give continuity to the original project, and complement courses in the area of structures.

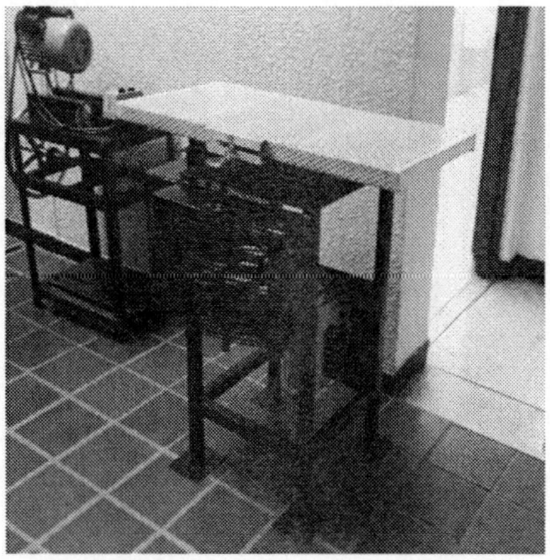

Figure 2: Preliminary shaking table.

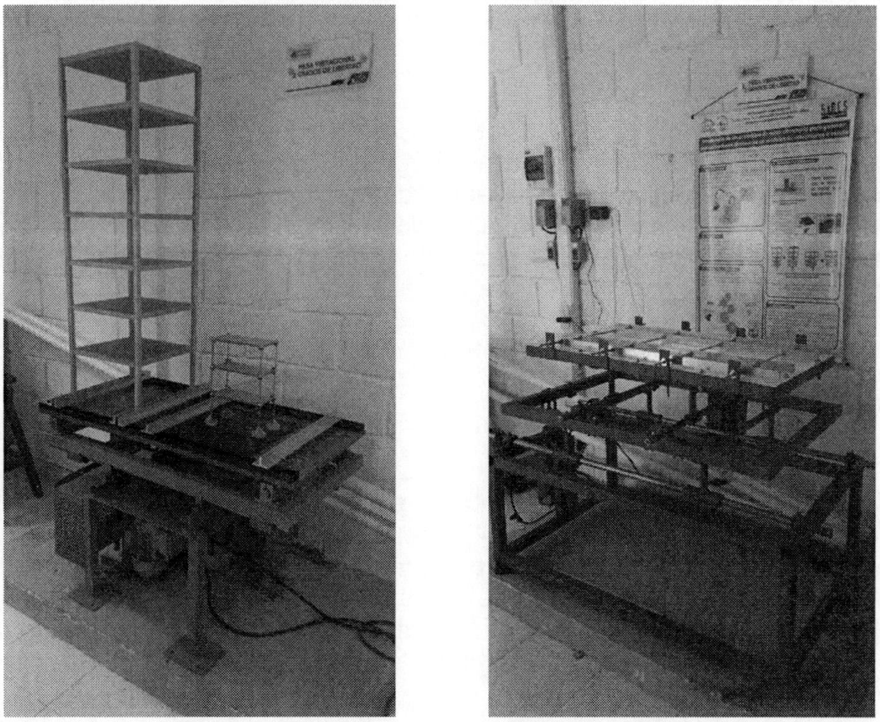

Figure 3: Improved versions of shaking tables.

Equipment

The two low-cost technological tools developed by students and teachers reproduce simple harmonic motion in two horizontal directions that work independently, and able to produce dynamic loading on structures models up to 500 N weight, over a plane platform of size 1.0 m x 0.5 m sliding on lineal bearings.

The motion is generated with a crank mechanism with two different engines that allow independence of both motion directions. Vibration frequencies of motion are controlled for each direction by electronic devices, within a range between 0 - 30 Hz.

Applications of bench-scale shaking tables in Civil Engineering lead to teachers in structural analysis generate a proper environment for experimentation from which students get closer to real-life situations in earthquake engineering. In addition, measurements of dynamic response of structures models in both directions allow analyzing torsional effects in buildings.

Table 2 presents a general comparison of costs for shaking tables designed in other institutions, and with different mechanisms. Advantages of the devices developed in University of Ibagué are directly related with the mechanism used to reproduce the motion, and the construction process performed by students.

Table 2: Costs of shaking tables.

Institution	(Dyke, et al., 2000)	(Carrillo et al., 2013)		**University of Ibagué (Mechanical system)**	
		Pneumatic system	Hydraulic system	High frequency	Low frequency
Total USD	$ 25.260,52	$ 2.733,91	$ 12.028,69	$ 1.356,23	$ 678,12

Implementation and Applications

The use of technological tools presented in this research is in process to be implemented. Results for these applications are showed in terms of the progress in developing documents to support experimental activities using the seismic simulators. A first approximation to such activities has been developed in a free vibration class, where students used experiments guides to analyze the dynamic response of a system in free vibration. Results are presented in next section.

Results and discussion

Preliminary experimental activities

One first approach to experimental experience for students has been developed during recent structural dynamics courses (Figure 4), in which they were asked through an experiment guide document, to establish hypothesis and suppositions about possible response of vibrating systems, based on previous lectures about dynamic properties of systems. Results for two different groups show that by first implementing the theory, the students in addition of applying the learned theoretical concepts begin to construct new ideas from the obtained experiences.

Figure 4: Development of the preliminary activity in structural dynamics course.

Additionally, experimental guidelines generated qualitative evidences that allow understanding the importance of developing practice strategies in courses with numerical complexity. The answers of students to questions like "What did you learned about dynamic behavior of buildings analyzed as a SDOF system?" showed that during and after the experiments, they got good interpretation of fundamental concepts studied in the class.

Some of the representative answers are presented here:

"Observing real effects of changes in physics characteristics in SDOF systems on dynamic properties is useful to analyze mechanical behavior of real-life buildings with variations in geometry and mass."

"Observing simulations of a vibrating SDOF system helps to understand possible behavior in structures under seismic loading. Making predictions was useful to imagine the structure response and then, after practice observations, to strengthen fundamental concepts."

Experimental activities

The first experimental activity implemented in the laboratory was the simulation of harmonic movement in the base of simple oscillators using a shaking table, in order to reproduce resonance conditions (Figure 5). Guidelines allowed the students to make a previous exercise of analyzing, interpreting and predicting the dynamic response of a SDOF system under harmonic base loading.

Methodological process was proposed based on three principal activities:

1. Preliminary hypothesis formulation, where students have a space to predict possible behavior of the vibrating systems as a response to harmonic loading.
2. Experimental process, where they make simulations of bodies under dynamic conditions and measure response parameters.
3. Analytical and concluding stage, where proposed predictions and real observations are compared to clarify the behavior of principal parameters that define the mechanical response of the systems.

During the first activity, students showed some difficulties integrating conceptual relations taken from the class with the real situations in the laboratory, before observations of the experiment. In this activity, students have already studied basic concepts of SDOF systems under harmonic loading at the base. However, most of the predictions at the first time did not fit with real observations, for analysis of variables depending on implicit parameters in movement equations.

Most of the students made predictions based on the mechanical properties of structural elements with regard to static conditions. Later, after observations of real movement, they realized about the contribution of additional dynamic variables to the response of the oscillators.

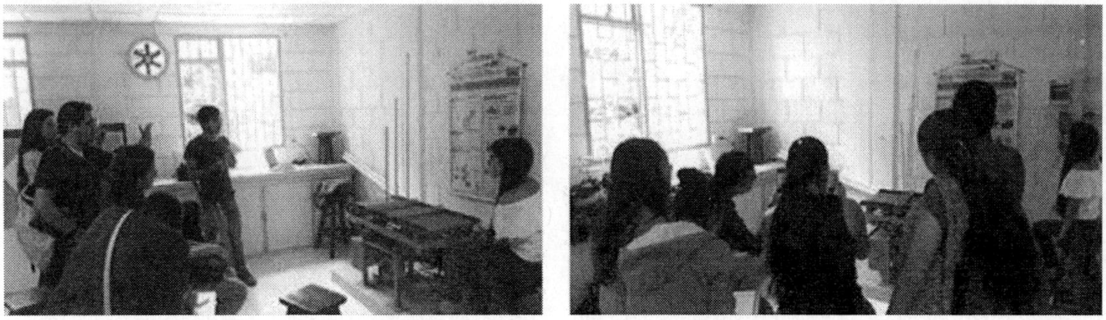

Figure 5: Development of the first experimental practice.

Scaled Structures Models

The scaled models for testing with the tables were defined from comparative analysis of buildings in different materials and dimensions, in order to identify the proper dynamic properties that show similarity with those from real-life structures, and that coincide with those of the vibration frequency of the base motion, with the intention of getting closer to resonance effects.

Future research

Future research is intended to continue with the implementation of practical experiments in structural dynamics courses that leads to strengthen teaching-learning processes in earthquake engineering, and the improvement of the technological tools by automation of motion and instrumentation of models response to dynamic loading. Both approaches are in progress now, and will lead students to quantify variations in structural response of building and other systems due to changes in mass, stiffness and damping, which initially represented the principal source of difficulties in students for learning mechanics.

References

Avitabile, P. (2007). Teaching Experimental Structural Dynamics Applications. *SOUND AND VIBRATION*, (November), 14–21.

Briede, B. (2013). A constructivist approach in engineering education. In *ENGINEERING FOR RURAL DEVELOPMENT* (pp. 584–589). Jelgava: May.

Carrillo, J., Bernal Ruíz, N. M., & Porras, P. (2013). DESIGN EVALUATION OF A SMALL SHAKE TABLE FOR TESTS. *Ciencia E Ingeniería Neogranadina*, *23(1)*, 89–105.

Dwight, R. A., & Carew, A. L. (2006). Investigating the causes of poor student performance in basic mechanics. In *Proc. 17th AAEE Annual Conf.* Auckland, New Zealand.

Dyke, S. J., Gould, P., & Truman, K. Z. (2001). *University Consortium of Instructional Shake Tables : Final Report to NSF*.

Dyke, S. J., Jiang, Z., Christenson, R., Gao, X., & Courter, S. (2007). Teleoperation and Teleparticipation of Instructional Shake Tables Using the NEES Cyberinfrastructure.

Dyke, S. J., Nepote, B., Caicedo, J. M., Johnson, S. M., & Oware, E. A. (2000). Earthquake Engineering Education : A Modern Approach. In *Proceedings of the American Society of Engineering Educators Conference*. St. Louis, Missouri: June.

Dyke, S. J., Truman, K. Z., & Gould, P. L. (2000). Current Directions in Earthquake Engineering Education : The University Consortium on Instructional Shake Tables. In *Proceedings of the American Society of Engineering Educators Conference*. St. Louis, Missouri: June.

Fukuwa, N., Tobita, J., Mori, M., Koide, E., & Hanai, T. (2008). DEVELOPMENT OF VIBRATION EXPERIMENT EDUCATION MATERIALS FOR STRUCTURAL AND SOIL DYNAMICS. In *The 14th World Conference on Earthquake Engineering*. Beijing, China: October.

Garrison, C., & Ehringhaus, M. (2007). Formative and Summative Assessment in the Classroom (pp. 1–3). Retrieved from http://www.amle.org/Publications/WebExclusive/Assessment/tabid /1120/Default.aspx

Goldfinch, T., Carew, A. L., Gardner, A., Henderson, A., & McCarthy, T. J. (2008). Cross-Institutional Comparison of Mechanics Examinations : A Guide for the Curious. In *Proc. 19th AAEE Annual Conf.* (pp. 1–8). Yeppoon, Australia.

Karim, R. (2011). Teaching and learning of fundamentals of mechanics in an innovative way to maximise students ' understanding. In *2nd WIETE Annual Conference on Engineering and Technology Education* (pp. 29–34). Pattaya, Thailand: January.

McHenry, A. L., Depew, D. R., Dyrenfurth, M. J., Dunlap, D. D., Keating, D. A., Stanford, T. G., … Deloatch, G. (2005). Constructivism : The Learning Theory That Supports Competency Development of Engineers For Engineering Practice and Technology Leadership Through

Graduate Education. In *Proceedings of the 2005 American Society for Engineering Education Annual Conference & Exposition*.

Muhlenkamp, M. (1997). *Analysis, design and construction of a shaking table facility*.

Acknowledgements

This project was funded by the Agreement 1026-2013 "Desarrollo de una cultura científica en niños y jóvenes del Tolima – Línea B: Fortalecimiento de Semilleros de Investigación del Tolima" between the Gobernación del Tolima and the Universidad de Ibagué. It was also funded by University of Ibagué, through investment resources for laboratories. The authors are grateful for all the students and teachers researchers that contributed to help reach all objectives, especially to students from Mechanical Engineering Program.

Copyright statement

Use of Free Hardware and Software in Engineering Educational Process of UNICATOLICA

J. E. Ordoñez[1,2], Ch. Burbano[1,2], C. E. Gamboa[1,2], and C. -W. Sánchez[1,2]

[1]Semillero Métodos Físicos Aplicados a la Industria - MeFAI

[2]Grupo de Investigación KIMSA, Facultad de Ingeniería

Fundación Universitaria Católica "Lumen Gentium", Cali, Colombia

jordonez@unicatolica.edu.co

Abstract: *The Fundación Universitaria Católica Lumen Gentium, UNICATÓLICA, is a higher education institution in Santiago de Cali, Valle del Cauca with twenty years of service, providing quality instruction to vulnerable communities in the region. The training process corresponding to the engineering area presents weaknesses regarding the fundamentals of electronics, particularly when trying to address issues of industrial automation and robotics. Due to this, we have taken advantage of the free hardware and software tools available as well as collaborative work applications, which provide a wide range of devices that allow us to create functional and quality developments in this area, to contribute to the professional training of UNICATOLICA engineers and supporting the processes of open science and knowledge dissemination. In this paper we gather practical experiences at a research level in UNICATOLICA, describing the tools used and the structure of the projects, their scope and perspectives. This way, we set an institutional precedent in the learning process of our engineering students.*

Introduction

The Fundación Universitaria Católica "Lumen Gentium" is an institution with 20 years of service to the community in the City of Cali, with two main sites in Pance a Meléndez offering professional programs and three satellites corresponding to the Ceres in Alfonso Lopez (Cali), Yumbo and Jamundi, the latter providing education at the technological level. Archbishop Isaías Duarte Cansino founded the institution with the purpose of favoring vulnerable communities in the city of Santiago de Cali, offering high quality education. Although the incursion into the research topic is relatively new, great strides have been made, especially in the humanities. UNICATOLICA counts with 5 investigation groups, of which 1 is recognized before Colciencias with C category. The faculty of engineering counts with one research group that covers the available engineering and technology programs, currently in the process of gathering requisites for its categorization before Colciencias. In Colombia, we use the term "Hotbed of Research", this to describe a group of students that through working in research activities support research lines previously stablished within the projects of the main groups. The engineering research group is called KIMSA, a term that means, "Work in community" and that carries this appellation because it gathers the research efforts of actors from different programs. Within KIMSA, lines of research are established, supported by formational subgroups of students, "research hotbeds", working on these lines. Physical Methods Applied to Engineering MeFAI, by its acronym in Spanish, is a new research hotbed with two years of operation, this one, has allowed the students to develop their degree works through activities within its domain. The investigation structure in UNICATOLICA is explained in the diagram of Figure 1.

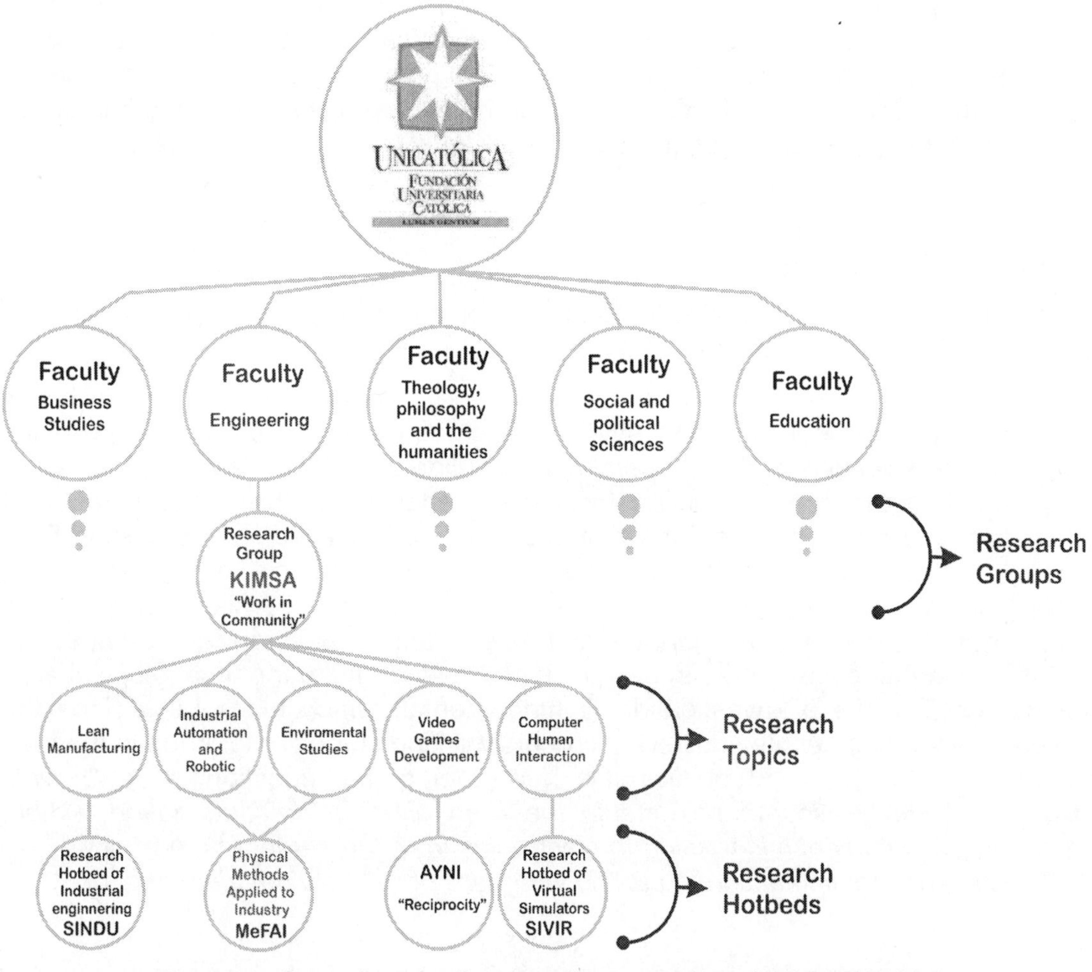

Figure 1. Organigram of research structure at UNICATOLICA

At the moment three students have been directly graduated and six more are in the process of achieving this. Similarly, about sixteen students work in different projects thought by themselves or the research teachers.

The fundamental initiative of this research hotbeds is that the work is framed in the application of free hardware and free software, seeking the dissemination of information in the media by pointing to open science initiatives. Below, we will present the training experience that has positively impacted students and researchers in the exercise of applying science to solve problems generated within the productive field. We will list the most transcendent projects and give a perspective of what has been achieved in each case.

Context

The *Fundación Universitaria Católica "Lumen Gentium"* is a young institution when it comes to the engineering area, now at days it counts with two programs, industrial and systems engineering. For the faculty it has been clear that even though this programs are a great improvement in the incursion of the Science Technologies Engineering Math - STEM area, we still have a serious disadvantage when it come to the specific field of electronics. With the interest of venturing in the automation and robotics fields, we opted for tools of free software and hardware, which offer the possibility of developing activities using the academic bases offered by the current academic programs. This, in order to create new professional perspectives for students who are members of KIMSA. This work brings together the experiences around the immersion process and the results obtained from it.

Research Questions

When we approach the challenge to educate people of different programs in just one group, we raised the research question.

The *Fundación Universitaria Católica "Lumen Gentium"* offers two programs within its engineering faculty, industrial and systems engineering to be precise. Due to the training of the scientists and engineers who coordinate the activities of the research group and hotbeds, lines of action have been adopted in the area of industrial automation and robotics. This lines demand high expertise in the electronics field. Although the institution has a technological program regarding this line, Technology in Electronics, the willingness of students interested in research activities is very low. As a consequence, we face a shortage of trained staff in electronics. For both the institution and the researchers, an approach of the students to the technological area that frames this line of investigation is fundamental, which leads us to face the challenge of forming people of different careers working in the same group in the area of Industrial automation and robotics. It is then, when certain questions arise, *"How to advance common research projects in the area of industrial automation without requiring a robust academic training in electronics from the student body?"* and *"How to offer research training according to the vanguard of national development?"*.

Theoretical frameworks

Currently, the research group KIMSA has members belonging to the programs of Systems Engineering and Industrial Engineering. With the formation of both programs it was possible to conceive, at a theoretical level, different projects in which industrial processes and technology are combined. However, it was necessary to use tools that allowed the physical implementation of such projects, as well as the idea of putting to use different free hardware initiatives. With the pass of time, these tools were manifested as a great help for the elaboration of research with a higher level of application in the industry, allowing us to obtain a closeness through abstractions of the real processes. In addition, these free hardware tools are easily accessible and free to use, which makes them feasible for learning processes. Taking as reference the study of Bitner, N., & Bitner, J. O. E. (2002) and Gil-flores, J., & Rodríguez-santero, J. (2017) where they treat the application of technologies in education processes specifically in the classroom, they recommend introducing these elements to students with care to get better results than traditionally obtained. Our research group considers the use of these technologies to offer special skills not covered by traditional curriculum to students. Among the tools used, we find the mini PC Raspberry Pi and Arduino board. The first one is a mini computer that, within the research process, is usually used as a server for the management of communications between software and hardware, while the second is a programmable circuit board, used for the control of hardware devices. The application of this tools (free hardware and software) as is evidenced on the investigations by Case et al (2015), Jamieson et al (2015), Moreno et al (2015), Paper et al (2013), Pereira et al (2014), Rubenstein (2015) and Valderrama and Rojas (2014) who use adaptations of free hardware and software with results that impact positively the skills aquired by the engineering students, while providing them with research concerns in the area of desing and development, forging an inquisitive attitude that will lead them to propose solutions in the labor sphere in the future.

Methodology

Due to the formation in sciences of the advisors and engineers, all the MeFAI projects have been directed to be approached from the singular parts towards the integration of a total system, from the particular towards the general, from simplicity to complexity, which demands of the student capacity of synthesis and organization. For the process of assigning tasks to students by researchers, we use tools, such as the Gantt diagram. This allow us to discretize the activities that take part in a project and to schedule the tasks during the set time, indicating the dates of beginning and ending, this methodology is based in concepts of project administrations how show in the work of Rivera and Hernandez (2010). In order to coordinate the activities of various research projects that have several students, the researchers use collaborative work tools such as the Zoho Projects platform[1], which lets users to load project progresses with contributions from different participants so that these can be integrated at any given time. This program makes use of a repository that stores information in a certain virtual space (cloud) that serves as backup of the work developed. We use these tools especially in projects that have to do with software development. In support of less rigorous activities, the Wunderlist application is convenient for assigning short activities in a very dynamic way, denoting the more important ones and the ones that have already been covered.

These tools have let us develop a large number of activities with students who can control the execution of their own tasks as they keep in contact with their assigned professors. Weekly meetings are planned in which the group of students compare their advances with their coordinators and director, so that they have a clear guide and a very dynamic adaptation process to changes.

Correspondingly the research methodologies applied in each project, constitute, at a general level, the following phases:

- Survey of requirements.
- Analysis of requirements.
- Design and deployment of research.
- Feedback

[1] Zoho Project is a website that offer support to collaborative work (https://www.zoho.com/projects/)

Purpose

At the outset, we tried to introduce the student to the paradigm of engineering: There is a problem and based on the acquired knowledge, we seek a solution. This supports an exercise in critical thinking, and assumes diverse points of view provided by the work team. The object for this was automation and robotics, the medium was the free hardware tools that today give us the possibility to make the jump without strong bases in the field of electronic, a clear weakness in our academy as aforementioned.

The purpose of the activities developed in the research hotbeds is to train in the technical and investigative level, preparing students with a high ethical quality, with strengths in the expression and communication of ideas, to make them later on invaluable elements in the business field. For this purpose, students are encouraged to participate in different events, congresses and symposia during the training period so that they not only lose the fear of speaking to a scenario, but build tolerance to criticism and the capacity for critical self-analysis. With this work, we have position students into international events after attending local, regional and national competitions. This has strengthened the quality of the research group we are part of, bringing us near the Colciencias classification with a contribution to the number of products required for this purpose.

Approach

The development of research activities around these tools has been based on approaching situation from a simple perspective to then build complex systems through the incorporation of concepts, requirements and solutions that lead us to generate knowledge and tangible products.

Accordingly, all projects begin with an approach to the inherent theoretical concepts, given, once assimilated these concepts we proceed to complement the bibliographic review with the applications of the concepts to specific problems. From a scientific perspective, we adapt and complement the information available in databases and online consultations from different sources; the coordinators verify this process. This way, studies related to the objectives set in the research hotbed are proposed as prototypes that usually constitute thesis dissertations, as example production process and marketing of products. These results also is possible obtain book chapters or journal publications. Finally, to culminate each activity, this is consigned in a free access book for the university community and the country.

Results

Using technologies of free hardware and software, engineering students have developed projects such as the following:

- A robot that carry out surveillance and inspection activities inside sewer pipes.
- A three-dimensional scanner that allows the user model the physical structure of an object in a digital setting.
- A system for monitoring gaseous variables (CO_2, CH_4, CO, Relative Humidity and Temperature) to evaluate the impact the institution has when it comes to greenhouse gases emissions.
- A system for monitoring environmental noise in common areas.

The projects that have already produced substantial results are in process of publication in several engineering magazines. Furthermore, the analysis and problem solving skills that the students have acquired by being involved in this kind of exercise signifies a benefit to their academic grow.

The project "Mobile robotic platform for inspection of sewerage pipelines with diameters equal or greater than 1 meter" (PRIDA) was the first one inside the MeFAI research hotbed to receive proper financial support, with an approximate $ 20.000.000 COP, which allowed us to design and build a robot that inspects the structural state of pipelines. One of the main purpose of this task is to reduce the loss of water due to illegal connections to the city system. This project looks up to answer the following question: “Can we carry the kind of inspections needed in city pipelines with a locally developed system that offers the economic advantage that lack the commercial systems, and can it be maintained through time opportunely?”. To answer this, it is appropriate to contextualize the situation; the project first began with the lifting of requirements for sewerage inspection processes, such as dimensions, sensors and mobility requirements. Afterwards we proceed to generate a design, done by one of the students in training, using the computer assisted design software provided by the institution. This design was set in meetings between advisors, students and experts, specifically consulted for this purpose. This project then continued with the construction phase in which the environmental variables registration, locomotion and location systems developed by students in direct consultation with the researchers were integrated. The electronics and coding components of the robot based on Arduino were develop within short projects among the students in the research hotbed, to be later integrated in a final prototype with a functioning operative system, and which manufacture process can be projected in a national level, with guaranties of being less expensive than the current robots used to this purpose in the country. This work allowed the university to attend the V Argentina Congress of Mechanical Engineering, V CAIM 2016, which also established collaborative links between universities, and was presented by Ordoñez et al (2016) under the title *“Robot de Exploración de Caños para Alcantarilla Desarrollo como Solución a un Problema Local”.* In addition, alternative degree works were developed with the economic support of the project in infrastructure and equipment, as well as a degree work titled *“Diseño del Modelo de Producción de un plataforma de inspección de tuberías”,* which places special emphasis on the production process that must be developed to market the product. In the figure 2 we have the function scheme of PRIDA where we can see free hardware elements integrated in this system.

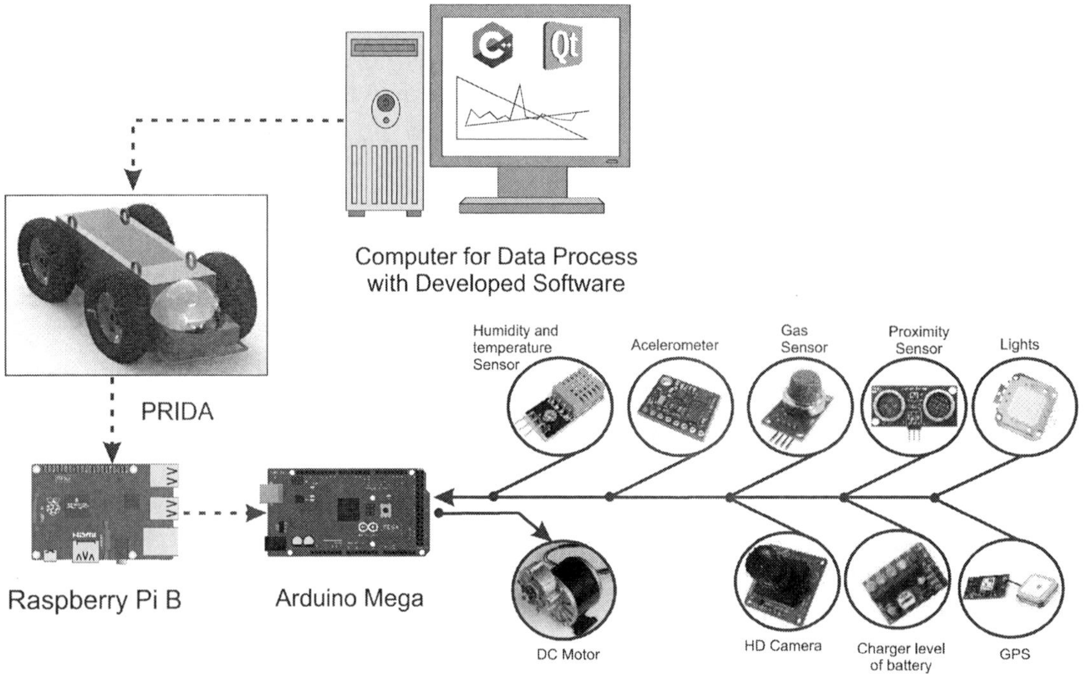

Figure 2. Function scheme of Mobile robotic platform for inspection of sewerage pipelines (PRIDA)

For the control of the remote platform via umbilical wire, this project has a fundamental part, which was done by a pair of engineering in systems students, the PRIDA robot control software, is made entirely in the C ++ language through the framework QT Creator, more information about that is enable in Lee (2016), which facilitates software development by implementing pre-designed windows and buttons that must be programmed in an object-oriented programming environment. It also makes use of the free use MySQL databases, which allows the system to carry registration of users and events of the system, supported by the work of Commeau (2016). This work is soon to be completed and constitutes the degree work of these two students.

One of the first secondary projects financed with the resources of the PRIDA project, was the development project of XKNR G, a 3D scanner developed in the KIMSA labs. This project aimed to explore the digital image processing so that in the future the PRIDA robot could be equipped with this technology as a further tool in the diagnosis of sewage pipes. The project was build up by a systems engineering student, approaching the problem first from the theoretical one and then expanding its knowledge to areas not explored in its academic formation, but the help of his adviser allowed him to acquire expertise in those. Management of medium-level languages such as C ++ and specialized libraries, such as openCV and openGL, was completely necessary for the processing of images, we supported this activities in work of Kaehler and Bradski (2008). This work generated a prototype that was worth the distinction of meritorious thesis to the student developer, who then continued with the institution, in the modality of young researcher and nowadays as a teacher in the "Teaching hotbed" the institution has, seeking to strengthen the line of research in this field. Further information on this work can be found in the article *"Construcción de un Escáner Tridimensional Usando C++ y Hardware Libre"*, which is approved for publication in the Journal of the Faculty of Sciences of the National University of Medellin. Currently, this work is in a stage of complementation, implementing algorithms and optimizing the cloud of data generated. This will let us use the device optimally with other programs such as Blender for 3D design, Freecad, Solid Work and AutoCAD. In the Figure 3, we have the function scheme of the 3D scan and the hardware elements involved in its operation.

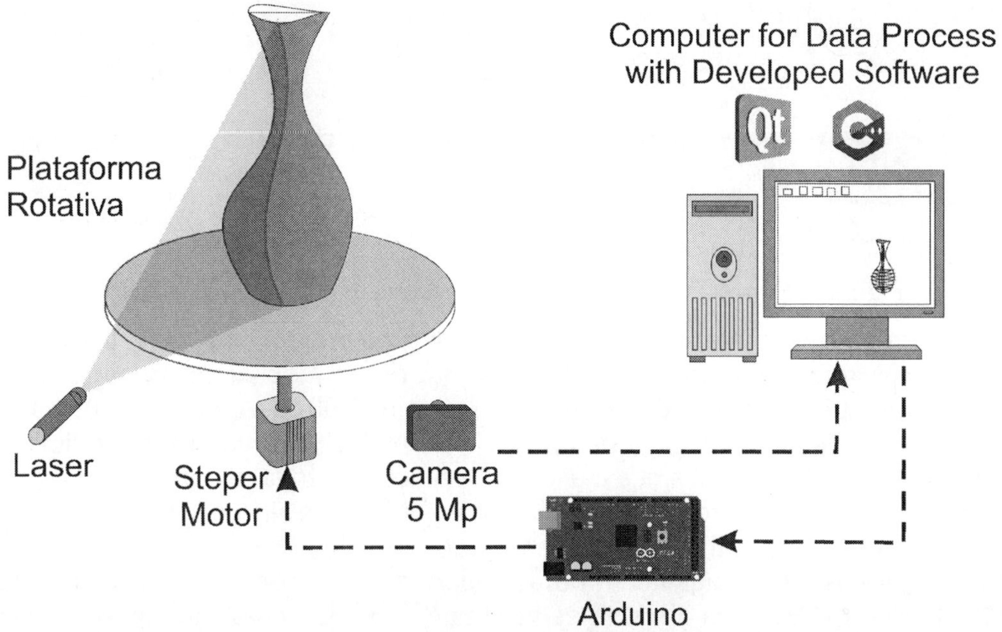

Figure 3.Function scheme of scan 3D developed at UNICATOLICA

Another project that was alternative to the PRIDA project and financed with its resources was the project: "Determinación de los níveles de CO_2 en la cafeteria de UNICATOLICA ". This project generated a gas monitoring system prototype based on the use of technologies of free hardware and software as those offered by Arduino and Raspberry Pi, that can be implemented by companies to make a CO_2 emissions diagnosis due to their industrial processes. This constitutes an environmental diagnosis tool that allows adjustments before the companies face control processes carried out by governmental entities. With this work the student represented the research hotbed at an institutional level, allowing her to participate in the local, regional and national events of the REDCOLSI and ACIET organizations, she was selected to represent Colombia in the international research hotbeds event to be held in the city of Santiago de Chile in August 2017. Two articles were produced and are currently under evaluation before the internal magazine of UNICATOLICA and one more in an indexed magazine of Colciencias. The Thesis dissertation was distinguished granting the engineer a scholarship as a young investigator and allowing her to initiate investigative activities as a professional. Her work as a young researcher led us to interfaculty articulation through her participation as a collaborator, representative of KIMSA, in a project consisting in Management of Sustainable campus, to be developed in 2017 by the Administration Faculty of the university. In the Figure 4, we have the scheme of function of the system with different elements of communications integrated with free hardware and software elements.

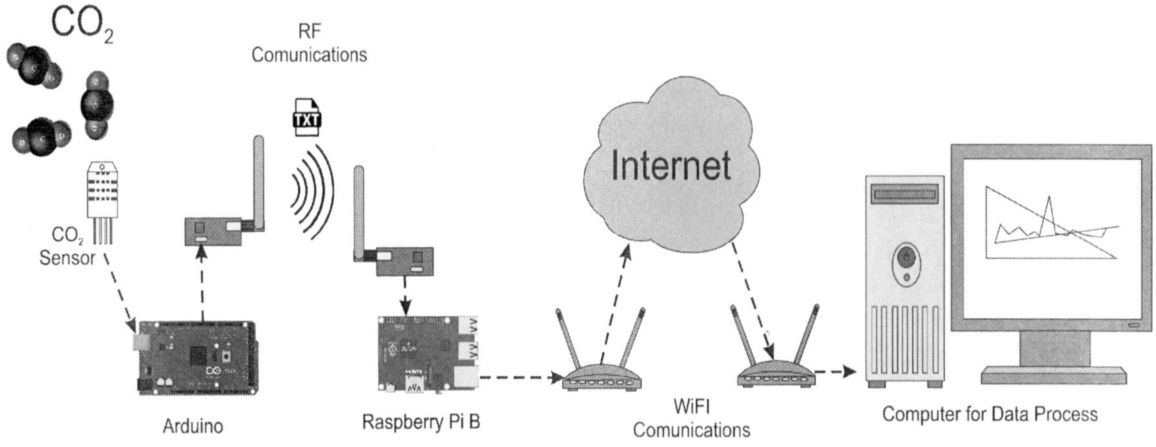

Figure 4. Function scheme of sensor system of gas measurement

Along with the environmental line, we develop a degree project that seeks to determine the level of environmental noise in the restaurant of the institution, a conglomerated place at hours of lunch (12:00 to 2:00 p.m. and 6:00 p.m. to 7:00 p.m.). The determination of the environmental noise that UNICATOLICA students and officials are subjected to is very important since it allows the institution to take corrective measures that seek to mitigate this problem that can cause injuries at the hearing level as well as situations of stress within the university community. This study is an ongoing initiative, but it has had a significant participation in the national research hotbeds of REDCOLSI and ACIET. The project has free hardware elements such as Arduino and microcomputers such as Raspberry PI, with free software elements like C ++, and open office suite, among others. The function scheme of this system in the figure 5, where we can appreciated a mixture of elements used in others projects.

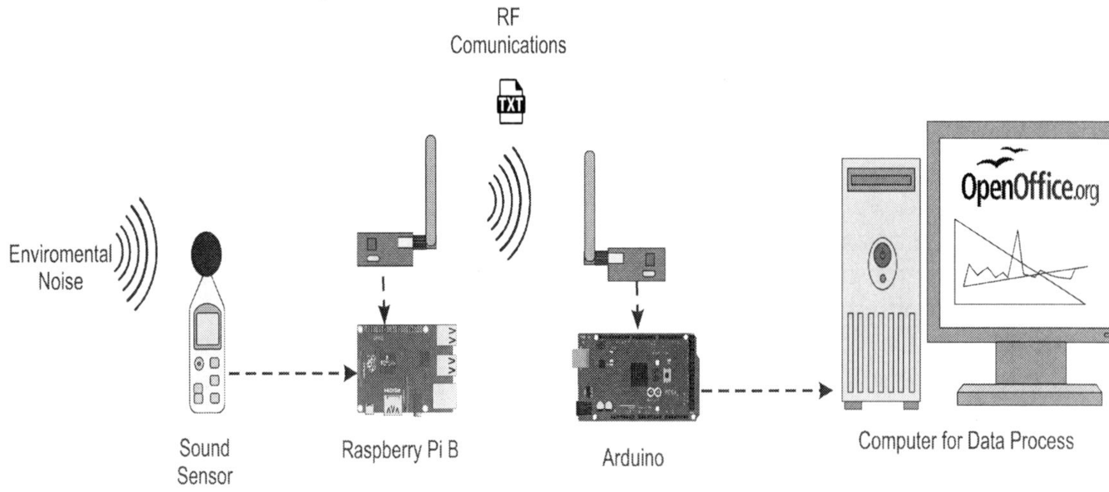

Figure 5. Function scheme of sensor system of noise measurement

Currently, we are working on a new project to develop a robotic arm that allows students a direct approach to advanced manufacturing techniques. This has funding that will allow us to work on alternative initiatives, as it was done with the PRIDA project. The first stage seeks the development of a small-scale robotic arm that can be used by students of industrial and systems engineering in the laboratory or classroom. Subsequently, a second phase will develop a robust robotic arm that allows students to immerse themselves in the real experience of working with industrial arms, their operation and implementation. Four students take part in this project, two from the industrial engineering program working on the hardware component and two from the systems engineering program working on the software component using C++ and MySQL, so, just as the PRIDA robot this software will allow to keep control of the users and the events. In addition this software will allow us to manipulate the arm remotely and program it to follow a specific sequence, this programming can be done by pseudocode and through autonomous learning of the same, and this constitutes the degree work of the students.

Today, through the use of the available tools of free hardware and software, the MeFAI research hotbed advances projects, with which it hopes to participate in the events programmed by REDCOLSI and ACIET as well as other events in the industrial engineering and systems field, such as congresses and symposia. Among this new concepts the research hotbed is working on, renewable energy projects are distinguished, still in the stage of implementation and study, but with a wide participation in events at national level. Works on recycling and processing of plastics aimed at the development of a pilot silver for processing polyethylene terephthalate (PET) in UNICATOLICA. In addition to this, we have been working the design and construction of a locomotion assistant for people with reduced mobility as well as the development of a cane that helps people with visual limitations in UNICATOLICA. The aforementioned works will have an important connotation when it comes to social and environmental pertinence in the Cali community. We cannot leave aside different projects developed by third parties with the support of the research hotbed, such as a “Pet Water Supply System” that uses free software and hardware tools in its implementation.

Conclusions

All the engineering projects that we advance in UNICATOLICA are in some way integrated to the community or with a focus on it (being part of the PIIC, by its acronym in Spanish Proyectos de Ingeniería Integrados a la Comunidad) this was intentionally fixed by the KIMSA group in all its lines of research . This kind of development seeks to benefit the local community with contributions such as monitoring systems to enhance environmental quality through its application. The investigation process is an inductive one in which the student in training turns to consult, inquiry, experiment, analyze and write to show their results.

For example, the project that focuses on the exploration of aqueduct and sewer lines rose in the students the need to research designs in search of the most successful one according to the requirements of the problem. This is one of the most important research topics in the literature.

Thought the entire research process carried in the KIMSA group students have shown improvement in and out of the hotbed activities, managing to use tools, attitudes and expertise previously acquired in their own projects. This results translate into better grades in the courses of each program and better abilities when it comes to communication of ideas to their teachers and peers in oral presentations.

Similarly, new projects such as an assistant for people with limited vision have an obvious social impact, seeking to improve the interaction of people who present these complications with their environment.

In general, the institutional orientation encourages the projects to provide the student, as a researcher in training, with a holistic vision encompassing not only the technique but also the impact of it in society, always in the interest of improvement. In addition, the trainees are looking for better writing and reading skills, while strengthening their ability to communicate ideas assertively, through active participation in regional and national outreach events.

The professionals trained in the MeFAI hotbed show concern about the care of the environment, solidarity with people with locomotive and visual limitations, that is to say, these processes have an impact on the integral formation of the individual, forming responsible professionals with ethical and moral characteristics that reflect in employability and investigative performance.

The use of different elements of hardware and software free has allowed students of careers such as industrial engineering and systems engineering to undertake projects ranging from industrial automation to environmental monitoring studies without having deep knowledge in the area of electronics. These technological tools grant us the collection of reliable and reportable experimental data with autonomy, so that serious studies are important in areas where automation is essential. The use of these tools encourages creativity in engineering students while actively contributing to team activities in the business field. From engineering, researchers contribute to the national scientific development with activities of applied science using hardware and software free tools, spreading information in means of free disclosure as open journals and placing at the disposal of codes development codes repositories, so that they serve to others as tools of advancement in their own projects, personal and academic growth.

Recommendations

Due to the research experience of our research group and hotbeds, we recommend that groups open themselves to the interdisciplinary of members in training that will allow them to approach the different problems to investigate from diverse perspectives that enrich the range of possibilities and solutions to a problem. Even more so, these will be approached from innovative points of view that will guarantee them to make unique contributions with new perspectives. The work presented here is a proof that it is possible to address technological challenges even with technical limitations supported by new technologies such as hardware and software free and find solutions to problems quickly and reliably. We are convinced of the open science as a way to the democratization of knowledge and our institution supports this type of initiatives, with a magazine that communicates in this area and with the research hotbeds and their publications in in the aforementioned magazine aiming always to the dissemination our activities in the community.

References

Bitner, N., & Bitner, J. O. E. (2002). Integrating Technology into the Classroom : Eight Keys to Success, 10, 95–100.

Case, L., García, D., Domínguez, M., Prada, M. A., Alonso, S., Alonso, S., & Reguera, P. (2015). A Low-Cost Open Source Hardware In Control Education . Open Source Source Hardware Hardware In In Control Control Study : Arduino-Feedback, 117–122.

Comeau, A.(2016) MySQL Explained: Your Step-by-step guide to databes design. OSTraining

Gil-flores, J., & Rodríguez-santero, J. (2017). Computers in Human Behavior Factors that explain the use of ICT in secondary-education classrooms : The role of teacher characteristics and school infrastructure. Computers in Human Behavior, 68, 441–449.

Jamieson, P., & Herdtner, J. (2015). More Missing the Boat - Arduino , Raspberry Pi , and Small Prototyping Boards and Engineering Education Needs Them.

Kaehler A. and Bradski G. (2008) Learning OpenCV 3: Computer Vision with the OpenCV Library. O'Reilly Media. Sebastopol CA.

Lee Z. E. (2016). Qt5 C++ GUI Programming Cookbook. Packt Publishing Ltd. Birmingham UK

Moreno, J. C., Berenguel, M., Guzm, L., & Berenguel, M. (2015). Teaching Control Engineering Concepts using Open Source tools on a Raspberry Pi board ** This work has been partially ... using Open Source tools on a Raspberry Pi, (March 2016).

Ordoñez, J. E., Meza, F. A., Gamboa, C. E. y Sánchez, C. -W. (2016) Robot De Exploración De Caños Para Alcantarilla "Desarrollo Como Solución A Un Problema Local". Universidad Nacional de Santiago del Esero Facultad de Ciencias exactas y Tecnologías (Eds.), Proceedings ofV Congreso Argentino de Ingenieria Mecánica (pp. 395-405). Santiago del Estero, República Argentina

Paper, C., Catarina, S., Sim, P. S., Catarina, S., Bento, J., & Catarina, S. (2013). Educational application of Remote Experimentation for mobile devices Educational application of Remote Experimentation for mobile devices, (February). http://doi.org/10.1109/REV.2013.6502905

Pereira, V., Fernandes, V. A., & Sequeira, J. (2014). Low Cost Object Sorting Robotic Arm using Raspberry Pi, 1–6.

Rivera, F., Hernández, G. (2010). Administración de Proyectos: guía de aprendizaje. México: Pearson Educación de México

Rubenstein, M., Cimino, B., & Nagpal, R. (2015). AERobot : An Affordable One-Robot-Per-Student System for Early Robotics Education, 6107–6113.

Salim, A. J., Salim, S. I., Samsudin, N. R., & Soo, Y. (2013). Educational Development Tools for Software and Hardware Processor Design, 622–627.

Valderrama R., O.C., Rojas G., A. A.(2014) Diseño, Elaboración E Implementación De Hardware Libre: Una Nueva Tendencia En La Enseñanza Para Ingenieros De Sistemas. Memorias del Encuentro Internacional de Eduación en Ingenierias ACOFI 2014, Nuevos escenarios en la enseñanza de la ingenieria. Ibagué, Colombia.Universidad Cooperativa de Colombia

Acknowledgements

This work is partially support by the internal proyect of UNICATOLICA titled *"Plataforma Robotica Móvil de Inspección de Ductos Superiores a 1m"* with code FI-010106 and Enginnering Faculty of UNICATOLICA. In addition, we would like to thank to MeFAI team for their help in each of the research projects.

Copyright statement

The following copyright statement should be included at the end of your paper. Substitute authors' names in final copy only.

Engaging students in a non-technical engineering course through active learning strategies: an exploratory case study

Maria Orjuela-Laverde
McGill University, Montreal, Canada
maria.orjuela-laverde@mcgill.ca

Viviane Yargeau
McGill University, Montreal, Canada
viviane.yargeau@mcgill.ca

Mikhail Smilovic
McGill University, Montreal, Canada
mikhail.smilovic@mail.mcgill.ca

***Abstract**: Active Learning strategies were implemented in a large enrolment Professional Practice engineering course. Short in-class group discussions and questions via a web based students' response system were used during the lecture portion of the course. Tutorials were designed using group presentations of current engineering case studies as its main instructional strategy. Students completed a peer and self-evaluation form that assessed performance on professional skills. This study presents results of a content analysis conducted on students' comments provided in the self-assessment form. The content analysis was done using Kearsley & Shneiderman (1998) Engagement Theory as a conceptual framework. Results suggest that the use of authentic case studies were the most valuable element to engage students. The results also provide insights for the implementation of activities that promote peer interactions during the case study presentations. Ideas and lessons learned on the use of group presentations in large enrolment classes will be provided.*

Introduction

The course used in this study is entitled - Engineering Professional Practice, which is offered by the Faculty of Engineering at a large, public, research-intensive North American university. This is a mandatory one-credit course for all last year Engineering students and the student population is comprised by learners from all disciplines to have a classroom representative of working environments, i.e. composed of engineers from various engineering specialities. It is worth noting that the student population in engineering at this University is very diverse and composed of about 50% of students from Quebec, 19% of students from the rest of Canada and 31% of international students. Such a diverse class in terms of disciplines, cultures and gender (28% of women), increases the challenge of finding common grounds to discuss ethical issues and professionalism in a way that keeps everyone's attention.

The course has two sections, with approximately 180 students each. Each section consisted of a once a week lecture and tutorial session (1.5 hours each). The course is offered every semester and on a rotation basis between the Faculty of Engineering departments implying that a different instructor teaches the course every semester. Since the creation of the course in 2009, students had reported that the tutorial sessions, which were based on the presentation by teaching assistants of ethical case studies, were not seen as of much benefit. In addition, instructors had reported that students' participation during lecture time and the tutorials was very little. Course evaluations also indicated that students were not able to grasp the importance of the course and did not feel engaged. Innovation was thus required to stimulate students' interest in this non-technical course and to promote their engagement.

This course is also of importance in the accreditation of our programs by the Canadian Engineering Accreditation Board. The accreditation criteria and procedures implies the qualitative and quantitative analysis of the curriculum content and the assessment of graduate attributes, including not only a knowledge base for engineering, problem analysis, investigation, design, use of engineering tools but also individual and team work, communication skills, professionalism, economics and project management, an understanding of the impact of engineering on society and the environment, ethics and equity, and the capacity for life-long learning.

The purpose of the present course re-design was to provide improved tools for the assessment of some key non-discipline specific attributes sporadically covered in technical courses (professionalism, impact of engineering on society and the environment, ethics and equity and life-long learning).

The Pedagogical Coordinator for the Faculty of Engineering together with a group of graduate students and instructors put together an initiative to re-design a more meaningful lecture and tutorial experience for students also providing opportunities to assess graduate attributes. It was hypothesized that the use of active learning strategies during lectures and tutorial sessions would make the course more attractive to students and result in increased engagement, attendance and participation.

Literature review and Theoretical Framework

Research in the field of student engagement in STEM disciplines have focused mainly on analyzing student's engagement in technical courses, and limited information is available on classes and topics similar to those covered in the course used for the present study.

Studies conducted in STEM classroom settings on the implementation of team work and project based learning strategies in fields such as software development (Miliszewska & Horwood, 2004), computer and video games (Dickey, 2005), Human-Computer Interaction (HCI) (Shneiderman, 1998), and materials heat transfer (Tlhoaele, Suhre, & Hofman, 2016) are available in literature. However, few studies explore student engagement in non-technical STEM courses where professional skills are introduced.

Engagement Theory (Kearsley & Shneiderman, 1998) is a conceptual framework used in STEM research to measure students' engagement in their learning experiences. Kearsley and Shneiderman (1998) define their theory as students being "meaningfully engaged in learning activities through interactions with others and worthwhile tasks (...) Engagement theory is based upon the idea of creating successful collaborative teams that work on ambitious projects that are meaningful to someone outside the classroom". The factors underlying the theory are that learning activities occur in collaborative teams (RELATE), are project-based (CREATE), and have an authentic focus (DONATE) (Kearsley & Shneiderman, 1998).

Research question

What factors within the Engagement Theory (Kearsley & Shneiderman, 1998) framework do students perceive to be most engaging in a non-technical engineering course?

Procedures / Background to the case study

Re-design of the lecture

The re-design of the lecture was based on two different approaches: 1. The use of active learning strategies and tools to promote student engagement during the lectures; and 2) The diversification of evaluation strategies used to obtain a better assessment of graduate attributes. To promote active learning, in-class activities were designed and a web-based student-response system (SRS) was used as an interactive tool during the lecture. In every lecture, students were polled 3 to 5 times and in almost every lecture, one of the polling involved discussion in small groups amongst the student (usually around 3-5 minutes) or discussion with the whole class (variable length). The questions asked were designed to stimulate the discussion on the topics relevant to each lecture but also to collect information to assess the graduate attributes as required by the accreditations process. Other evaluation methods such as a peer/self-evaluation of the group formed for the tutorials (as discussed in the next section) and a short essay on life-long learning were also created to better assess specific graduate attributes seldom evaluated in technical courses.

Re-design of the tutorials

The re-design of the tutorial section drastically changed the role of the teaching assistants (TAs) who became moderators during the tutorials. Rather than having a presentation by the TAs of the case studies related to concepts of ethics, responsibility, liability, sustainability, and environmental legislation, students were asked to present the result of their investigation of a case study. Examples of topics covered include Canadian Oil Sands, Genetically Modified Foods, Space Exploration, Drones, Artificial Intelligence, Nuclear Power, and 3D printing. These topics were selected to represent current hot topics in the different engineering disciplines and every week a different group of 22 students was presenting the results of their investigation of a case study in one of five presentation styles. Figure 1 presents an overview of the different presentation styles (blog post, video, panel discussion, debate, infographic and article) along with the number of students involved in each one. Three of these presentation styles (video, panel discussion and debate) were delivered during the 1.5-h weekly session while the other deliverables were made available online.

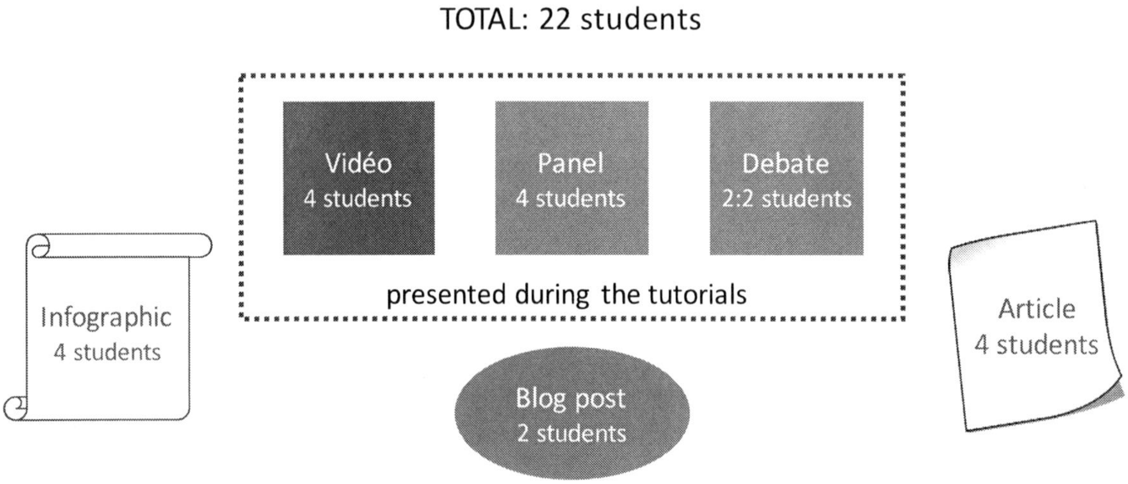

Figure 1 Diagram representing the various presentation styles used for each case studies.

Example of In-class activities

An example of in-class activity designed using active learning strategies is a discussion of a case study described below. Students were polled to determine if the decision made by the engineer was ethical or not.

> CASE STUDY: *Bob Baines was hired to work in the quality control department of a company that designs sensing devices for automobile air bags. About six weeks after joining the company, he was asked to sign off on a design that he felt very uncertain about. He checked with people involved in the design and found the situation*

ambiguous. Bob told his manager that he would not feel right about signing off, and asked that he not be required to do so. His manager kept putting pressure on him. Eventually, Bob decided that he wished neither to violate his values by doing something that he thought was wrong, nor to become involved in a battle in which his career might be a major casualty. He thus quietly resigned.

Results of the polling (Figure 2 – (A) Initial polling) were shared with the class and students were asked to discuss their arguments with their peers for about 5 minutes and to take into consideration the reversibility test previously discussed as a tool to evaluate if a decision is ethical. Students were then polled again (Figure 2 – (B) Re-polling) using the following question "*Reconsider your answer knowing that after leaving the company, other engineers continued to be quiet about defective design impacting the quality of sensors and putting your own safety at risk since the air bag was installed in your own car.*" Results of the re-polling clearly demonstrate that the approach used to introduce the concept of the reversibility test was a very efficient way to convince the students of the value of the test when considering ethics. Students were shocked to realize that so many of them were initially willing to consider the decision as ethical until they were asked to consider that they might personally be affected by the same decision, i.e. end up having the defective air bag in their own car.

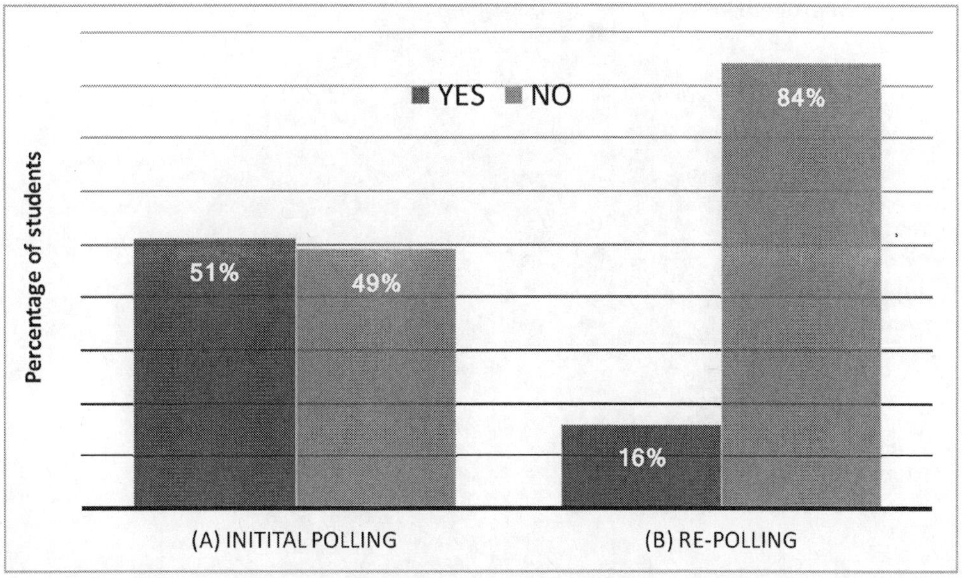

Figure 2 Impact on students' (*n* = 342*)* perception of an ethical question using an active learning strategy to teach a concept: (A) initial student responses, (B) students responses after initial polling and in-class discussion

Methodology

Data Collection

An online evaluation form was sent to the students to assess their performance on a set of indicators related to professional skills (e.g. cooperative work, consideration of others' perspectives). This form was divided into two sections: peer and self-evaluation. In addition, a comment box was included asking students to provide comments and suggestions for improvement of the tutorial sections, presentations, or case studies. The information compiled in the comment box was analyzed to address the research question for the present study.

Data analysis

A content analysis was employed to analyze data compiled in the comment box section of the online evaluation form. Content analysis is "one of the most important research techniques in the social sciences" (Krippendorff, 1989). In a 2004 publication, Krippendorff defines content analysis as a "research technique for making replicable and valid inferences from texts to the contexts of their use". Analysis of the data can be guided by existing theories or practices (Domas White & Marsh, 2006). In this study, the content analysis conducted was framed following Kearsley & Shneiderman's (1998) Engagement Theory principles.

A total of 290 comments were provided. Comments ranged from logistical suggestions to engagement with the course. Figure 3 illustrates the categories created after grouping students' comments. The data used in this study relates specifically to comments under the "Interesting and Engaging" category (119 comments). The coding scheme used in the results section to analyze these 119 comments followed the three components underlying the Engagement Theory: Relate – Create – Donate.

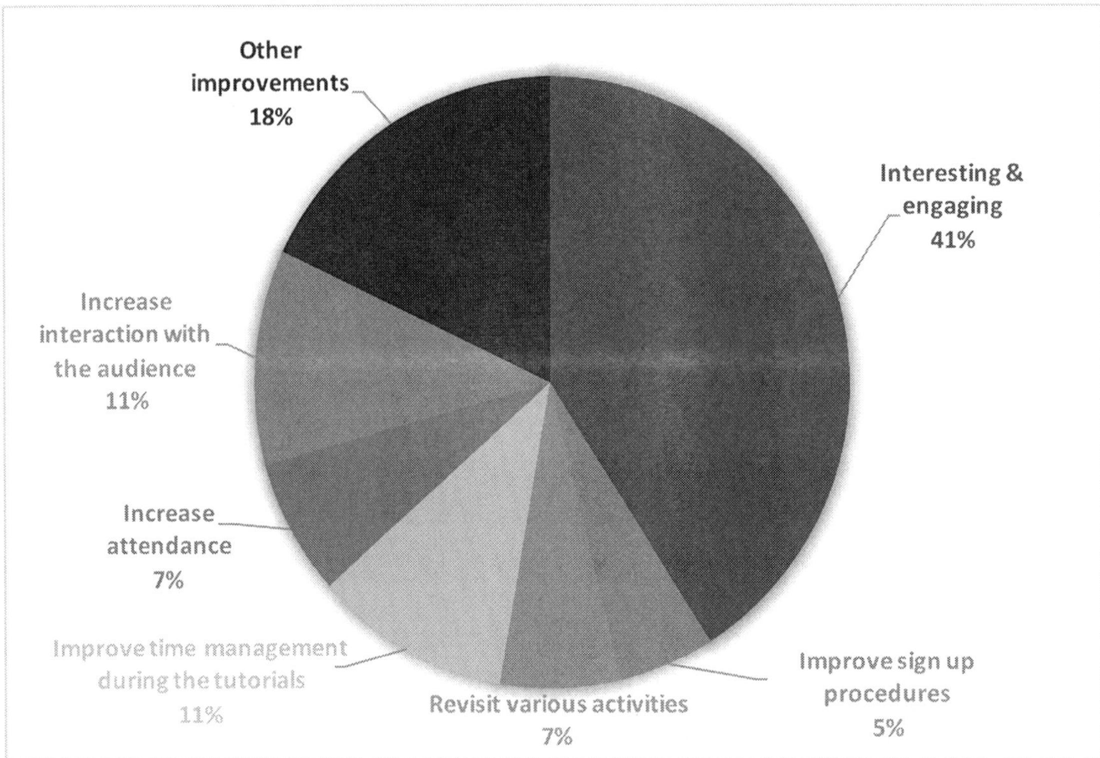

Figure 3. Number of comments per category after classification of the 290 comments provided as answers to the question "*Do you have any comments or suggestions for improvement of the tutorial sections, presentations, or case studies?*" after the re-design of the tutorial.

Results

A total of 119 comments were coded using the Relate – Create - Donate categories. Of these comments, 59% refer to the value of using authentic engineering case studies as a strategy to introduce course content (Professional knowledge and skills); 11% of the comments were related to students' experiences with team work; and, 30% referred to the creation of specific artifacts required in the group presentations (e.g. videos, infographics, blogs).

Examples of students' comments per category are as follow:

Relate:

"(...) there should be more discussion with the entire team working on the case study rather than just working only with either a partner or other members of the project they are working on (videos, debates, panels, etc.)".

"It was a great way to work as a team and listen to different opinions on a topic of choice".

"Great way to get the class involved and to work together in teams!"

Create:

"The tutorials and case study presentations are very dynamic. It is interesting to watch the case study presentations and different approaches to the subject, such as video, debate and panels during the same tutorial. The turning point questions are a great way to get students engaged. "

"I really enjoyed the whole process; I thought it was engaging and provides us with a solid understanding of the case study presented each week."

"I think that the case studies were a great way to educate the class on specific topics for two reasons. First, the case studies engaged students in a non-exam format, which was a welcome break from the standard evaluation format that we find in traditional engineering classes. Second, each group was given guidelines that structured the quality of the assignment submissions, while also giving each group flexibility on which topic to explore. Overall, I would love for this type of evaluation format to be incorporated into other engineering courses."

"I really like how we could pick the method used to do our case study. Worked out well for an introvert like me and I enjoyed the case study more"

Donate:

"I think the case studies are extremely interesting. "

"In general, I really enjoyed how the course was organized. I feel that it touched upon worthwhile topics. I feel as though a lot of effort was put in to making this an engaging experience."

"I gained knowledge about the various issues globally taking place."

"The case studies were on interesting current topics in engineering ethics, which I appreciated."

"The tutorials and case studies are helpful as they showcase real life examples in the field of engineering."

"It gives you experience for talk related events (public speaking). I realized that you have to be prepared for the most unlikely questions that may be posed by the audience."

"Tutorials are particularly interesting due to the important issues that are discussed by classmates"

"The case studies are a very useful learning tool to spark curiosity in engineering/global topics outside of typical classroom learning."

"All the case studies were relevant and concerned recent, important topics which made them interesting."

"The case studies are really well chosen. Each topic involves technical information and some controversies, which engage us to search and to learn. Also, these topics are not too far from our life. They are understandable and interesting."

Discussion

Results from the content analysis suggest that students perceive authentic tasks (real-life case scenarios) to be the most engaging component in a non-technical course, when compared to components such as work in collaborative teams and project based activities.

A possible explanation to the high value given to the authentic task component is the few experiences students have to relate content knowledge to applicable engineering scenarios. Because this course is taught at the end of the program, students have been frequently exposed to team work and project-based courses but much less to situations forcing them to consider their profession practice and the impact of their technical decisions on the environment and society.

In the book *How learning works: 7 research-based principles for smart teaching* (Ambrose, Bridges, DiPietro, Lovett, & Norman, 2010) the authors explain how motivation is one of the principles of learning. Establishing value to a task (e.g. assignment) is a factor that increases motivation to complete the task. A strategy suggested by the authors to increase the value of a learning activity is using authentic, real-world tasks in the classroom setting. This suggestion may also explain why students, in the current study, gave more value to authentic focus than to collaborative teams and project-based components.

In terms of students' attendance and participation during lecture time, Figure 4 presents comparison data between number of students responding to questions via a student response system (SRS) in Winter 2014 (pre re-design) and Winter 2016 (post re-design). Both sessions were taught by the same instructor. In addition to the differences in in-class activities and tutorials case studies, participation marks were given to students answering to SRS questions in the Winter-14 (pre-design). What is interesting to note is that attendance rates did not decrease noticeably in the post re-design semester during which attendance was not mandatory. This might be another indicator of improved student engagement in the course.

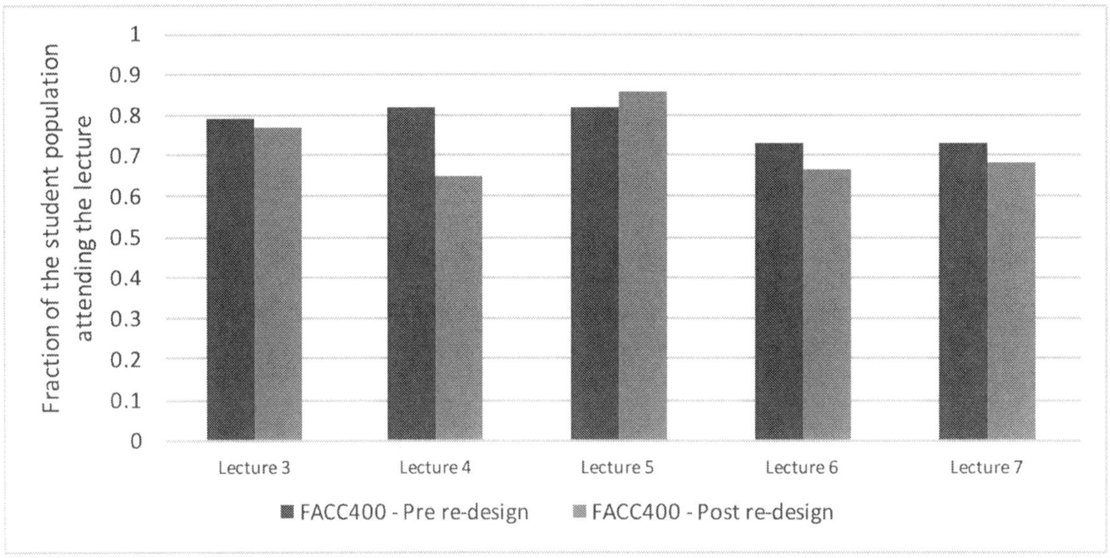

Figure 4. Comparison of lecture attendance before and after the re-design of the course for the same instructor, which also corresponds to the mandatory attendance (pre re-design) and without tracking attendance (post re-design). Pre re -design *n* = 341*;* Post re-design *n* = 342

Conclusions

Based on the feedback obtained from students' peer and self-assessment, as well as our personal experiences with the course, the use of active learning strategies is an efficient way to increase student engagement in non-technical courses. The use of authentic tasks is highly recommended to increase students' engagement with classroom content.

Additional to the conclusions related to this study, we would like to share lessons learned for those interested in implementing this type of instructional strategies:

- The success of such initiatives designed to improve students experience requires a close collaboration of all stakeholders (Instructor, Teaching Assistants, Education Developers) at the planning as well as the implementation phases.
- Feedback from students during and after the implementation phase provides valuable input for improvement and to identify the most efficient strategies.
- For courses taught by multiple instructors, the sustainability of the strategies implemented to enhance students' engagement requires collaboration between the instructors and the team of TAs. Classroom observation by instructors who will teach the course in the future is also strongly suggested.

References

Ambrose, S. A., Bridges, M. W., DiPietro, M., Lovett, M. C., & Norman, M. K. (2010). *How learning works: Seven research-based principles for smart teaching*. John Wiley & Sons.

Dickey, M. D. (2005). Engaging by design: How engagement strategies in popular computer and video games can inform instructional design. *Educational Technology Research and Development*, *53*(2), 67-83.

Kearsley, G., & Shneiderman, B. (1998). Engagement theory: A framework for technology-based teaching and learning. *Educational technology*, *38*(5), 20-23.

Krippendorff, K. (1989). Content analysis. In E. Barnouw, G. Gerbner, W. Schramm, T. L. Worth, & L. Gross (Eds.), International encyclopedia of communication (Vol. 1, pp. 403-407). New York, NY: Oxford University Press. Retrieved from http://repository.upenn.edu/asc_papers/226

Miliszewska, I., & Horwood, J. (2004). Engagement theory: A framework for supporting cultural differences in transnational education. *Higher Education Research Society of Australasia*.

Shneiderman, B. (1998). Educational Journeys on the Web Frontier. *Educom Review*, *33*(6).

Tlhoaele, M., Suhre, C., & Hofman, A. (2016). Using technology-enhanced, cooperative, group-project learning for student comprehension and academic performance. *European Journal of Engineering Education*, *41*(3), 263-278.

White, M. D., & Marsh, E. E. (2006). Content analysis: A flexible methodology. *Library trends*, *55*(1), 22-45.

Acknowledgements

We wish to thank the Faculty of Engineering at McGill University for the financial support provided to this project through a Gerald W. Farnell Teaching Scholar awarded to Prof. Viviane Yargeau.

Copyright statement

AUTHOR INDEX

AUTHOR INDEX

AUTHOR INDEX